Amphibians of Ecuador

Amphibians of Ecuador: Phyllomedusidae, Leptodactylidae, Ceratophryidae, Hemiphractidae, Eleutherodactylidae, Centrolenidae, Gymnophiona, and Caudata, Volume IV is the final installment in a four-volume series that provides comprehensive, well-illustrated, and authoritative insights into the amphibian species of Ecuador, making it an invaluable resource for biologists, conservationists, and others. The series explores, in comprehensive detail, the cultural history and the rich amphibian diversity of Ecuador, providing a thorough review of biogeography, amphibian declines, and conservation.

Volume IV specifically focuses on Phyllomedusidae, Leptodactylidae, Ceratophryidae, Hemiphractidae, Eleutherodactylidae, Centrolenidae, Gymnophiona, and Caudata. Characteristics of each species are listed, defined, and compared to similar other species. Reproductive behavior, where known, is described, as are data on vocalizations, larvae, and ontogenetic changes. Amphibian distributions are detailed and illustrated with physiographic maps with dots. This volume also addresses the declines, extinctions, and conservation status of each species, highlighting their occurrence in protected areas.

KEY FEATURES:

- Provides detailed and comprehensive accounts for all species from Ecuador of Phyllomedusidae, Leptodactylidae, Ceratophryidae, Hemiphractidae, Eleutherodactylidae, Centrolenidae, Gymnophiona, and Caudata. An Addendum includes accounts of 20 species described from 2023 to January 2025.
- New data are incorporated for many species.
- Describes with full color maps the distribution of all known taxa.
- Includes information on the ecology, reproduction, and behavior of all taxa.

Amphibians of Ecuador

Phyllomedusidae, Leptodactylidae, Ceratophryidae, Hemiphractidae, Eleutherodactylidae, Centrolenidae, Gymnophiona, and Caudata

Volume IV

Luis A. Coloma and William E. Duellman

CRC Press
Taylor & Francis Group
Boca Raton London New York

CRC Press is an imprint of the
Taylor & Francis Group, an **informa** business

Designed front cover image: Cover. Female of the gold-spotted glassfrog *Hyalinobatrachium aureoguttatum* with the green oocytes and other organs visible through its skin at Wikiri Sapoparque in Quito, Ecuador. Photo by Luis A. Coloma.

Back Cover. Male of the gold-spotted glassfrog *Hyalinobatrachium aureoguttatum* taking care of two clutches of eggs at Wikiri Sapoparque in Quito, Ecuador. Photo by Luis A. Coloma.

First edition published 2025
by CRC Press
2385 NW Executive Center Drive, Suite 320, Boca Raton FL 33431

and by CRC Press
4 Park Square, Milton Park, Abingdon, Oxon, OX14 4RN

CRC Press is an imprint of Taylor & Francis Group, LLC

ISBN: 978-1-032-89645-8 (hbk)
ISBN: 978-1-032-89646-5 (pbk)
ISBN: 978-1-003-54387-9 (ebk)

DOI: 10.1201/9781003543879

Typeset in Times LT Std
by Apex CoVantage, LLC

In memory of José Q. Coloma

Contents

About the Authors

Luis A. Coloma
Coloma is an amphibian biologist who was born in Guaranda, Provincia Bolívar, Ecuador, in 1962. He received his *licenciatura* in biological sciences from Pontificia Universidad Católica del Ecuador (PUCE) in 1987. In 1991, he was granted an MA in the Department of Systematics and Ecology by the University of Kansas, where he was mentored by William E. Duellman; his thesis was "Ecuadorian Frogs of the Genus *Colostethus* (Anura: Dendrobatidae)." For his PhD, he was under the guidance of Linda Trueb; in 1997, he completed his dissertation, "Morphology, Systematics, and Phylogenetic Relationships among Frogs of the Genus *Atelopus* (Anura: Bufonidae)." For 19 years, from 1991 to 2010, he was Professor and Senior Lecturer in charge of vertebrates and in charge of the herpetology section at the Escuela de Ciencias Biológicas at PUCE. He mentored 23 *licenciatura* students. Since 2011, he has been Director and Researcher at Centro Jambatu de Investigación y Conservación de Anfibios in Quito, Ecuador. Coloma has published 69 scientific papers, with 56 of these articles focused on amphibians and indexed in Scopus (as of 31 May 2023). Additionally, he has published 32 outreach publications, among the latter two coffee table photo books: *Ecuador Megadiverso* and *Sapos, Ecuador Sapodiverso.*

He has described or co-described 31 new species of frogs, among which, 28 are Ecuadorian frogs of the genera *Atelopus* (7), *Hyloxalus* (6), *Gastrotheca* (4), *Hyloscirtus* (4), *Engystomops* (3), *Pristimantis* (2), *Leucostethus* (1), and *Epipedobates* (1). Five species of frogs, one lizard, and one earthworm have been named in his honor by his colleagues. In 2007, he was awarded the Sabin Award for Conservation of Amphibians, presented by the World Conservation Union and Conservation International. In 2008, he received the Saint Louis Zoo Conservation Award in recognition of his extraordinary lifelong dedication to the conservation of Ecuadorian biodiversity. In 2009, he was appointed member of the Latin American Academy of Sciences. Additional information about Coloma's life appears in Chapter 2 of Volume I and references therein.

William E. Duellman (1930–2022)
Duellman was a herpetologist that was born in Dayton, Ohio, on 6 September 1930. He passed away on 25 February 2022. He earned three degrees from the University of Michigan—a BA (1951) in zoology with a minor in geography, an MS (1952) in zoology with a minor in botany, and a PhD (1956) in zoology with a minor in geology. His doctoral dissertation was on snakes of the genus *Leptodeira.* Since 1959, most of Duellman's academic life was at the University of Kansas, where he was a professor in the Department of Systematics and Ecology and Curator and Curator Emeritus of Herpetology in the Natural History Museum (now Biodiversity Institute). He published 386 titles (among them 12 books). Among these books are *Hylid Frogs of Middle America, An Equatorial Herpetofauna, Biology of Amphibians* (with Linda Trueb); *Cusco Amazónico: Lives of Amphibians and Reptiles in a Tropical Rainforest, Terrestrial Breeding Frogs (Strabomantidae) in Peru* (with Edgar Lehr); and *Marsupial Frogs and Allied Genera.* Duellman served as Major Professor for 12 master's and 34 PhD students, and he mentored seven post-doctoral scholars. During his time at the University of Kansas Natural History Museum, he built up the collection from 59,000 to more than 300,000 specimens through extensive field work by himself and with his students and by acquiring other collections, making it the fourth largest herpetological collection in the United States and by far the most significant collection of the herpetofauna of Latin America. He described or co-described 252 species of amphibians from the Neotropics, among which 93 are from Ecuador. Six species of frogs and five species of reptiles are named in his honor. Additional information about Duellman's life is provided in Chapter 2 of Volume I and references therein.

Contributors

Juan M. Guayasamin and Andrea Terán-Valdez assisted in the writing, review, and editing process of the Centrolenidae accounts. They are also credited alongside their contribution. Additional contributors include Juan C. Santos, who produced time-calibrated maximum likelihood phylogenetic trees illustrating the evolutionary relationships of amphibians in Ecuador, and Daniel J. Paluh, who assisted in the production of microCT images and high-resolution computed tomography reconstructions of the skeletons of Ecuadorian amphibian genera. The acknowledgments section further recognizes other contributions made by these individuals.

Juan M. Guayasamin
iBIOTROP, Colegio de Ciencias Biológicas y Ambientales COCIBA
Universidad San Francisco de Quito USFQ
Quito, Ecuador

Andrea Terán-Valdez
Centro Jambatu de Investigación y Conservación de Anfibios
Fundación Jambatu
Quito, Ecuador

ADDITIONAL CONTRIBUTORS

Juan C. Santos
Department of Biological Sciences, College of Liberal Arts and Sciences
St. John's University
NY, USA

Daniel J. Paluh
Department of Natural History, Florida Museum of Natural History
University of Florida
FL, USA

Foreword

This Encyclopedia of *Amphibians of Ecuador* is not only the outcome of a lifetime of work by William E. Duellman (1930–2022) and Luis A. Coloma but also a tribute to the thriving and dynamic field of amphibian biology in Ecuador over recent decades. Duellman's contributions helped establish the foundation for amphibian research in Ecuador, upon which independent teams of researchers continue to build. Importantly, through his vision and mentorship, Duellman also fostered generations of Ecuadorian herpetologists, many of whom became leading figures in the field, including Coloma (Duellman, 2015).

I first met Luis A. Coloma as an undergraduate student at the Museo de Zoología of Pontificia Universidad Católica del Ecuador in 2003, where he was a professor and Head of the Herpetology Division. Like Duellman, Coloma's work ethic and unwavering dedication to the study of amphibians have motivated additional generations of Ecuadorian herpetologists and produced impactful science and conservation outcomes. Coloma's contagious curiosity and fascination for frogs began early in life, first as a child, and later as a high school student, when he naively designed an experiment testing the effects of temperature changes on frogs (Coloma, 2024)—a line of questioning that has since become an important focus within herpetology in the face of climate change!

Luis' eagerness to explore the biology of amphibians found an ideal complement in his lifetime collaboration with William E. Duellman, which began during his graduate studies at the University of Kansas. Upon his return to Ecuador, Coloma mentored a new generation of biologists, many of whom now make critical contributions to the study and conservation of amphibians. Witnessing the devastating declines of amphibian diversity firsthand, Coloma expanded his focus beyond just the science, spearheading initiatives to protect these species while at the same time empowering Ecuadorian scientists and citizens. Observing this evolution from close-up has been both a privilege and an inspiration.

The Encyclopedia provides an in-depth assessment of the amphibians of Ecuador, accompanied by photographs of exceptional quality. Volume IV features detailed accounts of Ecuadorian amphibians from the anuran families Phyllomedusidae, Leptodactylidae, Ceratophryidae, Hemiphractidae, Eleutherodactylidae, and Centrolenidae and the orders Gymnophiona and Caudata. Ecuador's remarkable amphibian discovery and description rate is among the highest globally, a trend highlighted throughout the Encyclopedia and especially in Appendix 2, which documents 20 new taxa described between 2023 and January 2025. While the discovery of new species and their evolutionary relationships will continue, this contribution will undoubtedly serve as a springboard for the further advancement of amphibian biology.

The families featured in Volume IV showcase the incredible diversity of Ecuadorian amphibians and their unique biological traits. Many species are emblematic of Neotropical amphibians. For instance, the distinctive morphology of most phyllomedusids evokes iconic images of tropical forest frogs. The calls of some *Engystomops* (Leptodactylidae) species in the lowland regions of the Pacific Coast, or the calls of *Gastrotheca* (Hemiphractidae) in the Andean highlands, are common sounds in the natural background. In terms of reproductive strategies, many leptodactylids create foam nests to protect their eggs, while marsupial frogs of the genus *Gastrotheca* have pouches on their backs where eggs are kept until they are released as tadpoles or froglets, depending on the species. The mesmerizing glassfrogs (Centrolenidae) display parental care, with males of many species guarding eggs laid on leaves over streams, from which tadpoles drop into the water to develop. Gymnophiona species also vary in reproductive mode, with some being oviparous or viviparous and certain larvae feeding on maternal tissue. This volume also includes species that have contributed to advancements in other fields of research, such as *Engystomops* and *Gastrotheca*, providing new insights to developmental biology through the pioneering work of Eugenia Del Pino and collaborators.

Reading through the pages of Volume IV evokes a blend of awe for the remarkable biology of these organisms and concern for their looming threats. Many species within each family are classified under one of the IUCN's threatened categories. In certain cases, particularly with Caudata and Gymnophiona, our understanding of their systematics, biology, and ecology is so limited that we are unable to definitively assess their conservation status or the threats they face. Some species, like *Ceratophrys stolzmanni* and *Gastrotheca cornuta*, are now part of *ex situ* conservation programs at the Centro Jambatu of Amphibian Research and Conservation aimed at boosting population sizes, maintaining backup populations, and eventually reintroducing them to their natural habitats. Others, such as *Agalychnis spurrelli* and *Cryziohyla craspedopus*, are not endangered but are involved in programs designed to combat the illegal pet trade by offering legal and sustainable alternatives. Many species have not been seen in decades; hopefully, with the dedication of the numerous herpetologists now working in our country, these species will be rediscovered.

What does the future hold? The work presented here reflects decades of effort and passion, which merely mark the beginning of a monumental challenge ahead. We are living in a thrilling era of research, where our understanding of biodiversity is expanding and deepening, and new techniques are offering unprecedented tools to explore questions we never thought possible. Yet these are also critical times, as our biodiversity urgently needs these advancements to combat the accelerating decline and extinction rates of natural populations. As we continue to explore *what* species we discover, it is imperative that we also enquire *why* and *how* these species exist, coexist, survive, or vanish—all while developing multidisciplinary strategies to protect them. We must ask ourselves how we can harness this knowledge in a rapidly changing world that desperately needs our efforts. I am confident that the work presented here will inspire many and that, through our endeavors, we will uncover answers along the way.

Mónica I. Páez-Vacas

Preface

In the Preface of Volume I, there is a general recounting and reflections on the purpose and history of the complete opus (Volumes I–IV). Here, the focus is on a few events and taxa to highlight broader issues mostly relevant to the taxonomy and systematics of Ecuadorian taxa, for which we have compiled accounts for each species, genus, family, and order. This Volume IV includes the accounts of 132 species of Ecuadorian anurans of the families Phyllomedusidae, Leptodactylidae, Ceratophryidae, Hemiphractidae, Eleutherodactylidae, and Centrolenidae; of 25 species of Gymnophiona; and of seven species of Caudata. It also includes an addendum with 20 new taxa described between January 2023 and January 2025.

In 1758, Carl von Linnaeus formalized the modern system of binomial nomenclature to name species, while also coining the binomial name (scientific name) for a species now known from Ecuador. The species is *Pipa pipa,* and his taxonomic action was solely based on illustrations of this frog originally collected in Suriname. Nearly a century passed before the first description of a frog named from Ecuador was made: *Atelopus ignescens* (the Jambato), described by the Italian scientist Emilio Cornalia in 1849 in a 15-line morphological account, which was an extensive description for those times. Two centuries later, in 1958, about 200 amphibian species were known from Ecuador. By 1980, this number had increased to 332 species, half of the species currently known in Ecuador, and they were described based on morphological evidence, with a few accounts also including osteological features and phylogenetic analyses, especially the taxonomic works by the research group at the University of Kansas, USA, led by William E. Duellman and Linda Trueb. By the end of 2022, when we ceased adding species to this opus, 655 species were known from Ecuador, and in the recent four decades, this was mostly due to efforts by Ecuadorian taxonomists and systematists. From 2023 up to January 2025, 27 additional, mostly cryptic species have been described, and we included 20 of them in an addendum in Volume IV. One of the most recent descriptions is that of a cryptic species by the Ecuadorian Pablo Menéndez-Guerrero and four collaborators (2024), *Rhinella bella* (beautiful toad), in a detailed 27-page paper that includes data and analyses based on extensive specimen collections, a molecular phylogenetic hypothesis, and the combined evidence of mtDNA sequence variation, morphology, bioacoustics, and environmental information.

The particular cases of *Pipa pipa*, *Atelopus ignescens*, and *Rhinella bella* illustrate the trend of taxonomic knowledge about amphibians in Ecuador, as well as the current accelerating pace, dynamism, and increasing complexity involved in describing this diversity, which is expected to increase substantially in the coming years (but see Löbl *et al.,* 2023), especially in a country like Ecuador, where apparently no fewer than 230 species are awaiting discovery and description. However, this commendable progress is overshadowed by the poor or nonexistent knowledge in aspects of amphibian biology fields beyond morphology and DNA sequences (e.g., natural history data, calls, tadpoles, embryos, distributions, population dynamics, ethology) for most of the species, and even worse, by the possible extinction of nearly 50 of the known species, including an entire family of aquatic frogs, Telmatobiidae; a dozen *Atelopus* harlequin frogs; several monophasic marsupial frogs of the genus *Gastrotheca*; some craugastorids and *Hyloxalus* rocket frogs; and probably many other amphibians that were never documented.

The case of *Pipa pipa* helps to highlight the immense knowledge gaps that still need to be filled. For example, it is nearly unbelievable that a species such as *Pipa pipa* is still poorly known in Ecuador. In an era of genomics, paradoxically, its molecular phylogeny, its morphological and geographic variation, and its taxonomy are pending issues. This case highlights the missing pieces of basic systematic and taxonomic information and the still-inadequate samplings in museum collections. This is a generalized pattern also known for many other taxa (see, for example, Remarks for most species of the Microhylidae in Volume II or Gymnophiona in Volume IV).

Ironically, the case of the other mentioned species, *Atelopus ignescens*, whose taxonomy and systematics still have pending issues, serves to illustrate the catastrophe of amphibian extinctions and the desperate efforts to save a single population of an otherwise historically extremely abundant species having many populations. This has pushed the nascent Ecuadorian academic and conservationist community along with international collaborators to begin studying it in the contexts of population numbers, natural history, genomics, and ethnozoology (Vega-Yánez *et al.,* 2024). However, this species is somehow privileged given its cultural importance, while many hundreds of others suffering the same path of extinction are still awaiting attention.

The most recently described species, such as *Rhinella bella*, which is an extremely common species, are based on integrative taxonomic practices, and this undoubtedly represents a step forward in gaining data and knowledge but also highlights the state of scientific developments, which are not free of controversy. This is especially true given contemporary discussions of still-contentious issues and fierce debates such as what defines a species, species boundaries, taxonomic anarchy, taxonomic inflation in the absence of appropriate geographic sampling, the discovery of cryptic diversity, genetic introgression, interspecific hybridization, and basic taxonomic practice-related topics (Garnett and Christidis, 2017; Cook *et al.,* 2023; Arias-Cárdenas *et al.,* 2024; Dubois *et al.,* 2024).

To advance knowledge in various biological fields and reverse the extinction trends of over 400 species, it is crucial to complete the description of Ecuadorian amphibians. This requires accelerating taxonomic research in Ecuador, significantly increasing governmental and international funding for museum collections and biobanks, digitizing collections, providing open access to information, and establishing legal frameworks that support the scientific exchange of taxonomic specimens and tissues to benefit biodiversity research.

REFERENCES

Arias-Cárdenas, A., L. S. Barrientos, C. Pardo-Diaz, A. Paz, A. J. Crawford, and C. Salazar. 2024. Taxonomic inflation and a reconsideration of speciation in the Andes: the case of the high-elevation tree frog *Dendropsophus molitor* (Anura: Hylidae). *Zoological Journal of the Linnean Society* 200: 763–775.

Coloma, L. A. 2024. Crying in the rain in the middle of the world. In *Lost Frogs and Hot Snakes: Herpetologists' Tales from the Field*, Vol. 6, ed. M. L. Crump, 134–139. Ithaca, NY: Cornell University Press.

Cook, C. N., K. H. Redford, and M. W. Schwartz. 2023. Species conservation in the era of genomic science, *BioScience* 73: 885–890.

Dubois, A., W. Denzer, O. M. Entiasupe-Neto, T. Frétey, A. Ohler, A. M. Bauer, and R. A. Pyron. 2024. Nomenclatural problems raised by the recent description of a new anaconda species (Squamata, Serpentes, Boidae), with a nomenclatural review of the genus *Eunectes*. *Bionomina* 37: 8–58.

Duellman, W. E. 2015. *Herpetology at Kansas: A Centennial History*. Ithaca: Society for the Study of Amphibians and Reptiles.

Garnett, S., and L. Christidis. 2017. Taxonomy anarchy hampers conservation. *Nature* 546: 25–27.

Löbl, I., B. Klausnitzer, M. Hartmann, and F. Krell. 2023. The silent extinction of species and taxonomists—an appeal to science policymakers and legislators. *Diversity* 15(10): 1053.

Menéndez-Guerrero, P. A., S. P. L. dos Santos, M.-J. Salazar-Nicholls, D. M. Green, and S. R. Ron. 2024. Cryptic diversity in toads of the *Rhinella marina* species group (Anura, Bufonidae) with a subjectively beautiful new species from Western Ecuador. *Zoological Journal of the Linnean Society* 197: 1–26.

Vega-Yánez, M. A, A. B. Quezada-Riera, B. Ríos-Touma, M. D. C. Vizcaíno-Barba, W. Millingalli, O. Ganzino, L. A. Coloma, E. E. Tapia, N. Dupérré, M. Páez-Vacas, D. Parra-Puente, D. Franco-Mena, G. Gavilanes, D. Salazar-Valenzuela, C. A, Valle, and J. M. Guayasamin. 2024. Path for recovery: An ecological overview of the jambato harlequin toad (Bufonidae: *Atelopus ignescens*) in its last known locality, Angamarca Valley, Ecuador. *PeerJ* 12: e17344. https://doi.org/10.7717/peerj.17344

Acknowledgments

While William E. Duellman is no longer with us, his indelible mark on the production of this opus and the near-completion of all species' accounts remains immeasurable. His unique perspective, wealth of experience, knowledge, and unwavering enthusiasm for this project, coupled with meticulous attention to detail, continue to serve as a source of inspiration. Hopefully, his passion for science and amphibians will forever be etched in these pages as a tribute to his titanic contribution to the field of amphibian biology, as well as to the enduring nature of his friendship and partnership that transcends time and space. Similarly, I must express my deepest gratitude to Linda Trueb (his partner), my teacher and mentor at KU, who has been by Bill's side in charting the progress of *Amphibians of Ecuador* since 2015, steadfastly supporting him through challenging times as he embarked on what would ultimately become his final Ecuadorian endeavor. Linda's unwavering commitment also extended to securing part of the funding and seeing this Encyclopedia through to its completion. In January 2023, her wise and practical advice was much appreciated. She said,

> I really would urge you to cut off adding additional species at a logical time—e.g., no frogs described in 2023! Otherwise, you'll never see the end—kind of like Sisyphus pushing that stone ball up the side of the mountain.

As detailed in Chapter 2 (under the subtitle "From Anfibios de Ecuador to Amphibiawebecuador and to Anfibios del Ecuador-Bioweb at QCAZ") and briefly mentioned in the preface, this opus is the descendant of earlier endeavors involving a vast collective of individuals and friends, encompassing students, collectors, field companions, scientists, administrators, and more. I extend my deepest gratitude for their invaluable contributions. Also, among them, I express my immense appreciation for their input while at Pontificia Universidad Católica del Ecuador goes to Giovanni Onore, Néstor Acosta-Buenaño, Ítalo G. Tapia, Elisa Bonaccorso, and Juan M. Guayasamin. Onore has always been a source of inspiration and drive, even in the most difficult times. Tapia (the Collection Manager at the QCAZ) aided, with great energy and empathy, in the field, in the museum, and with the photography of frogs and their development. Many of the photos in the field and in the lab would have not been possible without his collecting efforts and his ability to raise them. Bonaccorso and Guayasamin collaborated by leading the JRS and SENECYT projects, which, in 2008–2010, provided the funding necessary to enhance the electronic version of AmphibiaWebEcuador. Since 2011, all of these individuals have continued to offer their support in numerous ways during my tenure at Centro Jambatu of Amphibian Research and Conservation (CJ), and I am immensely grateful. In particular, Onore welcomed CJ under the auspices of the *Fundación Otonga* until 2019, when CJ became an independent NGO. Acosta-Buenaño has been instrumental in building the CJ electronic encyclopedia (SapopediaEcuador). He not only developed the conceptual and physical models of the database but also played a key role in its programming, design, and layout. His initiative, serenity, and drive are inspiring. During my time at the QCAZ and the establishment of CJ, I received generous support and encouragement from Mark Wanner (Saint Louis Zoo), Joseph Mendelson III, Ron Gagliardo, and Kevin Zippel of Amphibian Ark, Lauren A. O'Connell (Stanford University), and Kim L. Hoke (Colorado State University). Their actions spoke louder than words.

For the realization of *Amphibians of Ecuador*, first and foremost, I extend my gratitude to our contributing authors and coauthors who devoted relevant efforts to write some sections. They are (in alphabetical order): Ana Almendáriz-Cabezas, Ernesto Arbeláez Ortiz, Sebastián Bermúdez-Puga, David Brito-Zapata, Mateo Dávila-Játiva, María D. Guarderas, Juan M. Guayasamin, Andrés León-Reyes, Carlos E. Montalvo-Puente, Mónica I. Páez-Vacas, Emilia Peñaherrera-Romero, Carolina Proaño-Bolaños, Carolina Reyes-Puig, David Romo, Camilo Rosero-Gómez, Juan Carlos Sánchez-Nivicela, Andrea Terán-Valdez, Gregory O. Vigle, María Vizcaíno-Barba, and Mario H. Yánez-Muñoz. Many of these authors (and or coauthors) engaged directly with myself, fostering enlightening and fruitful discussions, especially Andrea Terán-Valdez, Mónica Páez-Vacas, Juan M. Guayasamin, María D. Guarderas, Juan C. Santos, Ana Almendáriz-Cabezas, Gregory O. Vigle, Carolina Reyes-Puig, and Carlos E. Montalvo-Puente. My gratitude goes also to Daniel J. Paluh, who enthusiastically (and supported by David C. Blackburn from the Florida Museum of Natural History) embarked on the production of microCT images of each of the genera of Ecuadorian amphibians with amazing results; Juan C. Santos, who, in 2021, performed a phylogenetic analysis and obtained a phylogeny and chronogram for the Ecuadorian amphibians; and to Pedro A. Coloma, Morley Read, and Francisco Velásquez Espín, who enthusiastically analyzed the calls of 13 species: *Pristimantis appendiculatus*, *P. carlosceroni*, *P. ecuadorensis*, *P. eugeniae*, *P. glandulosus*, *P. parvillus*, *P. pycnodermis*, *P. sirnigeli* (PAC and IGT), *Atelopus balios*, *A. coynei*, *A. exiguus*, *A. ignescens* (MR), and *Ranitomeya ventrimaculata* (FVE, MR), most of which were recorded by Ítalo G. Tapia. Carlos E. Montalvo-Puente introduced me to the amazing world of archeology after a visit to Museo El Alabado.

Some of the authors of subtitles in Volumes I and IV also helped with other aspects of the complete opus in various ways. For example, Terán-Valdez, the CJ Collections Manager, edited and revised all of the accounts; she also helped to write descriptions of the tadpoles of 29 species and of many ontogenetic series. Her input to the Encyclopedia in countless ways has been very valuable. She was always there (including non-working hours and vacations), ready to help with anything. Camilo Rosero-Gómez helped with refining maps and associated data with amazing efficiency. Guarderas helped with many administrative tasks and the acquisition of funds related to the Encyclopedia; Juan C. Santos reviewed the section about David C. Cannatella; Almendáriz provided some data, photos, and critically reviewed the sections titled "Peters and the Harlequin Frogs, a Rebirth for Amphibian Collections" in Chapter 2 of Volume I; Vigle also edited and reviewed the MECN section under the subtitle "The DHMECN Collection: From the Friends of Nature to the National Biodiversity Institute" in Chapter 2 of Volume I; Mónica Páez-Vacas edited part of the section that includes the information about calls of Ecuadorian frogs and also helped with the organization of the Red List table. Guayasamin aided in obtaining information of authors from Scopus.

Additional reviews of sections of Volumes I–IV were conducted by John E. Simmons and Joseph Mendelson III (Chapters 1 and 2), Giovanni Onore ("Amphibians at the QCAZ"), Andrew Gray (Phyllomedusidae), Santiago Castroviejo-Fisher and his PhD student Lourdes Echevarría (*Hemiphractus*), Ross Maynard (accounts of species from Reserva Manduriacu), and Martha L. Crump (part of "In Memoriam"). John E. Simmons drew my (LAC) attention to Jorge Icaza's mention of frogs in the Ecuadorian literature. Pedro A. Coloma helped with the organization of the Red List information. Sofía Carvajal-Endara and Gustavo Pazmiño-Otamendi helped with refining the geographic information and the production of many of the distribution maps. Néstor Acosta-Buenaño aided in part of the compilation, organization, and editing of the information about vocalizations. I also appreciate the editorial support from Crumly Chuck, Kara Roberts, Balaji Karuppanan, Shayla McKeown, and the rest of the team at CRC Press and Taylor & Francis Group.

I couldn't have compiled the biographies of many of the amphibian biologists included in Chapter 2 of Volume I without their positive response to my request for their CVs, photos, and biographical information. I thank them all (in alphabetical order): Ana Almendáriz-Cabezas, Alejandro Arteaga, Diego Batallas-Revelo, Jorge Brito Molina, Marcel Adrián Caminer Rodríguez, Sofía Carvajal-Endara, Glenn Flores, Diego F. Cisneros-Heredia, Jaime Culebras, Juan M. Guayasamin, María José Navarrete, Jhael Ortega, Lauren A. O'Connell, Nadia B. Páez, Mónica I. Páez-Vacas, Carolina Proaño-Bolaños, Juan C. Santos, Morley Read, Carolina Reyes-Puig, Juan P. Reyes-Puig, Juan Carlos Sánchez-Nivicela, Diana Székely, Paul Székely, Elicio E. Tapia, Ítalo G. Tapia, Andrea Terán-Valdez, Verónica L. Urgilés, Jorge H. Valencia, and Mario H. Yánez-Muñoz.

Numerous other people have contributed in various ways, and their efforts are greatly appreciated. Joseph Mendelson III wrote an insightful foreword for Volume I with great enthusiasm and generosity. He was the perfect person for such a task, given his deep knowledge and his lifelong commitment to amphibian systematics and conservation in the Neotropics. Moreover, his close bond with Bill Duellman, serving as both mentor and friend, and his consistent support of Coloma in various professional pursuits made him an ideal contributor. Many thanks also go to Juan C. Santos, Juan M. Guayasamin, and Mónica Páez-Vacas for writing the forewords for Volumes II, III, and IV, respectively. Our long-standing friendship, which began when they were my students at PUCE, has enriched our lives and grown into a productive and successful collaboration on amphibian research. Giovanni Onore has called my attention to ethnozoological aspects of amphibians. Tonnie Woeltjes stood out as a dedicated resource, diligently searching for and providing copies of hard-to-find publications, particularly those pertaining to amphibians in the context of hobbyists. Juan C. Santos and John E. Simmons also made significant contributions by actively seeking and sharing copies of numerous publications. Jaime Culebras kindly pointed out some of my misidentifications, particularly within the genus *Pristimantis* (as he also provided photos), and generously shared his expertise derived from extensive field experience. He has gathered rare frogs and specimens, contributing to our *ex situ* colonies and museum collections. Furthermore, he has been a constant source of lively discussions and passion for frogs. Thanks go to Connie Dickinson, who generously organized a visit to a Tsáchila community, where Alfonso Aguavil, Primitivo Aguavil, José Jacinto Aguavil, and Milton Aguavil shared with us their hospitality and knowledge of common names of frogs in Tsáfiqui. Leonardo Negrete assisted in compiling the information related to the calls of centrolenid frogs. Néstor Acosta-Buenaño and Sofía Ballaz contributed to editing the information of calls of Ecuadorian frogs included in the Appendix. Alexander Hirtz's openhandedly provided access for photographing his collection of archeological pieces of amphibians. Erika Ullauri P. granted entry to photograph archeological artifacts of the Museo Nacional del Ecuador. Amanda B. Quezada helped to edit thousands of bibliographic references and numerous other editorial tasks with great efficiency.

Our deepest thanks go to the people who collected specimens in our respective institutions (KU, QCAZ, CJ). All are cited in the databases of the museum collections, and the ones that made the most significant contributions are also mentioned in the history section. At Centro Jambatu, we especially thank Elicio E. Tapia; Mónica I. Páez-Vacas; Morley Read; Ítalo G. Tapia; Arturo Guasti; Patricio Vargas Mena; Mark Wanner; Jaime Culebras; Jan, José, and Thomas Verkade; Juan C. Santos; Amadeus Plewnia; Manuel Morales-Mite; Sofía Carvajal-Endara; Amanda B. Quezada; Carolina Proaño-Bolaños; H. Mauricio Ortega-Andrade; and about 150 other collectors who brought specimens and associated data to CJ. Some of them also brought live amphibians to be photographed in the lab. My particular gratitude goes to the Verkade family (Jan, José, and Thomas) for their searching for specific species of frogs (under my request), and their tadpoles, recordings of calls, and photography skills (all material deposited at CJ). Additionally, Thomas found a rare caecilian from southwestern Ecuador. Thanks go to Fernando Vaca and Nanto Huamoni, who enthusiastically supported our CJ scientific work and collections at the Shiripuno Lodge.

Most of the photographs of tadpoles, ontogenetic series, and reproduction data would have not been possible without the dedication to raising them in the lab while at the QCAZ (up to 2010) and at CJ (2011–2023) of the following amphibian keepers, vivaria managers, veterinarians, and several assistants. They are Ítalo G. Tapia (2001–2010 at the QCAZ, 2023 to date at CJ), Diego Almeida-Reinoso (2004–2010 at the QCAZ, 2011–2014 at CJ), Freddy Almeida-Reinoso (Insectarium, QCAZ 2007–2010, CJ 2011–2014), Gabriela Aguiar (2011–2016), Patricio Vargas Mena (2013 to date), Marco de Rossi (CJ vivaria, 2014 to date), Elena Naydenova (veterinarian, 2015 to date), Diana Carolina Salazar (2016–2019), Gabriela Manzano (2016–2017), Carlos Valle Piñuela (2016–2018), Renata Gavilánez (2018–2019), Leonardo Negrete (2018–2019), Berenice Benavides (2021–2022), Estefanía Erazo (2017–2020), Xavier Cisneros (2019–2021), Johanna Paz (2019–2021), Daniel Romero (2019 to date), Fernanda Gordon Domínguez (2021–2023), Alex Achig (2021 to date), Belén Morales (2022–2023), and Mateo Gómezdelatorre (2022–2023). Among them, Ítalo G. Tapia and Patricio Vargas Mena joined me on innumerable collecting field trips, and I value them as great field companions. Carlos Valle Piñuela generously taught me the use of QGIS for the production of maps. Esther M. Langan, Collection Manager at the National Museum of Natural History of the Smithsonian Institution, graciously provided information in response to several inquiries.

When compiling data included in this Encyclopedia during my time at the QCAZ and CJ, I relied heavily on the invaluable assistance provided by the collection managers, curatorial assistants, volunteers, and their dedicated efforts in managing preserved specimens and the living collection. I am grateful for the support and contributions at the QCAZ of Ítalo G. Tapia (2002–2010) and Diego Almeida-Reinoso (2004–2010) and at CJ of Sofía Carvajal-Endara (2011–2013), Diego Almeida-Reinoso (2011–2014), Elicio E. Tapia (2011–2016), Manuel Morales-Mite (2015–2018), Morley Read (2017–2018), Sebastián Hernández-Nieto (2018), Andrea Terán-Valdez (2018–to date), and Amanda B. Quezada (2021).

Several aspects of the production of *Amphibians of Ecuador* received partial funding from various grants. One of these grants was the PARG project, which provided support during my temporal association with Universidad Regional Amazónica Ikiam from 2015 to 2017 and subsequently at CJ until 2019. I extend my gratitude to Andrea Coloma Santos and Adriana Paulina Guarderas Valverde for their assistance in crafting the PARG project proposal. In 2019–2020, another grant that contributed to this endeavor was an award to Lauren A. O'Connell (No./FAIN No. 1845651) from the US National Science Foundation. This grant played a vital role in covering an important portion of salaries and printing costs. Lauren's aid and research initiatives are greatly acknowledged from the foundation of CJ from 2011 to date. During 2012–2018, research grants to Kim L. Hoke and her student Molly C. Womack from the US National Science Foundation (NSF IOS–1350346, OISE–1157779, and PRFB–1611752) funded activities specifically related to the documentation of ontogenies of Bufonidae; thus, this project became an important aid to CJ and these books.

Additional financial and miscellaneous support for CJ and this work has been provided by Wikiri, an Ecuadorian biotrade enterprise, and the Saint Louis Zoo. These institutions have consistently offered crucial support to CJ and this project, and I could not have accomplished this without their assistance. I wish to express my profound gratitude to María D. Guarderas, the Manager of Wikiri and CJ Coordinator, whose inspirational leadership and unwavering dedication have been instrumental in sustaining CJ. She has initiated numerous projects to advance our scientific work and adeptly managed the complex bureaucratic and administrative tasks at CJ. She was joined in recent years by Andrea Terán-Valdez, who plays an instrumental function in scientific coordination. Their support while I have been immersed in this opus, often setting aside many other responsibilities at my institution, did not waver.

I would also like to extend my heartfelt thanks to the Saint Louis Zoo, represented by Jeff Bonner, Eric Miller, Lissa Kelley, Mark Wanner, Lauren Augustine, and more recently by Justin Eiden. Their financial assistance has played a pivotal role in our work. In addition, I want to convey my deepest appreciation to Mark Wanner at the Saint Louis Zoo, with whom I have shared numerous field trips over the years in our pursuit of frogs. Our friendship, which began in the bed of a pickup truck while traveling for a nighttime frog search during a frog meeting in Panama in 2006, has endured and enriched both our personal and professional lives.

Finally, I dedicate this work with my profound and infinite gratitude to my extended family, and especially to Alexandra Quiguango Ubillús (my wife); to our sons, Pedro Antonio and José Miguel; and to my mother, Lida María Román. Their devoted emotional support surpasses any words I can write.

PHOTO AND ILLUSTRATION CREDITS

All photos and illustrations are appropriately credited in their respective figure captions. We have employed abbreviations for the primary authors of the majority of the photos, namely Luis A. Coloma (LAC), William E. Duellman (WED), Diego Acosta López (DAL) Steven and Guevara Salvador (SGS). The latter two worked at CJ focusing on photographic documentation and photo designing in 2016–2020 and from 2018 onwards, respectively. Steven generously and patiently assisted with the detective work of finding the data associated with thousands of photos that had not been completely labeled. Our thanks go to Rafe M. Brown and Ana Motta for facilitating access to the KU photographic archive and diligently providing scans. We extend our special thanks to Jaime Culebras, Amanda B. Quezada, Morley Read, Paul Székely, and Mario H. Yánez-Muñoz, who graciously shared numerous photos with us, many of which were crucial to expand the photographic documentation of Ecuadorian amphibians. Additionally, we are grateful to Jorge Brito Molina, Juan Carlos Sánchez-Nivicela, Gregory O. Vigle (for his frog photos and photos of Kenneth Miyata), Carolina Reyes-Puig, Elicio E. Tapia, Fernando Coloma Román, Jan C. Verkade, Diego Batallas-Revelo, Juan P. Reyes-Puig, Ross Maynard (The Biodiversity Group), Amadeus Plewnia, Andreas Nöllert, John E. Simmons, Alejandro Arteaga, Fernando Rojas-Runjaic, Joel Sartore, Cristian Paul Nieves, Diego Almeida-Reinoso, Erwin Patzelt, José Vieira-Fernandes, Mateo Vega-Yánez, Daniela Franco-Mena, Patricia A. Burrowes, Marinus S. Hoogmoed, J. S. Bermingham, Alex Achig, Frank Pichardo, Ted R. Kahn, Jazmín Buitrón, Jenna Welch (Division of Amphibians & Reptiles, Smithsonian Institution), Matthijs Hollanders, Mark-Oliver Rödel, Sarah Bock and Francesca Angiolani for their generous contibutions of photo sets and/or unique, important images.

Organization and Content of Accounts

Following are the accounts of each of the 132 species of Ecuadorian anurans of the families Phyllomedusidae, Leptodactylidae, Ceratophryidae, Hemiphractidae, Eleutherodactylidae, and Centrolenidae; of 25 species of Gymnophiona; and of seven species of Caudata we recognize up to the year 2022. The Addendum in Appendix 2 includes accounts of 20 new species published between January 2023 and January 2025 (with submissions accepted until November 6, 2024).

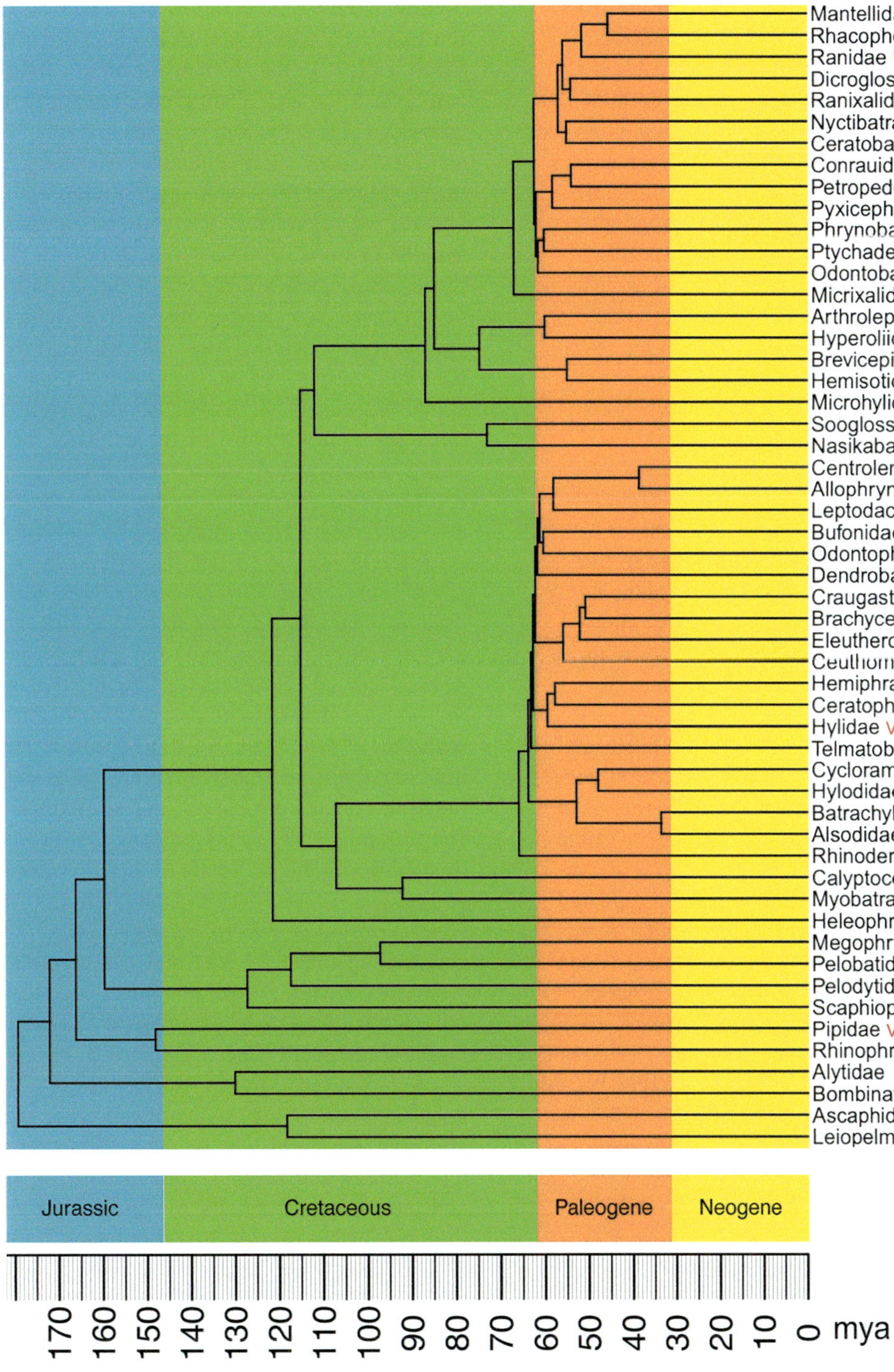

FIGURE 0.1 Topology of the families of Anura, extracted from a time-calibrated phylogenetic tree based on hundreds of loci. The families occurring in Ecuador are in red. The volume number in which the Ecuadorian families are included is indicated. Reprinted and modified from Portik *et al.* (2023) with permission from Elsevier Phyllomedusidae, which is not included in this figure, is covered in Volume IV.

Families of anurans are arranged phylogenetically from the oldest to the most recent lineages (see Figure 0.1). Accounts of genera and species of Anura are in alphabetical order within families. Within Caudata and Gymnophiona, accounts of families, genera, and species are arranged in alphabetical order.

Family and generic accounts include a section of contents. Each species account has the following structure:

(1) The currently recognized taxon name followed by the name of the author(s) and year of publication. The discoverers of a species are the authors who describe it in accordance with the rules of the International Code of Zoological Nomenclature. The parentheses indicates that the original name has suffered changes through its taxonomic history. The scientific nomenclature is based on Frost (2023), with a few exceptions noted under Remarks of the corresponding account. Sometimes, the publication date that appears in a publication is different from the valid date when the work actually was published. In such cases, both dates are indicated in the references.
(2) The common name. The English standard common name, chosen by us, is provided. A compilation of these names is also given in Table 0.1. The Spanish standard common name, chosen by us, is indicated in Table 0.1. The Spanish name is not necessarily a translation of the English standard common name (see "Standardized Names in Spanish and English" in Chapter 1 of Volume I).

The original name, author(s), year of publication, and page number. This is followed by type(s) information with museum number(s) and locality data. Type locality data were ordered from specific locality to country. In some cases, the spelling or description of the locality was corrected. Additionally, locality information was updated when necessary, such as current provincial allocations. Coordinates were excluded from this section, except when the type locality was not in Ecuador. For the original citations of locality data see Frost (2023 and onward).

(3) Synonym. If nomenclatural changes occurred to the original name, only the most recent modification to that name is included. The new name is followed by the author(s), year of publication, and page number. The synonymy information was primarily sourced from Frost (2023). For comprehensive and detailed lists of synonyms, please refer to his website hosted by the American Museum of Natural History. We opt not to incorporate subgenus names such as *Duellmania* and *Huicundomantis.*
(4) Distribution color bar. It represents the altitudinal regions of Ecuador. The bold line below the bar indicates the region of occurrence. From left to right: western tropical (0–1000 m), western subtropical (1001–2000 m), western temperate (2001–3000 m), high Andean (3001–4800 m), eastern temperate (2001–3000 m), eastern subtropical (1001–2000 m), eastern tropical (0–1000 m).
(5) Etymology. When available, the etymological description in the original description is included. Otherwise, the intended meaning is interpreted, usually based on the Greek or Latin derivation. In few cases, the vernacular name is also discussed.
(6) Identification. In the accounts of orders, families, and genera we use Definition (instead of Identification). This section (diagnosis *sensu* Wiley and Lieberman, 2011) is intended to convey more than synapomorphies or autapomorphies and may include any characters useful in demonstrating that the taxon is different from other taxa. Thus, readers can identify members of the taxon in the most concise manner possible without having to consult more detailed descriptions. It provides the characteristics of the taxa, which alone or in combination help to distinguish the taxon from other taxa. Descriptions of the color are in life, unless otherwise indicated. When available, phenotypic synapomorphies for families and genera are described. This identification information is generally derived from the literature, which is indicated under Key references. The features used to characterize, diagnose, and distinguish vary according to the taxa, and we have nearly transcribed the characters used by the authors in the original descriptions or posterior taxonomic revisions. Nonetheless, in a few cases, and if necessary, we did some rewriting, made corrections, or added information. For several species, we offer information on known masses primarily derived from CJ data. Many of the characteristics of families are taken from Duellman and Trueb (1994), Frost *et al.* (2006), and especially from the compilation made by Vitt and Caldwell (2014), AmphibiaWeb (2022), and references cited therein, although in some cases, we updated the information. For nearly all genera, a microCT scan of the skeleton of a representative species is provided in dorsal and ventral views. Identifications of species are usually accompanied by plates depicting photos of males and females, intraspecific variation, and morphological details. We use the following morphological abbreviations throughout the text: SVL = snout–vent length, TL = total length for tadpoles, LTRF = labial tooth row formula.
(7) Comparisons. Similar species are listed at the beginning of the paragraph, followed by comparisons to each of them. The comparisons were primarily conducted with Ecuadorian taxa; however, in some instances, comparisons were extended to include species from neighboring countries.

(8) Tadpoles. Tadpole descriptions were transcribed from the literature, and their source is cited under Key references and also provided under Table 0.2. We present plates and, in some cases, descriptions for 44 species that were previously undescribed, along with images depicting ontogenetic changes for 135 species. The ontogenetic variation was photographically documented under laboratory conditions either at the QCAZ (from 1991 up to 2010) by LAC or at CJ (from 2011 to 2024), mostly by Coloma, Diego Acosta López, and Steven Guevara Salvador. In the figure legends of plates depicting tadpoles and ontogenetic changes, we include, when available, information on stage, size, and time of development. Color descriptions are based on the tadpoles in life, unless otherwise indicated.

(9) Natural history. A synthesis of the most relevant aspects of these topics is provided, and the original sources or those that summarize them are mentioned under Key references. When relevant, collectors natural history data from specimens deposited at CJ were added. The information of CJ number and collector information can be found in the CJ species and specimens database. CJ laboratory notes were also included for species-maintained *ex situ*. Vocalizations are briefly described in this section, but a more complete account of vocalizations for each species is provided under the subtitle "Vocalizations of Ecuadorian Anurans" in Appendix 1. Sources of call descriptions are also compiled in Table 0.2.

(10) Distribution. The distribution range in Ecuador and in other countries is outlined. Then, the species is located in the Ecuadorian provinces, and the altitude ranges in which it has been recorded are indicated, along with the area of extent of occurrence.

Sometimes, the altitudes measured in the field do not exactly match those obtained from the coordinates in the geographic layers we used (WorldClim. Ver 2.1., Fick and Hijmans, 2017); thus, we generally report the altitudes provided in the original publication. In a separate paragraph, we specify the ecosystems and biogeographic sectors (Ministerio del Ambiente del Ecuador, 2013) where the species occurs, along with the average annual rainfall and temperature in their distributional ranges. We provide a compilation of the distribution of each of the species in the ecosystems (Tables 0.3 and 0.4) and biogeographic sectors (Table 0.5). Dot maps for each of the species are presented. We incorporated new distributional records, especially from the CJ collections, corrected errors, and made refinements to the level of precision of distribution records previously provided by Ortega-Andrade *et al.* (2021) (see more details under "The Red List of Ecuadorian Amphibians" in Chapter 3 of Volume I). The locality data used to build the maps are available at CJ.

(11) Conservation. The IUCN category, source of species categorization, and its justification are specified, accompanied by supporting details regarding the conservation status (refer also to "The Red List of Ecuadorian Amphibians" in Chapter 3 of Volume I). Additionally, a comprehensive compilation of categorizations for all species can be found in Table 0.2. Information is provided regarding the occurrence in state (public) protected reserves, protected forests, biosphere reserves, and some private reserves. It is explicitly mentioned whether the species is listed in CITES.

(12) Content (order, families, genera accounts). The number of families, genera, and species worldwide is provided based on Frost (2023) up to December 2023, and the number of species in Ecuador follows this Encyclopedia.

(13) Remarks. Any pertinent information and comments not included in other sections are provided herein. Additionally, if available, phylogenetic information and sources are provided in accounts for all taxa. Figures of phylogenetic trees including taxa not occurring in Ecuador (taken from the literature) are also given for many genera and some groups of species. Phylogenies published after December 2022, as well as recent molecular analyses and the phylogeny by Portik *et al.* (2023), are not included or discussed in the Encyclopedia, given that they were published after we ceased updating information, except where explicitly indicated (e.g., Figure 0.1). For example, we incorporated Strabomantidae within Craugastoridae, as suggested by Portik *et al.* (2023); we mentioned the papers by Franco-Mena *et al.* (2023) and Székely *et al.* (2023b) only when strictly necessary. The family accounts include a time-calibrated maximum likelihood phylogenetic tree depicting the evolutionary relationships of Ecuadorian species inferred from DNA sequences of mitochondrial and nuclear genes. The updated phylogeny of the Ecuadorian amphibian taxa up to the end 2021 was derived initially from alignments of the two largest phylogenetic reconstructions of amphibians to date (Pyron, 2014; Jetz and Pyron, 2018). The master alignments of both studies were appended, and taxa not found in Ecuador were removed; duplicate taxa were also removed by choosing the one with more sequence data, and taxonomic nomenclature was updated following AmphibiaWeb (2022). The final sequence alignment was realigned using DECIPHER (Wright, 2016) and manually reviewed and adjusted, particularly the mitochondrial ribosomal gene sequences. Multiple alignment programs were used to provide a good starting point, but adjustments were made by eye (Felsenstein, 2004; Baum and Smith, 2013). The optimized alignment was then used for the

phylogenetic estimation of sections of the global phylogeny; a maximum likelihood tree was estimated using IQ-TREE2 2.1.3 (Minh *et al.,* 2020), and node time constraints followed those of previous studies (Pyron, 2014; Jetz and Pyron, 2018). In some sections of the tree of Ecuadorian amphibians, the topology among the families of Hyloidea was constrained to match those of more comprehensive analyses with 100–1000s of nuclear genes and no mitochondrial data (Feng *et al.,* 2017; Hime *et al.,* 2021).

(14) Key references. This section presents references relevant to the Ecuadorian amphibian orders and species accounts. It is not intended to be an exhaustive list of publications but rather a selection of the most relevant ones. For a comprehensive list of citations related to taxonomy, please refer to the most recent version of *Amphibian Species of the World* (Frost, 2023 and onward) or other revisionary works in specific fields. The references are presented in chronological order and include a brief description of the main topic covered. Additionally, authors of tadpole and call descriptions are acknowledged. This section includes the references corresponding to information provided in most sections of the account, with the exception of the Conservation and Remarks sections, which have citations integrated throughout the text.

INSTITUTIONAL AND MUSEUM ABBREVIATIONS

Throughout the text, we use acronyms for institutions and museums. For museum abbreviations, see Frost (2023 and onward). The most commonly used museum or institutional abbreviations are:

CJ	Centro Jambatu de Investigación y Conservación de Anfibios, Ecuador.
DHMECN	División de Herpetología, Museo Ecuatoriano de Ciencias Naturales (part of INABIO, Instituto Nacional de Biodiversidad, Ecuador since 2017).
EPN	(museum) Escuela Politécnica Nacional, Museo de Zoología, Ecuador.
KU	University of Kansas, Museum of Natural History, Division of Herpetology, USA.
MCZ	Museum of Comparative Zoology, Harvard University, USA.
MHNG	Muséum d´histoire Naturelle, Geneva.
QCAZ	Museo de Zoología, Pontificia Universidad Católica del Ecuador.
ZSFQ	Universidad San Francisco de Quito, Museo de Zoología, Ecuador.
USNM	National Museum of Natural History, Division of Amphibians and Reptiles, USA.
MZUTI	Universidad Tecnológica Indoamérica, Museo de Zoología, Ecuador.

TABLE 0.1
List of Suggested Common Names in English and Spanish for Species of Ecuadorian Amphibians
See "Standardized Names in Spanish and English" in Chapter 1 in Volume I. * = Caudata, ** = Gymnophiona.

Centrolenidae	*Centrolene ballux*	Gold-dusted glassfrog	Rana de cristal dorada
Centrolenidae	*Centrolene buckleyi*	Buckley's glassfrog	Rana de cristal de Buckley
Centrolenidae	*Centrolene charapita*	Charapita glassfrog	Rana de cristal charapita
Centrolenidae	*Centrolene condor*	Cóndor glassfrog	Rana de cristal de El Cóndor
Centrolenidae	*Centrolene geckoidea*	Gecko glassfrog	Rana de cristal gecko
Centrolenidae	*Centrolene heloderma*	Warty glassfrog	Rana de cristal verrugosa
Centrolenidae	*Centrolene huilensis*	Huila glassfrog	Rana de cristal de Huila
Centrolenidae	*Centrolene lynchi*	Lynch's glassfrog	Rana de cristal de Lynch
Centrolenidae	*Centrolene medemi*	Medem's glassfrog	Rana de cristal de Medem
Centrolenidae	*Centrolene peristicta*	Dappled glassfrog	Rana de cristal jaspeada
Centrolenidae	*Centrolene pipilata*	Peeping glassfrog	Rana de cristal espía
Centrolenidae	*Centrolene sanchezi*	Sanchez's glassfrog	Rana de cristal de Sánchez
Centrolenidae	*Chimerella mariaelenae*	María Elena's glassfrog	Rana de cristal de María Elena
Centrolenidae	*Cochranella granulosa*	Granular glassfrog	Rana de cristal granular
Centrolenidae	*Cochranella litoralis*	Coast glassfrog	Rana de cristal litoral
Centrolenidae	*Cochranella mache*	Mache glassfrog	Rana de cristal de Mache
Centrolenidae	*Cochranella resplendens*	Resplendent glassfrog	Rana de cristal resplandeciente
Centrolenidae	*Espadarana audax*	Daring glassfrog	Rana de cristal atrevida
Centrolenidae	*Espadarana callistomma*	Pretty-eyed glassfrog	Rana de cristal de ojos bellos
Centrolenidae	*Espadarana durrellorum*	Durrell's glassfrog	Rana de cristal de Durrell
Centrolenidae	*Espadarana prosoblepon*	Variable glassfrog	Rana de cristal variable
Centrolenidae	*Hyalinobatrachium adespinosai*	Adela's glassfrog	Rana de cristal de Adela
Centrolenidae	*Hyalinobatrachium aureoguttatum*	Gold-spotted glassfrog	Rana de cristal de puntos dorados
Centrolenidae	*Hyalinobatrachium chirripoi*	Chirripó glassfrog	Rana de cristal Chirripó
Centrolenidae	*Hyalinobatrachium iaspidiense*	Jaspe's glassfrog	Rana de cristal de Jaspe
Centrolenidae	*Hyalinobatrachium mashpi*	Mashpi glassfrog	Rana de cristal de Mashpi
Centrolenidae	*Hyalinobatrachium munozorum*	Muñoz's glassfrog	Rana de cristal de Muñoz
Centrolenidae	*Hyalinobatrachium nouns*	Nouns' glassfrog	Rana de cristal de Nouns
Centrolenidae	*Hyalinobatrachium pellucidum*	Crystal-clear glassfrog	Rana de cristal transparente
Centrolenidae	*Hyalinobatrachium tatayoi*	Tatayo's glassfrog	Rana de cristal de Tatayo
Centrolenidae	*Hyalinobatrachium valerioi*	Valerio's glassfrog	Rana de cristal de Valerio
Centrolenidae	*Hyalinobatrachium yaku*	Yaku glassfrog	Rana de cristal yaku
Centrolenidae	*Nymphargus anomalus*	Anomalous glassfrog	Rana de cristal anómala
Centrolenidae	*Nymphargus balionotus*	Spotted glassfrog	Rana de cristal salpicada
Centrolenidae	*Nymphargus buenaventura*	Buenaventura glassfrog	Rana de cristal de Buenaventura
Centrolenidae	*Nymphargus cariticommatus*	Unadorned glassfrog	Rana de cristal desadornada
Centrolenidae	*Nymphargus cochranae*	Cochran's glassfrog	Rana de cristal de Cochran
Centrolenidae	*Nymphargus colomai*	Coloma's glassfrog	Rana de cristal de Coloma
Centrolenidae	*Nymphargus garciae*	García's glassfrog	Rana de cristal de García
Centrolenidae	*Nymphargus grandisonae*	Red-spotted glassfrog	Rana de cristal de puntos rojos
Centrolenidae	*Nymphargus griffithsi*	Griffiths' glassfrog	Rana de cristal de Griffiths
Centrolenidae	*Nymphargus humboldti*	Humboldt's glassfrog	Rana de cristal de Humboldt
Centrolenidae	*Nymphargus lasgralarias*	Las Gralarias glassfrog	Rana de cristal de Las Gralarias
Centrolenidae	*Nymphargus laurae*	Laura's glassfrog	Rana de cristal de Laura
Centrolenidae	*Nymphargus lindae*	Linda's glassfrog	Rana de cristal de Linda
Centrolenidae	*Nymphargus manduriacu*	Manduriacu glassfrog	Rana de cristal de Manduriacu
Centrolenidae	*Nymphargus mariae*	María's glassfrog	Rana de cristal de María
Centrolenidae	*Nymphargus megacheirus*	Large-handed glassfrog	Rana de cristal de manos grandes
Centrolenidae	*Nymphargus megistus*	Greater Andean glassfrog	Rana andina grande
Centrolenidae	*Nymphargus posadae*	Posada's glassfrog	Rana de cristal de Posada
Centrolenidae	*Nymphargus siren*	Siren glassfrog	Rana de cristal sirena
Centrolenidae	*Nymphargus sucre*	Sucre's glassfrog	Rana de cristal de Sucre

(*Continued*)

TABLE 0.1 (*Continued*)

List of Suggested Common Names in English and Spanish for Species of Ecuadorian Amphibians
See "Standardized Names in Spanish and English" in Chapter 1 in Volume I. * = Caudata, ** = Gymnophiona.

Centrolenidae	*Nymphargus wileyi*	Wiley's glassfrog	Rana de cristal de Wiley
Centrolenidae	*Rulyrana flavopunctata*	Yellow-spotted glassfrog	Rana de cristal de manchas amarillas
Centrolenidae	*Rulyrana mcdiarmidi*	McDiarmid´s glassfrog	Rana de cristal de McDiarmid
Centrolenidae	*Sachatamia albomaculata*	White-spotted glassfrog	Rana de cristal de puntos blancos
Centrolenidae	*Sachatamia ilex*	Holly's glassfrog	Rana de cristal de Holly
Centrolenidae	*Sachatamia orejuela*	Orejuela's glassfrog	Rana de cristal de Orejuela
Centrolenidae	*Teratohyla amelie*	Amélie's glassfrog	Rana de cristal de Amélie
Centrolenidae	*Teratohyla midas*	Midas' glassfrog	Rana de cristal de Midas
Centrolenidae	*Teratohyla pulverata*	Dusty glassfrog	Rana de cristal polvorienta
Centrolenidae	*Teratohyla spinosa*	Dwarf glassfrog	Rana de cristal enana
Centrolenidae	*Vitreorana ritae*	Rita's glassfrog	Rana de cristal de Rita
Ceratophryidae	*Ceratophrys cornuta*	Horned Pac-Man frog	Sapo pacman de cuernos
Ceratophryidae	*Ceratophrys stolzmanni*	Stolzmann's Pac-Man frog	Sapo pacman de Stolzmann
Eleutherodactylidae	*Adelophryne adiastola*	Confused hammer frog	Ranita martillo confusa
Eleutherodactylidae	*Diasporus gularis*	Orange little hammer frog	Ranita martillo naranja
Hemiphractidae	*Gastrotheca cuencana*	Cuenca marsupial frog	Rana marsupial de Cuenca
Hemiphractidae	*Gastrotheca elicioi*	Elicio's marsupial frog	Rana marsupial de Elicio
Hemiphractidae	*Gastrotheca litonedis*	Busa marsupial frog	Rana marsupial de Busa
Hemiphractidae	*Gastrotheca lojana*	Lojana marsupial frog	Rana marsupial Lojana
Hemiphractidae	*Gastrotheca plumbea*	Plumbeus marsupial frog	Rana marsupial plúmbea
Hemiphractidae	*Gastrotheca pseustes*	Liar marsupial frog	Rana marsupial mentirosa
Hemiphractidae	*Gastrotheca psychrophila*	Cold marsupial frog	Rana marsupial de frío
Hemiphractidae	*Gastrotheca riobambae*	Gualataco marsupial frog	Rana marsupial gualataco
Hemiphractidae	*Gastrotheca turnerorum*	Turner's marsupial frog	Rana marsupial de Turner
Hemiphractidae	*Gastrotheca yacuri*	Yacuri marsupial frog	Rana marsupial de Yacuri
Hemiphractidae	*Gastrotheca andaquiensis*	Andaqui marsupial frog	Rana marsupial Andaquí
Hemiphractidae	*Gastrotheca angustifrons*	Thin-snouted marsupial frog	Rana marsupial de nariz estrecha
Hemiphractidae	*Gastrotheca cornuta*	Horned marsupial frog	Rana marsupial con cuernos
Hemiphractidae	*Gastrotheca dendronastes*	Arboreal marsupial frog	Rana marsupial arbórea
Hemiphractidae	*Gastrotheca espeletia*	Espeletia marsupial frog	Rana marsupial de espeletia
Hemiphractidae	*Gastrotheca guentheri*	Lower-Toothed marsupial frog	Rana marsupial dentada
Hemiphractidae	*Gastrotheca longipes*	Long-footed marsupial frog	Rana marsupial de pies largos
Hemiphractidae	*Gastrotheca orophylax*	Guardian marsupial frog	Rana marsupial guardiana
Hemiphractidae	*Gastrotheca testudinea*	Tortoise marsupial frog	Rana marsupial tortuga
Hemiphractidae	*Gastrotheca weinlandii*	Weinland's marsupial frog	Rana marsupial de Weinland
Hemiphractidae	*Hemiphractus bubalus*	Crested casque-headed frog	Rana de casco crestada
Hemiphractidae	*Hemiphractus fasciatus*	Banded casque-headed frog	Rana de casco bandeada
Hemiphractidae	*Hemiphractus helioi*	Helio's casque-headed frog	Rana de casco de Helio
Hemiphractidae	*Hemiphractus proboscideus*	Long-snouted casque-headed frog	Rana de casco narizona
Hemiphractidae	*Hemiphractus scutatus*	Helmeted casque-headed frog	Rana de casco protector
Leptodactylidae	*Adenomera andreae*	Göldi's thin-toed frog	Rana dedilarga de Göldi
Leptodactylidae	*Adenomera hylaedactyla*	Dark-spotted thin-toed frog	Rana dedilarga de manchas oscuras
Leptodactylidae	*Edalorhina perezi*	Perez's foam-nest frog	Rana de espuma de Pérez
Leptodactylidae	*Engystomops coloradorum*	Colorado bully foam toad	Sapito bullanguero Colorado
Leptodactylidae	*Engystomops guayaco*	Guayaco bully foam toad	Sapito bullanguero guayaco
Leptodactylidae	*Engystomops montubio*	Montubio bully foam toad	Sapito bullanguero montubio
Leptodactylidae	*Engystomops petersi*	Peters' bully foam toad	Sapito bullanguero de Peters
Leptodactylidae	*Engystomops pustulatus*	Blistered bully foam toad	Sapito bullanguero ampollado
Leptodactylidae	*Engystomops puyango*	Puyango bully foam toad	Sapito bullanguero de Puyango
Leptodactylidae	*Engystomops randi*	Rand's bully foam toad	Sapito bullanguero de Rand
Leptodactylidae	*Leptodactylus discodactylus*	Disc-shaped thin-toed frog	Rana dedilarga discada
Leptodactylidae	*Leptodactylus knudseni*	Knudsen's thin-toed frog	Rana dedilarga de Knudsen
Leptodactylidae	*Leptodactylus labrosus*	Thick-lipped thin-toed frog	Rana dedilarga de labio grueso

TABLE 0.1 (*Continued*)
List of Suggested Common Names in English and Spanish for Species of Ecuadorian Amphibians
See "Standardized Names in Spanish and English" in Chapter 1 in Volume I. * = Caudata, ** = Gymnophiona.

Leptodactylidae	*Leptodactylus leptodactyloides*	Thin-toed frog	Rana dedilarga
Leptodactylidae	*Leptodactylus melanonotus*	Black-back thin-toed frog	Rana dedilarga dorsinegra
Leptodactylidae	*Leptodactylus mystaceus*	Mustache thin-toed frog	Rana dedilarga con bigote
Leptodactylidae	*Leptodactylus pentadactylus*	Gualag thin-toed frog	Rana dedilarga gualag
Leptodactylidae	*Leptodactylus peritoaktites*	Coastal dweller thin-toed frog	Rana dedilarga de la costa
Leptodactylidae	*Leptodactylus petersii*	Peters' thin-toed frog	Rana dedilarga de Peters
Leptodactylidae	*Leptodactylus rhodomerus*	Rose-thigh thin-toed frog	Rana dedilarga de muslos rosados
Leptodactylidae	*Leptodactylus rhodomystax*	Rose-lipped thin-toed frog	Rana dedilarga de labios rosados
Leptodactylidae	*Leptodactylus stenodema*	Popoco thin-toed frog	Rana dedilarga popoco
Leptodactylidae	*Leptodactylus ventrimaculatus*	Spotted-venter thin-toed frog	Rana dedilarga de vientre punteado
Leptodactylidae	*Leptodactylus wagneri*	Wagner's thin-toed frog	Rana dedilarga de Wagner
Leptodactylidae	*Lithodytes lineatus*	Striped ant-nest frog	Rana listada hormiguera
Phyllomedusidae	*Agalychnis buckleyi*	Buckley's leaf frog	Rana de hoja de Buckley
Phyllomedusidae	*Agalychnis hulli*	Hull's leaf frog	Rana de hoja de Hull
Phyllomedusidae	*Agalychnis psilopygion*	Smooth leaf frog	Rana de hoja lisa
Phyllomedusidae	*Agalychnis spurrelli*	Spurrell's gliding leaf frog	Rana de hoja de Spurrell
Phyllomedusidae	*Callimedusa ecuatoriana*	Ecuadorian leaf frog	Rana de hoja ecuatoriana
Phyllomedusidae	*Callimedusa perinesos*	Purple leaf frog	Rana de hoja púrpura
Phyllomedusidae	*Callimedusa tomopterna*	Tiger-striped leaf frog	Rana mono rayada tigre
Phyllomedusidae	*Cruziohyla calcarifer*	Splendid leaf frog	Rana de hoja espléndida
Phyllomedusidae	*Cruziohyla craspedopus*	Fringed leaf frog	Rana de hoja con flecos
Phyllomedusidae	*Phyllomedusa tarsius*	Tarsus leaf frog	Rana de hoja de tarso
Phyllomedusidae	*Phyllomedusa vaillantii*	White-lined leaf frog	Rana de hoja de línea blanca
Phyllomedusidae	*Pithecopus palliatus*	Mantle leaf frog	Rana de hoja de manto
*Plethodontidae	*Bolitoglossa chica*	Tiny mushroom tongue salamander	Salamandra apulmonada minúscula
*Plethodontidae	*Bolitoglossa equatoriana*	Ecuadorian mushroom tongue salamander	Salamandra apulmonada ecuatoriana
*Plethodontidae	*Bolitoglossa medemi*	Medem's mushroom tongue salamander	Salamandra apulmonada de Medem
*Plethodontidae	*Bolitoglossa palmata*	Webbed mushroom tongue salamander	Salamandra apulmonada con membranas
*Plethodontidae	*Bolitoglossa sima*	Snub-nosed mushroom tongue salamander	Salamandra apulmonada de nariz corta
*Plethodontidae	*Oedipina ecuatoriana*	Ecuadorian worm salamander	Salamandra gusano ecuatoriana
*Plethodontidae	*Oedipina villamizariorum*	Villamizar's worm salamander	Salamandra gusano de Villamizar
**Caeciliidae	*Caecilia abitaguae*	Abitagua caecilian	Ilulo de Abitagua
**Caeciliidae	*Caecilia albiventris*	White-bellied caecilian	Ilulo de vientre blanco
**Caeciliidae	*Caecilia attenuata*	Attenuated caecilian	Ilulo atenuado
**Caeciliidae	*Caecilia bokermanni*	Bokermann's caecilian	Ilulo de Bokermann
**Caeciliidae	*Caecilia buckleyi*	Buckley's caecilian	Ilulo de Buckley
**Caeciliidae	*Caecilia crassisquama*	Thick caecilian	Ilulo grueso
**Caeciliidae	*Caecilia disossea*	Bony caecilian	Ilulo óseo
**Caeciliidae	*Caecilia dunni*	Dunn's caecilian	Ilulo de Dunn
**Caeciliidae	*Caecilia guntheri*	Günther's caecilian	Ilulo de Günther
**Caeciliidae	*Caecilia leucocephala*	White-headed caecilian	Ilulo de cabeza blanca
**Caeciliidae	*Caecilia nigricans*	Black caecilian	Ilulo negro
**Caeciliidae	*Caecilia orientalis*	Eastern caecilian	Ilulo del oriente
**Caeciliidae	*Caecilia pachynema*	Giant caecilian	Ilulo gigante
**Caeciliidae	*Caecilia subterminalis*	Taylor's caecilian	Ilulo de Taylor
**Caeciliidae	*Caecilia tentaculata*	Yamba caecilian	Ilulo yamba
**Caeciliidae	*Caecilia tenuissima*	Guayaquil caecilian	Ilulo de Guayaquil
**Caeciliidae	*Oscaecilia bassleri*	Bassler's caecilian	Ilulo de Bassler
**Caeciliidae	*Oscaecilia equatorialis*	Ecuadorian caecilian	Ilulo ecuatoriano
**Rhinatrematidae	*Amazops amazops*	Amazonian-tailed caecilian	Ilulo colilarga amazónico
**Rhinatrematidae	*Epicrionops bicolor*	Bicolor caecilian	Ilulo bicolor
**Rhinatrematidae	*Epicrionops petersi*	Peters' caecilian	Ilulo de Peters
**Siphonopidae	*Microcaecilia albiceps*	Pale-headed caecilian	Ilulo de cabeza alba
**Siphonopidae	*Siphonops annulatus*	Ringed caecilian	Ilulo anillado
**Typhlonectidae	*Chthonerpeton onorei*	Onore's caecilian	Ilulo de Onore
**Typhlonectidae	*Potomotyphlus kaupii*	Kaup's caecilian	Ilulo de Kaup

TABLE 0.2
Summary of Endemism, Red List Categories, and Citations of References for Which Tadpole and Call Descriptions of 655 Amphibians from Ecuador Have Been Published.

Family	Species	End	Rel List	Tadpole References	Call References
Centrolenidae	*Centrolene ballux*		EN	Guayasamin *et al.* (this publication)	Márquez *et al.* (1996)
Centrolenidae	*Centrolene buckleyi*		CR	Guayasamin *et al.* (this publication)	Guayasamin *et al.* (2020b)
Centrolenidae	*Centrolene charapita*		CR	Guayasamin *et al.* (this publication)	Non-described
Centrolenidae	*Centrolene condor*	E	EN	Guayasamin *et al.* (2020b)	Almendáriz and Batallas (2012)
Centrolenidae	*Centrolene geckoidea*		CR (PE)	Lynch *et al.* (1983), Rueda-Almonacid (1994)	Non-described
Centrolenidae	*Centrolene heloderma*		EN	Guayasamin *et al.* (this publication)	Guayasamin *et al.* (2020b)
Centrolenidae	*Centrolene huilensis*		DD	Non-described	Non-described
Centrolenidae	*Centrolene lynchi*		EN	Non-described	Dautel *et al.* (2011)
Centrolenidae	*Centrolene medemi*		CR	Non-described	Non-described
Centrolenidae	*Centrolene peristicta*		NT	Guayasamin *et al.* (this publication)	Salgado and Guayasamin (2018)
Centrolenidae	*Centrolene pipilata*	E	CR (PE)	Non-described	Non-described
Centrolenidae	*Centrolene sanchezi*		EN	Guayasamin *et al.* (this publication)	Guayasamin *et al.* (2020b)
Centrolenidae	*Chimerella mariaelenae*	E	NT	Terán Valdez and Guayasamin (2014)	Batallas and Brito (2016); Guayasamin *et al.* (2020b)
Centrolenidae	*Cochranella granulosa*		DD	Starrett (1960); Hoffmann (2010)	Kubicki (2007)
Centrolenidae	*Cochranella litoralis*		EN	Non-described	Non-described
Centrolenidae	*Cochranella mache*		EN	Non-described	Ortega-Andrade *et al.* (2013)
Centrolenidae	*Cochranella resplendens*		LC	Terán-Valdez *et al.* (2009)	Non-described
Centrolenidae	*Espadarana audax*		NT	Guayasamin *et al.* (this publication)	Guayasamin *et al.* (2020b)
Centrolenidae	*Espadarana callistomma*		EN	Non-described	Guayasamin *et al.* (2020b)
Centrolenidae	*Espadarana durrellorum*		LC	Non-described	Non-described
Centrolenidae	*Espadarana prosoblepon*		LC	McCranie and Wilson (2002); Starrett (1960); Savage (2002)	Guayasamin *et al.* (2020b)
Centrolenidae	*Hyalinobatrachium adespinosai*	E	DD	Non-described	Guayasamin *et al.* (2020b)
Centrolenidae	*Hyalinobatrachium aureoguttatum*		EN	Ibañez *et al.* (1999a); Terán-Valdez *et al.* (2009)	Non-described
Centrolenidae	*Hyalinobatrachium chirripoi*		EN	Hoffmann (2010)	Guayasamin *et al.* (2020b)
Centrolenidae	*Hyalinobatrachium iaspidiense*		NT	Non-described	Señaris and Ayarzagüena (2005)
Centrolenidae	*Hyalinobatrachium mashpi*	E	EN	Non-described	Guayasamin *et al.* (2022)
Centrolenidae	*Hyalinobatrachium munozorum*		LC	Guayasamin *et al.* (this publication)	Guayasamin *et al.* (2020b)
Centrolenidae	*Hyalinobatrachium nouns*	E	EN	Non-described	Non-described
Centrolenidae	*Hyalinobatrachium pellucidum*	E	VU	Non-described	Guayasamin *et al.* (2020b)
Centrolenidae	*Hyalinobatrachium tatayoi*		NT	Guayasamin *et al.* (this publication)	Castroviejo-Fisher *et al.* (2007)
Centrolenidae	*Hyalinobatrachium valerioi*		EN	See Remarks in account	See Remarks in account
Centrolenidae	*Hyalinobatrachium yaku*	E	DD	Non-described	Guayasamin *et al.* (2017)
Centrolenidae	*Nymphargus anomalus*	E	EN	Guayasamin *et al.* (this publication)	Non-described
Centrolenidae	*Nymphargus balionotus*		EN	Non-described	Non-described

TABLE 0.2 (*Continued*)

Summary of Endemism, Red List Categories, and Citations of References for Which Tadpole and Call Descriptions of 655 Amphibians from Ecuador Have Been Published.

Family	Species	End	Rel List	Tadpole References	Call References
Centrolenidae	*Nymphargus buenaventura*		EN	Non-described	Non-described
Centrolenidae	*Nymphargus cariticommatus*	E	EN	Non-described	Non-described
Centrolenidae	*Nymphargus cochranae*	E	VU	Guayasamin *et al.* (this publication)	Non-described
Centrolenidae	*Nymphargus colomai*	E	EN	Non-described	Non-described
Centrolenidae	*Nymphargus garciae*		EN	Non-described	Non-described
Centrolenidae	*Nymphargus grandisonae*		LC	Ospina-Sarria *et al.* (2011)	Hutter *et al.* (2013)
Centrolenidae	*Nymphargus griffithsi*		VU	Guayasamin *et al.* (this publication)	Hutter and Guayasamin (2012)
Centrolenidae	*Nymphargus humboldti*	E	EN	Non-described	Non-described
Centrolenidae	*Nymphargus lasgralarias*	E	EN	Non-described	Hutter and Guayasamin (2012)
Centrolenidae	*Nymphargus laurae*	E	CR	Non-described	Non-described
Centrolenidae	*Nymphargus lindae*	E	EN	Non-described	Non-described
Centrolenidae	*Nymphargus manduriacu*	E	CR	Non-described	Guayasamin *et al.* (2019a)
Centrolenidae	*Nymphargus mariae*		LC	Non-described	Guayasamin *et al.* (2020b)
Centrolenidae	*Nymphargus megacheirus*		CR	Non-described	Non-described
Centrolenidae	*Nymphargus megistus*		EN	Non-described	Non-described
Centrolenidae	*Nymphargus posadae*		LC	Non-described	Non-described
Centrolenidae	*Nymphargus siren*		EN	Guayasamin *et al.* (this publication)	Guayasamin *et al.* (2020b)
Centrolenidae	*Nymphargus sucre*	E	CR	Non-described	Non-described
Centrolenidae	*Nymphargus wileyi*	E	DD	Non-described	Non-described
Centrolenidae	*Rulyrana flavopunctata*		VU	Non-described	Guayasamin *et al.* (2020b)
Centrolenidae	*Rulyrana mcdiarmidi*		VU	Non-described	Non-described
Centrolenidae	*Sachatamia albomaculata*		VU	Hoffmann (2010)	Non-described
Centrolenidae	*Sachatamia ilex*		EN	Hoffmann (2010)	Guayasamin *et al.* (2020b)
Centrolenidae	*Sachatamia orejuela*		EN	Non-described	Brunner and Guayasamin (2020)
Centrolenidae	*Teratohyla amelie*		LC	Non-described	Guayasamin *et al.* (2020b)
Centrolenidae	*Teratohyla midas*		LC	Guayasamin *et al.* (this publication)	Guayasamin *et al.* (2020b)
Centrolenidae	*Teratohyla pulverata*		VU	Hoffmann (2004)	Ibañez *et al.* (1999b); Kubicki (2007)
Centrolenidae	*Teratohyla spinosa*		VU	Savage (2002)	Guayasamin *et al.* (2020b)
Centrolenidae	*Vitreorana ritae*		LC	Guayasamin *et al.* (2020b); Schiesari *et al.* (2022)	Guayasamin *et al.* (2020b)
Ceratophryidae	*Ceratophrys cornuta*		LC	Duellman (1978, 2005); Schiesari *et al.* (2022)	Schlüter (1980); Duellman (2005)
Ceratophryidae	*Ceratophrys stolzmanni*		VU	Non-described	Non-described
Eleutherodactylidae	*Adelophryne adiastola*		LC	N/A	Heyer (1977)
Eleutherodactylidae	*Diasporus gularis*		EN	N/A	Non-described
Hemiphractidae	*Gastrotheca cuencana*	E	EN	Carvajal-Endara *et al.* (2019)	Carvajal-Endara *et al.* (2019)
Hemiphractidae	*Gastrotheca elicioi*	E	EN	Carvajal-Endara *et al.* (2019)	Carvajal-Endara *et al.* (2019)
Hemiphractidae	*Gastrotheca litonedis*	E	CR	Carvajal-Endara *et al.* (2019)	Carvajal-Endara *et al.* (2019)
Hemiphractidae	*Gastrotheca lojana*	E	CR	Carvajal-Endara *et al.* (2019)	Carvajal-Endara *et al.* (2019)
Hemiphractidae	*Gastrotheca plumbea*	E	EN	N/A	Sinsch and Juraske (2006)

(*Continued*)

TABLE 0.2 (*Continued*)
Summary of Endemism, Red List Categories, and Citations of References for Which Tadpole and Call Descriptions of 655 Amphibians from Ecuador Have Been Published.

Family	Species	End	Rel List	Tadpole References	Call References
Hemiphractidae	*Gastrotheca pseustes*	E	NT	Carvajal-Endara *et al.* (2019)	Carvajal-Endara *et al.* (2019)
Hemiphractidae	*Gastrotheca psychrophila*	E	CR (PE)	Non-described	Non-described
Hemiphractidae	*Gastrotheca riobambae*	E	EN	Duellman (2015)	Sinsch and Juraske (2006)
Hemiphractidae	*Gastrotheca turnerorum*	E	EN	Carvajal-Endara *et al.* (2019)	Non-described
Hemiphractidae	*Gastrotheca yacuri*	E	CR	Non-described	Carvajal-Endara *et al.* (2019)
Hemiphractidae	*Gastrotheca andaquiensis*		CR (PE)	N/A	Non-described
Hemiphractidae	*Gastrotheca angustifrons*		CR (PE)	N/A	Non-described
Hemiphractidae	*Gastrotheca cornuta*		EN	N/A	Duellman (2015)
Hemiphractidae	*Gastrotheca dendronastes*		CR (PE)	N/A	Non-described
Hemiphractidae	*Gastrotheca espeletia*		EN	Duellman (2015)	Sinsch and Juraske (2006)
Hemiphractidae	*Gastrotheca guentheri*		CR (PE)	N/A	Non-described
Hemiphractidae	*Gastrotheca longipes*		NT	N/A	Non-described
Hemiphractidae	*Gastrotheca orophylax*		EN	N/A	Sinsch and Juraske (2006)
Hemiphractidae	*Gastrotheca testudinea*		NT	N/A	Carvajal-Endara *et al.* (2019)
Hemiphractidae	*Gastrotheca weinlandii*		CR (PE)	N/A	Non-described
Hemiphractidae	*Hemiphractus bubalus*		NT	N/A	Non-described
Hemiphractidae	*Hemiphractus fasciatus*		CR (PE)	N/A	Non-described
Hemiphractidae	*Hemiphractus helioi*		CR (PE)	N/A	Non-described
Hemiphractidae	*Hemiphractus proboscideus*		NT	N/A	Non-described
Hemiphractidae	*Hemiphractus scutatus*		LC	N/A	Non-described
Leptodactylidae	*Adenomera andreae*		LC	Schiesari *et al.* (2022)	Marquez *et al.* (1995); Angulo *et al.* (2003)
Leptodactylidae	*Adenomera hylaedactyla*		LC	Heyer and Silverstone (1969); Schiesari *et al.* (2022)	Marquez *et al.* (1995); Angulo *et al.* (2003)
Leptodactylidae	*Edalorhina perezi*		LC	Schlüter (1990); Duellman (2005); do Nascimento *et al.* (2021)	Schlüter (1980, 1990); Duellman (2005)
Leptodactylidae	*Engystomops coloradorum*	E	EN	Cannatella and Duellman (1984)	Ron *et al.* (2005)
Leptodactylidae	*Engystomops guayaco*	E	VU	Ron *et al.* (2005)	Ron *et al.* (2005)
Leptodactylidae	*Engystomops montubio*	E	LC	Non-described	Ron *et al.* (2004)
Leptodactylidae	*Engystomops petersi*		LC	Duellman (1978)	Guerra and Ron (2008)
Leptodactylidae	*Engystomops pustulatus*		LC	Terán-Valdez and Coloma (this publication)	Ron *et al.* (2010)
Leptodactylidae	*Engystomops puyango*	E	NT	Non-described	Ron *et al.* (2004, 2010)
Leptodactylidae	*Engystomops randi*	E	LC	Terán-Valdez and Coloma (this publication)	Ron *et al.* (2004)
Leptodactylidae	*Leptodactylus discodactylus*		LC	Duellman (1978) (uncertain)	Straughan and Heyer (1976)
Leptodactylidae	*Leptodactylus knudseni*		LC	Hero (1990); Schiesari *et al.* (2022)	Heyer (2005)
Leptodactylidae	*Leptodactylus labrosus*		LC	Terán-Valdez and Coloma (this publication)	de Carvalho and Ron (2011)
Leptodactylidae	*Leptodactylus leptodactyloides*		LC	Heyer (1994)	Heyer (1994)
Leptodactylidae	*Leptodactylus melanonotus*		LC	McCranie and Wilson (2002)	Straughan and Heyer (1976)
Leptodactylidae	*Leptodactylus mystaceus*		LC	Heyer (1978); Schiesari *et al.* (2022)	Heyer *et al.* (1996)

TABLE 0.2 (*Continued*)

Summary of Endemism, Red List Categories, and Citations of References for Which Tadpole and Call Descriptions of 655 Amphibians from Ecuador Have Been Published.

Family	Species	End	Rel List	Tadpole References	Call References
Leptodactylidae	*Leptodactylus pentadactylus*		NT	Menin *et al.* (2010); Schiesari *et al.* (2022)	Heyer (2005)
Leptodactylidae	*Leptodactylus peritoaktites*	E	EN	Non-described	Heyer (2005)
Leptodactylidae	*Leptodactylus petersii*		LC	Duellman (2005); Schiesari *et al.* (2022)	Heyer (1994)
Leptodactylidae	*Leptodactylus rhodomerus*		VU	Non-described	Non-described
Leptodactylidae	*Leptodactylus rhodomystax*		LC	Rodrigues *et al.* (2007); Duellman (1978); Hero (1990); Schiesari *et al.* (2022)	Zimmerman and Bogart (1984); de Sá *et al.* (2014)
Leptodactylidae	*Leptodactylus stenodema*		NT	Non-described	Heyer (1979); Lescure and Marty (2000)
Leptodactylidae	*Leptodactylus ventrimaculatus*		LC	Terán-Valdez and Coloma (this publication)	Non-described
Leptodactylidae	*Leptodactylus wagneri*		LC	Non-described	de Carvalho *et al.* (2022)
Leptodactylidae	*Lithodytes lineatus*		LC	Regös and Schlüter (1984); Lamar and Wild (1995); Schlüter and Regos (1996); Hero (1990)	Marquez *et al.* (1995); Schlütter *et al.* (2009)
Phyllomedusidae	*Agalychnis buckleyi*		VU	Cannatella (1980)	Non-described
Phyllomedusidae	*Agalychnis hulli*		LC	Gray and Coloma (2022)	Non-described
Phyllomedusidae	*Agalychnis psilopygion*		EN	Cannatella (1980)	Cannatella (1980)
Phyllomedusidae	*Agalychnis spurrelli*		LC	Vargas-Salinas and Gutiérrez-Cárdenas (2005)	Ortega-Andrade (2008); Cossio and Medina-Barcenas (2020)
Phyllomedusidae	*Callimedusa ecuatoriana*	E	EN	Brito *et al.* (2014)	Brito *et al.* (2014)
Phyllomedusidae	*Callimedusa perinesos*		CR	Non-described	Non-described
Phyllomedusidae	*Callimedusa tomopterna*		LC	Duellman (1978); Schiesari *et al.* (2022)	Zimmerman (1983); Duellman (2005)
Phyllomedusidae	*Cruziohyla calcarifer*		NT	Gray *et al.* (2021)	Non-described
Phyllomedusidae	*Cruziohyla craspedopus*		LC	Hoogmoed and Cadle (1990)	Non-described
Phyllomedusidae	*Phyllomedusa tarsius*		LC	Duellman (1978); Hero (1990); Schiesari *et al.* (2022)	Duellman (1978)
Phyllomedusidae	*Phyllomedusa vaillantii*		LC	Duellman (1978); Hero (1990); Wild (1996); Schiesari *et al.* (2022)	Duellman (1978, 2005); Röhr *et al.* (2020)
Phyllomedusidae	*Pithecopus palliatus*		EN	Duellman (1978)	Duellman (1978, 2005)
*Plethodontidae	*Bolitoglossa chica*		CR	N/A	N/A
*Plethodontidae	*Bolitoglossa equatoriana*	E	LC	N/A	N/A
*Plethodontidae	*Bolitoglossa medemi*		EN	N/A	N/A
*Plethodontidae	*Bolitoglossa palmata*	E	CR	N/A	N/A
*Plethodontidae	*Bolitoglossa sima*		EN	N/A	N/A
*Plethodontidae	*Oedipina ecuatoriana*	E	CR	N/A	N/A
*Plethodontidae	*Oedipina villamizariorum*	E	CR	N/A	N/A
**Caeciliidae	*Caecilia abitaguae*	E	EN	N/A	N/A
**Caeciliidae	*Caecilia albiventris*		DD	N/A	N/A
**Caeciliidae	*Caecilia attenuata*		DD	N/A	N/A
**Caeciliidae	*Caecilia bokermanni*	E	DD	N/A	N/A
**Caeciliidae	*Caecilia buckleyi*	E	DD	N/A	N/A
**Caeciliidae	*Caecilia crassisquama*	E	DD	N/A	N/A
**Caeciliidae	*Caecilia disossea*		LC	N/A	N/A

(*Continued*)

TABLE 0.2 (*Continued*)

Summary of Endemism, Red List Categories, and Citations of References for Which Tadpole and Call Descriptions of 655 Amphibians from Ecuador Have Been Published.

Family	Species	End	Rel List	Tadpole References	Call References
**Caeciliidae	*Caecilia dunni*		CR	N/A	N/A
**Caeciliidae	*Caecilia guntheri*		EN	N/A	N/A
**Caeciliidae	*Caecilia leucocephala*		VU	N/A	N/A
**Caeciliidae	*Caecilia nigricans*		VU	N/A	N/A
**Caeciliidae	*Caecilia orientalis*		VU	N/A	N/A
**Caeciliidae	*Caecilia pachynema*	E	EN	N/A	N/A
**Caeciliidae	*Caecilia subterminalis*	E	DD	N/A	N/A
**Caeciliidae	*Caecilia tentaculata*		LC	N/A	N/A
**Caeciliidae	*Caecilia tenuissima*	E	DD	N/A	N/A
**Caeciliidae	*Oscaecilia bassleri*		VU	N/A	N/A
**Caeciliidae	*Oscaecilia equatorialis*	E	EN	N/A	N/A
**Rhinatrematidae	*Amazops amazops*	E	DD	Non-described	N/A
**Rhinatrematidae	*Epicrionops bicolor*		VU	Non-described	N/A
**Rhinatrematidae	*Epicrionops petersi*		EN	Non-described	N/A
**Siphonopidae	*Microcaecilia albiceps*		VU	N/A	N/A
**Siphonopidae	*Siphonops annulatus*		LC	N/A	N/A
**Typhlonectidae	*Chthonerpeton onorei*	E	DD	Non-described	N/A
**Typhlonectidae	*Potomotyphlus kaupii*		VU	Non-described	N/A

Note: Call descriptions are the ones that include quantitative data. For descriptions of tadpoles see accounts. Information on calls is given under accounts and in the section Vocalizations of Ecuadorian anurans. End = endemism, E = endemic, CR = Critically Endangered, CR (PE) = Critically Endangered, Possibly Extinct, EN = Endangered, DD = Data Deficient, LC = Least Concern, NT = Near Threatened, VU = Vulnerable. * = Caudata, ** = Gymnophiona.

TABLE 0.3

Ecuadorian Ecosystems *sensu* Ministerio del Ambiente del Ecuador (2013) and the Abbreviations Used Herein

Ecosystem	Abbreviation
Deciduous Forest of Cordillera Costera of Ecuatorial Pacific	DFCCEP
Deciduous Forest of Lowlands of Jama-Zapotillo	DFLJZ
Deciduous Low Montane Forest of Catamayo-Alamor	DLMFCA
Deciduous Montane Foothill Forest of Catamayo-Alamor	DMFFCA
Deciduous Shrub and Grasslands of Beaches of Litoral	DSGBL
Desert Shrub of Lowlands of Jama-Zapotillo	DSLJZ
Desert Shrub of South of the Valleys	DSSV
Evergreen Forest of Lowlands of Aguarico-Putumayo-Caquetá	EFLAPC
Evergreen Forest of Lowlands with Bamboo of Amazonia	EFLBA
Evergreen Forest of Lowlands of Ecuatorial Chocó	EFLEC
Evergreen Forest of Lowlands of Fan of Pastaza	EFLFP
Evergreen Forest of Lowlands of Napo-Curaray	EFLNC
Evergreen Forest of Lowlands of Tigre-Pastaza	EFLTP
Evergreen Forest of Páramo	EFP
Evergreen Forest on Sandstone Plateaus of Cordillera del Cóndor in Lowland Ecuadorian Amazonia	EFSPCCLEA
Evergreen Grassland and Shrub of Páramo of Volcán Sumaco	EGSPVS
Evergreen High Montane Forest of Catamayo-Alamor	EHMFCA
Evergreen High Montane Forest of Cordillera Occidental of the Andes	EHMFCOA
Evergreen High Montane Forest of North of Cordillera Oriental of the Andes	EHMFNCOA
Evergreen High Montane Forest of South of Cordillera Oriental of the Andes	EHMFSCOA
Evergreen Low Montane Forest of Catamayo-Alamor	ELMFCA
Evergreen Low Montane Forest of Cordilleras Cóndor-Kutukú	ELMFCCK
Evergreen Low Montane Forest of Cordillera Occidental of the Andes	ELMFCOA
Evergreen Low Montane Forest of Galeras	ELMFG
Evergreen Low Montane Forest of North of Cordillera Oriental of the Andes	ELMFNCOA
Evergreen Low Montane Forest of Cordillera Costera of Ecuatorial Chocó	ELMFCCEC
Evergreen Low Montane Forest of South of Cordillera Oriental of the Andes	ELMFSCOA
Evergreen Low Montane Forest on Sandstone Plateaus of Cordilleras Cóndor-Kutukú	ELMFSPCCK
Evergreen Montane Forest of Cordilleras Cóndor-Kutukú	EMCCCK
Evergreen Montane Forest of Catamayo-Alamor	EMFCA
Evergreen Montane Forest of Cordillera Occidental of the Andes	EMFCOA
Evergreen Montane Foothill Forest of Catamayo-Alamor	EMFFCA
Evergreen Montane Foothill Forest of Cordilleras Cóndor-Kutukú	EMFFCCK
Evergreen Montane Foothill Forest of Cordillera Occidental of the Andes	EMFFCOA
Evergreen Montane Foothill Forest of Galeras	EMFFG
Evergreen Montane Foothill Forest of North of Cordillera Oriental of the Andes	EMFFNCOA
Evergreen Montane Foothill Forest on Outcrops of Limestone Rock of Amazonian Cordilleras	EMFFOLRAC
Evergreen Montane Foothill Forest of South of Cordillera Oriental of the Andes	EMFFSCOA
Evergreen Montane Foothill Forest on Sandstone Plateaus of Cordilleras Cóndor-Kutukú	EMFFSPCCK
Evergreen Montane Forest of North of the Cordillera Oriental of the Andes	EMFNCOA
Evergreen Montane Forest of South of the Cordillera Oriental of the Andes	EMFSCOA
Evergreen Montane Forest on Sandstone Plateaus of Cordillera del Cóndor	EMFSPCC
Evergreen Montane Shrub from North of the Andes	EMSNA
Evergreen Montane Shrub from South of the Andes	EMSSA
Evergreen Riparian Shrub from Cordillera Oriental of the Andes	ERSCOA
Evergreen Seasonal Flooded Forest of Alluvial Plain Jama-Zapotillo	ESFFAPJZ
Evergreen Seasonal Forest of Lowlands of Ecuatorial Chocó	ESFLEC
Evergreen Seasonal Forest of Lowlands of Jama-Zapotillo	ESFLJZ
Evergreen Shrub and Grassland of Páramo	ESGP
Evergreen Subnivarean Grassland and Shrub of Páramo	ESGSP
Evergreen Seasonal Low Montane Forest of Catamayo-Alamor	ESLMFCA

(Continued)

TABLE 0.3 (*Continued*)
Ecuadorian Ecosystems *sensu* Ministerio del Ambiente del Ecuador (2013) and the Abbreviations Used Herein

Ecosystem	Abbreviation
Evergreen Seasonal Low Montane Forest of Cordillera Costera of Ecuatorial Pacific	ESLMFCCEP
Evergreen Seasonal Montane Foothill Forest of Catamayo-Alamor	ESMFFCA
Evergreen Seasonal Montane Foothill Forest of Cordillera Costera of Ecuatorial Chocó	ESMFFCCEC
Evergreen Seasonal Montane Foothill Forest of Cordillera Costera of Ecuatorial Pacífic	ESMFFCCEP
Evergreen Seasonal Montane Foothill Forest of Cordillera Occidental of the Andes	ESMFFCOA
Evergreen Shrub and Montane Grassland of Cordillera del Cóndor	ESMGCC
Evergreen High Montane Shrub from Southern Páramo	EHMSSP
Flooded Forest of Alluvial Plain of Amazonia	FFAPA
Flooded Forest of Alluvial Plain of Ecuatorial Chocó	FFAPEC
Flooded Forest of Alluvial Plain of Rivers of Andean and Amazonian Cordilleras Origin	FFAPRAAC
Flooded Forest of Alluvial Plain of Rivers of Amazonian Origin	FFAPRAO
Flooded Forest of Intertidal Plain of Ecuatorial Chocó	FFIPEC
Flooded Forest and Lacustrine-riparian Vegetation of Amazonian Black Waters	FFLVABW
Flooded Forest of Palms of Alluvial Plain of Amazonia	FFPAPA
Flooded Grassland of Páramo	FGP
Flooded Lacustrine-riparian of the Alluvial Plain of Amazonia	FLAPA
Flooded Lacustrine Grassland of Ecuatorial Pacific	FLGEP
Flooded Riparian Grassland of Lowlands of Ecuatorial Chocó	FRGLEC
Flooded Riparian Grassland of Lowlands of Jama-Zapotillo	FRGLJZ
Forest and Semideciduous Shrub of North of the Valleys	FSSNV
Forest and Semideciduous Shrub of South of the Valleys	FSSSV
Grassland of Páramo	GP
High Superior Humid Montane Grassland of Páramo	HSHMGP
Low Forest and Deciduous Shrub of Lowlands of Jama-Zapotillo	LFDSLJZ
Low Montane Lacustrine Grassland of South of Cordillera Oriental of the Andes	LMLGSCOA
Mangrove of Ecuatorial Chocó	MEC
Mangrove of Jama-Zapotillo	MJZ
Rosette Caulescent and Grassland of Páramo (Frailejones)	RCGP
Semideciduous Forest of Cordillera Costera of Ecuatorial Pacific	SFCCEP
Semideciduous Forest of Lowlands of Jama-Zapotillo	SFLJZ
Semideciduous Low Montane Forest of Catamayo-Alamor	SLMFCA
Semideciduous Montane Foothill Forest of Catamayo-Alamor	SMFFCA
Semideciduous Montane Foothill Forest of South of Cordillera Oriental of the Andes	SMFFSCOA
Semideciduous Shrub of South of the Valleys	SSSV
Subnivarean Humid Grassland of Páramo	SHGP
Ultra Humid Subnivarean Grassland of Páramo	UHSGP

TABLE 0.4

Distribution of 655 Species of Ecuadorian Amphibians in 89 Ecosystems (*sensu* Ministerio del Ambiente del Ecuador, 2013).

Centrolenidae	*Centrolene ballux*	ELMFCOA, EMFCOA
Centrolenidae	*Centrolene buckleyi*	EMSNA, ESGP, EHMFCOA, ELMFCOA, ELMFNCOA, ELMFSCOA, EMFCOA, EMFNCOA, RCGP
Centrolenidae	*Centrolene charapita*	EMFFOLRAC
Centrolenidae	*Centrolene condor*	EMCCCK, EMFSPCC
Centrolenidae	*Centrolene geckoidea*	ELMFCOA, EMFCOA
Centrolenidae	*Centrolene heloderma*	ELMFCOA, EMFCOA
Centrolenidae	*Centrolene huilensis*	I
Centrolenidae	*Centrolene lynchi*	ELMFCOA, EMFCOA
Centrolenidae	*Centrolene medemi*	ELMFNCOA
Centrolenidae	*Centrolene peristicta*	ELMFCOA, EMFCOA
Centrolenidae	*Centrolene pipilata*	ELMFNCOA
Centrolenidae	*Centrolene sanchezi*	ELMFNCOA, EMFNCOA
Centrolenidae	*Chimerella mariaelenae*	ERSCOA, EFLFP, ELMFCCK, ELMFNCOA, ELMFSCOA, EMFFNCOA
Centrolenidae	*Cochranella granulosa*	EFLEC
Centrolenidae	*Cochranella litoralis*	EFLEC
Centrolenidae	*Cochranella mache*	EFLEC, ESFLJZ, ESMFFCCEC, ELMFCCEC, EMFFCOA
Centrolenidae	*Cochranella resplendens*	FFAPA, EFLAPC, EFLNC, EFLTP, EMFFNCOA
Centrolenidae	*Espadarana audax*	ELMFNCOA, ELMFSCOA, EMFNCOA, EMFFCCK, EMFFNCOA
Centrolenidae	*Espadarana callistomma*	EFLEC, EMFFCOA
Centrolenidae	*Espadarana durrellorum*	EFLNC, ELMFNCOA, EMFFSCOA
Centrolenidae	*Espadarana prosoblepon*	FFAPEC, SFCCEP, EFLEC, ESMFFCCEC, ESMFFCA, ELMFCCEC, ELMFCOA, EMFFCOA
Centrolenidae	*Hyalinobatrachium adespinosai*	ELMFNCOA
Centrolenidae	*Hyalinobatrachium aureoguttatum*	EFLEC
Centrolenidae	*Hyalinobatrachium chirripoi*	EFLEC, EMFFCOA
Centrolenidae	*Hyalinobatrachium iaspidiense*	EFLAPC, EFLNC, EMFFNCOA
Centrolenidae	*Hyalinobatrachium mashpi*	EMFFCOA
Centrolenidae	*Hyalinobatrachium munozorum*	FFAPA, EFLAPC, EFLNC, EMFFNCOA
Centrolenidae	*Hyalinobatrachium nouns*	EMFFCOA
Centrolenidae	*Hyalinobatrachium pellucidum*	ELMFNCOA, EMFNCOA, EMFFCCK, EMFFSCOA
Centrolenidae	*Hyalinobatrachium tatayoi*	EFLEC, ESFLJZ, ESLMFCCEP, ELMFCOA, MEC
Centrolenidae	*Hyalinobatrachium valerioi*	I
Centrolenidae	*Hyalinobatrachium yaku*	EFLTP, EMFFNCOA
Centrolenidae	*Nymphargus anomalus*	ELMFNCOA, ELMFSCOA
Centrolenidae	*Nymphargus balionotus*	ELMFCOA, EMFFCOA
Centrolenidae	*Nymphargus buenaventura*	ESMFFCOA, ESMFFCA
Centrolenidae	*Nymphargus cariticommatus*	ELMFSCOA, EMFSCOA
Centrolenidae	*Nymphargus cochranae*	ELMFCCK, ELMFNCOA, EMCCCK, EMFFCCK, EMFFNCOA, EMFFSCOA
Centrolenidae	*Nymphargus colomai*	EMFFOLRAC
Centrolenidae	*Nymphargus garciae*	EMFNCOA
Centrolenidae	*Nymphargus grandisonae*	ELMFCOA, EMFCOA
Centrolenidae	*Nymphargus griffithsi*	ELMFCOA, EMFCOA
Centrolenidae	*Nymphargus humboldti*	ELMFCCK, ELMFNCOA
Centrolenidae	*Nymphargus lasgralarias*	ELMFCOA, EMFCOA
Centrolenidae	*Nymphargus laurae*	ELMFNCOA
Centrolenidae	*Nymphargus lindae*	EMFFOLRAC
Centrolenidae	*Nymphargus manduriacu*	EMFFCOA
Centrolenidae	*Nymphargus mariae*	EFLTP, EMFFNCOA
Centrolenidae	*Nymphargus megacheirus*	ELMFNCOA
Centrolenidae	*Nymphargus megista*	ELMFCOA

(*Continued*)

TABLE 0.4 (*Continued*)

Distribution of 655 Species of Ecuadorian Amphibians in 89 Ecosystems (*sensu* Ministerio del Ambiente del Ecuador, 2013).

Centrolenidae	*Nymphargus posadae*	ELMFNCOA, ELMFSCOA, EMFNCOA, EMFFSCOA
Centrolenidae	*Nymphargus siren*	ELMFNCOA, EMFNCOA
Centrolenidae	*Nymphargus sucre*	EMFSCOA
Centrolenidae	*Nymphargus wileyi*	ELMFNCOA, EMFNCOA
Centrolenidae	*Rulyrana flavopunctata*	EFLFP, ELMFNCOA, EMFNCOA, EMFFNCOA
Centrolenidae	*Rulyrana mcdiarmidi*	SMFFSCOA, EFLFP, ELMFCCK, EMFFCCK, EMFFSCOA, EMFFOLRAC
Centrolenidae	*Sachatamia albomaculata*	EFLEC, ELMFCCEC, ELMFCOA, EMFFCOA
Centrolenidae	*Sachatamia ilex*	EFLEC, EMFFCOA
Centrolenidae	*Sachatamia orejuela*	EMFFCOA
Centrolenidae	*Teratohyla amelie*	EFLNC, EFLTP, EMFFCCK, EMFFNCOA, EMFFOLRAC
Centrolenidae	*Teratohyla midas*	FFAPRAAC, FFAPA, EFLFP, EFLAPC, EFLNC, EFLTP, EMFFCCK, EMFFNCOA
Centrolenidae	*Teratohyla pulverata*	EFLEC
Centrolenidae	*Teratohyla spinosa*	EFLEC, EMFFCOA
Centrolenidae	*Vitreorana ritae*	FFAPA, EFLNC
Ceratophryidae	*Ceratophrys cornuta*	FFAPA, EFLAPC, EFLNC
Ceratophryidae	*Ceratophrys stolzmanni*	LFDSLJZ, DFLJZ, SFCCEP, ESMFFCCEP
Eleutherodactylidae	*Adelophryne adiastola*	FFAPA, EFLTP
Eleutherodactylidae	*Diasporus gularis*	EFLEC, EMFFCOA
Hemiphractidae	*Gastrotheca andaquiensis*	ELMFCCK, ELMFNCOA, EMFFCCK, EMFFNCOA
Hemiphractidae	*Gastrotheca angustifrons*	EFLEC
Hemiphractidae	*Gastrotheca cornuta*	EFLEC, EMFFCOA
Hemiphractidae	*Gastrotheca cuencana*	EHMFCOA
Hemiphractidae	*Gastrotheca dendronastes*	I
Hemiphractidae	*Gastrotheca elicioi*	EMSSA, EHMFSCOA, EMFSCOA
Hemiphractidae	*Gastrotheca espeletia*	RCGP
Hemiphractidae	*Gastrotheca guentheri*	ELMFCOA, EMFCOA, EMFFCOA
Hemiphractidae	*Gastrotheca litonedis*	EMFCOA
Hemiphractidae	*Gastrotheca lojana*	EMSSA, EMFCA, GP
Hemiphractidae	*Gastrotheca longipes*	FFAPRAAC, FFAPA, FFPAPA, EFLFP, EFLNC, EFLTP, EMFFCCK, EMFFNCOA
Hemiphractidae	*Gastrotheca orophylax*	EHMFNCOA, EMFNCOA
Hemiphractidae	*Gastrotheca plumbea*	EHMFCOA, ELMFCOA, EMFCOA
Hemiphractidae	*Gastrotheca pseustes*	EMSNA, ESGP, EHMFCOA, EHMFCA, EHMFNCOA, EMFCOA, GP, HSHMGP, UHSGP, ESGSP
Hemiphractidae	*Gastrotheca psychrophila*	I
Hemiphractidae	*Gastrotheca riobambae*	EMSNA, ESGP, EHMFNCOA, EMFCOA, FSSNV, GP
Hemiphractidae	*Gastrotheca testudinea*	ELMFCCK, ELMFNCOA, EMFNCOA, EMFSCOA, EMFFCCK, EMFFNCOA
Hemiphractidae	*Gastrotheca turnerorum*	ESGP, EHMFSCOA, EMFCA, FGP
Hemiphractidae	*Gastrotheca weinlandii*	ELMFCCK, ELMFNCOA, ELMFSCOA, EMFSCOA, EMFFCCK, EMFFSCOA
Hemiphractidae	*Gastrotheca yacuri*	EMFCA
Hemiphractidae	*Hemiphractus bubalus*	ELMFNCOA, EMFNCOA, EMFFNCOA
Hemiphractidae	*Hemiphractus fasciatus*	I
Hemiphractidae	*Hemiphractus helioi*	I
Hemiphractidae	*Hemiphractus proboscideus*	FFAPRAAC, FFAPA, FFPAPA, EFLNC, EFLTP, EMFFCCK
Hemiphractidae	*Hemiphractus scutatus*	FFAPRAAC, FFAPA, EFLFP, EFLNC, EFLTP, ELMFCCK, ELMFNCOA, ELMFSPCCK, EMFFNCOA
Leptodactylidae	*Adenomera andreae*	FFAPRAO, FFAPRAAC, FFLVABW, FFAPA, FFPAPA, EFLFP, EFLAPC, EFLNC, ELMFNCOA, EMFFCCK, EMFFNCOA, EMFFSCOA, FLAPA
Leptodactylidae	*Adenomera hylaedactyla*	FFAPRAO, FFAPA, EFLNC, EFLTP, EHMFSCOA, ELMFNCOA, EMFNCOA, EMFFNCOA, EMFFOLRAC
Leptodactylidae	*Edalorhina perezi*	FFAPRAO, FFAPRAAC, FFAPA, FFPAPA, EFLFP, EFLNC, EFLTP, EMFNCOA,
Leptodactylidae	*Engystomops coloradorum*	EMFFCOA

TABLE 0.4 (*Continued*)

Distribution of 655 Species of Ecuadorian Amphibians in 89 Ecosystems (*sensu* Ministerio del Ambiente del Ecuador, 2013).

Leptodactylidae	*Engystomops guayaco*	DFLJZ, SFLJZ
Leptodactylidae	*Engystomops montubio*	LFDSLJZ, DFCCEP, SFLJZ, ESFLJZ
Leptodactylidae	*Engystomops petersi*	FFAPRAO, FFAPRAAC, FFLVABW, FFAPA, FFPAPA, EFLFP, EFLAPC, EFLNC, EMFFCCK, EMFFNCOA
Leptodactylidae	*Engystomops pustulatus*	DSGBL, LFDSLJZ, DFLJZ, SFLJZ, ESFLJZ, ESMFFCCEP
Leptodactylidae	*Engystomops puyango*	SSSV, LFDSLJZ, DFLJZ, DMFFCA, SLMFCA, SMFFCA, ESMFFCA
Leptodactylidae	*Engystomops randi*	LFDSLJZ, DFLJZ, ESFLJZ, ESMFFCA
Leptodactylidae	*Leptodactylus discodactylus*	FFAPRAO, FFAPRAAC, FFAPA, FFPAPA, EFLFP, EFLAPC, EFLNC, EFLTP, FLAPA
Leptodactylidae	*Leptodactylus knudseni*	FFAPRAO, FFAPA, EFLAPC, EFLNC, EFLTP, EMFFOLRAC
Leptodactylidae	*Leptodactylus labrosus*	SSSV, LFDSLJZ, DFCCEP, DFLJZ, DMFFCA, SFLJZ, SLMFCA, SMFFCA, ESFLEC, ESFLJZ, ESMFFCCEP, ESMFFCA, ELMFCCEC, EMFFCOA, MEC
Leptodactylidae	*Leptodactylus leptodactyloides*	FFAPRAAC, FFPAPA, EFLFP, EFLAPC, EFLNC, ELMFCCK, EMFFCCK, MEC
Leptodactylidae	*Leptodactylus melanonotus*	EFLEC
Leptodactylidae	*Leptodactylus mystaceus*	FFAPRAO, FFAPRAAC, FFAPA, FFPAPA, EFLFP, EFLAPC, EFLNC, EMFFNCOA
Leptodactylidae	*Leptodactylus pentadactylus*	FFAPRAO, FFAPRAAC, FFLVABW, FFAPA, FFPAPA, EFLFP, EFLAPC, EFLNC, EMFFG, EMFFNCOA, EMFFOLRAC
Leptodactylidae	*Leptodactylus peritoaktites*	ESFLJZ
Leptodactylidae	*Leptodactylus petersii*	FFAPA, EFLFP, EFLAPC
Leptodactylidae	*Leptodactylus rhodomerus*	EFLEC, ESMFFCCEC, ELMFCCEC
Leptodactylidae	*Leptodactylus rhodomystax*	FFAPRAO, FFAPRAAC, FFAPA, FFPAPA, EFLFP, EFLAPC, EFLNC, EFLTP, EMFFNCOA
Leptodactylidae	*Leptodactylus stenodema*	FFAPA, FFPAPA, EFLNC, EFLTP, ELMFNCOA
Leptodactylidae	*Leptodactylus ventrimaculatus*	DFLJZ, SFCCEP, SFLJZ, EFLEC, ESFLJZ, ESLMFCCEP, ELMFCCEC, ELMFCOA,
Leptodactylidae	*Leptodactylus wagneri*	ERSCOA, FFAPRAO, FFAPRAAC, FFAPA, FFPAPA, EFLFP, EFLAPC, EFLNC, ELMFCCK, ELMFNCOA, ELMFSCOA, ELMFSPCCK, EMFFCCK, EMFFN-COA, EMFFSCOA, EMFFOLRAC
Leptodactylidae	*Lithodytes lineatus*	FFAPRAO, FFAPRAAC, FFLVABW, FFAPA, FFPAPA, EFLFP, EFLAPC, EFLNC, ELMFCCK, EMFFG, EMFFCCK, EMFFNCOA, EMFFSCOA, EMFFOLRAC, EMFFSPCCK, FLAPA
Phyllomedusidae	*Agalychnis buckleyi*	ELMFNCOA
Phyllomedusidae	*Agalychnis hulli*	FFAPRAO, EFLFP, EFLNC, EFLTP, EMFFCCK, EMFFNCOA, EMFFOLRAC
Phyllomedusidae	*Agalychnis psilopygion*	EFLEC, ESFLJZ
Phyllomedusidae	*Agalychnis spurrelli*	SFLJZ, EFLEC, ESMFFCCEP, ESMFFCA, ELMFCCEC, EMFFCOA
Phyllomedusidae	*Callimedusa ecuatoriana*	ELMFCCK, ELMFSPCCK, EMCCCK, EMFSPCC, EMFFSCOA
Phyllomedusidae	*Callimedusa perinesos*	ELMFNCOA
Phyllomedusidae	*Callimedusa tomopterna*	FFAPRAO, FFAPA, FFPAPA, EFLFP, EFLAPC, EFLNC, EFLTP, EMFFCCK, EMFFOLRAC
Phyllomedusidae	*Cruziohyla calcarifer*	EFLEC, EMFFCOA
Phyllomedusidae	*Cruziohyla craspedopus*	FFAPRAO, FFAPRAAC, FFLVABW, FFAPA, FFPAPA, EFLAPC, EFLNC, EFLTP, EMFFOLRAC
Phyllomedusidae	*Phyllomedusa tarsius*	FFAPRAAC, FFAPA, EFLFP, EFLAPC, EFLNC, EFLTP, EMFFCCK, EMFFNCOA,
Phyllomedusidae	*Phyllomedusa vaillantii*	FFAPRAO, FFAPRAAC, FFAPA, FFPAPA, EFLFP, EFLAPC, EFLNC, EFLTP, EMFFNCOA, EMFFOLRAC
Phyllomedusidae	*Pithecopus palliatus*	EFLAPC
*Plethodontidae	*Bolitoglossa chica*	FFAPEC, EFLEC, ESFLJZ, EMFFCOA
*Plethodontidae	*Bolitoglossa equatoriana*	FFAPRAO, FFLVABW, FFAPA, FFPAPA, EFLAPC, EFLNC, EMFFNCOA
*Plethodontidae	*Bolitoglossa medemi*	EFLEC, EMFCOA, EMFFCOA
*Plethodontidae	*Bolitoglossa palmata*	ELMFNCOA, EMFNCOA

(*Continued*)

TABLE 0.4 (*Continued*)

Distribution of 655 Species of Ecuadorian Amphibians in 89 Ecosystems (*sensu* Ministerio del Ambiente del Ecuador, 2013).

*Plethodontidae	*Bolitoglossa sima*	EFLEC, ELMFCCEC, EMFFCOA, MEC
*Plethodontidae	*Oedipina ecuatoriana*	I
*Plethodontidae	*Oedipina villamizariorum*	I
**Caeciliidae	*Caecilia abitaguae*	ELMFNCOA
**Caeciliidae	*Caecilia albiventris*	I
**Caeciliidae	*Caecilia attenuata*	I
**Caeciliidae	*Caecilia bokermanni*	FFAPRAAC
**Caeciliidae	*Caecilia buckleyi*	ELMFCOA, EMFCOA
**Caeciliidae	*Caecilia crassisquama*	EMCCCK
**Caeciliidae	*Caecilia disossea*	FFAPRAAC, FFAPA, EFLNC, EFLTP, EMFFNCOA
**Caeciliidae	*Caecilia dunni*	I
**Caeciliidae	*Caecilia guntheri*	I
**Caeciliidae	*Caecilia leucocephala*	EFLEC, ESMFFCCEC, EMFFCOA
**Caeciliidae	*Caecilia nigricans*	EFLEC, ESFLJZ, ESLMFCCEP, ELMFCCEC, EMFFCOA
**Caeciliidae	*Caecilia orientalis*	ELMFNCOA, EMFNCOA, EMFFNCOA
**Caeciliidae	*Caecilia pachynema*	DSSV
**Caeciliidae	*Caecilia subterminalis*	I
**Caeciliidae	*Caecilia tentaculata*	FFAPA, EFLNC, EFLTP, EMFFNCOA
**Caeciliidae	*Caecilia tenuissima*	I
**Caeciliidae	*Oscaecilia bassleri*	FFAPA, EMFFNCOA
**Caeciliidae	*Oscaecilia equatorialis*	ESMFFCCEC, EMFFCOA
**Rhinatrematidae	*Amazops amazops*	I
**Rhinatrematidae	*Epicrionops bicolor*	ELMFCOA
**Rhinatrematidae	*Epicrionops petersi*	ELMFNCOA, EMFNCOA
**Siphonopidae	*Microcaecilia albiceps*	FFAPA, EFLTP
**Siphonopidae	*Siphonops annulatus*	FFAPA, FFPAPA, EFLAPC, EFLNC, EFLTP, EMFFNCOA
**Typhlonectidae	*Chthonerpeton onorei*	ELMFNCOA
**Typhlonectidae	*Potomotyphlus kaupii*	FFAPRAAC

Note: Water (W) and intervened (I) areas are also included. * = Caudata, ** = Gymnophiona. Abbreviations are given in Table 0.3.

TABLE 0.5
Distribution of 655 Species of Ecuadorian Amphibians in 15 Biogeographic Sectors (*sensu* Ministerio del Ambiente del Ecuador, 2013).

Family	Species	AC	APC	CA	CCC	CCEP	COA	EC	FP	JZ	NC	NCOA	P	SCOA	TP	V	# sectors
Centrolenidae	*Centrolene ballux*						X										1
Centrolenidae	*Centrolene buckleyi*						X					X	X	X			4
Centrolenidae	*Centrolene charapita*	X															1
Centrolenidae	*Centrolene condor*	X															1
Centrolenidae	*Centrolene geckoidea*						X										1
Centrolenidae	*Centrolene heloderma*						X										1
Centrolenidae	*Centrolene huilensis*											X					1
Centrolenidae	*Centrolene lynchi*						X										1
Centrolenidae	*Centrolene medemi*											X					1
Centrolenidae	*Centrolene peristicta*						X										1
Centrolenidae	*Centrolene pipilata*											X					1
Centrolenidae	*Centrolene sanchezi*	X										X					2
Centrolenidae	*Chimerella mariaelenae*	X							X			X		X			4
Centrolenidae	*Cochranella granulosa*						X	X									2
Centrolenidae	*Cochranella litoralis*						X	X									2
Centrolenidae	*Cochranella mache*				X		X	X		X							4
Centrolenidae	*Cochranella resplendens*		X								X	X		X	X		5
Centrolenidae	*Espadarana audax*	X										X		X			3
Centrolenidae	*Espadarana callistomma*						X	X									2
Centrolenidae	*Espadarana durrellorum*	X									X	X		X			4
Centrolenidae	*Espadarana prosoblepon*			X	X	X	X	X		X							6
Centrolenidae	*Hyalinobatrachium adespinosai*											X					1
Centrolenidae	*Hyalinobatrachium aureoguttatum*						X	X									2
Centrolenidae	*Hyalinobatrachium chirripoi*						X	X		X							3
Centrolenidae	*Hyalinobatrachium iaspidiense*	X									X	X					3
Centrolenidae	*Hyalinobatrachium mashpi*						X										1
Centrolenidae	*Hyalinobatrachium munozorum*	X	X								X	X					4
Centrolenidae	*Hyalinobatrachium nouns*						X										1
Centrolenidae	*Hyalinobatrachium pellucidum*	X										X		X			3
Centrolenidae	*Hyalinobatrachium tatayoi*					X	X	X		X							4
Centrolenidae	*Hyalinobatrachium valerioi*																0
Centrolenidae	*Hyalinobatrachium yaku*										X	X			X		3
Centrolenidae	*Nymphargus anomalus*											X		X			2
Centrolenidae	*Nymphargus balionotus*						X										1

(*Continued*)

TABLE 0.5 *(Continued)*

Distribution of 655 Species of Ecuadorian Amphibians in 15 Biogeographic Sectors (*sensu* Ministerio del Ambiente del Ecuador, 2013).

Family	Species	AC	APC	CA	CCC	CCEP	COA	EC	FP	JZ	NC	NCOA	P	SCOA	TP	V	# sectors
Centrolenidae	*Nymphargus buenaventura*			X			X										2
Centrolenidae	*Nymphargus cariticommatus*													X			1
Centrolenidae	*Nymphargus cochranae*	X										X		X			3
Centrolenidae	*Nymphargus colomai*	X															1
Centrolenidae	*Nymphargus garciae*											X					1
Centrolenidae	*Nymphargus grandisonae*						X										1
Centrolenidae	*Nymphargus griffithsi*						X										1
Centrolenidae	*Nymphargus humboldti*	X										X					2
Centrolenidae	*Nymphargus lasgralarias*						X										1
Centrolenidae	*Nymphargus laurae*										X						1
Centrolenidae	*Nymphargus lindae*	X															1
Centrolenidae	*Nymphargus manduriacu*						X										1
Centrolenidae	*Nymphargus mariae*										X	X			X		3
Centrolenidae	*Nymphargus megacheirus*											X					1
Centrolenidae	*Nymphargus megista*						X										1
Centrolenidae	*Nymphargus posadae*											X		X			2
Centrolenidae	*Nymphargus siren*											X					1
Centrolenidae	*Nymphargus sucre*													X			1
Centrolenidae	*Nymphargus wileyi*											X					1
Centrolenidae	*Rulyrana flavopunctata*								X		X	X					3
Centrolenidae	*Rulyrana mcdiarmidi*	X							X					X			3
Centrolenidae	*Sachatamia albomaculata*				X		X	X									3
Centrolenidae	*Sachatamia ilex*						X	X									2
Centrolenidae	*Sachatamia orejuela*						X										1
Centrolenidae	*Teratohyla amelie*	X									X	X			X		4
Centrolenidae	*Teratohyla midas*	X	X						X		X	X			X		6
Centrolenidae	*Teratohyla pulverata*						X	X									2
Centrolenidae	*Teratohyla spinosa*						X	X									2
Centrolenidae	*Vitreorana ritae*		X								X						2
Ceratophryidae	*Ceratophrys cornuta*		X								X			X	X		4
Ceratophryidae	*Ceratophrys stolzmanni*					X				X							2
Eleutherodactylidae	*Adelophryne adiastola*														X		1
Eleutherodactylidae	*Diasporus gularis*						X	X									2
Hemiphractidae	*Gastrotheca andaquiensis*	X															1
Hemiphractidae	*Gastrotheca angustifrons*							X									1
Hemiphractidae	*Gastrotheca cornuta*						X	X		X							3

TABLE 0.5 (*Continued*)
Distribution of 655 Species of Ecuadorian Amphibians in 15 Biogeographic Sectors (*sensu* Ministerio del Ambiente del Ecuador, 2013).

Family	Species	AC	APC	CA	CCC	CCEP	COA	EC	FP	JZ	NC	NCOA	P	SCOA	TP	V	# sectors
Hemiphractidae	*Gastrotheca cuencana*						X						X	X			3
Hemiphractidae	*Gastrotheca dendronastes*						X										1
Hemiphractidae	*Gastrotheca elicioi*			X									X	X			3
Hemiphractidae	*Gastrotheca espeletia*						X					X	X				3
Hemiphractidae	*Gastrotheca guentheri*						X										1
Hemiphractidae	*Gastrotheca litonedis*						X										1
Hemiphractidae	*Gastrotheca lojana*			X			X						X			X	4
Hemiphractidae	*Gastrotheca longipes*	X	X						X		X	X			X		6
Hemiphractidae	*Gastrotheca orophylax*											X					1
Hemiphractidae	*Gastrotheca plumbea*						X					X					2
Hemiphractidae	*Gastrotheca pseustes*			X			X					X	X	X		X	6
Hemiphractidae	*Gastrotheca psychrophila*													X			1
Hemiphractidae	*Gastrotheca riobambae*						X					X	X			X	4
Hemiphractidae	*Gastrotheca testudinea*	X									X	X		X			4
Hemiphractidae	*Gastrotheca turnerorum*												X				1
Hemiphractidae	*Gastrotheca weinlandii*	X										X		X			3
Hemiphractidae	*Gastrotheca yacuri*												X				1
Hemiphractidae	*Hemiphractus bubalus*											X		X			2
Hemiphractidae	*Hemiphractus fasciatus*						X										1
Hemiphractidae	*Hemiphractus helioi*											X					1
Hemiphractidae	*Hemiphractus proboscideus*	X	X								X				X		4
Hemiphractidae	*Hemiphractus scutatus*	X	X						X		X	X			X		6
Leptodactylidae	*Adenomera andreae*	X	X						X		X	X		X	X		7
Leptodactylidae	*Adenomera hylaedactyla*	X	X								X	X	X	X	X		7
Leptodactylidae	*Edalorhina perezi*		X						X		X	X			X		5
Leptodactylidae	*Engystomops coloradorum*						X										1
Leptodactylidae	*Engystomops guayaco*							X		X							2
Leptodactylidae	*Engystomops montubio*					X				X							2
Leptodactylidae	*Engystomops petersi*	X	X						X		X	X		X	X		7
Leptodactylidae	*Engystomops pustulatus*			X	X	X		X		X							5
Leptodactylidae	*Engystomops puyango*			X						X						X	3
Leptodactylidae	*Engystomops randi*			X						X							2
Leptodactylidae	*Leptodactylus discodactylus*	X	X						X		X	X			X		6
Leptodactylidae	*Leptodactylus knudseni*	X	X								X	X		X	X		6

(*Continued*)

TABLE 0.5 (*Continued*)

Distribution of 655 Species of Ecuadorian Amphibians in 15 Biogeographic Sectors (*sensu* Ministerio del Ambiente del Ecuador, 2013).

Family	Species	AC	APC	CA	CCC	CCEP	COA	EC	FP	JZ	NC	NCOA	P	SCOA	TP	V	# sectors
Leptodactylidae	*Leptodactylus labrosus*			X	X	X	X	X		X						X	7
Leptodactylidae	*Leptodactylus leptodactyloides*	X	X						X		X				X		5
Leptodactylidae	*Leptodactylus melanonotus*							X		X							2
Leptodactylidae	*Leptodactylus mystaceus*		X						X		X	X			X		5
Leptodactylidae	*Leptodactylus pentadactylus*	X	X						X		X	X			X		6
Leptodactylidae	*Leptodactylus peritoaktites*						X	X		X							3
Leptodactylidae	*Leptodactylus petersii*		X						X						X		3
Leptodactylidae	*Leptodactylus rhodomerus*				X		X	X									3
Leptodactylidae	*Leptodactylus rhodomystax*		X						X		X	X			X		5
Leptodactylidae	*Leptodactylus stenodema*		X						X		X	X			X		5
Leptodactylidae	*Leptodactylus ventrimaculatus*				X	X	X	X		X							5
Leptodactylidae	*Leptodactylus wagneri*	X	X						X		X	X		X	X		7
Leptodactylidae	*Lithodytes lineatus*	X	X						X		X	X		X	X		7
Phyllomedusidae	*Agalychnis buckleyi*											X					1
Phyllomedusidae	*Agalychnis hulli*	X	X						X		X	X			X		6
Phyllomedusidae	*Agalychnis psilopygion*							X		X							2
Phyllomedusidae	*Agalychnis spurrelli*			X	X	X	X	X		X							6
Phyllomedusidae	*Callimedusa ecuatoriana*	X												X			2
Phyllomedusidae	*Callimedusa perinesos*											X					1
Phyllomedusidae	*Callimedusa tomopterna*	X	X						X		X	X		X	X		7
Phyllomedusidae	*Cruziohyla calcarifer*						X	X									2
Phyllomedusidae	*Cruziohyla craspedopus*	X	X								X				X		4
Phyllomedusidae	*Phyllomedusa tarsius*	X	X						X		X	X		X	X		7
Phyllomedusidae	*Phyllomedusa vaillantii*	X	X						X		X	X		X	X		7
Phyllomedusidae	*Pithecopus palliatus*		X														1
*Plethodontidae	*Bolitoglossa chica*						X	X		X							3
*Plethodontidae	*Bolitoglossa equatoriana*		X								X	X			X		4
*Plethodontidae	*Bolitoglossa medemi*						X	X									2
*Plethodontidae	*Bolitoglossa palmata*											X					1
*Plethodontidae	*Bolitoglossa sima*				X		X	X									3
*Plethodontidae	*Oedipina ecuatoriana*						X										1
*Plethodontidae	*Oedipina villamizariorum*						X										1
**Caeciliidae	*Caecilia abitaguae*	X					X										2
**Caeciliidae	*Caecilia albiventris*		X														1
**Caeciliidae	*Caecilia attenuata*											X					1
**Caeciliidae	*Caecilia bokermanni*														X		1

TABLE 0.5 (*Continued*)

Distribution of 655 Species of Ecuadorian Amphibians in 15 Biogeographic Sectors (*sensu* Ministerio del Ambiente del Ecuador, 2013).

Family	Species	AC	APC	CA	CCC	CCEP	COA	EC	FP	JZ	NC	NCOA	P	SCOA	TP	V	# sectors
**Caeciliidae	*Caecilia buckleyi*						X										1
**Caeciliidae	*Caecilia crassisquama*	X															1
**Caeciliidae	*Caecilia disossea*		X								X	X			X		4
**Caeciliidae	*Caecilia dunni*											X					1
**Caeciliidae	*Caecilia guntheri*						X										1
**Caeciliidae	*Caecilia leucocephala*					X	X	X									3
**Caeciliidae	*Caecilia nigricans*				X	X	X	X		X							5
**Caeciliidae	*Caecilia orientalis*											X					1
**Caeciliidae	*Caecilia pachynema*			X			X									X	3
**Caeciliidae	*Caecilia subterminalis*																0
**Caeciliidae	*Caecilia tentaculata*		X								X	X			X		4
**Caeciliidae	*Caecilia tenuissima*									X							1
**Caeciliidae	*Oscaecilia bassleri*		X								X	X					3
**Caeciliidae	*Oscaecilia equatorialis*						X	X									2
**Rhinatrematidae	*Amazops amazops*											X					1
**Rhinatrematidae	*Epicrionops bicolor*			X			X										2
**Rhinatrematidae	*Epicrionops petersi*											X		X			2
**Siphonopidae	*Microcaecilia albiceps*		X												X		2
**Siphonopidae	*Siphonops annulatus*		X						X		X	X			X		5
**Typhlonectidae	*Chthonerpeton onorei*											X					1
**Typhlonectidae	*Potomotyphlus kaupii*														X		1

Note: * = Caudata, ** = Gymnophiona. Abbreviations are as follows: Amazonian Cordilleras = AC; Aguarico-Putumayo-Caquetá = APC; Catamayo-Alamor = CA; Cordillera Costera of Chocó = CCC; Cordillera Costera of Equatorial Pacific = CCEP; Cordillera Occidental of the Andes = COA; Equatorial Chocó = EC; Fan of Pastaza = FP; Jama-Zapotillo = JZ; Northern Cordillera Oriental of the Andes = NCOA; Napo-Curaray = NP; Páramo = P; Southern Cordillera Oriental of the Andes = SCOA; Tigre-Pastaza = TP; Valleys = V.

1 Anura: Family Phyllomedusidae

LEAF FROGS

Phyllomedusidae Günther, 1858: 346. **Type genus.**—*Phyllomedusa* (Wagler, 1830).

Etymology.—The familial name is derived from the Greek words *phyll*, meaning leaf, and *Medousa*. The name alludes to the gelatinous egg masses deposited on leaves of trees.

Definition.—Phyllomedusid frogs are a monophyletic group diagnosed on the basis of molecular evidence and phenotypic features. Putative myological synapomorphies are: (1) submandibular intermandibularis muscle bearing a wide aponeurosis, and (2) posterolateral supplementary elements of the submandibular intermandibularis present, inserted in the ventral surface of the principal element of the intermandibularis.

The SVL varies from 45 mm in *Callimedusa atelopoides* to 119 mm in *Phyllomedusa bicolor*. The iris is vertically elliptical. The pectoral gridle is arciferal. They have eight procoelous presacral vertebrae. The calcaneum and astragalus are separated. There are intercalary cartilages between the terminal and penultimate phalanges. The terminal phalanges are claw-shaped. The chromosome complement is 2n = 26.

Distribution.—Phyllomedusids occur in the Neotropics, from Mexico to northern Argentina.

Content.—The family contains eight genera with 67 species. Five genera (*Agalychnis*, *Callimedusa*, *Cruzyohyla*, *Phyllomedusa*, *Pithecopus*) occur in Ecuador.

Remarks.—Figure 1.1. Phyllomedusids have been treated as a family by Günther (1858), Bossuyt and Roelants (2009), Duellman *et al.* (2016), Schmid *et al.* (2018). We adopt the family category for this monophyletic group instead of the subfamily rank (under Hylidae) that is chosen by other authors (see Frost, 2023). Faivovich *et al.* (2010) performed a parsimony analysis of 45 species and Duellman *et al.* (2016) a maximum likelihood analysis of 46 species, 78% of the known members of the family. The same lineages emerge. The latter authors recovered a major lineage of phyllomedusids with 100% support. Within this group, there are large frogs of the genus *Phyllomedusa* (100% support) with vomerine teeth and another lineage with 100% support of smaller frogs lacking vomerine teeth (100% support). Within the latter are two lineages with significant support. One lineage, *Pithecopus* (100% support), is characterized by having opposable thumbs and no bright flash colors. In the second lineage, *Callimedusa* (78% support), the thumb is not opposable, and with one exception, all have bright flash colors.

GENUS *AGALYCHNIS* (LEAF FROGS) COPE, 1864

Agalychnis Cope, 1864: 181. **Type species.**—*Hyla callidryas* (Cope, 1862) by original designation.

Etymology.—The generic name is from the Greek *aga*, an intensive prefix, and the Greek *lychnis*, a plant with scarlet flowers. Presumably, the name refers to the red-eyed treefrog, *Agalychnis callidryas*, the type species of the genus. The gender is feminine.

Definition.—Figure 1.2. *Agalychnis* is a group supported by molecular evidence. No morphological synapomorphies are known. The genus is composed of a range of similar-looking frogs, from the slender *A. lemur*, reaching a maximum SVL of 47 mm, to the larger-sized *A. spurrelli* at 110 mm; others such as those in the *A. callidryas* group are medium-sized, and others such as *A. dacnicolor* are more heavily built. This genus only includes phyllomedusids that have a single rather than bicolored iris (a characteristic seen in *Cruziohyla* and *Phrynomedusa*). The parotid glands in *Agalychnis* are absent or poorly developed. Extensive webbing is present on both the hands and feet of many *Agalychnis* species but is lacking or well reduced in those species previously named *Hylomantis* and *Pachymedusa*. Vomerine teeth are present, and most species that do have extensive webbing (except *A. dacnicolor*) also have a reticulated palpebral membrane to cover the eye. Egg-laying occurs on open leaves rather than folded, as seen in *Phasmahyla* and *Phyllomedusa*, or on tree trunks and lianas. The tadpoles of *Agalychnis* all have moderately small oral discs that are directed anteroventrally.

Distribution.—The genus occurs in the Pacific lowlands of Mexico from southern Sonora south, including the Balsas Depression and the State of Mexico, to the Isthmus of Tehuantepec; tropical southern Mexico and Central America; the Pacific lowlands of Colombia and northwestern Ecuador; and the upper Amazon Basin and lower Andean slopes in Colombia, Venezuela, and northeastern Peru, likely into eastern Ecuador.

Content.—Currently, 14 species are recognized in this genus. Four species are known in Ecuador.

Remarks.—Figure 1.3. Gomez-Mestre *et al.* (2008), Faivovich *et al.* (2010), and Pyron and Wiens (2011) provided a molecular phylogeny of species currently under *Agalychnis*. According to Duellman *et al.* (2016), the inclusion of *Agalychnis hulli* and *A. lemur* in this genus is problematic, even though their respective support values (88% and 75%) are relatively high. *Agalychnis hulli* is a geographic enigma inasmuch as it is a member of the

DOI: 10.1201/9781003543879-1

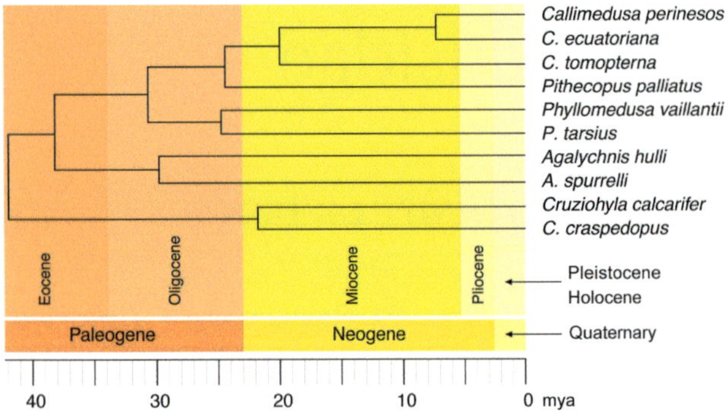

FIGURE 1.1 Time-calibrated maximum likelihood phylogenetic tree depicting evolutionary relationships of some of the Ecuadorian species of Phyllomedusidae, inferred from DNA sequences of mitochondrial and nuclear genes. Tree by Juan C. Santos (this publication).

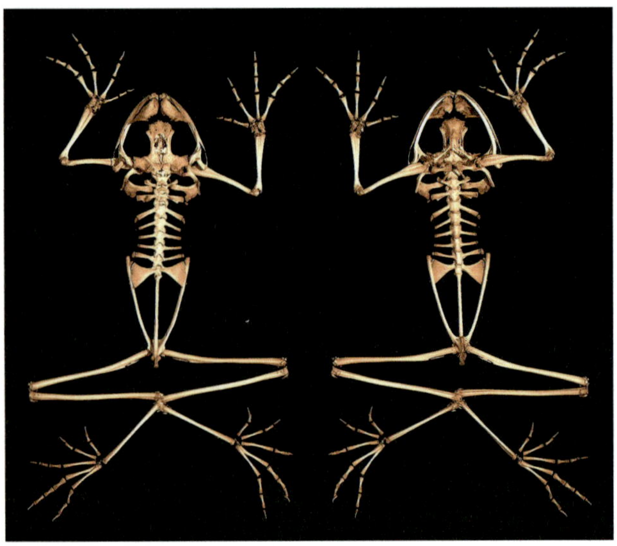

FIGURE 1.2 High-resolution computed tomography reconstructions of the skeleton of *Agalychnis spurrelli*, CJ 2224, Reserva Otokiki, Alto Tambo, Provincia Esmeraldas. Left: dorsal view of the skeleton. Dorsal (left) and ventral (right) views. Daniel J. Paluh.

"*Phyllomedusa buckleyi* group" as defined by Cannatella (1980) and exists in the Amazon Basin (Duellman and Mendelson, 1995), whereas other members of the group live in cloud forests on the slopes of the Andes and Central American highlands. *Agalychnis lemur* is the only species in the so-called "*Phyllomedusa buckleyi* group" in their analysis. Molecular data are needed from *A. buckleyi*, *A. danieli*, *A. medinae*, and *A. psilopygion* before the phylogenetic relationships can be clarified for a suitable classification.

Gomez-Mestre *et al.* (2008) studied the evolution of hatching plasticity in *Agalychnis* in a phylogenetic context. According to them, premature hatchlings of *Agalychnis* are more vulnerable to fish predation than are full-term hatchlings, indicating a conserved risk trade-off across hatching that would make plasticity advantageous. In contrast, the hatching response to snake attack has undergone major changes at least

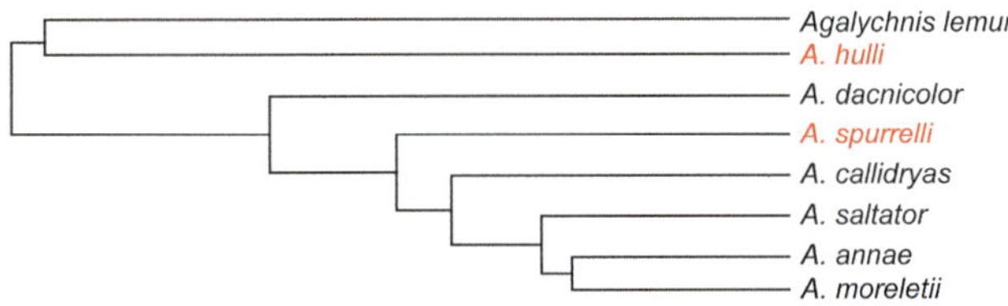

FIGURE 1.3 Topology of a tree analysis of *Agalychnis* based on molecular data, showing species occurring in Ecuador (in red). Taken and modified from Pyron and Wiens (2011).

twice in the *Agalychnis* lineage, with two species showing substantially lower escape success than the others. Responses to different threats have thus evolved independently.

In the 2010s, the United States of America alone imported 221,960 *Agalychnis* frogs, according to the Species Survival Network (SSN). In a resolution adopted on 2 March 2010 by the Convention on International Trade in Endangered Species (CITES) of Wild Fauna and Flora, all species belonging to the genus *Agalychnis* at that time were listed in Appendix II.

AGALYCHNIS BUCKLEYI (BUCKLEY'S LEAF FROG) (BOULENGER, 1882a)

Phyllomedusa buckleyi (Boulenger, 1882a): 425. **Holotype.**—BMNH 1947.2.22.35 from Sarayacu, Provincia Pastaza, Ecuador.

Agalychnis buckleyi (Faivovich *et al.*, 2010): 259.

Etymology.—The specific name is a patronym that honors Clarence Buckley (1868–1878), who collected the holotype and many other specimens for the British Museum (Natural History).

Identification.—Figure 1.4. In these frogs, males attain an SVL of 44.5 mm, and females of 54.7 mm. The head is slightly wider than the body. The snout is short, acutely rounded in dorsal view; in lateral view, it is sloping and rounded from lip to nostril in both sexes. The canthus rostralis is rounded, distinct. The loreal region is concave. The lips are thin and not flared. The nostrils are not protuberant, directed laterally. The internarial region is flat. The eyes are large and protuberant. The pupil is vertically elliptical. The eyelid is clear. The parotoid glands are not differentiated. The supratympanic fold is thin, indistinct, obscuring the dorsal edge of the tympanum and up to one-half of the tympanum in some; the fold extends posteroventrally to a point above the insertion of the arm. The tympanum is barely distinct, round, separated from eye by distance equal to the horizontal diameter of the tympanum. The upper arm is slender; the forearm is robust. The ulnar fold is low,

FIGURE 1.4 *Agalychnis buckleyi*. QCAZ 680, subadult, km 10, road Hollín-Loreto, Provincia Napo. LAC.

indistinct. The fingers are moderately long. The relative length of fingers from shortest to longest is 1-2-4-3. The digital discs are of moderate size, rounded. The subarticular tubercles are large and subconical. The distal tubercle on finger IV is bifid in most specimens. The palmar tubercle is low, diffuse, rounded. The prepollex is slightly enlarged and bears thin, horny nuptial excrescences in breeding males. There is slight rudimentary webbing between the last three fingers. The legs are moderately long and slender. There is a small, blunt calcar on the heel. There are a few small tubercles below the calcar. The inner tarsal fold is barely distinct; the outer tarsal fold is barely distinct, appearing merely as a row of low tubercles in some. The toes are of moderate length; the relative length from shortest to longest is 1-2-3-5-4; the discs are rounded, smaller than those on the fingers. The inner metatarsal tubercle is flattened, low, and elliptical. The outer metatarsal tubercle is small and rounded. Webbing is absent between toes I and II. There is rudimentary webbing between toes II and III and basal webbing between other toes. The anal opening is directed ventrally at the midlevel of the thighs. There is a short anal flap. The para-anal region is tubercular. The skin on the dorsum is minutely granular. There are white dorsal warts on some specimens and small granules in the loreal and temporal regions. The skin on throat, belly, and posteroventral surface of the thighs is distinctly granular; the skin elsewhere is smooth. The tongue is lanceolate, notched posteriorly, free for about one-half of its length. Vomerine teeth are present in most individuals; the dentigerous processes of the prevomers are small, moderately separated medially, oriented posteromedially at the anterior level of the choanae. Vocal slits are present in males; they are short, parallel to the jaw, extending from the posterolateral corner of the tongue to the corner of the mouth. The vocal sac is single, median, and subgular. The quadratojugal is present. The sacro-coccygeal articulation is bicondylar.

The dorsum is lavender–brown by night; by day, the dorsum is pale green to yellow–green with purple flecks. The dorsal surfaces of the toe and finger discs are yellowish-white with green wash. The ventral surfaces of hands and feet are flesh-colored. The venter is cream with an orange wash; the throat is off-white with a pale orange wash. The flanks and concealed surfaces are bright orange, fading to cream on the edges. The ulnar and tarsal folds and the anal tubercles are off-white. The dorsal warts are pale yellow. Finger IV is green dorsally, fading to pale yellow. Finger III is dirty yellow. Fingers I and II are orange dorsally. The eyelid is clear; the upper border is slightly pigmented. The iris is dirty cream. In a newly metamorphosed individual, the dorsal surfaces of the body and limbs are pale green with a yellow wash. The dorsal surface of the toe pads and concealed surfaces of the toes are dull orange; the dorsal wart is light yellow. The venter is fleshy orange. The upper lip is dirty yellow. The iris is off-white.

Comparisons. *Agalychnis hulli*, *A. psilopygion*. *Agalychnis buckleyi* differs from *A. hulli* by having a smaller tympanum (mean tympanum/eye = 0.352) that is barely evident, whereas the larger tympanum (mean tympanum/eye = 0.489) is distinct in *A. hulli*. Furthermore, *A. buckleyi* differs from *A. hulli* by having a small calcar and usually haying a small outer metatarsal tubercle. The dorsal skin is finely granular in *A. buckleyi* and coarsely granular in *A. hulli*. *Agalychnis psilopygion*, from the Pacific lowlands, has a large calcar and no para-cloacal tubercles.

Natural history.—Most individuals have been found at night on low vegetation overhanging pools of water. From field observations, the call is known to be a short "cluck." A female in amplexus deposited 98 eggs in a plastic bag. A similar sized clutch was found on vegetation above water. The eggs are quite large, having a total egg capsule diameter of approximately 4.5 mm at stage 8; that of the egg yolk alone is about 3.1 mm in size. The yolk is green, and the jelly is clear. Tadpoles have been found in streamside pools, temporary ponds, and puddles. *Agalychnis buckleyi* is sympatric with *Phyllomedusa vaillantii* and *P. tarsius* in the Río Pastaza Valley, Provincia Pastaza, and with *Callimedusa perinesos* in the Río Quijos Valley, Provincia Napo. At Río Salado, Provincia Napo, on 17–19 March 1975, 12 *C. perinesos* were taken from low vegetation around the pond and

FIGURE 1.5 Tadpole and metamorph of *Agalychnis buckleyi*, Top four: El Reventador, La Virgen, Provincia Napo. LAC. Bottom: CJ 13068, stage 46 = 28.6 mm SVL, trail Cascadas de Baeza, Provincia Napo. SGS.

cutover forest at the same locality. Seven *A. buckleyi* were collected in low vegetation adjacent to the pond; none were taken from the cutover forest. At this same site, about 60 *C. perinesos* and five *A. buckleyi* were collected on the evening of 17 July 1977 following a light rain. All of the *A. buckleyi* were taken from low vegetation overhanging a small pond; the specimens of *C. perinesos*, including amplectant pairs, were also found around the same pond, but the vast majority were in the surrounding forest clearing, several meters from water. These limited data suggest spatial segregation for the two species.

Tadpoles.—Figure 1.5. The tadpole was described based on 45 tadpoles from the Río Azuela and Río Salado, Provincia Napo, Ecuador, ranging in development from stage 25 to 41. They have body lengths of 8.4 – 24.1 mm and a total length of 22.1 – 58.0 mm The following description is based on two of these specimens. The body is as wide as deep, deepest and widest at two-thirds the length of the body. The top of the head is slightly convex. The snout is acutely rounded in lateral profile; in dorsal view, it is almost truncate. The nostrils are dorsolateral, directed anterolaterally. The internarial distance is less than the width of the oral disc. The eyes are dorsolateral and directed laterally. The spiracle is a flap-like tube, ventral and sinistral to the midline. The spiracular opening is at about the midlength of the body. The chondrocranial elements are not visible through the skin. The mouth is anteroventral and directed anteriorly. The cloacal tube is short and dextral to the caudal fin. The slender caudal musculature tapers gradually to the posterior end of the fin. The myomeres are moderately developed. At the midlength of the tail, the depth of the caudal musculature is slightly less than the depth of the ventral fin but greater than that of the dorsal fin. The musculature extends to the tip of the tail. The dorsal fin is shallow anteriorly, not extending onto the body; the fin is deepest at two-thirds its length from the anterior portion; the ventral fin is deepest at the midlength. The mouth is small, with a shallow lateral fold. The medial portion of the upper lip lacks papillae; elsewhere, papillae are present in two to three rows along the border; a few papillae are present medially to the border in the region of the lateral fold. The upper beak is moderately deep, with about 12 coarse serrations medially, otherwise finely serrate. The lower beak is shallow and finely serrate. There are two upper and three lower rows of denticles. The upper rows are of the same length; the first upper row is uninterrupted; the second row is narrowly broken medially. The three lower rows are of equal length, unbroken; the denticles of the first upper and third lower rows are smaller than those of other rows.

The body is olive brown above. The venter is silver–yellow to creamy green with greenish-gold iridescence laterally. The caudal musculature is cream to tan. The fins are transparent. The iris is silver to bronze. Metamorphs' emerging forelimbs have a green dorsum.

Distribution.—Figure 1.6. *Agalychnis buckleyi* occurs in the upper Amazon Basin and lower Amazonian slopes of the Andes in Ecuador and Colombia (Caquetá and Santander). In Ecuador, it occurs in Sucumbíos, Napo, Pastaza, and Tungurahua provinces. Its elevational range is 1176–1674 m with an extent of occurrence of about 2950 km^2.

This species inhabits subtropical forests and intervening areas in the evergreen low montane forest ecosystem of the northern Cordillera Oriental of the Andes biogeographic region, where the average annual rainfall is 1964–3600 mm and the average annual temperature is 18.1–24 °C.

Conservation.—*Agalychnis buckleyi* is categorized as Least Concern throughout its entire range (Castro *et al.*, 2004b; in IUCN Red List of 2022). In Ecuador, Ortega-Andrade *et al.* (2021) categorized it as Vulnerable. According to the latter authors, in Ecuador, this species exhibits an extent of occurrence of about 12,917 km^2. Our calculations of the area of occurrence in Ecuador are much lower (2950 km^2); thus, we suggest including it in the Endangered category given that its extent of occurrence is less than 5000 km^2 and there is a continuing decline in the area, extent, and/or quality of habitat. This species is mostly threatened by agriculture, roads, oil pipelines, a hydroelectric plant, deforestation, and volcanism from the El Reventador volcano. It is an uncommon species. In Colombia, the Santander population overlaps with the Guanenta-Alto Río Fonce National Park. In Ecuador, it has been recorded in

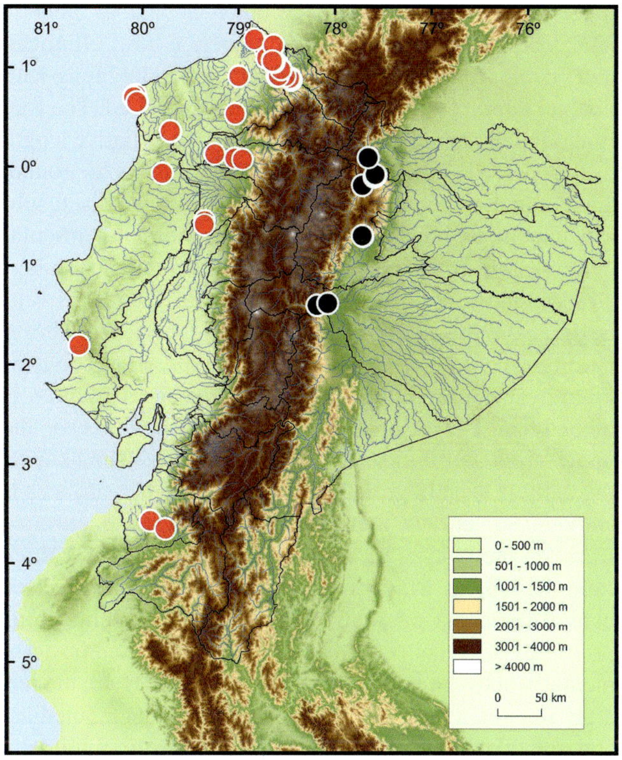

FIGURE 1.6 Localities of known occurrence of *Agalychnis buckleyi* (black dots) and *A. spurrelli* (red dots) in Ecuador.

the public reserves Parque Nacional Cayambe Coca and Parque Nacional Llanganates, the protected forests of Cerro Sumaco y Cuenca Alta del Río Suno, and the biosphere reserve Sumaco.

Remarks.—Its molecular phylogenetic relationships are unknown. It is in the *Agalychnis buckleyi* group *sensu* Faivovich *et al.* (2005). But Faivovich *et al.* (2010) excluded this species from any formal species group. The holotypes of *P. buckleyi* and its two junior synonyms (*Phyllomedusa loris* and *Hyla porifera*) are from the Río Pastaza drainage. Taxonomic research is needed to determine whether or not there are two different species in Colombia, considering the disjointed nature of the species distribution.

Key references.—(Boulenger, 1882a) (description, conservation); (Cannatella, 1980) (redescription, account, tadpole).

AGALYCHNIS HULLI (HULL'S LEAF FROG) (DUELLMAN AND MENDELSON, 1995)

Phyllomedusa hulli (Duellman and Mendelson, 1995): 348. **Holotype.**—URP-WED 60198 from 1.5 km north of Teniente López, 310 m, Provincia Loreto, Departamento Loreto, Peru.

Agalychnis hulli (Faivovich *et al.*, 2010): 259.

Etymology.—The specific name is a patronym that honors Clark Hull, former Environmental Manager of the Occidental Petroleum Company. His broad outlook on environmental issues set an outstanding example in Peru.

Identification.—Figures 1.7, 1.8, 1.9. In these frogs, males attain a maximum SVL of 37.1 mm, and females are unknown. The maximum known mass is 1.5 g in a male. The head is barely wider than the body. The snout is short, narrowly rounded in dorsal view, sloping anteriorly from the level of the nostrils to the margin of the lip in profile. The nostrils are moderately protuberant, directed dorsolaterally, at a level posterior to the anterior margin of the lower jaw. The internarial region is slightly depressed. The canthus rostralis is straight, acutely rounded in section. The loreal region is slightly concave. The lips are thin, not flared. The top of the head is flat; the interorbital distance is greater than the width of the upper eyelid. The eyes are large, protuberant; the pupil is vertically elliptical. The eyelid is clear. The parotoid glands are not differentiated. The supratympanic fold is thin, extending to a point above the angle of the jaw, not obscuring the upper edge of the tympanum. The tympanum is round, distinct, with a slightly elevated annulus, separated from

FIGURE 1.7 *Agalychnis hulli*. Left: CJ 6467 (top and middle), female. CJ 6490 (bottom), female. Right: CJ 2045, male, 33.5 mm SVL. Reserva Otoyacu, Provincia Pastaza. LAC.

FIGURE 1.8 Morphological variation in *Agalychnis hulli*. Top left: QCAZ 40924, Centro Ecológico Zanja Arajuno, Provincia Pastaza. Top right: CJ 2047, male, 48.9 mm SVL. Bottom left: CJ 6467, female. Reserva Otoyacu. Provincia Pastaza. Bottom right: Centro Ecológico Zanja Arajuno. LAC.

FIGURE 1.9 Morphological details of *Agalychnis hulli*. Top: CJ 4741, note color change in the same individual. Bottom: CJ 6467, female, Reserva Otoyacu, Provincia Pastaza. LAC.

the eye by a distance equal to two-thirds of the diameter of the tympanum. The upper arm is slender; the forearm is moderately robust. The dermal fold from the elbow to the base of the disc on finger IV is narrow, slightly crenulated. The fingers are moderately short, flattened in section, lacking webbing but having narrow lateral fringes on each side of each digit. The discs are round. The diameter of the disc on finger III is equal to the diameter of the tympanum. The relative length of the fingers is I < II < IV < III. The distal subarticular tubercles are conical; the proximal subarticular tubercles are round, barely elevated. The supernumerary tubercles are diffuse, present only on the proximal segments of the digits; the palmar tubercle is diffuse. The thenar tubercle is elliptical, poorly differentiated. There is a thin, keratinous nuptial excrescence on the dorsal surface of the basal part of the thumb. The hind limb is slender. The calcar and tubercles on the heel are absent. The outer tarsal fold is distinct, crenulated, extending from the heel to the base of the disc on toe V. The inner tarsal fold is absent. The outer metatarsal tubercle is absent. The inner metatarsal tubercle is low, elliptical, and visible from above. The toes are moderately long with distinct lateral fringes, about one-fourth webbed. The foot is about half webbed among toes III–V. The relative length of toes is I < II < III < V < IV. The digital discs are round, about equal in size to those on the fingers. The distal subarticular tubercles are conical; the proximal subarticular tubercles are round, elevated. The supernumerary tubercles are indistinct. The cloacal opening is directed ventrally at the lower level of the thighs; the cloacal sheath is long. The para-cloacal tubercles are present. The quadratojugal is present. The sacro-coccygeal articulation is bicondylar. The skin on the dorsal surfaces, except the upper arms and digits, is coarsely granular. The skin on the belly is weakly granular. The skin on other surfaces is smooth. Sometimes, there are one or two white, elevated warts on the body. The tongue is lanceolate, nearly three times as long as wide, distinctly notched posteriorly, free behind for about one-half of its length. The dentigerous processes of vomers each bear three teeth, angled posteromedially between large reniform choanae; they are moderately separated medially. The vocal sac is single, median, and subgular. The vocal slit extends from the midlateral base of the tongue nearly to the angle of the jaw.

At night, the dorsum is dark green with small, pale green flecks. By day, the dorsum is pale green; the flanks and hidden surfaces of the limbs are orange; the chin, belly, and labial, ulnar, tarsal, and infracloacal stripes are white. The iris is cream.

Comparisons.—*Agalychnis buckleyi*, *A. psilopygion*. *Agalychnis hulli* is unique in having coarsely granular skin on the dorsum; the dorsal skin is finely granular in *A. buckleyi*. Also, *A. buckleyi* has a smaller tympanum (mean tympanum/eye = 0.352) that is barely evident, whereas the larger tympanum (mean tympanum/eye = 0.489) is distinct in *A. hulli*. Furthermore, *A. buckleyi* differs from *A. hulli* by having a small calcar and usually having a small outer metatarsal tubercle. *Agalychnis psilopygion* has a calcar, no para-cloacal tubercles, and a truncate snout.

Natural history.—Figure 1.10. At Teniente López, Peru, an individual was perched on a leaf of a vine 4 m above the ground; the vine was attached to a tree at the edge of a forest pond. The frog was collected at 22:25 h. This pond was visited nightly for nine days (16–24 July 1993); no other individuals were observed. *Phyllomedusa tarsius* were found in trees near the pond, and four species of *Dendropsophus* (*D. koechlini*, *parviceps*, *rhodopeplus*, *sarayacuensis*) were calling from bushes and low herbs in and around the pond.

In Provincia Pastaza, amplectant pairs, egg clutches over leaves, and tadpoles were found in March 2008 and February 2014. Frogs were active at night on secondary vegetation up to 2.3 m, surrounding an artificial pond of about 4 m in length, and also around larger artificial and natural ponds. An egg clutch from Reserva Otoyacu contained 90 eggs. Amplexus is axillary, and multiple amplexus occurs.

FIGURE 1.10 *Agalychnis hulli.* Top right: male combat in terrarium. Top left: multiple amplexus. Middle left: amplectant pair. Middle right: egg clutch. Reserva Otoyacu. Bottom: QCAZ 37863, embryos. Centro Ecológico Zanja Arajuno. Provincia Pastaza. LAC.

FIGURE 1.11 Ontogenetic changes in *Agalychnis hulli* from tadpole to adult. Top to bottom, left to right: CJ 4741, stage 36, lateral, dorsal, and ventral views, TL = 43.2 mm; CJ 11821, undetermined stage, TL = 53 mm; stage 42, TL = 56 mm; stage 45; juvenile. CJ 4741, juvenile, 63 days after stage 36; adult 215 days. Reserva Otoyacu, Provincia Pastaza. LAC, SGS.

Additional reproductive biology data of pairs (wild and lab born) from Reserva Otoyacu, Provincia Pastaza, and Parque Nacional Yasuní, Provincia Orellana, have been obtained in the laboratory (CJ lab notes by Alex Achig, August 2021–February 2023). Male combats were observed. Amplexus is axillary and lasts 2–3 days. Oviposition events occurred at night. Clutches are irregularly ovoid and contain 36–65 eggs of 2.2–3.3 mm in diameter ($\bar{x} = 2.7$, n = 20); the animal and vegetal poles are clearly differentiated by gray and white coloration. The maximum diameter of the clutches varies between 22.6 and 44.8 cm. The clutches were deposited on the terrarium glass and on the apex of the underside of a leaf, 5–30 cm above water. Hatching required 12–25 days after egg deposition, and the tadpoles underwent complete development and metamorphosis in a span of 92–147 days at a water temperature of about 23.4 °C and a pH of 7.7. The development from froglet to adult in individuals of two clutches took around 231 and 319 days at about 22.5–25.2 °C.

Tadpoles.—Figure 1.11. The tadpole was described based on an individual at stage 36 from a pond near Río Pucayacu, Reserva Otoyacu, Provincia Pastaza. It has an ovoid body in dorsal view and is slightly triangular in lateral view, about equal in height and width, highest and widest at the midpoint of the body. The body length is 30.4% of the total length (BL = 10.5 mm; TL = 34.5 mm). The snout is rounded in dorsal view and subelliptical in lateral view. The eyes are large, located immediately above the midline of the body, directed laterally (eye diameter = 1. 9 mm), representing 25.7% of the interorbital distance and 14.7% of the body length, and visible from below. The nostrils are small, ovoid, and dorsolaterally positioned, with external openings directed anterolaterally, situated closer to the snout than the eyes (internarial distance = 2.1 mm representing approximately 28.4% of the interorbital distance). The spiracle is ventral, the opening scallop-shaped, flap-like, transparent, directed backwards, located at the midbody on the left side near the transversal midline, 1.4 mm long, and widest toward the opening (1.5 mm wide). The ventral surface has a silver covering; the internal organs, including the intestinal tube, are not visible through the ventral and lateral surfaces of the body. The anal tube is short, dextral, the right wall attached to the ventral fin, opening posteroventrally. The tail and body are of about equal height. The caudal musculature is well developed, slender, becoming progressively thinner caudally. The myosepta and caudal vein are defined until about the first half of the tail. The dorsal fin is considerably lower than the ventral fin (dorsal fin height = 1.5 mm, ventral fin height = 2.5 mm) and weakly arched; it does not extend onto the body, its origins are at the anterior third of the tail. The ventral fin is approximately 66.7% deeper than the dorsal; it is weakly arched distally; its origins are at the posterior quarter of the body; its widest point is at approximately the midpoint, corresponding with the dorsal fin; both tail fins graduate to a distal point. Caudal tip ends in a long flagellum.

The oral disc is positioned and directed anteroventrally, with the outer edge of the labium forming a short funnel-shaped structure, except for a median anterior labium gap in the upper labium being a continuation of the snout. Except for the median anterior labium gap, the labium is bordered by one continuous, uniserial row of about 80–100 small, conical, marginal papillae along the outer edge, and the inner surfaces have a uniform scattering of papillae, which are indiscreet, barely breaking the otherwise smooth surface (being clearly visible within the translucent labium when lit from beneath). The oral disc is large (oral disc width = 3.7 mm) representing 53.5% of the body width, having a small median margination in the lower labium. There are two anterior and three posterior rows of teeth; the LTRF is 2(2)/3(1). Teeth on the first anterior row (A-1) form a broad arch and are smaller in size than those on the second anterior row (A-2), which is interrupted medially; the anterior rows are longer than the first (P-1), second (P-2), and third (P-3) posterior rows, which decrease in length sequentially; the first posterior row (P-1) has a small medial gap, and the second (P-2) and third (P-3) rows are complete. The upper and lower jaw sheaths are robust, heavily pigmented, and both bear fine serrations; the upper jaw sheath is medially convex and bears moderately long lateral processes; the lower jaw sheath is V-shaped.

The background color of the body at stage 36 is flesh pink, having fine, dark brown speckling on the dorsal surfaces of the body, extending along the caudal musculature at the anterior of the tail. The tail musculature is flesh pink; the caudal vein is dark red until about the first half of the tail. The dorsal and ventral fins are transparent. The ventral surface of the body is silver. There is a silver covering on the internal organs; thus, the organs are not visible through the ventral and lateral surfaces of the body. The chondrocranial elements are visible dorsally, dark metallic brown and polygon-shaped. The iris silver–gray, the periphery with fine silver reticulations. The limb buds and anal tube follow the coloration of the body.

At stages 25–26, the distance between the nostrils and the eye are of about equal width, whereas from stage 26, the distance between the eye and nostril increases considerably compared to that between each nostril. Throughout development, the bilateral myotonic muscle masses and the caudal vein in the tail musculature remain clearly defined. At stage 36, the pupil has developed an elliptical shape, the spiracle opening is evident although not prominent, and a shallow groove from the posterior edge of the nostril to the eye is observed. Between stages 36 and 41, a notable increase in the growth of the ventral fin height is seen compared with that of the dorsal fin. This, and growth in the musculature, result in a notable overall change in tail height compared to tail length from about 26.8% at stage 36 to about 30.6% at stage 41. From stage 41, with the onset of metamorphosis, the spiracle and the anal tube disappear, the tail begins to shrink, and the nostrils become defined. The overall tail muscle height decreases, but the caudal vein remains prominent; the eye to nostril distance starts to decrease; the anterior limbs appear; and fine dorsal granulations seen in the adult also begin to appear. The nostrils become more pronounced, although the shallow groove between the nostril and the eye remains minimal, and a small notch appears at the bottom of the pupil to initiate the pupils' vertical development. One specimen in stage 44 showed a total length of 43.4 mm, including its partially absorbed tail.

Distribution.—Figure 1.12. *Agalychnis hulli* is known in the Amazonian region of southeastern Ecuador and northeastern Peru. In Ecuador, it occurs in Sucumbíos, Orellana, Napo, Pastaza, Morona Santiago, and Zamora Chinchipe provinces. Its elevational range is 233–1014 m with an extent of occurrence of about 57,160 km^2.

This species inhabits tropical rainforests and intervened areas in seven ecosystems (Table 0.4) within seven biogeographic regions (Table 0.5), where the average annual rainfall is 1711–4397 mm and the average annual temperature is 21.1–25.4 °C.

Conservation.—*Agalychnis hulli* is categorized as Least Concern throughout its entire range (IUCN SSC Amphibian Specialist Group, 2018f; in IUCN Red List of 2022) and in Ecuador by Ortega-Andrade *et al.* (2021) in view of its wide distribution and presumed large population. However, it is not an abundant species, and in Ecuador, it is considered

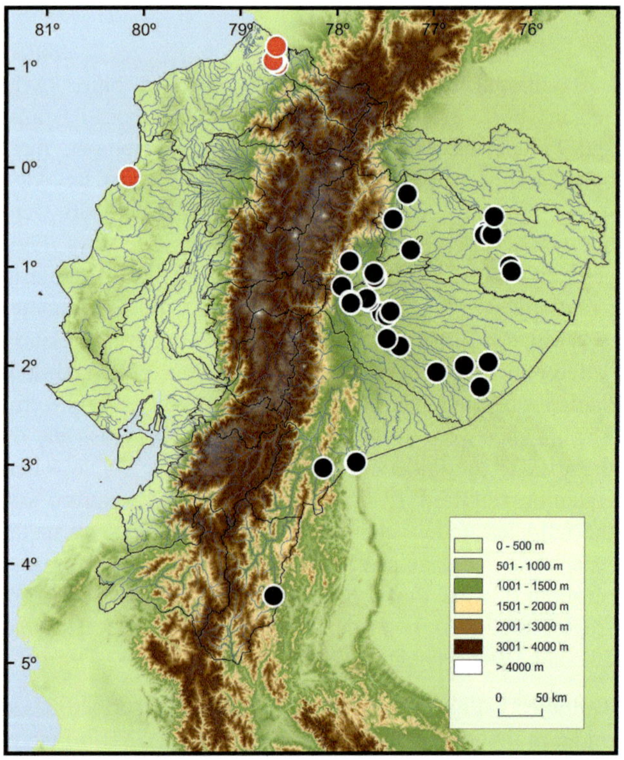

FIGURE 1.12 Localities of known occurrence of *Agalychnis hulli* (black dots) and *A. psilopygion* (red dots) in Ecuador.

uncommon to rare. It is mostly threatened by oil activities, mining, agriculture, deforestation, and the presence of roads. In Ecuador, it has been recorded in the public reserve Parque Nacional Yasuní; the protected forests of CEPLOA and Nacionalidad Zápara del Ecuador; the biosphere reserves Sumaco, Yasuní, and Podocarpus-El Cóndor; and some private reserves (e.g., Reserva Biológica Jatun Sacha, Reserva Biológica Río Bigal).

Remarks.—This species is in the *Agalychnis buckleyi* group *sensu* Duellman and Mendelson (1995) and Faivovich *et al.* (2005) but unassigned to a group by Faivovich *et al.* (2010). According to the latter authors, *Agalychnis hulli* is the sister taxon of *Hylomantis aspera* + *H. granulosa* from Brazil. Our calculations of the area of extent of occurrence in Ecuador are much larger (51,164 km^2) than those (27,806 km^2) provided by Ortega-Andrade *et al.* (2021).

Key references.—(Duellman and Mendelson, 1995) (description); (Gray and Coloma, 2022) (tadpole).

AGALYCHNIS PSILOPYGION (SMOOTH LEAF FROG) (CANNATELLA, 1980)

Phyllomedusa psilopygion (Cannatella, 1980): 32. **Holotype.**—KU 169608 from 8 km west of Danubio, 300 m, 3.61667° N, 76.78333° W, Río Anchicayá, Departamento del Valle, Colombia.

Agalychnis psilopygion (Faivovich *et al.*, 2010): 259.

FIGURE 1.13 *Agalychnis psilopygion*. QCAZ 29092, Durango, Provincia Esmeraldas. LAC.

Etymology.—The specific epithet is derived from the Greek *psilos*, meaning smooth or bare, and *pyge*, meaning rump; *pygion* is the gender diminutive form. It is applied in reference to the lack of para-cloacal tubercles, which are possessed by other species in the group.

Identification.—Figure 1.13. In these frogs, males attain an SVL of 42.0 mm, and females of 47.3 mm. The head is slightly wider than the body. The snout is short, truncate in lateral view, acutely rounded in dorsal view in both sexes. The nostrils are at the tip of the snout, not protuberant, directed laterally. The internarial region is flat. The canthus is rounded. The loreal region is barely concave. The lips are thin and not flared. The interorbital region is flat; the eyes are large and protuberant. The pupil is vertically elliptical. The eyelid is clear. The parotoid glands are not differentiated. The supratympanic fold is thin, obscuring the dorsal and posterior edge of the tympanum, extending posteroventrally to the angle of the jaw. The tympanum is round, distinct, separated from the eye by a distance equal to one-third of the diameter of the tympanum. The upper arm is slender; the forearm is moderately developed, with a weakly developed ulnar fold extending from the elbow to the disc of finger IV. The fingers are of moderate length, not webbed. The digital discs are rounded and small. The fingers are flattened in cross section. There is a slight fringe present on either side of the digits. The order of fingers from shortest to longest is I<II<IV<III. The distal subarticular tubercles are barely rounded and elevated. The tubercle on the fourth finger is slightly bifid. The supernumerary tubercles are small, flattened, present proximally. The palmar tubercle is diffuse, single. The prepollex is moderately developed, with a thin, horny nuptial excrescence in breeding males. The hind limbs are slender and moderately long. There is a weak outer tarsal fold extending from the small calcar and merging with toe fringe of toe V. The inner tarsal fold is very weak. The outer metatarsal tubercle is absent. The inner metatarsal tubercle is flat, elliptical, and not visible from

above. The toes are of moderate length and webbed basally. The discs are rounded, smaller than those on the fingers. The toe fringes are present, widening distally. The subarticular tubercles are large, rounded, flattened on the more distal segments. The supernumerary tubercles are small and indistinct; the order of toes from shortest to longest is I<II<III<V<IV. The anus is directed posteroventrally at the midlevel of the thighs, covered by a short anal flap. The tubercles are absent in the para-cloacal region. The skin on the belly and ventral surface of the groin is faintly granular, smooth elsewhere. White dorsal warts are present in some individuals. The tongue is lanceolate, slightly notched from behind, and free posteriorly for one-third of its length. Vomerine teeth are present; the dentigerous processes are oriented posteromedially, moderately separated at the level of the anterior margins of the elliptical choanae. The vocal sac is single, median, and subgular. Vocal slits are present in males, short, extending from the posterolateral corner of the tongue to the corner of the mouth. The quadratojugal is present. The sacro-coccygeal articulation is fused.

At night, the dorsum is dull green to reddish-brown with green flecks. By day, it is pale green with or without reddish-brown flecks. The dorsal warts are creamy yellow. The flanks, upper arms, anterior and posterior surfaces of the thighs, finger IV, toes I–IV, and ventral surfaces of the hind limbs are bright orange. The throat, belly, anal region, and ventral portion of the forearms are white. The tarsal stripe and elbow are pale gray. The iris is creamy white; the eyelid is clear.

Comparisons.—*Agalychnis buckleyi*, *A. hulli*. *Agalychnis psilopygion* differs from *A. buckleyi* and *A. hulli* by the lack of para-cloacal tubercles. Furthermore, *A. psilopygion* differs from *A. buckleyi* by lacking a small outer metatarsal tubercle.

Natural history.—In Río Anchicayá, Colombia, individuals were found in an area of steep slopes covered with rainforest. Some males were found calling from bushes and tangled vines at the mouth of a natural grotto at night. A clutch of eggs was found on the underside of a rock protruding from the side of the grotto; it was embedded in a single gelatinous matrix and contained 32 eggs at stage 16; the eggs had an average diameter of 4.3 mm. In life, the yolk is grayish-tan and the jelly clear. Tadpoles were collected from the same grotto. During the period of 10–14 June 1975, the range of temperatures at the locality was 21.0–32 °C; the average rainfall was 4.5 mm with a range of 2.0–9.5 mm.

The call has been recorded in Río Anchicayá, Departamento del Valle, Colombia. The call consists of a single chirp of 0.04–0.05 seconds. Additional data of the call are provided under the subtitle "Vocalizations of Ecuadorian Anurans" in Appendix 1.

Tadpoles.—Figure 1.14. The tadpole was described based on 15 individuals in stage 25 from Río Anchicayá, 8 km west of Danubio, Departamento del Valle, Colombia. The average body length is 11.8 mm (8.5–17.9), and the mean total length is 29.4 mm (24.1–40.4). The body is slightly wider than deep. The snout is acutely rounded in lateral profile; in dorsal view, the snout is almost truncate. The body is widest and deepest at two-thirds the length of the body. The nostrils are dorsolateral, directed anterolaterally, situated at the end of the snout. The internarial distance is equal to the width of the oral disc. The eyes are dorsolateral and directed laterally. The spiracle is a flap-like tube, ventral and sinistral to the midline. The spiracular opening is at a point slightly more than one-third the length of the body from the anterior end; the skin is transparent. The chondrocranial elements are visible through the dorsal skin of the head. The mouth is anteroventral and directed anteriorly. The cloacal tube is short and dextral to the caudal fin. The caudal musculature is slender, tapering distally to the end of fin. The myomeres are poorly differentiated. The ventral fin is deepest just posterior to the body and tapering gradually to the tail tip. The dorsal fin is narrow anteriorly, widening to its deepest depth at the midlength of the tail; the dorsal fin is never as deep as the ventral fin at the same vertical level. The dorsal fin does not extend onto the body. The mouth is small, without lateral folds. The upper lip lacks papillae medially; otherwise, papillae are present in one row laterally and two rows ventrally along the border; very few are present elsewhere. The upper and lower beaks are shallow, both with small, uniform serrations. There are two upper and three lower rows of denticles; the second upper row is interrupted medially; the upper rows are of equal length; the lower rows are unbroken and not as long as upper rows; the third lower row is shorter than the others. The denticles on the first upper and third lower rows are finer than those of other rows.

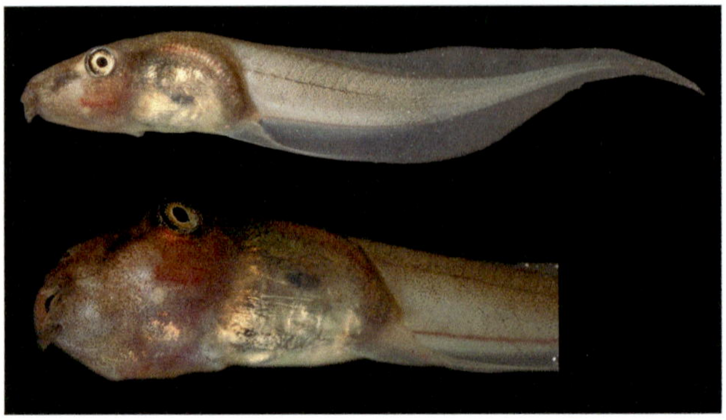

FIGURE 1.14 Tadpole of *Agalychnis psilopygion* in lateral (top), and ventral (bottom) views. QCAZ 29653, Durango, Provincia Esmeraldas. LAC.

The body is reddish-brown above with gold flecking below. The tail is bluish-gray. The smaller individuals are almost unpigmented. The iris is pale bronze.

Distribution.—See Figure 1.12. *Agalychnis psilopygion* occurs in the Pacific lowlands of southern Colombia (southern Valle del Cauca through Cauca to southern Chocó) into northeastern Ecuador. In Ecuador, it occurs in Esmeraldas and Manabí provinces. Its elevational range is 96–202 m with an extent of occurrence of about 1760 km^2.

This species inhabits tropical forests in the evergreen forest of lowlands of Equatorial Chocó and the evergreen seasonal forests of lowlands of Jama-Zapotillo ecosystems, where the average annual rainfall is 863–2796 mm and the average annual temperature is 24.5–25.3 °C.

Conservation.—*Agalychnis psilopygion* is categorized as Least Concern throughout its entire range (IUCN SSC Amphibian Specialist Group, 2019h; in IUCN Red List of 2022) in view of its wide distribution and presumed large population. In Ecuador, Ortega-Andrade *et al.* (2021) categorized it as Endangered given that its extent of occurrence is less than 5000 km^2 (1760 km^2, according to our data), its area of distribution is severely fragmented, and there is a continued decline in the area extent and habitat quality. It is a rare species, which is under constant pressure by palm plantations, deforestation by wood companies, mining activities, the presence of roads, the introduction of exotic species, and human settlements. The private reserve Jama-Coaque is one of the largest remnants of undisturbed coastal equatorial primary forest within the Tumbes-Chocó-Magdalena conservation corridor; therefore, the subpopulation occurring there is well-protected. However, the region surrounding the reserve is experiencing very high rates of deforestation. In Ecuador, it has been recorded in the protected forest of Asociación Agrícola Carchi-Imbabura.

Remarks.—Its molecular phylogenetic relationships are unknown. It is in the *Agalychnis buckleyi* group *sensu* Faivovich *et al.* (2005) but unassigned to a group by Faivovich *et al.* (2010). Interpopulation variation in the presence and absence of the small calcar has been noticed between two populations in Colombia. Our calculations of the area of extent of occurrence in Ecuador are much larger (1760 km^2) than those (7 km^2) provided by Ortega-Andrade *et al.* (2021).

Key references.—(Cannatella, 1980) (description, tadpole).

AGALYCHNIS SPURRELLI (SPURRELL'S GLIDING LEAF FROG) BOULENGER, 1913

Agalychnis spurrelli Boulenger, 1913: 1024. **Syntypes.**—BMNH 1947.2.24.24–25 from Peña Lisa, Condoto, 91 m, Departamento Chocó, Colombia.

Etymology.—The specific epithet is a patronym that honors Dr. Herbert George Flaxman Spurrell (1877–1918), who collected the type series in Colombia. He was a British biologist and physician whose work in South America and Africa led to the discovery of several new species.

Identification.—Figures 1.15, 1.16, 1.17. Males of this species attain a maximum SVL of 82.0 mm, and females of 110.0 mm (in Ecuadorian samples, males 72.9, females 89.5 mm). The upper surfaces are slightly rough to smooth. The venter is granular. The head is about as broad as long. The snout is subelliptical in dorsal outline, obtuse in profile. The eyes are large, and the upper eyelid is reticulate. The parotoid glands are indistinct. The tympanum is distinct; the diameter is equal to about one-half the diameter of the eye. There are short, paired vocal slits and a

FIGURE 1.15 *Agalychnis spurrelli.* Left: CJ alive, female, 75.3 mm. SVL Right: CJ alive, male, 62.7 mm SVL. Reserva Otokiki, Alto Tambo, Provincia Esmeraldas. LAC.

FIGURE 1.16 Morphological variation in juveniles of *Agalychnis spurrelli.* Top: Reserva Otokiki, Alto Tambo. Bottom left: CJ 11677, Reserva Otokiki. Bottom right: QCAZ 37744, Durango. Provincia Esmeraldas. LAC.

FIGURE 1.17 Morphological details of *Agalychnis spurrelli*. Note the nuptial pad of a male (top right). Reserva Otokiki, Alto Tambo, Provincia Esmeraldas. DAL.

slightly distensible single internal subgular vocal sac in adult males. Finger I is shorter than II. The finger discs are large; the disc on finger III is about equal to the diameter of the eye, larger than the tympanum. The distal subarticular tubercles under fingers III and IV are bifid in some individuals. The prepollex is elongated, flattened and ventrally elliptical; the nuptial pads are dark brown, large, horny, and not spiny; they can reach the thenar tubercle from the base of Finger II, covering the entire dorsal region of the prepollex. The fingers are extensively webbed. The thenar tubercle is greatly enlarged; a palmar tubercle, heel tubercle, and calcar are absent. There is a narrow dermal fold from the heel to the disc of toe V. The toes are extensively webbed. There are many plantar tubercles; the inner metatarsal tubercle is low, elliptical. There is no outer tubercle. A long, moderately developed tarsal fold is present. The cloacal opening is directed posteroventrally.

The dorsal surfaces are generally pale green during the day and darker green at night, often (80%) with white dorsal spots (up to 11) or a mix of single spots with clumped groups of spots that are usually outlined by black. The upper surface of the forearm, flanks, belly, anterior and posterior thighs, webbing, and most digits are either orange, yellow, or pale pink–purple, depending on which population they originate from. The outer edges of the tarsus and heel have bold yellowish-white folds. A similar fold extends along the outer edges of the forearm and finger IV. The surfaces of the belly, chest, throat, lower lip, and insertions of the arms and legs usually are white or cream. The iris is red with fine black reticulations. There is a golden reticulated nictitating membrane.

Juveniles are green during the day but reddish-brown at night. The general coloration of the dorsum in metamorphic juveniles (< 34.5 mm SVL, n = 244) varies from pale green to dark brown or reddish, usually with cream spots edged with reddish-brown. The webbing, fingers II and III, toes I–III, anterior and posterior surfaces of the legs, inner surfaces of the tarsus, arm, and forearm have an intense orange color. The surfaces of the flanks, belly, chest, gular region, lower lip, and insertions of the arms and legs are generally white. The eggs have a creamy green yolk and are covered by a transparent greenish gelatinous mass. The iris is light reddish, with a round pupil.

Comparisons.—*Agalychnis callidryas*, *A. saltator*. This medium to large species, having fully webbed hands and feet and lacking bars on the flanks, can be distinguished from all other *Agalychnis* by the presence of orange to pale rose or bright yellow flanks and thighs, dark red eyes, and a dorsum usually marked with black-bordered white spots. White pustular spots occur on the dorsal surfaces of many species of *Agalychnis*, but only in *A. spurrelli* are these spots always narrowly outlined in black. Other red-eyed *Agalychnis* from Central America (*A. callidryas* and *A. saltator*) have the flanks blue to purplish-brown rather than orange.

Natural history.—Figures 1.18, 1.19. It is a nocturnal, arboreal species of tropical humid forests. It moves toward the forest canopy (especially in the dry season) by climbing branches and lianas, although it is a generalist in the use of vertical substrates available in the habitat. Frogs glide by jumping in free fall. In planning, they extend legs and hands in such a way that they can be seen in their entire length. During the fall, they can be held at an angle of 45°; they can also turn. Groups of this species are commonly found in small pools, whether natural or artificial, and in torrential streams surrounded by bushy vegetation associated with the upper strata of the forest, from where they descend to reproduce. They are also found in primary and secondary forests and in intervened areas with some remaining trees and bodies of water.

In the period of 12 months, from 2006 to 2007, population sizes of 1400–6852 adult individuals were estimated in Laguna del Diablo, Provincia Esmeraldas, for a sampling area of about 2400 km^2. Their reproduction is explosive and occurs in permanent pools or in ephemeral ponds formed by rain during storms. Large numbers of animals gather in a small area after heavy rains; breeding occurs into the morning hours; males grapple for possession of females; several males may fertilize one clutch; frogs leave the swamp in processions up lianas, and egg clusters are laid up to 8 m above the water. The males are territorial, coexisting at an average distance of 40–50 cm or more, with fidelity to the vocalization sites. Vocalizing males perch on tree and shrub branches 1–3 meters above the ground. The amplexus is axillary, and the periods of amplexus occur between 19:00 and 23:00 h, while oviposition occurs between 4:00

FIGURE 1.18 *Agalychnis spurrelli.* Top: amplexus, Reserva Otokiki, Alto Tambo. Middle: metamorph at night. Bottom: adult in its microhabitat at night. Durango. Provincia Esmeraldas. LAC.

FIGURE 1.19 *Agalychnis spurrelli.* Egg clutches and embryos. Reserva Otokiki, Alto Tambo, Provincia Esmeraldas. Middle right: QCAZ 37740, Durango, Provincia Esmeraldas. LAC.

and 8:00 h. In Costa Rica, *Agalychnis spurrelli* breeds only a few times each rainy season.

Observations in Colombia show that before oviposition, the amplectant pair comes down from the vegetation and passes some minutes in the water, a time in which females fill their swim bladders with water to inject into the gelatinous cover of their eggs. Egg clutches are laid as irregularly shaped monolayers with little associated jelly. Clutches are attached to leaves, lianas, branches, or trunks suspended over the water and in the surroundings of streams, ponds or wetlands, although the use of flooded cavities in trunks has also been reported. Clutches are occasionally embedded in moss. Most of the eggs are laid communally. Communal egg-laying results in large aggregations of eggs spread over branches, leaves, trunks, mosses, and epiphyte roots in a continuous mass in which individual clutches (i.e., laid by individual females) are indistinguishable from one another. Eggs are laid in an irregularly shaped flat monolayer (4–5 mm deep) or rarely in a partial bilayer, with little associated jelly. Hatchlings fall into the water after eight days to continue their development to metamorphosis. Tadpoles are pelagic and forage on algae that grow on submerged surfaces. Embryos hatch simultaneously. Development is estimated to last 2–3 months (between February and April in Laguna del Diablo). The metamorphs begin their ascent to the middle and upper layers of the forest as soon as they emerge from the water. Most of the year, frogs prefer the upper layers of the forest, descending only to reproduce.

In the Osa Peninsula (Provincia Puntarenas, Costa Rica), in August 1970, there was a congregation of about 13,000 breeding individuals at an artificial pool. When the tadpoles hatched, they fell into the water in the pools below. Literal masses of tadpoles dripped from the foliage. Most of the spawns were 1.5–3 m above the pool, but they have been found a high as 8 m. Egg masses are generally deposited on the upper side of leaves. In 14 clutches, the number of eggs was 14–67. Males have been observed removing eggs from leaves with rotary movements of the feet. A 30 × 11 cm leaf contained a concentration of 533 eggs on its upper part and 83 on its back, that is, a total of 616 eggs on one leaf. Multiple clutches found together on another leaf in Colombia contained up to 322 eggs. Fifteen individual clutches from Guayacán, Provincia Limón, Costa Rica, had a mean of 46 eggs.

At Sirena Field Station in Corcovado National Park, Costa Rica, average egg mass size in *Agalychnis spurrelli* was 73 eggs (n = 20 clutches), and clutches were only 4.4 mm thick, only slightly thicker than the ovum diameter itself. They ranged from 12 to 231 eggs, with a mean of 77 eggs. They began hatching spontaneously during the fourth night after oviposition. Their peak of spontaneous hatching was during the night at age 5–6 days, and clutches were completely hatched the following night. At two days of age, they had a heartbeat and external gills (Gosner stage 19). Embryos became hatching competent in Gosner stage 23, the last stage with bilateral external gills. Embryos retained external gills and remained in stage 23 until they hatched, although other structures continued developing. Their gills regressed rapidly after hatching. At this site, egg predation was by snakes, mainly *Leptodeira septentrionalis* but also by *Leptophis ahaetulla* and *Imantodes inornatus*. Snakes did not always completely consume clutches, so that even for young clutches, mortality was not 100% in snake attacks. Escape success increased with development. The escape success in *A. spurrelli* is low and held even in clutches that had already begun hatching spontaneously. Embryos in attacked clutches remain largely motionless, even in eggs close to the snake's mouth. For instance, in one clutch attacked at the modal hatching age, over half of the eggs had hatched spontaneously prior to snake contact, and yet only 9–28% of the remaining embryos hatched when attacked, and consequently most were eaten. *Agalychnis spurrelli* embryos showed a stronger hatching response (57% hatched in 1 h) to submergence underwater than to snake attacks, even though submergence is a less frequent risk. This suggests they have a greater capacity for early hatching than is expressed in the context of snake attacks but have much lower sensitivity to snake cues than to flooding cues. In a hatchling predation experiment, younger hatchlings were more vulnerable than older hatchlings to fish predation.

In Colombia, the level of egg mortality was 25% due to predation by *Leptodeira*, ants, or fungal infection. In the water, *Astyanax* fish prey upon tadpoles. The *Leptodeira septentrionalis* snake is one of its main predators, especially of juveniles and eggs. Birds such as the swamphen (*Porphyrio porphyrio*) eat their eggs, while the lesser kingfisher (*Chloroceryle americana*) eat tadpoles.

Calls from Panama and Ecuador have been described. At Barro Colorado Island, Panama, males produced one type of advertisement call and two types of aggressive calls (A and B) that differed in temporal and spectral frequency. The advertisement call was a single note. Aggressive call A was a long, pulsed, single note, and call B was composed of more than two notes per call. Calling behavior was influenced by chorus size and male-to-male proximity, with males altering their advertisement calls with the increase in chorus size and eventually producing aggressive calls. In Ecuador, the described call is a simple and weak groan repeated at intervals of 4 seconds to several minutes. Additional data of calls are provided under the subtitle "Vocalizations of Ecuadorian Anurans" in Appendix 1.

Tadpoles.—Figure 1.20. The tadpole was described based on a series of 10 tadpoles in stages 34 to 37 from Rincón de Osa, Puntarenas Province, Costa Rica. They have body lengths of 14.5 to 17.0 mm ($\bar{x}$ = 15.8) and total lengths of 40.0 to 43.0 mm ($\bar{x}$ = 41.5 mm). The ratio of tail length to total length is 0.575 to 0.654 ($\bar{x}$ = 0.619). The largest tadpole in stage 41 has a body length of 19.5 mm and a total length of 52.3 mm; the ratio of tail length to total length is 0.589. Tadpoles in stage 36 have fully developed mouthparts and have undergone no external changes except for the development of small hind limbs. In these tadpoles, the body is as wide as deep; it is nearly of uniform width posterior to the eyes. The top of the head is flat. In lateral profile, the snout is rounded; in dorsal profile; it is truncate. The nostrils are dorsolateral and directed laterally; the diameter of the eye equals about one-third of the depth of the body. The spiracle is a flap-like tube ventral and sinistral to the midline; the spiracular opening is at a point about two-thirds of the distance from the snout to the posterior end of the body. The mouth is anteroventral and directed anteriorly. The cloacal tube is short and dextral to the caudal fin. The caudal musculature is rather weak; its depth at the midlength of the tail is about one-third of the depth of the tail. The musculature extends nearly to the tip of the tail; distally, the musculature is slender and curved

FIGURE 1.20 Ontogenetic changes in two individuals of *Agalychnis spurrelli* from tadpole to adult. Top to bottom, left to right. CJ 6064, stage 32, lateral, dorsal, and ventral views, TL = 43.2 mm. LAC. QCAZ 49608, stage 41, TL = 65.5 mm; stage 42, TL = 65.5 mm. LAC. CJ 6064, stage 46, 63 days after stage 32 = 23.3 mm SVL; froglet 75 days after stage 46 =31.2 mm SVL; juvenile 172 days = 43.4 mm SVL. Reserva Otokiki, Alto Tambo, Provincia Esmeraldas. DAL.

dorsally. The dorsal fin is shallow anteriorly and does not extend onto the body; the ventral fin is deepest at about one-third of its length. The mouth has a shallow lateral fold. The median two-thirds of the upper lip are bare; the rest of the mouth is bordered by two or three rows of papillae, and laterally, additional papillae are present. The upper beak is moderately shallow and forms a broad arch; the lateral processes are barely expanded distally. The lower beak is heavy. Small, blunt serrations are present on the upper beak and small, pointed serrations on the lower beak. There are two upper and three lower rows of teeth. The upper rows are about equal in length, and the second upper row is interrupted medially. The lower rows are much shorter than the upper rows; the lower rows are progressively shorter, so that the third lower row is the shortest. The snout and top of the head are olive brown; the sides of the body are dark bluish-gray, and the venter is pale bluish-gray. The caudal musculature is grayish-tan. Dark brown flecks are present on the sides of the body and on the proximal edges of the anterior one-half of the fins; dark brown reticulations are present on the anterior one-half of the caudal musculature. During development, the amount of dark pigment increases so that in tadpoles in stage 41, the entire tail is marked with bold reticulations.

In a population in Dagua, Departamento Valle del Cauca, Colombia, the embryos hatched from their eggs in stage 24 (*sensu* Kenny, 1968) with an average total length of 11.9 mm. Individuals completed their metamorphosis with a total length of about 44.2 mm in developmental stage 41. They exhibited white spots on their back. Their labial tooth row formula is 2(2)/3(1).

Distribution.—See Figure 1.6. *Agalychnis spurrelli* occurs from the Central western lowlands of Costa Rica to the Pacific lowlands of Colombia (Valle del Cauca and Chocó) and adjacent Ecuador. In Ecuador, it occurs in Esmeraldas, Carchi, Imbabura, Pichincha, Los Ríos, Manabí, Santa Elena, and El Oro provinces. Its elevational range is 0–820 m with an extent of occurrence of about 70,690 km^2.

This species inhabits tropical forests and intervened areas in six evergreen and semideciduous ecosystems (Table 0.4) within six biogeographic regions (Table 0.5), where the average annual rainfall is 588–2769 mm and the average annual temperature is 21.5–25.8 °C.

Conservation.—*Agalychnis spurrelli* is categorized as Least Concern throughout its entire range (Jungfer *et al.*, 2008; in IUCN Red List of 2022) and in Ecuador by Ortega-Andrade *et al.* (2021) in view of its wide distribution, presumed large population, and because it is unlikely to be declining fast enough to qualify for listing in a more threatened category. This species can present locally high population sizes, even in areas with alteration gradients. Its area of distribution is under constant pressure by palm plantations, mining activities, deforestation by wood companies, and the introduction of exotic species. These activities result in habitat reduction and fragmentation, as well as severe pollution. The habitats are patches of forest remnants, many of them surrounded by cultivated areas, grasslands, and human settlements. Museum specimens of this species were found infected with chytridiomycosis, but the current impact of this pathogen on the species in the wild is unknown. It has been recorded from a number of protected areas, including at least three in Panama and three in Costa Rica. In Ecuador, it has been recorded in the public reserve Reserva Ecológica Mache-Chindul; the protected forests of Asociación Agrícola Carchi-Imbabura, Cordillera Chongón Colonche, and Río Arenillas Presa Tahuín; and some private reserves (e.g., Reserva Río Manduriacu, Ross Maynard pers. comm., Reserva Otokiki, Reserva Biológica Canandé, Reserva Biológica Bilsa, Reserva Biológica Ayampe). Despite being part of the National System of Protected Areas, the Mache-Chindul Reserve is subject to selective timber extraction, the opening of vegetation cover for farms, and constant pressure from agricultural expansion, in addition to including human settlements. The other localities where it occurs are found outside the SNAP, in areas under strong environmental degradation and habitat loss, where the removal of vegetation cover is constant and the areas with forest remnants are mostly restricted to river basins, streams and buffer areas of the Parque Nacional Cotacachi-Cayapas, and the Mache-Chindul and Cayapas-Mataje Ecological Reserves. *Agalychnis spurrelli* was listed in CITES under Appendix II (on 21 March 2010). In the 2010s, the United States of America alone imported 221,960 *Agalychnis* frogs, according to the Species Survival Network (SSN). Ortega-Andrade *et al.* (2011) recommend it for management. Its reproduction and handling *in situ* and *ex situ* have been successful. Since 2011, it has been part of the biotrade program of Wikiri (Sinovas and Price, 2015).

Remarks.—It is in the *Agalychis callidryas* group *sensu* Faivovich *et al.* (2010). Differences in size have been noted between populations in southeastern Costa Rica and those in central and eastern Panama (Duellman, 1970). The largest female was recorded in Peña Lisa, Chocó, Colombia (SVL = 92.8 mm), whereas the largest male is from Tacarcuna, Provincia Darién, Panama (SVL = 75.6 mm) (Duellman, 2001). Also, according to Ortega-Andrade (2008), there are differences in SVL among five populations in northwestern Ecuador. The smaller ones are from Estero Aguacate, Provincia Esmeraldas. Additionally, specimens from Colombia and Panama have a larger mean SVL than Ecuadorian samples.

Key references.—(Boulenger, 1913) (description); (Duellman, 1970) (accounts, as *Agalychnis spurrelli* and *A. litodryas*); (Scott and Starrett, 1974) (natural history); (Gray, 1997) (natural history); (Savage, 2002) (account); (Ortega-Andrade, 2008) (morphology, systematics, natural history); (Köhler, 2011) (account); (Vargas-Salinas *et al.*, 2000) (reproductive biology); (Vargas-Salinas and Gutiérrez-Cárdenas, 2005) (tadpole); (Vargas-Salinas, 2007) (clutch); (Gomez-Mestre and Warkentin, 2007) (egg predation); (Stuart *et al.*, 2008) (account); (Jungfer *et al.*, 2008) (conservation); (Gomez-Mestre *et al.*, 2008) (phylogeny, predation); (Vargas-Salinas and Torres-Suárez, 2013) (account); (Cossio and Medina-Barcenas, 2020) (call, behavior); (Yánez-Muñoz *et al.*, 2020) (account).

GENUS *CALLIMEDUSA* (LEAF FROGS) DUELLMAN *ET AL.*, 2016

Callimedusa Duellman *et al.*, 2016: 33. **Type species.**—*Phyllomedusa perinesos* (Duellman, 1973).

Etymology.—The generic name is derived from the Greek *kalos*, meaning beautiful, and the Greek *Medousa*. The name alludes to the beautiful coloration of the flanks of members of this genus. The gender is feminine.

Definition.—Figure 1.21. *Callimedusa* is a group supported by molecular and morphological evidence. Phenetic synapomorphies include the non-opposable thumb, and with one exception, all have bright flash colors. They are small to medium-sized species (SVL of 44 mm in *Callimedusa atelopoides* to 62 mm in *C. tomopterna*); toe I is slightly longer than toe II; the vomerine teeth are present (except in *C. atelopoides*); the flash colors on the flanks and hidden surfaces of the thighs are purple or black. The palpebral membrane is reticulated or not; the tadpoles have moderately small oral discs directed anteroventrally.

Distribution.—The genus occurs in the Amazonian slopes of the Andes from Ecuador to central Peru, the upper Amazon Basin from Colombia to Bolivia, and the Guyanese Region.

Content.—Currently, six species are recognized in this genus. Three species are known in Ecuador.

Remarks.—Figure 1.22. The monophyly of this genus is rather well supported (78%), but that of the "*Phyllomedusa perinesos* group" of Cannatella (1982) within *Callimedusa* is only moderately supported (66%). The four species having allopatric ranges on the Amazon slopes of the Andes in Ecuador and Peru (*C. baltea*, *C. duellmani*, *C. ecuatoriana*, and *C. perinesos*) closely resemble one another. In the analysis by Duellman *et al.* (2016), the sister species is *C. atelopoides*, an inhabitant of the upper Amazon Basin. This species differs from all other phyllomedusids by being small, brown, and terrestrial; furthermore, it is unlike other *Callimedusa* in lacking vomerine teeth and clavicles in the pectoral girdle. The analysis places *C. tomopterna* as the sister species to the remaining members of the genus. This colorful frog that ranges throughout much of the Amazon Basin is widely allopatric to the Andean species.

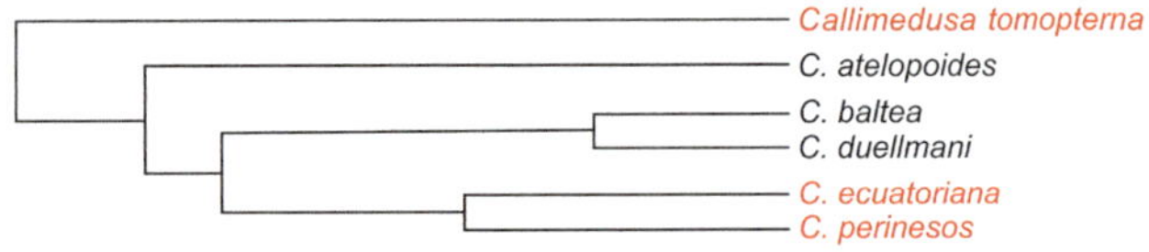

FIGURE 1.22 Topology of a tree analysis of *Callimedusa* based on molecular data, showing species occurring in Ecuador (in red). Taken and modified from Duellman *et al.* (2016).

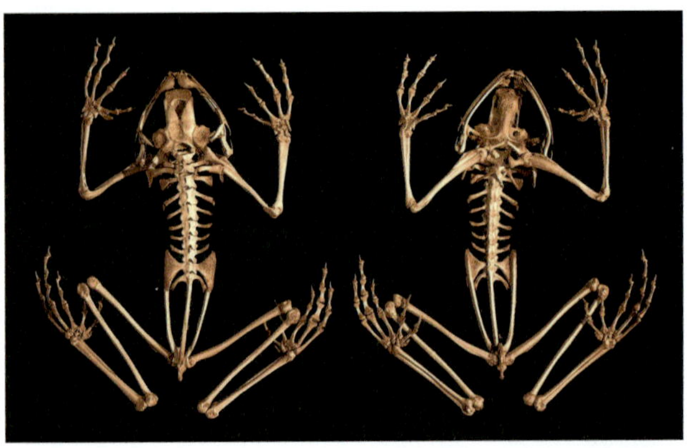

FIGURE 1.21 High-resolution computed tomography reconstructions of the skeleton of *Callimedusa perinesos*, MCZ 97778, 49.9 mm SVL, km 102 INECEL Station Cascadas de San Rafael, Río Quijos, Provincia Napo. Dorsal (left) and ventral (right) views. Daniel J. Paluh.

CALLIMEDUSA ECUATORIANA (ECUADORIAN LEAF FROG) CANNATELLA, 1982

Callimedusa ecuatoriana Cannatella, 1982: 507. **Holotype.**—USNM 215750 from Agua Rica, a one-house posada on the trail between Limón and Gualaceo, 1890 m, Provincia Morona Santiago, Ecuador.

Etymology.—The specific name is in honor of the country of its origin, the Republic of Ecuador.

Identification.—Figures 1.23, 1.24. Adults of this species attain a maximum SVL of 72.2 mm in females and of 55.4 in males. The enlarged para-anal tubercles are absent. The dorsal surfaces of the hands and feet are purple with irregular orange markings. The digital discs are uniform gray. A white stripe extends from the corner of the jaw to the flanks. There is a large orange blotch on the anterior and posterior surfaces of thighs. The concealed surfaces of shanks and tarsi are uniform purple, without discrete blotches or markings. The eyelid has scattered silvery flecks, not reticulated.

In life, the dorsal surfaces of the head, body, forearms, fourth fingers, finger and toe pads, thighs, shanks, tarsi, and fourth and fifth toes are green. The lower lip, ulnar, tarsal, and supra-anal folds are off-white. The dorsal surfaces of the hands and feet are purple with irregular carmine markings. The flanks are cream with large orange blotches, the posterior one extending onto the thigh. A creamy orange

FIGURE 1.23 *Callimedusa ecuatoriana*. Left: CJ 11808, female. Right: CJ 11807, male. Río Blanco, Provincia Morona Santiago. SGS.

band borders the ventrolateral side and extends from the corner of the jaw to the groin. The anal region is green with cream tubercles. The concealed surfaces of the hind limbs are purple; there is a single, large orange blotch on the anterior and posterior surfaces of the thighs and an orange blotch on the hidden surface of the arm. The belly, throat, ventral surfaces of the limbs, hands, and feet are purple and cream; there are distinct cream and light orange granules on the belly, anterior flanks, and anal region.

Comparisons.—*Callimedusa perinesos*, *C. tomopterna*. *Callimedusa ecuatoriana* differs from other tree frogs in Ecuador, except for *C. perinesos*, by the combination of a vertical pupil and the presence of large purple and orange spots on the flanks and on the hidden surfaces of the thighs. It differs from *C. perinesos* by the presence of a creamy orange band that borders the ventrolateral side and that extends from the corner of the jaw to the groin (absent in *C. perinesos*). Another similar species is *C. tomopterna*, from which it differs by having an elongated tubercle on the heel instead of developed calcars.

Natural history.—Figure 1.25. About eight males were vocalizing in Río Blanco, Provincia Zamora Chinchipe on 26 July 2011. One note and two to three pulses comprise the calls. Calls have an average duration of 30.87 ms. Detailed data and figures are provided under the subtitle "Vocalizations of Ecuadorian Anurans" in Appendix 1.

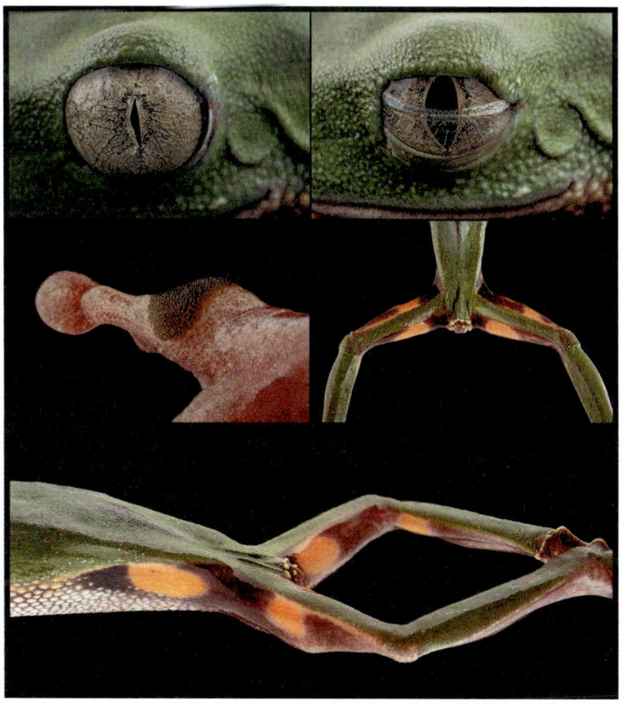

FIGURE 1.24 Morphological details of *Callimedusa ecuatoriana*. Top: CJ 12849, note the nictitating membrane. Middle and bottom: CJ 11807, male. Río Blanco, Provincia Morona Santiago. SGS.

FIGURE 1.25 Aspects of the natural history of *Callimedusa ecuatoriana*. Individual and tadpoles in their microhabitat at night, amplexus (in the lab), clutches of embryos (in the field), and eggs (in the lab). Río Blanco, Provincia Morona Santiago. LAC, SGS (top and middle right).

At Río Blanco, tadpoles in different stages were observed in April, May, and June 2009, as well as in July and October 2010, May 2012, and June 2013.

Two posts were recorded during visits to the study area; these were found deposited in the surrounding vegetation of a pool in semi-rolled leaves, and a gelatinous substance coated the eggs. In February of 2012, a female was collected in an advanced state of gravity, with large yellowish-cream oocytes.

A specimen was taken from a large swampy area strewn with logs and large rocks. Other adult specimens were sighted at night, perched on the riverside vegetation of small and/or medium seasonal pools, lentic waters, and tannin-stained water, either inside the forest or in open areas. The specimens were located between branches at heights up to 2.5 m; when captured they adopt a posture of inertia, possibly to evade predators.

Tadpoles were observed both day and night on the surface or shore of pools, with the head directed toward the surface, and when they detect the presence of a stranger, they swim quickly to the bottom.

Arachnids of the Argiopidae family were the most frequent prey in the stomachs of the specimens examined (20%), followed by other arthropods (28%), plant remains (16.6%), and Acari 2%. In the stomachs, nematodes also comprised 33.3% of the total volume. Overall, 80% of the total prey was consumed by males and 20% by females.

Reproductive biology data of pairs (wild and lab born) from Río Blanco, Provincia Zamora Chinchipe, have been obtained in the laboratory (CJ lab notes by Alex Achig, in May 2021 and January 2023) from two reproductive events. Amplexus is axillary. The clutches contained 31 and 27 unpigmented eggs. The clutches were deposited inside rolled leaves 5 cm above water. Hatching occurred 11 and 21 days after egg deposition. Tadpoles underwent complete development and metamorphosis in a span of 16 days (water temperature: 24.6 °C; pH: 7.1) and 153 days (water temperature: 20.6 °C; pH: 7.9). The development from froglets to adults in one clutch took around 530 days at about 21.4–23.4 °C.

FIGURE 1.26 Ontogenetic changes in *Callimedusa ecuatoriana* from tadpole to adult. Top to bottom, left to right. CJ 11806, lateral, dorsal and ventral views, stage 28, TL = 37.5 mm; undetermined stage, TL = 57.8 mm; stage 42, TL = 60.5 mm; stage 46, 22.2 mm SVL; 108 days after stage 46 = 38.9 mm SVL; juvenile 204 days = 43.6 mm SVL; 302 days = 55.2 mm SVL, 402 days = 60 mm SVL. Río Blanco, Provincia Morona Santiago. SGS.

Tadpoles.—Figure 1.26. The tadpole was described based on an individual at stage 26 from the Cordillera del Cóndor, Provincia Zamora Chinchipe, Ecuador. It has a total length of 55.9 mm and a body length of 21.6 mm, 38.6% of the total length. The width of the body is 12.6 mm; the body height is 14.3 mm. Dorsally, the snout is subovoid. The eyes are lateral, located 5.3 mm from the nostril. The tail length is 40.3 mm (72% of the total length). The muscle width of the tail is 5.0 mm; the height of the musculature of the tail is 7.4 mm. The dorsal and ventral fins begin at the base of the tail. The spiracle is sinistral, directed toward the left ventral region of the body; its anterior part is attached to the body; the tube length is 1.3 mm. The transverse width of the tube is 1.7 mm; the direction of thc spiracle is posteroventral. The cloacal tube is adhered to the ventral fin, open and with a posteroventral direction; the tube length is 5.2 mm. The oral disc is anterofrontal; its maximum diameter is 4.1 mm (32.1% of the body width). The anterior lip has a wide medial gap; it bears three rows of irregular marginal papillae on the anterior and posterior lips. The lateral line system is characterized by several lateral line canal openings, as a successive row from the postero-inferior region of the orbit to the lower medial part of the abdomen and turning in the direction of the groin. The dental formula is 2(2)/3(1). The A-1 row is complete, bearing 190 teeth. The A-2 row bears a gap in the middle of the row and has 62 teeth. The P-1 row has a small gap in the middle and has 117 teeth. The P-2 row is of similar length to P-1 and bears 146 teeth, which are shorter than the teeth of P-1. The P-3 row has 68 teeth that are smaller than the ones in rows P-1 and P-2. The variation in morphological characters in stages 26–42 is as follows: the total length ranges from 50.6 to 63.6 mm; the body length is 18.5–26 mm; the tail length varies from 36.6–45.2 mm. At stage 42, the dental rows disappear.

Distribution.—Figure 1.27. *Callimedusa ecuatoriana*, an endemic species to Ecuador, is distributed on the Amazonian slopes of the Cordillera Oriental of the Andes and the Cordillera del Condor. It occurs in Morona Santiago and Zamora Chinchipe provinces. Its elevational range is 1450–2354 m with an extent of occurrence of about 4920 km^2.

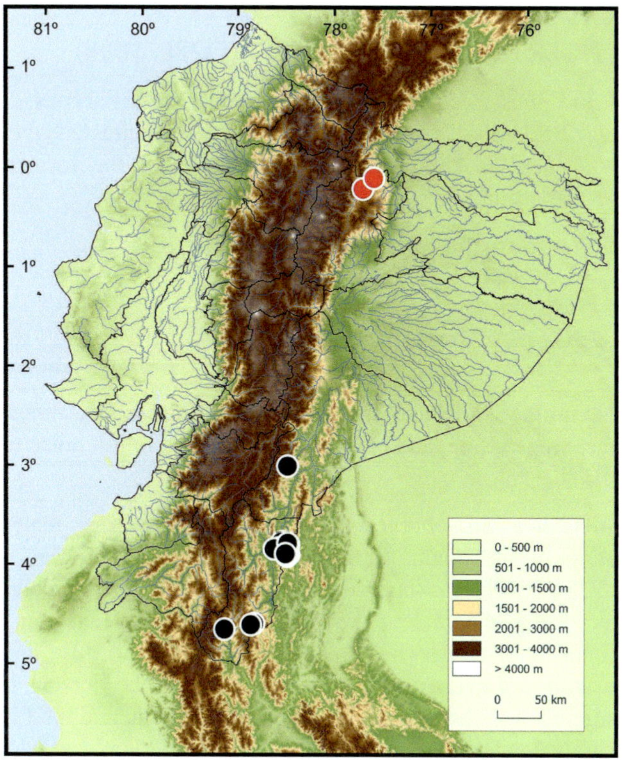

FIGURE 1.27 Localities of known occurrence of *Callimedusa ecuatoriana* (black dot) and *C. perinesos* (red dots) in Ecuador.

This nocturnal, arboreal species inhabits subtropical forests and nearby intervened areas in five ecosystems (Table 0.4) in the Amazonian Cordilleras and southern Cordillera Oriental of the Andes biogeographic sectors, where the average annual rainfall is 1120–1525 mm and the average annual temperature is 15.8–20.9 °C.

Conservation.—*Callimedusa ecuatoriana* is categorized as Vulnerable (IUCN SSC Amphibian Specialist Group, 2019i; in IUCN Red List of 2022). However, Ortega-Andrade *et al.* (2021) categorized it as Endangered, given that its extent of occurrence is less than 5000 km^2, and it is threatened by mining, agriculture, roads, pipelines, deforestation, and stochastic events. Populations of *C. ecuatoriana* in Tudayme, Río Blanco, and others in the Cordillera del Cóndor are in areas severely affected by mining operations. It is an uncommon species. In Ecuador, it has been recorded in the public reserve Reserva Biológica Cerro Plateado, the protected forests of Cordillera del Cóndor and Cuenca Alta del Río Nangaritza, the biosphere reserve Podocarpus-El Cóndor, and the private reserve Refugio de Vida Silvestre El Zarza.

Remarks.—According to Ron *et al.* (2013), it is closely related to *Callimedusa perinesos*.

Key references.—(Cannatella, 1982) (description); (Brito *et al.*, 2014) (morphology, tadpole, vocalization, natural history).

CALLIMEDUSA PERINESOS (PURPLE LEAF FROG) DUELLMAN, 1973

Callimedusa perinesos Duellman, 1973: 507. **Holotype.**—KU 146562 from Río Salado, about 1 km upstream from the Río Coca, 1410 m, Provincia Napo, Ecuador.

Etymology.—The specific name is from the Greek *perinesos*, meaning edged with purple, used in reference to the purple color of the flanks and hidden surfaces of the limbs.

Identification.—Figure 1.28. This small species of *Callimedusa* attains a maximum SVL of 65.2 mm in females and of 51.5 in a male. The snout is truncate in profile. The parotoid glands are diffuse, not forming elevated longitudinal ridges. The webbing is absent. Toe I is shorter than II. A short calcar is present.

The dorsum is green. The flanks are dark purple with white granules. The groin, hands, feet, and ventral and hidden surfaces of the limbs are purple with large,

FIGURE 1.28 *Callimedusa perinesos*. KU 164450, Río Salado, 1 km upstream from Río Coca, Provincia Napo. WED.

bright orange spots; the belly is pinkish-white. The ulnar, tarsal, and supra-anal stripes are cream; the iris is dull greenish-gray with fine black reticulations; the eyelid is unpigmented.

Comparisons.—*Callimedusa ecuatoriana, C. tomopterna.* It differs from other tree frogs in Ecuador, except *Callimedusa ecuatoriana*, by coloration features: the combination of a vertical pupil and the presence of large purple and orange spots on the flanks and on the non-visible surfaces of the thighs. It differs from *Callimedusa ecuatoriana* by the absence of a creamy orange band that borders the ventrolateral side and that extends from the corner of the jaw to the groin. Another similar species is *Callimedusa tomopterna*, from which it differs by having an elongated tubercle on the heel instead of developed calcars.

Natural history.—A frog was perched on the leaf of a bush in disturbed montane rainforest at 22:00 h.

Tadpoles.—Unknown.

Distribution.—See Figure 1.27. *Callimedusa perinesos*, and endemic species to Ecuador, is distributed on the Amazonian slopes of the northern Cordillera Oriental of the Andes. It is known from two localities (17 km apart) in Provincia Napo at an elevational range of 1382–1475 m.

This nocturnal, arboreal species inhabits cloud forests and nearby intervened areas in the evergreen low montane forests of the northern Cordillera Oriental of the Andes ecosystem, where the average annual rainfall is 2640–2730 mm and the average annual temperature is 19.1–19.3 °C.

Conservation.—*Callimedusa perinesos* is categorized as Endangered (IUCN SSC Amphibian Specialist Group, 2019j; in IUCN Red List of 2022). However, Ortega-Andrade *et al.* (2021) categorized it as Critically Endangered, given that its extent of occurrence is very small, and there is evidence of a population size reduction. It is mostly threatened by oil activities, hydroelectric power plants, agriculture, roads, pipelines, and volcanic activity from the El Reventador volcano. This is a rare species. Historical records come from the type locality in the Río Salado in Napo Province, Ecuador. Intensive searches in Río Azuela (Provincia Sucumbíos) between 2000 and 2003 did not register this species, even though it was there in the 1970s. Due to ongoing declines in the extent and quality of habitat, the population is suspected to be decreasing. The pools where the species was recorded in two of the three known localities in Ecuador, Río Salado and Río Azuela, have been destroyed by human activities. In Ecuador, it has been recorded in the public reserve Parque Nacional Cayambe Coca and in the biosphere reserve Sumaco.

Remarks.—According to Ron *et al.* (2013), it is closely related to *Callimedusa ecuatoriana*.

Key references.—(Cannatella, 1982) (description).

CALLIMEDUSA TOMOPTERNA (TIGER-STRIPED LEAF FROG) (COPE, 1868)

Pithecopus tomopternus (Cope, 1868): 112. **Syntypes.**—USNM 6651 (two specimens), now lost, from Río Napo, or Upper Amazon, below the mouth of the former, Departamento Loreto, Peru.

Callimedusa tomopterna Duellman *et al.*, 2016: 33.

Etymology.—The specific name is from the Greek *tomos*, meaning cutting, and *pterna*, meaning heel, in reference to the heel calcar.

Identification.—Figures 1.29, 1.30. In these frogs, males attain a maximum SVL of 49.1 mm, and females of 62.3 mm. The maximum known masses are 5.0 g in males and 9.8 g in females. The snout is truncate in males and rounded in females. The parotoid glands are diffuse and usually not evident externally. The skin on the dorsum is smooth. The discs are large; webbing is absent, and toe I is barely longer than and opposable to toe II. Large calcars are present. The heel bears a prominent triangular calcar. Breeding males have brown nuptial excrescences. The pupil is vertically elliptical.

FIGURE 1.29 *Callimedusa tomopterna*. Left: CJ 6473, male, 49.7 mm SVL, Reserva Otoyacu, Provincia Pastaza. Right: CJ 10952, juvenile, 37.0 mm SVL, San José de Payamino, Provincia Orellana. DAL.

FIGURE 1.30 Morphological details of *Callimedusa tomopterna*. Top: CJ 11918, female, 55.2 mm SVL, note the translucent nictitating membrane, Reserva Biológica Limoncocha, Provincia Sucumbíos. SGS. CJ 6473, male, 49.7 mm SVL, Reserva Otoyacu, Provincia Pastaza. LAC.

FIGURE 1.31 *Callimedusa tomopterna*. Top left: amplectant pair from Nangaritza, Provincia Zamora Chinchipe. Jaime Culebras. Top right and bottom: individual in its microhabitat at night, amplectant pair, and metamorph. Shiripuno Lodge, Provincia Orellana. Morley Read.

The dorsal surfaces and sides of the head are green. The flanks and hidden surfaces of the limbs are orange with vertical purplish-brown bars. The throat and chest are white, and the belly is pale orange. The iris is silvery gray.

Comparisons.—*Pithecopus palliatus*. *Callimedusa tomopterna* is readily distinguished by its coloration, bright green dorsum, and orange flanks and hidden surfaces of the thighs with bold black or dark brown vertical bars. The only other green phyllomedusid frog having brown markings on the flanks and thighs is *Pithecopus palliatus*, which differs from *C. tomopterna* in having the sides of the head and the flanks cream with brown flecks, smaller digital discs, and no calcars.

Natural history.—Figure 1.31. In Santa Cecilia, Provincia Sucumbíos, all individuals of this species were found at night on bushes and low trees in forests from March through November. Although *Callimedusa tomopterna* has an extensive breeding season, it is doubtful if it breeds throughout the year. Data are lacking for January, February, and December, usually the driest months in Santa Cecilia. Males call from bushes and trees over ponds. Five gravid females in April and July contained 60–89 unpigmented ovarian eggs 3.5 mm in diameter; no clutches were found. The call is a soft "cluck." Three of six stomachs contained a roach each, and the other three each contained an orthopteran; one of the latter also had a homopteran. *Callimedusa tomopterna*, despite its gaudy appearance, is the least conspicuous phyllomedusid in the area. It seems to prefer more dense vegetation than the other species. Half of our specimens were obtained from a palm-shrouded pond on 2 and 4 April 1972.

At Cusco Amazónico, Peru, all individuals were in terra firme forest; 89 of 93 adults were found in the rainy season. This is strictly an arboreal species; 90 adults were at heights of 0.3–10.0 m ($\bar{x}$ = 2.35 +/– 1.62) above the ground (73.3%) or in small pools of water (26.7%). Of 83 individuals with substrate data, 58 (69.9%) were on branches of trees, 20 (24.1%) were on leaves of trees or bushes, three (3.6%) were on palm fronds, and one each (1.2%) was on a vine and the trunk of a sapling. The individuals found in June (three) and July (one) consisted of a juvenile male (1.8 g), an adult male (4.9 g), and two adult females (6.6 and 7.2 g). Twenty males were observed calling from branches of trees at heights of 0.5–10.0 m ($\bar{x}$ = 2.74 + 2.03) in November–February; all but three were on perches above water. Eleven males were calling from vegetation above three different depressions resulting from fallen trees; such small pools of water were no more than 5 m in diameter and contained water no more than 40 cm deep. Six males were calling at a small (mean size 2.3 × 5.2 m) ephemeral pool, and three were calling above a large fallen tree trunk in which a cavity (0.3 × 3.0 m) contained water no more than 15 cm deep. Four amplectant pairs were found in January (one), February (two), and March (one) on vegetation at heights of 0.5–2.3 m ($\bar{x}$ = 1.25) above water; two pairs were on branches of trees and two were on leaves of trees. Apparently, large breeding congregations do not occur in this species, nor does the number of calling males necessarily coincide with heavy rainfall. The largest number of calling males at one site was seven on branches and leaves above a small pool on 7 November 1991, a day on which no rain fell. Males were also calling on three other dry nights. Two were calling on 28 November 1989, when 11.4 mm of rain fell, and only four were observed calling on 10 February, when 127.6 mm of rain fell. Seven females collected in November (one), December (one), January (two), and February (three) contained ovarian eggs. Five of these contained 56–67 ($\bar{x}$ = 61.4) mature eggs that were 2.5–3.5 mm in diameter in both oviducts, whereas two contained 28 and 33 eggs ($\bar{x}$ = 30.5) of equal size only in their left oviducts.

Both of these females and two that had mature eggs in both oviducts also contained numerous small eggs (+ 1 mm in diameter) in both oviducts. These data indicate that females only deposit the contents of one oviduct per clutch and have the potential of depositing at least four egg clutches each rainy season. No egg clutches were found in the field. The total number of ovarian eggs is slightly less than 60–89 eggs ($\bar{x}$ = 71.4, n = 5 females), having a maximum diameter of 3.5 mm found in Santa Cecilia, Ecuador. However, in French Guiana, clutches generally contain 50–100 eggs deposited on leaves 0.2–3.0 m above the ground. In 22 specimens examined, only six contained food. Roaches accounted for the largest proportion (37.5%) of the number of prey and the majority (51.3%) of the volume of prey, followed by two unidentified orthopterans (25.0%, 18.5%), a tettigoniid (12.5%, 13.0%), a spider (12.5%, 10.8%), and a lepidopteran (12.5%, 6.45%).

Tadpoles were found in November–March in small temporary ponds and in a small pool of water in a fallen tree trunk. At the temporary pond, tadpoles were present from 27 December until 23 March, during which time 653 tadpoles were sampled (Wild, 1996). Sampling in this pond revealed that tadpoles were captured about equally abundantly by day and at night and that 48.2% of the individuals were taken in bottom samples, whereas 43.2% were in midwater samples and 8.6% were in surface samples. Hatchling tadpoles are in stage 25. These small tadpoles have jaw sheaths but no labial teeth; the gut region is filled with yolk, and intestines are not visible. The progressive development of the labial tooth rows takes place during stage 25, during which stage tadpoles grow to a maximum total length of 29.6 mm.

At Cusco Amazónico, the call is a soft cluck or a pair of clucks, of which the second note is softer than the first. The notes are produced at a rate of 2–3/min. Notes are 0.05 s long. The dominant frequency is at about 1550 Hz in the first note and 610 Hz in the second note. Additional data of the call are provided under the subtitle "Vocalizations of Ecuadorian Anurans" in Appendix 1.

FIGURE 1.32 Ontogenetic changes in three individuals of *Callimedusa tomopterna* from tadpole to adult. Top to bottom, left to right. CJ 6434, lateral, dorsal and ventral views, stage 31, TL = 41.7 mm; stage 43, TL = 51.8 mm SVL; stage 45, 20.3 mm SVL; 28 days after stage 45 = 23.4 mm SVL; 438 days = 50.5 mm SVL. Reserva Otoyacu, Provincia Pastaza. LAC, DAL.

Tadpoles.—Figure 1.32. The tadpole was described based on an individual at stage 38 from 3 km from the Santa Cecilia turn-off toward Quito, Provincia Sucumbíos, Ecuador. It has a body length of 23 mm and a total length of 64 mm. The body is elongate, ovoid, as wide as deep, and deepest posteriorly. The snout is truncate. The large eyes are widely separated and directed laterally. The nostrils are dorsal and directed anteriorly at a point one-third the distance from the tip of the snout to the eyes. The opening of the sinistral spiracle is just off of the midline and directed posteriorly at about the midlength of the body. The anal tube is short and dextral. The caudal musculature is robust proximally and gradually diminishes to a terminal filament. The dorsal fin is very low; the ventral fin is slightly shallower than the musculature and extends to the tip of the tail. The tail is about 65% of the total length. The small terminal mouth lacks lateral folds. The upper lip is bare; elsewhere, the lips are bordered by two rows of small labial papillae. The beaks are moderately robust and bear fine serrations. The upper beak forms a broad arch, and the lower beak is in the form of a flared V. There are two upper and three lower rows of denticles; the second upper row is broadly interrupted medially. The body and tail are yellow or yellowish-tan; the belly is white. A diffuse orange spot is present on the posterior part of the ventral fin. The iris is silvery white.

Also, a tadpole from Cusco Amazónico in stage 39 from 15 km E of Puerto Maldonado, Departamento Madre de Dios, Peru, has been described. It has a body length of 14.6 mm and a total length of 50.1 mm. It has its greatest body width (8 mm) at the level of the eyes and the greatest body height (8.0 mm) at about four-fifths of the length of the body. In dorsal view, the body is ovoid. The eyes are positioned and directed laterally. In dorsal view, the snout is nearly truncate, whereas it is rounded in profile. The nostrils are situated at about one-fourth of the distance from the tip of the snout to the eyes and directed anterolaterally. The interorbital distance (5 mm) is less than twice the diameter of the eye (3 mm). Narial papillae are absent. The spiracle is sinistral just to the left of the midline; it originates at the level of the middle of the orbits. The spiracular tube is short and attached for its full length to the body; the spiracular opening is slightly narrower than the base and directed posteriorly. The vent tube is short and enclosed in the ventral fin. The caudal musculature is robust, highest at the junction with the body, and tapers gradually to a pointed tip. The dorsal fin originates on the proximal caudal musculature; the fin is shallow and highest posteriorly. The ventral fin originates on the posterior part of the body; it is highest anteriorly and throughout most of its length is higher than the dorsal fin, but at midlength of the tail, it is not as high as the caudal musculature. The oral disc is terminal, directed anteriorly, and lacks lateral folds. There is a single irregular row of small, blunt marginal papillae with a broad median gap on the anterior labium. The labial tooth row formula is 2(2)/3 or 3(2–3)/3. The jaw sheaths are slender and finely serrate; the anterior sheath is broadly arched, and the posterior sheath is broadly V-shaped. The body is pale tan, and the belly is white with a distinct golden tint. The caudal musculature is grayish-tan; the fins are nearly transparent, except for the edges of the dorsal and ventral fins in smaller tadpoles (stage 25), and only the ventral fins of larger tadpoles (stage 28 and later) are bright orange. The fins are translucent and not pigmented.

Distribution.—Figure 1.33. *Callimedusa tomopterna* occurs in the upper Amazon Basin in Brazil, Colombia, Ecuador, Peru, and Bolivia and in the Guiana region from southeastern Venezuela to French Guiana. In Ecuador, it occurs in Sucumbíos, Orellana, Napo, Pastaza, Morona Santiago, and Zamora Chinchipe provinces. Its elevational range is 177–1149 m with an extent of occurrence of about 71,390 km^2.

This nocturnal, arboreal species inhabits tropical rainforests and intervened areas in evergreen forests of lowlands, evergreen montane foothill forests, and flooded forests in 10 ecosystems (Table 0.4) within seven biogeographic sectors (Table 0.5), where the average annual rainfall is 1711–4397 mm and the average annual temperature is 21.2–25.9 °C.

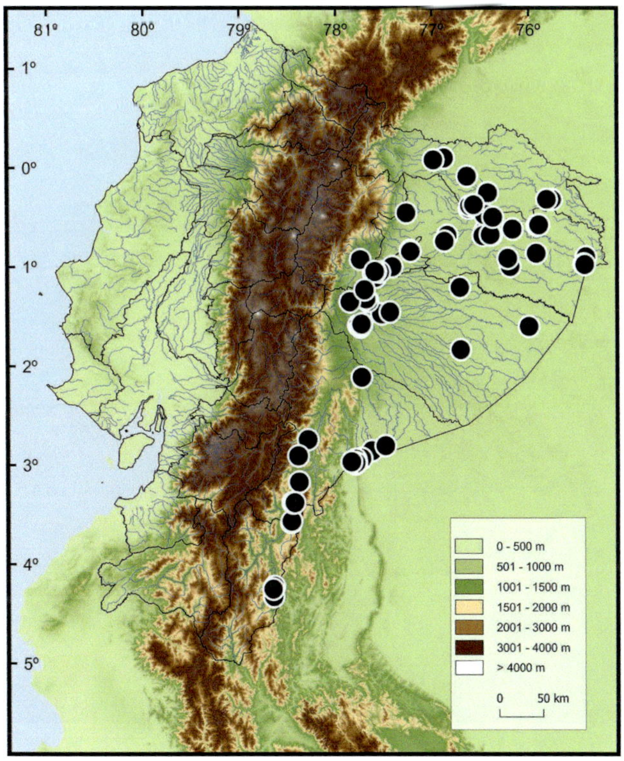

FIGURE 1.33 Localities of known occurrence of *Callimedusa tomopterna* in Ecuador.

Conservation.—*Callimedusa tomopterna* is categorized as Least Concern throughout its entire range (La Marca *et al.*, 2016; in IUCN Red List of 2022) and in Ecuador by Ortega-Andrade *et al.* (2021) in view of its wide distribution, presumed large population, and because it is unlikely to be declining to qualify for listing in a more threatened category. In Venezuela, the population occurring in Reserva Forestal Imataca faces a high risk due to logging practices and consequent depauperation of the original pristine forest vegetation. In Ecuador, it is mostly threatened by agriculture, roads, pipelines, and oil activities. It has been recorded in the public reserves Parque Nacional Yasuní, Reserva Biológica Limoncocha, and Reserva de Producción de Fauna Cuyabeno; the protected forests of CEPLOA, Cordillera Kutukú y Shaimi, Comuna Santa Elena, and Sacha Lodge; the biosphere reserves Sumaco, Yasuní, and Podocarpus-El Cóndor; and some private reserves (e.g., Estación de Biodiversidad Tiputini, Reserva Biológica Jatun Sacha).

Remarks.—Fouquet *et al.* (2007) provided molecular evidence that this is a species complex. According to Duellman *et al.* (2016), *Callimedusa tomopterna* is the sister taxon of the remaining members of the genus.

According to Duellman (2005), at Cusco Amazónico, Peru, *Callimedusa tomopterna* seems to breed at sites away from congeners. Although adults were found throughout the area, males called only from small ponds where other species of Phyllomedusidae were absent. However, this segregation is not consistent throughout the range of the species; for example, in Santa Cecilia, Ecuador, many males of *C. tomopterna* were calling from the same pond, as well as numerous *Phyllomedusa vaillantii*.

Key references.—(Cope, 1868) (description); (Crump, 1974) (natural history); (Duellman, 1974b) (account); (Lescure, 1975a) (reproduction); (Duellman, 1978) (account, call, tadpole); (Hero, 1990) (tadpole); (Rodríguez and Duellman, 1994) (account); (Lescure *et al.*, 1995) (account, natural history); (Barrio-Amorós, 1999) (notes); (Parmelee, 1999) (diet); (Lescure and Marty, 2000) (account); (Duellman, 2005) (account); (Barrio-Amorós, 2009) (notes); (Ouboter and Jairam, 2012) (account); (Villacampa *et al.*, 2017) (account); (Barrio-Amorós *et al.*, 2019) (notes); (Schiesari *et al.*, 2022) (tadpole).

GENUS *CRUZIOHYLA* (LEAF FROGS) FAIVOVICH *ET AL.*, 2005

Cruziohyla Faivovich *et al.*, 2005: 113. **Type species.**—*Agalychnis calcarifer* (Boulenger, 1902), by original designation.

Etymology.—The generic name is the Latinization of Cruz plus *Hyla*, a patronym for the Brazilian herpetologist Carlos Alberto Gonçalves da Cruz. The gender is feminine.

Definition.—Figure 1.34. The frogs in this genus are moderate in size and have large terminal discs on the fingers and

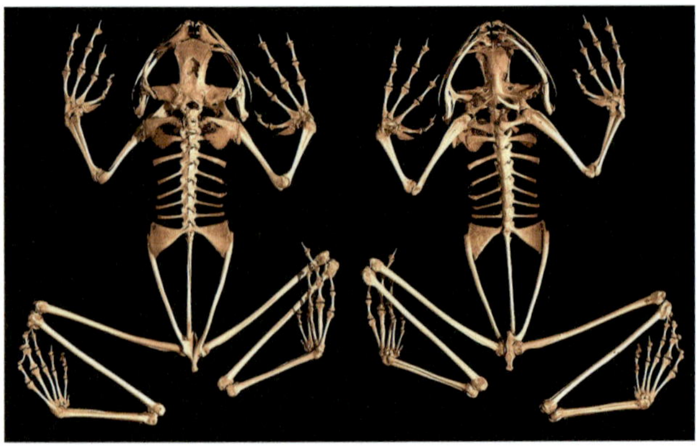

FIGURE 1.34 High-resolution computed tomography reconstructions of the skeleton of *Cruziohyla calcarifer*, UF 137291, 71.4 mm SVL, Bodega de Río Tapalwás, Gracias a Dios, Honduras. Dorsal (left) and ventral (right) views. Daniel J. Paluh.

toes, both of which are at least two-thirds webbed. Large calcars are present on the heels; the outer surfaces of the limbs bear a dermal fold or a scalloped fringe. The skin on the dorsum is smooth. The iris is pale gray medially and yellow peripherally. The eyelid is reticulated or not. The embryos have long, filamentous gills that are tightly pressed against the surface of the egg membrane and are resorbed before hatching. The tadpoles are unique among phyllomedusids in lacking a median gap in the marginal papillae.

Content.—The genus contains three species, two of which are found in Ecuador.

Distribution.—They occur on the Caribbean slopes of southeastern Nicaragua and eastern Honduras, through Costa Rica and Panama to the Chocoan lowlands of Colombia and northwestern Ecuador. One species occurs in the Amazonian lowlands in Colombia, Ecuador, Peru, and western Brazil.

Remarks.—Formerly, members of this genus were placed in *Agalychnis* until the systematic review of hylid frogs by Faivovich *et al.* (2005).

CRUZIOHYLA CALCARIFER (SPLENDID LEAF FROG) (BOULENGER, 1902)

Agalychnis calcarifer Boulenger, 1902: 52. **Holotype.**—BMNH 1747.2.24.22 (= 1901.4.27.25) from Río Durango, 350 feet, Provincia Esmeraldas, Ecuador.

Cruziohyla calcarifer Faivovich *et al.*, 2005: 113.

Etymology.—The specific epithet is derived from the Latin words *calcar* and *fero*, meaning spur or point. The name alludes to the triangular dermal flap on the heel.

Identification.—Figures 1.35, 1.36. Males of this species attain a maximum SVL of 81 mm, and females of 98 mm. In lateral view, the snout is bluntly rounded, and in profile, it is inclined anteroventrally. A dermal fold extends from the posterior corner of the orbit along the dorsal and posterior edges of the distinct tympanum to the angle of the jaw. A broad dermal fold extends from the elbow along the ventrolateral edge of the forearm to the disc on finger IV. The discs on the fingers are large, whereas those on the toes are somewhat smaller. The fingers are about three-fourths webbed. A large, triangular calcar is present on the heel. From the edge of the calcar, a narrow dermal fold extends along the outer edge of the tarsus to the disc on toe V. Another less conspicuous ridge extends the length of the dorsal edge of the tarsus onto the base of toe V. A weak inner tarsal fold is present on the distal one-third of the tarsus. The toes are about three-fourths webbed. The skin on the dorsal surfaces of the body and limbs is smooth. The skin on the belly and proximal posteroventral surfaces of the thighs is granular; other ventral surfaces are smooth.

The dorsal surfaces of the body and limbs are dark green; the flanks and hidden surfaces of the thighs are orange with vertical black bars. The narrow dorsal surfaces of the upper arm and throat are yellow. The belly, ventral surfaces of the limbs, hands, feet, and anterior and posterior surfaces of the thighs are deep orange. The vertical bars on

FIGURE 1.35 *Cruziohyla calcarifer.* Left: CJ kept alive, female. SGS. CJ 13011, male, LAC. Reserva Otokiki, Alto Tambo, Provincia Esmeraldas.

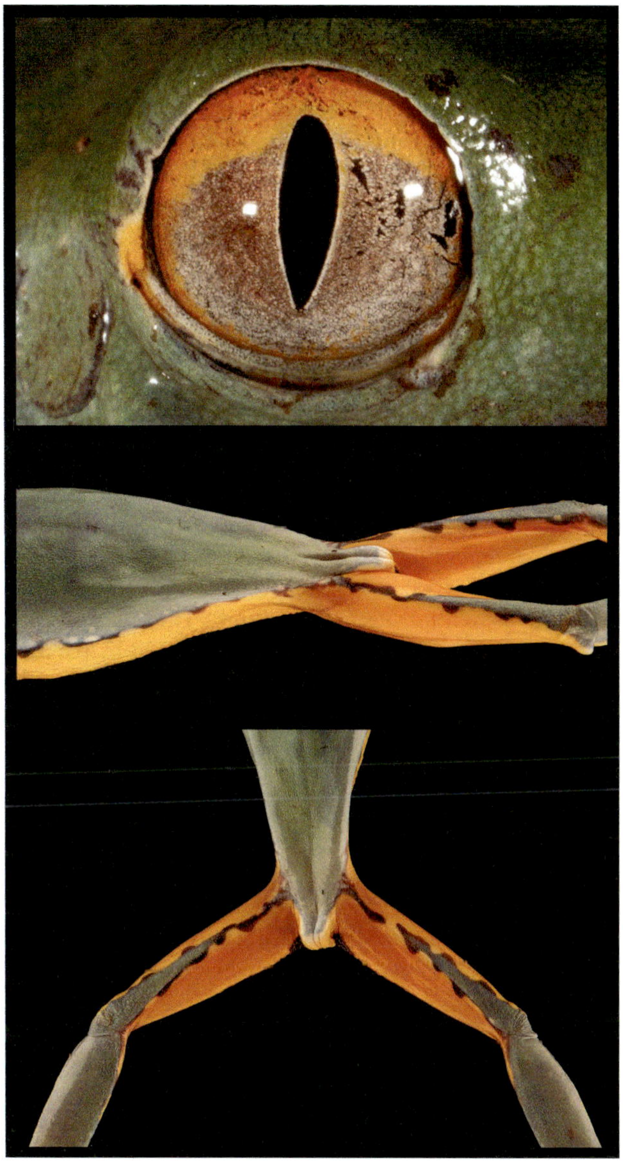

FIGURE 1.36 Morphological details of *Cruziohyla calcarifer*. CJ kept alive, Reserva Otokiki, Alto Tambo, Provincia Esmeraldas. SGS.

the flanks and transverse marks on the dorsal surfaces of the thighs are black. The stripes on the outer edge of the forearm and finger IV, on the inner edge of the shank, and on the outer edge of the tarsus and toe V are yellow. A yellow supracloacal stripe is present. The palpebral membrane is clear. The iris is pale grayish-lavender medially and dull yellow peripherally.

Comparisons.—*Agalychnis psylopygion*, *A. spurrelli*, *Boana pellucens*, *B. rubracyla*. There are no other frogs inhabiting the Pacific lowlands of Ecuador that have orange flanks with vertical black bars. Other green treefrogs in Chocoan Ecuador include two phyllomedusids—A. *psylopygion*, which has uniformly orange flanks, and *A. spurrelli*, which has pale yellow flanks and lacks a calcar. Also, there are two hylids, both of which have horizontally elliptical pupils and lack a triangular calcar—*B. pellucens*, which has a green dorsum with white flecks or small white spots, yellow to pale blue flanks, and a round tubercle on the heel, and *B. rubracyla*, which has reddish-brown canthal and dorsolateral stripes and green flanks.

Natural history.—Figure 1.37. In Changuinola, Panama, the species was locally common in the forest canopy and understory above a headwater streamlet at 170 m elevation on a forested ridge. The frogs were seen at night on leaves and trunks of trees at heights of 2–8 m above the ground. Only a few were low enough to reach with poles; most were unattainable or had to be knocked from their perches with hurled sticks. At Reserva Otokiki, Alto Tambo, Provincia Esmeraldas, Ecuador, a clutch contained 17 embryos.

Two kinds of vocalizations were heard at the Río Changuinola from specimens confined in plastic bags. Sounds emitted by the confined frogs included a low "whonk" and a "chuckle." The "whonk" call has a 0.1 s duration, with maximum energy output at about 800 Hz and a calculated pulse rate of about 140/s. The longer "chuckle" (possibly a release call) seems to be given at a similarly low frequency.

Reproductive biology data of a pair from Reserva Otokiki, Alto Tambo, Provincia Esmeraldas, have been obtained in the laboratory (CJ lab notes by Alex Achig, in April 2022) from one reproductive event that resulted in a non-viable clutch. Amplexus is axillary. The clutch is cluster-shaped contained 31 eggs of 4.5–5.7 mm ($\bar{x}$ = 5.0, n = 11) in diameter. The animal and vegetal poles are very evident, pale turquoise and white colored. The maximum diameter of the clutch is 91.0 cm; it was deposited on the glass wall of the aquarium, 5 cm above water.

Tadpoles.—Figure 1.38. The tadpoles and metamorphs of this species were described from lab-born and raised

FIGURE 1.37 Clutch of eggs of *Cruziohyla calcarifer* from Reserva Otokiki, Provincia Esmeraldas. LAC.

individuals from parents from Alto Tambo, Provincia Esmeraldas, Ecuador. At stage 37, the body is ovoid in lateral view, elliptical in dorsal view, depressed (Body Width = 9.5 mm; Body Height = 8.2 mm), highest and widest at about the midpoint of the body. The body length is approximately 30% of the total length (Body Length = 16.2 mm; Tail Length = 53.4 mm). In dorsal and lateral profiles, the overall snout shape is rounded. The rounded profile of the snout extends anteriorly, medial to distinct nares, which are minimally raised and situated on a shallow fold on either side of the mouth; the shallow fold slopes anteroventrally from the nostril to the outer margin of the oral disc. The nares are small (0.4 mm) yet well defined, positioned dorsolaterally, directed anteriorly, located on the same lateral plane as the center of the pupil. The distances of the nostrils are: from upper labia (1.5 mm), from the eye (4.2 mm), from each other (Inter Narial Distance = 3.6 mm). The eyes are dorsolateral, directed laterally; Eye Diameter = 2.0 mm; the interorbital distance is over twice that of the internarial distance (8.7 mm). The spiracle is ventral, sinistral to the midline, short (1.0 mm); the spiracular opening is oblique (1.3 mm wide), situated approximately midway and at the posterior edge of the body, an equal distance from the eye and mouth (5.0 mm). The ventral and dorsal tail fins curve outward distally, narrowing toward the terminus; the tail is highest at the midpoint (13.0 mm). The caudal musculature is moderately high and gradually tapers, not quite extending to the tip of the tail. The dorsal fin emerges at the junction of the body and tail musculature; the ventral fin does not extend onto the body. The height of the dorsal fin and caudal musculature at the midpoint of the tail are approximately equal (4.2 mm); the ventral fin is slightly higher (4.6 mm). The anal tube is short (1.5 mm), dextral to the caudal fin.

FIGURE 1.38 Ontogenetic changes in three individuals of *Cruziohyla calcarifer* from tadpole to metamorph. Reserva Otokiki, Alto Tambo, Provincia Esmeraldas. LAC.

The mouth is moderately small (Oral Disc Width = 3.8 mm wide), anteroventral, directed anteriorly. The oral disc is not emarginate; the labia are completely bordered by papillae. Anteriorly, a single row of marginal papillae is joined by very short length (n = 4 papillae) forming a second papilla row at the middorsal margin. Additional papillae are present submarginally, diminishing to a double row ventrally, having a short tertiary row and convex-shaped cluster of papillae midventrally. The upper jaw sheath has fine serrations and is convex medially, forming a smooth, broad arch with long, slender lateral processes extending distally. The lower jaw sheath is V-shaped, with distinct, sharply pointed serrations. The labial teeth are present in two anterior (upper) and three posterior (lower) rows. The labial tooth row formula is 2(2)/3(1). The anterior (upper) rows are long, extending laterally nearly to the submarginal papillae; the first row forms a distinct dip medially; the second upper row is narrowly interrupted medially. The posterior rows become progressively shorter posteriorly; the upper posterior labial tooth row is narrowly interrupted medially, and the lower two rows are complete.

The background color of the body at stage 37 is tan–brown, having fine pale-yellow marbling on the dorsal and ventral surfaces of the body, the spiracle, and the anterior part of the tail. The tail musculature is pale tan; there are pale gray reticulations on both the dorsal and ventral fins extending three-fourths the length of the tail. A series of fine gold speckles on the body surface represents clearly visible lateral line configurations. The internal organs are not clearly visible through the ventral surface of the body. The iris is silver–gray, its periphery having gold reticulations.

Changes in color are as follows: In stage 24, the body and tail are very dark brown; the tail musculature is lighter brown. The iris is brown with a golden inner edge. Some internal organ coloration is visible through the ventral surface of the body, primarily the intestinal tube, which is turquoise blue. In stages 25–26, the overall body coloration lightens considerably; the tail musculature is pink, and the redness of the heart muscle is visible through the ventral surface of the body. Externally, the orientation of neuromasts forming the lateral line system is clearly visible as fine, gold-colored speckles on the body surface and thus able to be detailed. In stages 25–30, the body is tan–brown with fine yellow marbling on the lateral body surfaces and anterior part of the tail. The pupil is round; the iris is silver with reticulations; its periphery has gold reticulation. Internally, the intestinal tube loses its blue coloration to become dark gray, and externally, a pink hind limb bud develops. In stages 30–39, the spiracle is evident but transparent; the brown/yellow marbling coloration on the body and at the junction of the caudal musculature becomes progressively more extensive and more contrasted in the later stages as the background body coloration becomes a darker brown. In later stages, the undersides of the toes develop dark brown pigmentation; the tubercles and toe pads are pale yellow. In stages 40–43, the dorsal median, upper labium, and tail are tan–brown. The dorsal surfaces of the arms, legs, flanks, and eye–snout are light gray–brown; the concealed surfaces are yellow. Distinct dark brown markings are present in the ventral thigh region. In stage 43, in contrast to the brown tail musculature, the dorsal and lateral fins turn distinctly black as the tadpole emerges and the tail atrophies.

Notable morphological changes include a distinct change in body shape. In the early stages (stages 25–26), the body is higher than wide, becoming dorsally depressed (wider than high) from stage 26. At stages 25–26, the tail musculature and both dorsal and ventral fins at the midpoint are of equal width, whereas from stage 26, the ventral fin is marginally higher than the caudal musculature and dorsal fin. At stage 30, the bilateral myotonic muscle masses in the tail musculature are more clearly defined; the nostrils become more prominent on either side of the snout. At stage 37, the pupil has developed an elliptical shape, the spiracle is still evident, and the dorsal lateral lines remain. Between stages 37 and 41, a notable increase in the growth of the tail length is seen compared with that of the body length; the hind limbs become well developed, and in later stages, further widening of the body is evident, attributable to the developing forearms that have not yet emerged. Between stages 39 and 40, the distance between the anterior labium and medial aspect of the snout reduces as the snout becomes truncate in the lateral plane. From stage 41, with the onset of metamorphosis, the tail begins to shrink; viewed from above, the snout becomes more pointed, and the nostrils are further defined. A small notch also appears at the bottom of the pupil, initiating the pupil's development to becoming vertical in shape. The outward-facing calcar on the heel becomes apparent in some specimens.

Distribution.—Figure 1.39. *Cruziohyla calcarifer* occurs through the Chocoan lowlands of northwestern Ecuador and Colombia, Panama, and the most southern part of Costa Rica. In Ecuador, it is known from the Pacific lowlands in Provincia Esmeraldas. Its elevational range is 82–710 m in an area of extent of occurrence of about 320 km².

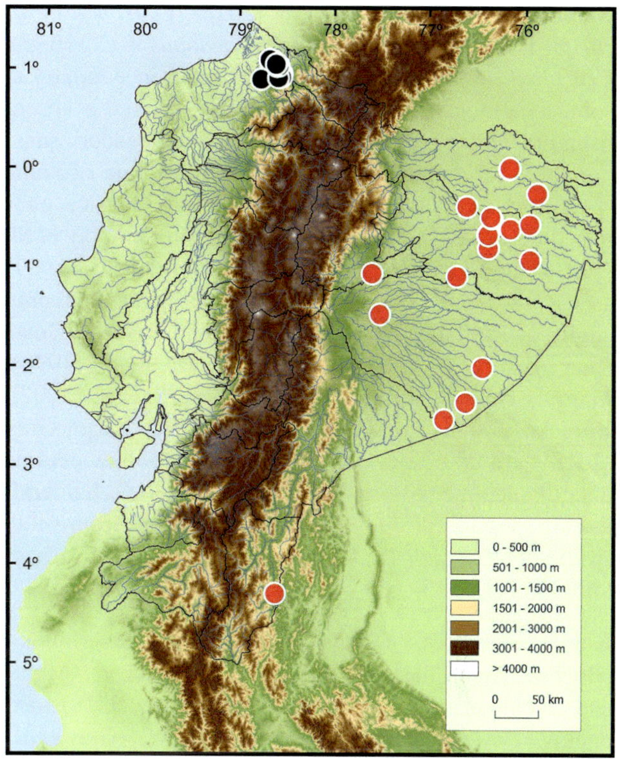

FIGURE 1.39 Localities of known occurrence of *Cruziohyla calcarifer* (black dots) and *C. craspedopus* (red dots) in Ecuador.

In Ecuador, this nocturnal canopy species inhabits rainforest habitats in the evergreen forests of lowlands of Equatorial Chocó and the evergreen montane foothill forests of the Cordillera Occidental of the Andes, where the average annual rainfall is 2359–2624 mm and the average annual temperature is 22.1–25.4 °C.

Conservation.—*Cruziohyla calcarifer* is categorized as Least Concern throughout its entire range (IUCN SSC Amphibian Specialist Group, 2020p, in the IUCN Red List of 2022). Although this species is known from isolated records in a large area, it is almost certainly severely under-recorded since it is a canopy species, and it has a wide distribution with a presumed large population. This is often considered a particularly rare frog, and subpopulations are highly restricted in all countries in which it is known to occur. Ortega-Andrade *et al.* (2021) categorized it as Near Threatened in Ecuador, where oil activities, agriculture, roads, pipelines, and deforestation mostly threaten it. Its area of distribution is under constant pressure by wood companies, palm plantations, mining activities, the introduction of exotic species, and human settlements. These activities result in habitat reduction and fragmentation, as well as severe pollution. Only occasional individuals are seen from time to time. In Ecuador, the species is recorded very infrequently. Due to ongoing declines in the extent and quality of its habitat, the population is suspected to be decreasing. This species is part of sustainable biotrade in Costa Rica and Ecuador. In Ecuador, it has been recorded in the protected forest of Asociación Agrícola Carchi-Imbabura and some private reserves (e.g., Reserva Otokiki).

Remarks.—According to Gray (2018), *Cruziohyla calcarifer* is the sister taxon of *C. craspedopus* + *C. sylviae*. *Cruziohyla calcarifer* and *C. sylviae* occur in sympatry at some localities in Central America, but the latter is absent from Ecuador. The description of *C. sylviae* renders questionable the identifications and associated data of some specimens of *C. calcarifer* from Central America published before 2018. For example, according to Gray *et al.* (2021), the tadpole described by Donnelly *et al.* (1987) as *C. calcarifer* belongs to *C. silviae*, given that only adults of the latter have been found at La Selva, Costa Rica. Also, data (as *C. calcarifer*) by Duellman (1970), Marquis *et al.* (1986), Caldwell (1994), Savage (2002), Kubicki (2004a), and others (see synonymy list in Gray (2018)) are of *C. sylviae*. Gray (2018) provides peptide data of specimens from Alto Tambo, Provincia Esmeraldas. Its peptide profile is unique and differs from other species of *Cruziohyla*. *Cruziohyla calcarifer* has a haploid number of 13.

Key references.—(Boulenger, 1902) (description); (Duellman and Cole, 1965) (chromosome number); (Myers and Duellman, 1982) (natural history); (Gray, 2018) (taxonomy, revision, phylogeny, peptides, account); (Gray *et al.*, 2021) (tadpole); (Yánez-Muñoz *et al.*, 2020) (account).

CRUZIOHYLA CRASPEDOPUS (FRINGED LEAF FROG) (FUNKHOUSER, 1957)

Phyllomedusa craspedopus Funkhouser, 1957: 23. **Holotype.**—CAS-SU 10310 from Chichirota, Río Bobonaza, Provincia Pastaza, Ecuador.

Cruziohyla craspedopus (Faivovich *et al.*, 2005): 113.

Etymology.—The specific name is from the Greek *kraspedon*, meaning edge or border, and refers to the crenulated dermal fringe on the outer surfaces of the limbs.

Identification.—Figures 1.40, 1.41. Males of this species attain a maximum known SVL of 57 mm, and females of 73 mm. In lateral view, the snout is bluntly rounded, and in profile, truncate. The weak supraorbital fold does not cover the dorsal and posterior edges of the distinct tympanum. A broad, scalloped dermal fold extends from the elbow along the ventrolateral edge of the forearm to the disc on finger IV. The discs on the fingers are large, whereas those on the toes are somewhat smaller. The fingers are about two-thirds webbed. Males have a brown nuptial pad consisting of numerous rounded keratinous denticles on the prepollex. A large, triangular calcar is present on the heel. From the edge of the calcar, a broad, scalloped dermal fold extends along the outer edge of the tarsus to the disc on toe V. An inner tarsal fold is absent. A scalloped dermal fringe

FIGURE 1.40 *Cruziohyla craspedopus*. Left: CJ 12850, female, 66.2 mm SVL. Right: CJ 12851, male, 57.6 mm SVL. Shiripuno Lodge, Provincia Orellana. SGS.

FIGURE 1.41 Morphological details of *Cruziohyla craspedopus*. Top: CJ 12851, male, note the nictitating and palpebral membranes on the top right. Middle left: flanks. Middle right: note the nuptial pad. SGS. Bottom: groin and thighs. DAL. Shiripuno Lodge, Provincia Orellana. SGS.

is present on the outer edge of the shank. The toes are about two-thirds webbed. The skin on the dorsal surfaces of the body and limbs in smooth. The skin on the belly and proximal posteroventral surfaces of the thighs is granular; other ventral surfaces are smooth.

The dorsal surfaces of the body and limbs are dark green with numerous grayish-white lichenous marks. The flanks, hidden surfaces of the limbs, and ventral surfaces are bright yellow–orange with narrow, wavy brown–black bars. On the flanks and limbs, these bars bifurcate and fuse dorsally to form a narrow black stripe separating the green dorsum and yellow–orange venter. A supracloacal stripe is absent. The palpebral membrane is dark green with pale green and tallow reticulations, and the nictitating membrane is mostly translucent, except at its upper margin, which is green and cream. Medially, the iris is grayish-white with fine black reticulations; peripherally, the iris is yellow.

Comparisons.—*Callimedusa tomopterna*, *Phyllomedusa tarsius*. Two other treefrogs in Amazonian Ecuador have coloration faintly resembling that of *Cruziohyla craspedopus*, but both lack scalloped dermal fringes on their limbs. Of these, *C. tomopterna* also differs by having a silver iris and a short calcar on the heel; the flanks and hidden surfaces of the limbs are yellow to orange with vertical black or dark brown bars. Subadults of *P. tarsius* have narrow vertical black bars on the flanks; these bars may be interconnected; *P. tarsius* differs from the others by lacking dermal adornment on the heels.

Natural history.—Figure 1.42. This species inhabits tall trees in primary and secondary forests and descends to low branches in order to breed. Observations of breeding individuals of *Cruziohyla craspedopus* have been made in Ecuador and Peru. Males congregate in the vicinity of water-filled cavities in logs, hollows, or depressions at ground level (pitfall buckets). One study noted that eggs were observed being deposited in water-filled cavities located in hollow logs approximately 1.0 m and 1.8 m above ground level. Additionally, eggs have been found in muddy pools on exposed ground within undisturbed primary forests of the Amazonian region of Peru. These frogs have also been reported to utilize water cavities in the buttresses of standing plank trees. At Las Piedras Biodiversity Station, Provincia Tambopata, Departamento Madre De Dios, Peru, tadpoles were reared in artificial breeding habitats.

Once in amplexus, the male continues to call. Amplexus takes place on vegetation over water-filled cavities and also above small puddles on the ground. At least some females during amplexus enter the water, presumably thereby absorbing water into their bladders. Egg clutches are attached to roots and vegetation structures above the water; leaves are not folded around the eggs. The clutch size is 14–21 eggs with a mean of 17 eggs, plus numerous empty egg capsules. The ovum is about 4 mm in diameter, and the diameter of the egg capsule is 10–12 mm. Hatching occurs 11–15 days after deposition; hatchling tadpoles have total lengths of 18–20 mm. A recently metamorphosed young has an SVL of 27 mm and a brown dorsum with many pale green flecks and small blotches. Adults display defense postures after being captured, contracting their body combined with eye protection.

FIGURE 1.42 *Cruziohyla craspedopus*. Top: amplectant pair laying eggs under laboratory conditions. SGS. Bottom left: clutches of eggs in a pond located at the root of a fallen tree. Bottom right: embryo. Shiripuno Lodge, Provincia Orellana. Morley Read.

Additional reproductive biology data of pairs (wild and lab born) from Shiripuno Lodge, Provincia Orellana, have been obtained in the laboratory (CJ and Wikiri lab notes by Alex Achig, January 2021–March 2022). Amplexus is axillary and lasts 2–21 hours ($\bar{x}$ = 13, n = 7). All the oviposition events occurred during the day (not at night) and lasted 15 minutes, except for one that lasted 25 minutes. Groups of about four eggs are released. Clutches are cluster-shaped and contain 4–25 light turquoise-colored eggs ($\bar{x}$ = 18, n = 6) of 3.9–7.8 mm in diameter ($\bar{x}$ = 5.5, n = 52). The maximum diameter of the clutches varies between 45.5 and 66.1 mm ($\bar{x}$ = 58.7, n = 6). The clutches were deposited in different places in the terrarium, on the glass, on the apex of a leaf, on the water sprinkler, and over leaves. Hatching requires 10–42 days ($\bar{x}$ = 22, n = 5) following egg deposition. Tadpoles undergo complete development and metamorphosis in a span of 24–1039 days

FIGURE 1.43 Ontogenetic changes in five individuals of *Cruziohyla craspedopus* from tadpole to adult. Top to bottom, left to right. CJ 6373a, stage 27, lateral, dorsal, and ventral views, TL = 49.2 mm. CJ 6373b, stage 41, TL = 70.5 mm. CJ 6373c, stage 42, TL = 61 mm. CJ 6373d, stage 46 = 29 mm SVL. DAL. CJ 6373e, juvenile = 53.6 mm SVL; adult = 65 mm SVL. Shiripuno, Provincia Orellana. SGS.

($\bar{x}$ = 65.8, n = 5) at a water temperature of about 24.8 °C and a pH of 7.7. The development from froglet to adult in individuals from one clutch took around 543 days.

The call is a soft cluck having a duration of 0.06–0.07 s and a dominant frequency of 600–800 Hz.

Tadpoles.—Figure 1.43. The tadpole was described based on an individual at stage 39, collected from Shiona, Provincia Pastaza, Ecuador. Its body length is 25 mm, and the total length is 65 mm. The body is ovoid, wider than high. The greatest width is just behind the eyes and is greater than the depth of the body. The snout is bluntly rounded in dorsal view and in profile. The eyes are moderately large (2.4 mm in diameter), situated dorsolaterally, directed laterally, and separated by a distance of 7.7 mm. The nostrils are situated dorsally and directed anteriorly 3.3 mm behind the snout. The spiracle is just sinistral of the midline. The short spiracular tube is attached to the body wall; the slit-like opening is 2.4 mm wide. The vent tube is dextral to the ventral fin. The caudal musculature is moderately slender and highest at its junction with the body; the musculature gradually tapers posteriorly to a slender tip. At the midlength of the tail, the musculature is less than one-third of the height of the tail. The dorsal fin originates on the posterior edge of the body; at the midlength of the tail, the dorsal and ventral fins are about equal in height and higher than the caudal musculature. The fins terminate posteriorly in a blunt tip. The oral disc is directed anteroventrally just posterior to the tip of the snout and has shallow lateral folds. Marginal papillae completely encircle the oral disc—one row anteriorly and two rows laterally and ventrally. The labial tooth row formula is 2(2)/3(1). The anterior rows of denticles are about equal in length, extending to the papillae, slightly longer than the ventral rows. The serrated jaw sheaths are moderately robust and form a broad arch.

The dorsum of tadpoles varies from dark olive gray-green to creamy yellow, with white canthal lines. The venter is bluish-gray to creamy yellow. The tail musculature varies from creamy yellow to brown and black with bluish-gray spots. The fins are translucent with dense bluish-gray stippling. The iris is grayish-white.

Distribution.—See Figure 1.39. *Cruziohyla craspedopus* occurs on the Amazonian lowlands from southern Colombia and Ecuador southward to southern Peru and eastward to Estado Pará, Brazil. In Ecuador, it is known in the Amazonian lowlands and foothills of Sucumbíos, Orellana, Napo, Pastaza, and Zamora Chinchipe provinces. Its elevational range 200–973 m with an extent of occurrence of about 52,760 km^2.

In Ecuador, it inhabits rainforest habitats in evergreen forests of lowlands, flooded forests, and evergreen montane foothill forests in 10 ecosystems (Table 0.4) within four biogeographic sectors (Table 0.5), where the average annual rainfall is 1776–4266 mm and the average annual temperature is 23.3–25.6 °C.

Conservation.—*Cruziohyla craspedopus* is categorized as Least Concern throughout its entire range (Angulo *et al.*, 2004b; in IUCN Red List of 2022) and in Ecuador by Ortega-Andrade *et al.* (2021) in view of its wide distribution, presumed large population, and because it is unlikely to be declining fast enough to qualify for listing in a more threatened category. In Ecuador, local populations are threatened by agriculture, roads, pipelines, stochastic events, and oil activities. It is rarely encountered due to its preference for habitats in the canopy. In Peru, it is known to occur in Parque Nacional Manu. In Colombia, it is present in the El Pure protected area. In Ecuador, it occurs in the public reserves Reserva de Producción de Fauna Cuyabeno, Reserva Biológica Limoncocha, and Parque Nacional Yasuní; the biosphere reserves Yasuní, Sumaco, and Podocarpus-El Cóndor; and some private areas (e.g., Shiripuno Lodge).

Remarks.—According to Gray (2018), *Cryziohyla craspedopus* is the sister taxon of *C. sylviae* from Panama and Costa Rica. This species is part of the biotrade program carried out by Wikiri in Ecuador.

Key references.—(Funkhouser, 1957) (description); (Hoogmoed and Cadle, 1990) (natural history, tadpole); (Block *et al.*, 2003) (reproduction); (Meneghelli *et al.*, 2011) (natural history); (Rodrigues *et al.*, 2011) (distribution, reproduction); (Núñez *et al.*, 2015) (natural history); (Turrell *et al.*, 2016) (natural history); (Villacampa *et al.*, 2017) (account); (Glaw *et al.*, 2021) (natural history).

GENUS *PHYLLOMEDUSA* (LEAF FROGS) WAGLER, 1830

Phyllomedusa Wagler, 1830: 201. **Type species.**—*Rana bicolor* Boddaert, 1772.

Etymology.—The generic name is from the Greek *phyll*, meaning leaf, and *Medousa*. The name alludes to the gelatinous egg masses deposited on leaves of trees. The gender is feminine.

Definition.—Figure 1.44. *Phyllomedusa* is a group supported by molecular evidence. The genus is composed of large frogs (SVL up to 130 mm in *P. bicolor*) with little or no webbing on the feet and none on the hand. The vomerine teeth are present. The palpebral membrane is not reticulated. The tadpoles have moderately small oral discs directed anteroventrally.

Distribution.—The genus occurs in the Amazon Basin, the Chacoan Region of Paraguay, and eastern Brazil southward to Uruguay, northeastern Argentina, the Guianas, Trinidad, northern Venezuela, Cordillera Occidental and Magdalena Valley in Colombia, and extreme eastern Panama.

Content.—Currently, 16 species are recognized in this genus. Two species are known in Ecuador.

Remarks.—Figure 1.45. The molecular phylogenetic analysis by Duellman *et al.* (2016) shows a 100% support for the monophyly of this genus, within which are three strongly supported lineages. The first of these contains only *Phyllomedusa bicolor* and *P. vaillantii*, which are unique among phyllomedusids by having osteoderms in the skin on the dorsum. Although recognition of the *P. burmeisteri* and *P. tarsius* groups, as proposed by Faivovich *et al.* (2010) is strongly supported in their analysis, there are no known morphological features that distinguish the two groups. Röhr *et al.* (2020) studied the evolution and causes of variability in the advertisement call in the Brazilian species, at different levels: intra-individual, intra-population, inter-population, intra-species, and inter-specific. Their analyses showed that the variation was attributed to the inter-specific level, while call parameters at the intra-individual level varied the least. The call interval was the most variable across all levels. Most temporal acoustic parameters were affected by environmental temperature, while pulse rate and dominant frequency were strongly influenced by body size. Only pulse rate was correlated with the geographic distance between populations, while all parameters presented a significant phylogenetic signal.

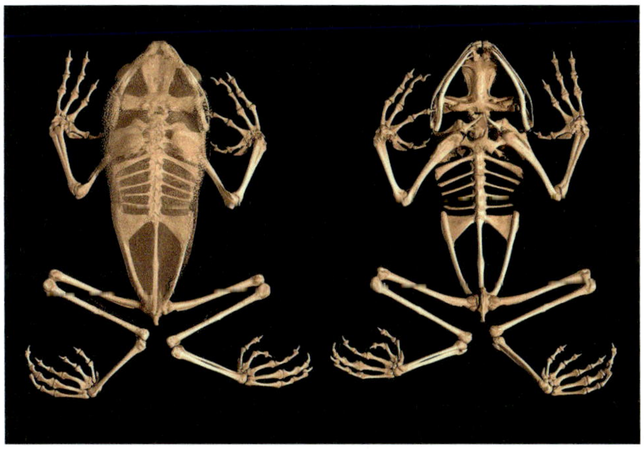

FIGURE 1.44 High-resolution computed tomography reconstructions of the skeleton of *Phyllomedusa bicolor*, CAS 257760, 106.7 mm SVL, Crabwood Creek, Guyana. Dorsal (left) and ventral (right) views. Daniel J. Paluh.

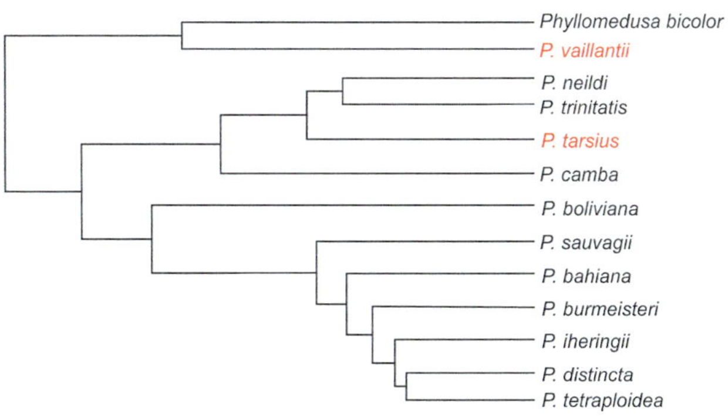

FIGURE 1.45 Topology of a tree analysis of *Phyllomedusa* based on molecular data, showing species occurring in Ecuador (in red). Taken and modified from Duellman *et al.* (2016).

PHYLLOMEDUSA TARSIUS (TARSUS LEAF FROG) (COPE, 1868)

Pithecopus tarsius (Cope, 1868): 113. **Type(s).**—USNM 6652, now lost, from Río Napo or upper Amazon, below the mouth of the former, Departamento Loreto, Peru.

Phyllomedusa tarsius Boulenger, 1882a: 428.

Etymology.—The specific name derives from the Latin *tarsus* in reference to the length of the tarsal bones.

Identification.—Figures 1.46, 1.47, 1.48. In these frogs, males attain an SVL of 97.0 mm, and females of 111.0 mm. The snout is inclined anteroventrally in both sexes and is angular above in males and round in females. The parotoid glands are rounded and extend to the midbody or sacrum. The skin on the dorsum is shagreen; that on the hind limbs is tuberculate in large individuals. The discs are moderately large; webbing is absent, and the first toe is longer than and opposable to the second. Calcars are absent.

The dorsal surfaces and sides of the head are green. The flanks are dull green or bluish-green with discreet or confluent cream, pink, or pale orange spots. The anterior and posterior surfaces of the thighs are brown or dull green with cream, pale orange, or bluish-gray spots. The throat, chest, and ventral surfaces of the limbs are dark brown or grayish-brown; the belly is gray or pale orange–brown. Ventral markings consist of a pair of white spots on the throat, a median white spot on the chest, and usually a pair of para-anal white spots. The iris is bright orange with bold black reticulations. Breeding males have grayish-brown nuptial excrescences.

Comparisons.—*Phyllomedusa vaillantii*. *Phyllomedusa tarsius* differs from other members of the genus by having rounded parotoid glands extending to the midbody or sacrum; flanks marked by cream, pink, or pale orange spots; and an orange iris with black reticulations. The species is

FIGURE 1.46 *Phyllomedusa tarsius*. Left: CJ 6485, female. Right: CJ 1115, male, SVL 79.2 mm. Reserva Otoyacu, Provincia Pastaza. LAC.

most easily confused with *P. vaillantii*, which has angular parotoid glands bearing a dorsolateral row of white tubercles, flanks that are green above and red below a row of longitudinally elliptical pale spots, and a pale gray iris. Other phyllomedusids have orange or cream flanks with brown markings and green frogs in other genera in the area have webbing and the first toe shorter than the second (webbing absent and first longer than second in *P. tarsius*).

Natural history.—Figure 1.49. In Santa Cecilia, Provincia Sucumbíos, individuals were found in March through October, with most individuals being observed in April through June. All were found in forests at night. Non-breeding individuals usually are perched on limbs 2–6 m above the ground. This species sporadically breeds after rain. Males have been heard calling from the limbs of trees and bushes over ponds at various times throughout the year. Several hundred of these large frogs congregated at a forest pond bordered by spiny palms on the night of 2 April 1972. As many as four individuals were observed on one palm frond. Two nights later, only a few individuals were found. The unpigmented eggs are deposited on or between leaves over ponds. Usually, the frogs fold leaves over the clutch so as to conceal the eggs. Eight clutches contained 426–650 ($\bar{x}$ = 551) eggs having diameters of 3.5–4.0 ($\bar{x}$ 3.7) mm. These clutches hatched in 7.0–8.3 ($\bar{x}$ = 7.4) days; hatchlings had total lengths of 11.5–13.0 ($\bar{x}$ = 12.6) mm. Fifteen females contained 410–750 ($\bar{x}$ = 549) mature ovarian eggs; thus, it seems as though this species usually deposits the entire ovarian complement as a single clutch.

FIGURE 1.47 Morphological variation in *Phyllomedusa tarsius*. Top: QCAZ 33526, subadult, Bocana del Suno, Provincia Orellana. Middle: QCAZ 40864. Bottom: QCAZ 36925. Centro Ecológico Zanja Arajuno, Provincia Pastaza. LAC.

Tadpoles orient head-up at an angle of about 45° to the surface of the water. Although some have been found singly or in small numbers, two large concentrations were found in open water. Each congregation consisted of at least 100 individuals; all were oriented in the same way. Fourteen metamorphosing young had snout–vent lengths of 20–29 ($\bar{x}$ = 25.3) mm.

Twenty of the 21 stomachs examined contained only Orthoptera, most of which were large, with only one

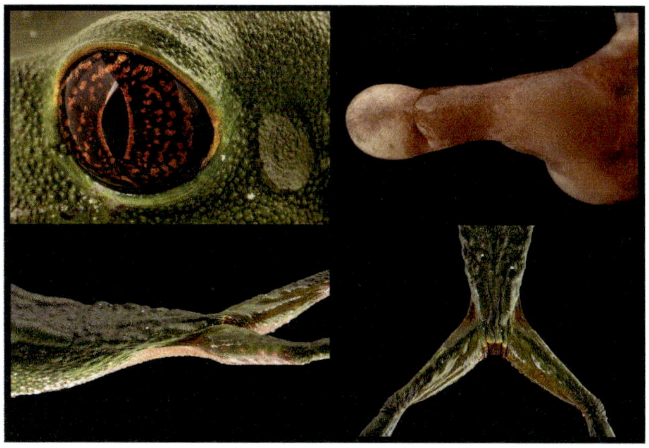

FIGURE 1.48 Morphological details of *Phyllomedusa tarsius*. Top: CJ 11917, male, 85.0 mm SVL, note the nictitating membrane. Middle left. CJ 11917, male, note the nuptial pad. Reserva Biológica Limoncocha, Provincia Sucumbíos. SGS. Middle right and bottom: CJ 6485, female, Reserva Otoyacu, Provincia Pastaza. LAC. Ecuador.

FIGURE 1.49 *Phyllomedusa tarsius*. Aspects of its natural history. Top left: QCAZ 40864, individual in its microhabitat at night, Centro Ecológico Zanja Arajuno, Provincia Pastaza. LAC. Top right: individual in its microhabitat at night, note color variation, Parque Nacional Yasuní, Provincia Orellana. Morley Read. Bottom left: recently laid clutch of eggs. Bottom right: clutch of eggs in a nest made of fern leaves. Centro Ecológico Zanja Arajuno, Provincia Pastaza. LAC.

specimen per stomach. The largest orthopteran was 43 mm in length. The other stomach contained a spider. The exclusive occurrence of gravid females of *Phyllomedusa tarsius* and *Pithecopus palliatus* is discussed in the account of *P. palliatus*.

FIGURE 1.50 Ontogenetic changes in *Phyllomedusa tarsius* from tadpole to adult. Top to bottom, left to right. QCAZ 78837, stage 27, dorsal and lateral views; undetermined stage; stage 42 = 33. 7 mm; stage 46, 15.4 mm SVL; juvenile 43 days after stage 46, 26.6 mm SVL; juvenile 82 days, 32.2 mm SVL; adult 257 days, 46.7 mm SVL. Centro Ecológico Zanja Arajuno. LAC.

The call is a moderately loud, low-pitched "cluck" with 2.0–3.2 notes per minute. Additional data of the call are provided under the subtitle "Vocalizations of Ecuadorian Anurans" in Appendix 1.

Tadpoles.—Figure 1.50. The tadpole was described based on an individual at stage 36 from Santa Cecilia, Provincia Sucumbíos, Ecuador. It has a body length of 18 mm and a total length of 50 mm. The body is ovoid, as wide as deep, and deepest posteriorly. The snout is truncate. The eyes are large, widely separated, and directed laterally. The nostrils are directed anteriorly nearly at the tip of the snout. The opening of the sinistral spiracle is directed posteriorly nearly on the midline and at a point at about the midlength of the body. The dextral anal tube is short. The caudal musculature is robust proximally, gradually tapering to a terminal point. A shallow dorsal fin is present only on the posterior two-thirds of the tail. The ventral fin is shallower than the musculature and extends to the tip. The tail is about 64% of the total length. The small terminal mouth lacks lateral folds. The upper lip is bare; there is one row of small labial papillae laterally and two rows ventrally. The jaw sheaths are moderately robust with coarse serrations. The upper jaw sheath is a narrow arch, and the lower beak is V-shaped. There are two upper and three lower rows of denticles; the second upper row is broadly interrupted medially.

The dorsal surfaces of the body and tail are pale tan; the sides and venter are pale metallic green. The iris is bronze.

Distribution.—Figure 1.51. *Phyllomedusa tarsius* occurs in the Amazon Basin in southern Colombia, Ecuador, Peru,

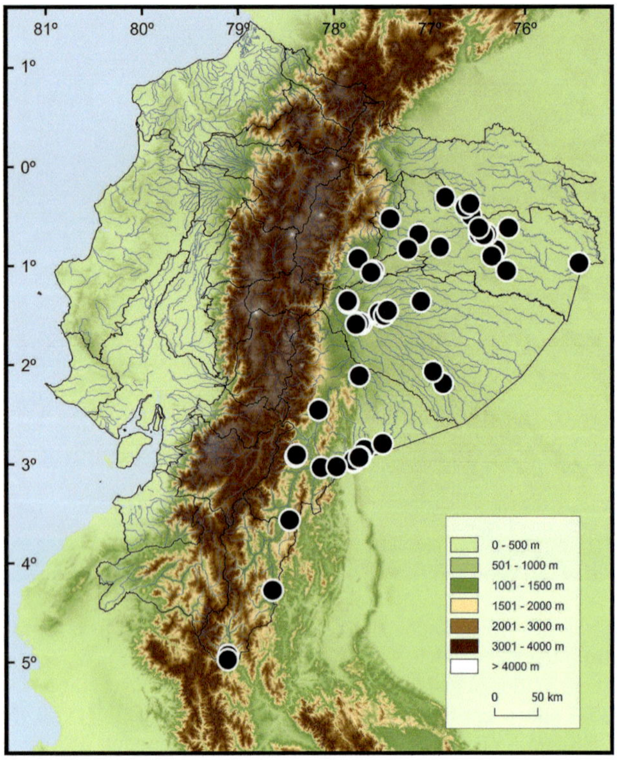

FIGURE 1.51 Localities of known occurrence of *Phyllomedusa tarsius* in Ecuador.

and Brazil, with seemingly isolated populations in the Cordillera Occidental of Colombia and along the lowlands on the eastern side of the Mérida Andes in western Venezuela and central Guyana (see Remarks). In Ecuador, it occurs in Sucumbíos, Orellana, Napo, Pastaza, Morona Santiago, and Zamora Chinchipe provinces. Its elevational range is 196–1036 m in an area of extent of occurrence of about 68,920 km^2.

This nocturnal, arboreal species inhabits tropical rainforests and intervened areas in nine ecosystems (Table 0.4) within seven biogeographic sectors (Table 0.5), where the average annual rainfall is 1104–4392 mm and the average annual temperature is 21.1–25.9 °C.

Conservation.—*Phyllomedusa tarsius* is categorized as Least Concern throughout its entire range (La Marca *et al.*, 2004b; in IUCN Red List of 2022) in view of its wide distribution, tolerance of a degree of habitat modification, presumed large population, and because it is unlikely to be declining fast enough to qualify for listing in a more threatened category. In Ecuador, Ortega-Andrade *et al.* (2021) also categorized it as Least Concern, although local populations are threatened by oil activities, agriculture, roads, pipelines, and deforestation. It is common throughout its range. In Colombia, the Santander population overlaps with the Guanenta-Alto Río Fonce National Park. In Ecuador, it has been recorded in the public reserve Parque Nacional Yasuní; the protected forests of Cordillera Kutukú y Shaimi, Comuna Santa Elena, and Nacionalidad Waorani del Ecuador; the biosphere reserves of Yasuní, Sumaco, and Podocarpus-El Cóndor; and some private reserves (e.g., Estación de Biodiversidad Tiputini, Reserva Biológica Jatun Sacha).

Remarks.—*Phyllomedusa tarsius* is closely related to *P. trinitatis* plus *P. neildi*, according to the molecular phylogeny of Duellman *et al.* (2016). Barrio-Amorós *et al.* (2019) questioned the presence of *Phyllomedusa tarsius sensu stricto* in Venezuela; these populations are suggested to be of an unnamed species. Székely *et al.* (2016) provided an altitudinal record extension in Ecuador.

Phyllomedusa coelestis (Cope, 1874) was described in Moyabamba, Departamento San Martín, Peru. Subsequently, it was recorded in Jatun Sacha, Provincia Napo, Ecuador, by Vigle (2008) based on two specimens found (not collected), one by Morley Read and another one by Gregory O. Vigle. We examined a color photo identified as *P. coelestis* of one of the specimens found by Morley Read at Jatun Sacha, and it exactly fits the coloration pattern of a subadult of *P. tarsius* (see Figure 1.50), which has orange spots on the lower flanks. Thus, we removed this species from the Ecuadorian fauna. When recording *P. coelestis* in Departamento Loreto, Peru, Duellman and Mendelson III (1995) redescribed the holotype (ANSP 11384) and provided data for an additional male specimen (KU 221944). They failed to find major differences between both specimens, except for dermal features that were attributed to preservation. They also compared these specimens of *P. coelestis* with *P. tarsius* and found size differences (53.3–61.8 mm SVL vs. 81–90 mm SVL, respectively), vocal slits present in *P. coelestis* (absent in *P. tarsius*), and small irregular pale marks (absent in *P. tarsius*). Except for the absence of vocal slits in *P. tarsius* (present in *P. coelestis*), the size and flank color pattern might be part of the intraspecific variation. Thus, *P. coelestis* might be a junior synonym of *P. tarsius*.

Key references.—(Cope, 1868) (description); (Duellman, 1974b) (account); (Duellman, 1978) (account, call, tadpole, natural history); (Hero, 1990) (tadpole); (Rodríguez and Duellman, 1994) (account); (Barrio-Amorós, 2009) (biology); (Schiesari *et al.*, 2022) (tadpole).

PHYLLOMEDUSA VAILLANTII (WHITE-LINED LEAF FROG) BOULENGER, 1882a

Phyllomedusa vaillantii Boulenger, 1882a: 427. **Holotype.**—BMNH 1947. 2. 22. 34 from Santarem, Estado Pará, Brazil.

Etymology.—The specific name is a patronym that honors the French herpetologist Léon Louis Vaillant (1834–1914). He became a professor at the Museum of Natural History in Paris.

Identification.—Figures 1.52, 1.53. In these frogs, males attain an SVL of 58.0 mm, and females of 84.0 mm. The maximum known masses are 9.0 g in males and 35.5 g in females. Although females are much larger than males, no sexual dimorphism is evident in the proportions. In profile, the snout is truncate in males and inclined anteroventrally in females. The parotoid glands are slightly elevated, angular, and extend to the sacrum; there is a longitudinal row of white granules along the angle of the gland. The skin on the dorsum is rough; it is tubercular on the hind limbs in large individuals. The discs are moderately large; webbing is absent, and toe I is longer than and opposable to toe II. Calcars are absent.

The dorsum and side of the head are green; the flanks are green above and reddish-brown below a row of longitudinally elliptical cream to pale orange spots. The anterior and posterior surfaces of the thighs are reddish-brown or lavender with small cream to pale orange spots. The venter is dull grayish-brown with a pair of lavender-bordered cream spots on the throat and a pale green spot on the chest. The iris is pale gray with a greenish tint. Breeding males have grayish-brown nuptial excrescences. Juveniles are dichromatic; the dorsum can change from pale green to orange–brown.

Comparisons.—*Phyllomedusa tarsius*, *Pithecopus palliatus*, *Callimedusa tomopterna*. *Phyllomedusa vaillantii* is a large green tree frog without webbing and characterized by angular parotoid glands bearing a dorsolateral row of small white tubercles. It is most similar to *P. tarsius*, which has round parotoid glands, no white tubercles, and an orange iris with black reticulations (pale gray in *P. vaillantii*). Other phyllomedusids in Ecuador (*P. palliatus*, *C. tomopterna*) have cream or orange flanks with brown markings; in *P. vaillantii*, the flanks are green above and red below a row of pale spots.

Natural history.—Figure 1.54. In Santa Cecilia, Provincia Sucumbíos, Ecuador, individuals were found in primary and secondary forests at night in all months, except November and December. Adults were on bushes and trees up to a height of 4 m. Males call from bushes over ponds, but no breeding congregations were encountered. The call consists of a short, harsh "cluck." Five gravid females in March, April, June, and August contained 912–1250 ($\bar{x}$ = 1114) mature ovarian eggs having diameters of 2.0 mm. One female deposited a clutch of 645 unpigmented eggs 2.0 mm in diameter. The eggs are deposited on vegetation above ponds. *Phyllomedusa vaillantii* apparently breeds at sites

FIGURE 1.52 *Phyllomedusa vaillantii* depicting variation. Left: CJ 6355, female, SVL 54.6 mm, Shiripuno Lodge. Right: QCAZ 43138, km 10 on road Pompeya-Iro, Parque Nacional Yasuní. Provincia Orellana. LAC.

FIGURE 1.53 Morphological details of *Phyllomedusa vaillantii*. Top: an individual depicting color change, Parque Nacional Yasuní, Provincia Orellana. LAC. Middle: CJ 6355, female, groin and thighs, Shiripuno Lodge, Provincia Orellana. LAC. Bottom: CJ 11922, female, 77.0 mm SVL, note the nictitating and palpebral membranes, Reserva Biológica Limoncocha, Provincia Sucumbíos. SGS.

FIGURE 1.54 *Phyllomedusa vaillantii* in its microhabitat at night. Top left: CJ 6253, female subadult, 51.3 mm SVL. LAC. Top right: Morley Read. Bottom left: amplectant pair. Morley Read. Bottom right: juvenile. Morley Read. Shiripuno Lodge, Provincia Orellana.

distinct from those used by the other three species of phyllomedusids in Santa Cecilia. At two ponds where the other species called commonly at various times throughout the year, *P. vaillantii* was never found. Instead, individuals of *P. vaillantii* were scattered throughout the forest or in small groups at temporary ponds. Eleven metamorphosing young from February (2), June (1), and July (8) have SVLs of 21–25 ($\bar{x}$ = 23.8) mm. Sixteen of 18 stomachs contained orthopterans up to 34 mm in length. The other two contained spiders.

At Cusco Amazónico, Peru, although individuals were found throughout the terra firme forest, all breeding congregations were at permanent ponds in inundated forest; 162 of 174 adults were found in the rainy season. This is primarily an arboreal species; of 162 adults found in the rainy season, 129 calling males were at heights of 0.2–5.0 m ($\bar{x}$ = 1.72), whereas non calling individuals were at heights of 0.0–4.0 m ($\bar{x}$ = 1.02). In the dry season, 12 individuals were at lower heights, 0.1–2.0 m ($\bar{x}$ = 0.76). Of 240 individuals with substrate data, 200 (83.3%) were on branches of trees or bushes and 33 (13.8%) were on leaves (31 on trees or bushes and two on *Heliconia* leaves). Two others were walking on the ground, and two were perched on vines; one was found on a tree root, one on a tree trunk, and one on a palm frond. All 129 calling males were above water; only four were on leaves, and the others plus two amplectant pairs were on branches. In the rainy season, 23 non-calling adults were on perches above land and 10 were above water, whereas 23 young were on perches above land and three above water. In contrast to adults, 20 (90%) of the juveniles were on leaves of bushes, and only three (10%) were on branches. Individuals found in June (13) and July (2) consisted of nine females, three males, and three juveniles; the males and females are notably smaller than those found in the rainy season, whereas the juveniles are larger. Most breeding activity takes place in the early part of the rainy season; 140 calling males were recorded in November, 38 in December, and two in January. The response of males to the first heavy rain of the rainy season is illustrated by observations at a permanent pond. On 17 November 1986, rainfall amounted to 62 mm; on that night and the following night, the pond was dominated by calling *Osteocephalus taurinus* and *Trachycephalus coriaceus*. Neither of these species was present in the pond on subsequent nights when *Phyllomedusa vaillantii* appeared in numbers: 49 on 19 November, 15 on the 20th, 20 on the 21st, and 20 on the 22nd. No males were calling on 23 November when 20 mm of rain fell; an additional 19 mm fell on 24 November when 40 males were calling, but only one male was calling on 25 November. A similar response was observed at Camp Cocha, Cusco Amazónico, in December 1989, when rainfall amounted to 87.7 mm on 13–15 December; the first *C. vaillantii* called on the 15th. Numerous males called from then through 26 December, during which time an additional 53.1 mm of rain fell. After 22.4 mm of rainfall on 27 January 1990, only one male called, but no males called on the night of 10 February 1990 when 127.6 mm of rain fell. Thus, it seems that this species reaches the height of its breeding activity during the early part of the rainy season and that no breeding takes place at times of heavy rains in January and February.

The call is a moderately loud, a single cluck-like sound, sometimes followed by a soft chuckle. Two or three notes are emitted per minute. Each note is about 0.1 s long. The analysis of one recording from Pilcopata, Departamento Cusco, Peru, shows 13.3 notes per minute having a duration of 0.5 s. Also, a call from Brazil was described. Detailed data and figures on calls are provided under the subtitle "Vocalizations of Ecuadorian Anurans" in Appendix 1.

Despite the abundance of this frog, only two amplectant pairs were found, both in November. One pair was on a branch 1.5 m above the water, and another was observed to ascend vines and branches 0.2–4.0 m above the water. The left and right oviducts in two gravid females contained 980 and 1020 ovarian eggs about 1.5–2.0 mm in diameter. No egg clutches were observed. The number and sizes of ovarian eggs corresponded to the number and sizes (912–1250 eggs, $\bar{x}$ = 1111.4, with a maximum diameter of 2.0 mm) reported for five females from Santa Cecilia, Ecuador. However, apparently, only about half of the eggs are deposited in a clutch. A clutch of 645 eggs was deposited on a leaf over a pond at Santa Cecilia, Ecuador, and clutches of 415–532 embryonated eggs wrapped in leaves were found in French Guiana. The existence of only one size and 912–1020 ovarian eggs in a given female suggest that females deposit two clutches consecutively and may not produce eggs for subsequent clutches later in the rainy season.

At Cusco Amazónico, Peru, tadpoles were found in permanent ponds from December through March and in a permanent pond in November. Of 2318 tadpoles sampled in Camp Cocha, 1332 were at midlevel in the water, 621 were in bottom samples, and 365 were in surface samples;

corresponding numbers for 55 tadpoles in another pond were 7, 44, and 4. At Camp Cocha, the majority (1388) of the tadpoles were found at night, whereas in the other pond, the majority (54) were found by day. At all three ponds, some, but not all, tadpoles were observed in schools containing > 50 individuals by day. Schooling behavior occurs by day and night. Five adults contained food; each had one item. Of the total volume, a cicada made up 55.86%, and an unidentified orthopteran 43.23%. Other food items (0.9% of volume) consisted of an ant, an isopod, and an unidentified larva.

Tadpoles.—Figure 1.55. The tadpole was described based on an individual at stage 38 from 3 km from the Santa Cecilia turn-off toward Quito, Provincia Sucumbíos, Ecuador. It has a body length of 19 mm and a total length of 52 mm. The body is elongate, ovoid, slightly deeper than wide. The snout is truncate. The large eyes are widely separated and directed laterally. The nostrils are directed anteriorly nearly at the tip of the snout. The spiracular opening is directed posteriorly just to the left of the midline at a point at about the midlength of the body. The anal tube is short and dextral. The caudal musculature is robust proximally and gradually narrows to a slender tip. The dorsal fin is very shallow, and the ventral fin is shallower than the musculature and extends to the tip of the tail, which is about 63% of the total length. The small terminal mouth lacks lateral folds. The upper lip is bare; elsewhere the mouth is bordered by two rows of small labial papillae. The jaw sheaths are moderately robust with fine serrations. The upper jaw sheath is a broad arch with slender alary processes; the lower haw sheath is broadly V-shaped. There are two upper and three lower rows of denticles; the second upper row is broadly interrupted medially.

The dorsum is olive green, paler on the tail than on the body. The belly is white, and the ventrolateral surfaces are iridescent yellowish-green. The iris is pale bronze.

At Cusco Amazónico, Peru, tadpoles hatch at stage 25 and have body lengths of 2.3–2.5 mm ($\bar{x}$ = 2.41, n = 10) and total lengths of 8.9–9.2 mm ($\bar{x}$ = 9.02, n = 10). The hatchlings have no keratinized mouthparts and a large quantity of yolk in the gut. Slightly larger tadpoles at stage 25 have body lengths of 4.0–5.5 mm ($\bar{x}$ = 4.73, n = 9) and total lengths of 11.0–15.0 mm ($\bar{x}$ = 12.17, n = 9); these tadpoles have keratinized jaw sheaths but no labial teeth. Two tadpoles in stage 26 have body lengths of 5.0 and 5.6 mm and total lengths of 12.1 and 15.2 mm; these and all those in later stages have a labial tooth row formula of 2(1)/3. Corresponding measurements of tadpoles in later stages are as follows: stage 28, body length 10.0 mm and total length 23.5 mm; stage 29, body lengths 10.5–11.0 mm ($\bar{x}$ = 10.8, n = 3) and total lengths 28.5–32.3 mm ($\bar{x}$ = 29.9, n = 3); stage 33, body length 14.4 mm and total length 39.0 mm; stage 37, body length 19.2 mm and total length 51.0 mm; stage 38, body length 18.8 mm and total length 54.0 mm; stage 41, body length 21.0 and total length 53.3 mm. Recently metamorphosed young in January and February have SVLs of 26.3–35.0 mm ($\bar{x}$ = 28.1, n = 10). A typical tadpole from Cusco Amazónico, Peru, in stage 37 has a body length of 19.2 mm and a total length of 51.0 mm. The tadpole has a greatest body width of 11.3 mm at the midlength of the body; the greatest height is 10.8 mm

FIGURE 1.55 Ontogenetic changes in *Phyllomedusa vaillantii* from tadpole to adult. Top to bottom, left to right. CJ 6389, stage 34, lateral, dorsal, and ventral views, TL = 43.1 mm; stage 42, TL = 54.3; stage 46, 34 days after stage 34 = 22.6 mm SVL; juvenile, 91 days after stage 46, 27.8 mm SVL. DAL. Adult, 294 days, 51.0 mm SVL. Shiripuno Lodge, Provincia Orellana. SGS.

at the posterior end of the body. In dorsal view, the body is ovoid. The snout is truncate in dorsal view and bluntly rounded, most protuberant dorsally, in profile. The eyes are positioned dorsolaterally and directed laterally. The interorbital distance is about twice the diameter of the eye. The nostrils are moderately small, positioned laterally, and directed anterolaterally at about one-third of the distance from the snout to the orbits. Narial papillae are absent. The spiracle is just sinistral to the midline. The spiracular tube originates just posterior to the orbit and throughout its length is attached to the body; the spiracular opening is slightly smaller than the base and is directed posteriorly. The vent tube is moderately long and attached for its entire length medially to the ventral fin. The caudal musculature is moderately robust, highest at its junction with the body, and tapers gradually to a pointed tip. The dorsal fin originates on the proximal caudal musculature; the fin is shallow throughout its length and highest posteriorly. The ventral fin originates on the posterior end of the body and is highest anteriorly; throughout its length, it is much higher than the dorsal fin, but at the midlength of the tail, it is not as high as the caudal musculature. The oral disc is anteroventral, is directed anteriorly, and lacks lateral folds. The disc has a single row of small, blunt marginal papillae laterally and two irregular rows ventrally; the median two-thirds of the anterior labium lack papillae. The labial tooth row formula is 2(2)/3. The jaw sheaths are slender and finely serrate. The anterior sheath is broadly arched; the posterior sheath is broadly V-shaped.

In living tadpoles in stage 25, the body is pale tan anteriorly and black posteriorly; the caudal musculature is pale tan, and the fins are translucent except for black on the proximal margin of the dorsal fin. In tadpoles beyond stage 33, the dorsum of the body is dull green, and the caudal musculature is paler green; the belly is white, and the ventrolateral surfaces of the body have a yellowish tint in some individuals.

Distribution.—Figure 1.56. *Phyllomedusa vaillantii* is known from Guyana, Suriname, French Guiana, and northeastern Brazil to the Amazon Basin of Colombia, Ecuador, Peru, and northern Bolivia. It has been reported in Cerro Neblina, extreme southern Amazonas, Venezuela. In Ecuador, it occurs in Sucumbíos, Orellana, Napo, Pastaza, Morona Santiago, and Zamora Chinchipe provinces. Its elevational range is 192–1027 m with an extent of occurrence of about 72,270 km^2.

This nocturnal, arboreal species inhabits tropical rainforests and intervened areas in flooded forests, evergreen forests of lowlands, and evergreen montane foothill forests in 11 ecosystems (Table 0.4) within seven biogeographic sectors (Table 0.5), where the average annual rainfall is 1763–4397 mm and the average annual temperature is 20.9–25.8 °C.

Conservation.—*Phyllomedusa vaillantii* is categorized as Least Concern throughout its entire range (Azevedo-Ramos *et al.*, 2010; in IUCN Red List of 2022) and in Ecuador by Ortega-Andrade *et al.* (2021) in view of its wide distribution, presumed large population, and because it is unlikely to be declining to qualify for listing in a more threatened category. In Ecuador, local populations are threatened by oil activities, agriculture, roads, pipelines, and deforestation. It is common to abundant throughout its range. Venezuelan populations are protected within Sierra de La Neblina National Park. In Ecuador, it has been recorded in the public reserves Parque Nacional Yasuní, Reserva Biológica Limoncocha, and Reserva de Producción de Fauna Cuyabeno; several protected forests (e.g., CEPLOA, Cordillera Kutukú y Shaimi, Cuenca Alta del Río Nangaritza, Pañacocha); and some private reserves (e.g., Estación de Biodiversidad Tiputini, Reserva Biológica Jatunsacha).

Remarks.—*Phyllomedusa vaillantii* is the sister taxon of *P. bicolor*, according to the molecular phylogeny of Faivovich *et al.* (2005), and corroborated by the most recent analyses by Duellman *et al.* (2016). The calls described from Santa Cecilia (Ecuador), Cusco Amazónico (Peru), and Brazil differ in several parameters such as the note rate and note duration. This variability in call parameters needs to be evaluated at different levels: intra-individual, intra-population, interpopulation, intra-species, and inter-specific.

Key references.—(Boulenger, 1882a) (description); (Duellman, 1974b) (account); (Crump, 1974) (natural history); (Duellman, 1978) (account, call, tadpole, natural history); (Hero, 1990) (tadpole); (Rodríguez and Duellman, 1994) (account); (Lescure *et al.*, 1995) (natural history); (Wild, 1996) (tadpole, natural history); (Lescure and Marty, 2000) (account); (Duellman, 2005) (account, tadpole, call);

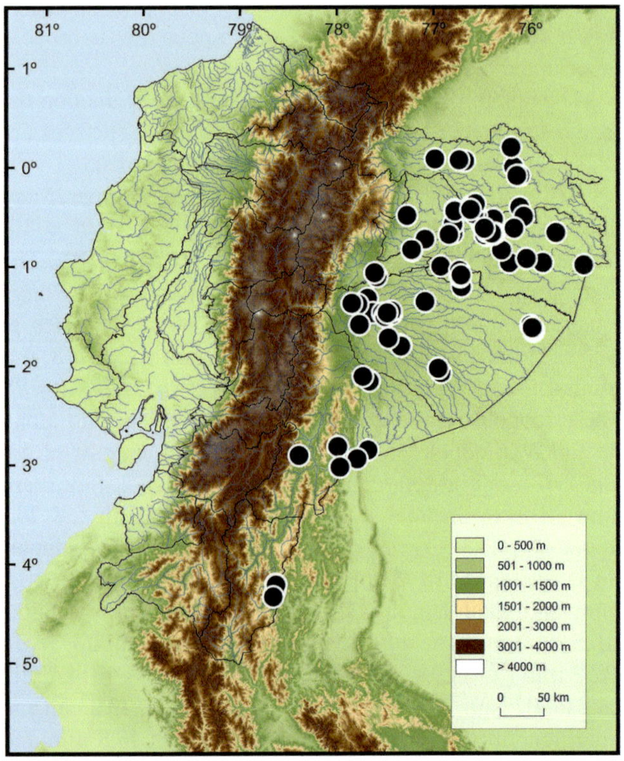

FIGURE 1.56 Localities of known occurrence of *Phyllomedusa vaillantii* in Ecuador.

(Kok and Kalamandeen, 2008) (account); (Barrio-Amorós, 2009) (biology); (Ouboter and Jairam, 2012) (account); (Cole *et al.*, 2013) (account); (Villacampa *et al.*, 2017) (account); (Barrio-Amorós *et al.*, 2019) (notes); (Röhr *et al.*, 2020) (call); (Schiesari *et al.*, 2022) (tadpole).

GENUS *PITHECOPUS* (LEAF FROGS) COPE, 1866

Pithecopus Cope, 1866: 86. **Type species.**—*Phyllomedusa azurea* (Cope, 1862), by original designation.

Etymology.—The generic name is derived from the Greek *pithekodes*, meaning ape-like. The gender is masculine.

Definition.—Figure 1.57. *Pithecopus* is a group supported by molecular evidence. No morphological synapomorphies are known. The genus is composed of medium-sized species (SVL ± 45 mm). Toe I is much longer than and opposable to toe II. The vomerine teeth are absent. The tadpoles have a moderately small oral disc directed anteroventrally.

Distribution.—The genus occurs in tropical South America east of the Andes from southern Venezuela to northern Argentina.

Content.—Currently, 12 species are recognized in this genus. One species is known in Ecuador.

Remarks.—Figure 1.58. The genus *Pithecopus* contains species that were designated as the former *Phyllomedusa hypochondrialis* group (e.g., Faivovich *et al.*, 2010). The molecular phylogenetic analysis by Duellman *et al.* (2016) shows a 100% support for the monophyly of this genus. Röhr *et al.* (2020) studied the evolution and causes of variability in the advertisement call of the Brazilian species at different levels: intra-individual, intra-population, interpopulation, intra-species, and inter-specific. Their analyses showed that the variation was attributed to the inter-specific level, while call parameters at the intra-individual level varied the least. Call interval was the most variable across all levels. Most temporal acoustic parameters were affected by environmental temperature, while pulse rate and dominant frequency were strongly influenced by body size. Only pulse rate was correlated with the geographic distance between populations, while all parameters presented a significant phylogenetic signal.

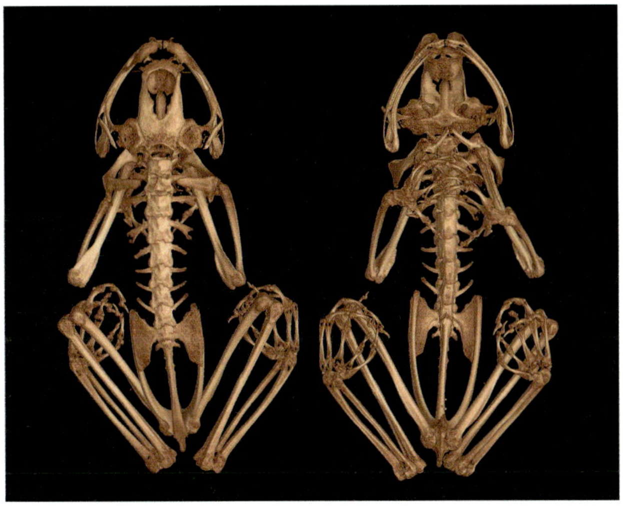

FIGURE 1.57 High-resolution computed tomography reconstructions of the skeleton of *Pithecopus azureus*, MCZ 29952, 33.6 mm SVL, El Carmen, Bolivia. Dorsal (left) and ventral (right) views. Daniel J. Paluh.

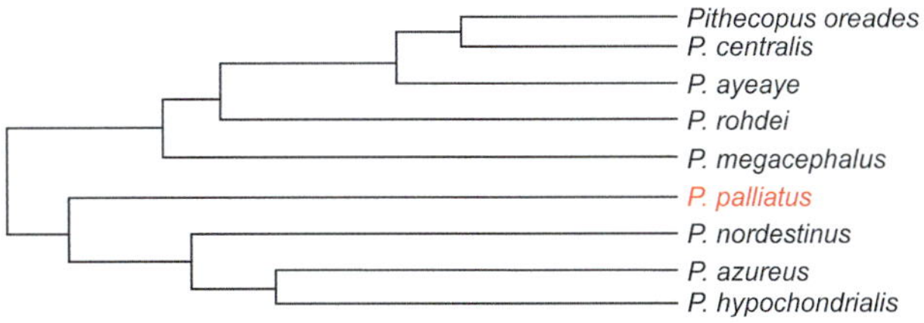

FIGURE 1.58 Topology of a tree analysis of *Pithecopus* based on molecular data, showing species occurring in Ecuador (in red). Taken and modified from Duellman *et al.* (2016).

PITHECOPUS PALLIATUS (MANTLE LEAF FROG) (PETERS, 1873b)

Phyllomedusa palliata (Peters, 1873b): 772. **Holotype.**—ZMB 7181 from Departamento Ucayali, Peru.

Pithecopus palliatus Duellman *et al.*, 2016: 32.

Etymology.—The specific name derives from the Latin *palliatus*, meaning cloak, presumably in reference to the contrasting color between the green dorsum and light brown flanks, which gives the frog the appearance of bearing a mantle.

Identification.—Figure 1.59. In these frogs, males attain an SVL of 39.1 mm, and females of 41.5 mm. The maximum mass is 3.4 g in males and 4.0 g in females. No sexual dimorphism is evident in proportions. In profile, the snout is rounded above; in males, it is truncate terminally, whereas it is slightly inclined anteroventrally in females. The parotoid glands are low and rounded; they extend only to the scapular region. The skin on the dorsum is smooth. The discs on the digits are only slightly wider than the digits; webbing and calcars are absent.

The dorsum and the sides of the head above the nostrils and middle of the orbits are dark green. The flanks and the sides of the head below the nostrils and orbits are creamy tan with irregular dark brown flecks and dashes on the flanks. The groin and hidden surfaces of the thighs are pale orange with small, irregular, dark brown marks. The venter is dull cream with brown flecks. Breeding males have

FIGURE 1.59 *Pithecopus palliatus*. KU 126661, male, 44.0 mm SVL, Santa Cecilia, Provincia Sucumbíos. WED.

brown nuptial excrescences. The iris is pale bronze with fine black reticulations.

Comparisons.—*Callimedusa tomopterna*, *C. atelopoides*, *Phyllomedusa camba*, *P. vaillantii*. *Pithecopus palliatus* differs from other phyllomedusids by having small digital discs, and the sides of the head and flanks are cream with brown flecks. All other green frogs in Ecuador have the sides of the head green. Only *C. tomopterna* has patterned flanks (orange with vertical brown bars); it differs further from *P. palliatus* by having moderately large discs and large calcars (absent in *P. palliatus*). With the exception of the much smaller, brown *C. atelopoides*, other phyllomedusids in Ecuador and Peru have proportionately large discs on the digits. The sides of the head and flanks are green in the larger *camba* and *P. vaillantii*; the former has a dark red iris and the latter a pale gray iris. In the slightly larger *C. tomopterna*, the flanks and hidden surfaces of the thighs are orange with black or dark brown vertical bars; furthermore, prominent calcars are present on the heels (absent in *P. palliatus*).

Natural history.—In Manu, Peru, it has been found in forest clearings near buildings. In Santa Cecilia, Provincia Sucumbíos, individuals were found about equally abundantly in primary and secondary forests and in forest-edge habitats, where they usually were associated with temporary or permanent ponds. At most breeding sites, the frogs were most common in low (< 1 m) herbaceous growth above the water. Individuals were observed in March through August and in October and December. Breeding occurs sporadically throughout the year. Males call from low, emergent vegetation in ponds in and at the edge of the forest. Eggs are deposited on the upper edges of leaves, usually in a teardrop shape on the tip of the leaf. Six clutches contained 38–71 ($\bar{x}$ = 49.3) unpigmented eggs having diameters of 3.5–4.0 ($\bar{x}$ = 3.6) mm. Fifteen gravid females contained 38–106 ($\bar{x}$ = 60.8) mature ovarian eggs. Apparently, the entire ovarian complement is not necessarily deposited in one clutch. The tadpoles occur in open ponds, where they orient head-up at an angle of about 45° to the surface. Usually, they are found singly or in groups of fewer than 10. Upon being disturbed, they seek refuge amid vegetation or detritus on the bottom. Five metamorphosing young have SVLs of 20–21 ($\bar{x}$ = 20.4) mm. In 20 stomachs, orthopterans and homopterans were the most frequently taken prey (40% and 35% of the stomachs, respectively). Spiders were present in one-fourth of the stomachs, whereas each of three frogs had eaten a fly, a beetle, and a caterpillar. Although *Pithecopus palliatus* and *Phyllomedusa tarsius* were commonly found at the same breeding sites at the same time, observations at one site throughout one year revealed that gravid females of only one species were present on a given night. The presence of one or the other was sporadic, not seasonal.

The call from Santa Cecilia was described as a soft click repeated at intervals of several seconds to three minutes. One recording in Tambopata, Peru revealed notes produced every 20–25 seconds; each note was about 0.12 seconds. Additional data of the call from Santa Cecilia and figures of the call from Manu are provided under the subtitle "Vocalizations of Ecuadorian Anurans" in Appendix 1.

At Cusco Amazónico, eight individuals were found at night in February in three different years. One was on a bush 1 m above the ground in the camp, one was on a leaf 1 m above the water in a temporary pond, and six were on leaves and branches of bushes 0.4–1.0 m ($\bar{x}$ = 0.58) above water in a temporary pond in a clearing just west of the Quebrada Madama. Individuals also were seen in terra firme forest in the rainy season. Two females were gravid. An analysis of the contents of five stomachs revealed that orthopterans made up half of the prey items and 93.3% of the volume of prey. Other prey items were one insect larva, one wasp, and one mite. The apparent scarcity and restricted spatial distribution of this species at Cusco Amazónico has not been explained. In comparison with other sites where this species is abundant, the diversity of temporary and permanent ponds seems to be the same.

Tadpoles.—The description of the tadpole was based on eight individuals in stages 36 and 37 from Santa Cecilia, Provincia Sucumbíos, Ecuador. They have body lengths of 13.1–13.8 mm ($\bar{x}$ = 13.43) and total lengths of 41.5–44.0 mm ($\bar{x}$ = 42.70); two specimens in stage 42 have body lengths of 15.2 and 16.9 mm. The following description is based on the largest individual in stage 37. The tadpole has a body length of 13.8 mm and a total length of 44.0 mm. In dorsal view, the body is elongate, ovoid, and widest (7.4 mm) at the level of the eyes; the body is slightly wider than high, and the greatest height (6.5 mm) is at the posterior part of the body. The eyes are positioned dorsolaterally and directed laterally; the interorbital distance is about twice the diameter of the eye. The nares are moderately small, positioned laterally at about one-third the distance from the snout to the orbit, and directed anteroventrally; narial papillae are absent. The short, broad spiracular tube is just sinistral to the midline; it originates at the level of the posterior borders of the orbits and is attached to the belly for its entire length. The vent tube is short, dextral, and attached for its entire length to the ventral fin. The caudal musculature is moderately robust, highest at its junction with the body and gradually tapering to a fine tip. The dorsal fin originates near the base of the tail; it is shallow throughout its length and highest posteriorly. The ventral fin originates on the posterior part of the belly; it is highest anteriorly and throughout most of its length is much higher than the dorsal fin. At the mid-length of the tail, the ventral fin is much shallower than the caudal musculature. The oral disc is terminal and not emarginate. One row of moderately small, blunt papillae is present on the anterior labium, except for a broad median gap; two irregular rows of marginal papillae are present laterally and on the posterior labium. The labial tooth row formula is 2(2)/3(1). The jaw sheaths are moderately slender and finely serrate. The anterior sheath is in the form of a broad arch, and the posterior sheath is broadly V-shaped.

The body and tail are bluish-gray, with an iridescent bluish-green sheen on the belly.

Distribution.—Figure 1.60. *Pithecopus palliatus* occurs in the upper Amazon Basin in Ecuador, Peru, Bolivia, and Brazil. In Ecuador, it occurs in Sucumbíos and Orellana provinces. Its elevational range is 269–322 m with an extent of occurrence of about 890 km^2.

This nocturnal, arboreal species inhabits tropical rainforests and secondary forests in the evergreen forests of lowlands ecosystem of the Aguarico-Putumayo-Caquetá biogeographic sector, where the average annual rainfall is 3417–3570 mm and the average annual temperature is 24.7–24.9 °C.

Conservation.—*Pithecopus palliatus* is categorized as Least Concern throughout its entire range (Angulo *et al.*, 2016; in IUCN Red List of 2022) in view of its wide distribution, tolerance of a degree of habitat modification, presumed large population, and because it is unlikely to be declining fast enough to qualify for listing in a more threatened category. However, it is important to note that in Ecuador, *P. palliatus* has been categorized as Endangered by Ortega-Andrade *et al.* (2021), given that its extent of occurrence is less than 5000 km^2 (890 km^2 according to our data), and there is a suspected population reduction based on a decline in the area of occupancy, extent of occurrence, and habitat quality. According to the latter, it is mostly threatened by agriculture, transport, deforestation, and stochastic events. In Peru, this species is known to occur in Parque Nacional Manu and Tambopata National Reserve. It is present in many Brazilian protected areas. In Ecuador, it has not been recorded in any public or private reserves.

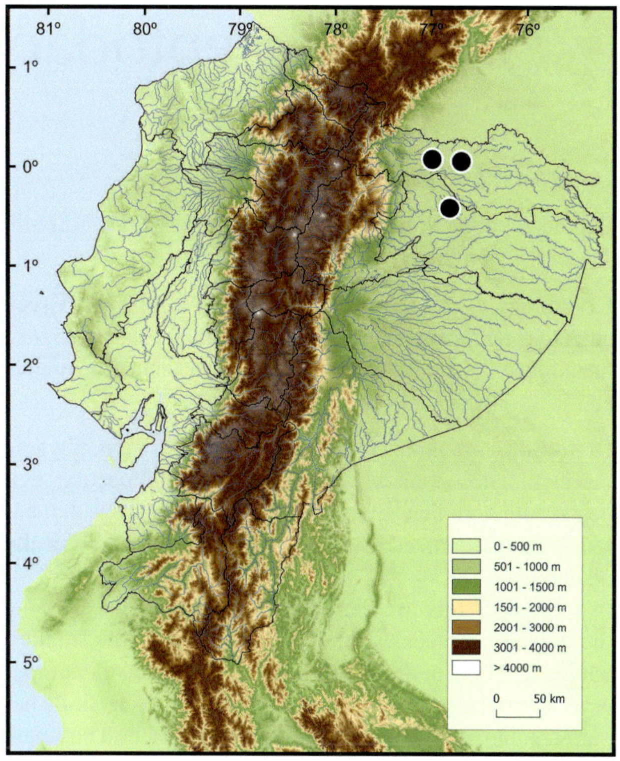

FIGURE 1.60 Localities of known occurrence of *Pithecopus palliatus* in Ecuador.

Remarks.—According to Duellman *et al.* (2016), *Pithecopus palliatus* is the sister taxon of a lineage that includes *P. nordestinus* and *P. azureus* + *P. hypocondrialis*.

Key references.—(Peters, 1873b) (description); (Duellman, 1974b) (account, taxonomy); (Duellman, 1978) (account, natural history, call, tadpole); (Rodríguez and Duellman, 1994) (account); (Duellman, 2005) (account, call); (Villacampa *et al.*, 2017) (account).

2 Family Leptodactylidae

FOAM-NESTING FROGS

Leptodactylidae Werner, 1896: 357. **Type genus.**—*Leptodactylus* Fitzinger, 1826.

Etymology.—The familial name is derived from the Greek adjective *lepto*, meaning thin, fine, slender, and *dactylos*, meaning finger, toe. The name refers to their thin fingers and toes.

Definition.—Leptodactylid frogs are a monophyletic group diagnosed mainly on the basis of molecular evidence, although they are mostly foam-nesting frogs that has long been recognized as behaviorally and morphologically distinctive.

Leptodactylids are extremely variable in size. Among the Ecuadorian species, *Engystomops randi* is the smallest having an SVL of 18.7 mm, and *Leptodactylus pentadactylus* is the largest, reaching 195 mm SVL. The skull has paired palatines and frontoparietals. The vertebral has eight presacral holochordal vertebrae, and all are procoelous. The transverse processes of the sacral vertebra are cylindrical, and this vertebra has a bicondylar articulation with the urostyle. Postmetamorphs have no dorsal ribs on the presacral vertebrae. The pectoral girdle is arciferal, rarely pseudofirmisternal, with a distinct sternum. The fibulare and tibiale are fused at their proximal and distal ends. No intercalary cartilage occurs between the terminal and penultimate phalanges of the digits, and the tips of the terminal phalanges are variable. There is a distinct tympanum. The tadpoles have keratinized mouthparts, and the left and right branchial chambers fuse behind the heart and are emptied by a spiracle on the left side at midbody.

Leptodactylids have a diverse array of life histories. Most species of the large genus *Leptodactylus* are terrestrial. Eggs are deposited in a foam nest produced from cloacal secretions in almost all species. The location of foam nests varies among species. Foam nests may be placed on the surface of the water, and tadpoles drop into the water below, or they may be deposited in depressions close to water, and tadpoles are washed into ponds during rain. In *Lithodytes lineatus*, eggs are deposited in foam nests in the underground galleries of ant nests. Some species of *Adenomera* deposit eggs in foam nests, but some species have nonfeeding, endotrophic tadpoles (e.g., *A. marmoratus*), whereas others have aquatic tadpoles. Species of *Paratelmatobius* have brightly colored venters that they display by turning ventral-side-up if disturbed. These frogs deposit large eggs either on pond bottoms (*P. cardosoi*) or adhering to rock surfaces by small rivulets (*P. poecilogaster*). Tadpoles are aquatic, living on the bottom of ponds until metamorphosis. Tadpoles of some species of *Leptodactylus* form large schools.

Distribution.—Leptodactylids are found in a wide distribution range stretching from the southernmost regions of Texas in the USA and southern Sonora in Mexico to the northern Antilles and further south to Brazil, Argentina, and Chile.

Content.—The family contains 13 genera with 233 species (Frost, 2022). In Ecuador, there are five genera and 25 species.

Remarks.—Figure 2.1. The nomenclatural and systematic history is detailed in Frost (2023 and onward). We follow Pyron and Wiens' (2011) arrangement as a family.

GENUS *ADENOMERA* (THIN-TOED FROGS) STEINDACHNER, 1867

Adenomera Steindachner, 1867: 37. **Type species.**—*Adenomera marmorata* Steindachner, 1867.

Etymology.—The generic name is derived from the Greek word *adeno*, meaning gland, and the Latin *merus*, meaning part or portion. This name likely refers to the prominent glandular structures or glands present on the body of these frogs, particularly notable in some species of this genus. The genus is feminine.

Definition.—*Adenomera* is a group supported by molecular evidence and life history traits; it has facultative endotrophic larvae that complete their development in the terrestrial nest.

Distribution.—The genus occurs in South America to the east of the Andes.

Content.—Currently, 30 species (but see Remarks) are recognized in this genus. Two species are present in Ecuador.

Remarks.—The *Leptodactylus marmoratus* group was transferred to *Adenomera* by Heyer (1974). Frost *et al.* (2006), Pyron and Wiens (2011), Fouquet *et al.* (2013), and de Sá *et al.* (2014) recovered *Adenomera* as the sister taxon of *Lithodytes*. Fouquet *et al.* (2013) estimated phylogenetic relationships among 320 *Adenomera* samples. These results were integrated with other lines of evidence to propose a conservative species delineation. Their delineation identified 31 confirmed candidate species (four remaining unconfirmed) representing a 94% increase in species richness. Angulo *et al.* (2003) described the calls of four species in southeastern Peru. Giaretta *et al.* (2011) described the larval morphology of *Adenomera* sp. from Uberlândia, Minas Gerais, Brazil, and suggested that the nominal *Adenomera* is a subtaxon of the *Leptodactylus fuscus* group, but this hypothesis was rejected by de Sá *et al.* (2014). de Carvalho

DOI: 10.1201/9781003543879-2

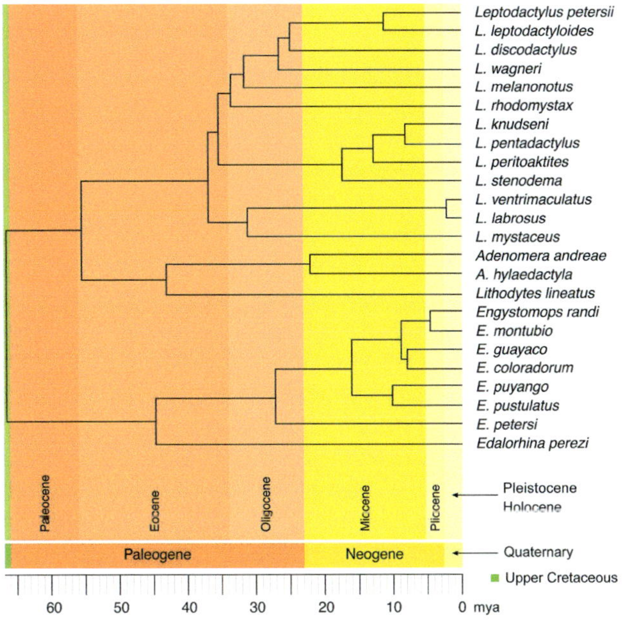

FIGURE 2.1 Time-calibrated maximum likelihood phylogenetic tree depicting evolutionary relationships of some of the Ecuadorian species of Leptodactylidae, inferred from DNA sequences of mitochondrial and nuclear genes. Tree by Juan C. Santos (this publication).

et al. (2020) provided a tree of relationships of the *Adenomera andreae* lineage, noting at least two unnamed species (*Adenomera* sp. D, the sister taxon of *Adenomera chicomendesi*, and *Adenomera* sp. T, the sister taxon of *Adenomera guarayo*).

ADENOMERA ANDREAE (GÖLDI'S THIN-TOED FROG) (MÜLLER, 1923)

Leptodactylus andreae (Müller, 1923): 40. **Neotype.**—ZSM 145/1911 from Peixeboi, Bragançabahn, Estado Pará, Brazil.

Adenomera andreae Heyer, 1974: 94, by implication.

Etymology.—The specific epithet is a patronym that honors André Göldi, a botanist and Head of the Estação Experimental de Agricultura of Peixeboi, in grateful memory of the endured hospitality and support of the collecting activity done by Müller.

Identification.—Figures 2.2, 2.3, 2.4. In these frogs, males attain a maximum SVL of 25.0 mm, and females of 27.0 mm. Fingers I and II are about equal in length. There are elongate tubercular ridges laterally. A prominent ridge extends diagonally from the scapular region to the groin. The tips of the digits are not expanded. Webbing is absent.

FIGURE 2.2 *Adenomera andreae* (but see Remarks). Left: CJ 6157, female, 25.6 mm SVL. Right: CJ 6158, male, 19 mm SVL. Shiripuno Lodge, Provincia Orellana, DAL.

The dorsum is dull olive brown to grayish-brown with darker brown markings consisting of an interorbital triangle with a posterior apex; short, broad longitudinal bars behind tympanum, in the groin, and posteriorly from the scapular region; a diagonal bar in the groin; round spots on the flanks; and transverse bars on the hind limbs. The upper lip is marked by two dark brown suborbital bars diverging from the eye. A short, creamy white bar extends from the tympanum to the base of the arm, which is orange–tan. The posterior surfaces of the thighs are brown. The venter is creamy white with brown flecks peripherally on the chin and laterally on the belly and thighs in some individuals. Nuptial pads and spines are absent in breeding males. The iris is dull bronze with black reticulations.

Comparisons.—*Adenomera hylaedactylus*, *Leptodactylus discodactylus*, *L. leptodactyloides*, *L. wagneri*, *Pristimantis nigrovittatus*. This small species is indistinguishable morphologically from *A. hylaedactylus*. It is most easily confused with small individuals of *L. discodactylus*, *L. leptodactyloides*, and *L. wagneri*, which have finger I longer than II and have lateral fringes on the toes. *Adenomera andreae* superficially resembles *P. nigrovittatus*, which has smooth skin dorsally, has a dark anal patch, and lacks an interorbital triangular mark.

FIGURE 2.3 Morphological variation in *Adenomera andreae* (but see Remarks). Left to right, top to bottom: CJ 11770, subadult female, 26.7 mm SVL, Palora, Provincia Morona Santiago. CJ 12728, female juvenile, 20.8 mm SVL, Puyo, Provincia Pastaza. SGS. QCAZ 47671, La Selva Amazon Ecolodge, Provincia Sucumbíos. QCAZ 29585, Misahuallí Amazon Lodge, Provincia Napo. QCAZ 33008, Panguitza, Provincia Zamora Chinchipe. QCAZ 37767, Reserva de Producción de Fauna Cuyabeno, Provincia Sucumbíos. CJ 898, female, Río Pucayacu, Reserva Otoyacu, Provincia Pastaza. CJ 13065, female, 25.7 mm SVL, road to Dolorosa, Provincia Morona Santiago. LAC.

FIGURE 2.4 Morphological details of *Adenomera andreae* (but see Remarks). Left: CJ 12728, female juvenile, Puyo, Provincia Pastaza. SGS. Right: CJ 6157, female, Shiripuno Lodge, Provincia Orellana, DAL.

Natural history.—In Santa Cecilia, Provincia Sucumbíos, Ecuador, this terrestrial frog is active by night and day throughout the year, although about 60% of the frogs were found at night. Ten individuals were hidden beneath logs by day, and two were resting on low vegetation at night. Eighty percent of the frogs were in leaf litter in primary or secondary forests; only one individual was in water. Eleven individuals were found in clearings at night. No males were heard calling. Females having mature ovarian eggs were found in February, March, May, July, August, and September; females having immature ovarian eggs were found in February–May and July–October. Five or more females were found monthly from July to August, and the percentage of females having mature ovarian eggs in each of those months was 62% in May (n = 5), 70% in July (n = 20), and 40% in August (n = 5). The data are insufficient to determine with certainty the seasonal nature of reproduction in this species. A foam nest, presumably produced by *Adenomera andreae*, was found beneath a log in primary forest in March 1967. Twelve females contained 6–16 unpigmented eggs ($\bar{x}$ = 8.7) having a maximum diameter of 3.0 mm. Of the three smallest juveniles (SVLs of 13–14 mm), one is from June and two from July.

In Brazil, foam nests are constructed in underground chambers by the males or inside decomposing fallen logs in terra firme forests. An egg clutch contained seven unpigmented eggs. Endotrophic tadpoles complete development in the nest.

Of 11 specimens containing food, ants were present in five; roaches (Blattidae) and beetles were present in two stomachs, whereas each of the following food items was present in only one individual: Dermaptera, Isoptera, Hemiptera, Homoptera, Chilopoda, Aranea, and Annelida.

The call was described in Tambopata National Reserve and Bahuaja Sonene National Park, Departamento Madre de Dios, Peru. It consists of a short call with no perceptible pulses, with a mean duration of 0.049 s and a mean call rate of 52.2 calls per minute. Detailed data and figures are provided under the subtitle "Vocalizations of Ecuadorian Anurans" in Appendix 1.

Tadpoles.—This description was based on a single tadpole at stage 39 from Central Amazonia. The total length is 13.1 mm. The body is elongate, oval in dorsal view and globular/depressed in lateral view. The snout is truncate in dorsal and lateral views. The eyes are small, positioned and directed laterally. The nostrils are small, circular, laterally positioned near the snout, with openings anterolaterally directed, without a projection on the marginal rim. The oral disc is anteroventral, non-emarginate; marginal papillae are elongate, uniseriate, with a wide anterior labium gap and a narrow ventral gap. Submarginal papillae are absent. The LTRF is 0/0, dermal ridges are present. Jaw sheaths are narrow, finely serrated; the anterior jaw sheath is arch-shaped; the posterior jaw sheath is U-shaped. The spiracle is absent. The vent tube is medial, fused to the ventral fin, with a medial opening. The caudal musculature is of moderate width; in lateral view, it gradually tapers to a pointed tip. The dorsal fin is of moderate height, originating at the tail–body junction, convex; the ventral fin

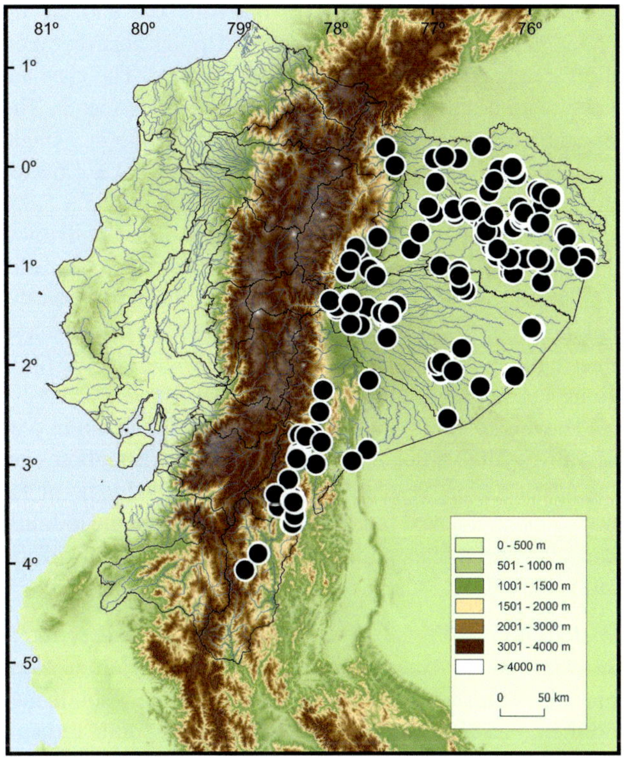

FIGURE 2.5 Localities of known occurrence of *Adenomera andreae* in Ecuador (but see Remarks).

is of moderate height and convex. The tail tip is pointed. The dorsum is brown; the venter is translucent yellow; the fins are translucent; the tail is whitish. Metamorphs have a brown dorsum with numerous dark marks in most individuals; the dorsum is brown with two light dorsolateral stripes; the venter is whitish.

Distribution.—Figure 2.5. *Adenomera andreae* is known in the lowlands of northern South America, to the east of the Andes. Its distribution is limited to the Amazonian region, extending to the south and east. In Ecuador, it occurs in Sucumbíos, Napo, Orellana, Pastaza, Morona Santiago, and Zamora Chinchipe provinces. Its elevational range is 177–1429 m in an area of extent of occurrence of about 86,020 km^2.

This nocturnal, terrestrial species inhabits tropical rainforests and secondary areas in 14 ecosystems (Table 0.4) within seven biogeographic sectors, where the average annual rainfall is 1549–4423 mm and the average annual temperature is 18.6–25.8 °C.

Conservation.—*Adenomera andreae* is listed as Least Concern throughout its entire range (La Marca *et al.*, 2004a; in IUCN Red List of 2022) and in Ecuador by Ortega-Andrade *et al.* (2021) in view of its wide distribution, tolerance of a range of habitats, presumed large population, and because it is unlikely to be declining fast enough to qualify for listing in a more threatened category. This is a common species. Clear cutting is a major threat to this species. In Ecuador, it has been recorded in the public reserves Reserva de Producción de Fauna Cuyabeno, Reserva Biológica Limoncocha, Parque Nacional Yasuní, and Parque Nacional Llanganates; in nine protected forests (e.g., Pañacocha, Río Aguarico, Cerro Sumaco y Cuenca Alta del Río Suno); and in the biosphere reserves Sumaco, Yasuní, and Podocarpus-El Cóndor.

Remarks.—Most of the specimens from the Santa Cecilia region were utilized by Heyer (1973) in his systematic analysis of the *Leptodactylus marmoratus* group. Heyer resurrected the name *L. andreae* for an Amazonian species that had been confused with *L. hylaedactylus*. Fouquet *et al*. (2007) provided molecular evidence that this is a species complex. The frequent sympatric occurrences of various species of *Adenomera* have caused identification problems. This is the case between *A. andreae* and *A. hylaedactylus*, which are indistinguishable based on morphological features according to Rodríguez and Duellman (1994). Thus, our plates and distribution maps may include a mix of more than one species.

Key references.—(Müller, 1923) (description); (Heyer, 1973) (account, as *Leptodactylus andreae*); (Crump, 1974) (natural history); (Duellman, 1978) (account); (Hero, 1990) (tadpole); (Rodríguez and Duellman, 1994) (account); (Marquez *et al*., 1995) (call); (Angulo *et al*., 2003) (call, taxonomy); (Ouboter and Jairam, 2012) (account, as *Leptodactylus andreae*); (Lima *et al*., 2012) (account); (Menin and Rodrigues, 2013) (tadpole); (Cole *et al*., 2013) (account); (Villacampa *et al*., 2017) (account); (Barrio-Amorós *et al*., 2019) (notes); (de Carvalho *et al*., 2019) (taxonomy); (Schiesari *et al*., 2022) (tadpole).

ADENOMERA HYLAEDACTYLA (DARK-SPOTTED THIN-TOED FROG) (COPE, 1868)

Cystignathus hylaedactylus (Cope, 1868): 115. **Holotype.**—ANSP 2240 from the Napo or upper Marañon River, Peru. de Carvalho *et al*. (2019) suggested that the holotype is unidentifiable, and the name may not apply to the species but to *Adenomera andreae* or *Adenomera simonstuarti*.

Adenomera hylaedactyla Heyer, 1974: 43.

Etymology.—The specific epithet derives from the Greek *hyla*, meaning forest, and *dactylos*, meaning finger, toe.

Identification.—In these frogs, males attain a maximum SVL of 26.2 mm, and females of 31.0 mm. The dorsum has several longitudinal rows of low tubercles. The head is longer than wide. The snout is moderately long and subovoid to pointed, more pointed in males. The canthus rostralis is indistinct. The tympanum is distinct. The male vocal slits are elongate. The vocal sac is single. The dorsum is smooth or with ridges and warts, rarely with dorsolateral folds. Most males exhibit white-tipped tubercles on the posterior dorsum, while females lack these tubercles. The females do not have distinctive glands; the males lack distinctive glands or bear paired, small, oval glands at the angle of the jaw, just in the back of the axillae, in the groin, and on either side of the anus on the hind limbs. Fingers I and II are equal in length, and the toes lack webbing and lateral fringes; all digits end in small, round discs that tend to be rounded. The toe tips are nonexpanded or slightly expanded, not developed into flattened discs. The subarticular tubercles are moderately developed or pronounced. There is a weak, sinuous or almost straight tarsal fold.

The dorsum varies from dull gray to tan with brown spots, with or without cream to pink dorsolateral stripes or a cream middorsal stripe posteriorly on the body; a dark brown triangular mark is present on the back of the head. There is a light middorsal pin (rarely broad) stripe from the snout to above the anus, or it is absent. Some males have black vocal folds. The posterior faces of the thighs are mottled to uniform. In many individuals, the forelimbs are creamy orange; the venter is creamy white. The iris is bronze with minute black flecks.

Comparisons.—*Adenomera andreae*, *Leptodactylus discodactylus*, *L. leptodactyloides*, *L. wagneri*, *Pristimantis nigrovittatus*. This small species is indistinguishable morphologically from *A. andreae*. It is most easily confused with small individuals of *L. discodactylus*, *L. leptodactyloides*, and *L. wagneri*, which have finger I longer than the II and have lateral fringes on the toes. *Adenomera andreae* superficially resembles *P. nigrovittatus*, which has smooth skin dorsally, has a dark anal patch, and lacks an interorbital triangular mark.

Natural history.—*Adenomera hylaedactyla* usually is active in open areas at night. In Iquitos, Peru, it excavates flask-shaped cavities, in which females deposit 9–14 unpigmented eggs about 3 mm in diameter in a foam nest. The tadpoles develop in the nest; they subsist entirely from the yolk provided in the eggs and emerge from the nest as miniature replicas of the adults. In Regina, French Guiana, at the beginning of the main French Guianese dry season, the tadpoles were found in two foam nests partially roofed over by soil and under pieces of metal and wood. The nests were shallow, circular cups bearing a distinct raised rim; they were 35 mm in diameter and consisted of a thin layer of foam resting directly over the soil. Large numbers of adults of *A. hylaedactyla* were observed during the daylight hours beneath the runway markers. When the sheltering markers were lifted, these frogs attempted to escape into the surrounding grass.

The call was described by Angulo *et al.* (2003) from Tambopata National Reserve and Bahuaja Sonene National Park, Departamento Madre de Dios, Peru. It consists of a short multi-pulsed call, with a mean duration of 0.046 s and a mean call rate of 210 calls per minute. Detailed data and figures are provided under the subtitle "Vocalizations of Ecuadorian Anurans" in Appendix 1.

Tadpoles.—This description is based on individuals from Regina, French Guiana. The total length of the largest specimen in stage 34 is 13.2 mm, that in stage 40 is 12.6 mm. The body length is 36.7 ± 3.1% of the total length. The nostril is positioned midway between the eye and the tip of the snout or closer to the snout. The distance between the nares is equal to the interorbital distance. The eyes are large, with a diameter 15.2 ± 1.6% of the body length. The mouthparts are subterminal. The oral papillae are in a single row. The oral disc is complete; its width is 18.0 ± 1.6% of the body length. The anterior oral papilla gap is 82.5 ± 5.8% of the oral disc width. There are no denticles on the tooth rows. The jaw sheaths are weakly developed. There is no spiracle. The dorsal fin originates at the body–tail junction or on the tail, posterior to the body–tail junction. The tail height is less than, equal to, or greater than the body height. The tail tip is elongate and rounded to pointed. The anal tube is median.

The dorsum has a uniform pattern, gray to brown in preservative. The venter has a suffusion of melanophores on the throat region, few melanophores on the belly, no melanophores on the anal tube, no melanophores on the tail fins, and scattered melanophores on the upper two-thirds of the tail musculature.

Distribution.—Figure 2.6. *Adenomera hylaedactyla* is known from the Amazonian forests of southeastern Colombia and Venezuela; east through the Guianas; south into northern, eastern, and central Brazil; and south through Amazonian Ecuador and Peru to Bolivia. In Ecuador, it occurs in Sucumbíos, Napo, Orellana, Pastaza, Morona Santiago, and Zamora Chinchipe provinces. Its elevational range is 223–1949 m in an area of extent of occurrence of about 75,190 km^2.

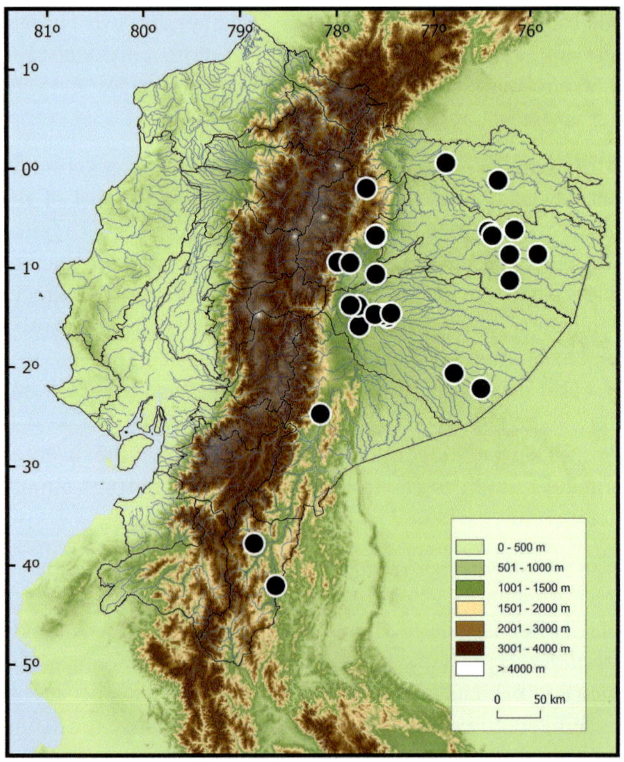

FIGURE 2.6 Localities of known occurrence of *Adenomera hylaedactyla* in Ecuador (but see Remarks).

This nocturnal, terrestrial species inhabits tropical rainforests and secondary areas in 10 ecosystems (Table 0.4) within seven biogeographic regions, where the average annual rainfall is 967–4423 mm and the average annual temperature is 10.3–25.8 °C.

Conservation.—*Adenomera hylaedactyla* is listed as Least Concern throughout its entire range (La Marca *et al.*, 2010a; in IUCN Red List of 2022) and in Ecuador by Ortega-Andrade *et al.* (2021) in view of its wide distribution, tolerance of a broad range of habitats, presumed large population, and because it is unlikely to be declining to qualify for listing in a more threatened category. This is a common species. There are no threats to the overall survival of the species. In Ecuador, it has been recorded in the public reserves Parque Nacional Cayambe Coca, Parque Nacional Yasuní, and Reserva Biológica Colonso Chalupas; the protected forests of Cerro Sumaco y Cuenca Alta del Río Suno, Subcuenca Alta del Río Leon y Microcuencas de Los Ríos San Felipe de Ona y Shincata, and Comunidad Kichwa Wamani Cokiwa; and the biosphere reserves Yasuní, Sumaco, and Podocarpus-El Cóndor.

Remarks.—This is a complex of species; thus, some of the populations have been referred to as *Adenomera aff. hylaedactylus* or *Adenomera* sp. (e.g., Dubeux *et al.*, 2020; Kokubum and Giaretta, 2005). The frequent sympatric occurrences of more than one species of *Adenomera* have caused identification problems. This is the case between *A. hylaedactyla* and *A. andreae*, which are indistinguishable based on morphological features according to Rodríguez and Duellman (1994). Therefore, caution should be exercised when interpreting its distribution map, as it may encompass a mixture of more than one species.

Key references.—(Cope, 1868) (description); (Heyer and Silverstone, 1969) (tadpole); (Heyer, 1973) (account); (Zimmerman, 1983) (call); (Hödl, 1990) (natural history); (Rodríguez and Duellman, 1994) (account, as *Adenomera andreae* and *A. hylaedactyla*); (Murphy, 1997) (account); (Köhler, 2000) (account); (Ouboter and Jairam, 2012) (account, as *Leptodactylus hylaedactyla*); (Cole *et al.*, 2013) (account); (de Sá *et al.*, 2014) (account); (Villacampa *et al.*, 2017) (account); (Barrio-Amorós *et al.*, 2019) (notes); (Schiesari *et al.*, 2022) (tadpole).

GENUS *EDALORHINA* (FOAM-NEST FROGS) JIMÉNEZ DE LA ESPADA, 1870

Edalorhina Jiménez de la Espada, 1870: 58. **Type species.**—*Edalorhina perezi* Jiménez de la Espada, 1870.

Etymology.—The generic name is derived from the Greek words *Edalos*, meaning visible, and *rhinos*, meaning nose. Presumably, it refers to the distinctive tubercles of the nasal region of these frogs. The genus is feminine.

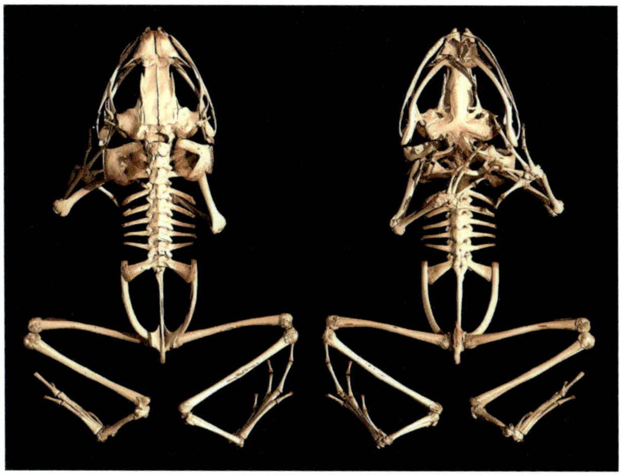

FIGURE 2.7 High-resolution computed tomography reconstructions of the skeleton of *Edalorhina perezi*, CAS-SU 11446, 34.2 mm SVL, Payamino, Napo-Pastaza provinces, Ecuador. Dorsal (left) and ventral (right) views. Daniel J. Paluh.

Definition.—Figure 2.7. These frogs attain a maximum SVL of 38 mm in females. *Edalorhina* is a group supported by molecular and tadpole anatomy features, but these features are known for only one of the two species currently recognized within *Edalorhina*. The combination of two lingual papillae, a filiform median ridge, and the lack of buccal roof papillae are diagnostic of *E. perezi* and putative autapomorphies of *Edalorhina*. In the chondrocranial anatomy, the presence of a uniquely shaped processus pseudopterygoideus and cartilago suprarostralis with the corpora and alae joined by dorsal and ventral connections readily differentiates this genus from other Leiuperinae genera. Eggs are deposited in foam nests.

Distribution.—The genus occurs in Colombia, Ecuador, Peru, and adjacent Brazil.

Content.—Currently, two species are recognized in this genus. One is present in Ecuador.

Remarks.—Faivovich *et al.* (2012) recovered *Edalorhina* as the sister taxon of *Engystomops* in the large group that also includes *Pleurodema* and *Physalaemus*. However, Ferraro *et al.* (2020) and do Nascimento *et al.* (2021) corroborate Pyron and Wiens' (2011) hypothesis that *Edalorhina* is the sister taxon of a group that includes *Engystomops*, *Eupemphix*, and *Physalaemus*. Ferraro *et al.* (2020) studied the evolution of macroglands on the flank and inguinal regions and its defensive mechanisms such as bright coloration.

References.—(Dunn, 1949) (notes); (Duellman and Morales, 1990) (account); (Faivovich *et al.*, 2012) (foam nest evolution).

EDALORHINA PEREZI (PEREZ'S FOAM-NEST FROG) (JIMÉNEZ DE LA ESPADA, 1870)

Edalorhina Perezi (Jiménez de la Espada, 1870): 58.
Holotype.—MNCN 1673 from Río Napo, Ecuador.
Edalorhina perezi Dunn, 1949: 2.

Etymology.—The specific name is a patronym that honors Laureano Pérez Arcas (1824–1894), a Spanish zoologist who was the first teacher of zoology of Marcos Jiménez de la Espada (1831–1898).

Identification.—Figures 2.8, 2.9. The maximum known SVL is 30.2 mm in males and 36.7 mm in females. The maximum mass is 2.5 g in males and 4.4 g in gravid females. It has a depressed body and tubercles on the snout. All individuals have a cluster of prominent, pointed tubercles on the upper eyelid; usually, the three largest tubercles are on the outer margin of the eyelid. A prominent dorsolateral fold extends posteroventrally from the posterior edge of the orbit to the groin. The dorsal skin texture is polymorphic in this species. Most individuals (70%) have many tubercles on the dorsum, whereas about 20% have longitudinal ridges between the dorsolateral folds, and less than 10% have a smooth dorsum with four or five tubercles. The skin on the venter is smooth.

The dorsal coloration consists of a gray or tan ground color; in those individuals having longitudinal dermal ridges on the back, the ridges commonly are reddish-brown to black. The dorsolateral fold is gray to orange–brown, and the flanks are black. The upper lips are pale gray to orange–brown with at least one dark gray or dark brown diagonal mark anteroventral to the orbit. The throat is white with irregular black spots or mottling; this pattern also is present on the chest in some individuals, but usually the chest and always the belly are white and black. The proximal ventral surfaces of the shanks, the axillae, and the proximal parts of the arms are orange to red; a large black spot in the inguinal region is narrowly bordered by cream, orange, or red. The iris is grayish-tan with a reddish-copper ring around the pupil.

Comparisons.—*Engystomops petersi.* This species is unique by having a depressed body, prominent dorsolateral folds, a cluster of pointed tubercles on the eyelid, tubercles on the snout, and a smooth white venter with large black spots laterally; the chest and belly are black. *Engystomops petersi* is about the same size, but it has a rugose dorsum and black and white mottling on the belly; it lacks dorsolateral folds and a pointed tubercle on the eyelids.

Natural history.—*Edalorhina perezi* is an inhabitant of terra firme forest, where it is primarily diurnal in shade. At Santa Cecilia and Lago Agrio, Provincia Sucumbíos, Ecuador, individuals were found active by day amid leaf litter on the ground in primary forest. The frogs moved by series of short hops and hid among dead leaves, where they were

FIGURE 2.8 *Edalorhina perezi.* CJ 9463, male, 34.6 mm SVL. Parque Nacional Yasuní, Provincia Orellana. DAL.

FIGURE 2.9 Morphological details of *Edalorhina perezi*. Top: CJ 9464, female, 41.1 mm SVL. SGS. Bottom: CJ 9463, male. DAL. Parque Nacional Yasuní, Provincia Orellana.

effectively camouflaged. One female collected in May contained 69 ovarian eggs; a second showed no evidence of reproductive activity. Three juveniles from April and May have SVLs of 16 and 17 mm. Four individuals contained two Diptera, one Hymenoptera, and one Orthoptera.

At Cusco Amazónico, Peru, of 115 adults recorded by day, 62 (54%) were observed between 9:00 and 12:00 h. During the day, 96 individuals were in or on leaf litter, 13 were on the ground, and three were at the edge of a pool of water in a fallen log. Also, during the day, one amplectant pair was in the water in the aforementioned pool, and another amplectant pair was in water in a small depression. Twenty individuals were found at night; five were in leaf litter and seven were sitting adjacent to a pool in a fallen log; four amplectant pairs were in water. The frogs seem to be equally abundant throughout the rainy season; 22 individuals were recorded in November, 29 in December, 25 in January, 30 in February, and 14 in March. Only 20 individuals were observed in the dry season (May–July). Males call from the leaf litter throughout the rainy season, especially on mornings after rain. Amplectant pairs were observed in November–February, but reproduction may occur throughout most of the year, as recently metamorphosed young were found in January, February, June, July, November, and December. Amplexus is axillary, and during oviposition, the male kicks the water and eggs into a foam nest in a constrained body of water. No amplectant pairs or nests were observed in permanent ponds or in the many temporary ponds where many other species of frogs breed. Of 10 foam nests observed, seven were floating on the water in a pool (0.3 × 3.0 m) in a cavity in a log about 1 m above the ground. Two others were in small, water-filled depressions in the ground, and one was on the ground at the edge of a water-filled depression. The two nests in the depressions were constructed by amplectant pairs at night. Foam nests also were found in palm bracts on the forest floor at Panguana, Peru. The pool in the log was utilized as a breeding site on many occasions between November 1989 and December 1991. This is the only place where more than one male was calling synchronously. As many as three males were observed at the edge of this pool by day, and on four occasions, two to four males were sitting at the edge of the pool at night. No individuals were observed climbing from the ground to this breeding site. To reach the pool, it would be necessary for the frog to climb up the root system of the tree and hop along the log to the pool. Development time is 4–5 days from deposition to hatching. There are reports of 3–6 days of egg development in individuals under terraria conditions, the former at 26 °C. Feeding commences on the second day after hatching. Four females having SVLs of 32.0–36.7 mm ($\bar{x}$ = 35.2) contained 78–98 ovarian eggs ($\bar{x}$ = 86.4) about 1.9 mm in diameter. *Edalorhina perezi* feeds by day on a wide range of small, soft- and hard-bodied arthropods in the leaf litter on the forest floor. It is uncertain if the species is an active forager or a sit-and-wait strategist, but the variety of prey consumed (17 prey categories) suggests that it is an active forager. Of 53 prey items contained in 28 individuals, spiders made up 11.32% of the number of prey and 21.77% of the volume of prey; small orthopterans (Blattaria, Gryllidae, Tetrigidae, and Tettigoniidae) accounted for 28.29% of the number and 13.45% of the volume, whereas dipterans made up 24.52% of the number and 16.60% of the volume. Less frequent items of prey were ants, beetles, and isopods. These data are in accordance with the recovery of 10 kinds of prey (0.3–15 mm) from 18 individuals at Panguana, Peru; one of those individuals contained a single gastropod.

Although males of *Edalorhina perezi* produce a variety of calls, there is one typical call that can be heard from long distances and may be interpreted as an advertisement call. Another kind of call (presumably a courtship call) is produced immediately before mating; this call consists of a series of notes having double pulses. The advertisement calls recorded at Cusco Amazónico, Peru, consist of three to five low whistles with two pulses/note. The calls are repeated four or five times at intervals of 0.13–0.20 s with periods of a minute or more between call groups. They differ from the advertisement calls described from Panguana, Peru. A recording from Pakitza, Peru, has series of 10–12 notes. Detailed data are provided under the subtitle "Vocalizations of Ecuadorian Anurans" in Appendix 1.

Tadpoles.—Figure 2.10. The tadpole was described based on individuals in stages 35–36 from Parque Nacional Yasuní, Provincia Orellana, and Campo Villano, Provincia Pastaza, Ecuador. They have a total length of 27.6–29.7 mm. The body is depressed, elliptical in dorsal and lateral views. The ventral contour of the body is convex in the peribranchial region and flat to slightly convex in the abdominal region. The body wall has a constriction at the level of the spiracle.

FIGURE 2.10 Ontogenetic changes in two individuals of *Edalorhina perezi* from tadpole to adult. Top to bottom, left to right. CJ 9464a, stage 27, lateral, dorsal, and ventral views, TL= 24.8 mm. CJ 9464b, stage 41; stage 42; stage 46, juvenile 50 days after stage 46; juvenile 98 days; adult 414 days = 34.9 mm. Parque Nacional Yasuní, Provincia Orellana. SGS.

The body length is about 45% of the total length; the maximum body width is at the middle third of the body and the maximum body height at the posterior third. The snout is truncated in dorsal view and rounded to slightly acuminate in lateral view. The eyes are located dorsally and directed dorsolaterally, representing nearly 50% of the intraorbital distance. The nostrils are dorsal and directed dorsolaterally; the external opening is circular, surrounded by a continuous and light marginal rim; each nostril represents about 23% of the internarial distance, 4% of the maximum body width, and they are located closer to the snout than to the eyes. The oral disc is about 36% of the maximum body width, almost terminal and laterally emarginated. The labium bears a single row of marginal papillae (~13 papillae/mm), with a wide gap on the upper labium. The papillae are overall conical, elongated (i.e., length greater than basal width), with rounded tips, their bases slightly offset and directed in alternating directions on the lower labium. The lateral papillae on the lower labium may be poorly differentiated. There are few submarginal papillae (2–3) found laterally on the lower labium. The labial tooth row formula is 2(2)/3(1); the length of each row is A-1 = A-2 > P-1 = P-2 > P-3. The P-1 and P-2 row lengths are about 90% of the anterior rows. The P-3 row length is about 70% of the other posterior rows. A-2 and P-1 have medial gaps (about 30% of the A-2 length and 10% of the P-1 length, respectively). The jaw sheaths are pigmented; the upper jaw sheath is arc-shaped and finely serrated (about 48 serrations/mm), the serrations are straight medially. The lower jaw sheath is broad, V-shaped, and finely serrated (about 45 serrations/mm); the serrations are dorsally oriented. The spiracle is sinistral, located below the body midline; its opening is on the middle third of the body and directed posterodorsally, forming an angle of about 25° with the longitudinal body axis; the inner wall is a slight ridge (i.e., mostly fused with the body), and the external wall ends are anterior to the insertion of the inner wall. The vent tube is about equal in length and width; the right wall is displaced more dorsally and anteriorly than the left wall, which is attached directly to the fin margin, resulting in a large dextral opening; the margin of the aperture is smooth. The tail length is about 55% of the total length; the maximum height is greater than the body height; it is highest on the first third of the tail. The tail tip is rounded. The tail musculature maximum height is less than half of the tail maximum height, becoming progressively thinner posteriorly and extending to the tail tip. The myosepta are distinct anteriorly and partially visible on the posterior half of the tail. The dorsal fin height is less than half the tail height and slightly lower than the ventral fin. The dorsal fin begins at the body–tail junction with an overall convex contour; its maximum height is reached in the first third of the tail. The ventral fin begins at the base of the tail; its origin is concealed by the vent tube and has a convex contour; its maximum height is reached at the middle third of the tail.

Tadpoles in stages 25 to 45 also have been described from Cusco Amazónico, Peru. Hatchling tadpoles are in stage 25. These small tadpoles have jaw sheaths but no labial teeth; the intestines are filled with visible yolk. Progressive development of the labial tooth rows takes place during stage 25, during which stage tadpoles grow to a maximum total length of 20.2 mm. The body and caudal musculature are pale tan, the belly is creamy white, and the fins are translucent with scattered melanophores.

Distribution.—Figure 2.11. *Edalorhina perezi* is known in Amazonian Peru, Ecuador, Colombia, Bolivia, and adjacent Brazil. In Ecuador, it occurs in Sucumbíos, Napo, Orellana, Pastaza, and Morona Santiago provinces. Its elevational range is 178–2501 m in an area of extent of occurrence of about 59,610 km^2.

This nocturnal, terrestrial species inhabits tropical rainforests and secondary areas in nine ecosystems (Table 0.4) in five biogeographic regions (Table 0.5), where the average annual rainfall is 2454–4412 mm and the average annual temperature is 14.2–25.6 °C.

Conservation.—*Edalorhina perezi* is listed as Least Concern throughout its entire range (Angulo *et al.*, 2004a; in IUCN Red List of 2022) and in Ecuador by Ortega-Andrade *et al.* (2021) in view of its wide distribution, tolerance of a degree of habitat modification, presumed large population, and because it is unlikely to be declining fast enough to qualify for listing in a more threatened category. In Ecuador, it has been recorded in the public reserves Reserva de

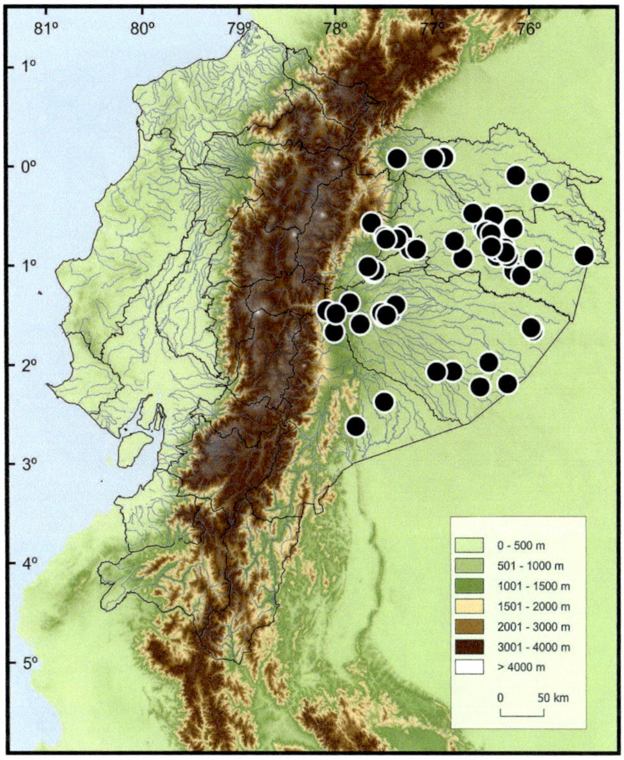

FIGURE 2.11 Localities of known occurrence of *Edalorhina perezi* in Ecuador.

Producción de Fauna Cuyabeno, Parque Nacional Sumaco Napo-Galeras, and Parque Nacional Yasuní; the protected forests of Cerro Sumaco y Cuenca Alta del Río Suno, Comuna Kichwa Añangu, Nacionalidad Zápara del Ecuador, and Nacionalidad Shiwiar del Ecuador Nashie; and the biosphere reserves Yasuní and Sumaco.

Remarks.—*Edalorhina perezi* is the sister taxon of *E. nasuta* (the only other species in the genus *Edalorhina*), but molecular data of the latter have not yet been analyzed. Duellman and Morales (1990) coded four dorsal skin textures and noted the presence of three types among 33 specimens from Departamento de Madre de Dios, Peru: type 2—smooth with four or five tubercles (9.1%); type 3—many tubercles (69.7%); type 4—longitudinal ridges (21.2%). Additional specimens that bring the total to 57 individuals show percentages of 7.1%, 70.1%, and 22.8%, respectively. do Nascimento *et al.* (2021) provide a description of the tadpole in Gosner stages 35–36 including internal morphology data (i.e., buccopharyngeal cavity and larval skeleton) based on samples from two populations from Ecuador. They did not compare these tadpoles to the ones described by Duellman (2005).

Key references.—(Jiménez de la Espada, 1871) (description); (Dunn, 1949) (notes); (Duellman, 1978) (account); (Schlüter, 1980) (call); (Duellman and Morales, 1990) (account); (Schlüter, 1990) (reproduction, tadpole); (Verhaagh, 1993) (natural history, diet); (Rodríguez and Duellman, 1994) (account); (Parmelee, 1999) (diet); (Duellman, 2005) (account, call, tadpole); (Villacampa *et al.*, 2017) (account); (Ferraro *et al.*, 2020) (evolution of defenses); (do Nascimento *et al.*, 2021) (tadpole Ecuador).

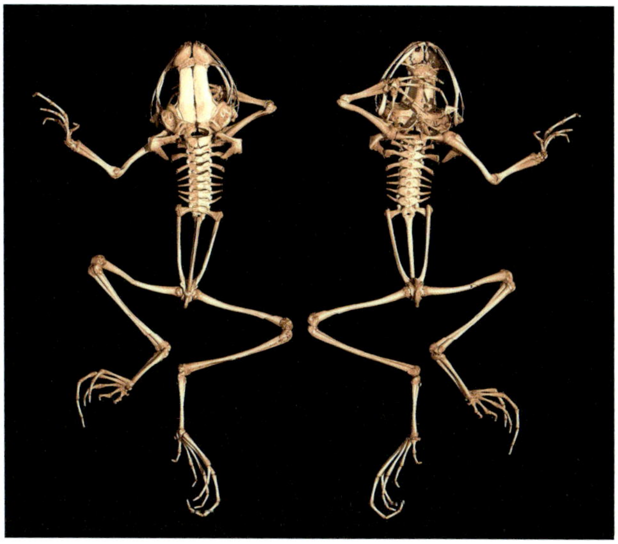

FIGURE 2.12 High-resolution computed tomography reconstructions of the skeleton of *Engystomops pustulatus*, UF 98141, 14.9 mm SVL, 40 km east of Durán, Provincia Guayas, Ecuador. Dorsal (left) and ventral (right) views. Daniel J. Paluh.

GENUS *ENGYSTOMOPS* (BULLY FOAM TOADS) JIMÉNEZ DE LA ESPADA, 1872

Engystomops Jiménez de la Espada, 1872: 86. **Type species.**—*Engystomops petersi* Jiménez de la Espada, 1872.

Etymology.—The generic name is derived from the Greek adjective *engyst*, meaning narrow, and the Greek noun *stoma*, meaning mouth. The name presumably refers to their slender mouth. The genus is masculine.

Definition.—Figure 2.12. These frogs attain a maximum SVL of 31.3 mm in males and 39.1 mm in females. The texture of the dorsal skin is tuberculated; the vomerine teeth are absent; the hypertrophied antebrachial tubercles are absent; there are parotoid glands and flank glands; the tympanic membrane is evident; the skull–mandible articulation is anterior to the transversal plane of the intersection between the alae and cultriform process of the parasphenoid; the frontoparietals overlap the anterior margin of the exoccipitals; the quadratojugals are present; the maxillary processes of the quadratojugals are present and developed; the eggs are deposited in foam nests.

Content.—The genus contains nine species, seven of which occur in Ecuador.

Distribution.—*Engystomops* has an extensive distribution throughout the mainland tropics in the Neotropics. It occurs

FIGURE 2.13 Topology of the species of *Engystomops*, extracted from a phylogenetic tree inferred from analysis of mitochondrial DNA sequences by Ron *et al.* (2010). Species occurring in Ecuador are in red.

throughout the lowlands of southern and eastern Mexico, Central America, and South America, where it occurs throughout the Pacific lowlands southward to southwestern Ecuador and in the Amazonian and Guianese regions (including Trinidad and Tobago) southward to Bolivia.

Remarks.—Figure 2.13. Ron *et al.* (2006) coined two lineage names for groups of species within *Engystomops*: *Duovox* (seven species) and *Edentulus* (three species). Within the *Duovox* lineage, two lineage names (*Brevivox* and *Vivavox*) were proposed. Ryan and Rand (1993a, 1993b) and Ron (2008) examined mate selection in females and studied the evolution of calls.

FIGURE 2.14 *Engystomops coloradorum.* Left: CJ 13066, female, 19.2 mm SVL. Right: CJ 13067, male, 17.7 mm SVL. Sapoparque La Florida, Provincia Santo Domingo de los Tsáchilas. SGS.

ENGYSTOMOPS COLORADORUM (COLORADO BULLY FOAM TOAD) (CANNATELLA AND DUELLMAN, 1984)

Physalaemus coloradorum (Cannatella and Duellman, 1984): 904. **Holotype.**—KU 178271 from Santo Domingo de los Tsáchilas, 580 m, Provincia Santo Domingo de los Tsáchilas, Ecuador.

Engystomops coloradorum Nascimento *et al.*, 2005: 312.

Etymology.—The specific name is a patronym for the Tsáchilas, also known as the Colorados, an indigenous group inhabiting the area of Santo Domingo de los Tsáchilas, the type locality of the species.

Identification.—Figure 2.14. The maximum known SVL is 24.1 mm in males and 27.7 mm in females. It has teeth on the maxilla and premaxilla and an elongate gland posteriorly on the flank; the snout is subacuminate, with the anterior margin of the lip at the level of the margin of the nostrils; the tympanic annulus is granular; the tympanum is recessed and smooth; the tarsal tubercle is absent; there is a pattern of tubercles on the dorsum consisting of scattered small tubercles and larger tubercles, some coalescing to form ridges.

The dorsum is brown, with or without an orange–brown middorsal stripe posteriorly; the forelimbs are orange–brown with dark brown markings; the hind limbs are orange–brown with dark brown to black bars; the posterior surfaces of the thighs are brown with few cream flecks; the venter is black, heavily mottled with white or bluish-white, becoming dark brown on the flanks and the hind limbs; the iris is pale grayish-brown or dull bronze with fine black reticulations.

Comparisons.—*Engystomops montubio*, *E. guayaco*, *E. randi*, *E. pustulatus*, *E. petersi*. *Engystomops coloradorum* is most similar to *E. montubio*, *E. guayaco*, and *E. randi* but can be distinguished by its more heavily tuberculated skin and larger dorsal tubercles, some of which coalesce into ridges. Additionally, *E. coloradorum* has a vertical loreal region, while *E. guayaco* and *E. montubio* have an oblique and sloping loreal region, respectively.

The absence of a tarsal tubercle and the presence of maxillary and premaxillary teeth immediately distinguish *Engystomops coloradorum* from the larger (31 mm SVL) *E. petersi*, which also has finger I noticeably longer than II. Finger I is equal in length to II in *E. coloradorum* and *E. pustulatus*. *Engystomops pustulatus* differs from *E. coloradorum* by its larger size (males 29.9 mm, females 36.5 mm), elongate row of small glands on the flank, round snout with the anterior margin of the lip at the level of the anterior edge of the nostril, non-tuberculate tympanic annulus, and pattern of tubercles on the dorsum consisting of many small tubercles with large tubercles forming a pair of rows diverging posteriorly in the scapular region.

FIGURE 2.15 *Engystomops coloradorum*. Amplexus and foam nests in La Florida Sapoparque, Provincia Santo Domingo de los Tsáchilas. LAC.

FIGURE 2.16 Ontogenetic changes in four individuals of *Engystomops coloradorum* from tadpole to adult. Top to bottom, left to right. CJ 13033a, stage 29, lateral, dorsal, and ventral views, stage 31, TL = 17.8 mm. CJ 13033b, undetermined stage, TL = 21.3 mm. CJ 13033c, stage 42, TL = 19.8 mm. CJ 13033d, stage 45, 8 mm SVL; stage 46, 8.1 mm SVL; juvenile 79 days after stage 46, 13.4 mm SVL; adult 265 days, 21.4 mm SVL. La Florida, Provincia Santo Domingo de los Tsáchilas. SGS.

Natural history.—Figure 2.15. Foam nests and tadpoles were found in temporary ponds and on shallow puddles at the edge of a field in Santo Domingo de los Tsáchilas on 4 March 1968 in shallow roadside ditches. Breeding occurred at least from 4 March to 9 June. In Sapoparque La Florida, Provincia Santo Domingo de los Tsáchilas, Ecuador, amplectant pairs and other individuals were found inside an artificial (cement) pond filled with water in an intervened area next to a house. Foam nests were found in temporary and shallow ponds filled with rain water, over pastures on the side of the road.
Under lab conditions, during amplexus, the eggs are enclosed in a foam nest that derives from the egg jelly and floats in the water. Foam nests are 5 cm in diameter and 4–5 cm high. Nests contained 130 ± 60 eggs of 1.3 mm in diameter (21 foam nests with a total of 2072 eggs). Upon hatching, the tadpoles moved into the underlying water. Development to hatching required four days. The white eggs and embryos of these frogs were effectively camouflaged in the white foam of the nest. In later stages, the embryos developed pigment around the nuclei.

Males call from secluded sites in or at the edge of temporary pools, on roots of partially submerged clumps of grass, and under leaves and debris partially submerged in water. The call of this species has been described in Santo Domingo de los Tsáchilas and in Sapoparque La Florida. It consists of a well-modulated note in which the first two harmonics are emphasized more than the others. Detailed data and figures are provided under the subtitle "Vocalizations of Ecuadorian Anurans" in Appendix 1.

Tadpoles.—Figure 2.16. The tadpole was described based on an individual at stage 38 from Provincia Santo Domingo de los Tsáchilas. It has a body length of 7.2 mm and a total length of 19.3 mm. The body is widest at midbody (at the level of the spiracle), wider than deep; the snout is short, rounded in lateral view; the eyes are small, narrowly separated, oriented dorsolaterally. The nostrils are closer to the eyes than to the tip of the snout, directed almost dorsally. The spiracle is sinistral, directed posterodorsally, its opening at about the midlength of the body. At the midlength of the tail, the caudal fins are slightly deeper than the slender caudal musculature; the tail is rounded at the tip; the dorsal fins do not extend onto the body. The oral disc is anteroventral, barely wider than the internarial distance; the upper lip is bare medially; there is one row of labial papillae laterally; the lower lip bears one continuous row of papillae; there is a moderately deep lateral fold bearing two or three rows; the jaw sheaths are slender, broad, serrate; there are two upper and three lower rows of denticles; the second upper row is interrupted medially. In life, the body is pale gray; the tail is transparent, flecked with light brown. Tadpoles in stages 37–39 have a total length of 15.6–20.3 mm and a body length of 7.2–8.3 mm.

Distribution.—Figure 2.17. *Engystomops coloradorum*, an endemic species to Ecuador, is known in the Pacific lowlands and foothills in Pichincha, Santo Domingo de los Tsáchilas, and Bolívar provinces. Its elevational range is 412–1051 m in an area of extent of occurrence of about 3770 km^2.

This nocturnal, terrestrial species inhabits disturbed habitats in the evergreen montane foothill forest ecosystem

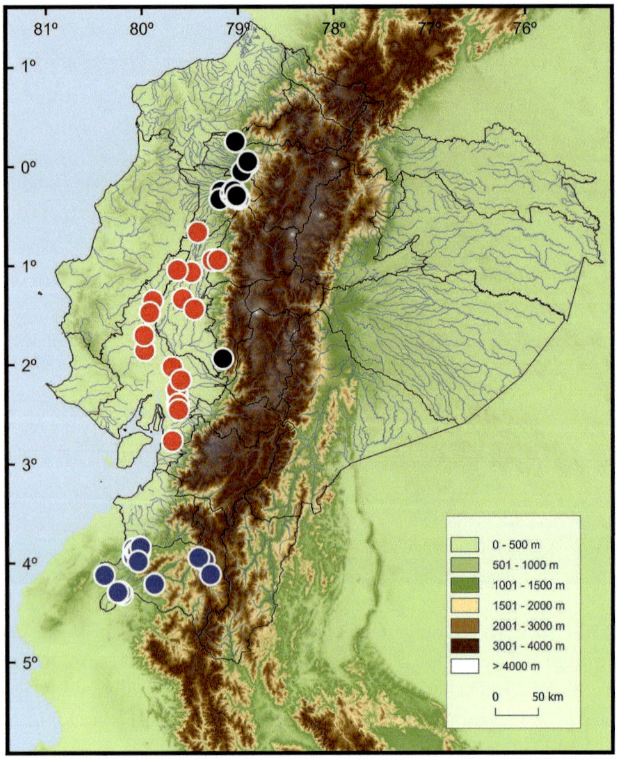

FIGURE 2.17 Localities of known occurrence of *Engystomops coloradorum* (black dots), *E. guayaco* (red dots), and *E. puyango* (blue dots) in Ecuador.

in the Cordillera Occidental of the Andes biogeographic region, where the average annual rainfall is 1781–3272 mm and the average annual temperature is 20.5–23.1 °C.

Conservation.—*Engystomops coloradorum* was categorized as Endangered by Ortega-Andrade *et al.* (2021) because they inferred a reduction in the population greater than or equal to 50% of the population in the next 100 years and a reduction in the extent of occurrence, which is estimated to be less than 5000 km^2 (3770 km^2, according to our data), its distribution is severely fragmented, and there is a continuing decline in the extent and quality of its habitat. Although this is an adaptable and abundant species during the rainy season, it is being affected by the very severe habitat degradation that is taking place within its range, mostly due to livestock grazing, pesticide use, clear cutting, roads, fires, and urban development. It occurs in the protected forest of Milpe Pachijal, the biosphere reserve Chocó Andino de Pichincha, and some private reserves (e.g., Sapoparque La Florida).

Remarks.—In a genetic dataset analysis by Ron *et al.* (2010), *Engystomops coloradorum* is the sister taxon to *E. guayaco* and is closely related to *E. montubio* and *E. randi*. These four species are members of the *Brevivox* sublineage that is part of the *Duovox* lineage (as defined by Ron *et al.* (2006)), which includes small species having calls of short duration compared to most other members of *Engystomops*. Targueta *et al.* (2012) described its karyotype (2n = 20) and described a case of triploidy. del Pino *et al.* (2007) described its early development and analyzed the gastrulation mode of *E. coloradorum* and found that notochord elongation overlapped with involution at the blastopore lip, as in *X. laevis* embryos. Ryan and Rand (1993a, 1993b) and Ron (2008) examined mate selection in females of this species and studied the evolution of their calls. Ryan and Drewes (1990) compared the laryngeal morphology of *E. coloradorum* with that of *E. petersi*.

Key references.—(Cannatella and Duellman, 1984) (description, tadpole, call); Ron *et al.* (2005) (call).

ENGYSTOMOPS GUAYACO (GUAYACO BULLY FOAM TOAD) (RON *ET AL.*, 2005)

Physalaemus guayaco (Ron et al., 2005): 180. **Holotype.**—QCAZ 23521 from private reserve Cerro Masvale, 92 m, Provincia del Guayas, Ecuador.
Engystomops guayaco Nascimento *et al.*, 2005: 312.

Etymology.—The specific name *guayaco* is a coloquial term that refers to the inhabitants of Guayaquil and Provincia del Guayas, Ecuador.

Identification.—Figure 2.18. The maximum known SVL is 19.4 mm in males and 21.0 mm in females. The skin on the dorsum bears scattered tubercles; the snout varies between truncate and subacuminate in dorsal view and round in lateral view; the vomerine teeth and odontophores are absent; the maxillary and premaxillary teeth are present; the parotoid and flank glands are present, the tarsal tubercle is absent; the nuptial pads are present; finger I is shorter than II; the tympanic annulus is evident, concealed dorsally; the tympanic membrane is not tuberculate.

The dorsum is grayish-brown with dark marks; the subocular bars are cream; the flanks are gray; the venter is reddish-cream posteriorly; the anterior half of the venter is cream with gray blotches; the ventral surfaces of the head are dark gray with a cream midventral stripe from the jaw tip to the scapular region; the thighs are salmon red ventrally; the iris is bronze.

Comparisons.—*Engystomops montubio*, *E. randi*, *E. petersi*, *E. pustulosus*, *E. pustulatus*, *E. coloradorum*. *Engystomops guayaco* is very similar to *E. montubio* and *E. randi*; the feet of both species have less extensive lateral fringes and basal webbing than *E. guayaco*. *Engystomops guayaco* has smaller and less abundant dorsal tubercles than *E. randi*. *Engystomops montubio* exhibits a larger size compared to *E. guayaco*, with male the mean SVL measuring 20.6 and 16.85 mm, respectively. The size differences become particularly noteworthy when examining specific populations. Notably, there is no significant disparity in size between the population of *E. montubio* with the lowest average SVL (located 52 km west of El Carmen on the road to Pedernales, Provincia Manabí) and that of *E. guayaco*, with the highest average SVL (15 km south of Naranjal, Provincia

FIGURE 2.18 Morphological variation in *Engystomops guayaco*. Left: QCAZ 35261, male. Right: QCAZ 35271. Buena Fé, Provincia Los Ríos. Bottom left: Cerro Masvale, Provincia Guayas. Bottom right: QCAZ 29639, Cerro Masvale. LAC.

del Guayas). However, despite the occurrence of significant size differences, these populations are geographically distant, separated by a distance of 630 km, at the opposite latitudinal extremes within the species' distribution ranges. *Engystomops guayaco* is smaller (non-overlapping SVL) than *E. petersi*, *E. pustulosus*, and *E. pustulatus*. The absence of a tarsal tubercle and the presence of teeth in the maxilla and premaxilla further distinguish it from *E. petersi* and *E. pustulosus*. The tadpole of *E. guayaco* differs from that of *E. petersi* in lacking paired elliptical paravertebral glands and in dorsal coloration. *Engystomops coloradorum* has bigger dorsal tubercles, some coalescing into ridges (tubercles are more scattered and never coalesce into ridges in *E. guayaco*). *Engystomops coloradorum* further differs from *E. guayaco* in having a vertical loreal region (oblique in *E. guayaco*).

Natural history.—At Cerro Masvale, Provincia Guayas, in 20 February 2002, *Engystomops guayaco* were calling from ditches and small ponds in pastures near the buildings of the Fundación Andrade field station. On 23 March 2003, males were calling from a nearby flooded rice field. At 11 km north of Cerro Masvale, a large chorus of *E. guayaco* was calling from a flooded rice field. At 15 km south of Naranjal, males were calling from ponds, less than 20 m away from the highway. *Scinax quinquefasciatus* and *E. randi* were calling syntopically. Observations of reproductive activity took place in February 2002 and March 2003 during the rainy season. Males began calling shortly after dusk. Males call from small ponds (down to 10 cm diameter) and ditches while floating in a few centimeters of water, usually concealed by herbaceous vegetation. Amplexus and egg deposition take place at the same sites where choruses call. *Engystomops guayaco* constructs floating foam nests during amplexus. While the female discharges the egg masses, the male beats them with his legs to produce the foam. On 20 March 2003, a submerged pair of amplectant frogs was found among vegetation. Two hours after capture, they were placed on a plastic container with water, and they began to build a foam nest. The male kicked the eggs with rapid movements in outbursts that lasted 2–3 seconds with intervals of approximately 10 seconds (time was measured from the digital video). During the intervals, the male rested with his legs semi-extended in the foam (angle between the thigh and shank).

The call of this species was recorded at Cerro Masvale, Provincia Guayas, and described as a single, short call that begins with a pulsed component followed by a whine-like second component. Detailed data and figures are provided under the subtitle "Vocalizations of Ecuadorian Anurans" in Appendix 1.

Tadpoles.—Figure 2.19. The tadpole was described based on an individual at stage 28 raised under laboratory conditions from a clutch of eggs that was laid in captivity collected 15 km south of Naranjal, Provincia Guayas. The tadpole shows, in dorsal view, an elliptical body, widest between the eye and spiracle, with a rounded snout. The eyes are relatively large (Body Length about 7.5 times larger than Eye Diameter), separated by a distance equal to about 1.9 times the internarial distance, directed and positioned dorsolaterally, not visible in ventral view. The external nares are oval, located dorsolaterally, at about one-third the distance between the anterior margin of the snout and anterior margin of the eye. In profile, the body is depressed (BW/BH = 1.4), flattened ventrally; the snout is rounded. The oral disc is posteriorly emarginated. The spiracle is sinistral, cylindrical inside the body wall, but protruding laterally, its tip is closer to the vent than to the eye; it is positioned ventrolaterally and oriented posterodorsally. The spiracle opening is rounded, situated at the level of the hind limbs. The tail musculature is conspicuous, decreasing in size toward the tip of the tail. The tail dorsal fin does not extend onto the body; it is slightly convex and reaches its highest size at about midtail. The tail tip is nearly rounded. The ventral fin is almost straight, beginning just posterior to the vent, about equal in size along its length. The anal tube is dextral, tubular, linked to the caudal muscle. The limb buds are of about equal length to the diameter. There are no lateral lines or glands. The oral disc is ventral, transversely elliptical; the anterior labium bears marginal papillae interrupted by a medial anterior labium gap extending two-fourths of the anterior margin of the labium; the posterior labium is emarginated, bordered by two rows of small rounded papillae. The labial tooth row formula is 1/3; A-2 = 7 + 3 labial teeth, interrupted by a medial gap; P-1 = 14 + 17 labial teeth, interrupted by a right gap; P-2 = 5 labial teeth; and P-3 = 32 labial teeth. The jaw sheaths are thin, serrated on the inner side; the upper is nearly straight, the lower is angular and shorter than the upper.

Stage 29 and stage 31 tadpoles are similar in shape and color pattern to the one described previously. Differences

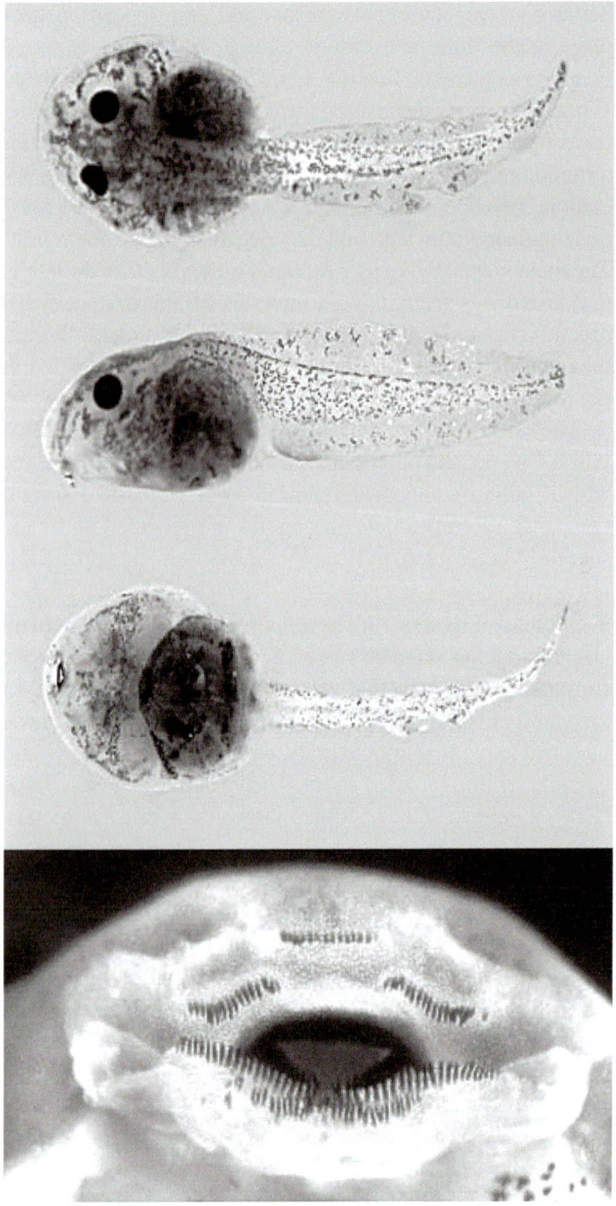

FIGURE 2.19 Tadpole and oral disc of *Engystomops guayaco*. Tadpole: QCAZ 24006a, TL = 13.8 mm. Oral disc: QCAZ 24006b, 1.58 mm. 15 km south of Naranjal, Provincia Guayas. Taken from Ron *et al.* (2005).

are noted on the labial tooth row formula, which varies from 2/2 to 2/3 in the number of labial teeth and their distribution. The number of labial teeth and their distribution in stage 28 are A-1 = 14, A-2 = 17 + 18, P-1 = 23, P-2 = 28, whereas those in stage 29 are A-1 = 3 + 21+ 3, A-2 = 16 + 20, P-1 = 21+ 24, P-2 = 39, P-3 = 5 and in stage 31 are A-1 = 49; A-2 = 12 + 14; P-1 = 44, P-2 = 36. A-2 and P-1 bear larger teeth than the outer rows. A brown tail tip mark is present in all of them and varies in its conspicuity.

Distribution.—See Figure 2.17. *Engystomops guayaco*, an endemic species to Ecuador, is known from the Pacific lowlands in Los Ríos, Cotopaxi, and Guayas provinces. Its elevational range is 4–229 m in an area of extent of occurrence of about 9540 km^2.
This nocturnal, terrestrial species inhabits open areas highly modified by human intervention in the deciduous forests of lowlands and semideciduous forests of lowlands ecosystems in the Equatorial Chocó and Jama-Zapotillo biogeographic regions, where the average annual rainfall is 830–2704 mm and the average annual temperature is 23.9–25.5 °C.

Conservation.—*Engystomops guayaco* was categorized as Vulnerable by Ortega-Andrade *et al.* (2021) because its extent of occurrence is estimated to be less than 20,000 km^2 (9540 km^2, according to our data); its distribution is severely fragmented, and there is a continuing decline in the extent and quality of its habitat. Although this species is common, adaptable, and does not appear to be in decline, the human impact along its distribution is severe. More than half of the land cover in this region has been converted to agriculture, cattle pasture, and urban areas. It has been recorded in the biosphere reserve Reserva de la Biósfera Macizo del Cajas (which extends form the Andes to the Pacific lowlands), the public reserve Reserva Ecológica Manglares Churute, and the private reserve Cerro Masvale.

Remarks.—In a genetic dataset analysis by Ron *et al.* (2010), *Engystomops guayaco* is the sister taxon to *E. coloradorum* and closely related to *E. montubio* and *E. randi*. These four species are members of the *Brevivox* sublineage that is part of the *Duovox* lineage (as defined by Ron *et al.*, 2006), which includes small species having calls of short duration compared to most other members of *Engystomops*. Calls of *E. guayaco* resemble those of *E. montubio*, except for a small difference in frequency. Differences may be explained by significant differences in SVL between both samples (*E. guayaco* mean SVL = 17.7 mm, *E. montubio* mean SVL = 20.1 mm). Targueta *et al.* (2012) described its karyotype (2n = 20).

Key references.—(Ron *et al.*, 2005) (description, tadpole).

ENGYSTOMOPS MONTUBIO (MONTUBIO BULLY FOAM TOAD) (RON *ET AL.*, 2004)

Physalaemus montubio (Ron *et al.*, 2004): 269. **Holotype.**—QCAZ 19524 from Cerro Puerto Rico, 30 m, Provincia Manabí, Ecuador.
Engystomops montubio Nascimento *et al.*, 2005: 312.

Etymology.—The specific name alludes to the Ecuadorian word *montubio*, which refers to the people who inhabit the countryside of the lowlands of western Ecuador. Because of its tolerance of habitat disturbance, *Engystomops montubio*

is well known by *montubios*, who refer to them as "*ranas bullangueras*" due to their loud advertisement calls.

Identification.—Figures 2.20, 2.21. The maximum known SVL is 22.6 mm in males and 24.2 mm in females. The skin on the dorsum bears numerous round or subconical tubercles; the snout is subacuminate in dorsal view, rounded in profile; the vomerine odontophores are absent; the maxillary teeth are present; the parotoid and flank glands are present; the tarsal tubercle is absent; the nuptial pads are present; finger I is shorter than II; the tympanic annulus is evident, concealed dorsally; the tympanic membrane is not tuberculate; the dentigerous processes of the vomers are thin and spikelike; the stalk of the alary process of the hyoid is very narrow; the anterior process of the hyale is well developed and prominent.

The dorsum is light gray to light and dark brown with darker blotches on the sides and on the limbs. There is a dark brown stripe posterior to eye. The upper lip is white. The lateral head coloration varies extensively between light gray and dark gray or dark brown. In individuals with a darker coloration, the suborbital cream stripes are short and ill defined. In frogs with a lighter coloration, a single, wide, cream band is present below the orbit. One or two additional cream to light brown labial bars may be present anteriorly. The venter is off-white near the abdomen with light brown speckling; the pelvic region, throat, and undersides of the legs are not pigmented. The distribution of ventral markings can be almost completely restricted to the head or the chest or present on all of the venter. On the abdomen, dark marks can be arranged in well-defined large spots or in diffuse speckled patterns; in all individuals, a midventral cream to light brown stripe (discrete to ill defined), starting on the head, is present. The extension of the stripe varies from being restricted to the head to reaching up to two-thirds of the SVL. The iris is bronze.

FIGURE 2.20 Morphological details of *Engystomops montubio*. CJ 10020, male, 21.6 mm, Salango, Provincia Manabí. SGS.

FIGURE 2.21 Morphological variation in males of *Engystomops montubio*. Top left: CJ 10034, 21.4 mm SVL. DAL. Top right. CJ 10033, 23.9 mm SVL. Bottom left: 10032, 21.7 mm SVL. Bottom right: CJ 10019, 21.8 mm SVL. Salango, Provincia Manabí. LAC.

Comparisons.—*Engystomops randi*, *E. guayaco*, *E. coloradorum*, *E. pustulosus*, *E. petersi*, *E. pustulatus*. *Engystomops montubio* is mostly similar to *E. randi*, *E. guayaco*, and *E. coloradorum*. *Engystomops montubio* has a greater SVL and proportionally shorter flank and parotoid glands than *E. randi*. *Engystomops montubio* is larger than *E. guayaco*, with a mean SVL in males of 20.6 and 16.85 mm, respectively, but see Remarks under *E. guayaco*. *Engystomops coloradorum* has more tuberculated skin than *E. montubio* and a nearly vertical loreal region (sloping gradually toward the lips in *E. montubio*). *Engystomops montubio* differs from the larger *E. pustulosus* and *E. petersi* by the absence of a tarsal tubercle and the presence of teeth on the maxilla and premaxilla. The ranges of adult size in *E. montubio* and *E. pustulatus* do not overlap (male SVL range 18.67–22.56 and 25.17–29.88 mm, respectively).

Natural history.—Individuals have been found by night in artificial open areas. Reproductive activity is nocturnal. In Puerto Rico, Provincia Manabí, frogs were breeding near buildings, pastures, and in other cleared areas. Choruses were heard in March 1999, February 2002, and March 2003 during the rainy season. Most individuals were congregated around small temporary pools that filled after heavy rains. Males called while floating in a few centimeters of water. In Puerto Rico, a large aggregation was found in a stream pool 50 m away from the Pacific Ocean. Dozens of males were calling among aquatic plants. Interestingly, the chorus was synchronized into two antiphonals "voices." Other species calling and breeding in the same pond were *Rhinella* *horribilis* and the hylids *Trachycephalus jordani* and *Scinax quinquefasciatus*. Amplexus and egg deposition take place at the same sites where choruses call. Amplectant pairs were observed to make floating foam nests. To do so, the male beats the egg masses with his legs as the female discharges them. In 1999 and 2002, *P. montubio* was breeding syntopically with *P. pustulatus* in Puerto Rico, Provincia Manabí. In some pools, males of both species were calling simultaneously.

The call of this species was recorded in Puerto Rico, Provincia Manabí, Ecuador. *Engystomops montubio* emits a single, short call that begins with two to five amplitude-modulated pulses followed by a nearly pure tone with a downward frequency sweep. Detailed data and figures are provided under the subtitle "Vocalizations of Ecuadorian Anurans" in Appendix 1.

Tadpoles.—Unknown.

Distribution.—Figure 2.22. *Engystomops montubio*, an endemic species to Ecuador, is known from along the coastal Pacific lowlands in Manabí, Guayas, and Santa Elena provinces. Its elevational range is 0–286 m in an area of extent of occurrence of about 18,500 km^2.

This nocturnal, terrestrial species inhabits open areas highly modified by human intervention in the deciduous forests of the Cordillera Costera of the Equatorial Pacific, low forests and deciduous shrub of lowlands of Jama-Zapotillo, evergreen seasonal forests of lowlands of Jama-Zapotillo, and the semideciduous forests of lowlands of Jama-Zapotillo ecosystems, where the average annual rainfall is 307–1897 mm and the average annual temperature is 23.6–25.5 °C.

Conservation.—*Engystomops montubio* was categorized as Least Concern by Ortega-Andrade *et al.* (2021) because it is common, adaptable, and does not appear to be in decline. This species is seasonally abundant. This is an adaptable species that is unlikely to be significantly threatened. It has not been recorded in any protected area.

Remarks.—In a genetic dataset analysis by Ron *et al.* (2010), *Engystomops montubio* is the sister taxon to *E. randi* and closely related to *E. coloradorum* and *E. guayaco*. These four species are members of the *Brevivox* sublineage that is part of the *Duovox* lineage (as defined by (Ron *et al.*, 2006), which includes small species having calls of short duration compared to most other members of *Engystomops*. *Engystomops montubio* has a genetic distance (pairwise uncorrected *p*-distance) from other species of the Brevivox lineage of *Engystomops* of at least 2.7% (*E. randi*) in fragments of aligned 12S and 16S mitochondrial DNA sequences. Targueta *et al.* (2012) described its karyotype (2n = 20).

Key references.—(Ron *et al.*, 2004) (description).

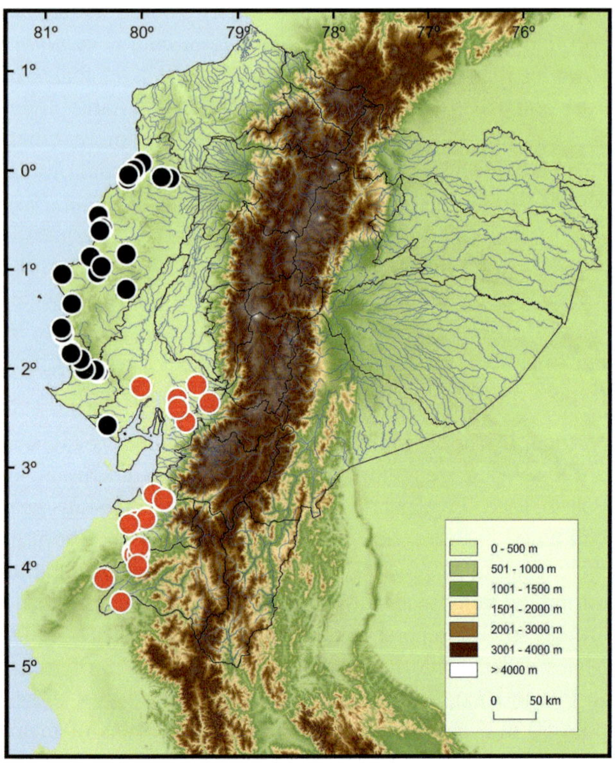

FIGURE 2.22 Localities of known occurrence of *Engystomops montubio* (black dots) and *E. randi* (red dots) in Ecuador.

ENGYSTOMOPS PETERSI (PETERS' BULLY FOAM TOAD) (JIMÉNEZ DE LA ESPADA, 1872)

Physalaemus petersi (Jiménez de la Espada, 1872): 86.
Types.—Lost, presumably originally in MNCN (two specimens cited in Jiménez de la Espada, 1875). **Type locality.**— "Oriente en el Ecuador;" "Archidona, provincia Oriental del Ecuador, cerca de las orillas del rio Napo" [Provincia Napo], according to Jiménez de la Espada (1875).
Engystomops petersi Nascimento *et al.*, 2005: 312.

Etymology.—The specific name honors Wilhelm Carl Hartwig Peters (1815–1883), former Director of the Zoology Museum of Berlin, and whose wise lessons helped to both direct and illustrate Espada's work.

Identification.—Figures 2.23, 2.24, 2.25. The known SVL is 24.5 mm in a male and 24.9 mm in a female from Payamino, Provincia Orellana, Ecuador. The vomerine, maxillary, and premaxillary teeth are absent; parotid and flank glands are elliptical; the snout is short and pointed; in lateral view, the snout is protruding, with the anterior margin of the lip located posteriorly to the nostril; the tympanic ring is generally well defined, at least ventrally; the tympanic membrane is smooth; the skin of the dorsum has small tubercles with scattered larger tubercles, sometimes forming divergent rows in the occipital and scapular regions; the skin of the ventral surfaces is smooth. The dentigerous

FIGURE 2.23 *Engystomops petersi*. Left: CJ 10926, female, 24.9 mm SVL. Right: CJ 10927, male, 24.5 mm SVL. Payamino, Provincia Orellana. DAL.

FIGURE 2.24 Morphological variation in *Engystomops petersi*. Top left: QCAZ 36298, Bobonaza, Provincia Morona Santiago. LAC. Top right: CJ 10928. Middle left: CJ 10930. Payamino, Provincia Orellana. DAL. Middle right: QCAZ 39589, Centro Ecológico Zanja Arajuno, Provincia Pastaza. Bottom left: CJ 2773, Reserva Otoyacu, Provincia Pastaza. Bottom right: QCAZ 36930, Centro Ecológico Zanja Arajuno, Provincia Pastaza. LAC.

processes of the vomers are thin and pointed (like a spine); the males have a large, medium subgular sac; nuptial pads are present; there is no webbing on the hands or feet; finger I is longer than II; the tarsal tubercle is present.

Individuals from Mera, Provincia Pastaza, Ecuador, have a brown dorsum with the tips of the tubercles being red; the limbs are brown with dark brown bars edged with black. The flanks are orange. The throat and chest are dull yellowish-brown; the belly and hidden surfaces of the thighs are black with white or bluish-white spots. About 10 to 20% of the frogs exhibit faint interorbital, occipital, or scapular rows of tubercles. A "pale morph" is characterized by having a pale snout and flanks, and it is more frequent in females than in males.

Comparisons.—*Engystomops coloradorum*, *E. montubio*, *E. guayaco*, *E. pustulatus*, *E. puyango*, *E. randi*. *Engystomops petersi* differs from the trans-Andean *Engystomops* (*Duovox* lineage) (*E. coloradorum*, *E. montubio*, *E. guayaco*, *E. pustulatus*, *E. puyango*, *E. randi*) by lacking maxillary teeth, having a tarsal tubercle, and having finger I longer than II (finger I shorter than or equal to finger II in species of *Duovox*).

Natural history.—Figure 2.26. Amplectant pairs were collected at night from April to November 2004 in Puyo, Provincia Pastaza. A study of sexual selection based on the advertisement call in *E. petersi* demonstrated that this parameter plays an important role in speciation.

The advertisement call of males from 5 km northeast of Puyo, Provincia Pastaza, Ecuador, consists of two components: a short, low-amplitude "prefix" followed by a longer, frequency-modulated "whine." Detailed data and figures are provided under the subtitle "Vocalizations of Ecuadorian Anurans" in Appendix 1.

Tadpoles.—Figure 2.27. The tadpole was described based on specimens from Mera, Provincia Pastaza, Ecuador. A striking feature is the paired elliptical glands located at midlength of the body. A typical tadpole in developmental stage 28 has a body length of 13.7 mm and a total length of 31.5 mm. The body is robust, widest posteriorly, and about as deep as wide. The snout is narrow and rounded. The eyes are small, close together, and directed dorsolaterally. The nostrils are closer to the eyes than to the tip of the snout. The spiracle is directed posterodorsally; the spiracular opening is at a point just below the midline at about one-third the length of the body. The caudal fins are deeper than the moderately robust caudal musculature, deepest at the midlength of the tail, and rounded at the tip; the dorsal fin does not extend onto the body. The oral disc is anteroventral, about as wide as the anterior part of the body. The upper lip is bare; elsewhere, there is one row of blunt labial papillae. The jaw sheaths are broad, slender, and weakly serrate.

FIGURE 2.25 Morphological variation and details of *Engystomops petersi*. Top left: CJ 12160, Puyo, Provincia Pastaza. SGS. Middle left: Reserva Río Bigai, Provincia Orellana. Amanda B. Quezada. Ecuador. Right and bottom: CJ 10932, note color of groin and dorsal surfaces of thighs, Payamino, Provincia Orellana. DAL.

FIGURE 2.26 *Engystomops petersi*. Top left: amplectant pair. Top right and bottom left: foam nests, Shiripuno. Provincia Orellana. Bottom right: QCAZ 40862, juvenile, Centro Ecológico Zanja Arajuno, Provincia Pastaza. LAC.

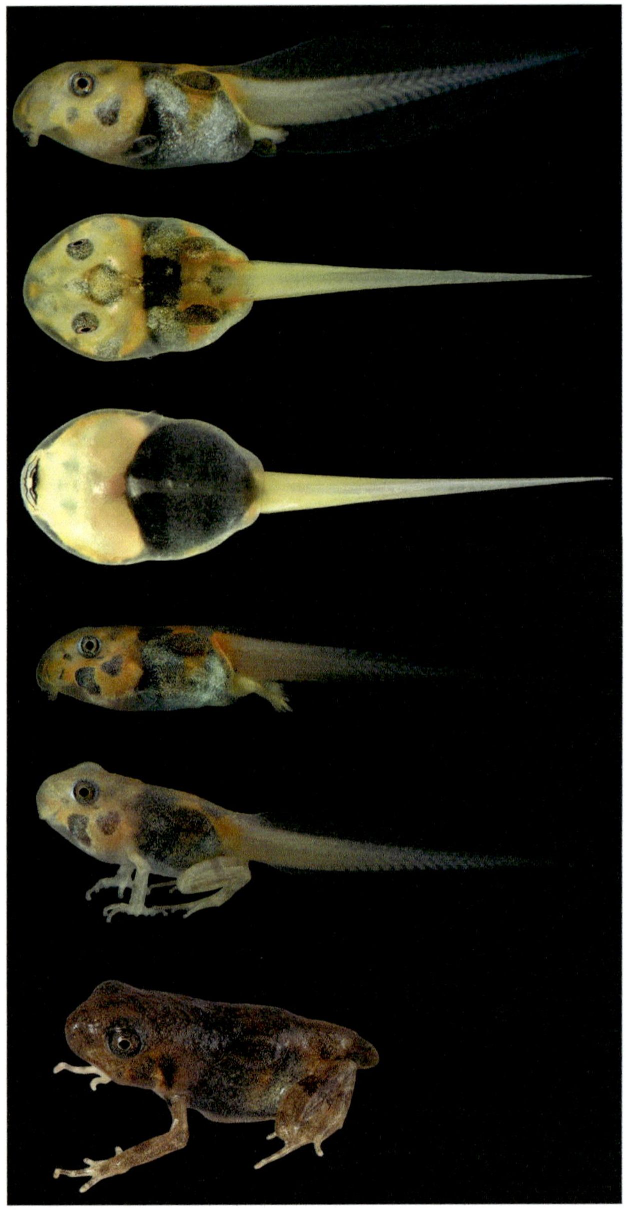

FIGURE 2.27 Ontogenetic changes in four individuals of *Engystomops petersi* from tadpole to froglet. Top to bottom. CJ 6409, stage 35, lateral, dorsal, and ventral views, TL = 25.5 mm; undetermined stage; stage 42, TL = 31 mm; stage 45 = 11.3 mm SVL. Reserva Otoyacu, Provincia Pastaza. DAL.

There are two upper and three lower rows of denticles; the upper rows are slightly longer than the lower rows, and the second upper row is narrowly interrupted medially.

The body and caudal musculature are pale brown with dark brown markings on the dorsum of the body and large brown spots on the tail. The dorsal markings usually consist of a pair of spots on the snout, a broad blotch posterior to the eyes, and a narrow transverse bar posteriorly on the body. The caudal fins are translucent, with brown flecks on the dorsal fin; flecks are absent in small tadpoles and most numerous in larger ones. In life, the body and musculature of the tail are black with cream spots on the tail; the fins are gray.

Distribution.—Figure 2.28. *Engystomops petersi* (see Remarks), an endemic species to Ecuador, is known on the Amazonian lowlands and eastern slopes of the Cordillera Oriental of the Andes. It occurs in Sucumbíos, Napo, Orellana, Pastaza, and Morona Santiago provinces. Its elevational range is 202–1310 m in an area of extent of occurrence of about 71,120 km^2.

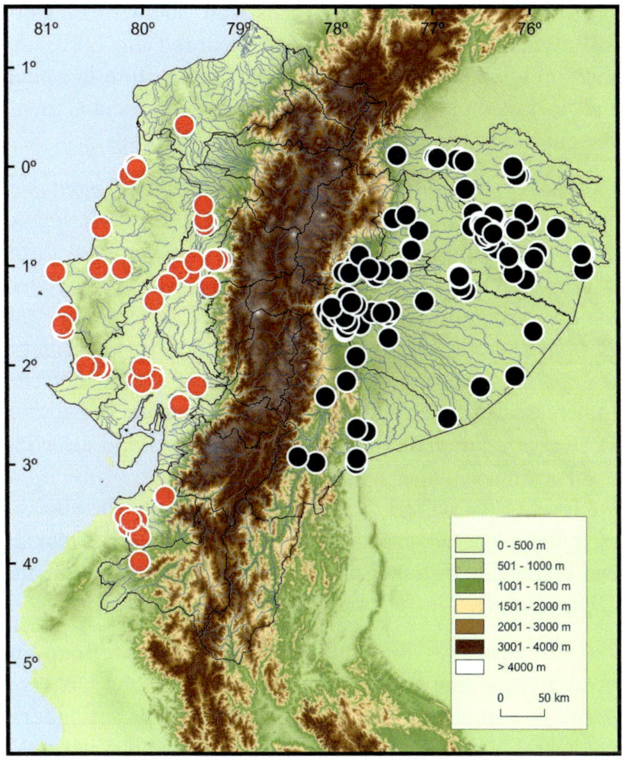

FIGURE 2.28 Localities of known occurrence of *Engystomops petersi* (black dots) and *E. pustulatus* (red dots) in Ecuador.

This nocturnal, terrestrial species inhabits primary and disturbed habitats in flooded forests, evergreen forests of lowlands, and evergreen montane foothill forests in 11 ecosystems (Table 0.4) within seven biogeographic sectors (Table 0.5), where the average annual rainfall is 1880–4434 mm and the average annual temperature is 19.7–25.7 °C.

Conservation.—*Engystomops petersi* is categorized as Least Concern throughout its entire range (IUCN SSC Amphibian Specialist Group, 2018d; in IUCN Red List of 2022) in view of its wide distribution, tolerance of a degree of habitat modification, presumed large population, and because it is unlikely to be declining to qualify for listing in a more threatened category. However, the four species recognized by Funk *et al.* (2012) in Amazonian Ecuador have relatively small ranges, which is one factor that can put them in a higher risk category. Thus, instead of four species of Least Concern, this complex may consist of several species of higher conservation concern, depending on additional factors such as evidence of population declines and threats. We suggest categorizing *E. petersi* as Vulnerable (following IUCN criteria B1ab) because its extent of occurrence is estimated to be less than 20,000 km^2, it is severely fragmented, and there is a continuing decline in the quality of its habitat. Local populations, especially in the Andean foothills, of *E. petersi sensu stricto* are impacted by habitat loss due to clear cutting, agricultural expansion, and urban development. It has been recorded in the public reserves Parque Nacional Yasuní and Reserva de Producción de Fauna Cuyabeno, the protected forests of Comuna Kichwa Añangu and Nacionalidad Waorani del Ecuador, the biosphere reserves Yasuní and Sumaco, and some private reserves (e.g., Centro Ecológico Zanja Arajuno, Reserva Otoyacu).

Remarks.—*Engystomops petersi*, as currently recognized, is a complex of species. Funk *et al.* (2007, 2012) provide morphological, acoustic, and molecular information, indicating that the Amazonian *Engystomops*, which were previously considered two species (*E. petersi* and *E. freibergi*), are actually five to seven species, four of which occur in Ecuador. Among them, populations from the eastern slopes of the Cordillera Oriental of the Andes of Ecuador and adjacent foothills and Amazonian lowlands were assigned to *E. petersi sensu stricto*. We follow the latter assignment, pending further systematic studies. Ron *et al.* (2006), Pyron and Wiens (2011), and Funk *et al.* (2007, 2009, 2012) analyzed genetic datasets. There are three well-defined lineages: one in the upper drainage of the Río Napo in Ecuador and adjacent Peru, another in the drainages of the Juruá and Madre de Díos rivers in Peru and adjacent Brazil, and a third from the State of Pará, Brazil. Gascon *et al.* (1998) found high interpopulation genetic diversity along the Juruá River, Brazilian Amazon, suggesting a certain degree of differentiation. However, when testing the Río Juruá hypothesis as a biogeographic barrier to gene flow, they found little support because both sides of the river showed genetic similarity for this species. Jiménez de la Espada (1875) provided external morphological and osteological data of two females (presumably the types) from Archidona. Within *Engystomops*, based on molecular data, *E. petersi* is within the Edentulus lineage (*sensu* Ron *et al.*, 2006). Duellman (1978) provided morphological, call, and natural history information based on specimens from Santa Cecilia and Puerto Libre, Provincia Sucumbíos. Ryan and Drewes (1990) studied the morphology of the vocal apparatus and its relationship with the advertisement call using specimens from the latter localities. Nonetheless, it is still unknown if the specimens from Santa Cecilia and Puerto Libre belong to *E. petersi* or any of its related undescribed species. Ryan and Drewes (1990) compared the laryngeal morphology of *E. petersi* with that of *E. coloradorum*.

Key references.—(Jiménez de la Espada, 1872) (description); (Jiménez de la Espada, 1875) (morphology, osteology); (Duellman, 1978) (account, tadpole); (Funk *et al.*, 2007) (phylogeny, biogeography); (Boul *et al.*, 2007) (natural history); (Funk *et al.*, 2009) (phylogeny, natural history); (Guerra and Ron, 2008) (natural history, call).

ENGYSTOMOPS PUSTULATUS (BLISTERED BULLY FOAM TOAD) (SHREVE, 1941)

Edalorhina pustulata (Shreve, 1941): 80. **Holotype.**—MCZ 7666 from Guayaquil, Provincia Guayas, Ecuador.
Engystomops pustulatus Nascimento *et al.*, 2005: 312.

Etymology.—The specific epithet is Latin, the past participle of *pustulare* (to blister); the name refers to the glandular skin characteristic of this species.

Identification.—Figure 2.29. The maximum known SVL is 29.9 mm in males and 36.5 mm in females. *Engystomops pustulatus* is characterized by having the skin on the dorsum highly tubercular, with elongated tubercles in the scapular region forming an inverted V-shape; the belly is smooth; there is an elongated row of glands on the flank; the snout is rounded; the canthus rostralis is visible; there is a rounded tympanic ring, about half the diameter of the eye; the maxillary and premaxillary teeth are present; the vomerine teeth are absent; the fingers do not have expanded digital discs and interdigital membranes; finger II is longer than I; the tarsal tubercle is absent; there are small metatarsal tubercles, the internal being a little larger than the external.

The dorsal coloration varies between light coffee to grayish or dark coffee, with or without dark gray and cream marks. Irregular dark marks may be present in diverse patterns, which include diffuse interorbital, sacral, and limb bands. A cream or light brown middorsal line usually extends along the posterior half of the body. There are usually rounded or oval cream marks on the middorsum at the anterior level of the body. The flanks are cream, brown, or gray, usually with a suffusion of green. The loreal region is cream, gray, or light brown and bears ill-defined darker bands. The belly is white or cream with an irregular marbling of gray or dark brown markings and spots. The undersides of the limbs are flesh-colored with pink. The throat is predominantly gray or brown, sometimes with a suffusion of green in the mandible margins. An ill-defined cream midventral stripe can be present from near the tip of the snout to the gular region or continue to the midventer. The iris is cream, clear brown, and gray with abundant fine, dark brown reticulations. In specimens from southern Ecuador in Provincia El Oro, there is a fine, clear red ring bordering the pupil, whereas in northern populations from la Maná, Provincia Cotopaxi, and Cerro Blanco, Provincia Guayas, the pupil is bordered by a creamy white fine ring.

Comparisons.—*Engystomops puyango*, *E. randi*, *E. montubio*, *E. guayaco*, *E. coloradorum*, *E. petersi*, *E. pustulosus*. *Engystomops pustulatus* is mostly similar to *E. puyango*. They differ in dorsal skin texture (less tuberculate in *E. puyango*). *Engystomops pustulatus* is larger than *E. randi*, *E. montubio*, and *E. guayaco* (non-overlapping SVL in adult males). *Engystomops coloradorum* has more prominent dorsal tubercles (tubercles are smaller and scattered in *E. pustulatus*). *Engystomops coloradorum* further differs from *E. pustulatus* in having a subacuminate snout with the anterior margin of the lip at the level of the posterior margin of the nostrils. The absence of a tarsal tubercle and the presence of teeth in the maxilla and premaxilla distinguish *E. pustulatus* from *E. petersi* and *E. pustulosus*.

Natural history.—Figure 2.30. Individuals were found in open areas where the original natural vegetation has been partly or completely removed and replaced by agricultural lands and human settlements. The diet of adults and juveniles of *Engystomops pustulatus* from three populations in the Ecuadorian Chocó was studied. Analysis of the items recovered from the gastrointestinal tracts of 95 individuals showed that *E. pustulatus* has a diverse diet, identifying 52

FIGURE 2.29 *Engystomops pustulatus*. Left: QCAZ 31273, female, Río Quaque, Pedernales, Provincia Manabí. Right: QCAZ 33232, male, Reserva Natural Laipuna, Macará, Provincia El Oro. LAC.

FIGURE 2.30 Aspects of the natural history of *Engystomops pustulatus*. Top left: calling male, Palenque, El Carmen, Provincia Los Ríos. LAC. Top right: amplectant pair, Reserva Fundación para la Conservación de los Andes Tropicales, Provincia Esmeraldas. Amanda B. Quezada. Middle left: Macul, Provincia Los Ríos. Juan Carlos Sánchez Nivicela. Middle and bottom right: Reserva Ecológica Arenillas, Provincia El Oro. Bottom left: Palenque, El Carmen. LAC.

prey types among 1211 items. Among juveniles at Cerro Blanco, Provincia Guayas, there was a high proportion of small prey items, including termites and ants. Larger items, including snails, coleopterans, and isopods, were most frequent in adults from Patricia Pilar, Provincia Los Ríos, and La Maná, Provincia Cotopaxi. *Engystomops pustulatus* is abundant during the rainy season in habitat areas modified by human activities. Couples in amplexus build foam nests in the water, which are formed by repetitive movements of the hind legs of the males. The nests contain an average of 320 eggs, of which almost all are fertile. It breeds in sympatry with *Engystomops montubio* and *E. randi*. Reproduction in *E. pustulatus* is characterized by high maternal investment (15% egg mass relative to body mass). There is evidence of size-assortative mating, with a tendency for larger females to mate with larger males. The clutch size is correlated with female weight, female condition, and male size, and larger females show a tendency to lay larger foam nests, which contain more eggs. In reproductive choruses, there is a male-biased operational sex ratio, indicative of high variance in male reproductive success. An observation of an amplectant couple spawning while an additional male was embedded in the foam could be evidence of an alternative mating strategy, where a small, non-amplectant male attempts to fertilize the eggs that are extruded by the amplectant female. A study on the spawning behavior of nesting couples from the lowlands of western Ecuador showed that the nest structure is influenced by the male physical effort and the female behavior (steady position during spawning and oviposition site selection). Nest building has two phases. In the first phase, the duration of resting periods, the duration of mixing events, and the number of kicks increase and are highly variable. During the second phase, the three variables stabilize until the end, and the volume of the nest mainly increases during this phase. In four nesting events, observations of kicking movements by the female were recorded. It seems that this behavior may be triggered by insufficient male effort.

The males sing at night from the water, and their vocalization is a very loud moan. The call of this species was recorded in Ecuador in Arenillas, El Empalme, Huaquillas (Provincia El Oro), La Maná (Provincia Cotopaxi), Patricia Pilar (Provincia Los Ríos), and Puerto Rico (Provincia Manabí). *Engystomops pustulatus* emits a call with two obligatory and one facultative components. The first component is characterized by a weak amplitude modulation. The second component is a whine-like note. A facultative third component (exclusive of La Maná population) is similar to the "chuck" of the call of *E. pustulosus*, but it is less discrete and has less acoustic complexity. Detailed data and figures are provided under the subtitle "Vocalizations of Ecuadorian Anurans" in Appendix 1.

Tadpoles.—Figure 2.31. Andrea Terán-Valdez and Luis A. Coloma (this publication). This description is based on an individual (series CJ 10042) from Salango, Provincia Manabí, at stage 25. It was collected by Luis A. Coloma, Chauna Schuette, Patricio Vargas Mena, and Mark Wanner on 14 February 2019. It has a body length of 6.3 mm and a total length of 18.5 mm. The body is ovoid and depressed, wider anteriorly than posteriorly, with a width surpassing its depth. The snout is rounded but slightly pointed in dorsal view and rounded in lateral view, and the eyes are directed dorsolaterally. The spiracle is sinistral, with the inner wall free from the body and its opening directed posteriorly. The vent tube is medial. The oral disc has an anteroventral position. The tail constitutes approximately 66% of the total length, reaching its maximum depth in the second third of its length and ending in an acutely rounded tip. The dorsal fin does not extend onto the body, maintaining a relatively constant depth along the tail, with the deepest point around the midlength, tapering gradually toward the end. The ventral fin, thinner than the dorsal fin, maintains a more or less constant depth along the tail, tapering gradually toward the end. The caudal musculature is deepest at its anterior part.

FIGURE 2.31 Ontogenetic changes in four individuals of *Engystomps pustulatus* from tadpole to adult. Top to bottom, left to right. CJ 10042abc, lateral, dorsal, and ventral views, stage 25, TL = 18.5 mm; undetermined stage, TL = 27.5 mm; stage 42, TL = 22 mm. CJ 10042d, stage 45 = 7. 5 mm SVL; froglet 49 days after stage 45 = 13.9 mm SVL; juvenile 82 days = 14.8 mm SVL; juvenile 271 days = 18.4 mm SVL; adult 406 days = 21.9 mm SVL. Salango, Provincia Manabí. SGS.

Dorsally, the tadpole displays brown and golden flecks, while ventrally, the skin is translucent with silvery flecks ventrolaterally, revealing visible organs. The tail has abundant brown flecks; there are tiny white flecks at the margins of the fins. The iris displays a blend of black and gold hues with a hint of pink.

Distribution.—See Figure 2.28. *Engystomops pustulatus*, an endemic species to Ecuador, is known along the Pacific coastal regions, lowlands, and foothills. It occurs in Esmeraldas, Manabí, Santo Domingo de los Tsáchilas, Cotopaxi, Los Ríos, Santa Elena, Guayas, El Oro, and Loja provinces. Its elevational range is 0–984 m in an area of extent of occurrence of about 51,590 km^2.

This nocturnal, terrestrial species inhabits open areas highly modified by human intervention in six ecosystems (Table 0.4) within five biogeographic sectors (Table 0.5), where the average annual rainfall is 205–2748 mm and the average annual temperature is 22.3–25.7 °C. The geographic range of *E. pustulatus* is characterized by seasonal precipitation, and it is abundant during the rainy season.

Conservation.—*Engystomops pustulatus* was categorized as Least Concern by Ortega-Andrade *et al.* (2021) in view of its relatively wide distribution, presumed large population, and because it is unlikely to be declining fast enough to qualify for listing in a more threatened category. It can be abundant during reproductive bouts. The main threats are habitat loss owing to agricultural expansion and human settlements. It is reasonably adaptable but probably cannot withstand very serious habitat degradation. It has been recorded in the public reserves Refugio de Vida Silvestre Pacoche, Parque Nacional Machalilla, and Reserva Ecológica Arenillas; the protected forest of Cerro Blanco; the biosphere reserve Bosque Seco; and some private reserves (e.g., Centro Científico Río Palenque).

Remarks.—In a genetic dataset analysis by Ron *et al.* (2010), *Engystomops pustulatus* is the sister taxon of *E. puyango* plus an undescribed species from 8.5 km N of Motupe, Departamento Lambayeque, Peru. These three species are members of the *Vivavox* sublineage that is part of the *Duovox* lineage (as defined by Ron *et al.*, 2006), which includes species having flank and parotoid glands and loud calls compared to other members of *Engystomops*. Within *E. pustulatus*, two lineages are strongly supported; one is distributed north of the central Costa Region (Cerro Blanco and Patricia Pilar), and the other is in the southern Costa Region (Huaquillas + Arenillas). The pairwise uncorrected *p*-distance between both lineages ranges from 0.019 to 0.021. This genetic differentiation is paralleled by differentiation in morphometric and acoustic space between lineages. *Engystomops pustulatus* has a genetic distance (pairwise uncorrected *p*-distance) from other species of the *Duovox* lineage of *Engystomops* of at least 6.5% (*E. puyango*) in fragments of aligned 12S and 16S mitochondrial DNA sequences. Targueta *et al.* (2012) described its karyotype (2n = 20). Narváez and Ron (2013) hypothesized that dietary niche breadth is correlated with the presence of teeth in *Engystomops*. *Engystomops pustulatus* has teeth and has a more generalist diet than the termite- and ant-specialist species *E. pustulosus*, *E. petersi*, and *E. freibergi*, which lack teeth.

Key references.—(Shreve, 1941) (description); (Ron *et al.*, 2004) (natural history); (Ron *et al.*, 2010) (phylogeny, natural history); (Ron *et al.*, 2014) (natural history); (Narváez and Ron, 2016) (natural history, diet, spawning behavior).

ENGYSTOMOPS PUYANGO (PUYANGO BULLY FOAM TOAD) RON *ET AL.*, 2010

Engystomops puyango Ron *et al.*, 2010: 28. **Holotype.**—QCAZ 26978 from Bosque Protector Puyango, near the bridge over the Río Puyango on the road Alamor Arenillas, 321 m, Provincia El Oro, Ecuador.

Etymology.—The specific epithet refers to the type locality, the Puyango Protected Forest. The 2658 ha reserve has one of the world's largest deposits of petrified tree trunks. It is also one of the largest reserves of tropical dry forest in Ecuador.

Identification.—Figure 2.32. The maximum known SVL is 30.5 mm in males and 32.7 mm in females. *Engystomops puyango* is characterized by having skin on the dorsum bearing a mixture of scattered pustules and minute tubercles; the snout varies between rounded and protruding in dorsal view and is rounded to subacuminate in lateral view; the vomerine teeth and odontophores are absent; the maxillary and

FIGURE 2.32 *Engystomops puyango*. Male, Bosque Protector Puyango, Provincia El Oro. LAC.

premaxillary teeth are present; the parotoid and flank glands are present, usually fused, with a mean length of 16.6 mm; the tarsal tubercle is absent; the nuptial pads are present; finger I is shorter than II; the tympanic annulus is evident, concealed dorsally and posteriorly; the tympanic membrane is typically non-tuberculate.

The dorsal coloration varies between coffee and grayish-coffee, with or without dark spots. Irregular dark marks may be present in diverse patterns. A clear middorsal line extends along the posterior half of the body, but in some, it only extends along the posterior one-fifth. The area below the eye and tympanum is cream in most specimens but can be restricted to a thin, light longitudinal stripe from the jaw articulation to the parotoid gland. Two or three vertical dark bars can be present in the loreal region and below the orbit. The belly is white or cream with irregular gray spots. The throat is predominantly gray. Marks are arranged in diverse patterns and vary from being restricted to the anterior half of the body (darker on folded vocal sacs) to being present over the entire venter (less abundant posteriorly). In a few specimens, the ventral marks are arranged in well-defined large spots. A midventral cream stripe can be present from near the tip of the snout to the gular region or continue to the midventer. In most individuals, the stripe is absent or ill defined.

Comparisons.—*Engystomops pustulatus*, *E. randi*, *E. montubio*, *E. guayaco*, *E. petersi*, *E. pustulosus*, *E. coloradorum*. *Engystomops puyango* is mostly similar to *E. pustulatus*. They differ in dorsal skin texture (less tuberculate in *E. puyango*). *Engystomops puyango* is larger than *E. randi*, *E. montubio*, and *E. guayaco* (non-overlapping SVL in adult males) and has less extensive to absent lateral fringes on the toes (fringes are prominent in *E. randi*, *E. montubio*, and *E guayaco*). The absence of a tarsal tubercle and the presence of teeth in the maxilla and premaxilla distinguish *E. puyango* from *E. petersi* and *E. pustulosus*. *Engystomops coloradorum* has more prominent dorsal tubercles (tubercles are smaller and scattered in *E. puyango*). *Engystomops coloradorum* further differs from *E. puyango* in having a vertical loreal region (oblique in *E. puyango*).

Natural history.—Individuals were found in open to sparsely forested areas, where the original natural vegetation has been partly or completely removed. At Bosque Protector Puyango, choruses were found in February 2004 and February 2005. Males were calling from water in ponds and ditches. Amplexus and egg deposition take place at the same sites where choruses call. *Engystomops puyango* constructs floating foam nests during amplexus. While the female discharges the egg masses, the male beats them with his legs to produce the foam. Males *E. puyango* and the smaller *E. randi* were frequently calling at the same ponds, sometimes at distances less than 20 cm, and might produce hybrids occasionally (see Remarks). Other species calling in syntopy with *E. puyango* were *Scinax quinquefasciatus* and *Trachycephalus quadrangulum*.

The call of *Engystomops puyango* was recorded in Bosque Protector Puyango, Provincia El Oro, Ecuador. It emits a call with two obligatory and one facultative components. The first component is characterized by a slight increase in frequency with decreasing amplitude modulation. The second component is a whine-like note. A facultative third component can be added after the whine and is characterized by a gradual increase in amplitude and a switch in dominant frequency from the first to the fifth or sixth harmonic. Detailed data and figures are provided under the subtitle "Vocalizations of Ecuadorian Anurans" in Appendix 1.

Tadpoles.—Unknown.

Distribution.—See Figure 2.17. *Engystomops puyango*, an endemic species to Ecuador, is known along the southwestern Pacific lowlands and foothills in El Oro and Loja provinces. Its elevational range is 222–1674 m in an area of extent of occurrence of about 3900 km^2.

This nocturnal, terrestrial species inhabits open areas highly modified by human intervention in seven ecosystems (Table 0.4) within the Catamayo-Alamor, Jama-Zapotillo, and valley biogeographic sectors, where the average annual rainfall is 451–1000 mm and the average annual temperature is 20.1–24.8 °C.

Conservation.—*Engystomops puyango* was categorized as Near Threatened by Ortega-Andrade *et al.* (2021). This species has a restricted area of extent of occurrence (3900 km^2) and is locally abundant during the breeding season (February), but the overall population trend is unknown. It occurs in disturbed areas, including the Puyango population, which occurs in the highly polluted Río Puyango. Although this is an adaptable species that lives in modified open areas, it is probably affected by the high deforestation rate, clear cutting, and very severe habitat degradation that is taking place within its range mostly due to agriculture, livestock grazing, and pesticide use. It has been recorded in the protected forest of Bosque Petrificado de Puyango and the biosphere reserves Podocarpus-El Cóndor and Bosque Seco.

Remarks.—In a genetic dataset analysis by Ron *et al.* (2010), *Engystomops puyango* is the sister taxon of an undescribed species from 8.5 km north of Motupe, Departamento Lambayeque, Peru. It is also closely related to *E. pustulatus*. These three species are members of the *Vivavox* sublineage that is part of the *Duovox* lineage (as defined by Ron *et al.*, 2006), which includes species having flank and parotoid glands and loud calls compared to other members of *Engystomops*. Ron *et al.* (2010) report a putative hybrid case between *Engystomops puyango* and *E. randi*. In a dense chorus of both species at Bosque Protector Puyango, on 19 February 2005, a male with an unusual advertisement call was found. The average call duration was below the range for *E. puyango* and close to the average of *E. randi* based on recordings from seven males from the same locality. The call interval was closer to that of *E. randi*. However, similarly to *E. puyango*, the first component of the call lacked the well-defined pulses present in calls of *E. randi*, and its SVL (28.31 mm) was outside the range of adult males of *E. randi*. Because the characteristics of the advertisement call of this individual are intermediate between *E. puyango* and *E. randi*, Ron *et al.* hypothesized that it is a hybrid individual. A sequence of 750 bases of gene 16S mtDNA is identical to that of *E. randi* from the same population. Thus, the hybrid could be the product of the cross of a female *E. randi* with a male *E. puyango*. According to Ron *et al.* (2010), *Engystomops puyango* has a chromosome number of 2n = 20, which subsequently was found to be characteristic of the *Duovox* lineage by Targueta *et al.* (2012). All species analyzed of *Eladorhina*, *Eupemphix*, *Pleurodema*, *Pseudopaludicola*, *Physalaemus*, and *Engystomops* have the karyotype 2n = 22. It has a genetic distance (pairwise uncorrected *p*-distance) from other species of the *Duovox* lineage of *Engystomops* of at least 6.5% (*E. pustulatus*) in fragments of aligned 12S and 16S mitochondrial DNA sequences.

Key references.—(Ron *et al.*, 2010) (description).

ENGYSTOMOPS RANDI (RAND'S BULLY FOAM TOAD) (RON *ET AL.*, 2004)

Physalaemus randi (Ron *et al.*, 2004): 263. **Holotype.**—QCAZ 19563 from Cerro Masvale, 92 m, Provincia Guayas, Ecuador.
Engystomops randi Nascimento *et al.*, 2005: 312.

Etymology.—The specific name honors A. Stanley Rand (1932–2005), who has contributed to *Engystomops* collections in western Ecuador and, most importantly, was a pioneer in behavioral studies of *Engystomops*. His extensive research, in collaboration with Michael J. Ryan, has greatly enriched the understanding of animal communication systems.

Identification.—Figures 2.33, 2.34. The maximum known SVL is 18.7 mm in males and 19.7 mm in females. The skin on the dorsum bears numerous tubercles; the snout is rounded in dorsal and lateral views; the vomerine teeth and odontophores are absent; the maxillary teeth are present; the parotoid and flank glands are present; the tarsal tubercle is absent; the nuptial pads are present; finger I is shorter than II; the tympanic annulus is evident, concealed dorsally; the tympanic membrane is not tuberculate; the dentigerous processed of the vomers are thin and spikelike.

The dorsum varies from light gray to dark gray or dark gray–brown with dark marks; the tubercles are all lighter than the background; there is an orange middorsal line, which is continuous or interrupted and varies between being restricted to the sacral region and almost reaching the head. In some individuals, there is a lighter orange area in the occipital region, limited by rows of tubercles forming an inverted V toward the scapular region, its margins are black, light gray inside; the interorbital and intercanthal regions are brown; there is a white labial line below the eye and tympanum; the tympanum is dark gray; the flanks are dark gray; the arms are tan–orange with darker forearms; the exposed surfaces of the thighs are brown; there are dark brown transversal bars on the exposed surfaces of the forelimbs and hind limbs. The ventral surfaces have a white cream background with dark brown and gray markings. In some, the markings are restricted to the head or absent only on the posterolateral margins of the abdomen. The variation between the extremes is continuous. On the abdomen, dark markings may be arranged in well-defined large spots or in diffuse speckled patterns. The relative length of the cream midventral stripe, beginning at the tip of the snout, varies from being restricted to the head to extending to two-thirds of the SVL. The lateral head coloration varies extensively between light gray and dark gray to dark brown. The suborbital cream stripes are thin, well defined, with the posterior stripe extended below the tympanum and the parotoid gland. In some individuals, the two suborbital cream stripes are fused into a single, wide band. In addition, two ill-defined brown labial bars are present anteriorly. Some have black canthal stripes. The iris is dark brown.

FIGURE 2.33 *Engystomops randi*. Top: QCAZ 19563, male, 17.6 mm SVL, Cerrro Masvale, Provincia Guayas. Bottom: CJ 9940, male, 90.2 mm SVL, Reserva Ecológica Manglares Churute, Provincia Guayas. DAL.

FIGURE 2.34 Morphological variation in *Engystomops randi*. Top: CJ 9989, male, 26.1 mm SVL. Middle: CJ 9994, male, 21.2 mm SVL. Bottom: CJ 9995, male, 22.0 mm SVL, Puyango, Provincia Loja. SGS.

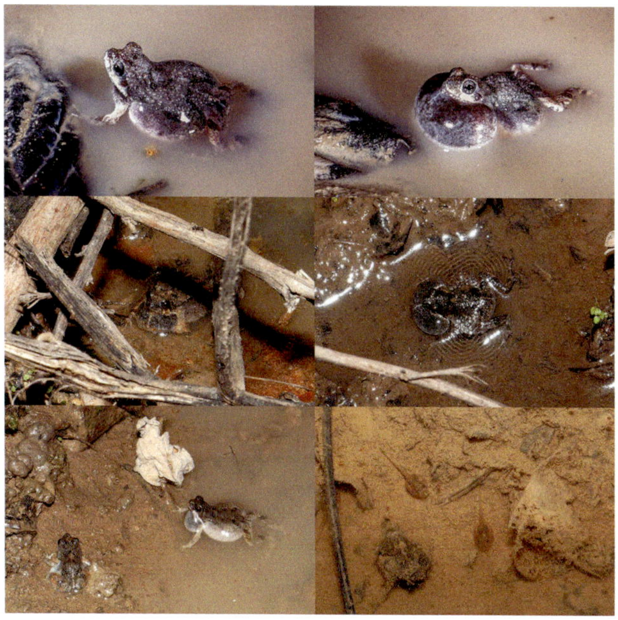

FIGURE 2.35 *Engystomops randi*. Calling males and tadpoles in their microhabitats. Top: Cerro Masvale, Provincia Guayas. Middle and bottom: Puyango, Provincia Loja. LAC.

Comparisons.—*Engystomops guayaco*, *E. montubio*, *E. coloradorum*, *E. pustulosus*, *E. petersi*, *E. pustulatus*, *E. puyango*. *Engystomops randi* is mostly similar to *E. guayaco*, *E. montubio*, and *E. coloradorum*. *Engystomops randi* has larger and more abundant dorsal tubercles than *E. guayaco*. *Engystomops randi* has a shorter SVL, proportionally larger flank and parotoid glands, and a wider dorsum than *E. montubio*. *Engystomops randi* has less tuberculated skin than *E. coloradorum* and a loreal region that slopes gradually toward the labial region (nearly vertical in *E. coloradorum*). *Engystomops randi* differs from the larger *E. pustulosus* and *E. petersi* by the absence of a tarsal tubercle and the presence of teeth on the maxilla and premaxilla. *Engystomops randi* is smaller than *E. pustulatus* and *E. puyango* (non-overlapping adult male SVL 15.45–18.65 mm, 25.17–29.88 mm, 23.78–30.48 mm, respectively). Furthermore, *E. randi* differs from *E. puyango* by having less extensive lateral fringes on the toes (extensive in *E. puyango*) and by the advertisement call.

Natural history.—Figure 2.35. Individuals are found in artificial open areas. At Cerro Masvale, Provincia Guayas, frogs were breeding in the vicinities of buildings, pastures, and agricultural lands. At 11 km north of Cerro Masvale, males were calling from a flooded rice field. At Puerto Inca, Provincia Guayas, males were calling from flooded grassland next to a banana plantation. Reproductive activity is nocturnal. Choruses of males were found in Cerro Masvale, 11 km north of Cerro Masvale, and in Puerto Inca (20 February 2002, 21–25 March 2003) during the rainy season. Males call from small ponds and ditches while floating in a few centimeters of water, usually concealed by vegetation. Amplexus and egg deposition take place at the same sites where choruses call. *Engystomops randi* constructs floating foam nests. The nest is constructed during amplexus; while the female deposits the egg masses, the male beats them with his legs. At some ponds in Cerro Masvale, calling males were abundant (1 individual/m^2), and some of them called within 20 cm of each other.

Under lab conditions, during amplexus, the eggs became enclosed in a foam nest that derived from the egg jelly and floated in the water. Foam nests are 5 cm in diameter and 4–5 cm high. Within the foam nest, embryonic development synchronously advanced until the tadpole stage. Nests contained 110 ± 69 eggs of 1.1 mm in diameter (17 foam nests with a total of 1952 eggs). Upon hatching, the tadpoles moved into the underlying water. Development to hatching required four days. The white eggs and embryos of these frogs were effectively camouflaged in the white foam of the nest. In later stages, the embryos developed pigment around the nuclei.

The call of *Engystomops randi* was recorded at Cerro Masvale, Provincia Guayas. It emits a call with two components: a pulsed sequence of amplitude-modulated pulses followed by a nearly pure-tone whine with a downward frequency sweep. Detailed data and figures are provided under the subtitle "Vocalizations of Ecuadorian Anurans" in Appendix 1.

FIGURE 2.36 Ontogenetic changes in two individuals of *Engystomops randi* from tadpole to adult. Top to bottom, left to right. CJ 10000, lateral, dorsal, and ventral views, stage 41, TL = 19.9 mm; stage 43 = 14.2 mm; froglet; juvenile 103 days after froglet = 18.4 mm SVL; 197 days = 19.1 mm SVL; 293 days = 22.7 mm SVL. Puyango, Provincia Loja. DAL.

Tadpoles.—Figure 2.36. Andrea Terán-Valdez and Luis A. Coloma (this publication). This description is based on an individual (not preserved, series CJ 10000) from Puyango, Provincia Loja, Ecuador, in stage 41. It was collected by Luis A. Coloma, Mark Wanner, Patricio Vargas Mena, and Chauna Schuette on 13 February 2019. It measures 7.3 mm in body length and 19.9 mm in total length. The body is ovoid and depressed, wider anteriorly than posteriorly, slightly deeper anteriorly, with a width surpassing its depth. The snout is rounded in both dorsal and lateral perspectives, and the eyes are directed dorsolaterally. In this stage, the spiracle is no longer present. The vent tube is medial. The oral disc has an anteroventral position. The tail constitutes approximately 64% of the total length, reaching its maximum depth at the second third of its length and ending in an acutely rounded tip. The dorsal fin does not extend onto the body, maintaining a relatively constant depth along the tail, with the deepest point occurring around the midlength, tapering gradually toward the end. The ventral fin, thinner than the dorsal fin, maintains a more or less constant depth along the tail. The caudal musculature is deepest at its anterior part.

Dorsally, the tadpole displays brown and golden flecks, while ventrally, the skin is translucent with silvery flecks on the sides diminishing toward the middle, revealing visible organs. The posterior legs and tail have scarce flecks, and the tip of the tail is characterized by a characteristic black coloration.

Distribution.—See Figure 2.22. *Engystomops randi*, an endemic species to Ecuador, is known along the coastal Pacific lowlands in Guayas, El Oro, and Loja provinces. Its elevational range is 11–984 m in an area of extent of occurrence of about 13,100 km^2.

This nocturnal, terrestrial species inhabits open areas highly modified by human intervention in four ecosystems (Table 0.4) within the Jama-Zapotillo and Catamayo-Alamor biogeographic sectors, where the average annual rainfall is 450–1781 mm and the average annual temperature is 22.3–25.7 °C.

Conservation.—*Engystomops randi* was categorized as Least Concern by Ortega-Andrade *et al.* (2021) because, although its currently known extent of occurrence is less than 20,000 km^2, it is common and does not appear to be in decline. This species is seasonally abundant and is an adaptable species that is unlikely to be significantly threatened. It has been recorded in the public reserves Reserva Ecológica Manglares Churute and Reserva Ecológica Arenillas, the protected forest of Bosque Petrificado del Puyango, the biosphere reserves Macizo del Cajas and Bosque Seco, and the private reserve Cerro Masvale.

Remarks.—In a genetic dataset analysis by Ron *et al.* (2010), *Engystomops randi* is the sister taxon to *E. montubio* and closely related to *E. coloradorum* and *E. guayaco*. These four species are members of the *Brevivox* sublineage that is part of the *Duovox* lineage (as defined by Ron *et al.* (2006), which includes small species having calls of short duration compared to most other members of *Engystomops*. It has a genetic distance (pairwise uncorrected *p*-distance) from other species of the Brevivox lineage of *Engystomops* of at least 2.7% (*E. montubio*) in fragments of aligned 12S and 16S mitochondrial DNA sequences. del Pino *et al.* (2007) analyzed the gastrulation mode of *E. randi* and found that notochord elongation overlapped with involution at the blastopore lip, as in *X. laevis* embryos. Romero-Carvajal *et al.* (2009) described also the early development and provided information for their maintenance and breeding in the laboratory.

Key references.—(Ryan and Rand, 2001) (call, under *pustulatus*); (Ron *et al.*, 2004) (description); (del Pino *et al.*, 2007) (embryology); (Romero-Carvajal *et al.*, 2009) (embryology, maintenance and breeding in the laboratory).

GENUS *LEPTODACTYLUS* (THIN-TOED FROGS) FITZINGER, 1826

Leptodactylus Fitzinger, 1826: 38. **Type species.**—*Rana typhonia* Latreille in Sonnini and Latreille, 1801 (= *Rana* Schneider, 1799) by subsequent designation of Fitzinger, 1843: 31.

Etymology.—The generic name is derived from the Greek adjective *lepto*, meaning thin, fine, slender, and *dactylos*, meaning finger, toe. The name refers to their thin fingers and toes. The genus is masculine.

Definition.—Figures 2.37, 2.38. These frogs range from medium to very large relative to most other Neotropical

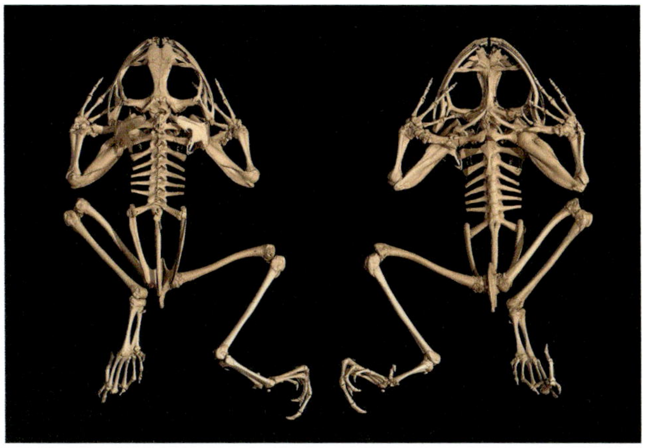

FIGURE 2.37 High-resolution computed tomography reconstructions of the skeleton of *Leptodactylus savagei* UF 103788, 144.4 mm SVL, ca. 3 km east of Quepos, Provincia Punta Arenas, Costa Rica. Dorsal (left) and ventral (right) views. Daniel J. Paluh.

anurans; they attain a maximum SVL of 195.0 mm in males and 174.2 mm in females. According to de Sá *et al.* (2014), the monophyly of *Leptodactylus* is supported by molecular characters and the following 15 synapomorphic phenetic characters: a dark supratympanic stripe extends dorsally above the tympanum and continues posterolaterally behind the tympanum; the nuptial excrescences on the thumb are two lateral keratinized spines; the termination of the pars facialis of the maxilla is at the level of the palatines; the posterior termination of the maxillary teeth reaches or surpasses the anterior tip of the quadratojugal; the position of the tectum nasi is posterior to the alary processes of the premaxillae; the posterior epiotic eminence is prominent and extends laterally beyond the otic capsule body; the anterior expansion of the nasals is present, the anterior apex forms a distinct protuberance; the vomerine teeth are arranged in a shallowly arched series; the number of vomerine teeth is > 8; the ridge on the ventral surface of the palatine is present and is superficial and hardly noticeable; the anterior extension of the anterior ramus of the pterygoid does not reach the palatine; the anteromedial process of the hyoid is absent; the depth of the hyoglossal sinus is deep, extending 2 mm past the anterior borders of the alary process; the neural spine of vertebrae I–V is imbricate; parental care is present, encompassing the provision of nest care.

Content.—The genus contains 84 species, 14 of which occur in Ecuador.

Distribution.—*Leptodactylus* has an extensive distribution throughout Mexico in North America to Central and South America and the West Indies. In South America, it occurs from Argentina, Uruguay, and Brazil (east of the Andes) to Ecuador and northern Peru (west of the Andes). *Leptodactylus* is predominantly a lowland lineage with most species occurring below 2000 m, but one species (*L. colombiensis*) reaches 2600 m.

FIGURE 2.38 Putative synapomorphic characters of *Leptodactylus*. Top: foam nest of *Leptodactylus ventrimaculatus*, CJ 3087, Reserva Otokiki, Provincia Esmeraldas. Bottom: lateral keratinized spines on thumb of male of *L. wagneri*, CJ 4586, 55.2 mm SVL, Comunidad Alto Tena, Provincia Napo. LAC.

Remarks.—Figure 2.39. Heyer (1969) redefined the genus. de Sá *et al.* (2014) provided a total evidence phylogenetic analysis of molecular (mitochondrial and nuclear markers) and non-molecular (adult and larval morphological and behavioral) characters for the genus *Leptodactylus*. They included > 80% of 75 recognized species until 2014. Their results support the monophyly of *Leptodactylus sensu stricto*, with *Hydrolaetare* placed as its sister group.

de Sá *et al.* (2014) summarize and provide citations of some important features of *Leptodactylus*. Their reproductive behavior is often complex and includes parental care.

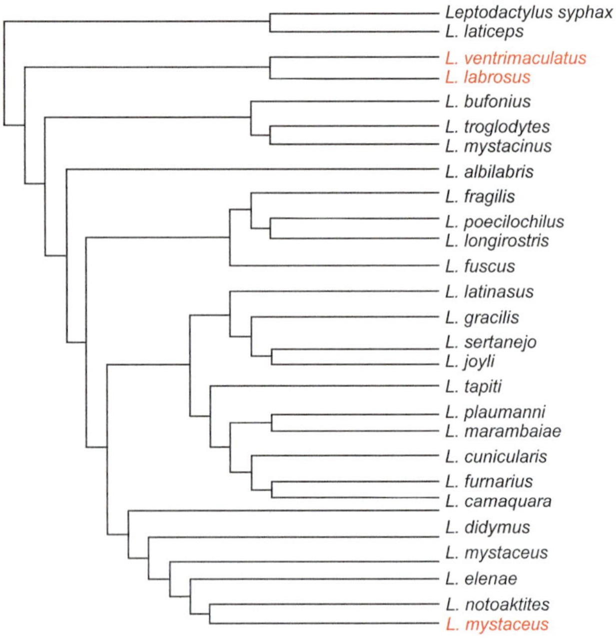

FIGURE 2.39 Topology of the species of *Leptodactylus*, extracted from a phylogenetic tree inferred from a total evidence analysis of mitochondrial DNA sequences, nuclear markers, and adult and larval morphological and behavioral characters. Taken from de Sá *et al.* (2014). Species occurring in Ecuador are in red.

Complex parental care has been reported in the genus, such as the construction of distinct underground burrows and foam nests in which embryos develop; various forms of protection by the parents of the foam nest and subsequently the larvae are exhibited, including adults actively attacking predators and emitting unique vocalizations and tadpole schooling. Furthermore, some species have evolved larval oophagy and carnivory. Some species are active foragers in the leaf litter, whereas some of the larger species are sit-and-wait predators found around ponds or outside of burrows that they occupy permanently. *Leptodactylus* species usually consume small invertebrate prey but can also prey on anurans as well as other small vertebrates. Some *Leptodactylus* species are a source of human food, and the dermal glands of several species are known to produce defensive skin secretions, including antimicrobial peptides and other protein toxins found in *L. fallax*, *L. labyrinthicus*, *L. laticeps*, *L. latrans*, and *L. pentadactylus*. Similarly, the foam nests of *L. vastus* possess a novel surfactant protein. Although some *Leptodactylus* species occur at considerable distances from water sources throughout the year, most species are strongly associated with water bodies; some species are associated primarily or exclusively with lotic systems (e.g., streams and creeks of riparian environments) and others with lentic systems (e.g., lakes, permanent and annual ponds, smaller ephemeral pools, and artificial water sources). Until recently, the fossil record for the genus was limited to remains of extant or indeterminate species from the Quaternary in Mexico, Bolivia, Brazil, Uruguay, Argentina, and the West Indies. However, the history of the genus was extended back to the early Pliocene in Argentina (Farola Monte Hermoso, Buenos Aires Province) based on fossil remains considered close to *L. latrans*.

LEPTODACTYLUS DISCODACTYLUS (DISC-SHAPED THIN-TOED FROG) BOULENGER, 1884

Leptodactylus discodactylus Boulenger, 1884: 637.
Holotype.—BMNH 1947.2.17.40 from Yurimaguas, Huallaga River, Departamento Loreto, Peru.

Etymology.—The specific name is from the Greek *diskos*, meaning disc, and *daktylos*, meaning finger/toe, characterizing the species as having disc-shaped toe tips.

Identification.—Figure 2.40. *Leptodactylus discodactylus* has a maximum known SVL of 34.2 mm in males and 39.9 mm in females; the male lacks a spatulate snout; the males lack spines on the thumb and chest; a light upper lip stripe is absent; there are no dorsal, dorsolateral, or lateral folds; the posterior thigh is usually mottled, and some individuals have dark and/or light stripes; the upper shank is barred; the belly is mottled; the toes are fringed.

The dorsum is dark brown, and the flanks are grayish-brown or reddish-tan, usually with a brown spot in the groin and one or two brown spots on the flank. A narrow pinkish-cream or gray interorbital bar usually is evident. The side of the head is dark brown, usually with four narrow diagonal or vertical cream bars (two below the orbit, one below the nostril, and one between the nostril and orbit); another cream bar is present on the tip of the snout. The limbs are dark brown, with darker transverse bars barely evident in some specimens. The posterior surfaces of the thighs are brown; black flecks are evident distally in some specimens. The throat is dark brown with cream flecks, and the belly and ventral surfaces of the limbs are cream with brown flecks laterally. Some adults of both sexes have bright yellow glandular areas laterally on the body. The iris is bronze with brown reticulations.

Comparisons.—*Leptodactylus leptodactyloides*, *L. petersii*, *L. wagneri*. *Leptodactylus discodactylus* is mostly similar to *L. leptodactyloides*, *L. petersii*, and juveniles of *L. wagneri*. The dorsal toe discs of *L. discodactylus* have 1–5 longitudinal grooves (visible under magnification), whereas the other listed similar species lack dorsal longitudinal grooves (on the dorsal toe tips). In Amazonian Ecuador, *Leptodactylus discodactylus* differs from the other species of *Leptodactylus* (except *L. wagneri* and *L. leptodactyloides*) by having lateral fringes on the toes. These three species differ in that *L. discodactylus* has expanded, round, terminal discs on the digits.

Natural history.—At night, individuals have been found resting on low vegetation, in a swamp, and on the forest floor. By day, they also have been found in swampy areas. Stomachal analysis of eight individuals revealed that five

FIGURE 2.40 Morphological details of *Leptodactylus discodactylus*. Top and middle left: CJ 6187, male, 39.2 mm SVL, Shiripuno Lodge, Provincia Orellana. Middle right and bottom: El Edén, Bloque 15, Provincia Orellana. LAC.

stomachs contained eight beetles, comprising 41% of the bulk. Two contained one homopteran each, representing 8.7% of the bulk. Other food items were represented by one individual each and included Diptera, Orthoptera, formicid, Hemiptera, Isopoda, and Annellida.

In Limoncocha, Provincia Sucumbíos, Ecuador, males were observed calling near a stream seepage within a forested area. The seepage area was 2–3 m in width, about 15 m in length, and the water was 1–2 cm deep. The males were calling along the seep's edges, under roots or leaves at the water's edge. They remained concealed in naturally occurring chambers. The calls were relatively weak, and males called day and night following heavy rainfall. No calling was observed during rainless periods or periods of weak rain. Two females were found with 226 unpigmented eggs measuring 1.5 mm in diameter and 1177 unpigmented eggs measuring 1.0 mm in diameter.

In Santa Cecilia, Provincia Sucumbíos, Ecuador, six females having 148–420 mature ovarian eggs ($\bar{x}$ = 234.8) were found in April (2), June, July (2), and October; the maximum diameter of the eggs was 1.2 mm. The smallest juveniles (7.3–7.7 mm SVL) were found in July. These limited data suggest that *L. discodactylus* may be a sporadic opportunistic breeder whose breeding activity is associated with periods of heavy rainfall.

Heyer (1997) described the call from four localities in Brazil and one in Limoncocha, Provincia Sucumbíos, Ecuador.

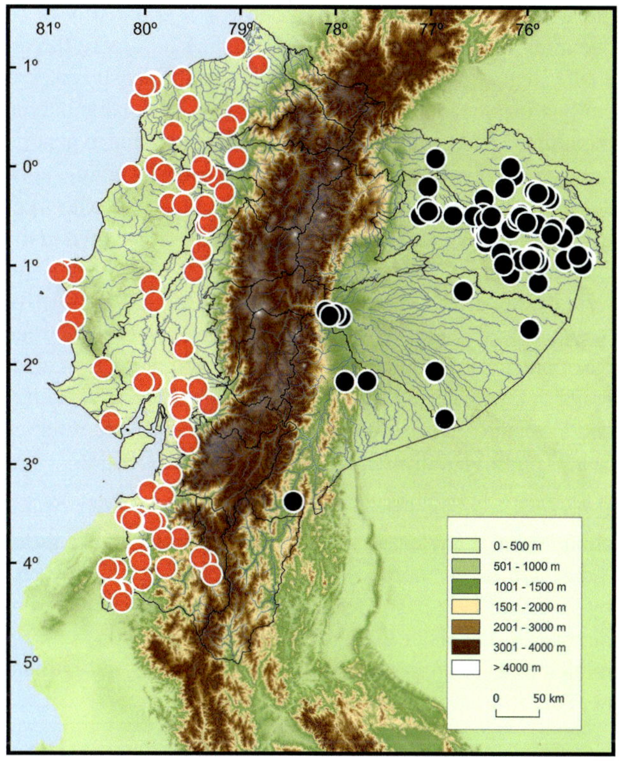

FIGURE 2.41 Localities of known occurrence of *Leptodactylus discodactylus* (black dots) and *L. labrosus* (red dots) in Ecuador.

The call from Limoncocha consists of a series of short notes, "wheeet, wheeet, wheeet," produced at a rate of 60 notes per minute and was described by Straughan and Heyer (1976). Detailed data and figures are provided under the subtitle "Vocalizations of Ecuadorian Anurans" in Appendix 1.

Tadpoles.—The description was based on a tadpole in stage 30 from Mera, Provincia Pastaza, Ecuador. However, Duellman's (1978) description and Heyer's (1998) illustration may or may not correspond to *L. discodactylus*. According to field notes by John D. Lynch (cited by Heyer, 1997), its specific assignment was a guess on the basis of habitat selection.

Distribution.—Figure 2.41. *Leptodactylus discodactylus* is known from Amazonian Brazil (states of Acre and Amazonas), Colombia, Ecuador, and Peru. In Ecuador, it occurs in Sucumbíos, Orellana, Pastaza, and Morona Santiago provinces. Its elevational range is 180–1251 m m in an area of extent of occurrence of about 67,080 km^2.

This nocturnal, terrestrial species inhabits intervened areas in the Amazonian region and swamp forest habitats in 10 ecosystems (Table 0.4) within six biogeographic sectors (Table 0.5), where the average annual rainfall is 1934–4422 mm and the average annual temperature is 20.1–25.6 °C.

Conservation.—*Leptodactylus discodactylus* is categorized as Least Concern throughout its entire range (Coloma *et al.*, 2004c; in IUCN Red List of 2016) and in Ecuador by Ortega-Andrade *et al.* (2021) in view of its wide distribution, presumed large population, and because it is unlikely to be declining fast enough to qualify for listing in a more threatened category. It is a reasonably common species. It is a generally widespread species with large areas of suitable habitat remaining. There is some localized habitat loss due to general human activities such as the collection of wood, logging, agriculture (crops, livestock, etc.), oil exploration, bauxite mining (Rhondonia), and colonization. In Peru, it is known in the Gueppi Reserved Zone. In Brazil (Acre), it is known in Floresta Nacional em Rhondonia and Chico Mendes Extractive Reserva. In Ecuador, it occurs in the public reserves Parque Nacional Yasuní and Reserva de Producción de Fauna Cuyabeno; in the protected forests of Río Aguarico, Sacha Lodge, and Comuna Kichwa Añangu; and in some private reserves (e.g., Tiputini Biodiversity Station, Shiripuno Lodge).

Remarks.—A total evidence analysis of molecular (mitochondrial and nuclear markers) and non-molecular features (adult and larval morphological and behavioral characters) by de Sá *et al.* (2014) revealed that *Leptodactylus discodactylus* is the sister taxon to *L. grisegularis* and is closely related to *L. podicipinus* and *L. diedrus*. Heyer and Diment (1974) reported the karyotype 2n = 22.

Key references.—(Boulenger, 1884) (description); (Heyer and Bellin, 1973) (reproduction, natural history); (Straughan and Heyer, 1976) (call); (Duellman, 1978) (reproduction, habitat, food); (Heyer, 1997) (call); (de Sá *et al.*, 2014) (phylogeny, account).

FIGURE 2.42 *Leptodactylus knudseni*. Top: QCAZ 31687, juvenile. Bottom: QCAZ 35362. Parque Nacional Yasuní, Provincia Orellana. LAC.

LEPTODACTYLUS KNUDSENI (KNUDSEN'S THIN-TOED FROG) HEYER, 1972

Leptodactylus knudseni (Heyer, 1972): 3. **Holotype.**—LACM 72117 from Limoncocha, Provincia Napo, Ecuador.

Etymology.—The specific name honors Dr. Jens W. Knudsen (1928–1994), whose mentoring influenced W. Ronald Heyer's pursuit of biology as a profession.

Identification.—Figure 2.42. *Leptodactylus knudseni* has a maximum known SVL of 170.0 mm in males and 154.0 mm in females; it lacks a spatulate snout; the adult males have a black spine on each thumb; the adult males have a pair of large black chest spines; the upper lip lacks a distinct light stripe; the dorsal folds are absent; dorsolateral folds are present, usually complete, sometimes interrupted, originating behind the eye; the lateral folds are interrupted or absent; the posterior thigh is usually dark with small to large light vermiculation or spots; the upper shank is barred; the toes have or lack weak lateral ridges.

The dorsum in adults is tan to reddish-brown or green, with dark brown markings consisting of an irregularly shaped spot on the top of the head anterior to the orbits, a transverse bar between the orbits, three to five broad transverse marks on the body at the level of the sacrum, and several small spots postsacrally. All markings on the body are between the orange–brown dorsolateral folds. The dorsal surfaces of the limbs are tan to reddish-brown with narrow brown transverse bars. The flanks are dull orange, and a pale orange–brown longitudinal stripe is present on the upper surface of the forearm. The upper lip is tan with a brown margin and brown spots, at least posteriorly, and a dark brown canthal stripe is present. The venter is cream with diffuse brown mottling, except for the throat, which is dark brown with creamy white flecks. In juveniles (from Ecuador), the head is mostly yellowish-green, the dorsum has greenish-yellow bands enclosing brownish-green areas that are black bordered, the chin has yellow marks along the edge, the posterior thigh is jet black, the belly is gray with lighter punctations, and the iris is gold–yellow above and rusty gold below. Notable ontogenetic changes occur in the color pattern. Juveniles tend to have a greenish-yellow dorsum and gray flanks, and the markings on the face and body are distinct; in the largest individuals, the pattern is barely discernible. Also, in juveniles, the posterior surfaces of the thighs are black, whereas in large adults, they are black to dark brown with a few creamy white spots dorsally.

Comparisons.—*Leptodactylus pentadactylus*, *L. stenodema*. Juveniles of *L. knudseni* often have a green dorsal coloration in life and solid black posterior thighs, whereas juveniles of *L. pentadactylus* are not green in life, nor do they have solid black posterior thigh patterns. Sexually active males of *L. knudseni* have a pair of chest spines, which are absent in *L. pentadactylus*. The dorsolateral folds of *L. knudseni* originate behind the eye; the dorsolateral folds of *L. stenodema* originate above the posterior edge of the tympanum.

Natural history.—At Yasuní, Provincia Orellana, individuals were observed in terra firme forest at the entrance of their burrows. A photograph taken therein reveals that the amplexus is axillary. At Cusco Amazónico, Peru, a few juvenile individuals were observed within the camp compound and in terra firme forest, where one male was sitting at the edge of a small pond, a juvenile and a small female were situated in the leaf litter, and the others remained on the ground. All the records are from nights in the rainy season (November–February). *Leptodactylus knudseni* has the largest average prey size of all terrestrial frogs at Cusco Amazónico. Beetles, ants, and millipedes accounted for 14 (46.7%) of the 30 prey items found in seven individuals, whereas tettigoniid orthopterans and millipedes made up 75.4% of the volume. Other food items included a small gastropod and unidentified coleopterans, hemipterans, and insect larvae.

Apart from an uncollected male, presumably of this species, calling on 11 November 1989 and a female obtained on 30 January 1986 containing numerous small ovarian eggs, no reproductive activity was noted at Cusco Amazónico, Peru, during the period from 1983 to 1991. Observations made at other sites in the Amazonian region (Manaus, Brazil; Panguana, Peru) reveal that at the beginning of the rainy season, amplectant pairs of *Leptodactylus knudseni* deposit foam nests in depressions or excavated basins along the periphery of temporary ponds or near to ponds located on clay soil, at least 50 m away from a stream. Tadpole development commences within the foam nests, with access to the pond facilitated by rainwater flooding the basins. Examined clutches showed an absence of trophic eggs, and tadpoles did not produce foam.

Heyer (2005) and Heyer and Heyer (2006) described the call from nine localities in Brazil, Peru, and French Guiana. *Leptodactylus knudseni* emits a rising whoop consisting of single notes almost entirely or partially pulsed at a rate of 16–66 calls/min. Detailed data and figures are provided under the subtitle "Vocalizations of Ecuadorian Anurans" in Appendix 1.

Tadpoles.—Figure 2.43. In the Reserva Florestal Adolpho Ducke, north of Manaus, Brazil, tadpoles are present in isolated and streamside ponds throughout the year. The tadpole was described based on an individual at stage 39. It has a body length of 17.0 mm and a total length of 59.3 mm. In dorsal view, the body is ovoid, widest (11.0 mm) at about midlength and wider than high; the greatest height (8.5 mm) is at about two-thirds of the length of the body. The snout is bluntly rounded in dorsal view and in profile. The nares are situated about midway between the tip of the snout and the eyes, which are situated and directed dorsolaterally. The interorbital distance is about twice the diameter of the eye. The base of the sinistral spiracle is at about the midlength of the body; the short spiracular tube has a small opening directed posterodorsally at the midline. The vent

FIGURE 2.43 Ontogenetic changes in *Leptodactylus knudseni* from tadpole to juvenile. Top to bottom, left to right. CJ 9417, stage 28, lateral, dorsal, and ventral views, TL = 51.3 mm; stage 40, TL = 76.7 mm; stage 42, TL = 80.2 mm; stage 46, 75 days after stage 28, 25.5 mm SVL; juvenile, 67 days, 49.1 mm SVL; juvenile, 132 days 49.7 mm SVL. Parque Nacional Yasuní, Provincia Orellana. SGS.

tube is short, median, and incorporated into the ventral fin. The caudal musculature is moderately slender, about equal in height throughout the proximal one-third of the tail and tapering gradually to a slender tip. The dorsal fin originates on the proximal caudal musculature; it reaches its greatest height at about two-thirds of the length of the tail and then gradually narrows to a rounded tip. The ventral fin originates on the body wall and is about equal in height throughout most of its length. At the midlength of the tail, the height of the ventral fin is equal to the height of the caudal musculature, whereas the dorsal fin is slightly higher. The oral disc is anteroventral and lacks lateral folds. With the exception of the median one-half of the anterior labium, the disc bears a single row of moderately large, blunt, marginal papillae; submarginal papillae are absent. The jaw sheaths are moderately slender; both sheaths are in the form of broad arches. The labial tooth row formula is 2(2)/2–3, with P-2 and P-3 with or without median gaps. In life, the dorsum of the body and the caudal musculature are gray or dark olive. The belly is silver, and the fins are translucent.

Distribution.—Figure 2.44. *Leptodactylus knudseni* is known in the Gran Sabana of Venezuela and adjacent Lavrado in northern Brazil and from southern Venezuela south to Bolivia and Brazil, extending eastward from Amazonian Ecuador, Colombia, and Peru, through Guyana, Suriname, and French Guiana to Trinidad. In Ecuador, it occurs in Sucumbíos, Orellana, Napo, Pastaza, Morona Santiago, and Zamora Chinchipe provinces. Its elevational range is 196–992 m in an area of extent of occurrence of about 71,470 km^2.

This nocturnal, terrestrial species inhabits primary rainforests, secondary forests, and open and intervened areas in the Amazonian region in evergreen forests of lowlands,

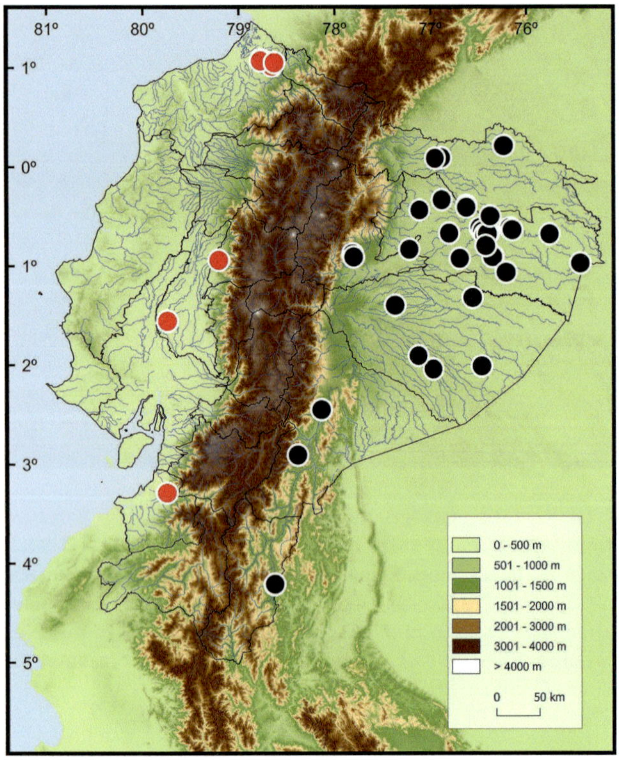

FIGURE 2.44 Localities of known occurrence of *Leptodactylus knudseni* (black dots) and *L. melanonotus* (red dots) in Ecuador.

evergreen montane foothill forests, and flooded forests in six ecosystems (Table 0.4) within six biogeographic sectors (Table 0.5), where the average annual rainfall is 1808–4278 mm and the average annual temperature is 22.1–25.6 °C.

Conservation.—*Leptodactylus knudseni* is categorized as Least Concern throughout its entire range (Heyer *et al.*, 2010; in IUCN Red List of 2022) and in Ecuador by Ortega-Andrade *et al.* (2021) in view of its wide distribution, tolerance of a degree of habitat modification, presumed large population, and because it is unlikely to be declining to qualify for listing in a more threatened category. It is common throughout its range. However, Duellman (2005) reported an apparent decline in their abundance in an undisturbed forest at Cusco Amazónico, Peru. Also, selective logging may affect larval growth and survival (Ernst *et al.*, 2007). The species occurs in many protected areas. In Venezuela, it occurs in the Canaima National Park and in biosphere reserves like the Orinoco-Casiquiare. In Ecuador, it occurs in the public reserves Parque Nacional Yasuní and Reserva Biólogica Limoncocha; the protected forests of Cuembí and Comuna Kichwa Añangu; the biosphere reserves Sumaco, Yasuní, and Podocarpus-El Cóndor; and in some private reserves (e.g., Tiputini Research Station).

Remarks.—In a total evidence analysis of molecular (mitochondrial and nuclear markers) and non-molecular features (adult and larval morphological and behavioral characters) by de Sá *et al.* (2014), *Leptodactylus knudseni* is the sister taxon to *L. fallax* and three species of the *L. labyrinthicus* complex. Heyer (1972) reported the karyotype 2n = 22, fundamental number 44, commonly with three pairs of metacentrics, four pairs of submetacentrics, and four pairs of subtelocentrics.

Key references.—(Heyer, 1972) (description, karyotype, ecology); (Hödl, 1990) (reproduction); (Hero, 1990) (tadpole); (Parmelee, 1999) (food, as *Leptodactylus pentadactylus*); (Heyer, 2005) (morphology, tadpole, call); (Duellman, 2005) (reproduction); (Heyer and Heyer, 2006) (call, tadpole); (Ernst *et al.*, 2007) (larval growth, survival); (de Sá *et al.*, 2014) (account); (Pinto and Menin, 2017) (reproduction); (Read, 2018) (notes); (Schiesari *et al.*, 2022) (tadpole).

LEPTODACTYLUS LABROSUS (THICK-LIPPED THIN-TOED FROG) JIMÉNEZ DE LA ESPADA, 1875

Leptodactylus labrosus Jiménez de la Espada, 1875: 36. **Lectotype.**—MNCN 3524 from Pimocha, shores of Río Daule, Provincia del Guayas, Ecuador.

Etymology.—The specific name is from the Latin *labrosus*, meaning thick-lipped.

Identification.—Figures 2.45, 2.46, 2.47. *Leptodactylus labrosus* has a maximum known SVL of 67.4 mm in males and 71.2 mm in females; the males and females have spatulate snouts; the males lack thumb spines and chest spines; the light upper lip stripe is indistinct or absent; the dorsal folds are absent; the dorsolateral folds are indistinct or absent; the lateral folds are interrupted or absent; the toes lack lateral fringes, and they may have vestigial basal ridges and basal web between toes I, II, and III.

The dorsum is spotted or rarely uniform; the spots are rarely fused. The dark suborbital bar is present or absent. A light stripe on the posterior face of the thigh is typically absent in almost all specimens (94%) or, in a few specimens, indistinct (6%). The upper shank is barred. The tibia is barred; its dorsal surface usually has many or scattered white tubercles, sometimes absent. The posterior surface of the tarsus usually has scattered white tubercles (78%); sometimes, they are absent (22%). The plantar surface of foot usually lacks white tubercles (91%); rarely, they are present (9%). The belly has uniform light to small melanophore blotches scattered across the anterior belly and extending centrally.

Comparisons.—*Leptodactylus ventrimaculatus. Leptodactylus labrosus* can be distinguished from *L. ventrimaculatus* by the typically smooth plantar surface of the foot observed in 91% of specimens, lacking tubercles. In contrast, *L. ventrimaculatus* exhibits distinct, scattered, or very few white tubercles on the plantar surface. Additionally, *L. labrosus* is characterized by a spatulate snout, a feature absent in *L. ventrimaculatus*.

FIGURE 2.45 *Leptodactylus labrosus.* Left: CJ 5473, female. Right: CJ 5474, female, 57.3 mm SVL. 6 km north of Huaquillas, Provincia El Oro. LAC.

FIGURE 2.46 Morphological variation in *Leptodactylus labrosus.* Top left: CJ 8801, male, 45.4 mm SVL, Bosque Protector Siete Cascadas, Provincia Esmeraldas. SGS. Top right: CJ 9983, male, 57.5 mm SVL. Bottom left: CJ 9985, male, 57.4 mm SVL. Puyango, Provincia Loja. Bottom right: CJ 9943, Reserva Ecológica Manglares Churute, Provincia Guayas. DAL.

FIGURE 2.47 Morphological details of *Leptodactylus labrosus.* Top and middle: CJ 9983, male, Puyango, Provincia Loja. Bottom: CJ 8801, male, Bosque Protector Siete Cascadas, Provincia Esmeraldas. DAL.

Natural history.—Figures 2.48, 2.49. Individuals have been found mainly on leaf litter at stream margins and in streams, rivers, swamps, drainage ditches, and disturbed areas. One individual was observed on a low bush 20 cm above the ground. They are active at night, but some were observed active late in the afternoon. In evergreen habitats, they were found as far as 100 m away from streams. At Arenillas, Provincia El Oro, they were active only during the rainy season; it is one of the first species to begin to breed, and males call during the entire rainy season from crevices underground. *Leptodactylus labrosus* is a sit-and-wait predator, and its diet includes ground-level, fossorial

FIGURE 2.48 *Leptodactylus labrosus*. Top and middle: individuals in their microhabitats, Reserva Ecológica Arenillas. Bottom: CJ 5483, foam nests, 6 km north of Huaquillas. Provincia El Oro. LAC.

FIGURE 2.49 Tadpoles and juveniles of *Leptodactylus labrosus* in their microhabitats. Puyango, Provincia Loja. LAC.

invertebrates (earthworms, gastropods, elaterid and scarabaeoid beetles, tiger beetle larvae, antlion larvae, terrestrial crickets, mole crickets) and flying insects (chrysomelid beetles, dipterans, cicadellids, winged ants, lepidopterans, tettigonid orthopterans), among which ants numerically predominate (up to 69.4%). Volumetrically, the diet is dominated by earthworms, crickets, grasshoppers, mole crickets (Orthoptera), slugs (Gastropoda), and caterpillars (Lepidoptera larvae). A juvenile *Bothrops asper* (322 mm SVL) has been recorded as a predator of an individual measuring 52.2 mm in SVL. *Leptodactylus labrosus* is sympatric with *L. ventrimaculatus* in seasonal evergreen forests at Centro Científico Río Palenque, Provincia Santo Domingo de los Tsáchilas (see also Remarks under *L. ventrimaculatus*).

de Carvalho and Ron (2011) described the advertisement calls from three localities in western Ecuador: Río Cuaque, Provincia Manabí; 20 km west of El Carmen, Provincia Santo Domingo de los Tsáchilas; and 10 km north of Zapotillo, Provincia Loja. The males call buried in underground chambers in open areas, adjacent to seasonal lentic water bodies. The advertisement call is a single, very low-pitched note with absent or barely defined pulses (completely overlaid) emitted at a rate of 118–135 calls/minute. A second type of call (probably aggressive call) from 10 km north of Zapotillo is broadcasted either alone or in groups of 2–4 simultaneous calls. It has a pulsed structure with 2–8 well-defined pulses. Detailed data and figures are provided under the subtitle "Vocalizations of Ecuadorian Anurans" in Appendix 1.

Tadpoles.—Figure 2.50. Andrea Terán-Valdez and Luis A. Coloma (this publication). Chondrocranial morphology was described from Centro Científico Río Palenque, Provincia Los Ríos, Ecuador. This description is based on an individual (series CJ 5472) from Centro Científico Río Palenque, Provincia Los Ríos, Ecuador, in stage 29. They were collected by Elicio E. Tapia and Patricio Vargas Mena on 14 February 2015. It has a body length of 6.2 mm and a total length of 18.4 mm. The body is ovoid and depressed, slightly wider anteriorly than posteriorly, with the deepest point posteriorly and a width surpassing its depth. The snout is rounded in both dorsal and lateral views, and the eyes are directed dorsolaterally. The spiracle is sinistral, with its inner wall free from the body, opening posterolaterally near the midlength of the body. The vent tube is wide and medial. The oral disc has an anteroventral position, featuring marginal papillae present laterally and ventrally, with a broad anterior labium gap. The labia are detached from the body. The

FIGURE 2.50 Ontogenetic changes in two individuals of *Leptodactylus labrosus* from tadpole to adult. Top to bottom, left to right. CJ 5472, stage 29, lateral, dorsal, and ventral views, TL = 18.4 mm. CJ 5473, stage 36, TL = 42.1 mm; stage 41, TL = 50.2 mm; stage 45; froglet five days after stage 45; adult, 218 days. 6 km north of Huaquillas, Provincia El Oro. LAC.

upper jaw sheath has a smooth arch with a sinous margin on the lower edge. The tail constitutes approximately 66% of the total length, reaching its maximum depth around mid-length and ending in a rounded tip. The dorsal fin does not extend onto the body, reaching its deepest point at the mid-length of the tail and tapering gradually toward the end. The caudal musculature is deepest at its anterior part. Golden dots forming lines are visible on the dorsum of body and anterior part of tail as part of the lateral line system.

The dorsum is nearly uniformly brown. Ventrally, the skin is translucent with scarce brown flecks toward the flanks. The fins are translucent, and the dorsal fin exhibits gray and white stippling, while the caudal musculature has gray stippling distributed along the tail. The iris has a golden ring.

Distribution.—See Figure 2.41. *Leptodactylus labrosus* is known from the coastal Pacific lowlands and midelevations of Ecuador and Peru south to the Peruvian departments of Cajamarca, Libertad, and Ancash. In Ecuador, it occurs in Esmeraldas, Pichincha, Santo Domingo de los Tsáchilas, Manabí, Los Ríos, Santa Elena, Guayas, Cañar, El Oro, and Loja provinces. Its elevational range is 0–1674 m in an area of extent of occurrence of about 96,170 km^2.

This nocturnal, terrestrial species inhabits mostly secondary forests and open areas highly modified by human intervention in 16 ecosystems (Table 0.4) within seven biogeographic sectors (Table 0.5), where the average annual rainfall is 231–2783 mm and the average annual temperature is 19.6–25.9 °C.

Conservation.—*Leptodactylus labrosus* is categorized as Least Concern throughout its entire range (IUCN SSC Amphibian Specialist Group, 2018c; in the IUCN Red List of 2022) and in Ecuador by Ortega-Andrade *et al.* (2021) in view of its wide distribution, tolerance of a degree of habitat modification, presumed large population, and because it is unlikely to be declining fast enough to qualify for listing in a more threatened category. It is abundant at Reserva Ecológica Arenillas, Provincia El Oro, Ecuador. Habitat destruction and degradation are threats to this species. Given its geographical distribution, this species might occur in Parque Nacional Cerros de Amotape and El Angolo Game Preserve in Peru. In Ecuador, it has been recorded in the public reserves Parque Nacional Machalilla, Refugio de Vida Silvestre Pacoche, Reserva Ecológica Arenillas, Reserva Ecológica Mache Chindul, Reserva Ecológica Manglares Cayapas Mataje, and Reserva Ecológica Manglares Churute; the protected forests of Bosque Petrificado del Puyango, La Ceiba, and Molleturo y Mollepungo; the biosphere reserves Macizo del Cajas, Podocarpus-El Cóndor, and Bosque Seco; and in some private reserves (e.g., Centro Científico Río Palenque, Reserva Biológica Bilsa).

Remarks.—In a total evidence analysis of molecular (mitochondrial and nuclear markers) and non-molecular features (adult and larval morphological and behavioral characters) by de Sá *et al.* (2014), *Leptodactylus labrosus* is the sister taxon to *L. ventrimaculatus*.

Key references.—(Jiménez de la Espada, 1875) (description, osteology); (Heyer and Peters, 1971) (natural history); (Larson and de Sá, 1998) (tadpole chondrocranial morphology); (Cisneros-Heredia, 2006) (natural history, diet); (Ortega-Andrade *et al.*, 2010b) (natural history); (de Carvalho and Ron, 2011) (call); (de Sá *et al.*, 2014) (phylogeny, account); (Székely *et al.*, 2016) (reproduction, natural history).

LEPTODACTYLUS LEPTODACTYLOIDES (THIN-TOED FROG) (ANDERSSON, 1945)

Eleutherodactylus leptodactyloides (Andersson, 1945): 43.

Holotype.—NHRM 1945 from Río Pastaza, between the Río Puyo and Río Copataza, Provincia Pastaza, Ecuador.

Leptodactylus leptodactyloides Heyer, 1994: 88.

Etymology.—The specific name refers to the similarity of this species to *Leptodactylus* because Andersson considered the type specimen to be a new species in the genus *Eleutherodactylus* that resembled frogs of the genus *Leptodactylus*.

Identification.—Figures 2.51, 2.52. *Leptodactylus leptodactyloides* has a maximum known SVL of 47.9 mm in males and 56.2 mm in females. Adult males do not have a spatulate snout; they have a pair of medium-sized, black, keratinized thumb spines and lack chest spines. A light upper lip stripe is absent between the tip of the snout to under the eye; the light stripe from under the eye to the jaw commissure varies from very distinct to indistinct; the dorsal folds are absent; the dorsolateral folds are absent or extend partially or totally from the eye to the groin; the lateral folds are absent or interrupted; the upper shank is barred to rarely uniform; the belly is mottled; the toes have lateral fringes.

The dorsal ground color varies from gray to tan and reddish-brown. In darker-colored individuals, the snout is paler than the rest of the dorsum. The dorsolateral fold usually is dark brown and delimits the dorsal coloration from that of the flanks, which usually are somewhat paler than the dorsum; in some individuals, the color of the flanks is different from that of the dorsum. For example, some individuals that have a reddish-brown dorsum have gray flanks. A narrow interorbital bar, dark brown or black in color, extends onto the eyelids, creating the base of a paler brown or gray, roughly triangular mark. The apex of this mark typically diffuses posteriorly. Small, irregularly shaped dark marks (brown or dark gray) may be present on the dorsum; most individuals have a few irregular black spots on the flanks. The upper lip is dark brown with two to four vertical cream dashes. In some individuals, the lateral aspect of the head appears dark brown, while in others, it matches the hue of the snout. In the latter scenario, distinct, dark brown canthal and supratympanic stripes are

FIGURE 2.51 Morphological details of *Leptodactylus leptodactyloides*. CJ 12155, female subadult, 31.9 mm SVL. CJ 12154, female subadult, 29.4 mm SVL. Puyo, Provincia Pastaza. SGS.

FIGURE 2.52 Morphological details of *Leptodactylus leptodactyloides*. CJ 12154, female subadult, note coloration of groin and thighs, Puyo, Provincia Pastaza. SGS.

present. In most individuals, a pale cream stripe extends from the posteroventral margin of the orbit onto the anterior base of the upper arm. The light posterior lip stripes rarely are very distinct (1%), sometimes are distinct (11%), usually are indistinct (52%), and often are not discernible (36%). The posterior surfaces of the thighs are mottled black or dark brown with pale tan to cream; there is a tendency for a creamy tan stripe on the distal part of the thighs in about one-half of the individuals. The throat is brown or gray with cream flecks; the rest of the venter is white with brown or dark gray reticulations on the hind limbs, chest, and, in some individuals, the anterior part of the belly. The iris is bronze with fine black reticulations.

Comparisons.—*Leptodactylus petersii*, *L. wagneri*. *Leptodactylus leptodactyloides* has more intense belly patterns anteriorly, and most individuals are moderately mottled; *L. petersii* specimens have more uniformly patterned bellies, often in an anastomotic pattern, and most individuals have extensively mottled bellies. *Leptodactylus leptodactyloides* is smaller than *L. wagneri* (*L. leptodactyloides* females 35–56 mm SVL, males 28–48 mm SVL; *L. wagneri* females 52–82 mm SVL, males 39–61 mm SVL). Very few *L. leptodactyloides* specimens have long dorsolateral folds; most *L. wagneri* have long dorsolateral folds. The bellies of *L. leptodactyloides* characteristically are finely mottled; many *L. wagneri* have boldly mottled bellies.

Natural history.—*Leptodactylus leptodactyloides* is widespread at Cusco Amazónico, Peru, but it is more common in inundated forests; individuals have been observed in the immediate vicinity of temporary and permanent ponds throughout the rainy season and at permanent ponds in the dry season. Three individuals were found on the banks of the Río Madre de Dios and six at the edge of the water in the Quebrada Madama in the rainy season; one was at the edge of the latter stream in the dry season. At night, the majority of adults (72 individuals) were observed either at the edges of shallow ponds, in shallow water, or perched on logs floating in the water. An additional 17 individuals were found on the ground or on leaf litter, away from the ponds. During the day, 16 individuals were located at the edge of ponds or in shallow water, while four others were observed on the ground away from the ponds. Additionally, one individual was found sheltered under a stump. During the day, 13 juveniles were found in the leaf litter. At night, six were spotted on leaves of herbs up to 20 cm above the ground. Also, four juveniles were found at the edges of ponds at night, and three were observed during the day. Between November and January, males of *L. leptodactyloides* were observed calling from concealed locations within shallow water, both in temporary and permanent ponds. On 3 June 1989, around 20 individuals were noted calling from a permanent pond. Gravid females were found in October to January; among them, four with SVLs of 40.7–44.9 mm ($\bar{x}$ = 42.2) contained 724–1053 eggs ($\bar{x}$ = 899.0) with diameters of 0.6–1.0 mm ($\bar{x}$ = 0.85).

In Limoncocha, Provincia Sucumbíos, Ecuador, numerous individuals were found in pastures, while others were located near ponds, puddles, or along stream and river banks within clearings. An analysis of the digestive tracts of 33 individuals revealed that *L. leptodactyloides* exhibited a broader range of prey consumption compared to other frogs at Cusco Amazónico. The frogs demonstrated a diverse diet, including relatively large prey such as snails, annelid worms, and crustaceans, as well as smaller prey like drosophilid flies, ants, and small beetles. Beetles constituted the highest volume in their diet (25.88%), followed by orthopterans (22.80%) and hemipterans (19.32%). Notably, ants were the most commonly consumed prey (26.53%), followed by beetles (20.41%) and insect larvae (19.83%). Males call at the edges of small, grassy ponds within the camp and pasture. Calling patterns were described during the night of 11 July 1971 in a roadside ditch puddle. The observations spanned from 20:00 to 20:40 h, where four males were actively calling, with the closest two positioned 1.5 m apart. Throughout the observation period, the frogs exhibited minimal movement. There was a discernible shift in dominance, both in terms of loudness and calling frequency, from the third male to the second male. Notably, individuals appeared to call randomly, lacking a specific order in their vocalizations. The call of *L. leptodactyloides* is relatively weak; males called actively both day and night following heavy rainfall. If there was a lack of rainfall for two or three days, *L. leptodactyloides* ceased calling. The weak call of *L. leptodactyloides* appears to serve the purpose of attracting nearby females to individual males. *Leptodactylus leptodactyloides* presents as a riparian species, with males and females typically found at or near the margins of grassy ponds. Given that the foam nest of *L. leptodactyloides* is deposited on the water's surface and hatchlings drop into the water a few days after fertilization, the survival of larvae is enhanced when adults reproduce during or immediately following rainstorms. Five females exhibited the following pigmented eggs: (1) 2793 eggs measuring 1.0 mm in diameter, (2) 3483 eggs measuring 1.0 mm in diameter, (3) 3375 eggs measuring 1.0–1.25 mm in diameter, (4) 3602 eggs measuring 1.0–1.25 mm in diameter, and (5) 2813 eggs measuring 1.0–1.5 mm in diameter.

During the morning of 27 July 1971, Heyer observed tadpoles in a 3 × 20 m grassy pond in the pasture. The larvae were associated in tight schools. At the beginning of the observation period, four schools were present. Each school was marked with a stake, and movements were recorded. The larvae within a school were extremely active, so much that a school could be seen from a considerable distance because of the boiling water and black mass of tadpoles. On close inspection, the larvae were moving their mouthparts extremely rapidly. Some larvae were also scraping blades of grass in a frantic manner. In the roiling school, detritus was stirred up in the water. The individual schools appeared ameboid; usually, the whole school would slowly shift position in the pond. In 20 min, one school moved 1.5 m. Some of the schools fragmented, and at one time during the observation period, two schools fused. An experiment with these larvae and aquatic insect predators revealed that the larvae apparently are not distasteful for them.

Heyer (1994) described the advertisement call of males from Limoncocha, Provincia Sucumbíos, Ecuador. The call consists of a long continuous series of wet popping notes. The advertisement call is a single note, unpulsed or with 3–5 partial pulses emitted at a rate of 0.3–3.3 calls/second. Detailed data and figures are provided under the subtitle "Vocalizations of Ecuadorian Anurans" in Appendix 1.

Tadpoles.—The following description is based on specimens from Limoncocha, Provincia Sucumbíos, Ecuador, and from Tambopata, Madre de Dios, Peru. The maximum total length at stage 40 is 28.3 mm; the body length constitutes 35–51% ($\bar{x}$ = 42.6) of the total length; the maximum tail depth is 32–51% ($\bar{x}$ = 41.6) of the body length. The nostrils are located closer to the tip of the snout than to the eyes or midway between the snout and the eyes; the internarial distance is slightly greater than the interorbital distance. The eye diameter in stages 25–31 represents 7–9% ($\bar{x}$ = 7.5) of the body length; in stages 36–40, it is 8–12% ($\bar{x}$ = 9.4). The spiracle is sinistral, the anal tube medial. The oral disc is located anteroventrally; the oral papilla formula is 1–2; the oral disc width in stages 25–31 represents 23–28% ($\bar{x}$ = 24.9) of the body length, whereas in stages 36–40, it represents 18–27% ($\bar{x}$ = 20.6). The oral papilla gap constitutes 24–43% ($\bar{x}$ = 35.0) of the oral disc width; the labial tooth row formula in stages 25–27 is 2(2)/3, and in stages 28–40, 2/3. Row A-2 at stages 25–31 has 62–75 labial teeth; in stages 36–40, 72–113.

The dorsal body pattern exhibits a brown hue, characterized by a not entirely uniform profusion of melanophores. The ventral body pattern displays moderate to heavy mottling anteriorly; a light spot behind the oral disc can be present. There is a very light to moderately scattered distribution of melanophores over the guts. The melanophores are present on the oral disc, while the anal tube bears scattered or very few melanophores. The tail is brown with scattered small flecks, and the pattern is most concentrated over the musculature. The ventral-most edge of the fin may or may not exhibit melanophores.

Distribution.—Figure 2.53. *Leptodactylus leptodactyloides* is known from the Amazonian lowlands and the Guianas. In Ecuador, it occurs in Sucumbíos, Orellana, Pastaza, Morona Santiago, and Zamora Chinchipe provinces. Its elevational range is 193–1480 m in an area of extent of occurrence of about 51,750 km^2.

This nocturnal, terrestrial species inhabits intervened areas in the Amazonian region and forest habitats in seven ecosystems (Table 0.4) within five biogeographic sectors (Table 0.5), where the average annual rainfall is 1451–3708 mm and the average annual temperature is 20.7–25.9 °C.

Conservation.—*Leptodactylus leptodactyloides* is categorized as Least Concern throughout its entire range (Coloma *et al.*, 2004b; in IUCN Red List of 2022) and in Ecuador by Ortega-Andrade *et al.* (2021) in view of its wide distribution, tolerance of a broad range of habitats, presumed large population, and because it is unlikely to be declining fast enough to qualify for listing in a more threatened category. It is common and not seriously threatened. It is possibly impacted locally by fire and habitat degradation, though it is tolerant to a high degree of habitat modification. It occurs in many protected areas, including Canaima National Park in Venezuela. In Ecuador, it has been recorded in the public reserves Reserva de Producción de Fauna Cuyabeno and Parque Nacional Yasuní and the biosphere reserves Sumaco and Yasuní.

Remarks.—In a total evidence analysis of molecular (mitochondrial and nuclear markers) and non-molecular features (adult and larval morphological and behavioral characters)

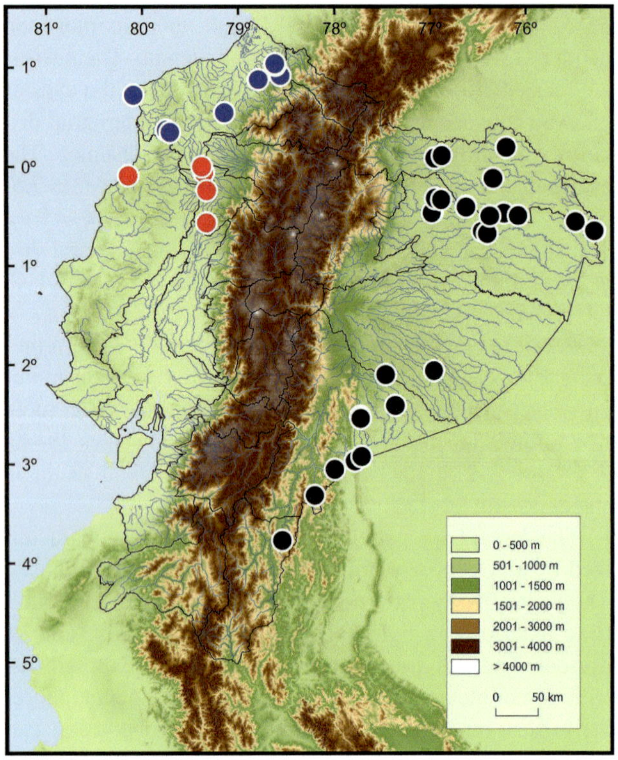

FIGURE 2.53 Localities of known occurrence of *Leptodactylus leptodactyloides* (black dots), *L. peritoaktites* (red dots), and *L. rhodomerus* (blue dots) in Ecuador.

FIGURE 2.54 Morphological details of *Leptodactylus melanonotus*. CJ 10772, male, 33.3 mm SVL, Durango, Provincia Esmeraldas. DAL.

by de Sá *et al.* (2014), *Leptodactylus leptodactyloides* is the sister taxon to *L. petersii* and closely related to *L. natalenis* and *L. pustulatus*.

Key references.—(Andersson, 1945) (description); (Heyer and Bellin, 1973) (reproduction, larval behavior, as *Leptodactylus wagneri*); (Heyer, 1994) (call, tadpole from Limoncocha, Ecuador, and Tambopata, Peru, natural history); (Parmelee, 1999) (food); (Duellman, 2005) (account, tadpole); (de Sá *et al.*, 2014) (account).

LEPTODACTYLUS MELANONOTUS (BLACK-BACK THIN-TOED FROG) (HALLOWELL, 1861)

Cystignathus melanonotus (Hallowell, 1861): 485. **Neotype.**—KU 184848 from Bonanza, Zelaya, Nicaragua.
Leptodactylus melanonotus Brocchi, 1881: 20.

Etymology.—The specific name is from the Greek *melanos*, meaning black, and *notos*, meaning back. The name alludes to the specimen on which the name was based being black on the dorsum.

Identification.—Figures 2.54. *Leptodactylus melanonotus* has a maximum known SVL of 43.4 mm in males and 48.1 mm in females. The males lack a spatulate snout, have a pair of keratinized thumb spines, and lack chest spines. The light upper lip stripe is absent, or there is a light stripe extending from the eye to the tympanum or arm; the dorsal folds are absent; the dorsolateral folds are weak or absent; the lateral folds are interrupted; the toes are fringed.

The dorsum is brown to gray with indistinct darker spots, blotches, bands, or stripes or without pattern; the interorbital area has a dark, light-outlined triangle, a light and dark bar, or a light triangle reaching the tip of the snout anteriorly and bordered posteriorly by a dark triangle. The upper lip is barred to uniform. The upper limb surfaces are barred to uniform. The ventral region shows a range of melanophores, from sparse, creating a light appearance, to a more profuse distribution, especially anteriorly, occasionally extending to a heavy profusion across the entire venter. The posterior thigh exhibits a mottled pattern.

Comparisons.—*Leptodactylus insularum*. *Leptodactylus melanonotus* is mostly similar to *L. insularum*, which is not present in Ecuador. *Leptodactylus melanonotus* has weak or absent dorsolateral folds, and adult males have a pair of keratinized thumb spines, whereas *L. insularum* has well-defined dorsolateral folds extending from the eye to the groin, and adult males have a single black spine on each thumb.

Natural history.—Individuals have been collected from choruses in temporary ponds, inundated areas, and along small streams. Tadpoles have been found in temporary ponds, inundated areas, and isolated pools in stream beds. Both males and females congregate at the margins of shallow bodies of water. Males are extremely shy and call from near the shore, often from concealed places, hiding in vegetation mats, the bases of grass clumps, or slight depressions. Male frogs exhibit territorial behavior, responding to intruders by raising and lowering the posterior portion of their bodies through the movement of the hind legs, exposing the ventrolateral glands. In most cases, the non-resident male retreats or cowers in response. If the intruder persists, the resident male leaps onto his back, leading to a confrontation involving biting, either from the defender or mutual biting between both individuals. Typically, the challenging male retreats at this stage. Amplexus takes place in the shallow waters, with males utilizing their thumb spines to facilitate clasping. During the deposition of eggs, the male employs short, quick strokes of his hind legs to beat glandular secretions and the gelatinous envelope of the eggs into a foam. This foam, resembling the consistency and color of meringue, then floats on the water's surface. The nest is usually deposited in shallow, often isolated, small pools or puddles or in the matted roots of aquatic vegetation. Nests contain 1000–2000 small black-and-yellow eggs, which are 1.0–1.5 mm in diameter. The nests are exceptionally robust, and having the eggs above the water level offers significant protection against aquatic predators. As the season progresses, heavy rain events wash the larvae in advanced development out of the nest. In some cases, the body of water may dry up without the foams decomposing, and heavy rain may release the tadpoles as the breeding site refills. The dark-colored tadpoles of this species form dense schools containing hundreds of individuals. In Honduras, males not only call at night but also exhibit frequent calling thorough the day, particularly at the peak of the breeding season.

Heyer (1970a) and Straughan and Heyer (1976) described the advertisement call from several localities in western Mexico. The advertisement call consists of one or two pulsed notes emitted at a rate of 150–160 calls/minute. Detailed data and figures are provided under the subtitle "Vocalizations of Ecuadorian Anurans" in Appendix 1.

Tadpoles.—The following description is from Costa Rica. Tadpoles in stage 40 have a total length of 40 mm, and the body is ovoid. The nares and eyes are located dorsally. The spiracle is lateral and sinistral. The tube vent is medial, and the caudal fin is rounded at its tip. The oral disc is small, positioned anteroventrally and directed ventrally; the labial tooth row formula is 2/3, with A-2 usually with a narrow median gap; jaw sheaths are small, blunt, and with serrations; there are double or triple rows of moderate papillae laterally, one or two rows ventrally, none above. The body and tail are mostly black with light areas lateral to the eyes; scattered small, light spots on the tail; and a dark venter.

Distribution.—See Figure 2.44. *Leptodactylus melanonotus* is known along the Atlantic and Pacific coasts of Mexico through Panama and the Pacific lowlands of Colombia and Ecuador. In Ecuador, it occurs in Esmeraldas, Cotopaxi, Los Ríos, and El Oro provinces. Its elevational range is 17–229 m in an area of extent of occurrence of about 14,300 km^2.

This nocturnal, terrestrial species inhabits forests and open areas modified by human intervention in the evergreen forests of lowlands ecosystem of the Equatorial Chocó and Jama-Zapotillo biogeographic sectors, where the average annual rainfall is 682–2704 mm and the average annual temperature is 23.9–25.7 °C.

Conservation.—*Leptodactylus melanonotus* is categorized as Least Concern throughout its entire range (IUCN SSC Amphibian Specialist Group, 2020o; in IUCN Red List of 2022) and in Ecuador by Ortega-Andrade *et al.* (2021) in view of its wide distribution, tolerance of a broad range of habitats, presumed large population, and because it is unlikely to be declining to qualify for listing in a more threatened category. It is common to abundant throughout its range. It is very adaptable and is not significantly threatened, except possibly by pollution resulting from the aerial spraying of pesticides. It occurs in many protected areas throughout its range. In Ecuador, it has been recorded in the protected forest of Asociación Agrícola Carchi-Imbabura and the biosphere reserve Macizo del Cajas.

Remarks.—In a total evidence analysis of molecular (mitochondrial and nuclear markers) and non-molecular features (adult and larval morphological and behavioral characters) by de Sá *et al.* (2014), *Leptodactylus melanonotus* is the sister taxon to *L. riveroi*. The image depicted in Yánez-Muñoz *et al.* (2020: Fig. 110) as *L. melanonotus* is *L. ventrimaculatus*.

Key references.—(Hallowell, 1861) (description); (Orton, 1951) (tadpole, Honduras); (Fouquette, 1960) (call); (Brattstrom and Yarnell, 1968) (reproduction); (Heyer, 1970a) (call); (Heyer, 1970b) (call, tadpole, Costa Rica); (Straughan and Heyer, 1976) (call); (Lee, 1996) (Lee, 2000) (tadpole, Honduras); (McCranie and Wilson, 2002) (reproduction, tadpole, natural history, Honduras); (Savage, 2002) (reproduction, tadpole from Costa Rica); (de Sá *et al.*, 2014) (account); (Yánez-Muñoz *et al.*, 2020) (account).

LEPTODACTYLUS MYSTACEUS (MUSTACHE THIN-TOED FROG) (SPIX, 1824)

Rana mystacea (Spix, 1824): 27. **Lectotype.**—ZSM 2504/0 from Solimões, Estado Amazonas, Brazil.
Leptodactylus mystaceus Fitzinger, 1826: 219.

Etymology.—The specific name is from the Greek *mystax*, *mystakos*, meaning upper lip, mustache, referring to the obvious light facial stripe.

Identification.—Figure 2.55. *Leptodactylus mystaceus* has a maximum known SVL of 52.2 mm in males and 56.1 mm in females. Males do not have a spatulate snout and lack thumb spines and chest spines. The light upper lip stripe is distinct (56% of specimens) or indistinct (44%); the dorsal folds are absent; the dorsolateral folds are distinct, complete; the lateral folds are absent; the posterior thighs have a light, transverse stripe (93%); the upper shank is barred; the toes lack lateral fringes.

The dorsum is gray, olive–tan, or reddish–tan marked by darker brown or grayish–brown blotches. These blotches typically include a triangular mark with the apex positioned toward the posterior part of the head, along with usually three crude chevron-shaped marks middorsally. The hind limbs are colored like the body and have dark transverse bars; the forelimbs are paler than the body and usually have faint transverse bars on the forearms. The posterior surfaces of the thighs are mottled brown and black with a longitudinal cream stripe distally. The dorsolateral fold is dark brown with a cream streak on the medial edge posteriorly. The flanks are pale gray or tan. The side of the head, including the tympanic region, is black, sharply defined above by a narrow bronze–tan canthal stripe. A broad, silvery white or creamy white labial stripe extends from the tip of the snout to the base of the arm. The venter is creamy white with brown flecks laterally on the throat and belly. The iris is bronze above and dark metallic reddish-brown below.

Comparisons.—*Leptodactylus fuscus*, *L. rhodomystax*. *Leptodactylus mystaceus* distinguishes from *L. fuscus* (not recorded for Ecuador) by lacking a pair of dorsal folds, which are present in the latter. In Amazonian Ecuador, the only other *Leptodactylus* having a pale labial stripe is *L. rhodomystax*, in which the stripe is pinkish-tan. Also, *L. rhodomystax* has a blunt snout and a dark venter; the venter in *L. mystaceus* is creamy white.

Natural history.—Figure 2.56. Frogs were found by day in leaf litter or under logs. In Santa Cecilia, Provincia Sucumbíos, *Leptodactylus mystaceus* was found throughout the year in primary forests (15%), secondary forests (50%), forest edges (5%), and clearings (30%). At night, they were on the ground in forests or clearings, but some were found on low bushes (< 1.5 m) and in water within swamps. Most (84%) individuals were not near water. Tadpoles have been found in shallow grassy pools in clearings, in a *Heliconia*-choked swamp in forest, and in a partially shaded pond at the edge of the forest. All tadpoles were observed in shallow water (< 25 cm), where they took refuge in grasses or among leaves on the bottom. In Montalvo, Provincia Pastaza, mature females and juveniles were found in the dry and wet seasons. In Santa Cecilia, 16 of 26 frogs examined contained orthopterans, and 10 of these contained only orthopterans, which made up 81% of the total bulk of food. Individual beetles were present in six stomachs and represented 8% of the bulk. The other food items were single examples of a variety of small arthropods. In Montalvo, two stomach contents contained orthopterans: a tettigoniid and a gryllotalpid each.

FIGURE 2.55 Morphological details of *Leptodactylus mystaceus*. Left: QCAZ 33137, Pañacocha, Provincia Sucumbíos. Top right: QCAZ 47789, La Selva Amazon Ecolodge, Provincia Sucumbíos. LAC. Bottom right: CJ 7115, juvenile, 48.8 mm SVL, Universidad Regional Amazónica Ikiam, Provincia Napo.

FIGURE 2.56 *Leptodactylus mystaceus*. Top: Reserva Biológica Jatun Sacha, Provincia Napo. Gregory O. Vigle. Bottom: Tiputini, Provincia Orellana. LAC.

In Limoncocha, Provincia Sucumbíos, males of *L. mystaceus* were calling from subterranean chambers located beneath logs or other types of debris. On the night of 19 June 1971, *L. mystaceus* exhibited active calling behavior. The first male was heard at 17:15 h, but calling was sporadic for the subsequent hour. At 18:11 h, while it was still light, numerous *L. mystaceus* were heard from a distance, with fewer in close proximity. The calls appeared to originate from an area with denser ground cover, and the calling sites may have darkened earlier than those in the open pasture. By 18:20 h, *L. mystaceus* was engaged in full calling activity. Even at 18:30 h, nearly dark, their calling remained as active as during the remainder of the observation period. By 18:45 h, it was completely dark. Calling patterns were recorded until 21:30 h and may be summarized as follows: (1) The number of calls within a call group varies. One individual exhibited the following series of calls per call group: 2, 5, 5, 19, 9, 4, 9, 12, 8, 14, 14. The highest number of calls recorded per call group was 43. (2) Individual males engage in calling for a few minutes, pause, and then resume calling. Some males were not heard for 10-minute intervals between calling activity, with one individual remaining quiet for half an hour and another for an entire hour. (3) There is no discernible chorus structure. While some individuals called simultaneously, call groups were initiated by different individuals. Three distinct individuals were responsible for the majority of the calling, each at different intervals throughout the evening. The call of *L. mystaceus* is loud, and males called every night irrespective of the rainfall pattern. In this species, the call serves to attract a female over a considerable distance to the calling site of the male. Five foam nests containing unpigmented eggs or larvae were collected, and their contents were as follows: (1) 77 eggs measuring 2.5 mm in diameter, (2) 156 eggs measuring 2.5 mm in diameter, (3) 246 larvae in stage 18, (4) 233 larvae in stage 25, and (5) 240 larvae in stage 27. Additionally, two females each had 285 and 266 unpigmented eggs measuring 2.0 mm in diameter.

In Santa Cecilia, Provincia Sucumbíos, males call throughout the year from secluded sites such as under logs, tree roots, and in ground holes. Females bearing mature ovarian eggs were found in every month except April, June, September, and December; no females were found in June. The unpigmented eggs are deposited in a foam nest; one nest about 100 mm in diameter and 50 mm in depth was in a cavity under a rock in a marshy clearing; others were found in weedy shallow water. Twenty-two females having SVLs of 52.0–60.0 mm ($\bar{x}$ = 54.8) contained 171–425 eggs ($\bar{x}$ = 280.5), with the largest having a diameter of 2.4 mm. Nine recently metamorphosed young have SVLs of 14.0–15.5 mm ($\bar{x}$ = 14.7). In Montalvo, Provincia Pastaza, one adult female with an SVL of 51 mm contained approximately 200 unpigmented eggs measuring 1.4 mm.

Duellman (1978) analyzed one recording from Santa Cecilia, Provincia Sucumbíos. The call consists of a series of notes, "oit, oit, oit;" the note repetition rate is 48 notes/minute. Heyer (1978) and Heyer *et al.* (1996) described the advertisement calls of individuals from Venezuela, Brazil, Ecuador (Limoncocha, Provincia Sucumbíos), and Peru. The advertisement call is a single-pulsed note emitted at a rate of 1.8 calls/second. The call from Limoncocha is emitted at a rate of 1–2 calls/second. Additional data and figures are provided under the subtitle "Vocalizations of Ecuadorian Anurans" in Appendix 1.

Tadpoles.—The description is based on a tadpole from Santa Cecilia, Provincia Sucumbíos, Ecuador (as *Leptodactylus amazonicus*). The maximum total length in stage 40 is 36.2 mm. The eye diameter constitutes 9–14% of the body length. The oral disc is positioned anteroventrally, with a width corresponding to 22–26% of the body length; the labial tooth row formula is 2(2)/3(1); the anterior oral papilla gap represents 50–70% of the oral disc width; there are 50–70 denticles on one side of the split tooth row anterior to the jaw sheaths. The body length comprises 33–40% of the total length. The tail displays a mottled pattern.

The following description is based on a tadpole from Santa Cecilia in stage 27, having a body length of 9 mm and a total length of 26 mm. The body is ovoid, widest medially and flat ventrally. In dorsal view, the snout is bluntly rounded, whereas in profile, it is inclined anteroventrally and rounded below. The eyes are moderate in size, well separated from one another, and directed dorsolaterally. The nostrils are midway between the tip of the snout and the eyes. The spiracle is sinistral; the opening is directed posterodorsally at a point just below the midline at about two-thirds of the length of the body. The oral disc is ventral and has lateral folds. The medium part of the upper lip is bare; elsewhere, there is one row of labial papillae. There are two upper and three lower rows of denticles; the second upper row is interrupted medially. The jaw sheaths are slender and finely serrated. The dorsal and ventral fins are equal in depth and shallower than the caudal musculature. The dorsal fin does not extend onto the body and is deepest at the midlength of the tail. The tip of the tail is pointed.

The body is olive tan; in large individuals (past stage 37), a dark brown dorsolateral stripe and an interorbital bar are evident. The venter is gray with gold flecks anteriorly. The tail is olive tan with dull green and orange–yellow flecks and dark brown reticulations. The iris is golden bronze.

Distribution.—Figure 2.57. *Leptodactylus mystaceus* has a broad distribution in the Amazon Basin, extending as far as the interior portions of the Brazilian states of Minas Gerais and São Paulo, as well as the northern Atlantic forests of Brazil. In Ecuador, it occurs in Sucumbíos, Orellana, Napo, Pastaza, and Morona Santiago provinces. Its elevational range is 177–639 m in an area of extent of occurrence of about 54,680 km^2.

This nocturnal, terrestrial species inhabits forests and intervened areas in the Amazonian region in flooded forests, evergreen forests of lowlands, and evergreen montane foothill forests in eight ecosystems (Table 0.4) within five biogeographic sectors (Table 0.5), where the average annual rainfall is 2629–4290 mm and the average annual temperature is 22.8–25.9 °C.

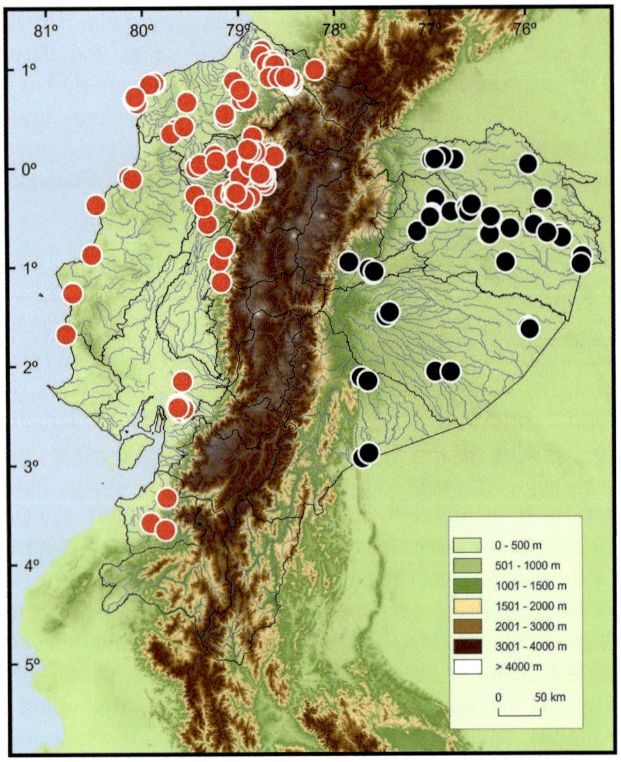

FIGURE 2.57 Localities of known occurrence of *Leptodactylus mystaceus* (black dots) and *L. ventrimaculatus* (red dots) in Ecuador.

Conservation.—*Leptodactylus mystaceus* is categorized as Least Concern throughout its entire range (Heyer and Rodrigues, 2010a; in IUCN Red List of 2022) and in Ecuador by Ortega-Andrade *et al.* (2021), in view of its wide distribution, tolerance of a broad range of habitats, presumed large population, and because it is unlikely to be declining to qualify for listing in a more threatened category. It is a fairly common species. There are no major threats to this widespread and adaptable species. Fire, clear cutting, logging, infrastructure development, and agricultural expansion might impact local populations. Its range includes several protected areas. In Ecuador, it has been recorded in the public reserves Parque Nacional Yasuní, Reserva de Producción de Fauna Cuyabeno, and Reserva Biológica Limoncocha; in the protected forests of Cordillera Kutukú y Shaimi, Río Aguarico, Comuna Santa Elena, and Comuna Kichwa Añangu; the biosphere reserves Sumaco and Yasuní; and in some private reserves (e.g., Reserva Biológica Jatun Sacha).

Remarks.—In a total evidence analysis of molecular (mitochondrial and nuclear markers) and non-molecular features (adult and larval morphological and behavioral characters) by de Sá *et al.* (2014), *Leptodactylus mystaceus* is a complex of at least three species occurring in Brazil. No molecular data are available for Ecuadorian populations.

Key references.—(Spix, 1824) (description); (Heyer and Bellin, 1973) (reproduction); (Heyer, 1978) (call, tadpole); (Duellman, 1978) (reproduction, natural history, food); (Almendáriz, 1987) (natural history, food); (Heyer *et al.*, 1996) (call); (de Sá *et al.*, 2014) (account); (Schiesari *et al.*, 2022) (tadpole).

LEPTODACTYLUS PENTADACTYLUS (GUALAG THIN-TOED FROG) (LAURENTI, 1768)

Rana pentadactyla (Laurenti, 1768): 32. **Neotype.**—RMNH 29559 from Suralcokamp V, Lelygebergte, Marowijne, Suriname.

Leptodactylus pentadactylus Boulenger, 1882a: 241.

Etymology.—The specific name is from the Greek *penta*, meaning five, and *dactylos*, meaning finger, toe. The illustration in Seba (1734: plate 75, figure 1) shows five fingers. No known frogs have five fingers. The Quechua Indian name for *Leptodactylus pentadactylus* is "*gualag*."

Identification.—Figure 2.58. *Leptodactylus pentadactylus* has a maximum known SVL of 195 mm in males and 174.2 mm in females. Males do not have a spatulate snout, and the thumb typically lacks spines, occasionally having one moderate spine; moreover, males do not have chest spines. The light upper lip stripe is absent; the dorsal folds are absent; the dorsolateral folds usually are complete, extending from the eye to the groin, complete from the eye to the sacrum, or interrupted between the sacrum and groin; the lateral folds are absent. The posterior thigh lacks a light stripe; the upper shank is barred; the toes lack flexible lateral fringes; the toes may have non-flexible ridges.

The dorsum is tan to reddish-brown with broad, reddish-brown marks on the body between the yellowish-tan dorsolateral folds. The dorsal surfaces of the limbs are tan to reddish-brown with narrow brown transverse bars. The flanks are creamy tan, and a narrow black canthal stripe is continuous with a narrow black stripe that parallels the fold from the tympanum onto the flank. The upper lip is tan with a brown margin and dark brown triangular spots. The posterior surfaces of the thighs are black with creamy tan flecks or mottling. The venter is cream with bold dark brown to black mottling, especially on the belly and hind limbs. The iris is bronze.

Comparisons.—*Leptodactylus knudseni*, *L. stenodema*. *Leptodactylus pentadactylus* can be distinguished from *L. knudseni* (characters in parenthesis) by its dark belly with conspicuous white mottling and vermiculation (uniformly white or dark) and the dorsolateral folds extending to the sacrum and groin. Sexually active males of *L. pentadactylus* usually lack a large black spine on each thumb and lack chest spines, whereas sexually active males of *L. knudseni* have a large black spine on each thumb and bear a pair of black chest spines. The dorsolateral fold originates from the

FIGURE 2.58 Morphological details of *Leptodactylus pentadactylus*. CJ 6254, male, 86.4 mm SVL, note color of groin and posterior surfaces of thighs, Shiripuno Lodge, Provincia Orellana. DAL.

posterior eye in *L. pentadactylus*, while in *L. stenodema*, the dorsolateral fold originates from above the posterior edge of the tympanum.

Natural history.—Individuals were found on the ground at night. By day, they shelter in burrows, under logs, or beneath tree roots. During axillary amplexus, the male creates a large foamy mass around the eggs by rapidly moving his hind legs through the jelly surrounding the fertilized eggs. This foam nest comprises a frothy mixture of sperm, skin secretions, water, and air. Females lay approximately 1000 light gray eggs, and the foam nests are constructed in dry cavities or depressions. Tadpoles can complete their development within the foam nest. A nest found at the Reserva Florestal Adolpho Ducke, Manaus, Estado Amazonas, Brazil, was in a burrow, approximately 50 cm deep and 20 cm wide, in clay soil. The nest contained 10 tadpoles at stage 40, immersed in foam within an excavated basin on the bottom of the burrow, which had no water. The burrow was about 50 m away from the water body. A second nest was found in a burrow, 13 cm in diameter and 25 cm deep, in a small remnant of paleo floodplain forest, about 200 m from the margin of a small river, the Paraná do Castanho. The nest contained more than 20 tadpoles at stage 39, immersed in foam. These tadpoles have a dorsal region with two well-defined, narrow transverse bands and irregular quadrangular markings. Eggs were observed in the gut of a tadpole at stage 29. Another nest was found in clay soil within the burrow of a fallen trunk, approximately 25 cm deep; this nest contained five juveniles (SVL = 28.04 ± 3.02 mm; 23.10–30.50 mm). A female (SVL = 146.0 mm) was found around this nest. Juveniles were similar to adults in color patterns and body shape.

Heyer (2005) described the advertisement call of individuals from Brazil, Colombia, and Peru. The advertisement call is a rising whoop consisting of a single-pulsed note emitted at a rate of 4–37 calls per minute. Additional data and figures are provided under the subtitle "Vocalizations of Ecuadorian Anurans" in Appendix 1.

Tadpoles.—Tadpoles have been described from several locations (see Key references). This description is based on individuals in stage 36 from Reserva Florestal Adolpho Ducke, Manaus, Estado Amazonas, Brazil. The body is ovoid and slightly compressed in lateral view and ovoid in dorsal view. The body comprises 23.8% of the total length. The snout is rounded in lateral and dorsal views. The eyes are small, directed dorsolaterally. The nostrils are small, rounded, positioned dorsolaterally. The spiracle is single, sinistral, lateroventral, directed posteriorly, with the opening below the midlevel of the body; it is visible in dorsal view. The vent tube is broad, medial, fused to the ventral fin and opening posteriorly. The tail is long, about 76.2% of the total length. The maximum tail height is two times greater than the tail muscle height. The caudal muscle is robust, higher than the dorsal and ventral fins along the anterior two-thirds of the tail. The dorsal fin originates at the tail–body junction, gradually increasing for one-third of the tail length and gradually diminishing after that toward the tip. The ventral fin, originating at the posterior ventral terminus of the body, is slightly arched and maintains the same height throughout the proximal two-thirds of the tail, diminishing gradually toward the tip. The oral disc is terminal, 3.0 mm wide; the marginal papillae are uniseriate, with a gap on the anterior labium; submarginal papillae are absent. The marginal papillae of the anterior labium are greater than those on the posterior labium. The labial tooth row formula is 1/2(1). The P-1 row is slightly interrupted medially, and the P-2 row is shorter than the others. The upper jaw sheath is wide, arch-shaped; the lower jaw sheath is wide, V-shaped; both finely serrated. In life, the color of the body and tail is dark gray, the belly is silver, and the fins are translucent; the intestines are visible through the skin.

Distribution.—Figure 2.59. *Leptodactylus pentadactylus* is known throughout the Amazon Basin from Colombia south to Peru, northern Bolivia, and northern Brazil and to Guyana, Suriname, and French Guiana. In Ecuador, it occurs in Sucumbíos, Orellana, Napo, Pastaza, Morona Santiago, and Zamora Chinchipe provinces. Its elevational range is 17–229 m in an area of extent of occurrence of about 73,630 km^2.

This nocturnal, terrestrial species inhabits forests and intervened areas in the Amazonian region in flooded forests, evergreen forests of lowlands, and evergreen montane foothill forests in 12 ecosystems (Table 0.4) within six biogeographic sectors (Table 0.5), where the average annual rainfall is 1799–4412 mm and the average annual temperature is 20.5–25.8 °C.

Conservation.—*Leptodactylus pentadactylus* is categorized as Least Concern throughout its entire range (Heyer *et al.*, 2008; in IUCN Red List of 2022) in view of its wide distribution, tolerance of a degree of habitat modification, presumed large population, and because it is unlikely to be declining fast enough to qualify for listing in a more threatened category. It seems to be an abundant species, and there are no major threats. However, it is important to note that in Ecuador, *Leptodactylus pentadactylus* has been categorized as Near Threatened by Ortega-Andrade *et al.* (2021), given localized threats such as deforestation, agriculture, and oil extraction and spills. In Ecuador, it has been recorded in the public reserves Parque Nacional Sumaco Napo-Galeras, Parque Nacional Yasuní, and Reserva de Producción de Fauna Cuyabeno; in the protected forests of Cordillera Kutukú y Shaimi and Comuna Kichwa Añangu; the biosphere reserves Sumaco, Yasuní, and Podocarpus-El Cóndor; and in some private reserves (e.g., Tiputini Biodiversity Station, Yachana Reserve, Shiripuno Lodge).

Remarks.—In a total evidence analysis of molecular (mitochondrial and nuclear markers) and non-molecular features (adult and larval morphological and behavioral characters) by de Sá *et al.* (2014), *Leptodactylus pentadactylus* is the sister taxon to *L. savagei* and closely related to *L. peritoaktites* and *L. myersi*.

Leptodactylus pentadactylus and *L. knudseni* have been historically treated as a single species, *L. pentadactylus*, until the taxonomic clarification by Heyer (2005), where he recognized two sympatric species. Consequently, any identifications, natural history observations, or other data provided prior to 2005 (e.g., Duellman, 1978; Almendáriz, 1987) should be approached with caution, as they may pertain to a combination of both species.

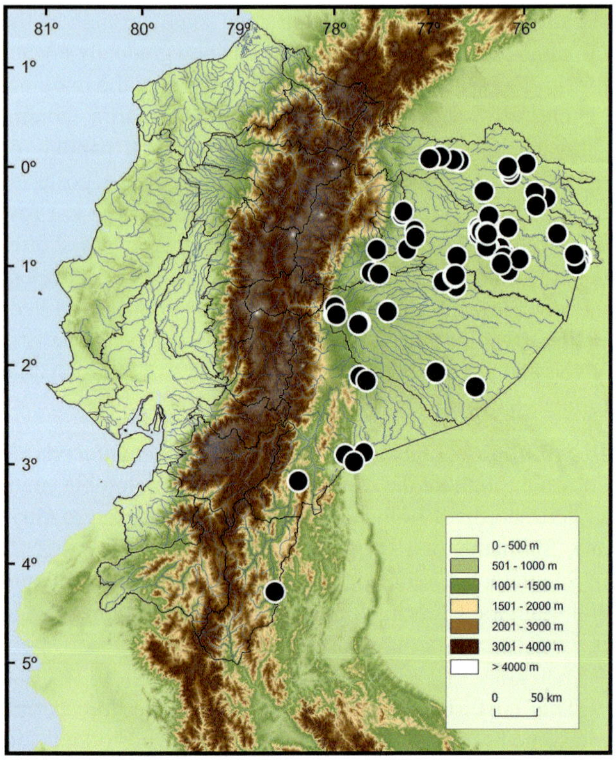

FIGURE 2.59 Localities of known occurrence of *Leptodactylus pentadactylus* in Ecuador.

Key references.—(Laurenti, 1768) (description); (Hero, 1990) (tadpole); (Lima *et al.*, 2012) (account); (Heyer, 2005) (taxonomy, call); (Duellman, 2005) (account, tadpole); (Menin *et al.*, 2010) (reproductive biology, tadpole); (Beirne and Whitworth, 2011) (natural history); (Heyer *et al.*, 2011) (tadpole); (de Sá *et al.*, 2014) (phylogeny, account, call); (Schiesari *et al.*, 2022) (tadpole).

LEPTODACTYLUS PERITOAKTITES (COASTAL DWELLER THIN-TOED FROG) HEYER, 2005

Leptodactylus peritoaktites Heyer, 2005: 325. **Holotype.**—USNM 196739 from Hacienda Equinox, 38 km north–northwest of Santo Domingo de los Colorados, 1000 m, Provincia Esmeraldas, Ecuador.

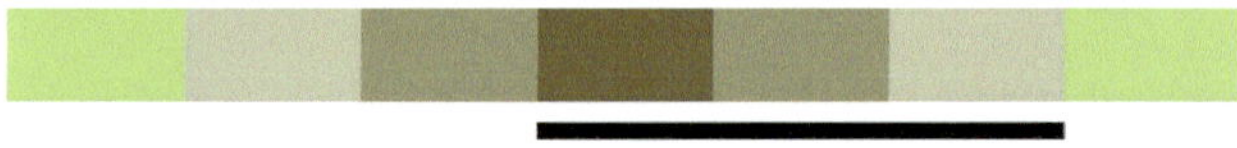

Etymology.—The specific name is from the Greek *peritos*, meaning west, and *aktites*, meaning coast dweller, in allusion to the geographic distribution of the species.

Identification.—Figure 2.60. *Leptodactylus peritoaktites* has a maximum known SVL of 146.3 mm in males and 133.1 mm in females. Males do not have a spatulate snout, typically have a moderate to large thumb spine, and lack chest spines. The light upper lip stripe is absent (upper lip usually has dark triangular marks). The dorsal folds are absent; the dorsolateral folds usually extend completely from the eye to the sacrum or, at a minimum, from the eye to a point between the sacrum and the groin; the lateral folds may be interrupted or entirely absent. The sides of the toes are weakly ridged (not fringed).

The dorsal surface exhibits varying shades of darker and lighter browns, as well as gray and brown–gray hues on the limbs. The dorsal pattern typically features irregular quadrangular or rectangular markings in dark brown, with equal or alternating intensity, either merging laterally or remaining distinct. Alternatively, the dorsal pattern may consist of a uniformly light brown background with one or two well-defined, narrow, dark brown transverse bands. In rare instances, the dorsal pattern may include no more than two broad, dark transverse bands of equal intensity, in addition to an interorbital band, which may or may not merge laterally. The canthal and postorbital stripes are dark brown to bold black, with the latter extending approximately to the level of the sacrum. The lip pattern typically displays dark brown to black triangular marks, with one or two elongated marks that may approach or extend into the lower

FIGURE 2.60 Morphological details of *Leptodactylus peritoaktites*. QCAZ 35265, Buena Fé, Recinto El Descanso, Dam 2 of the reservoir of the Río Baba, Provincia Los Ríos. LAC.

eye region. Additionally, the lip pattern frequently includes dark triangular marks along the edge of the upper lip only. The flanks are tan and brown. The ventral surface is usually dark, with large, distinct light spots. Alternatively, the belly may exhibit a mottled or uniformly dark appearance or feature small, light vermiculation on a dark background. The posterior thighs exhibit vibrant red markings. Typically, the pattern on the posterior thighs is dark, featuring large, contrasting, irregular red blotches or spots that extend from light dorsal transverse bars. Alternatively, the thigh pattern may be relatively uniform and dark or indistinctly mottled. It could also appear dark with small, distinct red vermiculation or spots or dark with large, distinct red vermiculation, some of which may coalesce. Another variation includes a dark pattern with distinct, discrete red spots. The iris is golden cream with fine black reticulations and a red suffusion covering the three-fourths of the lower portion.

Comparisons.—*Leptodactylus labrosus*, *L. rhodomerus*, *L. ventrimaculatus*. In life, juvenile *L. peritoaktites* have bright red coloration on the posterior thighs and groin; juveniles and adults of *L. labrosus* and *L. ventrimaculatus* lack red coloration on their posterior thighs. Sexually active males of *L. peritoaktites* have a single moderate to large black spine on each thumb and never exhibit the vermiculated belly pattern of *L. rhodomerus*; sexually active *L. rhodomerus* males have one tiny to small white or black spine on the thumb and a dark belly with large, light vermiculation. The outer tarsal surface of *L. peritoaktites* is smooth; the outer tarsal surfaces of *L. rhodomerus* and *L. ventrimaculatus* present white tubercles; the outer tarsal surfaces of *L. labrosus* usually have scattered white tubercles (78% of specimens). Adults of *L. peritoaktites* are larger (female SVL 115–133 mm, male SVL 124–146 mm) than *L. labrosus* (female SVL 50–71 mm, male SVL 48–67 mm).

Natural history.—Calls have been heard after heavy rains.

Heyer (2005) described the advertisement call of individuals from Ecuador. The call is a single-pulsed note emitted at a rate of 34–37 calls/minute. Additional data and figures are provided under the subtitle "Vocalizations of Ecuadorian Anurans" in Appendix 1.

Tadpoles.—Unknown.

Distribution.—See Figure 2.53. *Leptodactylus peritoaktites*, an endemic species to Ecuador, is known from the lowlands of Santo Domingo de los Tsáchilas and Manabí provinces. Its elevational range is 179–330 m in an area of extent of occurrence of about 2820 km^2.

This nocturnal, terrestrial species inhabits forests and adjacent intervened areas in the evergreen seasonal forests of lowlands ecosystem in the Cordillera Occidental of the Andes, Equatorial Chocó, and Jama-Zapotillo biogeographic sectors, where the average annual rainfall is 863–2828 mm and the average annual temperature is 23.8–24.5 °C.

Conservation.—*Leptodactylus peritoaktites* was categorized as Endangered by Ortega-Andrade *et al.* (2021) because its extent of occurrence is less than 5000 km^2 (2,820 km^2, according to our data), and there is a suspected population decline as a result of extensive habitat destruction in the lowlands of coastal Ecuador. Furthermore, its distribution is severely fragmented, and there is a continuing decline in the extent and quality of its forest habitat. The severe decline in habitat extent and quality is a result of industrial agriculture, logging, and urbanization in the lowlands of coastal Ecuador. It is a very rare species. There is an urgent need to conserve remaining areas of suitable forest for this species in western Ecuador. It has not been recorded in any public or private reserves.

Remarks.—In a total evidence analysis of molecular (mitochondrial and nuclear markers) and non-molecular features (adult and larval morphological and behavioral characters) by de Sá *et al.* (2014), *Leptodactylus peritoaktites* is the sister taxon to *L. savagei* and *L. pentadactylus*.

Key references.—(Heyer, 2005) (description, call); (de Sá *et al.*, 2014) (phylogeny, account, call).

LEPTODACTYLUS PETERSII (PETERS' THIN-TOED FROG) (STEINDACHNER, 1864)

Platymantis petersii (Steindachner, 1864): 254. **Holotype.**—NHMW, lost, from "Marabitanas," upper Río Negro, Amazonas, Brazil.

Leptodactylus (*Platymantis*) *petersii* Lutz, 1930: 1, 21, by implication.

Etymology.—The specific name honors Dr. Wilhelm Carl Hartwig Peters (1815–1883), a prolific herpetologist who was the Director of the Zoologisches Museum in Berlin for over 25 years.

Identification.—Figure 2.61. *Leptodactylus petersii* has a maximum known SVL of 41.1 mm in males and 51.3 mm in females. Males do not have a spatulate snout; they have a pair of black spines on each thumb and lack chest spines. The light upper lip stripes extend from the posterior corner of the eye and range from distinct to not discernible. The dorsal and lateral folds are absent; the dorsolateral folds are absent, short, or of moderate length, interrupted. The light posterior thigh stripes are typically absent (90% of specimens); the upper shank is barred; the belly usually is extensively mottled (66%); the toes have lateral fringes.

In frogs from Cusco Amazónico, Peru, the dorsum ranges from tan to reddish-brown, with a prominent dark brown or olive-brown triangular mark. This mark originates at the upper eyelids with its base and extends to the apex, which exhibits irregular margins posteriorly. Additionally, irregularly shaped dark spots are present on both the body and limbs. Distinct, narrow, cream bars separate two or three dark brown or black spots on the upper lip. The posterior surfaces of the thighs are dark brown with small cream flecks. The entire venter is mottled black and white. The iris is a dull bronze with radiating brown bars anteriorly, posteriorly, and ventrally.

Comparisons.—*Leptodactylus leptodactyloides*, *L. wagneri*. *Leptodactylus petersii* is similar to *L. leptodactyloides* and *L. wagneri* by having fringes on the toes. However, *L. petersii* can be distinguished from *L. leptodactyloides* by the ventral coloration patterns. In *L. petersii*, individuals typically exhibit relatively uniform and extensively patterned bellies, characterized by mottled black and white markings, often forming an anastomotic pattern. In contrast, *L. leptodactyloides* displays more developed belly patterns anteriorly, with most individuals showing moderate mottling;

FIGURE 2.61 *Leptodactylus petersii*. CJ 882, female juvenile, 25.5 mm SVL, Reserva Otoyacu, Provincia Pastaza. LAC.

the throat is brown, and the belly is white with brown mottling on the chest but without the characteristic anastomotic pattern observed in *L. petersii*. Most (90%) *L. petersii* lack distinct light posterior thigh stripes, whereas most *L. leptodactyloides* have distinct light posterior thigh stripes. In *L. petersii*, there is no distinct dorsolateral fold; instead, there are short, irregular, longitudinal ridges on the dorsolateral surfaces of the body; in *L. leptodactyloides*, low dorsolateral folds are continuous or interrupted and extend to the groin. *Leptodactylus petersii* is smaller than *L. wagneri* (*L. wagneri* females 52–82 mm SVL, males 39–61 mm SVL), and *L. petersii* lacks dorsolateral folds extending from the eye to groin, in contrast to most *L. wagneri*.

Natural history.—*Leptodactylus petersii* is an inhabitant of inundated forests, where it is commonly syntopic with *L. leptodactyloides*. At Cusco Amazónico, Peru, individuals were only found during the rainy season (November–March). Of 56 individuals with ecological data, 16 were observed by day on the ground at the edge or in shallow water in ponds. Of the 40 individuals found at night, 11 were in shallow water, 10 were on logs floating in water, eight were calling from under leaves at the edges of ponds, five were on the ground adjacent to ponds, and four were on logs on land near ponds. At night, two individuals were on leaves of herbs, one 0.2 m above the ground and one 0.1 m above the surface of a pond. Males were heard calling in February 1986 and in February and early March 1989. One female collected in February 1990 contained many small ovarian eggs.

Heyer (1994) described two types of calls of individuals from Brazil and Peru. The call type 1 is a single, partially pulsed note emitted at a rate of 2.9–4.2 notes/second. The call type 2 has two juxtaposed notes, the first note of 2–4 partial pulses and the second note not noticeably pulsed; the call rate is 0.6–1.3 call/second. Additional data and figures are provided under the subtitle "Vocalizations of Ecuadorian Anurans" in Appendix 1.

Tadpoles.—The following description is of tadpoles from Cusco Amazónico, Departamento Madre de Dios, Peru. The maximum total length in stage 36 is 20.8 mm; the oral disc is positioned anteroventrally; the labial tooth row formula is 2(2)/3; the tail is almost uniformly brown, or the musculature is moderately mottled brown. A typical tadpole in stage 37 from Cusco Amazónico has a body length of 8.0 mm and a total length of 19.3 mm. In dorsal view, the body is ovoid, widest (5.0 mm) at midlength; in lateral view, the body is shallowest anteriorly; the height at the level of the orbits is 3.0 mm, and the greatest height is 4.3 mm at about two-thirds of its length. The snout is acutely rounded in dorsal view and bluntly rounded in profile. The nostrils are situated dorsolaterally and directed anterolaterally at about midway between the tip of the snout and the orbits; narial papillae are absent. The eyes are small (0.7 mm in diameter), situated and directed dorsolaterally, and not visible from below; the interorbital distance is 1.1 mm. The spiracle is sinistral, with its base at about the midlength of the body. The spiracular tube is short and attached for its full length to the body wall; the circular opening is directed posterodorsally on the midline. The vent tube is short, dextral, and incorporated into the ventral fin; the aperture is diagonal. The caudal musculature is moderately robust, highest at its junction with the tail and tapering gradually to a slender tip. The dorsal fin originates on the caudal musculature, well behind the body, and reaches its greatest height at about two-thirds of the length of the tail before diminishing gradually to an acutely rounded tip. The ventral fin originates on the posterior body wall and is about the same height throughout the proximal three-fourths of the tail. At the midlength of the tail, the caudal musculature is slightly higher than the dorsal fin and much higher than the ventral fin. The oral disc is directed anteroventrally and lacks lateral folds. The greatest width of the disc is about one-half the greatest width of the body. The medial third of the anterior labium is bare; elsewhere, there is a single row of small subconical papillae. Submarginal papillae are absent. The jaw sheaths are slender and finely serrate. The anterior sheath is nearly straight, with the ends curved posteriorly; the posterior sheath is broadly V-shaped. The labial tooth row formula is 2(2)/2; A-2 is broadly interrupted medially, and P-1 is continuous or nearly interrupted. A weak third posterior row is present or absent. All rows are about equal in length; the posterior rows are nearly straight. In preservative, the body is brown, and the belly is transparent. The caudal musculature is cream, and the fins are translucent.

Distribution.—Figure 2.62. *Leptodactylus petersii* is known from the Guianas, the Amazon Basin, and isolated gallery forests in Cerrado open formations in central Brazil. In Ecuador, it occurs in Sucumbíos, Orellana, Pastaza, and Morona Santiago provinces. Its elevational range is 190–282 m in an area of extent of occurrence of about 24,800 km^2.

This nocturnal, terrestrial species inhabits forests and intervened areas in the Amazonian region in the flooded forests of alluvial plains of Amazonia, evergreen forests of lowlands of Aguarico-Putumayo-Caquetá, and evergreen forests of lowlands of Fan of Pastaza ecosystems, where the average annual rainfall is 2799–3563 mm and the average annual temperature is 24.8–25.7 °C.

Conservation.—*Leptodactylus petersii* is categorized as Least Concern throughout its entire range (Heyer and Rodrigues, 2010b; in IUCN Red List of 2022) and in Ecuador by Ortega-Andrade *et al.* (2021) in view of its wide distribution, tolerance of a broad range of habitats, presumed large population, and because it is unlikely to be declining to qualify for listing in a more threatened category. It is a common species. There are no major threats to this species; however, according to Ernst *et al.* (2007), the occurrence of logging can influence larval growth and survival by inducing changes in the thermal environment. Its range includes several protected areas. In Ecuador, it has been recorded in the public reserve Parque Nacional Yasuní, and the protected forest of Sacha Lodge, the biosphere reserve Yasuní, and some private reserves (e.g., Reserva Otoyacu).

Remarks.—In a total evidence analysis of molecular (mitochondrial and nuclear markers) and non-molecular features

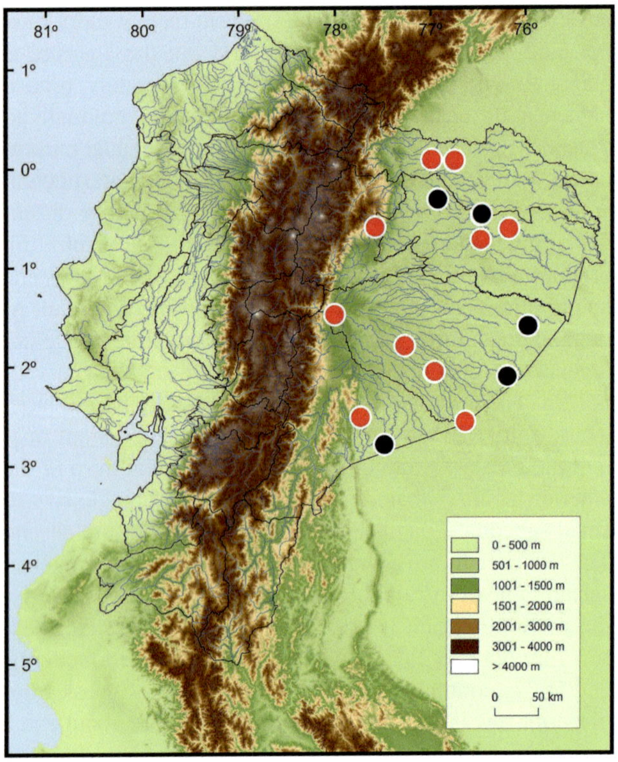

FIGURE 2.62 Localities of known occurrence of *Leptodactylus petersii* (black dots) and *L. stenodema* (red dots) in Ecuador.

(adult and larval morphological and behavioral characters) by de Sá *et al.* (2014), *Leptodactylus petersii* is the sister taxon to *L. leptodactyloides* and closely related to *L. natalensis* and *L. pustulatus*. According to Motta *et al.* (2018), who analyzed the16S rDNA sequence divergence of populations from Bolivia and central Amazonia, a probable case of a cryptic species may be suggested for the species they identified as *Leptodactylus* cf. *petersii*. Populations from Ecuadorian Amazonia have not been analyzed at the molecular level, and their taxonomic status is pending a correct neotype designation.

Key references.—(Steindachner, 1864) (description); (Heyer, 1994) (call, tadpole); (Duellman, 2005) (account, tadpole); (Ernst *et al.*, 2007) (larval growth, survival); (de Sá *et al.*, 2014) (phylogeny, account); (Schiesari *et al.*, 2022) (tadpole).

LEPTODACTYLUS RHODOMERUS (ROSE THIGH THIN-TOED FROG) HEYER, 2005

Leptodactylus rhodomerus Heyer, 2005: 327. **Holotype.**—ICN 13322 from campamento Chancos, 460 m, 3.96667° N, 76.73333° W, Vereda Campo Alegre, Municipio de Restrepo, Departamento Valle de Cauca, Colombia.

Etymology.—The specific name is from the Greek *rhodon*, meaning rose, and *meros*, meaning thigh, in reference to the red coloration on the posterior thighs in life.

Identification.—Figure 2.63. *Leptodactylus rhodomerus* has a maximum known SVL of 143.8 mm in males and 157.8 mm in females. Males do not have a spatulate snout; the thumb has one tiny to small black spine; they lack chest spines. The light upper lip stripe is absent; the dorsal folds are absent; the dorsolateral folds usually are complete from the eye to the groin; the lateral folds are interrupted or absent; the posterior thigh lacks a light stripe; the upper shank is barred; the toes are ridged (not fringed).

The dorsal pattern commonly exhibits a nearly uniform brown with irregular, dark brown quadrangular or rectangular markings. These markings vary in intensity, alternating between lighter and darker shades, and may either converge laterally or remain distinct. In addition, the dorsal pattern usually presents a uniform light brown color with one or two well-defined, narrow, dark brown transverse bands. The lip pattern has dark brown to black triangular marks, one or two elongate and approaching or entering the lower area of the eye. A dark brown to black canthal stripe is present, along with a dark brown to black postorbital stripe that extends to different levels from the posterior arm to the sacrum. The ventral surfaces are dark to light gray with small, well-defined white spots and vermiculation. The posterior thigh is gray with irregular bright red markings. The iris is golden cream with fine black reticulations and a horizontal red stripe that in some cases extends to the lower portion of eye.

Comparisons.—*Leptodactylus pentadactylus*, *L. peritoaktites*. *Leptodactylus rhodomerus* is mostly similar to *L. pentadactylus* and *L. peritoaktites*, which have allopatric or parapatric distributions with *L. rhodomerus*. *Leptodactylus rhodomerus* occurs in western Colombia and adjacent Ecuador; *L. pentadactylus* occurs in Amazonia. In life, *Leptodactylus rhodomerus* has bright red markings on the posterior thigh surfaces, whereas *L. pentadactylus* does not have red on the thighs. *Leptodactylus rhodomerus* has a parapatric distribution with *L. peritoaktites* to the south along Pacific coastal South America. Sexually active males of *L. rhodomerus* have one tiny to small spine on the thumb and a dark belly with large light vermiculation, whereas sexually active males of *L. peritoaktites* have a single moderate to large thumb spine and never exhibit the vermiculated belly pattern of *L. rhodomerus*.

Natural history.—Individuals have been found associated with forests, rivers, streams, and swamps. An adult male was found within a ground burrow; a female was observed near a burrow from April to September at the margin of a small pond.

Tadpoles.—Unknown.

Distribution.—See Figure 2.53. *Leptodactylus rhodomerus* is known from Departamento Antioquia in the Chocoan region of Pacific coastal Colombia to northern coastal

FIGURE 2.63 Morphological details of *Leptodactylus rhodomerus*. CJ 8296, female, note color of groin and posterior surfaces of thighs, 132.6 mm SVL. Reserva Biológica Bilsa, Provincia Esmeraldas. DAL.

Ecuador. In Ecuador, it occurs in Esmeraldas and Manabí provinces. Its elevational range is 0–699 m in an area of extent of occurrence of about 5820 km^2.

This nocturnal, terrestrial species inhabits forests and intervened areas in northwestern Ecuador in the evergreen forests of lowlands of Equatorial Chocó, evergreen seasonal montane foothill forests of the Cordillera Costera of Equatorial Chocó, and evergreen low montane forests of the Cordillera Costera of Equatorial Chocó ecosystems in the Cordillera Occidental of the Andes, Cordillera Costera of Chocó, and Equatorial Chocó biogeographic sectors, where the average annual rainfall is 1759–2566 mm and the average annual temperature is 22.2–25.4 °C.

Conservation.—*Leptodactylus rhodomerus* was categorized as Least Concern throughout its entire range (Heyer and Cisneros-Heredia, 2008; in IUCN Red List of 2022) in view of its wide distribution, presumed large overall population, tolerance of a degree of habitat modification, and because it is unlikely to be declining fast enough to qualify for listing in a more threatened category. However, it is important to note that in Ecuador, *Leptodactylus rhodomerus* has been categorized as Vulnerable by Ortega-Andrade *et al.* (2021). It is uncommon in Ecuador. Its extent of occurrence is 5820 km^2 (not 54 km^2, as indicated by Ortega-Andrade *et al.*, 2021), its distribution is severely fragmented, and there is continuing decline in the extent and quality of its forest habitat. Much of the forest within the known range of this species has been cleared for timber or through the conversion of land to agricultural use, with remaining forest remnants being greatly threatened by ongoing deforestation and mining activities. In Ecuador, it has been recorded in the public reserve Reserva Ecológica Mache Chindul, the protected forest of Asociación Agrícola Carchi-Imbabura, and some private reserves (e.g., Reserva Biológica Bilsa and Reserva Biológica Canandé).

Remarks.—The molecular phylogenetic relationships of *Leptodactylus rhodomerus* are unknown.

Key references.—(Heyer, 2005) (description); (Ortega-Andrade *et al.*, 2010a) (Ortega-Andrade *et al.*, 2010b) (natural history); (de Sá *et al.*, 2014) (account); (Yánez-Muñoz *et al.*, 2020) (account).

LEPTODACTYLUS RHODOMYSTAX (ROSE-LIPPED THIN-TOED FROG) BOULENGER, 1884

Leptodactylus rhodomystax Boulenger, 1884: 637.
Lectotype.—BMNH 1947.12.17.81 from Yurimaguas, Huallaga River, Departamento Loreto, Peru.

Etymology.—The specific name is from the Greek *rhodon*, meaning rose, and *mystax*, meaning upper lip.

Identification.—Figure 2.64. *Leptodactylus rhodomystax* has a maximum known SVL of 89.6 mm in males and 91.4 mm in females. Males do not have a spatulate snout; they have one black thumb spine and a pair of black chest spines. The upper lip has a distinct light stripe; the dorsal folds are absent; there is a pair of distinct, complete dorsolateral folds; the lateral folds are interrupted or absent; the posterior thigh has distinct light spots on a dark background, there is no light thigh stripe; the upper shank is barred or uniform brown or gray; the belly is gray with large irregular spots at least anteriorly; the toes have weak or smooth lateral ridges.

The dorsum is reddish-brown to grayish-brown; the dorsolateral folds are dark brown. The flanks are orange–tan. The groin and posterior surfaces of the thighs are dark grayish-brown with small, pale greenish-yellow spots. The side of the head, including the tympanic region, is dark brown. A pale, pinkish-tan labial stripe extends from the snout to the base of the arm. The throat is dark brown; the chest, belly, and ventral surfaces of the limbs are cream, heavily suffused with brown. The iris is dull bronze above and reddish-bronze below.

Comparisons.—*Leptodactylus knudseni*, *L. mystaceus*, *L. pentadactylus*, *L. stenodema*. *Leptodactylus rhodomystax* is mostly similar to species that lack fringes on toes and have complete dorsolateral folds in some or all individuals. All these species lack the characteristic posterior thigh pattern of distinct light spots on a dark background of *L. rhodomystax*. *Leptodactylus rhodomystax* lacks a light longitudinal stripe on the posterior thigh, while *L. mystaceus* has light posterior thigh stripes. *Leptodactylus rhodomystax* has a distinct light upper lip stripe (brilliant white or reddish in life), whereas *L. knudseni*, *L. pentadactylus*, and *L. stenodema* lack light upper lip stripes.

Natural history.—In Provincia Sucumbíos, Ecuador, three adult males were found at night in swamps during March (Santa Cecilia). One adult male was amid leaf litter within a mature forest in May (Lago Agrio), and one small female was beneath a stone at the edge of a clearing in July (Puerto Libre). Three individuals from these sites contained two roaches (Blattidae), one grasshopper (Orthoptera), and one millipede (Diplopoda). At Yachana Reserve, Provincia Napo, Ecuador, juveniles and adults are frequently found around a permanent swamp within primary forest. They are occasionally found in small temporary pools away from streams. The tadpoles are found in temporary ponds near streams.

In Brazil, *Leptodactylus rhodomystax* called from December to May 1984 and 1985 in the Manaus reserves, but it was not heard during the same periods or at any time in 1980 and 1983. Individuals are not particularly selective about the type of pools in which they breed, as both *baixões* (stream valleys that are wetter than the surrounding upland) and terra firme temporary and semi-permanent pools were used. Calling males were extensively distributed over several hundred hectares of forest at suitable breeding habitat sites. They were heard calling in or near puddles along the banks of two large streams measuring 4–6 m in width. Froglets and juveniles were found in these areas as well. Likewise, calling males and juveniles were found in both the terra firme rainwater pools frequented by breeding *Chiasmocleiss hudikarensis* and the semi-permanent pools used by Phyllomedusidae. Males called most frequently between 19:00 and 23:00 h. Males of *L. rhodomystax* were not observed in choruses, and often a single calling male was present at a site. During the January–February 1984 vocal peak, up to six calling individuals could be heard from a single location beside a large stream. In contrast, no more than two individuals were heard at the two terra firme pool sites, probably due to the smaller extent of pools in those areas compared to streams. Males have been observed calling from inside ground holes. Females deposit about 250 eggs in foam nests in dry ponds. These ponds may not fill for a month after egg laying. Many tadpoles survive in the foam until the pond fills, but many clutches are lost due to desiccation. Other observations revealed that ponds containing *L. rhodomystax* tadpoles had a mean size of 1.67 ± 0.47 m^2 (n = 21). Two foam nests were found deposited between shrubs, roots, and fallen tree trunks near a pond (51 and 41 cm away from the pond edge). Tadpoles reached the pond after rainfall flooded the area. The clutches contained 290 and 247 eggs. Tadpoles do not aggregate; they are free-swimming, benthic, and found in the leaf litter of ponds between January and May. Tadpoles are intra-specific predators of tadpoles and inter-specific predators of eggs and tadpoles of *Osteocephalus taurinus* and *Allobates* spp. On three occasions, tadpoles of *L. rhodomystax* were observed aggregated and buried in dry soil. Between October 2002 to September 2003, no tadpoles of *L. rhodomystax* were found in ponds, but in 2004 and 2005, tadpoles were located in 11 of 31 sites sampled.

Zimmerman and Bogart (1984) described the advertisement call of individuals from the Manaus reserves, Brazil.

FIGURE 2.64 *Leptodactylus rhodomystax*. CJ 6420, female subadult, 89.2 mm SVL, Shiripuno Lodge, Provincia Orellana. DAL.

The call is a single pulsed note emitted at a call rate of 0.23 per second. Additional data and figures are provided under the subtitle "Vocalizations of Ecuadorian Anurans" in Appendix 1.

Tadpoles.—Figure 2.65. The tadpole was described by Duellman (1978) from Santa Cecilia, Provincia Sucumbíos, Ecuador. Hero (1990) described a few external features (e.g., labial tooth row formula and body coloration) for an identification key of tadpoles of the Brazilian central Amazon. Rodrigues *et al.* (2007) provided data on a series of tadpoles from Reserva Florestal Adolpho Ducke, Manaus, Amazonas, Brazil: the maximum total length of stage 37 is 46.7 mm; the oral disc is anteroventral; the labial tooth row formula is 2(2)/3; the body and tail are black or dark brown.

The following description is based on an individual in stage 25 from Santa Cecilia. The body length is 6 mm, and the total length is 21.3 mm. The body is ovoid, widest anteriorly and deepest posteriorly. The snout is truncate in dorsal view and round in profile. The eyes are large, widely separated, and directed laterally. The nostrils are about one-third the distance from the tip of the snout to the eyes. The spiracle is sinistral; the opening is directed posterodorsally at the midline and about two-thirds of the length of the body. The mouth is anteroventral and has no lateral folds. The median part of the upper lip is bare; elsewhere, there is a single row of labial papillae. There are two upper and three lower rows of denticles; the second upper row is widely separated medially, and the third lower row is much shorter than the others. The tail is 72% of the total length and terminates in a filamentous tip. The dorsal fin is shallow, one-third the depth of the ventral fin, deepest at the midlength of the tail, where it is about equal to the depth of the caudal musculature and does not extend onto the body. In preserved tadpoles, the body and caudal musculature are brownish-gray, and the tail is marked with irregular darker flecks. The SVL of metamorphosed froglets from Reserva Florestal Adolpho Ducke was 18.0 mm (n = 27).

FIGURE 2.65 Ontogenetic changes in *Leptodactylus rhodomystax* from tadpole to froglet. Top to bottom. QCAZ 39943, stage 34, lateral and dorsal views; stage 41; stage 42; stage 45; stage 46; froglet three days after stage 46. Estación Biológica Wisui, Provincia Morona Santiago. LAC.

Distribution.—Figure 2.66. *Leptodactylus rhodomystax* is known in the Amazonian Basin and Guyana Shield in Bolivia, Brazil, Colombia, Ecuador, French Guiana, Peru, and Suriname. In Ecuador, it occurs in Sucumbíos, Orellana, Napo, Pastaza, and Morona Santiago. Its elevational range is 190–1390 m in an area of extent of occurrence of about 60,630 km^2.

This nocturnal, terrestrial species inhabits forests and intervened areas in the Amazonian lowlands in 10 ecosystems (Table 0.4) within five biogeographic sectors (Table 0.5), where the average annual rainfall is 2799–4402 mm and the average annual temperature is 18.9–25.7 °C.

Conservation.—*Leptodactylus rhodomystax* is categorized as Least Concern throughout its entire range (La Marca *et al.*, 2010b; in IUCN Red List of 2022) and in Ecuador by Ortega-Andrade *et al.* (2021) in view of its wide distribution, tolerance of a degree of habitat modification, presumed large population, and because it is unlikely to be declining to qualify for listing in a more threatened category. It is common throughout its range of distribution. There are no major threats. Many populations are within protected areas. In Ecuador, it has been recorded in the public reserves Parque Nacional Llanganates, Parque Nacional Yasuní, and Reserva de Producción de Fauna Cuyabeno; the protected forests of Cordillera Kutukú y Shaimi, Comuna Kichwa Añangu, and Nacionalidad Andwa de Pastaza del Ecuador; the biosphere reserves Sumaco and Yasuní; and in some private reserves (e.g., Tiputini Biodiversity Station, Yachana Reserve).

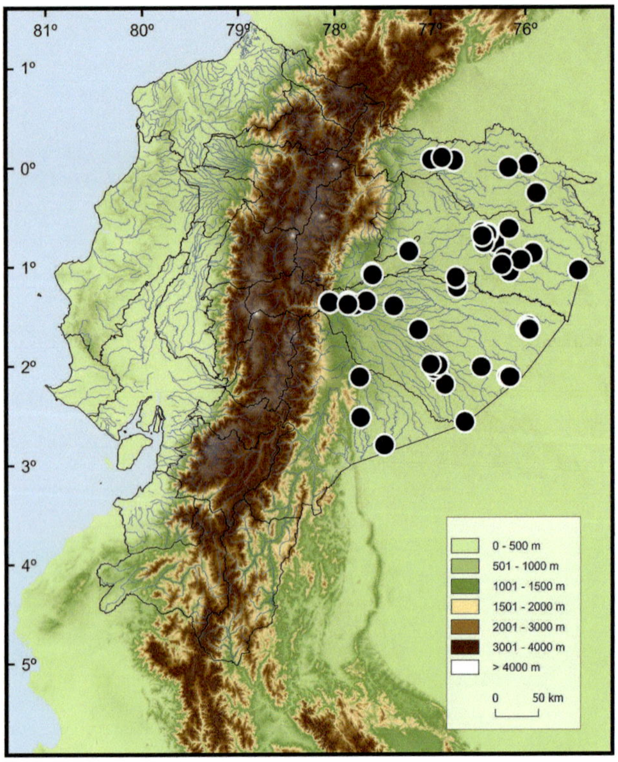

FIGURE 2.66 Localities of known occurrence of *Leptodactylus rhodomystax* in Ecuador.

Remarks.—In a total evidence analysis of molecular (mitochondrial and nuclear markers) and non-molecular features (adult and larval morphological and behavioral characters) by de Sá *et al.* (2014), *Leptodactylus rhodomystax* is the sister taxon to *L. rhodonotus* and closely related to *L. rugosus* and *L. lithonaetes.*

The tadpoles from Reserva Florestal Adolpho Ducke, Brazil, differ from the ones described in Santa Cecilia, Ecuador, because the snout is not truncate in dorsal view but rounded, and the tail tip is not filamentous but bluntly pointed.

Key references.—(Boulenger, 1884) (description); (Duellman, 1978) (tadpole, natural history); (Zimmerman and Bogart, 1984) (reproduction, call); (Hero, 1990) (tadpole); (Lima *et al.*, 2012) (account); (Rodrigues *et al.*, 2007) (reproduction, tadpole); (Beirne and Whitworth, 2011) (natural history); (de Sá *et al.*, 2014) (account, call); (Schiesari *et al.*, 2022) (tadpole).

LEPTODACTYLUS STENODEMA (POPOCO THIN-TOED FROG) JIMÉNEZ DE LA ESPADA, 1875

Leptodactylus stenodema Jiménez de la Espada, 1875: 64.
Lectotype.—MNCN 1687 from San José de Moti, Canton Quijos, Provincia Napo, Ecuador.

Etymology.—The specific name is from the Greek *stenos*, meaning narrow, and *demas*, meaning body. The Cauca Indian name for *Leptodactylus stenodema* is "*popoco*" (Pyburn and Heyer, 1975).

Identification.—*Leptodactylus stenodema* has a maximum known SVL of 99.7 mm in males and 105.0 mm in females. Males do not have a spatulate snout, and they lack thumb spines and chest spines. The light upper lip stripe is absent. The dorsal and lateral folds are absent; there are distinct dorsolateral folds. A light stripe on the posterior thigh is absent; the upper shank has wide or narrow bars; the belly is dark with various kinds of light marks. The lateral surface of toes is smooth or weakly ridged (not fringed).

The dorsum is dull reddish-brown, becoming orange–brown just medial to the black dorsolateral fold. The dorsal surfaces of the limbs are dark brown with indistinct narrow black transverse stripes. The posterior surfaces of the thighs are black. The side of the head is brown, and the lip is orange–brown with four triangular black marks. The flanks are dull brown above and orange-brown below. The throat is dark brown, and the other ventral surfaces are grayish-tan. The iris is bronze with a red median horizontal streak.

Comparisons.—*Leptodactylus pentadactylus*, *L. knudseni*. *Leptodactylus stenodema* is the only species in the genus in which the dorsolateral fold originates just posterior to the tympanum rather than the eye. *Leptodactylus pentadactylus* and *L. knudseni* are similar in size but differ by the presence of dark brown transverse bands between the dorsolateral folds and a black band from the nostril to the eye.

Natural history.—In Montalvo, Provincia Pastaza, Ecuador, two juvenile females were found during day on the leaf litter in May and August. In Santa Cecilia, Provincia Sucumbíos, Ecuador, an individual was found on the ground in secondary forest at night. In the valley of the Vaupes River, Colombia, it lives in burrows (of an armadillo in one case) in upland forests; the burrow is a branching system of tunnels, probably not excavated by the frog, that may pass 1 m or more below the surface and have several entrances that pass beneath roots and large stones. The stomach contents of two females contained Ortoptera and Coleoptera. Also, the stomach of a juvenile female with an SVL of 78 mm contained a hemipteran, a curculionid beetle, and remains of another beetle, probably a tenebrionid. Another juvenile female contained a chrysomelid beetle.

At Reserva Florestal Adolpho Ducke, Brazil, reproduction occurs throughout the year in burrows. The males are seen sitting or calling at the entrances to burrows, which they enter when threatened. Most calling occurs between 15 h and dusk.

Pyburn and Heyer (1975), Heyer (1979), and Lescure and Marty (2000) described the advertisement call of individuals from Wacará, Departamento Vaupés, Colombia, and from French Guiana. The call is a "whoop" emitted while the frog sits in a burrow, a single multi-pulsed note emitted at a call rate of 30 per minute. Additional data and figures are provided under the subtitle "Vocalizations of Ecuadorian Anurans" in Appendix 1.

Tadpoles.—Unknown.

Distribution.—See Figure 2.62. *Leptodactylus stenodema* is known from the Amazonian Basin and Guyana Shield in Brazil, Colombia, Ecuador, Peru, French Guiana, and Suriname. In Ecuador, it occurs in Sucumbíos, Orellana, Pastaza, and Morona Santiago provinces. Its elevational range is 231–1791 m in an area of extent of occurrence of about 41,100 km^2.

This nocturnal, terrestrial species inhabits forests and intervened areas in the Amazonian region in flooded forests of palms of alluvial plain of Amazonia, evergreen forests of lowlands, and evergreen low montane forest ecosystems (Table 0.4) within five biogeographic sectors (Table 0.5), where the average annual rainfall is 2953–4412 mm and the average annual temperature is 17.7–25.7 °C.

Conservation.—*Leptodactylus stenodema* is categorized as Least Concern throughout its entire range (Coloma *et al.*, 2010; in IUCN Red List of 2022) in view of its wide distribution, tolerance of a degree of habitat modification, and presumed large population. It is common in the Guianas and Brazil but uncommon in Colombia, Ecuador, and Peru. However, it is important to note that in Ecuador, *Leptodactylus stenodema* has been categorized as Critically Endangered by Ortega-Andrade *et al.* (2021) because they infer a general reduction in the population greater than 80%. Although this species has been rarely seen, we disagree with their inference because there is no evidence of sudden declines in lowland species, and there are many suitable habitats for the species within its extent of occurrence of 41,100 km^2 (not 27,956 km^2, as indicated by Ortega-Andrade *et al.*, 2021). Thus, we categorize it as Near Threatened. Habitat destruction and degradation, in particular agricultural expansion, logging, oil extraction and spills, and fires, are threats to this species. It occurs in many protected areas throughout its range. In Ecuador, it has been recorded in the public reserves Parque Nacional Sumaco Napo-Galeras and Parque Nacional Yasuní, the biosphere reserves Sumaco and Yasuní, and in some private reserves (e.g., Tiputini Biodiversity Station).

Remarks.—In a total evidence analysis of molecular (mitochondrial and nuclear markers) and non-molecular features (adult and larval morphological and behavioral characters) by de Sá *et al.* (2014), *Leptodactylus stenodema* is the sister taxon to a lineage that includes *L. vastus*, *L. paraensis*, and at least nine other species.

Key references.—(Jiménez de la Espada, 1875) (description, osteology, diet); (Pyburn and Heyer, 1975) (call, natural history); (Duellman, 1978) (natural history); (Heyer, 1979) (call); (Almendáriz, 1987) (natural history); (Lescure and Marty, 2000) (call); (Lima *et al.*, 2012) (account); (de Sá *et al.*, 2014) (account, call).

LEPTODACTYLUS VENTRIMACULATUS (SPOTTED-VENTER THIN-TOED FROG) BOULENGER, 1902

Leptodactylus ventrimaculatus Boulenger, 1902: 53. **Lectotype.**—BMNH 1947.2.17.78 from Pulún, 49 m, Provincia Esmeraldas, Ecuador.

Etymology.—The specific name is from the Latin *venter*, meaning belly, and *maculatus*, meaning spotted, variegated, full of spots.

Identification.—Figures 2.67, 2.68, 2.69. *Leptodactylus ventrimaculatus* has a maximum known SVL of 60.3 mm in males and 63.2 mm in females. Males do not have a spatulate snout, and they lack thumb spines and chest spines. The light upper lip stripe is absent. The dorsal and lateral folds are absent; the dorsolateral folds usually are present. The

FIGURE 2.67 *Leptodactylus ventrimaculatus*. Left: CJ 2844, female, 52.0 mm SVL, Reserva Otokiki, Provincia Esmeraldas. LAC. Right: CJ 8824, male, 47.2 mm SVL, Bosque Protector Siete Cascadas, Provincia Esmeraldas. DAL.

FIGURE 2.68 Morphological variation in *Leptodactylus ventrimaculatus*. Top left: CJ 10758, female, 50.0 mm SVL, Comunidad Awa Sabaleta, Provincia Carchi. DAL. Top right: QCAZ 31263, female, El Cinto, Provincia Pichincha. LAC. Middle left: CJ 8726, male, Bosque Protector Siete Cascadas, Provincia Esmeraldas. DAL. Middle right: CJ 1023, female, 57.2 mm SVL, Guayabillas, Provincia Pichincha. LAC. Bottom left: CJ 2832, male, 51.7 mm SVL, Sapoparque La Florida, Provincia Santo Domingo de los Tsáchilas. LAC. Bottom left: CJ 3088, Reserva Otokiki, Provincia Esmeraldas. LAC.

FIGURE 2.69 Morphological variation and details of *Leptodactylus ventrimaculatus*. Top left: CJ 966, Comunidad Mashpi, Provincia Pichincha. LAC. Top right: CJ 904, female, 43.6 mm SVL, Reserva Otokiki, Provincia Esmeraldas. LAC. Bottom: CJ 8824, Bosque Protector Siete Cascadas, Provincia Esmeraldas. DAL.

light stripe on the posterior thigh is almost always absent (97% of specimens) and rarely indistinct (3%); the upper shank is barred; the belly is slightly to heavily mottled; the toes lack lateral fringes.

The dorsum exhibits a range from dark to pale brown, adorned with irregular brown marks and subtle cross bars. It may be spotted, striped, or occasionally uniform. The dark suborbital bar is nearly always present. The two dorsolateral folds are dark brown or slightly orange. The anterior and posterior surfaces of the hind limbs are speckled with dark brown on a cream background; dark brown stripes run across the dorsal surfaces of the tibia and arms, interspersed with light brown. The dorsal surface of the tibia and the posterior surface of the tarsus have many white tubercles. The plantar surface of the foot has scattered or very few white tubercles. The belly is light brown or dark with subtle light spots. The ventrolateral region exhibits opaque yellow tones against a cream background. The throat is either white or has opaque brown spots. The upper lip is predominantly brown with cream spots. The iris is dark brown with fine reticulations, transitioning to golden brown in the upper half and silver in the lower half.

Comparisons.—*Leptodactylus labrosus*, *L. rhodomerus*. *Leptodactylus ventrimaculatus* is mostly similar to species that lack a distinct light posterior thigh stripe and lack toe fringes. *Leptodactylus ventrimaculatus* has distinct, scattered, or very few white tubercles on the plantar surface of the foot, whereas *L. labrosus* usually (91% of specimens) lacks white tubercles on the plantar surface of the foot. *Leptodactylus ventrimaculatus* is smaller (63.2 maximum SVL) than *L. rhodomerus* (157.8 mm maximum SVL), and their dorsum color patterns are different.

Natural history.—Figure 2.70. *Leptodactylus ventrimaculatus* is sympatric with *L. labrosus* in the Seasonal evergreen forests at Centro Científico Río Palenque.

Tadpoles.—Figure 2.71. Andrea Terán-Valdez and Luis A. Coloma (this publication). This description is based on an individual (series CJ 8767) from Bosque Protector Siete Cascadas, Provincia Esmeraldas, in stage 36. It was collected by Paola L. Tituaña Males. It has a body length of 12.2 mm and a total length of 34.0 mm. The body is ovoid and depressed, slightly wider anteriorly than posteriorly, with the deepest point posteriorly and a width surpassing its depth. The snout is rounded in both dorsal and lateral views, and the eyes are directed dorsolaterally. The spiracle is sinistral, with its inner wall free from the body, opening posterolaterally near the midlength of the body. The vent tube is wide and medial. The oral disc has an anteroventral position, featuring marginal papillae present laterally and ventrally, with a broad anterior labium gap. The labia are detached from the body. The upper jaw sheath has a smooth arch with a central V on the lower edge. The tail constitutes approximately 69% of the total length, reaching its maximum depth around midlength and ending in a rounded tip. The dorsal fin does not extend onto the body, reaching its deepest point at the midlength of the tail and tapering

FIGURE 2.70 *Leptodactylus ventrimaculatus.* Individuals in their microhabitats in Playón de San Francisco. Provincia Esmeraldas. LAC.

gradually toward the end. The caudal musculature is deepest at its anterior part. Golden dots forming lines are visible on the dorsum as part of the lateral line system.

The dorsum is nearly uniformly brown, with numerous golden flecks on the lateral areas. Ventrally, the skin is translucent with scarce golden flecks. The fins are translucent, and the dorsal fin exhibits abundant brown and golden flecks, while the caudal musculature has brown flecks distributed along the tail with very few golden spots. The iris has a golden coloration.

Distribution.—See Figure 2.57. *Leptodactylus ventrimaculatus* is known in the Chocó of Colombia and lowlands in Ecuador. In Ecuador, it occurs in Esmeraldas, Carchi, Imbabura, Pichincha, Santo Domingo de los Tsáchilas, Cotopaxi, Manabí, Guayas, and El Oro provinces. Its elevational range is 8–1796 m in an area of extent of occurrence of about 81,410 km^2.

This nocturnal, terrestrial species inhabits rainforests, cloud forests, and intervened areas in northwestern Ecuador in nine ecosystems (Table 0.4) within five biogeographic sectors (Table 0.5), where the average annual rainfall is 335–3141 mm and the average annual temperature is 17.0–25.7 °C.

FIGURE 2.71 Ontogenetic changes in *Leptodactylus ventrimaculatus* from tadpole to adult. Top to bottom, left to right. CJ 8767, lateral, dorsal, and ventral views, stage 36, TL = 34.0 mm; stage 41 = 43.6 mm; stage 43 = 40.6 mm; stage 45 = 15.1 mm SVL; juvenile 44 days after stage 45 = 19.1 mm SVL; juvenile 111 days = 20.7 mm SVL; juvenile 267 days = 32.0 mm SVL; adult, 359 days = 44.3 mm SVL. Bosque Protector Siete Cascadas, Provincia Esmeraldas. DAL.

Conservation.—*Leptodactylus ventrimaculatus* is categorized as Least Concern throughout its entire range (IUCN SSC Amphibian Specialist Group, 2019g, in IUCN Red List of 2022) in view of its wide distribution, tolerance of a broad range of habitats, presumed large population, and because it is unlikely to be declining fast enough to qualify for listing in a more threatened category. It is common in Ecuador. It is threatened by agrochemicals, but persists in severely deforested areas. In Ecuador, it has been recorded in the public reserves Reserva Ecológica Mache Chindul and Reserva Ecológica Manglares Churute; the protected forests of Asociación Agrícola Carchi-Imbabura, Mashpi, Milpe Pachijal, Predio Pacay, Río Arenillas Presa Tahuín, and Toachi Pilatón; and some private reserves (e.g., Centro Científico Río Palenque).

Remarks.—In a total evidence analysis of molecular (mitochondrial and nuclear markers) and non-molecular features (adult and larval morphological and behavioral characters) by de Sá *et al.* (2014), *Leptodactylus ventrimaculatus* is the sister taxon to *L. labrosus*.

According to Cisneros-Heredia (2006), McDiarmid and Miyata (cited in Heyer and Maxson, 1982) found habitat segregation between *Leptodactylus ventrimaculatus* and *L. labrosus* at Centro Científico Río Palenque; *L. labrosus* was present in pastures and other parts of cleared or highly disturbed forests, while *L. ventrimaculatus* was only found in the closed canopied forest. Specimens collected some years ago from the *L. labrosus* and *L. ventrimaculatus* contact zone at the surroundings of Centro Científico Río Palenque were reviewed by Heyer, who noted:

> There were quite a few specimens that I was unsure which species they were. I wonder if with the amount of deforestation that [*L.*] *labrosus* is coming into contact with [*L.*] *ventrimaculatus* much more frequently and perhaps there is some gene exchange going on that was not happening before. It would be worth getting a lot of advertisement call recordings and molecular data from a lot of localities of these two species to find out what is going on.

Yánez-Muñoz *et al.* (2020) incorrectly labeled their figure 113 as *Leptodactylus rhodomerus* (instead of *L. ventrimaculatus*).

Key references.—(Boulenger, 1902) (description); (Heyer and Maxson, 1982) (habitat); (de Sá *et al.*, 2014) (phylogeny, account); (Yánez-Muñoz *et al.*, 2020) (account).

FIGURE 2.72 *Leptodactylus wagneri*. Left: CJ 6255, female, 70.7 mm SVL. Right: CJ 6405, male, 62.5 mm SVL. Shiripuno Lodge, Provincia Orellana. DAL.

LEPTODACTYLUS WAGNERI (WAGNER'S THIN-TOED FROG) (PETERS, 1862a)

Plectromantis wagneri (Peters, 1862a): 232. **Type(s).**—ZSM 1080/0, lost, from "an den Westseite der Anden in Ecuador," data associated with lost type specimen was "Pastassathal" (= Pastaza Valley), Ecuador. **Neotype.**—QCAZ-A 26219, outskirts of Puyo, 1.443070°S, 77.996660°W; 954 m, Provincia Pastaza, Ecuador.
Leptodactylus wagneri Nieden, 1923: 479.

Etymology.—The specific name honors the collector Dr. Moritz Wagner (1813–1887). He was a German explorer, collector, geographer, and natural historian who developed the concept that geographical isolation could play a key role in speciation.

Identification.—Figures 2.72, 2.73. *Leptodactylus wagneri* has a maximum known SVL of 60.7 mm in males and 84.3 mm in females. The dorsolateral fold is long. The toe tips are unexpanded. The upper lip stripe extends from the posterior corner of the eye. The belly is heavily mottled, forming an anastomotic pattern. There are two keratinized spines on the base of the thumb in adult males. The advertisement call has a single note. The call is formed by incomplete pulses. The call has modest to pronounced frequency modulation.

Comparisons.—*Leptodactylus discodactylus*, *L. leptodactyloides*, *L. petersii*. *Leptodactylus wagneri* is distinguished from all other species of *Leptodactylus* in the *melanonotus* group by the combination of large SVL (a maximum SVL of 60.7 mm in males and 84.3 mm in females) and a dorsolateral fold extending from the eye to the groin. The species is also distinguished from *L. discodactylus* by having the toe tips unexpanded (toe discs expanded, with dorsal grooves in *L. discodactylus*). *Leptodactylus wagneri* is distinguished by an upper lip stripe extending from the posterior corner of the eye, extending from under the mid-eye or over the snout in *L. melanonotus*. *Leptodactylus wagneri* has an anastomotic belly pattern differing from the ventral coloration of almost all species of the group except *L. petersii*. *Leptodactylus wagneri* is also distinguished by the presence of keratinized thumb spines in adult males, as well as frequency-modulated calls formed by incomplete pulses. When combined, these two acoustic traits distinguish *L. wagneri* from all other species of the *L. melanonotus* group. *Leptodactylus wagneri* is larger than *L. leptodactyloides* (*L. leptodactyloides* females 35–56 mm SVL, males 28–48 mm SVL). Many *L. wagneri* have moderately to boldly mottled bellies, whereas the bellies of *L. leptodactyloides* characteristically are finely mottled. *Leptodactylus wagneri* is larger than *L. petersii* (females 31–51 mm SVL, males 27–41 mm SVL); the dorsolateral folds of *L. wagneri* commonly extend from behind the eye to the groin, and no *L. petersii* individuals have dorsolateral folds extending from the eye to the groin.

Natural history.—Figure 2.74. *Leptodactylus wagneri* occurs in a wide variety of habitats, including river edges, forests, forest clearings, and shallow swamps.

The calls from near Pacto Sumaco, Provincia Napo, Ecuador, and from Ichigkat, Eastern Cordillera, Peru, consist of a single, partly pulsed note given at a rate of 50 to 55 per minute. Additional data on vocalizations are provided under the subtitle "Vocalizations of Ecuadorian Anurans" in Appendix 1.

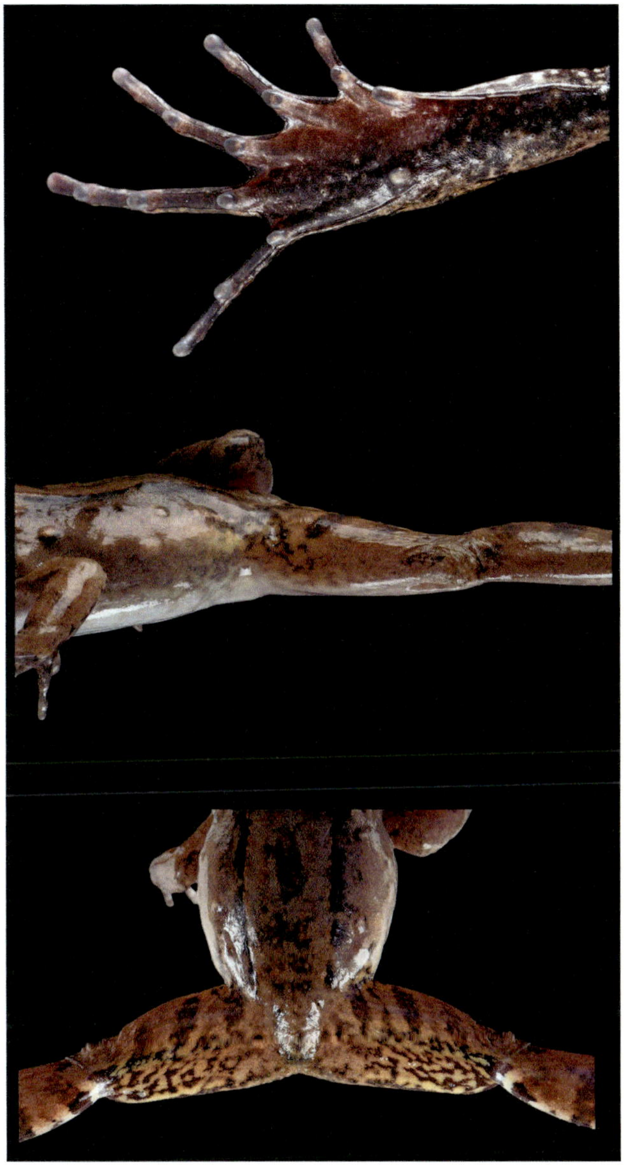

FIGURE 2.73 Morphological details of *Leptodactylus wagneri*. Top: CJ 6252, female, 82.3 mm SVL, plant. Middle and bottom: CJ 6313, juvenile, 39.7 mm SVL, groin and dorsum of thighs. Shiripuno Lodge, Provincia Orellana. DAL.

FIGURE 2.74 *Leptodactylus wagneri*. CJ 6252, female, active individual on a ground trail at Shiripuno Lodge, Provincia Orellana. LAC.

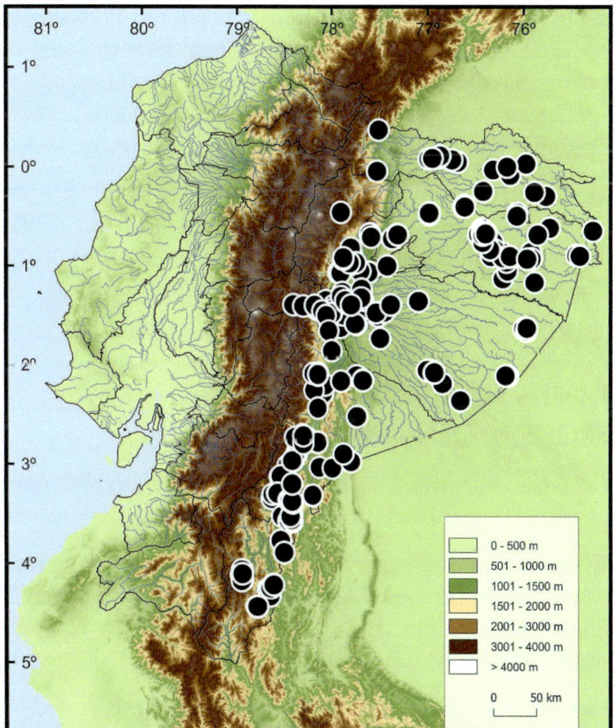

FIGURE 2.75 Localities of known occurrence of *Leptodactylus wagneri* in Ecuador.

Tadpoles.—Unknown.

Distribution.—Figure 2.75. *Leptodactylus wagneri* is known on the Amazonian slopes of the Andes in southern Colombia, Ecuador, and northern Peru, with a few records from lowland Amazonia (Brazil, Colombia, Ecuador, Peru). In Ecuador, it is known from Sucumbíos, Orellana, Napo, Tungurahua, Pastaza, Morona Santiago, and Zamora Chinchipe provinces. Its elevational range is 177–1859 m in an area of extent of occurrence of about 107,340 km^2.

This nocturnal, terrestrial species inhabits rainforests, flooded forests, foothill forests, and intervened areas in Amazonian Ecuador in 17 ecosystems (Table 0.4) within seven biogeographic sectors (Table 0.5), where the average annual rainfall is 1080–4434 mm and the average annual temperature is 17.1–25.7 °C.

Conservation.—*Leptodactylus wagneri* is categorized as Least Concern throughout its entire range (Silvano *et al.*, 2004; in IUCN Red List of 2022) and in Ecuador by Ortega-Andrade *et al.* (2021) in view of its wide distribution, tolerance of a broad range of habitats, presumed large population, and because it is unlikely to be declining fast enough to qualify for listing in a more threatened category. Its abundance is variable throughout its range. There are no serious threats to this species because it can adapt to modified habitats. It occurs in several protected areas throughout its range. In Ecuador, it has been recorded in the public reserves Parque Nacional Cayambe Coca, Reserva

Biológica Limoncocha, Reserva de Producción de Fauna Cuyabeno, Parque Nacional Yasuní, Parque Nacional Llanganates, Parque Nacional Sangay, and Área Ecológica de Conservación Municipal Siete Iglesias; in several protected forests such as Abanico, CEPLOA, Cerro Sumaco y Cuenca Alta del Río Suno, Cordillera Kutukú y Shaimi, and Cuenca Alta del Río Nangaritza; the biosphere reserves Sumaco, Yasuní, and Podocarpus-El Cóndor; and in some private reserves (e.g., Tiputini Biodiversity Station).

Remarks.—In a total evidence analysis of molecular (mitochondrial and nuclear markers) and non-molecular features (adult and larval morphological and behavioral characters) by de Sá *et al.* (2014), *Leptodactylus wagneri* is the sister taxon to *L. colombiensis* and closely related to three species of the *L. validus* complex. *Leptodactylus wagneri* and *L. leptodactyloides* have been historically treated as a single species, *L. wagneri*, until Heyer's (1994) taxonomic revision, in which he recognized two sympatric species. Thus, identifications, natural history, and other data provided prior to 1994 should be taken with caution because they might refer to a mix of both species (Duellman, 1978; Toft and Duellman, 1979; Aichinger, 1987; Zimmerman 1983; Almendáriz, 1987; Duellman and Salas, 1991; Rodríguez and Duellman, 1994).

Key references.—(Peters, 1862a) (description); (Heyer, 1994) (taxonomy); (Duellman and Mendelson, 1995) (account); (de Sá *et al.*, 2014) (phylogeny, account); (de Carvalho *et al.*, 2022) (account).

GENUS *LITHODYTES* (ANT-NEST FROGS) FITZINGER, 1843

Lithodytes Fitzinger, 1843b: 31. **Type species.**—*Hylodes lineatus* (Duméril and Bibron, 1841), by original designation.

Etymology.—The generic name is derived from the Greek words *lithos*, meaning stone, and *dytes*, meaning diver. It might refer to a presumed association with rocky environments. The genus is masculine in gender.

Definition.—Figure 2.76. *Lithodytes* is a group supported by molecular evidence. It is a monotypic genus with medium-sized species (SVL to 56 mm in females). The skin on the dorsum is finely spiculate, and the skin on the venter is smooth. The snout is rounded in dorsal view and in profile, and the tympanum is distinct and about the same diameter as the eye. Finger I is longer than finger II, and webbing is absent. *Lithodytes lineatus* uses large *Atta cephalotes* ant nests as calling and breeding sites. Eggs are deposited in foam nests in underground galleries of the ant nests. The tadpoles are translucent red.

Distribution.—The genus occurs in Eastern Peru, Ecuador, Bolivia, and Brazil through southern Venezuela to the Guianas.

Content.—Currently, one species is recognized in this genus. It is present in Ecuador.

Remarks.—Frost *et al.* (2006), Pyron and Wiens (2011), and de Sá *et al.* (2014) recovered *Lithodytes* as the sister taxon of *Adenomera*.

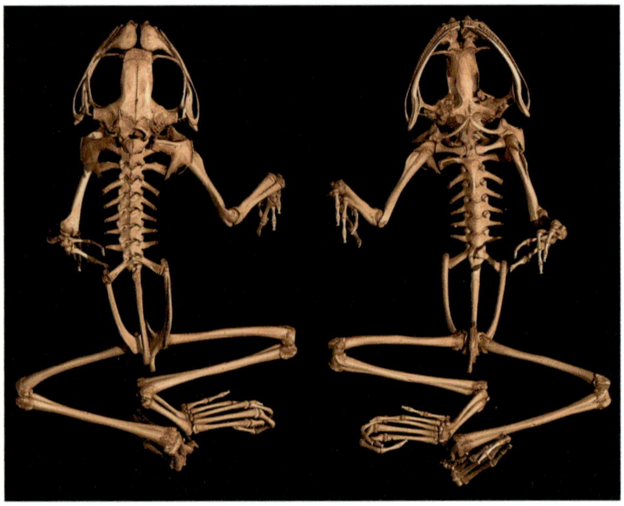

FIGURE 2.76 High-resolution computed tomography reconstructions of the skeleton of *Lithodytes lineatus*, MNHN 1978.335, 42.6 mm SVL, Crique Grégoire Sinnamary, Guyane Française. Dorsal (left) and ventral (right) views. Daniel J. Paluh.

LITHODYTES LINEATUS (STRIPED ANT-NEST FROG) (SCHNEIDER, 1799)

Rana lineata (Schneider, 1799): 105, and Daudin, 1803 "An. XI," Hist. Nat. Gen. Part. Rept., 8: 188). **Type locality.**—Not stated.

Lithodytes lineatus Fitzinger, 1843b 43.

Etymology.—The specific epithet derives from the Latin *lineatus*, meaning stripe or line, presumably named in reference to its bright dorsolateral stripes.

Identification.—Figures 2.77, 2.78. In these frogs, males attain a maximum SVL of 47.0 mm, and females of 56.0 mm. The maximum known masses are 5.8 g in males and 12.5 g in a female. The skin on the dorsum is finely spiculate, and the skin on the venter is smooth. The snout is rounded in dorsal view and in profile, and the tympanum is distinct and about the same diameter as the eye. Finger I is longer than the II, and webbing is absent. All dorsal surfaces and the flanks are black, except the upper arms, which are tan. A broad, bright yellow stripe extends from the tip of the snout, along the margin of the upper eyelid, and onto the groin. The stripes are creamy yellow in juveniles and golden tan in adults.

The dorsal surfaces of the limbs are pale brown with dark brown transverse marks. A large red spot is present in the groin, and a slightly smaller red spot is present on the posterior surface of each thigh. The throat and chest are brown, and the belly and ventral surfaces of the hind limbs are gray with creamy yellow flecks laterally. The iris is a dark coppery bronze with fine black reticulations.

Comparisons.—*Ameerega hahneli*, *Allobates femoralis*. This brightly colored frog with a black dorsum and

FIGURE 2.77 *Lithodytes lineatus.* Left: CJ 6352, female, 45.6 mm SVL, Shiripuno Lodge, Provincia Orellana. Right: QCAZ 33138, Pañacocha, Provincia Sucumbíos. LAC.

FIGURE 2.78 Morphological details of *Lithodytes lineatus.* CJ 6352, female, Shiripuno Lodge, Provincia Orellana. LAC.

yellow to tan dorsolateral stripes can be confused only with some dendrobatids, all of which have scutes on the dorsal surfaces of the terminal segments of each digit. Of these dendrobatids, *Ameerega hahneli* and *Allobates femoralis* resemble *L. lineatus* in having dorsolateral light stripes, but both have mottled venters. In addition, a lateral white stripe is present in *A. femoralis*. *Lithodytes* has a large red inguinal spot and another red spot on the posterior surface of the thigh, whereas red spots are lacking in the dendrobatids.

Natural history.—In Santa Cecilia, Provincia Sucumbíos, frogs of *Lithodytes lineatus* have been found in primary forests, secondary forests, and at the edges of forests. It is active by day and night. Eight individuals were found active on the ground by day and four by night; of the latter, one was sitting on a log. Two females found in May and July contained 195 and 230 ovarian eggs. Nine individuals contained a variety of small insects. Ants were present in seven stomachs; fly larvae in three; coleopterans, hemipterans, and orthopterans in two; and termites in one.

At Montalvo-Pucayacu, Provincia Pastaza, individuals were found on the ground of a secondary forest and under vegetation at the forest edge during day and night. Juveniles and adults were found in August and December. An adult female contained pigmented oocytes of synchronic development of 1 mm in diameter. The diet of two frogs revealed two coleopterans (scolitids) and two ants. On 15 August 1983 the mating call of *Lithodytes lineatus* was heard at 21:30 h. Males were observed calling from the entrances of nests of ants of the genus *Atta* in secondary forests. The area from which the chorus sounded was a little raised and covered by a leaf-cutter ant nest (*Atta* sp.). The ground was bare, without the usual layer of fallen leaves, and there were several piles of earth. The trees in this area were quite stunted, and the branches in the lower area had no more leaves. At several points, ants were strolling out of the nest in the surrounding forest. The entrances to the nest were situated at the base of recently elevated mounds of earth, predominantly on the south side, and led into tubes several centimeters wide. The ants were constantly busy filling up new earth by depositing small globules in large numbers on the piles. The freshly poured earth was easy to recognize by its light-yellow color. There, the males of *L. lineatus* sat and called. Each entrance had one or two animals. In total, seven specimens were captured, and another was observed. Not all the exits of the nest were located, and the total area examined within the nest was 30–50 m^2. The stomach contents of three animals were examined. One stomach was empty, one contained an unidentifiable white mass, and one contained the head of an ant. Apparently, the animals hardly eat any food during their calling activity.

At Jaburi Farm, Estado Rondônia, northwestern Brazil, from April 2001 to March 2002, information about *Lithodytes lineatus* was gathered using pitfall traps with drift fences as a survey method. The results indicated a higher capture frequency compared to that observed in other studies conducted in different Amazonian localities. This may be attributed to the species' secretive habits, including calling and breeding from nests of leaf-cutting ants, making them challenging to detect through visual encounter surveys. The breeding period occurs between September and March. Calling males and egg-bearing females were found between

September and February, and juvenile recruitment occurred mainly from the end of the rainy season to the beginning of the dry season (February to June). The number of ovarian eggs per female varies from 110 to 328, and analyses indicate that there is a significant correlation with SVL.

At Cusco Amazónico, Peru, one adult female was under the edge of a large log in terra firme forest at 22:20 h on 17 November 1986; on the following day, at 19:20 h, a male was calling from the leaf litter at the edge of the camp clearing. Two additional adults were on the ground (one by day and one at night) in terra firme forest in April. A female having an SVL of 51.0 mm obtained on 17 November contained unpigmented eggs 1.5 mm in diameter. No breeding was observed or tadpoles found at Cusco Amazónico. Neither of the individuals from Cusco Amazónico for which habitat data are available was near ant nests. The call is a continuous series of short whistles produced at a rate of about 90 notes/min. There were 35 prey items in nine prey categories in the digestive tracts of two individuals: one of these had regurgitated two earthworms that were not included in the analysis. Nineteen ants made up 54.3% of the number of prey but accounted for only 8.1% of the volume, whereas the greatest volumes came from a cricket (33.9%), an earthworm (22.0%), isopods (12.6%), and insect larvae (12.4%).

At Tambopata, Peru, *Lithodytes lineatus* uses large *Atta cephalotes* ant nests as breeding sites. Four males called synchronously from the interior of the same nest. In another *Atta* nest in the same area, one adult of *L. lineatus* and numerous tadpoles were found in a water-filled depression at the end of a system of underwater burrows. A vertical tunnel directly above the pool measured around 25 cm in diameter, and foam nests were adhered to the roots protruding from the otherwise smooth walls. The channel was frequented by a stream of descending ants. The tadpoles were at different developmental stages, reaching from individuals without legs to one metamorphosed juvenile frog. Adult and subadult ant frogs have also been observed along pathways in the forest or sitting in a hollow log without any obvious association with a visible ant nest. It is not clear whether these frogs represent juveniles that have dispersed from the nests or whether they are females or males that are temporarily sexually inactive. The stomach contents of one *L. lineatus* specimen caught at the entrance of an *Atta* nest contained only two nymphs of the predatory insect genus *Heza* (Reduviidae). The call is a continuous series of short whistles repeated at a rate of about 80–90 notes/min. Additional data on vocalizations are provided under the subtitle "Vocalizations of Ecuadorian Anurans" in Appendix 1.

At Villavicencio, Colombia, eggs were deposited at the mouth of a partially submerged burrow.

Eggs originating from captive-bred parents from Panguana, Peru, were deposited in a foam nest in water beneath a piece of bark within a terrarium. A total of 83 tadpoles hatched from the foam nest. The tadpoles left the foam nest 7–15 days after its construction, and they commenced feeding approximately three days after hatching.

FIGURE 2.79 Tadpole of *Lithodytes lineatus*. QCAZ 29257, Centro Experimental Fátima, Provincia Pastaza. LAC.

Tadpoles.—Figure 2.79. There are partial descriptions of tadpoles from captive-bred parents from Panguana, Peru. The tadpole was described based on an individual in stage 34 from Villavicencio, Departamento del Meta, Colombia. The tadpole has a body length of 18.1 mm and a total length of 47.4 mm. In dorsal view, the body is elongate, ovoid, widest at about one-third of its length and wider than high. The snout is truncate in dorsal view and bluntly rounded in profile. The eyes are small, situated dorsally, and directed laterally and slightly anterodorsally. The interorbital distance is about four times the diameter of the eye. The nares are small and situated about midway between the tip of the snout and the eyes. The spiracle is sinistral, well below the midline of the body; the origin of the spiracular tube is slightly posterior to the midlength of the body; the entire length of the tube is attached to the body, and its opening is directed posterodorsally. The vent tube is medial, moderately long, and attached to the ventral fin, but its tip is free. The caudal musculature is moderately robust, highest at about one-third the length of the tail. It gradually tapers toward the posterior end, culminating just before the rounded tip of the tail. The dorsal fin originates at the base of the tail and is highest at about the midlength of the tail. The ventral fin originates at the base of the tail and it is highest at about one-third of the length of the tail, where it is equal in height to the dorsal fin. The fins gradually narrow to a rounded tip. At the midlength of the tail, both fins are shallower than the caudal musculature. The oral disc is anteroventral and not emarginate. The anterior labium is bare; elsewhere, there is a single row of long marginal papillae. Submarginal papillae are absent. The jaw sheaths are slender, broadly arched, and finely serrate. The labial tooth row formula is 0/1(1). The presence of jaw sheaths and the labial tooth row formula are highly variable, 0/0 to 2(2)/3(1). In life, the body is bright pink with a short, metallic white middorsal stripe. The caudal musculature is colored like the body, but the color is less intense distally. The fins are translucent pink.

Distribution.—Figure 2.80. *Lithodytes lineatus* is known in eastern and northwestern Venezuela, south along the eastern side of the Andean foothills to northern Bolivia, and east in the Amazonian drainage to the states of Tocantins and Maranhão. In Ecuador, it occurs in Sucumbíos, Napo, Orellana, Pastaza, Morona Santiago, and Zamora Chinchipe provinces. Its elevational range is 211 in an area of extent of occurrence of about 86,190 km^2.

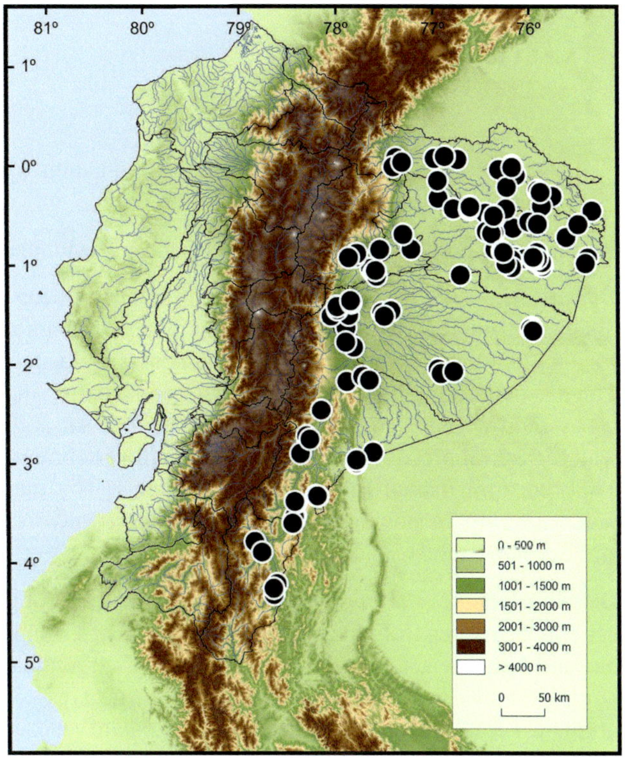

FIGURE 2.80 Localities of known occurrence of *Lithodytes lineatus* in Ecuador.

This diurnal and nocturnal terrestrial species inhabits tropical rainforests and secondary areas in 17 ecosystems (Table 0.4) within seven biogeographic sectors (Table 0.5), where the average annual rainfall is and the average annual temperature is 20.4–25.8 °C.

Conservation.—*Lithodytes lineatus* is listed as Least Concern throughout its entire range (La Marca *et al*., 2010c; in IUCN Red List of 2022) and in Ecuador by Ortega-Andrade *et al.* (2021) in view of its wide distribution, tolerance of a degree of habitat modification, presumed large population, and because it is unlikely to be declining to qualify for listing in a more threatened category. It is locally common throughout its range. There are no major threats to this species. In Ecuador, it has been recorded in the public reserves Parque Nacional Cayambe Coca, Reserva Biológica Limoncocha, Reserva de Producción de Fauna Cuyabeno, Parque Nacional Sumaco Napo-Galeras, and Parque Nacional Yasuní; several protected forests such as Pañacocha, Sacha Lodge, Río Aguarico, Cordillera Kutukú y Shaimi, and Cuenca Alta del Río Nangaritza; the biosphere reserves Sumaco, Yasuní, and Podocarpus-El Cóndor; and some private reserves (e.g., Shiripuno Lodge).

Remarks.—de Sá *et al.* (2014) suggest the existence of at least two species under the name *Lithodytes lineatus*. The coloration of *Lithodytes* sufficiently resembles that of the poisonous *Allobates femoralis* that the former could be considered a mimic and the latter the model. The resemblance is particularly striking in juvenile *Lithodytes*. Both species are terrestrial forest inhabitants from the Guianas to Peru; however, *Lithodytes* is active by day and night, whereas *Allobates* is diurnal.

This species has been reported to live in association with the large leaf-cutter ant nests (*Atta* spp.) during the breeding period at several sites. The association of *L. lineatus* with the leaf-cutter ant *Atta cephalotes* was first mentioned by Schlüter (1980), who recorded advertisement calls emanating from the interior of ant nests at the biological field station Panguana, Rio Llullapichis drainage, Departmento Huanuco, Peru. Based on additional observations on the frog's association with *A. cephalotes*, Schlütter and Regös (1981) suggested a mutual relationship ("true symbiosis"?) between the ants and frogs. Furthermore, additional biological aspects of *L. lineatus* such as different call types, foam nest-building behavior in captivity, and tadpoles of captive-raised individuals were reported by Regös and Schlüter (1984). Subsequently, Hoogmoed (1986), De la Riva (1993), Marquez *et al.* (1995) and Lamar and Wild (1995) confirmed the frog's association with leaf-cutter ants, the former in Montalvo, Provincia Pastaza, Ecuador.

Chemical-based mimicry and camouflage is used by *Lithodytes lineatus* to avoid being detected by social insects. Experiments demonstrate that the skin of *L. lineatus* has chemicals that prevent the attack of two species of leaf-cutting ants, *Atta laevigata* and *A. sexdens*. Individuals of *L. lineatus* were never attacked by the leaf-cutting ants of the genus *Atta*, while 100% of four other anuran species were. In addition, none of the 10 individuals of *Rhinella major* coated with skin extracts of frog *L. lineatus* were attacked, whereas controls (coated with ultrapure water) were attacked on each occasion. Cintra *et al.* (2014) describe and provide images of a defensive deimatic behavior, consisting in lowering the head and lifting the rear of the body by stretching the legs vertically. In addition, the nictitating membrane may be closed during such deimatic behavior for protection.

Key references.—(Schneider, 1799) (description); (Duellman, 1978) (account); (Schlütter and Regös, 1981) (natural history); (Regös and Schlüter, 1984) (tadpole); (Hoogmoed, 1986) (natural history); (Hero, 1990) (tadpole); (Rodríguez and Duellman, 1994) (account); (Lamar and Wild, 1995) (tadpole from Villavicencio, Colombia); (Marquez *et al*., 1995) (call); (Schlüter and Regos, 1996) (tadpole); (Parmelee, 1999) (diet); (Lescure and Marty, 2000) (account); (Duellman, 2005) (account); (Kok and Kalamandeen, 2008) (account); (Schlütter *et al*., 2009) (natural history); (Bernarde and Kokubum, 2009) (natural history); (Ouboter and Jairam, 2012) (account); (Cole *et al*., 2013) (account); (Cintra *et al*., 2014) (natural history); (de Lima Barros *et al*., 2016) (mimicry); (Villacampa *et al*., 2017) (account); (Barrio-Amorós *et al*., 2019) (notes).

3 Family Ceratophryidae

HORNED FROGS

Ceratophryidae Tschudi, 1838: 26.

Etymology.—The familial name is derived from the Latin noun *cerastes*, meaning horned men of Crete. The name refers to the prominent supraciliary processes, characteristic of some species, which have been termed eyelid horns.

Definition.—Figure 3.1. Ceratophryid frogs are a monophyletic lineage diagnosed on the basis of molecular evidence and morphological features of tadpoles and adults. Among them are a continuous row of papilla on the upper lip in larvae and the skull shape.

Adult ceratophryids are megalophagous and even cannibalistic. They are terrestrial or aquatic. Eggs are deposited in ponds, and the tadpoles are carnivorous. They are moderately large to immense frogs with broad heads, large mouths, and robust bodies. In some species, a bony shield lies over the presacral vertebra and fuses with the dermis. The transverse processes of the anterior presacral vertebra are long and expanded distally. These frogs possess non-pedicellate, fang-like teeth. They lack the pars palatina of the maxilla and premaxilla; the sternum is cartilaginous, and the tips of the terminal phalanges are knobbed.

Distribution.—Ceratophryids occur from the Caribbean lowlands in Colombia and Venezuela to central Argentina, including the arid coastal areas of northern Peru and southern Ecuador, the Amazon Basin, the Caatinga in northeastern Brazil, the Cerrado in central Brazil, the Atlantic forest, the Gran Chaco, and the Pampean grasslands.

Content.—According to Faivovich *et al.* (2014) the family contains three genera with 12 recognized species. Only one genus (*Ceratophrys*) occurs in Ecuador. The other genera (*Chacophrys* and *Lepidobatrachus*) are restricted to the Gran Chaco.

Remarks.—Figure 3.2. The nomenclatural and systematic history of this lineage is detailed in Frost (2023 and onward). It has been ranked either as a family, subfamily, or tribe. Lynch (1971) and Fabrezi (2006) discussed the morphological evolution of Ceratophryidae. We follow Pyron and Wiens' (2011) arrangement as a family. The familial common name is taken from that of the genus *Ceratophrys*. Eyelid "horns" are absent in *Chacophrys*, *Lepidobatrachus*, and *Ceratophrys stolzmanni*.

GENUS *CERATOPHRYS* (HORNED FROGS) WIED-NEUWIED, 1824

Ceratophrys (Wied-Neuwied, 1824): 672. **Type species.**—*Ceratophrys varius* (Wied-Neuwied, 1824) (= *Bufo auritus* (Raddi, 1823), by subsequent designation by Fitzinger, 1843).

Etymology.—Same as for the family. The gender is feminine.

Definition.—*Ceratophrys* is a monophyletic lineage diagnosed on the basis of cranial characters, morphological features of tadpoles, and molecular evidence. The dermal roofing bones are fused with one another and co-ossified with the overlying dermis. They are sit-and-wait terrestrial predators of insects and small vertebrates. They perform nuchal (gular, neck) amplexus. The tadpoles develop in ponds; they are voracious carnivores and develop rapidly.

Distribution.—*Ceratophrys* is widely distributed in tropical South America from the Caribbean lowlands of Colombia and Venezuela to the Pampean region of Argentina, Uruguay, and Brazil, including the xeric coastal areas of northern Peru and southern Ecuador, Amazonian Venezuela, Colombia, Ecuador, Peru, Brazil, and the Guianas; the Caatinga in northeastern Brazil; the Cerrado in central Brazil, the Atlantic coastal forests of Brazil; and the Chacoan region of Argentina, Bolivia, Brazil, and Paraguay.

Content.—As presently recognized, the genus contains eight species (but see synonym under Remarks of *Ceratophrys cornuta*), two of which occur in Ecuador.

Remarks.—Figure 3.3. Phylogenetic analyses of molecular data subjected to parsimony analysis (Faivovich *et al.*, 2014) corroborated the monophyly of *Ceratophrys* and provided evidence that *Ceratophrys* is the sister taxon of *Chacophrys* + *Lepidobatrachus*. It also corroborated previous hypotheses among species of *Ceratophrys* based on morphology (e.g., Lynch, 1971, 1982; Fabrezi, 2006).

CERATOPHRYS CORNUTA (HORNED PAC-MAN FROG) (LINNAEUS, 1758)

Rana cornuta (Linnaeus, 1758): 212. **Type.**—Frog figured by Seba (1734: pl. 72, figs. 1–2). **Type locality.**—"Virginia"; in error.

Ceratophrys cornuta Schlegel, 1837: 29.

Etymology.—The specific name is derived from the Latin adjective *cornutus*, meaning "horned." The name refers to the prominent supraciliary processes, which have been termed eyelid horns.

DOI: 10.1201/9781003543879-3

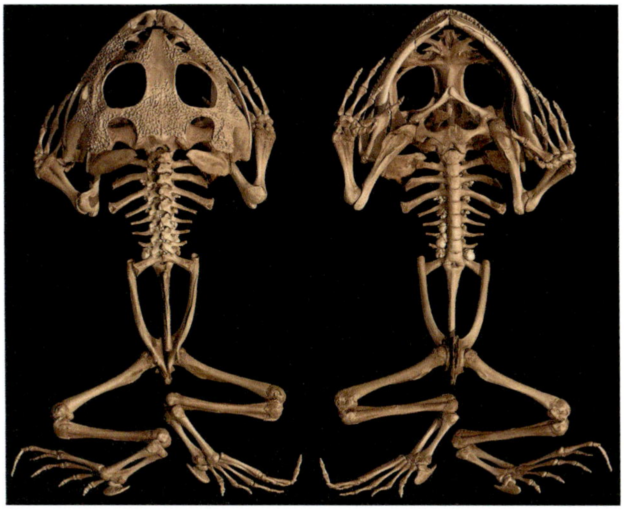

FIGURE 3.1 High-resolution computed tomography reconstructions of the skeleton of *Ceratophrys stolzmanni*, USNM 160970, male, SVL = 138.9 mm. Cuatro Hermanitos, 4 km west–northwest of Guayaquil, Provincia Guayas. Dorsal (left) and ventral (right) views. Daniel J. Paluh.

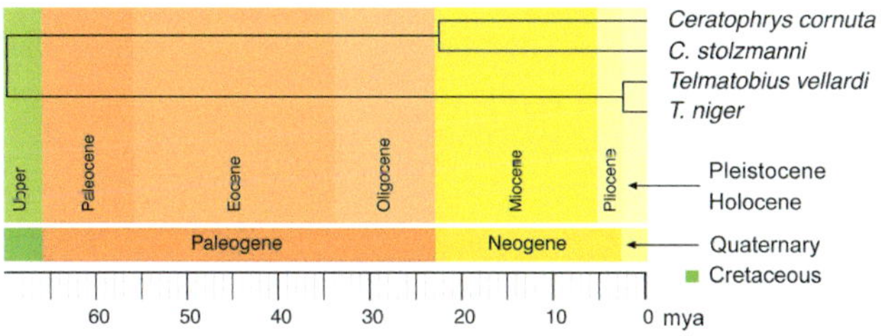

FIGURE 3.2 Time-calibrated maximum likelihood phylogenetic tree depicting evolutionary relationships of some of the Ecuadorian species of *Ceratophryidae* and *Telmatobiidae*, inferred from DNA sequences of mitochondrial and nuclear genes. Tree by Juan C. Santos (this publication).

FIGURE 3.3 Topology of the species of *Ceratophrys*, extracted from a phylogenetic tree inferred from parsimony analysis of DNA sequences by Faivovich *et al.* (2014).

Identification.—Figure 3.4. The maximum SVL in males is 80 mm, and females are notably larger, with a maximum known SVL of 117.5 mm. Weight (mass) is 56 g in males and 130 g in females. The head is immense; the large mouth is approximately 1.6 times wider than the length of the robust body. The supratympanic fold is poorly defined. Each mandible bears a medial odontoid directed toward the interior of the buccal cavity. The hind limbs are moderately short and robust. Finger I is longer than finger II. The fingers are unwebbed, and the toes are about one-half webbed. In dorsal view, the snout is rounded; in profile, it is steeply inclined anteroventrally from the anterior edges of the orbits. The skin on the dorsum and flanks is finely rugose with conical tubercles. A pair of paravertebral rows of laterally flattened tubercles extends from the occiput to the postsacral region. Long, pointed supraciliary processes give the impression of "horns" on the eyelids. A transverse row of low tubercles is present between the supraciliary processes, and a middorsal row of low tubercles extends from that interocular row to the end of the coccyx. The skin on the venter is nearly smooth, with small tubercles in the gular region. Breeding males have weakly developed, tan nuptial excrescences. A dorsal dermal shield is absent. A metatarsal keratinized dark brown spadix is present at the tarsus.

FIGURE 3.4 *Ceratophrys cornuta*. Top left: brown color morph, Parque Nacional Yasuní, Provincia Orellana. LAC. Top right: green color morph, KU 207699, Cusco Amazónico, Peru. Bottom left: burrowed into leaf litter at Cusco Amazónico, Peru. WED. Bottom right: gular amplexus in pair from Kaw Mountain, French Guiana. Jean-Pierre Vacher.

Color patterns are variable greens, browns, and beige with darker streaks of greens and browns. In life, about 50% of the individuals are green with brown markings, whereas the others are tan with brown markings. The snout and sides of the head usually are unicolor pale green or pale brown, but some individuals have dark brown diagonal marks in the labial region. In brown individuals, the region demarcated by the interocular and paravertebral rows of tubercles is pale tan. Brown markings consist of irregular spots lateral to the paravertebral rows of tubercles and transverse bars on the limbs. The belly and ventral surfaces of the limbs are dull cream, and the throat is brown (black in calling males).

Comparisons.—*Ceratophrys stolzmanni*. No other frog in the Amazon Basin can be confused with *C. cornuta*, which differs from *C. stolzmanni* by its distinctive long, pointed supraciliary processes; these are missing in the smaller *C. stolzmanni*.

Natural history.—Figure 3.4. *Ceratophrys cornuta* is widely distributed in terra firme and inundated rainforest areas in the Amazon Basin. These frogs prefer the leaf litter habitat associated with their cryptic coloration and burrowing habits. *Ceratophrys cornuta* seems to be far more abundant at some sites such as Cusco Amazónico, Peru, than elsewhere in its range. Sampling there during four rainy seasons from 1986 to 1991 provided an estimated minimal abundance of 1.8–2.1 individuals/ha, and it was not uncommon to find five or six non-calling individuals on any given night during the rainy season. Also, there was a highly significant correlation between the numbers of individuals observed and the average amount of rainfall for the two preceding 24-hour periods. Adult activity is mostly nocturnal, but juveniles were active during both day and night in leaf litter. The cryptic color pattern and the habit of burrowing into the leaf litter with only the head exposed makes these frogs extremely difficult to see. Adults found in the dry season were partially buried in the leaf litter; presumably, the frogs bury deep in the leaf litter during the dry season and thus are seldom observed. Daily movements of 18 individuals were monitored using tracking devices during a 45-day period in November and December 1991. It was found that the frogs moved only at night; the distances moved per night were 0–49 m; most movement occurred on rainy nights; and a significant positive correlation existed between SVL and the distance moved.

Ceratophrys cornuta is a major predator of other amphibians, both as larvae and as adults. Stomach content analysis found that large prey could account for more than half the volumetric content of an adult's diet. However, in contrast to observations in captivity that species of *Ceratophrys* feed only on large prey, *C. cornuta* feeds on diverse sizes of prey. If large orthopterans or small vertebrates do not pass within the feeding range of an individual, the frog will feed on smaller and more abundant prey, such as ants, until larger prey is available. Duellman and Lizana (1994) likened this behavior to the habit of surviving on numerous small snacks between infrequent feasts. An individual from Santa Cecilia, Ecuador, contained an adult frog, *Edalorhina perezi*; one spider; and four orthopterans. An analysis of the contents of the digestive tracts of 72 individuals, including juveniles, subadults, and adults, revealed that *C. cornuta*, a sit-and-wait predator, preys on a wide variety of organisms that inhabit the leaf litter on the forest floor, as well as two crab species that usually occur in water. Ants, beetles, and vertebrates were found in individuals of all size classes, but the proportion of ants decreased with the size of the frogs; the proportion of ants in the diet ranged from 91% in juveniles to 4% in adult females. Numerically, ants were the most common prey (70.22% of the prey), followed by beetles (8.28% of the prey). Vertebrates made up 53.05% of the volume, followed by grasshoppers at 22.22%. Other prey items included earthworms, millipedes, spiders, and various other insects. Vertebrate prey included six species of frogs (*Rhinella* sp. (*margaritifer* complex), *Pristimantis toftae*, *Hamptophryne boliviana*, *Dendropsophus parviceps*, *Boana punctata*, and *Scinax ruber*), a lizard (*Ptychoglossus brevifrontalis*), an unidentified snake, and three mice. The extremely large gape allows this species to consume proportionately large prey. One juvenile with an SVL of 28.5 mm was inadvertently kept overnight in a plastic bag containing other frogs: during the night, the juvenile *Ceratophrys* consumed an *H. boliviana* (SVL 23 mm) and an *S. ruber* (SVL 23 mm).

The tadpoles are predatory carnivores and are also cannibalistic. Tadpoles raised in individual containers fed voraciously on slightly smaller tadpoles of *Dendropsophus koechlini*, *D. leucophyllatus*, *Chiasmocleis ventrimaculata*, *Ctenophryne geayi*, and *Elachistocleis bicolor*, but not on those of *Trachycephalus coriaceus*. Feeding seems to occur by a gape-and-suck mechanism followed by a rapid gulping associated with the constriction of the jaw muscles, during which the medial spine on the posterior jaw sheath seems to puncture and deflate the body of the prey; the entire feeding bout requires < 5 s.

Males were observed calling at night in November 1986 and 1988 at ponds in Cusco Amazónico, Peru, but the presence of tadpoles in November and in February suggests that this species does not breed only at the beginning of the rainy season. Males call while sitting on the ground or in shallow water at the edges of ponds. The call is a loud "baaaaa." The call of a Bolivian frog from Panguana was emitted at irregular intervals and had a duration of about 1 s. Additional call data are provided under the subtitle "Vocalizations of Ecuadorian Anurans" in Appendix 1.

Tadpoles.—Figures 3.5, 3.6. The tadpole of this species was described from Santa Cecilia, Provincia Sucumbíos,

FIGURE 3.5 Tadpole of *Ceratophrys cornuta*. KU 207809, stage 36, TL = 44.8 mm, oral disc maximum diameter = 6.4 mm. Note the continuous row of large, pointed papillae on the anterior labium. Taken from Duellman (2005).

FIGURE 3.6 Metamorphic sequence in *Ceratophrys cornuta* at Cusco Amazónico, Peru. Top to bottom: tadpoles nearing metamorphosis, metamorphosing young (KU 215699, 26.3 mm SVL), and metamorphosed young (KU 205061, 28.5 mm SVL). Top, Erik R. Wild; middle and bottom, WED.

Ecuador, and from Cusco Amazónico, Peru. The following description is based on an Ecuadorian tadpole. A tadpole in stage 35 has a body length of 19 mm and a total length of 56 mm. The body is elongate, ovoid, wider than deep, and widest just anterior to the eyes. The snout is round. The eyes are small, close together, and directed anterolaterally. The nostrils are directed laterally at a point about one-third of the distance between the eyes and the tip of the snout. The opening of the sinistral spiracle is directed posterolaterally on the ventrolateral surface of the body at about the midlength of the body. The cloacal tube is short and dextral. The caudal musculature is moderately robust proximally and gradually tapers to a slender tip. The fins are equal in height and highest at the midlength of the tail, where each fin is slightly higher than the caudal musculature. The fins extend to the tip of the pointed tail, and the dorsal fin does not extend onto the body. The tail comprises about 66% of the total length. The large oral disc is directed anteriorly with expanded protruding lips, each bearing a single row of large, pointed papillae. The jaw sheaths are massive; the lower one has a medial spine that fits into a notch in the upper sheath. There are nine upper and nine lower rows (LTRF 9/9) of labial teeth, plus some short rows lateral to the jaw sheaths. The three innermost rows of both series are interrupted medially. The body is dull brown; the tail is tan with a brown stripe laterally on the proximal one-third of the tail. The entire caudal musculature and dorsal fin bear fine brown reticulations; the ventral fin is unpigmented.

A tadpole in stage 36 from Peru has a body length of 19.6 mm and a total length of 46.5 mm. The LTRF is 13(6)/8(2). A-1, A-2, and P-8 are shorter than the other rows, and the teeth on A-1 are shorter than those in the other rows.

The body and caudal musculature are tan; the belly is greenish-yellow, and the fins are translucent tan. The dorsal and lateral surfaces of the body and upper one-half of the caudal musculature have numerous melanophores. There are fewer melanophores on the ventral half of the caudal musculature and fewer still on the caudal fins. The skin on the venter is transparent with scattered guanophores. The massive jaw musculature is visible through the skin on the anterior one-third of the venter.

The adult color pattern, a green or tan snout and middorsal region and an interocular bar, becomes evident in stage 44. In metamorphosing young, the head and flanks are tan with reddish-brown spots. Distinct transverse bars are evident on the limbs. The supraciliary process is only a round tubercle.

Distribution.—Figure 3.7. This species has been reported in the Amazonian lowlands up to 832 m in Colombia, Ecuador, Peru, Brazil, and the Guianas. In Ecuador, it occurs in Sucumbíos, Orellana, Pastaza, and Morona Santiago provinces, but there are no records from Provincia Napo. Its elevational range is 183–832 m in an area of extent of occurrence of about 45,590 km^2.

This nocturnal, terrestrial species inhabits primary forests in the flooded forests of alluvial plains of Amazonia, evergreen forests of lowlands of Aguarico-Putumayo-Caquetá, and evergreen forests of lowlands of Napo-Curaray ecosystems (Table 0.4) in four biogeographic regions (Table 0.5), where the average annual rainfall is 2249–4004 mm and the average annual temperature is 22.2–25.7 °C.

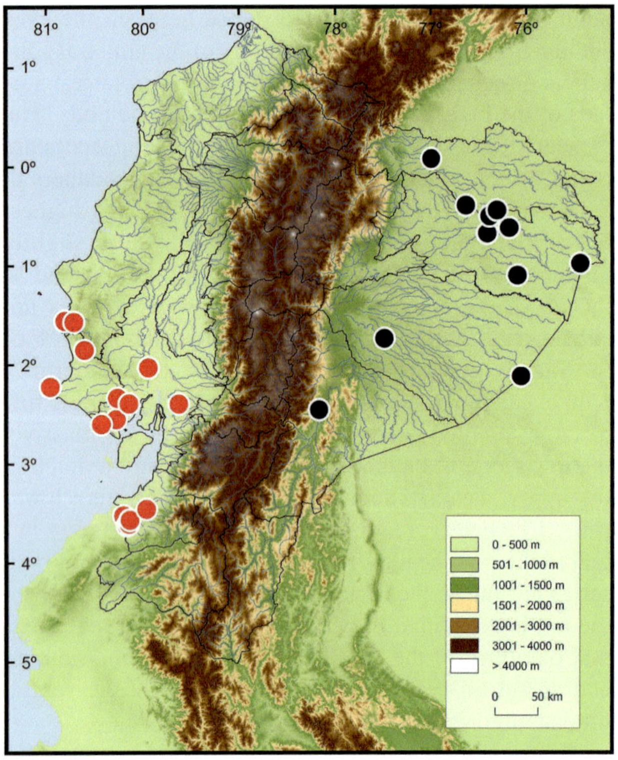

FIGURE 3.7 Localities of known occurrence of *Ceratophrys cornuta* (black) and *C. stolzmanni* (red) in Ecuador.

Conservation.—*Ceratophrys cornuta is* categorized as Least Concern throughout its entire range (La Marca *et al.*, 2004c; in IUCN Red List of 2022). Although it is rare (Aichinger, 1987) to locally common in Puerto Ayacucho, Venezuela (Barrio-Amorós, 1999), and Cusco Amazónico, Peru (Duellman, 2005), the species is still reasonably widely distributed, and its range includes several protected areas. Nonetheless, it is only found in primary forests and seems to be intolerant of any modification to its habitat. In Ecuador, it is a rarely collected frog. Within its range of distribution, its habitat is being lost to deforestation for agricultural development, logging, and human settlement. Another threat this species faces is pollution resulting from pesticide use. This species, known commercially as the horned Pac-Man frog, is also of interest to the pet trade, which breeds it under terraria conditions. In Ecuador, it has been recorded in the public reserves Reserva de la Biósfera Yasuní, Parque Nacional Yasuní, and Reserva Biológica Limoncocha; the community area Comunidad Sani Isla; and some private reserves (e.g., Estación de Biodiversidad Tiputini, La Selva Lodge).

Remarks.—Wild (1997) described in detail the adult skeleton and tadpole chondrocranium, including the ontogenetic development of the chondrocranium and the ossification sequence of the skeleton. The chondrocranium of the carnivorous larvae is unique in lacking a frontoparietal fontanelle and possessing a complete dorsal roof of cartilage. Furthermore, the chondrocranium is extremely robust, particularly those elements involved in the feeding mechanism. The large adult skull is hyperossified; most elements are fused into a single unit, and nearly all dermal elements are ornamented, casqued, and co-ossified. Calcification is present in nearly every cartilaginous element of the skeleton in larger (older) adults. With few exceptions, the ossification sequence relative to metamorphosis is consistent with those that are known for other anurans. Changes in the timing of development, or heterochrony, are involved with the evolution of the unusual skull and skeleton of ceratophryine frogs.

Andersson (1945) described *Ceratophrys testudo* from "Río Pastaza Watershed" in eastern Ecuador. The holotype (NHRM, 1912) is a juvenile male with an SVL of 32 mm. Duellman cursorily examined the specimen in 1969 and considered it the same as *C. cornuta*, as did Lynch (1982). Subsequent authors (Mercadal, 1988; Peri, 1993) recognized *C. testudo* as a valid species. Mercadal (1988) made comparisons of morphometric features of the holotype and specimens of *C. cornuta* (70.0 and 76.6 mm SVL). Peri's (1993) morphometric comparisons were made with a juvenile of *C. cornuta*; she concluded that the morphometric differences were not sufficient to distinguish the two species from one another. Both of these authors considered the shape of the supraciliary "horn" to be a valid character distinguishing *C. cornuta* from *C. testudo*. This minor difference probably is an artifact of preservation and storage. Other supposed distinguishing characters (nature of tubercles on the flanks and transverse interorbital crest) are subject to ontogenetic change and do not distinguish the species. Faivovich *et al.* (2014) remarked that the status of *C. testudo* should be carefully reassessed. We follow Lynch (1982) and do not consider *C. testudo* a valid species until further evidence of its distinctiveness becomes available. Therefore, it has no account herein.

The type locality is imprecise. Other species reported by Andersson (1945) from the Río Pastaza Watershed include *Ceratohyla cristata* (= *Hemiphractus proboscideus*), *Eleutherodactylus bufonius* (= *Pristimantis diadematus*), *Eleutherodactylus brevicrus* (= *Pristimantis brevicrus*), *Lithodytes cornutus* (= *Strabomantis cornutus*), *Leptodactylus tuberculosis* (= *Oreobates quixensis*), *Hyla lanciformis* (= *Boana lanciformis*), *Hyla marmorata* (= *Dendropsophus marmoratus*), *Hyla quadrangulum* (= *Trachycephalus quadrangulum*), *Hyla maxima* (= *Boana boans*), *Hyla rubra* (= *Scinax ruber*), *Hyla depressiceps* (= *Scinax funereus*), *Hyla macrotis* (= *Trachycephalus macrotis*), *Hyla depressa* (= *Osteocephalus taurinus*), *Hyla capitocarinata* (= *Gastrotheca longipes*), *Hyla membranacea* (= *Dendropsophus reticulatus*), and *Phyllomedusa edentula* (= *Phyllomedusa tarsius*). With the exception of *Strabomantis cornutus*, which occurs at elevations of 1100–1830 m on the eastern slopes of the Andes, all of these species inhabit the upper Amazon Basin, as does *Ceratophrys cornuta*. Thus, concern about the distribution of *C. testudo* at a higher elevation than *C. cornuta* is unwarranted.

Key references.—(Rivero, 1961) (account); (Cochran and Goin, 1970) (account); (Crump, 1974) (natural history); (Duellman, 1978) (account, tadpole); (Schlüter, 1980) (call); (Lynch, 1982) (phylogeny); (Obst *et al.*, 1988) (natural history); (Hero, 1990) (tadpole); (Rodríguez and Duellman, 1994) (account); (Duellman and Lizana, 1994) (natural history); (Marquez *et al.*, 1995) (call); (Duellman, 1995) (natural history); (Wild, 1997) (osteology); (Lescure and Marty, 2000) (account); (Duellman, 2005) (account, tadpole); (Lima *et al.*, 2006) (account); (Pyke and Ray, 2006) (natural history); (Valencia *et al.*, 2008) (account); (Chávez *et al.*, 2011) (diet); (Ouboter and Jairam, 2012) (account); (Cole *et al.*, 2013) (account); (Villacampa *et al.*, 2017) (account); (Barrio-Amorós *et al.*, 2019) (notes); (Schiesari *et al.*, 2022) (tadpole).

CERATOPHRYS STOLZMANNI (STOLZMANN'S PAC-MAN FROG) STEINDACHNER, 1882

Ceratophrys Stolzmanni Steindachner, 1882: 188. **Syntypes.**—NHMW 4631.1–2 and 4632 from Tumbes, Peru.

Etymology.—The specific name is a patronym for Jean Stanislaus Stolzmann (1854–1928), a Polish ornithologist who collected the frogs used in the description of this species.

Identification.—Figures 3.8, 3.9. Males attain an SVL of 67.9 mm, and females of 82.0 mm. The head and mouth are wide but not as wide as the body. In dorsal view, the snout is rounded; in profile, it is steeply inclined anteroventrally from the anterior edges of the orbits. The skin on the dorsum, flanks, and venter is finely rugose, but the gular region is smooth. The interocular region and paravertebral marks bear enlarged, laterally flattened tubercles. The flanks have small, scattered warts. Long, pointed supraciliary processes are absent. The tympanum is distinct, oval shaped, and about half the size of the eye; the supratympanic fold is poorly defined. The mandible is partially covered by the maxilla and bears an odontoid directed toward the buccal cavity. The fore and hind limbs are moderately short and robust. Finger I is longer than II; the fingers are unwebbed, and the toes are basally webbed. Toe IV is almost twice the size of toe V. Breeding males have dark brown nuptial excrescences at the bases of their thumb and a paired vocal sac. A dorsal dermal shield is absent. A metatarsal keratinized dark brown spadix is present distally on the tarsus.

The color pattern is variable; it ranges from pale brown and dark brown to pale green, dark green, and gray–green. The dorsal surfaces have elongate, rounded, well-defined, irregular marks darker than the ground color (e.g., brown, dark brown, black); these marks are narrowly bordered by cream. Markings on the head consist of a dark interorbital mark and an elongated dark mark from the tip of the snout along the canthus rostralis to the orbit. Dark brown transverse bars are present on the dorsal surfaces of the limbs. The belly is grayish-white with small brown spots laterally; irregular dark brown marks of variable size are present on the throat.

FIGURE 3.8 *Ceratophrys stolzmanni.* Top and middle left: female. Top and middle right: male. LAC. Bottom: ventral views of female and male. DAL. Reserva Ecológica Arenillas, Provincia El Oro.

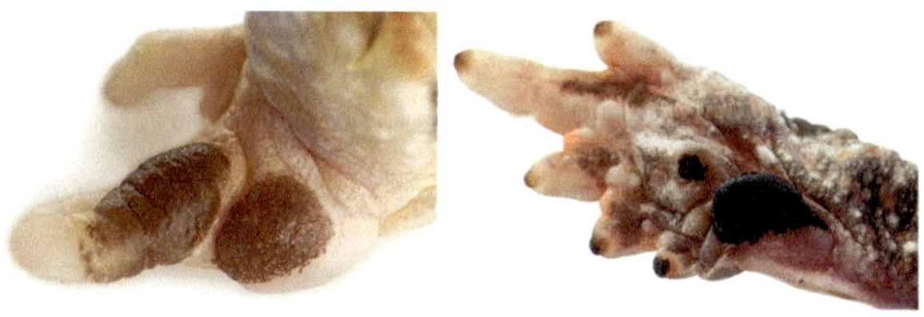

FIGURE 3.9 Morphological details of *Ceratophrys stolzmanni.* Left: nuptial excrescence at the base of the thumb of a male. DAL. Right: CJ 11999, keratinized metatarsal spadix, tubercle, and tips of toes, SGS. Reserva Ecológica Arenillas, Provincia El Oro.

Comparisons.—*Ceratophrys cornuta.* No other frog in the Pacific lowlands of Ecuador can be confused with *Ceratophrys stolzmanni*, which differs from the Amazonian

Ceratophrys cornuta by lacking distinctive, long, pointed supraciliary processes.

Natural history.—Figures 3.10, 3.11. *Ceratophrys stolzmanni* is a nocturnal frog. Some individuals have been found in open, sandy deserts with low, sparse vegetation without trees except in the vicinity of temporary streams. In western Ecuador, the species has been found only during the rainy season (January through March). Presumably, it lives underground in the dry season. In 2007, adults were collected between 21:00 and 00:00 h in and near temporary ponds along a path in the forest. The forest was surrounded by pastures and deforested areas. Subadults were collected between 00:00 and 00:30 h in a disturbed area with few trees and plowed soil. On average, the daily air temperature was ~30 °C, and the temperature of the water in the pond was ~28 °C. In 2011, juveniles were found at night in a swampy area and on the margin of a duck pond near human habitations.

FIGURE 3.10 *Ceratophrys stolzmanni.* Top left: in its microhabitat at night. Top right and middle: calling males and amplexus under lab conditions. Bottom: CJ 9980, tadpoles, note a conspecific predation event. Reserva Ecológica Arenillas, Provincia El Oro. LAC.

FIGURE 3.11 *Ceratophrys stolzmanni.* Clutch of eggs and embryos of *Ceratophrys stolzmanni* under laboratory conditions. 12 December 2012 13:58:01; 13 December 11:49:46, 11:54:11; 14 December 2012 14:24:14. Reserva Ecológica Arenillas, Provincia El Oro. LAC.

Reproduction is explosive and happens during the rainy season. The amplectant position differs from the typical axillary amplexus of other anurans; it is termed nuchal amplexus and may be the result of the unique morphology of Ceratophryidae. Reproductive biology data have been obtained in the laboratory (CJ and Wikiri lab notes in 2011–2016 by Gabriela Aguiar and in January 2023 by Italo G. Tapia). Reproduction takes place at night, and the deposition of eggs occurs on the night of courtship. Clutches contain up to 3459 eggs. Development time (~27 °C) from fertilization to hatching is 40 hours, and metamorphosis is completed in 20–32 days.

Tadpoles.—Figure 3.12. The tadpoles are described elsewhere but are depicted here.

Green metamorphosing tadpoles have dull brown dorsal marks in the paratoid region in the form of two ovoid ridges that become brown dorsal marks in stage 46. The venter and oral apparatus are creamy white, whereas the flanks and cheeks are pale yellow with profuse dark brown spots. The iris is yellow with pale brown reticulations. The caudal fins are translucent with green marbling, which becomes more profuse on the proximal part of the dorsal fin near its junction with the body. The caudal musculature is pale orange in the paler areas, with several dark brown blotches at the lower and distal sections of the tail. The upper section of the tail is profusely pigmented with dark brown and scattered pale (orange) areas. At metamorphosis, the arms and shanks have dull brown transverse bars, which become darker in postmetamorphic individuals.

In the brown morph, the dorsal coloration of the tadpole has a tan and brown pattern. The flanks are creamy silver with copious black mottling. The venter is translucent, revealing a predominantly dark gray gut. The iris is pale yellow with a brownish reticulation. The caudal fins are translucent and well-marked by red capillaries. The caudal musculature appears to be bicolored because of a greater melanophore concentration in the dorsal part (from midline up) that leaves a basically brown hue in the ventral part (from midline down). Melanophores are generally scattered, forming some larger blotches near the tail–body junction, where the general coloration is salmon. The myomere insertion point at the midline is

FIGURE 3.12 Ontogenetic changes in *Ceratophrys stolzmanni*. Top to bottom, left to right. stage 36, lateral, dorsal, and ventral views, TL = 38.3 mm; undetermined stage, TL = 91.5 mm; stage 42, TL = 93.5 mm; stage 46, 14 days after stage 36 = 37.2 mm SVL. Reserva Ecológica Arenillas, Provincia El Oro. SGS.

marked by a black stripe that diffuses toward the tail terminus. The hind limbs are green with diffuse transverse brown bars.

Distribution.—See Figure 3.7. *Ceratophrys stolzmanni* is found in the Pacific region of southern Ecuador and northern Peru (Tumbes department). In Ecuador, it is known from in Pacific lowlands of Manabí, Santa Elena, Guayas, and El Oro provinces. Its elevational range is 0–402 m in an area of extent of occurrence of about 18,420 km^2.

In Ecuador, this nocturnal and terrestrial burrowing species inhabits four ecosystems (Table 0.4) within the Cordillera Costera of Equatorial Pacific and Jama-Zapotillo biogeographic sectors, where the average annual rainfall is 155–1255 mm and the average annual temperature is 22.4–25.7 °C.

Conservation.—*Ceratophrys stolzmanni* is categorized as Vulnerable throughout its entire range (IUCN SSC Amphibian Specialist Group, 2018s; in IUCN Red List of 2022) and in Ecuador by Ortega-Andrade *et al.* (2021). Its conservation status is in need of revision. Ortiz *et al.* (2013) listed as Vulnerable because its extent of occurrence is less than 20,000 km^2, its distribution is severely fragmented, and there is a continuing decline in the extent and quality of its forest habitat in Peru and Ecuador. Although it seems to be a rare species, it has been regularly recorded in Ecuador since 2005. There is no information on its population status in Peru. It occurs in the Zona Reservada de Tumbes in Peru. The major threat is habitat loss because of agriculture, logging, and human settlement. Agrochemical soil pollution also is a threat.

In Ecuador, it has been recorded in the public reserves Parque Nacional Machalilla and Reserva Ecológica Arenillas; the protected forests of Bosque Protector Cerro Blanco, Reserva Ecológica Manglares Churute, Loma Alta y Ampliacíon, and community area Comuna Loma Alta; and some private reserves (e.g., Cerro Masvale). This species also is known in the pet trade, where it is bred under laboratory conditions. Its *ex situ* management and reproduction have been successful, and it currently is part of conservation programs (e.g., Arca de los Sapos of CJ) and biocommerce by the hobbyist industry (e.g., see Chapter 1 of Volume I).

Remarks.—Lynch (1982) published a table of morphological characters with the states of the characters and a figure of the eyelid. Since the species' original description, and for about 130 years, little was known about its biology, in part because of the habits of the species: these frogs stay burrowed underground most of the year, only emerging with the first heavy rains in the winter for breeding and feeding before burrowing again. More recently (since 2012 to date), a wealth of data on morphology, life history traits, demography, behavior, burrowing, sexual selection, husbandry, reproduction, ecology, embryology, and development were published (see Key references). Ortiz *et al.* (2013) provided pictures depicting the ontogenetic variation. Sáenz-Ponce *et al.* (2012) described the morphology of the gastrocoele, the cavity in the gastrula of embryos, and compared its development with that of other frogs. The gastrocoele ceiling plate is ciliated and probably originates from the mesoderm.

Key references.—(Steindachner, 1882) (description); (Peters, 1967) (natural history); (Valencia *et al.*, 2008) (account); (Stuart *et al.*, 2008) (account); (Sáenz-Ponce *et al.*, 2012) (development); (Ortiz *et al.*, 2013) (husbandry, natural history, tadpole, development); (Santillana-Ortiz and Del Pino, 2017) (development); (Székely *et al.*, 2016a) (morphology); (Székely *et al.*, 2016b) (natural history); (Székely *et al.*, 2017) (natural history); (Székely *et al.*, 2018a) (natural history); (Székely *et al.*, 2018b) (natural history); (Székely *et al.*, 2018c) (natural history); (Székely *et al.*, 2019) (diet); (Székely *et al.*, 2020a) (natural history).

4 Family Hemiphractidae

MARSUPIAL FROGS, HORNED TREEFROGS, AND EGG-BROODING FROGS

Hemiphractidae Peters, 1862b: 146. **Type genus.**—*Hemiphractus* Wagler, 1828.

Etymology.—The familial name is derived from the Greek inseparable prefix *hemisys*, meaning half, and the Greek *phraktos*, meaning fenced in or protected. Wagler (1828) apparently applied the name in allusion to the transverse crest seemingly fencing in the head posteriorly.

Definition.—Hemiphractid frogs are a monophyletic group diagnosed on the basis of molecular evidence and morphological features. Putative phenetic synapomorphies are (1) the habit of brooding eggs on the back or in a pouch and (2) embryos with large, sheet-like gills derived from branchial arches 1 and 2.

Hemiphractid frogs have an arciferal pectoral girdle—overlapping left and right halves ventrally without articulation. They have a full complement of four fingers and five toes. The terminal phalange on each digit is claw-shaped and offset from the next most distal (penultimate) phalange by a cartilaginous intercalary element. The embryos have one or two pairs of large, sheet-like gills that have one or two gill stalks derived from branchial arches I and II.

Species of *Hemiphractus* are robust-bodied and terrestrial, whereas other genera are treefrog-like. *Fritziana*, *Cryptobatrachus*, and *Stefania* are arboreal. *Flectonotus fitzgeraldi* is 19–24 mm in length, whereas species of *Gastrotheca* are often larger (e.g., *Gastrotheca cornuta*, SVL 66–81mm; *G. orophylax*, SVL 59–74 mm). Hemiphractids exhibit a wide range of sizes in Ecuador, spanning from the relatively small *G. espeletia*, with a maximum known female SVL of 52.6 mm, to the larger *G. angustifrons*, which can reach a maximum SVL of 91.8 mm.

Natural history.—Most species of hemiphractids are arboreal and have direct development, whereas others have a biphasic mode of reproduction. Amplexus and mating take place out of water. Females of hemiphractids transport eggs in specialized sacs, depressions, or pouches on the dorsa. Eggs of *Hemiphractus*, *Cryptobatrachus*, and *Stefania* are brooded openly on the dorsum of the female, not in pockets or brood chambers, and hatch as froglets. *Flectonotus*, *Fritziana*, and *Gastrotheca* have a specialized dorsal pouch in which eggs are carried. Developing embryos are partially (*Cryptobatrachus*, *Flectonotus*, *Fritziana*) or entirely (*Gastrotheca*, *Hemiphractus*, *Stefania*) enclosed in large, membranous, bell-shaped gills. The gills develop an extensive capillary net that acts as a placenta for the maternal transfer of gases, water, and nutrients. In *Flectonotus*, *Fritziana*, and some species of *Gastrotheca*, eggs hatch as advanced tadpoles. In other *Gastrotheca*, eggs remain in the pouch throughout the entire developmental period, and froglets hatch and emerge from the pouch after several months. Vocal slits are absent in *Cryptobatrachus*, *Hemiphractus*, and *Stefania*, although *Hemiphractus* produces a sound, whereas *Cryptobatrachus* is not known to call. Males of *Flectonotus*, *Fritziana*, and *Gastrotheca* have vocal slits and produce calls.

Distribution.—Hemiphractids are distributed in the Neotropics in Costa Rica, Panama, Venezuela, the islands of Tobago and Trinidad, and above the tree line in Colombia, Ecuador, Peru, Bolivia, and northwestern Argentina. Also, they are present in the Atlantic coastal forests in eastern and southeastern Brazil.

Content.—The family contains six genera with 119 species (Frost, 2022). In Ecuador, there are two genera and 25 species.

Remarks.—Figure 4.1. The nomenclatural and systematic history is detailed in Frost (2022 and onward). Throughout most of the 20th century, all of the hemiphractid genera were placed in Hylidae, the family of New World tree frogs. Some authors (e.g., Duellman, 2001) recognized hemiphractids as a subfamily of Hylidae. Early analyses of limited molecular data and small taxon sampling (e.g., Faivovich *et al.*, 2005) had mixed results and did not solidify the relationships of hemiphractid frogs. Further analyses of molecular data showed that the hemiphractid genera formed a monophyletic group (Guayasamin *et al.*, 2008b, Blackburn and Duellman, 2013; Castroviejo-Fisher *et al.*, 2015; Duellman, 2015). Heinicke *et al.* (2009) recognized five families of terrestrial breeding frogs (Brachycephalidae, Ceuthomantidae, Craugastoridae, Eleutherodactylidae, and Strabomantidae) in the unranked taxon Terrarana; in their phylogenetic analysis, Terraranae and Hemiphractidae are sister taxa.

GENUS *GASTROTHECA* (MARSUPIAL FROGS) FITZINGER, 1843

Gastrotheca Fitzinger, 1843: 30. **Type species.**—*Hyla marsupiata* (Duméril and Bibron, 1841).

Etymology.—The generic name is derived from the Greek *gastros*, meaning belly, and the Greek *theke*, meaning container; presumably, Fitzinger (1843) was referring to the brood pouch on the dorsum in females. Günther (1859: 115) noted that Fitzinger (1843) had erred in his choice of a generic name and stated: "Fitzinger appears to have believed in the existence of a pouch on the belly; his denomination . . . [was] used by the old anatomists for the brain-pan; therefore, I think myself justified in not accepting his name." Günther therefore proposed the name *Nototrema* (from the Greek *notos*, meaning "back"), in reference to the position

DOI: 10.1201/9781003543879-4

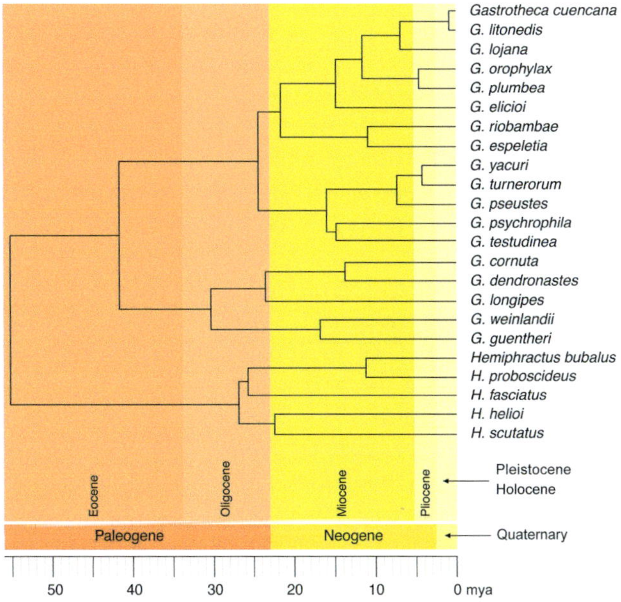

FIGURE 4.1 Time-calibrated maximum likelihood phylogenetic tree depicting evolutionary relationships of some of the Ecuadorian species of *Hemiphractidae*, inferred from DNA sequences of mitochondrial and nuclear genes. Tree by Juan C. Santos (this publication) for *Hemiphractus fasciatus* (see Remarks under its account).

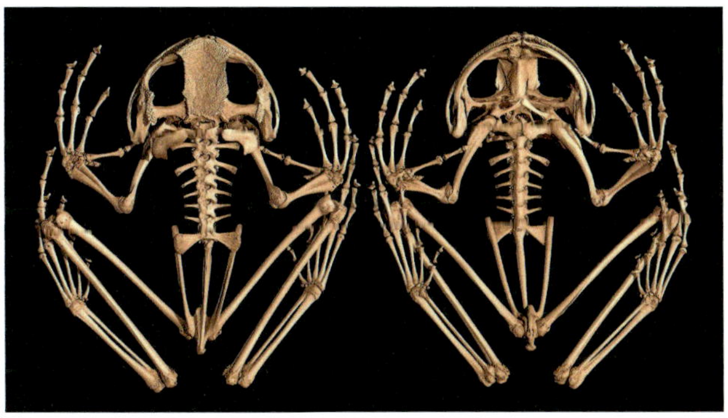

FIGURE 4.2 High-resolution computed tomography reconstructions of the skeleton of *Gastrotheca guentheri*, KU 164226, female, 69.8 mm SVL, 5 km east–southeast of Chiriboga, Provincia Pichincha. Dorsal (left) and ventral (right) views. Daniel J. Paluh.

of the brood pouch. Even though Günther's substitute name is more appropriate, the Law of Priority necessitates the use of the misleading generic name *Gastrotheca*.

Definition.—Figure 4.2. In these frogs, the skull is well ossified; in some species, the dermal roofing bones are exostosed and are co-ossified with the overlying skin. In most species, the frontoparietals articulate medially throughout their lengths, and the nasals are large and in contact anterior to the sphenethmoid. The neopalatines are edentate and serrated or not. A ventral spur of the neopalatine is absent, and the dentigerous processes of the vomers are anterior to the neopalatines. The teeth on the maxillaries, premaxillaries, and vomers are bicuspid. The mandible lacks serrations; odontoids are present in one species, and true teeth are present on the dentary in one species. The presacral vertebrae are procoelous and lack elongate neural spines. Vocal slits and a vocal sac usually are present. A fleshy proboscis on the tip of the snout is absent. Finger I is longer or shorter than II; nuptial excrescences are either present or absent. The SVL varies from 26 to 86 mm in males and from 32 to 110 in females. The eggs undergo development in a pouch on the back of the female. In some species, eggs complete their development therein and hatch as froglets (monophasic development); in others the eggs hatch as free-swimming, feeding tadpoles (biphasic development). The embryonic gills entirely cover the developing embryo and are derived from branchial arches I and II; the gill stalks remain separate, but the gills are fused to form a bell-shaped structure. The tadpoles belong to type IV tadpoles of Orton (1953) and the exotroph, benthic ecomorphological guild.

Distribution.—*Gastrotheca* is broadly distributed on the Caribbean versant in Costa Rica and Panama and the Pacific versant from eastern Panama to Ecuador and then in the Andes from Venezuela and Colombia to central Argentina. It also occurs in the upper Amazon Basin in Ecuador and northern Peru, the Cordillera de la Costa in northern Venezuela, and the Atlantic coastal forests in eastern Brazil. The greatest number of species is in the Andes of Colombia (16 species, six endemic), Ecuador (19 species, eight endemic), and Peru (16 species, 11 endemic).

Content.—Seventy-seven species are recognized; 20 species occur in Ecuador, and 10 of these are endemic.

Remarks.—Duellman (2015) recognized seven subgroups (under subgenera). Three of these lineages occur in Ecuador. The lineage *Amphignathodon* contains 10 long-legged arboreal species; their eggs undergo direct development; seven of these species (*Gastrotheca andaquiensis*, *G. angustifrons*, *G. cornuta*, *G. dendronastes*, *G. guentheri*, *G. longipes*, and *G. weinlandii*) are known from Ecuador. The lineage *Duellmania* is widespread in the Andes of Colombia, Ecuador, and extreme northern Peru. Eight species are known from Ecuador (*G. cuencana*, *G. elicioi*, *G. espeletia*, *G. litonedis*, *G. lojana*, *G. orophylax*, *G. plumbea*, and *G. riobambae*). The lineage *Gastrotheca* ranges from northern Ecuador to central Argentina; only four species (*G. pseustes*, *G. psychrophila*, *G. testudinea*, and *G. turnerorum*) occur in Ecuador. In the following species accounts, only the generic name is used.

GASTROTHECA ANDAQUIENSIS (ANDAQUI MARSUPIAL FROG) RUIZ-CARRANZA AND HERNÁNDEZ-CAMACHO, 1976

Gastrotheca andaquiensis Ruiz-Carranza and Hernández-Camacho, 1976: 11. **Holotype.**—IND-AN 401 from near Cueva de los Guácharos, ca. 2000 m, Departamento de Huila, Colombia.

Etymology.—The specific name refers to the Andaqui people, whose members inhabit the southern part of the Cordillera Oriental in Colombia.

Identification.—Figure 4.3. The maximum SVL is 76.8 mm in females and 62.0 in males. The skin on the dorsum is finely granular and not adherent to or co-ossified with the skull. A granular interocular ridge is present, as are conspicuous supraciliary processes. The tympanic annulus is granular. The snout is broadly rounded in dorsal view, deep, and abruptly inclined from the nostrils to the lip in profile. The interorbital distance is slightly greater than the width of the upper eyelid. Fingers I and II are equal in length. The fingers and toes have large terminal discs that are much wider than the adjacent digits. The fingers are webbed basally, and the toes are about three-fourths webbed. The hind limbs are long; the length of the tibia is more than 60% of the SVL. An elongate, triangular calcar is present on the heel; tarsal folds are absent.

The dorsum is green or tan, with dark to bright green or brown mottling and blotches. Markings on the head consist solely of a pale suborbital bar. A narrow cream dorsolateral stripe is present. The flanks are uniformly dark brown. The venter is uniformly cream to chartreuse. The iris varies from copper colored to bright bronze with brown reticulations. The tongue and the lining of the mouth are bluish-green. Specimens from 3.5 km west of Mera, Provincia Pastaza, Ecuador, are colored as follows: the dorsum is olive green to reddish-brown with indistinct tan spots. The side of the head and the flanks are dark olive brown. Various pale markings include a pale tan dorsolateral stripe, a cream suborbital bar, and tan webbing. The anterior and posterior surfaces of the thighs are pale orange–tan, with or without dull olive-green mottling. The discs, throat, and belly are green, with or without a brown suffusion or brown spots on the posterior part of the belly; the ventral surfaces of the thighs are dull tan; the tongue and buccal lining are green.

FIGURE 4.3 *Gastrotheca andaquiensis*. KU 106040, female, 62 mm SVL, 3,5 km west of Mera, Provincia Pastaza. WED.

Comparisons.—*Gastrotheca angustifrons*, *G. cornuta*, *G. dendronastes*, *G. guentheri*, *G. longipes*, *G. testudinea*, *G. weinlandii*. The presence of a granular interocular ridge is unique to *G. andaquiensis*. Of all of the other species having supraciliary processes (*G. cornuta*, *G. dendronastes*, and *G. guentheri*) all, except *G. guentheri*, differ from *G. andaquiensis* by having transverse dermal ridges on the dorsum of the body, a character also existing in *G. angustifrons*. Furthermore, *G. dendronastes* differs by having finger I longer than finger II and by having longitudinal cream stripes on brown flanks. The presence of mandibular teeth in *G. guentheri* separates that species from all other *Gastrotheca*. Only four species of *Gastrotheca* inhabit the Amazonian slopes of the Andes in Ecuador. Of these, *G. andaquiensis* is unique in having supraciliary processes and a granular interocular ridge. Furthermore, *G. longipes* and *G. testudinea* lack calcars; in *G. weinlandii*, the skin is co-ossified with the dermal roofing bones of the skull, and finger I is longer than II.

Natural history.—Specimens were found at night, 1–2 m above the ground on branches of bushes and trees adjacent to or overhanging streams. By day, the frogs take shelter in bromeliads; one was found on a leaf in a thicket of *Heliconia*.

Males call at night from trees. The call is a loud "wronk," followed by two or three shorter notes, "tonk-tonk." No recordings are available. The eggs hatch into froglets. The only datum regarding reproduction in this species is that a female having an SVL of 76.0 mm from Departamento Huila, Colombia, gave birth to 14 young on 14 March 1978. The young were colored like their green mother and had blunt supraciliary processes; six of the young had SVLs of 20.7–21.5 mm.

Distribution.—Figure 4.4. *Gastrotheca andaquiensis* is known from the Andes of southern Colombia and Ecuador, at an elevation of 1700 m in the Cordillera de Kutukú in southern Ecuador, and at an elevation of 2000 m at Cueva de los Guácharos on the western slopes of the Cordillera Oriental in the upper Río Magdalena Valley in Colombia. In Ecuador, it occurs in Sucumbíos, Napo, Pastaza, and Morona Santiago provinces. Its elevational range is 1072–2066 m in an area of extent of occurrence of about 7600 km^2.

This nocturnal, arboreal species inhabits lower montane and cloud forests in the evergreen low montane forests of the Cordilleras Cóndor-Kutukú, evergreen low montane forests of the northern Cordillera Oriental of the Andes, evergreen montane foothill forests of the Cordilleras Cóndor-Kutukú, and evergreen montane foothill forests of the northern Cordillera Oriental of the Andes ecosystems in the Amazonian Cordilleras biogeographic sector, where the average annual rainfall is 1454–3508 mm and the average annual temperature is 15.7–20.7 °C.

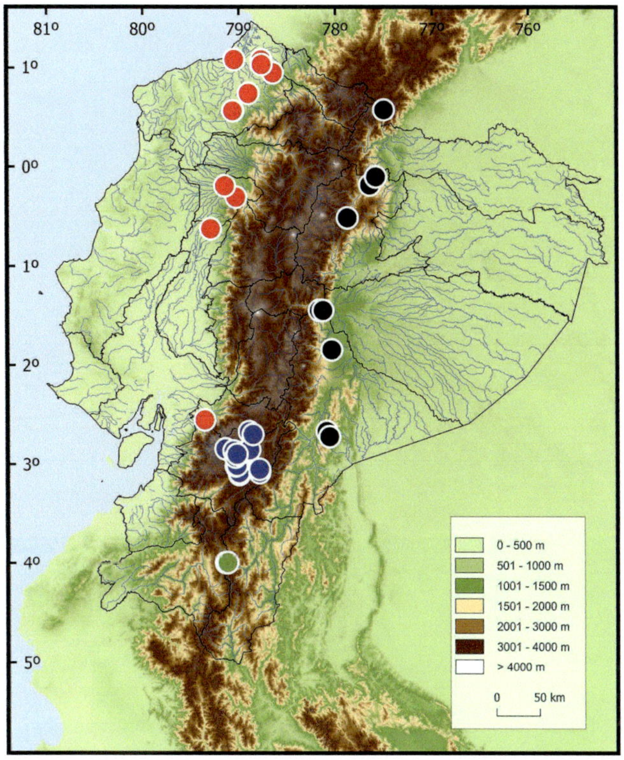

FIGURE 4.4 Localities of known occurrence of *Gastrotheca andaquiensis* (black dot), *G. cornuta* (red dots), *G. cuencana* (blue dots), and *G. psychrophila* (green dot) in Ecuador.

Conservation.—*Gastrotheca andaquiensis* is categorized as Least Concern throughout its entire range (IUCN SSC Amphibian Specialist Group, 2020m, in IUCN Red List of 2022). However, it is important to note that in Ecuador, *G. andaquiensis* has been categorized as Endangered by Ortega-Andrade *et al.* (2021). We categorize it as Critically Endangered following A2ac IUCN criteria because we infer a general reduction in the population greater than 90%. Frequent searches at its historical localities revealed no specimens, and it is likely that climate change and emerging pathogens have affected this population. Also, its distribution is severely fragmented, and there is a continuing decline in the extent and quality of its habitat. We suggest that it should additionally be tagged as Possibly Extinct given that in Ecuador, there have been no sightings for about four decades; the most recent record of this species is from 17 November 1984 in La Alegría, Provincia Sucumbíos. Threats to this species include deforestation and habitat fragmentation resulting from agricultural expansion, cattle ranching, timber extraction, wildfires, landslides, volcanic activity, and human settlements. Additionally, mining activities and pollution pose significant risks to its survival. In Ecuador, it has been recorded in the public reserve Parque Nacional Cayambe Coca, in the protected forests of Cordillera Kutukú y Shaimi, and within the biosphere reserve Sumaco.

Remarks.—Its molecular phylogenetic relationships are unknown. Although its range encompasses many protected areas, the lack of thorough collecting in these areas cannot substantiate the occurrence of *Gastrotheca andaquiensis* therein.

Key references.—(Ruiz-Carranza and Hernández-Camacho, 1976) (description); (Duellman and Lynch, 1988) (brief report); (Duellman, 2015) (account, cranial osteology, advertisement call).

GASTROTHECA ANGUSTIFRONS (THIN-SNOUTED MARSUPIAL FROG) (BOULENGER, 1898)

Nototrema angustifrons (Boulenger, 1898): 124. **Syntypes.**—BMNH 1947.2.22.38–39 from San Javier de Cachabí, Provincia Esmeraldas, Ecuador.
Gastrotheca angustifrons Peters, 1955: 339.

Etymology.—The specific name is derived from the Latin *angustus*, meaning narrow, and *frons*, meaning forehead.

Identification.—Figure 4.5. Adults of this species attain a maximum SVL of 91.8 mm in females and of 73.2 in males. The skin on the dorsum is smooth and not adherent to or co-ossified with the skull. Transverse dermal ridges are present on the dorsum of the body. Supraciliary processes are absent. The tympanic annulus is granular. The snout is rounded in dorsal view and in profile. The interorbital distance is slightly greater than the width of the upper eyelid. Fingers I and II are equal in length. The fingers and toes have large terminal discs that are much wider than the adjacent digits. The fingers are webbed basally, and the toes are about four-fifths webbed. The hind limbs are long; the length of the tibia is more than 60% of the SVL. There is no calcar on the heel; tarsal folds are absent.

FIGURE 4.5 *Gastrotheca angustifrons*. AMNH 88200, male 67.8 mm SVL, Quebrada Guangui, Departamento Cauca, Colombia. Charles W. Myers.

In life, the dorsum is gray, tan, olive brown, or yellowish-brown, with or without darker transverse bars. Markings on the head consist of a dark canthal stripe, a dark interorbital bar, and a white or pale yellow suborbital bar. A pale dorsolateral stripe is absent. The flanks are uniformly tan or yellowish-brown; the venter is white with or without small brown marks. The webbing is brown. The iris is pale bronze brown to pale olive.

Comparisons.—*Gastrotheca andaquiensis*, *G. cornuta*, *G. dendronastes*, *G. weinlandii*. There are three species of *Gastrotheca* with transverse dermal ridges on the dorsum of the body. Of these, *G. cornuta* and *G. dendronastes* have conspicuous supraciliary processes; furthermore, *G. cornuta* also differs by having a triangular calcar on the heel, whereas *G. dendronastes* differs by having finger I longer than II. The only other species of *Gastrotheca* in the Chocoan region is *G. guentheri*, which differs from *G. angustifrons* by having the skin on the top of the head co-ossified with the skull, supraciliary processes, calcars, and teeth on the mandible and by lacking transverse dermal ridges on the dorsum. All of the long-legged arboreal *Gastrotheca* on the Amazonian slopes of the Andes differ from *G. angustifrons* by lacking transverse dermal ridges on the dorsum. Of these, *G. andaquiensis* differs by having a granular interocular ridge and calcars; *G. weinlandii* also has calcars, finger I longer than II, and the skin on the top of the head co-ossified with the skull. *Gastrotheca longipes* and *G. testudinea* are like *G. angustifrons*, but *G. longipes* has a green dorsum, and *G. testudinea* has vertical black bars on the flanks.

Natural history.—Individuals have been found on bushes and trees near streams in lowland tropical rainforests at night.

Distribution.—Figure 4.6. In Ecuador, *Gastrotheca angustifrons* is known from the Pacific lowlands in Provincia Esmeraldas at an elevation of 34 m.

This nocturnal, arboreal species inhabits natural vegetation in the evergreen forests of lowlands of Equatorial Chocó, where the average annual rainfall is 2431 mm and the average annual temperature is 25.7 °C.

Conservation.—*Gastrotheca angustifrons* is categorized as Critically Endangered (Possibly Extinct) throughout its entire range (IUCN SSC Amphibian Specialist Group, 2019b; in IUCN Red List of 2022) because the species has not been recorded since the 1990s despite intensive surveys throughout its range. Ortega-Andrade *et al.* (2021) categorized this species as Endangered; however, its categorization seems to be an error given that they provide data and justification to be included in the Critically Endangered category. This species has not recuperated from a drastic population decline in the 1990s, inferred from its apparent disappearance or near disappearance from known sites. The last Colombian locality was reported on 29 May 1983, and in Ecuador, it has not been reported since 1898. The lack of recent records is perplexing given that the sympatric species *G. cornuta* is still present in northwestern Ecuador. The major suspected threats for its sudden declines are climate change and pathogens (such as chytridiomycosis), whereas it is also affected by habitat loss and degradation. If a population still remains, significant threats to its existence include habitat reduction and fragmentation, logging, deforestation for agricultural development (particularly palm plantations), fires, human settlements, and pollution. Furthermore, the area where it is distributed faces constant pressure from mining and wood companies. In Ecuador, it has not been recorded in any public or private reserves.

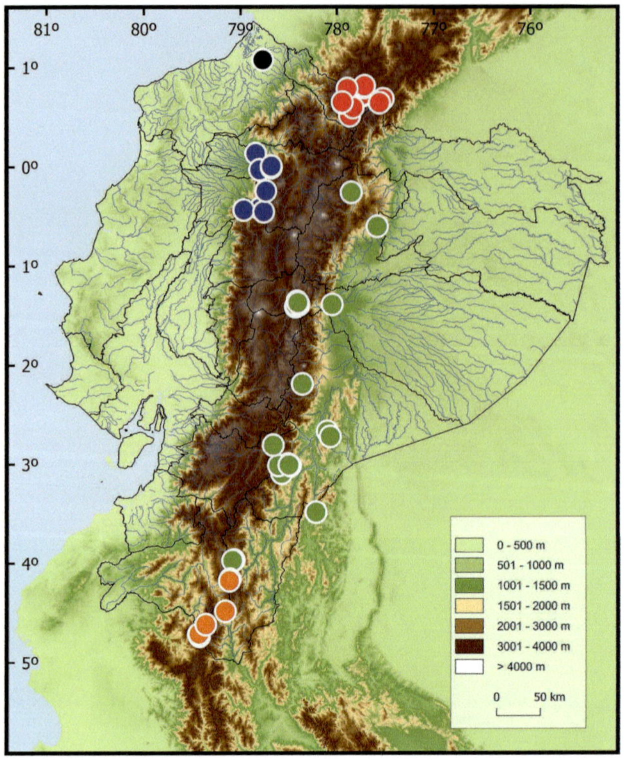

FIGURE 4.6 Localities of known occurrence of *Gastrotheca angustifrons* (black dot), *G. espeletia* (red dots), *G. guentheri* (blue dots), *G. testudinea* (green dots), and *G. turnerorum* (orange dots) in Ecuador.

Remarks.—Its molecular phylogenetic relationships are unknown. Duellman (2015) noted that two specimens from the Río Calima drainage in Colombia had a minute tubercle on each heel, and they differ somewhat from other specimens of *Gastrotheca angustifrons*.

Key references.—(Boulenger, 1898) (description); (Duellman, 2015) (account, cranial osteology, advertisement call).

GASTROTHECA CORNUTA (HORNED MARSUPIAL FROG) (BOULENGER, 1898)

Nototrema cornutum (Boulenger, 1898): 124. **Holotype.**—BMNH 1947.2.22.49 from San Javier de Cachabí, Provincia de Esmeraldas, Ecuador.

Gastrotheca cornuta Duellman, 1983: 872.

Etymology.—The specific name is derived from the Latin adjective *cornutus*, meaning horned. The name refers to the prominent supraciliary processes, which have been referred to as eyelid horns.

Identification.—Figures 4.7, 4.8. Adults of this species attain a maximum SVL of 77.7 mm in females and of 81.1 in males. The skin on the dorsum is smooth and not adherent to or co-ossified with the skull. Transverse dermal ridges are present on the dorsum of the body. Prominent supraciliary processes are present. The tympanic annulus is granular. The snout is rounded in dorsal view and nearly truncate in profile. The interorbital distance is much greater than the width of the upper eyelid. Fingers I and II are equal in length. The fingers and toes have large terminal discs that are much wider than the adjacent digits. The outer fingers are webbed basally, and the toes are about two-thirds webbed. The hind limbs are long; the length of the tibia is more than 60% of the SVL. A small triangular calcar is present on the heel; tarsal folds are absent.

In life, the dorsum is tan or brown with darker brown transverse markings on the body and limbs. Markings on the head consist of a dark brown interorbital bar and usually a suborbital bar. A cream suborbital spot usually is evident only by day. A pale dorsolateral stripe is absent. The flanks are creamy tan with dark brown flecks or small irregular marks. The venter is pinkish-tan or pale brown; the throat is brown. The iris is bronze brown peripherally and olive green medially.

FIGURE 4.7 *Gastrotheca cornuta*. Left: CJ 10525, female, 77.5 mm SVL. Right: CJ 10524, male, 75.1 mm SVL. Reserva Canandé, Provincia Esmeraldas. SGS.

Comparisons.—*Gastrotheca andaquiensis*, *G. angustifrons*, *G. dendronastes*, *G. longipes*, *G. testudinea*, *G. weinlandii*. Of the three species of *Gastrotheca* with transverse dermal ridges on the dorsum of the body *G. cornuta* is most like *G. dendronastes*, which differs from *G. cornuta* by having finger I longer than II and lacking a triangular calcar on the heel. *Gastrotheca angustifrons* differs from *G. cornuta* by lacking calcars and supraciliary processes. The only other species of *Gastrotheca* in the Chocoan region is *G. guentheri*, which differs from *G. cornuta* by having the skin on the top of the head co-ossified with the skull, supraciliary processes, calcars, teeth on the mandible, and by lacking transverse dermal ridges on the dorsum. All of the long-legged arboreal *Gastrotheca* on the Amazonian slopes of the Andes differ from *G. cornuta* by lacking transverse dermal ridges on the dorsum. Of these, *G. andaquiensis* differs by having a granular interocular ridge and calcars; *G. weinlandii* also has calcars, finger I longer than II, and the skin on the top of the head co-ossified with the skull. *Gastrotheca longipes* and *G. testudinea* are like *G. cornuta*, but *G. longipes* has a green dorsum, and *G. testudinea* has vertical black bars on the flanks.

Natural history.—Figure 4.9. The species seems to live primarily in the canopy. Although many individuals can be heard calling from the upper branches (estimated to be 20–35 m above the ground) of trees, few individuals are observed on lower branches.

Males call from the canopy. The call is a loud "bob." Usually, the call consists of one note, but sometimes, a frog

FIGURE 4.8 *Gastrotheca cornuta*. Left: axillary amplexus. Right: brooding female. Bottom: color variation in juveniles. Reserva Canandé, Provincia Esmeraldas. DAL.

FIGURE 4.9 *Gastrotheca cornuta*. Calling male in the laboratory, Reserva Canandé, Provincia Esmeraldas. LAC.

produces two or three notes in quick succession. In a call recorded at a temperature of 18.3 °C, the duration of each note is 0.08 s, and the interval between notes is 0.6 s. Each note noticeably diminishes in pitch and has three harmonics at about 800, 1600, and 2400 Hz, of which the lowest is the dominant frequency.

In laboratory breeding, Gagliardo *et al.* (2010) completely documented one breeding event that required three hours from the time amplexus was initiated until the last egg was placed in the pouch. Shortly after amplexus occurred, the male inflated and deflated his lungs several times before angling his hind legs upward so that his feet were between the female's cloaca and the opening of the brood pouch. Using his toes, the male rubbed a gel-like substance (presumably from the male's cloaca and containing sperm) over the lower back and pouch of the female. He continued cycles of lung inflation and deflation for about 45 min. The female arched her back so that her cloaca was directed upward, nearly in contact with the male's cloaca. Nearly two hours after amplexus was initiated, the first egg appeared, and five eggs emerged in the following 31 min. The male used his feet to position the eggs and push them into the brood pouch. The eggs undergo direct development in the brood pouch. Five to 10 eggs in the pouch were 9.2–17.0 mm in diameter. A female with an SVL of 72.5 mm contained 10 hatchling embryos having SVLs of 16.5 mm. Females that bred in captivity produced 23 and 14 young after gestation periods of 60–80 days. Brooding females have been found in January and July in Panama, in May and August in Colombia, and in November in Ecuador. Additional reproductive biology data of a pair from Reserva Canandé, Provincia Esmeraldas (male JMG 2260, 28 g and female JMG 2261, 36 g) have been obtained in the laboratory (CJ and Wikiri lab notes, by Elena Naydenova, 2019). In one reproduction event, the axillary amplexus lasted about one night. One clutch contained 11 eggs (10.82 mm diameter, n =1) that were placed in the pouch of the female. The gestation period was 63 days, with 10 froglets (1.46 g, 24.9 mm SVL, n =10) born with ambient temperature at about 22.1 °C.

Distribution.—See Figure 4.4. *Gastrotheca cornuta* has a wide distribution, encompassing the Caribbean slopes (90–1000 m) of eastern Costa Rica and Panama and the Pacific lowlands and slopes (100–1000 m) of southern Colombia and northwestern Ecuador. A hiatus of about 500 km in central and northern Chocoan Colombia exists between the known occurrence of this species in Central America and in western South America. This gap may be an artifact of sampling, as relatively little collecting has been done in that part of Colombia. In Ecuador, *G. cornuta* occurs in the Pacific lowlands in Esmeraldas, Santo Domingo de los Tsáchilas, and Azuay provinces. Its elevational range is 30–767 m in an area of extent of occurrence of about 11,700 km^2.

This nocturnal, arboreal species inhabits natural vegetation and partially intervened areas in the evergreen forest of lowlands of Equatorial Chocó and the evergreen montane foothill forests of the Cordillera Occidental of the Andes ecosystems in the Equatorial Chocó, Jama-Zapotillo, and Cordillera Occidental of the Andes biogeographic sectors, where the average annual rainfall is 1555–2649 mm and the average annual temperature is 21.4–25.8 °C.

Conservation.—*Gastrotheca cornuta* is categorized as Endangered throughout its entire range (Coloma *et al.*, 2008; in IUCN Red List of 2022) because of an observed past and projected future decline, estimated to be at least 50% over a 10-year period. However, Ortega-Andrade *et al.* (2021) categorized it as Vulnerable in Ecuador. We suggest placing it in the Endangered category in Ecuador because its area of occupancy is suspected to be less that 500 km^2, its distribution is severely fragmented, and there is a continuing decline in the extent and quality of its habitat. The precipitous decline of this species has been associated with the disease chytridiomycosis. This disease is also present in the lowlands of western Ecuador. There is no information on the population of this species. In Ecuador, other threats to its existence include habitat reduction and severe fragmentation, logging, deforestation for agricultural development (particularly palm plantations), fires, mining, the introduction of exotic species, human settlements, and pollution. Furthermore, the area where it is distributed faces constant pressure from wood companies. Areas of forest close to the type locality in Cachabí, Provincia Esmeraldas, Ecuador, have now been deforested, and no *G. cornuta* are present there. Also, most of the forests in the southern portion of its distribution (Santo Domingo de los Tsáchilas, Los Ríos and Cañar provinces) have been cleared, and records from these areas are historical. It occurs, or occurred, in several protected areas in Colombia, Panama, and Costa Rica. In Ecuador, it has been recorded in the protected forest of Centro Chachi Gualpi del Cayapas and within the biosphere reserve Macizo del Cajas. *Ex situ* conservation programs for this species are being carried out at the El Valle Amphibian Conservation Center in Panama and CJ de Investigación y Conservación de Anfibios in Ecuador.

Remarks.—A gap of nearly 500 km exists between records from Ecuador and southern Colombia and those in Central America. According to Duellman (2015), there are some differences between the frogs in these areas. Costa Rican specimens have the most transverse dermal ridges, 11–14. The differences in iris color are perplexing because geographic variation in this character is quite unusual. Should they be shown to be distinct species, the populations in Ecuador would still be *Gastrotheca cornuta*, whereas *Hyla ceratophrys* Stejneger (1911) is an available name for the Central American populations.

Key references.—(Boulenger, 1898) (description); (Duellman, 2015) (account); Yánez-Muñoz *et al.* (2020) (account).

GASTROTHECA CUENCANA (CUENCA MARSUPIAL FROG) CARVAJAL-ENDARA *ET AL.*, 2019

Gastrotheca cuencana Carvajal-Endara *et al.*, 2019: 17. **Holotype.**—CJ 1391 from Cuenca, 2579 m, Provincia Azuay, Ecuador.

FIGURE 4.10 *Gastrotheca cuencana*. Left: CJ 1940, female, 50.2 mm SVL. Right: CJ 1396, male, 52.2 mm SVL. Cuenca, Provincia Azuay. LAC.

Etymology.—The specific name is derived from an adjective referring to people born in Cuenca, the capital city of Provincia Azuay, where this species occurs. This name is also used to highlight the pioneer efforts of the people of Cuenca during a project to protect urban and suburban species of frogs through the biorestoration and protection of watersheds, river banks, wetlands, and frog ponds.

Identification.—Figures 4.10, 4.11, 4.12. Adults of this species attain a maximum SVL of 61.6 mm in females and of 53.7 in males. The skin on the dorsum is finely granular, not co-ossified with the skull, and lacks transverse ridges. The supraciliary processes are absent. The tympanic annulus is weakly granular to smooth. The interorbital distance is slightly larger than the width of the upper eyelid. Fingers I and II are about equal in length; the width of the discs is greater than digits. The fingers are unwebbed. The webbing between the external toes extends to the antepenultimate subarticular tubercle on toe IV and to the penultimate subarticular tubercle on toe V. The tibia length is 39–50% of the SVL, slightly shorter than the length of the foot. The heel lacks a calcar or tubercle.

The dorsum is green, tan, brown, or reddish-brown, with or without darker paravertebral marks. When present, the paravertebral markings usually are well defined; at the level of the sacrum, they become two rows of irregular blotches on each side of the vertebral axis. The head markings consist of a cream labial stripe that extends from the posterior margin of the lip to the insertion of the forelimb and, in some individuals, an inconspicuous canthal stripe, which is absent in most individuals. There is a pale white or cream dorsolateral stripe in most individuals. The flanks are cream, tan, brown, or green with cream flecks or spots. The tympanum is brown, tan, or olive green. The groin and anterior and posterior surfaces of the thighs are translucent cream, without marks and with a faint, pale blue tinge in some. The venter is uniform creamy white. The gular surface varies from white to creamy gray. When present, supracloacal and heel stripes are cream or tan. The ventral surfaces of the shanks have a faint tinge of pale blue. The iris is reddish-bronze with black reticulations.

Comparisons.—*Gastrotheca elicioi*, *G. litonedis*, *G. lojana*, *G. pseustes*, *G. riobambae*. In comparison with similar species in Ecuador, *G. cuencana* is most like *G. litonedis*, *G. lojana*, *G. elicioi*, and *G. riobambae*. It differs from these species by having a uniform creamy white venter; whereas the venter is pale gray in *G. litonedis* and cream with dark flecks, spots, or mottling in *G. lojana*, *G. elicioi*, and *G. riobambae*. In *G. cuencana*, the dorsal surfaces of the fingers are cream, whereas in the other species, the dorsal surfaces of the fingers are of the same color as the rest of the body (tan, brown, or green). *Gastrotheca cuencana* also differs from *G. litonedis*, *G. elicioi*, and *G. riobambae* by lacking a dark canthal stripe. Moreover, in *G. cuencana*, the groin and anterior surfaces of the thighs are mostly translucent cream without marks, whereas in *G. litonedis*, the groin and anterior

FIGURE 4.11 Color variation in *Gastrotheca cuencana*. Top left: CJ 1940, female, 50.2 mm SVL. Top right: CJ 1393, male, 46.7 mm SVL. Middle left: CJ 1395, female, 62.7 mm SVL. Middle right: CJ 1396, male, 52.2 mm SVL. Bottom left: CJ 1392, female, 49.8 mm SVL. Bottom right: CJ 1397, female, 55.1 mm SVL. Cuenca, Provincia Azuay. LAC.

FIGURE 4.12 Morphological details of *Gastrotheca cuencana*. Top: groin. Bottom: thighs. CJ 1940, Cuenca. Provincia Azuay. LAC.

surfaces of the thighs are usually bronze–brown or tan with cream or darks flecks and spots. In *G. lojana* and *G. riobambae*, the groin and anterior surfaces of the thighs are usually cream with black mottling; in *G. elicioi*, the groin and anterior surfaces of the thighs usually have dark and cream flecks and spots. In profile, the snout in *G. cuencana* is rounded above and inclined anteroventrally, whereas the snout in G. *elicioi*, *G. lojana*, and *G. litonedis* is bluntly rounded. Furthermore, *G. cuencana* has a conspicuous, elevated row of dorsolateral warts, whereas these are barely raised in *G. litonedis*; a narrow, cream supracloacal stripe is present in *G. litonedis*, *G. lojana*, and *G. riobambae*, whereas the stripe is fragmented in *G. elicioi* and absent in most *G. cuencana*. *Gastrotheca lojana* and *G. riobambae* also differ from *G. cuencana* and *G. litonedis* by having a complex call, whereas the call in the latter two species is a simple call. Additionally, the call of *G. cuencana* has a lower note rate, a shorter note duration, a longer inter-note interval, and a lower dominant and 90% bandwidth frequency compared with *G. litonedis*. Finally, *G. cuencana* differs from its sister species (*G. litonedis*) by having a genetic distance of 1.2% (in a DNA dataset of 438 bp, 16S gene). *Gastrotheca cuencana* occurs in syntopy with *G. pseustes* (*sensu lato*) throughout most of its range. *Gastrotheca pseustes* differs from *G. cuencana* by having coarsely granular skin on the dorsum, in contrast to the usually finely granular dorsum in *G. cuencana*. In *G. cuencana*, the venter is uniform creamy white, whereas it is cream or white with dark flecks or spots in *G. pseustes*; also, the call of *G. cuencana* is simple, whereas in *G. pseustes*, the call is complex.

Natural history.—Figure 4.13. This species is found in pasture lands and urban grasslands where lentic water is available for tadpoles; thus, the frogs occur close to irrigation ditches, water reservoirs, temporary ponds, or wetlands. The holotype and eight other individuals were collected in grassland in an urban area (Cuenca). Males were calling from small shrubs at approximately 80–100 cm above the ground. Females were perched on branches of shrubs at approximately 40 cm above the ground and also among the grass. Eight tadpoles were found in a small temporary pond. *Gastrotheca cuencana* is syntopic with *G. pseustes* (*sensu lato*) throughout most of its range.

Males call at night. Calls from two locations in Provincia Azuay (three individuals from Cuenca and three from Río Mazán) and from one location in Provincia Cañar (two individuals from Papaloma de la Nube), Ecuador, were described as a simple call composed of a series of short, single-pulsed notes (three and 19 notes per call) emitted at regular intervals of about 0.016 s and a mean inter-note interval of 2.783 s. Detailed data and figures are provided under the subtitle "Vocalizations of Ecuadorian Anurans" in Appendix 1.

FIGURE 4.13 *Gastrotheca cuencana*. Top: calling male. Middle and bottom: brooding females. Cuenca, Provincia Azuay. LAC. Bottom: Juan C. Sánchez Nivicela.

FIGURE 4.14 Ontogenetic changes in *Gastrotheca cuencana* from tadpole to froglet. Top to bottom. CJ 1945, stage 37, lateral, dorsal, and ventral views, TL = 47.5 mm; stage 41, lateral and dorsal views; stage 46. Cuenca, Provincia Azuay. LAC. Taken from Carvajal-Endara *et al.* (2019).

Tadpoles.—Figures 4.14, 4.15. The following description is based on a free-swimming tadpole in stage 38 from south Cuenca, 2579 m, Provincia Azuay, Ecuador. The total length is 44.6 mm and the body length is 17.3 mm (39% of total length). The body is ovoid in dorsal and lateral views and slightly depressed. The throat is slightly concave in lateral profile, sloping from the tip of the snout to the belly. The body width at the level of the spiracle is 11.5 mm, and the height at same position is 9.3 mm. The head width at level of the eyes is 8.5 mm. The lateral line system is present but barely visible. The supraorbital and infraorbital lines both rise at the level of the tip of the snout, extend parallel to the eye, and make contact immediately behind the eye; the angular line descends vertically just posterior to the eye toward the throat; it dorsally contacts the post infraorbital line. The anterior oral line descends vertically from the oral disc level and behind the naris level to the throat; it makes a curve that parallels the infraorbital line, forming a circuit continuous with angular and loreal lines. The dorsal body and middle body lines are not visible. The nostril is medium-sized (in proportion to body length), ovoid, protruding, having a fleshy annulus, its opening is

FIGURE 4.15 Oral disc of *Gastrotheca cuencana*. CJ 1945, stage 38, Cuenca (south control), Provincia Azuay. Taken from Carvajal-Endara *et al.* (2019).

directed anterolaterally. The snout–nostril distance is 2.3 mm; the internarial distance is 2.4 mm. The eye is directed dorsally. The eye length is 1.8 mm; the eye width is 1.8 mm; the interorbital distance is 4.3 mm. The spiracle is sinistral, located at the midbody level. The spiracular opening is oriented posteriorly; the distance from the tip of the snout to the spiracular opening is 11.0 mm. The end of the spiracular tube is rounded, attached to the body wall; the inner wall of the spiracular tube is not evident. The tube length is 3.1 mm, and the tube transverse width is 2.0 mm. The vent tube is dextral; its opening is oriented posteriorly. The tube length is 1.7 mm, and the tube transverse width is 1.8 mm. The tail length is 27.2 mm. The caudal musculature is slender, narrowing gradually until the tail terminus. The caudal muscle height is 3.7 mm, and its width is 3.2 mm. The caudal fins are well developed and proportional, arising near the tail–body junction, forming a low hump. The dorsal fin height is 3.0 mm; the ventral fin height is 2.8 mm; the maximum height of the tail is 9.3 mm; the tail terminus is rounded; the caudal musculature does not reach the fin terminus.

The oral disc is small, ventral, located near the tip of the snout, not protruding laterally beyond the body. The transverse width is 4.7 mm. It is surrounded by a uniserial row of marginal papillae, interrupted medially on the upper lip. The lower lip papillae alternate in orientation, giving the appearance of two rows. The upper lip has 29 papillae on the right side and 27 papillae on the left side. The lower lip bears 62 marginal papillae. The upper jaw sheath is medium-sized, forming a smooth arch, and finely serrated; its height is 0.36 mm; its transverse width is 2.8 mm (60% of width of oral width). The lower jaw sheath is V-shaped, open and finely serrated; its width is 1.92 mm; its height is 0.7 mm. The labial tooth row formula is 2/3(1). The tooth rows lengths are A-1: 3.6 mm, A-2: 3.5 mm, P-1 right row: 1.5 mm, P-1 left row: 1.6 mm, P-1 gap: 0.2 mm, P-2: 3.4 mm, and P-3: 3.0 mm.

The total length varies between 42.3 mm (stage 37) and 48.7 mm (stage 39), and the tail length proportion remains about 61%. The labial tooth row formula is 2/3(1). The number of marginal papillae varies among specimens and Gosner stages; variation in the number of ventral papillae on the lower lip is high (39–64).

In dorsal view, the body is olive cream; the snout and flanks paler olive cream. In ventral view, guts are dark gray in the belly area; the gills are reddish and visible throughout the throat. A nearly continuous line of cream marks covers the caudal musculature. The proximal half of the musculature is reddish-cream, with the myomeres and medial line well defined; the distal half is translucent. The dorsal and ventral fins are translucent, with scattered cream dots, heavily suffused near the tail–body junction. The spiracle, oral apparatus, and legs are olive cream. The iris is gold.

Changes in coloration during ontogenetic development have been documented in one mostly brown individual. At stage 41, the dorsum and flanks are diffuse brown with a well-defined pattern of brown–gray paravertebral stripes and a couple of blotches on the posterior dorsolateral body. A dark, gray–black stripe borders the canthal and dorsolateral body extending to the level of the midbody, bordered dorsally by a creamy white stripe that becomes wider at the distal end. The iris has a red suffusion. By stage 46, the markings on the dorsum and flanks are better defined, and the black–gray stripe on the flanks becomes a nearly black stripe; the fingers are yellowish-cream. There is scattered green pigmentation in the anterior portion of the paravertebral stripes.

Distribution.—See Figure 4.4. *Gastrotheca cuencana*, an endemic species to Ecuador, is known in the Paute Basin, which includes the eastern and western Cordilleras, in Cañar and Azuay provinces. Its elevational range is 2381 – 3329 m in an area of extent of occurrence of about 1180 km^2.

This nocturnal, semiarboreal species inhabits disturbed areas and a few forests in the evergreen high montane forests of the Cordillera Occidental of the Andes ecosystem in the Cordillera Occidental of the Andes and southern Cordillera Oriental of the Andes, where the average annual rainfall is 660–1192 mm and the average annual temperature is 8.8–15.2 °C.

Conservation.—*Gastrotheca cuencana* is categorized as Endangered (IUCN SSC Amphibian Specialist Group, 2022a; in the IUCN Red List 2022) because its extent of occurrence is 1494 km^2 (1180 km^2 according to our calculations), it occurs in five threat-defined locations, and there is a continuing decline in the extent and quality of its habitat, largely due to urbanization and the draining of wetlands in the Paute Basin in the provinces of Azuay and Cañar, Ecuador. For example, the city of Cuenca is rapidly growing and occupying previous rural areas, where swampy areas are being drained. It has been recorded in the public reserve Parque Nacional Cajas, the protected forests of Mazán, and within the biosphere reserve Macizo del Cajas. Some efforts (*in situ* and *ex situ*) to protect this species are underway by parties of the Parque Nacional Cajas, Amaru Zoo (Arbeláez Ortiz, 2014, as *G. litonedis*), and the city council of Cuenca (Siavichay-Pesántez *et al.*, 2016a, 2016b). However, these efforts urgently need to be reinforced.

Remarks.—*Gastrotheca cuencana* was confused with *G. litonedis* until 2018. The recognition of both species emphasizes the discrete morphological and acoustic differences between them. *Gastrotheca cuencana* belongs to the subgenus *Duellmania* (*sensu* Duellman, 2015). Note that the figure legends of Figs. 3 and 4 in Carvajal-Endara *et al.* (2019) have been incorrectly positioned. The figure legend intended for Fig. 4 has been mistakenly assigned to Fig. 3, and vice versa.

Key references.—(Carvajal-Endara *et al.*, 2019) (description, call, tadpole).

GASTROTHECA DENDRONASTES (ARBOREAL MARSUPIAL FROG) DUELLMAN, 1983

Gastrotheca dendronastes Duellman, 1983: 868. **Holotype.**—KU 169381 from Río Calima, 1.5 km [by road] west of Lago Calima, 1230 m, Departamento de Valle del Cauca, Colombia.

Etymology.—The specific name is derived from the Greek words *dendron*, meaning tree, and *nastes*, meaning inhabitant. The name was used in reference to the arboreal habits of this species.

Identification.—Figure 4.16. Adults of this moderately large species attain a maximum SVL of 79.5 mm in females and of 73.5 in males. The skin on the dorsum is finely granular with scattered tubercles; it is not adherent to or co-ossified with the skull. Nine to 12 tubercular transverse dermal ridges are present on the dorsum of the body. Small supraciliary processes are present. The snout is rounded in dorsal view and abruptly inclined anteroventrally in profile. The tympanic annulus is granular. The interorbital distance is barely greater than the width of the upper eyelid. Finger I is longer than II. The fingers and toes have large terminal discs that are much wider than the adjacent digits. The fingers are unwebbed, and the toes are about two-thirds webbed. The hind limbs are long; the length of the tibia is more than 60% of the SVL. A conical tubercle is present on the heel; tarsal folds are absent.

At night, the dorsum is tan with slightly darker transverse markings on the body and limbs. By day, the dorsum is pale brown to olive tan and olive brown with reddish-brown transverse marks and green flecks in some individuals. Markings on the head consist of a creamy white suborbital bar and a dark brown interorbital bar and canthal

FIGURE 4.16 *Gastrotheca dendronastes*. Top: USNM 258936, La Palma, 14.3 km northeast of Río Faisanes (old road Santo Domingo-Chiriboga), Provincia Pichincha. Roy W. McDiarmid. Bottom: KU 169381, holotype, male, 63.5 mm SVL, Río Calima, 1.5 km west of Lago Calima, Departamento Valle del Cauca, Colombia. WED.

stripe. A pale dorsolateral stripe is absent. The flanks are olive green to dark brown with cream longitudinal stripes bordered by dark brown. The posterior surfaces of the thighs are pale brown. The labial bars, supracloacal stripe, and heel stripes are cream. The webbing is reddish-brown. The venter is creamy yellow with reddish-brown flecks. The iris is dull bronze with fine black reticulations.

Comparisons.—*Gastrotheca andaquiensis*, *G. angustifrons*, *G. cornuta*, *G. longipes*, *G. testudinea*, *G. weinlandii*. *Gastrotheca angustifrons* and *G. cornuta* also have transverse dermal ridges on the dorsum of the body; *G. dendronastes* is most like *G. cornuta*, which differs from *G. dendronastes* by having fingers I and II equal in length and having a triangular calcar on the heel. *Gastrotheca angustifrons* differs from *G. dendronastes* by having fingers I and II equal in length and by lacking supraciliary processes. The only other species of *Gastrotheca* in the Chocoan region is

G. guentheri, which differs from *G. dendronastes* by having the skin on the top of the head co-ossified with the skull, calcars, and teeth on the mandible and by lacking transverse dermal ridges on the dorsum. All of the long-legged arboreal *Gastrotheca* on the Amazonian slopes of the Andes differ from *G. dendronastes* by lacking transverse dermal ridges on the dorsum. Of these, *G. andaquiensis* differs by having a granular interocular ridge and calcars; *G. weinlandii* has calcars, and the skin on the top of the head is co-ossified with the skull. *Gastrotheca longipes* and *G. testudinea* are like *G. dendronastes*, but *G. longipes* has a green dorsum, and *G. testudinea* has vertical black bars on the flanks.

Natural history.—All individuals have been found at night in ravines containing streams. Most adults have been observed 2–5 m above the ground on branches of trees and on stems of giant *Heliconia*.

The presence of juveniles in January, March, May, and December (in January and May at Río Faisanes in Provincia Pichincha, Ecuador) suggests that the breeding season may be long, perhaps throughout the year. The advertisement call is a loud, chuckle-like sound followed by three or four "clucks." Little else is known about the reproductive biology except that one female contained 11 ovarian eggs 9 mm in diameter.

Distribution.—Figure 4.17. *Gastrotheca dendronastes* is known on the Pacific slopes of the Cordillera Occidental in southern Colombia and northern Ecuador and in the northern part of the Cordillera Central in Colombia. In Ecuador, it is known in Provincia Santo Domingo de los Tsáchilas. Its elevational range is 1343 – 1405 m in two localities 0.5 km apart.

This nocturnal, arboreal species inhabits cloud forests of the evergreen low montane forests of the Cordillera Occidental of the Andes in the Cordillera Occidental of the Andes biogeographic sector, where the average annual rainfall is 2484 mm and the average annual temperature is 18.7 °C.

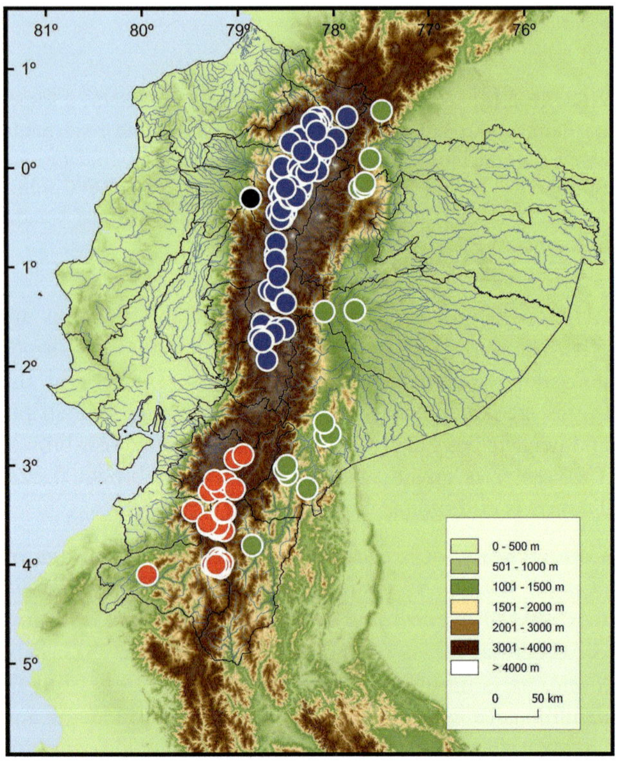

FIGURE 4.17 Localities of known occurrence of *Gastrotheca dendronastes* (black dot), *G. lojana* (red dots), *G. riobambae* (blue dots), and *G. weinlandii* (green dots) in Ecuador.

Conservation.—*Gastrotheca dendronastes* is categorized as Endangered throughout its entire range (IUCN SSC Amphibian Specialist Group, 2019c; in IUCN Red List of 2022) because a drastic population decline is suspected to have occurred in the 1990s, inferred from its apparent disappearance from known sites. However, it is important to note that in Ecuador, *G. dendronastes* has been categorized as Critically Endangered by Ortega-Andrade *et al.* (2021). In Ecuador, there have been no sightings for more than four decades; the most recent record of this species is 25 October 1979 from La Palma, Río Faisanes, Provincia Pichincha. The major suspected threats for its sudden declines are climate change and pathogens (such as chytridiomycosis), whereas it is also affected by habitat loss and degradation due to logging, mining activities, agriculture, human settlements, fires, and the use of pesticides. No individuals were found during the search efforts in Río Faisanes (Bustamante *et al.*, 2005). The most recent record in Colombia is from 2004 (Gutiérrez-Cárdenas and Barrientos, 2007). This species occurs in several protected areas in Colombia, including Parque Nacional Natural Farallones de Cali. In Ecuador, it has been recorded in the protected forests of Toachi–Pilatón.

Remarks.—Although *Gastrotheca dendronastes* is distinguished from other long-legged arboreal *Gastrotheca* species on the Pacific versant of South America by several characters of external morphology and coloration, it is similar in size and proportions to the other species. A stepwise discriminant analysis of morphometric data distinguished *G. dendronastes* from *G. angustifrons* and *G. cornuta* only by the nostril–jaw distance.

Key references.—(Duellman, 1983) (description); (Duellman, 2015) (account, cranial osteology, call).

GASTROTHECA ELICIOI (ELICIO'S MARSUPIAL FROG) CARVAJAL-ENDARA *ET AL.*, 2019

Gastrotheca elicioi Carvajal-Endara *et al.*, 2019: 34. **Holotype.**—CJ 1402 from Cajanuma, entrance to Parque Nacional Podocarpus in the Loja-Vilcabamba road, 2456 m, Provincia Loja, Ecuador.

Etymology.—The specific name is a patronym that honors Elicio E. Tapia, an Ecuadorian biologist and member of the Tapia family who has made significant contributions to the expansion of biodiversity collections in Ecuador. The authors highlight Elicio's intense field collecting efforts and enormous contribution to the museums that hold Ecuadorian amphibians, including many tissues, call recordings, ecological data, and specimens of *Gastrotheca*.

Identification.—Figures 4.18, 4.19. Adults of this moderately large species attain a maximum SVL of 67.5 mm in females and of 67.8 in males. The skin on the dorsum is finely granular; it is not co-ossified with the skull and lacks transverse ridges. The supraciliary processes are absent. The tympanic annulus is smooth to slightly granular. The interorbital distance is larger than the width of the upper eyelid. Fingers I and II are about equal in length; the width of the discs is about twice the width of the digits proximal to the discs. The fingers are unwebbed. The webbing between the external toes extends nearly to the antepenultimate subarticular tubercle on toe IV and proximal to the penultimate subarticular tubercle on toe V. The tibia length is 42–50% of the SVL, slightly longer than the foot. The heel lacks a calcar or tubercle. Each dentigerous vomerine process has 6–8 teeth.

The dorsum is green, tan, brown, or reddish-brown, with or without dark paravertebral marks. When present, these paravertebral marks usually are narrow, curved, and fragmented; these marks are adherent at the scapular level and may be connected with the interorbital mark. The head markings consist of a pale labial stripe, a black canthal stripe, and a dark triangular interorbital mark in some individuals. A short cream labial stripe is present on the margin of the lip; the stripe continues posteriorly to the insertion of the forelimb. The dark brown canthal stripe is always present, but it is inconspicuous in some individuals with a dark ground color. The tympanum is brown, tan, or olive green. A bronze, highly fragmented dorsolateral stripe is present in most individuals. The flanks are brown, bronze–brown, or green. The groin has bluish coloration with slight mottling. The posterior surfaces of the thighs usually are pale green with black flecks, and some specimens show black fringes. When present, supracloacal and heel stripes are cream and fragmented. The ventral surfaces usually are cream with dark marks, but some specimens lack dark marks. The gular surface varies from brownish-gray to white with black spots. The iris is usually copper, almost without black reticulations.

Comparisons.—*Gastrotheca cuencana*, *G. elicioi*, *G. lojana*, *G. litonedis*, *G. monticola*, *G. pseustes*, *G.*

FIGURE 4.18 *Gastrotheca elicioi*. Left: CJ 1940, female, 50.2 mm SVL. Right: CJ 1396, male, 52.2 mm SVL. Loja, Provincia Loja. LAC.

FIGURE 4.19 Color variation in metamorphs (stage 46 of Gosner) of *Gastrotheca elicioi* from Ecuador. Left: CJ 4313 from Loja-Abra de Zamora. Right: CJ 1950 from Loja, Provincia Loja. LAC.

turnerorum. In comparison with similar species, *Gastrotheca elicioi* is most like *G. lojana*, *G. cuencana*, *G. litonedis*, *G. turnerorum*, and *G. pseustes* in Ecuador and *G. monticola* in Peru. *Gastrotheca elicioi* differs from *G. lojana*, *G. cuencana*, and *G. litonedis* in color pattern as follows. *Gastrotheca elicioi* and *G. litonedis* have distinct dark canthal stripes, which are absent in *G. lojana* and inconspicuous or absent in *G. cuencana*. The groin and anterior and posterior surfaces of the thighs are slightly mottled in *G. elicioi*, whereas in *G. lojana*, they are heavily mottled; these surfaces are translucent cream without marks in *G. cuencana* and usually brown in *G. litonedis*. If any marks are present on the dorsum in *G. elicioi*, they usually consist of a triangular interorbital mark, which is connected or not with two narrow and curved paravertebral marks. In contrast, in *G. lojana*, the interorbital mark is a transverse bar usually connected with two broad paravertebral marks; in *G. cuencana* and *G. litonedis*, an interorbital mark is absent. *Gastrotheca lojana* and *G. cuencana* have a conspicuous, elevated row of dorsolateral warts, whereas they are barely raised in *G. elicioi* and *G. litonedis*. Additionally, in *G. elicioi*, the dark bars on the limbs, when present, are shorter, thinner, and less defined than in *G. lojana*, whereas in *G. cuencana* and in some *G. litonedis*, the bars are replaced by irregular blotches. *Gastrotheca elicioi* has a cream venter with dark flecks or marks, whereas the venter is uniform creamy white in *G. cuencana* and pale brownish-gray in *G. litonedis*. *Gastrotheca cuencana* also differs from *G. elicioi* by having cream dorsal surfaces of the fingers, whereas they are brown, green, or tan in *G. elicioi*. *Gastrotheca turnerorum* and *G. pseustes* differ from *G. elicioi* by having the skin on the dorsum areolate and weakly areolate, respectively, whereas in *G. elicioi*, the skin on the dorsum is finely granular. Also, they differ by lacking an interorbital mark and bars on the limbs, which can be present in *G. elicioi*. *Gastrotheca pseustes* is sympatric with *G. elicioi*, from which it differs notably by having smaller digital discs (larger in *G. elicioi*) and lacking pale dorsolateral and supracloacal stripes, which are fragmented in *G. elicioi*. *Gastrotheca monticola* is the sister species of *G. elicioi*, and their minimum genetic divergence is 3.0% (in a DNA dataset of 438 bp, 16S gene). *Gastrotheca monticola* differs from *G. elicioi* by having the axilla, groin, concealed surfaces of the thighs, and dorsal surfaces of the feet green with well-defined black spots on a green or tan background, whereas in *G. elicioi*, the groin and anterior and posterior surfaces of the thighs are lightly mottled with tinges of green or blue. Finally, the advertisement call of *G. elicioi* is unique among the calls of *Gastrotheca* in southern Ecuador in that long, pulsed notes are produced after the short ones, whereas the other species tend to produce long, pulsed notes before the short ones.

Natural history.—Figure 4.20. At the locality 5.5 km west of Loja, many *Gastrotheca elicioi* were found in a roadside agave fence row, where they were sitting on the agave leaves. At the locality 5.2 km west of Loja, an individual of *G. elicioi* was under an overhanging bank at the edge of small pond in a pasture by day. At Zamora Huayco, in the vicinity of Loja, two individuals were sitting on branches of shrubs at approximately 4–5 m above the ground; they were about 7 m from a river within moderately dense mountain forest. The holotype was one of a series of 16 tadpoles that were at the edge of a pond. At the type locality, *G. elicioi* is syntopic with *G. pseustes*; *G. elicioi* is also syntopic with *G. lojana*.

FIGURE 4.20 *Gastrotheca elicioi*. Calling male and gravid female. Loja and surroundings, Provincia Loja. LAC.

Males call at night. Calls from three locations in Provincia Loja (one individual from the old Loja-Catamayo road; three from Loja, Parque Universitario de Educación Ambiental y Recreación (PUEAR); and one from Loja, Quebrada El Salado) and two individuals in *ex situ* outdoor enclosures of CJ in San Rafael, Quito, Ecuador, were described as a complex call composed of up to six short, single-pulsed notes and followed by up to seven longer pulsed notes. It can be characterized as a series of short notes, which become gradually longer, pulsated notes by the end of the call. The long notes have a mean duration of 0.188 s and consist on average of 3.47 distinct pulses, partly fused, without silent intervals. The short notes have a mean duration of 0.027 s, and the inter-note interval is 0.452 s on average. Detailed data and figures are provided under the subtitle "Vocalizations of Ecuadorian Anurans" in Appendix 1.

This species has a biphasic mode of reproduction. On 5 December 2017, a pregnant female *Gastrotheca elicioi* was caught (Quebrada Shucos, Loja, Provincia Loja) and brought to the laboratory. After 20 days, the female deposited 434 tadpoles in Gosner stage 34. This is the largest number of tadpoles deposited by a female of the genus *Gastrotheca*. The female (SVL = 71.7 mm) weighed 43.06 g when caught and 24.38 g after she deposited the tadpoles.

Tadpoles.—Figures 4.21, 4.22. The following description is based on a specimen in stage 36 from a series of 50 tadpoles from Puntzará Alto, near Loja, 2311 m, Provincia Loja, Ecuador. The total length is 51.4 mm, and the body length is 19.4 mm (38% of total length). The body is ovoid in dorsal and lateral views, slightly depressed. The throat is slightly concave in lateral profile, sloping from the tip of the snout to the belly. The body width at the level of the spiracle is 11.9 mm, and the height at same position is 9.9 mm. The head width at the level of the eyes is 11.1 mm. The lateral line system is present but barely visible. The supraorbital line is not evident at the level of the snout; the infraorbital line is present at the level of the eye, touching the inferior portion of the orbit and making contact with the supraorbital line immediately behind the eye. The inferior oral line is visible at the eye level, where it contacts the angular line, which descends from the eye level. The supraorbital line is represented by scattered neuromasts and does not make contact with the orbit. The postorbital line forms a circle of neuromasts just behind the eye. The anterior oral line and loreal lines are not visible. The dorsal body and middle body lines are not visible. The nostril is medium-sized (in proportion to body length), ovoid, protruding; it has a fleshy annulus, and its opening is directed anterolaterally. The snout–nostril distance is 3.5 mm; the internarial distance is 2.9 mm. The eye is directed dorsally; the eye length is 2 mm, the eye width is 1.8 mm, and the interorbital distance is 5 mm. The spiracle is sinistral, located at the midbody level, and the spiracular opening is oriented posteriorly. The distance from the tip of the snout to the spiracular opening is 13.0 mm. The end of the spiracular tube is rounded, attached to the body wall; the inner wall of the spiracular tube is not evident. The spiracle length is 2.9 mm; the tube transverse width is 2.9 mm. The vent tube is dextral; its opening is oriented posteriorly; the tube length is 3.4 mm, and the vent tube transverse width is 3 mm. The tail length is 32.4 mm. The caudal musculature is slender, narrowing gradually until the tail terminus. The caudal muscle height is 4.4 mm; its width is 3.4 mm. The caudal fins are well developed and proportional, arising abruptly near the tail–body junction and forming a notorious hump, which makes an arc in the body plane. The dorsal fin height is 4.18 mm; the ventral fin height is 3.7 mm; the maximum height of the tail is 11.9 mm. The tail terminus is rounded, and the caudal musculature does not reach the fin terminus.

The oral disc is small, ventral; it is located near the tip of the snout and does not protrude laterally beyond the body. The transverse width is 4.9 mm. It is surrounded by a uniserial row of marginal papillae, interrupted medially on the upper lip. The lower lip papillae alternate in orientation, giving the appearance of two rows. The upper lip has 23 papillae on the right side and 20 papillae on the left side. The lower lip bears 52 marginal papillae. The upper jaw sheath is medium-sized, forming a smooth arch, and is finely serrated. Its transverse width is 2.6 mm (53% of oral disc width); its height is 0.4 mm. The lower jaw sheath is V-shaped, open, and finely serrated; its width is 4.1 mm; its height is 1.3 mm. The labial tooth row formula is 2/3(1); the tooth rows lengths are A-1: 4.15 mm, A-2: 4.0 mm, P-1 right row: 1.75 mm, P-1 left row: 1.75 mm, P-1 gap: 0.1 mm, P-2: 3.75 mm, and P-3: 3.70 mm.

FIGURE 4.21 Ontogenetic changes in two individuals of *Gastrotheca elicioi* from tadpole to adult. Top to bottom, left to right. CJ 1950a, stage 35, lateral, dorsal, and ventral views, TL = 41 mm; CJ 1950b, stage 40, lateral and dorsal views, TL = 64 mm; stage 46, lateral and dorsal views, 19.6 mm SVL; adult male, 76 days after stage 46. Barrio La Palmera, Loja, Provincia Loja. LAC.

FIGURE 4.22 Oral disc of *Gastrotheca elicioi*. CJ 4311, stage 36, Barrio Puntzará Grande, Loja, Provincia Loja. Taken from Carvajal-Endara *et al.* (2019).

In life (based on a specimen in stage 37) in dorsal and ventral views, the body is dark brown. The snout and flanks are dark brown. The guts are gray, and red gills are visible through the throat. The venter is gray. The caudal musculature is reddish-brown with brown stippling; its distal portion lacks pigment. The dorsal and ventral fins are gray with minute cream stippling. The vent tube is translucent. The iris is reddish-gold.

In stages 31–40, the total length varies between 15.8 mm (stage 31) and 73.2 mm (stage 39), and the tail length proportion varies from 57 to 70% until stage 38. The number of marginal papillae varies among specimens and Gosner stages; variation in the number of ventral papillae on the lower lip is high (43–64).

Changes in coloration during the ontogenetic development of one mostly brown individual are as follows. In stage 40, the dorsum and flanks are cream and brown with a diffuse pattern of brown–gray paravertebral marks and a dark gray stripe bordering the canthus and body dorsolaterally, extending to about level of the midbody, bordered dorsally by a diffuse cream area. The caudal musculature is pink proximally, with a gray suffusion distally. The caudal fins are pale gray. The iris is pale red. By stage 46, the markings on the dorsum of the body and limbs are diffuse gray and green on a dark brown background; the fingers and toes are yellowish-cream dorsally. The flanks have a wide, brown–gray band. There are well-defined, creamy white dorsolateral and labial stripes and supracloacal white marks. Six additional metamorphs vary from plain dark brown to plain green and have a complete supracloacal stripe.

Distribution.—Figure 4.23. *Gastrotheca elicioi*, an endemic species from Ecuador, is known from the vicinity of the city of Loja, Provincia Loja, in the Loja Basin and on the slopes of the adjacent cordilleras. Its elevational range is 2010–2914 m in an area of extent of occurrence of about 320 km^2.

This nocturnal, semiarboreal species inhabits mostly disturbed areas and a few forests in the evergreen montane forest of the southern Cordillera Oriental of the Andes, evergreen high montane forest of the southern Cordillera Oriental of the Andes, and evergreen montane shrub of the south of the Andes ecosystems in the Catamayo-Alamor, Páramo, and southern Cordillera Oriental of the Andes biogeographic sectors, where the average annual rainfall is 764 – 1128 mm and the average annual temperature is 12.6–17.4 °C.

Conservation.—*Gastrotheca elicioi* was categorized as Endangered (Carvajal-Endara *et al.*, 2019) according to criteria B1ab(i,ii,iii,iv) of the IUCN Red List. The authors suggested this conservation status because of its small known area of occurrence (103 km^2), which is extremely fragmented. Its habitats are in heavily human-populated areas. Loja and environs have been deforested and modified for human activities. The species has been recorded in the public reserves Parque Nacional Podocarpus, in the protected forests Cuenca del Río Malacatos en Loja and Cuencas de los Ríos: San Francisco, San Ramón y Sabanilla, and the private reserve Reserva Natural Madrigal del Podocarpus within the biosphere reserve Podocarpus-El Cóndor.

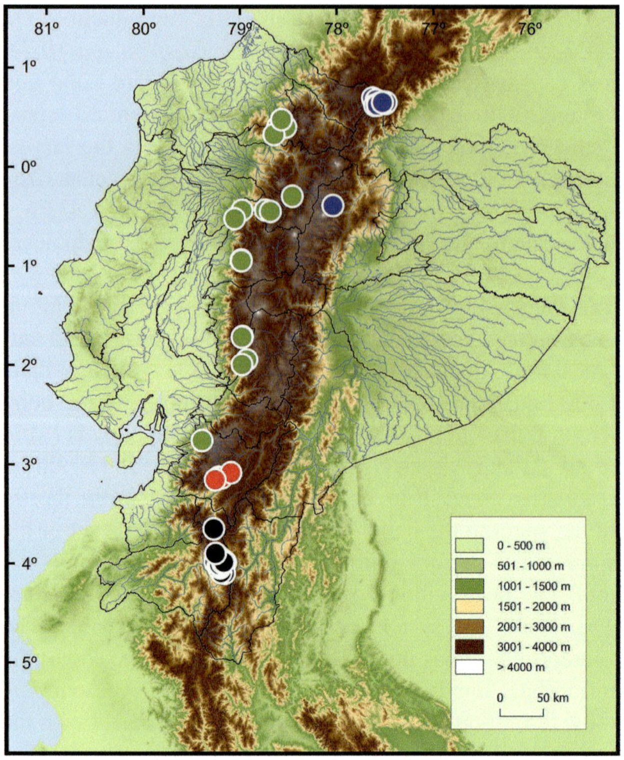

FIGURE 4.23 Localities of known occurrence of *Gastrotheca elicioi* (black dots), *G. litonedis* (red dots), *G. orophylax* (blue dots), and *G. plumbea* (green dots) in Ecuador.

Remarks.—In the past, *Gastrotheca elicioi* was confused with *G. lojana* and *G. monticola*. It belongs to the subgenus *Duellmania* and is reported as species B in Duellman (2015: Fig 11.1).

Key references.—(Duellman, 1974c) (coloration in life, as *Gastrotheca lojana*); (Carvajal-Endara *et al.*, 2019) (description, call, tadpole); (Székely *et al.*, 2020b) (account).

GASTROTHECA ESPELETIA (ESPELETIA MARSUPIAL FROG) DUELLMAN AND HILLIS, 1987

Gastrotheca espeletia Duellman and Hillis, 1987: 146.

Holotype.—KU 169401 from north shore of Lago de la Cocha, 2790 m, Departamento de Nariño, Colombia.

Etymology.—The specific name is a noun in apposition, referring to the genus *Espeletia*, a characteristic plant in the paramos inhabited by this species.

Identification.—Figure 4.24. Adults of this medium-sized species attain a maximum SVL of 51.9 mm in females and of 52.6 in males. The skin on the dorsal surfaces is smooth and not co-ossified with the skull, but that on the flanks is areolate; transverse dermal ridges are absent. There are no tubercles or supraciliary processes on the upper eyelid. The distinct tympanic annulus is smooth. The snout is acutely rounded in dorsal view and rounded in profile and protrudes well beyond the margins of the jaws. The interorbital distance is about equal to the width of the upper eyelid. Finger I is longer than II. The fingers are unwebbed, and the discs on the fingers and toes are small, round, and barely wider than the digits. The hind limbs are short and robust; the tibia length is less than 40% of the SVL. The length of the foot is greater than that of the tibia. The toes are long and about one-third webbed; the digital discs are small and round. Calcars and tarsal folds are absent.

The dorsal coloration is highly variable, ranging from dull dark brown to tan with brown markings or to green with dark green markings. Some individuals are uniformly bronze brown with green flecks on the posterior part of the body and hind limbs. The dark marks on the dorsum usually consist of paravertebral marks. Markings on the head usually consist of a pale labial stripe and a dark canthal stripe. A pale dorsolateral stripe is absent. The axillae, groin, anterior and posterior surfaces of the thighs, and inner surfaces of the shanks are pale greenish-blue, usually with small brown spots. The color of the venter varies from white to dull bluish-gray; the vocal sac is gray. The iris is reddish-brown with fine black reticulations.

FIGURE 4.24 *Gastrotheca espeletia*. Top: QCAZ 38417b, male, La Libertad, Provincia Carchi. LAC. Middle: KU 169402, paratype, male. Bottom: KU 169401, holotype, male. North shore of Lago de la Cocha, Departamento Nariño, Colombia. WED.

Comparisons.—*Gastrotheca pseustes*, *G. riobambae*. Among the high Andean short-legged *Gastrotheca* with small digital discs, *G. espeletia* is most like *G. pseustes* and *G. riobambae*, in both of which fingers I and II are equal in length, whereas finger I is longer than finger II in *G. espeletia*. Furthermore, *G. pseustes* differs from *G. espeletia* by having a broader snout and in coloration (see account of that species). *Gastrotheca riobambae* additionally differs from *G. espeletia* by having bold dark spots or mottling on the ventral surfaces of the shanks and a snout that does not protrude beyond the jaws.

Natural history.—Although this species descends to subparamos and cultivated areas, the species seems to be primarily associated with paramos supporting *Espeletia pycnophylla* (Asteraceae). With the exception of an outlying population in the Cordillera de los Llanganati in Provincia Tungurahua, central Ecuador, *Espeletia* reaches it southernmost distribution in the Páramo el Angel in northern Ecuador (Cuatrecasas, 1979), which also is the southern limit of the range of *Gastrotheca espeletia*. By day, these frogs inhabit bunch grass. At night, individuals of *G. espeletia* are on the ground or on low (< 50 cm) herbs.

This is a biphasic species. Males call by day and night from bunch grass in or at the edge of permanent marshy areas. The call consists of a single, multi-pulsed call. An analysis of 12 calls from two individuals showed that call energy usually increases gradually over time; the peak in amplitude is around the end of the call. There is a periodic amplitude modulation in most of the calls that has a duration of slightly less than one second. The call consists of two or three short, cluck-like notes; the pulsed notes have a dominant frequency of 1260 Hz. Detailed data and figures are provided under the subtitle "Vocalizations of Ecuadorian Anurans" in Appendix 1.

Most tadpoles have been found in grassy ponds in paramos. Tadpoles in nearly all stages of development have been found in February, May, and July; recently metamorphosed young were found in July. *Gastrotheca espeletia* presumably has a long breeding season; tadpoles may require many months to develop in the cold ponds.

Tadpoles.—Figure 4.25. The tadpole was described based on an individual at stage 35 from ca. 30 km southwest of Tulcán, Provincia Carchi. It has a body length of 25 mm and a total length of 67 mm. In dorsal view, the body is ovoid, narrowed anteriorly, and widest at three-fourths the length of the body; the snout is bluntly rounded. In lateral profile, the body is highest at three-fourths the length of the body but is not as high as wide; the snout is inclined from the nostrils to a truncate tip. The throat is convex. The eyes are small and are directed dorsolaterally. The interorbital distance is more than 33% of the width of the body at the level of the orbits. The nostrils are midway between the eyes and the tip of the snout and are directed anterolaterally. The spiracular tube is short and sinistral, with a U-shaped opening directed posterodorsally on the ventrolateral surface at the midlength of the body. The cloacal tube is short, conical, and median, with a transverse opening. The caudal musculature is moderately robust; its height is slightly less than that of the dorsal fin at the midlength of the tail. The musculature terminates gradually in a pointed tip well within the margin of the caudal fin. The dorsal fin arises abruptly from the posterior edge of the body; the dorsal and ventral fins are equal in height and terminate in an acutely rounded tip. The width of the oral disc is equal to about half the width of the body. The disc is bordered laterally and ventrolaterally by two alternating rows of marginal papillae and ventromedially by a single row of blunt papillae. There are two upper and three lower rows of labial teeth; the rows are about equal in length; the first lower row is narrowly interrupted medially. The upper jaw sheath is massive and forms a continuous broad arch; the lower jaw sheath is broadly V-shaped; the sheaths have barely noticeable serrations.

The body is grayish-green with yellow flecks on the tail. In preservative, the dorsum of the body and the caudal musculature are dull grayish-brown; minute brown flecks are present on the caudal fins.

FIGURE 4.25 Ontogenetic changes in *Gastrotheca espeletia* from tadpole to adult. Top to bottom, left to right. QCAZ 38417, undetermined stage: lateral, and dorsal views; stage 41, stage 42, stage 45; froglet 71 days after stage 45; juvenile 220 days; adult 338 days. La Libertad, Provincia Carchi. LAC.

Distribution.—See Figure 4.6. *Gastrotheca espeletia* occurs at elevations of 2530 – 3400 m in the southern part of the Cordillera Central in Colombia and in the Nudo de Pasto in southern Colombia and northern Ecuador. The southern end of the range in the western cordillera in Ecuador seems to be limited by the dry valley of the Río Chota. In the Cordillera Oriental in Ecuador, the range overlaps that of *G. riobambae*. *Gastrotheca espeletia* occurs on the high Amazonian slopes of the Cordillera Oriental at elevations of 2650–2710 m in the upper valley of the Río Chingual; there, it occurs sympatrically with the larger *G. orophylax*. The range of *G. espeletia* overlaps that of *G. argenteovirens* in the vicinity of Pasto in southern Colombia; both species have been recorded in Pasto, but they have not been found in microsympatry. In Ecuador, *Gastrotheca espeletia* is known from Pasto massif of the Andes in Provincia Carchi. Its elevational range is 2625 – 3700 m in an area of extent of occurrence of about 930 km^2.

This nocturnal, arboreal species inhabits natural vegetation and intervened areas in the caulescent rosette and herbazal of Páramo (frailejones) ecosystem in the Páramo, Cordillera Occidental of the Andes, and northern Cordillera Oriental of the Andes biogeographic sectors, where the average annual rainfall is 857 – 1566 mm and the average annual temperature is 7.1–13.2 °C.

Conservation.—*Gastrotheca espeletia* is categorized as Endangered (IUCN SSC Amphibian Specialist Group, 2019d; in Red List of 2019) because its extent of occurrence is 4995 km^2, its population is severely fragmented, and there is a continuing decline in the extent and quality of its habitat, as well as an ongoing decline in the number of mature individuals, mainly due to agricultural development, human settlement, and logging activities. This species was previously reported to be common and frequently encountered. However, there are no recent records in Ecuador, except for one in 2008 from La Libertad, Provincia Carchi, despite parts of the historical range in Ecuador having been well surveyed. There are several records from Colombia, including a northern range extension to Belen (Guerrero-Vargas *et al.*, 2007) in the department of Nariño (close to Cauca, where three individuals were found). It was recorded in 2014, 2015, and 2016 in Pasto and Aldana, but was missing from Laguna de El Encano, the type locality. Extensive agricultural activities have severely fragmented subpopulations of this species in thc paramos of southern Colombia and northern Ecuador, such that there is little or no dispersal between subpopulations. The population and the total number of mature individuals in the population, are inferred to be decreasing due to ongoing declines in the extent and quality of habitat. The major threats are deforestation for agricultural development (potato farming, cattle grazing), human settlement, and logging. Fire is also a problem in some places such as in the Páramos del Angel in Ecuador. The type locality (Laguna El Encano) is a Ramsar site but does not have formal protection. However, there are several small private reserves surrounding the site. In Ecuador, it has been recorded in the protected forest of Comuna Agropecuaria La Esperanza.

Remarks.—Morphometrically, this species falls within the range of *Gastrotheca riobambae*, from which *G. espeletia* was distinguished allozymically by Duellman and Hillis (1987). Careful field work is needed in the Cordillera de los Llanganati to determine if there is an outlying population of *G. espeletia* occurring with the disjunct population of *Espeletia pycnophylla*.

Key references.—(Duellman and Hillis, 1987) (description); (Sinsch and Juraske, 2006) (call); (Duellman, 2015) (account, tadpole); (Yánez-Muñoz *et al.*, 2020) (account).

GASTROTHECA GUENTHERI (LOWER-TOOTHED MARSUPIAL FROG) (BOULENGER, 1882a)

Amphignathodon guentheri (Boulenger, 1882a): 450. **Holotype.**—BMNH 1946.9.7.28 from Intag, Provincia de Imbabura, Ecuador.

Gastrotheca guentheri Duellman *et al.*, 1988: 531.

Etymology.—The specific name is a patronym in honor of Albert C. L. G. Günther, an English herpetologist of German origin. In 1856, he went to the British Museum, where he worked in ichthyology. In 1875, he became Director of the Department of Zoology of the Natural History Museum of England until 1895.

Identification.—Figure 4.26. Adults of this species attain a maximum SVL of 82.0 mm in females and of 76.0 in males. The skin on the dorsum is finely granular with scattered small tubercles; it is co-ossified with the skull. Transverse dermal ridges are absent on the dorsum of the body. Large, elongate, triangular supraciliary processes are present. The tympanic annulus is indistinct and tubercular. The snout is rounded in dorsal view and inclined anteroventrally in profile. The interorbital distance is much greater than the width of the upper eyelid. Fingers I and II are equal in length. The fingers and toes have large terminal discs that are much wider than the adjacent digits. The fingers are unwebbed, and the toes are about one-half webbed. The hind limbs are long; the length of the tibia is more than 60% of the SVL.

FIGURE 4.26 Color variation in *Gastrotheca guentheri* from Ecuador. Top right: from San Francisco de Las Pampas, Provincia Cotopaxi. LAC. Top left: KU 164224. Bottom left: KU 164226, female, 74.7 mm SVL. Bottom right: KU 164228, female, 82.0 mm SVL, Quebrada Zapadores, Provincia Pichincha. WED.

A triangular calcar or large triangular tubercle is present on the heel; tarsal folds are absent.

The dorsum of adults is tan with brown markings, dark brown to grayish-black with pale flecks, or green with or without an indistinct brown dorsolateral line or pale flecks on the flanks. All juveniles are green. A brown female from La Planada, Departamento Nariño, Colombia, gave birth to 14 young, all of which were uniformly green. The venter is brown, palest on the throat. The iris is pearly white above with dark brown triangles anteriorly and posteriorly and pink to reddish-brown below.

The variation in color is documented by the following descriptions of three adults from Quebrada Zapadores, 5 km east–southeast of Chiriboga, Provincia de Pichincha, Ecuador. KU 164224.—The dorsum is sepia umber with small bright red spots; the flanks are dark brown. The anterior surfaces of the thighs are brown, indistinctly mottled with cream; the posterior and ventral surfaces of the thighs are olive brown. The inner surfaces of the shanks and tarsi are brown, mottled with greenish-cream. The chin and pectoral region are greenish-gray, with brown and gray mottling and flecks; the abdomen is brown, with a few minute white flecks. The plantar surfaces are brown; the palmar surfaces are olive brown. The anal tubercles are white; the tongue is dull orange; the buccal lining is pale green. KU 164226.—The dorsum (except the thighs) is reddish-brown, with dark brown mottling on the back, a chevron-shaped interorbital mark, labial bars, and transverse bars on the limbs. The flanks are a paler reddish-brown, with prominent dark brown marks from the canthus along the upper eyelid and over the tympanum diagonally to the groin, as well as a broad stripe from the axilla to the groin, bordered above by bright deep yellow. The dorsal surfaces of the thighs are reddish-brown with bold, dark brown transverse bars extending onto the dull yellow anterior surfaces of the thighs. The posterior and ventral surfaces of the thighs, inner surfaces of the limbs, and palmar and plantar surfaces are dull gray brown. The throat is grayish-brown and paler than the abdomen, which has a few minute white flecks. KU 164228.—At night, the dorsum is pale emerald green; the belly is pale greenish-white. By day, the dorsum is darker green with cream flecks. The labial margin is dark brown. There are cream flecks along the outer margins of the forearms and hind limbs and cream spots and flecks along the flanks between the dorsal and ventral coloration. The venter is palest on the throat, where it is suffused with brown flecks and blotches that are interspersed with cream flecks; the venter is darkest on the belly, the ventral surfaces of the tarsi, and the plantar surfaces. The tongue and lining of the mouth are ochre. MCZ 108616 from Quebrada La Plata, Provincia de Pichincha, Ecuador.—This adult male is somewhat different. The dorsum is maroon, with indistinct black mottling. Between the folds of the brood pouch, the back is black; behind this, it is red. The head is cinnamon-colored, mottled with black and maroon; the flanks are yellowish-brown, with burnt umber longitudinal stripes. The thighs are yellowish-brown with burnt umber bars; other parts of the limbs are chestnut colored with burnt umber bars. Some yellowish-green patches are evident below and in front of the eyes. The venter is grayish-green, somewhat paler on the throat. The fingers are greenish-olive underneath. The tongue and margins of the jaws are bright yellow.

A brown female from La Planada, Departamento de Nariño, Colombia, gave birth to 14 young, all of which were uniformly green.

Comparisons.—*Gastrotheca andaquiensis*, *G. angustifrons*, *G. cornuta*, *G. dendronastes*, *G. longipes*, *G. testudinea*, *G. weinlandii*. *Gastrotheca guentheri* is unique among anurans by having true teeth on the dentary. The three other species of *Gastrotheca* on the Pacific versant (*G. angustifrons*, *G. cornuta*, and *G. dendronastes*) differ from *G. guentheri* by having transverse dermal ridges on the dorsum of the body and by not having the skin co-ossified with the skull. Moreover, *G. angustifrons* differs from *G. guentheri* by lacking calcars and supraciliary processes. Among the long-legged arboreal *Gastrotheca* species on the Amazonian slopes of the Andes, *G. weinlandii* is like *G. guentheri* in having the skin on the top of the head co-ossified with the skull, but it differs from *G. guentheri* by having a transverse occipital ridge and lacking prominent supraciliary processes. *Gastrotheca andaquiensis* differs by having a granular interocular ridge. *Gastrotheca longipes* and *G. testudinea* are like *G. guentheri*, but *G. longipes* has a green dorsum, and *G. testudinea* has vertical black bars on the flanks.

Natural history.—Figure 4.27. At night, most adults have been found at heights of 5–10 m on branches of trees. Juveniles were found 1–2 m above the ground on leaves of bushes. An Erotylidae and an Ortoptera were found in a stomach. Individuals are known to feed on large prey. An orthopteran found in the stomach of a female with a snout–vent length of 71.5 mm measured approximately 40 mm, making it over half the SVL.

FIGURE 4.27 *Gastrotheca guentheri*. Females from San Francisco de Las Pampas, Provincia Cotopaxi. LAC. Note one feeding on a lizard (*Pholidobolus montium*) under terrarium conditions.

The presence of brooding females in May, June, and November suggests that the reproductive season is long, and possibly is continuous throughout the year. Males call from branches of trees, mostly in the canopy. The advertisement call is a single loud "bop," usually repeated at intervals of several minutes, but some individuals have been heard to produce two or three notes in quick succession. The clutch size is small. Three females were brooding 12, 12, and 15 eggs; the diameters of the yolks were 11.0, 11.3, and 13.2 mm. A brooding female at La Planada, Departamento Nariño, Colombia, gave birth to 14 young on 28 May 1986; two juveniles having SVLs of 20.5 and 21.0 mm were found in March.

Distribution.—See Figure 4.6. In Ecuador, *Gastrotheca guentheri* is known on the western versant of Cordillera Occidental of the Andes in Pichincha and Cotopaxi provinces. Its elevational range is 1145–2100 m in an area of extent of occurrence of about 1270 km^2.

This nocturnal, arboreal canopy species inhabits natural vegetation and partially intervened areas in the evergreen low montane forests of the Cordillera Occidental of the Andes, evergreen montane foothill forests of the Cordillera Occidental of the Andes, and evergreen montane forests of the Cordillera Occidental of the Andes, where the average annual rainfall is 1684–2675 mm and the average annual temperature is 15.7–20.0 °C.

Conservation.—*Gastrotheca guentheri* is categorized as Data Deficient throughout its entire range (IUCN SSC Amphibian Specialist Group, 2019e; in IUCN Red List of 2022). However, it is important to note that in Ecuador,

G. guentheri was categorized as Critically Endangered by Ortega-Andrade *et al.* (2021). Herein, we categorize it in Ecuador as Critically Endangered (Possibly Extinct) following A2c IUCN criteria because we infer a general reduction in the population greater than 90% that may be reaching 100%. In Ecuador, there have been no sightings for more than three decades; the most recent record of this species is from early 1991 near San Francisco de las Pampas, Provincia Cotopaxi. The species has not been recorded since 1990 in Colombia, where there have been some surveys in limited parts of the range. All of the historical sites in Ecuador have been surveyed on multiple occasions. Data from Bosque Integral Otonga (Provincia Cotopaxi) suggest that it is now absent from where it once occurred. Surveys in Parque Nacional Natural Munchique (2014, 2015, and 2016) and La Planada (2005) in Colombia were unsuccessful at locating the species. Threats to this species are severe habitat destruction and degradation due to agriculture, clear cutting, grazing, pollution, mining, wildfires, and urbanization, but this does not account for the observed decline. Climate change and pathogens might be implicated as the likely causes of most of its decline. In Colombia, it has been recorded in Parque Nacional Natural Munchique. In Ecuador, it has not been recorded in public reserves, but there are records from the private reserve Bosque Integral Otonga and within the biosphere reserve Chocó Andino de Pichincha.

Remarks.—This species is unique among anurans in having true teeth on the lower jaw. Thus, Boulenger (1882a) erected a monotypic genus, *Amphignathodon*. Immunological data (Duellman *et al.*, 1988) and analysis of DNA sequences (Wiens *et al.*, 2007) clearly showed the species was nested within the long-legged arboreal *Gastrotheca*, a monotypic group for which Duellman (2015) applied the subgeneric name *Amphignathodon*.

The cranial osteology, external morphology, reproductive biology, and advertisement call also support this placement. Wiens (2011: 1294) used a time-calibrated phylogeny to show "strong support for the re-evolution of lost mandibular teeth in frogs." In his analysis, mandibular teeth were lost in the ancestor of modern frogs at least 230 million years ago. According to Wiens, these teeth "re-evolved" with the divergence of *Gastrotheca guentheri* and *G. weinlandii* 4.64 million years ago; presumably, the developmental constraint against mandibular teeth in frogs was breached in *G. guentheri*.

Although the external structure and coloration distinguish *Gastrotheca guentheri* from other long-legged arboreal *Gastrotheca* on the Pacific versant of South America, all of the species are similar in size and proportions. A stepwise discriminant analysis distinguishes *G. guentheri* on the first canonical axis: head width, interorbital distance, and thumb length. All known juveniles of *G. guentheri* are essentially a uniform green dorsally, a coloration known in only 10% of the adults. As noted by Duellman and Ruiz-Carranza (1986), ontogenetic polymorphism evidently occurs in this species: some adults develop a pattern of brown streaks; others are dark gray with red flecks, and some remain green.

Key references.—(Boulenger, 1882a) (description); (Duellman and Ruiz-Carranza, 1986) (morphology, reproduction); (Wiens, 2011) (teeth evolution); (Duellman, 2015) (account, cranial osteology, advertisement call); (Paluh *et al.*, 2019) (diet).

GASTROTHECA LITONEDIS (BUSA MARSUPIAL FROG) DUELLMAN AND HILLIS, 1987

Gastrotheca litonedis Duellman and Hillis, 1987: 149.
Holotype.—KU 202690 from 10 km (by road) northeast of Girón, 2750 m, Provincia Azuay, Ecuador.

Etymology.—The specific name is derived from the Greek *litos*, meaning plain, and *nedys*, meaning belly; the name is applicable to this species that characteristically has an unmarked venter.

Identification.—Figures 4.28, 4.29. Adults of this species attain a maximum SVL of 62.4 mm in females and of 57.4 in males. The skin on the dorsum is smooth, not co-ossified with skull, and lacks transverse ridges. The supraciliary

FIGURE 4.28 *Gastrotheca litonedis*. Left: CJ 1940, female, 50.2 mm SVL. Right: CJ 1396, male, 52.2 mm SVL. Laguna de Busa, Provincia Azuay. LAC.

FIGURE 4.29 *Gastrotheca litonedis.* Brooding females, Laguna de Busa, Provincia Azuay. LAC.

processes are absent. The tympanic annulus is distinct, smooth. The interorbital distance is about 1.5 times the width of the upper eyelid. Fingers I and II are about equal in length; the width of the discs is notably wider than the digits. The fingers are unwebbed. The webbing between the external toes extends to the penultimate subarticular tubercle on toe IV and to the distal subarticular tubercle on toe V. The tibia length is 40–54% of the SVL, slightly longer than the foot. The heel lacks a calcar or tubercle.

The dorsum varies from uniform green to brown; in some individuals, dark paravertebral marks and dark brown flecks are present. The ventral surfaces are pale gray in most specimens; some have evenly distributed dark flecks and spots, and some have a faint bluish tinge on the ventral surfaces. A narrow, dark brown canthal stripe is present; a cream supralabial stripe extends from the posterior margin of the lip to the insertion of the forelimb. The tympanum is brown, tan, or olive green. The posterior surfaces of the thighs are dark brown or cream with orange flecks. A cream dorsolateral stripe is evident in most individuals. The flanks, groin, and anterior surfaces of the thighs usually are bronze–brown or tan with cream or dark flecks and spots. In some individuals, a pink–orange tinge is present in the groin. The supracloacal and heel stripes are cream or tan, and the ventral surfaces of the shanks have a faint pale blue tinge. The iris is reddish-bronze with dense black reticulations.

The skin on the dorsum is mostly smooth in most specimens; however, in some specimens, it is weakly areolate. All adults have a moderate supratympanic fold, which usually extends from the upper part of the tympanum to a point above the insertion of the forelimb. Some individuals have several small, rounded tubercles anterolateral to the tympanum. Each dentigerous vomerine process has 2–10 teeth.

Comparisons.—*Gastrotheca cuencana, G. elicioi, G. plumbea, G. pseustes, G. turnerorum. Gastrotheca litonedis* most closely resembles five other species in southern Ecuador—*G. cuencana, G. plumbea, G. pseustes, G. turnerorum,* and *G. elicioi. Gastrotheca litonedis* differs from all of them by having a pale brownish-gray venter, whereas the venter is uniform creamy white in *G. cuencana*, cream in *G. plumbea*, cream with dark flecks or spots in *G. pseustes* and *G. elicioi*, and bronze or dark brown in *G. turnerorum. Gastrotheca litonedis*, like *G. plumbea*, has smooth skin on the dorsum, whereas it is finely granular in *G. elicioi* and *G. cuencana*, weakly areolate to smooth in *G. pseustes*, and areolate in *G. turnerorum.* A pale supracloacal stripe is present in *G. litonedis, G. pseustes*, and *G. turnerorum*, but it is absent in *G. cuencana* and *G. plumbea. Gastrotheca pseustes* differs further from *G. litonedis* by having a complex call structure (long and short pulsed notes), which is simple in *G. litonedis* (only short pulsed notes). The sister species of *G. litonedis* is the parapatric *G. cuencana*. Distinctive morphological and coloration features are as follows (character states for *G. cuencana* in parentheses): *Gastrotheca litonedis* is larger than *G. cuencana* (SVL Student's t-test: $t = -6.725$, $df = 99$, $p < 0.001$) and has a bluntly rounded snout in profile (rounded snout inclined anteroventrally in profile); a narrow dark brown canthal stripe (absent or inconspicuous); a row of barely elevated dorsolateral warts (conspicuous, elevated); green, tan, or brown dorsal surfaces of the fingers (fingers cream dorsally); and dark brown posterior surfaces of the thighs (translucent cream with a pale blue or green tinge). The call of *G. litonedis* has a higher note rate, longer note duration, shorter inter-note interval, higher dominant frequency, and higher 90% bandwidth frequency compared with that of *G. cuencana*. Finally, these two species also differ by having a genetic distance of 1.2% (in a DNA dataset of 438 bp, 16S gene). *Gastrotheca litonedis* is syntopic with *G. lojana* and *G. pseustes* (*sensu lato*), from which it differs notably by the texture of skin on the dorsum, ventral coloration, and call structure. The skin on the dorsum is smooth in *G. litonedis*, whereas it is finely granular in *G. lojana* and coarsely granular in *G. pseustes* (*sensu lato*); furthermore, both species have cream venters with dark flecks, spots, or mottling and a complex call structure with long and short pulsed notes.

Natural history.—Frogs were found among totora reeds (*Schoenoplectus californicus*) and shrubs at the border of the Laguna de Busa up to 2.5 m above the ground. Females were perched on totora reeds approximately 40 cm above the ground and along small streams that were close to the lagoon. During the day, a male and a female were basking on leaves of calla lilies (*Zantedeschia aethiopica*). In San Gerardo, most individuals were found on the border of an irrigation ditch in a pasture.

This species has a biphasic mode of reproduction. Males call mostly at night. Calls from two locations (six individuals from San Gerardo and three from San Fernando, Laguna de Busa), Provincia Azuay, Ecuador, were described as simple calls composed of a series of short, single-pulsed notes emitted at regular intervals. The calls have between one and 21 notes per call. The notes have a mean duration of 0.075 s and a mean inter-note interval of 1.768 s. Detailed data and figures are provided under the subtitle "Vocalizations of Ecuadorian Anurans" in Appendix 1.

In several visits made to Laguna de Busa (October and December 2007, August 2008, June 2011), gravid females were found, as well as tadpoles, in small streams, temporary ponds, and at the edge of the lagoon. Males were calling from totora reeds approximately 1–2 m above the ground. Females were perched on totora reeds approximately 40 cm above the ground and along small streams that were close to the lagoon. One of the gravid females collected on 9 June 2011 was transported to the laboratory, where it deposited 96 tadpoles. At San Gerardo, males were calling from the grass at approximately 50 cm above the ground. A brooding female of *G. litonedis* showed heliophilic behavior during most of the incubation time under captive conditions in an outdoor enclosure.

FIGURE 4.30 Ontogenetic changes in *Gastrotheca litonedis* from tadpole to froglet. Top 3 rows: CJ 6813, stage 37, lateral, dorsal, and ventral views, TL = 53.8 mm. Manuel Morales Mite. Bottom 3 rows: CJ 1947, stages 39, 41, 45 = 20.5 mm SVL. Laguna de Busa, San Fernando, Provincia Azuay. LAC.

Tadpoles.—Figures 4.30, 4.31. The tadpole was described based on tadpoles from Laguna de Busa, Provincia Azuay, Ecuador. The following description is based on a specimen in stage 37 from a series of 79 tadpoles born and reared under laboratory conditions from a mother collected while gravid in the field at Laguna de Busa, 2780 m. The total length is 52.8 mm, and the body length is 19.4 mm (37% of total length). The body is ovoid in dorsal and lateral views, slightly depressed. The throat is slightly concave in lateral profile, sloping from the tip of the snout to the belly. The body width at the level of the spiracle is 12.5 mm, and its height at same position is 10.3 mm; the head width at the level of the eyes is 10.0 mm. The lateral line system is present but barely visible. The supraorbital and infraorbital lines are not evident at the level of the snout, and the infraorbital line is present at the level of the eye, touching the inferior portion of the orbit and making contact with supraorbital line immediately behind the eye. The postorbital line is represented by three neuromasts arranged in line and forming a 45-degree angle with the body plane. The angular line, anterior oral line, and loreal lines are not visible. The dorsal body and middle body lines are not visible. The nostril is medium-sized (in proportion to the body length), ovoid, and protruding, having a fleshy annulus; its opening is directed anterolaterally. The snout–nostril distance is 4.2 mm; the internarial distance is 2.8 mm. The eyes are dorsally positioned; the eye length is 1.9 mm; the eye width is 1.8 mm. The interorbital distance is 5.2 mm. The spiracle is sinistral, located at midbody level; the spiracular opening is oriented

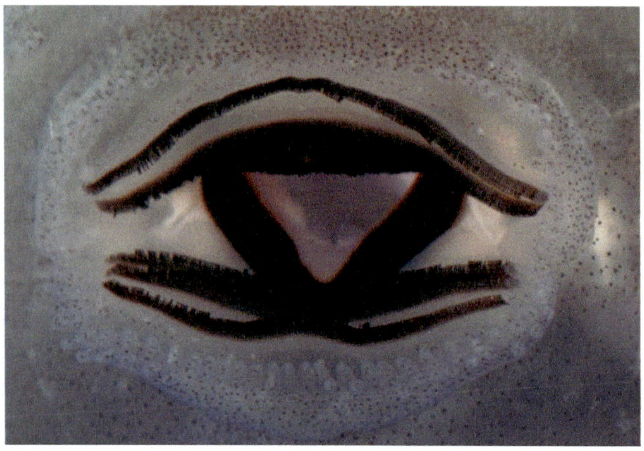

FIGURE 4.31 Oral disc of *Gastrotheca litonedis*. CJ 5292, stage 35, Laguna de Busa, San Fernando, Provincia Azuay. Taken from Carvajal-Endara *et al.* (2019).

posteriorly. The distance from the tip of the snout to the spiracular opening is 12.9 mm. The end of the spiracular tube is rounded, attached to the body wall; the inner wall of the spiracular tube is not evident; the spiracle length is 2.7 mm, and the spiracle transverse width is 3.5 mm. The vent tube is dextral; its opening is oriented posteriorly; its tube length is 2.5 mm, and the tube transverse width is 2.2 mm. The tail length is 33.3 mm. The caudal musculature is slender, narrowing gradually until the tail terminus. The caudal muscle height is 4.1 mm; its width is 3.6 mm. The caudal fins are well developed and proportional; the dorsal fin arises abruptly near the tail–body junction. The dorsal fin height is 4.1 mm; the ventral fin height is 3.7 mm; the maximum tail height is 9.2 mm. The tail terminus is rounded, and the caudal musculature does not reach the fin terminus.

The oral disc is small, ventral, located near the tip of snout, not protruding laterally beyond the body. The transverse width is 5.8 mm. It is surrounded by a uniserial row of marginal papillae, interrupted medially on the upper lip. The lower lip papillae alternate in orientation, giving the appearance of two rows. The upper lip has 23 papillae on the right side and 23 papillae on the left side. The lower lip bears 57 marginal papillae. The upper jaw sheath is medium-sized, forming a smooth arch, finely serrated; the transverse width is 2.9 mm (50% of width of oral width), and the height is 0.3 mm. The lower jaw sheath is V-shaped, open and finely serrated; the width is 2.6 mm, and the height is 0.7 mm. The labial tooth row formula is 2/3(1). The tooth rows lengths are A-1: 4.5 mm, A-2: 4.4 mm, P-1 right row: 1.65 mm, P-1 left row: 2.0 mm, P-1 gap: 0.2 mm, P-2: 4.1 mm, and P-3: 3.95 mm.

In life, in a specimen in stage 34 from a series from San Fernando, Laguna de Busa, Provincia Azuay, 2834 m, in dorsal view, the body is olive gray, and the head and tail are suffused with pink and cream. The flanks are cream anteriorly to olive gray posteriorly. In ventral view, it has cream anteriorly in the gular region, followed posteriorly by red gills visible throughout the throat and olive-gray guts. The caudal musculature is pink proximally, gradually becoming gray posteriorly. The dorsal and ventral fins are translucent, with olive stippling distributed regularly along the length of the tail except distally, where stippling diminishes. The spiracle and oral apparatus are nearly translucent. The iris is gold.

In stages 33–40, the total length varies between 31.4 (stage 33) and 61.9 mm (stage 40), and the tail length proportion varies from 58 to 69% until stage 40. The number of marginal papillae varies among specimens and Gosner stages; the variation in the number of ventral papillae on the lower lip is high (39–66).

Changes in the coloration of an individual during ontogenetic development are as follows: At stage 39, the dorsum and flanks are olive gray with a diffuse pattern of gray paravertebral marks. By stage 41, dorsum of the body and limbs is green with well-defined dark green paravertebral marks; a cream dorsolateral stripe is bordered below by a fine black line; a creamy white stripe is present on the face where the upper lip will be formed. The flanks are greenish-gray. At stage 46, the dorsum is nearly uniform green; the paravertebral marks are barely evident. The flanks are brown, and the iris is red. Another individual is similar to the above described but has scattered cream flecks on a green dorsum.

Distribution.—See Figure 4.23. *Gastrotheca litonedis*, an endemic species to Ecuador, is known from three localities (ca. Girón, San Fernando (Laguna de Busa), San Gerardo) on the eastern slope of the Cordillera Occidental in Provincia Azuay. Its elevational range is 2781–2839 m in an area of extent of occurrence of about 20 km^2.

This nocturnal, semiarboreal species inhabits forests and disturbed areas in the evergreen montane forests of the Cordillera Occidental of the Andes, where the average annual rainfall is 611–662 mm and the average annual temperature is 12.0–12.6 °C.

Conservation.—*Gastrotheca litonedis* was categorized as Endangered by Ortega-Andrade *et al.* (2021). We categorize it as Critically Endangered following B1ab(iii) IUCN criteria because (1) its extent of occurrence is estimated to be less than 100 km^2 (20 km^2 according to our data), (2) it is only known from the type locality, and (3) there is a projected decline in the extent and quality of its habitat. Currently, the known area of occurrence of *G. litonedis* is extremely small, and its area of occupancy is even smaller, although we suspect that its distribution might increase as further searches are conducted. Its populations are in heavily human-populated areas. For example, the surroundings of Laguna de Busa have been deforested and modified for human activities. The lake contains two introduced predatory fish (trout and carp), and its surroundings are greatly altered by the introduction of exotic eucalyptus and conifers. Additional threats at this site are unregulated tourism activities. It has not been recorded in the public and private reserves.

Remarks.—In the original description of *Gastrotheca litonedis* by Duellman and Hillis (1987), only the holotype, KU 202690, is *G. litonedis*, whereas the other specimens are *G. cuencana* and *G. lojana*. Genetic sequences published by Wiens *et al.* (2007) as *G. litonedis* are of *G. cuencana*.

Key references.—(Duellman and Hillis, 1987) (description); (Duellman, 2015) (account, advertisement call); (Carvajal-Endara *et al.*, 2019) (redescription, call, tadpoles).

GASTROTHECA LOJANA (LOJANA MARSUPIAL FROG) PARKER, 1932

Gastrotheca lojana Parker, 1932: 25. **Holotype.**—BMNH 1947.2.31.13 from Loja, Provincia Loja, Ecuador.

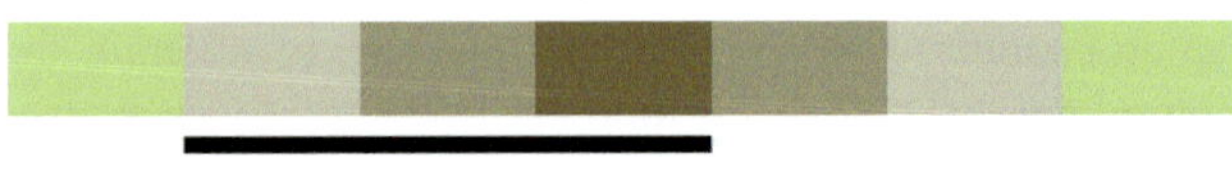

Etymology.—The specific name is derived from an adjective referring to Loja, the type locality.

Identification.—Figures 4.32, 4.33, 4.34. Adults of this species attain a maximum SVL of 76.1 mm in females and of 61.0 in males. The skin on the dorsum is finely granular; it is not co-ossified with the skull, and it lacks transverse ridges. The supraciliary processes are absent. The tympanic annulus is distinct and smooth. The interorbital distance is about 1.6 times the width of the upper eyelid. Fingers I and II are about equal in length; the discs on the fingers are about twice the width of digits proximal to the discs. The fingers are unwebbed. The webbing between the external toes extends to the antepenultimate subarticular tubercle on toe IV and to a point midway between the penultimate and distal subarticular tubercles on toe V. The tibia length is 41–56% of the SVL, longer than the foot. The heel lacks a calcar or tubercle.

The color of the dorsum is green, brown, or tan with darker interorbital and paravertebral marks. The interorbital mark is usually a bar that connects the two eyelids; however, in some individuals, it is divided into two blotches. Paravertebral marks are broad and are usually connected with the interorbital mark, or, in some individuals, they are fused at the level of the scapular region. The venter is white in most specimens; some have dark flecks and spots evenly distributed, and others have a heavily mottled pattern. On the head, there is a cream supralabial stripe, which is broader at the posterior margin of the lip and extends from the posterior margin of the lip to the insertion of the forelimb. The tympanum is brown, tan, or olive green. The flanks are brown, tan, or green with black and white spots. The posterior part of the flanks, groin, and anterior and posterior surfaces of thighs have a heavily mottled pattern; some individuals have green or purple tinge on the groin and posterior surfaces of the thighs. Supracloacal and heel stripes are cream or tan. The ventral surfaces of the shanks have a faint tinge of pale blue. The iris is reddish-bronze with a few black reticulations.

Comparisons.—*Gastrotheca cuencana*, *G. elicioi*, *G. litonedis*. *Gastrotheca lojana* most closely resembles three other species in southern Ecuador (*G. elicioi*, *G. cuencana*, and *G. litonedis*) and one species in northern Peru (*G. monticola*). *Gastrotheca lojana* differs from *G. elicioi* in some coloration patterns. A black canthal stripe is absent in *G. lojana* but present in *G. elicioi*. The anterior and posterior surfaces of the thighs and groin are heavily mottled in *G. lojana* but slightly mottled in *G. elicioi*. The interorbital mark in *G. lojana* is a transverse bar that sometimes is divided into two blotches and that is usually connected with two broad paravertebral marks. In *G. elicioi*, dorsal

FIGURE 4.32 *Gastrotheca lojana*. Left: CJ 407, female, 53.6 mm SVL. Right: CJ 408, male, 44.9 mm SVL. Oña, Provincia Azuay. LAC.

FIGURE 4.33 Color variation in *Gastrotheca lojana*. Top left: CJ 4302, male, Oña, Provincia Azuay. DAL. Top right: CJ 405, male. Middle left: CJ 408, male, 44.9 mm SVL. Middle right: CJ 409, subadult male, 45.5 mm SVL. Bottom left: CJ 411, female, 47.4 mm SVL. Bottom right: CJ 412, female, 55.1 mm SVL. Laguna de Busa, San Fernando, Provincia Azuay. LAC.

FIGURE 4.34 Color variation in metamorphs (Gosner stage 46) of *Gastrotheca lojana*. CJ 4304 from Oña, Provincia Azuay. LAC.

marks, when present, consist of a triangular interorbital mark, which may or may not be connected to two narrow and curved paravertebral marks. *Gastrotheca lojana* has a conspicuous, elevated row of dorsolateral warts, whereas it is barely raised in *G. elicioi*. In addition, in *G. elicioi*, the dark bars on limbs, when present, are shorter and less defined than in *G. lojana*. *Gastrotheca lojana* differs from its most related species, *G. cuencana* and *G. litonedis*, by having a cream venter with dark flecks or spots (uniform creamy white in *G. cuencana* and pale brownish-gray in *G. litonedis*), an interorbital bar or blotches usually connected to paravertebral marks (absent in *G. cuencana* and *G. litonedis*), dorsal surfaces of the limbs with dark transverse bars (irregular dark blotches in *G. cuencana* and usually unmarked in *G. litonedis*), anterior and posterior surfaces of the thighs and groin heavily mottled (translucent cream without marks in *G. cuencana* and usually brown and slightly molted in some *G. litonedis*), dorsal surfaces of the fingers cream or with dark brown spots (cream in *G. cuencana* and of same color of the rest of the body in *G. litonedis*), a fragmented dorsolateral stripe (continuous *in* *G. cuencana* and *G. litonedis*), and a complex call (simple call in *G. cuencana* and *G. litonedis*). The genetic divergence between *G. lojana* and *G. litonedis* is 2.5%, and that between *G. lojana* and *G. cuencana* is 2.8% (in a DNA dataset of 438 bp, 16S gene). *Gastrotheca lojana* differs from *G. monticola* by having the concealed surfaces of the thighs, groin, and dorsal surfaces of the feet with white mottling on a dark background, whereas in *G. monticola*, the axillae, groin, concealed surfaces of the thighs, and dorsal surfaces of the feet are green with black spots on a green or tan background. A distinctly different species, *G. pseustes*, occurs sympatrically with *G. lojana* in part of its range. *Gastrotheca lojana* differs from *G. pseustes* by having finely granular skin on the dorsum (smooth to coarsely granular in *G. pseustes*), larger digital discs, and pale dorsolateral and supracloacal stripes. Also, the call of *G. lojana* can be easily distinguished from the call of *G. pseustes* by the amplitude modulation of the longer note, shorter call duration, higher short-note rate, shorter interval between notes, fewer pulses, lower pulse rate, much lower dominant frequency, and lower 90% bandwidth frequency.

Natural history.—Figure 4.35. Frogs have been found up to 1.3 m above the ground in shrubs and among totora reeds (*Schoenoplectus californicus*). Throughout most of

FIGURE 4.35 *Gastrotheca lojana* in its microhabitat at night. Top: calling male, Pucará, Provincia Azuay. Juan Carlos Sánchez Nivicela. Bottom: Oña, Provincia Azuay, LAC.

its distribution, *G. lojana* is syntopic with *G. pseustes*; in San Fernando, it is syntopic with *G. pseustes* and *G. litonedis*, and in the southern part of its range, it is also syntopic with *G. elicioi*.

Reproductive biology data of pairs (wild and lab born) from Oña, Provincia Azuay, have been obtained in the laboratory (CJ and Wikiri lab notes by Daniel Romero Dueñas in September 2023) from two reproductive events. Amplexus is axillary and lasts around 3–6 days. Two females contained 194 and 229 eggs, with the latter having 43 non-viable eggs. Tadpoles underwent complete development and metamorphosis in a span of 80 days at a water temperature of 18.5 °C and a pH of 7.2.

Males call mostly at night. In Oña, Provincia Azuay, several *Gastrotheca lojana* were calling from agave leaves 50–130 cm above the ground and among grasses near the ground on 8 August 2003. A male from San Fernando, Provincia Azuay, was calling from approximately 1 m above the ground on a large stone. Calls from one location in Provincia Azuay (one individual from San Fernando, Laguna de Busa) and from one location in Provincia Loja (four individuals from Oña), Ecuador, were described as a complex call composed of one long pulsed note and followed (or not) by one or two short, single-pulsed notes. The long note had a mean duration of 0.562 s and consisted on average of 17.59 distinct pulses (pulse series) separated by silent intervals. The short notes have a mean duration of 0.064 s, and the inter-note interval is 0.356 s on average. Detailed data and figures are provided under the subtitle "Vocalizations of Ecuadorian Anurans" in Appendix 1.

FIGURE 4.36 Ontogenetic changes in three individuals of *Gastrotheca lojana* from tadpole to adult. Top to bottom, left to right. CJ 6556, lateral, dorsal, and ventral views of tadpole, stage 35, TL = 28.2 mm. Laguna de Busa, San Fernando, Provincia Azuay. LAC. CJ 9818a, stage 41. CJ 9818b, stage 42. Oña, Provincia Azuay. SGS. CJ 6556, juvenile 152 days after stage 35 = 47.5 mm SVL, juvenile 257 days = 50.1 mm SVL; adult 424 days = 54.4 mm SVL. LAC.

FIGURE 4.37 Oral disc of *Gastrotheca lojana*. CJ 4309, stage 37, El Tablón, Provincia Loja. Taken from Carvajal-Endara *et al.* (2019).

Tadpoles.—Figures 4.36, 4.37. The following description is based on a specimen in stage 36 from Oña, 2272 m, Provincia Azuay, Ecuador. The total length is 55.7 mm, and the body length is 19.8 mm (36% of total length). The body is ovoid in dorsal and lateral views, and it is slightly depressed. The throat shape is concave in lateral profile. The body width at the level of spiracle is 12.3 mm, and the height at the same position is 10.2 mm. The head width at the level of the eyes is 10.5 mm. The lateral line system is barely visible. The supraorbital and infraorbital lines both originate at the tip of the snout, running in parallel to the eye and making contact immediately behind the eye. The angular line descends vertically from just posterior to the eye toward the throat. The anterior oral line descends vertically from level of the oral disc and just anterior to level of the nares to throat, where it curves parallel to the infraorbital line, thereby forming a continuous circuit with the angular and loreal lines. The dorsal body and middle body lines are present. The nostril is medium-sized (in proportion to body length); it is ovoid and protruding, with a fleshy annulus; its opening is directed anterolaterally. The snout–nostril distance is 2.8 mm. The internarial distance is 3.2 mm. The eye is positioned and directed dorsally, not visible from below. The eye length is 2.1 mm, and the eye width is 1.9 mm. The interorbital distance is 5.7 mm. The spiracle is sinistral, and it is located ventrolaterally. The spiracular opening is directed posteriorly; the distance from the tip of the snout to the spiracular opening is 13.7 mm. The spiracle end is rounded, attached to the body wall, and the inner wall of the spiracle is not evident. The tube length is 4.0 mm, and the tube transverse width is 2.9 mm. The vent tube is dextral, its opening directed posteriorly. The tube length is 3.7 mm, and the tube transverse width is 2.1 mm. The tail length is 37.0 mm. The caudal musculature is slender, narrowing gradually until the tail terminus. The caudal muscle height is 5.2 mm, and the caudal muscle width is 4.3 mm. The caudal fins are well developed and about the same size, originating near the tail–body junction. The maximum height of the tail is 13.6 mm. The tail terminus is rounded, and the caudal musculature does not reach the fin terminus.

The oral disc is small, ventral, located near the tip of the snout, not protruding laterally beyond the body, and not visible dorsally. The transverse width is 4.1 mm. It is surrounded by a uniserial row of short, marginal papillae, interrupted medially on the upper lip. The lower lip papillae alternate in and out, giving the appearance of two series. The upper lip has 21 papillae on the right side and 19 papillae on the left side. The lower lip has 63 marginal papillae. The upper jaw sheath is medium-sized, forming a finely serrated, smooth arch; the height is 0.4 mm; the transverse width is 3.0 mm (48% of oral disc width). The lower jaw sheath is V-shaped, open, and finely serrated; the width is 2.1 mm; the height is 0.7 mm. The labial tooth row formula is 2/3(1). The tooth rows lengths are: A-1: 4.2 mm, A-2: 4.1 mm, P-1 right row: 1.9 mm, P-1 left row: 1.6 mm, P-1 gap: 0.3 mm, P-2: 3.2 mm, and P-3: 3.5 mm.

In dorsal and lateral views, the body is olive cream; the loreal and snout areas are the palest. Ventrally, the guts are not visible; the belly is white with gray marks; the gills are visible through the throat, with a reddish hue. The caudal musculature is cream–gray. The proximal third is reddish with myomeres and the medial line well defined, and the distal two-thirds have small cream flecks. The dorsal and ventral fins are translucent, suffused with minute cream flecks, clustering in rounded patches near the borders of the fins; they are most dense near the tail–body junction. The spiracle, oral apparatus, and legs are olive cream. The iris is gold.

Tadpoles in stages 34–41 vary in total length between 40.9 (stage 34) and 61.2 mm (stage 41), and tail length proportion varies from 59.1 to 63.7 mm until stage 41. The labial tooth row formula is 2/3(1). The number of marginal papillae varies among specimens and Gosner stages; variation in the number of ventral papillae on the lower lip is moderate (60 and 69). The color of the venter varies from nearly plain gray to gray flecks on a white background.

Changes during ontogenetic development in two individuals are as follows: At stage 46, the dorsum is nearly uniform clear brown; the paravertebral marks and interorbital marks are green and vary from nearly absent to well defined.

Distribution.—See Figure 4.17. *Gastrotheca lojana*, an endemic species to Ecuador, is known from basins in Azuay, Loja, and El Oro provinces. Its elevational range is 1676 – 3279 m in an area of extent of occurrence of about 6010 km^2.

This nocturnal, semiarboreal species inhabits mostly disturbed areas and a few forests in the evergreen montane forests of Catamayo-Alamor, evergreen montane shrub of the south of the Andes, and herbazal of Páramo ecosystems within four biogeographic sectors (Table 0.5), where the average annual rainfall is 566 – 1066 mm and the average annual temperature is 10.0–18.8 °C.

Conservation.—*Gastrotheca lojana* was categorized as Critically Endangered by Ortega-Andrade *et al.* (2021) according to criteria B1ab(iii,iv) of the IUCN Red List. Although its area of extent of occurrence is estimated to be 6010 km^2 (according to our data), the area is severely fragmented, and its habitats are in heavily human-populated areas. Search efforts at Loja and surroundings revealed no specimens, although it used to be abundant therein, as inferred from historical collections. The environs of Laguna de Busa and Oña have been deforested and modified for human activities. The Busa lake is populated by introduced fish (trout and carp) and its surroundings by introduced eucalyptus and conifers. Additional threats at Laguna de Busa are unregulated tourism activities. In Ecuador, it has not been recorded in public or private reserves, but there are records within the Macizo del Cajas, Podocarpus-El Cóndor, and Bosque Seco biosphere reserves.

Remarks.—*Gastrotheca elicioi* was confused under *G. lojana*, *G. marsupiata*, *G. monticola*, and *G. litonedis* before its description in 2019.

Key references.—(Parker, 1932) (description); (Duellman, 2015) (account, tadpoles, advertisement call); (Carvajal-Endara *et al.*, 2019) (description).

GASTROTHECA LONGIPES (LONG-FOOTED MARSUPIAL FROG) (BOULENGER, 1882a)

Nototrema longipes (Boulenger, 1882a): 418. **Syntypes.**—BMNH1947.1.31.4 from Canelos and 1947.2.31.5 from Sarayacu, Provincia de Pastaza, Ecuador.

Gastrotheca longipes Peters, 1955: 343.

Etymology.—The specific name is derived from the Latin words *longus* and *pes*, meaning long and foot, respectively. The name refers to the long foot of this species.

Identification.—Figure 4.38. Adults of this moderately large species attain a maximum SVL of 88.0 mm in females and of 71.0 in males. The skin on the dorsum is smooth, and that on the flanks is faintly granular. The skin is not adherent to or co-ossified with the skull. Transverse dermal ridges on the dorsum of the body are absent. Supraciliary processes are absent. The tympanic annulus is granular. The snout is rounded in dorsal view, and in profile, it is abruptly inclined from the nostrils to the tip. The interorbital distance is slightly greater than the width of the upper eyelid. Fingers I and II are equal in length. The fingers and toes have large terminal discs that are much wider than the adjacent digits. The fingers are about one-fourth webbed, and the toes are about three-fourths webbed. The hind limbs are long; the length of the tibia is about 56% of the

FIGURE 4.38 *Gastrotheca longipes*. Left: male, Curaray medio, Provincia Pastaza. José Vieira-Fernandes. Top and middle: male, Parque Nacional Yasuní. Alex Achig. Bottom right: gravid female, Sunka, Provincia Morona Santiago. Ana Almendáriz C.

SVL. Calcars and large tubercles are absent on the heels; a weak inner tarsal fold is present on the distal one-third of the tarsus.

The dorsum and flanks are green with brown marks on the snout and between the eyes and green with or without small white spots, respectively. Markings on the head consist of a dark interocular bar and a dark canthal stripe or not. The flanks are uniformly green. The venter is uniformly cream or green. Usually, there are narrow cream labial, ulnar, tarsal, heel, and supracloacal stripes. A pale dorsolateral stripe is absent. The iris is cream with brown reticulations. In many individuals, the dorsum is nearly a uniform green.

In a juvenile from 3.1 km south of Río Tiputini, Provincia de Orellana, Ecuador, the dorsum and flanks are green, with small white spots that are most numerous on the flanks. There are diffuse brown spots above the eye and three white spots bordered by brown between the eyes. The supratympanic ridge is white and brown; three white spots, bordered by brown, are present on the anterior part of the body. The limbs are green with white spots; the upper lip is white; the lower lip is green. The throat and chest are creamy yellow; the belly is green with creamy yellow spots; the ventral surfaces of the limbs are green. The limb bones also are green. The cloaca is bordered above by white and brown stripes and laterally by white spots. There are brown spots on the knee; the outer edge of the heel to the tip of the outer toe is white, bordered by brown. There is a white spot below the eye and another spot at the posterior edge of the jaw; the iris is white with brown reticulations.

Comparisons.—*Gastrotheca albolineata*, *G. andaquiensis*, *G. angustifrons*, *G. carinaceps*, *G. cornuta*, *G. dendronastes*, *G. orophylax*, *G. weinlandii*. In Amazonian Ecuador, there are no other large green frogs. Of the other *Gastrotheca* species on the Amazonian slopes of the Andes, the brown *G. weinlandii* differs from *G. longipes* by having the skin on the head co-ossified with the skull and prominent calcars on the heels. In contrast, *G. andaquiensis* differs from *G. longipes* by having granular skin on the dorsum, supraciliary processes, elongate calcars, and dark mottling on the dorsum. Of the four species on the Pacific versant in Ecuador, *G. guentheri* differs from *G. longipes* by having elongate supraciliary processes and the skin on the head co-ossified with the skull. The other three species (*G. angustifrons*, *G. cornuta*, and *G. dendronastes*) differ from *G. longipes* by being brown and having transverse dermal ridges on the body. The smaller (snout–vent length to 60 mm) *G. albolineata* differs by having a tibia length less than 50% of the snout–vent length, the first finger longer than the second, the outer fingers more than one-half webbed, a pale labial stripe, and a broad pale dorsolateral stripe. *G. orophylax* and *G. carinaceps* also are uniformly green; both have a proportionately wider interorbital distance (about 200% of the eyelid width) and unwebbed fingers. *Gastrotheca orophylax* also differs by having a truncate snout in profile, and *G. carinaceps* differs by having prominent cranial crests.

Natural history.—Individuals have been found at night on branches of bushes and trees. Two females contained 15 and 17 embryos in their brood pouches. The largest eggs were 10.9 mm in diameter. The small number of large eggs indicates that the eggs undergo direct development.

Distribution.—Figure 4.39. *Gastrotheca longipes* has a rather limited distribution in the upper Amazon Basin and on the lower slopes of the Andes in southern Ecuador and extreme northern Peru. In Ecuador, *G. longipes* is known in the Amazonian lowlands of Orellana, Pastaza, and Morona Santiago provinces. Its elevational range is 185 – 954 m in an area of extent of occurrence of about 32,390 km^2.

This nocturnal, arboreal species inhabits natural vegetation and partially intervened areas in eight ecosystems (Table 0.4) within six biogeographic sectors (Table 0.5), where the average annual rainfall is 2693 – 4424 mm and the average annual temperature is 21.0–25.6 °C.

Conservation.—*Gastrotheca longipes* is categorized as Least Concern throughout its entire range (IUCN SSC Amphibian Specialist Group, 2018a; in IUCN Red List of 2022) in view of its wide distribution and presumed large population. However, it is important to note that in

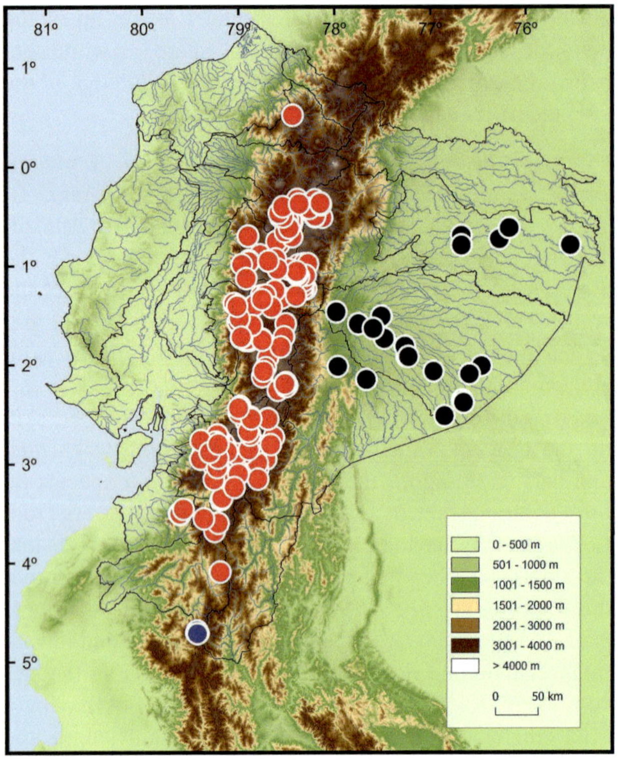

FIGURE 4.39 Localities of known occurrence of *Gastrotheca longipes* (black dots), *G. pseustes* (red dots), and *G. yacuri* (blue dots) in Ecuador.

Ecuador, *G. longipes* has been categorized as Near Threatened by Ortega-Andrade *et al.* (2021). This species is rare, possibly an artifact of its arboreal lifestyle. This species faces no major threats; it is a widespread species with large areas of suitable habitat remaining. There is some localized habitat loss to selective logging and agricultural activities. It is potentially susceptible to chytrid infections, like most monophasic marsupial frogs. It is present in the Santiago-Comaina Reserved Zone of Peru and Ecuador. In Ecuador, it has been recorded in the public reserves Parque Nacional Yasuní, the protected forests of Río Aguarico, and some private reserves (e.g., Tiputini Biodiversity Station).

Remarks.—The description of sexual dimorphism in coloration by Almendáriz and Cisneros-Heredia (2005) was based on two adults. The male had a brown spot on the snout and a brown interorbital mark, whereas the female was uniform green. The coloration of two other adults confirms their statements about color polymorphism. As seen in Figure 4.38, males also have brown spots on the body, and females have small white flecks, especially on the flanks.

Key references.—(Boulenger, 1882a) (description); (Almendáriz and Cisneros-Heredia, 2005) (morphology, distribution, natural history); (Duellman, 2015) (account, cranial osteology).

GASTROTHECA OROPHYLAX (GUARDIAN MARSUPIAL FROG) DUELLMAN AND PYLES, 1980

Gastrotheca orophylax Duellman and Pyles, 1980:5. **Holotype.**—KU 164243 from 11 km (by road) east–southeast of Papallacta, 2660 m, Provincia Napo, Ecuador.

Etymology.—The specific name is derived from the Greek *oros*, meaning mountain, and *phylax*, meaning guard or watchman. Duellman and Pyles (1980) proposed the name in allusion to frogs, probably of this species, witnessed in 1539 a band of Spanish conquistadors passing by the type locality on their way to the Amazon.

Identification.—Figures 4.40, 4.41. Adults of this species attain a maximum SVL of 74.0 mm in females and of 66.4 in males. The skin on the dorsum is smooth or weakly areolate; it is not co-ossified with the skull (co-ossified with the frontoparietals in large females). Supraciliary processes and calcars are absent. In profile, the snout is truncate. The interorbital distance is about twice the width of the upper eyelid. Finger I is shorter than II; the fingers lack webbing,

FIGURE 4.40 *Gastrotheca orophylax*. Left: CJ 6119, female, 68.8 mm SVL. Right: CJ 6122, male, 66.4 mm SVL. SGS.

and the toes are about one-half webbed. The tibia length is less than 55% of the SVL.

The dorsum and flanks are uniform green; the flanks and venter are paler green. A narrow cream labial stripe and brown supratympanic stripe are present, and the tympanum is golden yellow; the iris is bronze with black reticulations.

Comparisons.—*Gastrotheca agramma*, *G. elicioi*, *G. litonedis*, *G. lojana*, *G. longipes*, *G. plumbea*, *G. pseustes*, *G. psychrophila*, *G. riobambae*, *G. turnerorum*. At least some individuals of two other Ecuadorian monophasic *Gastrotheca* species are uniformly green dorsally. *Gastrotheca plumbea* differs from *G. orophylax* by usually having pale dorsolateral and supracloacal stripes and brown flanks. *Gastrotheca longipes* has an anteroventrally inclined snout in profile; fingers I and II are equal in length, and the tympanum is green. Some individuals of the biphasic *G. agramma*, *G. elicioi*, *G. litonedis*, *G. lojana*, *G. pseustes*, *G. psychrophila*, *G. riobambae*, and *G. turnerorum* are uniformly green dorsally, but they usually have flanks with contrasting colors, usually brown or green mottled with brown (uniform pale green in *G. orophylax*).

Natural history.—Figure 4.42. Nearly all individuals observed in Ecuador were perched on the large leaves of the herb *Gunnera* sp. (Gunneracea). One individual was under a stone by day; all others were on vegetation up to 2.5 m above the ground at night.

FIGURE 4.41 Color variation in metamorphs of *Gastrotheca orophylax*. Santa Bárbara, Provincia Sucumbíos. LAC.

FIGURE 4.42 *Gastrotheca orophylax*. Top: calling male. Bottom: brooding female. Santa Bárbara, Provincia Sucumbíos. LAC.

Males call at night. The advertisement call consists of a single, short note, "bonk," emitted in a series of up to five notes. Details of the call are given under the subtitle "Vocalizations of Ecuadorian Anurans" in Appendix 1. When individuals are disturbed, they exude an aromatic pale blue slime; presumably, this has a protective function, but the details are unknown.

The eggs of *Gastrotheca orophylax* undergo direct development in the pouch. Fourteen females having SVLs of 50.8–74.0 ($\bar{x}$ = 59.3) mm were brooding 12–37 ($\bar{x}$ = 23.4) eggs 5.2–7.5 ($\bar{x}$ = 6.2) mm in diameter. Twelve of these were collected in March, May, June, and July. Four females having an average snout–vent length of 60.0 mm were brooding an average of 28 eggs with diameters of 5.0 mm. Two brooding females collected on 21 July at El Carmelo, Provincia Carchi, gave birth to young on 31 August and 25 September. Ten of these young had snout–vent lengths of 15.7–17.2 ($\bar{x}$ = 16.1) mm.

Hatchings simply crawl out of the pouch. Usually, the gills slough off before emergence. During the final stages of parturition, part of the pouch lining protrudes as the female inserts one or both feet in the pouch to remove young, undeveloped eggs, and detritus. Parturition in the laboratory took place in water, and developing embryos were tadpole-like with pigmentation and well-developed mouthparts and tails. The tadpole-like embryos have many larval buccal structures.

Distribution.—See Figure 4.23. *Gastrotheca orophylax* is known from the Amazonian slopes of the Cordillera Oriental of the Andes in northern Ecuador and extreme southern Colombia. In Ecuador, it occurs in Carchi, Sucumbíos, and Napo provinces. Its elevational range is 2484 – 3308 m in an area of extent of occurrence of about 1030 km^2.

This nocturnal, arboreal species inhabits natural vegetation in cloud forests, subparamos, and partially intervened areas in the evergreen high montane forests of the northern Cordillera Oriental of the Andes and evergreen montane forests of the northern Cordillera Oriental of the Andes, where the average annual rainfall is 1181 – 1374 mm and the average annual temperature is 9.5–13.5 °C.

Conservation.—*Gastrotheca orophylax* is categorized as Vulnerable throughout its entire range (IUCN SSC Amphibian Specialist Group, 2018b; in IUCN Red List of 2022). In Ecuador, Ortega-Andrade *et al.* (2021) categorized this species as Near Threatened; however, their categorization seems to be an error given that they provide data and justification to be included in the Endangered category. We categorize it as Endangered following B1ab(iii) IUCN criteria because: (1) its extent of occurrence is estimated to be less than 5000 km^2 (1030 km^2, according to our data), (2) its distribution is severely fragmented, and (3) there is a continuing decline in the extent and quality of its habitat. Its population status is unknown, but it is a rare species, whereas it was formerly common. The few recorded localities in Ecuador are facing severe human-induced habitat deterioration. The major threat is deforestation for agriculture and cow raising. However, the decline in Ecuador remains unexplained and might perhaps be related to climatic change and diseases, since the species appears to have disappeared from pristine habitats. It is also possible that the use of agro-chemicals in agricultural areas (particularly for potato and tree tomato cultivation) could have led to declines in this habitat. In the Valle de Sibundoy, the species is threatened by timber extraction and intentional fires set for the burning of paramo habitats. Significant threats are deforestation due to agriculture, livestock, burning, the heavy use of pesticides on crops, forest plantations of pine and eucalyptus, urbanization and human settlements, etc. This species is part of the *ex situ* conservation program undertaken by CJ; however, no successful breeding has been reported yet. In Colombia, the species has been found in the Reserva Natural Privada Castelví. In Ecuador, it has been recorded in the public reserve Parque Nacional Cayambe Coca and in the Area Ecológica de Conservación Municipal La Bonita.

Remarks.—Duellman and Pyles (1980) provided a multivariate analysis showing distinctive differences between *Gastrotheca plumbea* and *G. orophylax*. Differences in morphometrics also are evident from univariate statistics given by Duellman (2015), who also provided data on cranial osteology. In the phylogenetic tree shown by Duellman (2015), *G. orophylax* and *G. plumbea* are sister species. These species, together with *G. nicefori* and *G. galeata*, are the only species in the *Duellmania* lineage that produce eggs that undergo direct development. The other 12 species in that lineage have tadpoles that develop in ponds. Del Pino and Escobar (1981) removed embryos from the pouch and raised the nonfeeding larvae in water to metamorphosis; they estimated that the duration of incubation is 88 days.

Yánez-Muñoz *et al.* (2020) incorrectly labeled their figure 90 as *Gastrotheca espeletia* (instead of *G. orophylax*).

Key references.—(Duellman and Pyles, 1980) (description); (del Pino and Escobar, 1981) (reproductive biology); (Wassersug and Duellman, 1984). (Duellman, 2015) (account); (Yánez-Muñoz *et al.*, 2020) (account).

GASTROTHECA PLUMBEA (PLUMBEUS MARSUPIAL FROG) (BOULENGER, 1882a)

Nototrema plumbeum (Boulenger, 1882a): 417. **Holotype.**—BMNH 1947.2.31.19 from Intag, Provincia Imbabura, Ecuador.

Gastrotheca plumbeum Peters, 1955: 346.

Etymology.—The specific name is derived from the Latin *plumbeus*, meaning of lead. Boulenger (1882a) apparently applied the name because of the gray color of the preserved holotype.

Identification.—Figures 4.43, 4.44, 4.45. Adults of this species attain a maximum SVL of 71.8 mm in females and of 66.4 mm in males. The skin on the dorsum is smooth or weakly areolate; it is not co-ossified with the skull. Supraciliary processes and calcars are absent. The interorbital distance is about twice the width of the upper eyelid. Finger I is shorter than II; the fingers lack webbing, and the toes are about one-half webbed. The tibia length is less than 50% of the SVL.

The dorsum is uniform green, and the venter is creamy tan. The flanks are uniform brown in most individuals but green in some; the axillae and groin usually are dull blue. A pale labial stripe and dorsolateral stripe are present. The tympanum is brown or green, and the iris is olive green to brown with black reticulations.

Comparisons.—*Gastrotheca agramma*, *G. elicioi*, *G. litonedis*, *G. lojana*, *G. longipes*, *G. orophylax*, *G. pseustes*, *G. psychrophila*, *G. riobambae*, *G. turnerorum*. The presence of pale labial and dorsolateral stripes and brown flanks separates *G. plumbea* from *G. orophylax* and *G. longipes*; the latter has an anteroventrally inclined snout in profile in contrast to a bluntly rounded snout in *G. plumbea*. The uniformly pale venter distinguishes *G. plumbea* from the biphasic *G. elicioi*, *G. litonedis*, *G. lojana*, *G. pseustes*, *G. psychrophila*, *G. riobambae*, and *G. turnerorum*. *Gastrotheca agramma* differs from *G. plumbea* by having bold black reticulations in the iris.

Natural history.—Figures 4.46, 4.47. Adults have been found by day in bromeliads throughout the year. In May, one individual was perched on a boulder in a stream by day. At night, some individuals have been found on branches of bushes in ravines, but most are perched on the large, hirsute leaves of the herb *Gunnera* sp. (Gunneraceae). Juveniles have been reported from the sheaths of bananas, *Musa ensete*.

Males call at night and have been heard calling in April, June, and July. Calling males were abundant at Pilaló, Provincia Cotopaxi, in June and July; several males were observed there in March, but none were calling. The advertisement call consists of a single, short note, "croak," emitted in a series. Detailed data and figures of calls from Pilaló, Provincia Cotopaxi, are provided under the subtitle "Vocalizations of Ecuadorian Anurans" in Appendix 1.

FIGURE 4.43 *Gastrotheca plumbea*. Left: CJ 6635, female, km 25, road Aloag-Tandapi, Provincia Pichincha. DAL. Right: QCAZ 62127, male, San Antonio, ca. Cuellaje, Provincia Imbabura. LAC.

FIGURE 4.44 Color variation in adults of *Gastrotheca plumbea*. Top left: CJ 348, male, 59.6 mm SVL. Top right: CJ 353, female, 71.8 mm DVL. Second row left: CJ 349, male, 66.4 mm SVL. Second row right: CJ 347, male. Third row left: CJ 357, male, 60.9 mm SVL. Third row right: CJ 346, male, 61.8 mm SVL. Dos Cascadas, road Aloag-Tandapi, Provincia Pichincha. LAC. Bottom row left: CJ 6637, km 25, road Aloag-Tandapi. DAL. Bottom row right: CJ 5277, male, Reserva Comunidad Junín, Provincia Imbabura. DAL.

FIGURE 4.45 Morphological details of *Gastrotheca plumbea*. CJ 6635, female, km 25, road Aloag-Tandapi, Provincia Pichincha. DAL.

FIGURE 4.46 *Gastrotheca plumbea*. Top left: calling male, La Victoria, Sigchos, Provincia Cotopaxi. Top right and middle left: brooding females, Aloag-Tandapi, Provincia Pichincha. Middle right: brooding female, Santiago, Provincia Bolívar. Bottom left: embryos just before hatching, Santiago, Provincia Bolívar. Bottom right: recently born metamorphs, Dos Cascadas, road Aloag-Tandapi. LAC.

FIGURE 4.47 Variation in hatchlings and ontogenetic changes in *Gastrotheca plumbea*. Top to bottom, left to right. CJ 13038a,b,c, stage 44, recently hatched, dorsolateral and ventral views of three hatchlings. CJ 8371a,b, froglets stage 46, TL = 17.3 mm. CJ 13038d, froglet 34 days after stage 44; adult 229 days = 47.2 mm SVL. Dos Cascadas, road Aloag-Tandapi, Provincia Pichincha. LAC.

A calling male from La Victoria, Sigchos, Provincia Cotopaxi, is depicted in Figure 4.46. Brooding females have been found in June and July in Pilaló. The eggs of *Gastrotheca plumbea* undergo complete development in the pouch and hatch as froglets. Two females with snout–vent lengths of 60.3 and 62.9 mm were brooding 14 and 34 eggs having diameters of 4.5 and 5.9 mm, respectively. Three females (mean SVL of 69 mm) had an average of 28 eggs 4 mm in diameter in their pouches. Other reported females were brooding 20–28 eggs 6.8 mm in diameter. Duration of incubation is 85–118 days. During parturition, the female uses her hind limbs to remove the young from the pouch. Ontogenetic changes from metamorph to adult are depicted in Figure 4.47.

Distribution.—See Figure 4.23. *Gastrotheca plumbea*, an endemic species from Ecuador, is known on the Pacific slopes of Cordillera Occidental of the Andes in Imbabura, Pichincha, Cotopaxi, Bolívar, Chimborazo, and Azuay provinces. Its elevational range is 1831–3028 m in an area of extent of occurrence of about 13,330 km^2.

This nocturnal, arboreal species inhabits natural vegetation, forest edges, and partially intervened areas in the evergreen high montane forests of the Cordillera Occidental of the Andes, evergreen low montane forests of the Cordillera Occidental of the Andes, and evergreen montane forests of the Cordillera Occidental of the Andes, where the average annual rainfall is 645–2265 mm and the average annual temperature is 11.8–17.1 °C.

Conservation.—*Gastrotheca plumbea* was categorized as Near Threatened by Ortega-Andrade *et al.* (2021). However, their categorization seems to be an error given that they provide criteria for it to be included under a threatened category. We categorize it as Endangered following A2ac IUCN criteria because we infer a general reduction in the population greater than 50%. Its general population status is unknown. However, it has declined, and some populations (e.g., from Santiago, Provincia Bolívar, and from Bosque Integral Otonga, Provincia Cotopaxi) have disappeared. Its apparent decline could be related to climate change and disease. The amphibian chytrid fungus has been reported in Cashca Totoras, Provincia de Bolívar (Ron and Merino-Viteri, 2000). Under laboratory conditions at CJ, this species has been shown to be highly susceptible to chytrid infections. Additional significant threats are severe habitat loss, fragmentation, deforestation, agriculture, livestock, fires, human settlements, and mining. It has been recorded in the public reserve Reserva Ecológica Los Illinizas, the protected forests of Cashca Totoras, and some private reserves (e.g., Bosque Integral Otonga). CJ (Quito, Ecuador) maintains a program of *ex situ* management.

Remarks.—Duellman (2015) provided an account including data on cranial osteology. According to Duellman (2015), when disturbed, these frogs exude an aromatic pale blue slime. Skins of three specimens were sent to the late Dr. Vittorio Erspamer (Department of Biochemistry, Universiti Roma) for an analysis of skin peptides and amines. Erspamer found no active substances in the skins, which were essentially the same as those of *G. pseustes* and *G. riobambae*, which do not exude an aromatic slime. See Remarks under *G. orophylax*.

According to del Pino and Escobar (1981), the embryos of *Gastrotheca plumbea*, like those of *G. orophylax*, develop through stages typical of tadpoles with pigmentation and well-developed mouthparts and tails. Furthermore, they removed embryos from the pouch and raised them in water until the nonfeeding tadpoles metamorphosed. A female gave birth to 26 young on 7 December, and another gave birth to 28 young on 30 September.

Key references.—(Boulenger, 1882a) (description); (del Pino and Escobar, 1981); (reproduction); (Auber-Thomay *et al.*, 1990) (morphology, reproductive biology, natural history); (Duellman, 2015) (account); (Yánez-Muñoz *et al.*, 2020) (account).

GASTROTHECA PSEUSTES (LIAR MARSUPIAL FROG) DUELLMAN AND HILLIS, 1987

Gastrotheca pseustes (Duellman and Hillis, 1987): 5. **Holotype.**—KU 203443 from 7.1 km by road north of San Lucas, 2940 m, Provincia Loja, Ecuador.

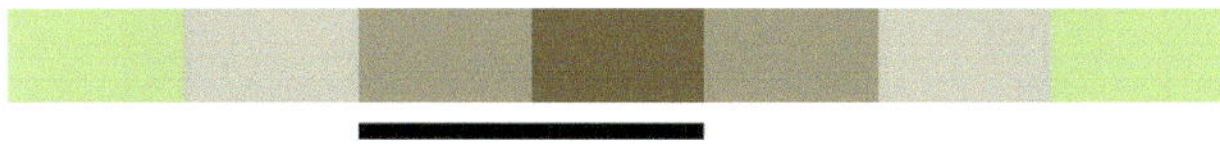

Etymology.—The specific name is derived from the Greek word meaning liar; the name was assigned in reference to the phenotypic similarity of the species to *Gastrotheca riobambae*, with which it had been confused previously.

Identification.—Figures 4.48, 4.49, 4.50, 4.51. Adults of this species attain a maximum SVL of 62.4 mm in females and of 54.4 in males. The skin on the dorsum is smooth to weakly pustular, not co-ossified with the skull, and lacks transverse ridges. Supraciliary processes are absent. The tympanic annulus is smooth. The interorbital distance is slightly greater than the width of the upper eyelid. Fingers

FIGURE 4.48 *Gastrotheca pseustes* (*sensu stricto*). Left: CJ (T 707), female. Right: CJ 202, male. Chilla, Provincia El Oro. LAC.

FIGURE 4.49 Color variation in metamorphs of *Gastrotheca pseustes*. CJ 551, Salado de Jimbura, Provincia Loja. LAC.

FIGURE 4.50 Color variation in metamorphs at stage 46 of *Gastrotheca pseustes* from Chillacocha, Provincia El Oro. LAC.

FIGURE 4.51 Color variation in subadults of a population from Papallacta, Provincia Napo. Top left: CJ 10508. Top right: CJ 12114. Middle left: CJ 12113. Middle right: CJ 10504. Bottom left: CJ 10502. Bottom right: CJ 10499. Papallacta-Laguna Baños, Provincia Napo. SGS.

I and II are equal in length and bear discs that are slightly wider than the digits. The fingers are unwebbed; on the foot, the webbing extends maximally to the antepenultimate subarticular tubercle on the fourth toe and the penultimate subarticular tubercle on the fifth toe. The tibia length is less than 50% of the snout–vent length and is slightly shorter than the foot. The heel lacks a calcar or tubercle.

The dorsum is green, usually with a darker green or brown middorsal blotch. Markings on the head usually consist of a pale labial stripe and a dark canthal stripe. A pale dorsolateral stripe is absent. The flanks are tan or brown with dark or pale spots. The venter is gray, usually with diffuse darker gray marks. Although most individuals of *Gastrotheca pseustes* have a green dorsum with darker green or brown markings, the colors are highly variable.

Nearly all specimens of *Gastrotheca pseustes* have at least a suffusion of blue in the axillae and groin and on the hidden surfaces of the thighs; in some individuals, this blue is intense. The venter is uniformly dull cream or creamy gray in some individuals, but in most individuals, diffuse gray flecks or spots are evident. These are not well-defined spots or mottling, as in *G. riobambae*, and the ventral surfaces of the limbs are never mottled, as in that species. Descriptions of the iris color in life vary from deep gold to bronze or copper, with black flecks or reticulations.

Comparisons.—*Gastrotheca espeletia*, *G. litonedis*, *G. plumbea*, *G. peruana*. *Gastrotheca pseustes* is like *G. espeletia* and *G. riobambae* in having short limbs and small digital discs. It differs from *G. espeletia* in having fingers I and II of equal length and by having a broader and blunter snout. It differs from *G. riobambae* by having less webbing on the feet and a more truncate snout and by lacking bold black spots or mottling on the ventral surface of the shank. Both *G. litonedis* and *G. plumbea* have color patterns like those exhibited by some specimens of *G. pseustes*, but both of the former species have larger digital discs and uniformly pale venters, as contrasted with the usually diffusely gray-spotted venter in *G. pseustes*. *G. pseustes* differs from Peruvian species in being larger than any except *G. peruana*. *G. pseustes* differs from that species by having a round instead of an acuminate snout in dorsal view, more webbing on the feet, and in coloration. In *G. pseustes*, the venter is gray, with or without diffuse dark spots; the posterior surfaces of the thighs are pale brown or blue, with or without dark spots. In *G. peruana*, the venter is tan; the posterior surfaces of the thighs are brown, with or without pale spots.

Natural history.—Figure 4.52. By day, individuals have been found under stones in pastures and in paramos, in crevices between leaves of agave plants, and in crevices in stone walls. *Gastrotheca pseustes* seems to be most common (or at least most easily found) in narrow irrigation ditches, where the frogs are on the banks of ditches overhung with vegetation. On cloudy or misty days, males commonly call from ditches or from beneath stones. At night, males have been found calling from open ground, branches of low bushes (e.g., *Baccharis*), margins of shallow ponds, and reeds in a lake. Tadpoles have been found in shallow grassy ponds.

FIGURE 4.52 *Gastrotheca pseustes* in their microhabitats. Top: Area Ecológica de Conservación Mazán, Provincia Azuay. Morley Read. Middle left: amplectant pair, Guanujo, Provincia Bolívar. Middle right: Páramo de Guamaní, Provincia Napo. LAC. Bottom left: Angamarca, Provincia Cotopaxi. María del Carmen Vizcaíno. Bottom right: Guaranda, Provincia Bolívar. LAC.

Males call at night. Calls from three locations in Provincia Loja (two individuals from Bosque Washapamba, one from Vía Urdaneta-Tutupali, and one from Cerro de Arcos), Ecuador, were described as complex calls composed of one to five long, pulsed notes and followed (or not) by one to six short, single-pulsed notes. The long note had a mean duration of 0.878 s and consisted on average of 32.72 distinct pulses, partly fused, without silent intervals. The short notes had a duration of 0.080 s, and the inter-note interval was on average 0.080 s. Detailed data and figures are provided under the subtitle "Vocalizations of Ecuadorian Anurans" in Appendix 1.

This species has a biphasic mode of reproduction.

Gastrotheca pseustes broods relatively small eggs, which hatch into tadpoles that complete their development in shallow ponds. Fourteen brooding females having snout–vent lengths of 47.4–60.2 mm ($\bar{x}$ = 54.2) were brooding 49–197 eggs ($\bar{x}$ = 109.1), in which the ovarian diameter was 2.6–6.3 mm ($\bar{x}$ = 4.5). Brooding females have been collected in January, February, March, June, and July. These months fall within the rainy season (November–August) in the Andes of central and southern Ecuador.

Reproductive biology data of pairs (wild and lab born) from Papallacta, Provincia Napo, have been obtained in the laboratory (CJ and Wikiri lab notes by Daniel Romero Dueñas, in April and May 2023) from two reproductive events. The amplexus is axillary and the clutches contained 57 and 96 eggs.

Tadpoles.—Figures 4.53, 4.54. Tadpoles belong to the type IV tadpoles of Orton (1953) and the exotroph, benthic ecomorphological guild. All measurements are expressed in millimeters. The following description is based on a specimen in stage 36, from a series (CJ 1949) obtained from a pond at Chillacocha, 8 km southwest Chilla, 3250 m, Provincia El Oro, Ecuador, by Elicio E. Tapia, Sofía Carvajal-Endara, and Henry Grefa on 10 June 2011. Total length 58.2 mm; body length 22.7 mm (39% of total length). The body is ovoid in dorsal and lateral views, slightly depressed; the throat is concave in lateral profile, sloping from the anterior margin of the snout to the belly; the body width at the level of the spiracle is 15.3 mm, and its height at the same position is 12.9 mm; the head width at the level of the eyes is 11.7 mm. The lateral line system is present but barely visible, with supraorbital and infraorbital lines both originating at the tip of the snout, running parallel to the eye and making contact immediately behind the eye; the angular line descends vertically from just posterior to the eye toward the throat; it dorsally contacts the post-infraorbital line; the post-supraocular line is present; the anterior oral line descends vertically from the oral disc level and behind the naris level to the throat, making a curve that parallels the infraorbital line, forming a circuit continuous with the angular and loreal lines; the ventral line dorsally surrounds the spiracle; the dorsal body and middle body lines are not visible. The nostrils are medium-sized (in proportion to body length), ovoid, protruding, with a fleshy annulus; the opening is directed anterolaterally. The snout–nostril distance 3.9 mm; the internarial distance is 3.6 mm. The eye is positioned and directed dorsolaterally; the eye length is 2.9 mm, eye width 2.3 mm, interorbital distance 7.1 mm. The spiracle is sinistral, located at midbody; the spiracular opening is oriented posteriorly; the distance from the tip of the snout to the spiracular opening 15.6 mm; the end of the spiracle is rounded, attached to the body wall; the inner wall of the spiracle is not evident; the tube length is 2.7 mm; the tube transverse width is 3.0 mm. The vent tube is dextral, with the opening directed posteriorly; the tube length is 2.9 mm; the tube transverse width is 2.8 mm. The tail length is 35.0 mm; the caudal musculature is robust, narrowing gradually until the tail terminus; the caudal muscle height is 4.5 mm; the caudal muscle width is 4.3 mm; the caudal fins are well developed and proportional; the dorsal fin originates near the tail–body junction, forming a low hump; the dorsal fin height is 4.2 mm; the ventral fin height is 3.8 mm; the maximum height of the tail is 12.3 mm; the tail terminus is rounded, with the caudal musculature not reaching the fin terminus.

FIGURE 4.53 Ontogenetic changes in two individuals of *Gastrotheca pseustes* from tadpole to adult. Top to bottom, left to right. CJ 11443, lateral, dorsal, and ventral views, stage 25, TL = 41.4 mm, Simiatug, Provincia Bolívar. CJ 12176, stage undetermined, TL = 59.1 mm; stage 44 = TL = 53.8 mm; stage 45 = 23.1 mm SVL; 37 days after stage 45 = 29.3 mm SVL; 126 days = 39.9 mm SVL; 137 days = 41.1 mm SVL; 371 days = 42.9 mm SVL. Parque Nacional Cajas, Provincia Azuay. SGS.

FIGURE 4.54 Oral disc of *Gastrotheca pseustes* (*sensu stricto*). CJ 1949, stage 37, Chillacocha, Provincia El Oro. Taken from Carvajal-Endara *et al.* (2019).

The oral disc is small, ventral, anteriorly reaching the level of the tip of the snout, not protruding laterally beyond the body or visible dorsally; the transverse width is 5.9 mm, surrounded by a uniserial row of small, marginal papillae, interrupted medially on the upper lip; the lower lip papillae alternate in orientation in and out, giving appearance of two rows; the upper lip has 17 papillae on the right side and 16 papillae on the left side; the lower lip has 52 marginal papillae; the upper jaw sheath is medium-sized, forming a finely serrated, smooth arch, with lateral processes of height 0.56 mm and transverse width 3.7 mm (63% of oral disc width); the lower jaw sheath is V-shaped, finely serrated, with a width of 2.9 mm and height of 0.68 mm. The labial tooth row formula is 2/3(1) with the following tooth row lengths: A-1: 4.2 mm, A-2: 4.5 mm, P-1 right row: 1.9 mm, P-1 left row: 2.0 mm, P-1 gap: 0.3 mm, P-2: 4.1 mm, P-3: 3.7 mm.

In dorsal view, the body is tan with black flecks. The flanks are reticulated with black and tan; the areas around the eyes and snout are paler tan. The venter is cream with black markings; the throat is translucent with small cream flecks; reddish gills are evident. The caudal muscles are reddish-pink, more evident on the proximal half; the myomeres are barely visible; the caudal muscles and fins have medium-sized cream dots, most dense on the upper side of the caudal muscles and near the tail–body junction, forming a nearly continuous stripe along the dorsolateral caudal musculature; otherwise, the caudal fins are translucent. The legs are cream with black markings on the toes. The oral apparatus is pale cream. The iris is copper–yellow with black reticulations.

The changes in coloration during ontogenetic development of one mostly pale brown individual are described next. At stage 41, the dorsum and flanks are yellowish-brown with a poorly defined pattern of brown-gray, elongated paravertebral marks, and a few blotches on the dorsum of the hind limbs. A diffuse brown–gray stripe borders the canthal and dorsolateral body and is bordered above by a creamy brown stripe. The iris has a reddish-gold suffusion. By stage 45, markings on the dorsum and flanks are better defined, with brown markings darkest peripherally; scattered green ill-defined flecks are present on the dorsum of the head, limbs, and canthus rostralis; the tips of the fingers are yellowish-cream. At stage 46, paravertebral green–brown markings on the dorsum of the body and green flecks on the limbs contrast with the surrounding light brown areas; the flanks have a dark brown stripe. The color variation in six metamorphs in stage 46 is depicted in Figure 4.50. They vary from dark brown to green with well-defined paravertebral marks or a uniformly colored dorsum.

Distribution.—See Figure 4.39. *Gastrotheca pseustes*, an endemic species from Ecuador, is known in the Cordilleras Oriental and Occidental of the Andes and inter-Andean basins in Imbabura, Pichincha, Napo, Cotopaxi, Tungurahua, Bolívar, Chimborazo, Cañar, Azuay, El Oro, and Loja provinces. Its elevational range is 2320–4708 m in an area of extent of occurrence of about 35,970 km^2.

This nocturnal, arboreal species inhabits natural vegetation in paramos, subparamos, and intervened areas in 11 ecosystems (Table 0.4) within six biogeographic sectors (Table 0.5), where the average annual rainfall is 525–1348 mm and the average annual temperature is 1.4–15.5 °C.

Conservation.—*Gastrotheca pseustes* was categorized as Least Concern by Ortega-Andrade *et al.* (2021) because, although declines have occurred in the past, the distribution is too large to qualify for a threatened category. They overestimated its area of extent of occurrence (48,593 km^2) and did not take into account some taxonomic problems and threats. We categorize this species as Near Threatened given that its area of extent of occurrence is 35,970 km^2, the population is severely fragmented, and many threats continue to affect the species. Additionally, *G. pseustes* seems to be a complex of species (see Remarks). The species still persists; however, many populations that were once abundant in the mid-1980s are now severely reduced, for instance, the populations of Parque Nacional Cajas and Reserva de Producción Faunística Chimborazo. This species is relatively tolerant of habitat destruction, although it requires some thickets to persist. The conversion of habitat for human settlements and agriculture has severely fragmented the habitat, and aggressive afforestation programs in paramo habitats using exotic pines are probably also a threat. Agrochemicals that enter the water bodies needed for reproduction are another threat. Chytridiomycosis has been confirmed in this species. It has been recorded in the public reserves Parque Nacional Cajas, Parque Nacional Cayambe Coca, Parque Nacional Cotopaxi, Parque Nacional Llanganates, Parque Nacional Sangay, Parque Nacional Cotacachi-Cayapas, Reserva Ecológica Antisana, and Reserva de Producción de Fauna Chimborazo and the protected forests of 15 Áreas del Interior de la Cuenca del Río Paute, Cashca Totoras, Mazán, Molleturo y Mollepungo, Sun Sun Yanasacha, Subcuenca del Río Tambo, Tamboyacu, Antisana, Pita, Cinto, Saloya, and Pichán y Quebrada San Juan.

Remarks.—*Gastrotheca pseustes* is highly variable in size, structure, and coloration. Accordingly, the variation in the species was analyzed by Duellman (2015), comparing morphometric and descriptive characters in samples from 11 areas in the Andes of Ecuador. *Gastrotheca pseustes* was described from 7.1 km by road north of San Lucas, 2940 m, Provincia Loja, Ecuador, in the southern Andean Cordillera. The type locality is located about midway between San Lucas and Saraguro. According to Carvajal-Endara *et al.* (2019), the current assignment of specimens to this taxon is constrained because: (1) There is a high phenotypic similarity between *G. pseustes* and *G. lateonota* from the Cordillera de Huancabamba in Peru. (2) There is no genetic information for topotypic *G. lateonota*, thereby precluding comparisons with specimens from southern Ecuador. (3) According to Duellman (2015), the variation in morphometrics, structural characters, and coloration within *G. pseustes* presents a geographic mosaic, except in the southernmost population from Saraguro. The frogs from Saraguro are somewhat more distinctive than the others in being larger in size and in having broader heads and some consistent characteristics of the color pattern, including dark flanks with pale spots and uniformly dark posterior surfaces of the thighs. (4) New molecular phylogeographic data suggest that there may be at least two species in what is now recognized as *G. pseustes*.

The tadpoles depicted as *Gastrotheca turnerorum* (Fig. 5E) and *G. pseustes* (Fig. 5F) in Carvajal-Endara *et al.* (2019) are erroneously labeled. Fig. 5E corresponds to *G. pseustes* and Fig. 5F to *G. turnerorum*.

Key references.—(Duellman and Hillis, 1987) (description); (Duellman, 2015) (account).

GASTROTHECA PSYCHROPHILA (COLD MARSUPIAL FROG) (DUELLMAN, 1974c)

Gastrotheca psychrophila Duellman, 1974c: 15. **Holotype.**—KU 120760 from ridge between Loja and Zamora, 2850 m, 13–14 km [by road] east of Loja, Provincia Loja, Ecuador.

Etymology.—The specific name is derived from the Greek words *psychros*, meaning cold, and *philos*, meaning "having an affinity for." The name was used in allusion to the cold climate at the type locality.

Identification.—Figure 4.55. Adults of this species attain a maximum SVL of 64.0 mm in females and of 50.0 in males. The skin on the dorsal surfaces is smooth, and that on the flanks is areolate; transverse dermal ridges are absent. In large females, the skin on the dorsum of the body is finely granular, and the skin on the head is co-ossified with the frontoparietals. There are no tubercles or supraciliary processes on the upper eyelid. The distinct tympanic annulus is smooth. The snout is broadly rounded in dorsal view and bluntly rounded in profile. The interorbital distance is much wider than the width of the upper eyelid. Finger I is shorter than II. The fingers are long and bear moderately large discs. The outer fingers are webbed basally. The hind limbs are moderately robust; the tibia length is about 50% of the SVL and barely longer than the foot. The toes are long and about two-fifths webbed; the digital discs are round and wider than the digits. Calcars and a distinct inner tarsal fold are present on the distal two-thirds of the tarsus; calcars are absent on the heels.

The dorsal coloration is rather dull, varying from dark brown to tan or green; dark markings are absent. Some individuals are dull green with scattered cream flecks, and others are uniform tan, dull brown, or dark bronze–brown. There are no markings on the head. A pale dorsolateral stripe is absent. Some individuals have orange spots in the groin and anterior surfaces of the thighs. The entire venter, axillae, groin, anterior and posterior surfaces of the thighs, and inner surfaces of the shanks may be blue or pale gray. The iris is coppery bronze with black reticulations.

FIGURE 4.55 Morphological variation in *Gastrotheca psychrophila*. Top left: KU 120760, holotype, female, 61 mm SVL. John D. Lynch. Middle left: KU 142635, paratype, male. WED. Top right: KU 164234, paratype, male. WED. Bottom left: KU 120761, paratype, female. John D. Lynch. Bottom right: KU 164233 (old number KU 142633), paratype, female, 63. 5 mm SVL. WED. Abra de Zamora, 13–14 km east of Loja, Loja-Zamora Chinchipe provinces. WED.

Comparisons.—*Gastrotheca ruizi*. *Gastrotheca psychrophila* is unique among Andean members of the genus by lacking a dorsolateral stripe and markings on the head. *Gastrotheca psychrophila* superficially resembles *G. ruizi*, which differs by having an acuminate snout, pale green paravertebral marks, and fingers I and II equal in length.

Natural history.—Most individuals have been found by day in large terrestrial bromeliads on the east slope of the ridge. Two adults and some juveniles were under stones by day, and one individual was on a bromeliad leaf at night.

This species produces tadpoles, but no brooding females are known. Females collected in March and June have small ovarian eggs. Males have been heard calling by day in March, May, June, and July.

Tadpoles.—Unknown.

Distribution.—See Figure 4.4. *Gastrotheca psychrophila*, an endemic species to Ecuador, is known in the Cordillera Oriental of the Andes on the Abra de Zamora ridge in Zamora Chinchipe and Loja provinces. Its elevational range is 2600 – 2757 m in two localities 2 km apart.

This nocturnal, arboreal species inhabits natural vegetation in the evergreen montane forest ecosystem in the southern Cordillera Oriental of the Andes biogeographic sector, where the average annual rainfall is 1133–1170 mm and the average annual temperature is 13.3–14.0 °C.

Conservation.—*Gastrotheca psychrophila* was categorized as Critically Endangered by Ortega-Andrade *et al.* (2021) because its extent of occurrence is less than 2 km^2. It occurs in a single threat-defined location, and there is a continuing decline in the extent and quality of its habitat due to pine plantations, agriculture, and urban expansion. Failure to find this species is intriguing and might be related to precipitous declines due to climate change and pathogens. It was originally collected in 1968–1975, but several subsequent search efforts at the type locality have been unsuccessful

(Carvajal-Endara *et al.*, 2019; Székely *et al.*, 2020b). We follow the suggestion of Székely *et al.* (2020b) and tag this species as Possibly Extinct given that in Ecuador, there have been no sightings for about five decades; the most recent record of this species is 8 March 1975 from 13 km east of Loja, Provincia Loja. It has been recorded in the northern margin of the public reserve Parque Nacional Podocarpus and the protected forests of Cuencas de los Ríos: San Francisco, San Ramón y Sabanilla.

Remarks.—Duellman and Hillis (1987) reported three convergent electromorphs in *Gastrotheca psychrophila* and *G. pseustes*, which otherwise are quite distinct allozymically, and they postulated a possible past hybridization between the two species. Phylogenetic analyses of molecular data (e.g., Wiens *et al.*, 2007; Blackburn and Duellman, 2013) placed *G. psychrophila* with Peruvian taxa (*G. peruana* and *G. stictopleura*). In the phylogenetic tree based on DNA sequences presented by Duellman (2015) and modified by Duellman and Venegas (2016), *G. psychrophila* is the sister to *G. oresbios* in a lineage containing three other species in the Northern lineage of the subgenus *Gastrotheca*.

Key references.—(Duellman, 1974c) (description); (Duellman and Hillis, 1987) (evolution); (Duellman, 2015) (account).

GASTROTHECA RIOBAMBAE (GUALATACO MARSUPIAL FROG) (FOWLER, 1913)

Hyla riobambae (Fowler, 1913): 157. **Holotype.**—ANSP 18161 from "Río Bamba" (= Riobamba), Provincia Chimborazo, Ecuador.

Gastrotheca riobambae Duellman and Fritts, 1972: 11.

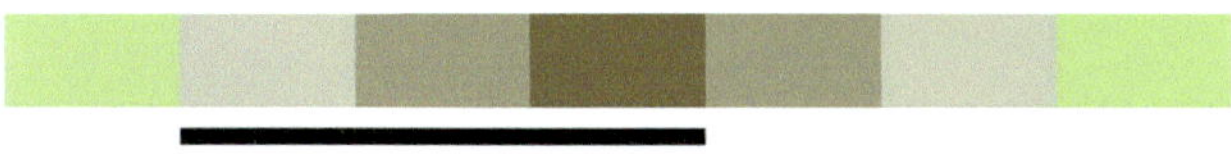

Etymology.—The specific name *riobambae* is latinized and refers to the type locality, the city of Riobamba, Provincia Chimborazo, Ecuador.

Identification.—Figures 4.56, 4.57, 4.58. Adults of this species attain a maximum SVL of 66.4 mm in females and of 56.8 in males. The skin on the dorsum is smooth, areolate, or granular; it is not co-ossified with the skull. Supraciliary processes and calcars are absent. The interorbital distance is barely greater than the width of the upper eyelid. Fingers I and II are equal in length; the terminal discs are only slightly wider than the digits. The fingers lack webbing, and the outer toes are about one-third webbed. The hind limbs are relatively short; the tibia length is less than 45% of the SVL and is slightly shorter than the foot.

The dorsum is brown or green, usually with darker brown longitudinal and paravertebral marks. Usually, there is no pale dorsolateral stripe. Dark spots are present on the pale gray or tan flanks and on the cream venter.

FIGURE 4.56 *Gastrotheca riobambae*. Left: CJ 12356a, female, 52.7 mm SVL. Right: CJ 12356b, male, 50.6 mm SVL. Hacienda Córdova, Valle de los Chillos, Provincia Pichincha. SGS.

There is considerable intrapopulation variation in the coloration of adults, metamorphs, and throughout its ontogeny. Intrapopulation color variation in adults from La Merced, Provincia de Pichincha, is shown in Figure 4.57. Intrapopulation color variation in metamorphs from Angamarca (near el Tingo), Provincia Pichincha, is shown in Figure 4.58.

Comparisons.—*Gastrotheca cuencana*, *G. elicioi*, *G. espeletia*, *G. litonedis*, *G. lojana*, *G. orophylax*, *G. plumbea*, *G. pseustes*, *G. psychrophila*, *G. turnerorum*. *Gastrotheca riobambae* is most like *G. espeletia* and *G. pseustes*; all three species have short limbs, small digital discs, and pale venters with dark spots or mottling. *Gastrotheca riobambae* differs from both of those species by more extensive webbing on the feet and by having bold dark spots or mottling on the ventral surfaces of the shanks. *Gastrotheca riobambae* lacks a bluish tinge on the thighs, axillae, and groin, which is characteristic of *G. pseustes*. Also, the snout is more rounded in *G. riobambae* than in *G. pseustes* and less protruding than in *G. espeletia*. *Gastrotheca lojana* also has dark spots on the belly, but it has a fragmented pale dorsolateral stripe. In other species of *Gastrotheca* in the Andes of Ecuador, the venter is uniform cream or creamy

FIGURE 4.57 Color variation in adults of *Gastrotheca riobambae*. La Merced, Valle de Los Chillos, Provincia Pichincha. LAC.

FIGURE 4.58 Color variation in metamorphs (Gosner stage 46) of *Gastrotheca riobambae*. CJ 13060, Barrio Angamarca, Valle de Los Chillos, Provincia Pichincha. LAC.

white (*G. cuencana*, *G. plumbea*), pale green (*G. orophylax*), pale gray (*G. litonedis*, *G. psychrophila*), cream with brown flecks (*G. elicioi*), or dark brown with cream spots (*G. turnerorum*). Furthermore, a pale dorsolateral stripe is present in *G. cuencana* and *G. litonedis*, is fragmented in *G. elicioi* and *G. litonedis*, and consists of a longitudinal series of bronze dots in *G. turnerorum*.

Natural history.—Figures 4.59, 4.60. It is often found near water—irrigation canals, ponds, streams, marshes, etc. It has been found in agave (*Agave* sp.), cornfields, and potato fields. On Volcán Pasochoa, its habitat consists of leaves, grasses, ferns, bushes (thorny or not), and small trees in a nearby farming area. Individuals prefer to take refuge in the undergrowth of bushes; they are also associated with vegetation around bodies of water. For perching, they seem to prefer microhabitats other than thorny shrubs and grass litter. Although they prefer low undergrowth (0–80 cm), they have been found up to 5 m above the ground. Their home range can reach 15 m. A study by Ramírez J. and Rodríguez B. (2011) on Volcán Pasochoa indicated the presence of four adults per km^2.

Throughout most of the year, males call on rainy or humid days and most usually at night from the branches of low bushes, clumps of grass, or the ground. The advertisement call consists of a low-pitched pulsed note; this primary note is followed by two or more shorter notes that form the territorial call. Details of the call are given under the subtitle "Vocalizations of Ecuadorian Anurans" in Appendix 1.

FIGURE 4.59 *Gastrotheca riobambae*. Top left: calling male. Top right: amplectant pair. Parque Metropolitano de Quito, Provincia Pichincha. LAC. Middle left: Quito, Provincia Pichincha. Middle right, bottom row: brooding females and tadpoles, basking behavior in a brooding female in the middle right, La Merced, Valle de los Chillos, Provincia Pichincha. LAC.

FIGURE 4.60 Sequences of egg insertion, fertilization (in an outdoor enclosure), and parturition (under laboratory conditions) of *Gastrotheca riobambae*. Left: amplectant male fertilizing and inserting eggs with his hind feet into the pouch, from top to bottom, 14:26:56, 14:27:03, 12:27:06, male from Quito, Provincia Pichincha, female from Hacienda Cobuendo, ca. Ibarra, Provincia Imbabura. LAC. Right: CJ 13061, parturition, from top to bottom, note the opening of the pouch using the outer three fingers of one or both hind feet, Hacienda Cobuendo. LAC.

Gastrotheca riobambae has a biphasic mode of reproduction. Females produce eggs that are transported as embryos in the pouch; the eggs hatch as tadpoles, which complete their development in shallow ponds. Brooding females have been found in January–October. Each female breeds only once each year. Mating takes place on land or slightly elevated vegetation. Subsequent to the initiation of amplexus, the female twists her hind limbs; in doing so, she elevates the cloacal region to a level equal to or above the opening of the pouch. The male produces fluid, which presumably contains sperm, from the cloaca. With his feet, he wipes this fluid over the female from her cloaca to the opening of the pouch. As eggs are extruded by the female, the male directs the eggs into the pouch with his feet. The eggs are inserted in the pouch one by one in a mean duration of 45 seconds (n = 8) (CJ video from 11 December 2023 by LAC). Recently fertilized eggs are uniformly pale yellow; the mean ovum diameter is 1.9 mm. Shortly after placement of the eggs in the pouch, the female seeks a secluded place, usually under a stone. She presses her back against the stone; this behavior results in the distention of the pouch and the arrangement of the eggs in one or two uniform layers in the pouch, wherein thin tissue develops between the layers of eggs. Incubation time is about three months (in 16 females, it was 70–108 days, $\bar{x}$ = 87). Near the end of incubation, the aperture of the pouch opens, and when the tadpoles hatch, the female moves to water. The parturition of tadpoles is partly spontaneous and partly assisted by the female, who distends the pouch opening with her hind feet and inserts one or both hind feet into the pouch to pull out the tadpoles. The largest number of tadpoles deposited by a female (CJ 13070) is 325 tadpoles (322 alive, three dead), at Hacienda Cobuendo, Imbaya, Provincia Imbabura, 64.7 mm in SVL (CJ laboratory notes of April 7, 2022, by Daniel Romero Dueñas). The mean clutch size is 125.4 eggs, n = 33. Twelve brooding females having SVLs of 44.3–62.2 mm contained 64–166 eggs, of which the diameter of the ova or embryos was 2.7–5.6 mm.

Newly released tadpoles are in stages 25–33; they are active swimmers and begin feeding immediately or after one or two days. The vitelline envelope and jelly capsule rupture during birth, and the tadpole straightens. Oral suckers are absent. Circulation through the external gills terminates almost immediately after the tadpole enters water, but the gills are not retracted into the opercular chamber until about a day after hatching. Within the opercular chambers, the gills reduce in size and finally disappear in about one week.

The period of larval development has been documented under laboratory conditions. The duration is highly variable and is mostly dependent on temperature and on the availability and quality of food. Normally, the duration varies from 30 to 66 days at temperatures of 17–26 °C. Tadpoles from Latacunga, Provincia Cotopaxi, maintained at 20–22 °C and fed *ad libitum* (with food available at all times) with boiled lettuce and fish food metamorphosed in 51–60 days after hatching. Under laboratory conditions, the dorsal pouch was formed after five months in individuals from San Rafael, Provincia Imbabura. Frogs from Ibarra, Provincia Imbabura, reached adulthood and reproduced after 10 months. Males raised in captivity began calling about 3–6 months after metamorphosis, and a female developed ovarian eggs about five months after metamorphosis. The ranges of snout–vent lengths of metamorphosing young are 22.1–24.5 mm ($\bar{x}$ = 23.1, n = 6) from Quito and 22.7–25.4 mm ($\bar{x}$ = 23.9, n = 13) from Latacunga.

Tadpoles.—Figure 4.61. The tadpole was described based on tadpoles from Quito, Provincia Pichincha. Some tadpoles of *Gastrotheca riobambae* grow to a large size (30 mm in body length and 90 mm in total length), but the maximum in most individuals is 25 mm in body length and 75 mm in total length. In dorsal view, the body is ovoid with a slight constriction just posterior to the orbits; the body is widest posteriorly and has a rounded snout. In profile, the body is highest posteriorly; the height is equal to about two-thirds of the width of the body. The throat is slightly concave, and the snout is bluntly rounded. The eyes are large and directed dorsolaterally. The interorbital distance is about one-third of the width of the body at the level of the orbits. The nostrils are midway between the eyes and the tip of the snout and are directed anterolaterally. The spiracular tube is short and sinistral; its opening is directed posterodorsally on the ventrolateral surface at a point slightly posterior to the midlength of

FIGURE 4.61 Ontogenetic changes in *Gastrotheca riobambae* from tadpole to metamorph. Top to bottom, left to right. CJ 10899, stage 31, TL = 20.3 mm; stage 41, TL = 57.2; stage 42, 115 days after stage 31, TL = 51.9; froglet, 11 days after stage 42, 19.6 mm SVL; juvenile, 98 days, 25.1 mm SVL; juvenile, 216 days, 34.8 mm SVL; adult, 466 days, 42.4 mm SVL. Tuquer, Provincia Carchi. SGS.

the body. The cloacal tube is short, conical, and medial, with a transverse opening. The caudal musculature is moderately slender, shallower than the dorsal fin at the midlength of the tail, and terminating gradually in a pointed tip well within the margin of the caudal fin. The dorsal fin arises gradually from the posterior edge of the body. The dorsal and ventral fins are equal in depth and terminate in a rounded tip.

The width of the oral disc is about 40% of the width of the body. The disc is bordered laterally and ventrally by two alternating rows of small marginal papillae. There are two upper and three lower rows of labial teeth. The tooth rows are about equal in length, and the first lower row is narrowly interrupted medially. The upper jaw sheath is slender, in the form of a continuous arch; the lower sheath is moderately robust and broadly V-shaped. Both sheaths have small pointed serrations. The body varies from light to dark gray. The caudal musculature and dorsum of the body have fine brown flecks. The belly and the fins have brown reticulations.

The morphology of the tadpole skull was described in detail by Haas (1996).

Living tadpoles and their habitat in La Merced, Provincia Pichincha, are depicted in Figure 4.59. We documented changes during larval ontogenetic development in individuals from Tuquer, Provincia Carchi.

Distribution.—See Figure 4.17. *Gastrotheca riobambae*, an endemic species to Ecuador, is widely distributed and known in the Cordilleras Oriental and Occidental of the Andes and inter-Andean basins in northern and central Ecuador in Carchi, Imbabura, Pichincha, Cotopaxi, Tungurahua, and Chimborazo provinces. From the Ambato Basin, it descends the valley of the Río Pastaza to a point below Baños at an elevation of 1590 m. Its elevational range is 1578 – 3449 m in an area of extent of occurrence of about 11,740 km^2.

This nocturnal, arboreal species inhabits natural vegetation as well as grasslands in subparamos, pastures, and cultivated fields in six ecosystems (Table 0.4) within four biogeographic sectors (Table 0.5), where the average annual rainfall is 525 – 1309 mm and the average annual temperature is 8.9–19.3 °C.

Conservation.—*Gastrotheca riobambae* was categorized as Near Threatened by Ortega-Andrade *et al.* (2021). However, their categorization seems to be an error, given that they provide criteria for it to be included under a threatened category. Also, their categorization was based on an overestimation of its area of extent of occurrence (11,740 km^2 according to our data). We categorize this species as Endangered because of a drastic population decline estimated to be more than 50% over the last three generations. It was once a common species but has undergone declines across much of its range. This species was also classified as Endangered by Ramírez J. and Rodríguez B. (2011), but those authors also suggested that it may be categorized as Critically Endangered. It has been reported in the following public protected areas: Refugio de Vida Silvestre Pasochoa, Reserva Geobotánica Pululahua, and Parque Nacional Cotacachi-Cayapas. Historically, it was a common species, as determined by the number of specimens and associated collection data in museums. For example, in the early 1950s, it was found in Quito at places like La Alameda, Calle Toledo, and Vicentina. It was also common in the 1970s (Duellman, 2015). Recently, numerous populations have disappeared while others have been severely fragmented and reduced. Such is the case of populations studied on Volcán Pasochoa, where they have formed metapopulations isolated from each other (Ramírez and Rodríguez, 2011). The species is especially endangered by the loss and fragmentation of habitats resulting from urban growth, deforestation, agriculture, livestock, burning, heavy use of pesticides, plantations of introduced trees (pine and eucalyptus), volcanic eruptions, etc. Their ranges have been affected by climate change (Ron *et al.*, 2003) and pathogens such as the chytrid fungus (e.g., Metropolitan Park of Quito, Pasochoa Wildlife Refuge) (Manzano, 2010) (Ramírez and Rodríguez, 2011), although there is no evidence that the chytrid fungus is affecting the populations studied. Despite being a tolerant species to changing habitat conditions, populations are restricted to areas where there are puddles, ponds, and water bodies with lentic or slow-flowing water and remnants of natural vegetation. Jiménez-Casalombo *et al.* (2023) studied the abundance of *G. riobambae* in an urbanization gradient, including conserved, rural, periurban, and urban areas in the province of Pichincha, Ecuador. The number of animals recorded was very low (e.g., 94 adults), and the authors concluded that *G. riobambae* occupies environments transformed by human activities in rural and peri-urban areas, but these habitats require specific suitable characteristics for their development and survival. In Ecuador, it has been recorded in the public reserves Refugio de Vida Silvestre Pasochoa and Parque Nacional Cotacachi-Cayapas and the protected forests of Flanco Oriental de Pichincha y Cinturón Verde de Quito. In 2012, this species was declared as an emblematic animal of the Metropolitan District of Quito, Provincia Pichincha (Consejo Metropolitano de Quito, 2012). Reproduction and *ex situ* management are part of conservation programs and public displays of Centro Jambatu, Vivarium in Quito, and Guayllabamba Zoo. Also, the species is being managed as part of the biotrade industry by Wikiri (see Chapter 1 of Volume I).

Remarks.—Duellman (2015) provided a detailed account including data and discussions on cranial osteology, morphometrics, and structural characters. *Gastrotheca riobambae* is the best-known anuran in Andean South America. Systematically, the species is placed in the subgenus *Duellmania* (Duellman, 2015). It is grouped with and is the sister taxon of a lineage that includes eight species: *G. argenteovirens*, *G. aureomaculata*, *G. dunni*, *G. nicefori*, *G. ruizi*, *G. trachyceps* (all endemic to Colombia), and the Ecuadorian *G. espeletia*. All of these species have a biphasic life history, but nested within this lineage is *G. nicefori*, the eggs of which undergo direct development.

From 1957 to 1971, numerous papers were published on observations of the mating, behavior, and developmental biology of *Gastrotheca marsupiata* (Harrison Matthews, 1958; Walker, 1957; Grenard, 1958; Oeser, 1959; Brenner, 1960; Weiss, 1960; Deckert, 1963; Kästle, 1963; Hoogmoed, 1967; Spannhof and Spannhof, 1971). In each of these publications, the source of specimens is stated as Quito. That identification is erroneous according to Duellman and Fritts (1972), who resurrected *G. riobambae* for the Ecuadorian species. *Gastrotheca marsupiata* occurs in the Andes in central Peru to Bolivia. In Ecuador, *G. pseustes*, which closely resembles *G. riobambae*, occurs sympatrically with *G. riobambae* in the outskirts of Quito in places such as Valle de los Chillos or Machachi, so possibly some of these works are based on *G. pseustes*. More recent studies (e.g., (Fitzgerald *et al.*, 1979) correctly referred Ecuadorian specimens to *G. riobambae*.

The cytogenetics of *Gastrotheca riobambae* have been studied exhaustively by Schmid *et al.* (2013) and references cited therein. This species has been the subject of many studies on embryology and development in the laboratory by del Pino and coauthors (1975–2007) (del Pino, 1975, 1977, 1980, 1983, 1989, 1996, 1998; del Pino and Sánchez, 1977; del Pino and Humphries, 1978; del Pino and Escobar, 1981; del Pino and Elinson, 1983; del Pino and Loor-Vela, 1990; del Pino and Alcocer, 1994; del Pino *et al.*, 1975, 1986, 2007). However, the identity of some of their specimens remains uncertain. For instance, *G. lojana* was misidentified as *G. riobambae* by Del Pino (1989: 111). Additionally, before the description of *G. pseustes* in 1987, it was frequently mistaken for *G. riobambae*. Del Pino and Escobar (1981) stated that most of their specimens came from the vicinity of Machachi, but that frogs from other localities in the provinces of Cotopaxi, Pichincha, and Tungurahua also were used. Both *G. riobambae* and *G. pseustes* occur at Machachi, and *G. pseustes* is found in the Andes in the provinces mentioned. Thus, the developmental work by del Pino and her collaborators may have occasionally included both species.

Key references.—(Fowler, 1913) (description); (Schmid *et al.*, 2013) (cytogenetics); (Duellman, 2015) (account, cranial osteology, tadpoles, call).

GASTROTHECA TESTUDINEA (TORTOISE MARSUPIAL FROG) (JIMÉNEZ DE LA ESPADA, 1871)

Nototrema testudineum (Jiménez de la Espada, 1871): 62.
Holotype.—MNCN 155 from San José de Moti, Provincia de Napo, Ecuador.
Gastrotheca testudinea Duellman, 1974a: 4.

Etymology.—The specific name is derived from the Latin *testudo*, meaning tortoise, and presumably was applied in reference to the robustness of the body.

Identification.—Figures 4.62, 4.63, 4.64. Adults of this species attain a maximum SVL of 82.6 mm in females and of 61.1 in males. The skin on the dorsum and flanks is areolate. The skin on the head is adherent to the skull. A transverse occipital ridge is present. Transverse dermal ridges on the dorsum of the body, supraciliary processes, and calcars are absent. In some adults of both sexes, folds of skin existing ventrolaterally in the pectoral region form pockets with the opening directed anteriorly. The tympanic annulus is tubercular. The snout is rounded in dorsal view and in profile is inclined from the nostrils to a rounded tip. The interorbital distance is about twice the width of the upper eyelid. Finger I is longer than II. The fingers and toes have large terminal discs that are much wider than the adjacent digits; breeding males have a brown nuptial excrescence on the thumb. The fingers are unwebbed, and the toes are about one-half webbed. The hind limbs are moderately long; the length of the tibia is about 54% of the SVL. A small tubercle usually is present on the heel; a low inner tarsal fold is present on the distal one-third of the tarsus.

The dorsum varies from tan to olive tan and pale or dark green, with or without brown transverse bars, a narrow mid-dorsal mark or flecks, or a narrow brown transverse occipital stripe. The dorsum may bear cream spots. Transverse bars on the limbs usually are brown, but in some individuals, they are dark green. Markings on the head usually consist of a pale labial stripe and a broad brown interorbital bar. The flanks are tan with dark brown flecks, irregular marks, or vertical bars. The venter is cream, tan, or pale gray with

FIGURE 4.62 *Gastrotheca testudinea*. Left: CJ 354, female, 69.1 mm SVL. Right: CJ 2180, male. Zuñag, Provincia Morona Santiago. LAC.

FIGURE 4.63 Morphological variation and details of *Gastrotheca testudinea* showing color variation and morphological details. Note the pectoral pockets of females (bottom left) and males (bottom right). Zuñag, Provincia Morona Santiago. DAL (dorsolateral views), LAC (frontal views).

FIGURE 4.64 *Gastrotheca testudinea*. Top left: axillary amplexus. Top right: brooding female. Bottom: froglet coloration variation. Zuñag, Provincia Morona Santiago. LAC.

or without brown spots; in breeding males, the vocal sac is brown. The iris varies from dull bronze with fine black reticulations to golden tan with reddish-brown reticulations.

Comparisons.—*Gastrotheca andaquiensis*, *G. ovifera*, *G. testudinea*. Individuals of *G. testudinea* that have a narrow middorsal dark mark are patterned like *G. ovifera*, which differs in having the skin co-ossified with the skull and a prominent transverse occipital ridge with dorsal projections. Superficially, *G. andaquiensis* is like *G. testudinea*, but the former has an elongate calcar on the heel and a granular interocular ridge, and it lacks dark vertical bars on the flanks.

Natural history.—At night, adults have been found on branches of bushes and trees in cloud forests; juveniles have been found on leaves and stems of herbs and low bushes. By day, one adult was found in an arboreal bromeliad, and another was under a log.

Detailed data and figures are provided under the subtitle "Vocalizations of Ecuadorian Anurans" in Appendix 1.

The eggs of *Gastrotheca testudinea* undergo direct development in the pouch. Brooding females have been collected in June, July, September, and October. *Gastrotheca testudinea* has the largest clutch size of all direct-developing species in the genus. Brooding females contained 30, 44, and 69 eggs 5.5–9.1 mm in diameter in their pouches. The induced parturition of a female having an SVL of 77.5 mm resulted in 59 young and three undeveloped ova. The young had SVLs of about 10.7–11.8 mm. The froglets with SVLs of 11.0 mm or more actively jumped immediately after leaving the pouch.

Males call at night. Calls from one location (five individuals from Zuñag) in Provincia Morona Santiago and one individual in *ex situ* facilities at CJ, Ecuador, were described as complex calls composed of one long, pulsed note followed (or not) by one to three short, single-pulsed notes. The long note had a mean duration of 0.746 s and consisted on average of 18.88 distinct pulses (pulse series) separated by silent intervals. Detailed data and figures of calls from Zuñag, Provincia Morona Santiago, are provided under the subtitle "Vocalizations of Ecuadorian Anurans" in Appendix 1.

Distribution.—See Figure 4.6. *Gastrotheca testudinea* is known from many localities at elevations of 1100 – 2775 m on the eastern slopes of the Andes in Ecuador, Peru, and Bolivia; it also occurs in three isolated mountain ranges to the east of the main Andean cordillera: Volcán Sumaco and Cordillera de Kutukú in Ecuador and Serranía de Sira in Peru. In Ecuador, *G. testudinea* is known on the Amazonian slopes of the Cordillera Oriental of the Andes in Orellana, Napo, Pastaza, Tungurahua, Morona Santiago, Azuay, and Zamora Chinchipe provinces. Its elevational range is 1091 – 2775 m in an area of extent of occurrence of about 26,770 km^2.

This nocturnal, arboreal species inhabits natural vegetation and partially intervened areas in six ecosystems (Table 0.4) within four biogeographic sectors (Table 0.5), where the average annual rainfall is788 – 4208 mm and the average annual temperature is 13.9–21.3 °C.

Conservation.—*Gastrotheca testudinea* is categorized as Least Concern throughout its entire range (IUCN SSC Amphibian Specialist Group, 2020n; in IUCN Red List of 2022) in view of its wide distribution, presumed large population, and because it is unlikely to be declining fast enough to qualify for listing in a more threatened category. However, it is important to note that in Ecuador, *G. testudinea* has been categorized as Near Threatened by Ortega-Andrade *et al.* (2021). This is a rare species. Important threats to its existence are habitat reduction and fragmentation, oil spills, deforestation, agriculture, mining, and volcanic eruptions.

It is known from three protected areas in Bolivia. In Ecuador, it has been recorded in the public reserves Parque Nacional Cayambe Coca, Parque Nacional Sumaco Napo-Galeras, Reserva Biológica El Cóndor, and Área Ecológica de Conservación Municipal Siete Iglesias; the protected forests of Cordillera Kutukú y Shaimi, Cuencas que Forman los Ríos: San Francisco, San Ramón y Sabanilla, and Tinajillas-Río Gualaceño; and within the Podocarpus-El Cóndor and Sumaco biosphere reserves.

Remarks.—As presently known, the presence of dermal pectoral pockets is unique to *Gastrotheca testudinea*; their function is unknown.

Key references.—(Jiménez de la Espada, 1871) (description); (Duellman and Chavéz, 2010) (reproduction); (Duellman, 2015) (account, cranial osteology, life history, etymology); (Székely *et al.*, 2020b) (brief account).

GASTROTHECA TURNERORUM (TURNER'S MARSUPIAL FROG) CARVAJAL-ENDARA *ET AL.*, 2019

Gastrotheca turnerorum Carvajal-Endara *et al.*, 2019: 48.

Holotype.—CJ 393 from Laguna Negra de Jimbura, Parque Nacional Yacuri, Amaluza, 3406 m, Provincia Loja, Ecuador.

Etymology.—The specific name is a latinized word that honors the Turner family. The Turner family stands out in their commitment to the conservation of global natural resources. From Ted Turner's pioneering efforts to protect very large land areas and restore them to their original and natural states of clean water and abundant fauna and flora to granddaughter Elizabeth's awareness-raising book, *Our Friends the Frogs*, there are multiple generations of active conservationists. Together, the Turner family is helping to make the planet a better place for the next generations.

Identification.—Figures 4.65, 4.66. Adults of this species attain a maximum SVL of 58.2 mm in females and of 55.6 in males. The skin on the dorsum is areolate, not co-ossified with the skull; it lacks transverse ridges. The supraciliary processes are absent. The tympanic annulus is smooth. The interorbital distance is slightly greater than width of upper eyelid. Finger I is slightly shorter than finger II; the width of the discs is much wider than the digits. The fingers are unwebbed. The foot webbing between the external toes extends to the antepenultimate subarticular tubercle on toe IV and to the penultimate subarticular tubercle on toe V. The tibia length is 37–42% of the SVL, shorter than the foot. The

FIGURE 4.65 *Gastrotheca turnerorum*. Left: CJ 1386, female, 58.2 mm SVL, Laguna Negra de Jimbura, Parque Nacional Yacuri, Provincia Loja. Right: CJ 7823, male, 46.7 mm SVL, Amaluza, Parque Nacional Yacuri. LAC.

FIGURE 4.66 Morphological details of *Gastrotheca turnerorum*. Top: groin. Bottom: thighs. CJ 7823, male, Amaluza, Parque Nacional Yacuri, Provincia Loja. DAL.

heel lacks a calcar or tubercle. The dorsum is uniform green or brown without dark paravertebral marks. The dorsal surfaces of the hands and feet have a pattern of brown dots over a lighter background. The head markings consist of pale labial and canthal stripes formed by a series of small bronze dots. The dorsolateral stripe is present, also consisting of a series of small bronze dots. The flanks are brown to bronze–brown, with or without white stippling, and the axillae and groin are green, both with or without a few cream spots, with a bluish suffusion in some specimens. The anterior surfaces of the thighs are dark brown, and the posterior surfaces of the thighs are bluish-brown with numerous cream tubercles proximal to the vent. The venter is dark brown with uniformly distributed cream spots, but some specimens have larger cream spots, thereby resulting in a paler venter. The gular region is dark brown. The iris is yellow–cream with brown reticulations and a brown horizontal bar across the eye.

Comparisons.—*Gastrotheca aguaruna*, *G. elicioi*, *G. litonedis*, *G. pseustes*. *Gastrotheca turnerorum* most closely resembles three other species in southern Ecuador, *G. pseustes*, *G. elicioi*, and *G. litonedis*. *Gastrotheca turnerorum* differs from all of them by having a different skin texture on the dorsum and a distinctive color pattern. The dorsal skin is areolate in *G. turnerorum*, whereas it is weakly areolate or smooth in *G. pseustes*, finely granular in *G. elicioi*, and smooth in *G. litonedis*. In *G. turnerorum*, the venter is dark brown with uniformly distributed cream spots, whereas it is cream with dark marks in *G. pseustes* and *G. elicioi* and pale brown–gray in *G. litonedis*. The labial stripe consists of a series of small bronze dots in *G. turnerorum* and is mostly uniform cream in *G. pseustes*, *G. elicioi*, and *G. litonedis*. *Gastrotheca pseustes* and *G. elicioi* have a dark brown canthal stripe, whereas *G. turnerorum* has a canthal stripe formed by a series of small bronze dots. *Gastrotheca turnerorum* is the sister species of *G. aguaruna* + *G. yacuri*, with a genetic distance of at least 2.85% (in a DNA dataset of 438 bp, 16S gene). The Peruvian *G. aguaruna* differs from *G. turnerorum* by having weakly granular to smooth dorsal skin (areolate in *G. turnerorum*), fingers I and II equal in length (finger I shorter in *G. turnerorum*), and a cream venter with dark fleck or spots (venter dark brown with uniformly distributed cream spots in *G. turnerorum*).

Natural history.—At the Lagunas del Compadre in Parque Nacional Podocarpus, a gravid female was sitting on moss approximately 2 m from the Río Sabanilla, which is about 1.5 km from Lagunas del Compadre. A female from Laguna Negra de Jimbura in the Parque Nacional Yacuri was found at the edge of a pond measuring about 3 × 2 m at the edge of a lagoon at 14:00 h. The water temperature was 16.4 °C, and the water pH was 5.3. Eighteen tadpoles were collected at this pond. Another group of tadpoles was found at 18:00 h; at the same locality in another lagoon, the water temperature was 7.4 °C. Strong winds were blowing across the lagoon, and only a small patch of forest remained at one border.

Gastrotheca turnerorum has a biphasic mode of reproduction. At the Lagunas del Compadre in Parque Nacional Podocarpus, a gravid female was collected on 1 December 1994. It deposited tadpoles on 12 December in the laboratory; some of them were raised, and others were preserved. On 20 September 2009, another female with eggs in her pouch was found at the same locality. Tadpoles were found on 12 June 2011.

Tadpoles.—Figures 4.67, 4.68. All measurements are expressed in millimeters. The following description is based on a specimen in stage 39 from Laguna Negra de Jimbura, Parque Nacional Yacuri, 3406 m, Provincia Loja, Ecuador. The total length is 61.3 mm; the body length is 23.5 mm (38% of total length). The body is robust, ovoid in dorsal and lateral views, and slightly depressed. The throat is slightly concave in lateral profile. The body width at the level of the spiracle is 14.7 mm, and the height at same position is 11.9 mm; the head width at the level of the eyes is 12.9 mm. The lateral line system is not evident. The nostril is medium-sized (in proportion to body length), ovoid, and protruding, with a fleshy annulus; its opening is directed anterolaterally. The snout–nostril distance is 3.7 mm. The eyes are positioned and directed dorsally; the eye length is 2.4 mm, and the eye width is 2.2 mm. The spiracle is sinistral, located at the midbody level. The spiracular opening is oriented posteriorly. The distance from the tip of the snout to the spiracle opening is 15.9 mm. The spiracle end is rounded, not free, attached to the body wall, and the

FIGURE 4.67 Ontogenetic changes in *Gastrotheca turnerorum* from tadpole to adult. Top to bottom, left to right. CJ 11386, stage 34, lateral, dorsal, and ventral views, TL = 38.9 mm; stage 40, TL = 63,7 mm; stage 42, TL = 51.4 mm; stage 46, 43 days after stage 34, 22.9 mm SVL; juvenile, 155 days after stage 46, 35.6 mm SVL; juvenile, 294 days, 42.3 mm SVL. Laguna Natosa, Parque Nacional Yacuri, Provincia Loja. LAC.

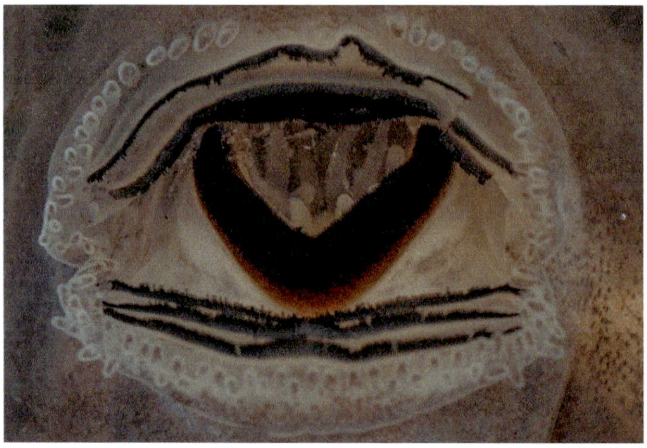

FIGURE 4.68 Oral disc of *Gastrotheca turnerorum*. CJ 1959, stage 39, Laguna Negra de Jimbura, Provincia Azuay. Taken from Carvajal-Endara *et al.* (2019).

inner wall of the spiracle is not evident. The tube length is 3.4 mm, and the tube transverse width is 1.4 mm. The vent tube is dextral, and the opening is oriented posteriorly. The tube length is 4.0 mm, and the tube transverse width is 2.6 mm. The tail length is 38.8 mm. The caudal musculature is robust in the proximal two-thirds of the body and narrows gradually until the tail terminus. The tail muscle height is 5.4 mm, and the tail muscle width is 4.8 mm. The caudal fins are thin on the proximal half of the tail, getting higher on the distal half and rising near the tail–body junction; the dorsal fin height is 3.5 mm, the ventral fin height is 3.7 mm, and the maximum height of the tail is 12.7 mm. The tail tip is rounded, and the tail musculature does not reach the fin terminus.

The oral disc is small, ventral, located near the tip of snout; it does not protrude laterally beyond the body; it is not visible dorsally, and the transverse width is 5.6 mm. It is surrounded by a uniserial row of marginal papillae, interrupted medially on the upper lip. The lower lip papillae alternate in and out, giving the appearance of two series. The upper lip has 18 papillae on the right side and 17 papillae on the left side. The lower lip has 42 marginal papillae. The upper jaw sheath is medium-sized, forming a finely serrated, smooth arch; the height is 0.4 mm, and the transverse width is 3.3 mm (59% of oral disc width). The lower jaw sheath is V-shaped, open, and finely serrated; the width is 2.4 mm, and the height is 0.7 mm. The labial tooth row formula is 2/3(1), and the tooth rows lengths are A-1: 4.7 mm, A-2: 4.0 mm, P-1 right row: 1.9 mm, P-1 left row: 2.1 mm, P-1 gap: 0.1 mm, P-2: 3.6 mm, P-3: 3.5 mm.

In life, in dorsal and lateral views, the body is tan, speckled with black. The areas around the snout are lighter. The venter is cream with black markings, semi-translucent. The guts are visible as a darker area, and the throat is translucent with small cream flecks. The gills are evident as a red hue. The caudal musculature is reddish-pink in the proximal half, decreasing in intensity toward the distal half. The myomeres and nerves are barely visible. The caudal musculature and fins have cream marks, which are clustered in a dorsal line on the caudal musculature; otherwise, the caudal fins are translucent. The legs are cream. The oral apparatus is light cream. The iris is copper–yellow with small black reticulations.

Tadpoles in stages 35–39 vary in total length between 61.3 (stage 39) and 75.0 mm (stage 35). The tail length proportion varies from 63.2 to 67.0 mm until stage 39. The labial tooth row formula is 2/3(1). The number of marginal papillae varies among specimens and Gosner stages, and variation in the number of lower lip papillae is high (42–57).

Changes in coloration during the ontogenetic development of two specimens are as follows: At stage 36, the dorsum and flanks are pale brown with dark brown areas on the posterior body and flanks. By stage 42, the dorsal surfaces of the body and limbs are uniform green; fine, cream, dorsolateral stripes are present, extending to the midbody. Cream stripes border the outer dorsal margins of the limbs. At stage 46, the dorsal surfaces of the body and limbs have more well-defined cream stripes and a cream labial stripe; the canthus rostralis is green, and remaining flanks are mostly dark brown with an upper diffuse black stripe. The dorsal surfaces of the fingers and toes are brown, except on the finger and toe discs, which are yellowish-cream. The iris is copper–yellow, lacking the brown horizontal band that is observed in adults.

Distribution.—See Figure 4.6. *Gastrotheca turnerorum*, an endemic species to Ecuador, is known from three localities in the Cordillera Oriental of the Andes in southern Ecuador in Loja and Zamora Chinchipe provinces. Its elevational range is 2899 – 3513 m in an area of extent of occurrence of about 540 km^2.

This mostly nocturnal, semiarboreal species mainly inhabits paramos and a few forests in four ecosystems (Table 0.4) in the Páramo biogeographic sector, where the average annual rainfall is 890 – 1278 mm and the average annual temperature is 10.1–13.1 °C.

Conservation.—*Gastrotheca turnerorum* is categorized as Endangered (Ortega-Andrade *et al.*, 2021) according to criteria B1 a(≤5), b(iii), b(iv) of the IUCN Red List. Although this species occurs in the Yacuri and Podocarpus National Parks, its small known area of occurrence (500 km^2) is vulnerable to improperly regulated tourism activities, introduced species such as trout, climate change, and pathogens. Laguna Negra de Jimbura has also suffered the synergistic effects of agriculture, cattle and sheep raising, fires, introduced species, and pesticide use.

Remarks.—*Gastrotheca turnerorum* belongs to the subgenus *Gastrotheca*, reported as species C in Duellman (2015: Fig 12.1). The tadpoles depicted as *G. turnerorum* (Fig. 5E) and *G. pseustes* (Fig. 5F) in Carvajal-Endara *et al.* (2019) are erroneously labeled, Fig. 5E corresponds to *G. pseustes* and Fig. 5F to *G. turnerorum.*

Key references.—(Carvajal-Endara *et al.*, 2019) (description, tadpole).

GASTROTHECA WEINLANDII (WEINLAND'S MARSUPIAL FROG) (STEINDACHNER, 1892)

Nototrema weinlandii (Steindachner, 1892): 837.
Holotype.—NMW 16481 from Ecuador.
Gastrotheca weinlandii Peters, 1955: 340.

Etymology.—The specific name is a patronym for the German zoologist David Friedrich Weinland (1829–1915), who was one of the authors of the description of *Notodelphys ovifera* (Lichtenstein and Weinland, 1854) and, in the same year, published a detailed study of embryonic stages of that species (Weinland, 1854).

Identification.—Figure 4.69. Adults of this robust species attain a maximum SVL of 75.7 mm in males and 86.0 mm in females. The skin on the dorsum is weakly granular, and that on the flanks is more granular. The skin on the head is completely co-ossified with the underlying dermal elements of the skull. A transverse occipital ridge is present. Transverse dermal ridges on the dorsum of the body are absent. Small triangular supraciliary processes are present. The tympanic annulus is indistinct and tubercular. The snout is broadly rounded in dorsal view, and in profile, it is slightly inclined anteroventrally and is nearly truncate. The interorbital distance is noticeably greater than the width of the upper eyelid. Finger I is longer than II. The fingers and toes have large terminal discs that are much wider than the adjacent digits. The fingers have basal webbing and prominent lateral fringes, and the toes are about two-fifths webbed. The hind limbs are long; the length of the tibia is more than 60% of the SVL. Large triangular calcars are present on the heels; a low inner tarsal fold is present on the distal one-fourth of the tarsus.

The dorsum and flanks are tan, grayish-tan, or brown with dark brown or olive-brown markings consisting of a narrow middorsal mark from the occiput to the sacrum with posterolateral diagonal marks extending from the middorsal mark onto the flanks and numerous transverse bars of the limbs. The venter varies from white to cream or pale green. A pale dorsolateral stripe is absent. The tongue and buccal lining are blue. The iris is reddish-bronze to coppery bronze with brown or black reticulations. Two juveniles were green or olive tan with brown marks dorsally. The venter is translucent yellowish-green.

FIGURE 4.69 *Gastrotheca weinlandii.* KU 103104, male, 85.7 mm SVL, 16,5 km northeast of Santa Rosa, Provincia Napo. WED.

Comparisons.—*Gastrotheca andaquiensis*, *G. angustifrons*, *G. cornuta*, *G. dendronastes*, *G. longipes*. Of the other *Gastrotheca* species on the Amazonian slopes of the Andes, the green *G. longipes* differs from *G. weinlandii* by not having the skin on the head co-ossified with the skull and by lacking prominent calcars on the heels. In contrast, *G. andaquiensis* differs from *G. weinlandii* by not having the skin on the head co-ossified with the skull, having finger I longer than II and a prominent interocular ridge, and by the absence of a tarsal fold. Of the four species on the Pacific versant in Ecuador, *G. guentheri* differs from *G. weinlandii* by having elongate supraciliary processes, entirely different coloration, and three teeth on the dentary. The other three species (*G. angustifrons*, *G. cornuta*, and *G. dendronastes*) differ from *G. weinlandii* by having transverse dermal ridges on the body.

Natural history.—Juveniles were collected at night on leaves of bushes and herbs, and adults were found 2–10 m above the ground on branches of trees. Males call at night.

The advertisements call is an explosive "wrock," usually followed by two shorter notes, "rock-rock." The eggs undergo direct development into froglets. One female having an SVL of 78.8 mm collected in September contained 14 eggs, about 7 mm in diameter each, in the brood pouch.

Distribution.—See Figure 4.17. *Gastrotheca weinlandii* is known on the Amazonian slopes of the Andes from southern Colombia to northern Peru. In Ecuador, it occurs in Sucumbíos, Napo, Pastaza, Morona Santiago, and Zamora Chinchipe provinces. Its elevational range is 1053 – 2389 m in an area of extent of occurrence of about 14,790 km^2.

This nocturnal, arboreal species inhabits cloud forests and partially intervened areas in six ecosystems (Table 0.4) in the Amazonian Cordilleras and northern and southern Cordillera Oriental of the Andes biogeographic sectors, where the average annual rainfall is 976–4353 mm and the average annual temperature is 15.4–21.4 °C.

Conservation.—*Gastrotheca weinlandii* is categorized as Least Concern throughout its entire range (IUCN SSC Amphibian Specialist Group, 2019f; in IUCN Red List of 2022) based on its relatively wide range and presumed large population. However, it is important to note that in Ecuador, *G. weinlandii* has been categorized as Critically Endangered by Ortega-Andrade *et al.* (2021) because of an observed and inferred reduction in the population (and its extent of occurrence) in the past for unknown causes. This species has not recuperated from a drastic population decline in the 1980s, inferred from its apparent disappearance or near disappearance from known sites. Frequent searches at historical localities revealed no specimens. Also, its distribution is severely fragmented, and there is a continuing decline in the extent and quality of its habitat. We categorize and tag it as Critically Endangered (Possibly Extinct), given that in Ecuador, there have been no sightings for about four decades; the most recent record of this species is from 22 February 1985 in Cascada San Rafael, Provincia Napo. In Ecuador, this species is threatened by deforestation, agriculture, the rearing of livestock, wood extraction, mining, landslides, wildfires, and volcanic eruptions. It is less seriously threatened in its Colombian and Peruvian ranges. It occurs in Parque Nacional Natural Alto Fragua-Indiwasi in Colombia. In Ecuador, it has been recorded in the public reserves Parque Nacional Cayambe Coca and Area Ecológica de Conservación Municipal Siete Iglesias, in the protected forests of Cordillera Kutukú y Shaimi and Tinajillas–Río Gualaceño, and within the Podocarpus-El Cóndor and Sumaco biosphere reserves.

Remarks.—In juveniles having SVLs of 25 mm or less, the skin is not co-ossified with the underlying cranial elements. Also, juveniles tend to be paler than adults and have a dark brown postorbital bar continuous to the midflank.

The cranial osteology was described by Linda Trueb in Duellman (2005).

Key references.—(Steindachner, 1892) (description); (Duellman, 2015) (account, cranial osteology, advertisement call).

GASTROTHECA YACURI (YACURI MARSUPIAL FROG) CARVAJAL-ENDARA *ET AL.*, 2019

Gastrotheca yacuri Carvajal-Endara *et al.*, 2019: 56.
Holotype.—CJ 7822 from El Salado de Jimbura, at about 2.5 km (by road) west from the entrance in the Parque Nacional Yacuri, 2914 m, Provincia Loja, Ecuador.

Etymology.—The specific name refers to the Yacuri National Park, where this species has been found. Yacuri takes its name from the Quechua words *Yacu*, meaning water, and *Quri*, which means gold. Ecuador's southernmost national park ranges from 2000 to 3700 meters in the Cordillera Oriental. This park of 43,090,60 hectares is unique by sheltering a series of 48 glacial lakes and many endemic species of flora and fauna. It is a Ramsar site (The Convention on Wetlands) and part of the Podocarpus-El Cóndor Biosphere Reserve declared by UNESCO.

Identification.—Figure 4.70. Adults of this species attain a maximum SVL of 58.9 mm in males (females unknown).

FIGURE 4.70 Morphological variation and details of *Gastrotheca yacuri*. Left, middle and bottom right: CJ 7822, holotype, male, 57.4 mm SVL, note color of groin, thighs, and shanks. DAL, Paul J. Székely (bottom right). Top right: Jaime Culebras. El Salado de Jimbura, Parque Nacional Yacuri, Provincia Loja.

The skin on the dorsum is weakly granular to smooth, co-ossified with the skull; it has a transverse ridge in the occipital region. The supraciliary processes are absent. The tympanic annulus is smooth. The interorbital distance is greater than the width of the upper eyelid. Finger I is shorter than II, and the width of the discs is much wider than the digits. The fingers are unwebbed. The toes are one-half webbed; webbing extends to the antepenultimate subarticular tubercle on toe IV and to the penultimate subarticular tubercle on toe V. The tibia length is 47–49% of SVL, larger than the foot. The heel lacks a calcar or tubercle.

The dorsum is uniform green or brown with brown paravertebral marks. The head markings consist of a brown labial stripe with cream glandular areas at the posterior end and black, brown, and bronze canthal and supraciliary stripes. The canthal stripe is diffuse; the labial stripe varies from brown anteriorly to white posteriorly until the insertion of the forelimb. The dorsolateral stripe is absent or present; when present above, it consists of a series of bronze and cream warts, bordered below with a black and brown line. The flanks are brown and green with a blue and black reticulum toward the lower flanks, axillae, groin, anterior and posterior surfaces of the thighs, shanks, and tarsus; the pelvic patch is brown. The venter is cream with nearly uniformly distributed, small, dark brown marks or brown with cream marks. The gular region is dark brown in males. The palmar and plantar surfaces are pinkish-gray; the ventral fingers and toes are fleshy. The iris is copper with abundant black reticulations.

Comparisons.—*Gastrotheca aguaruna, G. pseustes, G. psychrophila. Gastrotheca yacuri* differs from all other species of *Gastrotheca* in southern Ecuador by having a blue and black reticulated pattern on the flanks, axillae, groin, thighs, and shanks. It most closely resembles two other species, *G. pseustes* and *G. psychrophila*, by having blue colors, but they lack a blue and black reticulum. *Gastrotheca yacuri* differs from its sister species *G. aguaruna*, from the Cordillera Central in northern Peru, by lacking blue colors, having the skin co-ossified with the skull, and being smaller (SVL of males of *G. yacuri* 57.4–58.9 mm vs. 41.6–46.8 mm in *G. aguaruna*). The genetic distance between them is at least 2.8% (in a DNA dataset of 438 bp, 16S gene).

Natural history.—Figure 4.71. The male holotype was found near a road, calling from the branches of a bush about 3.5 m above the ground. Males were usually calling from the vegetation near the road. An additional subadult was found during the day in a large terrestrial bromeliad. In this location, in Parque Nacional Yacuri, *Gastrotheca yacuri* is syntopic with *G. turnerorum*.

Males call at night. Choruses were heard from April to late July 2016. Calls from El Salado, Provincia Loja, Ecuador, were described as complex calls composed of one to three (usually two) long, pulsed notes followed (or not) by one to three (usually two) short, single-pulsed notes. The long note had a mean duration of 0.634 s and consisted on average of 32.24 distinct pulses, partly fused, without silent intervals. The short notes had a mean duration of 0.072 s, and the inter-note interval was on average 0.549 s. Detailed data and figures are provided under the subtitle "Vocalizations of Ecuadorian Anurans" in Appendix 1.

FIGURE 4.71 *Gastrotheca yacuri* in its habitat at Parque Nacional Yacuri, Provincia Loja. Paul J. Székely.

Tadpoles.—Unknown.

Distribution.—See Figure 4.39. *Gastrotheca yacuri*, an endemic species from Ecuador, is only known from three nearby localities 2 km apart (maximum distance) in the Cordillera Oriental de los Andes in southern Ecuador, in Provincia Loja, at elevations of 2722 – 3470 m.

This nocturnal, semiarboreal species inhabits paramos and subparamos in the evergreen montane forest of Catamayo-Alamor ecosystem in the Páramo biogeographic sector, where the average annual rainfall is 890 – 1159 mm and the average annual temperature is 10.4–14.3 °C.

Conservation.—*Gastrotheca yacuri* is categorized as Endangered (Ortega-Andrade *et al.*, 2021). We categorize it as Critically Endangered following B1ab(iii) IUCN criteria because: (1) its extent of occurrence is estimated to be less than 100 km^2, (2) it is only known from the type locality and nearby areas, and (3) there is a continuing decline in the extent and quality of its habitat. Although it is reasonable to predict that further surveys conducted within the protected Parque Nacional Yacuri will reveal an expansion in its areas of occurrence and occupancy, and that its habitat currently appears to be relatively unaffected by major threats, it is important to acknowledge that wildfires pose a significant risk to its survival in the limited areas where it exists. In any case, the Lagunas Negras de Jimbura region has suffered the synergistic effects of agriculture, cattle and sheep raising, fires, introduced species such as trout, pesticide use, and unregulated tourism activities. Areas in the Salado de Jimbura that are outside of the Yacuri National Park have already been heavily modified by human actions.

Remarks.—*Gastrotheca yacuri* is reported as species D in Duellman (2015: Fig 12.1).

Key references.—(Duellman, 2015) (phylogeny, as species D); (Carvajal-Endara *et al.*, 2019) (description, call).

GENUS *HEMIPHRACTUS* (CASQUE-HEADED FROGS) WAGLER, 1828

Hemiphractus Wagler, 1828: 736. **Type species.**—*Hemiphractus spixii* Wagler, 1828 (= *Rana scutata* Spix, 1824).

Etymology.—The generic name is derived from the Greek prefix *hemi*, meaning half, and the Greek *phraktos*, meaning protected. The name refers to the helmet covering what appears to be about the anterior half of the body. The gender is masculine.

Definition.—Figure 4.72. The skull is triangular, extensively casqued, uniquely characterized by prominent, posterolaterally projecting paraoccipital processes; the dermal roofing bones are exostosed but not co-ossified with the overlying skin; a considerable amount of variation exists in the cranial characters. The maxillary teeth are fanglike, and anteriorly, each mandible has a large, fanglike odontoid. The vertebrae have enlarged neural processes that protrude dorsally. A fleshy proboscis and conical tubercles on the eyelids are present. Vocal slits and vocal sacs are absent, but large tympana are present. The eggs adhere to the female's back and undergo direct development. *Hemiphractus* cannot be confused with any other frogs in Ecuador.

Distribution.—The distribution of *Hemiphractus* is disjointed. Four species, *H. fasciatus*, *H. elitoi*, *H. kaylockae*, and *H. panamensis*, range from western Panama to the Pacific slopes and lowlands of Colombia and northwestern Ecuador. One species, *H. johnsoni*, is restricted to the Cordillera Central of Colombia. All other species of the genus are distributed on the forested slopes of the Andes and upper Amazon Basin in southern Colombia, Ecuador, Peru, northern Bolivia, and Brazil.

Content.—The genus currently contains nine species, five of which occur in Ecuador.

Remarks.—Figure 4.73. Phylogenetic analyses of molecular data subjected to maximum likelihood analysis and parsimony analysis revealed different topologies, which are sensitive to alignment, optimality criteria, taxon sampling, and indel coding (Duellman, 2015; Castroviejo-Fisher *et al.*, 2015; Echevarria *et al.*, 2021). Our topology also differs from them. Furthermore, the molecular and phenotypic analyses by Castroviejo-Fisher *et al.* (2015) revealed divergent lineages within *Hemiphractus scutatus*, as well as new species not yet described; more extensive analyses possibly will show that these species, as presently recognized, are composites.

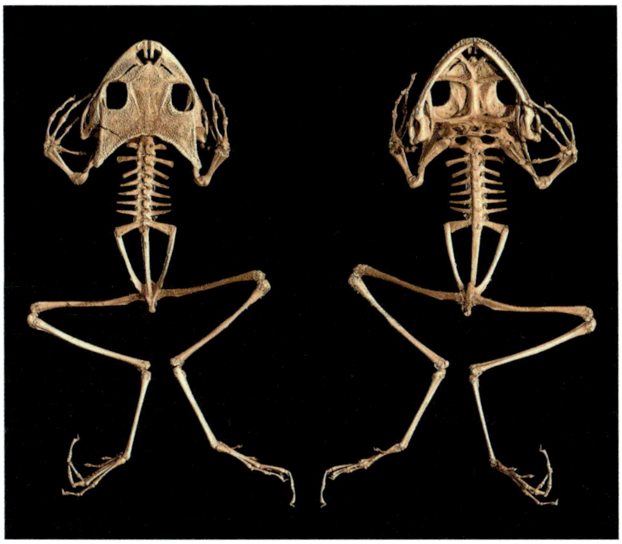

FIGURE 4.72 High-resolution computed tomography reconstructions of the skeleton of *Hemiphractus fasciatus*, CAS 122206, 59.8 mm SVL, near the lower Río Blanco, Río Esmeraldas watershed, Provincia Esmeraldas. Dorsal (left) and ventral (right) views. Daniel J. Paluh.

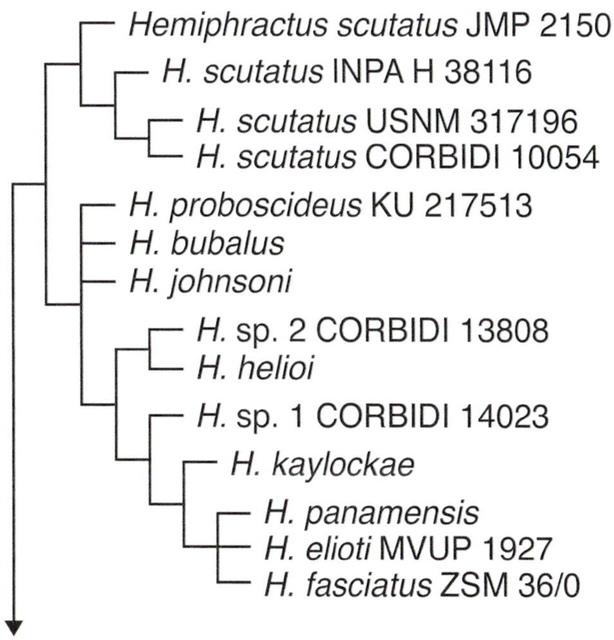

FIGURE 4.73 Topology of the species of *Hemiphractus* inferred from a tree-alignment + parsimony analysis of phenotypic characters and DNA sequences. Taken and modified from Echevarria *et al.* (2021).

HEMIPHRACTUS BUBALUS (CRESTED CASQUE-HEADED FROG) (JIMÉNEZ DE LA ESPADA, 1870)

Cerathyla bubalus (Jiménez de la Espada, 1870): 64. **Holotype.**—MNCN 1736 from Archidona, Provincia Napo, Ecuador.

Hemiphractus bubalus Noble, 1926: 19.

Etymology.—The specific name is derived from the Greek *boubalos*, a name referring to gazelles or antelopes. Presumably, the name refers to the bony pointed projections or horns, which makes them similar to such mammals.

Identification.—Figure 4.74. Considerable sexual dimorphism exists in this species. Males attain a maximum SVL of 50.5 mm, whereas females reach an SVL of 66.3 mm. The posterior projection of the paraoccipital process is relatively greater in males (18.4% of head length) than in females (4.0%); furthermore, the width between the distal tips of the processes is relatively greater in males (96.6% of head width) than in females (83.0%). The postorbital indentation is large; supraorbital, postorbital, and preoccipital ridges are present. The distinct transverse preoccipital ridge is partially continuous with other cranial ridges, which together form a pentagonal configuration on the top of the head; this cranial configuration is evident externally. Generally, the proboscis is small, triangular, and dorsoventrally depressed, except in one individual that has a large proboscis. The tubercles on the dorsal surfaces of the hind limbs are indistinct and are aligned transversely. A small, distinct calcar is present on the heel. Adhesive discs are present on the terminal digits of the fingers and toes.

The coloration of individuals varies between night and day. At night, the dorsum is yellowish-tan with dark brown markings, the venter is white with brown or gray markings, and the iris is cream. By day, the dorsum is darker brown, and the venter is brown with gray mottling.

FIGURE 4.74 *Hemiphractus bubalus*. Top left: KU 143107. Top right: KU 143108, Río Azuela, Provincia Napo. Middle left: KU 123160, note its large proboscis, which differs from KU 123161. Middle right: KU 123161, male. South slope Cordillera del Duc, above Río Coca, Provincia Napo. WED. Bottom: Volcán Reventador, Provincia Napo. LAC.

Comparisons.—*Hemiphractus fasciatus*, *H. helioi*, *H. scutatus*, *H. proboscideus*. *Hemiphractus bubalus* differs from all other species in the genus by its distinctive transverse preoccipital ridge, which, together with other cranial ridges, forms a single pentagonal configuration on the head. Besides this character, *H. bubalus* is readily differentiated from *H. fasciatus* and *H. scutatus* because it has a large postorbital indentation (absent in the former two species). *Hemiphractus bubalus* has only basal webbing between the toes, while it reaches the penultimate subarticular tubercle of toes IV and V in *H. helioi* and *H. proboscideus*. The latter species also differs from *H. bubalus* by having an elongate, laterally depressed proboscis (shorter and dorsoventrally compressed in *H. bubalus*).

Natural history.—Individuals camouflage in leaf litter or under moss in rocky areas by day; at night, they perch on the limbs of bushes and trees 0.3–2.5 m above the ground. Daytime activity became apparent through the register of a diurnal alopoglossid lizard, *Alopoglossus buckleyi*, in the stomachal contents of a female. It is not present in intervened areas.

This species reproduces by direct development, and the eggs are carried on the female's back.

Hemiphractus bubalus is like some other members of the genus in having a bright orange tongue. Trueb eloquently described the behavior associated with this structure:

> Several individuals (adult males and females) displayed what is described most appropriately as "gaping behavior"; when disturbed, these frogs typically open their mouths to their fullest extent exhibiting 1) the inordinately large size of the buccal cavity 2) a bright yellow or orange tongue, and 3) an impressive pair of mandibular odontoids.

Presumably, this behavior detracts potential predators.

Distribution.—Figure 4.75. *Hemiphractus bubalus* has been reported at elevations from 300 to 2000 m in the foothills and Amazonian slopes of the Andes from extreme southern Colombia to Peru. In Ecuador, it occurs in Sucumbíos, Orellana, Napo, and Pastaza provinces. Its elevational range is 562 – 2543 m in an area of extent of occurrence of about 5210 km^2.

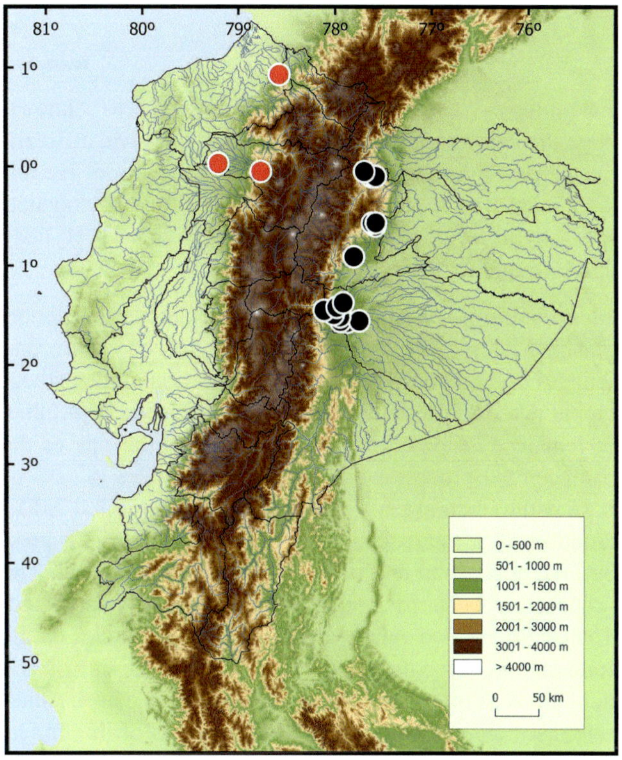

FIGURE 4.75 Localities of known occurrence of *Hemiphractus bubalus* (black dots) and H. *fasciatus* (red dots) in Ecuador.

This nocturnal, semi-arboreal species inhabits lower montane and cloud forests in the evergreen high montane forests of the northern Cordillera Oriental of the Andes, the evergreen montane forests of the northern Cordillera Oriental of the Andes, and the evergreen montane foothill forests of the northern Cordillera Oriental of the Andes in the northern Cordillera Oriental of the Andes and southern Cordillera Oriental of the Andes biogeographic sectors, where the average annual rainfall is 1267 – 4431 mm and the average annual temperature 14–23.1 °C.

Conservation.—*Hemiphractus bubalus* is categorized as Vulnerable throughout its entire range (IUCN SSC Amphibian Specialist Group, 2022b; in IUCN Red List of 2022). However, it is important to note that in Ecuador, *H. bubalus* has been categorized as Near Threatened by Ortega-Andrade *et al.* (2021). The main threat for the species is forest loss because of infrastructure development for human settlements, logging, agricultural development (including for the cultivation of crops and the rearing of livestock), and mining. Although much forest habitat remains within the range of this species, it is rarely registered, and it seems to be sensitive to slight habitat alteration. The rate of forest loss within the majority of its range in Ecuador is high. Populations at Volcán Reventador face volcanic eruptions. In Ecuador, it has been recorded in the public reserves Parque Nacional Cayambe Coca, Parque Nacional Sumaco Napo-Galeras, and Comuna Ancestral Kichwa de Canelos and the biosphere reserve Sumaco.

Remarks.—According to Castroviejo Fisher (pers. comm.), populations from Morona Santiago and Zamora Chinchipe provinces previously identified as *Hemiphractus bubalus* actually pertain to a yet-unnamed species. The lowland records for this species might not be as low as reported from the old specimens in collections, so those records should be approached with caution.

Key references.—(Jiménez de la Espada, 1870) (description); (Boulenger, 1903) (natural history); (Trueb, 1974) (external morphology, osteology, natural history); (Duellman and Lynch, 1988) (natural history).

HEMIPHRACTUS FASCIATUS (BANDED CASQUE-HEADED FROG) PETERS, 1862b

Hemiphractus fasciatus Peters, 1862b: 149. **Holotype.**—ZSM 36/0 from "Pastassa-Thal, an der Ostseite der Anden in Ecuador," in error according to Trueb, 1974: 28.

Etymology.—The specific name is Latin meaning band or belt. Presumably, the name refers to the transverse bars on the hind limbs.

Identification.—See Remarks.

Coloration in life is unknown.

Comparisons.—*Hemiphractus helioi*, *H. bubalus*, *H. proboscideus*, *H. scutatus*. *Hemiphractus fasciatus* differs from *H. helioi*, *H. bubalus*, and *H. proboscideus* by lacking a postorbital indentation and a supraorbital ridge. Also, it lacks the pentagonal configuration of ridges on the top of the head characteristic of *H. bubalus*. It is readily differentiated from *H. scutatus* because the digits of the latter are not expanded as discs (noticeably expanded as discs in *H. fasciatus*). The toes of *H. fasciatus* are like those in *H. bubalus* and *H. helioi* in having only basal webbing, whereas the toes are one-fourth webbed in *H. proboscideus* and about one-half webbed in *H. scutatus*. Furthermore, it differs from *H. proboscideus* by having a dorsoventrally compressed proboscis (laterally compressed and elongate in *H. proboscideus*). But see Remarks.

Natural history.—Unknown.

Distribution.—See Figure 4.75. This species occurs on the Pacific slopes of the Andes and adjacent lowlands of Colombia and northwestern Ecuador. In Ecuador, it occurs in Esmeraldas and Pichincha provinces. Its elevational range is 367 – 1283 m in an area of extent of occurrence of about 2750 km^2.

Hemiphractus fasciatus inhabits lowland tropical rainforests, lower humid montane forests, and cloud forests in evergreen montane foothill forests in the Cordillera Occidental of the Andes biogeographic sector, where the average annual rainfall is 2459 – 2574 mm and the average annual temperature is 19.5–23.6 °C.

Conservation.—*Hemiphractus fasciatus* is categorized as Vulnerable throughout its entire range (IUCN SSC Amphibian Specialist Group, 2022c; in IUCN Red List of 2022) because, although the species is still reasonably widely distributed, it is intolerant of any modification to its habitat, and much of its habitat is poorly protected. The population is suspected to be declining at a rate of at least 30% over a 10-year period (2015–2025) due to the high rate of habitat loss throughout much of this species' range in Colombia and Ecuador as a result of logging, agricultural expansion, and mining. However, it is important to note that in Ecuador, *H. fasciatus* has been categorized as Critically Endangered by Ortega-Andrade *et al.* (2021). In Ecuador, only four individuals have been collected despite abundant collecting efforts at the region where it was known (see also Remarks). We suggest that it should additionally be tagged as Possibly Extinct given that in Ecuador, there have been no sightings for more than seven decades; the most recent record of this species is from February 1949 in the upper area of the Río Cachabí watershed, Provincia Esmeraldas. If there is still a surviving population, the major threats to its existence include habitat reduction and fragmentation, logging, deforestation for agricultural development (particularly palm plantations), fires, mining, the introduction of exotic species, human settlements, and pollution. Furthermore, the area where it was distributed faces constant pressure from wood companies. In Ecuador, it has not been recorded in public or private reserves.

Remarks.—Descriptions of *Hemiphractus fasciatus* (e.g., Myers, 1966; Duellman, 1970) have been based on specimens from Panama, which now correspond to *H. elioti*, *H. panamensis*, and *H. kalyockae*. The proper assignment of specimens to the nominal *H. fasciatus* has been questioned by Hill *et al.* (2018), who became convinced that the taxon *H. fasciatus* does not occur in Panama. They point out taxonomic problems with this taxon, which are in need of revision. The type locality, "Pastaza Valley, on the east side of the Andes in Ecuador" (Peters, 1862b), is problematic and was considered an error because none of the specimens from east of the Andes examined by Trueb (1974) matched the morphology of the holotype.

Key references.—(Peters, 1862b) (description); (Crawford *et al.*, 2013) (cryptic species); (Hill *et al.*, 2018) (morphology, osteology).

HEMIPHRACTUS HELIOI (HELIO'S CASQUE-HEADED FROG) SHEIL AND MENDELSON, 2001

Hemiphractus helioi Sheil and Mendelson, 2001: 190.
Holotype.—KU 163304 from Río Kosñipata, 4 km southwest of Santa Isabel, 1700 m, Región Cusco, Peru.

Etymology.—The specific name is a patronym to acknowledge the contributions of Hélio Ricardo da Silva, a Brazilian herpetologist, friend of Sheil and Mendelson. He has significantly added to the understanding of the phylogenetics of hylid frogs, particularly within the families Hylidae and Hemiphractidae.

Identification.—Figure 4.76. There is slight sexual dimorphism. Males attain a maximum SVL of 42.7 mm, and females of 59.3 mm. The posterior projection of the paraoccipital process is less in males (19.5% of head length) than in females (27.3%). The distance between the tips of the paraoccipital processes is slightly less in males (82.1% of head width) than in females (90.1%). The postorbital indentation is narrow and distinct. The supraorbital ridge is present, but postorbital and preoccipital ridges are absent. The proboscis is small, triangular, and dorsoventrally depressed. The eyelid has two enlarged tubercles. The tubercles on the forearms are irregularly scattered; they are absent on the hind limbs. A prominent calcar is present on the heel. Adhesive discs are present on the terminal digits of the fingers and toes.

At night, the dorsum of the holotype was described as pale tan with barely discernible darker markings consisting of a narrow vertebral line and narrow transverse bars on the thighs and shanks; the venter was dull cream. By day, the dorsum was pinkish-tan with brown markings; the venter was grayish-white with grayish-brown flecks that were darkest on the limbs. The iris was pinkish-tan with small, irregular black markings.

Comparisons.—*Hemiphractus fasciatus*, *H. scutatus*, *H. proboscideus*, *H. bubalus*. *Hemiphractus helioi* differs from *H. fasciatus* and *H. scutatus* by having a postorbital indentation and a prominent supraorbital ridge (both absent in *H. fasciatus* and *H. scutatus*). The toes in the latter are about

FIGURE 4.76 *Hemiphractus helioi*. KU 163304, 52.7 mm SVL, Río Kosñipata, 4 km southwest of Santa Isabela, Cusco, Peru. WED.

one-half webbed, as opposed to only basal webbing in *H. helioi*. The skin on the dorsal surfaces of the hind limbs is smooth in *H. helioi* and granular in *H. proboscideus* and *H. bubalus*; the latter also differs from *H. helioi* by having a pentagonal configuration of ridges on the top of the head.

Natural history.—Individuals have been found in primary forests at night, perched on stems and branches of bushes at heights of 0.3–0.7 m above the ground. Two gravid females contained 17 and 18 ovarian eggs.

Distribution.—Figure 4.77. *Hemiphractus helioi* occurs in the upper Amazon Basin and lower Amazonian slopes of the Andes to 1700 m in Ecuador, Peru, and Estado Acre in western Brazil. In Ecuador, it is known from two nearby localities (Puyo and Canelos) in Provincia Pastaza at an elevation of 464 – 929 m.

This nocturnal species inhabits primary tropical rainforests in currently intervened areas in the northern Cordillera Oriental of the Andes biogeographic sector, where the average annual rainfall is 3905 – 4137 mm and the average annual temperature is 21.1–23.6 °C.

Conservation.—*Hemiphractus helioi* is categorized as Least Concern throughout its entire range (Icochea *et al.*, 2004; in IUCN Red List of 2022) due to its wide distribution and presumed large population. However, it is important to note that in Ecuador, *H. helioi* has been categorized as Critically Endangered by Ortega-Andrade *et al.* (2021). We suggest that it should additionally be tagged as Possibly Extinct given that in Ecuador, there have been no sightings for a century and a half; the most recent record of this species is from around 1880 in Canelos, Provincia Pastaza (but see Remarks). In Ecuador, it is only known from a couple of nearby historical localities, despite numerous collecting efforts near Canelos and Puyo, Provincia Pastaza. In Ecuador, it has not been recorded in any public or private reserves.

Remarks.—The presence of this species in Ecuador should be taken with caution, and specimens should be studied again. Trueb (1974) examined the specimen from Canelos but identified it as *Hemiphractus johnsoni*, and Sheil and Mendelson III (2001) assigned the specimens to *H. helioi* without studying them. The two specimens from Puyo have not been included in any study related to the systematics of *Hemiphractus*.

Key references.—(Trueb, 1974) (external morphology, osteology); (Sheil and Mendelson, 2001) (description).

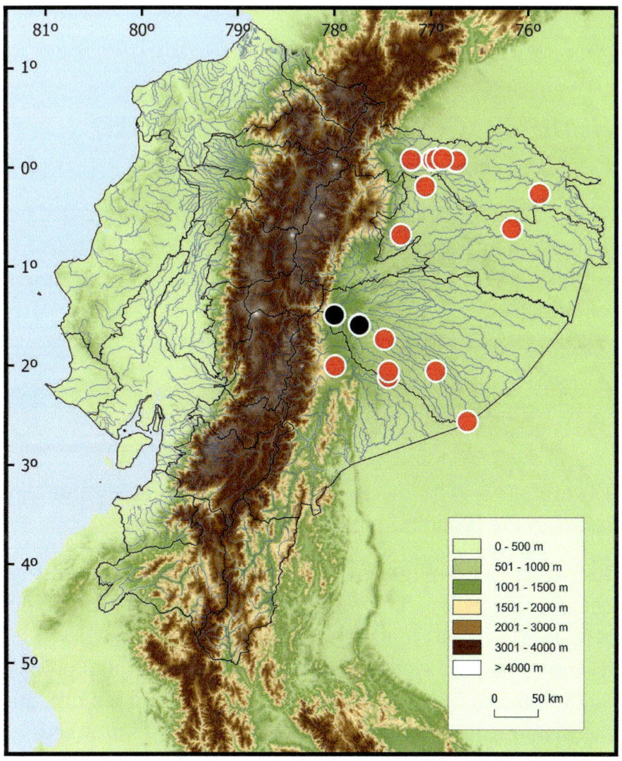

FIGURE 4.77 Localities of known occurrence of *Hemiphractus helioi* (black dot) and *H. proboscideus* (red dots) in Ecuador.

HEMIPHRACTUS PROBOSCIDEUS (LONG-SNOUTED CASQUE-HEADED FROG) (JIMÉNEZ DE A ESPADA, 1870)

Cerathyla proboscidea (Jiménez de la Espada, 1870): 64.
Holotype.—MNCN 1737 from Sumaco, Provincia Napo, Ecuador.
Hemiphractus proboscideus Noble, 1926: 19.

Etymology.—The specific epithet is the Latin *proboscis*, meaning snout. The name refers to the elongate, fleshy protuberance of this frog.

Identification.—Figure 4.78. Males attain a maximum SVL of 49.7 mm, and females are much larger, but there are no data available on their SVL. The posterior projection of the paraoccipital process is greater in males (24.5% of head length) than in females (19.8%). The distance between the tips of the paraoccipital processes is slightly proportionately

FIGURE 4.78 *Hemiphractus proboscideus*. KU 111727, 45.0 mm SVL, Santa Cecilia, Provincia Sucumbíos. WED.

greater in males (101.5% of head width) than in females (97.7%). The postorbital indentation is large, rounded, and about twice as long as wide. The supraorbital ridge is present; postorbital and preoccipital ridges are absent. The proboscis is large and laterally compressed; it has an irregular dorsal ridge. The eyelid has one large tubercle and many small tubercles. The forearm and shank have tubercles arranged in oblique rows. A small, triangular calcar is present on the heel. Adhesive discs are present on the terminal digits of the fingers and toes.

At night, the dorsum is yellowish-tan with brown transverse or diagonal marks on the body and transverse bars on the limbs; the venter is grayish-cream to white. By day, the dorsum is brown with darker brown markings, and the venter is brown to reddish-brown with orange or yellow blotches on the belly and yellow or white spots on the throat. The discs on fingers I and II are pale orange. The tongue is orange. The iris is pale gray, with a pinkish tint in some individuals, with diffuse darker gray triangular marks anteriorly and posteriorly.

Comparisons.—*Hemiphractus bubalus, H. scutatus. Hemiphractus proboscideus* differs from all other members of the genus by having an elongated, laterally compressed proboscis (shorter and dorsoventrally compressed in other species). Furthermore, it lacks the pentagonal arrangement of ridges on the top of the head present in *H. bubalus*. The toes are about one-fourth webbed in *H. proboscideus*, whereas webbing is basal in the other species, except *H. scutatus*, in which they are about one-half webbed

Natural history.—These frogs inhabit undisturbed tropical rainforest, where they live in leaf litter on the forest floor by day and, at night, ascend vegetation to heights of 1.5 m above the ground, where they most commonly perch on leaves and palm fronds. The presence of diurnal prey in their stomachs shows that *Hemiphractus proboscideus* feeds during daytime and nighttime. According to Duellman, "*Hemiphractus proboscideus* is a carnivore with a prodigious appetite." Contents of the stomachs of 10 individuals collected in Santa Cecilia, Provincia Sucumbíos, included 15 frogs of 12 species, ranging from dendrobatids to hylids, some nearly as large as the individual that had eaten them. Moderately large invertebrates also were eaten; these included an annelid worm, a leech, a homopteran, and a gryllotalpid cricket.

Reproduction aspects of this species are poorly known. One female collected in July contained 26 ovarian eggs. In Santa Cecilia, Provincia Sucumbíos, four juveniles having SVLs of 18–19 mm (possibly recently hatched) were found in an area of about 20 m^2 in March.

Duellman noted that some individuals, when disturbed, opened their cavernous mouths, exposing bright yellow tongues and emphasized, "Subsequent prodding by a careless observer can result in a quick snap of the jaws in a vice-like grip." He also mentions that apparently, this species does not have a mating call, but one male emitted a "grunt" when picked up.

Distribution.—See Figure 4.77. *Hemiphractus proboscideus* is known in the upper Amazon Basin in extreme southwestern Colombia and in the Amazon Basin and lower slopes of the adjacent Andes in Ecuador and Peru. In Ecuador, it occurs in Sucumbíos, Orellana, Pastaza, and Morona Santiago provinces. Its elevational range is 223 – 986 m in an area of extent of occurrence of about 41,480 km^2.

This species inhabits primary tropical rainforests in seven ecosystems (Table 0.4) within four biogeographic sectors (Table 0.5), where the average annual rainfall is 2292 – 4002 mm and the average annual temperature is 21.3–25.7 °C.

Conservation.—*Hemiphractus proboscideus* is categorized as Least Concern throughout its entire range (IUCN SSC Amphibian Specialist Group, 2018c; in IUCN Red List of 2022) in view of its wide distribution, presumed large population, and because it is unlikely to be declining fast enough to qualify for listing in a more threatened category. However, it is important to note that in Ecuador, *H. proboscideus* has been categorized as Near Threatened by Ortega-Andrade *et al.* (2021). Although this species is rarely seen, it is a widespread species with large areas of suitable habitat remaining. The major threats to its existence are habitat reduction and fragmentation, deforestation, and oil spills. In Ecuador, it has been recorded in the public reserve Reserva de Producción Faunística Cuyabeno and the biosphere reserve Sumaco.

Key references.—(Jiménez de la Espada, 1870) (description); (Duellman, 1978) (account, sound); (Rodríguez and Duellman, 1994) (account), (Gagliardi-Urrutia *et al.*, 2002) (account).

HEMIPHRACTUS SCUTATUS (HELMETED CASQUE-HEADED FROG) (SPIX, 1824)

Rana scutata (Spix, 1824): 28. **Types.**—Not designated but presumably specimen ZSM 37/0, now lost, illustrated in Spix, 1824: pl. 4, fig, 2, from Rio Solimões, Amazonas, Brazil.

Hemiphractus scutatus Wagler, 1830: 205.

Etymology.—The specific name is Latin meaning shield; it refers to the broad bony head of this frog.

Identification.—Figure 4.79. There is a considerable difference in size between the sexes; the maximum SVL in males is 62 mm, whereas females attain a maximum known SVL of 81 mm. The posterior projection of the paraoccipital process is the same in males and females (18% of head length), and the distance between the tips of the paraoccipital processes is the same in both sexes (93% of head width). There is no postorbital indentation. Supraorbital, postorbital, and preoccipital ridges are absent. The proboscis is small, triangular,

FIGURE 4.79 *Hemiphractus scutatus*. Top and second row: QCAZ 38917, female with 14 froglets on its back, ± 80 mm SVL. Third and fourth row: juvenile, ± 40 mm SVL. Bottom: recently hatched froglets from the back-riding brood of QCAZ 38917, ± 19 mm SVL. Nearby Villano, Comunidad Kurintza, Provincia Pastaza. LAC.

FIGURE 4.80 *Hemiphractus scutatus*. Top: CJ 6414 showing gaping behavior, 73.9 mm SVL. Bottom: CJ 6415, juvenile, 25.6 mm SVL. Shiripuno Lodge, Provincia Orellana. LAC.

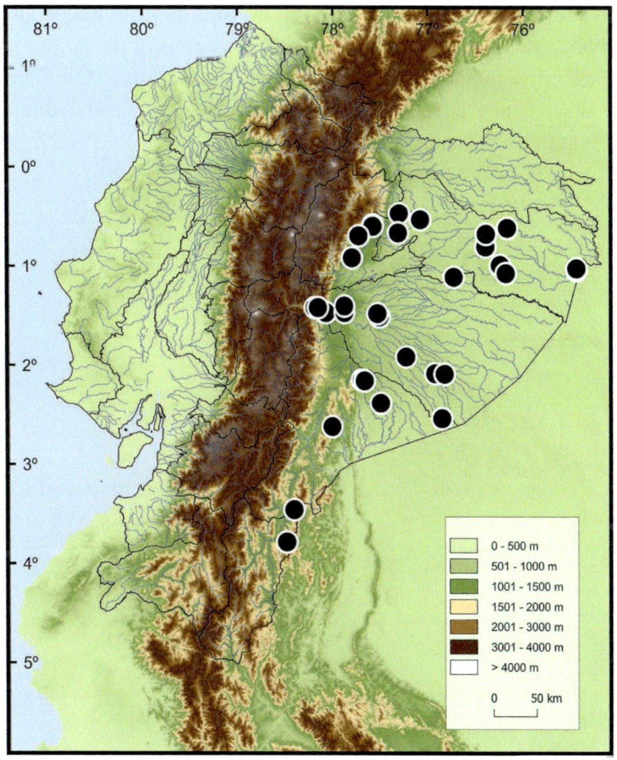

FIGURE 4.81 Localities of known occurrence of *Hemiphractus scutatus* in Ecuador.

and dorsoventrally depressed. The eyelid has one slightly enlarged conical tubercle. The tubercles on the forearm and shank are arranged in oblique rows corresponding to the dark bars. There is no calcar on the heel. Adhesive discs are absent on the terminal digits of the fingers and toes.

The dorsum is dull yellowish-tan to brown; markings consist of irregular, dark brown diagonal marks on the back, faint brown bars on the limbs, and usually a prominent black diagonal mark below the eye. Some individuals have a dark brown band posterior to the eye. Narrow brown bars on the upper lips become more prominent on the lower lips and fuse into the brown throat. The venter is dull cream. The tongue is orange. The iris is reddish-bronze.

Comparisons.—*Hemiphractus scutatus* differs from all other members of the genus by having narrow terminal digits that lack adhesive discs. Furthermore, the toes are one-half webbed, whereas the toes are one-fourth webbed in *H. proboscideus* and only basally webbed in the other species. Within the genus *Hemiphractus*, *H. scutatus* is unique in having prominent dark bars on the lips.

Natural history.—Figure 4.80. Individuals have a yellow tongue as part of a gaping behavior, a shared characteristic with other members of the genus. A juvenile was found being eaten by a spider.

Females have been found with 5–17 eggs, 7–10 mm in diameter, on their backs. One female, found in July, carried 14 recently hatched froglets on its back.

Distribution.—Figure 4.81. *Hemiphractus scutatus* occurs in the Amazon Basin in Brazil (as far east as Para), southeastern Colombia, and northern Bolivia, as well as in Ecuador and Peru, where it ascends the Amazonian slopes of the Andes to elevations of nearly 2000 m. In Ecuador, it occurs in Orellana, Napo, Tungurahua, Pastaza, Morona Santiago, and Zamora Chinchipe provinces. Its elevational range is 195 – 1963 m in an area of extent of occurrence of about 61,310 km^2.

This nocturnal, primarily terrestrial, leaf litter species inhabits primary and secondary lowland rainforests, lower montane forests, and cloud forests in 11 ecosystems (Table 0.4) within six biogeographic sectors (Table 0.5), where the average annual rainfall is 1427 – 4437 mm and the average annual temperature is 17.1–25.5 °C.

Conservation.—*Hemiphractus scutatus* is categorized as Least Concern throughout its entire range (Coloma *et al.*, 2004a; in IUCN Red List of 2022) and in Ecuador by Ortega-Andrade *et al.* (2021). It has a wide distribution and a presumed large population, and it is unlikely that it will be declining fast enough to qualify for listing in a more threatened category. It has no major threats and is a widespread species with large areas of suitable habitat remaining. There is some localized habitat loss because of different human activities, such as agriculture (crops, livestock, etc.) and selective logging. In Ecuador, it has been recorded in the public reserves Parque Nacional Sumaco Napo-Galeras, Parque Nacional Yasuní, and Reserva Biológica El Quimi; the protected forests of Cerro Sumaco y Cuenca Alta del Río Suno, Cordillera Kutukú y Shaimi, and Nacionalidad Andwa de Pastaza del Ecuador; and the biosphere reserves Sumaco and Yasuní.

Remarks.—There are two deeply divergent lineages (Castroviejo-Fisher *et al.*, 2015; Moraes and Pavan, 2018; Echevarria *et al.*, 2021). From a nomenclatural perspective, the situation is challenging. The holotype of *H. scutatus* is lost, and its type locality is described as the Solimões River, which corresponds to a huge area of the Amazon Basin between the borders of the neighboring towns of Leticia and Tabatinga in Colombia and Brazil, respectively, and Manaus.

Key references.—(Spix, 1824) (illustration); (Rodríguez and Duellman, 1994) (account); (Villacampa *et al.*, 2017) (account).

5 Family Eleutherodactylidae

FREE-TOED FROGS

Eleutherodactylinae (Part) Lutz, 1954: 157. Type Genus.—*Eleutherodactylus* Duméril and Bibron, 1841: 520.
Eleutherodactylidae Hedges *et al*., 2008: 47.

Etymology.—The familial and generic names are derived from the Greek *eleutheros*, meaning free, and the Greek *daktylos*, meaning finger.

Definition.—Frogs of the family Eleutherodactylidae have a cartilaginous sternum and lack a vertebral shield. They have eight presacral vertebrae; presacrals I and II are not fused. The sacral diapophyses are rounded or barely dilated. An omosternum is present. Cranial elements are not co-ossified with the overlying skin. The maxillary arch is complete. The nasals are large with a broad median contact and are not in contact with the frontoparietals; a frontoparietal fontanelle usually is absent. The pterygoid lacks a ventral flange, and its anterior ramus does not reach the neopalatine; a columella is present. The terminal phalanges are T-shaped. There are three phalanges on finger IV, and toe I is fully developed. The mandibular ramus of the trigeminal nerve passes lateral to the m. abductor mandibulae.

Distribution.—The family is distributed throughout the West Indies and occurs in peninsular Florida and southern Texas, USA; throughout much of Mexico and Central America; and southward to northwestern Ecuador. Some genera are discontinuously distributed in northeastern South America and the Amazon Basin.

Content.—As presently recognized, the family contains four genera and 217 species. Two genera each have one species in Ecuador.

Remarks.—Figure 5.1. The recognition of this family has varied with different phylogenetic analyses of molecular data (Hedges *et al*., 2008; Pyron and Wiens, 2011; Padial *et al*., 2014). Generally, two subfamilies are recognized. The subfamily Eleutherodactylinae consists of *Diasporus* and *Eleutherodacylus*, whereas the subfamily Phyzelaphryninae contains *Adelophryne* and *Phyzelaphryne*. In an analysis of molecular data, Fouquet *et al*. (2012) recognized four lineages of phyzelaphrynines, each containing multiple species in different regions in South America.

GENUS *ADELOPHRYNE* (HAMMER FROGS) HOOGMOED AND LESCURE, 1984

Adelophryne Hoogmoed and Lescure, 1984: 92. **Type species.**—*Adelophryne adiastola* Hoogmoed and Lescure, 1984.

Etymology.—The generic name is a combination of the Greek *adelos*, meaning unseen or unknown, and the Greek *phryne*, meaning toad. The name is in reference to the fact that these small toads had not been collected until the mid1900s. The gender is feminine.

Definition.—In these frogs, the maximum SVL in females is 17 mm. The head is no wider than the body, which is moderately robust and has relatively short limbs. The tympanic membrane is differentiated. Cranial crests are absent. The digits are flattened; the elongated discs are asymmetrically pointed, elongate, and have grooves laterally. Finger I is shorter than II; finger IV has two or three phalanges. Nuptial excrescences are absent, but males have a large subgular vocal sac. Toe III is longer than V; the toes are unwebbed. The terminal phalanges of the fingers and toes knobbed or T-shaped. The ilium lacks a dorsal crest. The frontoparietals meet medially, without exposing a fontanelle. Vomerine and maxillary teeth are present. The occipital condyles are widely separated. The last presacral vertebra is about as wide as the sacrum. The sacral diapophyses are rounded, and slightly wider distally. The omosternum lacks a bony element.

Distribution.—The members of this genus have disjunct distributions in the northern Amazon Basin and coastal forests in Brazil.

Content.—The genus contains nine species.

Remarks.—Fouquet *et al*. (2012) reported several undescribed species of *Adelophryne*.

ADELOPHRYNE ADIASTOLA (CONFUSED HAMMER FROG) HOOGMOED AND LESCURE, 1984

Adelophryne adiastola Hoogmoed and Lescure, 1984: 95. **Holotype.**—UTA A-4943 from Yapima, Río Vaupés, Departamento Vaupés, Colombia.

Etymology.—The specific name is derived from the Greek *adiastolos*, meaning not separated, confused, in reference to the fact that the type specimens were originally named as *Phyzelaphryne miriamae*.

Identification.—Figure 5.2. Adults of this minute species attain a maximum SVL of 13.9 mm in females and of 13.7 in males. The skin of dorsum and flanks is areolate or shagreened. The belly and limbs are smooth. The throat is

DOI: 10.1201/9781003543879-5

granular and wrinkled. There is a glandular series of warts from the eardrum to the insertion of the arm. The discoidal fold is absent. The head is longer than wide. There is an indistinct, rounded, and straight canthus rostralis. The loreal region is flat, inclined toward the upper lips. The snout is rounded in dorsal view and rounded to truncated in side view. The round eardrum is small but distinct. There is an incomplete tympanic ring and indistinct supratympanic fold. The vomerine processes are present with two rows of 6–8 teeth. The males have curved vocal grooves and a subgular vocal sac. The finger IV is reduced. The fingers are pointed, without an interdigital membrane. There are no skin flanges or discs. There are indistinct, long, and oval subarticular tubercles and a rounded supernumerary tubercle on the palm of the hand. The toes are asymmetric and pointed. The tarsal fold is absent. The internal metatarsal tubercle is long, flat, and oval. There is a rounded and protuberant external metatarsal tubercle and flat and oval subarticular tubercles.

The dorsum is evenly dark brown. The belly is black with cream spots. The iris is red.

Comparisons.—*Diasporus gularis*. Its rounded snout, small tympanum, and asymmetric and pointed fingers distinguish this species from other species from Ecuador, except *D. gularis*, which also has pointed fingers. However, it differs because the latter has a uniformly orange coloration and cream or yellow belly.

Natural history.—Individuals at Yapima, Colombia, were found during the rainy season at night associated with the leaf litter near streams or hills. In Ecuador, an individual was active during the day, moving among the leaf litter in a primary hill forest. An opilionid was found in the stomach of a female.

A male was found calling in the rainy season at Yapima, Colombia. Females can have white ovarian eggs. A call from Yapima was described (as *Phyzelaphryne miriamae*) as a simple call composed of a series of short, single-pulsed notes (three and 19 notes per call), emitted at regular intervals of about 0.016 s and a mean internote interval of 2.783 s. Detailed data and figures are provided under the subtitle "Vocalizations of Ecuadorian Anurans" in Appendix 1.

FIGURE 5.1 Topology of a phylogenetic reconstruction of a Bayesian analysis of three mitochondrial loci and three nuclear loci of Phyzelapryninae lineages by Fouquet *et al.* (2012). Regional abbreviations are SAMC = southern Amazonian clade; NAMC = northern Amazonian clade; NAFC = northern Atlantic forest clade; SAFC = southern Atlantic forest clade.

FIGURE 5.2 *Adelophryne adiastola*. UTA A-4943, female, 62 mm SVL, Yapima, Río Vaupés, Departamento Vaupés, Colombia. Taken from Hoogmoed and Lescure (1984).

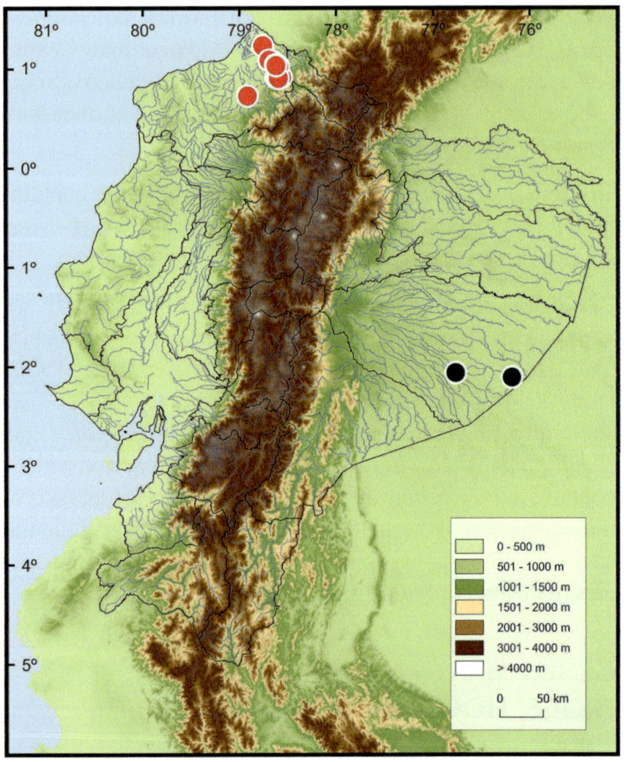

FIGURE 5.3 Localities of known occurrence of *Adelophryne adiastola* (black dots) and *Diasporus gularis* (red dots) in Ecuador.

Distribution.—Figure 5.3. *Adelophryne adiastola* is known in the Amazon Basin of Colombia (Amazonas and Vaupés), eastern Ecuador, northeastern Peru, and adjacent Brazil. In Ecuador, it occurs in Provincia Pastaza. Its elevational range is 190–339 m in a distance of 65 km between Kurintza and Juyuintza.

This leaf litter species inhabits primary rainforest forest in the evergreen forest of lowlands of Tigre–Pastaza and flooded forests of alluvial plains of Amazonia, where the average annual rainfall is 3227–3463 mm and the average annual temperature is 24.9–25.7 °C.

Conservation.—*Adelophryne adiastola* is categorized as Least Concern (Angulo *et al.*, 2004c; in IUCN Red List of 2022) in view of its relatively wide distribution, tolerance of a degree of habitat modification, presumed large population, and because it is unlikely to be declining fast enough to qualify for listing in a more threatened category. It is common in Colombia and uncommon in Peru and Ecuador. The conservation status of Ecuadorian populations is unknown. There are no major threats to the species. There is some localized habitat loss to different human activities such as agriculture (crops, livestock, etc.), but its range is in an area of low human impact. In Ecuador, it has not been recorded in public reserves, but it occurs in the private reserve Bosque El Chalpi-Saguangal.

Remarks.—Previously confused with *Phyzelaphryne miriamae*. Based on molecular data, it was assigned to the Eleutherodactylidae family, being closely related to species of the genus *Adelophryne* and *Phyzelaphryne*. Ortega-Andrade (2009) recorded it from Ecuador and compared it to *Noblella myrmecoides*.

Key references.—(Hoogmoed and Lescure, 1984) (description); (Lynch, 1986) (morphology); (Ortega-Andrade, 2009) (distribution, photos).

GENUS *DIASPORUS* (LITTLE HAMMER FROGS) HEDGES *ET AL.*, 2008

Diasporus Hedges *et al.*, 2008: 49. **Type species.**—*Lithodytes diastema* Cope, 1876: 155.

Etymology.—The generic name is from the Greek *diaspora*, meaning a dispersion from. It applies to the close relationships of this mainland genus to the West Indian *Eleutherodactylus*. The gender is masculine.

Definition.—Figure 5.4. The cranial elements are not co-ossified with overlying skin. The maxillary arch is complete. The nasals are large with a broad median contact and are not in contact with the frontoparietals; a frontoparietal fontanelle is absent. Morphologically members of this genus are quite similar. These are small frogs; the SVL varies from 11 mm in males of *Diasporus quidditus* to 26 mm in *D. hylaeformis*. The head is distinct from the body, which is moderately robust and has relatively short limbs. The snout is rounded to subacuminate in dorsal view. Cranial crests are absent. The skin on the dorsum is smooth to rugose; that on the belly is coarsely areolate. The tympanic membrane is usually differentiated; vocal sacs are present, but nuptial pads are absent. Finger I is shorter than II; toe V is much longer than III. The terminal digital discs are expanded, with or without lanceolate or papillate tips; circumferential grooves are present. The toes are unwebbed.

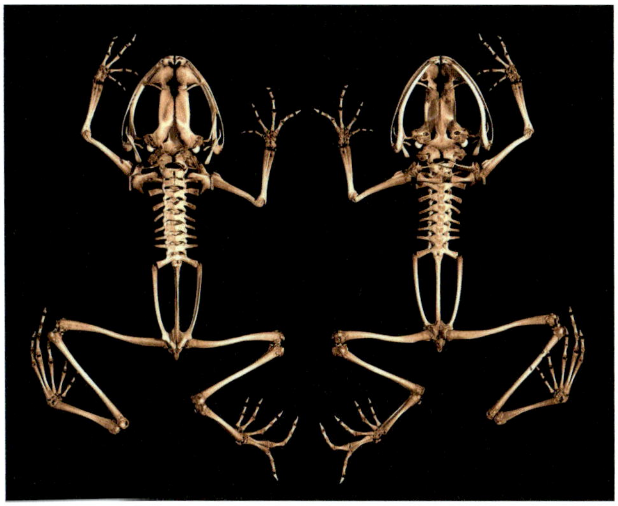

FIGURE 5.4 High-resolution computed tomography reconstructions of the skeleton of *Diasporus gularis*, CJ 2082, 21.7 mm SVL, Playón de San Francisco-Durango, Provincia Esmeraldas. Dorsal (left) and ventral (right) views. Daniel J. Paluh.

Distribution.—Species of *Diasporus* occur in forests from eastern Honduras through Panama to the Pacific versant of Colombia and northwestern Ecuador.

Content.—The genus contains 15 species.

Remarks.—Since Hedges *et al.* (2008) recognition of eight species in *Diasporus*, seven others have been described from Colombia and Panama.

DIASPORUS GULARIS (ORANGE LITTLE HAMMER FROG) (BOULENGER, 1898)

Hylodes gularis (Boulenger, 1898): 121. **Holotype.**—BMNH 1947.2.15.82 from San Javier de Cachabí, Provincia Esmeraldas, Ecuador.
Diasporus gularis Hedges *et al.*, 2008: 47.

Etymology.—The specific name is derived from the Latin *gula*, meaning throat.

FIGURE 5.5 *Diasporus gularis*. Top: Reserva Canandé, Provincia Esmeraldas. Amanda B. Quezada. Second row: CJ 418, Reserva Otokiki, Alto Tambo. LAC. Third row and bottom: CJ 2082, Playón de San Francisco, Provincia Esmeraldas. LAC.

Identification.—Figure 5.5. The maximum SVL is 22 mm in males and 25 mm in females. The skin on the dorsum is smooth; that on the venter is areolate. A discoidal fold and dorsolateral folds are absent. The snout is subacuminate in dorsal view and truncate in profile. A tympanic membrane is absent, but the tympanic annulus is visible through the skin. The upper eyelid lacks tubercles and is narrower than the interorbital distance. Finger I is shorter than II. The discs on the fingers are slightly wider than the digits; the disc covers are weakly pointed but lack papillae. The fingers bear weak lateral fringes and lack webbing. Tubercles are absent on the heel and tarsus; an inner tarsal fold is absent. The inner metatarsal tubercle is oval and about three times larger than the outer metatarsal tubercle. Toe V is much longer than III. The toes have basal webbing and lack lateral fringes. The discs are lanceolate and have minute papillae at their tips. The dorsum is orange and yellowish-tan to pale brown with faintly darker markings. The snout is pale. Markings on the head may consist of a dark brown interorbital bar, a narrow canthal stripe, and a broader postorbital stripe; the tip of the snout is paler than the dorsum. The ventral surfaces are cream with brown flecks on the throat. The posterior surfaces of the thighs are brown. The iris is golden bronze with a reddish-brown median horizontal streak.

Comparisons.—This is the only species of frog in western Ecuador that has lanceolate digital discs with papillae on their tips.

Natural history.—Usually, it is perched on leaves and branches close to streams. Males call from the leaves in dense vegetation at heights of 0.5–1.5 m above the ground. The advertisement call of *Diasporus gularis* was described from Colombia by Toro-Sánchez and Bernal-Bautista (2015). They determined that the call is a single whistle-like note with four harmonics ranging from 2.6 to 3.0 kHz, with a dominant frequency at 2.8 kHz. Egg deposition is unknown, but certainly it is like that in other members of the genus known to deposit their eggs amid leaf litter on the forest floor; their eggs undergo direct development.

Distribution.—See Figure 5.3. This species is distributed in the lowlands and mid-elevations up to 1770 m from northwestern Colombia (Departamento Antioquia) to northwestern Ecuador. In Ecuador, it occurs in Provincia Esmeraldas. Its elevational range is 60–699 m in an area of extent of occurrence of about 990 km^2.

This nocturnal species inhabits primary humid tropical rainforests, as well as disturbed areas in the evergreen forest of lowlands of Equatorial Chocó and evergreen montane foothill forests of the Cordillera Occidental of the Andes, where the average annual rainfall is 2299–2584 mm and the average annual temperature is 22.2–25.6 °C.

Conservation.—*Diasporus gularis* is categorized as Least Concern throughout its entire range (IUCN SSC Amphibian Specialist Group, 2019a; in IUCN Red List of 2022) in view of its wide distribution, tolerance of a broad range of habitats, presumed large population, and because it is unlikely to be declining to qualify for listing in a more threatened category. In some parts of its distribution, it is an apparently common species, not significantly threatened. However, it is important to note that in Ecuador, *D. gularis* has been categorized as Endangered by Ortega-Andrade *et al.* (2021), who did not provide IUCN criteria. In Ecuador, its current extent of occurrence is less than 5000 km^2 (990 km^2), its distribution is severely fragmented, and there is a continuing decline in the extent and quality of its habitat. In Ecuador, it has been found in low to moderate densities. It occurs in the public reserves Parque Nacional Cotacachi-Cayapas and Refugio de Vida Silvestre La Chiquita, in the protected forest of Asociación Agrícola Carchi-Imbabura, and in the private reserves Reserva Otokiki, Reserva La Chiquita, and Reserva Biológica Río Canandé.

Remarks.—Its molecular phylogenetic relationships are unknown. This species was treated in detail and illustrated by Lynch and Duellman (1997), who provided comparisons with other species of *Diasporus* known at that time. *Hyloxalus huigrae* was described by Fowler (1913), supposedly from "Huigra, at an elevation of 4000 ft, Ecuador." Most likely, the type locality is in error. This name (*H. huigrae*) was considered a synonym of *Eleutherodactylus diastema* by Dunn (1942a) and subsequently a synonym of *E. gularis* by Lynch (1976).

Key references.—Ortega-Andrade *et al.*, 2010a (account); (Morales-Mite, 2013) (account); (Yánez-Munoz *et al.*, 2020) (account).

6 Family Centrolenidae

GLASSFROGS

Juan Manuel Guayasamin, Luis A. Coloma, Andrea Terán-Valdez

Centrolenidae Taylor, 1951: 36. **Type genus.**—*Centrolene* (Jiménez de la Espada, 1872).

Etymology.—The familial name is derived from the type species *Centrolene geckoideum*, described by Jiménez de la Espada (1872). See etymology under *Centrolene*.

Definition.—Centrolenid frogs are a monophyletic group diagnosed on the basis of molecular evidence and phenotypic features. Putative morphological synapomorphies of the family are: (1) translucent dorsal and/or ventral skin; (2) forward-directed eyes; (3) a dilated medial process on metacarpal III; (4) an intercalary element between the distal and penultimate phalanges; (5) complete or partial fusion of the tibiale and fibulare; (6) a submandibular interhyoideus muscle with two distinct layers, with a criss-cross orientation in adult males; and (7) eggs deposited out of water. Centrolenids vary greatly in size, ranging from the tiny *Teratohyla spinosa* (< 21 mm adult SVL) to the large *Centrolene geckoidea*, which has a maximum SVL of 80.7 mm.

Distribution.—Centrolenids are distributed in the Neotropics in tropical Central America, through the tropical Andes, and in the Sierra Nevada de Santa Marta in Colombia, the Cordillera de la Costa of Venezuela, Tobago, the Guiana Shield, the Amazon Basin, and the Atlantic forests of Brazil.

Content.—The family contains 12 genera with 158 species (Frost, 2022). In Ecuador, there are 10 genera and 63 species.

Natural history.—Centrolenids are nocturnal, epiphyllous, and arboreal. They always deposit their eggs out of the water on vegetation (leaves, mosses, or branches) overhanging streams, or on rocks near streams; some species breed in ponds (*Centrolene buckelyi*, *Cochranella granulosa*). Glassfrogs are also remarkable by the variation in parental care. In species where this behavior is present, it can be divided into the following: (1) short-term maternal care, where brooding is provided for a few hours after oviposition; (2) prolonged male care, where paternal care is provided for several weeks; and (3) prolonged female care, where maternal care is provided for several weeks, only observed in *Ikakogi tayrona*.

Remarks.—Figure 6.1. The nomenclatural and systematic history is detailed in Frost (2023 and onward). The taxonomy presented herein follows the proposal by Guayasamin *et al.* (2009). Jetz and Pyron (2018) produced a large phylogenetic tree with all amphibian sequences available at the moment. Based on that tree, where *Espadarana* (*sensu* Guayasamin *et al.*, 2009) is paraphyletic with respect to *Sachatamia* + *Rulyrana*, Dubois *et al.* (2021) proposed the genus *Audaciella*. We note that the sequences used in the study by Jetz and Pyron (2018) are the same as those generated mainly by Guayasamin *et al.* (2008b), Castroviejo-Fisher *et al.* (2014), and Twomey *et al.* (2014). The relationships of species in the genus *Espadarana* in the Jetz and Pyron (2018) topology are at odds with all the studies on glassfrog systematics (see Twomey *et al.*, 2014; Castroviejo-Fisher *et al.*, 2014; Guayasamin *et al.*, 2020b; Mendoza-Henao *et al.*, 2023). All the aforementioned studies infer a monophyletic *Espadarana*. Therefore, we place *Audaciella* Dubois *et al.* (2021), under the synonymy of *Espadarana* Guayasamin *et al.* (2009).

One of the most outstanding traits of glassfrogs is their transparency. Recent studies have shown that translucent skin provides camouflage by disguising the frog's outline (edge diffusion, Barnett *et al.*, 2020). Also, transparency requires both see-through tissues and an active mechanism that clears pigments (red blood cell hemoglobin) from the circulatory system. Glassfrogs archive such a task by removing up to 90% of the red blood cells from circulation and packing them in the liver (Taboada *et al.*, 2022). Guayasamin *et al.* (2020b) provided accounts for all Ecuadorian species, including synonymy lists, common names, etymology, identification, characterization, color in life, color in ethanol, biology and ecology, call, tadpole, distribution, conservation status, evolutionary relationships, remarks, specimens examined, and localities from the literature. The data presented herein represent a comprehensive update and expansion of Guayasamin *et al.* (2020b).

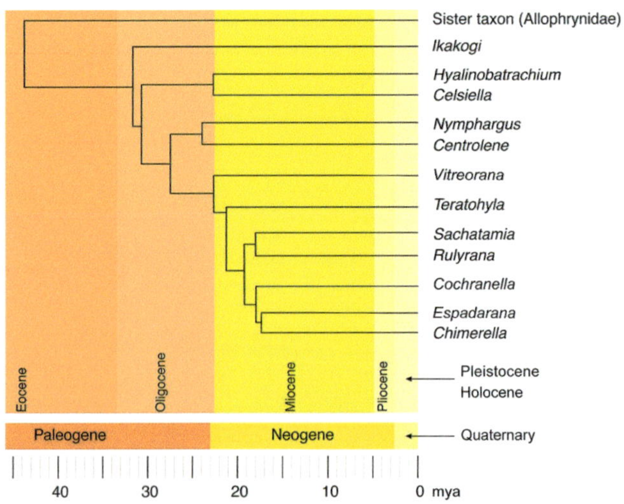

FIGURE 6.1 Time-calibrated maximum likelihood phylogenetic tree depicting evolutionary relationships of genera of Centrolenidae, inferred from DNA sequences of mitochondrial and nuclear genes. Phylogeny taken and redrawn from Guayasamin *et al.* (2020b).

DOI: 10.1201/9781003543879-6

Key references.—(Hayes and Starrett, 1980) (identification); (Duellman and Trueb, 1994) (overview); (Savage, 2002) (overview); (Frost *et al.*, 2006) (phylogeny, taxonomy); (Roelants *et al.*, 2007) (phylogeny); (Kubicki, 2007) (natural history); (Guayasamin and Trueb, 2007) (identification); (Guayasamin *et al.*, 2008b) (evolutionary relationships); (Guayasamin *et al.*, 2009) (generic taxonomy); (Pyron and Wiens, 2011) (phylogeny); (Vitt and Caldwell, 2014) (overview); (Pyron, 2014) (biogeography); (Bravo Valencia and Delia, 2016) (natural history); (Jetz and Pyron, 2018) (phylogeny); (Guayasamin *et al.*, 2020b, and references therein) (identification); (Dubois *et al.*, 2021) (phylogenetics, nomenclature).

GENUS *CENTROLENE* (GLASSFROGS) JIMÉNEZ DE LA ESPADA, 1872

Centrolene Jiménez de la Espada, 1872b: 62. **Type species.**—*Centrolene gekoideum* Jiménez de la Espada, 1872b.

Etymology.—The generic name is derived from the Greek *kentron*, meaning point or spur, and *olene*, meaning elbow, alluding to the humeral spine found in males of this genus. The genus is feminine in gender.

Definition.—Figure 6.2. Within the genus *Centrolene* size varies from a maximum SVL of 80.7 mm in *Centrolene geckoidea* to the tiny *C. peristicta* (adult SVL < 22 mm). Although there are no traits exclusive to *Centrolene*, all Ecuadorian species exhibit humeral spines in adult males and have some degree of webbing between fingers III and IV. Other glassfrog taxa found in Ecuador where males also present humeral spines are *Espadarana*, *Chimerella mariaelenae*, *Cochranella litoralis*, and *Nymphargus grandisonae*.

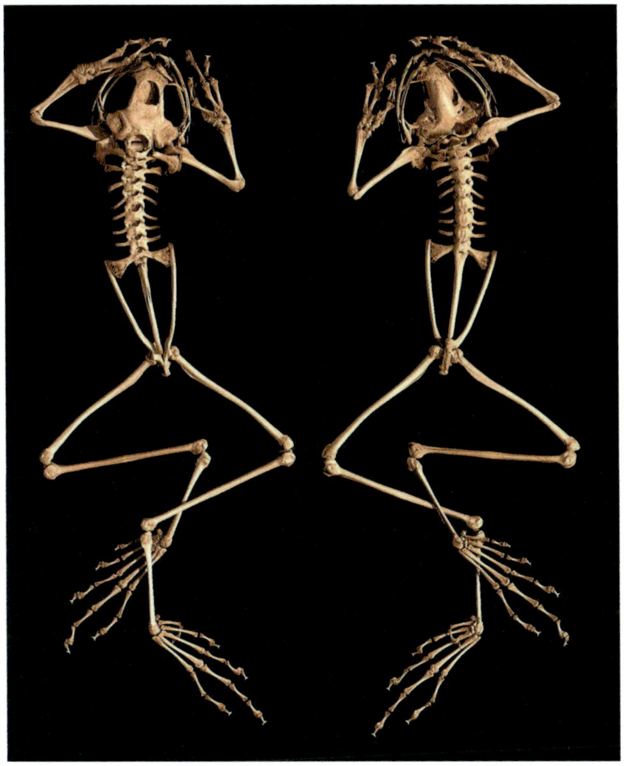

FIGURE 6.2 High-resolution computed tomography reconstructions of the skeleton of *Centrolene buckleyi*, UF 30578, 31.4 mm SVL, near Ambato, Provincia Tungurahua. Dorsal (left) and ventral (right) views. Daniel J. Paluh.

Content.—The genus contains 25 species, 12 of which occur in Ecuador.

Distribution.—Figure 6.3. *Centrolene* occurs in humid forests along the Andes from Colombia and Venezuela to Peru and on the Cordillera de la Costa of Venezuela and the Guiana region.

Remarks.—Figure 6.4. The phylogeny of species presented herein follows Guayasamin *et al.* (2020b).

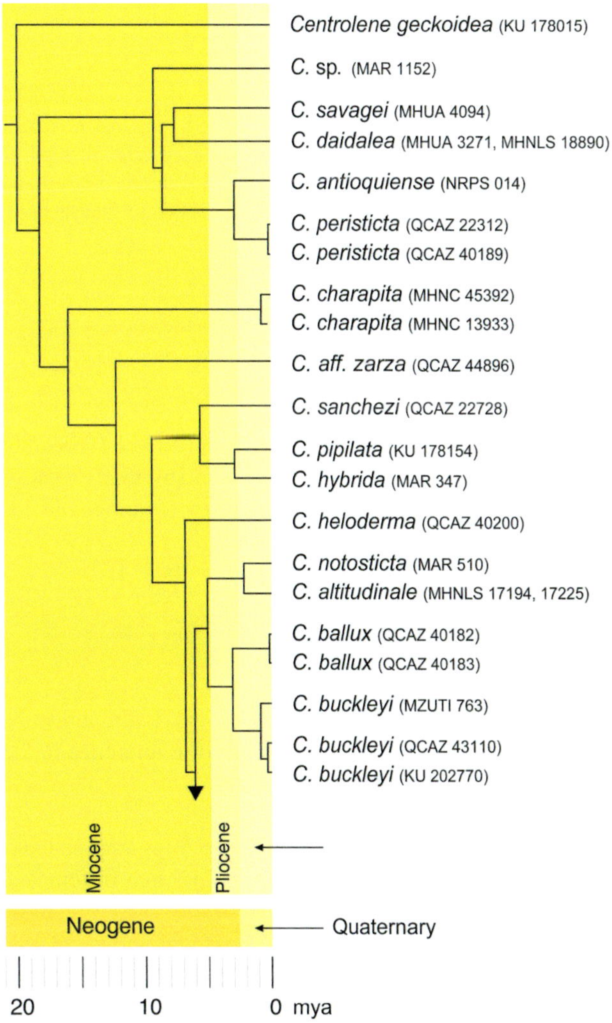

FIGURE 6.3 Time-calibrated maximum likelihood phylogenetic tree depicting the evolutionary relationships of *Centrolene*, inferred from DNA sequences of mitochondrial and nuclear genes. Phylogeny taken and redrawn from Guayasamin *et al.* (2020b). *Centrolene aff. zarza* follows Székely *et al.* (2023) cited in the Addendum in Appendix 2.

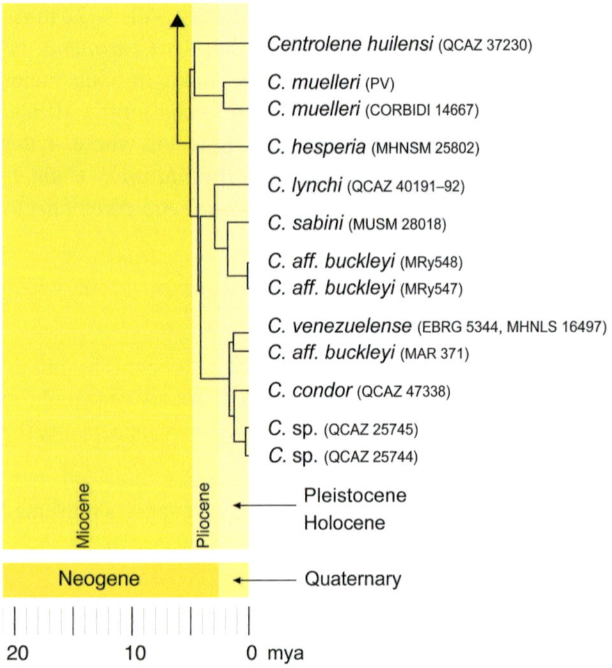

FIGURE 6.4 Continued. Time-calibrated maximum likelihood phylogenetic tree depicting evolutionary relationships of *Centrolene*, inferred from DNA sequences of mitochondrial and nuclear genes. Phylogeny taken and redrawn from Guayasamin *et al.* (2020b).

FIGURE 6.5 *Centrolene ballux*. Left: QCAZ 40196, female. Right: QCAZ 40199, male. Reserva Las Gralarias, Provincia Pichincha. LAC.

FIGURE 6.6 Morphological details of *Centrolene ballux*. QCAZ 48201, Reserva Las Gralarias, Provincia Pichincha. LAC.

CENTROLENE BALLUX (GOLD-DUST GLASSFROG) (DUELLMAN AND BURROWES, 1989)

Centrolenella ballux (Duellman and Burrowes, 1989): 80.
Holotype.—KU 164725 from 14 km (by road) west of Chiriboga, 1960 m, Provincia Santo Domingo de los Tsáchilas, Ecuador.
"Centrolene" ballux Guayasamin *et al.*, 2009: 1.

Etymology.—The specific epithet is the Latin adjective *ballux*; it means "gold dust" and is used in reference to the minute gold flecks on the dorsum.

Identification.—Figures 6.5, 6.6. These frogs attain a maximum SVL of 22.2 mm in males and 23.3 mm in females. The skin on the dorsum is shagreen with small greenish-white warts; the ventral surfaces of the thighs below the vent have a pair of enlarged tubercles. The anterior two-thirds of the ventral parietal peritoneum are white; the posterior third is transparent. There is a silvery white pericardium; no iridophores in the peritonea cover the intestines, stomach, testes, kidneys, gall bladder, and urinary bladder. The vomers lack teeth. The snout is bluntly rounded to truncate in dorsal and lateral profiles. The tympanum is oriented almost vertically, ED/TD = 31–34%; the tympanic annulus is visible, except for a dorsal border covered by the supratympanic fold; the tympanic membrane is partially pigmented, differentiated from the surrounding skin. The liver is tetralobed and lacks iridophores. Humeral spines are present in adult males. There is no webbing between fingers I and II; webbing between fingers II and III is basal or absent, and there is moderate webbing between the outer fingers. The webbing on the foot is moderate. The ulnar fold is low and white; the inner tarsal fold is low; the outer tarsal fold is absent, but small white tubercles are evident along the ventrolateral margin of the tarsus. The prepollex is clearly separated from finger I; in males, the nuptial pad is a large to medium-sized excrescence present on the dorsal, lateral, and/or ventral sides of the thumb. Finger II is slightly longer than finger I (finger I 91.4–98.0% of finger II). The disc of finger III is of moderate size, about 47.4–59.3% of the eye diameter.

The dorsum is lime green with small greenish-white to yellowish-white warts. The upper lip is yellowish-white to white. The fingers and toes are yellowish-green. The melanophores are mostly absent from the fingers and toes, except for a few on toes IV and V and on the base of finger IV. The bones are green. The iris is whitish-cream with fine black reticulations and a pale yellow hue around the pupil.

Comparisons.—*Nymphargus buenaventura*, *N. griffithsi*, *Centrolene peristicta*, *C. lynchi*, and *C. robledoi* (Colombia). On the Pacific versant of the Andes of Ecuador, only *N. buenaventura* and some populations of *N. griffithsi* resemble *C. ballux* in having a green dorsum with small greenish-white spots. These two species differ, however, by possessing basal webbing between the outer fingers (moderate webbing in *C. ballux*) and lacking humeral spines (present in males of *C. ballux*). "*Centrolene*" *robledoi*, endemic to Colombia, has small dark spots on its dorsum (absent in *C. ballux*), a prepollex not separated from finger I (prepollex separated from finger I in *C. ballux*), and is slightly larger than *C. ballux* (male SVL 19.9–24.4 mm in *C. robledoi*; 19.2–22.2 mm in *C. ballux*). *Centrolene peristicta* and *C. lynchi* can be differentiated from *C. ballux* by having dorsal dark and light minute spots (only light spots in *C. ballux*).

Natural history.—Figure 6.7. During the night, individuals are on the upper surfaces of leaves of bushes and trees 50–800 cm above and along small streams or on ferns over roadside ditches. *Centrolene ballux* is relatively abundant at Reserva Las Gralarias, Provincia Pichincha, where it maintains breeding populations along several streams (Ballux Creek, Five-Frog Creek, Heloderma Creek, Chalguayacu River, Kathy's Creek) that have been registered during the rainy seasons (December–April) of 2011–2018. Some reproductive observations are as follows: In Reserva Las Gralarias, one male was on a leaf 5 cm below another leaf on which there was a clutch of 18 eggs. Another clutch found at La Planada, Departamento Nariño, Colombia, had 13 eggs. Near Chiriboga, Provincia Pichincha, a gravid female with green eggs and five adult males were found along a stream on 8 May 1975. At Reserva Las Gralarias in March 2009, a clutch with 21 embryos was found on the upper side of a leaf, which was hidden by another leaf; an adult male was observed nearby (ca. 100 cm from the clutch). It is unknown if females provide short-term parental care to egg clutches; males do not exhibit parental care.

FIGURE 6.7 *Centrolene ballux*. Top left: male. Carl Hutter. Top right and bottom left: QCAZ 47924, clutches of embryos. LAC. Bottom right: metamorph. Carl Hutter. Reserva Las Gralarias, Provincia Pichincha.

Advertisement calls from Las Palmeras and Reserva Las Gralarias, Provincia Pichincha, Ecuador, were described as a single short note with duration of about 328.5–420.4 ms, with about 7–9 pulses. Detailed data and figures are provided under the subtitle "Vocalizations of Ecuadorian Anurans" in Appendix 1.

Tadpoles.—Figure 6.8. This description is based on an individual (not preserved, series QCAZ 47924) from Reserva Las Gralarias, Provincia Pichincha, Ecuador, in stage 28. It was collected by Diego Almeida-Reinoso on 3 March 2009. The tadpole had a body length of 11.2 mm and total length of 34.6 mm. The body is ovoid, uniformly wide; it is wider than deep. The snout is round both in dorsal view and in lateral profile. The eyes are narrowly separated and directed dorsally. The spiracle is sinistral, with the inner wall free from the body, and the spiracular opening is directed posterolaterally at approximately three-fourths the body's length. The vent tube is medial. The mouth has an anteroventral position. The tail comprises about 71% of the tadpole's total length. It is deepest at around its midlength and subacute at the tip. The dorsal fin does not extend onto the body; it is slightly deeper than the ventral fin and about half the depth of the caudal musculature at the midlength of the tail.

The body appears pinkish with brown blotches on the dorsum and the caudal musculature, except for a ventral unpigmented band that extends from the beginning of the tail to its midlength. The area surrounding the eyes lacks pigmentation, and the tail fin is translucent with a few scattered dark flecks.

In individuals at stages 38–40, the dorsum and posterior legs exhibit a light green coloration. By stage 41, the pinkish coloration has largely disappeared, and the dorsum and legs have a more pronounced light green hue. Blue bones become visible thought the skin in the legs and arms, and white dots appear on these limbs. There are brown flecks on the dorsum, but the tail maintains its dark brown coloration. At stage 44, the dorsum becomes predominantly green with brown flecks, and the arms and legs retain their white dots.

FIGURE 6.8 Ontogenetic changes in two individuals of *Centrolene ballux* from tadpole to froglet. Top to bottom. QCAZ 47924, stage 28, TL = 34.6 mm. CJ 12885, undetermined stage. QCAZ 47924, stage 42. CJ 12885, stage 45. QCAZ 47924, stage 46, 125 days after stage 27. Reserva Las Gralarias, Provincia Pichincha. LAC.

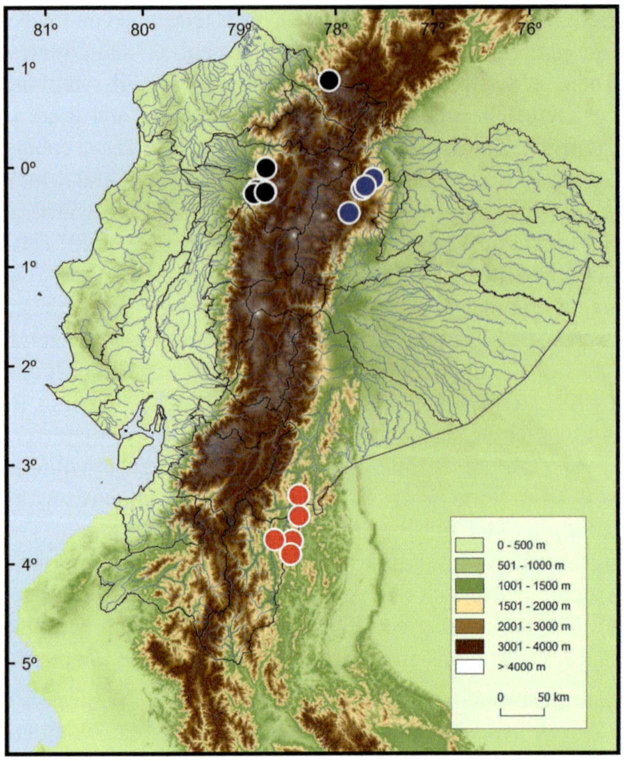

FIGURE 6.9 Localities of known occurrence of *Centrolene ballux* (black dots), *C. condor* (red dots), and *C. pipilata* (blue dots) in Ecuador.

Upon completing metamorphosis, which occurs around 125 days after stage 26, the tadpole adopts a greenish body color with brown flecks on the head. Scattered white dots are still visible on the body. The ventral vocal area takes on a light blue color, and the iris of the eye is golden brown.

Distribution.—Figure 6.9. *Centrolene ballux* is known from the Pacific slopes of the Cordillera Occidental of the Andes in southern Colombia (Departamento Nariño) and northern Ecuador. In Ecuador, it occurs in Carchi, Pichincha, and Santo Domingo de los Tsáchilas provinces, at elevations between 1843–2282 m in an area of extent of occurrence of about 1130 km^2.

This nocturnal, arboreal species inhabits natural forests and intervened areas in evergreen low montane forest and evergreen montane forest ecosystems in the Cordillera Occidental of the Andes biogeographic sector, where the average annual rainfall is 877–2151 mm and the average annual temperature is 15.0–16.9 °C.

Conservation.—*Centrolene ballux* is categorized as Endangered throughout its entire range (IUCN SSC Amphibian Specialist Group, 2019n; in IUCN Red List of 2022) and in Ecuador by Ortega-Andrade *et al.* (2021). In Ecuador, its extent of occurrence is less than 5000 km^2 (1130 km^2, according to our data), and there is continuing decline in the extent and quality of its habitat. According to Guayasamin *et al.* (2020b), it was apparently abundant when described in the 1970s; however, it is more difficult to find now. In Ecuador, it seems is no longer present in some of the localities. Recent sampling in historic locations in Ecuador (e.g., near Quebrada Zapadores and Chiriboga with a search effort of 58 hours per person, as well as in Las Palmeras) were unsuccessful in registering the species, and it had not been observed since 1989 until it was recorded in 2009 in Reserva Las Gralarias. In Colombia, the last published record is from Reserva La Planada in April–June 1986. This reserve is currently (as of 2016) being managed by an indigenous community, which appears to be successful at preventing further habitat deterioration within the reserve. The main threats include deforestation for agricultural activities, logging, and the introduction of predatory fish species in streams. Chytridiomycosis was listed as a potential threat in the 2004 assessment, but at Reserva Las Gralarias, although the amphibian chytrid fungus *Batrachochytrium dendrobatidis* infects *C. ballux* with a relatively high prevalence, no recent declines have been recorded (Guayasamin *et al.*, 2014).

In Ecuador, it has been recorded in the protected forests of Toachi-Pilatón, in the Reserva de la Biósfera Chocó Andino de Pichincha, and some private reserves (e.g., Reserva Las Gralarias).

Remarks.—The placement of *Centrolenella ballux* Duellman and Burrowes, 1989 in the genus *Centrolene* (*sensu* Guayasamin *et al*., 2009) is unambiguous. According to Amador *et al*. (2018), *Centrolene ballux* and *C. buckleyi sensu stricto* are sister species.

Key references.—(Duellman and Burrowes, 1989) (description); (Márquez and de la Riva, 1996) (call); (Hutter and Guayasamin, 2012) (habitat); (Arteaga *et al*., 2013) (natural history); (Delia *et al*., 2017) (natural history); (Guayasamin *et al*., 2020b) (account, call); Yánez-Muñoz *et al*. (2020) (account).

CENTROLENE BUCKLEYI (BUCKLEY'S GLASSFROG) (BOULENGER, 1882a)

Hylella buckleyi (Boulenger, 1882a): 88. **Neotype.**—KU 202770 from Isla Wolf of Laguna Cuicocha, 3070 m, Provincia Imbabura, Ecuador.
Centrolene buckleyi Ruiz-Carranza and Lynch, 1991a: 6.

Etymology.—The specific name honors Clarence Buckley (1832–1885 or 1886), who collected the type series of the species.

Identification.—Figures 6.10, 6.11. *Centrolene buckleyi* is a species complex that requires taxonomic clarification (see Remarks). Thus, the traits listed herein should be taken with caution. The maximum known SVL is 30.9 mm in males and 34.4 mm in females. The skin on the dorsum is finely shagreen; the males present spicules, and there is a pair of round tubercles posteroventral to the vent. The anterior half to two-thirds of the venter is covered by a white parietal peritoneum, and the posterior portion is translucent. There is a silvery white pericardium; there is a translucent peritoneum covering the intestines, stomach, testes, kidneys, gall bladder, and urinary bladder. The vomers lack teeth. The snout is round in dorsal view, slightly sloping to sloping in lateral profile. The tympanum is partially or completely hidden under the skin; when visible, it is oriented almost vertically; its diameter is 29.0–38.6% of eye diameter; the supratympanic fold is moderately heavy; the tympanic membrane is slightly thinner than the skin around the tympanum. The liver has four or five lobes and lacks iridophores. The humeral spines are present in males. There is no webbing between fingers I, II, and III, and there is reduced webbing between the outer fingers. The webbing on the foot is moderate. The ulnar fold is low; the ventrolateral margin of the arm is white; the inner tarsal fold is evident, and the outer tarsal fold is absent. The external ventrolateral margin of the tarsus is white. The prepollex is not separated from finger I; in males, the nuptial pad is a large to medium-sized excrescence present on the dorsal, lateral, and/or ventral sides of the thumb. Finger II is slightly longer than finger I (finger I 86.5–98.7% of finger II). The disc of finger III is of moderate size, about 54.2–64.5% of the eye diameter.

The dorsum is uniform bright to dark green. It is sharply demarcated laterally from the lower white flanks. The throat and most of the venter are pale green. The parietal peritoneum is yellowish-white. The edge of the upper lip is

FIGURE 6.10 *Centrolene buckleyi*. Left: QCAZ sc 27486, female, Morán, Provincia Carchi. Right: CJ 12889, male, Bosque Protector Cashca Totoras, Provincia Bolívar. LAC.

FIGURE 6.11 Variation and morphological details of *Centrolene buckleyi*. Top left: QCAZ sc 27486, Morán, Provincia Carchi. Top right: MZUTI 763, male, nearby Oyacachi, Provincia Napo. Middle left: QCAZ 40308, female, El Goatal, Provincia Carchi. Middle right: Guaranda, Provincia Bolívar. Bottom left: Chillogallo, Quito, Provincia Pichincha. Bottom right: Santiago, Provincia Bolívar. LAC.

white, usually continuing as a white line across the flanks. The ventrolateral borders of the arms and tarsus are white. There are small white warts posterior to the cloacal opening. Some individuals have whitish warts on the dorsum or small white spots on the legs and forearms. The melanophores are mostly absent from the fingers and toes, except for a few on toes IV, V, and on the base of the outer fingers. The bones are green. The iris is gray–white with a thin black reticulation and a horizontal brown stripe.

Comparisons.—*Centrolene heloderma*, *C. condor*. *Centrolene buckleyi* is one of the few species of glassfrogs found in the highlands of Ecuador. It is easily identified by having a green dorsal coloration that sometimes presents small white spots, a white upper lip, an inclined snout in lateral view, reduced webbing between the fingers, and, in males, a large humeral spine. *Centrolene heloderma* and *C. condor* can be confused with *C. buckleyi*; however, *C. heloderma* has pustular dorsal skin (shagreen in *C. buckleyi*), and *C. condor* has small light and dark spots on the dorsum (dark spots absent in *C. buckleyi*). Similar glassfrogs from other South American countries include *Centrolene hesperium*, *C. lemniscatum* (Peru), *C. altitudinale*, and *C. venezuelense* (Venezuela). See Remarks for comments on putative new species that are similar to *C. buckleyi*.

Natural history.—Figure 6.12. During the day, *Centrolene buckleyi* has been found in terrestrial and arboreal bromeliads near and away from streams in secondary forests and pastures. During the night, frogs are active on terrestrial bromeliads and vegetation 30–160 cm above streams, lakes, and marshes in primary and secondary forests. The males call from the upper sides of leaves. *Centrolene buckleyi* mainly breeds in rapid mountain streams, but it seems that it also breeds in bromeliads or in ponds. There was an observation of an egg clutch on the inner leaf of a bromeliad in an area where *C. buckleyi* was abundant. Males fight dangling upside down while holding onto vegetation with their hind limbs, grasping one another venter–venter. A male was found on the upper side of a leaf, approximately 160 cm above a stream. Females provide short-term parental care to egg clutches; males do not exhibit parental care.

The advertisement call from ca. Oyacachi, Provincia Napo, Ecuador is a single note, highly pulsed, with a duration of 0.248–0.288 s and a dominant frequency at 2816–2941 Hz. Calls are produced in series. Detailed data and figures are provided under the subtitle "Vocalizations of Ecuadorian Anurans" in Appendix 1.

FIGURE 6.12 *Centrolene buckleyi.* Top: Chillanes, Provincia Bolívar. Bottom: CJ 2173, clutch of embryos, Puente Río Unachi, Provincia Cotopaxi. LAC.

Tadpoles.—Figure 6.13. This description is based on an individual (not preserved, series CJ 2064) from Río Unachi, 2790 m, Provincia Cotopaxi, Ecuador, in stage 26. It was collected by Elicio E. Tapia, Nora Tapia, and Flavio Ernesto Gallo on 22 February 2014. Four tadpoles were found in marginal pools with a maximum diameter of 1 m and a depth of 0.5 m, with leaf litter and mud at the bottom. The water temperature was 11.2 °C at 11:00 h. The tadpole had a body length of 13.4 mm and total length of 42.6 mm. The body is ovoid and sinuous, widest posteriorly, and wider than deep. The snout appears round both in dorsal view and in lateral profile. The chondrocranial elements are visible through the dorsal skin. The eyes are narrowly separated and directed dorsally. The spiracle is sinistral, with the inner wall free from the body; its opening is directed posteriorly and is located at approximately two-thirds of the body's length. The vent tube is medial with a web between the tube and fin. The mouth is

FIGURE 6.13 Ontogenetic changes in *Centrolene buckleyi* from tadpole to juvenile. Top to fourth row: CJ 2064, TL (Top, stage 27) = 13.4 mm, Puente Río Unachi, Provincia Cotopaxi. Fifth and sixth rows: QCAZ 50393. Bottom: QCAZ 40308. Morán, Provincia Carchi. LAC.

medium-sized (around 46% of the body width) and is positioned anteroventrally. The oral disc is transversely elliptical and not emarginate. The upper labium is poorly developed, while the lower labium is larger and free from the body wall. Both the upper and lower jaw sheaths are completely keratinized, with the upper jaw sheath having a smooth arc. The tail comprises about 70% of the total length and is deepest at the posterior part (at the beginning of the third part of the tail) with a rounded tip. The dorsal fin does not extend onto the body and is slightly deeper than the ventral fin. Both fins are relatively narrow and approximately half the depth of the caudal musculature at the midlength of the tail.

The tadpole's overall appearance is characterized by an intensive red coloration, attributed to the translucent skin that reveals the extensive internal vascularization. Additionally, it exhibits dark brown blotches on its dorsal body surface. Dark brown flecks are present on the tail, both dorsally and laterally. The dorsal fin is mostly covered in dark brown blotches, while the ventral fin is translucent, with only a brown stripe adjacent to and running along the caudal musculature. The skin on the ventral side is translucent, allowing the internal organs to be visible.

As the tadpoles progress to stages 39–40, the red coloration is retained primarily on the lateral sides of the body and on the caudal musculature. The brown blotches disappear, leaving behind some scattered brown flecks on the tail. The body and legs take on a yellow coloration, with the anterior part of the body exhibiting a greenish tint. By stage 44, the tadpole's coloration begins to resemble that of adults, featuring light green on the body, while the feet and hands appear yellowish. Additionally, the bones appear blue, and the iris is brown with a golden ring surrounding the pupil.

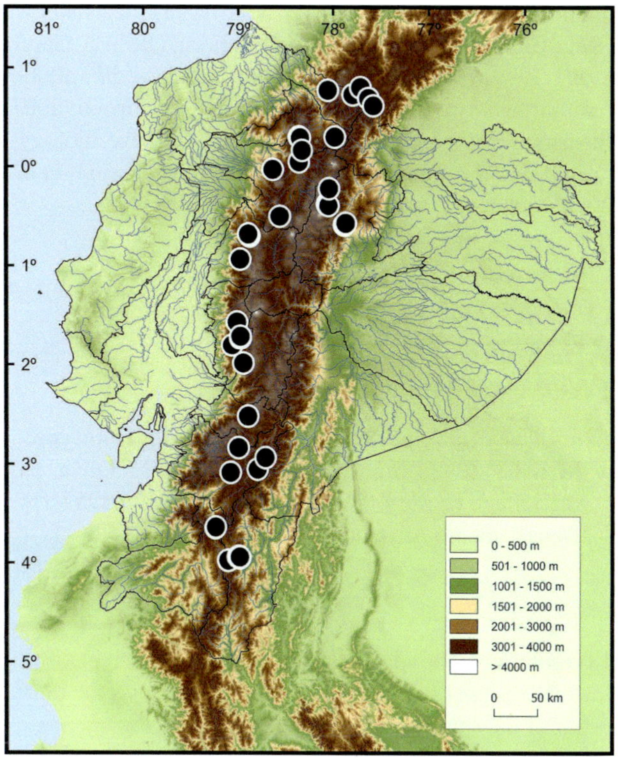

FIGURE 6.14 Localities of known occurrence of *Centrolene buckleyi* in Ecuador.

Distribution.—Figure 6.14. *Centrolene buckleyi* is known from the Andes of Colombia and Ecuador to Huacambamba in the Departamento de Piura in northern Peru. In Ecuador, it occurs along the Cordillera Oriental and Cordillera Occidental of the Andes and in inter-Andean valleys in Carchi, Sucumbíos, Imbabura, Pichincha, Napo, Cotopaxi, Bolívar, Chimborazo, Cañar, Azuay, Loja, and Zamora Chinchipe provinces at elevations between 1955 and 3408 m (but see Remarks) in an area of extent of occurrence of about 41,120 km^2.

This nocturnal, arboreal species inhabits primary and secondary forests and pasture areas in 10 ecosystems (Table 0.4) in three biogeographic sectors (Table 0.5), where the average annual rainfall is 557–2200 mm and the average annual temperature is 8.9–17.0 °C.

Conservation.—*Centrolene buckleyi* is categorized as Critically Endangered throughout its entire range (Guayasamin *et al.*, 2020b) and in Ecuador by Ortega-Andrade *et al.* (2021) because of a population decline, projected to be more than 80% over the next 10 years, in much of its Ecuadorian range. According to Guayasamin *et al.* (2020b) and citation therein, in Ecuador, *C. buckleyi* was abundant along the high Andes, but fieldwork demonstrates its absence or rarity at most of its historical localities (e.g., Cuicocha, Provincia Imbabura; Cashca Totoras, San Pablo de Atenas, Provincia Bolívar; Quito, Provincia Pichincha; Papallacta, Provincia Napo; Bosque Protector Mazán, Provincia Azuay). In Ecuador, at Yanayacu Biological Station, only three individuals of *C. buckleyi* were found after an intensive three-year inventory effort, suggesting that this species is quite rare at this locality. In Ecuador, almost half of the habitats suitable for this species have been deforested. In Ecuador, it has been recorded in the public reserves Área Ecológica de Conservación Municipal La Bonita and Parque Nacional Cotacachi-Cayapas and Parque Nacional Cayambe Coca; the protected forests of Toachi Pilatón, Cuenca Río Guayllabamba (Área 1-Área 2), Cashca Totoras, 15 Áreas al Interior de la Cuenca del Río Paute, Corazón de Oro, and Cuencas que Forman los Ríos: San Francisco, San Ramón y Sabanilla; and in the biosphere reserves Chocó Andino de Pichincha, Sumaco, Macizo del Cajas, and Podocarpus-El Cóndor.

Remarks.—Previous studies have shown that *Centrolene buckleyi* is a species complex that requires subdivision. Amador *et al.* (2018) propose at least three new candidate species within the species complex; therefore, the information presented in the current species account should be taken with caution, since upcoming taxonomic work on the *buckleyi* complex will redefine the distribution and conservation status of the "true" *C. buckleyi* (see also *Centrolene elisae* and *C. marcoreyesi* in the Addendum in the Appendix 2). Guayasamin *et al.* (2020b) suggests that locality records below 2400 m need further verification.

Key references.—(Boulenger, 1882a) (description); (Lynch and Duellman, 1973) (natural history, habitat, natural history); (Duellman and Wild, 1993) (distribution); (Ruiz-Carranza *et al.*, 1996) (distribution); (Bolívar-García *et al.*, 1999) (natural history); (Guayasamin *et al.*, 2006b) (natural history); (Cisneros-Heredia and Mcdiarmid, 2007) (distribution); (Delia *et al.*, 2017) (reproductive biology); (Yánez-Muñoz *et al.* (2020) (account); (Guayasamin *et al.*, 2020b) (account, description of neotype, call).

CENTROLENE CHARAPITA (CHARAPITA GLASSFROG) TWOMEY *ET AL.*, 2014

Centrolene charapita Twomey *et al.*, 2014: 12. **Holotype.**—MHNCP 13933 from near the village of La Oliva, past the village of Muyo, 682 m, 5.30107° S, 78.39571° W, Departamento Amazonas, Peru.

Etymology.—The specific name derives from a type of yellow chili pepper (ají charapita), which resembles the dorsal ocelli of the species.

Identification.—Figures 6.15, 6.16. The maximum known SVL is 37.9 mm in males; females are unknown. The skin on the dorsum has a complex texture with microspicules and low enameled warts on the ocelli; the venter and thighs are coarsely areolate; the other ventral surfaces are smooth. The cloacal ornamentation is conspicuous, formed by enameled folded skin surrounded by lower warts; the ventral surfaces of the thighs below the vent have a pair of enlarged tubercles. The anterior one-third to one-half of parietal peritoneum is white; the posterior portion is transparent. The pericardial, hepatic, renal, and gonadal peritonea are transparent; the visceral (stomach and intestines) peritoneum is white. The vomers have 4–10 vomerine teeth. In the Peruvian population, the snout is truncate in dorsal view and truncate to slightly round in profile; in the Ecuadorian population, the snout is round in dorsal view and sloping in profile. The tympanum is small and partially hidden under the skin; the supratympanic fold is present. The liver is trilobed and lacks iridophores. The humeral spines are absent. There is no webbing between fingers I and II; webbing is extensive on the outer fingers. The webbing on the foot is well developed. The ulnar and tarsal folds are scalloped and enameled, extending from the fringe on the postaxial edge of finger IV to the elbow and from the fringe on the postaxial edge of toe V to the ankle, respectively. The prepollex and prepollical spine are concealed by skin. In males, the nuptial pad is a large excrescence present on the dorsal, lateral, and/or ventral sides of the thumb and is not pigmented. When appressed, finger I is about equal in length to or slightly longer than finger II. The eye diameter is about 2.5 times the width of the disc on finger III.

FIGURE 6.15 Morphological variation in *Centrolene charapita*. Left: 73 11074, male. Right: Top: CJ 11072, male. Middle: CJ 11073, male. Bottom: CJ 11075, male. Reserva Maycu, Cordillera del Cóndor, Provincia Zamora Chinchipe. DAL.

FIGURE 6.16 Morphological details of *Centrolene charapita*. CJ 11074, male, Reserva Maycu, Cordillera del Cóndor, Provincia Zamora Chinchipe. DAL.

The dorsum and dorsolateral surfaces are green, with yellow or pale green spots of different sizes (ocelli). The upper lip and discs of the fingers and toes I, II, and III are white (enameled). The ventral surfaces, flanks, upper arms, fingers I and II, and toes I and II are not pigmented. The bones are green. The iris is delimited by a black ring, and the background coloration is white to light gray with yellow shades, black dots, and reticulations.

Comparisons.—*Centrolene charapita* is unique among Ecuadorian glassfrogs by having dorsal yellow ocelli of different sizes, scalloped and enameled ulnar and tarsal folds, and a relatively large size (adult male SVL = 34.7–37.0).

Natural history.—Figure 6.17. Individuals were found perched on riverine vegetation during the night. Males call from the upper side of leaves.

Tadpoles.—Figure 6.18. This description is based on an individual (not preserved) from Reserva Maycu, Provincia Morona Santiago, in stage 26. The clutch was collected from a palm leaf at approximately 2 m above the water by José Viera-Fernandes and Amanda B. Quezada on 9 September 2019. The tadpole had a body length of 9.4 mm and total length of 36.2 mm. The body is ovoid and flattened, wider than it is deep. The snout is round in dorsal view and appears subacute in lateral profile. The eyes are narrowly separated and directed dorsally. The spiracle is sinistral and situated at the posterior end of the body, where the tail begins; the spiracular opening is directed posteriorly. The vent tube is medial with a web between the tube and the fin. The mouth is of medium size and located anteroventrally. The lower labium is well developed and protrudes, while the upper labium is poorly developed. The tail constitutes approximately 74% of the total length, with its greatest depth at the beginning of the last third of its length, tapering to a subacute tip. Both the dorsal and ventral fins start at about the second third of the tail's length and exhibit similar depth, with the dorsal fin being slightly deeper. Both fins are about half the depth of the caudal musculature at the midlength of the tail.

The tadpole's body appears pinkish with brown blotches on the dorsal surface of the body. The area around the eyes lacks pigmentation. Brown flecks are present on the tail in both dorsal and lateral views, with the ventral caudal musculature being unpigmented on its anterior half. The dorsal and ventral fins display scattered brown flecks.

FIGURE 6.17 *Centrolene charapita.* CJ 11076, clutch of embryos, Reserva Maycu, Cordillera del Cóndor, Provincia Zamora Chinchipe. DAL.

FIGURE 6.18 Tadpole of *Centrolene charapita.* CJ 11076, TL (stage 26) = 36.2 mm, Reserva Maycu, Provincia Zamora Chinchipe. SGS.

Distribution.—Figure 6.19. *Centrolene charapita* is known in the Amazonian foothills and slopes of Departamento Amazonas, Peru, and southeastern Ecuador. In Ecuador, it has been only recorded in Reserva Maycu, Provincia Zamora Chinchipe, at elevations between 949 and 1322 m.

This nocturnal, arboreal species inhabits natural forests in the evergreen montane foothill forests on outcrops of limestone rock of the Amazonian Cordilleras ecosystem

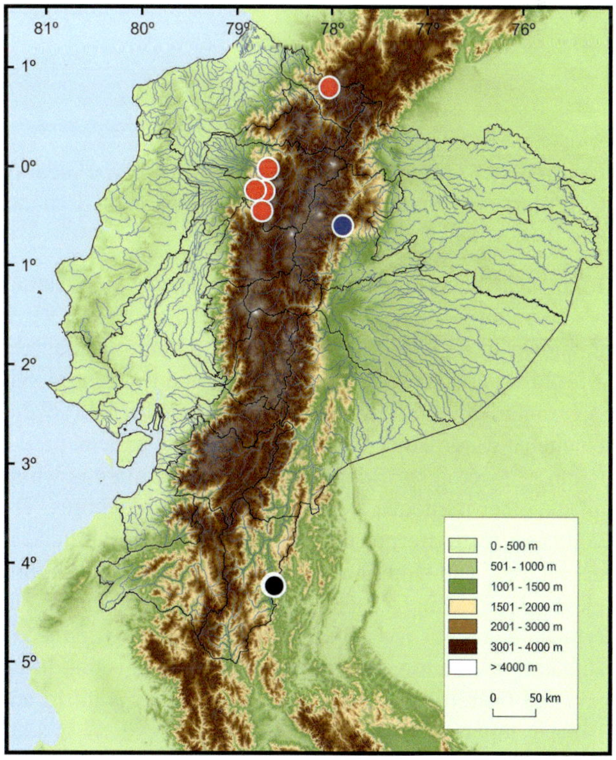

FIGURE 6.19 Localities of known occurrence of *Centrolene charapita* (black dots), *C. geckoidea* (red dots), and *C. huilensis* (blue dot) in Ecuador.

and the Amazonian Cordilleras biogeographic sector, where the average annual rainfall is 1431–1692 mm and the average annual temperature is 21.8 – 23.3 °C.

Conservation.—*Centrolene charapita* is categorized as Endangered throughout its entire range (IUCN SSC Amphibian Specialist Group, 2022e; in IUCN Red List of 2022). However, it is important to note that in Ecuador, *C. charapita* was categorized as Critically Endangered by Ortega-Andrade *et al.* (2021) because its extent of occurrence is estimated to be less than 100 km^2, it is only known from a single locality, and there is a projected decline in the extent and quality of its habitat. The whole distribution range of the species is within a mining concession. In Ecuador, it has not been recorded in public reserves, but there are records from the private reserve Reserva Natural Maycu.

Remarks.—According to Guayasamin *et al.* (2020b), *Centrolene charapita* is the sister species to a lineage formed by several *Centrolene* species.

Key references.—(Twomey *et al.*, 2014) (description, habitat, evolutionary relationships); (Guayasamin *et al.*, 2020b) (account).

FIGURE 6.20 *Centrolene condor.* Left: CJ 11685, male, 26.2 mm SVL, El Quimi, Provincia Morona Santiago. SGS. Right: CJ 7250, male, 29.2 mm SVL, Paquisha, Provincia Zamora Chinchipe. DAL.

FIGURE 6.21 Morphological details of *Centrolene condor.* Top: CJ 11685, El Quimi, Provincia Morona Santiago. SGS. Bottom: CJ 7249, Paquisha Alto, Provincia Zamora Chinchipe. DAL.

CENTROLENE CONDOR (CÓNDOR GLASSFROG) CISNEROS-HEREDIA AND MORALES-MITE, 2008

Centrolene condor Cisneros-Heredia and Morales-Mite, 2008: 50. **Holotype.**—QCAZ 37279 from Destacamento Militar Cóndor Mirador, western slope of the Cordillera del Cóndor, between 1750 and 1850 m, Provincia Zamora Chinchipe, Ecuador.

Etymology.—The specific name refers to the type locality, Cordillera del Cóndor, a mountain chain shared by Ecuador and Peru.

Identification.—Figures 6.20, 6.21. These frogs attain a maximum SVL of 27.6 mm in males, and the SVL is unknown in females. The skin on the dorsum is shagreen with low warts and abundant spicules, and the ventral skin is granular. The subcloacal area is also granular with several enameled warts. The upper two-thirds of the parietal peritoneum is covered by iridophores. The pericardium is white; all the other peritonea are not covered by iridophores. The vomerine teeth are present. The snout is subacuminate in dorsal view and sloping in profile; the nostrils are slightly elevated, producing a depression in the internarial area. The tympanic annulus is visible, vertical, and has a slight dorsolateral orientation. The tympanic membrane is not differentiated from the surrounding skin. The liver is lobed and lacks iridophores. Small humeral spines are present in adult males. There is no webbing between fingers I and II; the webbing between II and III is basal; the webbing between the outer fingers is moderate. The webbing on the foot is moderate. The ulnar fold is enameled; there is an enameled fringe on the edge of toe V that continues into a thin enameled metatarsal fold and then into a row of enameled tubercles along the outer edge of the tarsus; the nuptial excrescences are present on the dorsal, lateral, and/or ventral sides of the thumb. The prepollex is concealed. Finger I is shorter than II. The eye diameter is larger than the width of the disc on finger III. The melanophores are absent on the fingers and toes, except for a few on the dorsal surfaces of the outer fingers and outer toes.

The dorsum is green with abundant yellowish-white flecks and dark points. There are enameled spots and yellowish-white flecks on the arms and legs. A whitish-yellow labial line is present. The arm and leg folds are enameled. The bones are green. The iris is cream–yellow with fine, dark reticulations.

Comparisons.—*Centrolene pipilata.* The main traits that distinguish *C. condor* from other glassfrogs are a green dorsum with many small yellowish-white flecks and dark bluish-black/brown flecks and points, a sloping snout in lateral view, a light labial stripe, and, in adult males, humeral spines. Among Ecuadorian centrolenids, only *C. pipilata* shares a similar dorsal pattern, but *C. condor* has vomerine teeth (absent in *C. pipilata*) and a small, curved humeral spine (larger and straight in *C. pipilata*). Additionally, *C. condor* is only found in the Cordillera del Cóndor.

Natural history.—Males were found on the upper side of leaves of riverine vegetation from a few centimeters to about 4 m over water in mature elfin forest. At Loma Tigres Bajo, about 30 males were heard calling. At El Quimi, calling males were found over a river of lentic water with tannins that later formed a lagoon. Tadpoles are benthic.

Advertisement calls from Paquisha Alto, Provincia Zamora Chinchipe, Ecuador, were described as trills composed of two pulsed notes with a mean duration of about 650–700 ms. Detailed data and figures are provided under the subtitle "Vocalizations of Ecuadorian Anurans" in Appendix 1.

Tadpoles.—Figure 6.22. The tadpole was described based on an individual (not preserved) at stage 36 from Paquisha, Provincia Zamora Chinchipe, Ecuador. The tadpole has a body length of 8.4 mm and a total length of 29.9 mm. The body is elongated, postero-triangular, with the widest part anteriorly. In dorsal view, the snout is rounded, while laterally, it appears subacute. The body is flattened, wider than it is deep. The eyes are moderately large, narrowly separated, and directed dorsolaterally. Chondrocranial elements are visible. The sinistral spiracular opening is directed at approximately two-thirds of the body's length. A medial vent tube is present. The oral disc is medium-sized, located anteroventrally, and appears transversely elliptical. The upper labium is poorly developed, while the lower labium is larger and not attached to the body wall. Both the upper and lower jaw sheaths are entirely keratinized, with the upper jaw sheath having a smooth arc. The tail accounts for about 72% of the total length, reaching its maximum depth at the beginning of the last third of its length and tapering to a rounded tip. The dorsal fin is visible at the midlength of the tail, whereas the ventral fin is minimally developed in the anterior portion of the tail and widens as it approaches the tail's midlength. Both dorsal and ventral fins have approximately the same depth and are about half as deep as the caudal musculature at the midlength of the tail.

The skin is translucent, allowing for the visibility of the internal organs. The circulatory system is conspicuous, giving the tadpoles a bright red coloration. Some melanophores are present on the dorsal surfaces of the body and tail, with scattered melanophores along the tail midline. The legs are yellow.

In stage 42, which occurs 42 days after stage 36, an individual has a total length of 13.3 mm. The red coloration has disappeared, replaced by a light green dorsal coloration with darker green blotches and white dots. The arms and legs are light green with white dots, while the tail is yellowish with brown and white flecks. By stage 44, five days later, the dorsum becomes covered by brown and olive-green blotches with white dots. The bones in the arms and legs are visible through the skin, appearing blue. At stage 46, occurring 20 days later and featuring an SVL of 9.7 mm, most of the brown and green blotches have vanished, giving way to a more uniform greenish coloration in the body. The arms and feet take on a yellowish hue. As a froglet, with an SVL of 14.6 mm, and after the complete reabsorption of the tail, the coloration closely resembles that of the adult frogs.

FIGURE 6.22 Ontogenetic changes in *Centrolene condor* from tadpole to juvenile. Top to bottom. CJ 8618, stage 36, lateral, dorsal, and ventral views, TL = 29.9 mm; stage 42, TL = 13.4 mm; stage 45, 51 days after stage 36, TL = 10.4 mm; froglet, 20 days after stage 45 = 9.7 mm SVL; Paquisha, Provincia Zamora Chinchipe. DAL. CJ 11687, juvenile = 14.6 mm SVL, El Quimi, Provincia Morona Santiago. SGS.

Distribution.—See Figure 6.9. *Centrolene condor*, an endemic species to Ecuador, is known from the Cordillera del Cóndor from Morona Santiago and Zamora Chinchipe provinces. Its elevational range is 882–2254 m in an area of extent of occurrence of about 810 km^2.

This nocturnal, arboreal species inhabits natural forests and intervened areas in evergreen montane forest of Cordilleras Cóndor-Kutukú and evergreen montane forests on sandstone plateaus of Cordillera del Cóndor ecosystems in the Amazonian Cordilleras biogeographic sector, where the average annual rainfall is 1240–1633 mm and the average annual temperature is 16.4–23.6 °C.

Conservation.—*Centrolene condor* was categorized as Endangered by Ortega-Andrade *et al.* (2021). Its current extent of occurrence is less than 5000 km^2 (810 km^2, according to our data), and the species is known from few localities in an isolated mountain range (Cordillera del Cóndor). According to Guayasamin *et al.* (2020b), it was apparently fairly abundant, given the high number of calls attributed to this species that were heard from the canopy along the stream. Its major threats are extensive mining activities and associated deforestation, human colonization, agricultural expansion, and contamination. It has been recorded in the public reserve Reserva Biológica El Quimi and the protected forests of Cordillera del Cóndor and Pueblo Shuar Arutam.

Remarks.— See Remarks in the *Centrolene zarza* account included in the Addendum in Apendix 2.

Key references.—(Cisneros-Heredia and Morales-Mite, 2008) (description); (Almendáriz and Batallas, 2012) (natural history); (Guayasamin *et al.*, 2020b) (account, biology and ecology, call, evolutionary relationships).

CENTROLENE GECKOIDEA (GECKO GLASSFROG) (JIMÉNEZ DE LA ESPADA, 1872)

Centrolene geckoideum (Jiménez de la Espada, 1872b): 88. **Holotype.**—MNCN 1596 from the shores of Río Napo, Provincia Napo, Ecuador. Possibly in error; see Remarks.

Centrolene geckoidea Barrio-Amorós *et al.*, 2019: 19.

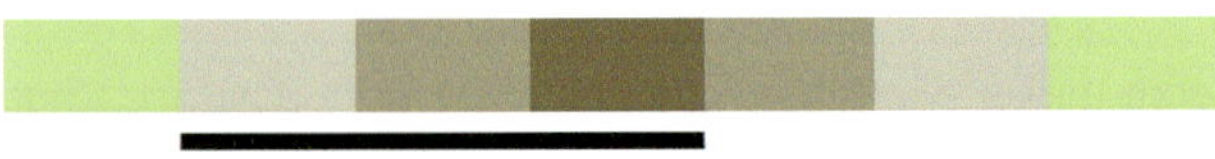

Etymology.—The specific epithet refers to the enormous size of the discs on fingers and toes of this species, which resemble those of gecko lizards (Gekkonidae).

Identification.—Figure 6.23. The maximum known SVL is 80.7 mm in males and 72.9 mm in females. The skin on the dorsum of males and females is covered with warts, and the spicules are evident only in males. The ventral surfaces of the thighs below the vent lack a pair of enlarged tubercles. The anterior two-thirds or the entire ventral parietal peritoneum are white. There is a silvery white pericardium; no iridophores in the peritonea cover the intestines, stomach, testes, kidneys, gall bladder, and urinary bladder. Each vomer has four or five teeth. The snout is truncate in dorsal and lateral profiles. The tympanum is visible, small, and oriented almost vertically; its diameter is 40.3–50.0% of the eye diameter; the supratympanic fold is moderate; the tympanic membrane is completely pigmented and differentiated from

FIGURE 6.23 *Centrolene geckoidea*. Top: KU 164490, male, 9 km southeast of Tandayapa, Provincia Pichincha. WED. Middle: ICN 41291, male, 61.0 mm SVL. Dagua Queremal, Departamento Valle del Cauca, Colombia. Bottom: male, Río La Plata, Provincia Carchi. Doug Wechsler.

the surrounding skin. The liver is tetralobed and lacks iridophores. The humeral spines are present in males. There is reduced webbing between fingers I, II, and III; there is extensive webbing between fingers III and IV. The webbing on the foot is extensive. The ulnar fold has small white tubercles; the inner tarsal fold is low and short; the outer tarsal fold is low with small white tubercles. The prepollex is not separated from finger I; in males, the nuptial pad is a large to medium-sized excrescence present on the dorsal, lateral, and/or ventral sides of the thumb. Finger I is about the same length as finger II (finger I 91.7–100.0% of finger II). The disc of finger III is wide, about 115–130% of the eye diameter.

The dorsum is dull green with enameled greenish or bluish warts and varies to dark gray. The throat is greenish-yellow. The margin of the upper lip is yellow. The venter is cream or yellow–cream. The ulnar, tarsal, and cloacal tubercles are creamy white. There are white flecks on the flanks. The bones are green. The males and females are dark gray to brownish-green during the night. The melanophores cover the dorsal surfaces of the hands and feet. The iris is pale greenish-gold with fine black reticulations. The eyelid is clear.

Comparisons.—*Centrolene paezorum*. *Centrolene geckoidea* is the largest known glassfrog species (body size in adult males: 70.2–80.7 mm; in adult females: SVL 61.8–72.9 mm). Males also have a conspicuous humeral spine, which in some individuals perforates the skin of the arm. Only *C. paezorum*, endemic to the high Andes of Colombia, could be confused with *C. geckoidea*. *Centrolene paezorum* is known only from a single female that is smaller (SVL 44.5 mm; see Remarks) and has less hand webbing than *C. geckoidea*.

Natural history.—Both male and female *Centrolene geckoidea* individuals exhibit an intriguing behavior of clinging to vertical or overhanging rock surfaces in the spray zone behind waterfalls. Males call from these areas or within the vicinity of spray zones found in fast-flowing streams. Males have call-site fidelity and have been observed vocalizing at the same site for up to a month. Egg clutches are deposited on rocks within the spray zone and consist of 89–112 black eggs, which are glued to the rocks by a unique brittle jelly not observed in other centrolenid species. Interestingly, each clutch exhibits an "empty space" at its center, adding to the distinctiveness of this species' reproductive biology. During the night, males have been observed sitting on or near up to four clutches of eggs while actively calling. During the day, males also exhibit a behavior of sitting in close proximity to or directly on the clutches located on the rockface behind waterfalls or concealed within crevices between rocks.

Advertisement calls from El Queremal, Departamento Valle del Cauca, Colombia, were described as loud, high-pitched, trilled whistles that lack any consistent pattern of amplitude modulation, with a duration of about 155–373 ms. Detailed data and figures are provided under the subtitle "Vocalizations of Ecuadorian Anurans" in Appendix 1.

Tadpoles.—Lynch *et al.* (1983) and Rueda-Almonacid (1994) described and provided illustrations of the tadpole of *Centrolene geckoidea* from Colombia. Lynch *et al.* (1983) described a tadpole collected in Vereda San Julián, Municipio Calarcá, Departamento Quindío, Colombia (2030–2100 m). The tadpole was characterized as having a typical centrolenid body form measuring 22.3 mm in total length. The tadpole's mouth lacked upper tooth rows, but two prominent lower rows (P-1) with a wide medial gap were evident. The upper jaw sheath appeared thick with minute serrations, while the lower jaw sheath was thin and poorly keratinized. Notably, a single row of large subconical marginal papillae bordered the oral disc ventrolaterally and posteriorly.

Rueda-Almonacid (1994) described tadpoles in Gosner stage 22, one month after hatching, as possessing two incomplete tooth rows on the anterior labium and three tooth rows on the posterior labium. However, three months after hatching, all the tooth rows were lost, and only dermal ridges remained visible. Additionally, the dorsal surfaces of the body and tail musculature displayed black pigmentation, while the venter lacked any pigmentation.

Distribution.—See Figure 6.19. *Centrolene geckoidea* is known from across the three Andean Cordilleras of Colombia (Cordillera Occidental, Central, and Oriental in the departments of Antioquia, Valle del Cauca, Caldas, Quindío, Risaralda, Tolima, Boyacá, and Caquetá), south to the Pacific Andean slopes of Ecuador. In Ecuador, it occurs in Carchi, Santo Domingo de los Tsáchilas, and Pichincha provinces. Its elevational range is 1843–3330 m in an area of extent of occurrence of about 1670 km^2.

This nocturnal, arboreal species inhabits natural forests in evergreen low montane forest and evergreen montane forest ecosystems in the Cordillera Occidental of the Andes biogeographical region, where the average annual rainfall is 920–2151 mm and the average annual temperature is 9.1–16.9 °C.

Conservation.—*Centrolene geckoidea* is categorized as Critically Endangered (Possibly Extinct) throughout its entire range (IUCN SSC Amphibian Specialist Group, 2019o; in IUCN Red List of 2022) and in Ecuador by Ortega-Andrade *et al.* (2021). This species appears to have undergone a drastic population decline in the 1990s, inferred from the apparent disappearance of most or all of the populations. Intensive fieldwork at historical localities such as Quebrada Zapadores, Tandayapa, and Dos Cascadas, Provincia Pichincha (Ecuador), has failed in finding the species. The last reports of *C. geckoidea* in Ecuador are from Río La Plata on the Maldonado–Tulcán road on 25 July 1988 and from Bosque Protector Río Guajalito in January to May between 1998 and 1999. A similar scenario exists with populations of *C. geckoidea* in Valle del Cauca, located in the Cordillera Occidental of Colombia, with no sightings reported for over a decade. The species is affected by continuous habitat loss and degradation due to agriculture, logging, and human settlements throughout its range. Other known threats include the introduction of exotic predatory fish in streams. In Ecuador, it has been recorded in the protected forest of Estación Científica Río Guajalito and the private area Bellavista.

Remarks.—*Centrolene geckoidea* is sister to all other species in the genus *Centrolene* (Guayasamin *et al.*, 2008b). Guayasamin *et al.* (2020b) discussed the problems associated with the type locality, which has been suggested to be in error, and that the holotype of *C. geckoidea* likely was collected in an Andean locality or in the headwaters of the Río Napo.

The external morphology, osteology, and myology of *Centrolene geckoidea* were studied in detail by Rueda-Almonacid (1994). Given the remarkable morphological similarity between *C. geckoidea* and *C. paezorum*, Guayasamin *et al.* (2020b) consider the possibility that *C. paezorum* represents a junior synonym of *C. geckoidea*. Differences between these species include the absence of vomerine teeth in *C. paezorum*, smaller size, and reduced webbing. Also, the two species occupy different elevations; *C. geckoidea* is found at elevations between 1750 and 2525 m, whereas *C. paezoroum* occurs at 3030 m. Additional specimens from the type locality of *C. paezorum* (Colombia: Departamento del Cauca: km 55–56 on the Popayán–Inza road, 3030 m) are necessary to clarify its status.

Key references.—(Jiménez de la Espada, 1872b) (description); (Goodman and Goin, 1970) (habitat); (Lynch *et al.*, 1983) (natural history); (Ruiz-Carranza *et al.*, 1986) (morphology); (Rueda-Almonacid, 1994) (morphology, osteology, myology); (Grant *et al.*, 1998) (natural history); Yánez-Muñoz *et al.* (2020) (account); (Guayasamin *et al.*, 2020b) (account).

FIGURE 6.24 *Centrolene heloderma.* Left: CJ 10834, female. Right: CJ 10835, male, 34.3 mm SVL. Road to Tandayapa, Provincia Pichincha. DAL.

CENTROLENE HELODERMA (WARTY GLASSFROG) (DUELLMAN, 1981)

Centrolenella heloderma (Duellman, 1981): 4. **Holotype.**—KU 164715 from Quebrada Zapadores, 5 km east–southeast of Chiriboga, 2010 m, 0.28333° S, 78.78333° W, Provincia Pichincha, Ecuador.

Centrolene heloderma Guayasamin *et al.*, 2009: 26.

Etymology.—The specific name is derived from the Greek words *helos*, meaning wart, and *derma*, meaning skin, and is used in allusion to the dorsal texture of the species.

Identification.—Figures 6.24, 6.25. The maximum known SVL is 31.5 mm in males and 32.3 mm in females. The skin on the dorsum of males and females is pustular; in males, minute spicules are evident only on the flanks and tympanic region. There is a pair of enlarged subclocacal warts. The anterior three-fourths of the ventral parietal peritoneum has white iridophores. The pericardium is white; there are no iridophores on the peritonea covering the intestines, stomach, kidneys, testes, gall bladder, and urinary bladder. The vomers lack teeth. The snout is subacuminate in dorsal profile, inclined anteriorly from the nostrils to the margin of the lip in lateral profile. The tympanum is completely visible, oriented almost vertically, relatively large; its diameter is 38.5–44.5% of the eye diameter; the supratympanic fold is moderate; the tympanic membrane is pigmented and clearly differentiated from the surrounding skin. The liver is tetralobed and lacks iridophores. In males, humeral spines are present. The webbing is absent between fingers I, II, and III and is moderate between fingers III and IV. The webbing on the foot is moderate. The external ulnar fold is evident and white; the inner tarsal fold is low and short; the outer tarsal fold is long, with low, white tubercles. The prepollex is concealed; in males, the nuptial pad is a large to medium-sized excrescence present on the dorsal, lateral, and/or ventral sides of the thumb. Finger II is slightly longer than finger I (finger I 91.0–95.5% of finger II). The disc of finger III is of moderate width, about 59.3–74.2% of eye diameter.

The dorsum is uniformly yellow–green to dark green with scattered green to bluish-white pustules. The margin of the lip is whitish-yellow, usually continuing as a white line across the flanks. The ventral parietal peritoneum is whitish-yellow. The throat is pale greenish-yellow. The cloacal, heel, and ulnar tubercles are white. The heart is not visible. The bones are green. The iris is yellow to pale golden yellow with fine black reticulations and a horizontal brown stripe.

Comparisons.—*Centrolene buckleyi*. *Centrolene heloderma* differs from all centrolenid species by having a green pustular dorsum and males with humeral spines. *Centrolene heloderma* can be confused with *C. buckleyi*; both species have humeral spines in males, the snout inclined in lateral profile, and a narrow white labial stripe. The most conspicuous difference between these species is that *C. buckleyi* lacks pustular dorsal skin.

Natural history.—Figure 6.26. During the rainy season (December–April), at night, *Centrolene heloderma* has been found active on vegetation near streams in primary and

FIGURE 6.25 *Centrolene heloderma* from Reserva Drácula, Provincia Carchi. Ross Maynard.

disturbed forests. Occasionally, males have been observed away from streams. At Reserva Las Gralarias, *C. heloderma* is abundant at Five-Frog Creek but has also been observed at Ballux Creek, Heloderma Creek, and Hercules Creek. Males call from the upper surfaces of leaves near streams. Males fight upside down, grasping one another venter–venter. Egg clutches, with 25–29 brownish eggs, are placed on the upper sides of leaves near streams. The females provide short-term parental care; prolonged parental care is absent. The egg clutches are deposited within male's territory, but parental care has not been reported.

Advertisement calls from Río Alambí and Reserva Las Gralarias, Provincia Pichincha, Ecuador, were described as relatively short, strongly pulsed, having two notes per call, with a mean duration of about 161 ms. Detailed data and figures are provided under the subtitle "Vocalizations of Ecuadorian Anurans" in Appendix 1.

Tadpoles.—Figure 6.27. This description is based on an individual (not preserved) from Tandayapa, Provincia Pichincha, Ecuador, in stage 26. A clutch was collected by Alejandro Arteaga Navarro on 20 May 2019. It has a

FIGURE 6.26 *Centrolene heloderma.* Top: male calling, Reserva Bellavista, Provincia Pichincha. Ross Maynard. Middle: CJ 2173, clutch of embryos, road to Tandayapa, Provincia Pichincha. DAL. Bottom: metamorph, Reserva Bellavista, Provincia Pichincha. Ross Maynard.

body length of 9.7 mm and total length of 31.3 mm. The body exhibits a postero-triangular elongated shape, with its widest part located anteriorly. The snout appears round both in dorsal view and in lateral profile. The eyes are narrowly separated and directed dorsally. The spiracle is sinistral with the inner wall free from the body; its opening is directed posteriorly and is situated at about two-thirds of the body's length. A medial vent tube is present. The mouth is medium-sized, positioned anteroventrally, and displays

FIGURE 6.27 Ontogenetic changes in *Centrolene heloderma* from tadpole to juvenile. Top to bottom. CJ 10837, stage 26, lateral, dorsal, and ventral views, TL = 31.3 mm; undetermined stage, TL = 39.4 mm; stage 42, TL = 35.6 mm; stage 46, 90 days after stage 26 = 12.1 mm SVL; juvenile 112 days after stage 46 = 12.6 mm SVL. Road to Tandayapa, Provincia Pichincha. DAL.

a transversely elliptical shape. The upper labium is barely developed, while the lower labium is more developed and not attached to the body wall. Both upper and lower jaw sheaths completely keratinized, with the upper jaw sheath having a smooth arc. The tail accounts for approximately 72% of the total length, reaching its maximum depth at the midlength and tapering to a subacute tip. The dorsal fin does not extend onto the body, has a similar depth to the ventral fin, and is about half as deep as the caudal musculature at the midlength of the tail.

The skin is translucent, allowing for the visibility of internal organs, resulting in a reddish coloration of the body. Brown flecks are present on the dorsum and along the tail, with a prominent brown stripe along its midline.

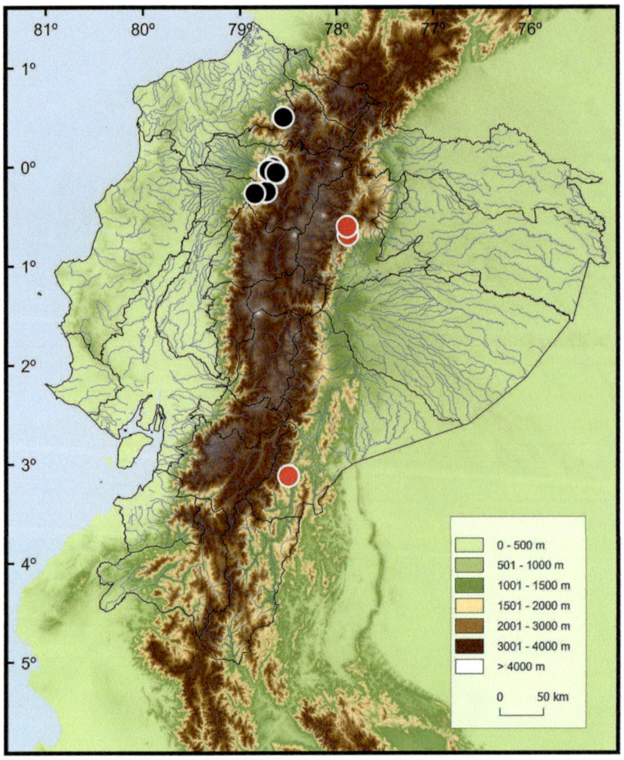

FIGURE 6.28 Localities of known occurrence of *Centrolene heloderma* (black dots), and *C. sanchezi* (red dots) in Ecuador.

In stage 39, which occurs 63 days after stage 26, the tadpole (total length of 39.4 mm) has a greenish coloration on both the body and the tail, with the brown flecks being less pronounced. By stage 43, 21 days later, with a total length of 35.6 mm, the body and extremities take on a light green hue with abundant white spots, and the still-translucent skin shows some red of the organs (in lateral view). The tail becomes green with brown and white flecks. In stages 45 (six days later, 12.1 mm SVL) and as juvenile (112 days after tail reabsorption, 12.6 mm SVL), the body assumes a uniform green coloration with white spots distributed along the body and extremities.

Distribution.—Figure 6.28. *Centrolene heloderma* occurs on the Pacific slopes of the Cordillera Occidental in Colombia (departments of Antioquia, Cauca, Valle del Cauca, and Risaralda) and south to northwestern Ecuador. In Ecuador, it occurs in Imbabura, Pichincha, and Santo Domingo de los Tsáchilas provinces. Its elevational range is 1859–2798 m in an area of extent of occurrence of about 850 km^2.

This nocturnal, arboreal species inhabits natural forests and intervened areas in evergreen low montane forest and evergreen montane forest ecosystems in the Cordillera Occidental of the Andes biogeographic sector, where the average annual rainfall is 1012–2107 mm and the average annual temperature is 12.4–16.8 °C.

Conservation.—*Centrolene heloderma* is categorized as Vulnerable through its entire range (IUCN SSC Amphibian Specialist Group, 2019p; in IUCN Red List of 2022). This categorization is based on its extent of occurrence, which is 19,495 km^2; it occurs in five to six threat-defined locations, and there is an ongoing decline in the extent and quality of its habitat and likely a continuing decline in the number of mature individuals. It was categorized as Endangered in Ecuador by Ortega-Andrade *et al.* (2021) due to its current extent of occurrence, which is less than 5000 km^2 (850 km^2, according to our data); its distribution is severely fragmented in fewer than five known localities, and there is a continuing decline in the extent and quality of its habitat. In Colombia, it is very rare and has not been recorded since 1996. According to Guayasamin *et al.* (2020b), due to ongoing declines in the extent and quality of habitat and the introduction of exotic predatory fishes, the population is suspected to be decreasing, and there is probably a continuing decline in the number of mature individuals. The type locality (Quebrada Zapadores, Provincia Pichincha), as well as nearby localities, has been visited numerous times between 2000 and 2015 (wet and dry seasons), with no new records of the species; in this area, the last record of the species dates from March 1979, when three individuals were collected at Quebrada Zapadores. In recent years, populations have been found through its entire range. Human activities have impacted approximately 34% of the potential distribution of the species in Ecuador, which corresponds to an area of 1067 km^2. According to Guayasamin *et al.* (2014), the amphibian chytrid fungus *Batrachochytrium dendrobatidis* has been found infecting *C. heloderma* at Reserva Las Gralarias, but no declines have been observed. In Ecuador, major threats include habitat loss, the introduction of exotic predatory fish (trout), climate change, fires, emerging diseases, expansion of the agricultural barrier, logging, and unplanned urban development. In Ecuador, it has been recorded in the protected forests of Cuenca Río Guayllabamba (Area 1–Area 2) and Toachi Pilatón and the private area Bellavista.

Remarks.—According to Guayasamin *et al.* (2020b), *Centrolene heloderma* is sister to a lineage formed by several *Centrolene* species.

Key references.—(Duellman, 1981) (description); (Hutter and Guayasamin, 2012) (habitat); (Arteaga *et al.*, 2013) (natural history); (Guayasamin *et al.*, 2014) (natural history, chytrid fungus); (Krynak *et al.*, 2018) (population); Yánez-Muñoz *et al.* (2020) (account); (Guayasamin *et al.*, 2020b) (account, call).

CENTROLENE HUILENSIS (HUILA GLASSFROG) (RUIZ-CARRANZA AND LYNCH, 1995a)

Centrolene huilense (Ruiz-Carranza and Lynch, 1995a): 10. **Holotype.**—ICN 7462 from 1 km northwest of Isnos, eastern versant of the Cordillera Central, 2190 m, Municipio de San José de Isnos, Departamento de Huila, Colombia.

Centrolene huilensis Barrio-Amorós *et al.*, 2019: 19.

Etymology.—The specific epithet refers to the type locality of the species, within the Huila Department, Colombia.

Identification.—Figure 6.29. The maximum known SVL is 23.8 mm in males and 28.7 mm in females. In Colombian populations, the adult males attain a maximum SVL of 26.7 mm. The skin on the dorsum is shagreen; males have low warts and spicules uniformly distributed; the ventral surfaces of the thighs below the vent have a pair of enlarged tubercles. The anterior 40–60% of the ventral parietal peritoneum is white; the posterior portion is translucent; there is a white pericardium; no iridophores are present on the peritonea covering the intestines, stomach, kidneys, gall bladder, and urinary bladder. The vomers lack teeth. The snout is short, round in dorsal aspect, and slightly sloping in lateral view. The tympanum is moderate and oriented almost vertically with slight lateral and posterior inclinations, and its diameter is 31–40% of the eye diameter. The tympanic annulus is completely visible, expect for the upper border; the supratympanic fold is evident, and the tympanic membrane is translucent and pigmented like the surrounding skin. The liver is lobed and lacks iridophores. The males have conspicuous humeral spines. There is no webbing between fingers I, II, and III; there is moderate webbing between the outer fingers. Webbing between

FIGURE 6.29 *Centrolene huilensis* from Yanayacu Biological Station, Provincia Napo. Mateo A. Vega-Yánez.

the toes is extensive; the feet are three/fourths webbed. The ulnar and tarsal folds have white tubercles. The prepollex is not exposed; the nuptial pad is a large to medium-sized excrescence present on the dorsal, lateral, and/or ventral sides of the thumb. Finger I is slightly shorter than finger II (finger I length 94–97% of finger II); the width of the disc of finger III is 51–63% of the eye diameter.

The dorsum is green with dark lavender and dark green spots of different sizes and smaller white spots. The upper lip is white. The region below the eye has small white warts. The bones are green. The upper flanks have the same color pattern as the dorsum. The lower flanks have numerous small white warts. The ulnar and tarsal folds have white tubercles. There are small white cloacal tubercles. The melanophores are absent from the dorsal surfaces of fingers and toes, except for finger IV and toes IV and V. The iris is cream with a slight yellow hue and thin black reticulations. The sclera is yellowish-cream to white.

Comparisons.—*Centrolene peristicta*, *C. daidaleum*, *C. condor*, *C. lynchi*, *C. muelleri*, *C. pipilata*, *C. savagei*, *C. solitaria*, *Nymphargus truebae*. *Centrolene huilensis* differs from other glassfrogs by having a green dorsum with a combination of dark green to dark lavender and white spots; white tubercles on the ventrolateral edges of finger V, the forearm, the elbow, toe V, the tarsus, and the heel; a humeral spine in adult males; and a relatively large body size (SVL = 23.6–26.7 mm in males; SVL = 28.7 mm in one female). The body size of *C. huilensis* is larger than that of *C. peristicta*, *C. pipilata*, *C. savagei*, and *C. solitaria* (in *C. peristicta*, male SVL = 17.9–22.0 mm, female SVL = 20.8–20.9 mm; in *C. pipilata*, male SVL 19.7–22.6 mm, female SVL 22.6–23.6 mm; in *C. savagei*, male SVL = 23.3–23.9 mm, female SVL = 19.8–22.6 mm; in *C. solitaria*, holotype male SVL = 19.3). Males of *C. huilensis* have humeral spines, which are absent in males of *C. daidaleum*, *C. savagei*, *C. solitaria*, and *N. truebae*. Additionally, some of the species are located in different biogeographic sectors; *C. peristicta* and *C. lynchi* are found on the Pacific slopes of the Andes of Ecuador and Colombia (*C. peristicta* is also found on the western slopes of the Cordillera Central, Colombia), and *C. muelleri* and *N. truebae* are only known from the Andes of Peru.

Natural history.—Males call from the upper sides of leaves. In Ecuador, individuals were found active during the night on the upper surfaces of leaves along slow-flowing streams.

Tadpoles.—Unknown.

Distribution.—See Figure 6.19. *Centrolene huilensis* is known from the type locality (2000–2190 m) in the Cordillera Central of Colombia (Departamento Huila Department) and from northeastern Ecuador on the Amazonian slopes of the Cordillera Oriental of the Andes. In Ecuador, it has been recorded at a single locality, Yanayacu Biological Station, in Provincia Napo at an elevation of 2119 m.

This nocturnal, arboreal species inhabits a patch of natural forest in the evergreen low montane forest of the northern Cordillera Oriental of the Andes ecosystem in the northern Cordillera Oriental de los Andes biogeographic sector, where the average annual rainfall is 2212 mm and the average annual temperature is 16.1 °C.

Conservation.—*Centrolene huilensis* is categorized as Endangered throughout its entire range (IUCN SSC Amphibian Specialist Group, 2019q; in IUCN Red List of 2022). However, it is important to note that in Ecuador, *Centrolene huilensis* is categorized as Data Deficient by Ortega-Andrade *et al.* (2021) given the absence of sufficient information for an evaluation. According to Guayasamin *et al.* (2020b), in Ecuador, assessing the conservation status of this species remains challenging because only a single additional locality (Yanayacu Biological Station) has been registered since its description in 1995. Although Yanayacu Biological Station is a relatively well-studied site, only a couple of individuals of *C. huilensis* have been observed, suggesting that it spends most of its time in the canopy and/or is extremely rare.

Remarks.—*Centrolene huilense* is most closely related to *C. muelleri* (Twomey *et al.*, 2014).

Key references.—(Ruiz-Carranza and Lynch, 1995a) (description); (Guayasamin *et al.*, 2020b) (account).

CENTROLENE LYNCHI (LYNCH'S GLASSFROG) (DUELLMAN, 1980)

Centrolenella lynchi (Duellman, 1980): 29. **Holotype.**—KU 164691 from 4 km northeast (by road) of Dos Ríos, 1140 m, Provincia Santo Domingo de los Tsáchilas, Ecuador.

Centrolene lynchi Ruiz-Carranza and Lynch, 1991a: 20.

Etymology.—The specific epithet honors Dr. John D. Lynch, who has made multiple and meaningful contributions to the field of amphibian systematics.

Identification.—Figures 6.30, 6.31. The maximum known SVL is 26.5 mm in males and 25.0 mm in females. The skin on the dorsum is shagreen in males and females; males have low, white warts, spicules, and spiculated warts on the sides of the head; the ventral surfaces of the thighs below the vent have a pair of enlarged tubercles. The anterior half to two-thirds of the ventral parietal peritoneum is white; the remaining posterior portion is transparent. There is a white pericardium; no iridophores are present in the peritonea covering the intestines, stomach, and kidneys; there is transparent peritoneum around the gall bladder and urinary bladder. There are no vomerine teeth. The snout is round or truncate in dorsal aspect and truncate to slightly inclined in lateral profile. The tympanic annulus is visible and oriented almost vertically, with slight lateral and posterior inclinations; its diameter is about 31.3–39.4% of the eye

FIGURE 6.30 *Centrolene lynchi.* CJ 10836, male, 24.8 mm SVL, road to Tandayapa, Provincia Pichincha. DAL.

FIGURE 6.31 Morphological variation and morphological details of *Centrolene lynchi*. Top left: male. Jaime Culebras. Top right: QCAZ 40191, male. Bottom left: QCAZ 40194, male. Bottom right: QCAZ 40192, male. Bottom: QCAZ 40192, male (note the humeral spines). Reserva Las Gralarias, Provincia Pichincha. LAC.

diameter; the supratympanic fold is evident; the tympanic membrane is translucent, partially pigmented and differentiated from the surrounding skin. The liver has four clearly defined lobes and lacks iridophores. The humeral spines are conspicuous in males. There is no webbing between fingers I and II; the webbing is absent or greatly reduced between fingers II and III, and the webbing is moderate between the outer fingers. The webbing between the toes is extensive. The ulnar fold is white; the external tarsal fold is absent, and the internal tarsal fold is short and low. The prepollex is usually not separated from finger I, except in a few individuals; the nuptial pad is a small circular or squarish excrescence present on the dorsal or dorsolateral face of the thumb. Finger I is slightly shorter than finger II or about equal to its length (finger I 94.5–100.0% of finger II). The width of the disc of finger III is about 48.4–57.0% of the eye diameter.

The dorsum is dull green to yellowish-green with minute yellowish-white warts (spiculated in males) and small, diffuse black spots. The bones are green. The throat and venter are creamy white. The fingers and toes are dull yellow. The upper lip is white. The ulnar and outer tarsal folds have a thin white line or low white tubercles. The flanks and venter are cream white. There are small white tubercles just posterior to the cloaca. There is a white parietal peritoneum covering the anterior half to two-thirds of the venter. The visceral and hepatic peritonea lack iridophores. The melanophores are mostly absent from the fingers and toes, except for a few on the base of fingers III and IV and along toes IV and V. The iris is grayish-white with black reticulations and a yellow hue surrounding the pupil.

Comparisons.—*Centrolene peristicta*, *C. antioquiense*, *C. quindianum*, *Nymphargus truebae*, *N. garciae*. *Centrolene lynchi* differs from most glassfrogs by having, in life, a green dorsum with minute yellowish-white flecks and diffuse small black spots. Among centrolenids found on the Pacific slopes of the Andes, only *C. peristicta* has a similar dorsal color pattern but is distinguished by being smaller (in

FIGURE 6.32 *Centrolene lynchi*. Top and bottom left: males in their microhabitat at night. Bottom right: QCAZ 48135, clutch of embryos. Reserva Las Gralarias, Provincia Pichincha. LAC.

males, SVL 23.3–26.5 mm in *C. lynchi*; 17.9–21.2 mm in *C. peristicta*). The Colombian *C. antioquiense* lacks black spots on the dorsum and has a white gastrointestinal peritoneum. *Centrolene lynchi* is also similar to the Colombian *C. quindianum*, which invariably has iridophores on the digestive tract (iridophores absent in *C. lynchi*). Other species, such as *Nymphargus truebae* (from the Amazonian slopes of the Peruvian Andes) and *N. garciae* (from the Cordillera Central of the Colombian Andes and the Amazonian slopes of the Ecuadorian Andes), have a coloration similar to that found in *C. lynchi* but lack humeral spines in males and webbing between fingers III and IV.

Natural history.—Figure 6.32. The males call from the upper sides of leaves. The males fight while hanging upside down, grasping one another venter–venter. At Reserva Las Gralarias (Lucy's Creek), Provincia Pichincha, females place clutches of 21–24 neon green eggs on vegetation in males' territories; most egg clutches are placed on top of large leaves or ferns above fast-flowing streams at a height of 165–600 cm. Only females provide parental care (short term) and prolonged parental care is absent in the species.

Advertisement calls from Reserva Las Gralarias, Provincia Pichincha, Ecuador were described as relatively short and consist of a tonal note followed by 1–3 peaked notes. The aggressive call is markedly different in structure from the advertisement call. It consists of a single short note containing two pulses, with a mean length duration of 150.4 ms. Detailed data and figures are provided under the subtitle "Vocalizations of Ecuadorian Anurans" in Appendix 1.

Tadpoles.—Unknown.

Distribution.—Figure 6.33. *Centrolene lynchi* is known on the Pacific slopes of the Cordillera Occidental of the Andes in Ecuador and southern Colombia. In Colombia, the species has been reported only in one locality (Reserva La Planada). In Ecuador, it occurs in Imbabura, Santo Domingo de los Tsáchilas, Pichincha, and Cotopaxi provinces. Its elevational range is 1447–2327 m in an area of extent of occurrence of about 1200 km².

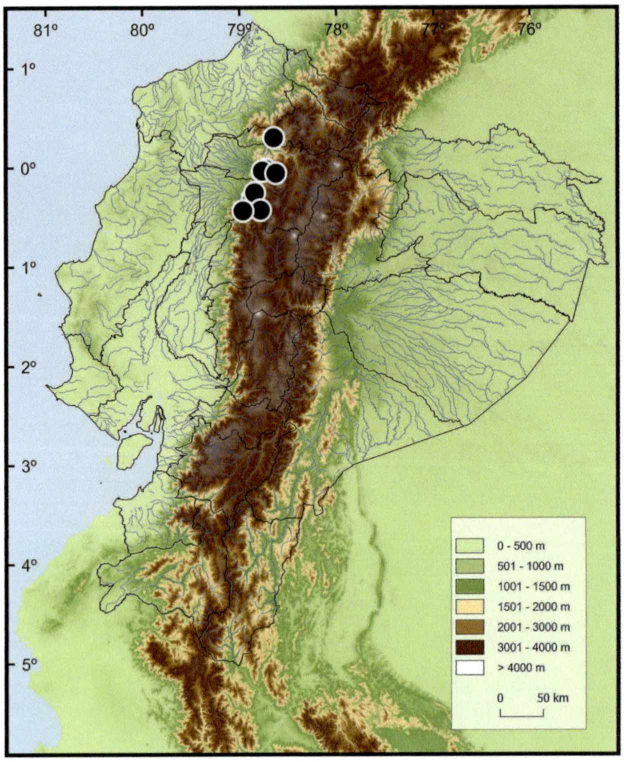

FIGURE 6.33 Localities of known occurrence of *Centrolene lynchi* in Ecuador.

This nocturnal, arboreal species inhabits natural forests and intervened areas in evergreen montane forest and evergreen low montane forest ecosystems in the Cordillera Occidental of the Andes biogeographic sector, where the average annual rainfall is 1185–2280 mm and the average annual temperature is 14.8–18.7 °C.

Conservation.—*Centrolene lynchi* is categorized as Endangered throughout its entire range (IUCN SSC Amphibian Specialist Group, 2022f; in IUCN Red List of 2022) and in Ecuador by Ortega-Andrade *et al.* (2021) because its entire extent of occurrence is 4268 km² (1200 km² in Ecuador according to our data), it occurs in five threat-defined locations, and there is a continuing decline in the extent and quality of its habitat. According to Guayasamin *et al.* (2020b), the species has not been found in some of its historic localities in the last 20 years, suggesting that the species has suffered population declines (e.g., San Francisco de Las Pampas). The amphibian chytrid fungus *Batrachochytrium dendrobatidis* has been found infecting *C. lynchi* at Reserva Las Gralarias, but no recent declines have been observed (Guayasamin *et al.*, 2014). About 33% of the potential distribution of the species is affected by human activities. Important threats to this species include deforestation due to agricultural activities, logging, fires, the introduction of exotic predatory fish species in streams, and other human-induced threats. In Ecuador, it has been recorded in the protected forests of Cuenca del Río Guayllabamba (Área 1–Área 2) and Toachi Pilatón and the biosphere reserve Chocó Andino de Pichincha.

Remarks.—Guayasamin *et al.* (2020b) place *Centrolene gemmatum* (Flores, 1985) in the synonymy of *Centrolene lynchi* given that several of the characteristics present in *C. gemmatum* fall into the variation observed in *Centrolene lynchi*, and others are the product of preservation artifacts. They also formally place *C. scirtetes* (Duellman and Burrowes, 1989), as defined by its holotype, in the synonymy of *Centrolene lynchi*, given that the holotype of *C. scirtetes* is undistinguishable from the holotype of *C. lynchi*. Additionally, they found that the paratypes of *C. scirtetes* are indistinguishable from individuals of *Nymphargus griffithsi*. Accordingly, they conclude that the Colombian material assigned to *C. scirtetes* is in fact representative of *N. griffithsi*.

Key references.—(Duellman, 1980) (description); (Flores, 1985) (description, as *Centrolene gemmatum*); (Duellman and Burrowes, 1989) (description of the holotype as *Centrolenella scirtetes*); (Dautel *et al.*, 2011) (natural history, call); (Arteaga *et al.*, 2013) (account, conservation); (Delia *et al.*, 2017) (natural history); (Guayasamin *et al.*, 2020b) (account).

"*CENTROLENE*" *MEDEMI* (MEDEM'S GLASSFROG) (COCHRAN AND GOIN, 1970)

Centrolenella medemi (Cochran and Goin, 1970): 511. **Holotype.**—USNM 152277 from Puerto Asís, upper Río Putumayo, Comisaría Putumayo, Colombia (apparently in error; see Distribution).

"*Centrolene*" *medemi* Guayasamin *et al.*, 2009: 57.

Etymology.—The specific epithet honors Dr. Federico Medem (Friedrich Johann Graf von Medem) (1912–1984), who found the specimens of the type series.

Identification.—Figure 6.34. The maximum known SVL is 30.8 mm in males and 44.3 mm in females. The skin on the dorsum of males and females is smooth to shagreen; the males have small spicules on the dorsum and flanks; the ventral surfaces of the thighs below have a pair of slightly enlarged tubercles. The anterior half of the ventral parietal peritoneum is white. There is a silvery white pericardium; no iridophores are present on the peritonea covering the intestines, stomach, testes, kidneys, gall bladder, and urinary bladder. The teeth on dentigerous process of the vomer are present or absent; each process bears 0–4 teeth. The snout is round in dorsal profile and truncate to slightly protruding in lateral profile. The tympanum is small, its diameter is 17.1–18.5% of the eye diameter; the dorsal third of tympanum is covered by the supratympanic fold; the tympanic membrane is pigmented and not clearly differentiated from the surrounding skin. The liver is tetralobed; the two large ventral lobes cover the two smaller lobes; the hepatic peritoneum lacks iridophores. In males, small humeral spines are present. The webbing between fingers I, II, and III is absent or basal; there is extensive webbing between fingers III and IV. The toes are fully webbed. The ulnar and tarsal folds are low. There prepollex is concealed in males; the nuptial pad is a large to medium-sized excrescence present on the dorsal, lateral, and/or ventral sides of the thumb. Finger I is slightly shorter than Finger II (Finger I is about 91–98% the length of Finger II). The disc of Finger III is large, 80–91% of the eye diameter.

The dorsal surfaces of the head, body, and limbs are olive green to grayish-brown with large (up to 2.8 mm) cream spots. The dorsum is, at night, black with bluish-green spots. The venter is dull blue. By day, the dorsum is dark olive brown with pale green spots. The flanks are cream. The lining of the mouth and tongue is pale blue. The bones are green. The webbing is pale yellowish-tan. The melanophores cover dorsal surfaces of the fingers and toes. The iris is dull grayish-bronze with minute black flecks or reticulation.

FIGURE 6.34 *Centrolene medemi*. KU 164493, female, 2 km southwest of Río Reventador, Provincia Napo. WED.

Comparisons.—*Centrolene medemi* has a unique coloration: dark olive green dorsum with large greenish-cream spots. Additionally, this species has a comparatively small tympanum (< 20% of eye diameter), and fully webbed foot. Adults are robust and relatively large (SVL 25.5–44.3 mm).

Natural history.—During the night, adults have been found on rocks, in rock crevices, and on rock cliffs along streams; juveniles were found on palm leaves or in rock crevices. The oviductal eggs of one female were dark brown. The eggs are deposited on rocks along streams.

Tadpoles.—Unknown.

Distribution.—Figure 6.35. *Centrolene medemi* is known from the Amazonian slopes of the Cordillera Oriental of Ecuador and on the eastern and western slopes of the Cordillera Oriental of the Colombian Andes. In Ecuador, it occurs in a single location in Provincia Napo, at an elevation of 1382 m.

This nocturnal, arboreal species inhabits natural forests in the evergreen low montane forest of North of Cordillera Oriental of the Andes ecosystem in the North of Cordillera Oriental of the Andes biogeographic sector, where the

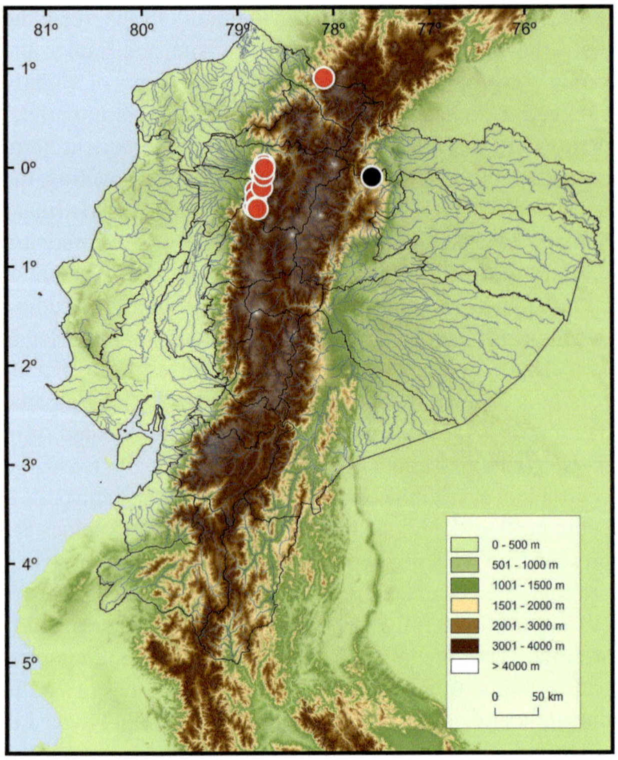

FIGURE 6.35 Localities of known occurrence of *Centrolene medemi* (black dot) and *C. peristicta* (red dots) in Ecuador.

average annual rainfall is 2580 mm and the average annual temperature is 19.3"C.

Conservation.—*Centrolene medemi* is categorized as Endangered throughout its entire range (IUCN SSC Amphibian Specialist Group, 2019r; in IUCN Red List of 2022). However, it is important to note that in Ecuador, *C. medemi* is categorized as Critically Endangered by Ortega-Andrade *et al.* (2021). We categorize and tag it as Critically Endangered (Possibly Extinct) following A1c IUCN criteria because we infer a general reduction of the population exceeding 90% and possibly nearing 100%. There are no sightings since about five decades; the most recent record of this species is 19 March 1975 from nearby Volcán Reventador, Provincia Napo. Exhaustive searches in the area have revealed no specimens. Threats to this species are severe habitat destruction and degradation, due to agriculture and clear cutting, grazing, pollution, fires, oil spills, landslide, and volcanic eruptions, but this does not account for the observed decline. Climate change and pathogens might be implicated as the likely causes of most of its decline. In Ecuador, it has been recorded in the public reserve Parque Nacional Cayambe Coca, and the biosphere reserve Sumaco.

Remarks.—Its molecular phylogenetic relationships are unknown.

Key references.—(Cochran and Goin, 1970) (description); (Lynch *et al.*, 1983) (natural history); (Ruiz-Carranza *et al.*, 1996) (distribution); (Suárez-Mayorga, 1999) (natural history); (Guayasamin *et al.*, 2006a) (distribution); (Guayasamin *et al.*, 2020b) (account).

CENTROLENE PERISTICTA (DAPPLED GLASSFROG) (LYNCH AND DUELLMAN, 1973)

Centrolenella peristicta (Lynch and Duellman, 1973): 45.
Holotype.—KU 118051 from Tandapi, 1460 m, Provincia Pichincha, Ecuador.
Centrolene peristicta Barrio-Amorós *et al.*, 2019: 19.

Etymology.—The specific epithet is derived from the Latin *peristiktos*, meaning dappled, and refers to the spotted color pattern of the species.

Identification.—Figures 6.36, 6.37. *Centrolene peristicta* is one of the smallest glassfrogs, reaching a maximum SVL of 21.2 mm in males and 20.9 mm in females. Also, it is one of the few species outside the genus *Hyalinobatrachium*, where males call mostly from the underside of leaves, and provide extended parental care. The skin on the dorsum is shagreen with small warts corresponding to

FIGURE 6.36 *Centrolene peristicta*. Left: QCAZ 40190, male. Right: QCAZ 40189, male. Reserva Las Gralarias, Provincia Pichincha. LAC.

FIGURE 6.37 Morphological details of *Centrolene peristicta*. CJ 11680, male, Santa Rosa, Provincia Pichincha. LAC.

yellowish-white spots, males and females lack spicules. The ventral surfaces of thighs below vent have a pair of enlarged tubercles. The anterior half of the ventral parietal peritoneum is white and the posterior half is transparent. There is a white pericardium; iridophores partially or completely cover the stomach and colon; no iridophores are present on the peritonea covering the kidneys, gall bladder, and urinary bladder. The vomers lack teeth. The snout is round in dorsal aspect and round or truncate in lateral profile. The tympanum is large and oriented almost vertically, with slight lateral and posterior inclinations; its diameter is 44.0–51.9% of the eye diameter; the tympanic annulus is visible; the supratympanic fold is evident; the tympanic membrane is translucent, partially pigmented and differentiated from surrounding skin. The liver is tetralobed and lacks iridophores. The males have conspicuous humeral spines. There is no webbing between fingers I and II; there is reduced webbing between fingers II and III; there is moderate to extensive webbing between the outer fingers. The webbing between the toes is extensive. The ulnar fold and the outer tarsal fold are present and have low white tubercles; the inner tarsal fold is low and short. The prepollex is exposed; the nuptial pad is a large to medium-sized excrescence present on the dorsal, lateral, and/or ventral sides of the thumb or a medium-sized spinous excrescence extending from the lateral side of the thumb to its dorsomedial surface. Finger I is slightly shorter than or as long as finger II (finger I length is 90.9–102.7% of finger II); the width of the disc of finger III is about 45.6–67.3% of the eye diameter.

The dorsum is green with minute yellowish-white flecks or spots and larger dark gray spots. The upper lip is white. The region below the eye has small white warts. The bones are green. The vocal sac is green. The upper flanks are green with minute white spots. The lower flanks are cream with minute white spots. The venter is yellowish-cream. The fingers and toes are dull yellow–green. The ulnar fold has a thin white line or low white ulnar tubercles. The outer tarsal fold has low white tubercles. There are small white tubercles just posterior to the cloaca. Melanophores are mostly absent from the fingers and toes, except for a few on the dorsal surfaces of finger IV and the proximal portion of finger III. The iris is white–gray with a yellow hue and thin black reticulations; a thin yellow line surrounds the pupil.

Comparisons.—*Centrolene antioquiense*, *C. lynchi*, *C. pipilata*, and *Nymphargus truebae*. *Centrolene peristicta* is distinguished from most glassfrogs by its minute body size (SVL < 21.3 mm) and having, in life, a yellowish-green dorsum with minute white flecks and small diffuse black spots. Additionally, adult males of *C. peristicta* have a small, curved humeral spine that is morphologically different from the spines present in most glassfrogs (i.e., not curved). *Centrolene antioquiense*, an endemic species to Colombia, is remarkably similar to *C. peristicta*, but *C. antioquiense* has less hand webbing. *C. lynchi* is larger than *C. peristicta* (in males, SVL 23.3–26.5 mm in *C. lynchi*) and has a very different call. *C. pipilata* occurs on the Amazonian slopes of the Andes, whereas *C. peristicta* is restricted to the Pacific slopes of the Andes. *N. truebae*, a species only known from the Andes of southern Peru, has the same color pattern as *C. peristicta*, but *N. truebae* lacks humeral spines and webbing between fingers III and IV (present in *C. peristicta*).

Natural history.—Figure 6.38. At Reserva Las Gralarias, Provincia Pichincha, Ecuador, *Centrolene peristicta* is active during the night on vegetation 40–600 cm above permanent streams in primary forests. The species is abundant at Lucy's Creek, but it is also found at Kathy's Creek and Río Santa Rosa. It breeds during the rainy season (December–April), but peaks occur in February–April. Males vocalize to advertise themselves and defend territories, in which females place clutches containing 6–41 eggs. Most of the time, egg clutches are placed on the underside of leaves and ferns 70–500 cm directly above streams or under dead leaves near streams. Males are polygynous and exhibit high site fidelity; some males were observed guarding egg clutches at different stages of development simultaneously. Parental care by males was demonstrated experimentally; unattended clutches have a significantly lower eclosion rate and longer hatching time than attended clutches. Clutch mortality was mainly because of desiccation, predation, and

FIGURE 6.38 *Centrolene peristicta*. Top: males. Jaime Culebras. Middle: males guarding clutches of embryos deposited under leaves. LAC. Bottom right: clutches of embryos. LAC. Reserva Las Gralarias, Provincia Pichincha. Bottom left: clutch of embryos, Mindo, Provincia Pichincha. Martín R. Bustamante.

FIGURE 6.39 Ontogenetic changes in *Centrolene peristicta* from tadpole to metamorph to juvenile. Top to bottom. QCAZ 30142, stage 33, lateral and dorsal views, TL = 31mm; undetermined stage, TL = 33 mm; stage 42; stage 45, 35 days after stage 33; juvenile 110 days after stage 45. Reserva Las Gralarias, Provincia Pichincha. LAC.

parasitism. Embryos develop for 17–27 days and then hatch as free-swimming larvae.

Advertisement calls of 35 individuals from Reserva Las Gralarias, Provincia Pichincha, Ecuador were described as one short and pulsed note that resembles the sound of a cricket chirp. The note has a mean duration of 0.062 s, and each call has one note. Detailed data and figures are provided under the subtitle "Vocalizations of Ecuadorian Anurans" in Appendix 1.

Tadpoles.—Figure 6.39. This description is based on an individual (not preserved) from Reserva Las Gralarias, Provincia Pichincha, Ecuador, in stage 33. It measures 8.7 mm in body length and 31.0 mm in total length. The body is ovoid, slightly wider posteriorly, and wider than it is deep. The snout appears round both in dorsal view and in lateral profile. Moderately large eyes are narrowly separated and directed dorsally. The chondrocranial elements are visible. The spiracle is sinistral, with the inner wall free from the body, and its opening is directed posterolaterally at about two-thirds of the body's length. The tail accounts for approximately 72% of the total length, with its greatest depth occurring at the midlength, and a subacute tip. Both dorsal and ventral fins are of similar depth, approximately half as deep as the caudal musculature at the midlength of the tail.

The skin is translucent, allowing for the visibility of internal organs and imparting a reddish coloration to the tadpole's body. Dorsally, there is an aggregation of melanophores forming a wide band along the body, starting at the level of the eyes and extending to the tail. Laterally, melanophores are present on the tail and concentrated along the midline. The rest of the body and tail remain unpigmented.

At stage 40, 24 days after stage 33, the tadpole acquires a light green coloration on the body and posterior legs, while the tail maintains its previous coloration. At stage 42, four days later, the green coloration intensifies, and the snout takes on a bluish-green hue. The bones are visible through the skin, appearing blue. Finally, at stage 45, seven days after stage 42, the individual's coloration closely resembles that of adults.

Distribution.—See Figure 6.35. *Centrolene peristicta* is known on the Pacific slopes of the Cordillera Occidental of the Andes in Ecuador and Colombia. In Ecuador, it occurs in Carchi, Santo Domingo de los Tsáchilas, and Pichincha provinces. Its elevational range is 1423–2396 m in an area of extent of occurrence of about 1490 km^2.

This nocturnal, arboreal species inhabits intervened areas and natural forests in evergreen low montane forest and evergreen montane forest ecosystems in the Cordillera Occidental of the Andes biogeographic sector, where the average annual rainfall is 1307–2211 mm and the average annual temperature is 14.6–18.9 °C.

Conservation.—*Centrolene peristicta* is categorized as Least Concern (IUCN SSC Amphibian Specialist Group, 2019s; in IUCN Red List of 2022) throughout its entire range. However, it is important to note that in Ecuador, *C. peristica* was categorized as Near Threatened by Ortega-Andrade *et al.* (2021), given the fragmented distribution of the species. According to Guayasamin *et al.* (2020b), several recent observations show that the species has a wider distribution than previously thought. Reproductive populations have been observed in Provincia Pichincha at Reserva Las Gralarias (December 2009–May 2018), Mindo Biology Station (February 2002; December 2014), and Río Pachijal (August 2001). The potential distribution of the species covers an area of 12,603 km^2, 44.6% of which is affected by human activities. The main threats to this species are deforestation and habitat fragmentation, mainly due to agricultural activities, including crops and livestock, pollution from agriculture, hydroelectric development, logging, fires, and mining. At Reserva Las Gralarias, the species is infected by the amphibian chytrid fungus *Batrachochytrium dendrobatidis*, but populations look healthy and there is no evidence of declines. It occurs in three protected areas in Colombia: La Planada and Rio Ñambí Natural Reserves and Munchique Natural National Park. In Ecuador, it has been recorded in the protected forests of Estación Científica Río Guajalito and Toachi Pilatón, the biosphere reserve Chocó Andino de Pichincha, and some private reserves (e.g., Reserva Las Gralarias, Mindo Biology Station).

Remarks.—*Centrolene antioquiense* and *C. peristicta* are nearly identical in morphology and could represent a single species. The subtle differences found between these species might be a consequence of incomplete sampling or intraspecific geographic variation. An analysis of the calls of the two taxa is needed to determine their species status.

Key references.—(Lynch and Duellman, 1973) (description); (Hutter and Guayasamin, 2012) (habitat); (Salgado and Guayasamin, 2018) (natural history); Yánez-Muñoz *et al.* (2020) (account); (Guayasamin *et al.*, 2020b) (account, call).

CENTROLENE PIPILATA (PEEPING GLASSFROG) (LYNCH AND DUELLMAN, 1973)

Centrolenella pipilata (Lynch and Duellman, 1973): 46.

Holotype.—KU 143278 from 16.5 km north–northeast of Santa Rosa, 1700 m, on Quito-Lago Agrio Road, Provincia de Napo, Ecuador.

Centrolene pipilata Barrio-Amorós *et al.*, 2019: 19.

Etymology.—The specific epithet is an adjectival derivative of the Latin verb *pipila*, meaning to peep, and refers to the call of this centrolenid frog.

Identification.—Figure 6.40. These frogs attain a maximum SVL of 22.6 mm in males and 23.6 mm in females. The skin on the dorsum is shagreen; the males have spicules coinciding with white spots; the ventral surfaces of the thighs below the vent have a pair of enlarged tubercles. The anterior 50–60% of the ventral parietal peritoneum is white, while the posterior portion is transparent. There is a white pericardium; no iridophores are present on the peritonea covering the intestines, stomach, kidneys, gall bladder, and urinary bladder. The vomers lack teeth. The snout is round in dorsal aspect and round or truncate in lateral profile. The tympanum is moderate, oriented almost vertically, with slight lateral and posterior inclinations, and its diameter 31.0–39.4% of the eye diameter; the tympanic annulus is completely visible; the supratympanic fold is evident; the tympanic membrane is translucent, partially pigmented, and clearly differentiated from surrounding skin. The liver is clearly tetralobed and lacks iridophores. The humeral spines of males are conspicuous. There is no webbing between fingers I, II, and III, and there is moderate webbing between outer fingers. The webbing between toes is extensive. The ulnar fold is present and white; the outer tarsal margin has low white tubercles; the internal tarsal fold is low. The prepollex is exposed; the nuptial pad is a small circular or squarish excrescence present on the dorsal or dorsolateral face of the thumb. Finger I is as long as finger II or slightly shorter (finger I 90.0–100% of finger II). The width of the disc of finger III is 44.8–56.5% of the eye diameter.

FIGURE 6.40 *Centrolene pipilata.* KU 143278, holotype, male, 22.9 mm SVL, 16.5 km northeast of Santa Rosa, Provincia Napo. WED.

The dorsum is green with minute yellowish-white flecks and larger diffuse black spots. The upper lip is white. The region below the eye presents small white warts that are spiculate in males. The upper flanks are green with minute white spots. The lower flanks are whitish-cream. The ulnar fold and the tubercles of the outer tarsal fold are white. There are small white tubercles just posterior to the cloaca. Melanophores are mostly absent from the fingers and toes, except for a few on the proximal half of finger IV and along toes IV and V. The bones are green. The iris is grayish-white with thin black reticulations and a yellow hue around the pupil.

Comparisons.—*Centrolene sanchezi*, *C. huilensis*, *Nymphargus truebae*. In life, *Centrolene pipilata* is differentiated from most glassfrogs by having a green dorsum with yellowish-white flecks, diffuse dark green marks, and a distinct prepollex. On the Amazonian slopes of the Ecuadorian Andes, only *C. sanchezi* and *C. huilensis* show similarities in their color patterns. However, *C. sanchezi* has a short, thin humeral spine (large, wide, and usually with the tip projected anteriorly in *C. pipilata*) and lacks a distinct prepollical spine. *Centrolene huilensis* is larger than *C. pipilata* (in *C. huilensis*, male SVL = 23.6–26.7 mm; SVL = 28.7 mm in one female; in *C. pipilata*, male SVL 19.7–22.6 mm, female SVL 22.6–23.6 mm).

Natural history.—All individuals have been found at night on vegetation along cascading mountain streams. At Río Azuela, *Centrolene pipilata* was found in sympatry with *Nymphargus anomalus*, *N. megacheirus*, *N. siren*, and *Hyalinobatrachium pellucidum*. At 16.5 km north–northeast of Santa Rosa, the species was found with *Espadarana audax*, *N. megacheirus*, and *N. siren*. Males call from the upper side of leaves. In the laboratory, a female deposited a clutch of 18 eggs. Eggs have clear jelly and pale green yolks.

Tadpoles.—Unknown.

Distribution.—See Figure 6.9. *Centrolene pipilata*, an endemic species to Ecuador, is known on the eastern slopes of the Cordillera Oriental of the Andes in Provincia Napo. Its elevational range is 1510–1898 m in an area of extent of occurrence of about 90 km^2.

This nocturnal, arboreal species inhabits natural forests and intervened areas in the evergreen low montane forests of the northern Cordillera Oriental of the Andes ecosystem in the northern Cordillera Oriental of the Andes biogeographic sector, where the average annual rainfall is 1982–2452 mm and the average annual temperature is 17 – 19 ºC.

Conservation.—*Centrolene pipilata* is categorized as Critically Endangered (Possibly Extinct) (IUCN SSC Amphibian Specialist Group, 2019t; in IUCN Red List of 2022) because it has a very restricted distribution (extent of occurrence = 90 km^2) in northeastern Ecuador, and it occurs in a single location where there is a continuing decline in the extent and quality of its habitat. Guayasamin *et al.* (2020b) also suggest that the species should be considered Critically Endangered based on the IUCN criteria A2c,e. According to Guayasamin *et al.* (2020b), the last confirmed report of *C. pipilata* was from 14.7 km northeast of Río Salado in February 1979. Recent surveys at Río Azuela and the type locality failed to find the species. The main threat to this species is the development of a hydroelectric project (Coca-Codo) that will affect the entire known distribution of the species. Other threats include habitat loss and degradation, mainly due to smallholder farming and livestock ranching, and the eruptions of the active volcano El Reventador. An oil pipe goes through this species' range, and it may present an impact if a spill occurs due to potential eruptions of the volcano. It has been recorded in the public reserve Parque Nacional Cayambe Coca and the biosphere reserve Sumaco.

Remarks.—Its molecular phylogenetic relationships are unknown.

Key references.—(Lynch and Duellman, 1973) (description); (Guayasamin *et al.*, 2020b) (account).

CENTROLENE SANCHEZI (SÁNCHEZ'S GLASSFROG) RUIZ-CARRANZA AND LYNCH, 1991b

Centrolene sanchezi Ruiz-Carranza and Lynch, 1991b: 22.

Holotype.—ICN 24293 from 3.1 km by road below Alto Gabinete, 2190 m, Vereda Gabinete, Municipio Florencia, Departamento Caquetá, Colombia.

Etymology.—The specific epithet honors Ricardo Sánchez, who, with John D. Lynch, found the species.

Identification.—Figures 6.41, 6.42. These frogs attain a maximum SVL of 22.3 mm in males and 20.9 mm in one female. The skin on the dorsum is shagreen with small spicules evident in most males; there is a pair of enlarged subcloacal tubercles. The anterior half of the ventral parietal peritoneum is white, while the posterior half is translucent. There is a white pericardium; there is a translucent visceral peritoneum. The vomerine teeth are absent. The snout is rounded in dorsal aspect and bluntly rounded or truncate in lateral profile. The tympanum is oriented almost vertically; its diameter is 31.4–37.8% of the eye diameter; the tympanic annulus is visible except for a dorsal border covered by the supratympanic fold; the tympanic membrane is partially pigmented but differentiated from the surrounding skin. The liver is tetralobed and lacks iridophores. The humeral spines are present in adult males. There is no webbing between fingers I and II, and it is basal among the others. The webbing on the foot is moderate. The ulnar and inner tarsal folds are low or absent; the outer tarsal fold is absent; the nuptial pad is a large to medium-sized excrescence present on the dorsal, lateral, and/or ventral sides of the thumb. The prepollex is not visible (concealed). Fingers I and II are about equal in length (FII/FI = 91.3–104.5%). The disc of finger III is of moderate size, and it is about 40.3–50.6% of the eye diameter.

FIGURE 6.41 *Centrolene sanchezi*. Left: CJ 11456, female, 20.7 mm SVL. Right: CJ 11455, male, 19.8 mm SVL. Yanayacu, Provincia Napo. SGS.

FIGURE 6.42 Morphological variation and details of *Centrolene sanchezi*. Top left: CJ 12885, male, Yanayacu, Provincia Napo. SGS. Top right: ICN 24293, holotype, 31 km by road, below the Alto Gabinete, Departamento de Caquetá, Colombia. Middle left: ICN 11685, km 84 road Popayán a Inzá, Río Guanacas, Departamento del Cauca, Colombia. Pedro Ruiz Carranza. Middle right and bottom: CJ 11455, male, Yanayaku, Provincia Napo. SGS.

The dorsum is dark green with small white spots. There is a series of white–cream tubercles under the eye and tympanum. The upper lip is white. The throat and ventral surfaces of the limbs are green. The digits are pale green. The cloacal region presents white warts. The ventral parietal peritoneum is white anteriorly and translucent posteriorly. The visceral peritoneum is translucent. The melanophores are mostly absent from the fingers and toes, except for a few on toes IV and V and on the base of the outer fingers. The bones are green. The iris is pale bronze with thin black reticulations.

Comparisons.—*Centrolene pipilata*. *Centrolene sanchezi* is easily differentiated from other glassfrogs by its minute body size (SVL 18.9–22.3 mm), green dorsum with small white spots, and the presence of white warts in an area that extends from below the eye to the insertion of the arm. Additionally, adult males have conspicuous humeral spines. The only other centrolenid from eastern Ecuador that can be confused with *C. sanchezi* is *C. pipilata*, which has a green dorsum with small, diffuse black spots and yellowish-white flecks and a prepollical spine. Additionally, the two species occupy different elevations on the Amazonian slope of the Andes; *C. sanchezi* is found at elevations of 1950–2350 m, whereas *C. pipilata* occurs at 1420–1910 m.

Natural history.—During the night, *Centrolene sanchezi* has been found on leaves approximately 130–300 cm above streams in primary and secondary forests. At Yanayacu Biological Station, *C. sanchezi* (reported as *C. bacatum*) is the most abundant centrolenid; here, males were heard calling during February (2013). Other sympatric species at Yanayacu include *Centrolene* aff. *buckleyi*, *C. huilensis*, *Nymphargus posadae*, *N. siren*, and *N. wileyi*. Males call from the upper sides of leaves. A male was found near two egg clutches, one on the underside and the other on the upper side of a single leaf; egg clutches have 14–18 pale yellowish-green eggs.

Advertisement calls of one individual from Reserva Yanayacu, Provincia Napo, Ecuador, were described as very short, with a mean duration of 11 ms. The call consists of one or two notes produced in series. Detailed data and figures are provided under the subtitle "Vocalizations of Ecuadorian Anurans" in Appendix 1.

Tadpoles.—Figure 6.43. This description is based on an individual (not preserved) from Estación Biológica Yanayacu, Provincia Napo, Ecuador, in stage 31. It was collected by Diego Almeida-Reinoso on 27 April 2008. It has a total length of 33.4 mm. The body exhibits an ovoid and sinuous shape, with equal width both posteriorly and anteriorly, widening in the midsection. The snout appears round in dorsal view. The eyes are moderately large, narrowly separated, and directed dorsally. The sinistral spiracular opening is directed posteriorly at about two-thirds

FIGURE 6.43 Ontogenetic changes in two individuals of *Centrolene sanchezi* from tadpole to juvenile. Top to bottom, left to right. QCAZ 38422a, stage 31, lateral, dorsal, and ventral views, TL = 33.4 mm. QCAZ 38422b, undetermined stage; stage 42; stage 45; froglet 52 days after stage 45; juvenile 112 days; juvenile 227 days. Yanayacu Biological Station, Provincia Napo. LAC.

the body's length. The tail represents approximately 70% of the total length, with the dorsal fin is not extending onto the body.

The skin is translucent, allowing for the visibility of the internal organs and giving the tadpole a red coloration. Brown blotches are present dorsally between the eyes and on the body and tail, while the rest of the body and tail remain unpigmented.

In individuals at stages 39–40, the body, legs, and tail exhibit a light green hue, while the snout takes on a greenish-blue color. The chondrocranial elements and vertebrae become visible dorsally through the skin. The brown blotches are transformed into scattered brown dots on the body, while those on the tail retain their original appearance. By stages 42–44, occurring seven days after stage 40, there is a uniform light green coloration with white dots along the entire dorsum. The brown blotches on the tail have transitioned into light brown dots. At stage 45, the green coloration becomes darker, turning into an olive green with white dots. As a froglet, the olive green becomes lighter and extremities take on a yellowish hue. Sixty days later, as a juvenile, the frog adopts a coloration similar to that of the adults.

Distribution.—See Figure 6.28. *Centrolene sanchezi* is known on the Amazonian slopes of the Andes of Ecuador and Colombia and from one locality on the eastern slopes of the Colombian Cordillera Central. In Ecuador, it occurs in Napo and Morona Santiago provinces. Its elevational range is 1553–2118 m in an area of extent of occurrence of about 430 km^2.

This nocturnal, arboreal species inhabits natural forests and intervened areas in evergreen low montane forest and evergreen montane forest ecosystems in the northern Cordillera Oriental of the Andes and Amazonian Cordilleras biogeographic sectors, where the average annual rainfall is 1346–2217 mm and the average annual temperature is 16.1–19.9 °C.

Conservation.—*Centrolene sanchezi* is categorized as Endangered through its entire range (IUCN SSC Amphibian Specialist Group, 2022g; in IUCN Red List of 2022) and in Ecuador (under *C. bacatum*) by Ortega-Andrade *et al.* (2021). In Ecuador, its current extent of occurrence is 430 km^2 (according to our data), its distribution is severely fragmented, and there is a continuing decline in the extent and quality of its habitat. The species' habitat is threatened by agricultural development, logging, and human settlement. The habitat of the species is fragmented by agriculture and pasture land and threatened by mining (mainly in southern Ecuador). In Ecuador, it has been recorded in the public reserve Parque Nacional Antisana, and the biosphere reserve Sumaco.

Remarks.—Its molecular phylogenetic relationships are unknown. Guayasamin *et al.* (2020b) placed *Centrolene bacatum* (Wild, 1994) and *Centrolene guanacarum* (Ruiz-Carranza and Lynch, 1995a) under the synonymy of *Centrolene sanchezi* (Ruiz-Carranza and Lynch, 1991b), given that the examination of the type material of *Centrolene sanchezi*, *C. bacatum*, and *C. guanacarum* revealed no morphological differences among them. Moreover, all specimens share two distinctive traits: a small, laminar humeral spine in males and the presence of a lateral row of enameled warts that extends from below the eye to just posterior to the insertion of the arm.

Key references.—(Ruiz-Carranza and Lynch, 1991b) (description); (Wild, 1994) (description as *Centrolene bacatum*); (Ruiz-Carranza and Lynch, 1995a) (description under *Centrolene guanacarum*); (Guayasamin *et al.*, 2006b) (natural history, under *Centrolene bacatum*); (Guayasamin and Funk, 2009) (natural history, under *Centrolene bacatum*); (Guayasamin *et al.*, 2020b) (account, call).

GENUS *CHIMERELLA* (GLASSFROGS) GUAYASAMIN *ET AL.*, 2009

Chimerella Guayasamin *et al.*, 2009: 31. **Type species.**—*Centrolene mariaelenae* Cisneros-Heredia and McDiarmid, 2006.

Etymology.—The generic name comes from the Greek *Chimaira*; the suffix *-ella* is a diminutive form. In Greek mythology, the Chimera is a creature composed of parts of multiple animals; the name refers to the peculiar combination of morphological characteristics of *Chimerella mariaelenae*. The genus is feminine.

Definition.—Figure 6.44. Species within the genus *Chimerella* are minute, with a body size reaching a maximum of 21 mm. All known species have a transparent parietal peritoneum and white pericardial, hepatic, and visceral peritonea. Adult males call from the upper surfaces of leaves and exhibit a small humeral spine.

Content.—*Chimerella* contains three species, of which one occurs in Ecuador.

Distribution.—*Chimerella* is endemic to the Amazonian slopes of the Andes of Ecuador and Peru at elevations between 610 and 1820 m.

Remarks.—Figure 6.45. The phylogeny of species presented herein follows (Guayasamin *et al.*, 2020b).

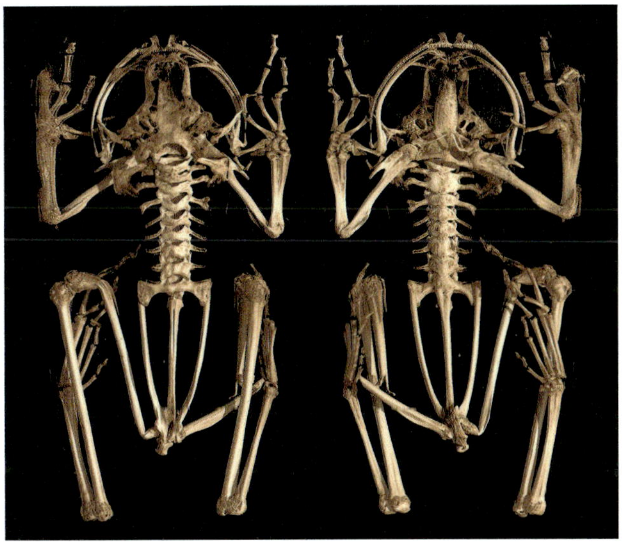

FIGURE 6.44 High-resolution computed tomography reconstructions of the skeleton of *Chimerella corleone*, CORBIDI 10475, male, 26.6 mm SVL, near the village of San José in the Cainarachi valley, Departamento San Martín, Peru. Dorsal (left) and ventral (right) views. Daniel J. Paluh.

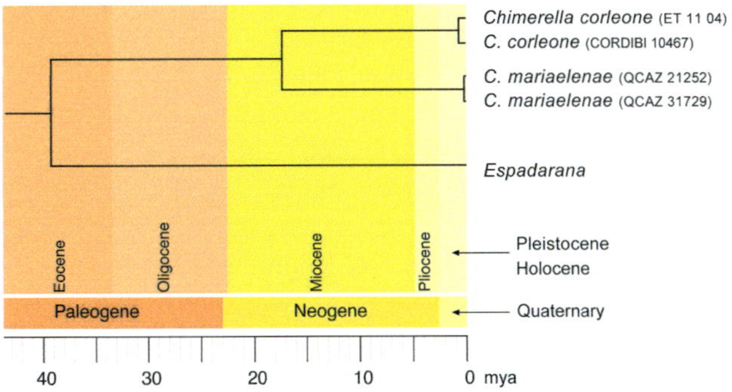

FIGURE 6.45 Time-calibrated maximum likelihood phylogenetic tree depicting evolutionary relationships of *Chimerella*, inferred from DNA sequences of mitochondrial and nuclear genes. Phylogeny taken and redrawn from Guayasamin *et al.* (2020b).

CHIMERELLA MARIAELENAE (MARÍA ELENA'S GLASSFROG) (CISNEROS-HEREDIA AND MCDIARMID, 2006)

Centrolene mariaelenae (Cisneros-Heredia and McDiarmid, 2006): 4. **Holotype.**—DFCH-USFQ D125 from tributary of the Río Jambue, ca. 16 km south from Zamora, Podocarpus National Park, 1820 m, Provincia Zamora Chinchipe, Ecuador.
Chimerella mariaelenae Guayasamin *et al.*, 2009: 31.

Etymology.—The specific epithet is a matronym for María Elena Heredia, Diego F. Cisneros-Heredia's mother.

Identification.—Figures 6.46, 6.47. These frogs attain a maximum SVL of 19.7 mm in males and 20.8 mm in one female. The skin on the dorsum is shagreen; males have minute dorsal spicules (only visible under 250×magnification);

FIGURE 6.46 Morphological variation in *Chimerella mariaelenae*. Top and second row. Left: QCAZ 49458, female, lab born, parents from Río Hollín, Provincia Napo. LAC. Right: CJ 7281, male, 19.5 mm SVL, Paquisha Alto, Provincia Zamora Chinchipe. DAL. Third row left: CJ 4386, male, Río Blanco, Provincia Zamora Chinchipe. Third row right: CJ 7290, male, Paquisha Alto, Provincia Zamora Chinchipe. LAC. Bottom left: male from Mera, Provincia Pastaza. Jaime Culebras. Bottom right: QCAZ 31729, male, Río Negro, Provincia Tungurahua. LAC.

FIGURE 6.47 Morphological details of *Chimerella mariaelenae*. QCAZ 49458, female, lab born, parents from Río Hollín, Provincia Napo. LAC.

FIGURE 6.48 *Chimerella mariaelenae*. Top right: amplectant pair, Reserva Tamandúa, Provincia Pastaza. Jaime Culebras. Top left: MZUTI 1680–81, amplectant pair, Quebrada Pangayacu, Provincia Napo. Eduardo Toral. Middle and Bottom left: CJ 4390, clutches on the upper sides of leaves, Río Blanco, Provincia Zamora Chinchipe. LAC. Bottom right: clutch of embryos, Río Hollín, Provincia Napo. LAC.

the ventral surfaces of the thighs below the vent have a pair of enlarged tubercles. The ventral parietal peritoneum is transparent; the iridophores in the pericardium and peritonea cover the digestive tract and testes; the kidneys and urinary bladder lack iridophores. The vomers lack teeth. The snout is truncate in dorsal aspect and truncate to slightly protruding in lateral profile. The tympanum is oriented posterolaterally with a slight dorsal inclination, its diameter is 23.0–27.4% of the eye diameter; the tympanic membrane is translucent and clearly differentiated from the surrounding skin. The liver is bulbous and covered by iridophores. Small humeral spines are present in males. Webbing is absent or basal between the inner fingers but moderate between outer fingers. The webbing between the toes is extensive. The ulnar and tarsal folds are absent. The prepollex is not separated from finger I; the nuptial pad is a large to medium-sized excrescence present on the dorsal, lateral, and/or ventral sides of the thumb. Finger I is slightly longer than finger II (finger II length is 95.6–97.3% of finger I). The disc of finger III is relatively narrow; its width is about 22.3–25.7% of the eye diameter.

The dorsum is yellowish-green with small gray to black spots. The venter is completely transparent (clear ventral parietal peritoneum), showing a white heart (covered by a white pericardium). The digestive tract, liver, testes, and gall bladder are white. There is a transparent peritoneum covering the urinary bladder. The bones are white to light green. The dorsal surfaces of the fingers and toes lack melanophores, except for some on the base of toe V. The iris is white with abundant spots or reticula forming a circular gray stripe on its periphery. The pupil is surrounded by a reddish-brown to orange line and a brick brown to black horizontal stripe.

Comparisons.—*Hyalinobatrachium* sp., *Teratohyla amelie*. On the Amazonian slopes of Ecuador, only species in the genus *Hyalinobatrachium* and *T. amelie* could be confused with *C. mariaelenae*. However, these species lack the small dark spots that characterize the dorsum of *C. mariaelenae*. Additionally, adult males of *C. mariaelenae* have small humeral spines, which are absent in all *Hyalinobatrachium* species and *T. amelie*.

Natural history.—Figure 6.48. All individuals have been found on the upper surfaces of leaves along small streams and ditches in cloud forests. Reproductively active individuals were observed at Quebrada Pangayacu in August 2012. The eggs are laid on the upper surfaces of leaves. Short-term maternal care present; parental care by males is absent.

Advertisement calls of one individual from Pangayacu creek, Provincia Napo, Ecuador, were described as variable from being short to a moderate length call. The typical call has two notes, but some calls can have up to 10 notes. Each note is extremely short at 4–7 ms. Detailed data and figures are provided under the subtitle "Vocalizations of Ecuadorian Anurans" in Appendix 1.

Tadpoles.—Figure 6.49. The tadpoles were described from Río Hollín, Provincia Napo, Ecuador. The description is based on a tadpole on stage 39 (TL = 35.5 mm;

FIGURE 6.49 Ontogenetic changes in three individuals of *Chimerella mariaelenae* from tadpole to juvenile. Top to bottom, left to right. QCAZ 49458, stage 28, lateral and dorsal views, TL = 28.9 mm; undetermined stage, lateral view, TL = 28.9 mm. Río Hollín, Provincia Napo. CJ 4390; stage 42, TL = 37 mm; stage 45; juvenile 107 days after stage 45; juvenile 261 days = 16.7 mm SVL; adult 339 days. Río Blanco, Provincia Zamora Chinchipe. LAC.

BL = 9.9 mm). The body is elongated, oval, and depressed, wider than higher. The snout is rounded in dorsal and lateral views. The lateral line system is visible. The eyes are dorsal; there are small narial apertures. The interorbital distance is shorter than the internarial distance. The short spiracle is single, sinistral, located in the posterolateral region of the body, 80.4% of the body length from the tip of the snout; the spiracular aperture has a dorsoposterior orientation, with the inner wall present as a low ridge. The vent tube is short and abdominal, free posteriorly, with the opening directed posteriorly. The tail length (26.1 mm) is 2.6 times the body length, with a subacute tip. Myotomes are visible throughout the entire length of the tail; a straight medial line is visible, separating the dorsal and ventral myotomes. The dorsal fin originates at about the midlength of the tail; its height (except for the beginning) is more or less uniform until the distal end, where it decreases abruptly. Proximally, the ventral fin originates at the base of the tail muscle, reaching its maximum height after the midlength of the tail. The medium-sized oral disc (width = 2.6 mm) is 51.2% of the body width, not emarginated, and anteroventral, surrounded by 49 marginal uniserial papillae. There are only ventral and lateral papillae, lacking dorsal papillae. The upper and lower jaw sheaths are nearly straight and completely keratinized, with a serrated edge. The lower jaw sheath is placed behind the upper jaw. The labial tooth row formula is 2(1)(2)/3. Tooth rows are about equal in length.

Dorsally, the tadpole is brownish with two areas without pigmentation at the anterolateral border of the eye. Some reddish pigmentation is seen in the anterior half of the dorsum because of its transparency. The most anterior part of the head is gray. Iridophore aggregations are present on the dorsum along the vertebral. The iris is bronze.

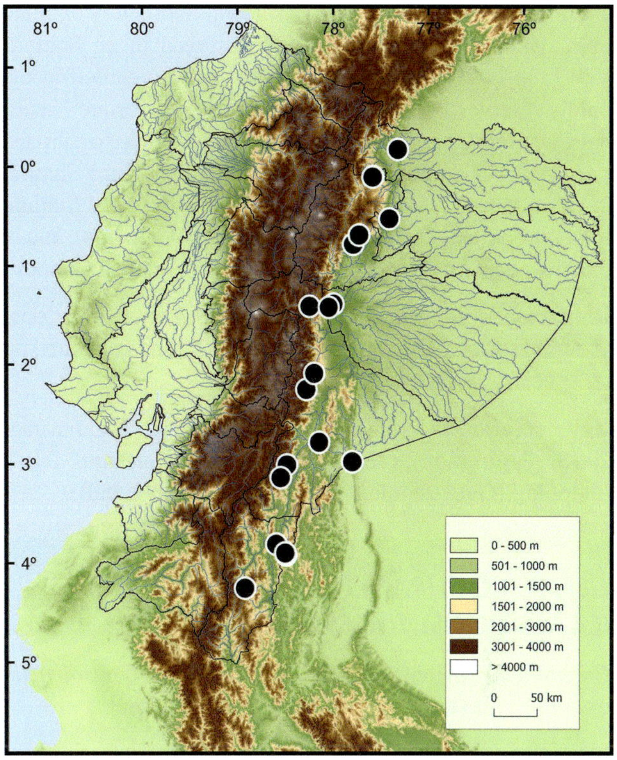

FIGURE 6.50 Localities of known occurrence of *Chimerella mariaelenae* in Ecuador.

Distribution.—Figure 6.50. *Chimerella mariaelenae* occurs from the Amazonian slopes of the Cordillera Oriental of the Andes in Ecuador to the Kampankis mountains between the Santiago and Morona rivers of northeastern Peru. In Ecuador, it occurs in Sucumbíos, Orellana, Napo, Tungurahua, Pastaza, Morona Santiago, and Zamora Chinchipe provinces. Its elevational range is 594–1932 m in an area of extent of occurrence of about 30,100 km^2.

This nocturnal, arboreal species inhabits natural forests and intervened areas in six ecosystems (Table 0.4) within four biogeographic sectors (Table 0.5), where the average annual rainfall is 1043–4015 mm and the average annual temperature is 17.2–24.3 °C.

Conservation.—*Chimerella mariaelenae* is categorized as Least Concern throughout its entire range (IUCN SSC Amphibian Specialist Group, 2018n; in IUCN Red List of 2022). This categorization is based on several factors: the species is widely distributed; it occurs in an area of extensive, suitable habitat, which appears not to be facing substantial threats; it is somewhat adaptable to habitat change; and it is presumed to have a large population. However, it is important to note that in Ecuador, *C. mariaelenae* has been categorized as Near Threatened by Ortega-Andrade *et al.* (2021). In Ecuador, major threats for the species include deforestation and mining. The distribution of *C. mariaelenae* covers an area of 30,100 km^2, and a portion of it (not less than 17%) is threatened by human activities. An apparently healthy subpopulation of this species occurs around

Cordillera Kampankis, Zona Reservada Santiago-Comaina, Peru, where 50 individuals were observed in an hour of searching in 2011. In Ecuador, it has been recorded in the public reserves Parque Nacional Sangay and Parque Nacional Cayambe Coca, in the protected forests of Cerro Sumaco y Cuenca Alta del Río Suno and Cordillera Kutukú y Shaimi, in some private reserves (e.g., Río Bigal Biological Station, Reserva Narupa, Zona Reservada Santiago-Comaina), and in the biosphere reserves Sumaco and Podocarpus-El Cóndor.

Remarks.—Phylogenetic trees based on molecular data show that *Chimerella mariaelenae* and *C. corleone* are sister species (Twomey *et al.*, 2014).

Key references.—(Cisneros-Heredia and McDiarmid, 2006) (description); (Terán Valdez and Guayasamin, 2014) (tadpole); (Guayasamin *et al.*, 2020b) (account, call).

GENUS *COCHRANELLA* (GLASSFROGS) TAYLOR, 1951

Cochranella Taylor, 1951: 62. **Type species.**—*Centrolenella granulosa* (Taylor, 1949) by original designation.

Etymology.—Taylor (1951) named the glassfrog genus *Cochranella* in honor of Doris M. Cochran, former curator at the Smithsonian Institution. Cochran's research focused on the herpetofauna of the West Indies and South America, particularly Brazil and Colombia. The genus is feminine in gender.

Definition.—Figure 6.51. *Cochranella* is diagnosed by the following combination of traits: (1) absence of humeral spines (small spine present in *C. litoralis*); (2) white digestive tract (translucent in *Cochranella nola*); (3) lobed liver covered by a transparent hepatic peritoneum; (4) ventral parietal peritoneum white anteriorly and transparent posteriorly; (5) moderate to extensive webbing between fingers III and IV; (6) green bones in life; (7) males that call from upper surfaces of leaves; (8) females that deposit eggs on the upper sides of leaves along streams.

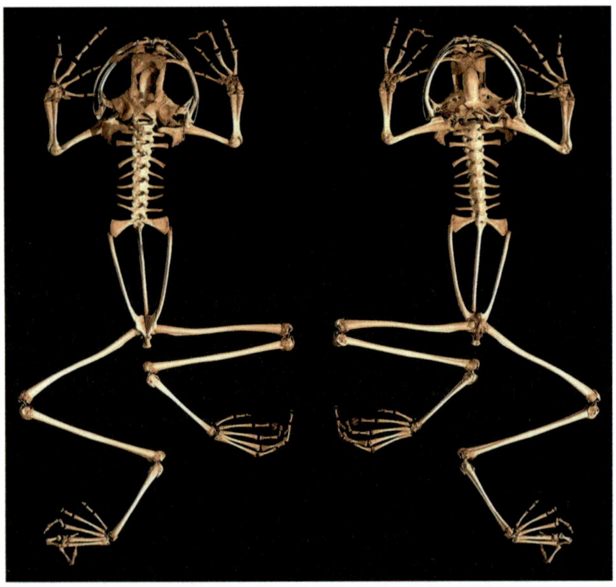

FIGURE 6.51 High-resolution computed tomography reconstructions of the skeleton of *Cochranella granulosa*, UF 137267, 26.5 mm SVL, Crique Wahatingni, Gracias a Dios, Honduras. Dorsal (left) and ventral (right) views. Daniel J. Paluh.

Content.—*Cochranella* contains eight species as of 13 October 2023 (Frost, 2023), four of which occur in Ecuador.

Distribution.—*Cochranella* occurs from Honduras to the Amazonian and Andean cloud forests of Colombia, Ecuador, Peru, and Bolivia.

Remarks.—Figure 6.52. The phylogeny of species presented herein follows Guayasamin *et al.* (2020b).

Key references.—(Guayasamin *et al.*, 2008b) (evolutionary relationships); (Guayasamin *et al.*, 2009) (generic taxonomy).

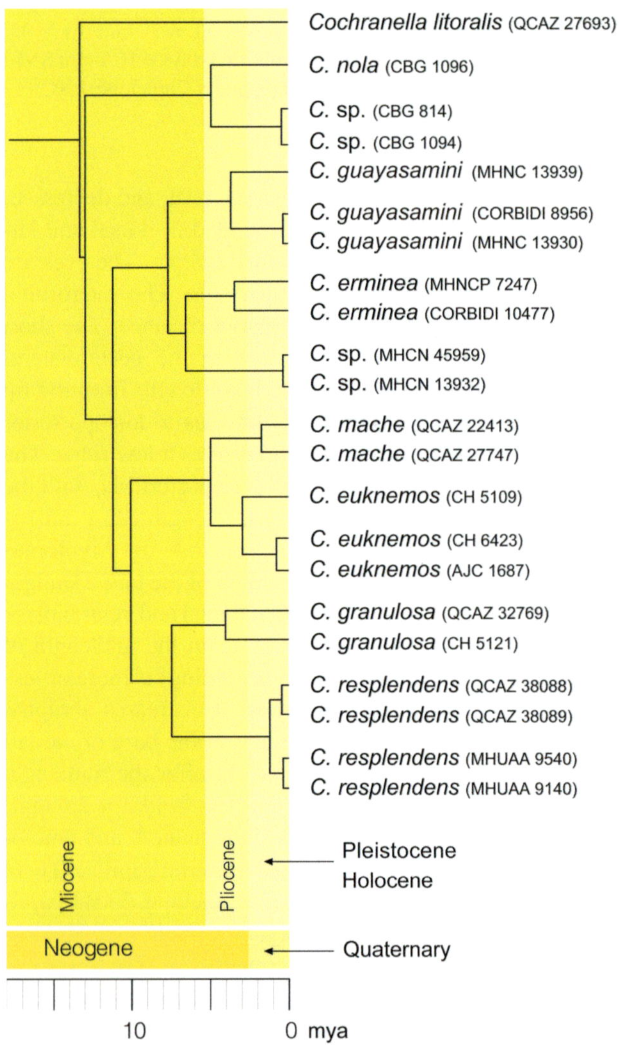

FIGURE 6.52 Time-calibrated maximum likelihood phylogenetic tree depicting evolutionary relationships of *Cochranella*, inferred from DNA sequences of mitochondrial and nuclear genes. Phylogeny taken and redrawn from Guayasamin *et al.* (2020b).

COCHRANELLA GRANULOSA (GRANULAR GLASSFROG) (TAYLOR, 1949)

Centrolenella granulosa (Taylor, 1949): 265. **Holotype.**—FMNH 178269 from Los Diamantes, one mile south of Guápiles, Cantón de Pococí, Provincia Limón, Costa Rica.

Cochranella granulosa Taylor, 1951: 34.

Etymology.—The specific epithet is derived from the Latin word *granum*, meaning grain, in allusion to the granular skin texture of the species.

Identification.—Figure 6.53. These frogs attain a maximum SVL of 29.0 mm in males and 32.0 mm in females. The skin on the dorsum is granular and lacks spicules; the venter is areolate and lacks a pair of enlarged tubercles on the ventral surfaces of the thighs. There is a white lining on the anterior 20–40% of the ventral parietal peritoneum; the peritoneum has white iridophores covering the digestive tract. The vomerine teeth are present; each vomer has 1–3 teeth. The snout is truncate in dorsal aspect and slightly sloping in lateral profile. The tympanum is oriented dorsolaterally with a slight posterior inclination, its diameter is about 35% of the eye diameter; the tympanic annulus is visible; a low supratympanic fold is evident, and the tympanic membrane is partially pigmented and clearly differentiated from the surrounding skin. The liver is tetralobed; the hepatic peritoneum lacks iridophores. The humeral spines are absent. There is no webbing between fingers I and II; there is vestigial to moderate webbing between fingers II and III; there is extensive webbing between III–IV. The webbing on the foot is extensive. The ulnar fold is low; the inner and outer tarsal folds are low and thin. The prepollex is not separated from finger I; the nuptial pad is a large to medium-sized excrescence present on the dorsal, lateral, and/or ventral sides of the thumb. Finger I is about same the length as finger II. The disc of finger III is narrow, about 40% of the eye diameter.

The dorsum is dark green to bluish-green, with yellowish or bluish granules and faint to clearly marked blue to black spots. The anterior 20–40% of the venter is white; the posterior portion is translucent. Melanophores are absent from the dorsal surfaces of hands and toes. Populations from Central America have a pale or whitish lip. One individual from Ecuador lacks dark dorsal spots. The bones are green. The iris is yellow to golden with few faint brownish streaks.

FIGURE 6.53 *Cochranella granulosa*. Left: QCAZ 32769, male, 4 km west of Durango, Provincia Esmeraldas. LAC. Right: MZUTI 4811, male, SVL 31.5 mm, El Jardín de los Sueños, Provincia Cotopaxi. Jaime Culebras (dorsolateral) and Javier Aznar (ventral).

Comparisons.—*Cochranella litoralis*, *Espadarana prosoblepon*. *Cochranella granulosa* is distinguished from most glassfrogs by having a granular dark green to bluish-green dorsum with whitish granules and faint to well-defined dark spots, yellowish-green hands and feet, a reduced white ventral parietal peritoneum (covering less than 40% of the venter), a white digestive tract, and by lacking humeral spines. Species that might have a sympatric distribution with *C. granulosa* and that present a similar coloration include *C. litoralis* and some populations of *E. prosoblepon*. However, *C. granulosa* differs by having granular skin (shagreen in *E. prosoblepon* and *C. litoralis*) and lacking humeral spines (present in males of *E. prosoblepon* and *C. litoralis*). Also, *C. litoralis* has an orange to red iris (yellow to golden in *C. granulosa*), and *E. prosoblepon* has a digestive tract that lacks white iridophores (present in *C. granulosa*).

Natural history.—The males call from the upper side of leaves. Males fight venter–venter, hanging from the vegetation by the tips of their toes. Reproduction usually occurs in riparian habitats, but it was also reported in a small forest pond about 10 m away from the nearest stream. Egg clutches consist of 40 to 104 dark eggs that are attached on masses on the upper distal half of leaves overhanging streams. Following hydration, the eggs form a pendent mass (i.e., drip tip) that hangs off the leaf and thereby ensures a constant flow of water over the developing larvae. There is no long-term parental care in this species. Females provide short-term parental care (brooding) to their egg clutches just after oviposition; this behavior, although brief (e.g., few hours), significantly reduces embryo mortality from dehydration. Parental care by males is absent. Males often call in a chorus, with one male initiating calling and then being answered by several males. In Los Laureles, Provincia Cotopaxi, males were actively calling on February 2016.

The call has been described from Costa Rica and Panama by Kubicki (2007), Savage (1967), and Ibañez *et al.* (1999b). The call consists of a rapid high-pitched pulsed trill, usually composed of three notes (range 3–5 notes). Usually, the first note is longer (average duration = 300 ms) than the subsequent notes (average duration = 150 ms). The mean duration for a three-note call is 800–900 ms. Detailed data and figures are provided under the subtitle "Vocalizations of Ecuadorian Anurans" in Appendix 1.

Tadpoles.—The tadpole of *Cochranella granulosa* was described from Costa Rica. Tadpoles of *C. granulosa* have a particularly slim shape. Recently hatched tadpoles (one to five days old) are markedly slender and dark brown. In the laboratory, older tadpoles are slender with a very long tail; they lack much body pigment, and their bright red coloration is from the highly vascularized internal anatomy anterior to the body cavity and visible through the body wall. The oral disc is not emarginate and is bordered laterally and along the posterior margin with 28–41 marginal papillae ($\bar{x}$ = 33 ± 4 papillae). The jaw sheaths are strongly serrate; the upper jaw forms a smooth m shape, and the lower sheath is V-shaped. Generally, all five tooth rows are well developed; the LTRF is 2(2)/3; the A-2 row has a large central gap, with the two halves extending as short wings on each side of the lateral descendent ends of the upper jaw sheath.

Distribution.—Figure 6.54. *Cochranella granulosa* is widely distributed; its range includes the humid lowland and premontane slopes from the Atlantic drainage in eastern Honduras to central Panama and the Pacific versant in humid upland or gallery forests from northern Costa Rica to northern Ecuador. In Ecuador, it occurs in Esmeraldas, Pichincha, and Cotopaxi provinces. Its elevational range is 201–709 m in an area of extent of occurrence of about 100 km^2.

This nocturnal, arboreal, riparian species inhabits natural tropical forests and streams in disturbed areas in the evergreen forests of lowlands of Equatorial Chocó ecosystem in the Cordillera Occidental of the Andes biogeographic sector, where the average annual rainfall is 2350–2410 mm and the average annual temperature is 21.8–24.8 °C.

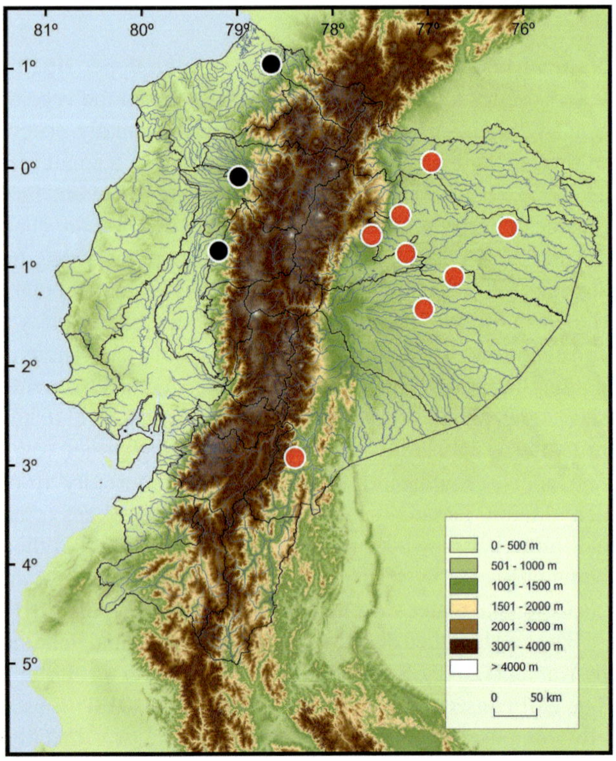

FIGURE 6.54 Localities of known occurrence of *Cochranella granulosa* (black dots) and *C. resplendens* (red dots) in Ecuador.

Conservation.—*Cochranella granulosa* is categorized as Least Concern (IUCN SSC Amphibian Specialist Group, 2020b; in IUCN Red List of 2022) due to its wide distribution, tolerance of a degree of habitat modification, and presumed large population. However, it is important to note that in Ecuador, *C. granulosa* was categorized as Data Deficient by Ortega-Andrade *et al.* (2021) because it only is known from two recent records. This species was common in Costa Rica but disappeared from the known sites around Monteverde between 1990 and 1994. It was also considered common in Panama, though it was extirpated in El Copé. There are few records from Honduras, and it is considered uncommon in El Quebracho, Nicaragua. Due to ongoing declines in the extent and quality of its habitat, the population is suspected to be decreasing. Generally, it is threatened by both severe habitat loss resulting from deforestation (from urban expansion, agriculture and logging), and water pollution. In Ecuador, it has not been recorded in any public reserve, but is it present in the protected forest of Asociación Agrícola Carchi-Imbabura.

Remarks.—The karyotype of *Cochranella granulosa* is 2N = 20. *Cochranella granulosa* is the sister species of *C. resplendens*, being an example of allopatric speciation mediated by the uplift of the Andes.

Key references.—(Taylor, 1949) (description); (Starrett, 1960) (natural history, tadpole); (Duellman, 1967) (karyotype); (Savage and Starrett, 1967) (call); (Mcdiarmid, 1978) (natural history); (Ibañez *et al.*, 1999b) (call); (Savage, 2002) (account); (Kubicki, 2007) (account, call); (Hoffmann, 2010) (natural history, tadpole); (Delia *et al.*, 2017) (natural history); (Guayasamin *et al.*, 2020b) (account).

COCHRANELLA LITORALIS (COAST GLASSFROG) (RUIZ-CARRANZA AND LYNCH, 1996)

Centrolene litoralis (Ruiz-Carranza and Lynch, 1996): 7.
Holotype.—ICN 13821 from La Guayacana, 100 m, Litoral Pacífico, Municipio de Tumaco, Departamento Nariño, Colombia.
Cochranella litoralis Guayasamin *et al.*, 2009: 32.

Etymology.—The specific epithet *litoralis* refers to the proximity of the type locality (La Guayacana, Colombia) to the Pacific Ocean.

Identification.—Figures 6.55, 6.56. These minute frogs attain a maximum SVL of 20.0 mm in males, and the SVL is unknown in females. The skin on the dorsum is smooth

FIGURE 6.55 *Cochranella litoralis*. Left: CJ 10774, male, 20.35 mm SVL, Durango, Provincia Esmeraldas. DAL. Top right: CJ 12588, male, Los Laureles, Provincia Cotopaxi. Jaime Culebras. Middle right: QCAZ 27693, male, Durango, Provincia Esmeraldas. LAC. Bottom right: CJ 12587, male, Cristóbal Colón, Quinindé, Provincia Esmeraldas. Ross Maynard.

FIGURE 6.56 Morphological details of *Cochranella litoralis*. Top left: CJ 12587, male, Cristóbal Colón, Quinindé, Provincia Esmeraldas. Ross Maynard. Top right and bottom: CJ 10774, male, Durango, Provincia Esmeraldas. DAL.

anteriorly and shagreen in the sacral region, it lacks spicules; the ventral surfaces of the thighs below the vent lack a pair of enlarged tubercles. The anterior half of the ventral parietal peritoneum has iridophores, and the posterior half is translucent. There are iridophores on the pericardium and gastrointestinal peritoneum; there are no iridophores on the testes, gall bladder, urinary bladder, and kidneys. The vomers lack teeth. The snout is truncate in dorsal and lateral profiles. The tympanum is of moderate size and oriented almost vertically, with slight lateral and posterior inclinations, and its diameter is about 30% of the eye diameter; the tympanic annulus is mostly visible, with the supratympanic fold covering its posterodorsal margin; the tympanic membrane is translucent, partially pigmented, and clearly differentiated from the surrounding skin. The liver is clearly lobed and covered by a transparent peritoneum. The humeral spines are small and pointy in males. There is no webbing between fingers I and II, the webbing varies from absent to vestigial between fingers II and III; there is moderate webbing between the outer fingers. The webbing between the toes is moderate. The ulnar and tarsal folds are absent or low and inconspicuous; they lack white coloration. The prepollex is exposed; the nuptial pad is a medium-sized diffuse excrescence formed by glandular clusters and individual glands. Finger I is as long as finger II. The width of the disc of finger III is about 33% of the eye diameter.

The dorsum is pale yellowish-green with dark spots that vary in size and conspicuousness. There are faint yellow–cream dorsolateral stripes. The anterior half of the venter is white, turning transparent posteriorly. The bones are pale green. The dorsal surfaces of the fingers and toes lack melanophores, except for some on the base of toe V. The iris is bright orange to red.

Comparisons.—*Espadarana prosoblepon*, *Nymphargus griffithsi*, *Cochranella guayasamini*. *Cochranella litoralis* is the only species in Centrolenidae that has a bright orange to red iris, a dorsum with dark spots, and adult males with humeral spines. In the Chocoan lowlands, only some populations of *E. prosoblepon* and *N. griffithsi* have a green dorsum with dark dots and humeral spines. Both *E. prosoblepon* and *N. griffithsi* are considerably larger, have a cream to white iris, and lack prepollical spines. The Peruvian *C. guayasamini* also has a red iris and humeral spines, but it lacks dark dorsal spots.

Natural history.—Figure 6.57. A male of *Cochranella litoralis* and an egg clutch, presumed to be from the same species, were found in riverine vegetation. The egg clutch was placed on the upper side of a leaf and contained 25 embryos.

Tadpoles.—Unknown.

Distribution.—Figure 6.58. *Cochranella litoralis* has been reported from the Pacific lowlands of northwestern Ecuador and the type locality in Departamento Nariño, Colombia. In Ecuador, it occurs in Esmeraldas and Cotopaxi provinces. Its elevational range is 14–414 m in an area of extent of occurrence of about 5420 km^2.

This nocturnal, arboreal species inhabits natural forests and intervened areas in the evergreen forests of lowlands of Equatorial Chocó ecosystem in the Cordillera Occidental of the Andes biogeographic sector, where the average

FIGURE 6.57 *Cochranella litoralis*. Top: CJ 12587, male, Cristóbal Colón, Quinindé, Provincia Esmeraldas. Ross Maynard. Bottom: clutch of embryos, Tungaloma Lodge, Provincia Esmeraldas. Lucas Bustamante.

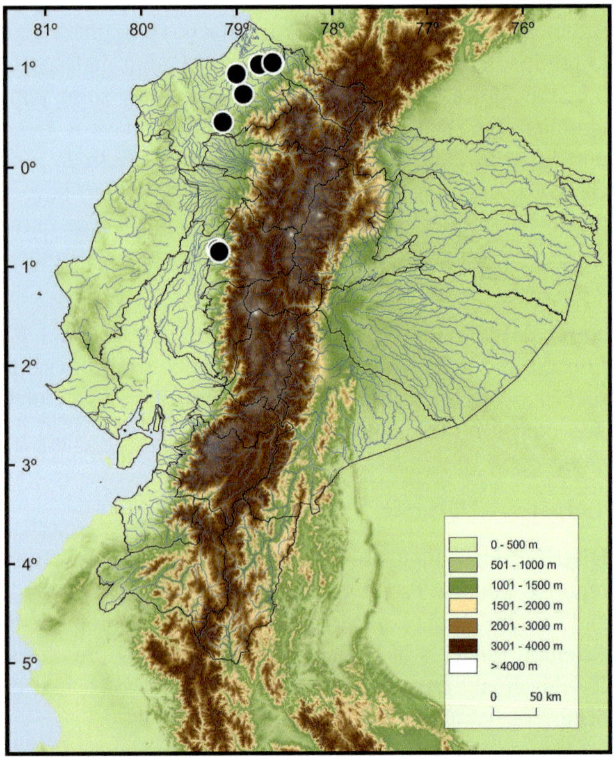

FIGURE 6.58 Localities of known occurrence of *Cochranella litoralis* in Ecuador.

annual rainfall is 2027–2518 mm and the average annual temperature is 23.3–25.8 °C.

Conservation.—*Cochranella litoralis* is categorized as Vulnerable throughout its entire range (IUCN SSC Amphibian Specialist Group, 2019u; in IUCN Red List of 2022). This categorization is based on several factors: the species' extent of occurrence was calculated as 7161 km^2, it occurs in approximately six threat-defined locations, the population is severely fragmented, and there is a continuing decline in the extent and quality of its habitat in the Colombian and Ecuadorian Andes. However, it is important to note that in Ecuador, *C. litoralis* has been categorized as Endangered by Ortega-Andrade *et al.* (2021). Although their calculations of the extent of occurrence (34 km^2) differ from ours (5420 km^2), it is clear that the area of occupancy is reduced and severely fragmented. Its major threats are habitat reduction and fragmentation, logging, deforestation for agricultural development (especially palm plantations), fires, mining, the introduction of exotic species, human settlements, and pollution. In Ecuador, it has been recorded in the protected forest of Asociación Agrícola Carchi-Imbabura and some private reserves (e.g., Jardín de los Sueños, Tundaloma Lodge).

According to Guayasamin *et al.* (2020b), *Cochranella litoralis* is sister to all other member of the genus *Cochranella*, but with low nodal support. Twomey et al. (2014) recovered a sister relationship between *C. litoralis* and *C. nola*.

Key references.—(Ruiz-Carranza and Lynch, 1996) (description); (Guayasamin *et al.*, 2020b) (account).

COCHRANELLA MACHE (MACHE GLASSFROG) GUAYASAMIN AND BONACCORSO, 2004

Cochranella mache Guayasamin and Bonaccorso, 2004: 486. **Holotype.**—QCAZ 22412 from Riachuelo La Ducha, 510 m, tributary of Río Aguacatal, Reserva Biológica Bilsa, Provincia Esmeraldas, Ecuador.

Etymology.—The specific epithet *mache* is a noun in apposition and refers to the Montañas de Mache, the type locality of the species, located in the Ecuadorian Chocó.

Identification.—Figures 6.59, 6.60, 6.61. These frogs attain a maximum SVL of 27.0 mm in males and 33.4 mm in females. The skin on the dorsum is shagreen with warts that usually correspond to light spots; the males have numerous minute spicules. The ventral skin is granular; there are several round and enameled warts around the cloaca; a cloacal fold is present; the ventral surfaces of the thighs below the vent lack a pair of enlarged tubercles. The anterior one-third of the parietal peritoneum is covered by

FIGURE 6.59 *Cochranella mache.* CJ 8320, 22.8 mm SVL, Reserva Biológica Bilsa, Provincia Esmeraldas. DAL.

iridophores; there are iridophores over the pericardium and visceral peritonea (digestive tract, gonads); the renal capsules are covered by iridophores in some individuals. Each vomer has three or four teeth on the dentigerous process.

FIGURE 6.60 Morphological variation in *Cochranella mache.* Top: Río Baltazar (left) and Alto Tambo (right), Provincia Esmeraldas. LAC. Bottom left: CJ 11856, male. Bottom right: CJ 11857. Itapoa, Provincia Esmeraldas. SGS.

FIGURE 6.61 Head in frontal view of *Cochranella mache.* CJ 8320, Reserva Biológica Bilsa, Provincia Esmeraldas. DAL.

The snout is subacuminate in dorsal aspect and gradually inclined in lateral profile. The tympanum is oriented almost vertically with slight lateral and posterior inclinations, and its diameter is about 33–37% of the eye diameter; the tympanic annulus is visible; the low supratympanic fold is evident; the tympanic membrane is translucent and pigmented like the surrounding skin. The liver is tri- or tetralobed; the hepatic peritoneum is clear or covered by small, isolated patches of iridophores. The humeral spines are absent. There is no webbing between fingers I and II; there is reduced webbing between fingers II and III; there is extensive webbing between the outer fingers. The webbing between the toes is extensive. The ventrolateral edges of finger IV, the forearm, the elbow, toe V, the tarsus, and the heel have dermal enameled folds and tubercles. The prepollex is not separated from finger I; the nuptial pad is large. Finger I is about the same length as finger II (finger I is about 94–98% of finger II). The width of the disc of finger III is about 44–54% of the eye diameter.

The dorsum varies from dark olive green to lavender–blue and pale blue, with numerous small yellow to orange spots. There is a large, dull to bright yellow patch on the top of the head. The upper lip has a thin white margin. There are white tubercles visible on the dermal folds of the hind and forelimbs. The innermost fingers and toes (fingers I and II, toes I and II) are white in males and partially white in females. The throat and ventral surfaces of the limbs are blue–green. There is white parietal peritoneum covering the anterior half of the venter, and the posterior portion is translucent. The pericardial and visceral peritonea are white. The hepatic peritoneum is transparent, but in some individuals, a small patch of white iridophores covers part of the liver. There are white warts surrounding the cloaca. The iris is whitish-cream to beige with thin brown reticulations. The bones are green. The melanophores cover the dorsal surfaces of fingers III and IV and toes IV and V. There is a white to golden ring surrounding the pupil.

Comparisons.—*Cochranella euknemos. Cochranella mache* is differentiated from most glassfrogs by having a snout gradually inclined in profile and dermal folds with white tubercles on the ventrolateral edges of finger V, the forearm, the elbow, toe V, the tarsus, and the heel. It also distinguished by its green to bluish-green dorsum, with numerous small yellow to orange spots. *Cochranella mache* can only be confused with *C. euknemos* (restricted to Costa Rica, Panama, and Colombia), which lacks the tubercles along the arms and legs that characterize *C. mache*.

FIGURE 6.62 *Cochranella mache*. Top: male calling, Reserva Biológica Bilsa, Provincia Esmeraldas. SGS. Bottom: amplexus, Río Baltazar, Alto Tambo, Provincia Esmeraldas. LAC.

Natural history.—Figure 6.62. *Cochranella mache* seems to be restricted to small streams and rivulets in primary and secondary forests in lowlands and piedmont forests. Although individuals have been observed on different strata of the forest up to 6 m high, *C. mache* prefers the midstory vegetation. The abundance of the species is correlated with rainfall, being higher during the rainy season. Males call from the upper surface of leaves or branches on trees and bushes; reproduction seems to be restricted to the rainy season. Amplexus is axillary, and a gravid female was observed to have about 30 eggs. The brown eggs are placed on the upper sides of leaves.

Advertisement calls of one individual (but see Remarks) from Estación Biológica Bilsa, Provincia Esmeraldas, Ecuador, were described as consisting of two notes separated by an interval of 107–130 ms. Each note has a mean duration of 29 ms. Detailed data and figures are provided under the subtitle "Vocalizations of Ecuadorian Anurans" in Appendix 1.

Tadpoles.—Unknown.

Distribution.—Figure 6.63. *Cochranella mache* has been reported from northwestern Ecuador and from the Pacific lowlands of Colombia (Departments of Antioquia and Valle del Cauca). In Ecuador, it occurs in Esmeraldas and Manabí provinces. Its elevational range is 82–654 m in an area of extent of occurrence of about 7290 km^2

This nocturnal, arboreal species inhabits natural forests and intervened areas in five ecosystems (Table 0.4) in four biogeographic sectors (Table 0.5), where the average annual rainfall is 833–2287 mm and the average annual temperature is 22.4–25.1 °C.

Conservation.—*Cochranella mache* is categorized as Near Threatened throughout its entire range (IUCN SSC Amphibian Specialist Group, 2020c; in IUCN Red List of 2022). However, it is important to note that in Ecuador, *C. mache* has been categorized as Endangered by Ortega-Andrade *et al.* (2021). Although its extent of occurrence is estimated to be more than 5000 km^2 (7290 km^2, according to our data), its distribution is severely fragmented, and there is a continuing decline in the extent and quality of its habitat. In Ecuador, more than half of the area of distribution is affected by human activities. The distribution of the species is likely to correspond to the remaining lowland forests in northwestern Ecuador and southwestern Colombia. Under scenarios of climate change, the potential distribution of *C. mache* suffers a reduction of 13–21% in its predicted range. When deforestation is taken into account, ~70% of the predicted distribution is reduced. Its major threats are habitat reduction and fragmentation due to logging and deforestation for agricultural development (especially palm plantations); its area of distribution is under constant pressure by wood and palm companies. Other threats are fires, the introduction of exotic species, human settlements, and pollution. Some of the larger fragments within its range are protected by a

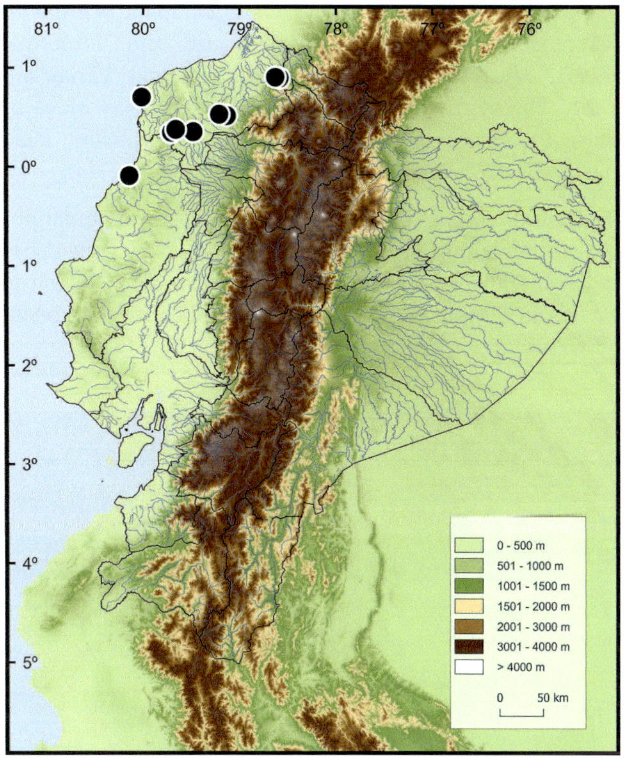

FIGURE 6.63 Localities of known occurrence of *Cochranella mache* in Ecuador.

public reserve and some private organizations, but many areas remain unprotected. Also, most conservation measures are ineffective because of institutional and funding restrictions and a lack of law enforcement. In Ecuador, it has been recorded in the public reserve Reserva Ecológica Mache Chindul and some private reserves (e.g., Reserva Biológica Bilsa, Reserva Biológica Canandé, Reserva Jama-Coaque).

Remarks.—*Cochranella mache* is the sister species of *C. euknemos* (Guayasamin *et al.*, 2020b). According to Jaime Culebras (pers. comm.), the call provided by Ortega-Andrade *et al.* (2013) was erroneously assigned to *C. mache*.

Key references.—(Guayasamin and Bonaccorso, 2004) (description); (Cisneros-Heredia *et al.*, 2008a) (definition); (Ortega-Andrade *et al.*, 2013) (natural history, call); (Guayasamin *et al.*, 2020b) (account).

COCHRANELLA RESPLENDENS (RESPLENDENT GLASSFROG) (LYNCH AND DUELLMAN, 1973)

Centrolenella resplendens (Lynch and Duellman, 1973): 51. **Holotype.**—KU 118053 from Santa Cecilia, 340 m, Provincia Sucumbíos, Ecuador.

Cochranella resplendens Ruiz-Carranza and Lynch, 1991a: 23.

Etymology.—The specific epithet is derived from the Latin verb *resplendo*, meaning to glitter, and is used in allusion to the jewel-like appearance of this frog.

Identification.—Figure 6.64. These frogs attain a maximum SVL of 26.6 mm in males, and the SVL is unknown in females. The skin on the dorsum is shagreen with elevated and spiculated warts corresponding to white spots. The venter is granular; the ventral surfaces of the thighs below the vent have a pair of slightly enlarged tubercles; an enlarged cloacal fold is present. There is a white lining (iridophores) on the anterior three-fourths of the ventral parietal peritoneum; the posterior one-fourth is transparent. There are iridophores on the pericardium and peritonea covering the intestines and stomach; the renal capsules, gall bladder, and urinary bladder lack iridophores. Each vomer has three or four teeth on the dentigerous process. The snout is round in dorsal aspect and gradually inclined in lateral profile. The tympanum is oriented almost vertically, with slight lateral and posterior inclinations, and its diameter is about 38% of the eye diameter; the tympanic annulus is partially visible; the low supratympanic fold is evident; the tympanic membrane is barely translucent and pigmented like the surrounding skin. The liver is lobate and covered by a transparent peritoneum. The humeral spines are absent. Webbing

FIGURE 6.64 Morphological variation and details of *Cochranella resplendens*. CJ 6176, male, 26.7 mm SVL, Shiripuno Lodge, Provincia Orellana. DAL.

is absent between the inner fingers, and there is extensive between the outer fingers. The webbing between the toes is extensive. The ventrolateral edges of finger IV, the forearm, the elbow, toe V, the tarsus, and the heel have dermal folds and enameled tubercles. The prepollex is not separated from finger I; the nuptial pad is not evident. Finger I is about the same length as finger II. The width of the disc of finger III is about 55% of the eye diameter.

The dorsum is dark green with numerous small, white to bluish-white spots. The ventral surfaces are mostly white, except for the posterior portion, which is translucent. The upper lip, tubercles on the forearm and foot, and cloacal warts are white. The bones are green. The dorsal surfaces of the fingers and toes have a few small and enameled spots; the melanophores are present only on the surfaces of finger IV and toe V. The iris is whitish-cream to beige, with fine gray to brown reticulations. The circumpupilar ring is pale yellow.

Comparisons.—*Cochranella mache*. *Cochranella resplendens* is distinguished from most glassfrogs by having a snout gradually inclined in profile and the ventrolateral edges of finger IV, the forearm, the elbow, toe V, the tarsus, and the heel with dermal folds and enameled tubercles. Additionally, the species lacks humeral spines and has a white venter, except in the posterior one-fourth, which is transparent. *C. mache* could be confused with *C. resplendens*; however, it occurs in the Pacific lowlands of Ecuador and not in the Amazon Basin.

Natural history.—Figure 6.65. *Cochranella resplendens* is a rare species. At the type locality (Santa Cecilia), fieldwork from 1967 through 1972 resulted in the collection of 7765 specimens of amphibians and reptiles but no additional specimens of *C. resplendens*. However, canopy surveys may reveal that its rarity is an artifact of microhabitat sampling. At the Tiputini Biodiversity Station, *Cochranella resplendens* was found to be sympatric with *Teratohyla midas*, *Hyalinobatrachium munozorum*, and *Vitreorana ritae*. The egg clutches are placed on the upper sides of leaves. Females provide short-term parental care; male parental care is absent.

Tadpoles.—Figure 6.66. The tadpoles have a non-emarginate oral apparatus; the tooth row formula is 2(2)/3; the upper jaw is slightly curved. The description is from Río Napinaza, Provincia Morona Santiago. The description is based on an

FIGURE 6.65 *Cochranella resplendens*. Top: amplexus and female guarding a clutch, Guagua Sumaco, Provincia Napo. Jesse Delia. Bottom left: clutch invaded by mosquito larvae. LAC. Bottom right: QCAZ 38088, clutch. Quebrada Río Napinaza, Provincia Morona Santiago. LAC.

FIGURE 6.66 Ontogenetic changes in *Cochranella resplendens* from tadpole to juvenile. Top to bottom. QCAZ 38088, stage 28, lateral, dorsal, and ventral views, TL = 26 mm; undetermined stage; stage 42; stage 45, 65 days after stage 28; juvenile 17 days after stage 45. Quebrada Río Napinaza, Provincia Morona Santiago. LAC.

individual in stage 36. The developmental variation was documented using 49 tadpoles—13 in stage 24, 20 in stage 25, five in stage 27, seven in stage 28, one in stage 31, two in stage 38, and one in stage 42. The total length measures 34.8 mm, while the body length is 10.5 mm, accounting for 30.2% of the total length. The body is elongated, oval, and depressed, being wider, with a body width of 4.8 mm, than its height, which is 3.2 mm. Chondrocranial elements are not visible. The snout appears rounded in both dorsal and lateral views. The lateral line system remains inconspicuous. The eyes are positioned dorsally, and the distance from the eye to the tip of the snout is 2.9 mm. The narial apertures are small, opening 1.7 mm from the tip of the snout, with an internarial distance of 1.8 mm. The distance from the narial opening to the anterior edge of the eye is 0.8 mm. The interorbital distance is 1.2 mm, which is less than the internarial distance.

A short spiracle is present, being single and sinistral, located at the posterolateral region of the body. The spiracular aperture has a posterodorsal orientation, with a slight ridge on the inner wall. The aperture diameter measures 0.7 mm, and the distance from the tip of the snout to the spiracular opening is 7.6 mm. The spiracle is situated at 72.4% of the body length from the tip of the snout. The vent tube is short and abdominal, free posteriorly, and its opening is directed backward. The tube's length is 0.4 mm, and its transverse width is 0.3 mm. The tail is long, measuring 24.2 mm, which is 2.3 times the body length, and it has a subacute tip. Myotomes are visible throughout the entire length of the tail, and a straight medial line separates the dorsal and ventral myotomes. The tail muscle width is 2.2 mm, while the tail muscle height is 2.3 mm. The maximum height of the tail is 4.0 mm, with the dorsal and ventral fin heights measuring 1.1 mm. The dorsal fin originates at approximately the midlength of the tail and maintains a nearly uniform height, except at its anterior and posterior ends. The ventral fin originates nearly at the base of the tail muscle and reaches its maximum height posterior to the midlength of the tail.

The medium-sized oral disc has a transverse width of 2.5 mm, which corresponds to 52% of the body width. It is non-emarginate and anteroventral. Around the oral disc, there are 38 marginal and uniserial papillae, mainly distributed on the lower labium, while the upper labium carries papillae only on its lateral extremes. These papillae are relatively large and acute. Among the lateral papillae (n = 15), lengths range from 0.04 to 0.16 mm, with a mean of 0.10 ± 0.03 mm, and widths range from 0.07 to 0.16 mm, with a mean of 0.09 ± 0.02 mm. The ventral papillae (n = 23) vary in lengths from 0.02 to 0.16 mm, with a mean of 0.11 ± 0.03 mm, and widths range from 0.07 to 0.18 mm, with a mean of 0.09 ± 0.03 mm.

The upper jaw sheath is completely keratinized and features a serrated edge. It is nearly straight, with a smooth arch in the middle. The lower jaw sheath is also keratinized, U-shaped, and serrated along its edge. The transverse width of the upper jaw sheath, including lateral processes, is 0.96 mm, accounting for 38.4% of the oral disc width. The lower jaw sheath measures 0.78 mm in width, which is 31.2% of the oral disc width. The serrations on both sheaths are broad-based and short, oriented straight medially. The labial tooth row formula is 2(2)/3, with tooth rows A-1 and P-3 exhibiting less pigmentation than the other rows. The anterior and posterior tooth rows are about equal in length, extending to the extremes of the anterior and posterior labia, respectively. Tooth row A-1 measures 2.3 mm in length, consisting of 208 teeth. Tooth row A-2 is interrupted medially, located at the sides of the upper jaw sheath, with each side measuring 0.6 mm in length and containing 86 teeth. Tooth row P-1 is 2.2 mm long, comprising 154 teeth. Tooth row P-2 measures 2.1 mm in length and contains 192 teeth. Tooth row P-3 is 2.2 mm long and contains 207 teeth.

On the dorsal side, the tadpole appears mostly red, with some clusters of iridophores forming an interorbital line and a middorsal band from behind the eyes to nearly the end of the body. The anteriormost part of the body dorsally is pale yellow. Laterally, the tadpole is translucent. The irises are mostly black, with a yellow ring encircling the pupil.

Distribution.—See Figure 6.54. *Cochranella resplendens* in known in the Amazon Basin of Colombia, Ecuador, Peru, and Bolivia, at elevations between 190 and 1100 m.

A discontinuous population of the species was recently found on the eastern slope of the central Andes in the Departamento Antioquia, Colombia, at much higher elevations (1309–1699 m). In Ecuador, it occurs in Sucumbíos, Orellana, Napo, Pastaza, and Morona Santiago provinces. Its elevational range is 231–1341 m in an area of extent of occurrence of about 25,620 km^2.

This nocturnal, arboreal species inhabits natural forests and intervened areas in five ecosystems (Table 0.4) in five biogeographic sectors (Table 0.5), where the average annual rainfall is 1848–3679 mm and the average annual temperature is 19.7–25.2 °C.

Conservation.—*Cochranella resplendens* is categorized as Least Concern (IUCN SSC Amphibian Specialist Group, 2018o; in IUCN Red List of 2022) in its entire range and in Ecuador by Ortega-Andrade *et al.* (2021) in view of its wide distribution, presumed large population, and occurrence in an area with extensive remaining suitable habitat. In Ecuador, according to Guayasamin *et al.* (2020b), although it is evident that *C. resplendens* is rare in collections, it is presumed that this is a consequence of inadequate sampling. The potential distribution of this species in Ecuador is 77,793 km^2, 14% of which is affected by human activities. The species is threatened by habitat fragmentation and habitat loss as a result of agricultural expansion, logging, and water pollution. In Ecuador, it has been recorded the protected forests of Cerro Sumaco y Cuenca Alta del Río Suno and Comunidad Kichwa Wamani Cokiwa, the biosphere reserves Sumaco and Yasuní, and some private reserves (e.g., Shiripuno Lodge).

Remarks.—*Cochranella resplendens* and *C. granulosa* are sister species, being a case of allopatric speciation mediated by the uplift of the Andes. Guayasamin *et al.* (2020b) placed *C. phryxa* (Aguayo and Harvey, 2006) in the synonymy of *Centrolenella resplendens*. The only difference that they found between the holotype of *Cochranella resplendens* and the Bolivian specimen is the presence of a small patch of iridophores on the hepatic peritoneum of the latter. However, it is possible that this patch is either an artifact of preservation (i.e., iridophores from the ventral parietal peritoneum can be attached to the liver) or intraspecific variation (i.e., the presence/absence of patches of iridophores on the peritoneum has been observed in other centrolenids, e.g., *Cochranella mache*).

Key references.—(Lynch and Duellman, 1973) (description); (Duellman, 1978) (account, natural history); (Cisneros-Heredia, 2005) (distribution); (Twomey and Brown, 2005) (distribution); (Aguayo and Harvey, 2006) (definition, distribution); (Terán-Valdez *et al.*, 2009) (tadpole); (Delia *et al.*, 2017) (natural history); (Guayasamin *et al.*, 2020b) (account).

GENUS *ESPADARANA* (GLASSFROGS) GUAYASAMIN *ET AL.*, 2009

Espadarana Guayasamin *et al.*, 2009: 33. **Type species.**—*Centrolenella andina* Rivero, 1968, by original designation.

Etymology.—The generic name is a patronym from Marcos Jiménez de la Espada, a Spanish zoologist who was part of the Comisión Científica del Pacífico that explored America between 1862 and 1865. Among his numerous contributions, Jiménez de la Espada described the first centrolenid frog, *Centrolene geckoideum*, in 1872. *Centrolene* is the type genus for the family Centrolenidae. In Spanish, the word *espada* means sword, in allusion to the humeral spines present in males of species in this genus. *Espadarana* is a combination of the words *espada* and *rana* (frog) and is feminine in gender.

Definition.—Figure 6.67. *Espadarana* is diagnosed by the following combination of traits: (1) adult males with conspicuous humeral spines; (2) lobed liver covered by a transparent hepatic peritoneum, digestive tract translucent; (3) ventral parietal peritoneum white anteriorly and transparent posteriorly; (4) moderate webbing between fingers III and IV; (5) bones green in life; and (6) males that call from the upper surfaces of leaves or branches and females that deposit eggs on the upper sides of leaves over streams. The presence of conspicuous humeral spines in the adult males of *Espadarana* distinguishes this lineage from most centrolenids. In Ecuador, the only other genus exhibiting this trait is *Centrolene*. Species in *Centrolene* usually have enameled pericloacal warts, which are absent in *Espadarana*. DNA data are needed for the unambiguous placement of species

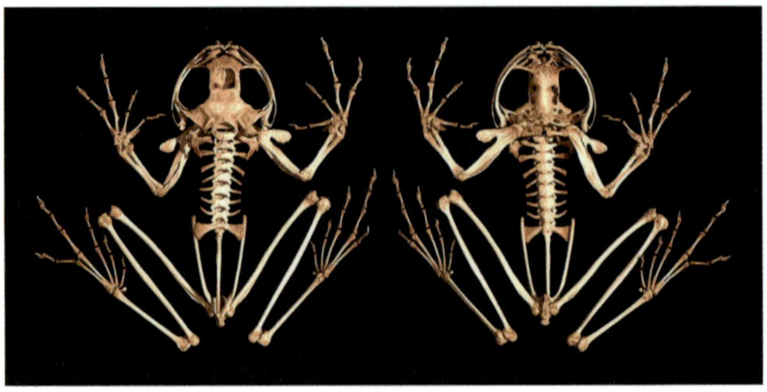

FIGURE 6.67 High-resolution computed tomography reconstructions of the skeleton of *Espadarana prosoblepon*, KU 164632, 24.1 mm SVL, 4 km northeast of Dos Ríos, Provincia Pichincha. Dorsal (left) and ventral (right) views. Daniel J. Paluh.

in either of these two lineages. *Chimerella* also has small humeral spines, but it has several distinguishing traits (e.g., transparent venter, white hepatic and visceral peritonea).

Content.—*Espadarana* contains five species, four of which occur in Ecuador (19 March 2020).

Distribution.—*Espadarana* occurs in the lowlands and mountains of Central America (Honduras, Nicaragua, Costa Rica, Panama), the Pacific lowlands of Colombia and Ecuador, and the cloud forests of the Andes in Colombia, Cordillera de Mérida, and Serranía del Perijá in Venezuela.

Remarks.—Figure 6.68. The phylogeny of species presented herein follows Guayasamin *et al.* (2020b).

ESPADARANA AUDAX (DARING GLASSFROG) (LYNCH AND DUELLMAN, 1973)

Centrolenella audax (Lynch and Duellman, 1973): 16. **Holotype.**—KU 146624 from Salto de Agua, 2.5 km north–northeast of Río Reventador on Baeza–Lago Agrio road, 1660 m, Provincia Napo, Ecuador.
Espadarana audax Twomey *et al.*, 2014: 1.

Etymology.—The specific epithet derives from the Latin *audax*, meaning daring, and is used in allusion to the precipitous regions inhabited by this species.

Identification.—Figures 6.69, 6.70. These frogs attain a maximum SVL of 25.5 mm in males and 28.8 mm in females. The skin on the dorsum of males and females is shagreened; the males have minute spicules; the ventral surfaces of the thighs below the vent have a pair of enlarged tubercles; other cloacal ornamentation is absent. The anterior two-thirds of the ventral parietal peritoneum are covered with white iridophores; the posterior third is transparent. There is a white pericardium; no iridophores are present on the peritonea covering the intestines, stomach, testes,

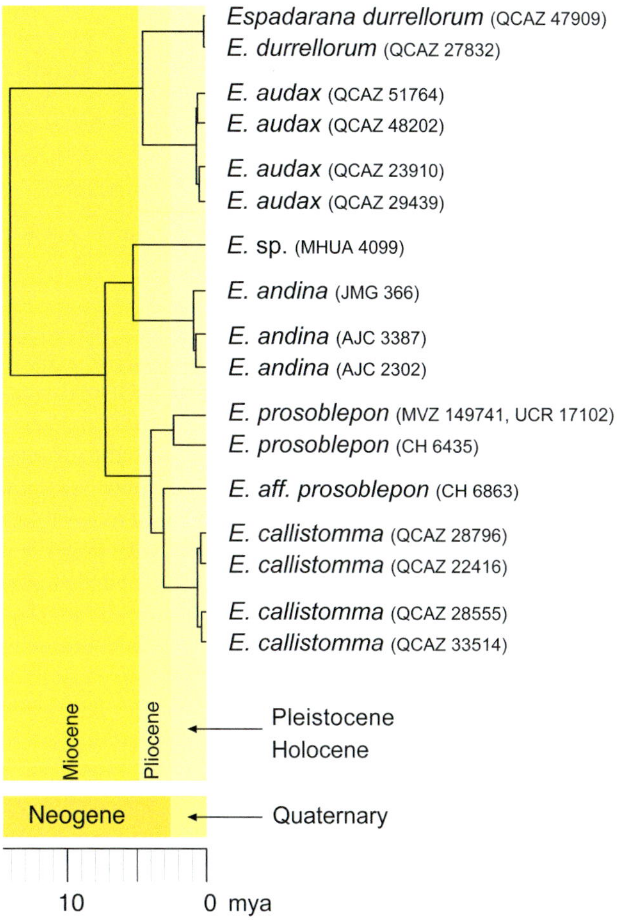

FIGURE 6.68 Time-calibrated maximum likelihood phylogenetic tree depicting evolutionary relationships of *Espadarana*, inferred from DNA sequences of mitochondrial and nuclear genes. Phylogeny taken and redrawn from Guayasamin *et al.* (2020b).

FIGURE 6.69 Morphological details of *Espadarana audax*. CJ 9509, male, 23.3 mm SVL, Hostería El Reventador, Provincia Napo. DAL.

kidneys, gall bladder, and urinary bladder. The dentigerous process of the vomer bears 2–4 teeth. The snout is rounded in dorsal profile and rounded to truncate in lateral profile. The tympanum is of moderate size, and its diameter is

FIGURE 6.70 *Espadarana audax*. Top: CJ 11469, male (note the humeral spines), Narupayaku, Provincia Orellana. SGS. Bottom: Río Negro, Provincia Tungurahua. DAL.

23–29% of the eye diameter; the low supratympanic fold is evident; the tympanic membrane is partially pigmented and clearly differentiated from the surrounding skin. The liver is tetralobed; the hepatic peritoneum lacks iridophores. In adult males, humeral spines are present. Webbing is absent between fingers I, II, and III and is basal between fingers III and IV. The foot is extensively webbed. The ulnar fold and the inner tarsal fold are present; the outer tarsal fold is absent. The prepollex is concealed; in males, the nuptial pad is a large to medium sized excrescence present on the dorsal, lateral, and/or ventral sides of the thumb. Finger I is about the same length as finger II (finger II is 95.6–102% the length of finger I). The disc of finger III is moderate, and its width is 43–56% of the eye diameter.

The dorsum is green with small yellow spots. The fingers and toes are pale yellow. The parietal peritoneum is white. The heart is not visible. The visceral peritoneum lacks iridophores. The ventral surfaces of the limbs are unpigmented. The bones are green. Melanophores are present on the outer fingers and outer toes. The iris is white or pale bronze to mustard, with a yellowish hue and thin black reticulations.

Comparisons.—*Espadarana durrellorum*. *Espadarana audax* is unique by having small yellow spots on the dorsum, short and distally curved humeral spines in males, and extensive webbing between fingers III and IV. In Ecuador, the species can only be confused with *E. audax*, but the latter differs by lacking yellow dorsal spots.

Natural history.—By day, one individual was found in a bromeliad. During the night, adults are found on the upper sides of leaves along streams. One female found in Río Salado had dark brown oviductal eggs visible through the body wall. The eggs are deposited in moss on branches. Females provide short-term parental care; male parental care is absent.

Advertisement calls of two individuals from Río Hollín, Provincia Napo, Ecuador, were described as follows: Calls

FIGURE 6.71 Ontogenetic changes in two individuals of *Espadarana audax* from tadpole to adult. Top to bottom, left to right. QCAZ 37871a, stage 27, lateral, dorsal, and ventral views. QCAZ 37871b, undetermined stage; stage 42; stage 46, four days after stage 42; froglet 24 days after stage 46; juvenile 189 days. Quebrada Río Napinaza, Provincia Morona Santiago. LAC.

are emitted in series; a typical call series has 4–5 calls and is relatively long (range = 2398–3285 ms). Each call has a single, pulsed note with a mean duration of 38 ms. Detailed data and figures are provided under the subtitle "Vocalizations of Ecuadorian Anurans" in Appendix 1.

Tadpoles.—Figure 6.71. This description is based on an individual in stage 27 (not preserved) from Quebrada Río Napinaza, Provincia Morona Santiago, Ecuador. It was collected by David Salazar-Valenzuela on 2 March 2008. The body is ovoid, widest posteriorly and slightly wider than it is deep. The snout is round in dorsal view and subacute in lateral profile. The eyes are moderately small, narrowly separated, and directed dorsally. The spiracle is sinistral, featuring an inner wall that forms a small ridge, with the opening directed posteriorly at about two-thirds of the body's length. The mouth is medium-sized, anteroventral, and transversely elliptical, with barely developed labia. Both the upper and lower jaw sheaths are completely keratinized, and the upper jaw sheath is narrow and M-shaped with a smooth arc. The tail represents approximately 73% of the total length, with its greatest depth occurring at the beginning of the last third of the tail's length, and it tapers to a rounded tip. The dorsal fin becomes visible at the midlength of the tail, while the ventral fin is minimally present on the anterior part of the tail and widens as it reaches the midlength. Both dorsal and ventral fins are equal in depth, about half the depth of the caudal musculature at the midlength of the tail.

The skin is translucent, making the internal organs visible. The circulatory system's conspicuity imparts a bright red coloration in the tadpoles. There are aggregated melanophores on the central area of the dorsum, starting in the middle of the body and extending to the tail, giving a brown coloration.

In individuals at stage 40, the red coloration is only present laterally on the anterior half of the body, having been replaced by a light brownish-green dorsal coloration. In stages 43–46, the dorsum adopts a uniform brownish-green hue, and the extremities become light green. Juveniles exhibit a brownish dorsum, light green extremities, and blue bones that are visible through the skin. As they mature into adults, the coloration evolves into a uniform light green, with some brownish areas remaining laterally on the body.

Distribution.—Figure 6.72. *Espadarana audax* occurs through the eastern montane region of southeastern Colombia in the departments of Huila (eastern slope of the Cordillera Central), Putumayo, Cauca, Caquetá, and Boyacá (eastern slope of the Cordillera Oriental); Ecuador; and northern Peru (department of San Martín) at elevations between 800 and 1900 m. In Ecuador, it occurs in Sucumbíos, Napo, Orellana, Pastaza, Morona Santiago, and Zamora Chinchipe provinces. Its elevational range is 311–1948 m in an area of extent of occurrence of about 26,150 km^2

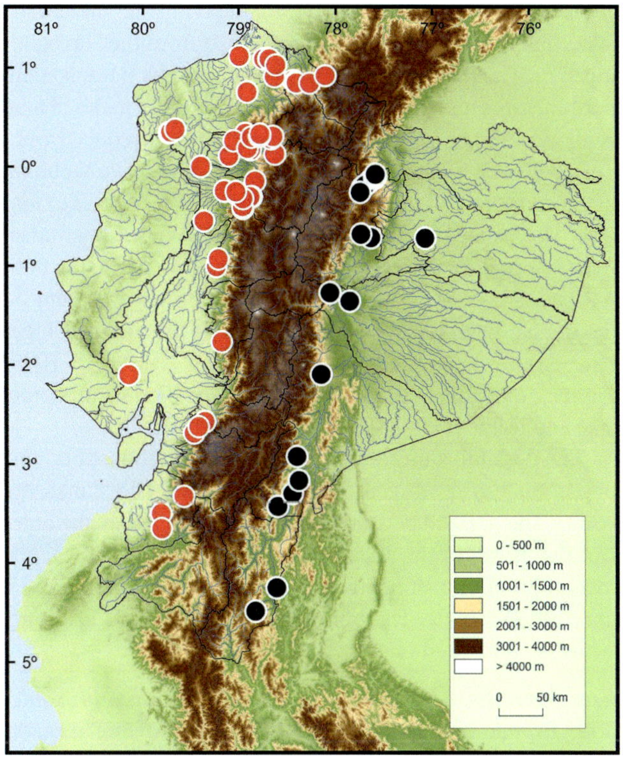

FIGURE 6.72 Localities of known occurrence of *Espadarana audax* (black dots) and *E. prosoblepon* (red dots) in Ecuador.

This nocturnal, arboreal species inhabits natural forests and intervened areas in five ecosystems (Table 0.4) in the Amazonian Cordilleras and northern and southern Cordillera Oriental of the Andes biogeographic sectors, where the average annual rainfall is 1209–4084 mm and the average annual temperature is 16.8–24.6 °C.

Conservation.—*Espadarana audax* is categorized as Least Concern throughout its entire range (IUCN SSC Amphibian Specialist Group, 2018p; in IUCN Red List of 2022) in view of its wide distribution and presumed large population. However, it is important to note that in Ecuador, *E. audax* was categorized as Near Threatened by Ortega-Andrade *et al.* (2021) due to habitat degradation in part of its range caused by human activities. According to Guayasamin *et al.* (2020b), the southern part of its distribution is threatened by mining activities, but these do not affect most of the species' distribution, and it is found in several protected areas. In Ecuador, *E. audax* has not been found in some of its historical localities but has been discovered recently at several new localities (General Leonidas Plaza Gutiérrez, Provincia Morona Santiago, Miazi Alto, Provincia Zamora Chinchipe). Although *Espadarana audax* has a much larger distribution than previously thought and occurs in protected areas, habitat destruction and degradation as a result of agriculture, logging, and mining activities are localized threats to this species in Ecuador. In Ecuador, it has been recorded in the public reserves Parque Nacional Cayambe Coca, Parque Nacional Llanganates, and Parque Nacional Sangay; the protected forests of Cerro Sumaco y Cuenca Alta del Río Suno, Cuenca Alta del Río Nangaritza, Comunidad Kichwa Wamani Cokiwa, and Pueblo Shuar Arutam; the biosphere reserves Sumaco and Podocarpus-El Cóndor; and some private reserves (e.g., Reserva Biológica Narupa).

Remarks.—The morphological differences (dorsal color pattern, humeral spine shape) that separate *Espadarana durrellorum* and *E. audax* are subtle. However, given the available genetic and morphological data, we consider them valid species.

Key references.—(Lynch and Duellman, 1973) (description); (Duellman and Schulte, 1993) (distribution); (Twomey *et al.*, 2014) (distribution); (Cisneros-Heredia and Guayasamin, 2014) (distribution); (Delia *et al.*, 2017) (natural history, habitat); (Guayasamin *et al.*, 2020b) (account, call).

ESPADARANA CALLISTOMMA (PRETTY-EYED GLASSFROG) (GUAYASAMIN AND TRUEB, 2007)

Centrolene callistommum (Guayasamin and Trueb, 2007): 29. **Holotype.**—QCAZ 25832 from a stream affluent of Río Bogotá, near San Francisco de Bogotá, 83 m, Provincia Esmeraldas, Ecuador.
Espadarana callistomma Guayasamin *et al.*, 2009: 33.

Etymology.—The specific epithet is derived from the Greek *kallistos*, meaning most beautiful, and *omma*, meaning eye, as a reference to the fantastic iris pattern in this species.

Identification.—Figures 6.73, 6.74. These frogs attain a maximum SVL of 29.6 mm in males and 31.8 mm in females. The skin on the dorsum of males and females is shagreen; minute spicules are evident in males (visible under magnification); the ventral surfaces of the thighs below the vent have a pair of enlarged tubercles. The anterior two-thirds of the ventral parietal peritoneum are covered with white iridophores; the posterior third is transparent. There is a white pericardium; a translucent peritoneum covers the intestines, stomach, gall bladder, testes, and urinary bladder; the kidneys are covered with a translucent peritoneum. The vomerine teeth are present; each vomer has 2–7 teeth. The snout is truncate in dorsal and lateral views. The tympanum is small (its diameter is 20–31% of the eye diameter), and it is oriented vertically with a lateral inclination; the tympanic annulus is visible except for the upper border, which is

FIGURE 6.73 *Espadarana callistomma*. Left: QCAZ 48340, female. Right: QCAZ 48341, male. 4 km on the train rail from Alto Tambo to El Placer, Provincia Esmeraldas. LAC.

FIGURE 6.74 Morphological details of *Espadarana callistomma*. QCAZ 48341, male, 4 km on the train rail from Alto Tambo to El Placer, Provincia Esmeraldas. LAC.

covered by the supratympanic fold; the tympanic membrane is partially pigmented and differentiated from the surrounding skin. The liver is tetralobed and covered by a transparent peritoneum. Humeral spines are present in adult males. There is no webbing between fingers I and II; there is reduced webbing between fingers II and III; there is extensive webbing between fingers III and IV. The webbing on the foot is extensive. The ulnar and inner tarsal folds are low; the outer tarsal fold is absent. The prepollex is concealed; in males, the nuptial pad is a large to medium-sized excrescence present on the dorsal, lateral, and/or ventral sides of the thumb. Finger II is slightly longer than finger III (finger III is 93.3–100% the length of finger II). The disc of finger III is of moderate size and is about 29–34% of the eye diameter.

The dorsum is uniformly olive green, without spots. The upper lip has a whitish-cream coloration. The flanks are white. The parietal peritoneum is white and covers the anterior two-thirds of the abdomen (the heart is not visible). The melanophores cover the dorsal surfaces of fingers III and IV and toes IV and V. The bones are green. The iris is silvery white with marked black reticulations.

Comparisons.—*Sachatamia ilex*, *Espadarana prosoblepon*. *Espadarana callistomma* has the following remarkable traits: a silvery white iris with black reticulations, a uniform green dorsal coloration, and a relatively large size (in males, SVL 26.7–29.6 mm; in females, SVL 29.5–31.8 mm). It differs from *S. ilex* by having, in males, an externally conspicuous humeral spine, whereas males of *S. ilex* have a small humeral spine that is embedded in the arm musculature. Some individuals of *E. prosoblepon* have the same color pattern as *E. callistomma*, but they have a significantly smaller body size (in males, SVL 23.2–27.5 mm; in females, SVL 25.3–27.8 mm) and a two-note call (3–4 notes in *E. callistomma*). See Remarks.

Natural history.—Figure 6.75. Males call from the upper sides of leaves, and females deposit pigmented eggs on the upper sides of leaves. Egg clutches obtained under laboratory breeding conditions from parents from Durango (Provincia Esmeraldas) contained an average of 35 eggs. The eggs measured approximately 2.1 mm in diameter. The animal hemisphere was dark brown, and the vegetal hemisphere was light brown. The developmental time was 12 days from the 32-cell stage until tadpole hatching. Archenteron elongation began in the early gastrula, and notochord elongation began in the mid to late gastrula phase. The development of the gastrocoel roof plate was precocious, in the late gastrula period.

Advertisement calls of one individual from Reserva Otokiki, Provincia Esmeraldas, Ecuador, were described as having 3–4 pulsed notes with a mean duration of 0.38 s. Notes have clearly defined pulses. Detailed data and figures are provided under the subtitle "Vocalizations of Ecuadorian Anurans" in Appendix 1.

Tadpoles.—Unknown.

FIGURE 6.75 Reproductive event of *Espadarana callistomma*. QCAZ 48340, female. QCAZ 48341, male, 4 km on the train rail from Alto Tambo to El Placer, Provincia Esmeraldas. LAC.

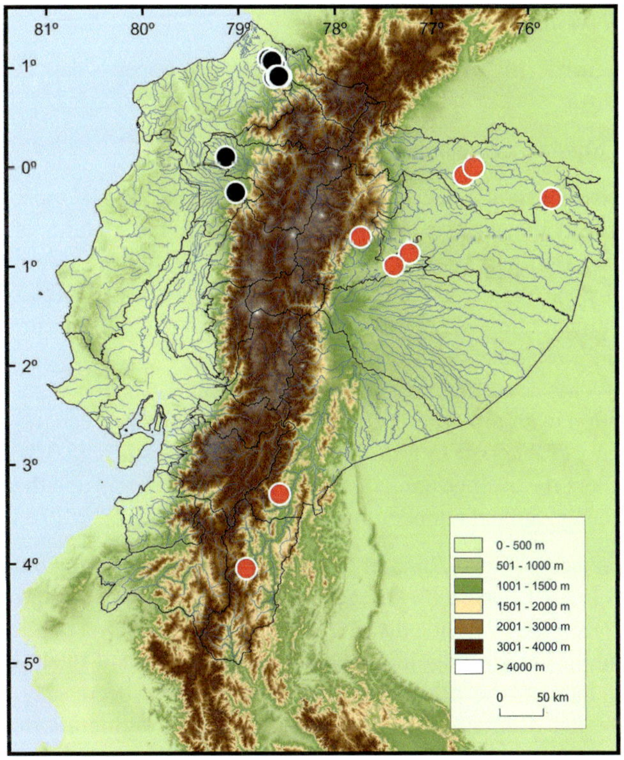

FIGURE 6.76 Localities of known occurrence of *Espadarana callistomma* (black dots) and *E. durrellorum* (red dots) in Ecuador.

Distribution.—Figure 6.76. *Espadarana callistomma* is known in the lowland rainforests of northwestern Ecuador and the Pacific lowlands of Colombia. In Ecuador, it occurs in Esmeraldas, Carchi, and Pichincha provinces. Its elevational range is 82–1117 m in an area of extent of occurrence of about 6460 km^2.

This nocturnal, arboreal species inhabits along streams and intervened areas in the evergreen forests of lowlands of Equatorial Chocó and evergreen montane foothill forests of the Cordillera Occidental of the Andes ecosystems, where the average annual rainfall is 2194–2412 mm and the average annual temperature is 20.7 – 25.4 °C.

Conservation.—*Espadarana callistomma* is categorized as Least Concern throughout its entire range (IUCN SSC Amphibian Specialist Group, 2019v; in IUCN Red List of 2022) in view of its wide distribution and presumed large population. However, it is important to note that in Ecuador, *E. callistomma* was categorized as Endangered by Ortega-Andrade *et al.* (2021) due to its current extent of occurrence and its fragmented distribution in fewer than five known localities. According to Guayasamin *et al.* (2020b), the most likely scenario is that *E. callistomma* is restricted to portions of the Chocó ecoregion, an area with a high rate of deforestation and mining. Also, even though the species is locally abundant at the localities where it has been found, the population is suspected to be decreasing due to ongoing declines in the extent and quality of its habitat. In Ecuador, it has been recorded in the protected forest of Asociación Agrícola Carchi-Imbabura and some private reserves (e.g., Reserva Otokiki, Reserva Río Canandé).

Remarks.—Guayasamin *et al.* (2020b) state that the specific status of *Espadarana callistomma* is uncertain because of the conflict among different sets of data. Genetically, *E. callistomma* and *E. prosoblepon* are not reciprocally monophyletic. On the other hand, the body size of males and females of *E. callistomma* is significantly larger than that of *E. prosoblepon* (t-test, $P < 0.001$), and the dorsal and iris color patterns of *E. callistomma* are different from most *E. prosoblepon*. In terms of vocalization, *E. callistomma* has a call with more notes (3–4 notes per call) than *E. prosoblepon* (two notes per call). Given that they found biologically meaningful differences among the species (i.e., body size, vocalizations), the authors recognize both species as valid. The lack of molecular divergence could be the product of a recent divergence, but this is a hypothesis that remains untested. Another pending taxonomic problem is that the type of *Hylella parabambae* matches the spotless coloration of *E. callistomma* and some populations of *E. prosoblepon*.

The embryonic development from cleavage to hatching was described by Salazar-Nicholls and del Pino (2015) based on samples from Ecuador.

Key references.—(Guayasamin and Trueb, 2007) (description); (Ospina-Sarria *et al.*, 2010) (distribution); (Salazar-Nicholls and del Pino, 2015) (development); Yánez-Muñoz *et al.* (2020) (account); (Guayasamin *et al.*, 2020b) (account, call).

ESPADARANA DURRELLORUM (DURRELL'S GLASSFROG) (CISNEROS-HEREDIA, 2007)

Centrolene durrellorum (Cisneros-Heredia, 2007a): 28.
Holotype.—DFCH-USFQ D131 from small rivulet tributary of the Jambue River, ca. 6 km south of Zamora, 1150 m, Provincia Zamora Chinchipe, Ecuador.
Espadarana durrellorum Twomey *et al.*, 2014: 1.

Etymology.—The specific epithet honors Gerald Durrell (1925–1995) and Lee McGeorge Durrell (Durrell Wildlife Conservation Trust) for their continuous contributions to the conservation of biodiversity.

Identification.—Figure 6.77. These frogs attain a maximum SVL of 26.1 mm in males and 27.0 mm in one female. The skin on the dorsum is shagreen with minute spicules uniformly distributed. The ventral skin is granular; the cloacal area is granular with two large, rounded, and flat subcloacal tubercles. In addition, there are two folds on the sides of the cloacal opening, forming a distinct cloacal sheath. The upper two-thirds of the parietal peritoneum are covered by iridophores. There is a white pericardium; all other peritonea lack iridophores. The vomerine teeth are present. The snout is rounded in dorsal view and bluntly truncate in lateral view. The tympanic annulus is evident and oriented dorsolaterally; there is a very weak supratympanic fold above the tympanum. The liver is lobed and lacks iridophores. In adult males, the humeral spines are short and straight with sharp points. Webbing is absent between fingers I and II; basal or absent webbing is between II and III; moderate webbing is present between the outer fingers. The webbing between the toes is extensive, more so in a female than in males. There are no dermal folds on the hands, forearms, feet, or tarsus. The nuptial excrescences are present, type I. The prepollex is concealed. Finger I is slightly longer than II. The eye diameter is larger than the width of the disc on finger III.

The dorsum is uniformly green without flecks or spots. The bones are green. The melanophores are widespread on the outer fingers and outer toes. The iris is golden with thin, dark reticulations.

FIGURE 6.77 Morphological variation and details of *Espadarana durrellorum*. QCAZ 47909, male, Comunidad Ñukanchi Allpa, Río Canoayacu, Provincia Napo. LAC.

Comparisons.—*Espadarana audax*. *Espadarana durrellorum* is distinguished from all glassfrogs by having a green dorsal coloration, short and straight humeral spines in adult males, a moderate body size (25.7–26.1 mm adult males), and by lacking iridophores on the visceral peritonea. *Espadarana durrellorum* is most similar to *E. audax* (see Remarks in *E. audax*). The main trait to differentiate between the two species is the absence of yellow dorsal spots in *E. durrellorum* (present in most *E. audax*).

Natural history.—Males call from the upper sides of leaves.

Tadpoles.—Unknown.

Distribution.—See Figure 6.76. *Espadarana durrellorum*, an endemic species to Ecuador, is known in the foothills of the Cordillera Oriental of the Andes in Sucumbíos, Napo, Morona Santiago, and Zamora Chinchipe provinces. Its elevational range is 212–1618 m in an area of extent of occurrence of about 43,430 km^2.

This nocturnal, arboreal species inhabits riverine vegetation in natural forests and intervened areas in four ecosystems (Table 0.4) in four biogeographic sectors (Table 0.5), where the average annual rainfall is 1233–3890 mm and the average annual temperature is 19.5–25.4 °C.

Conservation.—*Espadarana durrellorum* was categorized as Least Concern by Ortega-Andrade *et al.* (2021) in view of its relatively wide distribution and presumed large population. Due to ongoing declines in the extent and quality of habitat, the population is suspected to be decreasing. Habitat destruction in the foothills of the Cordillera Oriental is extensive; most areas are widely deforested and heavily populated. The northern distribution is affected by oil extraction and is heavily populated, the central area is affected by deforestation and cattle grazing, and the southern part is affected by mining activities. It has been recorded in the public reserve Reserva de Producción de Fauna Cuyabeno, the protected forests of Cerro Sumaco y Cuenca Alta del Río Suno, the biosphere reserves Sumaco and Podocarpus-El Cóndor, and some private reserves (e.g., Yachana Protected Forest).

Remarks.—See Remarks under *Espadarana audax*.

Key references.—(Cisneros-Heredia, 2007a) (description); (Guayasamin *et al.*, 2020b) (account).

ESPADARANA PROSOBLEPON (VARIABLE GLASSFROG) (BOETTGER, 1892)

Hyla prosoblepon (Boettger, 1892): 45. **Lectotype.**—SMF 3756 from La Junta, Cantón de Siquirres, Provincia de Limón, Costa Rica.

Espadarana prosoblepon Guayasamin *et al.*, 2009: 33.

Etymology.—The specific epithet is derived from the Greek words *proso*, meaning forward, onward, in front, and *blepo*, meaning look or see, probably referring to the forward orientation of the eyes in centrolenid frogs.

Identification.—Figures 6.78, 6.79, 6.80. These frogs attain a maximum SVL of 27.5 mm in males and 27.8 mm in females. The skin on the dorsum is shagreen; males and females may possess minute spicules on the flanks and tympanic region; the ventral surfaces of the thighs below the vent have a pair of enlarged tubercles. There are white iridophores on the anterior 50–70% of the ventral parietal

FIGURE 6.78 *Espadarana prosoblepon*. Right: CJ 974, female, 25.9 mm SVL, Comunidad Mashpi, Provincia Pichincha. LAC. Left: CJ 10880, male, 25.4 mm SVL, Chinambí, Provincia Carchi. SGS.

FIGURE 6.79 Morphological variation in *Espadarana prosoblepon*. Top left: CJ 10974, female, Otongachi, Provincia Santo Domingo de los Tsáchilas. DAL. Top right: CJ 8293, male, 1099 24.8 mm SVL, Reserva Biológica Bilsa, Provincia Esmeraldas. DAL. Middle left: CJ 10998, male, 27.8 mm SVL, Junín, Valle de Intag, Provincia Imbabura. DAL. Middle right: CJ 973, male, 23.2 mm SVL, Comunidad Mashpi, Provincia Pichincha. LAC. Bottom left: CJ 7771, male, Reserva Biológica Bilsa, Provincia Esmeraldas. DAL. Bottom right: CJ 10999, female, Junín, Valle de Intag, Provincia Imbabura. DAL.

FIGURE 6.80 Morphological details of *Espadarana prosoblepon*. Top left: CJ 8559, male. Top right: CJ 10880, male. Chinambí, Provincia Carchi. SGS. Bottom left: CJ 1043, Comunidad Mashpi, Provincia Pichincha. LAC. Bottom right: QCAZ 45992, Río Aguas Verdes, buffer zone of Parque Nacional Cotacachi-Cayapas, Provincia Imbabura. LAC.

peritoneum; the posterior portion is transparent. There is a white pericardium; translucent peritonea cover the intestines, stomach, kidneys, gall bladder, and urinary bladder. Each vomer has 2–7 teeth on the dentigerous process. The snout is usually truncate in dorsal and lateral profiles. The tympanum is small and oriented almost vertically, with slight lateral and posterior inclinations; its diameter is 20.5–30.8% of the eye diameter; the tympanic annulus is mostly visible, with a supratympanic fold covering its posterodorsal margin; the tympanic membrane is translucent, partially pigmented, and clearly differentiated from the surrounding skin. The liver is clearly tetralobed and covered by a transparent peritoneum. The humeral spines are conspicuous in males. There is no webbing between fingers I and II; the webbing between fingers II and III is basal or absent; there is moderate webbing between the outer fingers. The webbing between the toes is moderate. The ulnar and tarsal folds are absent or low and inconspicuous, lacking white coloration. The prepollex is concealed; the nuptial pad is a large to medium-sized excrescence present on the dorsal, lateral, and/or ventral sides of the thumb. Finger I is as long as finger II or slightly longer (finger II length is 93.8–101% of finger I). The disc on finger III is moderate, and its width is 28.6–49.0% of the eye diameter.

The dorsal coloration varies between populations. In Costa Rica, the dorsum is green with dark spots. About one-half of the Honduran specimens lack small, dark dorsal spots. In Ecuador, several color patterns have been reported: (1) uniform green dorsum; (2) green dorsum with dark spots; (3) green dorsum with yellow spots; (4) green dorsum with dark and yellow spots, separated from each other; and (5) green dorsum with dark spots surrounding yellow spots, forming false ocelli. The bones are green. The upper lip is whitish-cream. The white venter is transparent posteriorly. The ventrolateral margins of the forearm and tarsus lack white coloration. The melanophores cover the dorsal surfaces of fingers III and IV and toes IV and V. The iris varies from grayish-white or pale bronze with fine dark reticulations to silvery white with thick black reticulations.

Comparisons.—*Cochranella litoralis*, *Espadarana callistomma*, and *Sachatamia ilex*. *Espadarana prosoblepon* is distinguished from most glassfrogs from the Pacific versant of the Andes by having a large, flat, and projected humeral spine and a green dorsum with small black spots; however, *E. prosoblepon* has a highly variable dorsal coloration, including individuals that have both black and yellow spots, only yellow spots, only black spots, or the absence of spots. From similar species in the region, *Cochranella litoralis* is readily differentiated by having a bright orange iris. The unspotted *E. prosoblepon* can be confused with *E. callistomma*, which is distinguished by its larger body size (in males, SVL 26.7–29.6 mm; in females, SVL 29.5–31.8 mm), contrasting black-and-white iris, and a call with 3–4 notes (call with two notes in *E. prosoblepon*). Individuals of *E. prosoblepon* with a uniform dorsum and white iris with black reticulations resemble *S. ilex*; however, males of

FIGURE 6.81 *Espadarana prosoblepon*. Top: males in Sapoparque La Florida, Provincia Santo Domingo de los Tsáchilas. LAC. Bottom: amplexus and clutch of embryos, Junín, Valle de Intag, Provincia Imbabura. Gustavo Pazmiño.

E. prosoblepon have a conspicuous humeral spine, whereas males of *S. ilex* have a small humeral spine that is embedded in the arm musculature.

Natural history.—Figure 6.81. *Espadarana prosoblepon* feeds on a variety of small arthropods, including beetles, moths, spiders, and mites. When threatened, these frogs emit pungent odors as a mechanism defense. The mean annual survival probability is 0.46.

During the breeding season, which coincides with the rainy season, they are commonly found at night on vegetation along the edges of streams in primary to slightly degraded evergreen forests and sometimes even in pastures. Reproduction is characterized by increased male vocalization and territorial combats. Males call from the upper sides of leaves or small branches to establish their territory along the stream. If an intruder enters their territory, the resident male produces a rapid series of calls, and if the intruder approaches too closely, a fight ensues, with both males dangling upside-down while holding onto vegetation with their legs and grappling with their arms and humeral spines. The combat concludes when the loser drops from the fight site or signals submission by flattening its body against the leaf surface. Males initiate amplexus by jumping on the back of approaching gravid females. Females lay 20–52 black eggs on a variety of substrates, including the upper surfaces of leaves, moss-covered rocks, or branches directly above the water. Females provide short-term parental care; male parental care is absent. After hatching, tadpoles usually hide under leaf litter or sand at the bottom of streams.

Additional reproductive biology data from a wild pair from Junín, Provincia Imbabura, have been obtained in the laboratory from one reproductive event (Centro Jambatu lab notes by Alex Achig, in September 2021). Amplexus is axillary and lasts around one day. The oviposition event occurred at night. The clutch contained 32 light brown-colored eggs of 2.1–2.7 mm ($\bar{x}$ = 2.4, n = 11) in diameter. The clutch was ovoid–triangular in shape, and the maximum diameter is 31.0 cm; it was deposited on the water sprinkler nozzle 20 cm above the terrarium pool. Hatching required 21 days following egg deposition. Tadpoles undergo complete development and metamorphosis in a span of 78 days at water temperature of about 22.2 °C and a pH of 6.9.

The call of *Espadarana prosoblepon* from Costa Rica shows some differences from the Ecuadorian call. Advertisement calls of one individual from Río Sune Chico, Provincia Pichincha, Ecuador, were described as being produced every 4–7 minutes. Each call consists of two pulsed notes and lasts 207–223 ms. Each note has a mean duration of 40 ms. Detailed data and figures are provided under the subtitle "Vocalizations of Ecuadorian Anurans" in Appendix 1.

Males of *Espadarana prosoblepon* from Monteverde, Costa Rica, call at irregular intervals (17 calls/h). The advertisement call consists of two to five short "beeps." When encountering another frog within 15 cm, the resident males often produce a series of rapid short beeps. Males of *E. prosoblepon* call vigorously while in amplexus and also immediately after egg deposition. Other calls from Costa Rica are described as a rapid three-note series, but sometimes males emit 4–5 notes; each note has an average duration of 30 ms. Typical call series have a total duration lasting 500 ms (average three-note series) or 750 ms (average four-note series). Because males call from exposed sites on the upper surfaces of leaves and twigs, these periodic calling bouts, often synchronized among a few nearby males, may make it more difficult for predatory bats (i.e., *Trachops cirrhosus*) to locate and eat calling males of *E. prosoblepon*.

Tadpoles.—Figure 6.82. This description is based on an individual in stage 37 (not preserved) from Centro Científico Río Palenque, Provincia Los Ríos, Ecuador. The body is ovoid, widest posteriorly and slightly wider than it is deep. The snout is round both in dorsal view and in lateral profile. The eyes are moderately large, narrowly separated, and directed dorsolaterally. The sinistral spiracular opening is directed posteriorly, and it is located at about two-thirds of the body's length. The tail comprises approximately 70% of the total length. The dorsal fin emerges around the midlength of the tail, while the ventral fin is initially narrow in the anterior part of the tail and widens toward the midlength. Both fins, dorsal and ventral, are approximately equal in depth and about half the depth of the caudal musculature at the tail's midlength. The tadpoles from Costa Rica have a ventral mouth, a complete oral disc with medium-sized beaks, and 2–3 rows of denticles. A-2 exhibits a significant gap above the mouth; the lower beak is serrated laterally, and there is a single row of labial papillae, absent above the mouth.

The body and tail exhibit a light green coloration with clusters of melanophores that create a brownish appearance on the dorsal surface of the body and the caudal musculature. Due to their translucent skin, the organs are visible laterally on the body. The snout appears bluish, and the legs are light green. By stage 42, the melanophores have

FIGURE 6.82 Ontogenetic changes in *Espadarana prosoblepon* from tadpole to adult female. Top to bottom, left to right. CJ 4876, stage 37, lateral and dorsal views, TL = 31 mm; stage 42, TL = 28 mm; stage 46, 44 days after stage 37; juvenile 70 days after stage 46; juvenile 251 days; adult 436 days. Centro Científico Río Palenque, Provincia Los Ríos.

diminished, and the predominant coloration is light green, while the tail retains melanophore clusters. At stage 44, the dorsum and extremities exhibit light green and blue coloring, and the white parietal peritoneum is visible ventrally. As the tadpole transitions into a juvenile, it begins to acquire its adult coloration, which is typically green, with or without the presence of dots.

Distribution.—See Figure 6.72. *Espadarana prosoblepon* is found in eastern Honduras, Nicaragua, Costa Rica, and Panama, south along the Pacific slopes of the Andes of Colombia and Ecuador from sea level to about 1600 m elevation. In Colombia, it occurs on the northern and eastern flanks of the Eastern Cordillera south to Caldas and in the Magdalena Valley. In Ecuador, it occurs in Carchi, Esmeraldas, Manabí, Imbabura, Pichincha, Santo Domingo de los Tsáchilas, Cotopaxi, Los Ríos, Bolívar, Guayas, Azuay, and El Oro provinces. Its elevational range is 14–1789 m in an area of extent of occurrence of about 57,100 km^2.

This nocturnal, arboreal species inhabits natural forests and intervened areas in eight ecosystems (Table 0.4) in six biogeographic sectors (Table 0.5), where the average annual rainfall is 706–2606 mm and the average annual temperature is 17.1–25.8°C.

Conservation.—*Espadarana prosoblepon* is categorized as Least Concern through its entire range (IUCN SSC Amphibian Specialist Group, 2020d; in IUCN Red List of 2022) and in Ecuador by Ortega-Andrade *et al.* (2021) in view of its large distribution and presumed large populations. This species is abundant in many parts of its range in Costa Rica. It is present in La Selva and its surrounding secondary forests and is still present in Braulio Carrillo National Park.

However, it is rare in San Vito and in Monteverde, where the population crashed in 1987 but still persisted on upper slopes as of 1990–1994. Other Costa Rican populations appear stable. In some sites such as Zarcero, it is the most common glassfrog. The species is common in Nicaragua and moderately common in Honduras and Colombia. It is also common in Panama, though declining at El Copé. It is common in Ecuador. The major threats are deforestation for agricultural development, logging, and human settlement. Recorded declines in Costa Rica might be due to the disease chytridiomycosis, though that is uncertain given that the species exhibits strong predicted resistance to *Batrachochytrium dendrobatidis*, the causative agent of chytridiomycosis (Woodhams *et al.*, 2006). In Ecuador, it has been recorded in the public reserve Reserva Ecológica Mache Chindul; the protected forests of Los Cedros, Mashpi, Toachi Pilatón, Río Arenillas Presa Tahuín, and Comuna San Miguel Negro; the biosphere reserves Chocó Andino de Pichincha and Macizo del Cajas; and some private reserves (e.g., Reserva Biológica Bilsa; Reserva Comunitaria Ecojunín; Reserva Río Manduriacu, Ross Maynard pers. comm.).

Remarks.—While the general morphology of *Espadarana prosoblepon* is mostly conservative, its color pattern is extremely variable, and because of an insufficient understanding of the species' variation, different names have been assigned to distinctive morphs. For example, populations whose individuals have dorsal coloration consisting of black spots surrounding yellow spots and forming ocelli were called *ocellifera*, while others with the dorsum lacking spots have been variously named *parambae* and *callistomma*. Several authors have described some of the interspecific variation characteristic of *E. prosoblepon*, but much remains to be done. For example, most descriptions and color photographs of *E. prosoblepon* depict the iris of this species as gray with fine dark reticulations, but Kubicki (2007) reported individuals with a white iris with black reticulations. Unfortunately, most centrolenid descriptions have paid little attention to iris patterns; in fact, with the exception of the descriptions of *Sachatamia ilex*, *Hyalinobatrachium ignioculus*, and *H. eccentricum*, iris coloration, if reported at all, usually is based on an examination of photographs; thus, the inter- and intraspecific variation in iris color also is poorly known.

Genetic data suggest that *Espadarana prosoblepon* and *E. callistomma* might represent one evolving lineage. Morphological differences between *E. callistomma* and *E. prosoblepon* are in body size (*callistomma* being larger), dorsal coloration (without spots in *callistomma*; usually with spots in *prosoblepon*), and iris coloration (white with contrasting black reticulation in *E. callistomma*; less contrasting in *E. prosoblepon*). In terms of vocalization, *E. callistomma* has a call with more notes (3–4 notes per call) than *E. prosoblepon* (two notes per call). At the moment, Guayasamin *et al.* (2020b) consider that the morphological and acoustic differences justify the recognition of *E. callistomma* as a valid species.

Key references.—(Boettger, 1892) (description); (Lynch and Duellman, 1973) (account); (Jacobson, 1985) (natural history, call, habitat); (Hayes, 1991) (natural history, habitat); (Savage, 2002) (account, larvae); (McCranie and Wilson, 2002) (account); (Cisneros-Heredia and Mcdiarmid, 2007) (morphology); (Hoffmann, 2010) (natural history, habitat); (Kubicki *et al.*, 2010) (distribution); (Delia *et al.*, 2010) (natural history); (Arteaga *et al.*, 2013) (natural history); (McCaffery and Lips, 2013) (survival, abundance); (Delia *et al.*, 2017) (natural history); Yánez-Muñoz *et al.* (2020) (account); (Guayasamin *et al.*, 2020b) (account, call).

GENUS *HYALINOBATRACHIUM* (GLASSFROGS) RUIZ-CARRANZA AND LYNCH, 1991b

Hyalinobatrachium Ruiz-Carranza and Lynch, 1991b: 23.
Type species.—*Hylella fleischmanni* (Boettger, 1893), by original designation.

Etymology.—The generic name derives from the Greek words *hyalos* (glass) and *batrachion* (frog), alluding to the translucent and delicate appearance of the species in this genus. For nomenclatural purposes, the generic name is considered a genderless noun.

Definition.—Figure 6.83. The genus *Hyalinobatrachium* is unambiguously diagnosed by these main traits: (1) transparent venter; (2) white visceral and hepatic peritonea; (3) absent humeral spines; (4) most species with white bones; (5) males that usually vocalize from the undersides of leaves; (5) females that deposit one layer of eggs on the undersides of leaves; and (6) complete fusion of tibiale and fibulare.

Content.—*Hyalinobatrachium* contains 33 species, nine of which occur in Ecuador (19 March 2020).

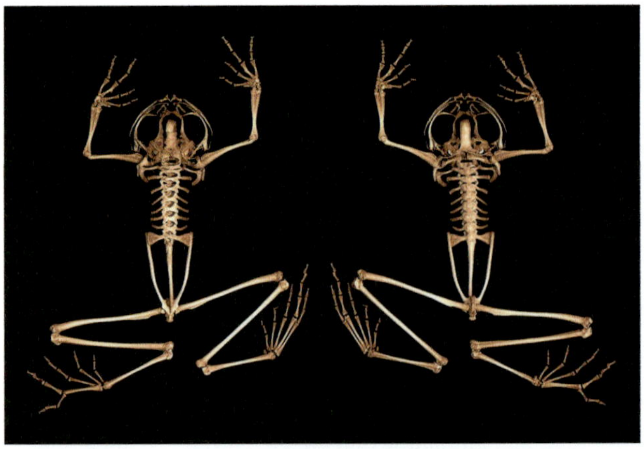

FIGURE 6.83 High-resolution computed tomography reconstructions of the skeleton of *Hyalinobatrachium adespinosai*, CJ 11563, male, 22.3 mm SVL, Guarumales, Provincia Zamora Chinchipe. Dorsal (left) and ventral (right) views. Daniel J. Paluh.

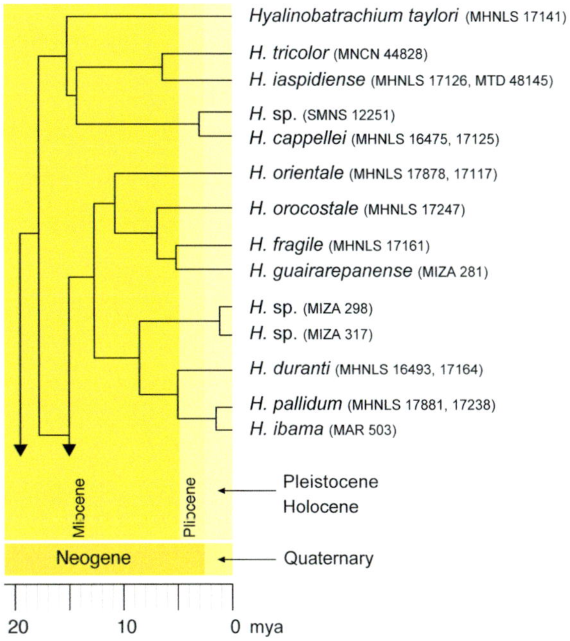

FIGURE 6.84 Time-calibrated maximum likelihood phylogenetic tree depicting evolutionary relationships of *Hyalinobatrachium*, inferred from DNA sequences of mitochondrial and nuclear genes. Phylogeny taken and redrawn from Guayasamin *et al.* (2020b).

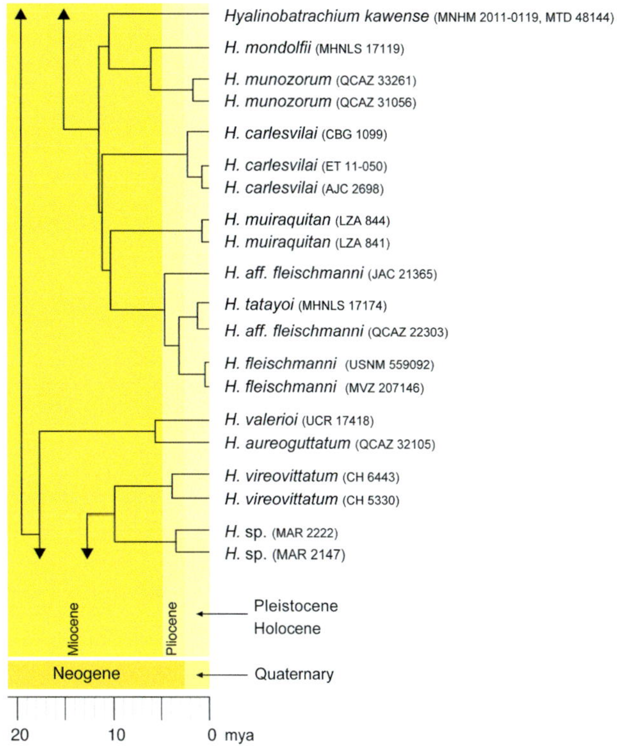

FIGURE 6.85 Continued. Time-calibrated maximum likelihood phylogenetic tree depicting evolutionary relationships of *Hyalinobatrachium*, inferred from DNA sequences of mitochondrial and nuclear genes. Phylogeny taken and redrawn from (Guayasamin *et al.*, 2020b).

Distribution.—*Hyalinobatrachium* occurs from tropical Mexico to southeastern Brazil and Argentina at elevations below 1900 m.

Remarks.—Figures 6.84, 6.85, 6.86. The phylogeny of species presented herein follows Guayasamin *et al.* (2020b).

HYALINOBATRACHIUM ADESPINOSAI (ADELA'S GLASSFROG) GUAYASAMIN *ET AL.*, 2019b

Hyalinobatrachium adespinosai Guayasamin *et al.*, 2019b: 135. **Holotype.**—ZSFQ 1648 from Río San Jacinto, 1795 m, Provincia Tungurahua, Ecuador.

Etymology.—The specific epithet is a matronym that honors Adela Espinosa, an Ecuadorian conservationist and board member of the Jocotoco Foundation. Adela has concentrated her work on the conservation of species and ecosystems. *Hyalinobatrachium adespinosai* is found only within the limits of a natural reserve owned by Adela and her husband, Antonio Páez.

Identification.—Figure 6.87. These small frogs attain a maximum SVL of 22.2 mm in males, and the SVL is unknown in females. The skin on the dorsum is shagreen and lacks tubercles; the ventral skin is areolate. Cloacal ornamentation is absent; the paired round tubercles below the vent are absent. The parietal peritoneum is transparent. The pericardium has a thin layer of iridophores (in life, a red heart is mostly visible ventrally); the liver, viscera, and testes are covered by iridophores. The dentigerous process of the vomer lacks teeth. The snout is truncate in dorsal and lateral profiles. The tympanum is barely visible and hidden under the skin, with a coloration similar to that of surrounding skin. The liver is white and bulbous. The humeral spines are absent. The hand webbing formula is basal. The foot webbing is moderate. The fingers and toes have thin lateral fringes; the ulnar and tarsal folds are present but difficult to distinguish; there is a thin layer of iridophores that extends to ventrolateral edge of finger IV and toe V. The nuptial excrescence is present as a small pad on finger I, a medium-sized, diffuse excrescence formed by glandular clusters and individual glands. The prepollex is not enlarged; the prepollical spine is not projecting (the spine is not exposed). When appressed, finger I longer than II. The diameter of the eye is about two times wider than the disc on finger III.

The dorsum is green to yellowish-green with small, diffuse yellow spots and minute gray to black melanophores. The melanophores are absent from the fingers and toes, except for finger IV and toes IV and V. Ventrally, the parietal peritoneum and pericardium are transparent, with a red

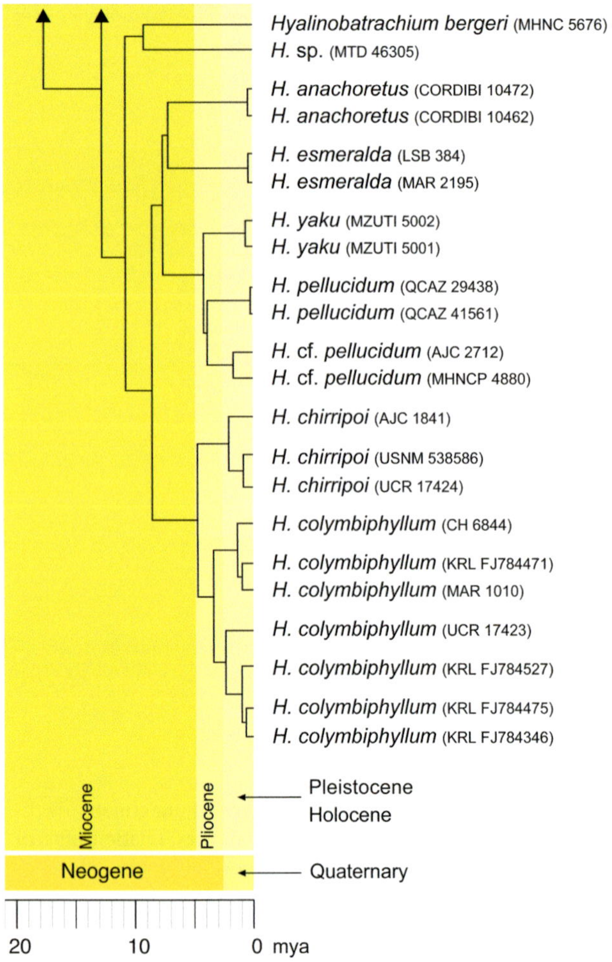

FIGURE 6.86 Continued. Time-calibrated maximum likelihood phylogenetic tree depicting evolutionary relationships of *Hyalinobatrachium*, inferred from DNA sequences of mitochondrial and nuclear genes. Phylogeny taken and redrawn from Guayasamin *et al.* (2020b).

heart always visible, even when a very thin layer of iridophores is present on the pericardium of some individuals. The visceral peritoneum of the gall bladder and urinary bladder are transparent. The hepatic and visceral peritonea are white. The ventral vein is red. The bones are white. The melanophores are absent from most fingers and toes but present on finger IV and toes IV and V. The iris is white with a pale yellow hue and minute lavender spots.

Comparisons.—*Hyalinobatrachium anachoretus*, *H. pellucidum*, *H. esmeralda*. Among glassfrogs, *Hyalinobatrachium adespinosai* is diagnosable mainly by having a transparent pericardium. However, it is morphologically cryptic, with three closely related taxa (*H. anachoretus, H. pellucidum, H. esmeralda*) that display a similar size and color pattern (pale green dorsum with yellow dots and a transparent venter and pericardium; red heart visible ventrally). However, calls have diverged noticeably; the major difference is the structure of the call, with two species (*H. adespinosai* and *H. anachoretus*) having pulsed calls and the others having tonal vocalizations. The call

FIGURE 6.87 *Hyalinobatrachium adespinosai*. ZSFQ 1648, holotype, male, 22.2 mm SVL, Río San Jacinto, Provincia Tungurahua. José Vieira-Fernandes.

of *H. adespinosai* is further differentiated from that of *H. anachoretus* by being longer, having more pulses per note, and being produced at a higher rate.

Natural history.—Males were found calling on the undersides of leaves of riverine vegetation along Río San Jacinto,

Provincia Tungurahua, Ecuador. The section of river was fast-flowing and had visible rapids. Although the population is locally abundant, individuals are difficult to observe because they are usually found at the canopy level (4–16 m above ground level).

Males were calling in the months of July and August. One male was apparently guarding an egg clutch containing 22 embryos; both the adult male and the egg clutch were on the same leaf most of the time, but the male also moved to nearby leaves.

Advertisement calls of nine individuals from Río San Jacinto, Provincia Tungurahua, Ecuador, were described as having a striking resemblance to the chirp of a cricket. Each call is composed of a single, high-pitched, pulsed note and has a mean duration of 380 ms. Detailed data and figures are provided under the subtitle "Vocalizations of Ecuadorian Anurans" in Appendix 1.

Tadpoles.—Unknown.

Distribution.—Figure 6.88. *Hyalinobatrachium adespinosai*, an endemic species to Ecuador, is known from a single location on the eastern slopes of the Cordillera Oriental of the Andes in Provincia Tungurahua at an elevation of 1880 m.

This nocturnal, arboreal species inhabits natural forests in the evergreen low montane forests of the northern Cordillera Oriental of the Andes ecosystem in the northern Cordillera Oriental of the Andes, where the average annual rainfall is 1768 mm and the average annual temperature is 17.2 ℃.

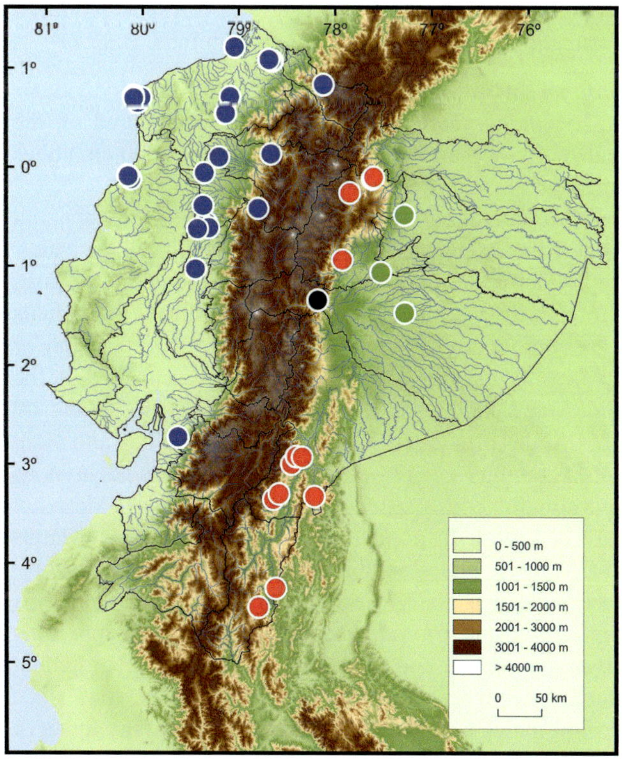

FIGURE 6.88 Localities of known occurrence of *Hyalinobatrachium adespinosai* (black dot), *H. pellucidum* (red dots), *H. tatayoi* (blue dots), and *H. yaku* (green dots) in Ecuador.

Conservation.—*Hyalinobatrachium adespinosai* was categorized as Data Deficient by Ortega-Andrade *et al.* (2021). It is only known from one locality, and there is little information about the population of this species.

Remarks.—Genetic distances for the 16S marker between *Hyalinobatrachium adespinosai* and closely related species (*H. pellucidum*, *H. esmeralda*, *H. yaku*) are > 2.5%. However, *H. adespinosai* and *H. anachoretus* only have a 1% genetic distance, although they are separated by a major biogeographic barrier (i.e., Río Marañon valley).

Key references.—(Guayasamin *et al.*, 2019b) (description).

HYALINOBATRACHIUM AUREOGUTTATUM (GOLD-SPOTTED GLASSFROG) (BARRERA RODRÍGUEZ AND RUIZ CARRANZA, 1989)

Centrolenella aureoguttata (Barrera Rodríguez and Ruiz Carranza, 1989): 77. **Holotype.**—ICN 17506 from km 23 carretera El Carmen–Quibdó, 1030 m, Municipio El Carmen de Atrato, Departamento del Chocó, Colombia.

Hyalinobatrachium aureoguttatum Ruiz-Carranza and Lynch, 1991b: 23.

Etymology.—The specific epithet is derived from the Latin words *aureus* (gold) and *guttatus* (dappled, speckled, spotted), referring to the dorsal coloration pattern of the species.

Identification.—Figures 6.89, 6.90. These frogs attain a maximum SVL of 24.0 mm in males and 23.9 mm in females. The skin on the dorsum is smooth to shagreen and lacks spicules. The venter is smooth; the ventral surfaces of the thighs below the vent lack a pair of enlarged tubercles. The ventral parietal peritoneum is transparent. There is a polymorphic pericardium, having or lacking iridophores. There is a white peritoneum covering the intestines, stomach, and testes; there is a transparent peritoneum covering the kidneys and urinary and gall bladders. The vomers lack teeth. The snout is truncate in dorsal and lateral profiles. The tympanum is of moderate size (the tympanum diameter is 30–34% of the eye diameter), with dorsolateral orientation and posterior inclination; the supratympanic fold is low; the tympanic membrane is clearly differentiated from the surrounding skin. The liver is bulbous and covered by a white peritoneum. The humeral spines are absent. There is basal webbing between fingers I and II and extensive webbing between the outer fingers. The foot webbing is extensive. The ulnar and tarsal folds are absent. The prepollex is

FIGURE 6.89 *Hyalinobatrachium aureoguttatum*. Left top and middle: CJ 11696, female. Left bottom: CJ 12510, female. Durango, Provincia Esmeraldas. SGS. Right: CJ 287, male, Río Mira, Provincia Esmeraldas. LAC.

FIGURE 6.90 Morphological variation and details of *Hyalinobatrachium aureguttatum*. Top left: QCAZ 45366. LAC. Top right: QCAZ 45364. LAC. Middle left: CJ 287, male, Río Mira, Provincia Esmeraldas. LAC. Middle right: lab-born female. Joel Sartore. CJ 11697, lab born. Durango, Provincia Esmeraldas. SGS.

not separated from finger I; in males, the nuptial pad is a medium-sized spinous nuptial excrescence extending from the lateral side of the thumb to its dorsomedial surface. Finger I is longer than finger II (finger II is about 87–95% the length of finger I). The disc of finger III is of moderate size, being 35–56% of the eye diameter.

The dorsum is greenish-yellow with 2–7 large yellow spots (diameter 0.5–2.1 mm) and with or without black flecks. The upper lip is unpigmented. The ventral parietal peritoneum is transparent. The pericardium is polymorphic, transparent (red heart visible ventrally), or white (heart not visible ventrally). There is a transparent peritoneum covering the kidneys and urinary and gall bladders. There is white peritoneum covering the liver, intestines, stomach, and testes. The fingers and toes lack melanophores, except for a few melanophores on the proximal portions of toes IV and V. The bones are white. The iris is white to yellow, with minute dark lavender flecks around the pupil or forming a horizontal stripe.

Comparisons.—*Hyalinobatrachium valerioi*. *Hyalinobatrachium aureoguttatum* is distinguished from most glassfrogs by having a completely transparent venter; a white, bulbous liver; a white visceral peritoneum; and by the absence of humeral spines. Within *Hyalinobatrachium*, it is characterized by its large yellow dorsal spots. *H. aureoguttatum* can only be confused with *H. valerioi*, from which is differentiated by having large yellow dorsal spots produced by xanthophores and iridophores (absent in *H. valerioi*).

Natural history.—Figures 6.91, 6.92. Most individuals of *Hyalinobatrachium aureoguttatum* were found on the undersides of leaves of *Heliconia* spp., 100–700 cm above small streams during the night.

Most males call from the undersides of leaves, but some have been observed calling from the upper sides. Reproductive season seems to be during July and August, although, in Durango, Provincia Esmeraldas, individuals were reproductively active in May. The mating behavior involves axillary amplexus, and eggs are deposited on the undersides of leaves. Clutches have between 27 and 49 eggs. Males provide parental care during the day and night until hatching ($\bar{x}$ = 17.1 days ± 1.8). Males are polygynous and simultaneously attend up to five clutches. Maternal care is absent. In populations from the Pacific lowlands of Colombia, males exhibit high fidelity to their territory; each male repeatedly uses the same leaf (*Heliconia* spp., *Anthurium* spp., *Philodendron* spp., *Cyclanthus* spp., *Calathea* spp., *Musa* spp.) for perching, calling, mating, and clutch attendance. Territoriality seems to be low, since several males were found in close proximity, and fights or aggressive behaviors were not observed in intrusion events by co-specific males.

Additional reproductive biology data of pairs (wild and lab born) from Durango, Provincia Esmeraldas, have

FIGURE 6.91 *Hyalinobatrachium aureoguttatum*. CJ 12513, clutch of eggs and sequence of embryonic development under laboratory conditions at CJ. Parental stock from Durango, Provincia Esmeraldas. Development time from egg releasing and fertilization (top left clutch) to tadpoles hatching (bottom right). The clutch was photographed daily for 15 consecutive days, except for the top right photo, which has a separation of two days from the picture to its left. LAC.

been obtained in the laboratory (CJ and Wikiri lab notes by Alex Achig, May 2021–September 2021). Ten reproductive events were observed. Amplexus is axillary and lasts from six hours to about one day. One oviposition event lasted about 10 minutes at night. Clutches contain 25–70 light green-colored eggs ($\bar{x}$ = 48.5, n = 6) of 1.53–2.2 mm in diameter ($\bar{x}$ = 1.73, n = 42). Clutches vary in shape between ovoid, circular, and irregularly ovoid. The maximum diameter of the clutches varies between 3.1 and 22.9 cm (mean =12.0, n = 5). In the terrarium, the clutches were deposited on the glass wall above water. Eclosion requires 12–14 days (n = 3) following egg deposition (air temperature = 21.6, 22.4 °C). Tadpoles undergo complete development and metamorphosis in a span of at least 51 days (water temperature: 20.1, 22.4, 21.4 °C, pH = 7.7). The development from froglet to adult takes around 66 days (air temperature: ~ 21 – 22°C).

Tadpoles.—Figure 6.93. The description includes illustrations of tadpoles from Estero Piedras, near Caimito-Tonchigue and Río Quingue, Provincia Esmeraldas. The description is based on an individual in stage 35. The ontogenetic variation was documented using 15 tadpoles—one in stage 23, one in stage 24, five in stage 25, three in stage 26, one in stage 27, three in stage 28, and one in stage 35. The total length measures 34.0 mm, while the body length is 10.4 mm, accounting for 30.6% of the total length. The body is elongate, oval, depressed, and wider (body width: 4.5 mm) than high (body height: 3.1 mm). The snout is rounded in dorsal and lateral views. The lateral line system is inconspicuous. The dorsally situated eyes are medium-sized and separated from each other by a distance corresponding

FIGURE 6.92 *Hyalinobatrachium aureoguttatum*. Males guarding clutches of eggs and embryos. Top: under laboratory conditions. Durango, Provincia Esmeraldas. LAC. Middle: Bosque Jardín de los Sueños, Provincia Cotopaxi. Jaime Culebras. Bottom: Bilsa, Provincia Esmeraldas. Jorge Brito Molina.

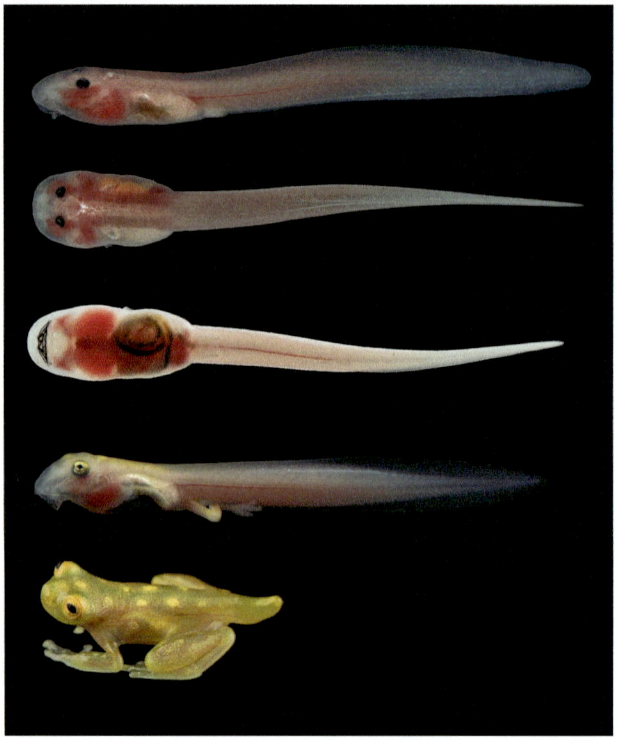

FIGURE 6.93 Ontogenetic changes in *Hyalinobatrachium aureoguttatum* from tadpole to metamorph. Top to bottom. CJ 12513, stage 25, lateral, dorsal, and ventral views, TL = 36.5 mm; undetermined stage, TL = 36.8 mm; stage 45 = 10.5 mm SVL. Durango, Provincia Esmeraldas. SGS.

to the 22% of the body width; the distance from the eyes to the tip of the snout is 2.6 mm. The interorbital distance is 1.0 mm. The narial apertures are very small; the openings are 1.8 mm from the tip of snout; the internarial distance is 1.5 mm; the distance from the narial opening to the anterior edge of the eye is 1.0 mm. The spiracle is short, single, sinistral, located in the posterolateral region of the body; the spiracular aperture has a posterodorsal orientation; its inner wall is free from the body. The tube length is 0.3 mm; the tube width is 0.4 mm; the aperture diameter is 0.4 mm. The distance from the tip of the snout to the spiracular opening is 7.0 mm; the spiracle is located at 67.3% of the body length from the tip of the snout. The vent tube is short and abdominal, free posteriorly, with the opening directed posteroventrally. The tube length is 1.0 mm; the tube transverse width is 0.6 mm. The tail is long (24.5 mm, 2.4× BL), with a subacute tip. The myotomes are visible throughout the entire length of the tail; there is a straight medial line visible separating the dorsal and ventral myotomes. The tail muscle width is 2.1 mm; the tail muscle height is 2.3 mm; the maximum tail height is 4.7 mm; the dorsal fin height is 1.3 mm; the ventral fin height is 1.4 mm. The dorsal and ventral fins originate at the base of tail. The dorsal fin reaches its maximum height posterior to the midlength of tail; the ventral fin reaches its maximum height in the last quarter of tail. The oral disc is large; its transverse width is 2.9 mm (65% of BW); it is emarginate and anteroventral; there are 101 marginal and uniserial papillae (ventral and lateral) distributed around the oral disc; four submarginal papillae are present on the upper labium. The papillae are short and subacute. The lateral papillae (n = 39) are 0.02–0.09 mm ($\bar{x}$ = 0.06 ± 0.01) in length, 0.02–0.11 mm (0.05 ± 0.02) in width; the ventral papillae (n = 62) are 0.04–0.11 mm ($\bar{x}$ = 0.07 ± 0.02) in length, 0.04–0.11 mm ($\bar{x}$ = 0.06 ± 0.01) in width. The upper jaw sheath is M-shaped; its medial part is thinner than the lateral parts; it has a serrated edge (serrations are broad-based and short, oriented straight and medially). The lower jaw sheath is U-shaped, with a serrated edge (serrations on the sides of the "U" have the same shape as the ones on the upper jaw; serrations in the proximal region are more acute and narrower at their base). The transverse width of the upper jaw sheath includes lateral processes, 1.2 mm (representing 41.4% of the ODW); lower jaw sheath, 0.7 mm (24.1% of ODW). The labial tooth row formula is 2(2)/3. The anterior and posterior tooth rows are long, reaching the extremes of the anterior and posterior labia, respectively. Tooth row A-1 is M-shaped, like the upper jaw sheath, 2.2 mm long; number of teeth = 259. The A-2 tooth row is interrupted medially and located on the sides of upper jaw sheath; the length of each side is 0.7 mm; number of teeth = 110. The P-1 tooth row is 2.0 mm long; number of teeth = 158. The P-2 tooth row is 2.1 mm long; number of teeth = 206. The P-3 tooth row is 2.1 mm long; number of teeth = 206; the teeth in this row are shorter and lighter than those in other rows.

Dorsally, the anterior end of the body is gray; the rest of the body and the anterior half of the tail are red; the posterior half of the tail changes to a gray coloration. Dorsally, there is an aggregation of iridophores that form an interorbital line and a middle band on the posterior part of the body. The irises are yellow. The posterior legs (not fully developed) are pale yellow. One individual in stage 23 bears a large yolk sac (length = 1.8 mm, 72% of body length); it does not have a developed spiracle, and the vent tube is closed. The oral apparatus is not fully developed; jaw sheaths are slightly pigmented, and they do not have the shape described in stage 35. Only tooth rows A-1 and P-1 are present; the latter has more teeth and pigmentation; the other tooth rows are visible as low ridges with no labial teeth. In stage 24, the yolk sac has almost disappeared. The spiracle is present, and the vent tube is still closed. All the structures of the oral apparatus are present, except for tooth row P-3, which is visible as a ridge. Larvae in stages 23–25 have black dots and marks along the body, dorsally and laterally. The eyes are C-shaped because of the lack of pigmentation. In stage 25, tooth row P-3 has almost no pigment, and P-2 has slight pigmentation. In stage 26, there is a considerable increase in total length, and the dark coloration has diminished; the posterior tooth rows are slightly more pigmented. In stage 27, coloration is very similar to that of stage 26; eyes are still C-shaped.

In stage 28, the total length is twice the size of stage 26, and the dark coloration has totally disappeared, except for some black dots dorsally on the tail. The eyes and oral apparatus are totally pigmented. Finally, from stage 28 to 35, there is no considerable change in size, coloration, or oral disc morphology. At stage 35, the vent tube is longer, and the cloacal aperture is oriented posteroventrally (not posteriorly, as in previous stages). The number of marginal papillae varies considerably ontogenetically and among individuals in the same stage; the general tendency, however, is an increase in the number of papillae as the tadpoles develop. One individual in stage 23 has 24 papillae; one in stage 24 has 31 papillae; individuals in stage 25 have 30–51 papillae (n = 5); in stage 26 there are 60, 77, and 69 papillae (n = 3); in stage 27, one individual has 87 papillae; and in stage 28, there are 92–105 (n = 3). Some individuals bear few submarginal papillae on the upper labium; in stage 26, one individual bears one submarginal papilla; in stage 27, one individual has one and other has two submarginal papillae; finally, two individuals in stage 28 have four submarginal papillae, and one has five submarginal papillae.

Distribution.—Figure 6.94. *Hyalinobatrachium aureoguttatum* is known in extreme southwestern Panama and the Pacific lowlands and western slopes of the Cordillera Occidental of Colombia and Ecuador at elevations below 1340 m. In Ecuador, it occurs in Carchi, Esmeraldas, and Cotopaxi provinces. Its elevational range is 62–665 m in an area of extent of occurrence of about 19,540 km^2

This nocturnal, arboreal species inhabits natural forests and intervened areas in the evergreen forest of lowlands of Chocó ecosystem in two biogeographic sectors (Table 0.5), where the average annual rainfall is 1329–2410 mm and the average annual temperature is 22.3–25.5 °C.

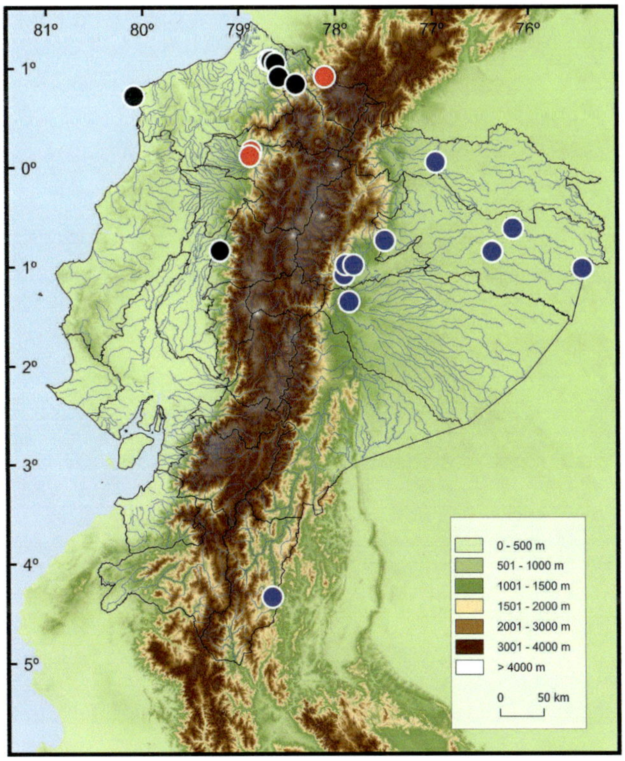

FIGURE 6.94 Localities of known occurrence of *Hyalinobatrachium aureoguttatum* (black dots), *H. mashpi* (red dots), and *H. munozorum* (blue dots) in Ecuador.

Conservation.—*Hyalinobatrachium aureoguttatum* is categorized as Least Concern throughout its entire range (IUCN SSC Amphibian Specialist Group, 2020e; in IUCN Red List of 2022). However, it is important to note that in Ecuador, *H. aureoguttatum* has been categorized as Endangered by Ortega-Andrade *et al.* (2021) because its distribution is severely fragmented and there is a continuing and projected decline in the extent and quality of its habitat. According to Guayasamin *et al.* (2020b), in Ecuador, 42% of the potential distribution of *H. aureoguttatum* is affected by human activities. It is a common species but it has localized threats, such as habitat loss and fragmentation due to the expansion of agriculture, including the cultivation of illegal crops, mining, and water pollution as a result of these activities. Its range includes Parque Nacional de Las Orquídeas and Parque Nacional Natural Farallones de Cali in Colombia and Parque Nacional Darién in Panama. In Ecuador, it has been recorded in the protected forests of Asociación Agrícola Carchi-Imbabura and some private reserves (e.g., Reserva Otokiki, Bosque Jardín de Los Sueños).

Remarks.—Some populations from Colombia usually have small (but visible) brown flecks on the dorsal surfaces of the head and body; Ecuadorian populations lack these flecks. Also, all Colombian specimens examined have a white pericardium, a character that is polymorphic in the Ecuadorian populations (white, partially white, or transparent pericardium). It is possible that these differences correspond to independent evolutionary lineages; however, Guayasamin *et al.* (2020b) chose to maintain these populations as one species until more data (e.g., acoustic, molecular) are available. Guayasamin *et al.* (2022) also found that within lowland populations of *H. aureoguttatum* in Ecuador, there are two groups, but they remarked that further analyses should determine if these genetic differences are the result of different evolutionary trajectories or retained ancestral polymorphisms.

Key references.—(Barrera Rodríguez and Ruiz Carranza, 1989) (description, habitat); (Ruiz-Carranza *et al.*, 1996) (distribution); (Ibañez *et al.*, 1999a) (distribution); (Acosta-Galvis, 2000) (distribution); (Bustamante *et al.*, 2007) (distribution); (Terán-Valdez *et al.*, 2009) (reproductive biology, tadpole); (Valencia-Aguilar *et al.*, 2012) (reproductive biology); Yánez-Muñoz *et al.* (2020) (account); (Guayasamin *et al.*, 2020b) (account).

HYALINOBATRACHIUM CHIRRIPOI (CHIRRIPÓ GLASSFROG) (TAYLOR, 1958)

Cochranella chirripoi (Taylor, 1958): 59. **Holotype.**—KU 36865 from Cocales Creek, Suretka, Cantón de Talamanca, Provincia Limón, Costa Rica.

Hyalinobatrachium chirripoi Ruiz-Carranza and Lynch, 1991a: 24.

Etymology.—The specific epithet refers to the Chirripó/Cabécar people, local inhabitants of the area where the species was first found, in Costa Rica. Chirripó also means "land of the eternal waters."

Identification.—Figures 6.95, 6.96. These frogs attain a maximum SVL of 25.5 mm (Ecuador) and 26 mm (Costa Rica) in males and about 30 mm in females. The skin on the dorsum is shagreen and lacks spicules; the venter is areolate; the ventral surfaces of the thighs below the vent lack a pair of enlarged tubercles; the ventral parietal peritoneum is transparent. There is a transparent pericardium; a white peritoneum covers the intestines and stomach, and the transparent peritoneum covers the urinary and gall bladders; the kidneys are covered by a mostly translucent peritoneum, except for some iridophores on their ventral portions. The vomers lack teeth. The snout is truncate in dorsal view and truncate to slightly protruding in lateral profile. The tympanum is almost completely concealed; when visible, it is small (tympanum diameter 2.3–2.8% of SVL); the tympanic membrane is not differentiated from the surrounding skin. The liver is bulbous and covered by a white peritoneum. The humeral spines are absent. There is basal to moderate webbing between fingers I and II; there is extensive webbing between the outer fingers. The foot webbing is extensive. The ulnar and tarsal folds are present and have minute iridophores. The prepollex is not separated from finger I; in males, the nuptial pad is a medium-sized, diffuse excrescence formed by glandular clusters and individual glands. Finger I is longer than finger II (finger II is about 85–90% the length of finger I). The disc of finger III is of moderate size; it is 38–47% of the eye diameter.

The dorsum is light green with several small yellow spots. The ventral parietal peritoneum is transparent. The pericardium is transparent (the red heart is visible ventrally). There is a transparent peritoneum covering the urinary bladder. There is a white peritoneum covering the liver, intestines, and stomach. There is a transparent peritoneum covering most of the kidneys. The bones are white. The fingers and toes lack melanophores, except for few on toes IV and V. The iris is yellowish-white, with minute dark lavender flecks.

Comparisons.—*Hyalinobatrachium chirripoi* is unique among glassfrogs by its light green dorsum with small yellow spots, a completely transparent ventral parietal peritoneum (red

FIGURE 6.95 *Hyalinobatrachium chirripoi.* Left: QCAZ 45380. Right: QCAZ 45379. Durango, Provincia Esmeraldas. LAC.

FIGURE 6.96 Morphological variation and details of *Hyalinobatrachium chirripoi.* Top left: CJ 12966, metamorph, Chinambí, Provincia Carchi. LAC. Top right and middle left: Canandé, Provincia Esmeraldas. Jaime Culebras. Middle right and bottom: QCAZ 45380, Durango, Provincia Esmeraldas. LAC.

heart visible ventrally), its relatively large size, and by lacking humeral spines. Additionally, it differs from most *Hyalinobatrachium* by having more webbing between fingers II and III.

Natural history.—Figure 6.97. Most males call from the undersides of leaves, but some have been observed calling from the upper sides. In Costa Rica, *Hyalinobatrachium chirripoi* deposits 53–80 greenish-white eggs in a single layer on the undersides of leaves overhanging streams. As development continues, the embryos become pale tannish-red. Males were most often seen calling from below palm fronds or other smooth leaves between 1 and 4 m above the water. One male was seen guarding eggs during the night, and another male was observed guarding eggs during daylight hours. At Reserva Itapoa, Provincia Esmeraldas, Ecuador, a male was observed calling and guarding an egg clutch on the underside of a leaf. Maternal care is absent; prolonged parental care is provided by males.

Advertisement calls of one individual from Reserva Itapoa, Provincia Esmeraldas, Ecuador, were described as a high-pitched insect-like buzz. Each call is composed by a single and pulsed note, and has a mean duration of 250 ms. Detailed data and figures are provided under the subtitle "Vocalizations of Ecuadorian Anurans" in Appendix 1.

Tadpoles.—The larvae were described from Costa Rica. Hatchlings (Gosner stages 25 and 26) have an unusual pattern on their dorsum: two rows of brown-pigmented, spider-like dots of melanophores stretch longitudinally over the entire body and join in a semicircle at the tip of snout; another line of pigmented spots extends dorsally over the tail musculature. Because of the general lack of pigment and high transparency, the body coloration is partially determined by the red color of the gills and heart. As development continues (Gosner stages 25–41), the tadpoles become very elongated and are among the most elongated centrolenid larvae known (ratio of body length to body width of 2.41 ± 0.22); they have completely transparent skin, and the dorsolateral lines are mostly lost; the body has a pale rose sheen, and the tail is cream or sometimes yellowish. The oral disc is emarginate and, except for a moderate anterior labium gap, is bordered by a single row of about 45 marginal papillae; the ventral ones being flatter than the lateral ones. The LTRF is 2(2)/3; the tooth rows are about of equal width (P-3 is slightly narrower) and extend across most of the oral disc. The A-2 gap is about as wide as the upper jaw, which is broadly arched; the lower jaw is V-shaped; the edges of both jaw sheaths are edged by many narrow serrations.

Distribution.—Figure 6.98. *Hyalinobatrachium chirripoi* is known from lowland forest localities in Central America in southeastern Costa Rica and eastern Panama south into South America to western Colombia and northwestern Ecuador. In Ecuador, it occurs in Esmeraldas and Manabí provinces. Its elevational range is 62–451 m in an area of extent of occurrence of about 10,910 km^2.

FIGURE 6.97 *Hyalinobatrachium chirripoi.* Males guarding clutches of embryos underside leaves, Reserva Itapoa, Provincia Esmeraldas. Jaime Culebras.

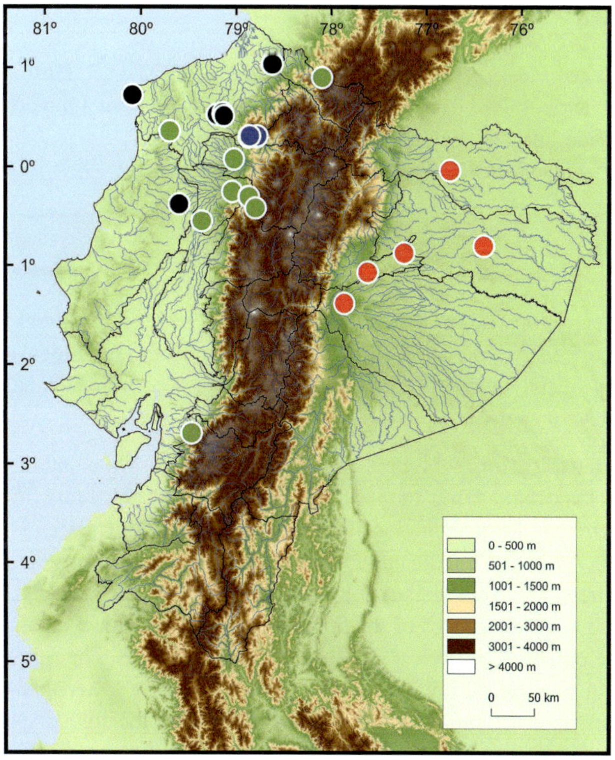

FIGURE 6.98 Localities of known occurrence of *Hyalinobatrachium chirripoi* (black dots), *H. iaspidiense* (red dots), and *H. nouns* (blue dots) in Ecuador.

This nocturnal, arboreal species inhabits natural forests and intervened areas in the evergreen forests of lowlands of Chocó and evergreen montane foothill forests of the Cordillera Occidental of the Andes ecosystems in the Cordillera Occidental of the Andes, Ecuadorian Chocó, and Jama-Zapotillo biogeographic sectors, where the average annual rainfall is 1329–2385 mm and the average annual temperature is 23.3–25.0 °C.

Conservation.—*Hyalinobatrachium chirripoi* is categorized as Least Concern throughout its entire range (IUCN SSC Amphibian Specialist Group, 2022h; in IUCN Red List of 2022). However, it is important to note that in Ecuador, *H. chirripoi* was categorized as Endangered by Ortega-Andrade *et al.* (2021). Although their calculations of the extent of occurrence (4728 km^2) differ from ours (10,910 km^2), it is clear that the area of occupancy is reduced and severely fragmented. In Ecuador, the species is threatened by habitat destruction because of agriculture, pasture lands, and mining. In Ecuador, it has been recorded in the protected forest of Asociación Agrícola Carchi-Imbabura.

According to Guayasamin *et al.* (2020b), molecular data support a sister relationship between *Hyalinobatrachium chirripoi and H. colymbiphyllum.*

Key references.—(Taylor, 1958) (description); (Kubicki, 2004b) (reproductive biology); (Kubicki, 2007) (reproductive biology); (Hoffmann, 2010) (tadpole); (Delia *et al.*, 2017) (reproductive biology); (Guayasamin *et al.*, 2020b) (account, call).

HYALINOBATRACHIUM IASPIDIENSE (JASPE'S GLASSFROG) (AYARZAGÜENA, 1992)

Centrolenella iaspidiensis (Ayarzagüena, 1992): 23. **Holotype.**—EBD 28803 from Quebrada Jaspe, San Ignacio de Yuruaní, Estado Bolívar, Venezuela.

Hyalinobatrachium iaspidiense Myers and Donnelly, 1997: 16.

Etymology.—The specific epithet is derived from the Greek word *iaspis*, meaning jasper, in reference to the type locality of the species, Quebrada de Jaspe, in the Gran Sabana of Venezuela.

Identification.—Figures 6.99, 6.100. These frogs attain a maximum SVL of 19.9 mm in males and 21.9 mm in females from Ecuador, whereas the maximum SVL is 22.7 mm in a female and 21.8 mm in males from Venezuela. The skin on the dorsum is shagreen; the venter is smooth; the ventral surfaces of the thighs below the vent have a pair of enlarged tubercles, which are difficult to see. The ventral parietal peritoneum is transparent; there is a transparent pericardium. A white peritoneum covers the intestines and stomach, and there is a transparent peritoneum on the

FIGURE 6.99 *Hyalinobatrachium iaspidiense.* CJ 12463, male, 21.0 mm SVL, Puyo, Provincia Pastaza. SGS.

FIGURE 6.100 Morphological details of *Hyalinobatrachium iaspidiense*. CJ 12463, male, Puyo, Provincia Pastaza. SGS.

urinary bladder. The vomerine teeth are absent. The snout is truncate in dorsal aspect and slightly protruding in profile. The tympanum is not visible; the supratympanic fold is absent. The liver is bulbous; the hepatic peritoneum is white. The humeral spines are absent. There is no webbing between fingers I, II, and III; there is moderate webbing between the outer fingers. The foot webbing is extensive. The ulnar fold is present and enameled; the external tarsal fold is present, enameled; the internal tarsal fold is short and low. The prepollex is not separated from finger I; the nuptial pad is a large excrescence formed by a granular pad that extends from the side of the thumb to its dorsomedial surface and on the proximal dorsolateral surface of finger II. Finger II is shorter than finger I (finger II is about 93% of finger I). The width of the disc of finger III is 39–57% of the eye diameter.

The dorsum is translucent, with a yellowish-green background coloration; large, light green blotches; and small black spots. The venter is transparent, the pericardium is transparent (red heart visible), and the visceral and hepatic peritonea are white. The fingers and toes lack melanophores, except for few present on finger IV and toes IV and V. The bones are white. The iris is yellowish to grayish-white.

Comparisons.—No other glassfrog in the Amazonian lowlands of Ecuador shares the dorsal color pattern of *Hyalinobatrachium iaspidiense* (green with light green blotches and small black spots).

Natural history.—Individuals have been found on vegetation 4–7 m above streams, and males call from the undersides of leaves. There is little information about females; however, an adult female from the Yachana Reserve (Ecuador) was found on a long blade of a tall grassy shrub overhanging water, located in the middle of a stream (4 m wide) that runs through the reserve. The stream habitat is variable along its length (~2500 m within reserve), but this frog was found in a shallow area (~20 cm deep), where the stream becomes a riffle and is more fast-flowing. The stream bottom is composed of about 80% pebbles and small rocks and 20% sand (close to the edges). The forest adjacent to the stream is near pristine and about 25–30 m high.

Advertisement calls of four individuals from Quebrada de Jaspe (Estado Bolívar) and Primer Raudal del Caño Acoima (Estado Delta Amacuro), Venezuela, were described as single notes with 16 pulses and have a duration of 62.7–75.1 ms. Detailed data and figures are provided under the subtitle "Vocalizations of Ecuadorian Anurans" in Appendix 1.

Tadpoles.—Unknown.

Distribution.—See Figure 6.98. *Hyalinobatrachium iaspidiense* is known in the Guiana region of Venezuela, central Guyana, Suriname (Sipaliwini District), and French Guiana, as well as in the Amazon Basin of Ecuador, Peru, and Brazil at elevations below 1000 m. In Ecuador, it occurs in Sucumbíos, Orellana, Napo, and Pastaza provinces. Its elevational range is 288–962 m in an area of extent of occurrence of about 8280 km^2.

This nocturnal, arboreal species inhabits natural forests in the evergreen forests of lowlands of Aguarico-Putumayo-Caquetá and Napo-Curaray and the evergreen montane foothill forests of the northern Cordillera Oriental of the Andes ecosystems in the Amazonian Cordilleras biogeographic sector, where the average annual rainfall is 3196–4095 mm and the average annual temperature is 21.1–24.9 °C.

Conservation.—*Hyalinobatrachium iaspidiense* is categorized as Least Concern throughout its entire range (IUCN SSC Amphibian Specialist Group, 2022i; in IUCN Red List of 2022), given the large distribution of the species and the absence of immediate threats. However, it is important to note that in Ecuador, *H. iaspidiense* has been categorized as Near Threatened by Ortega-Andrade *et al.* (2021). Localized threats to its habitat are agricultural development (cultivation of crops, rearing of livestock), as well as logging, road construction, and oil spills. In Ecuador, it has been recorded in the protected forest of Comuna Cofán Dureno, the biosphere reserves Yasuní and Sumaco, and some private reserves (e.g., Reserva Yachana).

Remarks.—*Hyallinobatrachium iaspidiense* is the sister taxa of *H. tricolor* (Guayasamin *et al.*, 2020b).

Key references.—(Ayarzagüena, 1992) (description); (Lescure and Marty, 2000) (distribution); (Cordeiro-Duarte *et al.*, 2002) (distribution); (Señaris and Ayarzagüena, 2005) (distribution, habitat, call); (Ernst *et al.*, 2005) (distribution); (Caldwell and Shepard, 2005) (distribution); (Kok and Castroviejo-Fisher, 2008) (distribution); (Barrio Amoros and Brewer-Carias, 2008) (taxonomy); (Yánez-Muñoz *et al.*, 2009) (taxonomy); (Guayasamin and North, 2009) (taxonomy, distribution, habitat); (Guayasamin *et al.*, 2020b) (account).

HYALINOBATRACHIUM MASHPI (MASHPI GLASSFROG) GUAYASAMIN *ET AL.*, 2022

Hyalinobatrachium mashpi Guayasamin *et al.*, 2022: 6.

Holotype.—CJ 11642 from Río San Vicente, Reserva Mashpi, 1040 m, Provincia Pichincha, Ecuador.

Etymology.—The specific epithet refers to the Mashpi area in northwestern Ecuador. There are several conservation efforts to preserve the last patches of forest remaining in Mashpi (e.g., Reserva Mashpi Lodge, Reserva Tayra, Reserva Amagusa, Biosphere Reserve Chocó Andino de Pichincha). Mashpi is a Yumbo word that means "friend of water," an apt description of this glassfrog, which depends on healthy streams for its reproduction.

Identification.—Figure 6.101. The maximum known SVL is 20.9 mm in males and unknown in females. The dentigerous processes of the vomers lack teeth. The snout is truncate in dorsal view and slightly protruding in lateral view. The tympanum is oriented almost horizontally. The tympanic annulus is barely visible, hidden under the skin; the tympanic membrane is differentiated, with coloration similar to that of surrounding skin. The dorsal skin is shagreened; the ventral skin is areolate. The cloacal ornamentation is absent. The paired round tubercles below the vent are absent. The parietal peritoneum and pericardium are translucent (in life, red heart visible in ventral view). The liver, viscera, and testes are covered by iridophores. The liver is bulbous. The humeral spines are absent. The hand webbing is basal. The foot webbing is moderate. The fingers and toes have thin lateral fringes; the ulnar and tarsal folds are absent. The nuptial excrescence is present as a small pad on finger I; the prepollex is not enlarged; the prepollical spine is not projecting (spine not exposed). When appressed, finger I is longer than II. The diameter of the eye is about two times wider than the disc on finger III.

The dorsal surfaces are light green with small, light yellow spots. The melanophores are absent from the fingers and toes, except from toes IV and V. The iris is yellow to golden yellow; the pupil is surrounded by a lavender ring.

FIGURE 6.101 *Hyalinobatrachium mashpi*. Left: CJ 12659, 20.6 mm SVL, Quebrada Naranjo, Provincia Carchi. Top right: CJ 12300, male, 22.0 mm SVL, Mashpi, Provincia Pichincha. SGS. Bottom right: Mashpi Lodge. Amanda B. Quezada.

Comparisons.—*Hyalinobatrachium aureoguttatum*, *H. nouns*, *H. talamancae*, *H. valerioi*, *H. vireovittatum*. *Hyalinobatrachium mashpi* is distinguished from other species in the genus mainly by its dorsal coloration and by its transparent pericardium (i.e., red heart visible in ventral view). Nevertheless, it can be confused with other *Hyalinobatrachium* species with similar dorsal coloration patterns. *H. mashpi* can be distinguished from *H. aureoguttatum* and *H. valerioi* by an interorbital bar, usually present in *H. mashpi* and absent in the other two species. Additionally, the heart is always visibly red in *H. mashpi*, whereas in *H. aureoguttatum* and *H. valerioi*, it can also be white or red and white. *H. mashpi* resembles the Central American species *H. talamancae* and *H. vireovittatum*, but the distributions do not overlap. *H. mashpi* is morphologically cryptic with *H. nouns*; however, they exhibit a considerable genetic distance (16S; 4.6–4.7%), which is particularly remarkable given that they are found in relatively close geographic proximity (straight distance = 18.9 km), but the Intag-Guayllabamba river valley acts as a barrier.

Natural history.—Figure 6.102. Most individuals of *Hyalinobatrachium mashpi* were found on the undersides of leaves among riverine vegetation, 3–14 m above the ground along steep creeks. This habitat placement poses challenges

FIGURE 6.102 *Hyalinobatrachium mashpi*. Top left: CJ 12300, male guarding a clutch of embryos. Jaime Culebras. Top right and bottom: CJ 12301, development of embryos depicted in the photo in the top right to the one in the bottom left lasted 6 days. SGS. Reserva Tayra, Provincia Pichincha.

for the observation of the species. Males have been observed calling in the months of April, May, June, August, September, October, and November. Egg clutches are located on the undersides of leaves overhanging streams. Males that were guarding egg clutches while calling were observed during the rainy season (October 2014, 18 February 2019; 7 May 2021) and dry season (June 2015 and August 2021). Examined egg clutches contain 31–34 eggs (n = 3). A female with mature eggs visible through the skin was observed on 27 May 2015.

The advertisement calls of two individuals from Reserva Mashpi and Reserva Tayra, Provincia Pichincha, Ecuador, were described as single pulsed (amplitude modulated) notes with a duration of 0.37–0.46 s. Detailed data and figures are provided under the subtitle "Vocalizations of Ecuadorian Anurans" in Appendix 1.

Tadpoles.—Unknown.

Distribution.—See Figure 6.94. *Hyalinobatrachium mashpi*, an endemic species to Ecuador, is known on the Pacific slopes of the Cordillera Occidental of the Andes of Ecuador in Carchi and Pichincha provinces. Its elevational range is 1020–1501 m in an area of extent of occurrence of about 160 km^2.

This species inhabits riverine areas in the evergreen montane foothill forests of the Cordillera Occidental of the Andes ecosystem in the Cordillera Occidental of the Andes biogeographic sector, where the average annual rainfall is 1518–2591 mm and the average annual temperature is 19–20.7 °C.

Conservation.—*Hyalinobatrachium mashpi* has not been evaluated by the IUCN.

Guayasamin *et al.* (2022), in their description of the species, suggest categorizing *Hyalinobatrachium mashpi* as Endangered following IUCN Red List criteria B1ab(iii). The extent of occurrence is less than 5000 km^2 (160 km^2 according to our data), it is only known from three localities, and there is a continued and projected decline in the extent and quality of its habitat. The main threats for this species are habitat loss and contamination due to cattle ranching and agriculture. An extremely troubling emerging threat within the species' distribution range is the establishment of large-scale metal mining concessions. These activities pose a significant danger to numerous species in the Andean region, primarily due to habitat degradation and contamination. It has been recorded in the protected forest of Mashpi, the biosphere reserve Chocó Andino de Pichincha, and some private reserves (e.g., Reserva Mashpi Lodge, Reserva Amagusa, Reserva Tayra).

Remarks.—According to Guayasamin *et al.* (2022), *Hyalinobatrachium mashpi* is the sister species of undetermined haplotypes from the Colombian Andes and *H. nouns* based on mitochondrial DNA (mitochondrial gene 16S). Other closely related taxa are endemic to Central America: *H. vireovittatum* and *H. talamancae*. Unidentified and closely related frogs from Colombia (Departamento de Risaralda; Departamento de Valle del Cauca) may prove to be conspecifics of *H. mashpi*.

Remarks.—*Hyallinobatrachium mashpi* is the sister taxa of *H.* sp. from Colombia (Guayasamin *et al.*, 2022).

Key references.—(Guayasamin *et al.*, 2022) (description, call).

HYALINOBATRACHIUM MUNOZORUM (MUÑOZ'S GLASSFROG) (LYNCH AND DUELLMAN, 1973)

Centrolenella munozorum (Lynch and Duellman, 1973): 40. **Holotype.**—KU 118054 from Santa Cecilia, 340 m, Provincia Sucumbíos, Ecuador.

Hyalinobatrachium munozorum Ruiz-Carranza and Lynch, 1991a: 24.

Etymology.—The specific epithet is a patronym for Ing. Ildefonso Muñoz B. and Sra. Blanca Muñoz, hosts of the extensive herpetological work carried out in Santa Cecilia by William E. Duellman and his students. Idelfonso Muñoz was a key Colombian leader in the development of Lago Agrio, Ecuador, and author of the text "*And the Jungle Did Not Devour Them.*"

Identification.—Figure 6.103. These frogs attain a maximum SVL of 21.9 mm in males and 23.6 mm in females. The skin on the dorsum is shagreen; the males lack spicules. The ventral skin texture is areolate; the ventral surfaces of the thighs below the vent lack a pair of enlarged tubercles. The ventral parietal peritoneum is transparent. White peritonea cover the intestines and stomach. The urinary bladder is transparent. The vomers lack teeth. The snout varies from truncate to rounded in dorsal aspect and round to truncate in lateral view. The tympanum is usually obscured by the skin; when evident, it is small, and its diameter is about 25% of the eye diameter; the supratympanic fold is low. The liver is bulbous and covered by a white peritoneum. The humeral spines are absent. There is no webbing between the inner fingers; the webbing between fingers III and IV is extensive. The feet are about three-fourths webbed. The ulnar and tarsal folds vary from inconspicuous to pronounced. The prepollex is not separated from finger I; in males, the nuptial pad is not evident. Finger I is slightly longer than finger II (finger II length is 93.7–97.6% of finger I). The disc of finger III is of moderate size, and its width is 37.9–56.2% of the eye diameter.

The dorsum of the head and body are green with small, diffuse, pale yellow spots and minute dark brown flecks. The parietal peritoneum is transparent. There are white peritonea covering the heart, liver, digestive tract, and testes. The urinary bladder is transparent. The dorsal surfaces of the fingers and toes lack melanophores. The bones are white. The iris is pale to bright gold with minute dark flecks.

FIGURE 6.103 *Hyalinobatrachium munozorum*. Top: KU 123225, paratype, female, 23.0 mm SVL, Santa Cecilia, Provincia Sucumbíos. WED. Middle and bottom: MZUTI 1616, male, near Tena, Provincia Napo. Eduardo Toral.

Comparisons.—*H. iaspidiense*, *Chimerella mariaelenae*, and *Teratohyla amelie*. *Hyalinobatrachium munozorum* is differentiated from most glassfrogs by having a green dorsum with pale yellow spots, a completely transparent ventral parietal peritoneum, a white liver and digestive tract, and by lacking humeral spines. Similar species with a completely transparent venter that inhabit the Ecuadorian Amazonian lowlands include *H. iaspidiense*, *C. mariaelenae*, and *T. amelie*. *Hyalinobatrachium munozorum* differs from all of these species by its pale yellow dorsal spots, which are absent in the aforementioned species. Also, *H. munozorum* lacks humeral spines, present in *C. mariaelenae*.

FIGURE 6.104 *Hyalinobatrachium munozorum*. CJ 12903, clutch of embryos on the underside of a leaf, Parque Nacional Yasuní, Provincia Orellana. SGS.

Natural history.—Figure 6.104. Individuals occupy diverse habitats, and they have been found in various locations on leaves of bushes and trees at night: over a pond, away from water in primary forests, 2 m above streams, and at a rivulet in a secondary forest. Also, an individual from Lago Agrio, Provincia Sucumbíos, was obtained from the foliage of a large tree that was felled during the clearing of a primary forest.

Males were found calling from riverine vegetation at a locality near Tena, Provincia Napo, Ecuador, on August 2012. The egg clutches are laid on the undersides of leaves.

Advertisement calls of one individual from Río Bigal, Provincia Orellana, Ecuador, were described as a single note and have a mean duration of 170 ms. Detailed data and figures are provided under the subtitle "Vocalizations of Ecuadorian Anurans" in Appendix 1.

Tadpoles.—Figure 6.105. This description is based on an individual (not preserved, series CJ 12903) from Parque Nacional Yasuní, Provincia Orellana, Ecuador, in stage 28. A clutch was collected from a leaf at approximately 3 m above the water of a small stream by Morley Read on 9 March 2022. It measures 8.9 mm in body length and 26.3 mm in total length. The body is ovoid, slightly wider anteriorly, and it is wider than it is deep. The snout appears round in both dorsal and lateral profiles, with narrowly separated eyes directed dorsally. The chondrocranium elements are visible. The spiracle is sinistral, with the inner wall free from the body, and the opening is directed posterolaterally on the deadline at about two-thirds of the body's length. The mouth is medium-sized, located anteroventrally, and exhibits a transversely elliptical shape. The upper labium is bare, while the lower labium is more developed and detached from the body wall. The upper jaw

FIGURE 6.105 Ontogenetic changes in two individuals of *Hyalinobatrachium munozorum* from tadpole to juvenile. Top to bottom: CJ 12903a, stage 29, lateral, dorsal, and ventral views, TL = 26.3 mm; stage 35, TL = 32 mm. CJ 12903b, stage 42, TL = 33.3 mm; stage 45 = 11.1 mm SVL; juvenile 126 days = 16.1 mm SVL. Parque Nacional Yasuní, Provincia Orellana. SGS.

sheath is completely keratinized, with a smooth arc interrupted medially and serrated on the inner side. The lower jaw sheath is partially keratinized. The tail constitutes about 66% of the total length, with the greatest depth at the midlength and a rounded tip. The dorsal fin does not extend onto the body, is slightly deeper than the ventral fin, and is approximately half the depth of the caudal musculature at the tail's midlength.

The skin is translucent, allowing the internal organs to be visible, resulting in a pinkish coloration of the body. White dots are present dorsally and ventrally on the tail; forming a line on the inner part of the fins in lateral view, outlining the caudal musculature.

By stage 40, occurring 83 days after stage 28, with a total length of 32 mm, the tadpole displays a uniform light green coloration, with the pink organs visible laterally on the body. At stage 42, 13 days later and with a total length of 27 mm, there are no significant changes in coloration, although the green becomes more vivid, and the snout takes on a bluish hue. At stage 45, the individual appears predominantly green but acquires a bronze coloration on the dorsum. The hands and feet exhibit a light blue shade, and the iris turns orange. After three months of tail reabsorption, the juvenile starts to resemble the coloration typical of an adult individual.

Distribution.—See Figure 6.94. *Hyalinobatrachium munozorum* is known from a few localities in the Amazonian lowlands and Andean foothills of Colombia, Ecuador, Peru, and Bolivia at elevations below 980 m. In Ecuador, it occurs in Sucumbíos, Orellana, Napo, Pastaza, Zamora Chinchipe provinces. Its elevational range is 195–939 m in an area of extent of occurrence of about 66,480 km^2.

This nocturnal, arboreal species inhabits natural forests and intervened areas in five ecosystems (Table 0.4) within four biogeographic sectors (Table 0.5), where the average annual rainfall is 1630–4084 mm and the average annual temperature is 21.2 – 25.5°C.

Conservation.—*Hyalinobatrachium munozorum* is categorized as Least Concern throughout its entire range (IUCN SSC Amphibian Specialist Group, 2022j; in IUCN Red List of 2022) and in Ecuador by Ortega-Andrade *et al.* (2021), in view of its wide distribution, presumed large population, and because it is unlikely to be declining to qualify for listing in a more threatened category. Although *H. munozorum* is rare in collections, it is presumed that this is a consequence of inadequate sampling at the canopy level. There is some localized habitat loss to general human activities such as agriculture (crops, livestock, etc.) and colonization; however, it seems that the species tolerates some degree of disturbance. In Ecuador, it has been recorded in the public reserve Parque Nacional Yasuní, the protected forests of Cerro Sumaco y Cuenca Alta del Río Suno and Cuenca Alta del Río Nangaritza, and the biosphere reserves Yasuní, Sumaco, and Podocarpus-El Cóndor.

Remarks.—Guayasamin *et al.* (2020b) place *Hyalinobatrachium ruedai* (Ruiz-Carranza and Lynch, 1998), in the synonymy of *H. munozorum* given that all individuals of the type series of *H. munozorum* and *H. ruedai* have a white heart, and the range of the webbing on the hand observed in *H. ruedai* falls within the variation in *H. munozorum*.

Key references.—(Lynch and Duellman, 1973) (description, habitat); (Ruiz-Carranza and Lynch, 1998) (description, distribution, under *H. ruedai*); (Cisneros-Heredia, 2005) (distribution); (Cisneros-Heredia and Mcdiarmid, 2007) (distribution, habitat); (Castroviejo-Fisher *et al.*, 2011) (distribution); (Guayasamin *et al.*, 2020b) (account, call).

HYALINOBATRACHIUM NOUNS (NOUNS' GLASSFROG) GUAYASAMIN *ET AL.*, 2022

Hyalinobatrachium nouns Guayasamin *et al.*, 2022: 19.
Holotype.—MZUTI 3299 from Bosque Protector Los Cedros, Cordillera de Toisán, 1420 m, Provincia Imbabura, Ecuador.

Etymology.—The specific epithet honors Nouns DAO, a global decentralized autonomous organization ("DAO") composed of owners of Nouns characters, which are digital art creations that live on the blockchain. The mission of Nouns DAO is to promote and build the Nouns brand throughout the physical and digital world. One of the ways Nouns DAO accomplishes this is by building public works and funding philanthropic projects that support the wonder of nature.

Identification.—Figures 6.106, 6.107. The maximum known SVL is 21.3 mm in males and unknown in females. The dentigerous processes of the vomers lack teeth. The snout is truncate in dorsal view and slightly protruding in lateral view. The tympanum is oriented almost horizontally; the tympanic annulus is barely visible, hidden under the skin. The tympanic membrane is differentiated, with coloration similar to that of the surrounding skin. The enameled tubercles are absent from the sides of the head. The dorsal skin is shagreened. The ventral skin is areolate. The cloacal ornamentation is absent; paired round tubercles below the vent are absent. The parietal peritoneum and pericardium are translucent (in life, the red heart is visible in ventral view); the liver, viscera, and testes are covered by iridophores. The liver is bulbous. The humeral spines are absent. The hand webbing is basal. The foot webbing is moderate. The fingers and toes have thin lateral fringes. The ulnar and tarsal folds are absent. The nuptial excrescence is present as a small pad on finger I. The prepollex is not enlarged. The prepollical spine is not projecting (spine not exposed). When appressed, finger I is longer than II. The diameter of the eye is about two times wider than the disc on finger III.

The dorsal coloration is light green with small, light yellow spots; melanophores are absent from the fingers and toes, except on toes IV and V. The iris is yellow to golden yellow; the pupil is surrounded by a lavender ring.

Comparisons.—*Hyalinobatrachium aureoguttatum*, *H. mashpi*, *H. talamancae*, *H. valerioi*, and *H. vireovittatum*. *Hyalinobatrachium nouns* is distinguished from other species in the genus mainly by its dorsal coloration and by its transparent pericardium. Nevertheless, it can be confused with other *Hyalinobatrachium* species with similar dorsal coloration patterns. *H. nouns* can be distinguished from *H. aureoguttatum* and *H. valerioi* by an interorbital bar, usually present in *H. nouns* and absent in the other two species. Additionally, the heart is always visibly red in *H. nouns*, whereas in *H. aureoguttatum* and *H. valerioi*, it can also be white or red and white. *H. nouns* resembles the Central

FIGURE 6.106 *Hyalinobatrachium nouns*. Left: CJ 12905, female, 21.9 mm SVL. Amanda B. Quezada. Right: CJ 12916, male. SGS. Sector Río Manduriacu, Provincia Imbabura.

FIGURE 6.107 Metamorph and morphological details of *Hyalinobatrachium nouns*. Top left: CJ 12904, metamorph. Amanda B. Quezada. Top right and middle left: CJ 12905, female. Middle right and bottom: CJ 12916, male. Sector Río Manduriacu, Provincia Imbabura. SGS.

FIGURE 6.108 *Hyalinobatrachium nouns*. CJ 12918, clutches of tadpoles, Sector Río Manduriacu, Provincia Imbabura. SGS.

American species *H. talamancae* and *H. vireovittatum*, but the distributions do not overlap. *H. nouns* is morphologically cryptic with *H. mashpi*, but they exhibit a considerable genetic distance (16S; 4.6–4.7%), which is remarkable given that they are found at a relatively close geographic proximity (straight distance = 18.9 km), but with the Intag-Guayllabamba river valley acting as a barrier.

Natural history.—Figure 6.108. At Bosque Protector Los Cedros and Reserva Río Manduriacu, individuals were found on the undersides of leaves, 1–6 m above a stream during the months of February, March, and November. Males call from the undersides of leaves and during the rainy season; in February, a male was observed calling next to an egg clutch. Metamorphs have been found perched on leaves 50–150 cm above streams in October and November. The parental care is unknown.

Tadpoles.—Unknown.

Distribution.—See Figure 6.98. *Hyalinobatrachium nouns*, an endemic species to Ecuador, is known from the Toisán mountain range on the Pacific slopes of the Cordillera Occidental of the Andes of Ecuador in Provincia Imbabura. Its elevational range is 1133–1501 m in an area of extent of occurrence of about 10 km^2.

This species inhabits the evergreen montane foothill forests of the Cordillera Occidental of the Andes ecosystem in the Cordillera Occidental of the Andes biogeographic sector, where the average annual rainfall is 1863–2082 mm and the average annual temperature is 18.3 – 20.0°C.

Conservation.—*Hyalinobatrachium nouns* has not been evaluated by the IUCN. In the description of the species, Guayasamin *et al.* (2022) suggest that *H. nouns* should be categorized as Endangered, following IUCN Species Survival Commission (2012) criteria B1ab(iii). Its extent of occurrence is extremely restricted, 10 km^2 according to our data; it is known to exist only in two localities; and there is a continuing and projected decline in the extent and quality of its habitat. The main threats for this species are habitat loss and contamination due to cattle ranching and agriculture. An extremely troubling emerging threat within the species' distribution range is the establishment of large-scale metal mining concessions. These activities pose a significant danger to numerous species in the Andean region, primarily due to habitat degradation and contamination. It has been recorded in the protected forest of Los Cedros and some private reserves (e.g., Reserva Río Manduriacu, Reserva Los Cedros).

Remarks.—According to Guayasamin *et al.* (2022), *Hyalinobatrachium nouns* is the sister to a lineage formed by *H. mashpi* and unidentified haplotypes from the Colombian Andes. However, this relationship has low support (bootstrap support = 60). Other closely related taxa are endemic to Central America: *H. vireovittatum* and *H. talamancae*.

Key references.—(Guayasamin *et al.*, 2022) (description).

HYALINOBATRACHIUM PELLUCIDUM (CRYSTAL-CLEAR GLASSFROG) (LYNCH AND DUELLMAN, 1973)

Centrolenella pellucida (Lynch and Duellman, 1973): 43.
Holotype.—KU 143298 from Río Azuela, 1740 m, Quito-Lago Agrio road, Provincia Sucumbíos, Ecuador.

Hyalinobatrachium pellucidum Ruiz-Carranza and Lynch, 1991a: 24.

Etymology.—The specific epithet is derived from the Latin word *pellucidus*, meaning clear, transparent, and refers to the transparent parietal peritoneum of this centrolenid frog.

Identification.—Figures 6.109, 6.110, 6.111. These frogs attain a maximum SVL of 21.4 mm in males and 21.6 mm in females. The skin on the dorsum is shagreen; the ventral skin texture is areolate; the ventral surfaces of the thighs below the vent lack a pair of enlarged tubercles. The ventral parietal peritoneum is transparent. A transparent pericardium and a white peritoneum cover the intestines and stomach. There is a transparent peritoneum on the urinary bladder. The vomerine teeth are absent. The snout is truncate in dorsal aspect and profile. The tympanum is partially hidden under the skin, and its diameter is about 30% of the eye diameter; the supratympanic fold is low. The liver is bulbous, the hepatic peritoneum is white. The humeral spines are absent. There is no webbing between fingers I and II; there is extensive webbing between the outer fingers. The webbing between the toes is extensive. The ulnar and tarsal folds are present and usually are white. The prepollex is not separated from finger I; the nuptial pad type is unknown. Finger I is slightly longer than finger II (finger II is about 95% of finger I). The width of the disc on finger III is about 50% of the eye diameter.

FIGURE 6.109 *Hyalinobatrachium pellucidum*. CJ 12991, male, 21.1 mm SVL, Quebrada Río Napinaza, Provincia Morona Santiago. SGS.

The dorsum is pale green with diffuse yellow spots. The venter and hidden surfaces of the limbs lack pigment, exposing the red heart. The fingers and toes are yellow. The parietal peritoneum is clear. The red heart is visible. The bones are white. The dorsal surfaces of fingers and toes lack melanophores. The iris is pale silvery bronze.

FIGURE 6.110 Morphological variation in *Hyalinobatrachium pellucidum*. Top left: CJ 11786, Runahurco, Provincia Morona Santiago. SGS. Top right: CJ 1193, Río Oyacachi, Provincia Napo. LAC. Second row left: QCAZ 31883. Second row right: QCAZ 42000. Third row and bottom left: QCAZ 32794, metamorph. Third row and bottom right: QCAZ 41910, froglet. Quebrada Río Napinaza, Limón Indaza, Provincia Morona Santiago. LAC.

Comparisons.—*Hyalinobatrachium iaspidiense*. *Hyalinobatrachium pellucidum* is distinguishable from most glassfrogs by having a completely transparent ventral parietal peritoneum, a white liver, and transparent pericardium (in life, red heart visible ventrally). On the Amazonian slopes of the Andes and the Amazon Basin, the only other species with a visible red heart in life is *H. iaspidiense*, which differs from *H. pellucidum* by having large, light green blotches and small black spots on the dorsum (both absent in *H. pellucidum*).

Natural history.—Figure 6.112. These frogs have been found active at night on leaves of herbs, shrubs, or trees 1 m above streams. Egg clutches are placed on the undersides of leaves, and males guard them. Short-term maternal care is absent; males provide prolonged parental care. Clutches from Río Napinaza, Provincia Morona Santiago, were found in August 2022 and December 2006; they contain up to 39 embryos.

The advertisement calls of one individual from Río Azuela, Provincia Napo, Ecuador, were described as relatively short single notes with a mean duration of 128 ms. Detailed data and figures are provided under the subtitle "Vocalizations of Ecuadorian Anurans" in Appendix 1.

Tadpoles.—Unknown.

FIGURE 6.111 Morphological details of *Hyalinobatrachium pellucidum*. CJ 12991, male, Quebrada Río Napinaza, Limón, Provincia Morona Santiago. SGS.

FIGURE 6.112 *Hyalinobatrachium pellucidum*. Males guarding clutches of embryos. Top left: QCAZ 32793, LAC. Top right: CJ 12991, SGS. Quebrada Río Napinaza, ca. Limón. Bottom left: Parque Nacional Sangay, Jorge Brito Molina. Bottom right: Runahurco, Provincia Morona Santiago. Jaime Culebras.

Distribution.—See Figure 6.88. *Hyalinobatrachium pellucidum* is known from the lower montane rainforest on the Amazonian Andean slopes of Ecuador and Peru at elevations between 1000 and 1897 m. In Ecuador, it occurs in Napo, Morona Santiago, and Zamora Chinchipe provinces. Its elevational range is 1036–1897 m in an area of extent of occurrence of about 15,710 km^2.

This nocturnal, arboreal species inhabits natural forests and intervened areas in four ecosystems (Table 0.4) within three biogeographic sectors (Table 0.5), where the average annual rainfall is 1233–2832 mm and the average annual temperature is 16.6 – 22.7°C.

Conservation.—*Hyalinobatrachium pellucidum* is categorized as Near Threatened throughout its entire range (IUCN SSC Amphibian Specialist Group, 2022k; in IUCN Red List of 2022). Although the species is still reasonably widely distributed, much of its habitat is poorly protected, and the population is suspected to be declining at a rate of at least 20–25% over a 10-year period (2015–2025) due to the high rate of habitat loss throughout much of this species' range on the Amazonian slopes of the Andes in Colombia, Ecuador, and Peru as a result of logging, agricultural expansion, and mining activities. However, it is important to note that in Ecuador, *H. pellucidum* has been categorized as Vulnerable by Ortega-Andrade *et al.* (2021). Its extent of occurrence is estimated at 15,710 km^2, and there is a continuing decline in the extent and quality of its habitat on the Amazonian slopes of both the Ecuadorian and Peruvian Andes. Habitat destruction, fragmentation, and degradation seem to be the main threat to this species, primarily due to smallholder farming, logging, and mining. A study aiming to predict glassfrog distributions and the impact of deforestation in western Ecuador suggests that, in this country, almost half of the habitats suitable for this species have been deforested (Cisneros-Heredia, 2008). In Ecuador, it has been recorded in the public reserves Parque Nacional Cayambe Coca and Reserva Biológica Colonso Chalupas and the protected forests of Cuenca Alta del Río Nangaritza and Pueblo Shuar Arutam.

Remarks: *Hyalinobatachium pellucidum* might represent a species complex. The species is closely related to *H. yaku* (Guayasamin *et al.* 2020b).

Key references.—(Lynch and Duellman, 1973) (description, habitat); (Castroviejo-Fisher *et al.*, 2009) (distribution); (Delia *et al.*, 2017) (reproductive biology); (Guayasamin *et al.*, 2020b) (account, call).

HYALINOBATRACHIUM TATAYOI (TATAYO'S GLASSFROG) CASTROVIEJO-FISHER *ET AL.*, 2007

Hyalinobatrachium tatayoi Castroviejo-Fisher *et al.*, 2007: 52. **Holotype.**—MHNLS 17174 from a stream near Tokuko, 9.84183° N, 72.82044° W, 301 m, Estado de Zulia, Venezuela.

Etymology.—The specific epithet honors Santiago Castroviejo Bolíbar "Tatayo" (1946–2009) for his work on plant taxonomy and systematics in the Neotropics and the Palearctic and for his help, support, and love to Santiago Castroviejo-Fisher.

Identification.—Figures 6.113, 6.114. Males attain a maximum SVL of 22.7 mm, and females of 22.8. The vomerine teeth are absent. The snout is semirounded to round in dorsal view and truncate to round in lateral profile. The tympanum is covered by skin. The dorsal skin is shagreen. The ventral skin is granular, transparent. The parietal peritoneum is transparent; the visceral and pericardium peritonea are covered in white iridophores. The liver is bulbous. The humeral spines are absent. Hand webbing is present between fingers III and IV; it is absent between fingers I and II and basal between fingers II and III. The webbing between toes is well developed. The enameled ulnar and tarsal folds are present. There are nuptial pads in adult males. Finger I is almost as long as finger II. The width of the disc of finger III is about 56% of the eye diameter.

The dorsum is dark apple green with small, pale yellow spots. The dorsal surface of the fingers lacks melanophores, but there are small melanophores reaching the last phalange of the fifth toe. The bones are white. The iris is yellow with black flecks more concentrated toward the pupil, creating a horizontal band that connects the pupil with the lateral edges of the eye.

Comparisons.—Among glassfrogs from the Ecuadorian Pacific lowlands and Andean foothills, *Hyalinobatrachium tatayoi* is unique in its transparent venter, white pericardium, white visceral peritonea, and lack of humeral spines.

Natural history.—Figure 6.115. The species is found on leaves over small and big streams, intermittent courses of water, and along big rivers (e.g., Río Atrato, Chocó, Colombia, 200 m width), calling in vegetation up to 10 meters above stream water.

In Ecuador, multiple males called from the upper surfaces of leaves at dusk, followed by a retreat to the undersides of the leaves after dark. During heavy rains, the species can be heard calling between 4:40 and 5:30 h. In Colombian populations (Chocó and Magdalena), calling activity began at dusk (19:10 h) and decreased near midnight. Only males are involved in the parental care of the eggs.

The territorial and reproductive behavior was studied in Lutz stream at Barro Colorado Island, Panama, where males were marked and monitored for 1.5 months. The results indicated that males were territorial, and most of them occupied up to two sites; calling males were spaced at distances of 0.9–2.1 m. Calling sites consisted of a single plant or a branch of a large shrub. Females are attracted to

FIGURE 6.113 *Hyalinobatrachium tatayoi*. Left: QCAZ 45385, female. San Lorenzo, Provincia Esmeraldas. LAC. Right: CJ 10545, male, Puerto Limón, Provincia Santo Domingo de los Tsáchilas. DAL.

FIGURE 6.114 Morphological variation and details of *Hyalinobatrachium tatayoi*. Top left: QCAZ 45386, Durango, Provincia Esmeraldas. LAC. Top right: QCAZ 32761, male, 4 km west of Durango, Provincia Esmeraldas. Middle left: CJ 5415, male, 19.9 mm SVL. Middle right: CJ 5417, male. Provincia Santo Domingo de los Tsáchilas. Bottom: CJ 5415, male, Provincia Santo Domingo de los Tsáchilas. LAC.

males' territories to lay eggs, and males exhibit 1–11 matings. The mating success of males was positively correlated with the number of days males occupied calling sites. Sites more than 0.6 m high were occupied on more days and used for oviposition more frequently than sites below 0.6 m. Males at high sites exhibited successful matings at a faster rate than those at low sites. There was relatively little predation on eggs, and the overall survival of eggs to hatching was 61–71%. In Magdalena populations in La Dorada municipality (Caldas, Colombia), males were found during dry (February) and humid (May) seasons, calling from the undersides of leaves (Araceae and Heliconiaceae), and have been observed with up to three clutches. Katydids and predatory wasps predate eggs, but males show defensive behavior, knocking the predator with their hind legs. Males show venter-to-venter combat, which starts with both males dangling upside-down while holding vegetation with their hind limbs and lasts for about 20 minutes until one frog is knocked from the leaf.

Egg clutches laid under laboratory conditions by parents from Durango, Provincia Esmeraldas, Ecuador, contained an average of 23 eggs. The eggs measured approximately 2.1 mm in diameter and were uniformly pale green. The developmental time was six days from the 32-cell stage until tadpole hatching.

FIGURE 6.115 Aspects of the natural history of *Hyalinobatrachium tatayoi*. Top: calling male and amplectant pair in their microhabitat at night, La Perla, Provincia Esmeraldas. Jaime Culebras. Middle left: clutch of eggs, Durango, Provincia Esmeraldas. LAC. Middle right: CJ 5409, embryos. LAC. Bottom left: CJ 10545, embryos. DAL. Puerto Limón, Provincia Santo Domingo de los Tsáchilas. Bottom right: male guarding a clutch with embryos, Durango, Provincia Esmeraldas. LAC.

Additional reproductive biology data of a pair from Santo Domingo de los Tsáchilas, Provincia Santo Domingo de los Tsáchilas, and ca. Puerto Quito, Provincia Pichincha, have been obtained in the laboratory (CJ lab notes by Alex Achig, in May 2021) from one reproductive event. Amplexus is axillary and lasts about one day. The oviposition event occurred at night. The clutch contained 26 unpigmented eggs of 1.7–2.2 mm in diameter. The clutch was ovoid–triangular in shape, and the average maximum diameter is 3 cm; it was deposited under a leaf 30 cm over the terrarium pool. Hatching lasted 13 days following egg deposition (air temperature = 22.6 °C), and tadpoles underwent complete development and metamorphosis in a span of at least 75 days (water temperature = 20.3 °C).

The call repertoire (advertisement, encounter, and courtship) was described from Lutz stream in Barro Colorado Island, Panama. The advertisement call was a peep approximately 0.1 s long and was delivered every 6–16 s (4–10 calls per min). During agonistic encounters, males produced advertisement calls interspersed with a call, which was termed a mew. This was a single note approximately 0.45 s long with a more gradual frequency change than the advertisement call. Males produced three types of calls during courtship. When a male first detected a female nearby, either visually or through tactile cues, he would begin interspersing mews with peeps. The mews recorded during courtship were shorter than mews recorded during aggressive encounters but had essentially the same frequency structure. Courting males also gave chirp calls about 0.1 s long. These were given most frequently when a female was on the same surface as a male. As courtship progressed, males usually stopped giving peeps and gave mostly chirps. Immediately before amplexus, each male gave a rapid series of brief chirps. Detailed data and figures are provided under the subtitle "Vocalizations of Ecuadorian Anurans" in Appendix 1.

Tadpoles.—Figure 6.116. The following description details embryos and early free-living tadpoles from Durango, Provincia Esmeraldas. Embryonic development showed that archenteron elongation began in the early gastrula and notochord elongation in mid to late gastrula phase. The development of the gastrocoel roof plate was precocious, in the late gastrula phase. When tadpoles hatch at stage 25, the size varies between 9 and 15 mm in length; the mouth is ventral; the oral disc is complete, surrounded by a row of papillae; they have 213 rows of denticles organized in the dental formula 2(2)/3; the infrakeratostome has a Λ-shape and is finely serrated, while the suprakeratostome has a slightly pronounced M-shape. The tadpoles of this species are characterized by being poorly pigmented, with fine osseous dots on the anterodorsal part of the tail, forming a row of dots on each side of the fin; the spiracle is located posteromedially with respect to the rostrum. The body is elongated and depressed, with narrow caudal fins and the tip of the tail rounded. Individuals have small eyes covered by skin.

FIGURE 6.116 Ontogenetic changes in three individuals of *Hyalinobatrachium tatayoi* from tadpole to adult. Top to bottom, left to right. CJ 10545a, stage 37, lateral, dorsal, and ventral views, TL = 26.4 mm. CJ 10545b, undetermined stage, TL = 28.2 mm. CJ 10545c, stage 42, TL = 26 mm; stage 45 = 10 mm SVL; froglet 26 days after stage 45; juvenile 90 days = 13.8 mm SVL; juvenile 189 days = 18.2 mm SVL; adult 284 days = 20 mm SVL. Puerto Limón, Provincia Santo Domingo de los Tsáchilas. DAL, SGS.

Distribution.—See Figure 6.88. *Hyalinobatrachium tatayoi* is known from the Venezuelan Cordillera de Perijá on the northern border between Colombia and Venezuela, through dry forests in the middle and upper Magdalena Valley, including the xeric scrub ecoregion in the Caribbean lowlands, up to the department of Tolima; in the Isthmian–Atlantic moist forests (Caribbean slope of Talamanca range) in Panama and Costa Rica; and from the Choco-Darien moist forests through Colombia to southern Ecuador. In Ecuador, it occurs in Esmeraldas, Carchi, Pichincha, Santo Domingo de los Tsáchilas, Manabí, Los Ríos, and Guayas provinces. Its elevational range is 0–1991 m in an area of extent of occurrence of about 51,081 km^2.

This nocturnal, arboreal species inhabits tropical rainforests and intervened areas in four evergreen ecosystems (Table 0.4) in four biogeographic sectors (Table 0.5), where the average annual rainfall is 833–2619 mm and the average annual temperature is 16.6 – 25.9°C.

Conservation.—*Hyalinobatrachium tatayoi* is categorized as Least Concern throughout its entire range (IUCN SSC Amphibian Specialist Group, 2022l; in IUCN Red List of 2022) and in Ecuador by Ortega-Andrade *et al.* (2021). However, we suggest that it be listed as Near Threatened; although its distribution is relatively large and the species is adapted to modified areas and is highly tolerant to water pollution (it has been observed even downstream of mining camps), current habitat degradation within its range is extensive. Significant threats to this species are deforestation, agriculture, the rearing of livestock, wood extraction, pollution, mining, and landslides.

In Ecuador, it has been recorded in the public reserves Reserva Ecológica Manglares Cayapas Mataje, the protected forests of Asociación Agrícola Carchi-Imbabura and Cerro Golondrinas, and some private reserves (e.g., Reserva Biológica Bilsa).

Remarks.—Until 2020, this species has been confused with *Hyalinobatrachium fleischmanni*, which is its sister taxa (Guayasamin et al., 2020b). The embryonic development from cleavage to hatching was described by Salazar-Nicholls and del Pino (2015) based on samples from Ecuador.

Key references.—(Starrett, 1960) (tadpole); (Savage, 2002) (tadpole); (Greer and Wells, 1980) (reproductive biology, tadpole, natural history); (Hayes, 1991) (reproductive biology, natural history); (Castroviejo-Fisher *et al.*, 2007) (description); (Delia *et al.*, 2010) (reproductive biology, natural history); (Salazar-Nicholls and del Pino, 2015) (development); (Rada *et al.*, 2019) (tadpole); (Mendoza *et al.*, 2019) (phylogeny); (Barrio-Amorós *et al.*, 2019) (notes); (Mendoza-Henao *et al.*, 2020) (phylogeny, taxonomy); (Guayasamin *et al.*, 2020b) (account, as *Hyalinobatrachium fleischmanni*).

HYALINOBATRACHIUM VALERIOI (VALERIO'S GLASSFROG) (DUNN, 1931)

Centrolene valerioi (Dunn, 1931): 397. **Holotype.**—MCZ 16003 from La Palma, 1372 m, Cantón de Coronado, Provincia San José, Costa Rica.

Hyalinobatrachium valerioi Ruiz-Carranza and Lynch, 1991a: 25.

Etymology.—The specific epithet is a patronym for Manuel Valerio, who, with Emmett R. Dunn (1894–1956), found the species in the field.

Identification.—Figure 6.117. These frogs attain a maximum SVL of 24.4 mm in males and 25.1 mm in females. The skin on the dorsum is finely shagreen, and the ventral skin is granular; there is no cloacal ornamentation. The parietal peritoneum is transparent. There are iridophores covering all the visceral peritonea. There is a variable pericardium from white to mostly translucent. The vomerine teeth are absent. The snout is truncate in dorsal view and truncate to protruding in lateral view. The tympanum has a dorsolateral orientation, and it is of moderate size (the tympanum diameter is 30–43% of the eye diameter); the tympanic membrane is evident; only the lower border of the tympanic annulus is visible. The liver is bulbous and covered by a white peritoneum. The humeral spines are absent. There is reduced webbing between fingers I and II; there is moderate webbing between fingers II and III; there

FIGURE 6.117 *Hyalinobatrachium valerioi*. Costa Rica. Bryan Kubicki.

is expanded webbing between the outer fingers. The webbing on the feet is moderate. The ulnar and tarsal folds are absent. The nuptial excrescences are medium-sized, diffuse excrescences formed by glandular clusters and individual glands. The prepollex is concealed. finger I is slightly longer than II. The eye diameter is larger than the width of the disc on finger III (the disc of finger III is 32–36% of the eye diameter).

The dorsum is yellowish-green with green reticulations (or green with numerous yellowish-green spots). The venter is transparent. The pericardium varies from completely white to almost transparent. The liver and digestive tract are white. The bones are white. The melanophores are absent from the fingers and toes, except for few on toes IV and V. The iris is golden with dark pigmentation scattered throughout, especially surrounding the pupil. Several individuals present dark pigmentation restricted to the lateral regions of the pupil.

Comparisons.—*Hyalinobatrachium aureoguttatum, H. mashpi, H. nouns*. Among Ecuadorian centrolenids from the Pacific lowlands and Andean foothills, only *H. aureoguttatum, H. mashpi, H. nouns* have similar color patterns. *H. aureoguttatum* differs by having large yellow to golden dorsal spots, which are produced by the interaction of xanthophores (yellow) and iridophores (white); the spots in *H. valerioi* are structurally different since they lack iridophores. The dorsal yellow spots of *H. valerioi* are larger than those of *H. mashpi* and *H. nouns*.

Natural history.—In Costa Rica, males are territorial and vigorously call from under leaves, where females deposit their egg clutches, but sometimes, they vocalize from the upper leaf surface. Physical combat is amplexus-like and occurs between males if another male intrudes into a male's territory despite the calls of the owner; both males may squeak during the fight; once one of the males is pinned venter-down, he is held for a while and then leaves. Pale greenish-white eggs (25–40 in number, average clutch size is 35) are deposited in a single layer on the underside of a leaf. Males continue to advertise and may end up with as many as seven clutches from different females. The males attend the eggs both during the day and at night. At night, the male continues to call and occasionally exhibits hydric brooding behavior, in which he apparently empties his bladder over the eggs. Only males are involved in parental care of the eggs. Egg masses are attacked by diurnal wasps, which remove eggs one at a time and carry them away, presumably to a nest, until the egg mass is depleted. Higher survivorship to hatching in this species than in the syntopic *Hyalinobatrachium colymbiphyllum* was attributed to the diurnal parental care. There is a positive effect of parental care on embryonic survivorship. Arthropod predation is the main cause for embryonic mortality.

In Costa Rica, the call is a high-pitched *seet* lasting 200 to 250 ms, repeated every 7–10 seconds. However, there is call variation among Costa Rican populations.

Tadpoles.—The following information refers to descriptions of larvae from Costa Rica mainly. The tadpole is similar to that of *Hyalinobatrachium fleischmanni*, but the tail is longer at hatching and the color is paler; the tail of older tadpoles is two- and three-fifths the length of the body; the dorsal part of the body has brown blotches; the tail musculature has brown blotches dorsally and laterally; pigment is present in depressions between myotomes posteriorly; the tail fins are clear. The mouth is ventral, nearly terminal; papillae surround all but the anterior portion of the disc. The LTRF is 2(2)/3; A-1 is complete, and the A-2 row is very short, occurring only lateral to the jaw sheaths; the outermost posterior row is not well developed at this stage (probably, in larger tadpoles, nearly as long as the inner two); the anterior jaw sheath is more heavily pigmented than the posterior one, which is doubly arched and weakly pigmented medially, often appearing broken; the posterior jaw sheath is armed with small, equal-length serrations. Tadpoles of *H. valerioi* are distinguished by their accented slenderness and by the noticeably longer tail, about three times longer than the body.

Distribution.—As currently recognized, *Hyalinobatrachium valerioi* occurs from Costa Rica to Ecuador, although there are no reliable records from Ecuador yet. See Remarks.

Conservation.—*Hyalinobatrachium valerioi* is categorized as Least Concern throughout its entire range (IUCN SSC Amphibian Specialist Group, 2020f; in IUCN Red List of 2022) in view of its wide distribution, tolerance of a degree of habitat modification, and presumed large population.

Remarks.—According to Guayasamin *et al.* (2022), the Ecuadorian populations previously assigned to *Hyalinobatrachium valerioi* belong to new taxa that they recently described: *H. nouns* and *H. mashpi*. No confirmed records of *H. valerioi* in Ecuador are available. They also found that the calls of the sister species *H. aureoguttaum* and *H. valerioi* are very similar. Given the importance of calls in species recognition, two scenarios explain the observed data: (1) the two species are fully allopatric, and the ancestral call traits have been retained or (2) the two species actually represent one evolutionary lineage. Based on their current dataset, the authors favor the first hypothesis, as there are color and genetic differences between *H. aureoguttaum* and *H. valerioi*. Nevertheless, full clarification would require more sampling (especially in Colombia) and studies in potential contact areas.

Key references.—(Dunn, 1931) (description); (Starrett, 1960); as *Cochranella reticulata*); (McDiarmid and Adler, 1974) (reproductive biology, habitat); (McDiarmid, 1975) (reproductive biology, habitat); (Mcdiarmid, 1978) (reproductive biology); (Hayes, 1991) (reproductive biology, habitat); (Savage, 2002) (account, larvae); (Kubicki, 2007) (account); (Vockenhuber *et al.*, 2008) (reproductive biology); (Hoffmann, 2010) (tadpole); (Guayasamin *et al.*, 2020b) (account).

HYALINOBATRACHIUM YAKU (YAKU GLASSFROG) GUAYASAMIN *ET AL.*, 2017

Hyalinobatrachium yaku Guayasamin *et al.*, 2017: 5. **Holotype.**—MZUTI 500 from a stream affluent of the Río Kallana, nearby the Kichwa community of Kallana, 325 m, Provincia Pastaza, Ecuador.

Etymology.—The specific epithet is the Kichwa word for water.

Identification.—Figures 6.118, 6.119. These frogs attain a maximum SVL of 22.3 mm in males and 21.1 mm in a female. The skin on the dorsum is shagreen, and the ventral skin is areolate. The cloacal area is glandular with tubercules and a slightly enameled patch on each side of the cloaca. A pair of round tubercles below the vent is absent. The parietal peritoneum, pericardium, kidneys, and urinary bladder are transparent (lacking iridophores). The hepatic, gastrointestinal, and testicular peritonea are covered by iridophores. The dentigerous process of the vomer lacks teeth. The snout is truncate in dorsal and lateral views. The lower half of the tympanic annulus is visible; the tympanic membrane is clearly differentiated and has coloration similar to that of the surrounding skin. The liver is white and bulbous. The humeral spines are absent. There is basal webbing between fingers I and II; there is moderate webbing between the external fingers. The foot webbing is moderate. The fingers and toes have thin lateral fringes. The ulnar and tarsal folds are present but difficult to distinguish, with a thin layer of iridophores that extends to the ventrolateral edge of finger IV and toe V. The nuptial excrescence is present as a small pad

FIGURE 6.118 *Hyalinobatrachium yaku*. Male from Reserva Río Bigai, Provincia Orellana. Amanda B. Quezada.

FIGURE 6.119 Morphological variation in *Hyalinobatrachium yaku*. Top left: QCAZ 53354, juvenile. Ross Maynard. Top right and bottom left: QCAZ 55628, male, Ryan Lynch. San José de Payamino, Provincia Orellana. Bottom right: MZUTI 5001, male, near the Kichwa community of Kallana, Provincia Pastaza. Carlos Morochz.

on finger I. The prepollex is not enlarged, and the prepollical spine is not projecting (the spine is not exposed). When appressed, finger I is longer than finger II. The diameter of the eye is about 2.1 times wider than the disc on finger III.

The dorsum is light green to yellowish-green with small yellow spots and minute gray to black melanophores. The posterior head and anterior half of the body present a few small and well-defined dark green spots placed middorsally. The anteriormost spot is generally the largest. The hands and feet are cream with a light green hue, and the tips of the fingers and toes are yellowish-green. There are no melanophores on the fingers and toes, except on finger IV and toes IV and V. Ventrally, the parietal peritoneum and pericardium are transparent (red heart fully visible). The visceral peritoneum of the gall bladder and urinary bladder are transparent. The hepatic and visceral peritonea are white. The ventral vein is red. The bones are white. Melanophores are present on finger IV and toes IV–V but absent on other fingers and toes. The iris is silver to yellow, with minute dark spots that encircle the pupil, giving the impression of diffuse rings.

Comparisons.—*Hyalinobatrachium yaku* is distinguishable from all other glassfrogs by having dark green middorsal spots on the anterior half of the body and a completely transparent venter and pericardium (red heart visible ventrally).

Natural history.—Figure 6.120. Males call from the undersides of leaves of riverine vegetation. One male was on the same leaf as two egg clutches, approximately 3 m above the stream. Observations suggest that prolonged parental care is provided by males.

At the locality of Ahuano, the species was found along a small stream, a tributary of the Río Arajuno; the stream was slow-flowing, very narrow (approximately 1 m wide), shallow (approximately 40 cm deep), and covered by

FIGURE 6.120 *Hyalinobatrachium yaku*. San José de Payamino, Provincia Orellana. Ross Maynard.

secondary forest. Individuals from San José de Payamino were found perched on leaves of small shrubs, ferns, and grasses (30–150 cm above ground) in a disturbed secondary forest. Additionally, all individuals recorded at San José de Payamino were found > 30 m from any stream.

The advertisement calls of one individual from Kallana, Provincia Pastaza, Ecuador, were described as single and high-pitched tonal notes with a mean duration of 300 ms and an average call rate of nine calls/minute. Detailed data and figures are provided under the subtitle "Vocalizations of Ecuadorian Anurans" in Appendix 1.

Tadpoles.—Unknown.

Distribution.—See Figure 6.88. *Hyalinobatrachium yaku*, an endemic species to Ecuador, is known from the Amazonian lowlands of Ecuador in Orellana, Napo, and Pastaza provinces. Its elevational range is 297–363 m in an area of extent of occurrence of about 1500 km^2.

This nocturnal, arboreal species inhabits natural forests in three ecosystems (Table 0.4) within three biogeographic sectors (Table 0.5), where the average annual rainfall is 3318–3897 mm and the average annual temperature is 24.1 – 24.6°C.

Conservation.—*Hyalinobatrachium yaku* is categorized as Endangered through its entire range (IUCN SSC Amphibian Specialist Group, 2022m; in IUCN Red List of 2022). However, it is important to note that in Ecuador, *H. yaku* was categorized as Data Deficient by Ortega-Andrade *et al.* (2021) because there is insufficient information about its populations to suggest a conservation category.

Remarks.—According to Guayasamin et al. (2020b), *Hyalinobatrachium yaku* is the sister taxa of *H. pellucidum*.

Key references.—(Guayasamin *et al.*, 2017) (description); (Guayasamin *et al.*, 2020b) (account).

GENUS *NYMPHARGUS* (GLASSFROGS) CISNEROS-HEREDIA AND MCDIARMID, 2007

Nymphargus Cisneros-Heredia and Mcdiarmid, 2007: 34.
Type species.—*Cochranella cochranae* (Goin, 1961).

Etymology.—The generic name derives is formed from the *Greek nymphae*, in allusion to the nymphs, beautiful goddesses in Greek mythology that personify the creative and fostering activities of nature, living in mountains, valleys, springs, and rivers, and *argus* in allusion to the mythological Greek *Argus*, nephew of the nymph Io, a giant with a hundred eyes, whose eyes became the ocelli in the peacock's tail. The name is masculine and alludes to the true ocelli found on the dorsum of some of the most peculiar species of the genus.

Definition.—Figure 6.121. Within the genus *Nymphargus*, the size of species from Ecuador varies from a maximum SVL of 37.4 mm in males to 40.0 mm in females. The most conspicuous trait shared by all species in *Nymphargus* is the reduced webbing between the outermost fingers (fingers III and IV). Additionally, the following character states are characteristic of *Nymphargus*. Humeral spines are absent, except in males of *N. grandisonae* and some populations of *N. griffithsi* and *N. balionotus* that have a small spine. The liver is tri- or tetralobed and covered by a transparent hepatic peritoneum. The ventral parietal peritoneum is white anteriorly and transparent posteriorly. The bones are green in life (white in *N. rosadus* and *N. anomalus*). The dorsum is lavender in preservative, with or without spots. The dorsum of males usually has conspicuous spinules during the breeding season. The type I nuptial pads are conspicuous in reproductive males. The males call from the upper sides of leaves, and females deposit egg masses on the upper sides of leaves along streams.

Content.—*Nymphargus* contains 41 species, 20 of which occur in Ecuador (4 March 2021).

Distribution.—*Nymphargus* occurs on the Western Andean slopes of Colombia and Ecuador and the eastern Andean slopes of Colombia, Ecuador, Peru, and Bolivia.

Remarks.—Figures 6.122, 6.123, 6.124. The phylogeny of species presented herein follows (Guayasamin *et al.*, 2020b).

NYMPHARGUS ANOMALUS (ANOMALOUS GLASSFROG) (LYNCH AND DUELLMAN, 1973)

Centrolenella anomala (Lynch and Duellman, 1973): 14.
Holotype.—KU 143299 from Río Azuela, 1740 m, Quito-Lago Agrio road, Provincia Napo, Ecuador.
Nymphargus anomalus Cisneros-Heredia and Mcdiarmid, 2007: 34.

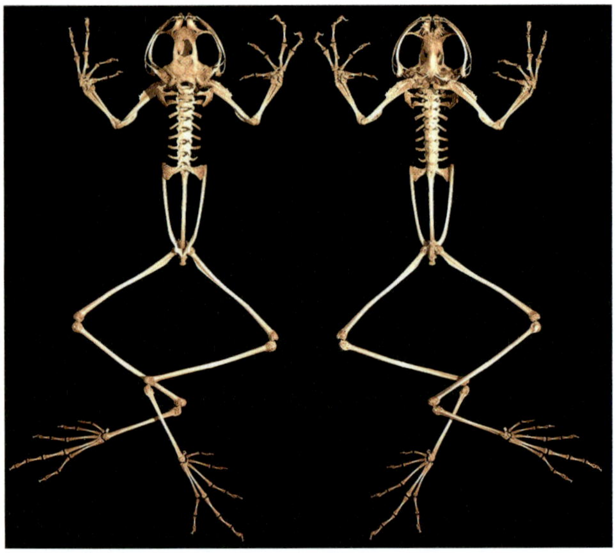

FIGURE 6.121 High-resolution computed tomography reconstructions of the skeleton of *Nymphargus griffithsi*, KU 118016, SVL = 26.4 mm, Tandapi, Provincia Pichincha. Dorsal (left) and ventral (right) views. Daniel J. Paluh.

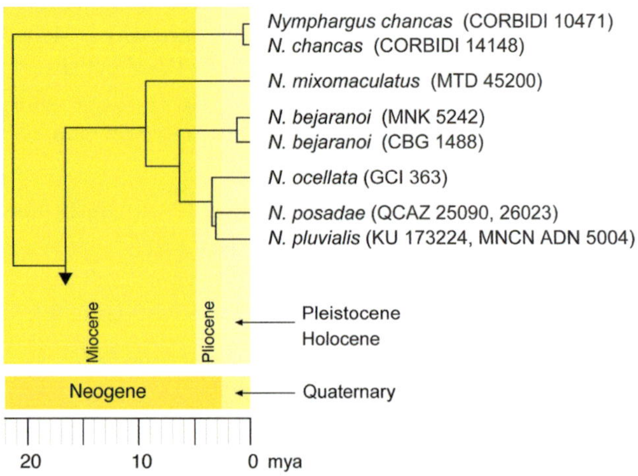

FIGURE 6.122 Time-calibrated maximum likelihood phylogenetic tree depicting evolutionary relationships of *Nymphargus*, inferred from DNA sequences of mitochondrial and nuclear genes. Phylogeny taken and redrawn from Guayasamin *et al.* (2020b).

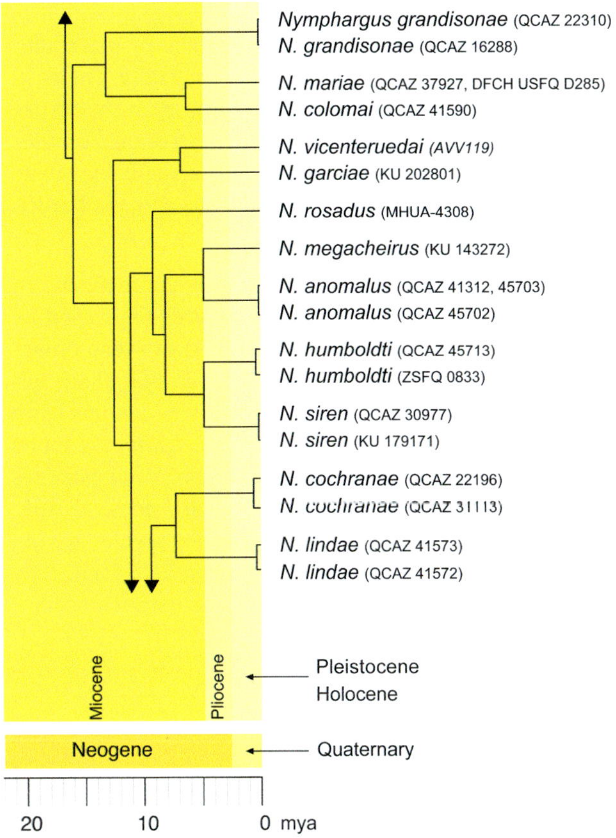

FIGURE 6.123 Continued. Time-calibrated maximum likelihood phylogenetic tree depicting evolutionary relationships of *Nymphargus*, inferred from DNA sequences of mitochondrial and nuclear genes. Phylogeny taken and redrawn from Guayasamin *et al.* (2020b)

Etymology.—The specific epithet refers to its unusual dorsal coloration.

Identification.—Figures 6.125, 6.126. These frogs attain a maximum SVL of 24.8 mm in males and 27.0 mm in females. The skin on the dorsum is shagreen, with or without minute spiculae; in females, the dorsal skin is shagreened without spiculae. The venter is areolate; the ventral surfaces of the thighs below the vent have a pair of enlarged tubercles. The anterior half of the ventral parietal peritoneum is white, and the posterior half is transparent. There is a white pericardium; the translucent peritoneum covers the intestines. The vomers lack teeth. The snout is truncate in dorsal and lateral profiles. The tympanum is relatively small, oriented almost vertically, with evident lateral and posterior inclinations; its diameter is about 20% of the eye diameter; the tympanic annulus is mostly visible, with a supratympanic fold covering its dorsal margin; the tympanic membrane is differentiated, translucent, and partially pigmented. The liver is tetralobed and covered by a transparent peritoneum. The humeral spines are absent. There is no webbing between fingers I and III; there is basal or moderate webbing between fingers III and IV. There is moderate webbing between toes. The ulnar fold is low; the inner and outer tarsal folds are present but low and difficult to distinguish. The prepollex is not separated from finger I. The nuptial pad is a large to medium-sized excrescence present on the dorsal, lateral, and/or ventral sides of the thumb. Finger I is slightly shorter than finger II (finger I length is 91.5–99.2% of finger II). The width of the disc of finger III is about 50% of the eye diameter.

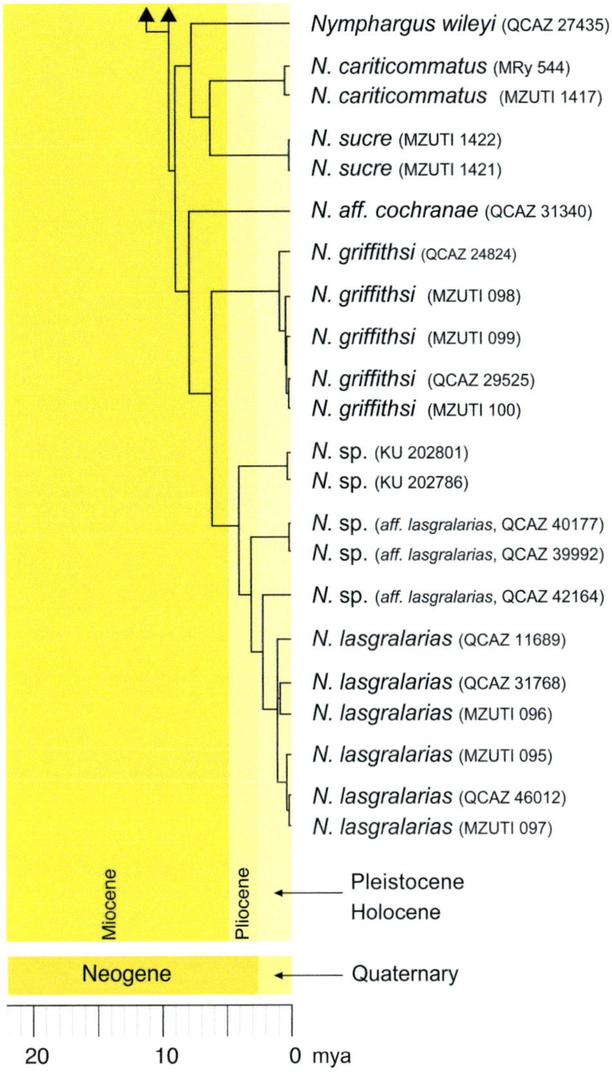

FIGURE 6.124 Continued. Time-calibrated maximum likelihood phylogenetic tree depicting evolutionary relationships of *Nymphargus*, inferred from DNA sequences of mitochondrial and nuclear genes. Phylogeny taken and redrawn from Guayasamin *et al.* (2020b).

The dorsum is pale brown or tan with a slight yellow to pink hue. The dorsal surfaces of the head, body, and limbs present small brown flecks and black ocelli that enclose brownish-orange spots. The anterior part of the venter is white. The posterior half of the venter is translucent. The dorsal surfaces of all fingers and toes usually have a few melanophores, except finger I, which usually lacks melanophores. The bones are white. *Nymphargus anomalus* exhibits ontogenetic variation in its dorsal coloration, being

FIGURE 6.125 *Nymphargus anomalus*. Left: CJ 8238, male, 24.4 mm SVL. Right: CJ 8245, male, 24.0 mm SVL. Río Cugusha Chico, Provincia Morona Santiago. DAL.

FIGURE 6.126 Morphological variation and details of *Nymphargus anomalus*. Top left: CJ 8240, male, 23.3 mm SVL. Top right: CJ 8242. Middle left: CJ 8246, male, 22.9 mm SVL. Middle right and bottom: CJ 8238, male. Río Cugusha Chico, Provincia Morona Santiago. DAL.

green in froglets and then gradually acquiring the tan to pale brown coloration that is characteristic of the adults. The spotted pattern also changes ontogenetically. The postmetamorphs have yellow spots that progressively change toward ocelli composed of black rings with orange centers. The iris is pale yellow with thin black reticulation; there is a bright yellow circumpupilar ring. The color of the iris becomes whitish toward the margins.

Comparisons.—*Nymphargus cochranae*, *N. ignotuss*, *N. laurae*. *Nymphargus anomalus* is readily distinguished by having, in life, a tan dorsum with black ocelli with orangish-tan centers, instead of the usual green that characterizes glassfrogs. Among centrolenids, only *N. ignotus* has a similar dorsal color pattern; however, *N. ignotus* is found on the Pacific slopes of the Andes, whereas *N. anomalus* is restricted to the Amazonian slopes of the Andes. *Nymphargus anomalus* further differs from other species having ocellated patterns (*N. cochranae*, *N. ignotus*, *N. laurae*) by having scattered black and lavender flecks between ocelli.

Natural history.—The holotype of *Nymphargus anomalus* was found on a mossy limb of a bush about 1.5 m above a cascading rivulet at night.

At Río Yana Challuwa Yaku, Provincia Pastaza, Ecuador, males were found calling on the upper surfaces of leaves and on branches during the night. Two egg clutches were on a mossy branch above the stream. Parental care is unknown.

Tadpoles.—Figure 6.127. This description is based on an individual in stage 41 (not preserved) from Reserva Ankaku, Provincia Pastaza, Ecuador. It was collected by Elicio E. Tapia on 15 October 2009. The body is ovoid, widest anteriorly,

FIGURE 6.127 Ontogenetic changes in *Nymphargus anomalus* from tadpole to juvenile. Top to bottom, left to right. QCAZ 49623, stage 35, lateral and dorsal views, TL = 43.6 mm; stage 41, lateral, dorsal, and ventral views, TL = 43.7 mm; stage 42, TL = 17.5 mm; stage 45, 37 days after stage 35; froglet, 44 days after stage 45; juvenile 75 days; adult 154 days. Río Challuwa Yaku, Reserva Ankaku, Provincia Pastaza. LAC. Note the color change from green froglet to brown subadult.

and wider than it is deep. The snout appears round in both dorsal view and lateral profile. The eyes are moderately large and directed dorsolaterally. The sinistral spiracular opening is directed posteriorly, positioned about halfway along the body's length. The mouth is located anteroventrally, with a poorly developed upper labium, while the lower labium is more developed and detached from the body wall. The jaw sheaths are fully keratinized. The tail constitutes approximately 72% of the total length, with the greatest depth occurring in the posterior third and a subacute tip. The dorsal fin does not extend onto the body; it is approximately equal in depth to the ventral fin and about half the depth of the caudal musculature at the midlength of the tail.

The species exhibits remarkable ontogenetic changes in coloration. In stage 41, the tadpole's dorsal surfaces are very pale green, with light blue coloring on the snout and posterior end of the tail. Scattered melanophores are present on the dorsal surfaces and laterally on the caudal musculature, more concentrated on the upper side of the tail. The translucent skin allows for the visualization of internal organs on the lateral and ventral aspects of the body. The ventral surfaces lack pigmentation, although a few scattered melanophores can be observed on the caudal musculature.

By stage 42, which occurs nine days after stage 41, the green coloration becomes more vivid, and the blue hue on the snout shifts to the anteroventral area of the body, particularly around the oral disc. Progressing to stage 45, 12 days later, the metamorph undergoes further changes. The dorsal surfaces display a darker green color with a few small yellow dots, while the blue coloration is now limited to the throat area. The hands and feet take on a yellowish hue. As the individual matures into a froglet, it acquires a deeper green dorsal coloration with scattered, larger yellow spots, and the blue coloration migrates to the scapular area. As the individual grows to an adult, the green and blue hues completely disappear, giving way to the typical adult coloration.

Distribution.—Figure 6.128. *Nymphargus anomalus*, an endemic species to Ecuador, is known on the Amazonian slopes of the Cordillera Oriental of the Andes in Napo, Pastaza, Tungurahua, and Morona Santiago provinces. Its elevational range is 1497–1770 m in an area of extent of occurrence of about 5230 km^2.

This nocturnal, arboreal species inhabits natural forests in the evergreen low montane forest ecosystems of the northern and southern Cordillera Oriental of the Andes biogeographic sectors, where the average annual rainfall is 1862–3107 mm and the average annual temperature is 15.7 – 18.9°C.

Conservation.—*Nymphargus anomalus* was categorized as Endangered by Ortega-Andrade *et al.* (2021) because of its extent of occurrence, which they estimated to be less than 5000 km^2. According to our data, the extent of occurrence of *N. anomalus* is 5230 km^2; therefore, it does not fit in the Endangered category according to the B1 criteria. Nevertheless, we agree with Ortega-Andrade *et al.* (2021) to categorize it as Endangered according to B2ab because its distribution is severely fragmented in few known localities, and there is a continuing decline in the extent and quality of its habitat.

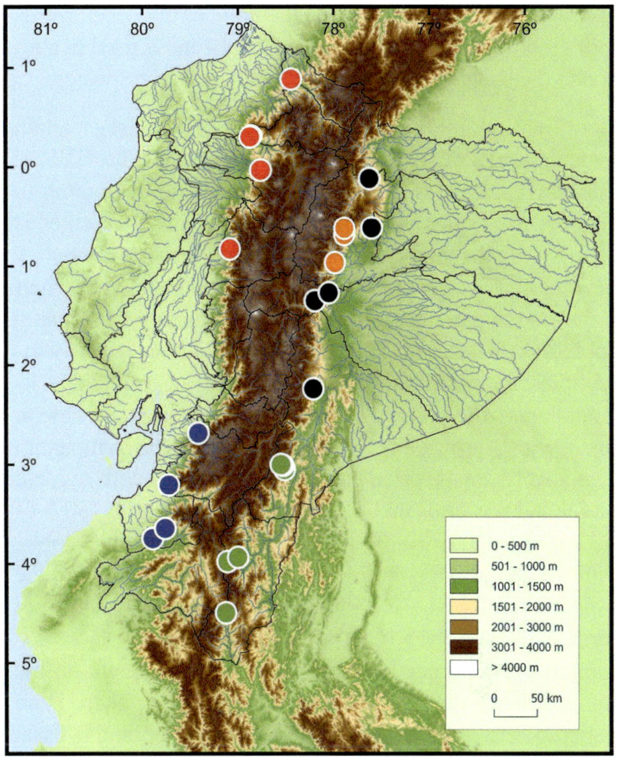

FIGURE 6.128 Localities of known occurrence of *Nymphargus anomalus* (black dots), *N. balionotus* (red dots), *N. buenaventura* (blue dots), *N. cariticommatus* (green dots), and *N. wileyi* (orange dots) in Ecuador.

Until recently, the species was known from a single individual collected on 23 October 1971 (Río Azuela). Surveys at the type locality have failed in finding this species, but other populations have been discovered. At the type locality, patches of forest have been cleared for agriculture and livestock farming. It has been recorded in the public reserves Parque Nacional Cayambe Coca, Parque Nacional Llanganates, and Parque Nacional Sumaco Napo-Galeras; the protected forest of Abanico; and the biosphere reserve Sumaco.

Remarks.—*Nymphargus anomalus* and *N. megacheirus* are sister species (Guayasamin *et al.*, 2020b).

Key references.—(Lynch and Duellman, 1973) (description); (Guayasamin *et al.*, 2020b) (account).

NYMPHARGUS BALIONOTUS (SPOTTED GLASSFROG) (DUELLMAN, 1981)

Centrolenella balionota (Duellman, 1981): 1. **Holotype.**—KU 164702 from 3.5 km (by road) northeast of Mindo, 1540 m, Provincia de Pichincha, Ecuador.

Nymphargus balionotus Guayasamin *et al.*, 2019a: 14.

Etymology.—The specific epithet combines the Greek words *balios* (spotted, dappled) and *notos* (back) and is used in allusion to the spots in the dorsal color pattern of the species.

Identification.—Figures 6.129, 6.130. These frogs attain a maximum SVL of 21.8 mm in males, and the SVL is unknown in females. The skin on the dorsum is smooth to shagreened and lacks spicules in the males. The venter is slightly granular, and the ventral surfaces of the thighs below the vent have a pair of enlarged tubercles. The anterior half of the ventral parietal peritoneum has iridophores; the posterior half is transparent. There is a pericardium covered by iridophores; the peritoneum covers the digestive tract and lacks iridophores. The vomers lack teeth. The snout is truncate in dorsal view, protruding in lateral profile. The tympanum is relatively small, oriented almost vertically, with evident lateral and posterior inclinations; its diameter is 23–30% of the eye diameter; the tympanic annulus is mostly visible, with the supratympanic fold covering its dorsal margin; the tympanic membrane is differentiated, translucent, and not pigmented. The liver is tetra- or pentalobed and lacks iridophores. The males have a bladelike ventral crest on the humerus that resembles a small humeral spine. There is no webbing between the inner fingers, and the webbing is basal on fingers IV and V. There is moderate webbing between the toes. The ulnar fold is low; the inner and outer tarsal folds are present but low and difficult to distinguish. The prepollex is not separated from finger I. The nuptial pad in males is variable, ranging from a large to medium-sized excrescence present on the dorsal, lateral, and/or ventral sides of the thumb to a medium-sized spinous nuptial excrescence extending from the lateral side of the thumb to its dorsomedial surface. Finger I is slightly longer than finger II (finger II length is 91.8–97.4% of finger I). The width of the disc of finger III is about 36–44% of the eye diameter.

The dorsum is pale green with reddish-brown dorsolateral stripes, small reddish-brown spots, and elevated larger yellow spots. The head has a reddish-brown interorbital bar and postorbital stripe that is continuous with the dorsolateral stripe. There is a large, elevated, yellow spot on the anteromedial edge of each eyelid. The dorsal surfaces of the fingers and

FIGURE 6.129 *Nymphargus balionotus.* Left: female. Ross Maynard. Right: CJ 12930, male, 21.0 mm SLV. SGS. Reserva Río Manduriacu, Provincia Imbabura.

FIGURE 6.130 Morphological variation in *Nymphargus balionotus.* Top: male. Jaime Culebras. Middle: male. Ross Maynard. Bottom: ZSFQ 533, male. José Vieira-Fernandes. Reserva Río Manduriacu, Provincia Imbabura.

toes lack melanophores. The bones are pale green. The iris is grayish-white with numerous minute brown spots.

Comparisons.—Among glassfrogs, *Nymphargus balionotus* has a unique coloration pattern, featuring a pale green dorsum with reddish-brown dorsolateral stripes, small reddish-brown spots, and larger yellow spots. Additionally, males are small (SVL < 23 mm) and possess a bladelike ventral crest on the humerus that can be confused with a humeral spine. See Remarks.

Natural history.—Figure 6.131. It usually is found on vegetation along clear-water streams during the night. At the type locality, males were calling less than 1 m above the water from the upper surfaces of leaves of herbs and ferns overhanging a trickling stream. A calling male was recorded in Manduriacu, Provincia Pichincha, on 31 July 2021.

Other stream-breeding frogs at the type locality included *Nymphargus grandisonae*, *N. griffithsi*, and *Hyloscirtus alytolylax*.

Tadpoles.—Unknown.

Distribution.—See Figure 6.128. *Nymphargus balionotus* occurs between 400 and 1540 m along the western slope of the Cordillera Occidental of Colombia (Departamento de Cauca) and south to Ecuador. In Ecuador, it occurs in Carchi, Cotopaxi, Imbabura, and Pichincha provinces. Its elevational range is 643–1620 m in an area of extent of occurrence of about 2480 km^2.

This nocturnal, arboreal, epiphyllous species inhabits pristine forests and intervened areas in the evergreen low montane forest and evergreen montane foothill forest ecosystems of the Cordillera Occidental of the Andes biogeographic sector, where the average annual rainfall is 1919–2207 mm and the average annual temperature is 18.1–22.6 °C.

FIGURE 6.131 *Nymphargus balionotus*. Top: male in its microhabitat and calling male. Ross Maynard. Middle and bottom: calling male, amplectant pairs, and clutch of embryos. Jaime Culebras. Reserva Río Manduriacu, Provincia Imbabura.

Conservation.—*Nymphargus balionotus* is categorized as Endangered throughout its entire range (IUCN SSC Amphibian Specialist Group, 2020g; in IUCN Red List of 2022) and in Ecuador by Ortega-Andrade *et al.* (2021) because its area of occupancy is less than 2000 km^2, its distribution is severely fragmented, and there is a continuing decline in the extent and quality of its habitat, the number of locations, and the number of mature individuals in Colombia and Ecuador. A single reproductive population is known in Ecuador in Reserva Río Manduriacu, which is threatened by mining activities. It is reasonably common in Colombia. *Nymphargus balionotus* was relatively abundant at the type locality (3.5 km northeast of Mindo), where 13 individuals were observed on two nights (7–8 April 1975). Significant threats to this species are habitat loss as a result of deforestation for agricultural development, the rearing of livestock, pollution, mining, landslides, wildfires, and human settlements. Other threats include the introduction of exotic predatory fish and pollution. Like some other centrolenids, this species might also be affected by the movement of the cloud layer up the mountain sides as a result of climate change, resulting in reduced humidity within the altitudinal range of the species (probably exacerbated by habitat fragmentation). In Ecuador, it has been recorded in the public reserve Reserva Ecológica Los Ilinizas, the biosphere reserve Chocó Andino de Pichincha, and some private reserves (e.g., Reserva Río Manduriacu).

Remarks.—Based on mitochondrial data, Guayasamin *et al.* (2019a) found that *Nymphargus balionotus* is sister to *N. manduriacu*. According to Guayasamin *et al.* (2020b), Ecuadorian individuals of *Nymphargus balionotus* have a small, pointed humeral spine, whereas the Colombian populations have a variable *crista ventralis* that, in some cases, takes the form of a pointed humeral spine. The polymorphism in this characteristic has created confusion in the generic placement of the species. It also has been suggested that at least two species are currently confused under the name *N. balionotus*.

Key references.—(Duellman, 1981) (description, habitat); (Cisneros-Heredia and Yánez-Muñoz, 2007) (distribution); (Arteaga *et al.*, 2013) (account); (Guayasamin *et al.*, 2019a) (distribution); Yánez-Muñoz *et al.* (2020) (account); (Guayasamin *et al.*, 2020b) (account).

NYMPHARGUS BUENAVENTURA (BUENAVENTURA GLASSFROG) (CISNEROS-HEREDIA AND YÁNEZ-MUÑOZ, 2007)

Cochranella buenaventura (Cisneros-Heredia and Yánez-Muñoz, 2007): 2. **Holotype.**—DHMECN 3563 from Reserva Buenaventura, 1200 m, Cantón Piñas, Provincia El Oro, Ecuador.

Nymphargus buenaventura Cisneros-Heredia and Mcdiarmid, 2007: 34.

Etymology.—The specific epithet refers to the type locality, Reserva Buenaventura, in Ecuador.

Identification.—Figure 6.132. These frogs attain a maximum SVL of 22.4 mm in males and 23.5 mm in a female. The skin on the dorsum is slightly shagreen, with scattered flat tubercles corresponding to light spots. The ventral skin is areolate; there is a pair of large and flat tubercles on the ventral surfaces of the thighs below the vent, and the subcloacal skin is granular and enameled. The upper half of the parietal peritoneum is covered by iridophores; all the other peritonea are transparent, except for the renal capsules and pericardium covered by iridophores. The vomerine teeth are absent. The snout is truncate in dorsal view and profile; the nostrils are elevated. The internarial area is slightly concave. The lower half of the tympanic annulus is evident and oriented dorsolaterally with dorsoventral inclination; the supratympanic fold is absent or very weak. The liver is tetralobed and uncovered by iridophores. The humeral spine is absent. There is no webbing between fingers I, II, and III; there is basal webbing between fingers III and IV. The webbing on the feet is moderate. The outer ventral edges of the forearms and tarsi have low folds. The unpigmented nuptial pad is a large to medium-sized excrescence present on the dorsal, lateral, and/or ventral sides of the thumb. The prepollex is concealed. Finger I is shorter than II. The eye diameter is larger than the width of the disc on finger III.

The dorsum is green to pale green with scattered pale yellow to cream spots on all the dorsal surfaces, including the arms and legs. The hands and feet are yellowish-green with pale to bright yellow discs. The venter is cream. Melanophores are present on the outer fingers and outer toes. The bones are green. The iris is yellowish-silver with fine dark maroon reticulations.

Comparisons.—*Nymphargus cariticommatus*, *N. griffithsi*, *N. lasgralarias*, and *N. wileyi*. *Nymphargus buenaventura* is distinguished from most glassfrogs by its dorsal coloration (green with small white spots), reduced webbing between the fingers, absence of a humeral spine, and absence of iridophores on the digestive visceral peritonea, but with iridophores covering the renal capsules. The last character is shared with at least three other species, *N. cariticommatus*, *N. griffithsi*, and *N. wileyi*. *Nymphargus cariticommatus* and *N. wileyi* occur only on the Amazonian Andean slopes of Ecuador, whereas *N. buenaventura* is restricted to the Pacific Andean slopes. *N. griffithsi* differs by having dark dorsal spots. Another species that could be confused with *N. buenaventura* is *N. lasgralarias*, which lacks the yellow spots characteristic of the former.

Natural history.—Figure 6.133. It has been found on the leaves of shrubs and low trees, 1–3 m above streams.

FIGURE 6.132 *Nymphargus buenaventura*. Reserva Buenaventura, Provincia El Oro. Mario Yánez-Muñoz.

FIGURE 6.133 *Nymphargus buenaventura*. Clutch of embryos from Reserva Buenaventura, Provincia El Oro. Mario Yánez Muñoz.

The males call from the upper sides of leaves along streams. The egg clutches are placed on the upper sides of leaves along streams. A male was found near an egg clutch with 38 embryos.

Tadpoles.—Unknown.

Distribution.—See Figure 6.128. *Nymphargus buenaventura*, an endemic species to Ecuador, is known in the southwestern Pacific lowlands and midelevations in Azuay and El Oro provinces. Its elevational range is 417–1925 m in an area of extent of occurrence of about 1040 km^2.

This nocturnal, arboreal species inhabits natural forests and intervened areas in the evergreen seasonal montane foothill forests of Catamayo-Alamor and the Cordillera Occidental of the Andes, where the average annual rainfall is 669–1019 mm and the average annual temperature is 22.7 – 23.8°C.

Conservation.—*Nymphargus buenaventura* is categorized as Endangered by Ortega-Andrade *et al.* (2021) because its extent of occurrence is estimated to be less than 5000 km^2 (1040 km^2, according to our data), its distribution is severely fragmented in fewer than five known localities, and there is a continuing decline in the extent and quality of its habitat. *N. buenaventura* is restricted to the southern part of the western Ecuadorian region (as suggested in the original description, Cisneros-Heredia and Yánez-Muñoz, 2007), an area with severe deforestation, the main threat for the species. It has been recorded in the protected forests of Río Arenillas Presa Tahuín and Uzchurrumi; La Cadena, Peña Dorada, Brazil; and some private reserves (e.g., Reserva Biológica Buenaventura).

Remarks.—Its phylogenetic relationships are unknown.

Key references.—(Cisneros-Heredia and Yánez-Muñoz, 2007) (description); (Guayasamin *et al.*, 2020b) (account).

NYMPHARGUS CARITICOMMATUS (UNADORNED GLASSFROG) (WILD, 1994)

Cochranella cariticommata (Wild, 1994): 304. **Holotype.**—KU 202806 from 11.2 km west–southwest of Plan de Milagro, 2350 m, Provincia Morona Santiago, Ecuador.

Nymphargus cariticommatus Cisneros-Heredia and Mcdiarmid, 2007: 34.

Etymology.—The specific epithet is an adjective formed from the Latin word *caritus*, meaning without, joined with *kommos*, a Greek word meaning decoration or embellishment. The name refers to the species' plain appearance and lack of distinctive features.

Identification.—Figures 6.134, 6.135. These frogs attain a maximum SVL of 23.3 mm in a male and 25.5 mm in a

FIGURE 6.134 *Nymphargus cariticommatus*. CJ 11569, male, 22.8 mm SVL, Guarumales, Provincia Zamora Chinchipe. SGS.

FIGURE 6.135 Morphological details of *Nymphargus cariticommatus*. CJ 11569, male, Guarumales, Provincia Zamora Chinchipe. SGS.

female. The skin on the dorsum is shagreen, with minute spicules in males. There is a pair of enlarged subcloacal tubercles. The anterior two-thirds of the ventral parietal peritoneum are white, and the posterior third is transparent. There is a silvery white pericardium; the translucent peritoneum covers the intestines, stomach, testes, gall bladder, and urinary bladder; the esophagus is white; the kidneys are white with unpigmented spots. The vomers lack teeth. The snout is round in dorsal profile and truncate in lateral profile. The tympanum is evident and of moderate size (the tympanum diameter is 26.7–27.1% of the eye diameter), the dorsal border of the tympanic annulus is covered by a supratympanic fold; the tympanic membrane is differentiated from the surrounding skin. The liver is tetralobed; three ventral lobes partially cover one smaller lobe; the hepatic peritoneum is transparent. The humeral spines are absent. There is no webbing between the inner fingers. There is moderate foot webbing. The ulnar fold is low in males and not visible in females; the inner tarsal fold is low in both sexes. The prepollex is concealed. The nuptial pad is a large to medium-sized excrescence present on the dorsal, lateral, and/or ventral sides of the thumb. Finger II is slightly longer than finger I (finger I is about 90–94% the length of finger II); the disc of finger III is of moderate size, 50–56% of the eye diameter.

The dorsum is green with small yellow spots. The dorsal surfaces of the fingers and toes are yellow. The dorsal surfaces of the fingers and toes lack melanophores. The iris is silvery white with thin black reticulations. There is a yellow ring around the pupil.

Comparisons.—*Nymphargus wileyi*, *N. griffithsi*, and *N. buenaventura*. *Nymphargus cariticommatus* has a green dorsum with small yellow dots (uniform green dorsum in *N. wileyi*), an esophageal peritonea covered by iridophores (translucent in *N. griffithsi* and *N. buenaventura*, and with a thin layer of iridophores in *N. wileyi*; but see Remarks), and slightly less hand webbing on the outer fingers than *N. buenaventura*. Additionally, *N. cariticommatus* is found on the Amazonian versant of the Andes, whereas *N. griffithsi* and *N. buenaventura* inhabit the Pacific versant.

Natural history.—*Nymphargus cariticommatus* was found on the upper sides of leaves and ferns within 2 m of cascading streams. Its call is described as a soft "zeet, zeet, zeet." The eggs are deposited on the upper sides of leaves.

Tadpoles.—Unknown.

Distribution.—See Figure 6.128. *Nymphargus cariticommatus*, an endemic species to Ecuador, is known on the Amazonian slopes of Cordillera Oriental of the Andes in Morona Santiago and Zamora Chinchipe provinces. Its elevational range is 1189–2419 m in an area of extent of occurrence of about 2180 km^2.

This nocturnal, arboreal species inhabits natural forests and intervened areas in the evergreen low montane forest and evergreen montane forest ecosystems in the southern Cordillera Oriental of the Andes biogeographic sector, where the average annual rainfall is 855–1602 mm and the average annual temperature is 15.4–21.0 °C.

Conservation.—*Nymphargus cariticommatus* was categorized as Endangered by Ortega-Andrade *et al.* (2021) because its extent of occurrence is estimated to be less than 5000 km^2 (2180 km^2, according to our data), its distribution is severely fragmented in fewer than five known localities, and there is a continuing decline in the extent and quality of its habitat. The main threats to the species are habitat destruction and degradation due to agricultural development, logging, and mining. It has been recorded in the protected forests of Cuencas que Forman los Ríos: San Francisco, San Ramón y Sabanilla, Corazón de Oro and Tinajillas-Río Gualaceño and in some private reserves (e.g., Reserva Biológica Tapichalaca).

Remarks.—As mentioned in the diagnosis, subtle morphological differences separate *Nymphargus cariticommatus* from *N. wileyi*. Guayasamin *et al.* (2020b) examined all the available material of the two species, including the type series, and observed that none of the specimens of *N. wileyi* have dorsal spots, whereas all individuals of *N. cariticommatus* have them. Also, in the preserved material, the esophageal peritoneum of *N. cariticommatus* is more pronounced and extended than that in *N. wileyi*, in which the iridophores are restricted to the anterior portion of the esophageal peritoneum. Considering those data, Guayasamin *et al.* (2020b) maintain the species status of *N. wileyi*.

Key references.—(Wild, 1994) (description, habitat); (Cisneros-Heredia and Yánez-Muñoz, 2007) (distribution); (Guayasamin *et al.*, 2020b) (account).

NYMPHARGUS COCHRANAE (COCHRAN'S GLASSFROG) (GOIN, 1961)

Cochranella cochranae (Goin, 1961): 97. **Holotype.**—BM 1912.11.1.68 from El Topo, 1280 m, Río Pastaza, Provincia Pastaza, Ecuador.

Nymphargus cochranae Cisneros-Heredia and Mcdiarmid, 2007: 34.

Etymology.—The specific epithet honors Doris M. Cochran (1898–1968), former Curator of the National Museum of Natural History, Smithsonian Institution.

Identification.—Figures 6.136, 6.137, 6.138. These frogs attain a maximum SVL of 26.2 mm in males and 30.3 mm in females. The skin on the dorsum of males and females is shagreen, with minute spicules usually present in males. The ventral surfaces of the thighs below the vent have a pair of enlarged tubercles. The anterior two-thirds of the ventral parietal peritoneum are white, and the posterior third is transparent. There is a white pericardium; there are translucent peritonea covering the intestines, stomach, testes, kidneys, gall bladder, and urinary bladder. The vomers usually have teeth. The snout is truncate in dorsal profile and truncate to slightly protruding in lateral profile. The tympanum is relatively small, and its diameter is 19.5–27.8% of the eye diameter; the dorsal border of the tympanic annulus is covered by a supratympanic fold; the tympanic membrane is pigmented like the surrounding skin. The liver is tetralobed; two large ventral lobes partially cover the two smaller lobes; the hepatic peritoneum is transparent. The humeral spines are absent. There is no hand webbing. The foot is about two-thirds webbed. The ulnar and tarsal folds are low. The prepollex is concealed. The nuptial pad is a large to medium-sized excrescence present on the dorsal, lateral, and/or ventral sides of the thumb. Finger I is about the same length as finger II (finger II is 96.3–101.8% of the length of finger I). The disc of finger III is small, 19.4–28.3% of the eye diameter.

The dorsum is green with small, dark blue to black ocelli enclosing bright orange spots. The ventral surfaces are mostly white, except for the posterior third, which is translucent. The dorsal surfaces of the hands and feet lack melanophores, except for the proximal portion of finger IV and toe V. The bones are green. The iris is grayish-white with a thin, dark gray reticulation. The area immediately around the pupil is pale brownish-yellow.

Comparisons.—*Nymphargus anomalus*, *N. lindae*, *N. laurae*, *N. megacheirus*, *N. ocellatus*. *Nymphargus cochranae*

FIGURE 6.136 *Nymphargus cochranae*. Left: CJ 7307, male, 26.9 mm SVL. Right: CJ 7309, male, 28.7 mm SVL. Paquisha Alto, Provincia Zamora Chinchipe. LAC.

FIGURE 6.137 Morphological variation and eye of *Nymphargus cochranae*. QCAZ 31113, Pacto Sumaco, Provincia Napo. LAC.

FIGURE 6.138 Morphological details of *Nymphargus cochranae*. CJ 9316, female, 29.3 mm SVL, trail to Volcán Sumaco, Provincia Napo. DAL.

is differentiated from all other glassfrogs by its green dorsum with small, dark blue to black ocelli enclosing orange dots; relatively large size (adult males, SVL 24.0–26.2 mm; adult females, SVL 27.8–30.3 mm); absence of webbing between the fingers; and lack of humeral spines. *Nymphargus cochranae* further differs from other species with ocellated dorsal patterns (*N. anomalus*, *N. lindae*, *N. laurae*, *N. ocellatus*) by having relatively smaller ocelli. *N. anomalus* further differs by having scattered black and lavender flecks between ocelli (absent in *N. cochranae*); *N. ocellatus* has a green dorsal coloration with large, dark rings that, in life, have greenish-white centers; *N. laurae*, known from a single adult male, is smaller (SVL = 19.7 mm) and has pointed papillae on toes I–IV (absent in *N. cochranae*) and larger ocelli. *Nymphargus cochranae* is most similar to its sister species, *N. lindae*, in female body size (*N. cochranae*, SVL = 27.8–30.3 mm; *N. lindae*, SVL 27.2–27.8 mm) and the relative size of the ocelli (minute in *N. cochranae* and clearly larger in *N. lindae*), which is the most conspicuous difference between the two species. The only other species similar to *N. cochranae* is *N. megacheirus*, the dorsal dots of which coincide with spicules and could be confused with ocelli. However, *N. megacheirus* is larger (SVL of males = 26.8–31.5 mm; SVL of females = 31.2–32.9 mm), and the males have conspicuous spicules and spiculated warts on the dorsum (minute spicules in males of *N. cochranae*).

Natural history.—Figure 6.139. In the reproductive season, males call at night from the upper surfaces of leaves or branches overhanging or near streams. At Estación Científica San Francisco, Provincia Zamora Chinchipe, Ecuador, a male was found next to a clutch of embryos, 1.2 m above a small stream in a secondary forest.

The advertisement call was described as a single, high-pitched note, but the specific spectral and temporal characteristics of the call are unknown.

Tadpoles.—Figure 6.140. This description is based on an individual (not preserved, series CJ 7310) from Paquisha Alto, Provincia Zamora Chinchipe, Ecuador, in stage 26. A clutch was collected at approximately 1.5 m above the water of a stream of about 1 m in diameter by Morley Read on 26 August 2017. The tadpole measures 8.5 mm in body length and 30.4 mm in total length. The body is elongated, postero-triangular shaped, widest anteriorly, and wider than it is deep. The snout appears round both in dorsal view and in lateral profile. The eyes are narrowly separated and

FIGURE 6.139 *Nymphargus cochranae*. Top: male calling from Domono, Provincia Morona Santiago. Jorge Brito Molina. Middle: Sumaco, Provincia Napo. DAL. Bottom: CJ 7310, clutch of embryos, Paquisha Alto, Provincia Zamora Chinchipe. LAC.

FIGURE 6.140 Ontogenetic changes in two tadpoles of *Nymphargus cochranae*. Top to bottom. CJ 7310a, stage 26, lateral, dorsal, and ventral views, TL = 30 mm. CJ 7310b, undetermined stage, TL = 38.4 mm. Paquisha Alto, Provincia Zamora Chinchipe. DAL.

directed dorsally. The spiracle is sinistral, with its inner wall free from the body wall. Its opening is directed posterolaterally at approximately the half of the body's length. The oral disc is medium-sized, positioned anteroventrally, non-emarginate, and transversely elliptical. The upper labium is bare, while the lower labium is more developed and detached from the body wall. Both upper and lower jaw sheaths are fully keratinized, with the upper jaw sheath M-shaped and forming a smooth arc. The tail accounts for approximately 73% of the tadpole's total length. It is deepest at about the midlength and tapers to a subacute tip. The dorsal fin does not extend onto the body, and it is slightly deeper than the ventral fin. It is about half the depth of the caudal musculature at the midlength of the tail.

The tadpole's skin is predominantly translucent, except for the presence of melanophores that form large brown blotches along the dorsum, starting from the snout and extending to the tail. The rest of the body and tail lack pigmentation, allowing for the visibility of the internal organs through the ventral skin. At stage 41, the body adopts a pale green hue, while the legs appear yellow.

Distribution.—Figure 6.141. *Nymphargus cochranae* occurs in Ecuador and Colombia (Serranía de Churumbelos) on the Amazonian slopes of the Andes at elevations between 1170 and 1960 m. In Ecuador, it occurs in Sucumbíos, Napo, Pastaza, Tungurahua, Morona Santiago, and Zamora Chinchipe provinces. Its elevational range is 929–2354 m in an area of extent of occurrence of about 22,310 km².

This nocturnal, arboreal species inhabits six evergreen montane forest ecosystems (Table 0.4) in the Amazonian Cordilleras and northern and southern Cordillera Oriental of the Andes biogeographic sectors, where the average annual rainfall is 1061–3679 mm and the average annual temperature is 15.8 – 23.5°C.

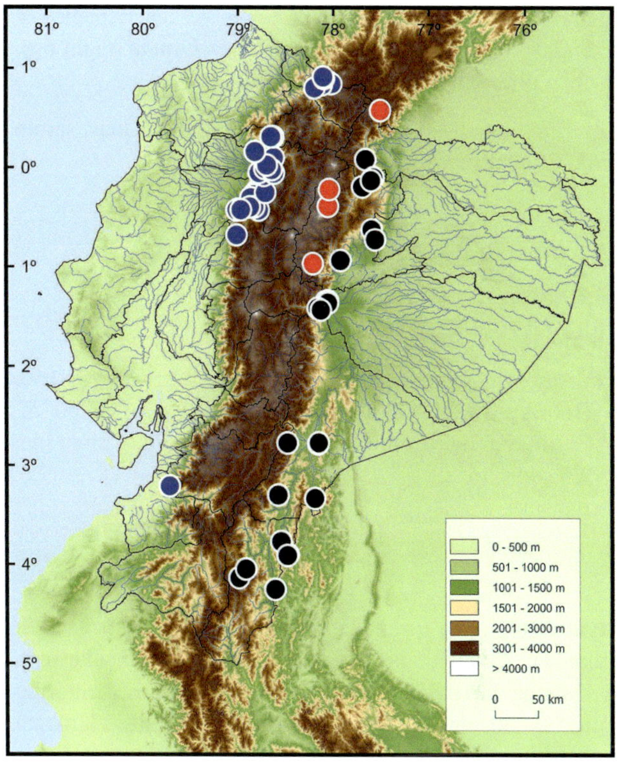

FIGURE 6.141 Localities of known occurrence of *Nymphargus cochranae* (black dots), *N. garciae* (red dots), and *N. grandisonae* (blue dots) in Ecuador.

Conservation.—*Nymphargus cochranae* was categorized is as Near Threatened by Ortega-Andrade *et al.* (2021). However, their categorization seems to be an error, given that they provide criteria for it to be included under the Vulnerable category. They infer a reduction in the population greater or equal to 30% of the population in the next 100 years and a reduction in the extent of occurrence, which was underestimated to be less than 20,000 km² (23,310 km², according to our data); its distribution is severely fragmented, and there is a continuing decline in the extent and quality of its habitat. We categorize it as Vulnerable following their inference. Almost half of the habitats suitable for this species have been deforested. According to Guayasamin *et al.* (2020b), the species remains abundant at several localities, including the Sumaco and Cayambe Coca national parks. However, part of the species' range is fragmented. Habitat destruction and degradation are threats to this species, mainly as a result of smallholder farming activities, clear-cutting of forests, and mining. In Ecuador, it has been recorded in the public reserves Parque Nacional Cayambe Coca, Reserva Biológica Colonso Chalupas, Parque Nacional Llanganates, Parque Nacional Río Negro Sopladora, and Parque Nacional Podocarpus; the protected forests of Cerro Sumaco y Cuenca Alta del Río Suno, Cuenca del Río Paute, Cordillera del Cóndor, and Cordillera Kutukú y Shaimi; the biosphere reserves Podocarpus and Podocarpus-El Cóndor; and some private reserves (e.g., Estación Científica San Francisco)

Remarks.—*Nymphargus cochranae* is the sister species of *N. lindae*. Southern populations of *N. cochranae* might represent a different species.

Key references.—(Goin, 1961) (description); (Guayasamin *et al.*, 2020b) (account).

NYMPHARGUS COLOMAI (COLOMA'S GLASSFROG) GUAYASAMIN AND HUTTER, 2020

Nymphargus colomai Guayasamin and Hutter, 2020: 160.
Holotype.—QCAZ 41590 from Miazi Alto, 1260 m, Cordillera del Cóndor, Provincia Zamora Chinchipe, Ecuador.

Etymology.—The specific epithet is a patronym for Luis A. Coloma in recognition of his pioneering and continual efforts in studying and protecting amphibians, as well as mentoring numerous students. Coloma is the Director of Centro Jambatu de Investigación y Conservación de Anfibios, a center that houses dozens of critically endangered amphibians, in an outstanding effort to conserve frogs and toads. Coloma received the Sabin Award for Amphibian Conservation in 2007.

Identification.—Figure 6.142. These frogs attain a maximum SVL of 25.7 mm in males, and the SVL is unknown in females. The skin on the dorsum is shagreened, with numerous spiculated warts and spicules in males. The venter is areolate; the ventral surfaces of the thighs have a pair of enlarged tubercles; there is a white parietal peritoneum covering about the anterior half of the venter. There is a white pericardium and translucent peritonea on the kidneys, intestines, stomach, and gall and urinary bladders. The dentigerous process of the vomer lacks teeth or has few teeth (up to three). The snout is truncate to slightly protruding in lateral profile and truncate in dorsal view. The tympanum is oriented almost vertically, with slight lateral and posterior inclinations, and its diameter is about 23–27% of the eye diameter; the upper half of the tympanic annulus is obscured by the supratympanic fold and warts; the tympanic membrane is pigmented like the surrounding skin. The liver is lobate and covered by a translucent peritoneum. Humeral spines are absent. There is no webbing between fingers I, II, and III; there is no or basal webbing between fingers III and IV. The feet are about two-thirds webbed. The ulnar and tarsal folds are low. The prepollex is not separated from finger I. The nuptial pad is a large to medium-sized excrescence present on the dorsal, lateral, and/or ventral sides of the thumb. Finger II is slightly longer than finger I. The width of the disc of finger III is about 46–50% of the eye diameter.

The dorsum can be dull yellowish-green, tangerine yellow, brown, or gray–olive, always with numerous small and diffuse yellow to orange spots. The anterior half of venter is white; the posterior portion is transparent. There are melanophores on the dorsal surfaces of fingers III and IV and toes IV and V. The bones are green. The iris is silvery white with a slight yellow hue and a clearly marked, black horizontal stripe.

FIGURE 6.142 *Nymphargus colomai.* Top left: QCAZ 41590, male, 25.0 mm SVL. Top right: QCAZ 41641, juvenile. Bottom: QCAZ 41591 or 41592, paratypes. Miazi Alto, Provincia Zamora Chinchipe. Holger Braun and Juan M. Guayasamin.

Comparisons.—*Nymphargus cariticommatus*, *N. chancas*, *N. humboldti*, *N. siren*, and *N. sucre*. *Nymphargus colomai* is unique in having a white iris with a contrasting black horizontal stripe and a dorsum that varies from dull yellowish-green to tangerine yellow, brown, or gray–olive, always with numerous small yellow to orange spots. Additionally, *N. colomai* lacks humeral spines and hand webbing. Among glassfrogs that inhabit the Amazonian slopes of the Andes, only *N. cariticommatus*, *N. chancas*, *N. humboldti*, *N. siren*, and *N. sucre* could be confused with *N. colomai*. However, *N. cariticommatus*, *N. chancas*, *N. siren*, and *N. humboldti* have a green dorsum with few and well-defined dorsal spots, whereas *N. colomai* has a non-green dorsum with numerous and diffuse yellow dorsal spots. Moreover, *N. siren* is smaller than *N. colomai* (*N. siren*, SVL < 23.3 mm; *N. colomai* SVL = 24.7–25.7 mm), and *N. cariticommatus* has a white esophagus and renal capsules (cream in *N. colomai*). Both *N. colomai* and the Peruvian *N. chancas* share the clearly defined black horizontal stripe on a silver background, but the two species are easily differentiated by their dorsal color patterns (*N. chancas*: yellowish-green dorsum with small yellow dots).

Natural history.—Individuals were found on vegetation 1.0–1.7 m above fast-flowing streams. Males were calling on 8–12 April 2009 from the upper surfaces of leaves.

Tadpoles.—Unknown.

Distribution.—Figure 6.143. *Nymphargus colomai*, an endemic species to Ecuador, is known from a single locality at 1361 m in the Cordillera del Cóndor, Provincia Zamora Chinchipe.

This nocturnal, arboreal species inhabits pristine forests in the evergreen montane foothill forests on outcrops of limestone rock ecosystem in the Amazonian Cordilleras biogeographic sector, where the average annual rainfall is 1423 mm and the average annual temperature is 21.6°C.

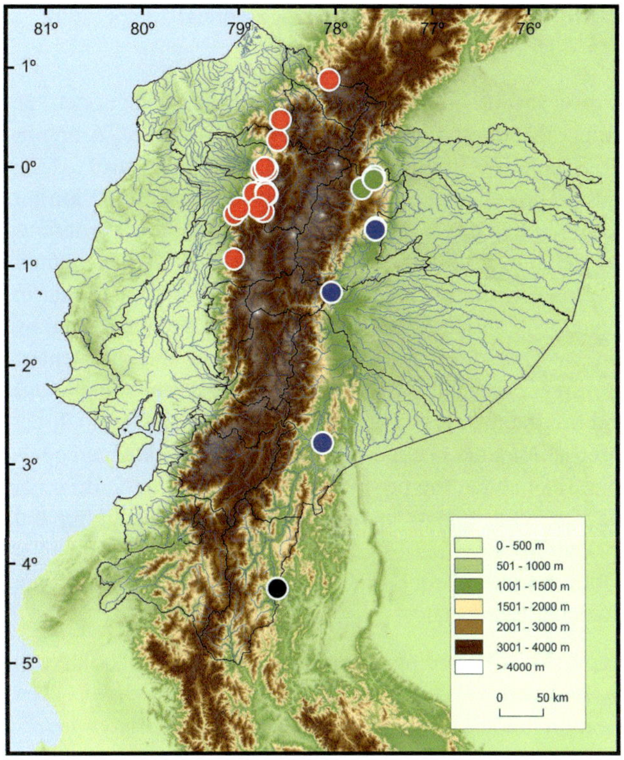

FIGURE 6.143 Localities of known occurrence of *Nymphargus colomai* (black dot), *N. griffithsi* (red dots), *N. humboldti* (blue dots), and *N. megacheirus* (green dots) in Ecuador.

Conservation.—*Nymphargus colomai* was categorized as Endangered by Ortega-Andrade *et al.* (2021) because it is known from fewer than three localities, and there is a continuing and projected decline in the extent and quality of its habitat. The main threats for this species in Cordillera del Cóndor are habitat destruction due to agricultural activities and water contamination associated with mining. It has been recorded in the biosphere reserve Podocarpus-El Cóndor.

Remarks.—*Nymphargus colomai* was previously confused with the Peruvian *N. chancas* (see Guayasamin *et al.*, 2011). However, the two species show several morphological differences and are not closely related *Nymphargus colomai* is sister to *N. mariae* (Guayasamin *et al.*, 2020b).

Key references.—(Guayasamin and Hutter, 2020) (description).

NYMPHARGUS GARCIAE (GARCÍA'S GLASSFROG) (RUIZ-CARRANZA AND LYNCH, 1995b)

Cochranella garciae (Ruiz-Carranza and Lynch, 1995b): 14. **Holotype.**—ICN 11752 from km 64–73 on road Popayán to Inzá, eastern versant of Cordillera Central, municipio de Inzá, 2590–2660 m, 2.56667° N, 76.06667° W, Departamento de Cauca, Colombia.

Nymphargus garciae Cisneros-Heredia and Mcdiarmid, 2007: 34.

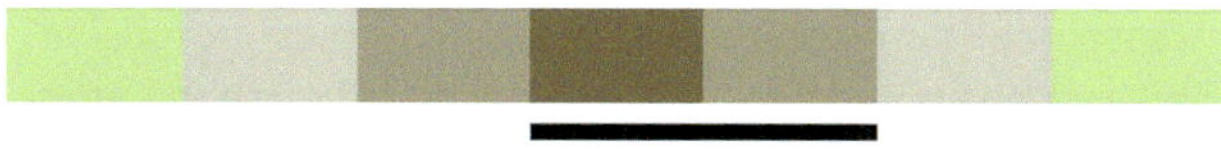

Etymology.—The specific epithet is a patronym for Dr. Evaristo García Piedrahita (1845–1921), author of the pioneer book *Los ofidios venenosos del Cauca*, published in 1896.

Identification.—Figure 6.144. These frogs attain a maximum SVL of 29.9 mm in males and 28.4 mm in females. The skin on the dorsum is shagreen, with spicules evident in males. There is a pair of enlarged subcloacal tubercles. The anterior half of the parietal peritoneum is white. There is a white pericardium, renal capsules, and a translucent visceral peritoneum. The vomerine teeth are absent. The snout is rounded in dorsal aspect and truncate to slightly sloping in lateral profile. The tympanum is oriented almost vertically and is small in size (the tympanum diameter is 22.9–30.6% of the eye diameter), and it has a slightly posterior inclination; the tympanic annulus is visible except for the dorsal border covered by supratympanic fold; the tympanic membrane is partially pigmented but differentiated from the surrounding skin. The liver is tetralobed and covered by

FIGURE 6.144 *Nymphargus garciae.* MZUTI 764, Oyacachi-El Chaco, Provincia Napo. Eduardo Toral.

a transparent peritoneum. The humeral spines are absent. There is no webbing between the fingers, although the junction of the lateral fringes resembles the basal webbing. The ulnar and tarsal folds are present and enameled. The nuptial pad is a large to medium-sized excrescence present on the dorsal, lateral, and/or ventral sides of the thumb. The prepollex is concealed. Finger II is slightly longer than finger I (FI/FII = 0.87–0.95). The disc of finger III is relatively large (3DW/ED = 0.56–0.77).

The dorsum is dark green with cream tubercles and with or without dark green to blue spots. Some of the photographs of Ecuadorian individuals of *Nymphargus garciae* show a uniform green dorsal coloration. The hands and feet are yellowish-green. The glandular pericloacal region is white. The pericardium and anterior half of the ventral parietal peritoneum is white. The visceral peritonea are translucent. The upper lip and ulnar and tarsal folds are white. The fingers lack melanophores dorsally; the melanophores are restricted to the proximal dorsal areas of toes IV and V. The bones are green. The iris is grayish-white with thin black reticulations.

Comparisons.—*Centrolene buckleyi*, *Nymphargus megacheirus*. *Nymphargus garciae* can be differentiated from other glassfrogs by its moderate body length (SVL 25.9–28.4 mm in adult females, SVL 25.1–29.9 mm in adult males); green dorsum, usually with dark green to blue spots; and absence of humeral spines. In Ecuador, *N. garciae* could be confused with *C. buckleyi* and *N. megacheirus*. However, *C. buckleyi* has basal webbing between fingers III and IV (absent in *N. garciae*), lacks dark spots on the dorsum (usually present in *N. garciae*), and has males with prominent humeral spines (absent in *N. garciae*). *Nymphargus megacheirus* has a similar dorsal coloration and hand webbing pattern to *N. garciae*, but the two species have allopatric distributions, with *N. garciae* inhabiting higher elevations (2152–3084 m) than *N. megacheirus* (1246–1769 m). Additionally, *N. megacheirus* is slightly larger than *N. garciae* (in *N. megacheirus*, SVL 26.8–31.5 mm in adult males, 31.2–32.9 mm in adult females).

Natural history.—Adults were found on vegetation and rocks along streams. The males call from the upper sides of leaves. One gravid female had numerous eggs, which had a dark brown animal and a cream vegetal pole.

Tadpoles.—Unknown.

Distribution.—See Figure 6.141. *Nymphargus garciae* is known on the eastern slopes of the Cordillera Central at elevations of 1900–2700 m in Colombia and in the Cordillera Oriental of the Ecuadorian Andes. In Ecuador, it occurs in Sucumbíos and Napo provinces. Its elevational range is 2152–3084 m in an area of extent of occurrence of about 1610 km^2.

This nocturnal, arboreal species inhabits natural forests and intervened areas in the evergreen montane forest ecosystem of the northern Cordillera Oriental of the Andes biogeographic sector, where the average annual rainfall is 1010–1320 mm and the average annual temperature is 10.5 – 15.2°C.

Conservation.—*Nymphargus garciae* is categorized as Vulnerable throughout its entire range (IUCN SSC Amphibian Specialist Group, 2019w; in IUCN Red List of 2022), as this species has an extent of occurrence of 11,745 km^2; it occurs in five threat-defined locations, and there is continuing decline in the extent and quality of its habitat mainly due to agriculture. However, it is important to note that in Ecuador, *N. garciae* has been categorized as Endangered by Ortega-Andrade *et al.* (2021), given that its area of extent of occurrence is small (1610 km^2 according to our calculation) and fragmented. It is a rare species. While there is no direct information, due to ongoing declines in the extent and quality of its habitat, the population is suspected to be decreasing. Its major threats include habitat loss due logging and agriculture activities. In Ecuador, it has been recorded in the public reserves Parque Nacional Cayambe Coca and Parque Nacional Llanganates.

Remarks.—According to Guayasamin *et al.* (2020b), *Nymphargus garciae* is the sister species of *N. vicenteruidai*. Given the morphological similarities between *Nymphargus garciae*, *N. nephelophilus*, and *N. oreonympha*, it is possible that they are actually one species with some phenotypic geographic variation. Complementary acoustic and genetic studies should be performed to establish the validity of these taxa.

Key references.—(Ruiz-Carranza and Lynch, 1995b) (description); (Ruiz-Carranza *et al.*, 1996) (distribution); (Guayasamin *et al.*, 2020b) (account).

NYMPHARGUS GRANDISONAE (RED-SPOTTED GLASSFROG) (COCHRAN AND GOIN, 1970)

Centrolenella grandisonae (Cochran and Goin, 1970): 513.
Holotype.—BM 1910.7.11.68 from Pueblo Rico, 1524 m, Departamento Caldas, Colombia.
Nymphargus grandisonae Guayasamin *et al.*, 2009: 28.

Etymology.—The specific epithet is a matronym for Alice G. C. Grandison (1927–2014), former curator of the British Museum.

Identification.—Figures 6.145, 6.146. These frogs attain a maximum SVL of 29.3 mm in males and 30.7 mm in females. The skin on the dorsum of males and females is finely pustular; there are minute spicules evident only in males. The ventral surfaces of the thighs below the vent have a pair of enlarged tubercles. There are white iridophores covering all or most of the ventral parietal peritoneum. There is a white pericardium; there is a translucent peritoneum covering the intestines, stomach, testes, and gall

FIGURE 6.145 *Nymphargus grandisonae*. Left: 10831 789, female, Reserva Las Gralarias, Provincia Pichincha. LAC. Right: CJ 10329, male, 29.6 mm SVL, Junín, Provincia Imbabura. DAL.

FIGURE 6.146 Morphological variation in males and morphological details of *Nymphargus grandisonae*. Top left: CJ 10833. Top right: CJ 10831. Middle left: CJ 10830. Near Tandayapa, Provincia Pichincha. DAL. Middle right: QCAZ 40001, 10 km east of Chilma Bajo, Provincia Carchi. LAC. Bottom: CJ 13017, male, ACUS Mashpi, Provincia Pichincha. SGS.

bladder; the kidneys are dorsally and laterally covered with a white lining; the urinary bladder is completely covered with white iridophores. The dentigerous process of vomer usually bears teeth; each process has 0–7 teeth. The snout is round in dorsal view and truncate in lateral profile. The tympanum is partially hidden under the skin and oriented almost vertically, and its diameter is 27.0–34.5% of the eye diameter; the supratympanic fold is moderate; the tympanic membrane is pigmented and barely differentiated from the surrounding skin. The liver is tetralobed and covered by a transparent peritoneum. In males, the humeral spines are small. There is no webbing between fingers I, II, and III; there is moderate webbing between fingers III and IV. The webbing on the foot is moderate between the outer toes. The ulnar fold is low and white; the inner tarsal fold is low, short; the outer tarsal fold is absent or low and inconspicuous. The prepollex is concealed. The nuptial pad is a large to medium-sized excrescence present on the dorsal, lateral, and/or ventral sides of the thumb. Finger I is about the same length as finger II (finger I is 92.0–102.2% of finger II). The width of the disc of finger III is moderate, about 55.2–64.0% of the eye diameter.

The dorsum is green with small red spots. The throat is pale green. The margin of the upper lip is white. The flanks and venter are white. The peritonea on the liver and digestive tract lack iridophores. The bones are green. The iris is yellowish-gray.

Comparisons.—*Nymphargus grandisonae* is unique in having a green dorsum with small, bright red spots. Another diagnostic trait is the presence of a white urinary bladder, a characteristic otherwise known only in glassfrogs from the Atlantic forests of Brazil and Argentina.

Natural history.—Figure 6.147. *Nymphargus grandisonae* is a nocturnal and epiphyllous glassfrog that prefers riparian vegetation in evergreen and cloud forests. The species is active on vegetation near streams in primary and secondary forests and on shrubs along the borders of pastures. In Reserva Las Gralarias, the species has been found at Ballux Creek, Five-Frog Creek, Heloderma Creek, Hercules Creek, Chalguayacu River, Kathy's Creek, Lucy's Creek, Waterfall Trail, and the Río Santa Rosa trail. Experimental studies have shown that tadpoles of *N. grandisonae* are susceptible to introduced trout, having a higher mortality and changing their morphology.

The males call from the upper sides of leaves near streams. *Nymphargus grandisonae* is a highly territorial species, with males having site fidelity for prolonged periods of time. Also, males have frequent fights and complex combat behaviors, including the following types: (1) dangling amplexus-like, where two males are dangling facing vent to dorsum, often grasping the head of the other male, which results from tumbling off the substrate in an amplexus-like position; (2) dangling venter–venter where two males dangle from the substrate by their hind limbs, positioned venter-to-venter, which may include males grasping the other around the head; (3) amplexus-like, where one male

FIGURE 6.147 *Nymphargus grandisonae.* Top left: male, Estación Experimental La Favorita, Provincia Pichincha. Carolina Reyes-Puig. Middle left: calling male, Chiriboga, Provincia Pichincha. Jaime Culebras. Bottom left: clutch of eggs deposited under lab-conditions in CJ, parents from Reserva Las Gralarias, Provincia Pichincha. LAC. Bottom right: QCAZ 40004, clutch of embryos, Chilma Bajo, Provincia Carchi. LAC. Top right: fighting behavior between two males under laboratory conditions in CJ, Provincia Pichincha. SGS.

is positioned upon the dorsum of another; accompanied by struggling and jumping; (4) a head–vent wrestle, where one male's head is positioned near the vent of the other and one or both males grasp their opponent around the waist with the forelimbs; (5) reverse dangling amplexus-like, where two males are dangling from the substrate with the head of one male positioned near the vent of another, resulting from tumbling off the substrate, where one male climbs the other; and (6) limb grasping, where one male immobilizes the other by grasping its limbs, which occurs while one male remains dangling from the substrate and the other attempts to exit onto the substrate surface. As a consequence of fighting, males usually present injuries, likely produced by humeral spines. Interspecific combat has also been observed between *N.* aff. *grandisonae* and *Espadarana prosoblepon*. *Nymphargus grandisonae* reproduces in the rainy season (December–April); amplexus occurs within males' territories, and egg clutches containing 30–71 eggs are deposited on the upper sides of leaves above streams. Short-term parental care is provided by females; no parental care is provided by males.

Additional reproductive biology data of a pair from either ca. Tandayapa or Área de Conservación y Uso Sostenible Mashpi, Provincia Pichincha, or Quebrada Naranjo, Provincia Carchi, have been obtained in the laboratory (CJ lab notes by Alex Achig, in July 2022) from one reproductive event. Amplexus is axillary and lasts around one day. The oviposition event occurred at night. The clutch contained 39 unpigmented eggs of 1.8–2.3 mm ($\bar{x}$ = 2.0, n = 12) in diameter. The clutch was ovoid–triangular in shape, and the maximum diameter was 21.4 cm; it was deposited over a leaf, on its apex, 0.7 cm above the terrarium pool. The clutch was not viable, and all the eggs decomposed.

Advertisement calls of 22 individuals from Reserva Las Gralarias, Provincia Pichincha, Ecuador, were described with a mean duration of 115 ms. The courtship, territorial, encounter, distress, and release calls also were described from the same locality. Detailed data and figures are provided under the subtitle "Vocalizations of Ecuadorian Anurans" in Appendix 1.

Tadpoles.—Figure 6.148. This description is based on an individual in stage 26 (not preserved) from Chilma Bajo, Provincia Carchi, Ecuador. It was collected by Omar Torres-Carvajal on 26 February 2009. The body is ovoid, widest anteriorly, and slightly wider than it is deep. The snout is round in both dorsal view and in lateral profile, with dorsally directed eyes. Chondrocranial elements are visible. The spiracle is sinistral, with its inner wall free from the body wall, and its opening is directed posteriorly at about two-thirds of the body's length. The vent tube is medial,

FIGURE 6.148 Ontogenetic changes in two individuals of *Nymphargus grandisonae* from tadpole to froglet. Top to bottom. QCAZ 40004, stage 27, lateral and dorsal views; Chilma Bajo, Provincia Carchi. CJ, stage 41, TL = 50 mm SVL; stage 42; stage 46, 27 days after stage 41. Reserva Las Gralarias, Provincia Pichincha. LAC.

with both walls attached directly to the ventral fin. The oral disc is anteroventral, non-emarginate, with developed upper and lower labia that are free from the body wall. The upper jaw sheath is M-shaped, and there is a small gap in row A-2. The tail comprises about 71% of the total length, with its greatest depth at the midlength and a rounded tip. The dorsal fin does not extend onto the body; it is equally as deep as the ventral fin and about half the depth of the caudal musculature at the midlength of the tail.

The skin is translucent, allowing the internal organs to be visible through the skin and giving rise to a pale pink coloration in certain areas of the body. Some scattered melanophores are present, but in general, the tadpole is unpigmented.

At stage 41, the high vascularization of the tadpole gives it a bright red coloration throughout the body and the tail, with green pigmentation on the snout. Progressing to stage 42, the metamorph exhibits a brown coloration on the body, legs and tail, with scattered yellow to orange spots. The tip of the snout and the end of the tail are green. As the metamorph matures into a froglet, it acquires an olive-green coloration with orange and red spots on the dorsal surfaces.

Distribution.—See Figure 6.141. *Nymphargus grandisonae* occurs from the western slopes of the Cordillera Occidental and Cordillera Central in Colombia (Departamento Antioquia) south to the Pacific versant of the Andes in northern and central Ecuador. In Ecuador, it occurs in Carchi, Imbabura, Pichincha, Cotopaxi, Santo Domingo de los Tsáchilas, and El Oro provinces. Its elevational range is 1246–2391 m in an area of extent of occurrence of about 14,230 km^2.

This nocturnal, arboreal species inhabits natural forests and intervened areas in three ecosystems (Table 6.4) within the Cordillera Occidental of the Andes biogeographic sector, where the average annual rainfall is 709–2299 mm and the average annual temperature is 14.1 – 23.8°C.

Conservation.—*Nymphargus grandisonae* is categorized as Least Concern throughout its entire range (IUCN SSC Amphibian Specialist Group, 2022n; in IUCN Red List of 2022) and in Ecuador by Ortega-Andrade *et al.* (2021). Although its extent of occurrence is less than 20,000 km^2 (14,230 km^2, according to our data), it is common species and tolerates certain levels of habitat disturbances. It is presumed to have a large population, and it is unlikely to be declining fast enough to qualify for listing in a more threatened category. According to Guayasamin *et al.* (2014), at Reserva Las Gralarias, Provincia Pichincha, this glassfrog was infected by the chytrid fungus *Batrachochytrium dendrobatidis*, but populations have not shown evidence of drastic declines, and the species is fairly common in the area. The major threats for the species are deforestation for agricultural development, logging, human settlements, and the introduction of exotic predatory fish in streams. In Ecuador, it has been recorded in the public reserves Reserva Ecológica Los Ilinizas; the protected forests of Cerro Golondrinas, Maquipucuna, Cuenca Río Guayllabamba (Area 1–Area 2), Estación Científica Río Guajalito, Toachi-Pilatón, and Uzchurrumi, La Cadena, Peña Dorada, Brazil; the biosphere reserves Chocó Andino de Pichincha and Macizo del Cajas; and some private reserves (e.g., Reserva Las Gralarias; Reserva Integral Otonga; Reserva Río Manduriacu, Ross Maynard pers. comm.).

Remarks.—Duellman (1980) redescribed this species. *Nymphargus grandisonae* is most closely related to *N. mariae* and *N. colomai* (Guayasamin *et al.*, 2020b).

Key references.—(Cochran and Goin, 1970) (description); (Duellman, 1980) (redescription); (Bolívar *et al.*, 2004) (distribution); (Arteaga *et al.*, 2013) (account); (Hutter and Guayasamin, 2012) (distribution); (Hutter *et al.*, 2013) (reproductive biology, habitat); (Vanegas *et al.*, 2014) (distribution); (Delia *et al.*, 2017) (reproductive biology); (Martín-Torrijos *et al.*, 2016) (predation); (Sorokin and Steigerwald, 2017) (reproductive biology); (Arteaga *et al.*, 2013) (account); (Yánez-Muñoz *et al.*, 2020) (account); (Guayasamin *et al.*, 2020b) (account).

NYMPHARGUS GRIFFITHSI (GRIFFITHS' GLASSFROG) (GOIN, 1961)

Cochranella griffithsi (Goin, 1961): 97. **Holotype.**—BM 1940.2.20.4 from Río Saloya, 1219 m, Provincia Pichincha, Ecuador.

Nymphargus griffithsi Cisneros-Heredia and Mcdiarmid, 2007: 34.

Etymology.—The specific epithet honors Dr. Ivor Griffiths of Birkbeck College, London, in recognition of his contributions to the understanding of relationships among higher categories of frogs.

Identification.—Figures 6.149, 6.150. These frogs attain a maximum SVL of 27.5 mm in males and 27.9 mm in females. The skin on the dorsum is slightly shagreen. The ventral surfaces of the thighs below the vent have a pair of enlarged tubercles. The upper half of the ventral parietal peritoneum is white, and the posterior half is transparent. There is a white pericardium; all other peritonea are clear. The vomers lack teeth. The snout is truncate in dorsal view and truncate to protruding in profile. The lower half of the tympanic annulus is evident and oriented dorsolaterally with a dorsoventral inclination; the tympanic membrane is pigmented like the surrounding skin; the supratympanic fold is absent or very weak. The liver is tetralobed; the hepatic peritoneum is transparent. The humeral spines are absent, although enlarged humeral crests are apparent in some specimens. There is no hand webbing. The foot is about two-thirds webbed. The ulnar and tarsal folds are enameled and low. The prepollex concealed. The nuptial

FIGURE 6.149 *Nymphargus griffithsi*. Left: CJ 11541, male, 23.3 mm SVL. Puranqui, Provincia Imbabura. SGS. Right: ZC 0699, male, 23.9 mm SVL, Quebrada Zapadores, Provincia Pichincha. SGS.

FIGURE 6.150 Morphological details of *Nymphargus griffithsi*. Note the nictitating membrane in the top left photo. ZC 0699, Quebrada Zapadores, Provincia Pichincha. SGS.

pad is a large to medium-sized excrescence present on the dorsal, lateral, and/or ventral sides of the thumb. Finger I is shorter than finger II. The disc of finger III is 20.5–26.5% of the eye diameter.

The dorsum varies from uniformly green to green with small dark flecks or some scattered yellow flecks or dots. The ventral surfaces are mostly greenish-white. The fingers and toes lack melanophores, except for some on toes IV and V. The bones are green. The iris varies from white to cream–beige, with brown to black points forming thin reticulations. In some individuals, a yellow hue is visible around the pupil.

Comparisons.—*Nymphargus buenaventura*, *N. lasgralarias*. Among glassfrogs found on the Pacific slope of the Andes, *N. griffithsi* can be distinguished by having a uniformly green dorsum with small dark flecks and/or small yellow spots and by lacking webbing between the fingers. The species also lacks humeral spines, although enlarged humeral crests are apparent in some specimens. Two species, *N. buenaventura*, a species restricted to southwestern Ecuador (Reserva Buenaventura, Provincia de El Oro), and *N. lasgralarias*, from Reserva Las Gralarias (Pichincha Province), can be confused with *N. griffithsi* because there are no distinctive morphological traits to distinguish them. However, *N. buenaventura* is slightly smaller and has a green dorsum with yellow spots, whereas most populations of *N. griffithsi* lack these yellow spots and have some dark flecks on the dorsum. The most reliable traits to differentiate between *N. lasgralarias* and *N. griffithsi* are the calls; *N. griffithsi* produces a single tonal or multi-pulsed (i.e., two or more pulses) call, while the calls of *N. lasgralarias* are always pulsed. Also, *N. griffithsi* emits its advertisement call as a single note, whereas *N. lasgralarias* emits its calls singly or in a series. In addition, *N. lasgralarias* has a shorter call duration than *N. griffithsi* (call duration in *N. lasgralarias* = 0.016–0.044 s; call duration in *N. griffithsi* = 0.103–0.148 s).

Natural history.—Figure 6.151. During the rainy season (December–April), at night, *Nymphargus griffithsi* is active on vegetation 0.5–9 m above streams, waterfalls and small rivers in primary and secondary cloud forests. In Reserva Las Gralarias, *N. griffithsi* has been observed at Five-Frog Creek, Heloderma Creek, Hercules Creek, and Kathy's Creek.

In Mindo, Reserva Las Gralarias, and throughout its range in Ecuador, reproductive activity peaks in February–April. The males call from the upper sides of leaves overhanging streams. Males are highly territorial and engage in aggressive interactions with conspecifics; when fighting, males have been observed dangling upside-down venter–venter holding onto a branch with their hind legs. Females lay clutches of 14–22 eggs on the tips of leaves overhanging fast-flowing water. These later expand into a hanging gelatinous mass upon the absorption of water; eventually, tadpoles hatch and fall into the stream below. Short-term parental care is provided by females; no parental care is provided by males.

The advertisement calls of 11 individuals from Reserva Las Gralarias, Provincia Pichincha, Ecuador, were described as tonal calls produced at a mean rate of 1.6 calls per minute. The call is composed of a single note with a mean duration of 122 ms. Detailed data and figures are provided under the subtitle "Vocalizations of Ecuadorian Anurans" in Appendix 1.

FIGURE 6.151 *Nymphargus griffithsi*. Clutches and embryos, Reserva Las Gralarias, Provincia Pichincha. LAC.

FIGURE 6.152 Ontogenetic changes in *Nymphargus griffithsi* from tadpole to froglet. Top to bottom. QCAZ 48136, stage 29, lateral and dorsal views; stage 42; stage 45, 165 days after stage 29. Reserva Las Gralarias, Provincia Pichincha. LAC.

Tadpoles.—Figure 6.152. This description is based on an individual in stage 26 (not preserved, series QCAZ 48136) from Reserva Las Gralarias, Provincia Pichincha, Ecuador. It was collected by Luis A. Coloma and Juan M. Guayasamin on 10 April 2010. The body is ovoid and globular, widest anteriorly, equally wide and deep. The snout appears round in dorsal view and subacute in lateral profile. The eyes are directed dorsally. The spiracle is sinistral, with the inner wall present as a small ridge, and its opening is directed posteriorly on the deadline at around half the length of the body. The oral disc is anteroventral, with a bare upper labium and a more developed lower labium that is detached from the body wall. The tail makes up approximately 71% of the total length, with its greatest depth at around the midlength and a rounded tip. The dorsal fin does not extend onto the body, shares an equal depth with the ventral fin, and is about half the depth of the caudal musculature at the tail's midlength. The skin is translucent, but the aggregation of melanophores results in brown blotches on the dorsum of the body, starting from the snout and extending to the tail. The rest of the body and tail remain unpigmented.

By stage 43, which occurs 145 days after stage 26, the tadpole's body, tail, and legs acquire a uniform yellowish-green coloration. The area surrounding the mouth takes on a bluish tint. At stage 45, 20 days later, the frog displays a consistent light green coloration on the dorsum and extremities, with a golden yellow iris.

Distribution.—See Figure 6.143. *Nymphargus griffithsi* occurs in Ecuador and Colombia on the western slopes of the Andes. In Ecuador, it occurs in Carchi, Imbabura, Pichincha, Cotopaxi, and Santo Domingo de los Tsáchilas provinces. Its elevational range is 1359–2790 m in an area of extent of occurrence of about 4960 km^2.

This nocturnal, arboreal species inhabits natural forests in the evergreen low montane forest and evergreen montane forest ecosystems of Cordillera Occidental of the Andes biogeographic sector, where the average annual rainfall is 877–2207 mm and the average annual temperature is 12.5–19.4°C.

Conservation.—*Nymphargus griffithsi* is categorized as Least Concern throughout its entire range (IUCN SSC Amphibian Specialist Group, 2018q; in IUCN Red List of 2022) in view of its wide distribution, tolerance of a degree of habitat modification, and presumed large population. However, it is important to note that in Ecuador, *N. griffithsi* has been categorized as Vulnerable by Ortega-Andrade *et al.* (2021) due to its small area of extent of occurrence (4960 km^2, according to our data), which is fragmented. This assessment is made on the basis that all subpopulations currently assigned to *N. griffithsi* form part of a single taxonomic group (but see Remarks). It is uncommon to locally common throughout its range. Due to ongoing declines in the extent and quality of its habitat, the population is suspected to be decreasing. The population is not severely fragmented. Localized threats include habitat loss as a result of deforestation for agricultural development, logging, and human settlement. Other threats include the introduction of exotic predatory fish and pollution. Specimens of this species have tested positive for *Batrachochytrium dendrobatidis (Bd)*, however, it remains uncertain whether chytridiomycosis is a threat. According to Guayasamin *et al.* (2014), at Reserva Las Gralarias, Provincia Pichincha, the species is not infected by the chytrid fungus *Bd*, although the number of individuals sampled was low (n = 5). This species occurs in several protected areas in Colombia. In Ecuador, it has been recorded in the protected forest of Toachi Pilatón, the biosphere reserve Chocó Andino de Pichincha, and some private reserves (e.g., Reserva Las Gralarias, Reserva Integral Otonga, Estación Experimental La Favorita).

Remarks.—According to Guayasamin *et al.* (2020b), *Nymphargus griffithsi*, as currently recognized, represents a species complex. Vocalization studies by Arcila-Pérez *et al.* (2017) on Colombian populations assigned to *N. griffithsi* also suggest that several species are concealed under this name. The former emphasizes that given the morphological similarity between *N. griffithsi*, *N. lasgralarias*, and potential cryptic species, extreme care is necessary when identifying these species. Genetic and acoustic identification are necessary tools for this species complex.

Key references.—(Goin, 1961) (description); (Duellman and Savitzky, 1976) (reproductive biology); (Hutter and Guayasamin, 2012) (morphology, call, reproductive biology, habitat); (Arteaga *et al.*, 2013) (distribution, reproductive biology); (Guayasamin *et al.*, 2014) (habitat); (Delia *et al.*, 2017) (reproductive biology); (Arcila-Pérez *et al.*, 2017) (call, natural history); Yánez-Muñoz *et al.* (2020) (account); (Guayasamin *et al.*, 2020b) (account).

NYMPHARGUS HUMBOLDTI (HUMBOLDT'S GLASSFROG) GUAYASAMIN *ET AL.*, 2020a

Nymphargus humboldti Guayasamin *et al.*, 2020a: 175. **Holotype.**—ZSFQ 0388 from Volcán Sumaco, 1738 m, Provincia Napo, Ecuador.

FIGURE 6.153 *Nymphargus humboldti*. QCAZ 47514, male, 25.2 mm SVL, Río Yan Challuwa, Provincia Pastaza. LAC.

Etymology.—The specific epithet honors Alexander von Humboldt (1769–1859) for his unparalleled contributions to biogeography, integrative perspective of the sciences and arts, humanism, and also for his devotion to mountains.

Identification.—Figure 6.153. These frogs attain a maximum SVL of 25.2 mm in males and 27.4 mm in females. The skin on the dorsum is shagreen, with minute spicules in males. The venter is areolate; the ventral surfaces of the thighs below the vent have a pair of enlarged tubercles. There is a white parietal peritoneum covering the anterior 50–60% of the venter. There is a white pericardium; there are translucent peritonea covering the intestines, stomach, kidneys, and gall and urinary bladders. The vomers have an edentate dentigerous process. The snout is truncate to bluntly rounded in lateral profile and truncate in dorsal view. The tympanum is oriented almost vertically, with slight lateral and posterior inclinations, and its diameter is about 20–27% of the eye diameter; the upper fourth of the tympanic annulus is obscured by the supratympanic fold; the tympanic membrane is pigmented like the surrounding skin. The liver is tetralobed, covered by a transparent peritoneum. Humeral spines are absent. There is no webbing between the inner fingers; there is no or basal webbing between fingers III and IV. The feet are about two-thirds webbed. The ulnar and tarsal folds are low. The prepollex is not separated from finger I. The nuptial pad is a large to medium-sized excrescence present on the dorsal, lateral, and/or ventral sides of the thumb. Finger I is about the same length as finger II (finger I length is 94–102% of finger II). The width of the disc of finger III is about 50–55% of the eye diameter.

The dorsum is green with small yellow spots. The upper lip is white. The anterior 60% of the venter is white. The posterior portion is transparent. The bones are green. The dorsal surfaces of the fingers and toes lack melanophores, although there are a few on toe V. The bones are green. The iris is grayish-cream with a slight yellow hue and fine black reticulations. The circumpupilary ring is yellow.

Comparisons.—*Nymphargus cariticommatus*, *N. siren*, *Espadarana audax*, *Rulyrana flavopunctata*, *R.*

mcdiarmidi, and *Teratohyla midas*. Among glassfrogs that inhabit the Amazonian slopes of the Andes and the Amazonian lowlands, *Nymphargus humboldti* is distinguished by having a green dorsum with small yellow spots and a moderate body size (male SVL 23.3–25.2 mm; female SVL 24.3–25.9 mm) and by lacking hand webbing and humeral spines. *Nymphargus humboldti* is most similar to *N. siren*, which is smaller (SVL in males, 19.8–22.6 mm; in females, 22.5–23.3 mm). Other species with a similar dorsal color pattern include *E. audax*, *N. cariticommatus*, *N. siren*, *R. flavopunctata*, *R. mcdiarmidi*, and *T. midas*. Differences among these species include the presence of hand webbing between fingers III and IV in *E. audax*, *R. flavopunctata*, *R. mcdiarmidi*, and *T. midas* (absent in *N. humboldti*), a white digestive tract in *T. midas* (opaque in *N. humboldti*), and humeral spines in males in *E. audax* (absent in *N. humboldti*). *Nymphargus cariticommatus* differs by having a white renal peritoneum and white esophagus (transparent renal pericardium and opaque esophagus in *N. humboldti*) and less foot webbing.

Natural history.—At Río Challuwa Yaku, Provincia Pastaza, Ecuador, *Nymphargus humboldti* was found in sympatry with *N. anomalus*. Individuals (including calling males) were found on 20–30 March 2009 on the upper surfaces of leaves near a fast-flowing stream.

Tadpoles.—Unknown.

Distribution.—See Figure 6.143. *Nymphargus humboldti*, an endemic species to Ecuador, is known on the Amazonian slopes of the Cordillera Oriental of the Andes in Napo, Morona Santiago, and Pastaza provinces. Its elevational range is 1497–1770 m in an area of extent of occurrence of about 3930 km^2.

This nocturnal, arboreal species inhabits natural forests in the evergreen low montane forests of the Cordillera Cóndor-Kutukú and northern Cordillera Oriental of the Andes ecosystems in the Amazonian Cordilleras biogeographic sector, where the average annual rainfall is 1782–3288 mm and the average annual temperature is 17.8 – 18.9°C.

Conservation.—*Nymphargus humboldti* has been categorized as Vulnerable by Ortega-Andrade *et al.* (2021). We categorize it as Endangered following B1ab(iii) IUCN criteria because: (1) its extent of occurrence is estimated to be less than 5000 km^2 (3930 km^2, according to our data), (2) its distribution is severely fragmented in fewer than five known localities, and (3) there is a continuing decline in the extent and quality of its habitat. Significant threats to its existence are habitat reduction and fragmentation, oil spills, deforestation, agriculture, fires, landslides, and volcanic eruptions. In Ecuador, it has been recorded in the public protected areas Parque Nacional Sumaco Napo-Galeras and Parque Nacional Llanganates, the protected forests Cerro Sumaco y Cuenca Alta del Río Suno and Cordillera Kutukú y Shaimi, and the biosphere reserve Sumaco.

Remarks.—*Nymphargus humboldti* is closely related to *N. siren* and *N. laurae* (Sánchez-Carvajal *et al.*, 2021).

Key references.—(Guayasamin *et al.*, 2020a) (description).

NYMPHARGUS LASGRALARIAS (LAS GRALARIAS GLASSFROG) HUTTER AND GUAYASAMIN, 2012

Nymphargus lasgralarias Hutter and Guayasamin, 2012: 5. **Holotype.**—MZUTI 096 from Five-Frog Creek, at Reserva Las Gralarias, 2150 m, Provincia Pichincha, Ecuador.

Etymology.—The specific epithet refers to the type locality of the new species, Reserva Las Gralarias. The species was dedicated to the reserve and the team of people led by Dr. Jane Lyons for their efforts to conserve Ecuadorian cloud forests.

Identification.—Figures 6.154, 6.155, 6.156. These frogs attain a maximum SVL of 26.5 mm in males and 27.2 mm in females. The skin on the dorsum is finely shagreened, with microspiculations. The ventral skin is areolate, with a

FIGURE 6.154 *Nymphargus lasgralarias.* Left: CJ 10690, male, 24.4 mm SVL, Reserva Integral Otonga, Provincia Cotopaxi. Right: CJ 2930, male, 27.5 mm SVL, Reserva Las Gralarias, Provincia Pichincha. LAC.

FIGURE 6.155 Morphological variation in *Nymphargus lasgralarias*. Top: QCAZ 40184, male. Middle: QCAZ 48132. Reserva Las Gralarias, Provincia Pichincha. LAC. Bottom: QCAZ 46012, Nanegal, Provincia Pichincha. LAC.

pair of large tubercles on the ventral surfaces of the thighs below the vent; the cloaca is surrounded by low warts. The upper half of the ventral parietal peritoneum is covered by iridophores, and all the other peritonea are translucent, except for a thin layer of iridophores covering the heart and renal capsules. The vomerine teeth are absent. The snout is truncate in dorsal profile and protruding in lateral profile. The tympanum is small; the supratympanic fold is present; the tympanic membrane is translucent and pigmented only on its upper half. The liver is lobed and lacks iridophores. The humeral spines are absent. There is no webbing between the fingers. The foot is about half webbed. The ulnar and tarsal folds are low and barely evident. The

FIGURE 6.156 Morphological details of *Nymphargus lasgralarias*. Top left: CJ 2930, male, Reserva Las Gralarias, Provincia Pichincha. Top right and bottom: CJ 10690, male, 24.8 mm SVL, Bosque Integral Otonga, Provincia Cotopaxi. LAC.

nuptial pad is a large to medium-sized excrescence present on the dorsal, lateral, and/or ventral sides of the thumb. The prepollex is not separated from finger I. Finger I is slightly shorter than II. The eye diameter is larger than the width of the disc on finger III.

The dorsum is light green and lacks dark spots. The flanks are yellowish-white. The fingers and toes are yellow with a faint green tint. The venter is white anteriorly and translucent posteriorly. The melanophores are absent on the fingers and toes or, when present, restricted to the dorsal surfaces of finger IV and toes IV and V. The bones are green. The iris background is golden with numerous small, black spots and very light reticulations.

Comparisons.—*Nymphargus buenaventura* and *N. griffithsi*. Among *Nymphargus* species found on the Pacific versant of the Ecuadorian Andes, *N. lasgralarias* can be confused only with *N. buenaventura* and *N. griffithsi*. The dorsal texture and color pattern readily separate *N. buenaventura*, which, in life, has a light green dorsum with warts corresponding to pale yellow spots, whereas the dorsum of *N. lasgralarias* is homogenously green and lacks warts. Additionally, *N. buenaventura* is smaller (male SVL in *N. lasgralarias* = 24.6–26.5 mm [n = 7]; male SVL in *N. buenaventura* = 20.9–22.4 mm [n = 3]). *Nymphargus lasgralarias* is most similar to *N. griffithsi*. The two species differ in the following traits: (1) dorsal color—pattern homogenously green in *N. lasgralarias*, green with very small black spots in *N. griffithsi*; (2) body size—male SVL in *N. lasgralarias* = 24.6–26.5 mm, male SVL in *N. griffithsi* = 22.5–24.2 mm; and (3) call—short, pulsed note lasting 0.016–0.044 s in *N. lasgralarias*, long tonal note lasting 0.103–0.148 s in *N. griffithsi*.

Natural history.—Figure 6.157. The species is active during the night and emits advertisement calls from the tops of small-sized ferns, small leaves, and long palm leaves 1–9 m above streams. At Reserva Las Gralarias, the species has been found at Ballux Creek, Five-Frog Creek, Heloderma Creek, Hercules Creek, Chalguayacu Creek, Kathy's Creek,

FIGURE 6.157 *Nymphargus lasgralarias*. Top, middle, and bottom left: individual in its microhabitat at night, calling male, clutches of embryos, and spider (Ctenidae) preying on egg clutch. Reserva Las Gralarias, Provincia Pichincha. Jaime Culebras. Bottom: clutch of embryos, Pululahua, Provincia Pichincha, Gustavo Pazmiño Otamendi.

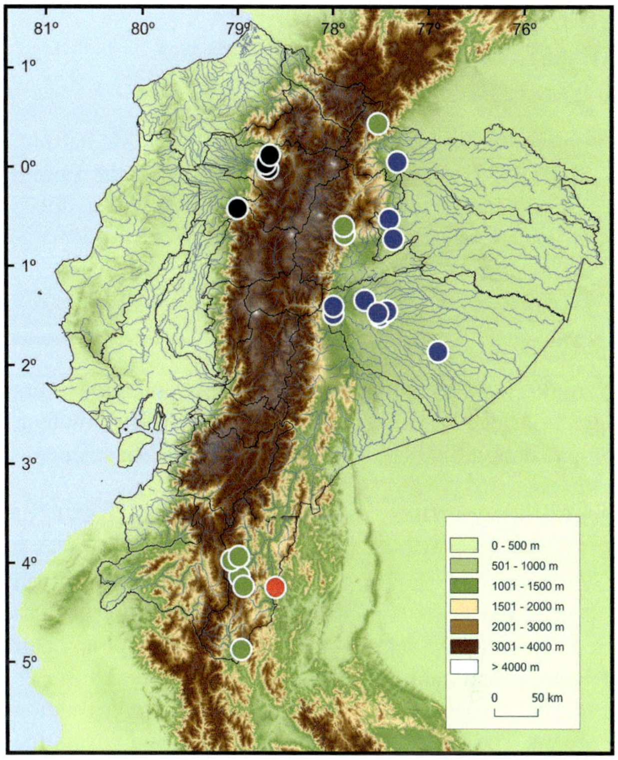

FIGURE 6.158 Localities of known occurrence of *Nymphargus lasgralarias* (black dots), *N. lindae* (red dot), *N. mariae* (blue dots), and *N. posadae* (green dots) in Ecuador.

and Lucy's Creek. *Nymphargus lasgralarias* occurs sympatrically with six other centrolenid species: *Centrolene ballux*, *C. heloderma*, *C. lynchi*, *C. peristicta*, *Nymphargus grandisonae*, and *N. griffithsi*.

Small clutches (12–36 eggs per clutch; n = 23) are deposited on the upper surfaces of leaves near the terminal margin and transition into hanging masses as the eggs develop. Reproductive activity was recorded between 5 April and 1 July 2011 but increased dramatically at the end of April and peaked in the middle of May. Females provide short-term parental care; male parental care is absent.

The advertisement calls of six individuals from Reserva Las Gralarias, Provincia Pichincha, Ecuador, were described as sounding like a "tick" or "click;" the mean call duration is 25.7 ms. Calls are pulsed and emitted in series of 1–4 calls per call series. Detailed data and figures are provided under the subtitle "Vocalizations of Ecuadorian Anurans" in Appendix 1.

Tadpoles.—Unknown.

Distribution.—Figure 6.158. *Nymphargus lasgralarias*, an endemic species to Ecuador, is known from the western slopes of the Cordillera Occidental of the Andes in Pichincha and Cotopaxi provinces. Its elevational range is 1414–2208 m in an area of extent of occurrence of about 230 km^2.

This nocturnal, arboreal species inhabits small, permanent streams (ca. 3 m width) within primary montane forests with minimal disturbance in the evergreen low montane forest and evergreen montane forest ecosystems of the Cordillera Occidental of the Andes biogeographic sector, where the average annual rainfall is 1285–2030 mm and the average annual temperature is 15.3 – 19.2°C.

Conservation.—*Nymphargus lasgralarias* was categorized as Endangered by Ortega-Andrade *et al.* (2021) because its extent of occurrence is estimated to be less than 5000 km^2 (230 km^2 according to our data), its distribution is severely fragmented, and there is a continuing decline in the extent and quality of its habitat. Significant threats to the species are habitat destruction and fragmentation due to agriculture and cattle; other likely threats are introduced species (trout), infectious diseases (*Batrachochytrium dendrobatidis*), and climate change. According to Guayasamin *et al.* (2014), at Reserva Las Gralarias, Provincia Pichincha, the species is infected by the chytrid fungus *B. dendrobatidis*, but no recent declines have been observed. It has been recorded in the public reserve Area Protegida Privada Bellavista, the biosphere reserve Chocó Andino de Pichincha, and some private reserves (e.g., Reserva Las Gralarias).

Remarks.—According to Guayasamin *et al.* (2020b), *Nymphargus lasgralarias* is the sister taxa of *N. griffithsi*.

Key references.—(Hutter and Guayasamin, 2012) (description); (Guayasamin *et al.*, 2014) (chytrid), (Delia *et al.*, 2017) (reproductive biology); (Guayasamin *et al.*, 2020b) (account).

NYMPHARGUS LAURAE (LAURA'S GLASSFROG) CISNEROS-HEREDIA AND MCDIARMID, 2007

Nymphargus laurae Cisneros-Heredia and Mcdiarmid, 2007: 37. **Holotype.**—USNM 288453 from the Loreto region near the town of Loreto, ca. 500 m, lower slopes of Volcán Sumaco, Provincia Orellana, Ecuador.

Etymology.—The specific epithet is a matronym for Laura Heredia, Diego Francisco Cisneros Heredia's grandmother, for her support of Diego's interest in animals and science.

Identification.—These frogs attain a maximum SVL of 22.1 mm in adult males; a single subadult female has a SVL of 22.1 mm. The skin on the dorsum is slightly shagreen with elevated warts corresponding to the ocelli and scattered spicules. The ventral skin is granular. There is a pair of large, round, and flat tubercles on the ventral surfaces of the thighs below the vent; other cloacal ornamentation is absent. The parietal peritoneum is white and covers the anterior two-thirds of the abdomen. There is a white pericardium; all the other peritonea are clear. The vomerine teeth are absent. The snout is truncate in dorsal and profile views; the nostrils are slightly elevated, producing a slight depression in the internarial area; the loreal region is concave. The tympanic annulus is evident and oriented dorsolaterally with a dorsal inclination; the supratympanic fold is weak from behind the eye to the insertion of arm. The liver is lobed; the hepatic peritoneum is clear. The humeral spine is absent. There is basal webbing between fingers I, II, and III. The foot is about two-thirds webbed. There are no dermal folds or tubercles on the hands, forearms, feet, or tarsi. The prepollex is concealed, and there is an unpigmented nuptial pad. Finger II is longer than I. The eye diameter is greater than the width of the disc on finger III.

The dorsum is green with black ocelli with yellow centers. Melanophores are absent on fingers and toes.

Comparisons.—*Nymphargus anomalus*, *N. lindae*, and *N. cochranae*. *Nymphargus laurae* is distinguished from most glassfrogs by the presence of dorsal ocelli (a trait shared only with *N. anomalus*, *N. lindae*, and *N. cochranae*), two papillae on each toe except toe V, and a relatively small size (SVL = 19.9 mm in male holotype). The presence of two papillae on each toe disc is a characteristic unique to Centrolenidae. *Nymphargus anomalus* differs by having smaller and more abundant ocelli and dark spots and points amid the ocelli. *Nymphargus lindae* differs from *N. laurae* mainly by being larger (in *N. lindae*, male SVL = 23.0–26.3 mm; female SVL = 27.2–27.8 mm), having vomerine teeth (absent in *N. laurae*), and lacking papillae on the toe discs. *Nymphargus laurae* is very similar to *N. cochranae*; both species share a similar color pattern and are likely to be sympatric; *N. cochranae*, however, differs by having much smaller ocelli, a larger body size (male *N. cochranae* SVL = 23.8–26.7 mm), and lacking papillae on the toes. See Remarks.

Natural history.—Unknown.

Tadpoles.—Unknown.

Distribution.—Figure 6.159. *Nymphargus laurae*, an endemic species to Ecuador, is known in the Amazonian lowlands of Orellana and Napo provinces at an elevational range of 442–1537 m.

This species inhabits natural forests in the evergreen low montane forests of the northern Cordillera Oriental of the Andes ecosystem in the Napo-Curaray biogeographic sector, where the average annual rainfall is 2832–3600 mm and the average annual temperature is 18.8 – 23.9°C.

Conservation.—*Nymphargus laurae* was categorized as Endangered by Ortega-Andrade *et al.* (2021). However, we categorize it as Critically Endangered following B1ab(iii) IUCN criteria because its extent of occurrence is estimated to be less than 100 km^2, its distribution is severely fragmented, and there is a projected decline in the extent and quality of its habitat. Collections during the 1950s on the Loreto Plateau were intensive and took place over several years because the collector (part of the Olalla family) lived in the area; however, only one specimen of *N. laurae* was ever collected (versus numerous specimens of other species of centrolenids), suggesting that even at the time of

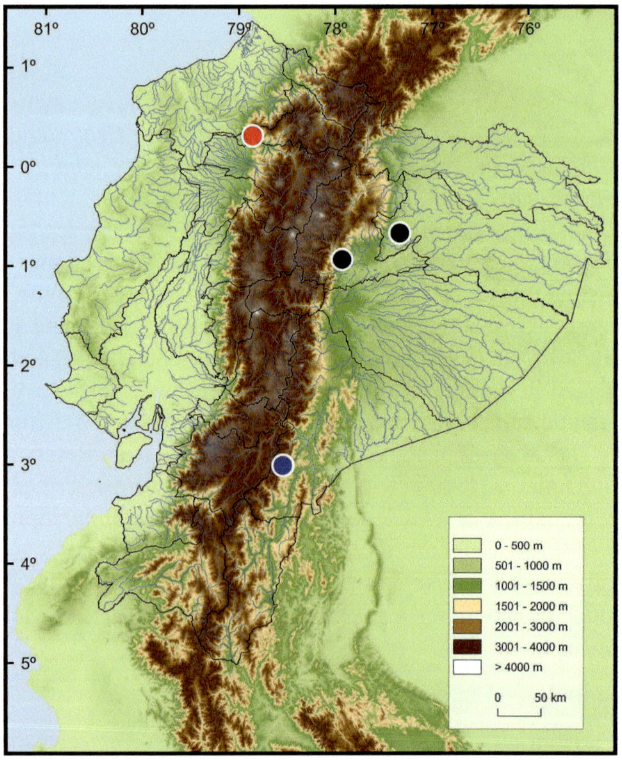

FIGURE 6.159 Localities of known occurrence of *Nymphargus laurae* (black dots), *N. manduriacu* (red dot), and *N. sucre* (blue dot) in Ecuador.

collection (when the quality of the habitats was better), the species was at least uncommon. Habitat destruction in the Loreto Plateau (the only known locality for the species) is extensive, and most forested areas in the plateau have already been replaced by agricultural and pasture lands. In general, the entire northern foothills of the Cordillera Oriental of Ecuador, where *N. laurae* occurs, are widely deforested. Wildfires also pose a significant risk to its survival in its small areas of occurrence and occupation.

Nymphargus laurae has been recorded in the public reserve Reserva Biológica Colonso Chalupas, the protected forest of Cerro Sumaco y Cuenca Alta del Río Suno, and the biosphere reserve Sumaco.

Remarks.—*Nymphargus laurae* is the sister species of *N. siren*, based on mitochondrial and nuclear data (Sánchez-Carvajal *et al.*, 2021).

Key references.—(Cisneros-Heredia and Mcdiarmid, 2007) (description); (Guayasamin *et al.*, 2020b) (account); (Sánchez-Carvajal *et al.*, 2021) (redescription).

NYMPHARGUS LINDAE (LINDA'S GLASSFROG) (GUAYASAMIN, 2020)

Nymphargus lindae Guayasamin, 2020: 184. **Holotype.**—QCAZ 41572 from Miazi Alto, 1200 m, Cordillera del Cóndor, Provincia Zamora Chinchipe, Ecuador.

FIGURE 6.160 *Nymphargus lindae*. Top: QCAZ 41597, paratype, female. Bottom: paratype, male. Miazi Alto, Provincia Zamora Chinchipe. Juan M. Guayasamin.

Etymology.—The specific epithet honors Linda Trueb, one of the most influential amphibian systematist of our time. Linda, as the Curator of the herpetological collection of the University of Kansas, has led one of the most prolific and solid research groups on amphibian biology, mentoring numerous students (including JMG) and generating an extraordinary body of work.

Identification.—Figure 6.160. These frogs attain a maximum SVL of 26.3 mm in males and 27.8 mm in females. The skin on the dorsum is shagreen with minute spicules in males. The venter is areolate; the ventral surfaces of the thighs below the vent have a pair of enlarged tubercles. There is a white parietal peritoneum covering about the anterior 60% of the venter. There is a white pericardium; there are translucent peritonea covering the intestines, stomach, kidneys, and gall and urinary bladders. The dentigerous process of the vomer has 2–4 teeth. The snout is truncate to bluntly rounded in lateral profile and truncate in dorsal view. The tympanum is oriented almost vertically with slight lateral and posterior inclinations, and its diameter is about 20–26% of the eye diameter; the upper fourth of the tympanic annulus is obscured by the supratympanic fold; the tympanic membrane is pigmented like the surrounding skin. The liver is tetralobed and covered by a transparent peritoneum. The humeral spines are absent. There is no webbing between the inner fingers; there is absent or basal webbing between fingers III and IV. The feet are about two-thirds webbed. The ulnar and tarsal folds are low. The prepollex is not separated from finger I. The nuptial pad is a large to medium-sized excrescence present on the dorsal, lateral, and/or ventral sides of the thumb. Finger I is about the same length as finger II or slightly shorter (finger I length is 94–100% of finger II). The width of the disc of finger III is about 46–50% of the eye diameter.

The dorsum is green with dark lavender to black ocelli enclosing yellow to orange spots. The upper lip is white. Anteriorly, about 60% of the ventral parietal peritoneum is white. The posterior portion is translucent. The dorsal surfaces of the fingers and toes lack melanophores, except for few on toe V. The bones are green. The iris is white with a slight, pale yellow hue and thin black reticulations.

Comparisons.—*Nymphargus anomalus*, *N. cochranae*, *N. laurae*. *Nymphargus lindae* can be distinguished from most other glassfrogs by having, in life, a green dorsum with dark lavender to black ocelli enclosing yellow to orange spots, and by lacking (or nearly lacking) webbing between the fingers. *Nymphargus lindae* is most similar to other Ecuadorian glassfrogs with dorsal ocellated patterns (*N. anomalus*, *N. cochranae*, *N. laurae*). *Nymphargus anomalus* differs by having, in life, a pale brown dorsum and scattered black and lavender flecks between ocelli (green dorsum lacking flecks in *N. lindae*). *N. laurae*, known from a single adult male, is smaller (SVL = 19.7 mm in *N. laurae*; male SVL = 23.0–26.3 mm in *N. lindae*), has pointed papillae on toes I–IV (absent in *N. lindae*), and lacks teeth on the vomers (present in *N. lindae*). *Nymphargus lindae* differs from its sister species *N. cochranae* by the relative size of the ocelli (minute in *N. cochranae* and conspicuously larger in *N. lindae*) and female body size (*N. cochranae*, SVL = 27.8–30.3 mm; *N. lindae*, SVL 27.2–27.8 mm).

Natural history.—*Nymphargus lindae* is the most abundant species in streams at Miazi Alto, Cordillera del Cóndor. Males were observed calling on the upper surfaces of leaves and on branches about 30–400 cm above a stream. An amplectant pair and two additional males were observed fighting on a branch; males that were not in amplexus continuously approached the amplectant male and tried to push him away with kicks.

Tadpoles.—Unknown.

Distribution.—See Figure 6.158. *Nymphargus lindae*, an endemic species to Ecuador, is known from a single locality at 1391 m in the Cordillera del Cóndor, Provincia Zamora Chinchipe.

This nocturnal, arboreal species inhabits streams of natural forests in the evergreen montane foothill forests on outcrops of limestone rock ecosystem in the Amazonian Cordilleras biogeographic sector, where the average annual rainfall is 1395 mm and the average annual temperature is 21.5 °C.

Conservation.—*Nymphargus lindae* was categorized as Endangered by Ortega-Andrade *et al.* (2021) following IUCN criteria B2a, B2(iii). It is known from fewer than three localities, and there is a continuing and projected decline in the extent and quality of its habitat. The main threats for the species in Cordillera del Cóndor are habitat destruction and contamination associated with mining activities. It has only been recorded in the biosphere reserve Podocarpus-El Cóndor.

Remarks.—*Nymphargus lindae* is the sister species to *N. cochranae* (Guayasamin *et al.*, 2020b).

Key references.—(Guayasamin, 2020) (description).

NYMPHARGUS MANDURIACU (MANDURIACU GLASSFROG) GUAYASAMIN *ET AL.*, 2019a

Nymphargus manduriacu Guayasamin *et al.*, 2019a: 7.
Holotype.—ZSFQ 0466 from Reserva Río Manduriacu, 1215 m, Provincia Imbabura, Ecuador.

Etymology.—The specific epithet refers to the type locality of the species, Río Manduriacu Reserve, Ecuador, a conservation area managed by Fundación EcoMinga.

FIGURE 6.161 *Nymphargus manduriacu.* Left: ZSFQ 462, paratype, female, 28.8 mm SVL. Right: ZSFQ 466, holotype, male, 25.3 mm SVL. Reserva Río Manduriacu, Provincia Imbabura. José Vieira-Fernandes.

Identification.—Figure 6.161. These frogs attain a maximum SVL of 25.7 mm in males and 28.8 mm in a female. The skin on the dorsum is shagreened, with microspicules in adult males. The ventral skin is granular; the subcloacal area has two large subcloacal warts. The parietal peritoneum is white; the iridophores cover one-third to one-half of the parietal peritoneum. There is a white pericardium (i.e., covered by iridophores); all the other visceral peritonea are clear. The dentigerous process of the vomer is low or absent, and it lacks vomerine teeth. The snout is truncate in dorsal view and truncate to slightly rounded in lateral view. The tympanic annulus is barely evident; the lower three-fourths are visible; the tympanic membrane is colored like the dorsal skin; the supratympanic fold is present. The liver is lobed, and the hepatic peritoneum is clear (lacking iridophore layer). The adult males lack humeral spines. There is no webbing between fingers I, II, and III; there is basal webbing between fingers III and IV. There is basal webbing between toes I and II. There are no dermal ornamentations in the form of tubercles, folds, or fringes on the hands, arms, feet, or legs. The prepollex is concealed. The nuptial pad varies from a large to medium-size excrescence present on the dorsal, lateral, and/or ventral sides of the thumb to nuptial excrescences formed by a combination of clustered and individual glands scattered along the flanks of the body. Finger I is slightly longer than finger II.

The diameter of the eye is greater than the width of the disc on finger III.

The dorsum is grayish-green to olive green with yellow spots, with melanophores concentrated around yellow spots, sometimes looking like false ocelli. The upper lip is unpigmented. The inner fingers and toes have a yellowish hue. The anterior half of the ventral parietal peritoneum is white. The posterior portion is translucent. There are abundant melanophores along fingers III and IV, less dense on finger II, rarely present on finger I, present and abundant along toes IV and V, less dense on toe III, and only at the base of toes I and II. The bones are green. The iris is light gray with thin, dark gray reticulations and a pale yellow hue around the pupil.

Comparisons.—*Nymphargus buenaventura*. *Nymphargus manduriacu* is easily differentiated from most glassfrogs by lacking webbing between the inner fingers and having, in life, a grayish-green dorsum with numerous yellow spots, which sometimes are surrounded by an ill-defined black ring (i.e., false ocelli). On the Pacific slopes of the Ecuadorian Andes, only *N. buenaventura* could be confused with *N. manduriacu*, but it differs by having a light green dorsum with diffuse, pale yellow spots.

Natural history.—Figure 6.162. It has only been found in Río Manduriacu Reserve, and although the reserve has been visited several times, *Nymphargus manduriacu* was only regularly detected during February 2018, with the site experiencing particularly heavy rains on a daily basis. The males call from the upper sides of leaves. The egg masses are deposited on the upper sides of leaves; the clutch size is 15–32 eggs (n = 4); no long-term parental care is provided by either males or females.

The advertisement call of one individual from Reserva Río Manduriacu, Provincia Pichincha, Ecuador, was described as a high-pitched "chirp" that consists of a pulsed single note with a mean duration of 100 ms. Detailed data and figures are provided under the subtitle "Vocalizations of Ecuadorian Anurans" in Appendix 1.

Tadpoles.—Unknown.

FIGURE 6.162 *Nymphargus manduriacu*. Individual at Reserva Río Manduriacu, Provincia Pichincha. Ross Maynard.

Distribution.—See Figure 6.159. *Nymphargus manduriacu*, an endemic species to Ecuador, is known from a single location at 1250 m on the Pacific slopes of Cordillera Occidental of the Andes in Provincia Imbabura.

This nocturnal, arboreal species inhabits natural forests in the evergreen montane foothill forest ecosystem of the Cordillera Occidental of the Andes biogeographic sector, where the average annual rainfall is 1951 mm and the average annual temperature is 19.3°C.

Conservation.—*Nymphargus manduriacu* was categorized as Critically Endangered by Ortega-Andrade *et al.* (2021) following A3cB1a(1)b(iii) IUCN criteria. This species is only known from a single locality, where there are important threats such as habitat destruction and contamination associated with cattle ranching, agriculture, and, most seriously, mining activities. In Río Manduriacu Reserve (the only known locality of the species), mining has become one the most dangerous threats to biodiversity, especially to species with very restricted distributions. A thorough explanation of current threats at Río Manduriacu Reserve is provided by Guayasamin *et al.* (2019a).

Remarks.—Based on mitochondrial data, Guayasamin *et al.* (2019a) found that *Nymphargus manduriacu* is sister to *N. balionotus*.

Key references.—(Guayasamin *et al.*, 2019a) (description, call, conservation).

NYMPHARGUS MARIAE (MARÍA'S GLASSFROG) (DUELLMAN AND TOFT, 1979)

Centrolenella mariae (Duellman and Toft, 1979): 66. **Holotype.**—KU 174713 from Serranía de Sira, approximately 1550 m, Departamento Huánuco, Peru.

Nymphargus mariae Guayasamin *et al.*, 2009: 28.

Etymology.—The specific epithet is a matronym for María Koepcke (1924–1971), who devoted her life to biological exploration in Peru.

Identification.—Figure 6.163. These frogs attain a maximum SVL of 31.7 mm in males and 30.1 mm in females. The skin on the dorsum is shagreen, with spiculated flat warts corresponding to light spots. The venter is areolate; the ventral surfaces of the thighs below the vent have a pair of enlarged tubercles. There is a white lining on the anterior 40% of the ventral parietal peritoneum; there are translucent peritonea covering the intestines, stomach, gall bladder, and urinary bladder. The vomerine teeth are present; each vomer has three to four teeth. The snout is truncate in dorsal aspect and round in lateral profile. The tympanum is oriented dorsolaterally with a slight posterior inclination, and its diameter is about 32–35% of the eye diameter; the tympanic annulus is visible; the supratympanic

FIGURE 6.163 Morphological variation and details of *Nymphargus mariae*. QCAZ 48529, male, Reserva Río Bigal, Provincia Napo. LAC.

fold is low but evident; the tympanic membrane is partially pigmented and clearly differentiated from the surrounding skin. The liver is tetralobed and covered by transparent peritoneum. The humeral spines are absent. There is no webbing between fingers I, II, and III; there is reduced or absent webbing between the outer fingers. The webbing on the foot is basal. The ulnar fold is low; the inner and outer tarsal folds are low and thin. The prepollex is not separated from finger I. The nuptial pad is a large excrescence present on the dorsal, lateral, and/or ventral sides of the thumb. Finger I is about the same length as finger II. The disc of finger III is narrow, about 34–38% of the eye diameter.

The dorsum is green to dark green, with minute and large yellow–green spots. The anterior half of the venter is white. The posterior part is translucent. Melanophores are absent from fingers I and II and toes I, II, and III; few are present on finger III; they are numerous on finger IV and toes IV and V. The bones are green. The iris is grayish-cream with a brown transverse bar and fine dark gray reticulations. The circumpupilar ring is pale yellow.

Comparisons.—*Centrolene medemi*, *Rulyrana flavopunctata*, *Teratohyla midas*, *Nymphargus cariticommatus*, *N. humboldti*, *N. siren*, *Sachatamia albomaculata*. *Nymphargus mariae* can be distinguished from most glassfrogs by having, in life, a green dorsum with relatively large yellow–green spots, a white parietal peritoneum covering the anterior half of the venter, and by lacking humeral spines and webbing between the fingers. *R. flavopunctata*, *T. midas*, *N. cariticommatus*, *N. humboldti*, and *N. siren* have a similar color pattern, but the yellow spots are conspicuously larger in *N. mariae*. *Sachatamia albomaculata* and *C. medemi* also have large yellow spots on the dorsum, but *S. albomaculata* has webbing between fingers III and IV and inhabits the Pacific lowlands (*N. mariae* lacks hand webbing and is found on the Amazonian slopes of the Andes and Amazonian lowlands). Also, *C. medemi* is larger than *N. mariae* (*C. medemi*: SVL 25.5–30.8 mm in adult males, 34.7–44.3 mm in adult females; *N. mariae*: SVL 23.4–30.2 mm in males, SVL 25.1–32.8 mm in females) and has more webbing between fingers III and IV, and males have humeral spines.

Natural history.—Males call from the upper surfaces of leaves. The advertisement call of one individual from Río Lliquino, Provincia Pastaza, Ecuador, was described as a single, non-frequency-modulated note; it has a duration of 15–18 ms. Detailed data and figures are provided under the subtitle "Vocalizations of Ecuadorian Anurans" in Appendix 1.

Tadpoles.—Unknown.

Distribution.—See Figure 6.158. *Nymphargus mariae*, an endemic species to Ecuador, is known on the Amazonian slopes of the Cordillera Oriental of the Andes in Sucumbíos, Orellana, and Pastaza provinces. Its elevational range is 283–1036 m in an area of extent of occurrence of about 12,180 km^2.

This nocturnal, arboreal species inhabits natural forests and intervened areas in the evergreen forests of lowlands and evergreen montane foothill forest ecosystems in the Napo-Curaray, Tigre-Pastaza, and northern Cordillera Oriental of the Andes biogeographic sectors, where the average annual rainfall is 2860–4103 mm and the average annual temperature is 20.6–25.0 °C.

Conservation.—*Nymphargus mariae* is categorized as Least Concern throughout its entire range (IUCN SSC Amphibian Specialist Group, 2020h; in IUCN Red List of 2022) and in Ecuador by Ortega-Andrade *et al.* (2021). It has a wide distribution and presumed large population. The population status of this species is unknown; however, it is suspected to be decreasing due to ongoing declines in the extent and quality of its habitat. In Ecuador, specimens have been recorded in 2008 and 2012 (Guayasamin *et al.*, 2014). Although the Sira area is relatively inaccessible, there is illegal logging, coca plantations, cattle farming, subsistence farming, and illegal mining taking place in the region, and there are concerns that these are also occurring within the boundaries of the Reserva Comunal El Sira (Exploration Sira), Peru. Presumably the Ecuadorian subpopulations are being locally affected by deforestation for agriculture, livestock, and wood extraction. In Ecuador, it has been recorded in the protected forest of Cerro Sumaco y Cuenca Alta del Río Suno and the biosphere reserve Sumaco.

Remarks.—*Nymphargus mariae* and *N. colomai* are sister species (Guayasamin *et al.*, 2020b).

Key references.—(Duellman and Toft, 1979) (description); (Flores and McDiarmid, 1989) (distribution); (Guayasamin *et al.*, 2020b) (account).

NYMPHARGUS MEGACHEIRUS (LARGE-HANDED GLASSFROG) (LYNCH AND DUELLMAN, 1973)

Centrolenella megacheira (Lynch and Duellman, 1973): 36.
Holotype.—KU 143245 from 16.5 km north–northeast of Santa Rosa, on Quito-Lago Agrio road, 1700 m, Provincia Napo, Ecuador.
Nymphargus megacheirus Cisneros-Heredia and Mcdiarmid, 2007: 34.

Etymology.—The specific epithet derives from the Greek words *megas*, meaning large, and *cheiros*, meaning hand; the name is used to refer to the exceedingly large hands of the species.

Identification.—Figure 6.164. These frogs attain a maximum SVL of 31.5 mm in males and 32.9 mm in females. The skin on the dorsum is shagreen to pustular in males and shagreen in females. Numerous spicules are present in males, whereas in females, spicules are present only on the head, tympanic region, and limbs. The skin of the venter is areolate; the ventral surfaces of the thighs below the vent have a pair of enlarged tubercles. The anterior two-thirds to three-fourths of the venter are covered by a white parietal peritoneum, and the posterior portion is transparent. There is a white pericardium; there is a translucent peritoneum covering the intestines, stomach, testes, kidneys, gall bladder, and urinary bladder. The vomers lack teeth. The snout is truncate in dorsal and lateral profiles. The tympanum is relatively small, and its diameter is 18.8–25.4% of the eye diameter; the dorsal border of the tympanic annulus is covered by the supratympanic fold; the tympanic membrane is pigmented like the surrounding skin. The liver is tetralobed; there are two large ventral lobes partially covering the two smaller lobes; the hepatic peritoneum is transparent. The humeral spines are absent. There is no hand webbing between the inner fingers; there is no or basal webbing between the outer fingers; the foot is about one-half webbed. The ulnar fold is conspicuous, and the tarsal folds are low. The prepollex is not separated from finger I. The nuptial pad is a large excrescence present on the dorsal, lateral, and/or ventral sides of the thumb. Finger I is usually slightly shorter than finger II (finger I is 88.9–101.3% the length of finger II). The disc of finger III is relatively large, 49.7–59.0% of the eye diameter.

FIGURE 6.164 *Nymphargus megacheirus*. Top: KU 143245, holotype, male, 16.5 km northeast of Santa Rosa, Provincia Napo. Bottom: KU 164614, 2 km southwest of Río Reventador, Provincia Napo. WED.

The dorsum is green with small blue to black spots. The ventral surfaces are mostly white, except for the translucent posterior third of the venter. The upper lip and ulnar folds are white. The dorsal surfaces of the fingers and toes lack melanophores. The bones are green. The iris is grayish-bronze with thin black reticulations.

Comparisons.—*Nymphargus cochranae*, *N. garciae*. *Nymphargus megacheirus* can be distinguished from other glassfrogs by its green dorsum with small blue to black spots, relatively large size (adult males, SVL 26.8–31.5 mm; adult females, SVL 31.2–32.9 mm), basal or no webbing among fingers, and lack of humeral spines. On the Amazonian slopes of the Ecuadorian Andes, only *N. cochranae* has similar characteristics; however, *N. cochranae* has small, dark ocelli enclosing orange dots on the dorsum, whereas *N. megacheirus* has solid dark spots. *Nymphargus megacheirus* has a similar dorsal coloration and hand webbing pattern as *N. garciae*, but the two species have allopatric distributions, with *N. garciae* inhabiting higher elevations (1900–2700 m) than *N. megacheirus* (1300–1740 m). Additionally, *N. megacheirus* is slightly larger than *N. garciae* (*N. garciae*, SVL 25.1–29.9 mm in adult males: SVL 25.9–28.4 mm in adult females).

Natural history.—At a site 16.5 km north–northeast of Santa Rosa, Provincia Napo, during the reproductive season, males were calling from the upper surfaces of leaves overhanging fast-moving streams in October 1971.

Tadpoles.—Unknown.

Distribution.—See Figure 6.143. *Nymphargus megacheirus* occurs on the Amazonian slopes of the Cordillera Oriental of the Andes of Ecuador and Colombia. In Ecuador, it occurs in Provincia Napo. Its elevational range is 1246–1769 m in an area of extent of occurrence of about 20 km^2.

This nocturnal, arboreal species inhabits streams of natural forests in the evergreen low montane forest ecosystem of the northern Cordillera Oriental of the Andes biogeographic sector, where the average annual rainfall is 2136–2706 mm and the average annual temperature is 17.9 – 19.9°C.

Conservation.—*Nymphargus megacheirus* is categorized as Critically Endangered through its entire range (IUCN SSC Amphibian Specialist Group, 2022o; in IUCN Red List of 2022) and in Ecuador by Ortega-Andrade *et al.* (2021) due to its restricted distribution. In Ecuador, its current extent of occurrence is 20 km^2, according to our data, and there are no recent records. The last records of *Nymphargus megacheirus* correspond to specimens collected at Río Azuela and Río Salado on 24 February 1979 and deposited at USNM. Surveys at Río Azuela have failed to find additional individuals. There is no recent information from Colombia. The major threats include deforestation for agricultural development, logging, human settlements, and the introduction of exotic predatory fish in streams. Like some other centrolenids, this species might also be affected by the movement of the cloud layer up the mountain slopes as a result of climate change, resulting in reduced humidity within the altitudinal range of the species (probably exacerbated by habitat fragmentation). In Ecuador, it has been recorded in the public reserve Parque Nacional Cayambe Coca and the biosphere reserve Sumaco.

Remarks.—*Nymphargus megacheirus* and *N. anomalus* are sister species (Guayasamin *et al.* 2020b).

Key references.—(Lynch and Duellman, 1973) (description, ecology); (Ruiz-Carranza *et al.*, 1996) (distribution); (Guayasamin *et al.*, 2020b) (account).

NYMPHARGUS MEGISTUS (GREATER ANDEAN GLASSFROG) (RIVERO, 1985)

Centrolenella megista (Rivero, 1985): 353. **Neotype.**— MHNCSJ 0291 from Urrao, Parque Las Orquídeas (Polo), 1700–1800 m, Departamento Antioquia, Colombia.
Nymphargus megistus Trageser *et al.*, 2021: 30.

Etymology.—The specific epithet derives from the Greek *megistos*, meaning greatest or largest.

Identification.—Figure 6.165. These frogs attain a maximum SVL of 37.4 mm in males and 40.0 mm in females. The skin on the dorsum of males has abundant conical spicules of uniform distribution; in females, the dorsum bears almost flat warts and few spicules. The ventral skin is granular; the subcloacal region is weakly granular. The cloacal ornamentation is present as fine whitish spots just in the area below the cloaca. There is a white parietal peritoneum covering the anterior 50% of the venter. There is a white parietal and pericardium peritoneum; the hepatic and the visceral peritonea are translucent. The vomers have a dentate dentigerous process. The snout is subovoid in dorsal view and rounded in lateral profile. There is a tympanum and tympanic annulus that is small, posterodorsally oriented. The liver is tetralobed. The humeral spines are absent. There is moderate webbing between fingers III and IV. The feet are about two-thirds webbed. There is a row of small tubercles on the ulnar and tarsal folds. The prepollex is not externally visible. The nuptial pad is a large to medium-sized excrescence present on the dorsal, lateral, and/or ventral sides of the thumb. Finger I is about as long

FIGURE 6.165 Morphological variation in *Nymphargus megistus*. Top left: Mindo, Provincia Pichincha. Amanda B. Quezada. Middle and bottom left: adults. Right: female. Reserva Río Manduriacu, Provincia Imbabura. Ross Maynard.

as II when appressed. The eye diameter is about 1.5 times greater than the size of the disc of finger III.

The dorsum is green with black spots. The pericardium and anterior half of the parietal peritoneum are white. The visceral peritoneum is translucent. There are melanophores on the fingers and toe discs. The bones are green. The iris is pale gray with a grayish reticulum.

Comparisons.—*Nymphargus megistus* can be distinguished from all other species of *Nymphargus* by its large size (SVL in adult males = 30.8–37.4 mm SVL; in adult females = 36.9–40.0 mm), green dorsum with black spots, moderate webbing between fingers III and IV, absence of humeral spines in adult males, and lack of iridophores on the gastrointestinal peritoneum. *Nymphargus megistus* is an atypical species of the genus *Nymphargus*, where all species have a reduced webbing between finger III and IV. In contrast, the hand webbing in *N. megistus* is more extensive. No additional variation or sexual dimorphism has been observed.

Natural history.—Figure 6.166. In Ecuador, individuals were found in primary cloud forest. Two individuals (male and female) were observed ca. 1.75 m from the nearest stream and within 25 m of one another; another male was two meters away from a creek. They were perched on a broad-leaf stalk (2 cm diameter) and on leaves between 0.2 and 1 m high. They were at temperatures of 14.8–15.6 °C and 95–100% relative humidity. A gravid female was found in December 2019.

In Colombia, individuals were found in the vegetation associated with streams. A gravid female was found where males vocalized near the clutches (close branches), mainly in leaves of Araceae (1.50–1.80 m high). However, some frogs were found far from bodies of water, interned in mature forest (0.5–2 m above ground). Observations in Colombia and Ecuador suggest a possible migration of adults during the non-breeding season into the forest.

Tadpoles.—Unknown.

Distribution.—Figure 6.167. *Nymphargus megistus* is known on the western slopes of the Cordillera Occidental of the Andes of Ecuador and the Cordillera Occidental in Antioquia, Risaralda, Departamento Valle del Cauca, Colombia. In Ecuador, it occurs in two localities in Imbabura and Pichincha provinces. Its elevational range is 1409–1760 m, and the straight-line distance between the two localities is around 40 km.

This nocturnal, arboreal species inhabits primary cloud forests in the evergreen low montane forest ecosystem of the Cordillera Occidental of the Andes biogeographic sector, where the average annual rainfall is 1894–2178 mm and the average annual temperature is 17.5 – 18.5°C.

Conservation.—*Nymphargus megistus* is categorized as Endangered throughout its entire range (IUCN SSC Amphibian Specialist Group, 2017; in IUCN Red List of 2022). In Ecuador, we also categorize it as Endangered because it is only known from two localities. Results from

FIGURE 6.166 *Nymphargus megistus*. Reserva Río Manduriacu, Provincia Imbabura. Ross Maynard.

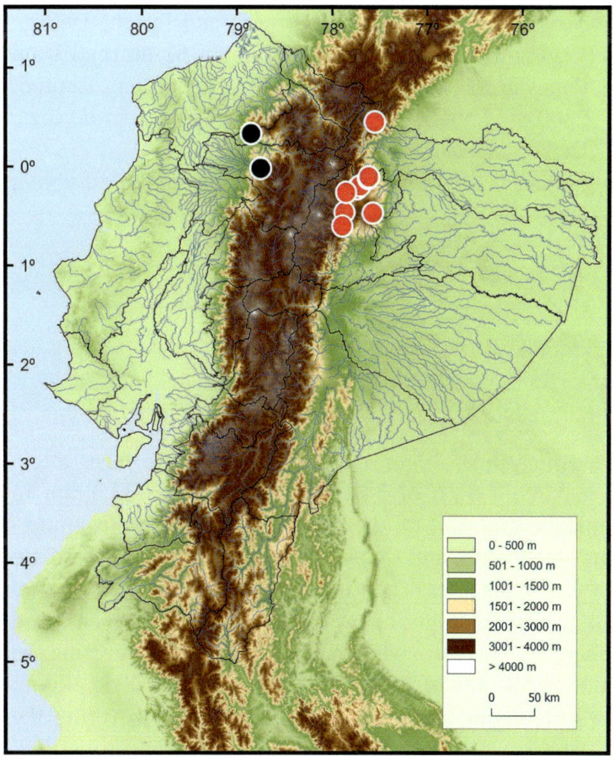

FIGURE 6.167 Localities of known occurrence of *Nymphargus megistus* (black dots) and *N. siren* (red dots) in Ecuador.

the Colombia Red List Assessment Workshop in 2016 indicate that the majority of the population resides in small subpopulations with no genetic exchange. Furthermore, severe fragmentation may already exist along specific corridors at the elevations necessary for this species to disperse, as Andean cloud forests in Colombia are disproportionately excluded from regional and national protected area networks. Additionally, in light of increasing mining activity in Ecuador and immediate threats facing Reserva Manduriacu, additional surveying of the vast area between the Colombian and Ecuadorian occurrences is necessary to better understand its distribution, natural history, and conservation status. In Ecuador, it has been recorded in the biosphere reserve Chocó Andino de Pichincha and in the private reserve Río Manduriacu.

Remarks.—According to Trageser *et al.* (2021), *Nymphargus megistus* and *N. grandisonae* are sister taxa; the genetic distance between them is 2.0%. Also, they state that it is unknown if the Ecuadorian populations of *N. megistus* are distinct from those in Colombia.

Key references.—(Rivero, 1985) (description); (Rada and Guayasamin, 2008) (redescription); (Stuart *et al.*, 2008) (notes, as *Cochranella megistra*); (Trageser *et al.*, 2021) (morphology, natural history).

FIGURE 6.168 *Nymphargus posadae*. Left: CJ 11383, female, 39.4 mm SVL. Right: CJ 11571, male, 35.3 mm SVL. Guarumales, Provincia Zamora Chinchipe. DAL.

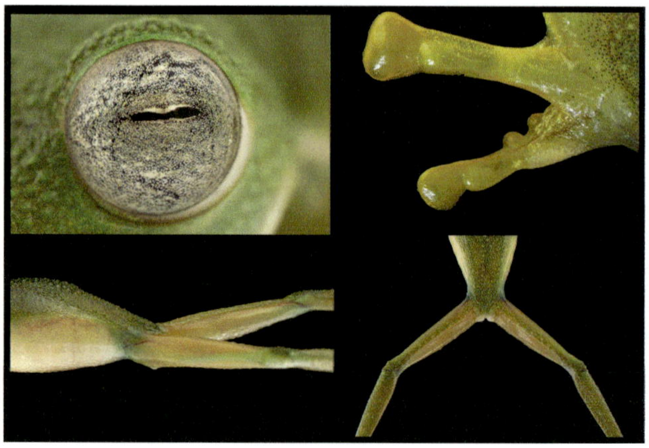

FIGURE 6.169 Morphological details of *Nymphargus posadae*. CJ 11571, male, Guarumales, Provincia Zamora Chinchipe. SGS.

NYMPHARGUS POSADAE (POSADA'S GLASSFROG) (RUIZ-CARRANZA AND LYNCH, 1995b)

Cochranella posadae (Ruiz-Carranza and Lynch, 1995b): 18. **Holotype.**—ICN 11307 from km 61 on road from Popayán to Inzá, eastern versant of Cordillera Central, 2800 m, 2.56667° N, 76.06667° W, Municipio de Inzá, Departamento de Cauca, Colombia.

Nymphargus posadae Cisneros-Heredia and Mcdiarmid, 2007: 35.

Etymology.—The specific epithet honors Dr. Andrés Posada Arango (1839–1922) for his work in the fields of zoology, botany, education, and conservation biology in Colombia.

Identification.—Figures 6.168, 6.169. These frogs attain a maximum SVL of 34.1 mm in males and 33.3 mm in females. The skin on the dorsum is covered with numerous small warts and some scattered larger warts; there are no visible spicules. The ventral skin is areolate; the ventral surfaces of the thighs below the vent have a pair of enlarged tubercles. There is a white parietal peritoneum covering the anterior 50–60% of the venter. There is a white pericardium; no iridophores are present on the peritonea covering the intestines, stomach, and kidneys; there is a translucent peritoneum around the gall and urinary bladders. The vomers have edentate dentigerous processes. The snout is truncate to round in dorsal aspect and truncate to slightly sloping in lateral profile. The tympanum is almost indistinguishable; it is small when visible, and its diameter is 21.2–26.5% of the eye diameter; the tympanic membrane is not differentiated from the surrounding skin; the supratympanic fold is present. The liver is lobate and covered by a transparent peritoneum. Humeral spines are absent. The webbing between the fingers is absent or greatly reduced. The feet are about two-thirds webbed. The ulnar and tarsal folds are low or absent. The prepollex is not separated from finger I. The nuptial pad is a large to medium-sized excrescence present on the dorsal, lateral, and/or ventral sides of the thumb. Finger II is longer than finger I (finger I length is 92.2–95.7% of finger II). The width of the disc of finger III is about 49.3–52.9% of the eye diameter.

The dorsal surfaces of the head, body, and limbs are bright green with small, scattered, and greenish-white warts. The upper lip is white. The lower lip has a thin white border. The ventrolateral border of the arm, finger IV, the tarsus, and toe V is white. The cloacal region presents numerous small, white warts. A white parietal peritoneum covers about the anterior half of the venter. The dorsal surfaces of the fingers and toes lack melanophores, except for the proximal portion of toes IV and V. The bones are green. The iris is white with thin, dark gray reticulations.

Comparisons.—*Centrolene buckleyi*. Among glassfrogs that inhabit the Amazonian slopes of the Ecuadorian Andes, *Nymphargus posadae* is unique in having a green dorsum with small greenish-white warts, a slightly sloping snout in lateral profile, and by lacking webbing between the fingers. The only species that could be confused with *N. posadae* is *C. buckleyi*, which has humeral spines in adult males (humeral spines absent in *N. posadae*).

Natural history.—Figure 6.170. In Reserva Yanayacu, Provincia Napo, Ecuador, males were found calling on ferns from the upper surfaces of leaves 110–220 cm above a stream.

In Colombia, the species has been observed on vegetation and rocks along creeks.

Tadpoles.—Unknown.

Distribution.—See Figure 6.158. *Nymphargus posadae* is known from the eastern flank of the Central Cordillera of the Andes of Colombia (Caldas, Cauca, and Huila departments), between 1100 and 2800 m, south to the Amazonian slopes of the Cordillera Oriental of the Andes of Ecuador and the Cordillera del Cóndor in northeastern Peru. In Ecuador, it occurs in Sucumbíos, Napo, and Zamora Chinchipe provinces. Its elevational range is 1382–2118 m in an area of extent of occurrence of about 11,260 km^2.

This nocturnal, arboreal species inhabits streams of natural forests in four ecosystems (Table 0.4) in the northern and southern Cordillera Oriental of the Andes biogeographic sectors, where the average annual rainfall is 750–2217 mm and the average annual temperature is 16.1–21.2 °C.

FIGURE 6.170 *Nymphargus posadae.* Top left: Palanda, Zamora Chinchipe. Amanda B. Quezada. Top right and bottom left: embryos from Estación Científica San Francisco, Provincia Zamora Chinchipe. Juan Carlos Sánchez-Nivicela. Bottom right: clutch of embryos from El Chaco, Provincia Napo. Jaime Culebras.

Conservation.—*Nymphargus posadae* is categorized as Least Concern throughout its entire range (IUCN SSC Amphibian Specialist Group, 2018r; in IUCN Red List of 2022) and in Ecuador by Ortega-Andrade *et al.* (2021) because it is widespread, and it is found in several large protected areas in Ecuador. The population is suspected to be decreasing due to ongoing declines in the quality and extent of its habitat. Threats to this species include habitat loss, fragmentation, contamination, and degradation due to the expansion of agriculture, mining activities, and timber extraction. In Ecuador, it has been recorded in the public reserves Parque Nacional Podocarpus and Parque Nacional Antisana, the protected forests of Corazón de Oro and Cuencas que Forman los Ríos: San Francisco, San Ramón y Sabanilla, and some private reserves (e.g., Estación Biológica Yanayacu).

Remarks.—*Nymphargus posadae* and *N. pluvialis* are sister species (Guayasamin *et al.*, 2020b).

Key references.—(Ruiz-Carranza and Lynch, 1995b) (description); (Guayasamin *et al.*, 2006b) (color in life, distribution, habitat); (Cisneros-Heredia and Mcdiarmid, 2007) (distribution); (Cisneros-Heredia *et al.*, 2008b) (distribution); (Guayasamin *et al.*, 2020b) (account).

NYMPHARGUS SIREN (SIREN GLASSFROG) (LYNCH AND DUELLMAN, 1973)

Centrolenella siren (Lynch and Duellman, 1973): 54. **Holotype.**—KU 146610 from a small tributary of the Río Salado, about 1 km upstream from the Río Coca, 1410 m, Provincia Napo, Ecuador.

Nymphargus siren Cisneros-Heredia and Mcdiarmid, 2007: 35.

Etymology.—The specific epithet derives from Greek mythology, where sirens were bird women who, with their sweet singing, enticed seafarers to destruction; the name is used in allusion to the calls of these frogs that entice biologists to the nocturnal perils of streams.

Identification.—Figures 6.171, 6.172. These frogs attain a maximum SVL of 22.6 mm in males and 23.3 mm in females. The skin on the dorsum is shagreened, usually

FIGURE 6.171 *Nymphargus siren*. Left: MZUTI 774, Oyacachi-El Chaco, Provincia Napo. Right: CJ 1198, male, 18.0 mm SVL, 12 km north of El Chaco, Provincia Napo. LAC.

with spicules in males. The ventral surfaces of the thighs below the vent have a pair of enlarged tubercles. There is a white parietal peritoneum covering the anterior half of the venter. There is a white pericardium; there are translucent to opaque peritonea covering the intestines, stomach, and kidneys; there is a translucent peritoneum around the gall and urinary bladders. The liver is lobate and covered by a transparent peritoneum. The vomers have edentate dentigerous process. The snout is usually truncate in dorsal aspect and truncate to slightly protruding in lateral profile. The tympanum is oriented almost vertically with slight lateral and posterior inclinations, and its diameter is about 20.5–30.5% of the eye diameter; the upper fourth of the tympanic annulus is obscured by the supratympanic fold; the tympanic membrane is clearly differentiated from the surrounding skin. The humeral spines are absent. There is no webbing between the inner fingers; there is no or basal webbing between the outer fingers. The feet are about two-thirds webbed. The ulnar and external tarsal folds are absent; the inner tarsal fold is low but evident. The prepollex is not separated from finger I. The nuptial pad is a large to medium-sized excrescence present on the dorsal, lateral, and/or ventral sides of the thumb. Finger II is slightly longer than finger I (finger I length is 91.5–99.6% of finger II). The width of the disc of finger III is about 43.3–58.5% of the eye diameter.

The dorsum is green with small yellow spots, which are narrowly bordered by a black ring in some individuals.

FIGURE 6.172 *Nymphargus siren*. Top: Cloud Forest Organics, Provincia Napo. Jaime Culebras. Middle and bottom: QCAZ 37971, Estación Científica Yanayacu, Provincia Napo. LAC.

The anterior half of the venter is white; the posterior half is transparent. Melanophores are absent on the dorsal surfaces of the fingers and toes, except for few on toe V. The bones are green. The iris is whitish-cream, with a yellow hue around the pupil and fine, dark gray reticulations.

Comparisons.—*Espadarana audax*, *Nymphargus humboldti*, *Rulyrana flavopunctata*, and *Teratohyla midas*. Among glassfrogs that inhabit the Amazonian slopes of the Andes of Ecuador, *N. siren* is differentiated by having a green dorsum with small yellow spots, a partially white venter, and a small size (male SVL < 22.6 mm; female SVL

< 23.3 mm) and by lacking humeral spines. Similar species from eastern Ecuador include *E. audax*, *N. humboldti*, *R. flavopunctata*, and *T. midas*. Differences among these species include the presence of humeral spines in adult males of *E. audax* (humeral spines absent in *N. siren*), moderate webbing between fingers III and IV in *E. audax* and *R. flavopunctata* (webbing absent in *N. siren* and basal in *T. midas*), and the white peritoneum that covers the digestive tract of *T. midas* (digestive tract opaque or translucent in *N. siren*, *E. audax*, and *R. flavopunctata*). *Nymphargus humboldti* and *N. siren* have non-overlapping body sizes (*N. humboldti*, male SVL = 23.3–25.2 mm; female SVL = 24.3–25.9 mm).

Natural history.—Figure 6.173. Relatively large numbers of individuals of *Nymphargus siren*, including calling males, were observed at a tributary of the Río Salado, Provincia Napo, Ecuador, on 7 April 1972 (13 individuals) and again on 18 March 1975 (15 individuals). Many frogs were also noted 3.2 km north–northeast of Oritoyacu, Provincia Napo, on 15 July 1977 (21 individuals). Males call from the upper surfaces of leaves.

The skin shedding of a female took 84 s.

The advertisement call of two individuals from Oyacachi and El Chaco, Provincia Napo, Ecuador, was described as short and composed of a single note. Note mean duration is 24 ms, and notes are generally pulsed. Detailed data and figures are provided under the subtitle "Vocalizations of Ecuadorian Anurans" in Appendix 1.

FIGURE 6.173 Sequence of *Nymphargus siren* shedding its skin. QCAZ 46666, La Bonita, Provincia Sucumbíos. LAC. Top left: 16:22:40 h. Top right: 16:24:04 h.

Tadpoles.—Figure 6.174. This description is based on an individual in stage 25 (not preserved, series QCAZ 50585) from Pacto Sumaco, Provincia Napo, Ecuador. It was collected by Elicio E. Tapia on 30 March 2009. The body is ovoid, slightly widest anteriorly, and wider than it is deep. The snout appears round both in dorsal view and in lateral profile, and the eyes are directed dorsolaterally. The spiracle is sinistral, with the inner wall free from the body wall, and its opening is directed posteriorly on the deadline at about half the body's length. The oral disc is anteroventral, with a narrow upper labium and a more developed lower labium. The tail constitutes about 72% of the total length, with the greatest depth at the midlength and subacute tip. The dorsal fin does not extend onto the body, shares the same depth as the ventral fin, and is about two-thirds as deep as the caudal musculature at the tail's midlength.

FIGURE 6.174 Ontogenetic changes in *Nymphargus siren* from tadpole to froglet. Top to bottom, QCAZ 50585, stage 25, lateral and dorsal views, TL 23.4 mm; stage 41, TL = 43 mm; stage 42, TL = 48.9 mm; stage 45, 275 days after stage 25; froglet 12 days after stage 45. Pacto Sumaco, Provincia Napo. LAC.

The skin is translucent, revealing the internal organs, and the body exhibits a pale pinkish coloration. Melanophores are present dorsally on the body, starting at the snout and extending to the tail, forming a pattern of brown flecks and blotches. Laterally, melanophores aggregate on the posterior half of the body and along the midline of the tail.

At stage 40, the tadpole takes on a uniform light green coloration across the body, tail, and legs, with a pale blue snout. By stage 42, the green becomes more vivid, and a few small white spots appear on the dorsum. The blue coloration migrates to the throat area. In stage 45, the green dorsal coloration remains relatively consistent, while the blue shifts to the scapular region and the groin. As a juvenile, the dorsal surfaces adopt a darker green hue, and the spots on the dorsum become yellow. The scapular region and groins maintain the blue coloration, and a white parietal peritoneum is evident on the ventral side.

Distribution.—See Figure 6.167. *Nymphargus siren* inhabits the Amazonian slopes of the Ecuadorian and Colombian Andes at elevations between 1410 and 2000 m. In Ecuador, it occurs in Sucumbíos, Orellana, and Napo provinces. Its elevational range is 1510–2119 m in an area of extent of occurrence of about 2280 km^2.

This nocturnal, arboreal species inhabits natural forests and intervened areas in the evergreen low montane forest and evergreen montane forest ecosystems of the northern Cordillera Oriental of the Andes biogeographic sector, where the average annual rainfall is 1393–3231 mm and the average annual temperature is 16.1 – 19°C.

Conservation.—*Nymphargus siren* is categorized as Endangered throughout its entire range (IUCN SSC Amphibian Specialist Group, 2022p; in IUCN Red List of 2022) and in Ecuador by Ortega-Andrade *et al.* (2021). There is a suspected population decline of more than 50% over a 10-year period (2015–2025) based on the high rate of habitat loss on the Amazonian slopes of the Ecuadorian and Colombian Andes as a result of expanding agriculture, cattle ranching, logging, and infrastructure development. No data are available on the population demography of the species, nor on its susceptibility to chytridiomycosis, climate change, and/or changes in UV radiation. The population status of the species at localities where it was historically abundant (e.g., Río Salado; 3.2 km north–northeast of Oritoyacu) is unknown. In Ecuador, it has been recorded in the public reserves Parque Nacional Cayambe Coca and Parque Nacional Sumaco Napo-Galeras, the biosphere reserve Sumaco, and some private reserves (e.g., Estación Científica Yanayacu).

Remarks.—*Nymphargus siren*, *N. laurae*, and *N. humboldti* are closely related species. Although Lynch and Duellman (1973) mentioned that some individuals have up to two teeth on the dentigerous process of the vomer, all the specimens examined by Guayasamin *et al.* (2020b) had edentate vomers.

Key references.—(Lynch and Duellman, 1973) (description); (Guayasamin and Funk, 2009) (presence data); (Guayasamin *et al.*, 2020b) (account).

NYMPHARGUS SUCRE (SUCRE'S GLASSFROG) GUAYASAMIN, 2013

Nymphargus sucre Guayasamin, 2013: 194. **Holotype.**—MZUTI 1421 from the creek on the Plan de Milagro–Gualaceo road, 2159 m, Provincia Morona Santiago, Ecuador.

Etymology.—The specific epithet is a noun in apposition and honors Antonio José de Sucre (1795–1830), who, with Simón Bolívar, led the independence of most Andean countries (Bolivia, Colombia, Ecuador, Peru, and Venezuela) from Spain. The epithet also makes reference to the national currency of Ecuador between 1884 and 2000; the sucre disappeared in the year 2000, when it was replaced by the US dollar after a disastrous economic policy that affected millions of Ecuadorians. As an analogy, the current destruction of habitats in southeastern Ecuador (as in many other regions) is likely to drive many species to extinction if activities such as mining, oil extraction, road building, cattle, and agriculture are promoted irresponsibly, and without assessing their effect on megadiverse areas and endangered species.

Identification.—Figure 6.175. These small glassfrogs attain a maximum SVL of 22.3 mm in males and 24.3 mm in a female. The skin on the dorsum is shagreened, with numerous minute spicules in males. The venter is areolate; there is a pair of enlarged tubercles below the level of the vent. There is white parietal peritoneum covering about the anterior half of the venter. There is a white pericardium; there are translucent peritonea covering the intestines, stomach, kidneys, and gall and urinary bladders. The dentigerous process of the vomer is present but lacks teeth. The snout is truncate in lateral and dorsal views. The tympanum is visible without magnification; it is oriented almost vertically with slight lateral and posterior inclinations, and its diameter is about 23–26% of the eye diameter. The upper fourth of the tympanic annulus is obscured by the supratympanic fold; the tympanic membrane is translucent but with melanophores like those on the surrounding skin. The liver is trilobed and covered by a transparent peritoneum. The humeral spines are absent. There is no webbing between the inner fingers; there is basal webbing between fingers III and IV. The feet are about two-thirds webbed. The ulnar and tarsal folds are present but low and inconspicuous and lack pigmentation. The prepollex is not separated from finger I. The nuptial pad is a large to medium-sized excrescence present on the dorsal, lateral, and/or ventral sides of the thumb. Finger I is slightly shorter than finger II (finger I length is 93–96% of finger II). The width of the disc of finger III is about 37–53% of the eye diameter.

The dorsum is brownish-yellow (sometimes with a greenish hue) with small yellow spots. The upper lip is unpigmented. The anterior half of the ventral parietal peritoneum is white; the posterior portion is translucent. The dorsal surfaces of the fingers and toes lack melanophores except for

FIGURE 6.175 *Nymphargus sucre*. Top and middle: MZUTI 1421, holotype, male, 21.6 mm SVL. Bottom: MZUTI 1422, female, 24.3 mm SVL. Plan de Milagro-Gualaceo, Provincia Morona Santiago. Alejandro Arteaga.

few on fingers III and IV and toes III, IV, and V. The color of the bones is unknown. The iris is white with minute dark spots, thin reticulations, and a yellow hue around the pupil.

Comparisons.—*Nymphargus anomalus*, *N. colomai*. *Nymphargus sucre* is distinguished from most glassfrogs by having, in life, a brownish-yellow dorsal surface with yellow spots and lacking webbing between the fingers. *Nymphargus sucre* is most similar to two other Ecuadorian species that lack the typical green dorsal coloration of centrolenids (*N. anomalus*, *N. colomai*). *Nymphargus anomalus* differs from *N. sucre* mainly by having, in life, a pale tan to brown dorsum with black ocelli (ocelli absent in *N. sucre*) and lacking yellow spots. *Nymphargus colomai* differs by having numerous diffuse yellow spots on the dorsum (fewer and well-defined spots in *N. sucre*) and a white iris with a contrasting horizontal black stripe (iris lacking horizontal stripe and having a yellow hue around the pupil in *N. sucre*).

Natural history.—Only one other species of glassfrog (*Nymphargus cariticommatus*) was found at the type locality, a creek on the Plan de Milagro–Gualaceo road, Provincia Morona Santiago, Ecuador. During the night, males were observed calling from the upper surfaces of leaves 90–130 cm above a stream on 6 June 2012.

Tadpoles.—Unknown.

Distribution.—See Figure 6.159. *Nymphargus sucre*, an endemic species to Ecuador, is known from a single locality on the Amazonian slopes of the Cordillera Oriental of the Andes in Provincia Morona Santiago at an elevational range of 2305–2331 m.

This nocturnal, arboreal species inhabits natural forests in the evergreen montane forests of the southern Cordillera Oriental of the Andes biogeographic sector, where the average annual rainfall is 1088–1092 mm and the average annual temperature is 15.6°C.

Conservation.—*Nymphargus sucre* is categorized as Critically Endangered by Ortega-Andrade *et al.* (2021) because its extent of occurrence is estimated to be less than 100 km^2, it only occurs at the type locality, and there is a continuing decline in the extent and quality of its habitat. Furthers surveys may reveal larger areas of occurrence and occupation; however, there are significant threats to this species, such as deforestation, agriculture, livestock, wood extraction, pollution, mining, landslides, and wildfires.

Remarks.—According to Guayasamin *et al.* (2020b), *Nymphargus sucre* is sister to *N. cariticommatus*.

Key references.—(Guayasamin, 2013) (description, habitat); (Guayasamin *et al.*, 2020b) (account).

NYMPHARGUS WILEYI (WILEY'S GLASSFROG) (GUAYASAMIN *ET AL.*, 2006b)

Cochranella wileyi (Guayasamin *et al.*, 2006b): 504. **Holotype.**—QCAZ 26028 from Yanayacu Biological Station, 2100 m, Provincia Napo, Ecuador.

Nymphargus wileyi Cisneros-Heredia and Mcdiarmid, 2007: 35.

Etymology.—The specific epithet is a noun in the genitive case and a patronym for Edward O. Wiley for his influential work on the development of phylogenetic systematics and use of the evolutionary species concept.

Identification.—Figures 6.176, 6.177. These frogs attain a maximum SVL of 26.2 mm in males and 27.1 mm in a female. The skin on the dorsum of males and females is shagreened, with small spicules evident in males. The venter is areolate; the ventral surfaces of the thighs below the vent have a pair of enlarged tubercles. The anterior 40–60% of the ventral parietal peritoneum is white, and the posterior portion is transparent. There is a white pericardium; there is a translucent to opaque peritoneum covering the intestines, stomach, testes, gall bladder, and urinary bladder; the peritoneum around the kidneys is white with unpigmented spots. The vomers lack teeth. The snout is truncate in dorsal aspect and truncate to protruding in lateral profile. The tympanum is oriented almost vertically with slight posterior and lateral inclinations; its diameter is 31.4–37.8% of the eye diameter; the tympanic annulus is visible anteroventrally; the tympanic membrane is unpigmented and clearly differentiated from the surrounding skin. The liver is tetralobed and covered by a transparent peritoneum. The humeral spines are absent. There is no webbing between the inner fingers; there is reduced webbing between the outer fingers. There is extensive webbing between the toes. The prepollex is not separated from finger I. The nuptial pad is a large to medium-sized excrescence present on the dorsal, lateral, and/or ventral sides of the thumb. Finger II is slightly longer than finger I (finger I is 92–97% of finger II). The disc of finger III is of moderate size, about 50–60% of the eye diameter. The ulnar fold is low or absent; the inner tarsal fold is low and thin; the outer tarsal fold is absent.

The dorsal surfaces are pale green and lacks spots. The lower venter is transparent. The parietal peritoneum is white and covers the anterior part of the abdomen (the heart is not visible). The fingers and toes lack melanophores on the dorsal surfaces. The bones are green. The iris is coppery white with black reticulations.

Comparisons.—*Nymphargus cariticommatus*, *Espadarana durrellorum*. *Nymphargus wileyi* differs from most species of glassfrogs by having a uniform green dorsum, white renal peritoneum, and by lacking any webbing between the fingers. On the Amazonian slopes of the Andes, only *N. cariticommatus* can be confused with *N. wileyi*. The differences between these two species are subtle; *N. wileyi* has a uniform dorsum (dorsum with small yellow dots in *N. cariticommatus*) and has a faint layer of iridophores covering the anterior portion of the esophagus (white peritoneum covering most of the esophagus in *N. cariticommatus*). The only other species with a uniform green dorsum on the Amazonian slopes of the Andes is *E. durrellorum*, which differs by having webbing between fingers III and IV and a humeral spine in males.

FIGURE 6.176 *Nymphargus wileyi*. Left: CJ 11458, male, 24.0 mm SVL. Right: CJ 11459, male, 24.4 mm SVL. Estación Científica Yanayacu, Provincia Napo. DAL. Note variation in the length of the peritonea between both individuals.

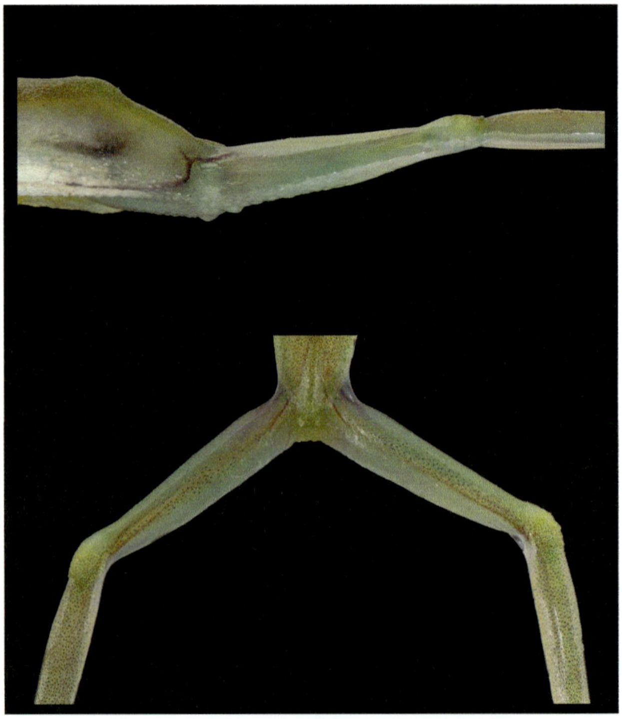

FIGURE 6.177 Morphological details of *Nymphargus wileyi*. CJ 11458, male. Estación Científica Yanayacu, Provincia Napo. LAC.

FIGURE 6.178 Egg clutches of *Nymphargus wileyi*. Estación Científica Yanayacu. Provincia Napo. Martín R. Bustamante.

Natural history.—Figure 6.178. Individuals have been found at night on leaves 120–220 cm above streams (five males) or above the ground (one female).

Males call from the upper surfaces of leaves; males are territorial and are usually found near egg clutches, which are on the tips of leaves or the upper surfaces near streams. The distances between egg clutches and males can be as close as 20 cm. Males were never observed on the same leaf as the egg clutch; also, several egg clutches had no male nearby; these observations suggest that prolonged parental care is absent.

Tadpoles.—Unknown.

Distribution.—See Figure 6.128. *Nymphargus wileyi*, an endemic species to Ecuador, is known on the Amazonian slopes of the Cordillera Oriental of the Andes in Provincia Napo. Its elevational range is 1519–2107 m in an area of extent of occurrence of about 40 km^2.

This nocturnal, arboreal species inhabits primary forests in the evergreen low montane forest and the evergreen montane forest ecosystems of the northern Cordillera Oriental of the Andes biogeographic sector, where the average annual rainfall is 2216–2740 mm and the average annual temperature is 16.1 – 18.7°C.

Conservation.—*Nymphargus wileyi* was categorized as Data Deficient by Ortega-Andrade *et al.* (2021), since there is still very little known about its extent of occurrence, area of occupancy, status, and ecological requirements. The species is only known from the type series, comprised of six specimens, which were collected during three years of inventory work at Estación Científica Yanayacu, suggesting that *N. wileyi* is a rare species. Habitat loss is a conspicuous threat for the species. It is thought to have a restricted range, although further field work is needed to test this. It has been recorded in the public reserves Parque Nacional Antisana and Reserva Biológica Colonso Chalupas, the biosphere reserve Sumaco, and some private reserves (e.g., Estación Científica Yanayacu).

Remarks.—Only one discrete morphological character separates *Nymphargus wileyi* from *N. cariticommatus*, the absence of pale spots in *N. wileyi*. Also, in preserved material, the esophageal peritoneum of *N. cariticommatus* is more pronounced and extensive than in *N. wileyi*, in which the iridophores are restricted to the anterior portion of the esophageal peritoneum. Surprisingly, the two species are not sister species to each other.

Key references.—(Guayasamin *et al.*, 2006b) (description); (Guayasamin *et al.*, 2020b) (account).

GENUS *RULYRANA* (GLASSFROGS) GUAYASAMIN *ET AL.*, 2009

Rulyrana Guayasamin *et al.*, 2009: 34. **Type species.**—*Centrolenella flavopunctata* (Lynch and Duellman, 1973), by original designation.

Etymology.—The generic name is in honor of Pedro Ruiz-Carranza and John D. Lynch, who have contributed enormously to the understanding of anuran diversity, biology, and evolution. The generic name *Rulyrana* comes from an arbitrary association of the two first letters of Ruiz and Lynch (Ruly) and the word *rana* (= frog). *Ruly* also is the nickname of Juan Manuel Guayasamin's good friend and colleague Martín R. Bustamante, whose work on biodiversity communication, animal conservation, and education is remarkable. The gender is feminine.

Definition.—Figure 6.179. The following characteristics are diagnostic of the genus *Rulyrana*: (1) humeral spines absent; (2) lobed liver covered by a transparent hepatic

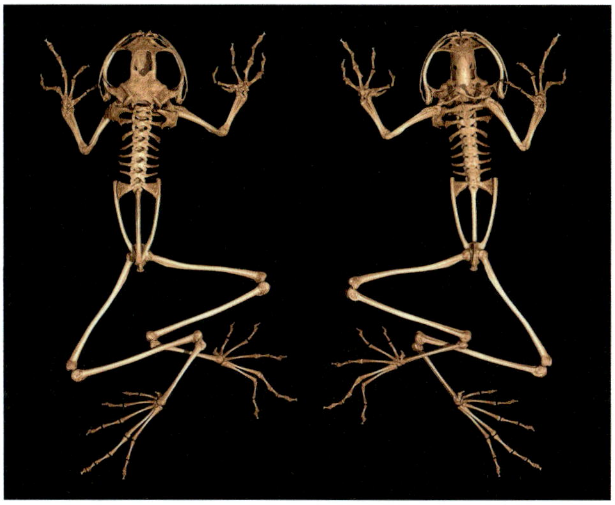

FIGURE 6.179 High-resolution computed tomography reconstructions of the skeleton of *Rulyrana flavopunctata*, CJ 4563, male, 22.4 mm SVL, Río Pachimbi, buffer zone of Reserva Ecológica Colonso Chalupas, Provincia Napo. Dorsal (left) and ventral (right) views. Daniel J. Paluh.

peritoneum, digestive tract translucent; (3) ventral parietal peritoneum white anteriorly and transparent posteriorly; (4) moderate to extensive webbing between fingers III and IV; (5) bones green in life; (6) dorsum lavender in preservative, with or without spots; (7) dentigerous process of the vomer present, vomerine teeth usually present; and (8) males that call from the upper surfaces of leaves or rocks, females that deposit eggs on the upper sides of leaves or rocks.

Content.—*Rulyrana* contains six species, two of which occur in Ecuador (December 2022).

Distribution.—*Rulyrana* occurs on the Amazonian slopes of the Andes in Ecuador and Peru, the western slopes of the Cordillera Oriental of the Andes in Colombia, and the eastern slopes of the Cordillera Central of the Andes in Colombia.

Remarks.—Figures 6.180, 6.181. There are no phenotypic differences between the genera *Rulyrana* and *Sachatamia*. Therefore, generic allocation at this level requires molecular data.

The phylogeny of species presented herein follows (Guayasamin *et al.*, 2020b).

RULYRANA FLAVOPUNCTATA (YELLOW-SPOTTED GLASSFROG) (LYNCH AND DUELLMAN, 1973)

Centrolenella flavopunctata (Lynch and Duellman, 1973): 25. **Holotype.**—KU 121048 from Mera, Provincia Pastaza, Ecuador.

Rulyrana flavopunctata Guayasamin *et al.*, 2009: 34.

Etymology.—The specific epithet is a combination of the Latin *flavus*, meaning golden yellow, and *punctatus*, meaning dotted, and is used in reference to the dorsal coloration of the species.

Identification.—Figures 6.182, 6.183, 6.184. These frogs attain a maximum SVL of 23.9 mm in males and 27.4 mm in females. The skin on the dorsum is shagreen; the males have spicules. The skin of the venter is areolate; the ventral surfaces of the thighs below the vent have a pair of slightly enlarged tubercles. The anterior half of the ventral parietal peritoneum is white, and the posterior half is translucent. There is a white pericardium; there are peritonea covering the intestines, stomach, and kidneys lacking iridophores. The urinary bladder is transparent. The teeth on the vomers are present or absent; each vomer has 0–3 teeth. The snout is usually rounded in dorsal and lateral views. The tympanum is visible, and its diameter is 27.9–38.7% of the eye diameter; the supratympanic fold is low. The liver is lobed and lacks iridophores. The humeral spines are absent. There is absent or basal webbing between fingers I and II; there is moderate webbing between fingers II and III; there is extensive webbing between fingers III and IV. The feet are about three-fourths webbed. The ulnar fold is low or absent; the inner tarsal fold is low and short; the outer tarsal fold is absent. The prepollex is not separated from finger I. The nuptial pad is a large to medium-sized excrescence present on the dorsal, lateral, and/or ventral sides of the thumb. Finger I is about the same length as finger II (finger I length is 95.5–104.0% of finger II). The disc of finger III is relatively large, and its width is 47.7–67.0% of the eye diameter.

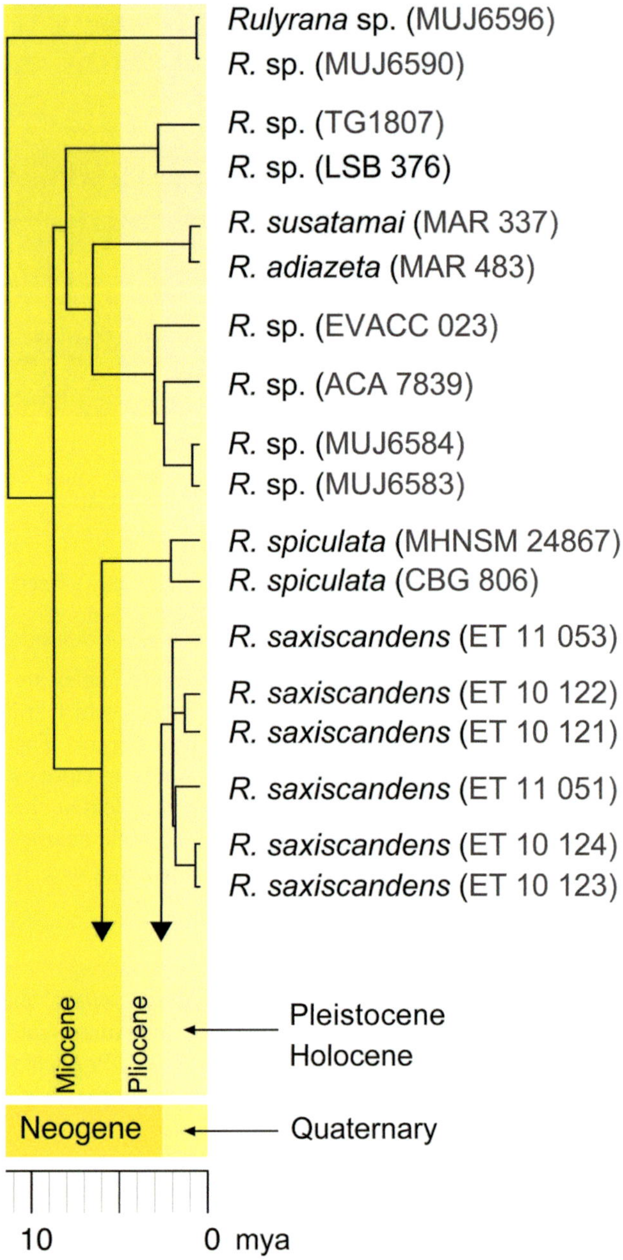

FIGURE 6.180 Time-calibrated maximum likelihood phylogenetic tree depicting evolutionary relationships of *Rulyrana*, inferred from DNA sequences of mitochondrial and nuclear genes. Phylogeny taken and redrawn from Guayasamin *et al.* (2020b).

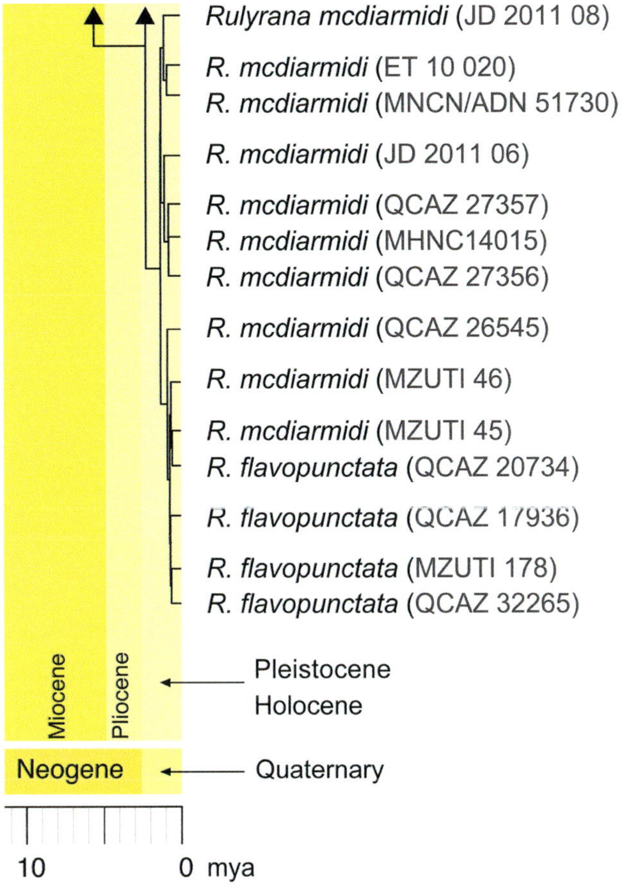

FIGURE 6.181 Continued. Time calibrated maximum likelihood phylogenetic tree depicting evolutionary relationships of *Rulyrana*, inferred from DNA sequences of mitochondrial and nuclear genes. Phylogeny taken and redrawn from Guayasamin *et al.* (2020b).

FIGURE 6.182 *Rulyrana flavopunctata*. Left: CJ 11466, male, 23 mm SVL. Right: CJ 11465, male, 25 mm SVL. Narupayaku, Provincia Orellana. LAC.

FIGURE 6.183 Morphological variation in *Rulyrana flavopunctata*. Top left: CJ 4563, male. Bottom left: CJ 4565, male, 21.8 mm SVL. Río Pachimbi, Provincia Napo. LAC. Top right: CJ 11467, male. Bottom right: CJ 11457, male. Narupayaku, Provincia Orellana. LAC.

Conspicuous variation has been observed in the dorsal and ventral coloration.

The dorsal surfaces of the head and body vary from green to very dark green, with small, well-defined yellow spots. The gular zone is translucent, contrasting with the yellow venter and white peritoneum. The dorsal surfaces of the fingers and toes usually have melanophores, but some individuals lack melanophores on finger I or on fingers I and II. The bones are green. At one locality (Río Tayuntza), the individuals showed considerable variation in the dorsal coloration, from green to almost black, with yellow spots, indicating that intraspecific variation in this species is not necessarily associated with geography. The iris is pale grayish-white to brown, with or without yellow to golden tint, and dark gray or brown flecks or fine reticulations.

Comparisons.—*Espadarana audax*, *Nymphargus siren*, *N. humboldti*, *N. mariae*, *Rulyrana mcdiarmidi*, and *Teratohyla midas*. Among glassfrogs that inhabit the Amazonian slopes of the Ecuadorian Andes, *R. flavopunctata* is distinguished by having a green to dark green dorsum with small yellow spots, a partially white venter, a small to moderate body size (male SVL 21.0–23.9 mm; female SVL 24.0–27.4 mm), and extensive webbing between fingers III and IV and by lacking humeral spines. Species with a similar color pattern from eastern Ecuador include *E. audax*, *N. siren*, *N. humboldti*, *N. mariae*, *R. mcdiarmidi*, and *T. midas*. None of the species in the genus *Nymphargus* have webbing

FIGURE 6.184 Morphological details of *Rulyrana flavopunctata*. Top left: QCAZ 37911, Centro Ecológico Zanja Arajuno, Provincia Pastaza. LAC. Top right: CJ 4563, male. Río Pachimbi, Provincia Napo. LAC. Bottom: CJ 11457, male, Narupayaku, Provincia Orellana. DAL.

FIGURE 6.185 *Rulyrana flavopunctata*. Top and bottom right: Río Pachimbi, Provincia Napo. Bottom left: QCAZ 37911, Centro Ecológico Zanja Arajuno, Provincia Pastaza. LAC.

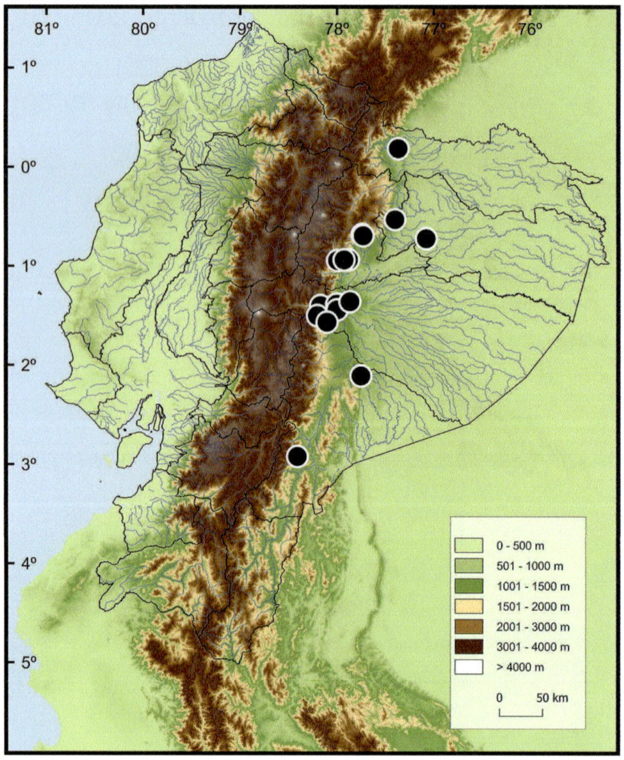

FIGURE 6.186 Localities of known occurrence of *Rulyrana flavopunctata* in Ecuador.

between fingers III and IV; males of *E. audax* have humeral spines; *T. midas* has very few yellow spots on the upper flanks and a white digestive tract; and *R. mcdiarmidi* has a slightly larger body size (male SVL 22.8–26.9 mm, female SVL 25.4–30.2 mm) and warts on its dorsum (absent in *R. flavopunctata*).

Natural history.—Figure 6.185. Most individuals were found in a deep ravine, where males were perched on small herbs in the spray zone of a small waterfall.

The males call from the upper side of leaves or rocks. At Río Tayuntza, Provincia Morona Santiago, egg clutches were found on rock walls during the rainy season.

The advertisement call of two individuals from Quebrada Pangayaku, Provincia Napo, Ecuador, was described as short and composed of a single note. Note mean duration is 4 ms, and notes are clearly pulsed. Detailed data and figures are provided under the subtitle "Vocalizations of Ecuadorian Anurans" in Appendix 1.

Tadpoles.—Unknown.

Distribution.—Figure 6.186. *Rulyrana flavopunctata* is known from the Amazonian slopes of the Andes of Colombia and Ecuador at elevations between 300 and 1715 m. In Ecuador, it occurs in Sucumbíos, Orellana, Napo, Tungurahua, Pastaza, and Morona Santiago provinces. Its elevational range is 311–1949 m in an area of extent of occurrence of about 19,420 km^2.

This nocturnal, arboreal species inhabits natural forests and intervened areas in four ecosystems (Table 0.4) within four biogeographic sectors (Table 0.5), where the average annual rainfall is 1780–4088 mm and the average annual temperature is 17.1–24.6 °C.

Conservation.—*Rulyrana flavopunctata* is categorized as Least Concern throughout its entire range (IUCN SSC Amphibian Specialist Group, 2022q; in IUCN Red List of 2022) in view of its wide distribution, presumed large population, and because it is unlikely to be declining fast enough to qualify for listing in a more threatened category. However, it is important to note that in Ecuador, *R. flavopunctata* has been categorized as Vulnerable by Ortega-Andrade *et al.* (2021). The latter infer a reduction in the population greater than or equal to 30% of the population in the next 100 years and a reduction in the extent of occurrence, which is estimated to be less than 20,000 km^2 (19,420 km^2, according to our calculations); its distribution is severely fragmented in fewer than 10 known localities, and there is a continuing decline in the extent and quality of its habitat. In Ecuador, there are no studies on populations of this species. The major threats to its existence are habitat reduction and fragmentation, oil spills, deforestation, agriculture, fires, landslides, and volcanic eruptions. In Colombia, it is known in two protected areas, Los Picachos Nature Reserve and Cueva de los Guácharos Nature Reserve. In Ecuador, it has been recorded in the public reserves Reserva Biológica Colonso Chalupas and Parque Nacional Sangay; the protected forests of Cerro Sumaco y Cuenca Alta del Río Suno, Cordillera Kutukú y Shaimi, and Comunidad Shuar Wisui; the biosphere reserve Sumaco; and some private reserves (e.g., Centro Ecológico Zanjarajuno).

Remarks.—Morphological differences between *Rulyrana flavopunctata* and *R. mcdiarmidi* are subtle (body size, skin texture, color pattern), and the two species are genetically indistinguishable. Therefore, it is possible that *R. mcdiarmidi* represents a geographic variation in *R. flavopunctata* or that the two species have recently diverged. Information on call variation and other traits (ecological requirements, morphology) are necessary to resolve the species status of *R. mcdiarmidi*.

Key references.—(Lynch and Duellman, 1973) (description); (Ruiz-Carranza *et al*., 1996) (distribution); (Castro *et al*., 2004a) (distribution); (Jimenez-Robles *et al*., 2015) (reproductive biology); (Guayasamin *et al*., 2020b) (account, call).

RULYRANA MCDIARMIDI (MCDIARMID'S GLASSFROG) (CISNEROS-HEREDIA *ET AL*., 2008b)

Cochranella mcdiarmidi (Cisneros-Heredia *et al*., 2008b): 342. **Holotype.**—DFCH-USFQ D132 from a small rivulet tributary of the Río Jambue, ca. 6 km south of Zamora, 1150 m, Provincia Zamora Chinchipe, Ecuador.
Rulyrana mcdiarmidi Guayasamin *et al*., 2009: 34.

Etymology.—The specific epithet *mcdiarmidi* honors Dr. Roy McDiarmid, in recognition of his contributions to the understanding of Neotropical herpetology, including reference studies on biological monitoring, tadpoles, and snakes.

Identification.—Figures 6.187, 6.188. These frogs attain a maximum SVL of 26.9 mm in males and 30.2 mm in females. The skin on the dorsum is smooth to shagreen with numerous warts. The ventral skin is coarsely granular; the subcloacal area is coarsely granular with abundant low and flat warts; the other cloacal ornamentation is absent. The ventral parietal peritoneum is white; there are iridophores entirely or almost entirely covering the abdomen to the level of the groin. There is a white pericardium; all the other visceral peritonea are clear. The vomerine teeth are present. The snout is rounded to subtruncate in dorsal and lateral views. The tympanic annulus is evident and oriented dorsolaterally; the supratympanic fold is weak. The liver is lobed and lacks iridophores. The humeral spine is absent. There is no webbing between fingers I and II; there is basal webbing between II and III; there is moderate to extensive webbing between the outer fingers. There is extensive webbing on the feet. The ulnar fold is thick, non-enameled, and non-crenulated; the inner tarsal fold is low and short. The prepollex is concealed. The nuptial pad is a large to medium-sized excrescence present on the dorsal, lateral, and/or ventral sides of the thumb. Finger I is slightly shorter than II. The eye diameter is larger than the width of disc on finger III.

FIGURE 6.187 Morphological variation in *Rulyrana mcdiarmidi*. Top: QCAZ 37915, male. LAC Middle left: QCAZ 27356. Middle right: QCAZ 41937. Bottom left: QCAZ41997. Bottom right: QCAZ27357. Quebrada Río Napinaza, Limón Indaza, Provincia Morona Santiago. LAC.

The dorsum is dark olive green with darker suffusions and yellowish-cream, orange, or light green spots. The males present smaller dorsal spots than the females. The venter is yellowish-cream. There are abundant melanophores widespread on all fingers and toes. The bones are green. The iris is olive brown to brown with fine dark reticulations.

Comparisons.—*Espadarana audax*, *Nymphargus siren*, *N. humboldti*, *N. mariae*, *Rulyrana flavopunctata*, and *Teratohyla midas*. *Rulyrana mcdiarmidi* is differentiated from other centrolenid species by having a moderate-sized body (male SVL 22.8–26.9 mm; female SVL 25.4–30.2 mm),

FIGURE 6.188 Morphological details of the head of *Rulyrana mcdiarmidi*. Top: QCAZ 41997. Bottom: QCAZ 41937. Quebrada Río Napinaza, Limón Indaza, Provincia Morona Santiago. LAC.

green dorsum with diverse darker shadows and pale yellow to green spots, dorsal skin with warts, the ventral parietal peritoneum completely white or with a translucent posterior portion, thick ulnar folds, and extensive hand and foot webbing. Among glassfrogs on the Amazonian slopes of the Ecuadorian Andes, species with a similar color pattern include *E. audax*, *N. siren*, *N. humboldti*, *N. mariae*, *R. flavopunctata*, and *T. midas*. None of the species in the genus *Nymphargus* have webbing between fingers III and IV; males of *E. audax* have humeral spines; *T. midas* has very few yellow spots on the upper flanks and a white digestive tract; *R. flavopunctata* differs by being slightly smaller (male SVL 21.0–23.9 mm; female SVL 24.0–27.4 mm), having yellow dorsal spots, and lacking dorsal warts.

Natural history.—Figure 6.189. Males call from rocks along streams and in the spray zone of small waterfalls; clutches are deposited on rocks in the same spray zone.

Adults were observed inside Cueva de los Tayos (78.6551 °W, 04.3556 °S, 965 m), Provincia Zamora Chinchipe, approximately 15 m from the main entrance. Within the cave an amblypygid, *Heterophynus* sp. (Phrynidae), was observed preying on an adult frog at around 11:00 h (Elicio E. Tapia and Nadine Dupérré, field notes of 1 August 2024).

FIGURE 6.189 *Rulyrana mcdiarmidi*. Individuals and clutch of embryos from La Canela, Provincia Zamora Chinchipe. Amanda B. Quezada.

Tadpoles.—Unknown.

Distribution.—Figure 6.190. *Rulyrana mcdiarmidi* is known on the Amazonian slopes of Cordillera Oriental of the Andes in southeastern Ecuador and northeastern Peru (Departamento de Cajamarca) at elevations between 1150 and 1500 m. In Ecuador, it occurs in Morona Santiago and Zamora Chinchipe provinces. Its elevational range is 508–1711 m in an area of extent of occurrence of about 23,810 km^2.

This nocturnal, arboreal species inhabits natural forests and intervened areas in six ecosystems (Table 0.4) within three biogeographic sectors (Table 0.5), where the average annual rainfall is 1049–2394 mm and the average annual temperature is 19–24.6 °C.

Conservation.—*Rulyrana mcdiarmidi* is categorized as Near Threatened throughout its entire range (IUCN SSC Amphibian Specialist Group, 2022r; in IUCN Red List of 2022) because of a population decline, suspected to be occurring at a rate of 20–25% over a 10-year period (2015–2025), due to widespread habitat loss throughout much of its range as a result of industrial agriculture, logging, and mining activities. However, it is important to note that in Ecuador, *R. mcdiarmidi* has been categorized as Vulnerable by Ortega-Andrade *et al.* (2021). The latter infer a reduction in the population greater than or equal to 30% of the population in the next 100 years and a reduction in the extent of occurrence, which is estimated to be less than 20,000 km^2 (23,810 km^2 according to our calculations); its distribution is severely fragmented in fewer than 10 known localities, and there is a continuing decline in the extent and quality of its habitat. In Ecuador, there are no studies on populations

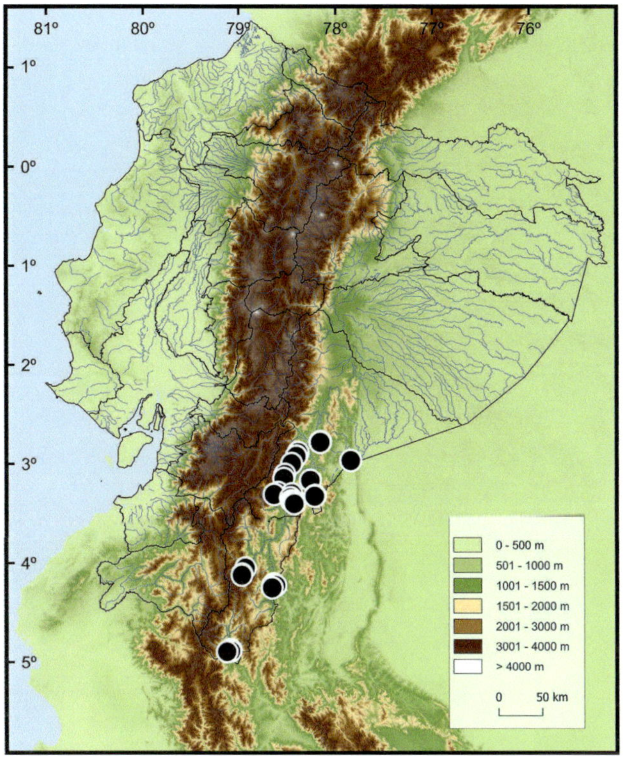

FIGURE 6.190 Localities of known occurrence of *Rulyrana mcdiarmidi* in Ecuador.

of this species. Its threats are habitat reduction and fragmentation, deforestation, agriculture, mining, landslides, and fires. It has been recorded in the public reserve Parque Nacional Podocarpus, the protected forests of Cordillera Kutukú y Shaimi and Pueblo Shuar Arutam, and the biosphere reserve Podocarpus-El Cóndor.

Remarks.—See Remarks under *Rulyrana flavopunctata*.

Key references.—(Cisneros-Heredia *et al.*, 2008b) (description); (Guayasamin *et al.*, 2020b) (account).

GENUS *SACHATAMIA* (GLASSFROGS) GUAYASAMIN *ET AL.*, 2009

Sachatamia Guayasamin *et al.*, 2009: 36. **Type species.**—*Centrolenella albomaculata* (Taylor, 1949) by original designation.

Etymology.—The generic name comes from the Quichua words *sacha*, meaning forest, and *tamia*, meaning rain, and refers to the tropical rainforest occupied by the species in this genus. *Sachatamia* is feminine in gender.

Definition.—Figure 6.191. The following combination of characteristics is diagnostic of the genus *Sachatamia*: (1) humeral spines present (*S. ilex*) or absent (*S. albomaculata*, *S. punctulate*); (2) lobed liver covered by a transparent hepatic peritoneum, translucent digestive tract; (3) ventral

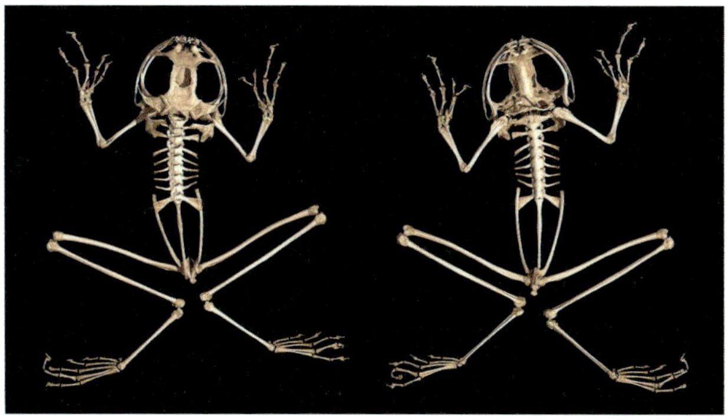

FIGURE 6.191 High-resolution computed tomography reconstructions of the skeleton of *Sachatamia ilex*, KU 221613, 30.1 mm SVL, Puerto Quito, Provincia Pichincha. Dorsal (left) and ventral (right) views. Daniel J. Paluh.

parietal peritoneum white anteriorly and transparent posteriorly; (4) moderate to extensive webbing between fingers III and IV; (5) green bones in life; (6) lavender dorsum in preservative, with or without spots; (7) dentigerous process of the vomer present, bearing teeth; (8) males that call from the upper surfaces of leaves or rocks, females that deposit pigmented eggs on the upper sides of leaves or rocks; and (9) quadratojugal articulating with the maxilla. We note that there are no discrete differences between the genera *Sachatamia* and *Rulyrana*; geographically, however, *Rulyrana* is found on the central and Amazonian slopes of the Andes, whereas *Sachatamia* occurs in Central America, the South American Chocó, and the central Andes. DNA data are necessary to unambiguously allocate species in one of these two lineages.

Content.—*Sachatamia* contains five species, two of which occur in Ecuador (31 December 2022).

Distribution.—*Sachatamia* occurs in Central and South America, from Honduras through Nicaragua, Costa Rica, and Panama to Colombia and northwestern Ecuador.

Remarks.—Figure 6.192. There are no phenotypic differences between the genera *Rulyrana* and *Sachatamia*. Therefore, generic allocation at this level requires molecular data. The phylogeny of species presented herein follows (Guayasamin *et al.*, 2020b).

SACHATAMIA ALBOMACULATA (WHITE-SPOTTED GLASSFROG) (TAYLOR, 1949)

Centrolenella albomaculata (Taylor, 1949): 267. **Holotype.**—KU 23814 from Los Diamantes, one mile south of Guápiles, Cantón de Pococí, Provincia Limón, Costa Rica.

Sachatamia albomaculata Guayasamin *et al.*, 2009: 36.

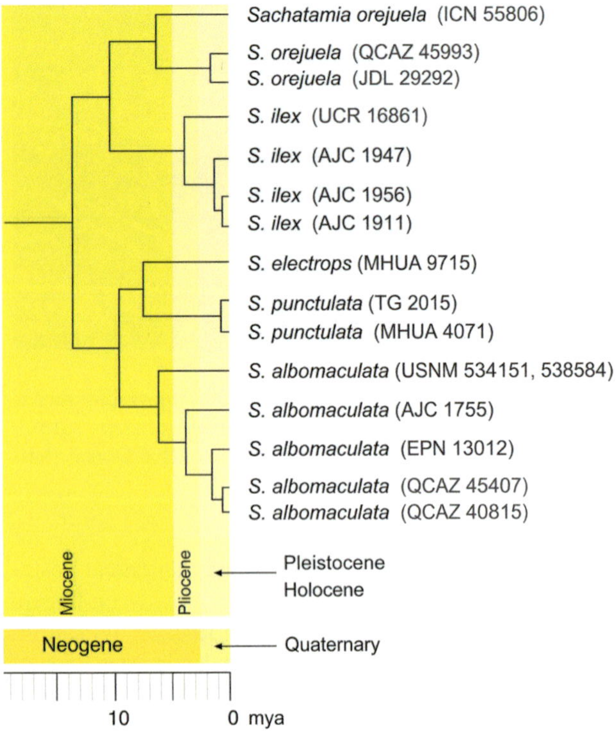

FIGURE 6.192 Time-calibrated maximum likelihood phylogenetic tree depicting evolutionary relationships of *Sachatamia*, inferred from DNA sequences of mitochondrial and nuclear genes. Phylogeny taken and redrawn from Guayasamin *et al.* (2020b).

FIGURE 6.193 *Sachatamia albomaculata*. Left: CJ 12511, female, 26 mm SVL. Right: CJ 12512, male, 23.4 mm SVL. 5 km east of Puerto Quito, Provincia Pichincha. SGS.

Etymology.—The specific epithet is a combination of the Latin words *albus*, meaning white, and *macula*, meaning spot, stain, mark, and refers to the dorsal color pattern of the species.

Identification.—Figures 6.193, 6.194, 6.195. These frogs attain a maximum SVL of 24.7 mm in males and 29.0 in females. The skin on the dorsum of males and females is shagreen; spicules are present in males. The skin on the venter is areolate; the ventral surfaces of the thighs below the vent have a pair of enlarged tubercles, which are more evident in females. The anterior half of the ventral parietal peritoneum is covered by white iridophores; the posterior half is transparent. There is a white pericardium; there is a translucent peritoneum covering the intestines, stomach, testes, kidneys, gall bladder, and urinary bladder. The teeth on the dentigerous process of the vomer are present; each process bears 4–6 teeth. The snout is round to truncate in dorsal profile and mostly truncate in lateral profile. The tympanum is evident, its diameter is 25.7–34.2% of the eye diameter; the dorsal border of the tympanic annulus is covered by a supratympanic fold; the tympanic membrane is clearly differentiated from the surrounding skin. The liver is tetralobed, where two large ventral lobes cover two smaller lobes; the hepatic peritoneum is transparent. The humeral spines are absent. There is no webbing between fingers I and II, basal webbing between fingers II and III,

FIGURE 6.194 Morphological variation in *Sachatamia albomaculata*. Top left: CJ 12514. Top right: CJ 12515. Middle left: CJ 12516. Middle right: CJ 12517. 5 km east of Puerto Quito, Provincia Pichincha. SGS. Bottom left: QCAZ 45409. Bottom right: QCAZ 45408. Durango, Provincia Esmeraldas. LAC.

FIGURE 6.195 Morphological details of *Sachatamia albomaculata*. Top left: CJ 12931, La Florida, Provincia Santo Domingo de los Tsáchilas. LAC. Top right and bottom: CJ 12511, 5 km east of Puerto Quito, Provincia Pichincha. SGS.

FIGURE 6.196 *Sachatamia albomaculata*. Top left: calling male, Reserva Manduriacu, Provincia Imbabura. Jorge Brito Molina. Top right and middle left: calling male and amplectant pair from Reserva Tayra, Provincia Pichincha. Jaime Culebras. Middle right and bottom left: amplectant pair and clutches of embryos, Lita, Alto Tambo, Provincia Esmeraldas. LAC. Bottom right: lab-born clutch, parents from Reserva Biológica Bilsa, Provincia Esmeraldas.

and moderate webbing between fingers III and IV. There is extensive foot webbing. The ulnar and tarsal folds are low. The prepollex is not separated from finger I. The nuptial pad is a large to medium-sized excrescence present on the dorsal, lateral, and/or ventral sides of the thumb. Finger I is about the same length as finger II (finger I is 96.5–103.9% the length of finger II). The disc of finger III is of moderate size, 41.9–45.0% of the eye diameter.

In populations from Central America, there are two clearly different dorsal coloration patterns; one has a green dorsum and small yellow spots, and the other has small and large yellow spots. The holotype of the species, as well as Ecuadorian populations, exhibit the latter coloration pattern. The venter is white on the anterior half and transparent posteriorly. There are usually melanophores on the dorsal surfaces of the outer fingers and toes. The bones are green. The iris varies from white to yellow with dark reticulations.

Comparisons.—*Sachatamia albomaculata* is the only glassfrog on the Pacific versant of the Andes that has, in life, a green dorsum with minute and large yellow spots, relatively extensive webbing between fingers III and IV, and no humeral spines.

Natural history.—Figure 6.196. Males typically call from the upper surfaces of leaves but have also been seen calling from plants growing on the trunks of fallen trees crossing streams, from vegetation and branches growing along rocky riparian walls, from rocky surfaces along streams, and in the spray zone of waterfalls. Females deposit eggs on the upper and undersides of leaves. At the Reserve of the Universidad de la Paz, Costa Rica, during the wet season, males and egg clutches were found throughout the stream channel, and females deposited their eggs on the upper surfaces of leaves hanging over the water. At the beginning of the dry seasons (December 2001 and 2002), they noted a shift in the reproductive behavior and oviposition sites of this species. A large aggregation of males and egg clutches (n = 53; December 2002) was found on rocks in the splash area of two waterfalls, and calling males and two egg clutches were also seen on a large boulder in the Río Jaris. These observations suggested that in some areas, *Sachatamia albomaculata* uses wet splash zones to prolong breeding during part of the dry season. Females provide short-term parental care; prolonged parental care is absent.

Reproductive biology data of pairs (wild and lab born) from Reserva Biológica Bilsa, Provincia Esmeraldas, have been obtained in the laboratory (CJ and Wikiri lab notes by Alex Achig, August 2021–January 2022). Amplexus is axillary and usually lasts two days (n = 5). Six oviposition events occurred at night. Clutches contain 41–69 eggs ($\bar{x}$ = 61, n = 6), light brown (1), brown (4), or black (1) colored, of 2.0–2.71 mm in diameter ($\bar{x}$ = 2.34, n = 40). Clutches vary in shape, between ovoid and triangular–ovoid. The maximum diameter of the clutches varies between 25.7 and 38.4 cm ($\bar{x}$ = 33.5, n = 5). In the laboratory, most of the clutches were deposited on the undersides of leaves, except one that was over one leaf, 2–10 cm above the water pool of the terrarium. Eclosion requires 12–21 days ($\bar{x}$ = 14.5, n = 6) following egg deposition. Tadpoles undergo complete development and metamorphosis in a span of 119–133 days (n = 2), at water temperature of about 21.2°C and a pH of 7.6. The development of individuals from froglet to adult from the same clutch took around 136 days.

The call consists of a rapid high-pitched single note, "tik;" the average note duration of the call is 0.001–0.002 s; the dominant frequency is 6.1–7.1 kHz.

Tadpoles.—Figure 6.197. This description is based on a lab-born individual (not preserved) from parents collected in Comunidad Mashpi, Provincia Pichincha, Ecuador, in stage 26 (29 April 2021). Parents were collected by Elicio E. Tapia. It measures 5.7 mm in body length and 20.4 mm in total length. The body is ovoid and flattened, slightly widest anteriorly and wider than it is deep. The snout appears round in dorsal view and subacute in lateral profile, with dorsally directed eyes. The spiracle is sinistral, with the inner wall free from body. Its opening is directed posterolaterally on the deadline at about two-thirds of the body's length. The oral disc is anteroventral, of medium size, transversely elliptical, and not emarginate. The upper labium is poorly developed, while the lower labium is more developed and free from the body wall. The upper jaw sheath is completely keratinized, moderately slender, serrated, and smoothly curved. The lower jaw is V-shaped, and the labial tooth row formula is 1–2/3. The tail constitutes about 72% of the total length, with relatively straight edges on the fins, maintaining a uniform depth along the entire medial area of the tail. The dorsal fin does not extend onto the body (it starts at the end of the first quarter of the tail), is equally as deep as the ventral fin, and is approximately half the depth of the caudal musculature at the tail's midlength.

The body and tail are brown on the dorsal and lateral surfaces, with aggregations of melanophores in certain areas (e.g., the posterolateral area of the body), creating darker blotches. Ventrally, melanophores are more sparsely distributed, leaving many areas without pigmentation, thereby allowing the translucent skin to reveal the internal organs, resulting in a ventral red appearance. On the tail, melanophores are situated just in the middle of the caudal musculature, forming a brown stripe along the central axis.

At stage 40, the tadpole displays an overall brown coloration with whitish dots and spots, except for the light green legs. In stage 42, which is 14 days later, the brown coloration takes on an olive-green tint on the dorsum, and the white spots become more prominent. The extremities are light green. By stage 46, the froglet exhibits a dark green dorsum with brown patches. Numerous white spots are present on all the dorsal surfaces, and the extremities are green. A month after tail reabsorption, the frog maintains the green and brown dorsal coloration, which now extends to the extremities. The yellow spots are larger and are surrounded by a light green halo, with white dots visible on the dorsal surfaces. After three months, the brown coloration has largely disappeared, leaving a dark green background with yellow spots and white dots. Seven months after tail reabsorption, the frog acquires the typical adult coloration.

Distribution.—Figure 6.198. *Sachatamia albomaculata* is known in humid lowlands and premontane slopes from

FIGURE 6.197 Ontogenetic changes in two individuals of *Sachatamia albomaculata* from tadpole to adult. Top to bottom, left to right. CJ 11463a, stage 26, lateral, dorsal, and ventral views, TL = 20.4 mm. CJ 11463b, undetermined stage, TL = 20.7 mm; stage 42, TL = 21 mm; stage 46, 10.9 mm SVL; froglet 34 days after stage 46 = 12.4 mm SVL; juvenile 101 days = 17.2 mm SVL; juvenile 240 days = 21.8 mm SVL. Comunidad Mashpi, Provincia Pichincha. SGS.

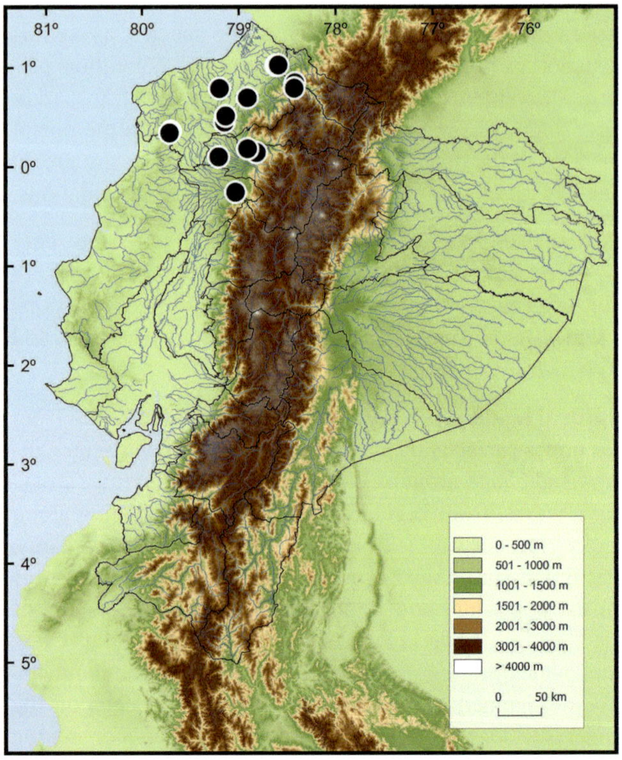

FIGURE 6.198 Localities of known occurrence of *Sachatamia albomaculata* in Ecuador.

north–central Honduras to western Colombia and Ecuador at elevations between 20 and 1500 m. In Ecuador, it occurs in Imbabura, Esmeraldas, Manabí, Pichincha, and Santo Domingo de los Tsáchilas provinces. Its elevational range is 61–1620 m in an area of extent of occurrence of about 9930 km^2.

This nocturnal, arboreal species inhabits natural forests and intervened areas in four ecosystems within three biogeographic sectors, where the average annual rainfall is 1518–2412 mm and the average annual temperature is 18.2 – 25.4°C.

Conservation.—*Sachatamia albomaculata* is categorized as Least Concern throughout its entire range (IUCN SSC Amphibian Specialist Group, 2020i; in IUCN Red List of 2022) in view of its wide distribution, presumed large population, and because it is unlikely to be declining to qualify for listing in a more threatened category (but see Remarks). However, it is important to note that in Ecuador, *S. albomaculata* is categorized as Vulnerable by Ortega-Andrade *et al.* (2021). The latter infer a reduction in the population greater than or equal to 50% of the population in the next 100 years and a reduction in the extent of occurrence, which is estimated to be less than 20,000 km^2; (9930 km^2, according to our calculations); its distribution is severely fragmented, and there is a continuing decline in the extent and quality of its habitat. The species is known from the Chocó, an area with the highest deforestation rate in Ecuador and spreading mining. Other significant threats to its existence include habitat reduction and fragmentation, logging, deforestation for agricultural development (particularly palm plantations), fires, human settlements, and pollution. This species has been recorded in many protected areas in Costa Rica and Panama. It was recently recorded in Parque Nacional Saslaya in Nicaragua. In Ecuador, it has been recorded in the public reserve Reserva Ecológica Mache Chindul, the biosphere reserve Chocó Andino de Pichincha, and some private reserves (e.g., Sapoparque La Florida, Reserva Biológica Bilsa).

Remarks.—*Sachatamia albomaculata* has two discrete dorsal color patterns: (1) dorsum with small yellow spots, (2) dorsum with small and large yellow spots. Thus, it is possible that two species occur within what is now recognized as *S. albomaculata* or, alternatively, the species has a natural polymorphic coloration. Acoustic and genetic data are needed to clarify the taxonomy of the species. Also, we note that the Colombian species *S. punctulate* has a coloration that matches pattern (1) of *S. albomaculata*.

The karyotype of *Sachatamia albomaculata* is 2N = 20.

Key references.—(Taylor, 1949) (description); (Duellman, 1967) (karyotype); (McCranie and Wilson, 2002) (account); (Savage, 2002) (account); (Puschendorf *et al.*, 2004) (natural history); (Kubicki, 2007) (account, call); (Hoffmann, 2010) (call); (Delia *et al.*, 2017) (reproductive biology); Yánez-Muñoz *et al.* (2020) (account); (Guayasamin *et al.*, 2020b) (account).

SACHATAMIA ILEX (HOLLY'S GLASSFROG) (SAVAGE, 1967)

Centrolenella ilex (Savage, 1967): 326. **Holotype.**—LACM 25205 from 14 km southwest of Amubri, Río Lari, 300 m, Cantón de Talamanca, Provincia Limón, Costa Rica.
Sachatamia ilex Guayasamin *et al.*, 2009: 36.

Etymology.—The specific epithet *ilex* honors Priscilla Hollister "Holly" Starrett for her contributions to centrolenid systematics. In botany, *Ilex* is the genus that contains holly trees.

Identification.—Figure 6.199. These frogs attain a maximum SVL of 29 mm in males and 35 mm in females. The skin on the dorsum of males and females is smooth to shagreen and lacks spicules; the venter is areolate; the ventral surfaces of the thighs below the vent lack a pair of enlarged tubercles. The anterior two-thirds of the ventral parietal peritoneum are white, the posterior third is transparent. There is a white pericardium; there is a translucent peritoneum covering the intestines, stomach, kidneys, testes, gall bladder, and urinary bladder. Each dentigerous process of the vomer bears 3–5 teeth. The snout is truncate in dorsal and lateral profiles. The tympanum is oriented almost vertically, relatively small, and its diameter is 26.1–28.7% of the eye diameter; the supratympanic fold is conspicuous, covering the dorsal margin of the tympanic annulus; the tympanic membrane is pigmented but clearly differentiated from the surrounding skin. The liver is tetralobed and covered by a transparent peritoneum. In males, the humeral spines are present but embedded in the arm musculature. There is no webbing between fingers I and II; there is reduced webbing between fingers II and III; there is extensive webbing between III and IV. The webbing on the foot

FIGURE 6.199 *Sachatamia ilex*. Male from Reserva Río Manduriacu, Provincia Imbabura. Ross Maynard.

is moderate. The ulnar fold is absent or low and inconspicuous; the inner tarsal fold is low and short; the outer tarsal fold is absent. The prepollex is not separated from finger I. The nuptial pad is a large to medium-sized excrescence present on the dorsal, lateral, and/or ventral sides of the thumb. Finger I is slightly shorter than or about the same length as finger II (finger I is 97.2–100.0% of finger II). The disc of finger III is relatively narrow; its width is about 34.0–45.9% of the eye diameter.

The dorsum is uniformly dark green. The throat and venter are creamy white. The hands and feet are dull yellow. The upper lip has a thin white line. There is a white parietal peritoneum covering the anterior two-thirds of the venter. The visceral and hepatic peritonea are clear. The melanophores are usually present on the fingers and toes, but sometimes they are absent from toes IV and V. The bones are green. The iris is silvery white to greenish-yellow with thick, contrasting black reticulations.

Comparisons.—*Espadarana prosoblepon*, *E. callistomma*, *Cochranella prasina*, and *Teratohyla spinosa*. *Sachatamia ilex* is distinguished from other glassfrogs by having, in life, a uniform dark green dorsum, a white to pale yellow or greenish-yellow iris with thick and contrasting black reticulations, and a moderate body size (male SVL 27.0–29.0 mm; female SVL 28.0–34.0 mm). Adult males have a pointed humeral spine that is embedded in the arm musculature (not easily visible externally). *Sachatamia ilex* closely resembles spotless populations of *E. prosoblepon* and *E. callistomma*, the males of which have externally visible humeral spines; most individuals of *E. prosoblepon* lack the thick black reticulation of the iris that is evident in *S. ilex*. Other species with uniform green dorsal coloration that could be confused with *S. ilex* are *C. prasina*, which lacks webbing between fingers II, III, and IV (present in *S. ilex*), and *T. spinosa*, which is much smaller in body size (SVL < 21 mm) and lacks humeral spines.

Natural history.—Figure 6.200. On 10 November 1999, at Hacienda La Joya, San Vicente de Andoas, Provincia Pichincha, one male was calling and two females were gravid, suggesting reproductive activity. The males fight upside down, grasping one another venter–venter. The combat behavior of *Sachatamia ilex* is as follows: males hang vertically from vegetation by the tips of their toes and grasp each other chest to chest in a type of arm lock; males may hang in this position for several minutes to hours and try to dislodge the other male's feet so that he loses grip of the vegetation. During this behavior, the two males produce softer "preep" calls. Males call from the upper surfaces of leaves near streams. The brown eggs are deposited on the upper side of leaves.

The advertisement call of one individual from Reserva Otokiki, Provincia Esmeraldas, Ecuador, was described as single, pulsed note. Note mean duration is 80 ms. Ten calls were emitted during a five-minute interval. Detailed data and figures are provided under the subtitle "Vocalizations of Ecuadorian Anurans" in Appendix 1.

FIGURE 6.200 *Sachatamia ilex*. Top left: amplectant pair from Reserva Jevon, Provincia Esmeraldas. Jaime Culebras. Top right: Reserva Río Manduriacu, Provincia Imbabura. Ross Maynard. Bottom: clutches of embryos in Reserva Tayra, Provincia Pichincha. Jaime Culebras.

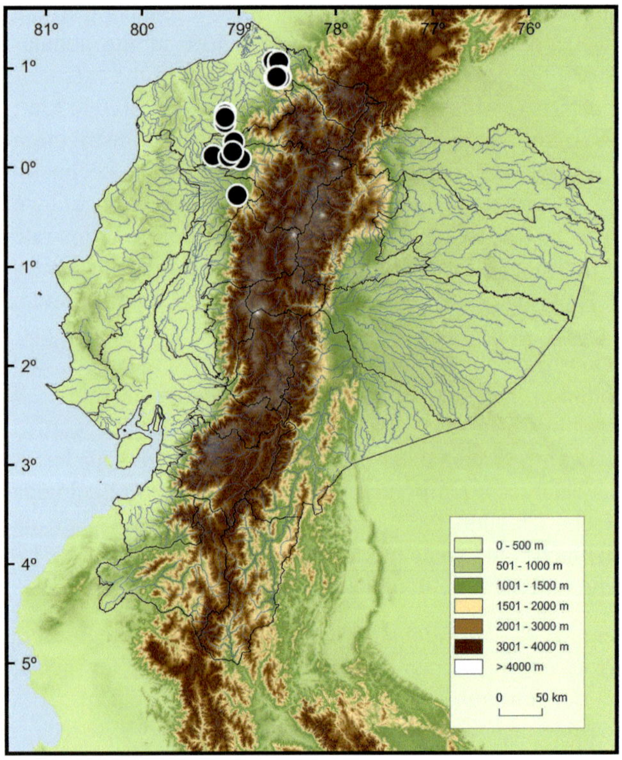

FIGURE 6.201 Localities of known occurrence of *Sachatamia ilex* in Ecuador.

Tadpoles.—The tadpoles have a flattened body shape and blackish-brown dorsal coloration; the venter is translucent; the upper jaw is in the form of a broad arch; the lower jaw is V-shaped; the labial tooth row formula is 0/2–3.

Distribution.—Figure 6.201. *Sachatamia ilex* is known in eastern Nicaragua, Costa Rica, western Panama, western Colombia, and western Ecuador at elevations from sea level up to 1420 m. In Ecuador, it occurs in Esmeraldas, Pichincha, and Santo Domingo de los Tsáchilas provinces.

Its elevational range is 133–869 m in an area of extent of occurrence of about 4920 km^2.

This nocturnal, arboreal species inhabits natural and intervened forests in the evergreen forests of lowlands of Equatorial Chocó and evergreen montane foothill forests of the Cordillera Occidental of the Andes ecosystems, where the average annual rainfall is 1897–2413 mm and the average annual temperature is 21.1 – 25.1°C.

Conservation.—*Sachatamia ilex* is categorized as Least Concern throughout its entire range (IUCN SSC Amphibian Specialist Group, 2020j; in IUCN Red List of 2022) in view of its wide distribution, presumed large population, and because it is unlikely to be declining fast enough to qualify for listing in a more threatened category. However, *S. ilex* was categorized as Vulnerable by Ortega-Andrade *et al.* (2021). We categorize it as Endangered following B1ab(iii) IUCN criteria because: (1) its extent of occurrence is estimated to be less than 5000 km^2 (4920 km^2), (2) its distribution is severely fragmented, and (3) there is a continuing decline in the extent and quality of its habitat. This species is difficult to find but is regularly recorded in suitable habitat. *Sachatamia ilex* is fairly common in much of its range along the Caribbean drainage in Costa Rica. In Ecuador, all known populations of *S. ilex* have been discovered recently. The species is known in the Chocó, an area with the highest deforestation rate in Ecuador and spreading mining. Other significant threats to its existence include habitat reduction and fragmentation, logging, deforestation for agricultural development (particularly palm plantations), fires, human settlements, and pollution. The species has been recorded in a number of protected areas in Costa Rica, Panama, and Colombia. In Ecuador, it has been recorded in the protected forest of Asociación Agrícola Carchi-Imbabura and the biosphere reserve Chocó Andino de Pichincha.

Remarks.—According to Guayasamin *et al.* (2020b), molecular evidence places *Sachatamia ilex* and S. *orejuela* as sister species.

Key references.—(Savage, 1967) (description); (Savage, 2002) (distribution); (Puschendorf *et al.*, 2004) (distribution); (Guayasamin *et al.*, 2006a) (reproductive biology, distribution); (Kubicki, 2007) (combat behavior); (Solís *et al.*, 2010) (distribution); Yánez-Muñoz *et al.* (2020) (account); (Guayasamin *et al.*, 2020b) (account).

SACHATAMIA OREJUELA (OREJUELA'S GLASSFROG) (DUELLMAN AND BURROWES, 1989)

Centrolenella orejuela (Duellman and Burrowes, 1989): 5.

Holotype.—KU 145081 from between El Tambo and La Costa, 800 m, Departamento de Cauca, Colombia.

Sachatamia orejuela Castroviejo-Fisher *et al.*, 2014: 73.

Etymology.—The specific epithet is a patronym for the Orejuela family (Jorge, Anamaría, and Tomás), who resided and administrated the Reserva La Planada, Colombia, at the time when the species was discovered.

Identification.—Figures 6.202, 6.203. These frogs attain a maximum SVL of 28.3 mm in males and 33.8 mm in females. The skin on the dorsum is finely shagreen, with small spicules evident in sexually active males. The ventral surfaces of the body are areolate. In males, the cloacal region has minute spinules. A pair of enlarged subcloacal tubercles are absent. The anterior half of the parietal peritoneum is white; there is a white pericardium; there is a translucent visceral peritoneum. Each vomer has 2–5 teeth. The snout is truncate in dorsal aspect, slightly protruding in lateral profile. The tympanum is small and visible, and its diameter is 18.5 25.6% of the eye diameter; the tympanic membrane is pigmented like the surrounding skin; the supratympanic fold is conspicuous. The liver is tetralobed and covered by a transparent peritoneum. The humeral spines are absent. There is basal webbing between fingers I and II; there is moderate webbing between fingers II and III; there is extensive webbing between fingers III and IV. There is extensive webbing on the foot. The ulnar and tarsal folds absent. The prepollex is concealed. The nuptial pad is a large to medium-sized excrescence present on the dorsal, lateral, and/or ventral sides of the thumb. Finger I is slightly

FIGURE 6.202 Morphological variation in *Sachatamia orejuela*. Left: CJ 11860, male, 26.6 mm SVL, Itapoa, Provincia Esmeraldas. SGS. Top right: CJ 1039, male, 25.8 mm SVL, Guayabillas, Provincia Pichincha. LAC. Middle right: CJ 978, male, 22.9 mm SVL, Comunidad Mashpi, Provincia Pichincha. LAC. Bottom right: CJ 8554, male, 27.3 mm SVL, Chinambí, Provincia Carchi. DAL.

FIGURE 6.203 Morphological details of *Sachatamia orejuela*. CJ 8554, male, Chinambí, Provincia Carchi. DAL.

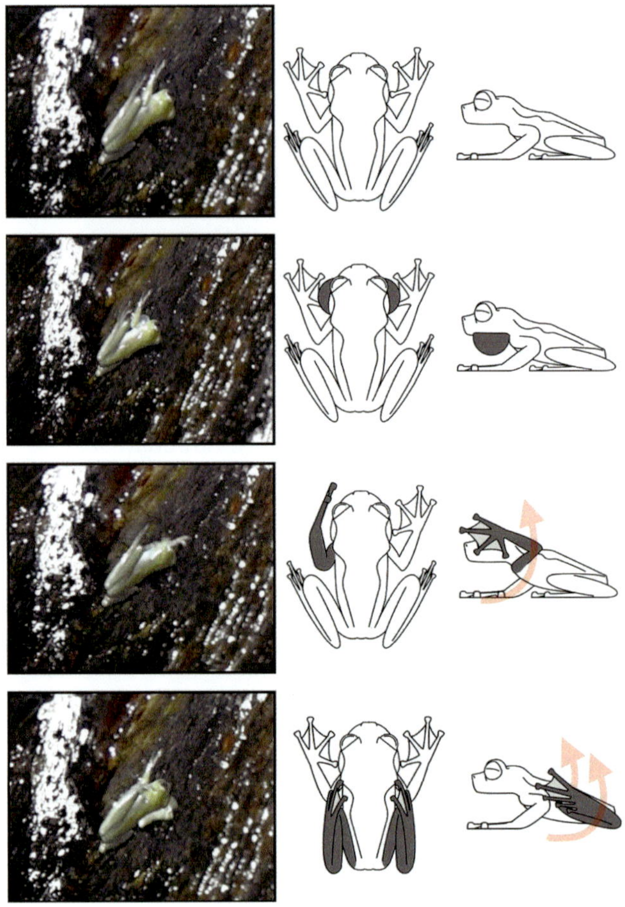

FIGURE 6.204 Visual signaling sequence of *Sachatamia orejuela*. Top to bottom: not calling or displaying, vocal sacs inflated while calling, arm-waving, double foot-flagging. Illustrations in dorsal and lateral views. Taken from Brunner and Guayasamin (2020).

longer than finger II (finger II is about 95–98% the length of finger I). The disc of finger III is large, and its width is about 62–70% of the eye diameter.

The dorsum is uniformly dark olive green. The flanks are grayish-green. The throat is translucent, with a light green hue. The venter is whitish-cream. There are melanophores on the dorsal surfaces of all fingers and toes. The bones are green. The iris is dark gray to brown with a yellow ring around the pupil.

Comparisons.—*Nymphargus prasinus*, *Sachatamia ilex*, *S. orejuela*, *Teratohyla spinosa*, and *E. prosoblepon*. *Sachatamia orejuela* has two traits that, as far as we know, are unique among glassfrogs: the species has a paired vocal sac, and its call has a dominant frequency (7.2–7.7 kHz) higher than that of any other centrolenid. Additionally, the species is mainly recognized by having a uniformly olive-green dorsum. In the Pacific versant of the Andes, only the following species have a uniformly green dorsum: *N. prasinus*, *S. ilex*, *S. orejuela*, *T. spinosa*, and some populations of *E. prosoblepon*. From these species, *S. orejuela* is distinguished by the absence of humeral spines (present in males of *E. prosoblepon* and *S. ilex*), relatively large body size (SVL 27.3–33.8 mm in *S. orejuela*; SVL < 22.0 mm in *T. spinosa*), and extensive webbing between fingers III and IV (webbing absent or basal in *N. prasinus*). Additionally, *S. orejuela* is found in a very specific microhabitat, on rocks nearby waterfalls.

Natural history.—Figures 6.204, 6.205. The males call from rocks along or within streams. During the night, *Sachatamia orejuela* is mostly found on rocks along steep stream banks or within the stream in the spray zone of waterfalls. Spiders of the genus *Clubiona* have been observed to prey on juveniles of *S. orejuela*. The call, emitted as a single "peep," has the highest frequency recorded in glassfrogs (7.2–7.7 kHz). The species also communicates through visual signaling. Detailed data and figures are provided under the subtitle "Vocalizations of Ecuadorian Anurans" in Appendix 1.

FIGURE 6.205 *Sachatamia orejuela*. Top left: Parque Nacional Cotacachi-Cayapas, Provincia Imbabura. LAC. Top right: juvenile from Reserva Tayra, Provincia Pichincha. Bottom: Reserva Manduriacu, Provincia Imbabura. Jaime Culebras.

FIGURE 6.206 Ontogenetic changes in two individuals of *Sachatamia orejuela* from tadpole to adult. Top to bottom, left to right. CJ 11863a, stage 26, lateral, dorsal, and ventral views, TL = 32.6 mm; undetermined stage, TL = 38 mm; stage 42, TL = 35.4 mm. CJ 11863b, stage 46 = 13. 4 mm SVL; froglet 43 days after stage 46 = 17.4 mm SVL; juvenile 124 days = 20 mm SVL; juvenile 215 days = 27.5 mm SVL; adult 317 days = 29.4 mm SVL. Río San Vicente, Reserva Mashpi, Provincia Pichincha. SGS.

Tadpoles.—Figure 6.206. This description is based on an individual (not preserved, CJ 11863) from Reserva Mashpi, Provincia Pichincha, Ecuador, in stage 26. It was collected by Juan Manuel Guayasamin on 6 July 2021. It measures 8.8 mm in body length and 32.6 mm in total length. The body is ovoid, equally wide anteriorly and posteriorly, and slightly wider than it is deep. The snout appears round in dorsal view and subacute in lateral profile, with dorsally directed eyes. The spiracle is sinistral, with the inner wall free from the body, and its opening is directed posteriorly on the deadline at the body's end. The vent tube is medial, attached to the ventral fin. The oral disc is medium-sized, anteroventral, transversely elliptical, and not emarginate. The upper labium is poorly developed, while the lower labium is more developed and not attached to the body wall. The upper jaw sheath is completely keratinized, deeply serrated, with a smooth arc. The tail constitutes about 73% of the total length, with the greatest depth at the midlength and a subacute tip. The dorsal fin does not extend onto the body, commencing in the second third of the tail. The ventral fin is very narrow in the first third of the tail and becomes deeper in the second third. The dorsal fin is slightly deeper than the ventral fin and about half the depth of the caudal musculature at the tail's midlength.

Tadpoles of *Sachatamia orejuela* are distinctive due to their heavily melanized dorsum, resulting in a dark brown appearance on the dorsal and lateral surfaces of the body and tail. On the ventral side, there is no pigmentation on the body, allowing the organs to be visible through the translucent skin, while the tail features some dark flecks.

At stage 40, which occurs 99 days after stage 26 and with a total length of 38.0 mm, the body loses most of its brown coloration and becomes predominantly green. However, the tail retains brown melanophores but also acquires a green tint. By stage 43, six days later and with a total length of 35.4 mm, the green coloration becomes brighter, and light blue becomes visible on the throat and scapular region. At stage 46, with an SVL of 13.4 mm, the dorsum turns dark green, while the extremities take on a lighter green. About one month after tail reabsorption, with an SVL of 17.4 mm, the green becomes lighter, resulting in uniform light green dorsal surfaces of the body and extremities. As a juvenile, three months later and with an SVL of 20 mm, the frog displays dark green on the dorsum and legs, while the arms, hands, and feet have a lighter green appearance. The region above the mouth, below the eyes and nares, as well as the tympanic area, features a light green color. Seven months after tail reabsorption, with an SVL of 27.5 mm, all dorsal surfaces are uniformly green, albeit lighter than the green observed in the 3-month-old juvenile. Blue becomes visible in the armpits. At an age of 10 months, with an SVL of 29.4 mm, the coloration closely resembles that of an adult.

Distribution.—Figure 6.207. *Sachatamia orejuela* is known from the Pacific lowlands and midelevations of southern Colombia and northwestern Ecuador. In Ecuador, it occurs in Esmeraldas, Imbabura, Pichincha, and Santo Domingo de los Tsáchilas provinces. Its elevational range is 303–1250 m in an area of extent of occurrence of about 2060 km^2.

This nocturnal, arboreal species inhabits natural forests and intervened areas in the evergreen montane foothill forest ecosystem of the Cordillera Occidental of the Andes biogeographic sector, where the average annual rainfall

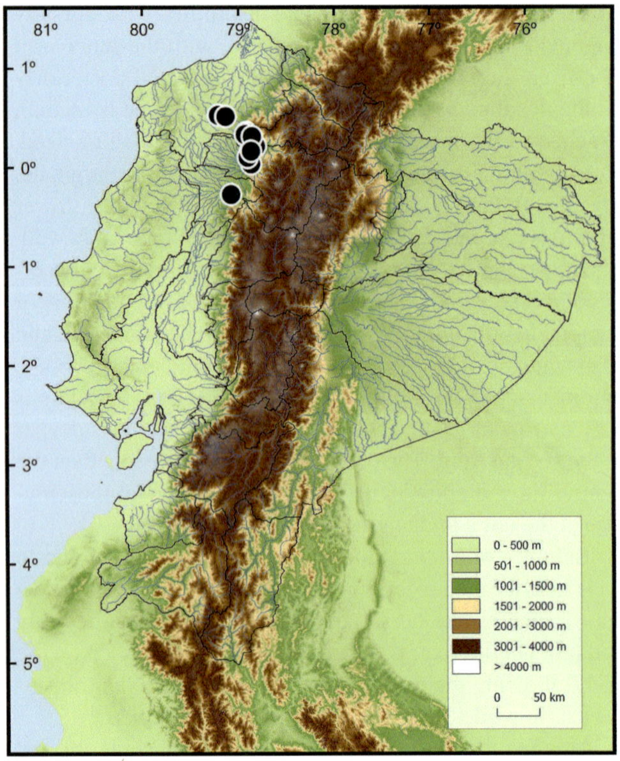

FIGURE 6.207 Localities of known occurrence of *Sachatamia orejuela* in Ecuador.

is 1390–2818 mm and the average annual temperature is 19.3 – 24.2°C.

Conservation.—*Sachatamia orejuela* is categorized as Least Concern throughout its entire range (IUCN SSC Amphibian Specialist Group, 2019x; in IUCN Red List of 2022) in view of its wide distribution and presumed large population. However, it is important to note that in Ecuador, *S. orejuela* is categorized as Endangered by Ortega-Andrade *et al.* (2021), given that there is a continuing and projected decline in the extent and quality of its habitat. Also, according to our data, its current extent of occurrence is less than 2060 km^2. In Ecuador, the habitat of the species is fragmented and threatened by logging and mining, which are serious threats for the species. Due to ongoing decline in the extent and quality of habitat, the population is suspected to be decreasing. In Ecuador, it has been recorded in the protected forest of Mashpi, the biosphere reserve Chocó Andino de Pichincha, and some private reserves (e.g., Reserva Ecológica Maquipucuna, Reserva Río Canandé, Reserva Río Manduriacu).

Remarks.—According to Guayasamin *et al.* (2020b), molecular evidence places *Sachatamia orejuela* and *S. ilex* as sister species.

Key references.—(Duellman and Burrowes, 1989) (description, habitat); (Ruiz-Carranza *et al.*, 1996) (distribution); (Almeida-Reinoso and Coloma, 2005) (predation); (Yánez-Muñoz and Cisneros-Heredia, 2008) (distribution, habitat); (Guevara-Molina and Rada, 2018) (distribution); Yánez-Muñoz *et al.* (2020) (account); (Guayasamin *et al.*, 2020b) (account); (Brunner and Guayasamin, 2020) (behavior, call).

GENUS *TERATOHYLA* (GLASSFROGS) TAYLOR, 1951

Teratohyla Taylor, 1951: 62. **Type species.**—*Centrolenella spinosa* (Taylor, 1949) by original designation.

Etymology.—The generic name is derived from the Greek *teras*, meaning monster, marvel, or wonder, and the word *Hyla*, traditionally associated with treefrogs. The origin of the frog name *Hyla* is based on the mythological Greek boy *Hylas*, and although the boy's name is masculine, it has been unambiguously treated as feminine by amphibian systematists.

Definition.—Figure 6.208. The genus *Teratohyla* is diagnosed mainly by the following combination of traits: (1) humeral spines absent; (2) relatively small body size, with SVL < 24.5 mm in males and < 28.3 mm in females; (3) moderate to extensive webbing between fingers III and IV; (4) bones pale to dark green in life; (5) restricted to lowlands (< 1000 m).

Content.—*Teratohyla* contains five species, four of which occur in Ecuador (31 December 2022).

Distribution.—*Teratohyla* occurs in the lowlands of Central America from northeastern Honduras south and in the Pacific and Amazonian wet tropical lowlands of South America below 1000 m, in a seemingly isolated population in French Guiana, and presumably in intervening Amazonian Colombia and Brazil.

Remarks.—Figure 6.209. The phylogeny of species presented herein follows Guayasamin *et al.* (2020b).

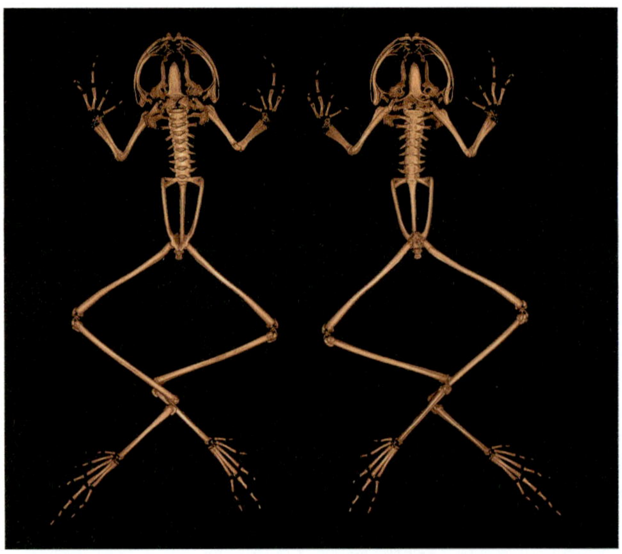

FIGURE 6.208 High-resolution computed tomography reconstructions of the skeleton of *Teratohyla midas*, KU 123221, 17.1 mm SVL, Puerto Libre, Río Aguarico, Provincia Sucumbíos. Dorsal (left) and ventral (right) views. Daniel J. Paluh.

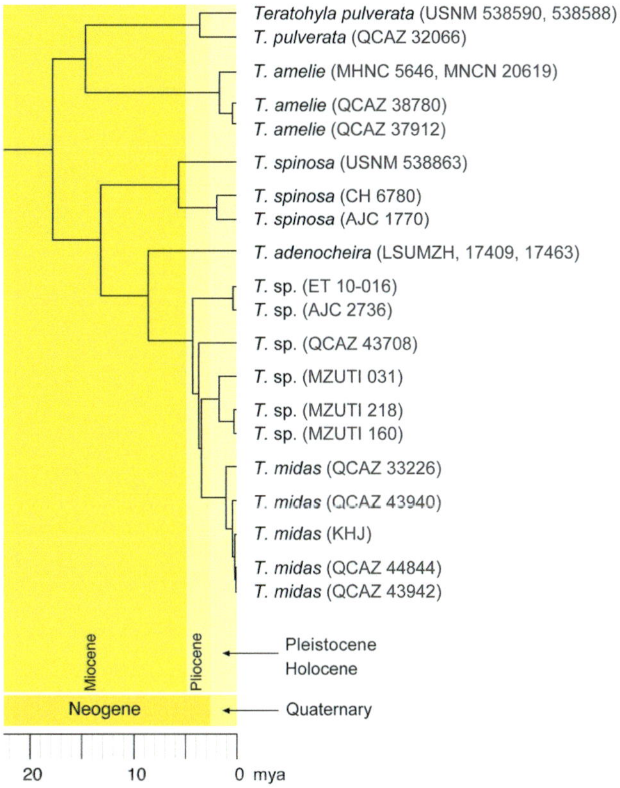

FIGURE 6.209 Time-calibrated maximum likelihood phylogenetic tree depicting evolutionary relationships of *Teratohyla*, inferred from DNA sequences of mitochondrial and nuclear genes. Phylogeny taken and redrawn from Guayasamin *et al.* (2020b).

FIGURE 6.210 *Teratohyla amelie*. CJ 10963, female, 19.6 mm SVL. CJ 10964, male, 19.1 mm SVL. San José de Payamino, Provincia Orellana. DAL.

TERATOHYLA AMELIE (AMÉLIE'S GLASSFROG) (CISNEROS-HEREDIA AND MEZA-RAMOS, 2007)

Cochranella amelie (Cisneros-Heredia and Meza-Ramos, 2007): 34. **Holotype.**—DHMECN 3066 from Comunidad de Oglán, 600 m, Cantón Arajuno, Provincia Pastaza, Ecuador.

Teratohyla amelie Guayasamin *et al.*, 2009: 37.

Etymology.—The specific epithet refers to the fictional character Amélie, the protagonist of the movie "Le Fabuleux Destin d'Amélie Poulain."

Identification.—Figure 6.210, 6.211. These frogs attain a maximum SVL of 19.3 mm in males, and females are unknown. The skin on the dorsum is shagreen; the ventral and subcloacal surfaces are granular; the ventral surfaces of the thighs below the vent have a pair of enlarged tubercles. The parietal peritoneum is completely transparent and lacks iridophores; the visceral peritonea are covered with iridophores. The vomerine teeth are absent. The snout is truncate in dorsal view and rounded in lateral view. The tympanic

FIGURE 6.211 Morphological variation and details of *Teratohyla amelie*. Left: QCAZ 37912. Right: QCAZ 48421. Centro Ecológico Zanja Arajuno, Provincia Pastaza. LAC. Middle left: CJ 10963, female. Middle right and bottom: CJ 10964, male. San José de Payamino, Provincia Orellana. DAL.

annulus is evident and oriented dorsolaterally with posterior inclination; there is a very weak supratympanic fold above the tympanum. The liver is white and bulbous. The humeral spine is absent. There is no webbing on the hand between fingers I and II; there is moderate to extensive webbing between the outer fingers. There is extensive webbing on the feet. There are low, non-enameled ulnar and inner tarsal folds. The prepollex is concealed. The nuptial pad is a large to medium-sized excrescence present on the dorsal, lateral, and/or ventral sides of the thumb. Finger I is longer than II. The eye diameter is greater than the width of the disc on finger III.

The dorsum is green to bluish-green and lacks flecks or spots. The throat is greenish; other ventral surfaces are transparent. The heart is white, and the viscera are visible ventrally. There are numerous melanophores on all fingers and toes. The bones are green. The iris is creamy brown.

Comparisons.—*Hyalinobatrachium* spp., *Chimerella mariaelenae*, *Teratohyla midas*, *T. pulverata*. *Teratohyla amelie* differs from most species inhabiting the Amazonian lowlands and Amazonian slopes of the Andes by having a completely transparent ventral parietal peritoneum. In Ecuador, only species in the genus *Hyalinobatrachium* and *C. mariaelenae* share the transparent parietal peritoneum character. *Chimerella mariaelenae* differs from *T. amelie* by having a green dorsum with many dark lavender points and scattered larger dark spots (dorsum uniformly green in *T. amelie*). Also, adult males of *C. mariaelenae* have a small humeral spine (absent in *T. amelie*). *Teratohyla amelie* differs from all species in the genus *Hyalinobatrachium* (characteristics in parentheses) by having a uniformly green dorsum in life (green dorsum with small yellow spots), eggs deposited on the upper surfaces of leaves (underside of leaves), and a nuptial pad present as a large to medium-sized excrescence on the dorsal, lateral, and/or ventral sides of the thumb (medium-sized diffuse nuptial excrescence formed by glandular clusters and individual glands in most *Hyalinobatrachium*).

Teratohyla midas is another species found in the Amazonian lowlands in Ecuador, but it has yellow dorsal spots and a venter that is white anteriorly and transparent posteriorly. *Teratohyla pulverata* also has a transparent ventral peritoneum and deposits eggs on the upper surfaces of leaves; however, it is restricted to the west of the Andes (*T. amelie* is endemic to the Amazonian lowlands of Ecuador and Peru). The main differences from *T. amelie* are a sloping snout in lateral view and a green dorsum with white spots in life.

Natural history.—Males call from the upper sides of leaves next to their clutches. Clutches of eggs are deposited in a single layer on the upper surfaces of *Heliconia* leaves about 80 cm above small streams.

The advertisement call of one individual from Río Liliquino, Provincia Pastaza, Ecuador, was described as a single, pulsed note. The mean note duration is 13 ms. Detailed data and figures are provided under the subtitle "Vocalizations of Ecuadorian Anurans" in Appendix 1.

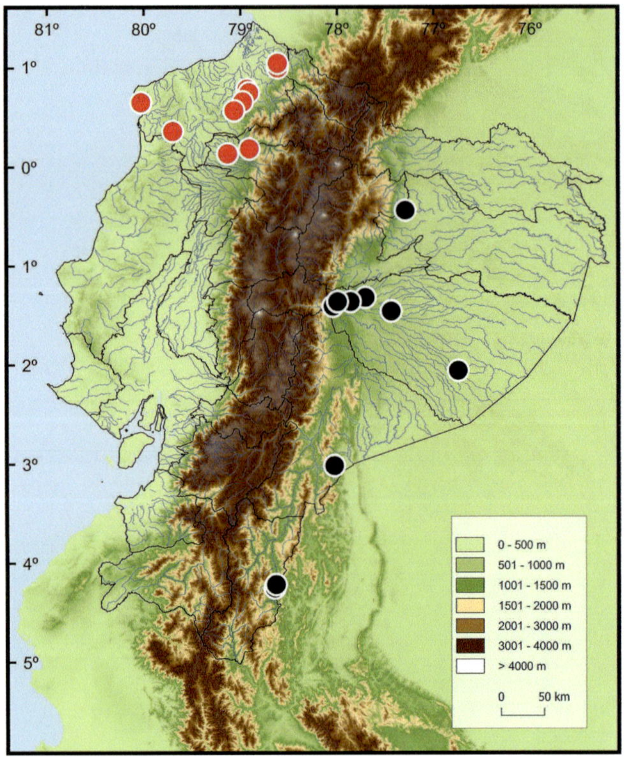

FIGURE 6.212 Localities of known occurrence of *Teratohyla amelie* (black dots) and *T. pulverata* (red dots) in Ecuador.

Tadpoles.—Unknown.

Distribution.—Figure 6.212. *Teratohyla amelie* is known in the foothills and Amazonian lowlands of Ecuador and Peru at elevations between 350 and 1037 m. In Peru, *T. amelie* is known from a single locality in Departamento de Cusco, km 10 from Quincemil toward Puerto Maldonado. In Ecuador, it occurs in Orellana, Pastaza, Morona Santiago, and Zamora Chinchipe provinces. Its elevational range is 291–1159 m in an area of extent of occurrence of about 35,840 km^2.

This nocturnal, arboreal species inhabits natural forests and intervened areas in five ecosystems (Table 0.4) within four biogeographic sectors (Table 0.5), where the average annual rainfall is 1679–4103 mm and the average annual temperature is 20.1–25.1 °C.

Conservation.—*Teratohyla amelie* is categorized as Least Concern throughout its entire range (IUCN SSC Amphibian Specialist Group, 2019y; in IUCN Red List of 2022) and in Ecuador by Ortega-Andrade *et al.* (2021). *Teratohyla amelie* has a presumed large population, wide distribution, and tolerates some degree of habitat degradation. No major threats are known for this species, although habitat destruction and degradation caused by agriculture occur in many of the areas of distribution. In Ecuador, it has been recorded in the biosphere reserves Sumaco and Podocarpus-El Cóndor.

Remarks.—*Teratohyla amelie* is a sister species to *T. pulverata*. These sister species represent a good example of allopatric speciation mediated by the uplift of the Andes.

Key references.—(Cisneros-Heredia and Meza-Ramos, 2007) (description); (Cisneros-Heredia and Meza-Ramos, 2007) (distribution); (Twomey *et al.*, 2014) (distribution); (Guayasamin *et al.*, 2020b) (account, call).

TERATOHYLA MIDAS (MIDAS' GLASSFROG) (LYNCH AND DUELLMAN, 1973)

Centrolenella midas (Lynch and Duellman, 1973): 38.
Holotype.—KU 123219 from Santa Cecilia, 340 m, Provincia Sucumbíos, Ecuador.
Teratohyla midas Guayasamin *et al.*, 2009: 37.

Etymology.—The specific epithet refers to King Midas in Greek mythology, at whose touch everything turned into gold. The epithet is associated with Río Aguarico (= rich water), near the type locality of the species, in reference to gold found in the river and in allusion to the golden spots that characterize this glassfrog.

Identification.—Figures 6.213, 6.214. These frogs attain a maximum SVL of 22.5 mm in males and 26.8 in females. The skin on the dorsum is smooth; the males have minute spicules visible only under magnification (× 250). The venter is areolate; the ventral surfaces of the thighs below the vent have a pair of enlarged tubercles. There is a white parietal peritoneum covering the anterior 40–50% of the venter; also, the pericardium and gastrointestinal peritoneum are white; there are translucent peritonea on the liver and the gall and urinary bladders. The vomers have dentigerous process bearing 0–4 teeth. The snout is truncate in dorsal aspect and truncate to bluntly rounded in lateral profile. The tympanum is moderate, oriented almost vertically, with slight lateral and posterior inclinations, and its diameter is 29.1–36.3% of the eye diameter; only the upper border of the tympanic annulus is obscured by the supratympanic fold; the tympanic membrane is clearly differentiated from the surrounding skin. The liver is lobate and covered by a transparent peritoneum. The humeral spines are absent. There is no webbing between fingers I and II; there is reduced webbing between fingers II and III; there is moderate webbing between the outer fingers. The feet are about two-thirds webbed. The ulnar and tarsal folds are absent. The prepollex is not separated from finger I. The nuptial pad is a large to medium-sized excrescence present on the dorsal, lateral, and/or ventral sides of the thumb. Finger I is slightly longer than finger II (finger II length is 94.1–98.4% of finger I). The width of the disc of finger III is about 34.4–39.7% of the eye diameter.

The dorsum is green with few small yellow spots, usually concentrated on the upper flanks. The upper lip is white. The venter is white anteriorly and transparent posteriorly.

FIGURE 6.213 *Teratohyla midas*. Left: CJ 12923, female, 20.2 mm SVL. CJ 12924, male, 19.2 mm SVL. Comuna 12 de Febrero, Provincia Sucumbíos. SGS.

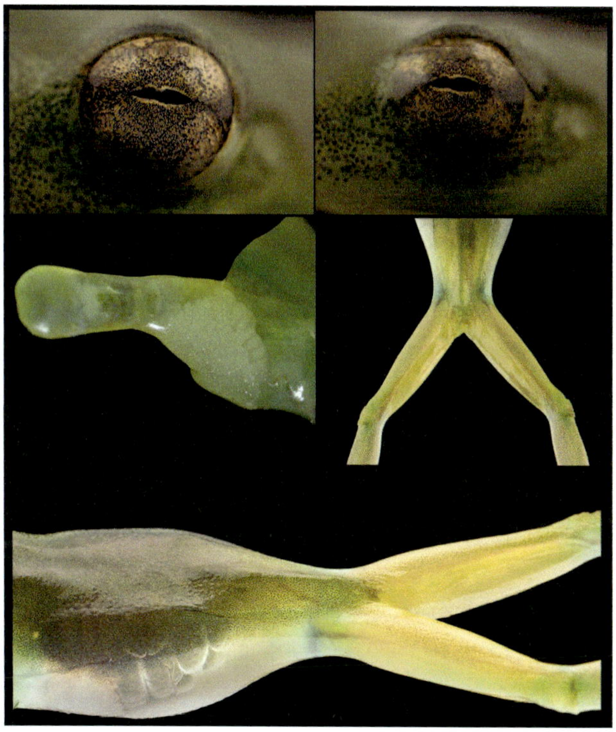

FIGURE 6.214 Morphological details of *Teratohyla midas*. CJ 12924, male. Comuna 12 de Febrero, Provincia Sucumbíos. Note the nictitating membrane. SGS.

Melanophores are usually covering the fingers and toes. The bones are green. The iris is silvery white with a pale yellow hue, faint brown horizontal stripe, and thin gray reticulations. There is a yellow ring around the pupil.

Comparisons.—*Nymphargus cariticommatus*, *N. siren*, *N. humboldti*, and *Rulyrana flavopunctata*. Among glassfrogs in the Amazonian lowlands of Ecuador, *Teratohyla midas* is unique by having a green dorsum with a few small yellow dots that are usually concentrated on the upper flanks; other diagnostic traits include a white digestive tract, small body size (male SVL < 22.5 mm), and the absence of humeral spines. Species with a similar dorsal coloration are *N. cariticommatus*, *N. siren*, *N. humboldti*, and *R. flavopunctata*, all of which have numerous yellow spots uniformly distributed on the dorsum (limited to the upper flanks of the body in *T. midas*). Also, none of these four species have a white digestive tract, and species of *Nymphargus* have less hand webbing than *T. midas*.

Natural history.—Figure 6.215. This species has been found throughout the year on the leaves of herbs, shrubs, and trees along small rivulets in the Amazonian rainforest. The males call from the upper surfaces of leaves along streams, and the eggs are deposited there as well. There is short-term maternal care; male parental care is absent.

In Reserva Natural y Comunitaria el Manantial, Departamento Caquetá, Colombia, the amplexus behavior of two pairs was observed. The axillary amplexus lasted for approximately 150 min; during this time, the females moved over leaves or branches looking for places for oviposition. While the females oviposited, the males made circular movement with their hind legs; once oviposition was finished, the males left, and the females covered the eggs with their bodies for approximately 60 min (brooding). Clutches of embryos contained on average 21.6 eggs (12–38 eggs, n = 21).

In Santa Cecilia, Provincia Sucumbíos, Ecuador, *Teratohyla midas* is sympatric with *Hyalinobatrachium munozorum* and *Cochranella resplendens*.

The advertisement call of one individual from Boanamo community, Provincia Orellana, Ecuador, was described as short and composed of three to four notes. Call mean duration is 82 ms. The dominant frequency measure at its peak is 7210–7390 Hz ($\bar{x}$ = 7285 ± 60.4). Calls are similar to those reported for Peruvian populations of the species, although they have a slightly lower dominant frequency (6760–7060 Hz). Detailed data and figures are provided under the subtitle "Vocalizations of Ecuadorian Anurans" in Appendix 1.

FIGURE 6.215 *Teratohyla midas*. Top and middle: QCAZ 32519, calling male, Bobonaza, Provincia Morona Santiago. LAC. Bottom: amplectant pair from Reserva Hyla, Provincia Pastaza. Jaime Culebras.

Tadpoles.—Figure 6.216. This description is based on an individual (not preserved, series QCAZ 41083) from Pacto Sumaco, Provincia Napo, Ecuador, in stage 31. A clutch of eggs was collected from the apex of an *Anthurium* leaf at approximately 5 m from the water by Elicio E. Tapia and Raúl E. Ruiz on 20 March 2009. The tadpole measures 9.8 mm in body length and 37.6 mm in total length. The body is ovoid, widest posteriorly, and wider than it is deep. It exhibits lateral expansions, bump-like structures in the anterior region, just before the spiracle. The snout appears round in both dorsal view and in lateral profile, and the eyes are directed dorsolaterally. Chondrocranial elements are visible. The spiracle is sinistral, with the inner wall free from the body, and its opening is directed posteriorly on the deadline at about two-thirds of the body's length. The mouth is anteroventral, with

FIGURE 6.216 Ontogenetic changes in *Teratohyla midas* from tadpole to juvenile. Top to bottom. QCAZ 41083, stage 31, lateral and dorsal views, TL = 37.6 mm; stage 42, lateral and ventral views, TL = 37.5 mm; stage 45, 105 days after stage 31; juvenile 31 days after stage 45. Pacto Sumaco, Provincia Napo. LAC.

a poorly developed upper labium and a more developed lower labium that is free from the body wall. The tail constitutes approximately 74% of the total length, with its greatest depth around the midlength and a subacute tip. The dorsal fin does not extend onto the body, is slightly deeper than the ventral fin, and is about two-thirds of the depth of the caudal musculature at the tail's midlength.

The skin is translucent, giving the body a pale pinkish appearance due to the visibility of internal organs and musculature. The tadpole lacks pigmentation, except for a few scattered light brown flecks on the body and tail.

By stage 42, which is 98 days after stage 31, and with a total length of 34.5 mm, the tadpole takes on a general pale

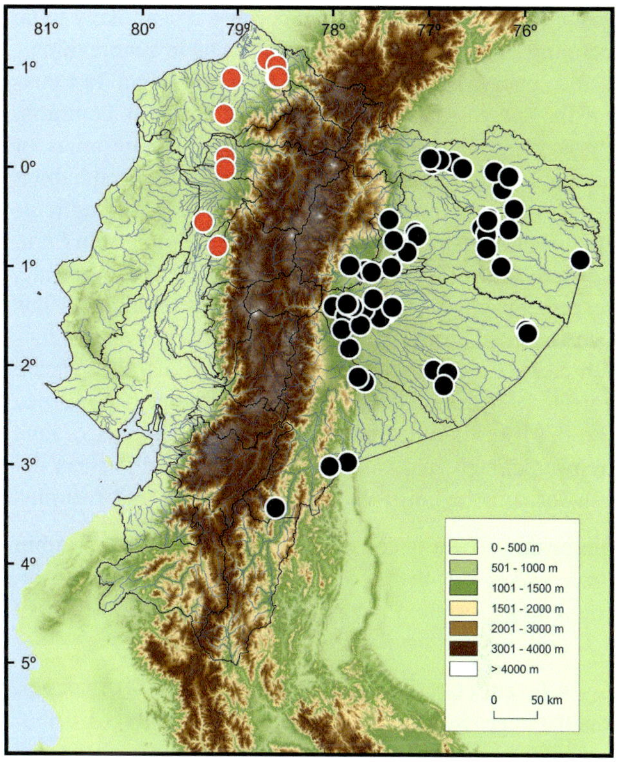

FIGURE 6.217 Localities of known occurrence of *Teratohyla midas* (black dots) and *T. spinosa* (red dots) in Ecuador.

green coloration, with the ventral side of the body having a bluish tint. The skin on ventral surface remains translucent, allowing the internal organs to be seen. At stage 44, seven days later, the green color becomes more vibrant, and there is a brownish hue dorsally on the head. The eyelids exhibit a bronze coloration. The blue coloration is confined to the throat and scapular area, with very few yellow spots appearing on the dorsum. As a froglet, 31 days later, the brownish hue extends along the dorsum, the blue color is limited to the scapular region, and the yellow spots become more distinct.

Distribution.—Figure 6.217. *Teratohyla midas* has been reported in the Amazonian lowlands (< 940 m) of Ecuador, Peru, and Brazil. In Ecuador, it occurs in Sucumbíos, Orellana, Napo, Pastaza, and Morona Santiago provinces. Its elevational range is 180–1040 m in an area of extent of occurrence of about 62,550 km^2.

This nocturnal, arboreal species inhabits natural forests and intervened areas in nine ecosystems (Table 0.4) within six biogeographic sectors (Table 0.5), where the average annual rainfall is 1677–4122 mm and the average annual temperature is 20.6–25.5 °C.

Conservation.—*Teratohyla midas* is categorized as Least Concern throughout its entire range (IUCN SSC Amphibian Specialist Group, 2022q; in Red List of 2020) and in Ecuador by Ortega-Andrade *et al.* (2021) in view of its wide distribution, tolerance of a degree of habitat modification, presumed large population, and because it is unlikely to be declining fast enough to qualify for listing in a more threatened category. The population status is unknown. In Ecuador, the species has a broad distribution and is common. It is a generally widespread species with large areas of suitable habitat remaining. There is some localized habitat loss to general human activities such as logging, agriculture (crops, livestock, etc.) and human settlements. In Peru, it occurs in the Manu Reserve Zone. Conservation units are present within its range in Brazil. In Ecuador, it has been recorded in the public reserves Reserva de Producción de Fauna Cuyabeno and Parque Nacional Yasuní; the protected forests of Pañacocha, Cerro Sumaco y Cuenca Alta del Río Suno, Cordillera Kutukú y Shaimi, Comuna Kichwa Añangu, and Nacionalidad Andwa de Pastaza del Ecuador; and the biosphere reserves Sumaco and Yasuní.

Remarks.—There is considerable genetic structure within *Teratohyla midas*; therefore, it should be considered a species complex. Among recognized taxa, the closest relative of *T. midas* is *T. adenocheira*.

Key references.—(Lynch and Duellman, 1973) (description, habitat); (Kok and Castroviejo-Fisher, 2008) (distribution); (Twomey *et al*., 2014) (distribution); (Diaz Ricaurte *et al*., 2016) (reproductive biology); (Araújo *et al*., 2018) (distribution); (Guayasamin *et al*., 2020b) (account).

TERATOHYLA PULVERATA (DUSTY GLASSFROG) (PETERS, 1873a)

Hyla pulverata (Peters, 1873a): 614. **Holotype.**—ZMB 7842 from Chiriqui, Atlantic and Pacific versants of extreme western Panama, Panama.

Teratohyla pulverata Guayasamin *et al*., 2009: 37.

Etymology.—The specific epithet originates from the Latin *pulvereus*, meaning dusty, and refers to the numerous minute white spots on the dorsum of the species.

Identification.—Figures 6.218, 6.219. These frogs attain a maximum SVL of 24.5 mm in males and 28.3 mm in females. The skin on the dorsum is shagreen; the males have small spicules (only visible under magnification); the ventral surfaces are granular; the thighs below the vent lack a pair of enlarged tubercles. The ventral parietal peritoneum is completely transparent. There is a white pericardium and a white gastrointestinal peritoneum. The dentigerous processes of the vomers have 2–4 teeth. The snout is rounded in dorsal view, sloping in lateral profile. The tympanum is visible, relatively small, and its diameter is 20.2–23.3% of the eye diameter; the tympanic annulus is visible except for the upper border, which is covered by the supratympanic fold; the tympanic membrane is differentiated, translucent, and pigmented like the surrounding skin. The liver is white and bulbous (i.e., covered by iridophores). The humeral spines

FIGURE 6.218 Morphological variation in *Teratohyla pulverata*. Left: CJ 290, female. Río Cachaví, Provincia Esmeraldas. LAC. Top right: QCAZ 45414, Durango, Provincia Esmeraldas. Middle right: CJ 12929, 25.0 mm SVL, Reserva Canandé, Provincia Esmeraldas. SGS. Bottom right: CJ 969, 23.0 mm SVL. Comunidad Mashpi, Provincia Pichincha. LAC.

FIGURE 6.219 Morphological details of *Teratohyla pulverata*. Top: CJ 12929, Reserva Canandé, Provincia Esmeraldas. SGS. Bottom: QCAZ 45414, Durango, Provincia Esmerladas. LAC.

are absent. There is absent or basal webbing between fingers I and II. The feet is about two-thirds webbed. The metacarpal, ulnar, metatarsal, and tarsal folds are present and enameled, with low enameled tubercles. The prepollex is concealed. The nuptial pad is a large to medium-sized excrescence present on the dorsal, lateral, and/or ventral sides of the thumb. Fingers I and II are about the same length (finger II length is 98–103% of finger I length). The disc of finger III is small; its width is 20.1–23.5% of the eye diameter.

The dorsum of the head and body is green with small white flecks and dots. Dorsally, the limbs are green and present pale yellow spots with white centers. The upper lip is white. There are small white dots visible below the eye. The ulnar and tarsal folds present low, white tubercles. The cloacal region presents numerous small, white warts. The parietal peritoneum is transparent (all the internal organs are visible ventrally). There are white peritonea covering the heart, liver, digestive tract, testes, and gall bladder; the peritonea covering the urinary bladder and kidneys is transparent. Melanophores partially cover the dorsal surface of finger IV, absent from fingers I–III. The bones are green. The iris is grayish-white with thin, dark gray reticulations and minute yellow flecks. There is a thin yellow to cream ring around the pupil.

Comparisons.—*Hyalinobatrachium* spp. Among glassfrogs that inhabit the Chocoan forest of Ecuador, *Teratohyla pulverata* is unique by having a green dorsum with small white spots, a completely transparent ventral parietal peritoneum, white hepatic and gastrointestinal peritonea, a sloping snout in lateral profile, and by lacking humeral spines. Only species in the genus *Hyalinobatrachium* could be confused with *T. pulverata*; however, *T. pulverata* has green bones (white in *Hyalinobatrachium*), deposits eggs on the upper sides of leaves (undersides of leaves in *Hyalinobatrachium*), and has conspicuous teeth on the vomers (teeth absent in *Hyalinobatrachium*).

Natural history.—Figure 6.220. Males of *Teratohyla pulverata* call from the upper surfaces of leaves along streams, and the clutches of embryos are deposited there too. There is short-term maternal care; prolonged parental care is absent. The short-term brooding provided by females of *T. pulverata* enhances the survival of clutches because it reduces mortality associated with dehydration, predation, and fungal infection.

In Ecuador, *Teratohyla pulverata* was found sympatrically with *Espadarana prosoblepon* and *Hyalinobatrachium fleischmanni* in Estero Aguacate, Provincia Esmeraldas, and with *Sachatamia ilex*, *S. albomaculata*, *E. prosoblepon*, and *H. aureoguttatum* in Río Bogotá, Provincia Esmeraldas.

The call has been described from Costa Rica (Kubicki, 2007) and Panama (Ibañez *et al.*, 1999b). It is a quick high-pitched "tik," normally repeated several times with roughly a one-second pause between calls. The call is emitted as a series of three notes (each note = 0.05 s). Detailed data and figures are provided under the subtitle "Vocalizations of Ecuadorian Anurans" in Appendix 1.

FIGURE 6.220 *Teratohyla pulverata*. Reserva Tayra, Provincia Pichincha. Jaime Culebras.

Tadpoles.—The tadpole of *Teratohyla pulverata* was described from Costa Rica. The ecomorphological guild of the tadpole is exotroph, lotic, and burrower. When hatching (Gosner stage 24–25), tadpoles have a dark reddish coloration due to the dorsal star-like pigmentation spots and the reddish visceral and muscular systems. Young larvae have a nearly circular shape in cross-section, whereas later larval stages change to a typical flat shape. The LTRF is 2/3; the A-2 row has a wide gap; P-3 is slightly shorter than the equally sized P-1 and P-2. The upper jaw sheath forms a smooth arc and has a pointed serration. The lower jaw sheath is smooth and V-shaped. Marginal papillae are present on both sides of the oral disc; the papillae are arranged in a single row but have a wide anterior labium gap. The eye position is dorsal.

Distribution.—See Figure 6.212. *Teratohyla pulverata* is known from the humid lowlands on the Atlantic versant in north–central Honduras and on the Pacific slopes in southwestern Costa Rica up to 960 m; it is also known in the lowland forests on the western slopes of the Andes of Colombia and Ecuador. In Ecuador, it occurs in Esmeraldas and Pichincha provinces. Its elevational range is 27–572 m in an area of extent of occurrence of about 7880 km^2.

This nocturnal, arboreal species inhabits natural forests and intervened areas in the evergreen forest of lowlands of Equatorial Chocó ecosystem in the Cordillera Occidental of the Andes biogeographic sector, where the average annual rainfall is 1490–2365 mm and the average annual temperature is 22.6–25.6 °C.

Conservation.—*Teratohyla pulverata* is categorized as Least Concern throughout its entire range (IUCN SSC Amphibian Specialist Group, 2020k; in IUCN Red List of 2022) in view of its wide distribution and presumed large population. However, in Ecuador, *T. pulverata* was categorized as Vulnerable by Ortega-Andrade *et al.* (2021). The latter inferred a reduction in the population greater than or equal to 30% of the population in the past and a reduction in the extent of occurrence, which is estimated to be less than 20,000 km^2 (7880 km^2, according to our calculations), its distribution is severely fragmented, and there is a continuing decline in the extent and quality of its habitat. The species is known in the Chocó, an area with the highest deforestation rate in Ecuador and spreading mining. Other significant threats to its existence include habitat reduction and fragmentation, deforestation for agricultural development (particularly palm plantations), cattle ranching, human settlements, and pollution. This species is moderately common in Honduras but uncommon in Costa Rica and Colombia. In Ecuador, it has been recorded in the protected forests of Comuna San Miguel Negro, Centro Chachi Corriente Grande, and Centro Chachi Gualpi del Cayapas; the biosphere reserve Chocó Andino de Pichincha; and some private reserves (e.g., Reserva Canandé, Reserva Tayra).

Remarks.—According to Guayasamin *et al.* (2020b), it is evident that iridophores (= white iridophores) usually are maintained in preserved specimens during long periods of time. For example, the specimen KU 85476 was collected in August 1964 and still maintains the white coloration in most of its internal organs (only a portion of the digestive tract has lost the iridophores and now is cream). However, small amounts of iridophores are diffused in ethanol. In *Teratohyla pulverata*, the white dots on the dorsal surfaces and ulnar and tarsal folds are not visible in some preserved specimens. Given the importance of the presence/absence of iridophores in centrolenid taxonomy, it is important that researchers take the time to carefully record and photograph the coloration in life.

Key references.—(Peters, 1873a) (description); (Ruiz-Carranza *et al.*, 1996) (distribution); (Acosta-Galvis, 2000) (distribution); (Köhler, 2001) (distribution); (Savage, 2002) (distribution); (McCranie and Wilson, 2002) (distribution); (Bustamante *et al.*, 2007) (distribution); (Hoffmann, 2004) (tadpole); (Guayasamin *et al.*, 2008a) (distribution); (Delia *et al.*, 2017) (reproductive biology); (Kubicki, 2007) (account, call); (Guayasamin *et al.*, 2020b) (account).

FIGURE 6.221 *Teratohyla spinosa*. Left: QCAZ 45412, female. Right: QCAZ 45410, male. Durango, Provincia Esmeraldas. LAC.

TERATOHYLA SPINOSA (DWARF GLASSFROG) (TAYLOR, 1949)

Centrolenella spinosa (Taylor, 1949): 259. **Holotype.**—KU 23809 from Los Diamantes, one mile south of Guápiles, Cantón Pococí, Provincia Limón, Costa Rica.
Teratohyla spinosa Guayasamin *et al.*, 2009: 37.

Etymology.—The epithet derives from the Latin *spina* (= thorn) and refers to the spine-like prepollex that is evident next to the thumb.

Identification.—Figures 6.221, 6.222. These frogs attain a maximum SVL of 19.7 mm in males and 20.7 mm in females. The skin on the dorsum is smooth and lacks spicules; the ventral skin is smooth to slightly areolate; the ventral surfaces of the thighs below the vent lack a pair of enlarged tubercles. There is a white parietal peritoneum covering the anterior 50–60% of the venter. There is a white pericardium; there are no iridophores in the peritonea covering the intestines, stomach, and kidneys; there is a translucent peritoneum around the gall and urinary bladders. The vomers have dentigerous processes that bear or lack teeth. The snout is truncate in dorsal aspect, truncate to slightly protruding in lateral profile. The tympanum has a pronounced lateral inclination, and its diameter is about 23.5–32.0% of the eye diameter; the upper border of the tympanic annulus is obscured by the supratympanic fold; the tympanic membrane is clearly differentiated from the surrounding skin. The liver is lobate and covered by a transparent peritoneum. The humeral spines are absent. There is no webbing between fingers I and II; there is reduced webbing between fingers II and III; there is extensive webbing between fingers III and IV.

FIGURE 6.222 Morphological variation and details of *Teratohyla spinosa*. Top left: QCAZ 48335. 4 km along the train track from Alto Tambo to El Placer, Provincia Esmeraldas. Top right: QCAZ 45411. Middle left: QCAZ 45413. Middle right: CJ 2841. Durango, Provincia Esmeraldas. Bottom: QCAZ 45413. Durango, Provincia Esmeraldas. LAC.

There is relatively extensive webbing on the feet. The prepollex is clearly separated from finger I. The nuptial pad is a large to medium-sized excrescence present on the dorsal, lateral, and/or ventral sides of the thumb. Finger I is slightly longer than finger II (finger II length is 90.1–92.5% of finger I). The ulnar fold is low; the tarsal folds are absent. The width of the disc of finger III is about 33.1–40.4% of the eye diameter.

The dorsum is uniform yellowish-green. The anterior half of the venter is white, and the posterior part is transparent. Melanophores are present only on the dorsal surfaces of finger IV (and sometimes finger III) and toes IV and V. The bones are green. The iris is silvery white to yellowish-white, with black contrasting reticulations.

Comparisons.—Among glassfrogs that inhabit the lowlands of western Ecuador (Chocó ecoregion), *Teratohyla spinosa* is unique in having a small body size (SVL < 21.0 mm), a uniform green dorsum, and a white to yellowish-white iris with contrasting black reticulations and lacking humeral spines. Also, this species is exceptional by having a spine (prepollex) that is distally separated from finger I.

Natural history.—Figure 6.223. In Ecuador, *Teratohyla spinosa* has been found breeding along streams. Individuals are active during the night, and males call from the upper surfaces of leaves. Females provide short-term parental care, whereas any form of prolonged parental care is absent. In Costa Rica, egg masses are attached on the undersides of leaves near their margins 0.5 to 2.5 m above the water and sometimes over the shore. Egg masses contain 15–25 eggs, with each egg and associated capsule measuring 5–7 mm in diameter; the clutch mass is small, measuring about 21 × 25 × 7 mm.

The call from a population from Panama was described as a series of harsh "creep-creeps" followed by a considerable pause. The advertisement call of one individual from Reserva Otokiki, Provincia Esmeraldas, Ecuador, was described as being composed of 1–3 pulsed notes but usually two notes. Calls with one note have a duration of 124–139 ms; calls with two notes last 431–594 ms; a single three-note call had a duration of 793 ms. Calls are emitted every 12–35 s. Detailed data and figures are provided under the subtitle "Vocalizations of Ecuadorian Anurans" in Appendix 1.

FIGURE 6.223 *Teratohyla spinosa*. Top and middle: Reserva Otokiki, Provincia Esmeraldas. LAC. Bottom: clutch of embryos from Reserva Canandé, Provincia Esmeraldas. Jaime Culebras.

Tadpoles.—The larvae were described from Costa Rica. They are tiny on hatching, with a total length of about 11 mm at stage 25; the body is elongate, depressed; the mouth is ventral; the nares and crescent-shaped eyes are dorsal; there is a midlateral, posterior, sinistral spiracle; the vent tube is median; the tail is long with a rounded tip; the caudal fins are reduced. The oral disc is complete, moderate, with a single row of marginal papillae (21–45) laterally and ventrally and a wide anterior labium gap above mouth; the LTRF is probably 2/3, but when rows are present, rows of teeth are weaker and sometimes irregular; the A-1 row is often not evident; A-2 is usually present but often visible only as traces of teeth on the lateral extremes of the tooth ridge; the jaw sheaths are robust and well developed, the upper broadly arched and the lower one V-shaped, both with sharp serrations along their entire lengths. The body and tail are grayish-brown dorsally, lighter ventrally; a bright yellow yolk mass colors the posterior half of the body and covers the liver and intestine; the tail musculature is mostly brown, except for a pale area anteroventrally; the fins are opaquely transparent with scattered dark spots.

Distribution.—See Figure 6.217. *Teratohyla spinosa* is distributed from eastern Honduras through Nicaragua to Costa Rica and Panama in Central America. In South America, it is found in the lowlands of the Chocó ecoregion of Colombia south to Río Palenque in Ecuador. In Ecuador, it occurs in Esmeraldas, Pichincha, Santo Domingo de los Tsáchilas, and Los Ríos provinces. Its elevational range is 28–691 m in an area of extent of occurrence of about 7520 km^2.

This nocturnal, arboreal species inhabits natural forests and intervened areas in the evergreen forests of lowlands of Equatorial Chocó and evergreen montane foothill forest ecosystems of the Cordillera Occidental of the Andes biogeographic sector, where the average annual rainfall is 1907–2564 mm and the average annual temperature is 22.2 – 25.8°C.

Conservation.—*Teratohyla spinosa* is categorized as Least Concern throughout its entire range (IUCN SSC Amphibian Specialist Group, 2020l; in IUCN Red List of 2022) in view of its wide distribution and presumed large population. However, in Ecuador, *T. spinosa* was categorized as Near Threatened by Ortega-Andrade *et al.* (2021). We categorize it as Vulnerable following B1ab(iii) IUCN criteria because: (1) its extent of occurrence is estimated to be less than 20,000 km^2 (7520 km^2, according to our data), (2) its distribution is severely fragmented, and (3) there is a continuing decline in the extent and quality of its habitat. In Ecuador, significant threats to its existence include habitat reduction and fragmentation, logging, deforestation for agricultural development (particularly oil palm plantations), human settlements, and pollution. Furthermore, the area where it is distributed faces constant pressure from mining and wood companies. It is a common species in Colombia but is uncommon in Ecuador. It is regularly encountered in Costa Rica, is abundant in the Caribbean lowlands, and was present at La Selva and surrounding secondary forests. Though the species has been extirpated at El Copé, Panama, it is regularly encountered at other sites in the country. In a lowland Costa Rican subpopulation, three individuals of 10 sampled were infected with chytrid fungus; however, the effects of the chytrid fungus on the population remain unknown. It occurs in many protected areas in Colombia and Central America. In Ecuador, it has not been recorded in any protected areas.

Remarks.—According to Guayasamin *et al.* (2020b), *Teratohyla spinosa* is sister species to a lineage formed by *T. midas* + *T. adenocheira. Teratohyla spinosa* is found in the Chocó, whereas *T. midas* and *T. adenocheira* are endemic to the Amazon basin. The most like speciation mechanism that separated *T. spinosa* from its closest relatives is vicariance through the uplift of the Andes.

Key references.—(Taylor, 1949) (description); (Starrett, 1960) (tadpole); (Duellman and Burrowes, 1989) (distribution); (Ruiz-Carranza *et al.*, 1996) (distribution); (McCranie and Wilson, 2002) (distribution); (Savage, 2002) (account); (Cisneros-Heredia, 2005) (distribution); (Sunyer and Köhler, 2007) (distribution); (Kubicki, 2007) (account); (Sunyer *et al.*, 2009) (distribution); (Hoffmann, 2010) (tadpole); (Köhler, 2011) (distribution); (Delia *et al.*, 2017) (account); Yánez-Muñoz *et al.* (2020) (account); (Guayasamin *et. al.* 2020b) (account).

GENUS *VITREORANA* (GLASSFROGS) GUAYASAMIN *ET AL.*, 2009

Vitreorana Guayasamin *et al.*, 2009: 38. **Type species.**—*Centrolenella antisthenesi* Goin, 1963, by original designation.

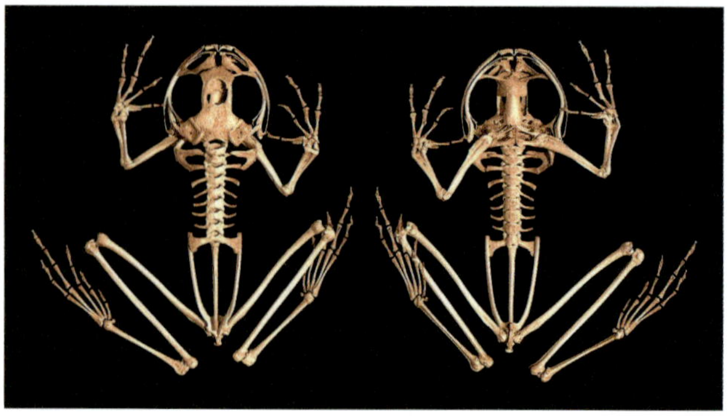

FIGURE 6.224 High-resolution computed tomography reconstructions of the skeleton of *Vitreorana antisthenesi*, KU 167350, 2Figure 4.0 mm SVL, km 29, road Maracay-Ocumare de la Costa road, Estado Aragua, Venezuela. Dorsal (left) and ventral (right) views. Daniel J. Paluh.

Etymology.—The generic name originates from the Latin words *vitreum*, meaning glass, and *rana*, meaning frog. The name refers to the ventral transparency of these frogs. The gender is feminine.

Definition.—Figure 6.224. The most conspicuous feature that diagnoses *Vitreorana* is the presence of a white hepatic peritoneum; this trait is a synapomorphy at this level of the phylogeny (also present in *Hyalinobatrachium*, *Chimerella*, and some species of *Teratohyla*). Additionally, most species in *Vitreorana* are small (SVL < 27 mm) and have a white gastrointestinal peritoneum and green bones.

Content.—*Vitreorana* contains 11 species, one of which occurs in Ecuador (22 October 2023).

Distribution.—*Vitreorana* occurs from the Cordillera de la Costa of Venezuela and the Guianas to French Guiana, Amazonia of Colombia and Ecuador, and in the Atlantic forests of Brazil and Argentina.

Remarks.—Figure 6.225. The phylogeny of species presented herein follows (Guayasamin *et al.*, 2020b).

VITREORANA RITAE (RITA'S GLASSFROG) (LUTZ IN LUTZ AND KLOSS, 1952)

Centrolene ritae (Lutz in Lutz and Kloss, 1952): 658. **Neotype.**—MCZ A96522 from headwaters of Río Caiwima, a tributary of the Río Amayaca-Yacu, ca. 70 km north–northeast of Puerto Nariño, Departamento Amazonas, Colombia.

Vitreorana ritae Cisneros-Heredia, 2013: B1.

Etymology.—The specific epithet honors Gertrud Rita Kloss, assistant of Bertha Lutz.

Identification.—Figure 6.226. These frogs attain a maximum SVL of 21 mm in males and 24 mm in females. The

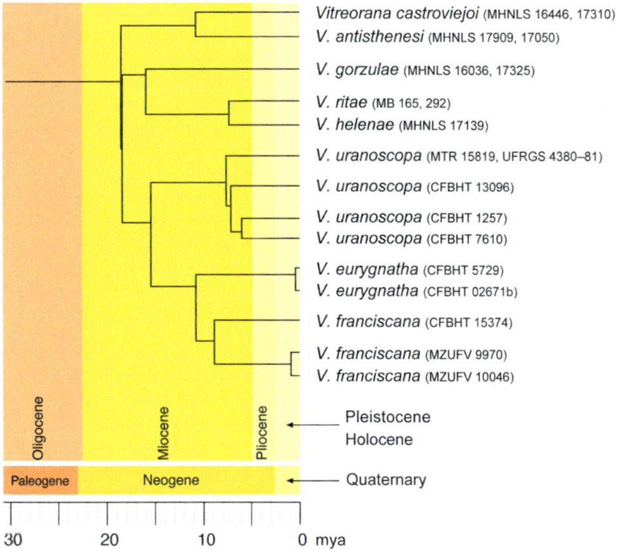

FIGURE 6.225 Time-calibrated maximum likelihood phylogenetic tree depicting evolutionary relationships of *Vitreorana*, inferred from DNA sequences of mitochondrial and nuclear genes. Phylogeny taken and redrawn from Guayasamin *et al.* (2020b).

skin on the dorsum is shagreen; spicules are absent. The ventral surfaces are granular; there is a pair of enlarged tubercles on the ventral side of the thighs. The anterior 25–40% of the ventral parietal peritoneum is white; the posterior portion is transparent. The pericardium and gastrointestinal peritoneum are white. The dentigerous process of the vomer have one tooth or lack teeth. The snout is rounded in dorsal and lateral views. The tympanum is visible and moderate in size, and its diameter is 25.8–35.4% of the eye diameter; the tympanic annulus is visible except for the upper border covered by the supratympanic fold; the tympanic membrane is differentiated, translucent, and pigmented like the surrounding skin. The liver is lobed and covered by a transparent peritoneum, except for the anterior border, which may be covered by a thin layer of iridophores. The humeral spines are absent. There is no webbing between fingers I–III; there is moderate webbing between fingers III and IV. There is moderate webbing between the toes. The ulnar fold is low and lacks iridophores. A low inner tarsal fold is present, lacking iridophores. The outer tarsal fold is absent. The prepollex is distinct (the distal end is separated from finger I). The nuptial pad is a large to medium-sized excrescence present on the dorsal, lateral, and/or ventral sides of the thumb. Finger I is slightly longer than finger II (finger II length is 84–92% of finger I). The disc of finger III is moderate; its width is 31.0–42.3% of the eye diameter.

The dorsum is green with small dark blue to black spots. The venter is white anteriorly, turning transparent posteriorly. The digestive tract is white. There are melanophores covering the dorsal surfaces of fingers III and IV. The bones are green. The iris is grayish-white with thin dark reticulations.

Comparisons.—*Vitreorana helenae*. Among glassfrogs, *V. ritae* is unique in having a green dorsum with small black spots, a small size (≤ 24 mm), a white gastrointestinal peritoneum, and a distinct prepollex. The only glassfrog that could be confused with *V. ritae* is *V. helenae*, which differs by having a yellow iris (grayish-white with a fine dark reticulation in *V. ritae*), a light greenish-yellow dorsum with dark points (green with dark points in *V. ritae*), and a mostly white hepatic peritoneum (hepatic peritoneum mostly transparent, showing the brown liver, except for some iridophores on the upper border in *V. ritae*). *Vitreorana helenae* has not been recorded in Ecuador.

FIGURE 6.226 *Vitreorana ritae*. Parque Nacional Yasuní, Provincia Orellana. Jaime Culebras.

Natural history.—During the night, *Vitreorana ritae* has been found on vegetation at about 1–5 m over streams and on the trunk of a *Ceiba*, about 7 m above the ground. The males call from the upper sides of leaves, and clutches of eggs are deposited there, too. Short-term maternal care is present; prolonged parental care is absent. In central Amazonia of Brazil, males of *V. ritae* were observed calling on vegetation above streams, about 0.5 to 3 m high on leaves of aquatic or stream-side plants. At Reserva Florestal Adolpho Ducke, Manaus, Departamento Amazonas, Brazil, breeding individuals and clutches of *V. ritae* were observed on vegetation above small streams in forests. Adults were found throughout the year, and vocalizations were heard throughout the night and were more common between January and May (from the middle to the end of the rainy season).

The advertisement call of two individuals from Boanamo community and Shiripuno Lodge, Provincia Orellana, Ecuador, were described as relatively short and composed of 2–5 notes. The mean note duration is 76 ms, and notes are strongly pulsed. The call described for the Ecuadorian populations matches the general information for Brazilian populations. Detailed data and figures are provided under the subtitle "Vocalizations of Ecuadorian Anurans" in Appendix 1.

Tadpoles.—The tadpoles in stage 25 have a labial tooth row formula of 0/1–2. The oral disc is small and ventral, with one row of large marginal papillae laterally and posteriorly. The upper jaw sheath is wide and robust; the lower jaw sheath is wide and V-shaped. The dorsum is reddish-brown. The venter is whitish. The tail muscle is reddish, and the tail fins are transparent.

Distribution.—Figure 6.227. *Vitreorana ritae* is found in Suriname, French Guiana, and Brazil and across western Amazonia in Colombia, Ecuador, Peru, and Bolivia. In Ecuador, it occurs in Sucumbíos, Orellana, Napo, and Pastaza provinces. Its elevational range is 237–355 m in an area of extent of occurrence of about 7440 km^2.

This nocturnal, arboreal species inhabits natural forests and intervened areas in the flooded forests of alluvial plains of Amazonia and evergreen forests of lowlands of Napo-Curaray ecosystems within the Aguarico-Putumayo-Caquetá and Napo-Curaray biogeographic sectors, where the average annual rainfall is 3008–3679 mm and the average annual temperature is 24.4–25.2 °C.

Conservation.—*Vitreorana ritae* is categorized as Least Concern throughout its entire range (IUCN SSC Amphibian Specialist Group, 2022s; in IUCN Red List of 2022) and in Ecuador by Ortega-Andrade *et al.* (2021) due to the broad distribution of the species in the Amazon Basin. This species is known only from a handful of localities but is likely to occur widely in the Amazon Basin. Further surveys with an emphasis on sampling bromeliads may provide a better understanding of the population dynamics of this species.

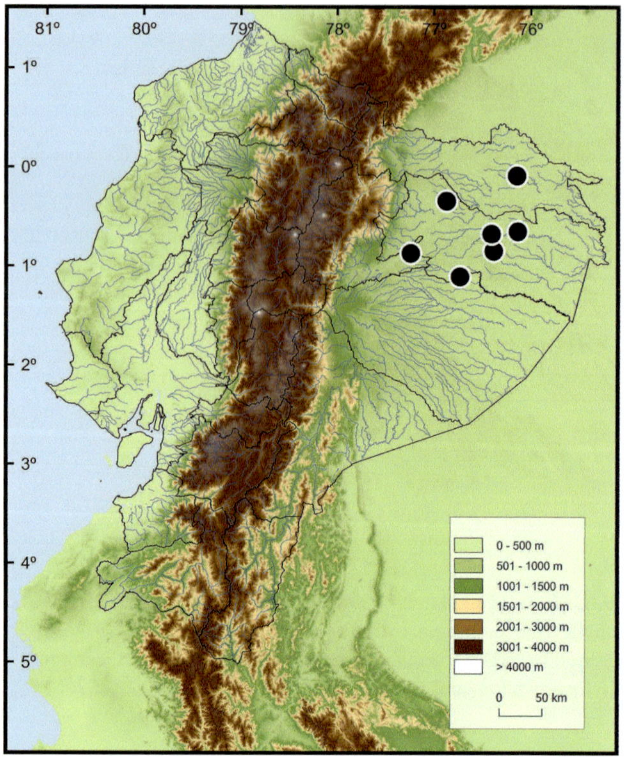

FIGURE 6.227 Localities of known occurrence of *Vitreorana ritae* in Ecuador.

There is no information on specific threats to this species, although the sites near Leticia, Colombia, are threatened by logging. In Ecuador, it has been recorded in the public reserves Reserva de Producción de Fauna Cuyabeno and Parque Nacional Yasuní and the biosphere reserves Sumaco and Yasuní.

Remarks.—According to Cisneros-Heredia (2013), *Vitreorana oyampiensis* is a synonym of *V. ritae*. Guayasamin *et al.* (2020b) cautiously adopt the change proposed by Cisneros-Heredia (2013) but note that a definitive solution to the taxonomic uncertainty involving *V. ritae* and *V. oyampiensis* requires the collection of new material of *ritae* from its type locality ("Benjamin Constant, Alto Solimões," Estado do Amazonas, Brazil).

Key references.—(Lutz in Lutz and Kloss, 1952) (description); (Lescure, 1975b) (description, under *Centrolenella oyampiensis*); (Flores, 1987) (distribution, habitat); (Lescure and Marty, 2000) (distribution); (Lima *et al.*, 2006) (distribution); (Cisneros-Heredia and Mcdiarmid, 2007) (distribution); (Kok and Castroviejo-Fisher, 2008) (distribution); (Guayasamin *et al.*, 2008b) (reproductive biology, distribution, habitat); (Menin *et al.*, 2009) (distribution); (Muñoz-Saravia and Aguayo-Vedia, 2009) (distribution); (Delia *et al.*, 2017) (reproductive biology); (Guayasamin *et al.*, 2020b) (account); (Schiesari *et al.*, 2022) (tadpole).

7 Order Gymnophiona

CAECILIANS

Gymnophia Rafinesque, 1814: 104.

Etymology.—The name of this order comes from the Greek words *gymno*, meaning naked, and *ophis*, meaning snake. Caecilians were initially believed to be closely related to snakes.

Definition.—Gymnophiona (the living caecilians) is a monophyletic group diagnosed on the basis of molecular evidence and numerous phenotypic synapomorphies. The body is greatly elongate and segmented by annular grooves; it is segmented by primary annuli, and usually, each segment contains a single vertebra. Some taxa have secondary and tertiary annuli. There are scales under the skin in many species. The smooth skin contains granular and mucous glands. Limbs are absent, and there is no remnant of the pectoral or pelvic girdles. The skull is solidly roofed, enabling it to withstand the jarring forces of digging or burrowing. Some elements, such as the maxillary and palatine, are fused as single bones. The teeth generally are curved and present in the premaxillae, maxillopalatines, vomers, dentaries, and usually on the splenials. The vertebral bears the atlas and 95–286 trunk vertebrae. The tail is usually absent; if present, it is short and pointed. The eyes are vestigial, small, covered by skin, and, in some cases, beneath the skull bones. The snout bears a retractable sensory tentacle on each side of the head between the nostril and the eye; the tentacle serves as an olfactory organ and aids in the location and identification of prey. The upper jaw protrudes beyond the lower jaw; this position allows for prey capture in narrow spaces yet does not interfere with the use of the head in burrowing. The jaw-closing mechanism of caecilians is unique in having a muscle that attaches to the retroarticular process on the dentary that, when contracted, causes the lower jaw to swing upward. External ear openings are absent, and the middle ear bears a massive columella, or it is absent. The left lung usually is rudimentary. Males have a phallodeum, a single, median, protrusible copulatory organ.

Caecilians are mostly fossorial, inhabiting moist soils adjacent to water bodies. Some species are aquatic. They are characterized by their blunt, bullet-shaped heads and cylindrical, limbless bodies for digging burrows. Three types of locomotion have been identified in caecilians: lateral undulation, whole-body concertina, and internal concertina. The latter two types depend on the skin and vertebral moving independently. Typhlonectids are exceptions to these generalities and use lateral undulation in water and whole-body concertina in wide tunnels. Presumably, all caecilians exhibit internal fertilization. Reproductive behavior in caecilians has been observed only a few times. Offspring develop internally or externally, depending upon the species, and if externally, development is indirect or direct.

Distribution.—This order of amphibians is pantropical, except for Madagascar, Oceania, and Australia.

Content.—The order contains 11 families, 33 genera, and 221 species (Frost, 2023). In Ecuador there are four families, eight genera, and 25 species.

Remarks.—For a detailed account of the nomenclatural and systematic history of caecilians, see Frost (2023 and onward). Comparative studies of their morphology have been summarized by Duellman and Trueb (1994). Due to their secretive nature, much of their biology is unknown, and our knowledge primarily comes from observations during heavy rains when they surface or from laboratory and museum specimens. Their conservation status is also mostly unknown.

Key references.—(Taylor, 1968) (taxonomic revision); (Nussbaum, 1991) (cytotaxonomy); (Hass *et al.*, 1993) (immunology, phylogeny); (Duellman and Trueb, 1994); (overview); (Wilkinson, 1997) (neuroanatomy); (Savage, 2002) (overview); (Frost *et al.*, 2006) (phylogeny, taxonomy); (Wilkinson *et al.*, 2011) (taxonomy); (Pyron and Wiens, 2011) (phylogeny); (Maddin *et al.*, 2012) (braincase); (Vitt and Caldwell, 2014) (overview); (San Mauro *et al.*, 2014) (life history evolution); (Bardua *et al.*, 2019) (caecilian skull); (Torres-Sánchez *et al.*, 2019) (genomics, burrowing); (Dubois *et al.*, 2021) (phylogenetics, nomenclature).

FAMILY CAECILIIDAE (COMMON CAECILIANS)

Cecilinia Rafinesque, 1814: 104. **Type genus.**—*Caecilia* (Linnaeus, 1758).

Etymology.—The family name is derived from the Latin *caecus*, meaning blind.

Definition.—Caeciliids are a monophyletic group diagnosed on the basis of molecular evidence and morphological features. Putative morphological synapomorphies are: (1) middle ear with an imperforate stapes; (2) inner mandibular teeth; (3) eyes surrounded or covered by the maxillopalatine; (4) monocuspid teeth. These species are generally slender, some quite large (up to 1.7 m in *Caecilia thompsoni*). All caeciliids have primary annuli; secondary annuli are present but fewer in number. Scales are typically present in the annular grooves, although *Oscaecilia elongata* lacks scales. The posterior end of the body is capped with a terminal shield but lacks a true tail. The eyes generally are not covered by bone, but there is intraspecific variation, and in

DOI: 10.1201/9781003543879-7

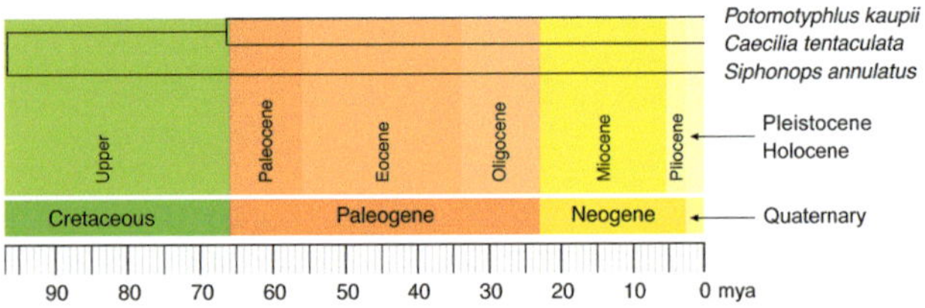

FIGURE 7.1 Time-calibrated maximum likelihood phylogenetic tree depicting evolutionary relationships of some of the Ecuadorian species of Gymnophiona, inferred from DNA sequences of mitochondrial and nuclear genes.

some species, the eyes are covered by bone. Tentacles are located below the nostrils.

Much of the natural history and biology of this fossorial family remains unknown. While *Caecilia orientalis* from Ecuador is known to be oviparous, many other aspects of their reproductive biology are still a mystery. Predators of these amphibians include hawks and snakes, and their diet may include earthworms.

Distribution.—This family of amphibians is found throughout Central and South America, ranging from Costa Rica and Panama in the north to Bolivia and southern Brazil in the south.

Content.—The family contains two genera (*Caecilia* and *Oscaecilia*) with 49 species (Frost, 2023). In Ecuador, there are two genera and 18 species.

Remarks.—Figure 7.1. The nomenclatural and systematic history is detailed in Frost (2023 and onward). The phylogeny of this group has been reconstructed using DNA sequences from both mitochondrial and nuclear genes. Evidence from molecular data suggests that the group originated at least as far back as 67 million years ago in the Upper Cretaceous period, as indicated by sequences from *Caecilia tentaculata*. The molecular phylogeny of most Ecuadorian species remains unpublished.

Key references.—(Nussbaum and Wilkinson, 1989) (phylogeny); (Frost *et al.*, 2006) (phylogeny); (Roelants *et al.*, 2007) (phylogeny); (Pyron and Wiens, 2011) (phylogeny); Wilkinson *et al.* (2011) (systematics); (Mauro *et al.*, 2012) (phylogeny); (Maddin *et al.*, 2012) (braincase, phylogeny); Vitt and Caldwell (2014) (account) Köhler (2011) (accounts species Central America).

GENUS *CAECILIA* (CAECILIANS) LINNAEUS, 1758

Caecilia Linnaeus, 1758: 229. **Type species.**—*Caecilia tentaculata* (Linnaeus, 1758) by subsequent designation by Dunn, 1942b: 494.

Etymology.—The generic name is derived from the Latin *Caecilia*, feminine form of *Caecilius*. The name refers to the byname *caecus*, meaning blind.

Definition.—Figure 7.2. These are caeciliids with the eye not covered with bone. They lack temporal fossae. The mesethmoid is exposed between frontals. The splenial teeth are present. The secondary grooves are present or absent. The scales are present or absent. The subdermal scales are present or absent. The tentacular opening is directly below the external naris, closer to the naris than

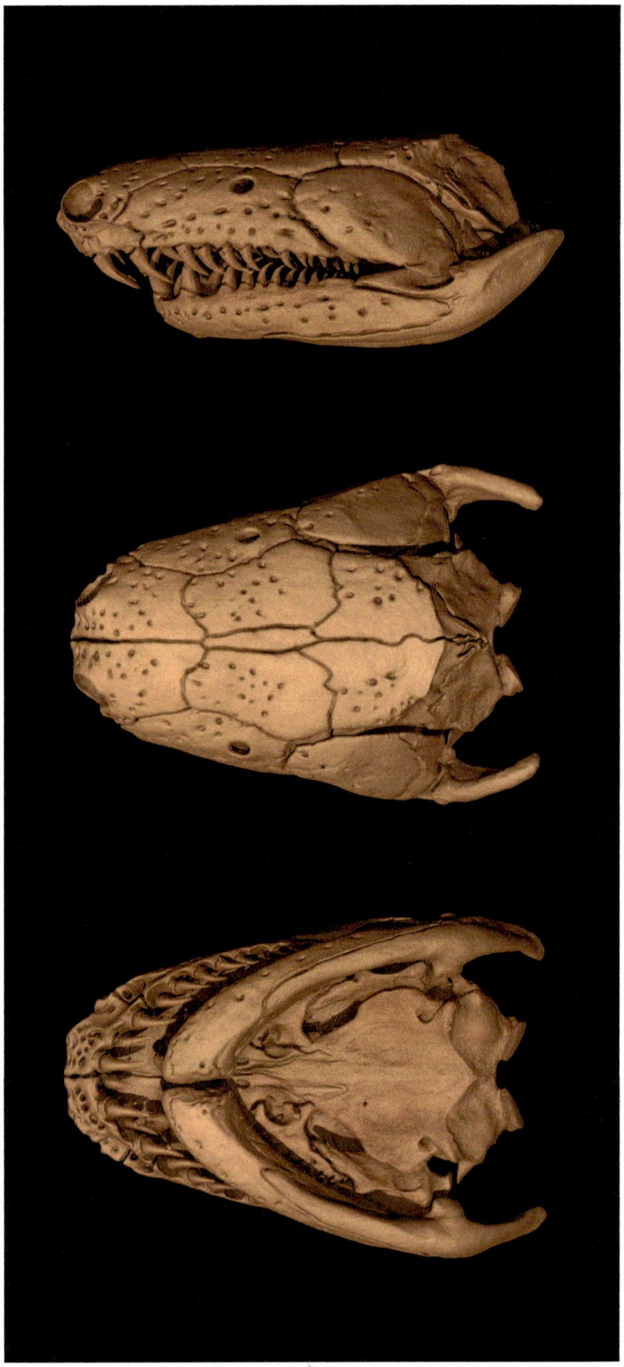

FIGURE 7.2 High-resolution computed tomography reconstructions of the cranial skeleton of *Caecilia tentaculata*, KU 175441, head length = 26.2 mm, Santa Cecilia, Provincia Sucumbíos. Lateral (upper), dorsal (middle), and ventral (lower) views.

to the eye. An unsegmented terminal shield is present or not. Narial plugs are present. There is not a diastema between the vomerine and palatine teeth. There is no terminal keel. The teeth are relatively few and large, usually placed alternately in groups; the vomeropalatine tooth row is displaced posteriorly, not parallel to the premaxillary–maxillary tooth row, diverging from the latter anteriorly and forming an angle where the two rows meet rather than a semicircle.

Content.—*Caecilia* is the largest genus of Gymnophiona in the Neotropics with 34 species, of which 15 occur in Ecuador.

Distribution.—*Caecilia* is most diverse in the lowlands up to elevations of 2689 m in Ecuador. It occurs throughout Panama and northern South America south to Bolivia.

Remarks.—Within the genus *Caecilia*, size might reach a maximum SVL of 1767 mm in males of *C. thompsoni* (Arredondo Salgar, 2007). For definition purposes, species with a maximum of total body length of 500 mm are labeled as small, the ones between 500 and 1000 mm as medium-sized, and species with a maximum size of more than 1000 mm are labeled as large. It should be noted that the individual body length of caecilians can vary according to how the animal was positioned at the time of preservation. This measurement is an estimate of the total length in life because animals might contract at death or stretch when preserved, given that the vertebrae have flexible joints, so the distance among vertebrae might vary.

CAECILIA ABITAGUAE (ABITAGUA CAECILIAN) DUNN, 1942b

Caecilia abitaguae Dunn, 1942b: 508. **Holotype.**—UMMZ 89930 from Abitagua, 1100 m, Provincia Pastaza, Ecuador.

Etymology.—The specific name refers to the type locality in the Cordillera Abitagua in Provincia Pastaza, Ecuador.

Identification.—Figure 7.3. It reaches a known length of 1303 mm. The body width is 22.1 mm. Its length is 43–59 times its width. It has 139–150 primary grooves and 3–6 secondary grooves, mostly incomplete. The tentacle is below the nostril. The eye is in a socket not covered by bone. The terminal shield is very small. Scales are present, beginning between the 40th and 50th grooves; these scales are small and few in number; at the middle of the body, the scales are transversely widened and thickened. In the posterior part of the body, the scales are large, strongly overlapping; all scales are enclosed in separate, very thin tissue pockets. There is group replacement in the teeth. The splenials are 3–3.

FIGURE 7.3 *Caecilia abitaguae*. KU 146973, western slope of the Cordillera del Cóndor, east of San José, Provincia Morona Santiago. John E. Simmons.

The dorsum is gray, dark lavender; it is slightly lighter laterally and lateroventrally. The head is bluish, contrasting with the gray body. The anterior part of the venter is somewhat brownish and somewhat violet posteriorly. The scales are often salmon colored.

Comparisons.—*Caecilia tentaculata*, *C. orientalis*, *C. disossea*, *Oscaecilia bassleri*. *Caecilia abitaguae* is similar to *C. tentaculata* in being a heavy-bodied and large caecilian. *Caecilia abitaguae* is more slender (the length is 43–59 times the diameter of the body) than *C. tentaculata* (the length is 30–35 times the diameter of the body). *Caecilia orientalis* is smaller (total length = 625) and thicker than *C. abitaguae*, its length is 21–40 times the diameter of the body. *Caecilia disossea* and *O. bassleri* have a length more than 80 times the diameter.

Natural history.—Individuals have been found dead on roads.

Distribution.—Figure 7.4. *Caecilia abitaguae*, an endemic species to Ecuador, is known from the eastern slopes of Cordillera Oriental of the Andes in Pastaza and Morona Santiago provinces. Its elevational range is 1054–1492 m in an area of extent of occurrence of about 190 km^2.

This fossorial species inhabits natural forests in the evergreen low montane forests of the northern Cordillera Oriental of the Andes ecosystem in the Amazonian Cordilleras biogeographic sector, where the average annual rainfall is 1921–3235 mm and the average annual temperature is 18.4–22.2 °C.

Conservation.—*Caecilia abitaguae* was categorized as Endangered by Ortega-Andrade *et al.* (2021). There is no information on the population status of this species. Its major threats are deforestation for agricultural development, habitat reduction and fragmentation, logging, and human settlements. Its area of distribution is under constant pressure by oil companies. It has been recorded in the public reserve Parque Nacional Llanganates.

Remarks.—Its molecular phylogenetic relationships are unknown. Records from near the type locality and from Cordillera del Cóndor (Taylor and Peters, 1974, also KU 146973, Fig, ABI1) seem to differ from the type series and require verification.

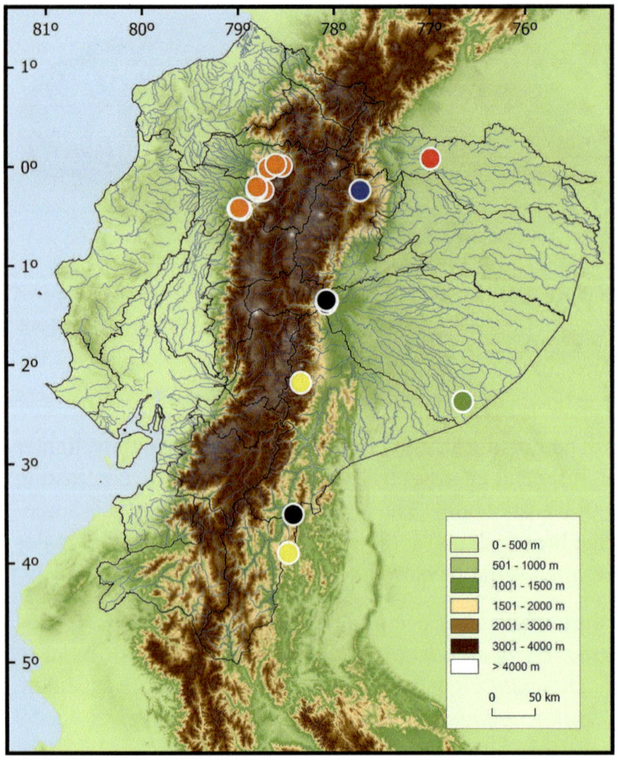

FIGURE 7.4 Localities of known occurrence of *Caecilia abitaguae* (black dots), C. *albiventris* (red dot), *C. attenuata* (blue dot), *C. bokermanni* (green dot), *C. buckleyi* (orange dots), and *C. crassisquama* (yellow dots) in Ecuador.

Key references.—(Dunn, 1942b) (description); (Taylor, 1968) (account); (Taylor and Peters, 1974) (brief account).

CAECILIA ALVIBENTRIS (WHITE-BELLIED CAECILIAN) DAUDIN, 1803

Caecilia albiventris Daudin, 1803: 423. **Holotype.**—MNHNP 840 from Suriname.

Etymology.—The specific epithet is the Latin *albus*, which means white, and *venter*, which means belly, referring to its ventral coloration.

Identification.—The species reaches a length of more than 600 mm and a body width of 18 mm in one female. The number of primary folds is approximately 144, and that of secondary folds is 45–53. The type specimen was reported as having 147 primary and 45 secondary folds. The scales begin to appear near the middle of the body, and soon, a single overlapping row is present in each succeeding fold. The scales are very large. The squamulae appear to be arranged in an orderly fashion.

Comparisons.—*Caecilia tentaculata. Caecilia albiventris* occurs in the same general area as *C. tentaculata*. It differs in color markings and, unlike *C. tentaculata*, has subdermal scales.

Natural history.—Unknown.

Distribution.—See Figure 7.4. *Caecilia albiventris* is known in Suriname, Amazonian Ecuador, and presumably in the intermediate regions, albeit not yet reported from there. In Ecuador, it occurs from Provincia Sucumbíos, at an elevation of 322.

This fossorial species inhabits natural forests in currently intervened areas in the Aguarico-Putumayo-Caquetá biogeographic sector, where the average annual rainfall is 3417 mm and the average annual temperature is 24.7 °C.

Conservation.—*Caecilia albiventris* is categorized as Data Deficient (IUCN SSC Amphibian Specialist Group, 2018h; in IUCN Red List of 2022) in view of continuing doubts as to its taxonomic validity, uncertain provenance, extent of occurrence, population status, and ecological requirements. There is no information on the population status of this species. In Ecuador, the only known locality has been deforested. It has not been recorded in any public or private reserves.

Remarks.—Its molecular phylogenetic relationships are unknown.

Key references.—(Daudin, 1803) (description); (Taylor, 1972) (taxonomy); (Taylor and Peters, 1974) (figure, range).

CAECILIA ATTENUATA (ATTENUATED CAECILIAN) TAYLOR, 1968

Caecilia attenuata Taylor, 1968: 358. **Holotype.**—NHMW 9117.1 from Peru.

Etymology.—The specific epithet *attenuata* is Latin and presumably refers to the decrease in coloration of specimens (because of the desiccation of the types) that appear lighter-colored than they would be otherwise.

Identification.—Figure 7.5. It reaches a known length of 396 mm. The body width is 6.3 mm. Its length is 62–66 times its width. It has 186–192 primary grooves and lacks secondary grooves. The tentacle is below the nostril, much closer to the nostril than to the eye. The eye is clearly visible under the skin in an open socket. There are no scales in the grooves. There are no subdermal scales. The splenial teeth are relatively large, only one by side. There is group replacement of the teeth.

The dorsum is generally light brown above and below; the grooves are darker brown. The head is somewhat lighter than the body; a seemingly distinct ventrolateral series of lighter arcs form a stripe, scarcely distinguishable, broken by darker grooves. There is a vaguely darker median ventral stripe.

Comparisons.—*Caecilia dunni, C. orientalis. Caecilia attenuata* occurs in the same general area as *C. dunni* and *C. orientalis*. They differ because *C. attenuata* is more slender than *C. dunni* and *C. orientalis* (its length is 62–66 times its width vs. about 33 times in *C. dunni* and 21–40 times in

FIGURE 7.5 *Caecilia attenuata*. KU 143666, 10.4 km north of Santa Rosa, Provincia Napo. WED.

C. orientalis). Also, *C. attenuata* has more primary grooves (186–192 in *C. attenuata* vs. 124 in *C. dunni* and 106–123 in *C. orientalis*). Furthermore, *C. attenuata* lacks secondary grooves (present in *C. dunni*).

Natural history.—Unknown.

Distribution.—See Figure 7.4. *Caecilia attenuata* is known in Ecuador and an imprecise type locality in Peru. In Ecuador, it occurs in Provincia Napo at an elevation of 1803 m.

This fossorial species inhabits natural forests in the evergreen low montane forests of the northern Cordillera Oriental of the Andes ecosystem in the Tigre-Pastaza biogeographic sector, where the average annual rainfall is 2247 mm and the average annual temperature is 17.8 °C.

Conservation.—*Caecilia attenuata* is categorized as Data Deficient (IUCN SSC Amphibian Specialist Group, 2018i; in Red List of 2019) in view of continuing uncertainties as to its extent of occurrence, status, and ecological requirements. There is no information on the population status of this species. Deforestation is a potential threat in Ecuador, although not enough information is known about this species to confirm this. It has been recorded in the public reserve Parque Nacional Cayambe-Coca and within the biosphere reserve Sumaco.

Remarks.—Its molecular phylogenetic relationships are unknown. Taylor (1973) reported the species from Ecuador based on a specimen from 10.4 km north of San Rosa, 1910 m, Provincia Napo. The large geographic gap between this specimen and the type locality (Peru) make their conspecificity suspicious.

Key references.—(Taylor, 1968) (description); (Taylor, 1973) (record from Ecuador).

CAECILIA BOKERMANNI (BOKERMANN'S CAECILIAN) TAYLOR, 1968

Caecilia bokermanni Taylor, 1968: 359. **Holotype.**—KU 200984 from Chichirota, Río Bobonaza, Provincia Pastaza, Ecuador.

Etymology.—The specific name is a patronym for Werner Carl August Bokermann (1929–1995), in the collection of whom the type specimen was originally deposited. Bokermann was a Brazilian herpetologist and ornithologist who worked at Fundación Parque Zoológico de São Paulo.

Identification.—It reaches a known length of 527 mm. The body width is about 10.2 mm. Its length is 52–68 times its width. It has 190–192 primary grooves and 15–16 secondary grooves. The eyes are vaguely visible through the skin, situated in a round socket, not or scarcely elevated. The collars are rather distinct, the second with a dorsal transverse groove. There is a very small, unsegmented terminal shield; the terminus is nearly as wide as body. The vent is nearly circular. The scales are present in grooves, somewhat irregular. There is group replacement of the teeth. The splenials are 2–2. The dorsum is a rather dark, dull violet generally; the shade on the underside is slightly lighter. The head is gray–white above; the lower side of the head is colored like the body. The area around the vent is a little lighter than the surroundings.

Comparisons.—*Caecilia dunni*, *C. orientalis*. *Caecilia bokermanni* is similar to *C. dunni* and *C. orientalis* in size. They differ in the number of primary (190–192 vs. 124 and 106–123, respectively) and secondary grooves (15–16 vs. 67 and 0 respectively).

Natural history.—Unknown.

Distribution.—See Figure 7.4. *Caecilia bokermanni* is known in the Amazonian lowlands of Ecuador and Colombia. In Ecuador, it occurs in Provincia Pastaza at an elevation of 262 m.

This fossorial species inhabits natural forests in the flooded forest of alluvial plain of Rivers of Andean and Amazonian Cordilleras Origin ecosystem in the Tigre-Pastaza biogeographic sector, where the average annual rainfall is 3168 mm and the average annual temperature is 25.5 °C.

Conservation.—*Caecilia bokermanni* is categorized as Least Concern (IUCN SSC Amphibian Specialist Group, 2019k; in IUCN Red List of 2022) in view of its relatively wide distribution, presumed large population, and lack of significant threats. There is no information regarding this species' conservation status; it is not regularly encountered. There is no information on specific threats to it, but it occurs in an area of low human impact and so is unlikely to be significantly threatened. It has not been recorded in any public or private reserves.

Remarks.—Its molecular phylogenetic relationships are unknown. It is known only from two specimens from the type locality and a specimen reported by Lynch (1999) from Amacayacu, Departamento Amazonas, Colombia.

Key references.—(Taylor, 1968) (description); (Taylor and Peters, 1974) (brief account).

CAECILIA BUCKLEYI (BUCKLEY'S CAECILIAN) BOULENGER, 1884

Caecilia buckleyi Boulenger, 1884: 398. **Holotype.**—BMNH 1946.9.5.15 from Intag, Ecuador.

Etymology.—The specific name is a patronym for Clarence Buckley (1832–1885 or 1886), who collected many specimens for the British Museum (Natural History).

Identification.—Figure 7.6. It reaches a known length of 898 mm. The body width is about 13.6 mm. Its length is 62–66 times its width. It has 149–177 primary grooves, almost all complete, and 0–8 secondary grooves. The terminal shield is present. The eye is always in a fully open socket. The scales are present from primary groove 110 to the terminus. The subdermal scales are present. The splenial teeth are present, 2–6.

The dorsum is uniformly bluish. The head is lighter and contrasts with the body color by having a bluish-rose appearance, except for the tip of the snout, which is bluish.

Comparisons.—*Caecilia guntheri*. The only similar sympatric species is *C. guntheri*, which is larger, purplish, and has fewer (108–133) body grooves.

Natural history.—At night, individuals have been found active at ground level, crossing roads, trails, or flooded areas. During the day, it is hidden under rocks, logs, or buried in the ground. As a defense mechanism, this species secrets a whitish and presumably noxious sticky substance. It feeds on termites and beetles. It is preyed upon by snakes, including *Bothrocophias campbelli*, and it is internally parasitized by nematodes.

Distribution.—See Figure 7.4. *Caecilia buckleyi*, an endemic species from Ecuador, is known on the western slopes of Cordillera Occidental of the Andes from Pichincha, Santo Domingo de los Tsáchilas, and Cotopaxi provinces. Its elevational range is 1526–2689 m in an area of extent of occurrence of about 570 km^2.

This fossorial and terrestrial species inhabits natural forests and occasionally adjacent pastures in the evergreen low montane forests of the Cordillera Occidental of the Andes and evergreen montane forests of the Cordillera Occidental of the Andes, where the average annual rainfall is 1104–2434 mm and the average annual temperature is 13.7–18.2 °C.

FIGURE 7.6 *Caecilia buckleyi*. QCAZ 47030, Nono-Tandayapa, Provincia Pichincha. LAC.

Conservation.—*Caecilia buckleyi* is categorized as Data Deficient (Ortega-Andrade *et al.*, 2021). It has been recorded in the protected forests of Cuenca Río Guayllabamba (Area 1–Area 2), Estación Científica Río Guajalito, Flanco Oriental de Pichincha y Cinturón Verde de Quito; some private reserves (e.g., Reserva Las Gralarias and Séptimo Paraíso); and within biosphere reserve Chocó Andino de Pichincha.

Remarks.—Its molecular phylogenetic relationships are unknown. This species was historically confused with *Caecilia pachynema*, from which synonymy was removed by Arteaga *et al.* (2013) following the taxonomic review done by Fletcher-Lazo (2002).

Key references.—(Boulenger, 1884) (description); (Fletcher-Lazo, 2002) (taxonomy, distribution, natural history); (Arteaga *et al.*, 2013) (account).

CAECILIA CRASSISQUAMA (THICK CAECILIAN) TAYLOR, 1968

Caecilia crassisquama Taylor, 1968: 369. **Holotype.**—AMNH 23434 from Zúñac, Río Upano, 1400–1800 m (eastern slope of the Andes), Provincia Morona Santiago, Ecuador.

Etymology.—The specific name derives from the Latin *crassis*, which means thick, and *squamea*, which means scaly, presumably referring to the thickened scales (inflexible scales) under the skin in the posterior grooves.

Identification.—Figure 7.7. It reaches a known length of 728 mm. The body width is about 11.0 mm. Its length is 66 times its width. It has 174 primary grooves. The grooves are incomplete above and below. It lacks secondary grooves. It has an unsegmented terminal shield. It has a few thickened, bony, inflexible scales in the posterior grooves. It lacks subdermal scales. The splenial teeth are present, 6.

The dorsum is olive slate. The head is lighter. The flanks are vaguely yellowish-olive with a median, not clearly defined, slate stripe, somewhat lighter than the dorsum. It is lighter ventrally under the head and neck. The anal region is yellowish-white; the grooves are faintly darker. There is a small light spot on the eye, nostril, tentacle, and vent.

FIGURE 7.7 *Caecilia crassisquama*. 9 de Octubre, Provincia Morona Santiago. Alex Achig.

Comparisons.—*Caecilia orientalis*. *Caecilia crassisquama* is similar to *C. orientalis* by lacking secondary grooves. They differ because *C. crassisquama* is slightly larger (total length = 728 vs. 625 mm) and more slender (its length is 66 times the diameter of the body vs. 31) than *C. orientalis*.

Natural history.—Unknown.

Distribution.—See Figure 7.4. *Caecilia crassisquama*, an endemic species from Ecuador, is known on the eastern slopes of the Cordillera Oriental of the Andes in Morona Santiago and Zamora Chinchipe provinces. Its elevational range is 2039–2354 m from two localities 190 km apart.

This fossorial species inhabits natural forests in the evergreen montane forests of the Cordilleras Cóndor-Kutukú ecosystem in the Amazonian Cordilleras biogeographic sector, where the average annual rainfall is 1314–1400 mm and the average annual temperature is 15.8–16.2 °C.

Conservation.—*Caecilia crassisquama* is categorized as Data Deficient (Cisneros-Heredia *et al.*, 2004a; in IUCN Red List of 2022) in view of the absence of recent information on its extent of occurrence, status, and ecological requirements. It is known only from the type specimen. Deforestation as a result of smallholder farming is a potential threat to it, but there is not enough information to confirm this. It has been recorded in the protected forests of the Cordillera del Cóndor.

Remarks.—Its molecular phylogenetic relationships are unknown.

Key references.—(Taylor, 1968) (description); (Taylor and Peters, 1974) (brief account).

CAECILIA DISOSSEA (BONY CAECILIAN) TAYLOR, 1968

Caecilia disossea Taylor, 1968: 375. **Holotype.**—AMNH 42832 from the mouth of the Río Santiago, Peru (rising into southern Ecuador, emptying into Río Marañón).

Etymology.—The specific name derives from the Latin *di-*, which means two, and *ossea*, which refers to bone or bony. The etymological intention is unknown.

Identification.—It reaches a known length of 1170 mm in females and 9661 mm in males. The body width is about 8.0 mm in the type specimen (total length 686). Its length is 86–132 times (less in young specimens) its width. It has 216 to 252 primary grooves and 16–34 secondary grooves. The tentacle is below and close to eye. The eye is in an open socket. The terminal shield is very small and unsegmented. The scales are present in grooves at about the beginning of the last third of the body. Subdermal scales are present, very small and circular. Splenial teeth are present, 2–2 or 3–3.

The dorsum is black with minute blue flecks, giving it a gray appearance. It is vaguely lighter on the venter and on the anterior 10 cm of the body. The vent area is light. The tip of the snout is lighter than the remainder of the head. The edge of the lower lip is pale gray.

Comparisons.—*Oscaecilia bassleri*. *Caecilia disossea* is similar to *Oscaecilia bassleri* by being extremely elongate, having a body length more than 80 times the diameter. In *C. disossea*, the eye is in a bony socket and is not covered by bone. In *O. bassleri*, the eye is covered by a thin layer of bone.

Natural history.—In Santa Cecilia and Lago Agrio, Provincia Sucumbíos, two individuals were moving in a clearing after heavy rains; one was unearthed from a stream bank in a clearing, and two were uncovered by bulldozers clearing a primary forest. Two females, both from June, contained small ova.

Distribution.—Figure 7.8. *Caecilia disossea* is known in Amazonian Ecuador and Peru. In Ecuador, it is known from Sucumbíos, Orellana, and Pastaza provinces. Its elevational range is 182–955 m in an area of extent of occurrence of about 39,770 km^2.

This fossorial species inhabits natural forests and adjacent intervened areas in five ecosystems (Table 0.4) within four biogeographic sectors (Table 0.5), where the average annual rainfall is 2896–4409 mm and the average annual temperature is 21.1–25.6 °C.

Conservation.—*Caecilia disossea* is categorized as Least Concern throughout its entire range (IUCN SSC Amphibian Specialist Group, 2018j; in IUCN Red List of 2022) and in Ecuador by Ortega-Andrade *et al.* (2021) in view of its wide distribution, tolerance of a degree of habitat modification, and presumed large population. It is a rare species, and the population status is poorly known but probably overlooked due to its fossorial habits. This species occurs in an area of low human impact and has been found in cleared areas; therefore, it is unlikely to be significantly threatened. In Ecuador, it has been recorded in the public reserve Parque Nacional Yasuní, the protected forests of CEPLOA, an the Reserva de la Biósfera Yasuní.

Remarks.—Its molecular phylogenetic relationships are unknown.

Key references.—(Taylor, 1968) (description); (Taylor and Peters, 1974) (records from Ecuador); (Duellman, 1978) (account).

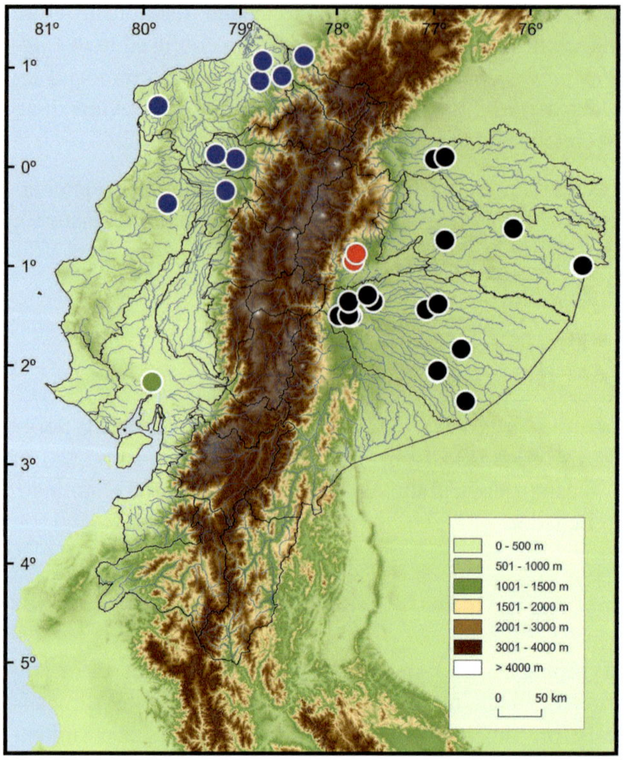

FIGURE 7.8 Localities of known occurrence of *Caecilia disossea* (black dots), *C. dunni* (red dots), *C. leucocephala* (blue dots), and *C. tenuissima* (green dot), in Ecuador.

CAECILIA DUNNI (DUNN'S CAECILIAN) HERSHKOVITZ, 1938

Caecilia dunni Hershkovitz, 1938: 370. **Holotype.**—UMMZ 82901 from the middle of the road between the settlements of Archidona and Tena (eastern slope), about 520 m, Provincia Napo, Ecuador.

Etymology.—The specific name is a patronym for Dr. Emmet Reid Dunn (1894–1956) in acknowledgment of his labors in the field of research on Gymnophiona, of his examination of the authors' specimens, and of his unselfish assistance in supplying necessary data difficult to procure otherwise.

Identification.—It reaches a known length of 464 mm. The body width is about 14.1 mm. Its length is about 33 times its width. It has 124 primary grooves and 67 secondary grooves. It has a very small, unsegmented terminal shield. The eye is small, in a socket, usually visible. The tentacle is below the nostril. The scales are present in grooves, and there are also scattered subdermal scales. The tongue has two blackish narial plugs (or valvules). It has splenial teeth, 6.

The dorsum is light blue; the ventral surface color is paler with grayish mottling.

Comparisons.—*Caecilia orientalis*, *C. bokermanni*. *Caecilia dunni* is similar to *C. orientalis* and *C. bokermanni* in size. It differs from *C. orientalis* by having secondary grooves (absent in *C. orientalis*), and from *C. bokermanni* by having fewer primary grooves (124 vs. 190–192).

Natural history.—Unknown.

Distribution.—See Figure 7.8. *Caecilia dunni*, an endemic species to Ecuador, is known on the eastern slopes of the Cordillera Oriental of the Andes in Provincia Napo. Its elevational range is 526–667 m in only two nearby localities.

This fossorial species inhabits natural forests in the evergreen montane foothill forests of the northern Cordillera Oriental of the Andes, where the average annual rainfall is 4038–4398 mm and the average annual temperature is 22.5–23.2 °C.

Conservation.—*Caecilia dunni* was categorized as Endangered by Ortega-Andrade *et al*. (2021). It is a rare species that is only known from two localities. Thus, we categorize it as Critically Endangered following B1ab(iii) IUCN criteria because: (1) its extent of occurrence is estimated to be less than 100 km², (2) its distribution is severely fragmented, and (3) there is a continuing decline in the extent and quality of its habitat. Its major threats are deforestation for agricultural development, habitat reduction and fragmentation, logging, and human settlements. It has been recorded in the Reserva de la Biósfera Sumaco.

Remarks.—Its molecular phylogenetic relationships are unknown.

Key references.—(Hershkovitz, 1938) (description).

CAECILIA GUNTHERI (GÜNTHER'S CAECILIAN) DUNN, 1942b

Caecilia guntheri Dunn, 1942b: 510. **Holotype.**—BMNH 1946.9.5.12 from western Ecuador.

Etymology.—The specific name is a patronym in honor of Albert C. L. G. Günther (1830–1914), an English herpetologist of German origin. In 1856, he went to the British Museum, where he worked in ichthyology. In 1875, he became the Director of the Department of Zoology of the Natural History Museum of England until 1895.

Identification.—Figure 7.9. This is the second-largest known species of *Caecilia*. It reaches a known length of 1602 mm in a female and 1010 mm in a male. It has 108–133 primary grooves and 7–22 secondary grooves. The terminal shield is present. The eye is always in a fully open socket. Scales begin at about the 26th fold, few at first, later forming a single complete row in each groove posteriorly (data from the holotype). The splenials are present, 4–6.

FIGURE 7.9 *Caecilia guntheri*. CJ 2786, Reserva Otokiki, Provincia Esmeraldas. LAC.

The body is purplish-gray, bluish-gray, and pale purple dorsally with a few scattered paler or white spots; the flanks and venter are creamy brown, lead gray, and creamy brown; the grooves are reddish-purple ventrally and creamy brown on the flanks.

Comparisons.—*Caecilia buckleyi*. The only similar sympatric species is *C. buckleyi*, which is bluish and has 149–177 body grooves (108–133 in *C. guntheri*).

Natural history.—Figure 7.10. Individuals have been found active on the ground after heavy rains and flooding or otherwise buried or under rotten logs. It is strong, slimy, and jittery. It will sometimes bite if handled. It feeds on slugs and earthworms in the wild and crickets and mice under laboratory conditions. Gravid females have been found to contain 22 eggs.

Distribution.—Figure 7.11. *Caecilia guntheri*, an endemic species from Ecuador, is known in the western lowlands and slopes of the Cordillera Occidental of the Andes in Esmeraldas, Pichincha, Santo Domingo de los Tsáchilas, and Cotopaxi provinces. Its elevational range is 665–1940 m in an area of extent of occurrence of about 1360 km².

FIGURE 7.10 *Caecilia guntheri*. QCAZ 11860, Reserva Integral Otonga, Provincia Cotopaxi. LAC.

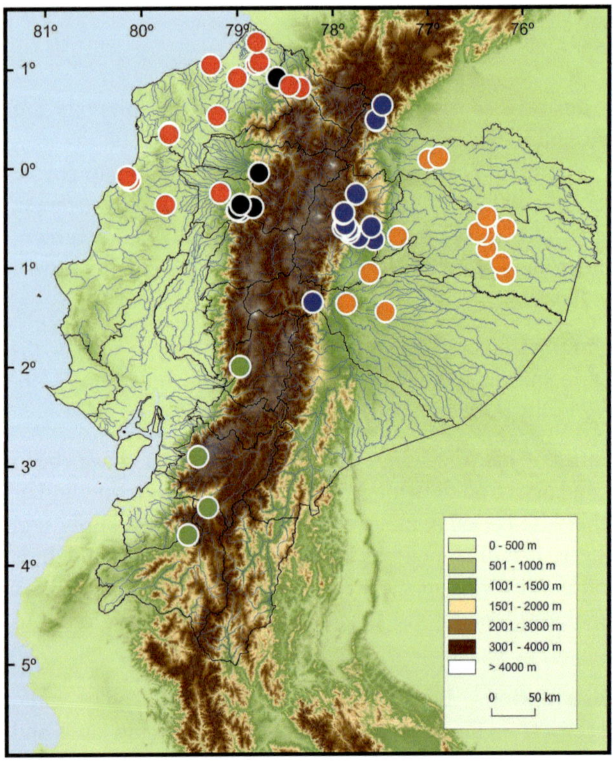

FIGURE 7.11 Localities of known occurrence of *Caecilia guntheri* (black dots), *C. nigricans* (red dots), *C. orientalis* (blue dots), *C. pachynema* (green dots), and *C. tentaculata* (orange dots) in Ecuador.

This cathemeral and fossorial species inhabits natural forests and pastures in the evergreen montane foothill forest, evergreen low montane forest, and evergreen montane foothill forest ecosystems in the Cordillera Occidental of the Andes biogeographic sector, where the average annual rainfall is 2209–2549 mm and the average annual temperature is 16.3–22.3 °C.

Conservation.—*Caecilia guntheri* is categorized as Least Concern throughout its entire range (IUCN SSC Amphibian Specialist Group, 2019l; in IUCN Red List of 2022). However, it is important to note that in Ecuador, *C. guntheri* has been categorized as Endangered by Ortega-Andrade *et al.* (2021) because its extent of occurrence is estimated to be less than 5000 km^2 (1360 km^2, according to our data), its distribution is severely fragmented, and there is a continuing decline in the extent and quality of its habitat. There is no information on its population status. Significant threats to its existence include habitat reduction and fragmentation, logging, deforestation for agricultural development, fires, mining, human settlements, and pollution. Furthermore, the area where it is distributed faces constant pressure from mining and wood companies. It has been recorded within the biosphere reserve Chocó Andino de Pichincha, and it occurs in some private reserves (e.g., Reserva Otokiki, Reserva Otonga).

Remarks.—Its molecular phylogenetic relationships are unknown. Fletcher-Lazo (2002) reviewed its taxonomy, but his thesis is not yet published. According to him, the population of *Caecilia guntheri* from Mindo differs in some morphometric features from the type, the locality of which in Ecuador was not specified. This species is one of the largest amphibians of the world and the second-largest known caecilian.

Key references.—(Dunn, 1942b) (description); (Fletcher-Lazo, 2002) (taxonomy, distribution, natural history); (Arteaga *et al.*, 2013) (account).

CAECILIA LEUCOCEPHALA (WHITE-HEADED CAECILIAN) TAYLOR, 1968

Caecilia leucocephala Taylor, 1968: 404. **Holotype.**—KU 200985 from Virology Field Station, Río Raposo, Valle del Cauca, south of Buenaventura, Colombia.

Etymology.—The species name comes from the Greek words *leuco*, which means white, and *kefali*, which means head and refers to the white head of this species.

Identification.—Figure 7.12. It reaches a known length of 189 mm in the type. The body width is about 6.0 mm in the type. Its length is about 32 times its width. It has 118 incomplete primary grooves and 42 secondary grooves, with eight or nine complete. The skull appears to be largely cartilaginous. The terminal shield is very small and unsegmented. The eye is in a socket and is vaguely visible externally.

FIGURE 7.12 *Caecilia leucocephala*. Top and middle: QCAZ 48290. Bottom: 48291. Reserva Otokiki, Provincia Esmeraldas. Ecuador. LAC.

The tentacle is below and in front of the nostril. The scales are in a single large row posteriorly. There is a median dentary tooth in the lower jaws. There are two slightly swollen areas of glandules on the sides of the occiput that are not in contact. The splenial teeth are very small, 4.

The body is bluish-gray, and the head is rose (under a lens, showing minute flecks of pigment except on the snout). The lower jaw is about the same color as the snout. The body anteriorly is rather brownish for 3 cm or so before gradually darkening, becoming blackish posteriorly on the sides of the neck and ventral part, which is much lighter. The chin is dark, and the remainder of the venter grows darker and also becomes blackish posteriorly. The area around the vent is whitish. The anal glands are yellowish (male).

Comparisons.—*Caecilia nigricans*. *Caecilia leucocephala* differs from *C. nigricans* (with which it is sympatric) by having fewer primary folds (*C. leucocephala* 118 and *C. nigricans* 156–192), being shorter (maximum body length for *C. leucocephala* 189 mm, for *C. nigricans* 1440 mm), and by having a rose head instead of a bluish-gray head in *C. nigricans*.

Natural history.—A female found in June 1985 in Puerto Quito, Provincia Pichincha, contained follicular ova reaching 1.5 mm.

Distribution.—See Figure 7.8. *Caecilia leucocephala* is known from eastern Panama through the Pacific lowlands of western Colombia (including Gorgona Island) to northwestern Ecuador. In Ecuador, it occurs in Esmeraldas, Carchi, Pichincha, Santo Domingo de los Tsáchilas, and Manabí provinces. Its elevational range is 39–699 m in an area of extent of occurrence of about 14,700 km^2.

This fossorial species inhabits natural forests and adjacent intervened areas in the evergreen forests of lowlands of Equatorial Chocó, evergreen montane foothill forests of the Cordillera Occidental of the Andes, and evergreen seasonal montane foothill forests of the Cordillera Costera of Chocó, where the average annual rainfall is 1522–3441 mm and the average annual temperature is 22.2–25.6 °C.

Conservation.—*Caecilia leucocephala* is categorized as Least Concern (Lynch *et al.*, 2008; in IUCN Red List of 2022) through its entire range in view of its wide distribution, presumed large population, and because it is unlikely to be declining fast enough to qualify for listing in a more threatened category. It was categorized as Vulnerable in Ecuador by Ortega-Andrade *et al.* (2021). In Ecuador, its current extent of occurrence is less than 20,000 km^2 (14,700 km^2), its distribution is severely fragmented in fewer than 10 known localities, and there is a continuing decline in the extent and quality of its habitat. It is a rarely encountered species. The major threats to its existence include habitat reduction and fragmentation, logging, deforestation for agricultural development (particularly palm plantations), fires, mining, the introduction of exotic species, human settlements, and pollution. Furthermore, the area where it is distributed faces constant pressure from wood companies. In Ecuador, it has been recorded in the public reserve Reserva Ecológica Mache Chindul.

Remarks.—Its molecular phylogenetic relationships are unknown. Fletcher-Lazo (2002) and Arroba (2017) reviewed its taxonomy, but their theses have not yet been published. According to the latter, the specimens identified as *Caecilia leucocephala* in museum collections need to be reidentified given that several cryptic species are involved.

Key references.—(Taylor, 1968) (description); (Fletcher-Lazo, 2002) (taxonomy, distribution, natural history); (Lynch, 1999) (distribution, morphology); (Ortega Andrade *et al.*, 2010b) (brief account); (Arroba, 2017) (systematics, phylogeny); (Yánez-Muñoz *et al.*, 2020) (account).

CAECILIA NIGRICANS (BLACK CAECILIAN) BOULENGER, 1902

Caecilia nigricans Boulenger, 1902: 51. **Holotype.**—BMNH 1946.9.5.52, Río Lita, 915 m, northwestern Ecuador.

Etymology.—The species name comes from the Latin *nigricans*, which means blackish.

Identification.—It reaches a known length of 1440 mm. Its length is about 44–83 times its width. It has 156–192 primary grooves and 35–69 secondary grooves. There is a tentacle below and slightly behind the nostril. The terminal shield is absent. The eye is in a fully open socket. The subdermal scales are present, as well as scales in the primary and secondary grooves. The splenials are present, 2–6. It has 175–196 vertebrae.

The dorsum is nearly uniform black. The head is bluish-gray. The flanks show a lighter, slightly lavender–slate shade. The median ventral part is slate, of a slightly grayer shade than the dorsum. The tip of the snout, the upper lip, the area around the tentacle, and the vent are whitish. In certain light, the grooves show a slight light line bordered by a darker color, this is scarcely discernible.

Comparisons.—*Caecilia leucocephala*. *Caecilia nigricans* differs from *C. leucocephala* by having more body grooves (156–189 vs. 142), by being larger (maximum body length for *C. nigricans* 1440 mm vs. 438 mm in *C. leucocephala*), and by having a bluish-gray head instead of rose in the latter.

Natural history.—Unknown.

Distribution.—See Figure 7.11. *Caecilia nigricans* is known from the eastern lowlands of southeastern Darien Province, Panama, south through the Pacific lowlands of Colombia (including Gorgona Island) to northern Ecuador and in the Caribbean drainage of Venezuela, Colombia, and Panama. In Ecuador, it occurs from Esmeraldas, Imbabura, Santo Domingo de los Tsáchilas, and Manabí provinces. Its elevational range is 22–844 m in an area of extent of occurrence of about 18,280 km^2.

This diurnal, fossorial species inhabits natural forests and adjacent intervened areas in five ecosystems (Table 0.4) within five biogeographic sectors (Table 0.5), where the average annual rainfall is 863–2666 mm and the average annual temperature is 21.8–25.8 °C.

Conservation.—*Caecilia nigricans* is categorized as Least Concern (Lynch *et al.*, 2010; in IUCN Red List of 2022) through its entire range in view of its wide distribution, presumed large population, and because it is unlikely to be declining to qualify for listing in a more threatened category. It was categorized as Vulnerable in Ecuador by Ortega-Andrade *et al.* (2021). In Ecuador, its current extent of occurrence is less than 20,000 km^2 (18,280 km^2), its distribution is severely fragmented, and there is a continuing decline in the extent and quality of its habitat. It is a rarely encountered species. In Colombia, it is a common species. In Panama, it is known only from a single locality. The major threats to its existence include habitat reduction and fragmentation, logging, deforestation for agricultural development (particularly palm plantations), fires, mining, the introduction of exotic species, human settlements, and pollution. Furthermore, the area where it is distributed faces constant pressure from wood companies.

In Ecuador, it has been recorded in the public reserve Reserva Ecológica Mache Chindul.

Remarks.—Its molecular phylogenetic relationships are unknown. Fletcher-Lazo (2002) reviewed its taxonomy, but his thesis has not yet been published.

Key references.—(Boulenger, 1902) (description); (Fletcher-Lazo, 2002) (taxonomy, distribution, natural history); (Lynch, 1999) (distribution, morphology); (Lynch *et al.*, 2010) (conservation); (Ortega-Andrade *et al.*, 2010b) (brief account); (Yánez-Muñoz *et al.*, 2020) (account).

CAECILIA ORIENTALIS (EASTERN CAECILIAN) TAYLOR, 1968

Caecilia orientalis Taylor, 1968: 417. **Holotype.**—USNM 159789 from La Bonita, 1920 m, Provincia Sucumbíos, Ecuador.

Etymology.—The specific name presumably refers to the region of the type locality, which is in the eastern region or Región Oriental of Ecuador.

Identification.—Figure 7.13. It reaches a known length of 625 mm. The body width (in the type) is about 12.1 mm at midbody. Its length is about 31 times its width. It has 106–123 primary grooves; they are incomplete above and below. There are no secondary grooves. The eyes are in a socket, visible externally. The tentacle is below and very slightly in front of the nostril, closer to the mouth than to the nostril. The scales are present in the topotype population or absent in the Baeza and surrounding population; there are no subdermal scales. There is a terminal unsegmented shield. There is group replacement of the teeth. The splenial teeth are present, 2–2. There are 120–128 vertebrae (n = 2).

Comparisons.—*Caecilia dunni*. *Caecilia orientalis* is similar to *C. dunni* in being stout. They differ mostly because *C. orientalis* lacks secondary grooves, whereas they are present in *C. dunni*.

Natural history.—Figure 7.14. Individuals were found during day underneath large, highly decomposed logs (see also Distribution under *Epicrionops bicolor*). In a study at the private reserve Cabañas San Isidro, Provincia Napo,

FIGURE 7.13 *Caecilia orientalis*. Top Right: El Reventador, Provincia Napo. Amanda B. Quezada. Bottom and right: Left: MZUTI 1373, road Oyacachi to El Chaco, 1800 m, Provincia Napo. LAC.

FIGURE 7.14 *Caecilia orientalis*. Top: QCAZ 21420, clutch of eggs. Bottom: embryo of the same clutch. Estación Científica Yanayacu, near Cosanga, Provincia Napo. LAC.

Caecilia orientalis (southern population) was the main prey item brought to the nest by two barred hawk parents (*Leucopternips princeps*) to feed a single chick brooded each year. Prey items delivered to nestlings included 50 caecilians (48.1% of the diet), and they are a dominant prey in the diet of the barred hawk. Barred hawks brought significantly more caecilians than snakes to the nest during hour-long time intervals when it was raining. During several years of research at this site and nearby Yanayacu Biological Station, caecilians were rarely encountered on the surface in the day, even when it was raining. Thus, it is unknown how barred hawks are able to consistently find these elusive amphibians. This species is oviparous and exhibits direct development with no aquatic larval stage. Observations were made at Yanayacu Biological Station, Baeza, and Cosanga, Provincia Napo (southern population, see Remarks). A clutch with seven eggs was found on 16 January 2001. The nest site was very wet and was several meters from a small, 1 m wide creek. The female was found approximately 5 cm from the clutch (presumably attending the clutch), and one of the males was approximately 5 cm from the female (presumably, biparental care might occur). The eggs were connected to each other by cords attached to both ends of each egg. The eggs were transparent, such that embryos could be seen inside. Embryos had distinct dark eyes, feathery external gills, and creamy yellow yolk sacs and moved when the eggs were touched. The lengths and diameters of eggs were 10.6, 10.3, and 10.2 mm, and the diameters in the same order were 8.8, 8.3, and 8.1 mm. Four females had large numbers of ova (mean ± SE = 129 ± 6 ova). Most ova were small (< 1.5 mm), although the female found with the clutch contained seven ova ranging from 3.5 to 5.7 mm long, and two of the other females had 30 and 22 medium-sized ova, respectively, ranging from 1.5 to 2.5 mm long. The different sizes of ova likely represent different stages of development. The temporal sequence of ovum maturation in the dissected females suggests that reproduction is seasonal in *C. orientalis*. Ovulation starts by January, the beginning of the rainy season at Yanayacu Biological Station. The female found in August had only very small ova, suggesting reproductive quiescence during this dry period of the year. The two females found in November have medium-sized ova, indicating that by this point, vitellogenesis is occurring. The observation that a female had mature ovarian ova at the same time she was guarding the reported clutch also suggests that *C. orientalis* is able to produce two clutches per rainy season. This conclusion is supported by the observation that three *C. orientalis* females found after the rainy season did not have mature ova. Thus, the female likely would have laid her mature ovarian ova later during the same rainy season in which she was caught.

Distribution.—See Figure 7.11. *Caecilia orientalis* is known from the Cordillera Occidental and Cordillera Central of Colombia and the Cordillera Oriental in northeastern Ecuador. In Ecuador, it occurs in Sucumbíos, Napo, and Tungurahua provinces. Its elevational range is 1041–2253 m in an area of extent of occurrence of about 5250 km^2.

This fossorial species inhabits natural forests and adjacent intervened areas in the evergreen low montane forest, evergreen montane foothill forest, and the evergreen montane forest ecosystems in the northern Cordillera Oriental of the Andes biogeographic sector, where the average annual rainfall is 1321–3896 mm and the average annual temperature is 14.6–21.0 °C.

Conservation.—*Caecilia orientalis* is categorized as Least Concern (IUCN SSC Amphibian Specialist Group, 2018k; in Red List of 2019) through its entire range in view of its relatively wide distribution, presumed large population, and tolerance of a degree of habitat modification. It was categorized as Vulnerable in Ecuador by Ortega-Andrade *et al.* (2021). In Ecuador, its current extent of occurrence is less than 20,000 km^2 (5250 km^2), its distribution is severely fragmented, and there is a continuing decline in the extent and quality of its habitat. There is no information on the population of this species. The major threats to its existence are habitat reduction and fragmentation, oil spills, deforestation, agriculture, and volcanic eruptions. It exhibits a degree of tolerance to habitat disturbance (secondary forest and pastures).

In Ecuador, it has been recorded in the public reserves Parque Nacional Cayambe Coca, Parque Nacional Sumaco Napo-Galeras, and Reserva Ecológica Antisana; the protected forests of Cerro Sumaco y Cuenca Alta del Río Suno, Cumandá; and some private reserves (e.g., Cabañas San Isidro).

Remarks.—Its molecular phylogenetic relationships are unknown. Taylor (1968) distinguish two populations of *Caecilia orientalis*, the northern (from La Bonita and 3 km north of Sebundoy, Provincia Sucumbíos) and the southern (from Baeza–Borja, Provincia Napo). Paratype specimens from the southern population were distinguished in the original description of the type (northern population) by lacking scales, which are generally present in the northern population. These differences suggest that two species might be included under this name. Also, the records from Colombia cited by Lynch (1999) need further revision, given that he noted some differences with Ecuadorian specimens.

The morphology of prehatching embryos of *Caecilia orientalis* (southern population) was described by Pérez *et al.* (2009). The state of development of advanced embryos of *C. orientalis* most closely resembles those of stage 47/48 *Gegeneophis ramaswamii*, an Indian caeciliid, and stage 47/48 *Hypogeophis rostratus*, a Seychellean caeciliid, both direct developers, in details of bone mineralization, chondrocranial degeneration, and vertebrogenesis. They are most like stage 45 *H. rostratus* in external features (gills, pigmentation, etc.). The similarity of developmental features among the direct developers suggests a correlation with the mode of life history. A noteworthy feature is that *C. orientalis* has an armature of multiple rows of teeth on the lower jaw with tooth crowns that resemble the "fetal" teeth of viviparous taxa and that are covered with a layer of oral mucosal epithelium until full development and eruption, but the upper jaw bears a single row of widely spaced, elongate, slightly recurved teeth that resemble those of the adult.

Key references.—(Taylor, 1968) (description); (Lynch, 1999) (distribution, morphology); (Funk *et al.*, 2004) (natural history); (Greeney *et al.*, 2008) (predation); (Pérez *et al.*, 2009) (development).

CAECILIA PACHYNEMA (GIANT CAECILIAN) (GÜNTHER, 1859)

Coecilia pachynema (Günther, 1859): 417. **Holotype.**—BMNH 1946.9.6.83 from western Ecuador.

Caecilia pachynema Cope, 1868: 118.

Etymology.—The specific name comes from the Greek *pachy*, which means thick, and *nema*, meaning thread or filament. Although the etymological intention is unknown, presumably, it refers to the body shape.

Identification.—Figure 7.15. (Based in the holotype only, see Remarks). It reaches a known length of 944 mm in a female. The body width is about 11.3 mm. Its length is about 84 times its width. There are 169 primary grooves; there are no complete rings dorsally, and the last eight grooves are complete ventrally. There are no secondary grooves. The terminal shield is conspicuous, and it interrupts four primary grooves dorsally. Elliptical and rectangular scales are present near the terminus, making complete rows around the body in the last grooves. The eyes are located dorsolaterally, covered by whitish skin and visible externally as two small black circles. The splenials are present, 2–2.

The dorsum and venter are dark blue, while the head is slightly lighter than the body. White stripes run along the ventrolateral side of the body.

Comparisons.—*Caecilia buckleyi*, *C. nigricans*, *C. tenuissima*. The species most similar to *C. pachynema* is *C. buckleyi*. They differ in coloration by the presence of light lateral stripes in *C. pachynema* (absent in *C. buckleyi*) and morphometry. *Caecilia nigricans* and *C. tenuissima* have primary grooves and ranges that overlap *C. pachynema*'s, but they differ from the latter by having secondary folds and not having a terminal shield or light-colored lateral stripes.

Natural history.—Figure 7.16. At Paccha, Provincia El Oro, three individuals were found (two live, one dead) on

FIGURE 7.15 *Caecilia pachynema*. QCAZ 46983, Río Chipla, near Molleturo, 2500 m, Provincia Azuay. LAC.

the road during heavy rain. The site has grassland on one side of the road and a steep hill on the other side (Thomas Verkade, field notes from 29 March 2021).

Distribution.—See Figure 7.11. *Caecilia pachynema*, an endemic species to Ecuador, is known from the Pacific drainage systems of western Ecuador in Chimborazo, Azuay, El Oro, and Loja provinces. Its elevational range is 1180–1950 m in an area of extent of occurrence of about 3810 km^2.

This fossorial species inhabits natural forests in the desert shrub of southern valleys ecosystem in the Catamayo-Alamor, Cordillera Occidental of the Andes, and valleys biogeographic sectors, where the average annual rainfall is 687–1267 mm and the average annual temperature is 16.5–21.9 °C.

Conservation.—*Caecilia pachynema* was categorized as Least Concern by Ortega-Andrade *et al.* (2021). However, we consider *C. pachynema* (*sensu* this publication, see Remarks) as Endangered following B1ab(iii) IUCN criteria because: (1) its extent of occurrence is estimated to be less than 5000 km^2 (3810 km^2), (2) its distribution is severely fragmented, and (3) there is a continuing decline in the extent and quality of its habitat. There are important threats to this species; they are deforestation, agriculture, the rearing of livestock, wood extraction, pollution, mining, landslides, roads, urbanization, and wildfires. In Ecuador, it has not been recorded in any public or private reserves.

FIGURE 7.16 *Caecilia pachynema*. Paccha, Provincia El Oro. Thomas and Jan C. Verkade.

Remarks.—Its molecular phylogenetic relationships are unknown. *Caecilia pachynema* was historically confused with *C. buckleyi*. Their synonymy was removed by Arteaga *et al.* (2013) following the taxonomic review done by Fletcher-Lazo (2002). The account by Yánez-Muñoz *et al.* (2020) seems to include misidentifications.

Key references.—(Günther, 1859) (description).

CAECILIA SUBTERMINALIS (TAYLOR'S CAECILIAN) TAYLOR, 1968

Caecilia subterminalis Taylor, 1968: 437. **Holotype.**—EHT-HMS 1732 from Ecuador.

Etymology.—The specific name derives from the Latin *sub-*, which means below or under, and *terminalis*, which means at the end or terminal. It presumably refers to the short terminal shield.

Identification.—It reaches a known length of 436 mm. It has 170 primary grooves (complete on venter) and 16 secondary grooves. The terminal shield is short and partly segmented. The eye is vaguely dark, forming a slightly raised elevation on the surface. Scales begin near the head (14th primary); the scales are small, dorsolateral, two to three on a side; in the middle of the body, there are more scales in a single lateral row. At beginning of the secondaries, they nearly surround the body in the primary grooves, posteriorly with one complete row in each fold. The secondaries have scales throughout their length; subdermal scales also are present in connective tissues, widespread. Splenial teeth are present, 4.

The dorsum is dark lavender–slate, the head grayish; the upper lip and the tip of the snout are yellowish, as is the chin and jaws to the middle of the second collar. There is a yellowish stripe or mark on the side of the head and along the flanks. The flank stripes widen and fuse in front of the vent. The venter is colored like the dorsum, with a vaguely lighter shade anteriorly.

Comparisons.—*Caecilia pachynema*. *Caecilia subterminalis* superficially resembles *C. pachynema*, but the latter has dark lines bordering the grooves.

Natural history.—Unknown.

Distribution.—*Caecilia subterminalis*, an endemic species to Ecuador, is known only from a general type locality, Ecuador.

Conservation.—*Caecilia subterminalis* is categorized as Data Deficient (Cisneros-Heredia *et al*., 2004b; in Red List of 2019) in view of continuing doubts as to its taxonomic validity, as well as absence of recent information on its extent of occurrence, status, and ecological requirements. There is no information on the population of this species. There is no information on threats to this species. It has not been recorded in any public or private reserves.

Remarks.—Its molecular phylogenetic relationships are unknown. Fletcher-Lazo (2002) suggests that this species could have been collected in western Ecuador and that it is closely related or belongs to the taxon *Caecilia occidentalis* (from near Popayán, Departamento Cauca, Colombia); nonetheless, more specimens and data are needed to confirm these suggestions.

Key references.—(Taylor, 1968) (description); (Fletcher-Lazo, 2002) (account).

CAECILIA TENTACULATA (YAMBA CAECILIAN) LINNAEUS, 1758

Caecilia tentaculata (Linnaeus, 1758): 229. **Type(s).**—NHRM (one lost specimen); animals figured by Linnaeus, 1749 and presumably Linnaeus, 1754, from America.

Etymology.—The specific name derives from the Latin *tentaculum*, presumably in reference to the tentacle below the nostril.

Identification.—It reaches a known length of 1025 mm. The body width is about 32.0 mm. Its length is 28–35 times its width. It has 123–124 primary grooves (complete on venter) and 57–62 secondary grooves. The grooves for the most part are incomplete. The terminal shield is small and unsegmented. The eye is in a socket, visible through the skin. The tentacle is below and slightly in front of the nostril; its distance from the eye is much greater than the distance from the nostril, close to the mouth and not visible from above the head. Some scales are present in grooves throughout most of the body; posteriorly in each groove, there is a single row of large, overlapping scales. Subdermal scales are absent. There are splenial teeth, 2–2.

The dorsum varies from uniform light blue to dark bluish-gray above, slightly paler below, and lacks any pale markings. The grooves on the sides are grayish-black. The eye usually is evident as a dark spot below the skin.

Comparisons.—*Caecilia abitaguae*. *Caecilia tentaculata* is similar to *C. abitaguae* in being a heavy-bodied and large caecilian. *Caecilia tentaculata* is stouter (the length is 30–35 times the diameter of the body) than *C. abitaguae* (the length is 43–59 times the diameter of the body).

Natural history.—In Lago Agrio and Santa Cecilia, Provincia Sucumbíos, individuals have been found in a wet grassy clearing by day, removed from a trench in a secondary forest, and under a log in a primary forest, and many were unearthed by bulldozers during the clearing of primary forests. In Veracruz, Provincia Pastaza, and individual was found lying beside a trail made of small logs laid crosswise, often with standing water between the logs, and had obviously just been hit with a machete a few moments before. Since the time of collection was shortly before noon, the animal had been out and crawling on the surface during the day. The area is one of heavy tropical rainforest. In five adult females from Lago Agrio and Santa Cecilia, Provincia Sucumbíos, three from May show no evidence of reproductive activity, whereas two from July have small ova.

Distribution.—See Figure 7.11. *Caecilia tentaculata* is known from lowlands in the Guiana shield and Amazonia of South America in Venezuela, French Guiana, Guyana, Suriname, Brazil, Colombia, Ecuador, and Peru. In Ecuador, it occurs in Sucumbíos, Orellana, Napo, and Pastaza provinces. Its elevational range is 231–950 m in an area of extent of occurrence of about 17,920 km^2.

This fossorial species inhabits natural forests and adjacent intervened areas in the flooded forests of alluvial plains of Amazonia, evergreen forests of lowlands of Napo-Curaray, evergreen forests of lowlands of Tigre-Pastaza, and evergreen montane foothill forests of the northern Cordillera Oriental of the Andes, where the average annual rainfall is 3225–4392 mm and the average annual temperature is 21.1–25.2 °C.

Conservation.—*Caecilia tentaculata* is categorized as Least Concern (Coloma *et al.*, 2004d; in Red List of 2019) in view of its wide distribution, presumed large population, and because it is unlikely to be declining fast enough to qualify for listing in a more threatened category. It is said to be uncommon throughout its range, but there is little information on its abundance. It occurs widely in an area of very limited human impact and so is unlikely to be facing any significant threats.

In Ecuador, it has been recorded in the public reserve Parque Nacional Yasuní, in the protected forest Comuna Kichwa Añangu, and within the biosphere reserve Sumaco.

Remarks.—The records from Colombia cited by Lynch (1999) and from other countries need further revision, given the uncertainties with the taxonomy of this species, the type locality of which (America) is imprecise.

Key references.—(Linnaeus, 1758) (description); (Taylor and Peters, 1974) (account); (Duellman, 1978) (account); (Lynch, 1999) (distribution, morphology); (Barrio-Amorós *et al.*, 2019) (brief account).

CAECILIA TENUISSIMA (GUAYAQUIL CAECILIAN) TAYLOR, 1973

Caecilia tenuissima Taylor, 1973: 219. **Holotype.**—USNM 12353 from Guayaquil, Provincia Guayas, Ecuador.

Etymology.—The specific name derives from the Latin *tenuis*, meaning thinnest, finest, slenderest. Presumably, it refers to the slender body shape.

Identification.—It reaches a known length of 390 mm (stretched to eliminate kinks), with a head width of 5.7 mm and an average height of the body of 4.8 mm. Its length is 81 times its width. It has 186 primary grooves, incomplete for the most part, with nine secondary grooves, four complete. The terminal shield is absent. The eye is in a socket. The scales are present, at least in the last 10 to 15 cm of the body. There are no subdermal scales. The narial plugs are pointed. The tongue is darkened. The splenial teeth are 2–2. There is no evidence of group loss or group replacement of the teeth. There are 191 vertebrae.

The dorsum is rather dark brown, vaguely lighter on the lower part of the head. The anterior part of the body a little lighter than the posterior. The top and sides of the head are whitish or yellowish in life, with blackish pigment around the tentacular area; there is some dark pigment on the front of both jaws. Cream or whitish flecking extends irregularly over the whole body.

Comparisons.—*Oscaecilia equatorialis*. The most similar species is *O. equatorialis*. They differ because *Caecilia tenuissima* has a visible eye not covered by bone, whereas *O. equatorialis* has an eye covered by bone (but see Remarks).

Natural history.—Unknown.

Distribution.—See Figure 7.8. *Caecilia tenuissima*, an endemic species from Ecuador, is known from the Pacific lowlands in Provincia Guayas at an elevation of 8 m.

This fossorial species in only known from the type locality in the semideciduous forests of lowlands of Jama-Zapotillo, which currently is urbanized and heavily populated, where the average annual rainfall is 1116 mm and the average annual temperature is 25.6 °C.

Conservation.—*Caecilia tenuissima* was categorized as Endangered by Ortega-Andrade *et al.* (2021). However, the categorization of this species requires the clarification of its taxonomic status (see Remarks). Thus, we include it in the Data Deficient category in view of the absence of information; it is known only from two specimens. However, we acknowledge that even with a larger range, this species could still be Endangered or Critically Endangered due to the presence of significant threats.

It has not been recorded in any public or private reserves.

Remarks.—Its molecular phylogenetic relationships are unknown. Fletcher-Lazo (2002) suggests that *Oscaecilia equatorialis* is a junior synonym of *Caecilia tenuissima*, given that the morphological variation among both taxa is overlapping.

Key references.—(Taylor, 1973) (description).

GENUS *OSCAECILIA* (CAECILIANS) TAYLOR, 1968

Oscaecilia Taylor, 1968: 598. **Type species.**—*Caecilia ochrocephala* (Cope, 1866).

Etymology.—The generic name is derived from the Latin *os*, meaning bone, and *Caecilia*, the feminine form of *Caecilius*, derived from the byname *caecus*, meaning blind. It refers to caecilians with the eye under bone.

Definition.—Figure 7.17. These are caeciliids with the eye under bone; no temporal fossae; the mesethmoid exposed dorsally; splenial teeth present; secondary grooves present; scales present; subdermal scales present or absent; tentacular opening directly below the external naris, closer to the naris than to the eye; no unsegmented terminal shield; narial plugs present; no diastema between the vomerine and palatine teeth; no terminal keel; relatively few and large teeth placed alternately in groups; vomeropalatine tooth row displaced posteriorly, not parallel to the premaxillary–maxillary tooth row, diverging from the latter anteriorly and forming an angle where the two rows meet rather than a semicircle.

Content.—It contains nine species, two of which occur in Ecuador.

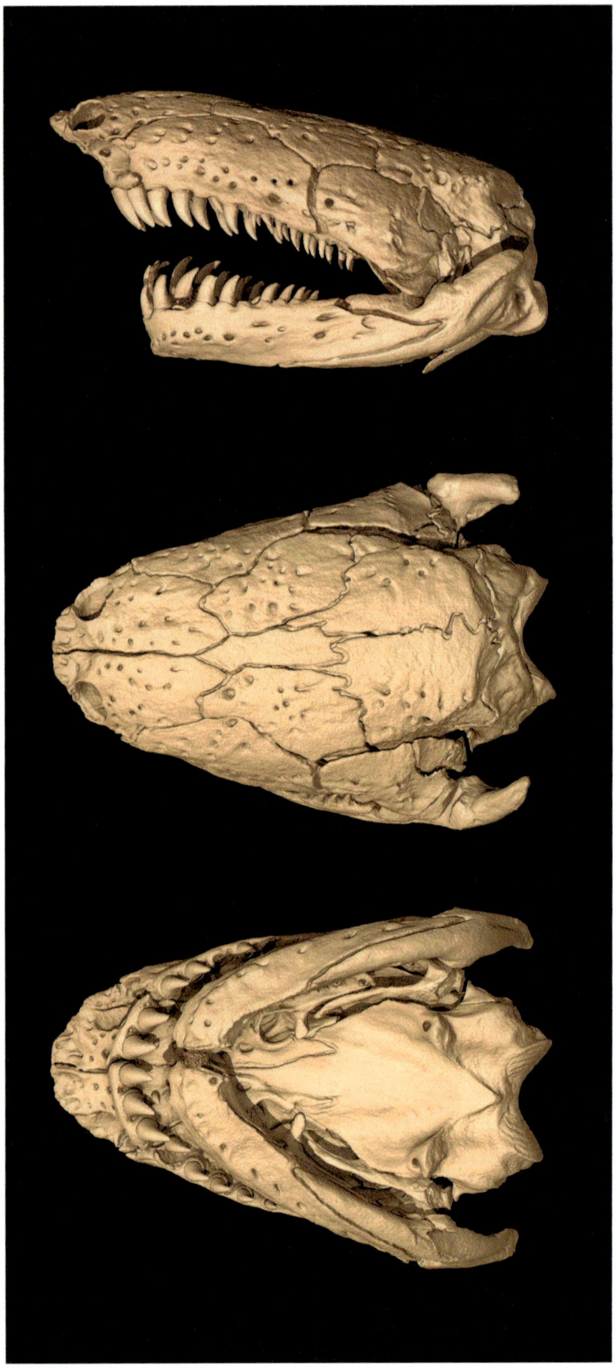

FIGURE 7.17 High-resolution computed tomography reconstructions of the cranial skeleton of *Oscaecilia ochrocephala*, CAS SUA 21864, head length = 11.1 mm, Albrook Air Force Base, Canal Zone, Panama. Lateral (upper), dorsal (middle), and ventral (lower) views.

Distribution.—Southern central and northern South America, Costa Rica, and Panama.

Remarks.—*Oscaecilia* was established by Taylor (1968) through the partitioning of *Caecilia* on the basis of a single difference (the eye covered with bone or not) of dubious value. Fletcher-Lazo (2002) considered *Oscaecilia* a junior synonym of *Caecilia* based on morphological data. Also, on the basis of molecular evidence, Pyron and Wiens (2011) suggested that *Caecilia* is paraphyletic with respect to *Oscaecilia*. Until the Fletcher-Lazo study is published and a more complete phylogeny and nomenclature issues are specifically addressed, we keep using both genera.

OSCAECILIA BASSLERI (BASSLER'S CAECILIAN) (DUNN, 1942b)

Caecilia bassleri (Dunn, 1942b): 518. **Holotype.**—MCZ 19401 from Río Pastaza (Canelos), Provincia Pastaza, Ecuador.
Oscaecilia bassleri Taylor, 1968: 598.

Etymology.—The specific name is a patronym that honors Harvey Bassler (1883–1951), whose collection of Peruvian caecilians included five species.

Identification.—It reaches a known length of 864 mm, with a body width of 6 mm. Its length is 90–140 times its width. It has 206–285 primary grooves and 14–41 secondary grooves. The eye is visible under a layer of bone. It has subdermal scales.

The dorsum is uniform dull bluish-black dorsally and slightly paler below.

Comparisons.—*Caecilia disossea*. *Oscaecilia bassleri* is similar to *C. disossea* in being slender. Other caecilians have much heavier bodies. In *O. bassleri*, the eye is invisible or visible under a thin layer of bone, whereas in *C. disossea*, the eye is in a socket and not covered by bone.

Natural history.—Individuals from Lago Agrio, Provincia Sucumbíos, and from Misión, Provincia Pastaza, were obtained during the clearing of primary forest by bulldozers. In Lago Agrio, Provincia Sucumbíos, one of two females obtained in May contained small ova.

Distribution.—Figure 7.18. *Oscaecilia bassleri* is known in Ecuador and Colombia. In Ecuador, it occurs in the Amazonian lowlands from Sucumbíos, Orellana, Napo, and Pastaza provinces. Its elevational range is 231–501 m in an area of extent of occurrence of about 12,520 km^2.

This fossorial species inhabits natural forests in the flooded forests of alluvial plains of Amazonia and evergreen montane foothill forests of the Cordillera Oriental of the Andes in the Aguarico-Putumayo-Caquetá, Napo-Curaray, and northern Cordillera Oriental biogeographic sectors, where the average annual rainfall is 3225–4264 mm and the average annual temperature is 23.4–25.2 °C.

Conservation.—*Oscaecilia bassleri* is categorized as Least Concern (IUCN SSC Amphibian Specialist Group, 2018l; in IUCN Red List of 2022) in view of its wide distribution,

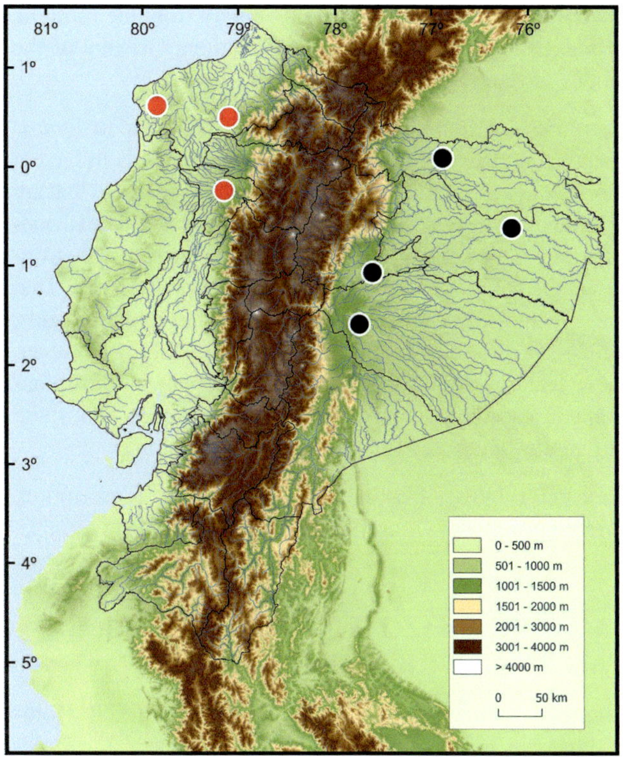

FIGURE 7.18 Localities of known occurrence of *Oscaecilia bassleri* (black dots) and *C. equatorialis* (red dots) in Ecuador.

presumed large population, and because it occurs in a region where there are still large tracts of suitable habitat remaining. However, it is important to note that in Ecuador, *O. bassleri* has been categorized as Vulnerable by Ortega-Andrade *et al.* (2021). In Ecuador, its area of extent of occurrence is of 12,520 km^2 (not 18,657 km^2, as indicated by Ortega-Andrade *et al.*, 2021), its distribution is severely fragmented, and there is a continuing decline in the extent and quality of its habitat. The major threats to its existence are habitat reduction and fragmentation, oil spills, deforestation, and agriculture. Population information is currently unavailable for this species. In Ecuador, it has not been recorded in any public or private areas, but it is present in the biosphere reserve Sumaco.

Remarks.—Its molecular phylogenetic relationships are unknown.

Key references.—(Dunn, 1942b) (description); (Duellman, 1978) (account); (Almendáriz, 1987) (brief account); (Lynch, 1999) (distribution, morphology).

OSCAECILIA EQUATORIALIS (ECUADORIAN CAECILIAN) TAYLOR, 1973

Oscaecilia equatorialis Taylor, 1973: 221. **Holotype.**—USNM 166421 from Dyott Farm, km 121 from Quito, 6 km east of Santo Domingo de los Colorados, Provincia Pichincha, Ecuador.

Etymology.—The specific name refers to the type locality in Ecuador.

Identification.—It reaches a known length of 432 mm, with a body width of 5 mm. Its length is 86 times its width. It has 180 primary grooves, mostly incomplete dorsally, and 10 secondary grooves, only one complete. The terminal shield is absent. The eye is covered by bone. Scales begin about 13 cm before the terminus. A single row of scales is present in the terminal grooves. Splenial teeth are present.

The head is grayish above and below and probably whitish or yellowish in life; the neck is lighter than the rest of the body, which is a blackish-slate, except for the terminus and subterminal area, which are whitish.

Comparisons.—*Caecilia tenuissima*. *Oscaecilia equatorialis* is similar to *C. tenuissima* by having a similar size and by being slender caecilians. They differ because the eye is covered by bone in *O. equatorialis*, whereas the eye is visible in a socket in *C. tenuissima*.

Natural history.—Unknown.

Distribution.—See Figure 7.18. *Oscaecilia equatorialis*, an endemic species to Ecuador, is known from the Pacific lowlands in Esmeraldas and Santo Domingo de los Tsáchilas provinces. Its elevational range is 125–548 m in an area of extent of occurrence of about 3440 km^2.

This fossorial species inhabits natural forests in the evergreen seasonal montane foothill forests of the Cordillera Costera of Chocó and evergreen montane foothill forests of the Cordillera Occidental of the Andes, where the average annual rainfall is 1522–2591 mm and the average annual temperature is 22.6–24.8 °C.

Conservation.—*Oscaecilia equatorialis* was categorized as Endangered by Ortega-Andrade *et al.* (2021), but they did not provide IUCN criteria. Its area of extent of occurrence is less than 5000 km^2 (3440 km^2), its distribution is severely fragmented in fewer than five known localities, and there is a continuing decline in the extent and quality of its habitat. The population status of this species is unknown. The major threats to its existence include habitat reduction and fragmentation, logging, deforestation for agricultural development (particularly palm plantations), the introduction of exotic species, human settlements, and pollution. Furthermore, the area where it is distributed faces constant pressure from wood companies.

It has been recorded in the public reserve Reserva Ecológica Mache Chindul.

Remarks.—Its molecular phylogenetic relationships are unknown. Fletcher-Lazo (2002) suggests that *Oscaecilia equatorialis* is a junior synonym of *Caecilia tenuissima*, given that the morphological variation among both taxa is overlapping.

Key references.—(Taylor, 1973) (description).

FAMILY RHINATREMATIDAE (NEOTROPICAL TAILED CAECILIANS)

Rhinatrematidae Nussbaum, 1977: 3. **Type genus.**—*Rhinatrema* (Duméril and Bibron, 1841)

Etymology.—The family named is derived from the Greek *Rhinatrema*: *rhin*, meaning nose, and *atrema*, meaning without holes, referring to the lack of tentacular apertures on the snout.

Definition.—Rhinatrematidae is a group supported by molecular evidence, and the only known putative phenotypic synapomorphy is the primary adductor musculature originating dorsomedially and passing through the upper temporal fenestrae. The body is strongly annulated. The primary annuli are divided by secondary and tertiary grooves, and numerous scales are present in the primary annular grooves. The body ends in a short tail. The eyes are visible externally and lie beneath the skin in bony sockets. A tentacle arises near or at the anterior edge of each eye. The middle ear contains a stapes. Most of the species are very poorly known. Free-living larvae are known for *Epicrionops*.

Distribution.—Species occur in northern South America, east and west of the Andes, throughout Colombia, Ecuador, Peru, Venezuela, French Guiana, and Suriname.

Content.—Rhinatrematidae contains 14 species in three genera; two genera (*Amazops* and *Epicrionops*) and three species occur in Ecuador.

Remarks.—It is considered the sister taxon of all other living caecilians.

Key references.—(Nussbaum, 1977) (description); (Nussbaum, 1979) (phylogeny); (San Mauro *et al.*, 2004) (phylogeny); (Frost *et al.*, 2006) (phylogeny); (Wilkinson and Nussbaum, 2006) (phylogeny); (Roelants *et al.*, 2007) (phylogeny); (Pyron and Wiens, 2011) (phylogeny); (Blackburn and Wake, 2011) (taxonomy); (Mauro *et al.*, 2012) (phylogeny); (Maddin *et al.*, 2012) (braincase, phylogeny); (Vitt and Caldwell, 2014) (account).

GENUS *AMAZOPS* (AMAZONIAN TAILED CAECILIANS) WILKINSON *ET AL.*, 2021

Amazops Wilkinson *et al.*, 2021: 28. **Type species.**—*Amazops amazops* (Wilkinson *et al.*, 2021).

Etymology.—The generic name is a portmanteau combining a reference to the Amazonian provenance of the type and only known species and the distinctive topological relationships of its eye and orbit, particularly the contribution of the squamosal to the bony margin of the orbit, which is unknown in any other rhinatrematid. The gender is masculine.

Definition.—These are caecilians with the squamosal contributing to the margin of the orbit.

Content.—*Amazops* contains one species that occurs in Ecuador.

Distribution.—The genus is known only from the foothills of the Amazonian versant of the Cordillera Oriental of the Andes of Ecuador.

Remarks.—Externally, *Amazops* is very similar to *Epicrionops*. *Amazops* differs from other rhinatrematids in having more elongate squamosals anteriorly occupying areas that are occupied by the maxillopalatine in other rhinatrematids, contributing to the bony orbits and making contact with the frontals at the anterior ends of the upper temporal fenestrae. That these features do not occur elsewhere within rhinatrematids, which suggests they might be derived conditions. However, these features are common (nearly universal) in non-rhinatrematid caecilians, implying either convergence with these or that they are instead ancestral to Gymnophiona.

Key references.—(Wilkinson *et al.*, 2021) (description, osteology).

AMAZOPS AMAZOPS (AMAZONIAN TAILED CAECILIAN) WILKINSON *ET AL.*, 2021

Amazops amazops (Wilkinson *et al.*, 2021): 28. **Holotype.**—USNM 320729 from Finca Virgen La Dolores, at the km 57 sign on Hollín–Loreto Road (E20), ca. 1000 m, Provincia Orellana, Ecuador.

Etymology.—As for the genus. For nomenclatural purposes, the specific epithet is considered a genderless noun.

Identification.—A female reaches 173.0 mm. The body length is 22 times the tail length. The body is subcylindrical, mostly somewhat dorsoventrally compressed, less so posteriorly, not compressed at the level of the vent, fairly uniform, narrowing slightly anteriorly and more notably posteriorly. The tail is moderately long, 7.6 mm, slightly laterally compressed at a level halfway between the vent and tail tip, tapering more strongly in dorsal than in lateral view; the tip is much more broadly blunt in lateral than in dorsal view; the dorsal and ventral margins are symmetrical in lateral view, the ventral surface is not flat. The head in dorsal view is slightly more V- than U-shaped; it is dorsoventrally compressed. The mouth is marginally subterminal. The circular eyes are small (diameter 0.4 mm). The tentacular apertures are small (0.6 mm), almost horizontal, arc-like slits; the tips of the paired tentacular ducts of each side (leading from the tentacle to the vomeronasal organ) are visible through the skin adjacent to the anterior end of the tentacular apertures. The nares are small (0.3 mm), visible in anterior but not ventral views. The teeth point posteriorly (at angles of 30° to 45° from the jaws); they are bicuspid, not strongly recurved. The anterior and posterior teeth of each series are smaller than those in between; none are hypertrophied. There are nine premaxillary teeth and 33 maxillary teeth, for a total of 42 premaxillary–maxillary teeth; 13 vomerine and 26 palatinal teeth, for a total of 39 vomeropalatine teeth; and 37 outer mandibular teeth and 22 symmetrically disposed inner mandibular teeth. The tongue has more or less longitudinal plicae over its entire surface. The choanae are elongate. The first groove in the collar region (first nuchal groove) is poorly marked on the dorsum and dorsolaterally; it is not visible in ventral view. The second nuchal groove is faint dorsally, pale and clearly marked laterally and ventrally. The third nuchal groove is complete; it bows slightly anteromedially on the dorsum and resembles subsequent annular grooves. The first nuchal collar is much shorter (1.2 mm) than the second (4.5 mm). There are four regularly spaced dorsal transverse grooves on the second collar. Behind the collars, there are 247 annular grooves (= folds). The first and last (~15–20) annuli are a little longer than the others. The annular grooves are complete ventrally, except for two interrupted by the vent and disc; the tail (area behind the vent) has 10 complete and one (the last) dorsoventrally incomplete annular groove. A single row of small subcircular scales is present in shallow pockets (about one-quarter of the length of the midbody annulus) below the dorsal transverse grooves on the second nuchal collar. At midbody and posteriorly, scales occur in two well-defined rows, a posterior row of larger scales and an anterior row of slightly smaller scales, in pockets about as deep as the length of an annulus at the midbody, about one and a half times the length of an annulus posteriorly. Additional scattered scales may lie below (posterior to) the row of larger scales. The vent is slightly longitudinal, bordered by an irregular array of partially subdivided denticulations. There is no disc other than what is delimited by the denticulations around the vent, approximately egg-shaped, with the narrow apex anteriorly formed by the unpigmented denticulations.

The body is almost uniformly dark, brownish-lavender, slightly paler on the head, more so on the throat. There are pale areas around the eyes and a slightly paler snout tip encompassing nares. There are pale, narrow, faint paramandibular stripes (inset from lips) on the ventral surface of the head and a distinctive pale second nuchal groove in the ventral collar region. It is paler around the vent, especially anteriorly. The annular grooves have a light posterior margin and a typically slightly longer, dark, anterior nonglandular area. It has 77 vertebrae, of which the last seven are entirely posterior to the vent. The bony ribs are absent from the last four of the caudal vertebrae.

Comparisons.—*Epicrionops colombianus*, *E. petersi*. *Amazops amazops* is readily distinguished from all other rhinatrematids by having very few (fewer than four) annular grooves interrupted by the vent. Further, the combination of a low number of annular grooves (< 275) and its uniform color distinguishes it from all other rhinatrematids except *Epicrionops colombianus*, the only known specimen of which has even fewer annular grooves (< 225). The similar *E. petersi* is mostly brown with scattered yellow dots and flecks, and its flanks are mostly yellow with irregular scattered brown flecks.

Natural history.—There are very small ovarian eggs (largest ca. 0.8 mm diameter) in an individual collected on 16 August 1990.

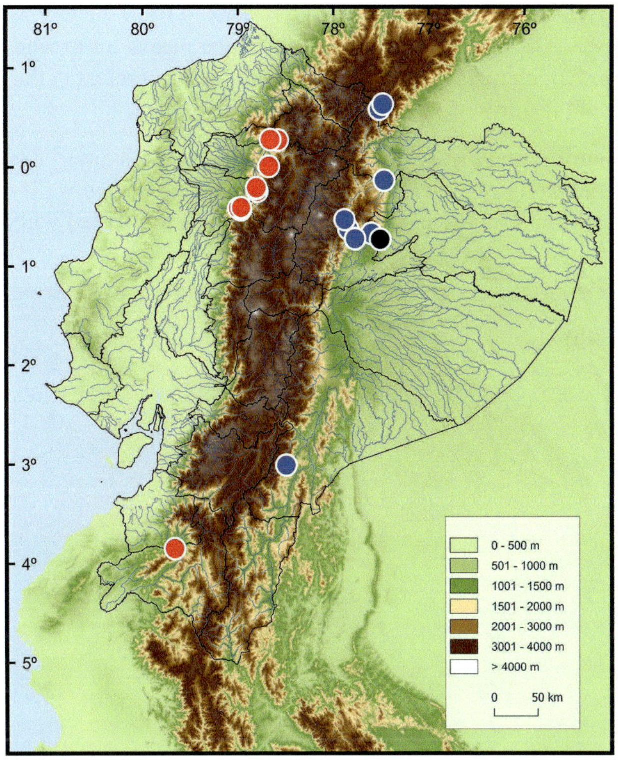

FIGURE 7.19 Localities of known occurrence of *Amazops amazops* (black dot), *Epicrionops bicolor* (red dots), and *E. petersi* (blue dots) in Ecuador.

Larvae.—Unknown.

Distribution.—Figure 7.19. *Amazops amazops*, an endemic species to Ecuador, is known only from the type locality on the Amazonian slopes of the Cordillera Oriental of the Andes in Provincia Napo at an elevation of 845 m.

This species has been recorded in currently intervened areas in the northern Cordillera Oriental of the Andes biogeographic sector, where the average annual rainfall is 4007 mm and the average annual temperature is 21.9 °C.

Conservation.—*Amazops amazops* is not listed yet on the IUCN Red List as of 31 December 2023. Wilkinson *et al.* (2021) suggest that the IUCN category of the conservation status of this species should be Data Deficient.

Remarks.—Its molecular phylogenetic relationships are unknown. Wilkinson *et al.* (2021) provided microCT scans of the cranial and caudal osteology and scales.

Key references.—(Wilkinson *et al.*, 2021) (description, osteology).

GENUS *EPICRIONOPS* (TAILED CAECILIANS) BOULENGER, 1883

Epicrionops Boulenger, 1883: 202. **Type species.**—*Epicrionops bicolor* Boulenger, 1883.

Etymology.—The generic name is derived from the Greek *epi*, meaning upon or over; *crion*, which means lump, and *ops*, which typically means face. Presumably, it refers to the facial features of these caecilians

Definition.—Figure 7.20. Derived features that support the monophyly of *Epicrionops* are unknown. *Epicrionops* differs from *Rhinatrema* primarily in having a longer tail with well-developed hemal arches and a more longitudinal vent and in retaining a small ceratobranchial III, all of which are probably ancestral character states.

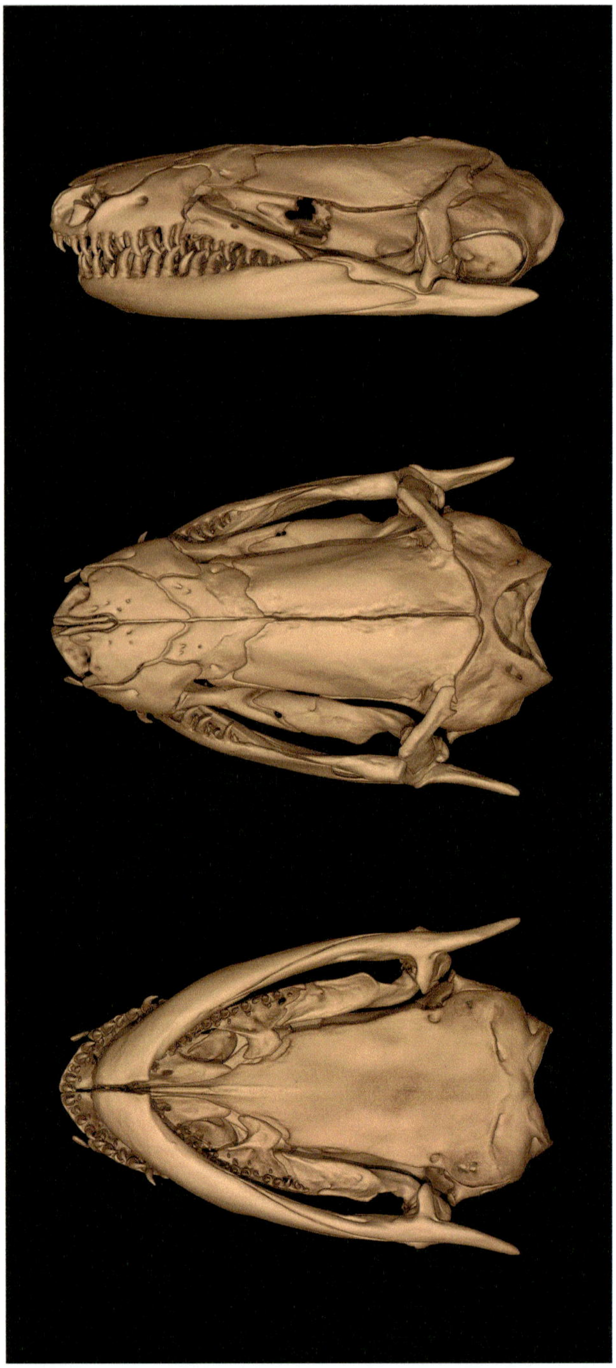

FIGURE 7.20 High-resolution computed tomography reconstructions of the cranial skeleton of *Epicrionops petersi*, CJ 11804, head length = 7.7 mm (from snout tip to os basale at midline), Narupayaku, Provincia Napo. Lateral (upper), dorsal (middle), and ventral (lower) views.

Content.—*Epicrionops* contains seven species, of which two occur in Ecuador.

Distribution.—It occurs throughout Colombia, Ecuador, Peru, and Guyana.

Remarks.—According to Pyron and Wiens (2011), *Epicrionops* and *Rhinatrema* are paraphyletic to each other; thus, a generic rearrangement is needed. Its conservation is a matter of concern given its rarity in recent herpetological surveys.

EPICRIONOPS BICOLOR (BICOLOR CAECILIAN) BOULENGER, 1883

Epicrionops bicolor Boulenger, 1883: 202–203. **Holotype.**—BMNH 1946.9.5.66 from Intag, Provincia Imbabura, Ecuador.

Etymology.—The specific epithet derives from the Latin *bicolor*, which refers to its color pattern of yellow and brown.

Identification.—Figure 7.21. It reaches a known length of 363 mm in females and 324 mm in males. The stripes are 4–5 mm in width. The body length is 21 times the tail length; the body width fits into the total length 25.5 times. It has 244 (240) annuli behind the collars, of which 14 are on the tail. It has three scale rows on the annuli of the posterior part of the body.

FIGURE 7.21 *Epicrionops bicolor.* Left: from Jardín de los Sueños, Provincia Cotopaxi. Jaime Culebras. Right: from Reserva Otonga, Provincia Cotopaxi. LAC.

The dorsum is marbled in a combination of brown over a bright yellow background that is plain on the flanks. The head is mostly brown dorsally. It has a wide brown ventral band that extends from the chin to the extreme of the tail. Sometimes, the band is interrupted in the area of the neck collars. The chin and jaws are light brown.

Natural history.—Individuals have been found active on the ground after heavy rains and under stones and thoroughly rotted logs.

Distribution.—See Figure 7.19. *Epicrionops bicolor* is known in Colombia, Ecuador, and Peru in the Pacific lowlands and western slopes of the Cordillera Occidental of the Andes. In Ecuador, it is known in Imbabura, Pichincha, Cotopaxi, Santo Domingo de los Tsáchilas, and Loja provinces. Its elevational range is 977–1984 m in an area of extent of occurrence of about 6390 km^2.

This semi-aquatic species inhabits streams or flooded areas in natural forests in the evergreen low montane forests of the Cordillera Occidental of the Andes in the Catamayo-Alamor and Cordillera Occidental of the Andes biogeographic sectors, where the average annual rainfall is 1150–2466 mm and the average annual temperature is 16.3–23.1 °C.

Conservation.—*Epicrionops bicolor* is categorized as Least Concern throughout its entire range (IUCN SSC Amphibian Specialist Group, 2019m; in IUCN Red List of 2022) in view of its wide distribution and presumed large population. However, it is important to note that in Ecuador, *Epicrionops bicolor* has been categorized as Vulnerable by Ortega-Andrade *et al.* (2021). In Ecuador, its current extent of occurrence is less than 20,000 km^2 (6390 km^2), its distribution is severely fragmented in fewer than 10 known localities, and there is a continuing decline in the extent and quality of its habitat. It is a rarely encountered species that is closely tied to watercourses in pristine forests. Its absence or rarity has been documented in the Mindo region, Provincia Pichincha, and especially in San Francisco de Las Pampas and the Reserva Otonga, Provincia Cotopaxi, where it was relatively abundant up until the 1980s but became very rare later. The most recent record of the species is of one individual at Reserva Otonga in 2010. It appears to have undergone a major decline due to causes associated with precipitous declines during the 1980s and 1990s (climatic abnormalities, pathogens). It should be noted that some other species of frogs that are associated with streams have undergone dramatic declines and disappearances, so the status of this species should be investigated further. Additional threats to this species throughout its range are pollution, and fires. These threats are linked to intensive agriculture, livestock, forestry, logging, mining, and human settlements. In Ecuador, it has been recorded in the biosphere reserve Chocó Andino de Pichincha and the private reserve Bosque Integral Otonga.

Remarks.—Its molecular phylogenetic relationships are unknown. Fletcher-Lazo (2002) reviewed its taxonomy, but his thesis is not yet published. According to him, the records of *Epicrionops bicolor* on the eastern slopes of the Andes of the Cordillera Oriental likely are *Epicrionops petersi*. Also, according to him, *E. marmoratus* is a junior synonym of *E. bicolor*. Accordingly, we follow his suggestions, excluding the Amazonian records from our account and including *E. marmoratus* record under *E. bicolor*. Also, Wake (1989) suggested that the populations from southwestern Ecuador and northwestern Peru are presumably a different species from the one in northwestern Ecuador and southern Colombia. However, to confirm the latter, further studies are needed.

Key references.—(Boulenger, 1883) (description); (Fletcher-Lazo, 2002) (taxonomy, distribution, natural history); (Arteaga *et al.*, 2013) (account); (Yánez-Muñoz *et al.*, 2020) (account).

EPICRIONOPS PETERSI (PETERS' CAECILIAN) TAYLOR, 1968

Epicrionops petersi petersi Taylor, 1968: 224. **Holotype.**—USNM 159794 from Agua Rica, between Limón and Gualaceo, Provincia Morona Santiago, Ecuador.

Etymology.—The specific name is a patronym in honor of James Arthur Peters, collector of the type. His main subject of research was herpetology and zoogeography of Latin America, especially Ecuador.

Identification.—Figures 7.22, 7.23. It reaches a known length of 286.5 mm. The width fits into the total length from 12 (in young) to 17 times (in adults). The tail length fits

FIGURE 7.22 *Epicrionops petersi*. Narupayaku, Provincia Napo. Jaime Culebras.

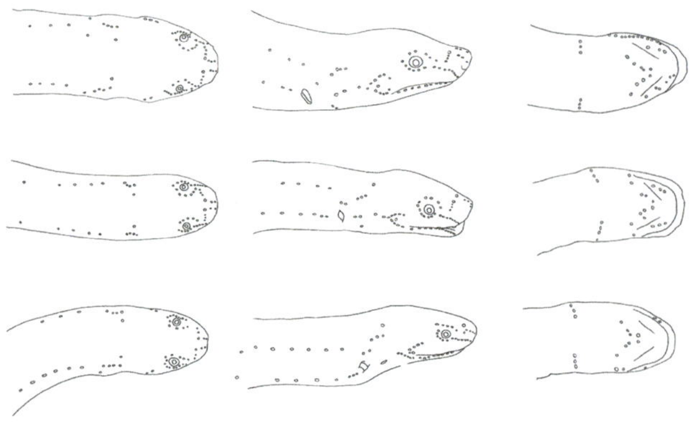

FIGURE 7.23 Distribution and variation in neuromasts of the lateral line system of three larvae of *Epicrionops petersi* from Narupayaku, Provincia Napo. Top: larger. Middle: mid-sized. Bottom: smaller. Illustrations by Francesca Angiolani based on photographs by Jaime Culebras.

into the body length from 21 to 23.6 times. The splenial teeth in large adults are 13–13 to 16–16; the premaxillary–maxillary are 17–17 to 21–21. It has total 269–316 folds (primary, secondary, and tertiary). It has five to seven short dorsal folds on the second collar. It has 78–81 vertebrae.

It is mostly brown with scattered yellow dots and flecks. The flanks are mostly yellow with irregular scattered brown flecks.

Comparisons.—*Amazops amazops*. *Epicrionops petersi* most closely resembles the macrosympatric *A. amazops*. *Epicrionops petersi* is mostly brown with scattered yellow dots and flecks, and its flanks are mostly yellow with irregular scattered brown flecks, whereas *A. amazops* is uniformly dark brownish-lavender, slightly paler on the head.

Natural history.—Individuals have been found sympatrically with *Caecilia orientalis*, living under the same environmental conditions. Most of the specimens in the type series were found under a thick layer of mold, roots, and dirt covering huge fallen trees. The logs had been in place for a considerable length of time and were heavily overgrown by vines and roots that were in turn covered by humus, detritus, and dirt, all forming a layer about four to six inches thick. The area in which the logs were found was swampy, with the entire layer across the log very wet. The caecilians live in the detritus layer. The technique used for collecting the animals was to chop through the layer at two points about three feet apart. Then, a horizontal cut was made in the layer low on one side, and the entire layer then rolled across the log like a carpet. The caecilians were exposed on the log surface as the layer rolled away from them and were easily picked up. On several occasions, there were three or four individuals on a single log, and *E. petersi* was found less frequently than *C. orientalis*.

Distribution.—See Figure 7.19. *Epicrionops petersi* is known on the Amazonian slopes of the Cordillera Oriental of the Andes in Ecuador and Peru. In Ecuador, it is known in Sucumbíos, Napo, and Morona Santiago provinces. Its elevational range is 1183–2284 m in an area of extent of occurrence of about 8010 km^2.

This semiaquatic species inhabits streams or flooded areas in natural forests in the evergreen low and montane forests of the Cordillera Oriental of the Andes in the northern and southern Cordillera Oriental of the Andes biogeographic sectors, where the average annual rainfall is 1346–3581 mm and the average annual temperature is 14.6–20.3 °C.

Conservation.—*Epicrionops petersi* is categorized as Least Concern throughout its entire range (IUCN SSC Amphibian Specialist Group, 2018m; in IUCN Red List of 2022) in view of its wide distribution, presumed large population, and because it is unlikely to be declining fast enough to qualify for listing in a more threatened category. However, it is important to note that in Ecuador, *Epicrionops petersi* has been categorized as Endangered by Ortega-Andrade *et al.* (2021). In Ecuador, its current extent of occurrence is 8010 km^2, its distribution is severely fragmented, and there is a continuing decline in the extent and quality of its habitat. It is a rarely encountered species that is closely tied to watercourses in pristine forests. The population status of this species is unknown. Its absence or rarity has been documented along several places on the eastern slopes of the Ecuadorian Andes, which have been sampled in recent decades and only revealed a few specimens. It appears to have undergone a major decline due to causes associated with precipitous declines during the 1980s and 1990s (climatic abnormalities, pathogens). It should be noted that some other species of frogs that are associated with streams have undergone dramatic declines and disappearances, so the status of this species should be investigated further. Additional threats to this species throughout its range are habitat loss and fragmentation, pollution, and fires. These threats are linked to intensive agriculture, livestock, forestry, logging, mining, and human settlement. In Ecuador, it has been recorded in the public reserve Parque Nacional Antisana, in the protected forests of La Cascada and Cerro Sumaco y Cuenca Alta del Río Suno, and in some private reserves (e.g., Reserva Narupayaku).

Remarks.—Its molecular phylogenetic relationships are unknown. Specimens reported as *Epicrionops bicolor* from eastern Ecuador and Peru might belong to *E. petersi*, as suggested by Fletcher-Lazo (2002).

Key references.—(Taylor, 1968) (description); (Taylor and Peters, 1974) (ecology).

FAMILY SIPHONOPIDAE (SOUTH AMERICAN CAECILIANS)

Siphonopina Bonaparte, 1850: 1 p. **Type genus.**—*Siphonops* Wagler, 1828.

Etymology.—The family named is derived from the Latin *sipho*, meaning siphon, tube, presumably referring to its shape.

Definition.—Siphonopids are a group of fossorial and oviparous caecilians that have been diagnosed as monophyletic based on molecular evidence. They are unique among oviparous caecilians in having a middle ear with an imperforate stapes and no inner mandibular teeth. Siphonopids range in size from 112 to 120 mm in total length and have primary annuli, with some having secondary annuli and none with tertiary. Some species have dermal scales, while all have a terminal shield at the caudal end of the body but lack a true tail. They are found burrowing in the moist humid forest floors of South America. Many siphonopid species exhibit maternal dermatophagy, a reproductive behavior in which the young feed on a specialized, lipid-rich outer layer of their mother's skin using a unique set of "fetal" multicuspid teeth.

Distribution.—They occur in South America from the Magdalena Valley of northwestern Colombia through southern Venezuela to the Guianas and from Amazonian Ecuador, Peru, Bolivia, and Brazil south to Misiones, Argentina, and Paraguay.

Content.—Siphonopidae contains 28 species in five genera; two genera (*Microcaecilia* and *Siphonops*) and two species occur in Ecuador.

Remarks.—The nomenclatural and systematic history is detailed in Frost (2023 and onward). The phylogeny of this group has been reconstructed using DNA sequences from both mitochondrial and nuclear genes; however, its phylogeny is still weakly supported. Evidence from molecular data suggests that the group originated at least as far back as 97 million years ago in the Upper Cretaceous period, as indicated by sequences from *Siphonops annulatus*.

Key references.—(Wilkinson *et al.*, 2008) (parental care) (Pyron and Wiens, 2011) (phylogeny); Wilkinson *et al.* (2011) (systematics); (Maddin *et al.*, 2012) (braincase, phylogeny); Vitt and Caldwell (2014) (account).

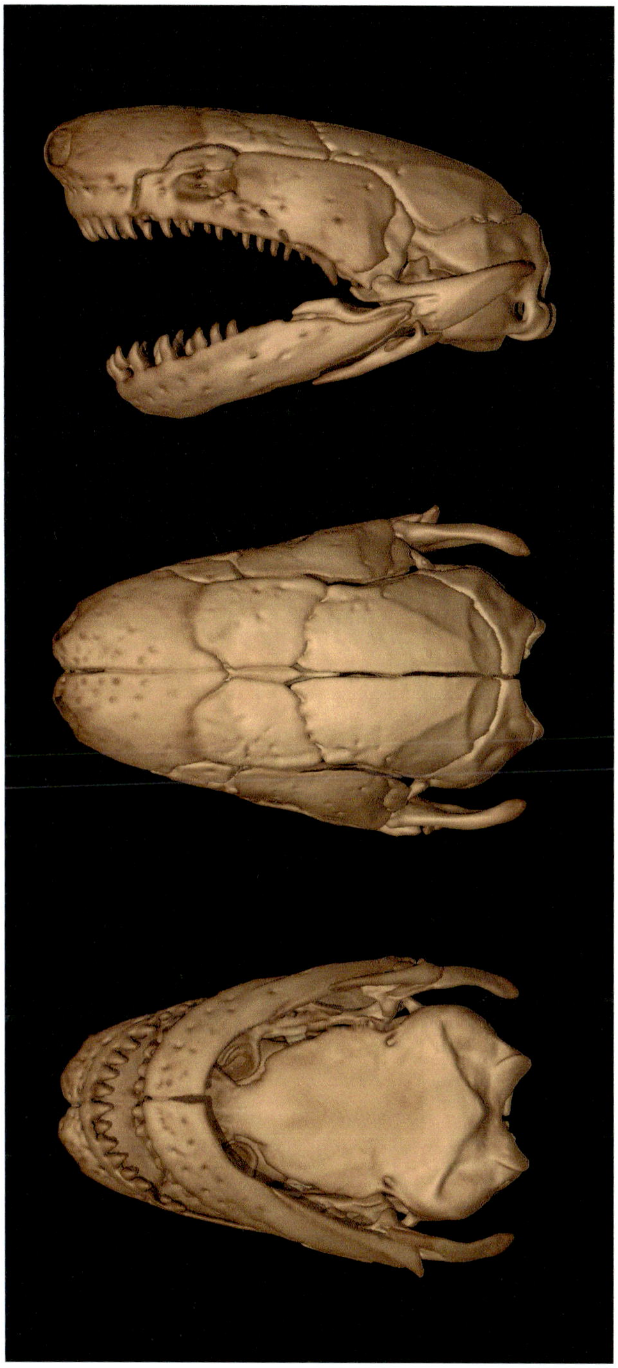

FIGURE 7.24 High-resolution computed tomography reconstructions of the cranial skeleton of *Microcaecilia albiceps*, CAS SUA 10363, head length = 5.5 mm (from snout tip to osbasale at midline), Río Cotapino, Provincia Napo. Lateral (upper), dorsal (middle), and ventral (lower) views.

GENUS *MICROCAECILIA* (TINY CAECILIANS) TAYLOR, 1968

Microcaecilia Taylor, 1968: 532. **Type species.**—*Dermophis albiceps* Boulenger, 1882b, by original designation.

Etymology.—The generic name is derived from the Latin *micro*, meaning small, and *Caecilia*, the feminine form of *Caecilius*. The latter refers to the byname *caecus*, meaning blind.

Definition.—Figure 7.24. These are the only siphonopids with scales and lungs and without substantial diastemata between the vomerine and palatine teeth.

Content.—*Microcaecilia* contains 16 species, of which only one occurs in Ecuador.

Distribution.—*Microcaecilia* occurs throughout Brazil, Colombia, Ecuador, French Guiana, Guyana, Suriname, and Venezuela.

Remarks.—Wilkinson *et al.* (2014) discussed the lung morphology of the genus.

MICROCAECILIA ALBICEPS (PALE-HEADED CAECILIAN) (BOULENGER, 1882b)

Dermophis albiceps (Boulenger, 1882b): 98. **Holotype.**—BMNH 1946.9.5.32 from Ecuador.
Microcaecilia albiceps Taylor, 1968: 532.

Etymology.—The specific name derives from the Latin *albus*, which means white, and *ceps*, which means headed, referring to the white head of this species.

Identification.—Figure 7.25. It reaches a known length of 227 mm. The width fits into the total length from 30 to 45.4 times. It has 115–123 primary folds and 48–62 secondaries. The scales are present. The splenial teeth are absent. The tentacle is distant from the nostril and very close to the eye. The eye is under bone, often visible externally.

Above, the body is generally grayish-brown, lighter anteriorly. The venter is vaguely lighter, the mid-ventral portion being slightly darker than the sides. The head and the anterior part of the neck are yellowish. The terminal region, vent, and a few folds preceding the vent are yellowish.

Natural history.—Unknown.

Distribution.—Figure 7.26. *Microcaecilia albiceps* is known in Ecuador and Colombia (Florencia, Departamento Caquetá). In Ecuador, it occurs in the Amazonian lowlands in Sucumbíos, Orellana, and Pastaza provinces. Its elevational range is 216–366 m in an area of extent of occurrence of about 12,890 km^2.

This fossorial species inhabits natural forests in the flooded forests of alluvial plains of Amazonia and the evergreen forests of lowlands of Tigre-Pastaza in the Aguarico-Putumayo-Caquetá and Tigre-Pastaza biogeographic sectors, where the average annual rainfall is 2991–4150 mm and the average annual temperature is 24.2–25.3 °C.

FIGURE 7.25 *Microcaecilia albiceps*. KU 110512, Santa Cecilia, Provincia Sucumbíos. WED.

Conservation.—*Microcaecilia albiceps* is categorized as Least Concern throughout its entire range (Castro *et al.*, 2004c; in IUCN Red List of 2022) in view of its relatively wide distribution, presumed large population, and because it is unlikely to be declining fast enough to qualify for listing in a more threatened category. However, it is important to note that in Ecuador, *Microcaecilia albiceps* has been categorized as Vulnerable by Ortega-Andrade *et al.* (2021) because its extent of occurrence is estimated to be less than 20,000 km^2 (12,890 km^2, not 119 km^2 as indicated by Ortega-Andrade *et al.*, 2021), its distribution is severely fragmented in fewer than 10 known localities, and there is a continuing decline in the extent and quality of its habitat. The population status of this species is unknown. Agriculture and livestock are most likely localized threats to the habitat of this species, but it occurs in a region of relatively low human impact. In Colombia, the species is not known in any protected areas. It has been recorded in the protected forest of Comuna Sinchi Chicta Caspi.

Remarks.—Its molecular phylogenetic relationships are unknown. According to Taylor and Peters (1974), the close similarity between the type and specimens from the Amazonian region of Ecuador suggests that the type originated in that area.

Key references.—(Boulenger, 1882b) (description); (Taylor, 1968) (account); (Lynch, 1999) (distribution, morphology).

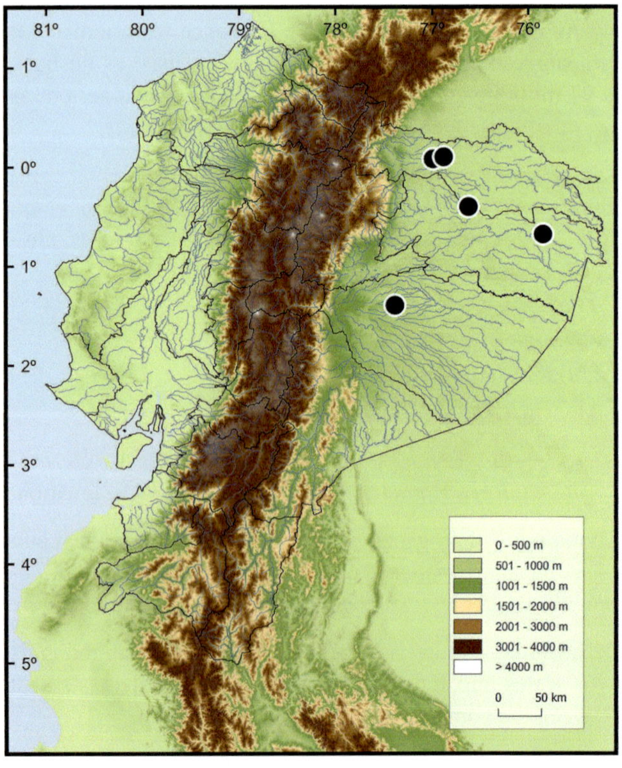

FIGURE 7.26 Localities of known occurrence of *Microcaecilia albiceps* in Ecuador.

GENUS *SIPHONOPS* (RINGED CAECILIANS) WAGLER, 1828

Siphonops Wagler, 1828: 742. **Type species.**—*Caecilia annulate* Mikan, 1820.

Etymology.—The generic name is derived from the Greek *siphon*, meaning tube, and *ops*, meaning appearance.

Definition.—Figure 7.27. These are caeciliids with the eye in a socket, not under bone. They lack temporal fossae. The mesethmoid is exposed dorsally. They lack narial plugs, splenial teeth, secondary grooves, scales, and a terminal keel. The tentacular opening is closer to the eye than to the external naris. It has an unsegmented terminal shield. There is no diastema between the vomerine and palatine teeth.

Content.—*Siphonops* contains four species, one of which occurs in Ecuador.

Distribution.—*Siphonops* occurs in South America east of the Andes in Argentina, Bolivia, Brazil, Colombia, Ecuador, Guyana, Paraguay, Peru, and Venezuela.

Remarks.—No phenotypic synapomorphies have been established for *Siphonops*. According to Wilkinson and Nussbaum (2006), the three species *annulatus*, *leucoderus*, and *paulensis* have a distinctive and presumably derived color pattern with a blue background and whitish annular ring, but this is shared with the species of *Mimosiphonops*, suggesting that *Siphonops* may be paraphyletic.

SIPHONOPS ANNULATUS (RINGED CAECILIAN) (MIKAN, 1822)

Caecilia annulate (Mikan, 1822): pl. 11 and associated text. **Type (lectotype).**—RMNH 2419 from Rio de Janeiro, Brazil.
Siphonops annulatus Wagler, 1828: 742.

Etymology.—The specific name derives from the Latin *anulus*, which means ring, referring to the unique white rings of this species.

Identification.—Figure 7.28. It reaches a known length of 539 mm. Its length is 17–32 times its width. It has 78–98 annuli (primary grooves), completely encircling the body except for except for an anterior primary groove and three or four near the vent. Secondary grooves are absent. The collars are very distinct. The tentacle is much closer to the eye than to the nostril. The narial plugs are indistinct. The eyes are visible in open sockets and slightly elevated above the surface of the head. It has a large, unsegmented terminal shield extending from anterior to the vent to the posterior tip of the body. The scales are absent. The splenials are absent. Dorsally, it varies from gray to dark ultramarine throughout, except that the dorsal part of the head is vaguely lighter, as is the terminus. A white spot covers the vent area, and small white spots are present at the tentacle and nostril. The grooves, except for the sides and underside of the collars, are bordered strongly with white. Females brooding young are often much paler than other adults.

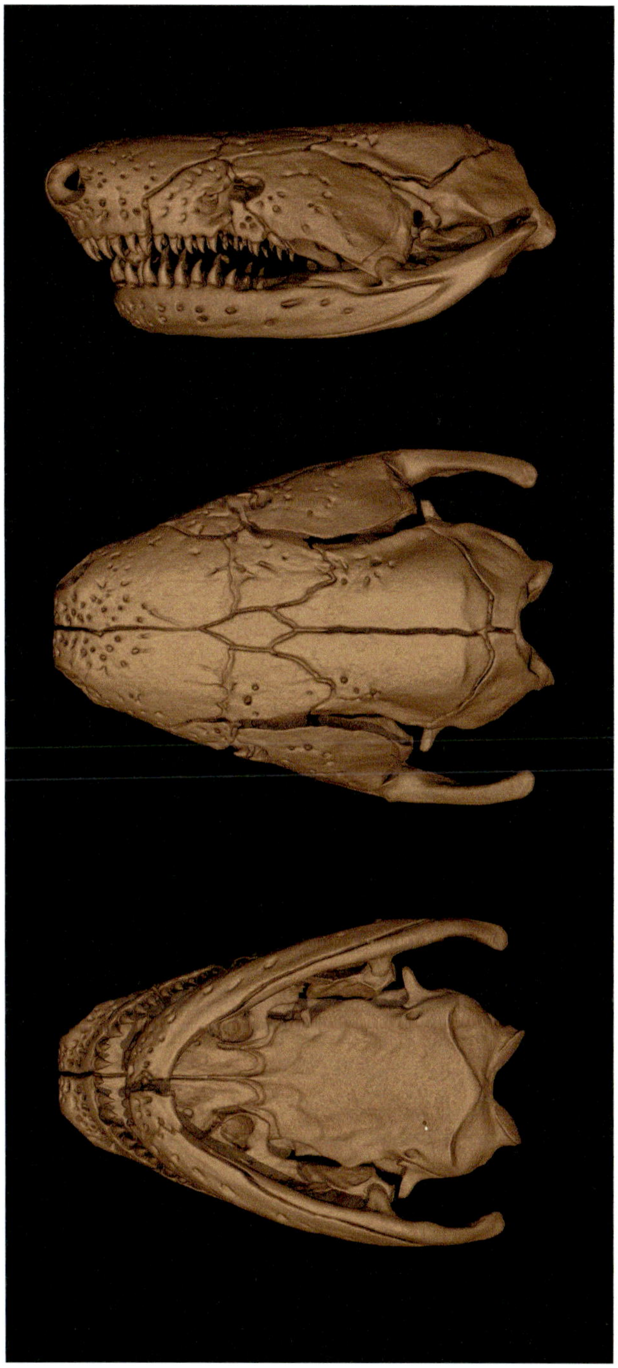

FIGURE 7.27 High-resolution computed tomography reconstructions of the cranial skeleton of *Siphonops annulatus*, UF Herp 168032, head length = 11.3 mm (from snout tip to osbasale at midline), Santa Rita, Paraiba, Brazil. Lateral (upper), dorsal (middle), and ventral (lower) views.

Natural history.—Caeciliids were found under decomposing trunks of African oil palms, even in grazed pastures. An

FIGURE 7.28 *Siphonops annulatus*. Bataburo Lodge, Provincia Pastaza. Andreas Nöllert.

individual was at the base of a banana plant in a secondary forest, another in a trench in early morning; presumably, it fell in during the night. Other individuals have been found in pitfall traps.

It is oviparous; a female with a clutch of eggs was found under a tree trunk moved during the ploughing of a pasture. The offspring feed on maternal skin (see Remarks).

Distribution.—Figure 7.29. *Siphonops annulatus* is known from cis-Andean South America in Argentina, Bolivia, Brazil, Colombia, Ecuador, French Guiana, Guyana, Paraguay, Peru, Suriname, and Venezuela. In Ecuador, it occurs in Sucumbíos, Orellana, Napo, Pastaza, and Morona Santiago provinces. Its elevational range is 209–508 m in an area of extent of occurrence of about 34,530 km^2.

This fossorial species inhabits natural forests and modified areas in six ecosystems (Table 0.4) within five biogeographic sectors (Table 0.5), where the average annual rainfall is 2999–4434 mm and the average annual temperature is 23.3–25.5 °C.

Conservation.—*Siphonops annulatus* is categorized as Least Concern (IUCN SSC Amphibian Specialist Group, 2014, in IUCN Red List of 2022) in view of its wide distribution, tolerance of a broad range of habitats, and presumed large population. It appears to be common in some parts of its range. Surveys in

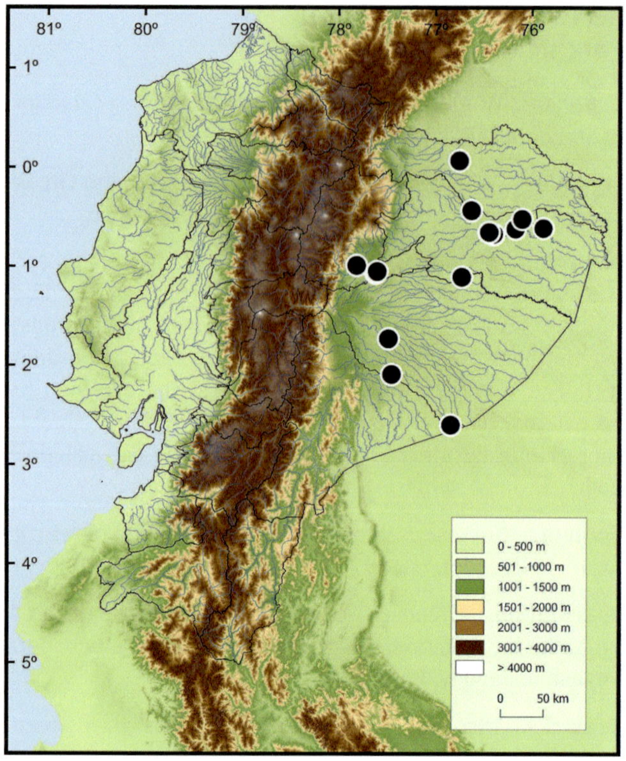

FIGURE 7.29 Localities of known occurrence of *Siphonops annulatus* in Ecuador.

the region of San Martín, Peru, found just two individuals after 90 and > 100 person/days in 2006 and 2008, respectively (von May *et al.*, 2008). There appear to be no major known threats to this species. It occurs in many protected areas. In Ecuador, it has been recorded in the public reserve Parque Nacional Yasuní and within the biosphere reserve Sumaco.

Remarks.—*Siphonops annulatus* is the caecilian with the largest reported distribution. Taylor (1968) linked its wide distribution to the fact that its moisture requirements are less than those of many other caecilians. Lynch (2006) also mentioned its tolerance of dry conditions. According to Maciel and Hoogmoed (2011), the ability of the species better to withstand desiccation seemingly is responsible for this distribution since it is reputedly found, at least occasionally, in relatively dry situations; also, it would appear to utilize a somewhat greater variety of food, which also might be a factor. Also, within its large distribution, *Siphonops annulatus* presents low variation, at least in external morphological characters.

Wilkinson *et al.* (2008) described maternal dermatophagy, the eating of maternal skin by offspring. This unusual form of parental investment co-evolved specializations of both the maternal skin and offspring dentition. The detailed similarities of skin feeding in different caecilian species provide strong evidence of its homology, implying its presence in the last common ancestor of these species, which has probably persisted in multiple lineages for more than 100 Myr.

In addition to skin feeding, young *S. annulatus* were observed to congregate around, and imbibe liquid exuded from, the maternal cloacal opening. Read (2018: 216) provided a photo of a specimen having an unusual coloration.

Key references.—(Mikan, 1820) (description); (Goeldi, 1899) (reproductive biology); (Taylor, 1968) (brief account); (Lynch, 1999) (morphology); (Taylor and Peters, 1974) (account); (Wilkinson *et al.*, 2008) (dermatophagy); (Maciel *et al.*, 2013) (distribution, morphology); (Barrio-Amorós *et al.*, 2019) (brief account).

FAMILY TYPHLONECTIDAE (AQUATIC CAECILIANS)

Typhlonectidae Taylor, 1968: 231. **Type genus.**—*Typhlonectes* (Peters, 1880).

Etymology.—The family named is derived from the Greek *thyphlos*, meaning blind, and *nectes*, meaning swimmer, referring to the aquatic habits of its species.

Definition.—Typhlonectidae is a group supported by morphological and molecular evidence. Individuals have fused, sheet- or sac-like embryonic gills. This unusual gill structure is found only in typhlonectids. Typhlonectids possess tracheal lungs (an expanded trachea used as an accessory gas exchange organ), narial plugs, and secondary zygokrotaphic skulls and lack annular scales as well as secondary annuli. They have only undivided primary annuli, although some may have pseudo secondary grooves. Dermal scales are absent in the grooves. The eyes are moderately well developed and always visible in bony sockets beneath the skin. The small sensory tentacles are nonprotrusible but functional and are usually closer to the nostrils than the eyes. They lack a true tail. Stapes are present in the middle ear. Typhlonectids are unique among caecilians in that they are secondarily aquatic. Members range from 140–750 mm in total length. The bodies of the aquatic taxa are laterally compressed and bear a middorsal fold or fin, which presumably assists their undulatory swimming.

All typhlonectids are viviparous, and embryonic development can be divided into three primary stages. During the first stage, the embryo relies on the yolk for nutrition. During the second stage, when the yolk is depleted, the embryo is released into the uterus, where it feeds on secretions and cells originating from the wall of the uterus. It may also feed on eggs or embryos of siblings. In the last stage, a pair of large, sac-like gills surrounds the embryo and comes in contact with the uterine wall, where it serves as both a respiratory structure and a functional placenta. Members may be completely aquatic or semi-aquatic; all make burrows either at water level or underwater. Lungs are present, but in some species, they are rudimentary, and one member of this family, *Atretochoana eiselti*, is the largest extant lungless tetrapod (reaching up to one meter) and is rarely seen. A 500 mm specimen of *Chthonerpeton viviparum* was eaten by the fish *Hoplias malabaricus*.

Distribution.—They occur in northern South America, east of the Andes.

Content.—Typhlonectidae contains 14 species in five genera; two genera (*Chthonerpeton* and *Potomotyphlus*) and two species occur in Ecuador.

Remarks.—Typhlonectidae is the sister taxon of Caeciliidae. The monophyly of this group is strongly supported through morphology, ecology, and geography. Maciel *et al.* (2017) analyzed relationships of the semiaquatic and aquatic South American caecilians (Typhlonectidae), including all five genera. They found that pulmonary reduction occurred convergently in *Atretochoana* and *Potamotyphlus*.

Key references.—(Wilkinson, 1989) (phylogeny); (Wilkinson, 1996) (key to the genera); (Wilkinson, 1997) (neuroanatomical morphology); (Wilkinson and Nussbaum, 1999) (phylogeny, morphology); (Frost *et al.*, 2006); (Roelants *et al.*, 2007) (phylogeny); Wilkinson *et al.* (2011) (phylogeny, taxonomy); (Pyron and Wiens, 2011) (phylogeny); (Blackburn and Wake, 2011) (taxonomy); (Maddin *et al.*, 2012) (braincase, phylogeny); Vitt and Caldwell (2014) (account); (Maciel *et al.*, 2017) (morphology, molecular phylogeny).

GENUS *CHTHONERPETON* (CAECILIANS) PETERS, 1880

Chthonerpeton Peters, 1880: 930. **Type species.**—*Siphonops indistinctus* (Reinhardt and Lütken, 1862).

Etymology.—The generic name is derived from the Greek words *chthon*, meaning the earth, and *-erpeton*, literally meaning creeping thing. The gender is masculine.

Definition.—Figure 7.30. Uniquely derived traits of *Chthonerpeton* are unknown. The tentacular aperture is in an intermediate imposition between the eye and the external naris. The tentacular groove is not covered with bone. The fetal gills attach laterally to the nuchal region. The two gill bases are well separated dorsally. There is no lateral compression of the body. There is no middorsal ridge or free fold (fin). The left lung is rudimentary. The external nares are ovate. The choanal valve aperture extends along the entire length of the valve. The cloacal disc is subcircular. They have a semiaquatic habit.

Content.—*Chthonerpeton* contains nine species, of which one occurs in Ecuador.

Distribution.—It occurs in Amazonian Ecuador, southern Brazil, Uruguay, and northern Argentina.

Remarks.—*Chthonerpeton* species may be found in wet soil far from permanent water.

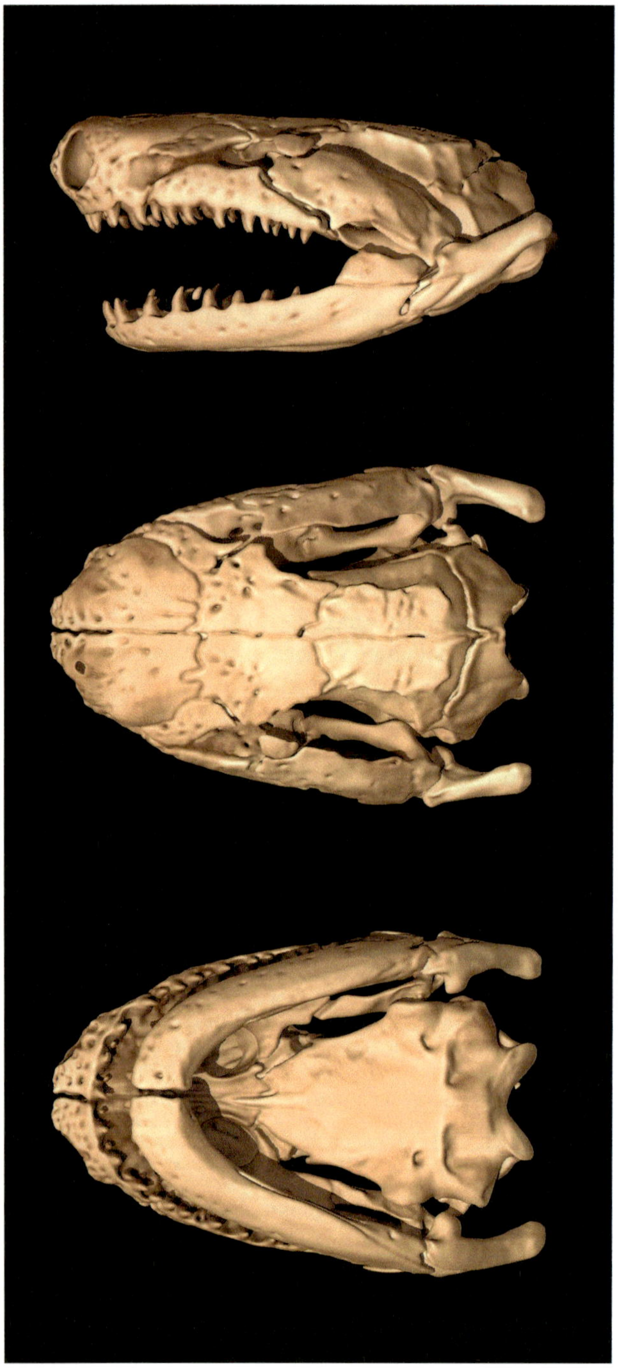

FIGURE 7.30 High-resolution computed tomography reconstructions of the cranial skeleton of *Chthonerpeton indistinctus*, CAS HERP 85521, head length = 14.1 mm (from snout tip to osbasale at midline), Porto Alegre, Rio Grande do Sul, Brazil. Lateral (upper), dorsal (middle), and ventral (lower) views.

CHTHONERPETON ONOREI (ONORE'S CAECILIAN) NUSSBAUM, 1986

Chthonerpeton onorei Nussbaum, 1986: 912. **Holotype.**—MHNG 2251.06 from El Reventador, 1500 m, Provincia Napo, Ecuador.

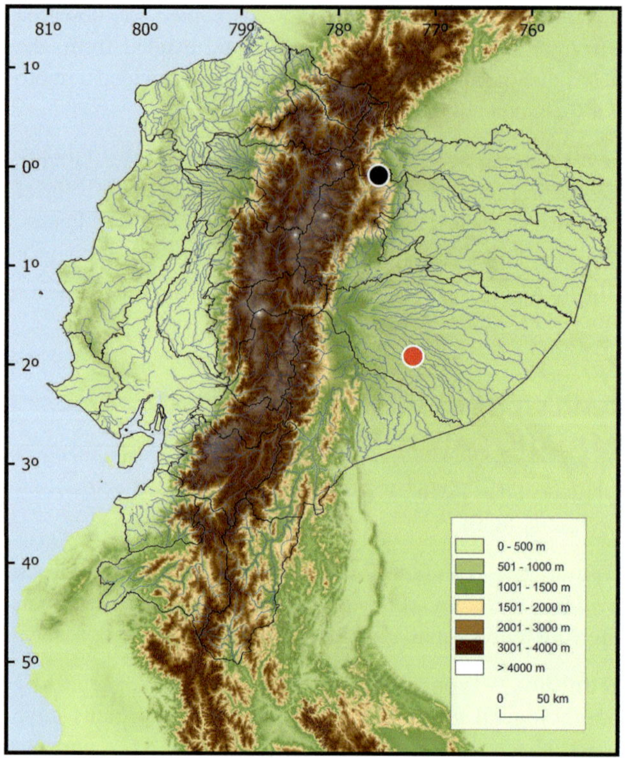

FIGURE 7.31 Localities of known occurrence of *Chthonerpeton onorei* (black dot) and *Potomotyphlus kaupii* (red dot) in Ecuador.

Etymology.—The species name is a patronym in honor of Giovanni Onore in recognition of his contributions to the herpetology of Ecuador.

Identification.—It reaches a known length of 519 mm with a body width of 8 mm. Its length is 65 times its width. It has 164 primary grooves, and there are no secondary grooves. The terminal shield is unsegmented. The eye is distinct. There are no annular scales. There are no subdermal scales. There is not a light-colored eye tentacle tract. There are 164 splenial teeth. There are 173 vertebrae.

Comparisons.—*Chthonerpeton braestrupi*, *C. indistinctum*, *C. viviparum*. *Chthonerpeton onorei* differs from the *C. braestrupi*, *C. hellmichi*, and *C. indistinctum* in having far more primary annuli (164 versus 73–104) and vertebrae (173 versus 111 or fewer) and from *C. viviparum* in having the tentacular aperture equidistant between the eye and nostril or slightly closer to the nostril; the tentacular aperture is much closer to the eye in *C. viviparum*.

Natural history.—Unknown.

Distribution.—Figure 7.31. *Chthonerpeton onorei*, an endemic species from Ecuador, is known only from the type locality on the eastern slopes of the Cordillera Oriental of the Andes in Provincia Sucumbíos at an elevation of 1589 m.

This fossorial species inhabits natural forests in the evergreen low montane forest ecosystem in the northern Cordillera Oriental of the Andes biogeographic sector, where the average annual rainfall is 2495 mm and the average annual temperature is 18.5 °C.

Conservation.—*Chthonerpeton onorei* is categorized as Data Deficient (Almendáriz *et al.*, 2004a; in Red List of 2019) in view of the absence of recent information on its extent of occurrence, status, and ecological requirements. There is no information on the population status of this species, which is known only from two specimens. There is no information on threats to this species. It has not been recorded in any public or private reserves.

Remarks.—Its molecular phylogenetic relationships are unknown.

Key references.—(Nussbaum, 1986) (description).

GENUS *POTOMOTYPHLUS* (AQUATIC CAECILIANS) TAYLOR, 1968

Potomotyphlus Taylor, 1968: 256. **Type species.**—*Caecilia kaupii* (Berthold, 1859).

Etymology.—The generic name is derived from the Greek *potamos*, meaning river, and *typhlós*, meaning blind or hidden.

Definition.—Figure 7.32. These are typhlonectids with an aquatic habit. The tentacular aperture is close behind the external naris. It has the tentacular groove partially roofed with bone in adults. Fetal gills are unknown. The body is laterally compressed. It has a middorsal free fold or ridge. The left lung is well developed, dilated, much wider than the right lung. The head is small relative to the body. The cloacal disc is subcircular posteriorly, with a narrower anterior portion. The external naris is subtriangular. The choanae are extremely large. The choanal valve has the aperture restricted to a small, funnel-like flap.

Content.—*Potomotyphlus* contains only one species, which occurs in Ecuador.

Distribution.—*Potomotyphlus* occurs throughout Brazil, Ecuador, French Guiana, Peru, Venezuela.

Remarks.—According to Wilkinson and Nussbaum (2006) *Potomotyphlus* appears to be the sister genus of the lungless *Atretochoana* on the basis of a number of features associated with a reduction in pulmonary respiration.

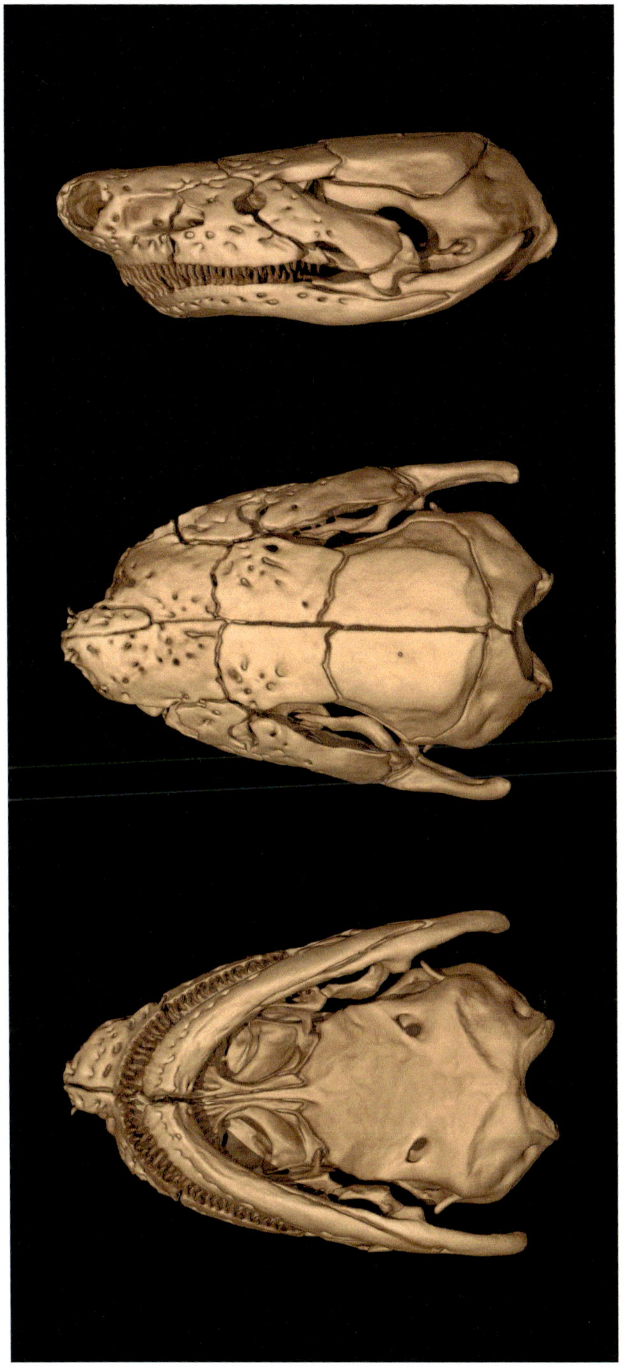

FIGURE 7.32 High-resolution computed tomography reconstructions of the cranial skeleton of *Potomotyphlus kaupii*, YPMA 23166, head length = 10.7 mm (from snout tip to osbasale at midline), unknown locality. Lateral (upper), dorsal (middle), and ventral (lower) views.

POTOMOTYPHLUS KAUPII (KAUP'S CAECILIAN) (BERTHOLD, 1859)

Caecilia kaupii (Berthold, 1859): 181. **Holotype.**—ZFMK 27684 from Angostura, Ciudad Bolívar, Estado Bolívar, Venezuela.
Potomotyphlus kaupii Taylor, 1968: 257.

Etymology.—The specific name is a patronym for Johann Jakob Kaup (1803–1873), a German naturalist and ichthyologist.

Identification.—It reaches a known length of 695 mm. There are 88–99 primary grooves behind the collars. The grooves usually are distinct and marked by a dark line. It has 108–114 vertebrae. It lacks secondary grooves. The scales are absent. The squamosals form a suture with the parietals. The body is compressed throughout. The dorsal fin is evident, beginning several centimeters behind the head and continuing around the terminus of the body to near the vent. The tail is absent. The eyes are visible, never covered by bone. The tentacle is near but posterior to and a little below the middle of the nostril. There is a more or less distinct copulatory disc around the vent. The males have a clasper. There is strong sexual dimorphism in the anal regions. The eyes in very young specimens (100–200 mm in length) are distinct, while in older specimens, they may be barely visible or entirely invisible below skin and flesh. The eye, however, is invariably in a socket and is never covered by the bones of the skull. There is some variation in the extent of the dorsal fin. In some specimens, it reaches forward to the first or second collar; in others, it may fail to reach the head by 6 or 7 cm. Large choanae with thickened valves are present. The tongues in all specimens seen have two anterior pad-like elevations (narial plugs) that correspond to the large choanal openings and may serve in part to control the movement of water through the choanae or prevent its movement.

The dorsum is gray–olive; the sides and venter are yellowish-olive; the head and neck are slightly lighter. The floor of the anal depression is cream. At Montalvo, Provincia Pastaza, the body color is dirty yellow-ivory on the lower sides and venter. It is grayish on the dorsum, the fin itself being a dark lavender–gray. The head is dark lavender. The snout is blackish below. The lower jaw is clouded with gray, while the grooves are blackish.

Natural history.—Caeciliids have been found in fishing nets or under stones in shallow water. One from the mouth of the Amazon was collected from docks in the river. A fisherman in San Vicente, Río Apure (a major tributary of the Orinoco), obtained one specimen fishing and used it as bait.

Distribution.—See Figure 7.31. *Potomotyphlus kaupii* is known from the Amazon and Orinoco drainage systems of South America in Brazil, Colombia, Ecuador, French Guiana, Peru, and Venezuela. In Ecuador, it occurs in a single locality in Provincia Pastaza at an elevation of 340 m.

This aquatic species inhabits the flooded forests of alluvial plains of rivers of Andean and Amazonian Cordilleras origin in the Tigre-Pastaza biogeographic sector, where the average annual rainfall is 3733 mm and the average annual temperature is 24.6 °C.

Conservation.—*Potomotyphlus kaupii* is categorized as Least Concern (Wilkinson *et al.*, 2010; in IUCN Red List of 2022) in view of its wide distribution, presumed large population, and because it is unlikely to be declining to qualify for listing in a more threatened category. It is poorly known, but it is occasionally seen in good numbers. However, it is important to note that in Ecuador, *Potomotyphlus kaupii* has been categorized as Vulnerable by Ortega-Andrade *et al.* (2021) because its small population has an area of extent of occurrence less than 20 km^2 and it is only known from a single locality. In Ecuador, it is a very rare species in collections. It has not been recorded in any public or private reserves.

Remarks.—According to Taylor (1968), the species is aquatic, retaining the dorsal fin throughout life. The only known specimen from Ecuador (USNM 166415) was collected in Río Bobonaza, Pucayacu, located between Montalvo and Sarayacu in Provincia Pastaza, by one member of the Olalla family on an unknown date (likely in the first half of the 1900s). Its absence from subsequent herpetological and ichthyological surveys in the Amazonian region of Ecuador is perplexing.

Key references.—(Berthold, 1859) (description); (Taylor, 1968) (account); (Lynch, 1999) (distribution, morphology); (Barrio-Amorós *et al.*, 2019).

8 Order Caudata

SALAMANDERS, NEWTS

Caudata Oppel, 1811: 409.

Etymology.—The name of this order comes from the Latin word *caudatus*, meaning tailed.

Definition.—Caudata is a monophyletic group diagnosed on the basis of molecular evidence and several synapomorphies: (1) the ossification sequence of the skull, including the late appearance of the maxillae; (2) a remodeling of the palate during metamorphosis; (3) the absence of a middle ear; (4) the origin of the jaw adductor muscle at the top and back of the skull; and (5) the presence of gill slits and external gills in aquatic larvae.

Salamanders are typically recognized by their elongated body, long tail, distinct head, and presence of limbs (although two groups have reduced or absent limbs). They lack a tympanum, and the middle ear consists of a large footplate and short stylus, which is absent in some taxa. When present, the aquatic larvae have true teeth on both jaws. Teeth are also present on the jaws and vomers of adults. Salamander skulls are reduced by the loss of many elements, and other cranial elements are partly or totally cartilaginous. Cartilaginous elements occur in the postcranial skeleton as well. Heterochrony involving pedomorphosis (interspecific) or pedogenesis (intraspecific) has occurred repeatedly in salamander evolution and is recognized by the retention of larval traits in adults, such as gill slits and gills, and the absence of eyelids.

Salamanders exhibit a diverse range of courtship behaviors, including external fertilization and internal fertilization without copulation. Males have specialized cloacal glands that produce spermatophores, which are gelatinous pyramidal structures capped with sperm. Females pick up the sperm cap in the cloaca and store it in a spermatheca. Fertilization typically occurs as the eggs pass through the cloaca. Salamanders typically deposit their eggs singly, in clumps, or in strings in water. The eggs hatch into aquatic larvae. However, some species are ovoviviparous or viviparous, and most plethodontids undergo direct development from terrestrial eggs.

Distribution.—This order of amphibians occurs from the northern hemisphere south to the Neotropical region in Central and South America (southwestern Brazil).

Content.—The order contains nine families, 62 genera, and 811 species. In Ecuador there is one family with two genera, seven species, and several undescribed candidate species.

Remarks.—For a detailed account of the nomenclatural and systematic history of salamanders, see Frost (2023 and onward).

Key references.—(Duellman and Trueb, 1994); (overview); (Selmi *et al.*, 1997) (sperm ultrastructure); (Savage, 2002) (overview); (Wiens *et al.*, 2005); (paedomorphy); (Frost *et al.*, 2006) (phylogeny, taxonomy); (Roelants *et al.*, 2007) (phylogeny); (Sessions, 2008) (karyological evolution); (Struck, 2007) (paedomorphy); (Zhang and Wake, 2009) (phylogenetics, time divergence); (Carroll, 2009) (paleontology); (Pyron and Wiens, 2011) (phylogeny); (Vitt and Caldwell, 2014) (overview); (Pyron, 2014) (biogeography); (San Mauro *et al.*, 2014) (life history evolution); (Dubois *et al.*, 2021) (phylogenetics, nomenclature); (Raffaëlli, 2022) (accounts).

FAMILY PLETHODONTIDAE (LUNGLESS SALAMANDERS)

Plethodontidae Gray, 1850: 31. **Type genus.**—*Plethodon* (Tschudi, 1838).

Etymology.—The family name is derived from the ancient Greek words *plethos*, meaning fullness or abundance, and *odont*, meaning tooth, referring to the numerous tiny teeth found on the jawbones of these salamanders.

Definition.—Plethodontids are a monophyletic group diagnosed on the basis of molecular evidence and morphological features. Plethodontids are characterized by a number of synapomorphies, including their lack of lungs and their specialized tongue projection mechanism. Plethodontids are typically recognized by having four limbs; some taxa are stocky and short limbed, and others are elongate and slender limbed; some have tails equal to the body length, and in others, the tails are twice the length of the body. The adult body size ranges from 25 to 30 mm TL in the diminutive *Thorius* to 320 mm TL in *Pseudoeurycea belli* (both from Mexico). The lower jaw has the angular and prearticular bones fused, and the upper jaw has both premaxillae and maxillae, while the lacrimal bone is absent. Adults lack gills and gill slits; have moveable eyelids, except in the pedomorphic taxa; and possess costal grooves on the trunk, as well as a pair of nasolabial grooves on the snout. Fertilization is internal. Adult females have spermathecae in the cloaca, and adult males possess six sets of cloacal glands. Most species of plethodontids have the ability to project their tongues ballistically, allowing them to capture rapidly moving prey. Plethodontids also display a diversity of reproductive modes. Some species have typical aquatic larvae, while others deposit eggs on land and undergo direct development.

Distribution.—This family has a wide distribution, ranging from extreme southern Alaska and Nova Scotia (Canada) in the north to eastern Brazil and central Bolivia in South America. They can also be found in southern Europe and Korea.

DOI: 10.1201/9781003543879-8

Content.—The family contains 29 genera with 510 species (Frost, 2023). In Ecuador, there are two genera (*Bolitoglossa* and *Oedipina*) and seven species.

Remarks.—The nomenclatural and systematic history is detailed in Frost (2023 and onward). The phylogeny of this group has been reconstructed using morphology and DNA sequences from both mitochondrial and nuclear genes.

Key references.—(Wake, 1966) (morphology, evolutionary relationships); (Larson, 1984) (phylogeny); (Duellman and Trueb, 1994); (account); (Sever, 1994) (cloacal morphology, phylogeny); (Savage, 2002) (overview); (Mueller *et al.*, 2004) (molecular phylogeny); (Chippindale *et al.*, 2004) (molecular phylogeny); (Frost *et al.*, 2006) (phylogeny, taxonomy); (Vieites *et al.*, 2011) (phylogeny and taxonomy); (Pyron and Wiens, 2011) (phylogeny); (Wake, 2012) (taxonomy); (Zheng *et al.*, 2011) (time of origin); (Vitt and Caldwell, 2014) (overview); (Shen *et al.*, 2016) (historical biogeography); (Herrboldt *et al.*, 2021) (pheromone gene diversification, evolution of courtship glands); (Raffaëlli, 2022) (accounts).

GENUS *BOLITOGLOSSA* (MUSHROOMTONGUE SALAMANDERS) DUMÉRIL *ET AL.*, 1854

Bolitoglossa Duméril *et al.*, 1854: 88. **Type species.**—*Bolitoglossa mexicana* Duméril *et al.*, 1854, by subsequent designation by Taylor, 1944: 218.

Etymology.—The generic name is derived from the Latin *boletus*, meaning mushroom, ultimately stemming from the ancient Greek word *bolos*, a lump or clod. Glossa stems from the Greek *glosso*, meaning tongue. The gender is feminine.

Definition.—Figure 8.1. This genus is diagnosed by transformations in nuclear and mitochondrial protein and ribosomal genes. They have a projectile boletoid tongue, as well as hands and feet with more or less developed webbing, which are adapted to an arboreal habitat. This is a derived characteristic that has evolved independently four times within the genus. They exhibit tail autotomy and varying levels of paedomorphosis. This genus differs from all other Neotropical genera by its hyobranchial apparatus and the absence of a sublingual fold. There are 13 costal grooves, as in all other tropical bolitoglossine genera, except *Oedipina*. High-altitude species are more generalized than lowland ones, the latter having diversified several times within this genus, which is characterized by a high level of homoplasy and a very large radiation. Individuals range in size from 5 to 26 cm. Males are smaller than females but have longer limbs and a more pronounced nasolabial region. The mechanism of tail autotomy differs among subgenera, with the first caudal vertebra bearing either branched (in *Nanotriton*, *Mayamandra*, *Magnadigita*, and *Pachymandra*) or unbranched transverse processes.

Distribution.—*Bolitoglossa* is broadly distributed from northeastern Mexico through central and southern Mexico to Central America, Colombia, Venezuela, Ecuador, Peru, northeastern Brazil, and central Bolivia.

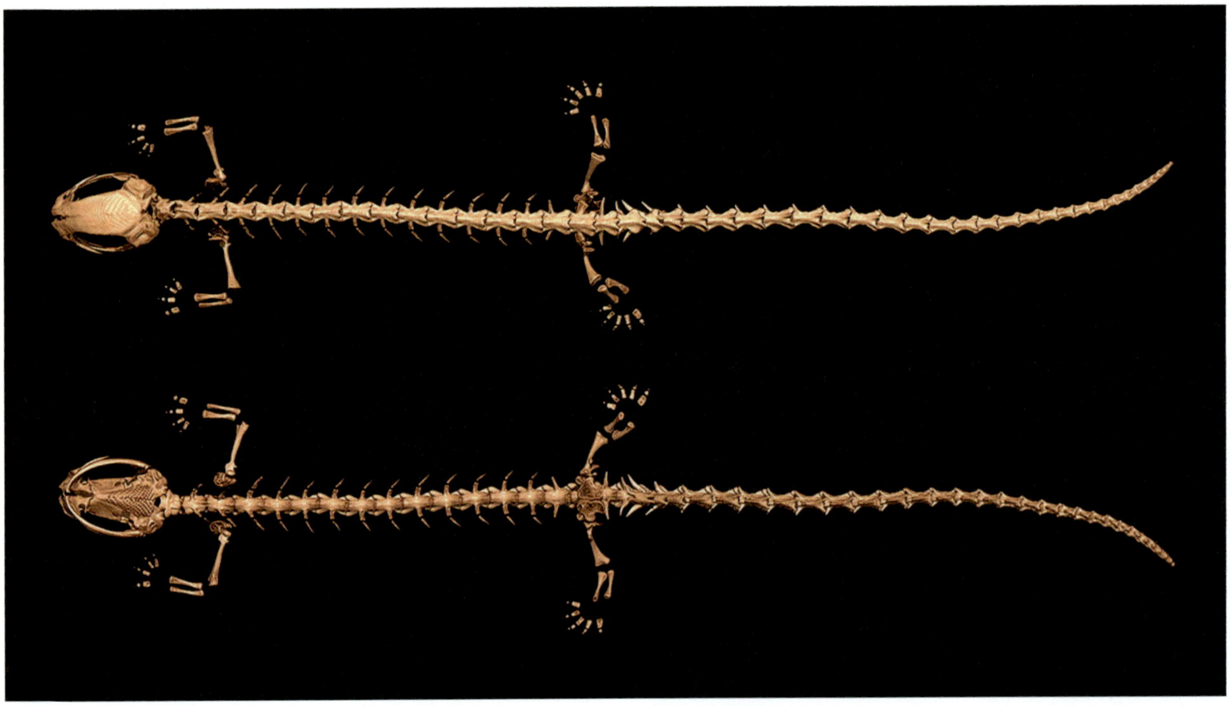

FIGURE 8.1 High-resolution computed tomography reconstructions of the skeleton of *Bolitoglossa* sp., CJ 6198, 74.2 mm SVL, Shiripuno Lodge, Provincia Orellana. Dorsal (upper) and ventral (lower) views.

Content.—132 species are recognized; five species occur in Ecuador (but see Remarks).

Remarks.—Figures 8.2, 8.3. Raffaëlli (2022) summarizes the phenetic genus features included above under definition. Elmer *et al.* (2013) provided a phylogenetic tree and inferred the lineage diversity and timing of divergence of the South American *Bolitoglossa* using mitochondrial and nuclear sequence data from specimens primarily from localities in the Andes and upper Amazon Basin. According to them, the ancestors of South American salamanders likely arrived at least by the Early Miocene, well before the completion of the Late Pliocene Panamanian land bridge (widely accepted as ca. 3 Mya). This date is in agreement with recent, controversial, arguments that an older, perhaps short-lived, land connection may have existed between South America and present-day Panama 23–25 Mya. Since its arrival in South America, *Bolitoglossa* has diversified more extensively than previously presumed and currently includes several cryptic species within a relatively small geographic area. Rather than two upper Amazonian species currently recorded for this region, *B. equatoriana sensu lato* and *B. peruviana sensu lato*, there are at least eight species awaiting formal descriptions. Stuart *et al.* (2008), Köhler (2011), and Raffaëlli (2013, 2022) provided accounts of the species, photographs, and range maps.

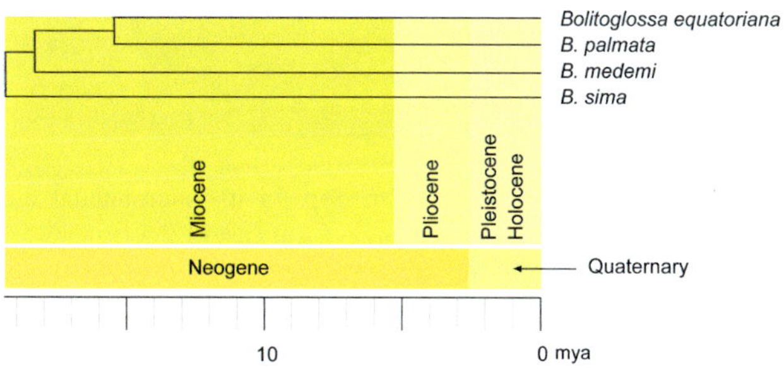

FIGURE 8.2 Time-calibrated maximum likelihood phylogenetic tree depicting evolutionary relationships of some of the Ecuadorian species of *Bolitoglossa*, inferred from DNA sequences of mitochondrial and nuclear genes.

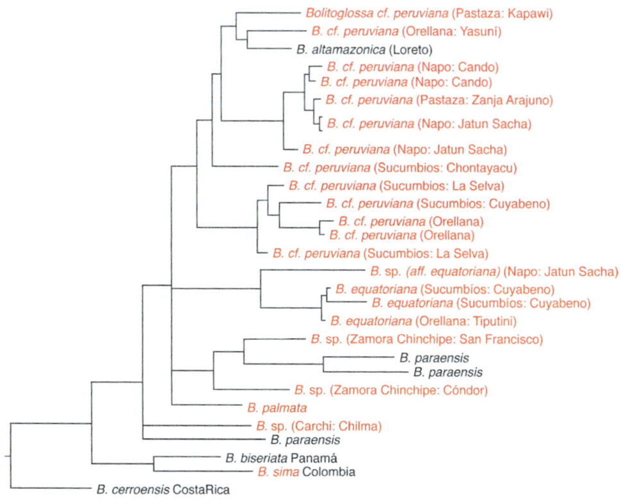

FIGURE 8.3 Topology of several species of South American *Bolitoglossa*, extracted (and modified) from a phylogenetic tree inferred from analysis of mitochondrial DNA and nuclear sequence data by Elmer *et al.* (2013). Species (described and undescribed) occurring in Ecuador are in red.

BOLITOGLOSSA CHICA (TINY MUSHROOMTONGUE SALAMANDER) BRAME AND WAKE, 1963

Bolitoglossa chica (Brame and Wake, 1963): 16. **Holotype.**—USNM (James A. Peters 4366) from near Hotel Zaracay, Santo Domingo de los Tsáchilas, 670 m , Provincia Santo Domingo de los Tsáchilas, Ecuador.

Etymology.—The specific epithet is derived from the Spanish word *chica*, meaning small.

Identification.—Figures 8.4, 8.5. The maximum known SVL is 40.3 mm in females and 38.7 mm in males. The snout is of moderate length, broad and truncate, slightly rounded at the tip. The nostril is small. The labial protuberances of the nasolabial groove are moderate. The canthus rostralis is indistinct, gently rounded. There is a relatively deep, distinct groove below the eye that extends the full length of the eye opening; it does not communicate with the lip. A poorly defined postorbital groove extends posteriorly from the eye as an irregular, indistinct depression. There are 16 vomerine teeth in slightly curved series that extend to the medial border of the internal nares. It has very few or no maxillary teeth (4 male, 0 female) and no premaxillary

FIGURE 8.4 *Bolitoglossa chica*. Left column: CJ 11598, female, 40.67 mm SVL, 5 km east of Puerto Quito, Provincia Pichincha. Right: CJ 13069, male, 43.14 mm SVL, Reserva Canandé, Provincia Esmeraldas. SGS.

FIGURE 8.5 *Bolitoglossa chica*. KU 119434, Santo Domingo de los Tsáchilas, Provincia Santo Domingo de los Tsáchilas. John D. Lynch.

FIGURE 8.6 *Bolitoglossa chica*. Top: in its microhabitat at night. Bottom: female with laid eggs. 5 km east of Puerto Quito, Provincia Pichincha. Morley Read.

teeth. The tail is relatively robust and round with no lateral compression, with a marked basal constriction. The post-iliac gland is not evident. The limbs are short and slender. The webbing of the hands and feet is nearly complete; the digits are fairly broad and well defined; the tips of the digits are free of webbing, triangular, and broadly rounded. The longest finger and longest toe are more pointed than others. The fingers, in order of decreasing length, are 3, 2, 4, 1; the toes in order of decreasing length are 3, 4, 2, 5, 1.

The dorsum is cream, and the sides and venter are brown. The limbs are dark, and the hands and feet are noticeably lighter in color.

Comparisons.—*Bolitoglossa sima*, *B. equatoriana*, *B. colonnea*. *Bolitoglossa chica* is mostly similar to *B. sima* in the Pacific region of Ecuador and to *B. equatoriana* in the Amazonian region. It differs from all of them by having very few or no maxillary teeth. It is further distinguished from *B. sima* by a highly significantly shorter adult snout–vent length and lighter dorsal coloration. It is different from *B. colonnea* (from Panama) by the absence of an interorbital dermal ridge.

Natural history.—Figures 8.6, 8.7. One individual was on a broad-leafed fern at night during moderately heavy rain. Individuals were found on banana leaves in banana plantations and on *Heliconia* leaves and elephant ear leaves in forest-edge habitats at night. The salamanders were found nearer to the ground in the early evening and higher later on. There seemed to be no difference in habitat preference between *B. chica* and *B. sima* in areas where they were found sympatrically.

Distribution.—Figure 8.8. *Bolitoglossa chica* is known in the coastal region of Ecuador and in Reserva Natural La Planada, 1650 m elevation, Nariño, Colombia, in the southwestern part of the western Andes. In Ecuador, it occurs in Esmeraldas, Pichincha, Manabí, Santo Domingo de dos

FIGURE 8.7 *Bolitoglossa chica*. Development of embryos (81 days from egg to hatchling). 5 km east of Puerto Quito, Provincia Pichincha. Morley Read.

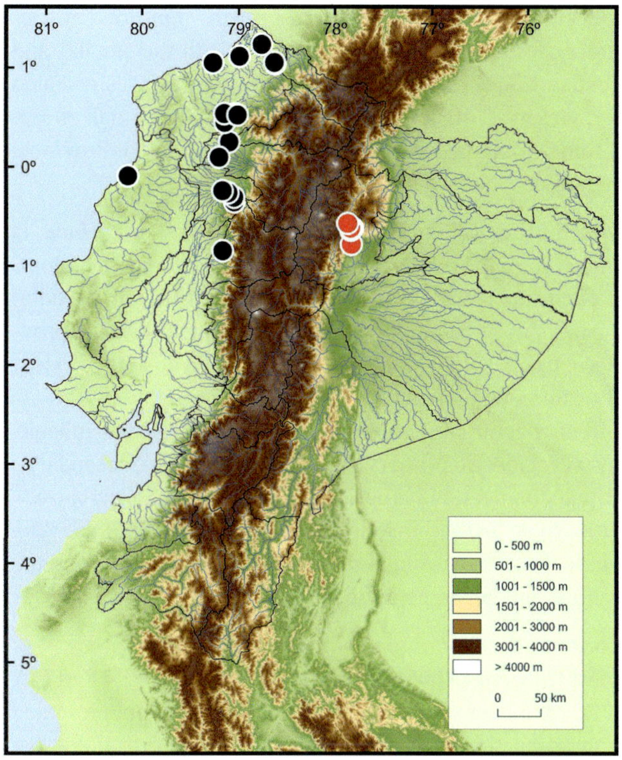

FIGURE 8.8 Localities of known occurrence of *Bolitoglossa chica* (black dots) and *B. palmata* (red dots) in Ecuador.

Tsáchilas, and Cotopaxi provinces. Its elevational range is 14–1082 m in an area of extent of occurrence of about 19,290 km^2.

This nocturnal, arboreal species inhabits the low vegetation of forest edges in the flooded forests of intertidal plains of Equatorial Chocó, evergreen forests of lowlands of Equatorial Chocó, evergreen seasonal forests of lowlands of Jama-Zapotillo, and evergreen montane foothill forests of the Cordillera Occidental of the Andes ecosystems within several biogeographic sectors (Table 0.5), where the average annual rainfall is 863–2663 mm and the average annual temperature is 20.1–25.8 °C.

Conservation.—*Bolitoglossa chica* is categorized as Critically Endangered (IUCN SSC Amphibian Specialist Group, 2022d; in IUCN Red List of 2022) because of a population decline, suspected to be occurring at a rate of at least 80% over the next 10 years, due to a severe loss of habitat as a result of industrial agriculture and commercial logging in the lowlands of northwestern Ecuador. It is considered quite rare. It is categorized as Critically Endangered in Ecuador, as well by Ortega-Andrade *et al.* (2021). There has been a dramatic reduction in its area of occupancy within its area of occurrence (19,290 km^2). Its major threats are habitat reduction and fragmentation, logging, deforestation for agricultural development (especially palm plantations), fires, mining, the introduction of exotic species, human settlements, and pollution. Its area of distribution is under constant pressure by wood companies.

It has been recorded in the protected the protected forests of Asociación Agrícola Carchi-Imbabura and some private reserves (e.g., Reserva Biológica Canandé, Reserva Jama Coaque, Reserva Tesoro Escondido, Tinalandia, Bosque Protector La Chiquita).

Remarks.—Its molecular phylogenetic relationships are unknown. According to Parra-Olea *et al.* (2004), *Bolitoglossa chica* is within the subgenus *Eladinea*.

Key references.—(Brame and Wake, 1963) (description); (Crump, 1978) (morphology, ecology); (Stuart *et al.*, 2008) (account); (Ortega-Andrade *et al.*, 2010b) (account); (Raffaëlli, 2013) (account).

BOLITOGLOSSA EQUATORIANA (ECUADORIAN MUSHROOMTONGUE SALAMANDER) BRAME AND WAKE, 1972

Bolitoglossa equatoriana (Brame and Wake, 1972): 23.
Holotype.—LACM 70550 from Limoncocha, 260 m, Provincia Sucumbíos, Ecuador.

Etymology.—The specific epithet refers to the geographic location of the type in Ecuador.

Identification.—Figures 8.9, 8.10. The maximum known SVL is 57.9 mm in females and 44.0 mm in males. It has low numbers of maxillary (mean 25) and moderate numbers of vomerine (mean 21) teeth. It has a broad head, extensive webbing, and long hands and feet.

The dorsum is reddish-brown; the proximal parts of the hind limbs are dull yellow–orange. The ventral coloration is sexually dimorphic; in males, the venter is gray with white flecks clumped into irregular spots, whereas in females, it is gray with white stippling. The iris is bronze.

Comparisons.—*Bolitoglossa walkeri*, *B. medemi*, *B. ramosi*, *B. altamazonica*. *Bolitoglossa equatoriana* is distinguished from *B. walkeri* by its broader head, more extensively webbed and slightly longer hands and feet, and spotted

FIGURE 8.9 *Bolitoglossa equatoriana*. Left: CJ 9402, female, 55.5 mm SVL, km 97, road Pompeya Sur-Iro, Parque Nacional Yasuní, Provincia Orellana. Right: CJ 9397, Pompeya Sur-Iro, Parque Nacional Yasuní. SGS.

FIGURE 8.10 Morphological details of *Bolitoglossa equatoriana*. Top: QCAZ 12338, Comunidad Samona Yuturi, Puerto Yuturi, Provincia Orellana. LAC. Bottom: CJ 9401, km 97, road Pompeya Sur-Iro, Parque Nacional Yasuní, Provincia Orellana. SGS.

ventral color pattern; from *B. medemi* by its less numerous maxillary teeth and somewhat shorter legs; from *B. ramosi* by its narrower head and less numerous maxillary teeth; and from *B. peruviana* by being larger and proportionately more robust. It has a broader head, larger hands and feet, and less numerous maxillary teeth, as well as a spotted ventral color pattern; the venter in *B. peruviana* is gray with small pale flecks. It differs from *B. altamazonica* by its broader head, larger hands and feet, and spotted ventral color pattern. *Bolitoglossa equatoriana* is distinguished from all other Panamanian and South American species of *Bolitoglossa* by the combination of extensively webbed hands and feet, coloration, and its size and dentitional features.

Natural history.—Individuals have been found in leaf litter on the forest floor (by day) and up to 2 m above the ground on broad leaves, palm leaves, and stems of low bushes or herbaceous vegetation along stream banks. Once, two individuals and at another time three individuals were found together on a single plant, less than 20 cm from each other. After capture by hand, some individuals adopt a defensive position with the body and tail coiled around the head. At Tiputini Biological Station, 60% of the specimens were collected in non-flooded forests (terra firme), while 40% were in flooded forests (varzea). Specimens from the Jatun Sacha Biological Station and from the Runa Huasi lodge were collected in closed-canopy, old-growth, non-flooded secondary forests. It occurs in sympatry with the smaller and slenderer *B. peruviana*. Stomach content analyses revealed the presence of ants, small beetles, and a small wasp. A female obtained in September in Santa Cecilia, Provincia Sucumbíos, had small ovarian eggs.

Distribution.—Figure 8.11. *Bolitoglossa equatoriana* is known from the western Amazon Basin of southern Colombia and Ecuador below 500 m (see Remarks). In Ecuador, it occurs in Sucumbíos, Orellana, Napo, and Pastaza provinces. Its elevational range is 177–434 m in an area of extent of occurrence of about 35,740 km^2.

This nocturnal, terrestrial, and arboreal species inhabits low vegetation in primary and secondary forests, secondary–primary borders, and in agricultural clearings in seven ecosystems (Table 0.4) within four biogeographic sectors (Table 0.5), where the average annual rainfall is 2732–4285 mm and the average annual temperature is 23.8–25.6 °C.

Conservation.—*Bolitoglossa equatoriana* is categorized as Least Concern (Almendáriz *et al.*, 2004b; in IUCN Red List of 2020) in view of its relatively wide distribution, presumed large population, and because it is unlikely to be declining fast enough to qualify for listing in a more threatened category.

It is locally common, but it appears to experience population fluctuations (being more abundant at some times of the year than at others). Agriculture, both crops and livestock, as well as logging are major threats to the species' habitat.

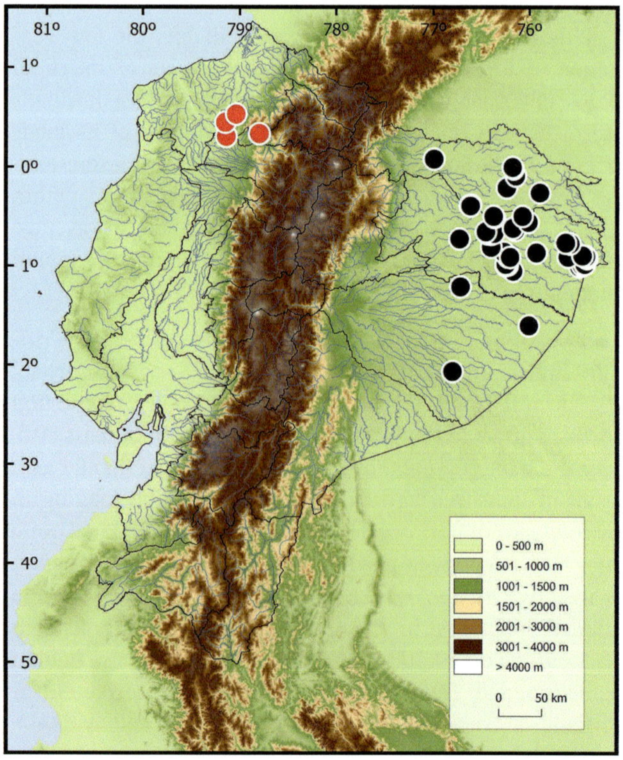

FIGURE 8.11 Localities of known occurrence of *Bolitoglossa equatoriana* (black dots) and *B. medemi* (red dots) in Ecuador.

It has been recorded in the public reserves Reserva de Producción de Fauna Cuyabeno, Reserva Biológica Limoncocha, and Parque Nacional Yasuní; the protected forests of Comuna Kichwa Añangu and Río Aguarico; some private reserves (e.g., La Selva Amazon Ecolodge, Biodiversity Station Tiputini, Bataburo Lodge); and within the biosphere reserve Sumaco.

Remarks.—According to Elmer *et al.* (2013), who analyzed Cytb and Rag1, *Bolitoglossa equatoriana* is within a lineage that includes at least one unnamed closely related cryptic species (from Jatun Sacha, Provincia Napo), plus a complex of species currently under *B. peruviana* and *B. altamazonica*.

Key references.—(Brame and Wake, 1972) (description); (Duellman, 1978) (account); (Cisneros-Heredia, 2007b) (distribution, ecology); (Elmer *et al.*, 2013) (phylogeny); (Raffaëlli, 2013, 2022) (accounts).

BOLITOGLOSSA MEDEMI (MEDEM'S MUSHROOMTONGUE SALAMANDER) BRAME AND WAKE, 1972

Bolitoglossa medemi (Brame and Wake, 1972): 2. **Holotype.**—LACM 42276 from Finca Chibiguí, on the Río Arquía, approximately, 300 m, Departamento Antioquia, Colombia.

Etymology.—The specific epithet honors Professor Fred (Federico) Medem, (Friedrich Johann Graf von Medem) (1912–1984) of the Universidad Nacional de Colombia (Villavicencio), who was a good friend of Arden H. Brame Jr. and David B. Wake and aided and encouraged their studies of Colombian salamanders.

Identification.—The maximum known SVL is 58.0 mm in females and 46.7 mm in males. It has 28–59 maxillary teeth and a mean of 31 vomerine teeth. Its head is relatively broad. The snout is truncate dorsally, lengthened in lateral profile. The eyes are large, slightly protuberant. The tail is laterally compressed. The tips of toe III and finger III are sharply pointed, abruptly protruding from the hand and foot, which are nearly completely webbed.

The dorsum is dark brown with disperse golden marks or irregular clear brown stains. The venter is grayish-brown, lighter than the dorsum color, with a few widely scattered, irregular, light yellow stains and spots. The iris is metallic golden or coppery. Specimens from Panamá (San Blas ridge and Río Tuquesa) vary from nearly uniform gray to light brown with irregular dark blotches dorsally. Scattered light blue stippling is present on the snout and flanks.

Comparisons.—*Bolitoglossa ramosi*, *B. walkeri*, *B. equatoriana*. *Bolitoglossa medemi* is distinguished from *B. ramosi* by having fewer maxillary teeth and a darker dorsal ground color; from *B. walkeri* by its broader head, more extensively webbed feet, and longer legs; and from *B. equatoriana* by its more numerous maxillary teeth and somewhat longer legs. *Bolitoglossa medemi* is distinguished from other Panamanian and South American species by the combination of large, extensively to completely webbed hands and feet, with the tips of toe III and finger III sharply pointed and abruptly protruding from the hand and foot, relatively broad head, long legs, distinctive coloration, and dentitional features.

Natural history.—In Reserva Biológica Canandé, Provincia Esmeraldas, one individual was found at night on the leaf of a bush 1.7 m above the ground. Other individuals (in Colombia) were collected during daylight hours, exposed on the surface, and also during the night, sitting under leaves between 0.7 and 2.5 m above the ground. In lowland rainforests of Ecuador, it is found in sympatry with *B. biseriata*.

Distribution.—Figure 8.11. *Bolitoglossa medemi* is known from the Darién and San Blas regions of Panama and the Chocó region through western Colombia to northwestern Ecuador. In Ecuador, it occurs in Esmeraldas, Imbabura, and Pichincha provinces. Its elevational range is 186–2338 m in an area of extent of occurrence of about 560 km^2.

This nocturnal, arboreal species inhabits mostly primary forests in the evergreen forests of lowlands of Equatorial Chocó and evergreen montane and foothill forests of the Cordillera Occidental of the Andes, where the average annual rainfall is 1568–2216 mm and the average annual temperature is 14.6–24.6 °C.

Conservation.—*Bolitoglossa medemi* is categorized as Least Concern throughout its entire range (IUCN SSC Amphibian Specialist Group, 2020a; in IUCN Red List of 2022) in view of its wide distribution and presumed large population. However, it is important to note that in Ecuador, *Bolitoglossa medemi* has been categorized as Endangered by Ortega-Andrade *et al.* (2021) because its extent of occurrence is estimated to be less than 5000 km^2 (560 km^2, according to our data), its distribution is severely fragmented in fewer than five known localities, and there is a continuing decline in the extent and quality of its habitat. Important threats to its existence include habitat reduction and fragmentation, logging, deforestation for agricultural development (particularly palm plantations), fires, mining, the introduction of exotic species, human settlements, and pollution. Furthermore, the area where it is distributed faces constant pressure from wood companies. It is a rare species. It occurs in Parque Nacional Darién in Panama. It is not known in any protected areas in Colombia. In Ecuador, it has not been recorded in public reserves, but there are records from the protected forest Los Cedros and from some private reserves (e.g., Reserva Biológica Canandé).

Remarks.—Ortega-Andrade *et al.* (2010b) provided information on specimens from Canandé, Provincia

Esmeraldas. According to Batista *et al.* (2014), who analyzed 16S RNA, *Bolitoglossa medemi* (samples from Provincia Darién, Panama) is the sister taxon of *B. taylori* from the same region.

Key references.—(Brame and Wake, 1972) (description); (Lynch and Suárez-Mayorga, 2004) (distribution); (Rivera-Correa and Gutiérrez-Cárdenas, 2006) (distribution); (Mueses-Cisneros *et al.*, 2007) (distribution); (Solís *et al.*, 2008) (conservation); (Stuart *et al.*, 2008) (account); (Ortega-Andrade *et al.*, 2010b; Köhler, 2011; Raffaëlli, 2013, 2022) (accounts).

BOLITOGLOSSA PALMATA (WEBBED MUSHROOMTONGUE SALAMANDER) (WERNER, 1897)

Spelerpes palmatus (Werner, 1897): 266. **Lectotype.**—NHMW 22862 from Cordillera, Ecuador.

Bolitoglossa palmata (Brame and Wake, 1962): 173.

Etymology.—The specific epithet is derived from the Latin *palmata*, meaning (hand-) palm-like, in reference to the shape of the hands.

Identification.—Figure 8.12. The maximum known SVL is 53.3 mm in females and 35.6 mm in males. It has s relatively short snout and limbs and has a distinctive foot shape. The number of vomerine teeth is 14–23. The number of costal folds not covered when the limbs are appressed to the sides of the body is 2, 1.5, 5, 4, and 4.5 from the smallest to largest specimens, respectively.

The dorsum is light reddish-brown to uniformly tan with indistinct dark lines. The venter is rich dark brown with a few tan dashes and spots, with dirty ochre yellow arising from the rear area. There is a sharp demarcation between the lighter dorsolateral color and darker ventral color along the ventrolateral line from forelimb to hind limb. The tail above is darker than the trunk; the uniform dorsal color breaks up into patches and spots ventrally and ventrolaterally. The legs are mottled, with the darker color of the venter prevailing over the lighter dorsum coloration.

FIGURE 8.12 *Bolitoglossa palmata.* Left: CJ 7540, dorsolateral, dorsal, and ventral views. Top right: head. Middle right: ventral view of plantar surface. Bottom right: ventral view of palmar surface. Cóndor-Paquisha, Provincia Zamora Chinchipe. DAL.

Comparisons.—*Bolitoglossa palmata* is mostly similar to *B. equatoriana*, from which it differs by lacking a spotted venter.

Natural history.—Figure 8.13.

Distribution.—See Figure 8.8. *Bolitoglossa palmata* is known from the Amazonian slopes of the Andes of Ecuador in Caquetá, Colombia. In Ecuador, it occurs in Provincia Napo. Its elevational range is 1250–2241 m in an area of extent of occurrence of about 50 km^2.

This nocturnal species has been recorded in primary forests in the evergreen low and montane forests of the northern Cordillera Oriental of the Andes, where the average annual rainfall is 2221–3492 mm and the average annual temperature is 15.1–20 °C.

Conservation.—*Bolitoglossa palmata* is categorized as Least Concern (IUCN SSC Amphibian Specialist Group,

FIGURE 8.13 *Bolitoglossa palmata.* Cordillera de Guacamayos, Provincia Napo. LAC.

2018g; in IUCN Red List 2022) through its entire range in view of its wide distribution and presumed large population. It was categorized as Endangered in Ecuador by Ortega-Andrade *et al.* (2021). We categorize this species is Critically Endangered in Ecuador following B1b(iii)c(ii) IUCN criteria because: (1) its extent of occurrence is estimated to be less than 100 km^2 (50 km^2) and (2) its habitat is currently affected by deforestation, fragmentation, and stochastic events such as volcanic eruptions, fires, and landslides. This species is rarely recorded. It is not known from any protected areas in Colombia. In Ecuador, it has been recorded in the public reserves Parque Nacional Antisana and within the biosphere reserve Sumaco.

Remarks.—Parra-Olea *et al.* (2004) hypothesized that *Bolitoglossa palmata* is the sister taxon of *B. peruviana* plus *B. paraensis* using Cytb and 16S RNA. Elmer *et al.* (2013) placed *B. palmata* as the sister taxon of *B. medemi* based on Rag1 and Cytb. However, according to the most recent phylogeny by Batista *et al.* (2014), who analyzed 16S RNA, *Bolitoglossa palmata* is the sister taxon of *B. peruviana* plus *B. altamazonica*.

This species was previously only known on the eastern slopes of the Andes in Ecuador. However, it has now also been recorded on the eastern slopes of the Andes in southern Colombia (Acosta-Galvis and Gutiérrez-Lamus, 2012).

Key references.—(Werner, 1897) (description); (Raffaëlli, 2007; Stuart *et al.*, 2008; Köhler, 2011; Raffaëlli, 2013, 2022) (accounts).

BOLITOGLOSSA SIMA (SNUB-NOSED MUSHROOMTONGUE SALAMANDER) (VAILLANT, 1911)

Spelerpes simus (Vaillant, 1911): 58. **Holotype.**—MNHNP 1906.284 from Ecuador.
Bolitoglossa sima (Wake and Brame, 1963): 386.

Etymology.—The specific epithet is derived from the Greek *simos*, meaning snub-nosed, alluding to one of the most striking features, the shape of the truncated snout, a peculiarity that Vaillant did not find cited or represented to the same degree in any animal of the genus.

Identification.—Figures 8.14, 8.15, 8.16. The maximum known SVL is 59.4 mm in females and 46.7 mm in males. The head is of moderate width (standard length 6.2–6.6 times the head width). It has 28–44 maxillary teeth and fairly short limbs (limb interval 3–4). The hands and feet are extensively webbed. The snout is of moderate length and width and is flattened and truncate anteriorly. The eyes are of moderate size. The maxillary teeth increase steadily with size, with the largest specimen (45.9 standard length) having the highest numbers of teeth (44). The maxillary teeth extend posteriorly to a point between the front and center of the eye. The vomerine teeth are in a single series in three individuals and are slightly patched laterally in two. The vomerine series extends to the medial border of the choanae in most. The tail is virtually round in cross-section, with little lateral compression. The webbing of the hands and feet is complete and is more extensive than in any other South American *Bolitoglossa*. The digits are very flattened and indistinct. The tips of the longest digits project as small obtuse angles from the

FIGURE 8.14 *Bolitoglossa sima.* Reserva Jevon. Jaime Culebras. Right: KU 142654, Río Baba, 10 km south, 4 km east of Santo Domingo de los Tsáchilas, Provincia Santo Domingo de los Tsáchilas. WED.

FIGURE 8.15 *Bolitoglossa sima.* Left: CJ 10773, female, Durango, Provincia Esmeraldas. DAL. Right: QCAZ 27752, Alto Tambo, Provincia Esmeraldas. LAC.

FIGURE 8.16 Morphological details of *Bolitoglossa sima*. Top: head. Middle: palmar surfaces. Bottom: plantar surfaces. CJ 10773, female, Durango, Provincia Esmeraldas. DAL.

pad-like webbing. The size of the hands and feet is greater in *B. sima* than in other Ecuadorian species.

The holotype has the following features: one premaxillary tooth; 24 maxillary teeth (plus several that have been knocked out), 20 vomerine teeth (in single rows), 13 costal grooves, limb interval 3V2, standard length 45.7 mm, axilla–groin length 26.0 mm, head width 6.9 mm, snout–gular fold length 10.9 mm, tail length 49.0 mm, hind limb length 9.8 mm. The maxillary teeth extend to the front of the eye on the left side and to the center of the eye on the right. The vomerine series extends to the lateral edge of the internal nares on both sides.

The dorsum is generally cream to brownish, sometimes with elongated dark brown flecks. Black stippling covers the dorsum of the head and body, denser on the head. The flanks are gray–brown, sometimes highly contrasting with the dorsal color and divided. All ventral surfaces are uniformly cream to tan covered with tiny white and black stippling. The white stippling is less homogeneous than the black stippling. The iris is golden cream with regular black stippling and a hint of red. A black ring borders the iris.

Comparisons.—*Bolitoglossa palmata*, *B. peruviana*, *B. equatoriana*, *B. altamazonica*, *B. chica*. *Bolitoglossa sima* is distinguished from *B. palmata*, *B. peruviana*, *B. equatoriana*, and *B. altamazonica* by its extensively webbed hands and feet and distinctive coloration (light dorsal ground color mottled with several shades of brown). It is distinguished further from *B. altamazonica* and *B. chica* by higher numbers of maxillary teeth and from *B. altamazonica* and *B. peruviana* by larger hands and feet (standard length 9.2–10.6 times the foot width versus 11.0–13.3 in *B. altamazonica* and 11.6–13.7 in *B. peruviana*).

Natural history.—At Santo Domingo de los Tsáchilas, Provincia Santo Domingo de los Tsáchilas, individuals were found in the nesting bases of the stems of giant elephant ear plants, which catch and hold water. All were found in plants that were shaded either by trees or by other elephant ear plants. The plants grow in large clumps, 3–4.6 m in diameter, and the salamanders were in the innermost plants, well down toward the base of the stem. The general area was dense wet forest, but the actual collection locality was a grassy field, cleared for cattle grazing, with scattered trees left for shade. No salamanders were found in the same plants growing in the forest proper. The individuals were collected at night in the same types of habitats as *B. chica*, that is, on banana leaves in banana plantations and on *Heliconia* leaves and elephant ear leaves in forest-edge habitats at night. They seemed to be found lower in the early evening and higher above the ground later in the evening. Specimens in Reserva Biológica Canandé, Provincia Esmeraldas were collected on leaves up to 3 m.

Distribution.—Figure 8.17. *Bolitoglossa sima*, an endemic species to Ecuador, is known from the northwestern Pacific lowlands of Ecuador in Carchi, Imbabura, Esmeraldas, Manabí, Pichincha, and Santo Domingo de los Tsáchilas provinces. Its elevational range is 6–1020 m in an area of extent of occurrence of about 14,420 km^2.

This nocturnal, arboreal species inhabits the low vegetation of forest edges in four ecosystems (Table 0.4) within the Cordillera Occidental of the Andes and Equatorial Chocó biogeographic sectors, where the average annual rainfall is 1782–2783 mm and the average annual temperature is 20.7–25.9 °C.

Conservation.—*Bolitoglossa sima* was categorized as Endangered by Ortega-Andrade *et al*. (2021). There is no information on the population status of this species. Its major threats are habitat reduction and fragmentation, logging, deforestation for agricultural development (especially palm plantations), fires, mining, the introduction of exotic

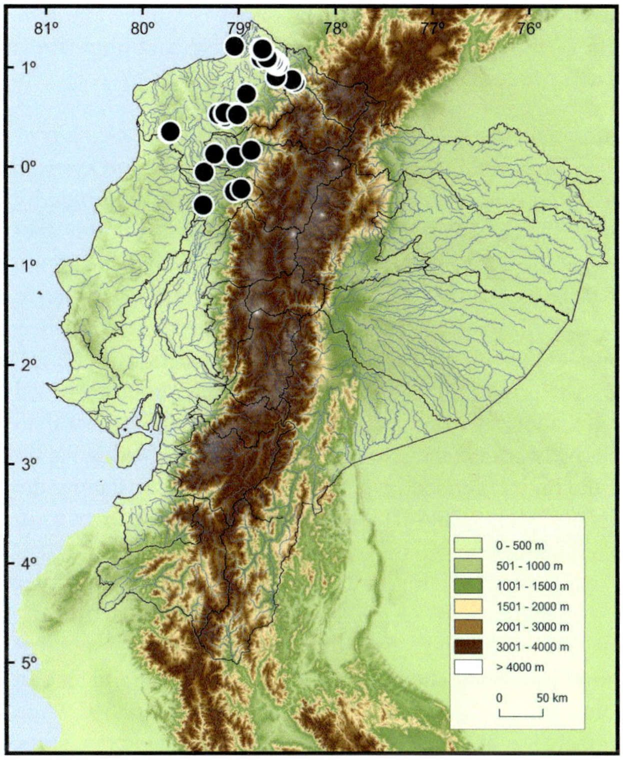

FIGURE 8.17 Localities of known occurrence of *Bolitoglossa sima* in Ecuador.

species, human settlements, and pollution. Its area of distribution is under constant pressure by wood companies. In Ecuador, it has been recorded in the public reserves Reserva Ecológica Reserva Ecológica Manglares Cayapas Mataje and Reserva Ecológica Mache Chindul, in the protected forests of Asociación Agrícola Carchi-Imbabura and Mashpi, in some private areas (e.g., Reserva Biológica Bilsa), and within the biosphere reserve Chocó Andino de Pichincha.

Remarks.—Vaillant (1911) described *Bolitoglossa sima* from a single Ecuadorian specimen. Brame and Wake, 1963) redescribed the holotype and added specimens from northwestern Ecuador (from Playa de Oro and Cachabí, Provincia Esmeraldas) to this taxon. According to Parra-Olea *et al.* (2004), *B. sima* (sample from Departamento Valle del Cauca, Colombia) is the sister taxon of *B. biseriata* (sample from Provincia Nusagandi, Panama) and is closely related to *B. adspersa* within the subgenus *Eladinea*, based on Cytb and 16S RNA. Cisneros-Heredia (2007b) commented on the close relationship between *B. biseriata* and *B. sima* and assigned Ecuadorian specimens to *B. biseriata* based on the coloration pattern. However, molecular data and analyses by Hervas *et al.* (2016) placed the latter populations under *B. sima*. Data and accounts by Crump (1978) (morphology, ecology), Stuart *et al.* (2008) (brief account), Ortega-Andrade *et al.* (2010a) (account, as *B. biseriata* and *B. sima*), Raffaëlli (2013, 2022) (account), and Yánez-Muñoz *et al.* (2020) (accounts as *B. biseriata* and *B. sima*) should be taken with caution until the taxonomy of this species is clarified.

Key references.—(Vaillant, 1911) (description); (Brame and Wake, 1963) (redescription).

GENUS *OEDIPINA* (WORM SALAMANDERS) KEFERSTEIN, 1868

Oedipina (Keferstein, 1868): 299. **Type species.**—*Oedipina uniformis* (Keferstein, 1868).

Etymology.—The generic name is a latinized word presumably derived from the Greek *Oedipus*, meaning swollen foot. *Oedipus* was originally coined by Tschudi, 1838, whose etymological intention is unknown. It might also derive from the Greek mythology to refer to the King of Thebes, son of Laius and Jocasta, who unwittingly killed his father and married his mother. The gender is feminine.

Definition.—Figure 8.18. This genus is diagnosed by molecular characters (16S and Cytb) *sensu* McCranie *et al.* (2008). No phenotypic synapomorphies have been noted.

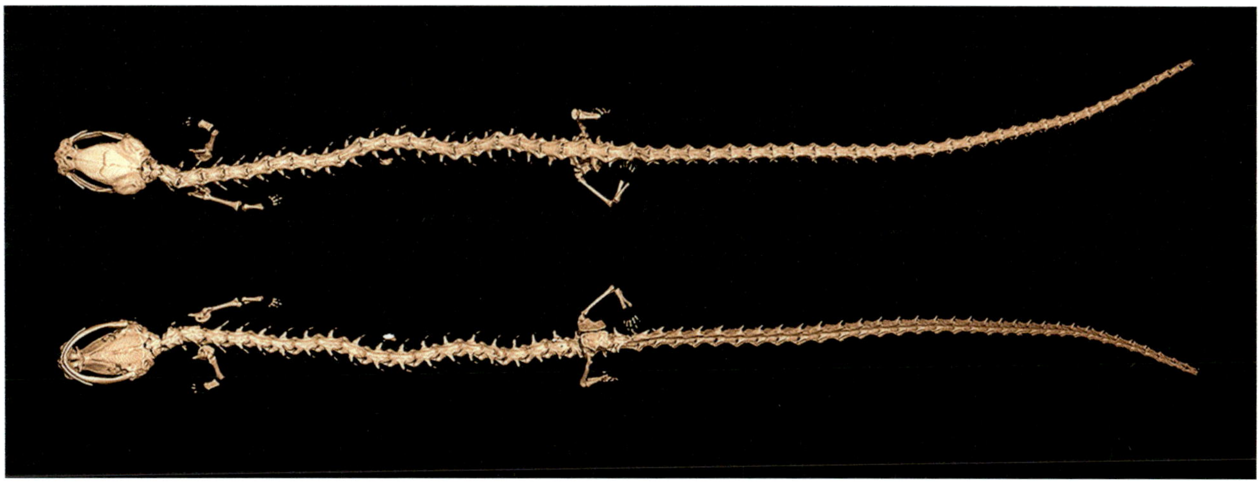

FIGURE 8.18 High-resolution computed tomography reconstructions of the skeleton of *Oedipina complex*, YPM 020155, total length = 23.4 mm, Cocobolo Nature Reserve (CREA Panama), Chepo, Panama. Dorsal (upper) and ventral (lower) views.

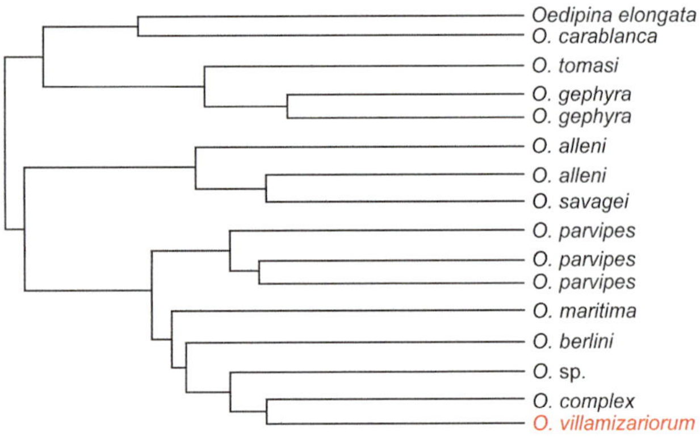

FIGURE 8.19 Topology of a subgroup of *Oedipina*, extracted from a phylogenetic tree inferred from a molecular analysis of partial sequences of mitochondrial DNA. Taken from Reyes-Puig *et al.* (2020). One of the species occurring in Ecuador is shown in red.

Distribution.—*Oedipina* is broadly distributed from Chiapas, Mexico, south through Central America to western Colombia, Magdalena Valley, and northwestern Ecuador.

Content.—Thirty-eight species are recognized; two species occur in Ecuador.

Remarks.—Figure 8.19. Phylogenies of *Oedipina*, based on molecular characters, have been provided by García-París and Wake (2000), McCranie *et al.* (2008), McCranie and Townsend (2011), and Reyes-Puig *et al.* (2020). According to the latter, *Oedipina* is the sister taxon of *Bradytriton*. Köhler (2011) and Raffaëlli (2013) provide accounts of the species.

OEDIPINA ECUATORIANA (ECUADORIAN WORM SALAMANDER) REYES-PUIG *ET AL.*, 2020

Oedipina ecuatoriana Reyes-Puig *et al.*, 2020: 17. **Holotype.**—BMNH 1901.3.29.115 from Paramba, 1067 m, Provincia Imbabura, Ecuador.

Etymology.—The specific epithet honors the Republic of Ecuador, an important center of amphibian biodiversity.

Identification.—Elongate body and tail with small and short legs; the maximum known SVL is 45.6 mm in a male. It has a narrow head (4.9 mm HW) and a broadly rounded snout. It has 18 trunk vertebrae and 17 costal grooves per side. The sublingual fold is present. The body is greatly attenuated. The limbs are small (nine costal folds exposed when limbs are appressed to the sides of the trunk). The tail is very long tail, more than twice the standard length.

In an Ecuadorian specimen (in preservative) from Paramba, Provincia Esmeraldas, the dorsum and flanks are dark brown. The venter is gray–brown. The head has a large white patch from the anterior edge of the eyes to the level of the gular fold. The snout is white. The dorsum of the proximal portions of the fore and hind limbs is whitish.

Comparisons.—*Oedipina villamizariorum*, *O. parviceps*, *O. complex*, *O. maritima*, *O. berlini*, *O. fortunensis*. *Oedipina ecuatoriana* differs from all known *Oedipina* by the presence of paired prefrontal bones unique to this species in the entire genus and a reduced phalangeal formula of 0-0-1-0; 0-1-2-1-1. *Oedipina ecuatoriana* is similar to *O. villamizariorum*, *O. parviceps*, *O. complex*, *O. maritima*, *O. berlini*, and *O. fortunensis*. It differs from *O. villamizariorum* (which is the only other *Oedipina* recorded in Ecuador) by having a broadly rounded snout rather than relatively pointed head with a blunt snout, by having the tip of the digits rounded instead of pointed, and by having the tips of fingers II and III rather than II, III, and IV free as in *O. villamizariorum*; the two species also differ in the extent of white pigmentation on the head, which is much greater in *O. ecuatoriana*. *Oedipina ecuatoriana* differs from *O. parvipes* by having a broadly rounded rather than pointed snout; by having short, blunt-tipped rather than longer and pointed digits; and by having much more extensive white pigment on the head. It differs from *O. complex sensu stricto* by its larger size and by having a longer snout that is rounded rather than blunt-tipped; shorter and more robust, rounded digits rather than weak and flattened; and often longer digits. *Oedipina ecuatoriana* was once assigned to *O. parvipes*, from which it differs by having a blunter and shorter snout, more extensive white pigmentation on the head, and shorter, blunter digits. *Oedipina ecuatoriana* is distinguished from *O. maritima* in having 17 versus 0–8 maxillary teeth. *Oedipina ecuatoriana* can be distinguished from *O. berlini* (characters in parentheses) by the presence of a narrow head and a broadly rounded snout (flat head, more than twice as wide as high, with relatively large protruding eyes). *Oedipina ecuatoriana* can also be distinguished from *O. fortunensis* (characters in parentheses) by the presence of a narrow head and a broadly rounded snout (narrow head and a relatively short rounded snout).

Natural history.—Unknown.

Distribution.—Figure 8.20. *Oedipina ecuatoriana*, an endemic species to Ecuador, is known only from the type locality, presumed to be Hacienda Paramba (see Remarks), in the western foothills of the Cordillera Occidental of the Andes in Provincia Imbabura at an elevation of 1067 m.

This species inhabits primary forests in currently intervened areas of the Cordillera Occidental of the Andes biogeographic sector, where the average annual rainfall is 2039 mm and the average annual temperature is 22.2 °C.

Conservation.—*Oedipina ecuatoriana* is suggested to be categorized as Critically Endangered (Reyes-Puig *et al.*, 2020) following the IUCN criteria (CR B1ab (i, iii) of IUCN). The authors inferred based on geographical data, deforestation statistics, and the distribution of

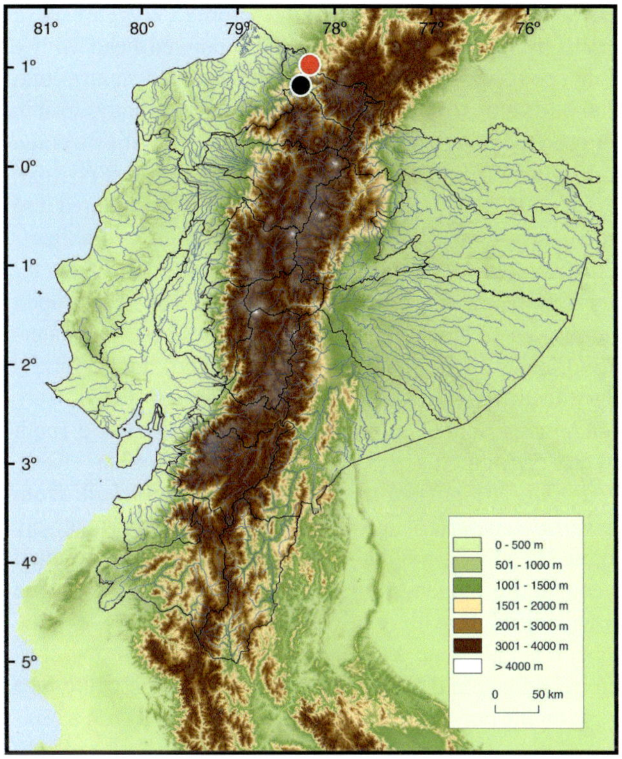

FIGURE 8.20 Localities of known occurrence of *Oedipina ecuatoriana* (black dot) and *O. villamizariorum* (red dot) in Ecuador.

current and projected mining concessions that a continuous decline in its habitat quality has occurred over the last 10 years and is expected to continue in the future. In Ecuador, its major threats are habitat reduction and fragmentation, logging, deforestation for agricultural development (especially palm plantations), fires, mining, the introduction of exotic species, human settlements, and pollution. Its area of distribution is under constant pressure by wood companies. It has not been recorded in public or private reserves.

Remarks.—Its molecular phylogenetic relationships are unknown. See Remarks under *Oedipina villamizariorum* for a discussion regarding taxonomy. Reyes-Puig *et al.* (2020) provide a detailed osteological description including microCT scan images. Data on the historic collecting locality of Paramba (the type locality of *O. ecuatoriana*) was incomplete; thus, Reyes-Puig *et al.* (2020) provided additional information, concluding that the exact location of the Hacienda Paramba of Rosenberg is difficult to establish with precision. The unique specimen of *Oedipina ecuatoriana* from Paramba was part of a collection sold to the British Museum (Natural History) (= The Natural History Museum, London) by the naturalist, collector, and natural history dealer William Frederick Henry Rosenberg (1868–1957), who collected in western Ecuador from November 1896 to September 1897, but his colleagues Georg Flemming and Rudolf Miketta kept sending him collections after his departure at least until 1903. Either Flemming or Miketta most probably collected the specimen of *O. ecuatoriana*. Boulenger (1898) and Hartert (1898) cited a description of Paramba provided by Rosenberg (the latter being more detailed): "a farm on the western bank of the Río Mira. Its elevation is 3500 ft altitude [= 1067 m], and it is still in the forest region, but the open country commences two or three miles higher up the Mira."

Key references.—(Brame and Wake, 1963) (account, as *Oedipina complex*); (Reyes-Puig *et al.*, 2020) (description).

OEDIPINA VILLAMIZARIORUM (VILLAMIZAR'S WORM SALAMANDER) REYES-PUIG *ET AL.*, 2020

Oedipina villamizariorum Reyes-Puig *et al.*, 2020: 8.
Holotype.—DHMECN 14489 from Quinshull, Chical, 921 m , Provincia Carchi, Ecuador.

Etymology.—The specific epithet honors Jorge and María Teresa Villamizar of Cumbayá, Ecuador, parents of Felipe Villamizar, who has been an important supporter of forest conservation in the area where this species was discovered. The noun is the genitive plural of the last name of the family.

Identification.—Figure 8.21. The maximum known SVL is 42.1 mm in a female. It has a narrow, relatively pointed head with a blunt snout. It has 18 trunk vertebrae and 17 costal grooves per side. The sublingual fold is present. The body is greatly attenuated. The limbs are very short (8.5 costal folds exposed when limbs are appressed to the sides of the trunk). The tail is very long, up to more than twice the standard length (the tail is broken but based on the size of the remaining vertebrae apparently was very long when complete).

The dorsal surfaces are chocolate, densely dotted with light brown, turquoise–cream, and brownish-orange. A white facemask is present, bounded with brownish-orange. The upper eyelid is slightly pigmented with turquoise. The loreal and rostral regions have scattered white marks. The flanks and dorsal surfaces of the hands and feet have distinctive irregular turquoise stains. The dorsal surfaces of the humerus and femur are orange. The ventral surfaces of the body are brownish-cream. The iris is coppery brown.

Comparisons.—*Oedipina ecuatoriana*, *O. parvipes*, *O. complex*, *O. maritima*, *O. elongata*, *O. carablanca*, *O. berlini*, *O. fortunensis*. *Oedipina villamizariorum* is most similar to *O. ecuatoriana*, *O. parvipes*, *O. complex*, *O. maritima*, *O. elongata*, *O. carablanca*, *O. berlini*, and *O. fortunensis*. It differs from *O. ecuatoriana* (which is the only other *Oedipina* recorded in Ecuador) by its apparently smaller size (the unique specimen of *O. ecuatoriana* is a male, typically the smaller of the two sexes, yet it is larger than the female holotype of *O. villamizariorum*), its

FIGURE 8.21 *Oedipina villamizariorum*. Left and two top right rows: Reserva Canandé, Provincia Esmeraldas. Two bottom right: rows: DHMECN 14489, female, 42.1 mm SVL, Quinshull, Chical, Provincia Carchi. Carolina Reyes Puig (left), Mario Yánez Muñoz (right).

narrower and more pointed snout, its pointed rather than round-tipped long digits, and much less white pigmentation on the head. The two species differ in several osteological characters (see herein). *Oedipina villamizariorum* is distinguished from *O. parvipes* (topotype) by having a somewhat more blunt-tipped snout, 26 maxillary teeth (19 in the holotype but only two in the topotypic individuals of *O. parvipes*), and dorsolaterally oriented eyes that are moderate in size (small in *O. parvipes*). It differs from *O. complex sensu stricto* by having a longer and more pointed head with a blunt snout (short and rounded to blunt-tipped snout), limb interval (vertical grooves on the sides of the body, visible between the points where the front and hind limbs meet the body when the limbs are positioned at their closest possible distance) of $8^{1/2}$ (7–$8^{1/2}$), and tips of the digits pointed and moderately long (blunt and short). *Oedipina villamizariorum* also can be distinguished from *O. maritima* (characters in parentheses) by having a pointed and blunt-tipped snout (pointed and elongate); moderately sized, dorsolaterally oriented eyes (small, laterally oriented); and 26 maxillary teeth (0–8). *Oedipina villamizariorum* differs from *O. elongata* and *O. carablanca* by its smaller size (both exceed 50 mm SL) and the presence of maxillary teeth. *Oedipina villamizariorum* can be distinguished from *O. berlini* (characters in parentheses) by a narrow, relatively pointed head with a blunt snout and dorsolaterally oriented eyes that are moderate in size (flat head, more than twice as wide as high, with relatively large protruding eyes); the dorsum is chocolate, densely dotted with light brown, turquoise–cream, and brownish-orange and with a white face mask on the dorsal surface of the head (colored with a mixture of pale earthy tones ranging from tan to dark reddish-brown, with finer white and dark brown to black spots and irregular markings scattered throughout). *Oedipina villamizariorum* differs from *O. fortunensis* by having a narrow, relatively pointed head with a blunt snout (a narrow head and a relatively short, rounded snout in *O. fortunensis*), and the dorsum chocolate in color, densely dotted with light brown, turquoise–cream, and brownish-orange (brown with white flecks in *O. fortunensis*).

Natural history.—The holotype was collected during excavations along the stony edges of a stream with bushy, thick canopy near forest clearings between 10:00 and 12:00 h in the morning. The specimen was found at approximately 15 cm depth inside moist, humus-rich soil covered with moderate leaf litter. One caecilian was found a few meters away.

Distribution.—Figure 8.20. *Oedipina villamizariorum*, an endemic species to Ecuador, is known only from the type locality in the western foothills of the Cordillera Occidental of the Andes in Provincia Carchi at an elevation of 912 m.

This species has been recorded in currently intervened areas in the Cordillera Occidental of the Andes biogeographic sector, where the average annual rainfall is 2978 mm and the average annual temperature is 21.1 °C.

Conservation.—*Oedipina villamizariorum* was suggested to be categorized as Critically Endangered by Reyes-Puig *et al.* (2020) following the IUCN criteria (CR B1ab (i, iii) of IUCN). They inferred based on geographical data, deforestation statistics, and the distribution of current and projected mining concessions that a continuous decline in its habitat quality has occurred over the last 10 years and is expected

FIGURE 8.22 *Oedipina* sp. (*ecuatoriana-villamizariorum*-aff. *villamimizariorum*?) (see Remarks under *O. villamizariorum*). Non-collected, San Lorenzo, Provincia Esmeraldas. Giovanni Onore.

to continue in the future. In Ecuador, its major threats are habitat reduction and fragmentation, logging, deforestation for agricultural development (especially palm plantations), fires, mining, the introduction of exotic species, human settlements, and pollution. Its area of distribution is under constant pressure by wood companies. It has not been recorded in public or private reserves.

Remarks.—Figure 8.22. Its molecular phylogenetic relationships are unknown. Reyes-Puig *et al.* (2020) provide a detailed osteological description including microCT scan images. Reyes-Puig *et al.* (2020) mention a possible unnamed species, *Oedipina* aff. *villamizariorum*, for which they also provide osteological data and images. They examined a specimen of *Oedipina* (FHGO 9642) from Padre Santo, Playa de Oro (0.867353° N, 78.804113° W, 130 m), Provincia Esmeraldas, Ecuador, collected by Manuel Morales-Mite in 2000 at 130 m of elevation. Both FHGO 9642 and DHMECN 14489 (holotype of *O. villamizariorum*) differ more from *Oedipina ecuatoriana* (BM 1901.3.29.115) than from each other. The authors think that the FHGO specimen represents a species closely related to *O. villamizariorum*, but they lack conclusive evidence given the small sample size, which precludes assessing variation. The phalangeal formulae of the hands of two osteological preparations differ. FHGO has one fewer phalanx on one digit. The two specimens are adults, as shown by the fully articulated frontals and parietals and absence of a fontanelle between these bones. However, the two specimens also differ with respect to whether the atlantal pedicels are fused (DHMECN 14489) or remain separated (FHGO 9642) over the nerve cord. The snout is more rounded, less "pinched" in FHGO 9642. There are some small differences in the shapes of the squamosals; the squamosal spur is a bit more evident in FHGO 9642, but it is still far less prominent than in Central American species. The authors strongly suspect FHGO 9642 is a third species from Ecuador but refrained from describing it.

An additional photographed individual (non-collected) from San Lorenzo, Provincia Esmeraldas, is similar in head coloration to both *O. villamizariorum* and *O. ecuatoriana*. This specimen is geographically closer to the specimen from Palo Santo identified as *O.* aff. *villamizariorum* by Reyes-Puig *et al.* (2020). The snout and body shape appear to be more similar to *O. ecuatoriana*. Thus, we cannot unambiguously assign this photo to any of the three putative taxa, and doubts arise regarding the validity of two of them. Thus, the claimed differences between the three putative taxa by Reyes-Puig *et al.* (2020) may be an artifact of the limited sample size (three specimens), and these differences could be interpreted as part of the variation (e.g., sex and geographic variation) within a single species, as well.

Key references.—(Reyes-Puig *et al.*, 2020) (description).

Appendix 1: Vocalizations of Ecuadorian Anurans: Phyllomedusidae, Leptodactylidae, Hemiphractidae, Eleutherodactylidae, and Centrolenidae

Introductory call information for each Ecuadorian anuran species is provided in individual species accounts. References for call descriptions are also listed under Key references in each account and compiled in Table 0.2. This section aims to enhance the information on the calls of Ecuadorian anurans found in the accounts, with a focus on presenting more technical details, especially when graphs of oscillograms, spectrograms, and related figures associated with the calls have been published. Additionally, we supply descriptions of the calls of 10 species that lacked prior documentation.

The knowledge of vocalizations among Ecuadorian amphibians remains incomplete and is currently limited to Anura. While there is evidence that caecilians emit sounds (LAC; pers. obs.), no formal descriptions are available at this time. Notably, no sounds have been recorded for Caudata. Presently, calls for 252 species (40%) out of 623 anurans occurring in Ecuador have been quantitatively described, with most of them from Ecuador and some from other countries. However, the calls of 371 anurans (60%) are still awaiting formal description. It is worth noting that the call repertoire an its ethological context is only known for a handful of species (e.g., *Nymphargus grandisonae* under field conditions, Hutter *et al.* (2013); *Hyloxalus toachi* under laboratory conditions, Quiguango-Ubillús and Coloma (2008)). Therefore, a world of acoustic and ethological discoveries lies ahead for scholars in the field of amphibian biology and evolution.

FAMILY PHYLLOMEDUSIDAE

Agalychnis psilopygion

Source: Cannatella (1980). Figure A1.1.
Voucher specimen and locality: KU 169611: Río Anchicayá, 8 km W of Danubio, Departamento del Valle, Colombia. Collected by William E. Duellman. KU Tape 1350.
Date and time: 13 June 1975, 21:30 h.
Temperature: 23 °C.
Call type: A single chirp.
Sample size: Five calls.
Call duration: 0.04–0.05 s.
Fundamental frequency: About 250 Hz.
Dominant frequency: 1900 Hz.
Harmonics: At about 250, 700, 1100, 1500, 1900, 2300, 2700, 3200, 3600, 4000, and 4800 Hz, with the most intense frequencies at 1500, 1900, and 2300 Hz.
Remarks: Recordings were analyzed with a Vibralizer (Kay Electric Company).

Agalychnis spurrelli 1

Source: Ortega-Andrade (2008). Figure A1.2.
Voucher specimen and locality: No voucher: Laguna del Diablo, Provincia Esmeraldas, Ecuador.
Recording equipment: Sony microcassette-corder M-670V.
Call type: A single, soft "groan" at intervals between 4 s and several mins.
Sample size: Four calls.
Call rate: Five calls per minute.
Number of notes: One.
Note duration: $\bar{x}$ = 187 ms; range = 156–215 ms.
Pulses per note: $\bar{x}$ = 12.8; range = 9–16.
Dominant frequency: $\bar{x}$ = 893.25 ± 73; range = 818–990 Hz.
Remarks: A stress call has 24 pulses lasting 0.248 milliseconds and a dominant frequency of about 990 Hz. Calls were analyzed with Adobe Audition.

Agalychnis spurrelli 2

Source: Cossio and Medina-Barcenas (2020). Figure A1.2.
Voucher specimen and locality: Not collected: Kingfisher pond (seasonal pond), Barro Colorado Island, Panama.
Date and time: May–December 2007; 21:30–06:00 h.
Temperature: 25 °C and 24 °C air temperature.
Recording equipment: Panasonic digital recorder-RRUS395PS. Sony ECM-C115 Microphone.
Sample size: 35 individuals.
  Advertisement call: 16 individuals, n = 96 calls.
  Aggressive call A: Seven individuals, n = 28 calls.
  Aggressive call B: 12 males, n = 33 calls.

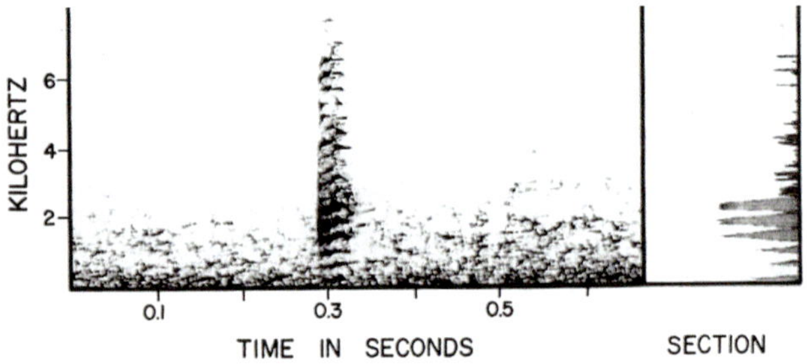

FIGURE A1.1 Advertisement call of *Agalychnis psilopygion*. Río Anchicayá, 8 km west of Danubio, Departamento del Valle, Colombia. Taken from Cannatella (1980).

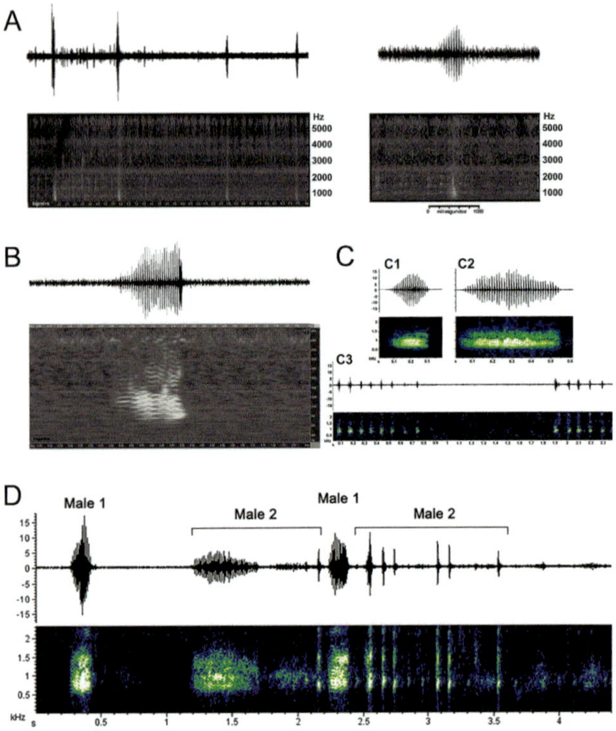

FIGURE A1.2 Call variation in *Agalychnis spurrelli*. A: advertisement call (left, section of a call; right a call). B: stress call. Laguna del Diablo, Provincia Esmeraldas. Taken from Ortega-Andrade (2008). C: waveform (top) and spectrogram (bottom) of calls produced by different males of *Agalychnis spurrelli*. C1: advertisement call. C2: aggressive call A. C3: aggressive call B. D: agonistic interaction between two males of *Agalychnis spurrelli*: male (1) and male (2). Waveform (top) and spectrogram (bottom). Male 1 produced two advertisement calls, while male 2 emitted a combination of aggressive calls: call A followed by call B. 25°C air temperature. Barro Colorado Island, Panama. Taken from Cossio and Medina-Barcenas (2020).

Call behavior: Aggressive calls (A and B) were produced by males during agonistic encounters.

Call type: Three clearly differentiated types of calls based on differences in call structure and social context: advertisement call, aggressive call A, and aggressive call B. The advertisement call consists of a single, simple note. Aggressive call A was produced singly but occasionally in combination with call B.

Notes per call:

Advertisement call: One note per call.

Aggressive call A: One note per call.

Aggressive call B: $\bar{x}$ = 9.3 ± 0.8; range = 2–22 notes per call.

Note duration:

Aggressive call B: $\bar{x}$ = 0.018 ± 0.001; range = 0.012–0.027 s.

Call duration:

Advertisement call: $\bar{x}$ = 0.20 0 ± 0.01; range = 0.13–0.36 s.

Aggressive call A: $\bar{x}$ = 0.59 ± 0.03; range = 0.50–0.69 s.

Aggressive call B: $\bar{x}$ = 0.97 ± 0.2; range = 0.17–1.62 s.

Dominant frequency:

Advertisement call: $\bar{x}$ = 884.2 ± 25.6; range = 750–1054.7 Hz.

Aggressive call A: $\bar{x}$ = 904.3 ± 28.2; range = 781.2–1062.5 Hz.

Aggressive call B: $\bar{x}$ = 809.2 ± 11.5; range = 750–847.7 Hz.

Call rate:

Advertisement call: $\bar{x}$ = 3.1 ± 0.2; range = 2–5 calls per minute.

Aggressive call A: $\bar{x}$ = 3.5 ± 0.6; range = 2–6 calls per minute.

Aggressive call B: $\bar{x}$ = 4.46 ± 0.7; range = 2–5 calls per minute.

Interval between calls:

Advertisement call: $\bar{x}$ = 16.1 ± 1.5; range = 4–26.9 s.

Aggressive call A: $\bar{x}$ = 5.1 ± 1.2; range = 1.4–18.0 s.

Aggressive call B: $\bar{x}$ = 4.05 ± 0.7; range = 1.4–15.4 s.

Note rate:

Aggressive call B: $\bar{x}$ = 5.1 ± 0.6; range = 2–11 notes per second.

Pulses per note:

Advertisement call: $\bar{x}$ = 17 ± 1.3; range = 14–30 pulses per note.

Aggressive call A: $\bar{x}$ = 45 ± 2.4; range = 28–65 pulses per note.

Aggressive call B: $\bar{x}$ = 14 ± 0.9; range = 8–26 pulses per note.

Pulse rate:

Advertisement call: $\bar{x}$ = 81.9 ± 2.9; range = 61–99 pulses per second.

Aggressive call A: $\bar{x}$ = 133.3 ± 9.4; range = 44–241 pulses per second.

Aggressive call B: $\bar{x}$ = 58.3 ± 7.5; range = 21–100 pulses per second.

Remarks: Microphone was at about 1 m from the individual. Vocalizations were digitized at 22 kHz with 16-bit resolution and analyzed using the software Raven Pro 1.5 software (Bioacoustics Research Program, 2011) through fast Fourier transform at a 512-point width, Hanning window type, and overlapping at 90%. Cossio and Medina-Barcenas (2020) describe the acoustic repertoire of *Agalychnis spurrelli* from Barro Colorado Island, Panama. There are three types of calls which are clearly differentiated on the basis of differences in call structure and social context. Duellman (2001) cites that the dominant frequency is between 435 and 700 Hz, the duration of the call ranges between 340 and 400 milliseconds, and each note consists of 19 to 24 pulses. Scott Jr. and Starrett (1974) reported what was probably a courtship call in Costa Rica.

Callimedusa ecuatoriana

Source: Brito *et al.* (2014). Figure A1.3.
Voucher specimen and locality: MEPN 14250: Río Blanco, 1630 m, Paquisha, Provincia Zamora Chinchipe, Ecuador,
Date: 26 July 2011.
Temperature: 17 °C.
Humidity: 90% RH.
Recording equipment: Olympus WS-750 digital recorder. Sennheiser K6-C microphone.
Sampling frequency: 44.1 kHz.
Resolution: 16 bits.
Sample size: One individual, eight calls,
Call duration: $\bar{x}$ = 30.87 ± 10.83 ms (n = 8).

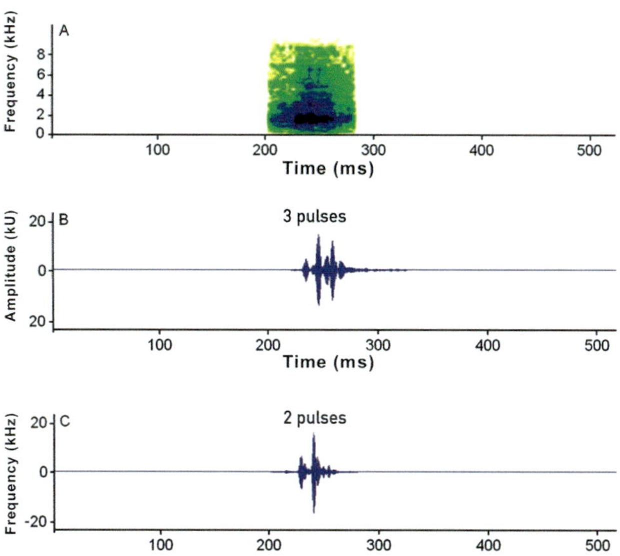

FIGURE A1.3 Call of *Callimedusa ecuatoriana*. Río Blanco, Paquisha, Provincia Zamora Chinchipe, Ecuador. A: spectrogram. B: oscillogram with three pulses. C: oscillogram composed of two pulses. Taken from Brito *et al.* (2014).

Interval between calls: $\bar{x}$ = 20.46 ± 11.06 s (n = 8).
Notes per call: One.
Pulses per call: Two (one of the analyzed calls presented three pulses, which could be considered a rare and infrequent pulse) (n = 8).
Pulse duration: First pulse = 0.0065 ± 0.00053 s; second pulse = 0.016 ± 0.00845 s (n = 8).
Interval between pulses: 0.0056 ± 0.00088 s (n = 8).
Dominant frequency: 1.53 ± 0.11 kHz (n = 8).
Fundamental frequency: 0.85 ± 0.06 kHz (n = 8).
Remarks: For acoustic analysis, Adobe Audition 3.0 was used at a sampling rate of 44.1 kHz and 16 "bits" of resolution; for the diagramming of the oscillogram and sonogram, Raven 1.4 software was used at 256 points of resolution of the fast Fourier transform (FFT).

Callimedusa tomopterna 1

Source: Zimmerman (1983).
Voucher specimen and locality: No voucher: Manaus, Brazil.
Date: August to October 1978.
Recording equipment: Uher 4000 Report L Tape recorder. Uher M517 microphone.
Call behavior: Chorus.
Call type: Advertisement call (Zimmerman, 1983). Mating call. The call is a soft "cluck" (Duellman, 1978).
Pulse rate: 205 pulses per second.
Dominant frequency: 1.79 kHz.
Sound type: Wide (broadband), amplitude modulated or pulsed.
Remarks: Sonograms were made on a Kay Electronics Model 7030A sonagraph, at normal tape-recording speed, and a narrow (45 Hz) and/or wide (300 Hz) filter bandwidth setting. Plots of instantaneous frequency versus amplitude (sections) were made with the sonagraph when emphasized (or dominant) frequencies in a call were not obvious from distinct contrast differences in the markings on the sonagram. Sonagrams were measured on a digitizer computer system with programs written by Dr. J. P. Bogartand J. R. Malcolm of the University of Guelph.

Callimedusa tomopterna 2

Source: Duellman (2005). Figure A1.4.
Voucher specimen and locality: Cusco Amazónico, Departamento Madre de Dios, Peru. KU cassette 101/31.
Temperature: 24 °C.
Call type: A soft cluck or a pair of clucks, of which the second note is softer than the first.
Note duration: 0.05 s.

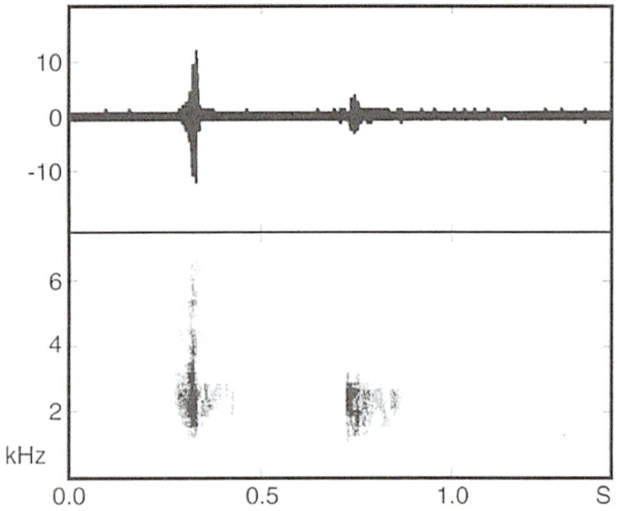

FIGURE A1.4 Oscillogram and spectrogram of the advertisement call of *Callimedusa tomopterna*. Tambopata, Peru. Taken from Duellman (2005).

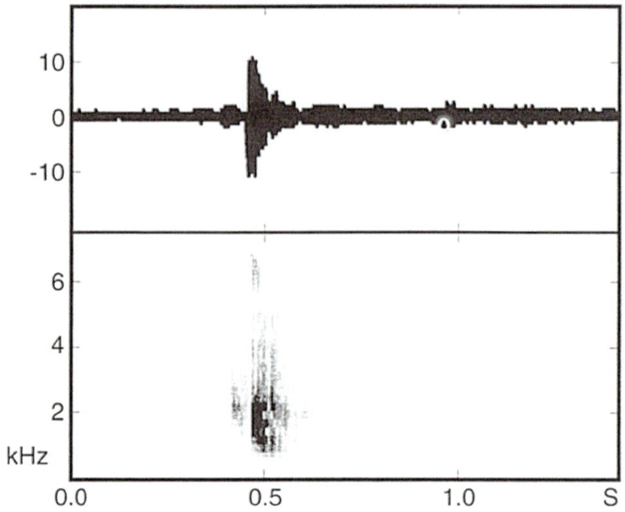

FIGURE A1.5 Oscillogram and spectrogram of the advertisement call of *Phyllomedusa vaillantii*. Cusco Amazónico, Peru. Taken from Duellman (2005).

Note rate: 2–3 notes per minute.

Dominant frequency: About 1550 Hz in the first note, 610 Hz in the second note.

Phyllomedusa tarsius

Source: Duellman (1978).

Voucher specimen and locality: No voucher: Santa Cecilia, Provincia Sucumbíos, Ecuador. Collected by W. E. Duellman.

Call type: Mating call. A moderately loud, low-pitched "cluck."

Sample size: Three recordings.

Note duration: $\bar{x}$ = 0.58; range = 0.5–0.7 s.

Note rate: $\bar{x}$ = 2.7; range = 2.0–3.2 notes per minute.

Pulse rate: 80 pulses per second.

Fundamental frequency: $\bar{x}$ = 375; range = 304–522 Hz.

Dominant frequency: $\bar{x}$ = 375; range = 304–522 Hz.

Phyllomedusa vaillantii 1

Source: Duellman (1978).

Voucher specimen and locality: Pilcopata, Departamento Cusco, Peru.

Call type: Mating call. Short, harsh, "cluck."

Sample size: One recording.

Note duration: 0.5 s.

Note rate: 13.3 notes per minute.

Pulse rate: 240 pulses per second.

Dominant frequency: At about 1800 Hz.

Phyllomedusa vaillantii 2

Source: Duellman (2005). Figure A1.5.

Voucher specimen and locality: Cusco Amazónico, Departamento Madre de Dios, Peru. KU Cassette 101/91.

Temperature: 23 °C.

Call type: The call is a single, moderately loud, short, cluck-like sound.

Sample size: One recording.

Note duration: 0.1 s.

Note rate: 2–3 notes per minute.

Dominant frequency: At about 1720 Hz.

Phyllomedusa vaillantii 3

Source: Röhr *et al.* (2020).

Voucher specimen and locality: Brazil.

Recording equipment: Marantz PMD 660 digital recorder with a sampling rate of 48 kHz and 16-bit resolution. Sennheiser ME66 directional microphone.

Temperature: Air $\bar{x}$ = 23.05 ± 1.37; range = 22–25 °C.

Call type: The call is a single, moderately loud, short, cluck-like sound.

Sample size: Four individuals, 12 calls.

Call duration: $\bar{x}$ = 0.04 ± 0.01; range = 0.03–0.053 s.

Interval between calls: $\bar{x}$ = 64.12 ± 65.69; range = 11.99–180.15 s.

Pulse number: $\bar{x}$ = 7.92 ± 1.31; range = 5–10.

Pulse rate: $\bar{x}$ = 178.12 ± 15.68; range = 155.56–212.12 pulses per second.

Pulse duration: $\bar{x}$ = 0.004 ± 0.001; range = 0.002–0.005 s.

Dominant frequency: $\bar{x}$ = 1609.4 ± 149; range = 1313–1875 Hz.

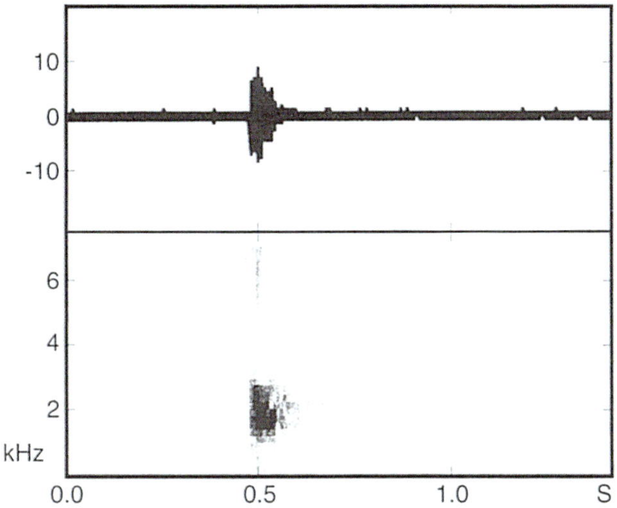

FIGURE A1.6 Oscillogram and spectrogram of the advertisement call of *Pithecopus palliatus*. Tambopata, Peru. Taken from Duellman (2005).

Remarks: Acoustic parameters were measured in Raven Pro 1.4 software and spectrograms produced as follows: FFT window width = 256; frame = 100; overlap = 50%.

Pithecopus palliatus

Source: Duellman (1978).

Voucher specimen and locality: Not vouchered; Santa Cecilia, Provincia Sucumbíos, Ecuador. Collected by W. E. Duellman.

Call type: Mating call. A soft "click" repeated at intervals of several seconds to three minutes.

Sample size: Two recordings.

Note duration: $\bar{x}$ = 0.135; range = 0.13–0.14 s.

Pulse rate: $\bar{x}$ = 190; range = 180–200 pulses per second.

Fundamental frequency: About 1500 Hz.

Dominant frequency: 3000 Hz.

Remarks: A call from Tambopata, Peru, was briefly described by Duellman (2005), who provided a figure. Figure A1.6.

FAMILY LEPTODACTYLIDAE

Adenomera andreae 1

Source: Marquez *et al.* (1995). Figure A1.7.

Voucher specimen and locality: Buenavista, Santa Cruz (17°21'S/63°40'W), Bolivia.

Date: 1987 to 1990. At night.

Recording equipment: Sony WM D6C or a Sanyo M 1120 tape recorder. Sennheiser Me 80 directional microphone.

Call type: Advertisement call. Composed of a single short note, repeated at a fast rate.

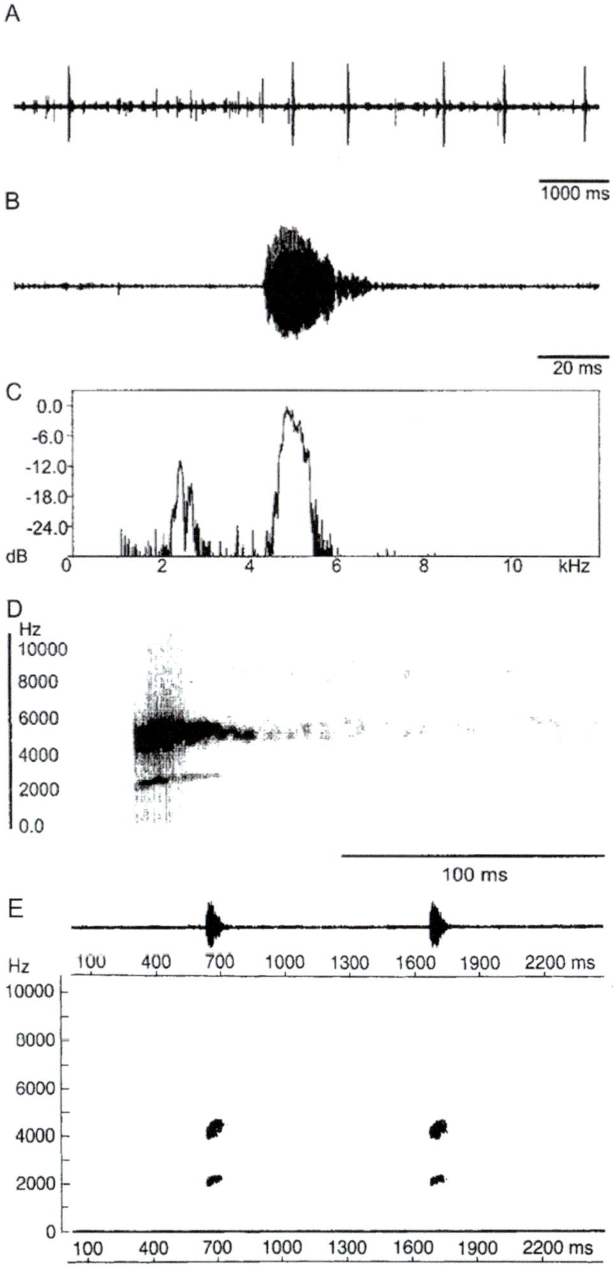

FIGURE A1.7 Advertisement call of *Adenomera andreae*. One of the four voucher specimens (not specified which), at Tambopata National Reserve and Bahuaja Sonene National Park, Departamento Madre de Dios, Peru. A: oscillogram depicting calling pattern. B: waveform of one advertisement call at higher resolution. C: power spectrum. D: spectrogram of one advertisement call. Taken from Angulo *et al.* (2003). E: audiospectrograms and oscillograms of a 2.5 s section of the advertisement call of *Adenomera andreae*. Taken from Marquez *et al.* (1995).

Call behavior: On the ground near flooded areas.

Sample size: 19 calls.

Call rate: $\bar{x}$ = 82.5 ± 23; range = 52.6–128.4 calls per min.

Notes per call: One.

Note duration: $\bar{x}$ = 66.5 ± 7.6; range = 56.6–88.3 ms.

Note rate: $\bar{x}$ = 82.5 ± 23; range = 52.6–128.4 notes per min.
Fundamental frequency: $\bar{x}$ = 2228.6 ± 35.7; range = 2180.8–2281.7 Hz.
Dominant frequency: $\bar{x}$ = 4322.2 ± 80; range = 4200–4502.9 Hz.
Change in DF: $\bar{x}$ = 269.2 ± 85.9; range = 80.8–424 Hz.
Remarks: Recordings were processed with an Apple Macintosh-based digital signal analysis system. Digitalization and edition were completed at a sampling frequency of 44.1 kHz and 16 bits of resolution with Sound Tools hardware and software. Signalyze software was used to obtain numerical information and to generate audiospectrograms and oscillograms. Frequency information was obtained through fast Fourier transform (FFT) (width, 1024 points).

Adenomera andreae 2

Source: Angulo *et al.* (2003) (as Forest Call Type III). Figure A1.7.
Voucher specimen and locality: USNM 343236, USNM 343232, USNM 343237: Tambopata National Reserve and Bahuaja Sonene National Park, Departamento Madre de Dios, Peru. Collected by Reginald B. Cocroft.
Date and time: 3 November 1990, 16:55 h (USNM 343236); 2 November 1990, 17:15 h (USNM 343232); 3 November 1990, 18:20 h (USNM 343237).
Temperature: 16.6 °C (USNM 343236); 26.2 °C (USNM 343232); 25.9 °C (USNM 343237).
Recording equipment: Sony TCD-5M tape recorder with a Sennheiser ME-80 microphone, or Sony Walkman Professional WM D6C tape recorder with a Sony ECM-907 microphone.
Sampling frequency: 44.1 kHz.
Resolution: 16 bits.
Sample size: Four individuals, 40 calls (10 calls per individual).
Call type: Advertisement call. Short call with no perceptible pulses.
Call duration: $\bar{x}$ = 48.61; range = 37.05–89.98 ms.
Call rise time: $\bar{x}$ = 10.14; range = 5.61–18.09 ms.
Call rate: $\bar{x}$ = 52.22; range = 48–60 calls per minute.
Pulses per call: One.
Fundamental frequency: $\bar{x}$ = 2506.70 Hz; range = 2208–2679.5 Hz.
Dominant frequency: $\bar{x}$ = 5106.58 Hz; range = 4654.6–5338.8 Hz.
Change in DF: $\bar{x}$ = 811.97; range = 62–1375 Hz.
Remarks: Calls were analyzed using the Macintosh-based digital signal analysis program Signalyze 3.12 (Keller, 1994). The sampling frequency rate was set at 44.1 kHz and 16 bits precision. Figures of temporal and spectral features of calls were produced using a combination of a PC-based sound analysis program, DADiSP, and Signalyze 3.12, and later edited with Corel Draw.

Adenomera hylaedactyla 1

Source: Marquez *et al.* (1995). Figure A1.8.
Voucher specimen and locality: Puerto Almach, Santa Cruz (15°46'S/62°15'W), Bolivia.

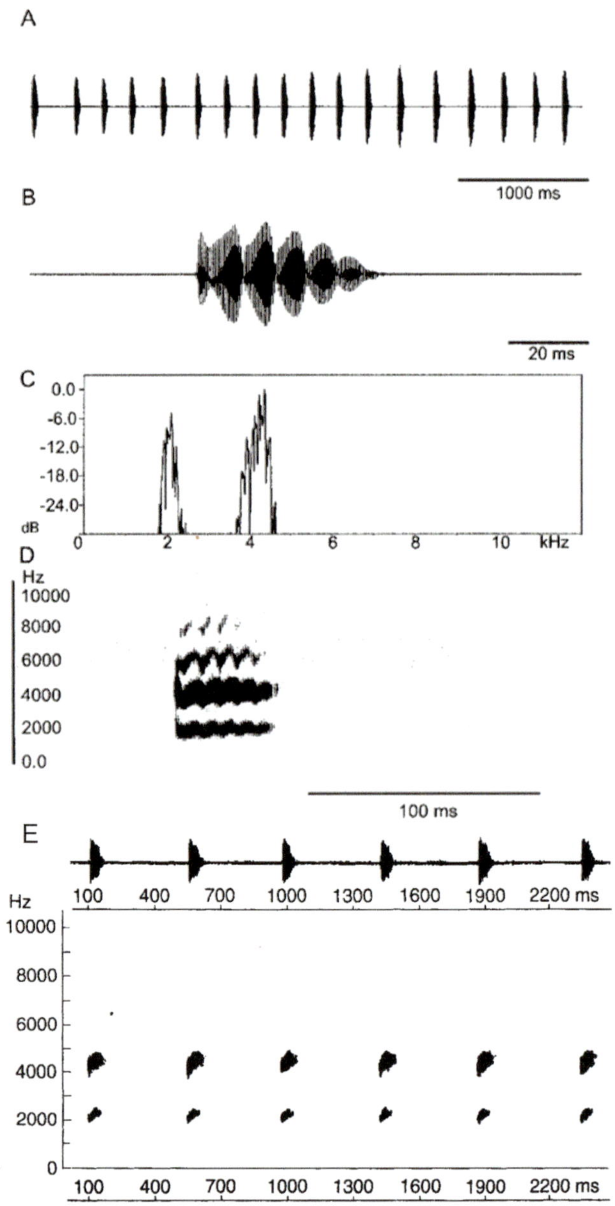

FIGURE A1.8 Advertisement call of *Adenomera hylaedactyla*. Tambopata National Reserve and Bahuaja Sonene National Park, Departamento Madre de Dios, Peru. A: oscillogram showing calling pattern. B: waveform of one advertisement call at higher resolution. C: power spectrum. D: spectrogram of one advertisement call. Taken from Angulo *et al.* (2003). E: audiospectrograms and oscillograms of a 2.5 s section of the advertisement call of *Adenomera hylaedactyla*. Taken from Marquez *et al.* (1995).

Date: 1987 to 1990. At day and night.
Recording equipment: Sony WM D6C or a Sanyo M 1120 tape recorder. Sennheiser Me 80 directional microphone.
Call type: Advertisement call. Single short note repeated regularly.
Call behavior: Perched at a low height or on the ground, always in or under a bush.
Sample size: 41 calls.
Call rate: $\bar{x}$ = 132.2 ± 6.7; range = 117.8–149.6 calls per min.
Notes per call: One.
Note duration: $\bar{x}$ = 56.8 ± 5.9; range = 43.9–69.2 ms.
Note rate: $\bar{x}$ = 132.2 ± 6.7; range = 117.8–149.6 notes per min.
Fundamental frequency: $\bar{x}$ = 2240.2 ± 101.8; range = 2079.8–2463.5 Hz.
Dominant frequency: $\bar{x}$ = 4448.2 ± 138.5; range = 4200–4785.6 Hz.
Change in DF: $\bar{x}$ = 249.2 ± 75.9; range = 121.2–424 Hz.
Remarks: Recordings were processed with an Apple Macintosh-based digital signal analysis system. Digitalization and edition were completed at a sampling frequency of 44.1 kHz and 16 bits of resolution with Sound Tools hardware and software. Signalyze software was used to obtain numerical information and to generate audiospectrograms and oscillograms. Frequency information was obtained through fast Fourier transform (FFT) (width, 1024 points).

Adenomera hylaedactyla 2

Source: Angulo *et al.* (2003). Figure A1.8.
Voucher specimen and locality: USNM 342986, USNM 342985: Tambopata National Reserve and Bahuaja Sonene National Park, Departamento Madre de Dios, Peru. ROM 40105, ROM 40106, ROM 40109; recorded by Ariadne Angulo. Recorded by R. B. Cocroft.
Date and time: 26 January 1999, 19:11 h (ROM 40105); 29 January 1999, 19:04 h (ROM 40106); 29 March 1999, 01:03 h (ROM 40109); 6 January 1989, 18:30 h (USNM 342986); 30 December 1988, 18:45 h (USNM 342985).
Temperature: 25.5 °C (ROM 40105); 25 °C (ROM 40106); 27 °C (ROM 40109); 24.2 °C (USNM 342986); 25 °C (USNM 342985).
Recording equipment: Sony TCD-5M tape recorder. Sennheiser ME-80 microphone or a Sony Walkman Professional WM D6C tape recorder and a Sony ECM-907 microphone.
Sampling frequency: 44.1 kHz.
Resolution: 16 bits.
Sample size: Five individuals, 10 calls (n = 50 calls).
Call type: Advertisement call. Short multi-pulsed call.
Call duration: $\bar{x}$ = 46.02; range = 34.77–62.27 ms.
Call rise time: $\bar{x}$ = 11.26; range = 00.84–22 ms.
Call rate: $\bar{x}$ = 210; range = 162–252 calls per minute.
Pulses per call: 4–6.
Pulse rate: $\bar{x}$ = 130; range = 80.4–153.5 per second.
Fundamental frequency: $\bar{x}$ = 2069.11; range = 1949.4–2208 Hz.
Dominant frequency: $\bar{x}$ = 4248.43 Hz; range = 3958.4–4475.6 Hz.
Change in DF: $\bar{x}$ = 1057.04 Hz; range = 515.62–1375 Hz.
Remarks: Calls were analyzed using the Macintosh-based digital signal analysis program, Signalyze 3.12 (Keller, 1994). Sampling frequency rate was set at 44.1 kHz and 16 bits of precision. Figures of temporal and spectral features of calls were produced using a combination of a PC-based sound analysis program, DADiSP, and Signalyze 3.12 and later edited with Corel Draw.

Edalorhina perezi 1

Source: Duellman (2005). Figure A1.9.
Voucher specimen and locality: Cusco Amazónico, Departamento Madre de Dios, Peru.
Call type: Advertisement. Low whistles.
Call rate: 4–5 at intervals of 0.13–0.20 s.
Interval between calls: A minute or more between call groups.
Number of notes: 3–5.

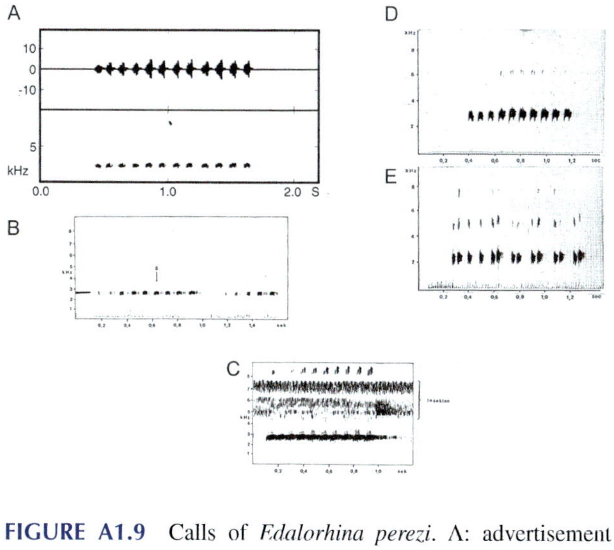

FIGURE A1.9 Calls of *Edalorhina perezi*. A: advertisement call. Pakitza, Departamento Madre de Dios, Peru. Taken from Duellman (2005). B: advertisement call. C: courtship calls. D: advertisement call. E: double-pulsed courtship call. Panguana, Departamento Loreto, Peru. Taken from Schlüter (1980).

Pulses per note: Two.
Fundamental frequency: 2700 Hz.
Dominant frequency: 2700 Hz.
Harmonics: At about 5400 and 8100 Hz.

Edalorhina perezi 2

Source: Schlüter (1980) and Schlüter (1990). Figure A1.9.
Voucher specimen and locality: Panguana field station on the lower Rio Yuyapichis (= Llullapichis), about 5 km upriver from its confluence with the Rio Pachitea (about 9°35'S, 74°56'W, elevation 220 m), Peru.
Sample size: Nine series, 1–26 calls.
Call type: Mating call. Series of notes having double pulses.
Call duration: 0.8–1.2 s.
Doble pulse rate: 6.4–6.6 calls per second.
First pulse duration: 0.025–0.030 second.
Second pulse duration: Always longer than the first.
Interval between first and second pulses: 0.020–0.045 s.
Dominant frequency: 1800–2700 Hz.
Harmonics: Concentrated at 4500–5600 Hz and 7000–7800 Hz.

Engystomops coloradorum

Source: Ron *et al*. (2005). Figure A1.10.
Voucher specimen and locality: QCAZ 19412, 19416–8, and 19441: Sapoparque La Florida, 5 km NW La Florida, Provincia Santo Domingo de los Tsáchilas, Ecuador.
Date and time: 15 February 2000.
Call type: Advertisement call. A single, short, pulsed component followed by a whine-like second component.
Sample size: Five individuals.
Pulses of the first component: Weakly defined.
Frequencies of the first component: The dominant frequency is approximately 0.149 kHz higher than that of the second. Although the dominant frequency of the first component is in the first harmonic, at 1.104 kHz, 45% of its sound energy is allocated above 1.2 kHz. The first component contains about 65% of the energy of the call and also is amplitude-modulated. The frequency with the most energy in the second harmonic of the first component is 2.227 kHz.

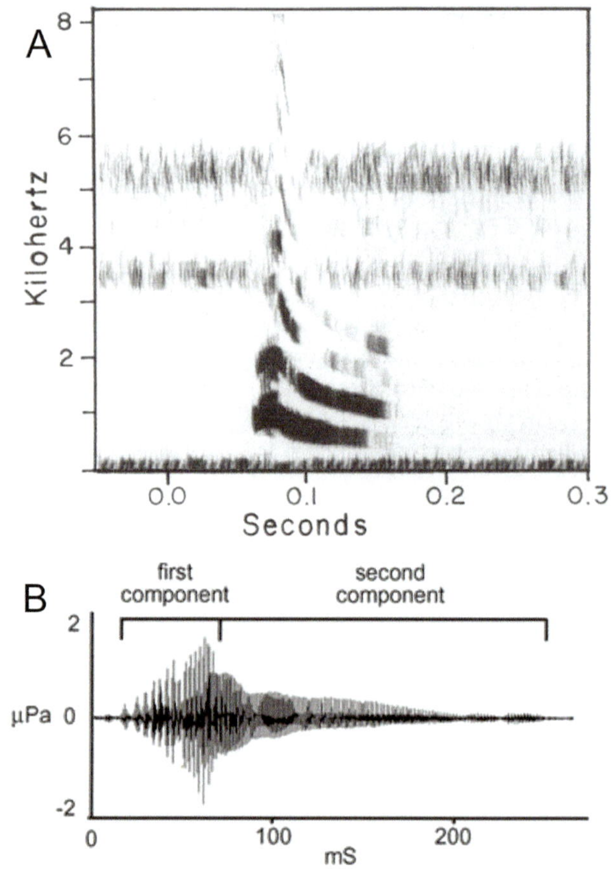

FIGURE A1.10 Advertisement call of *Engystomops coloradorum*. A: Santo Domingo de los Tsáchilas, Provincia Santo Domingo de los Tsáchilas, Ecuador. Taken from Cannatella and Duellman (1984). B: Sapoparque La Florida, 5 km NW of La Florida, Provincia Santo Domingo de los Tsáchilas, Ecuador. Taken from Ron *et al*. (2005).

Engystomops guayaco

Source: Ron *et al*. (2005). Figure A1.11.
Voucher specimen and locality: QCAZ 19561–2, 23510, 23512: Cerro Masvale, Provincia Guayas, Ecuador, 92 m. QCAZ 23652: 15 km south of Naranjal, Provincia Guayas, Ecuador, 74 m. QCAZ 23508: No locality. Collected by David C. Cannatella, Luis A. Coloma, Alisha Holloway, Santiago R. Ron, Martín R. Bustamante, and Ítalo G. Tapia.
Date and time: 19 February 2002, 23:00 h (QCAZ 19562); 20–25 March 2003 (QCAZ 23508, not collected).
Temperature: Water = 26.6 °C, air = 24.8 °C (QCAZ 19562); water = 27.8 °C, air = 28.0 °C, (QCAZ 23510); water = 27.7 °C, air = 27.7 °C (QCAZ 23512); water = 26.6 °C, air = 28.5 °C (QCAZ 23652); water = 27.8 °C, air = 27.2 °C (QCAZ 23508).
Recording equipment: Sony WM-D6C analog tape recorder. Sennheiser ME-67 directional microphone.
Call behavior: One male in a chorus at Cerro Masvale was recorded while calling partly submerged in 2 cm of water, 2 m from a ditch, in a grassy area.

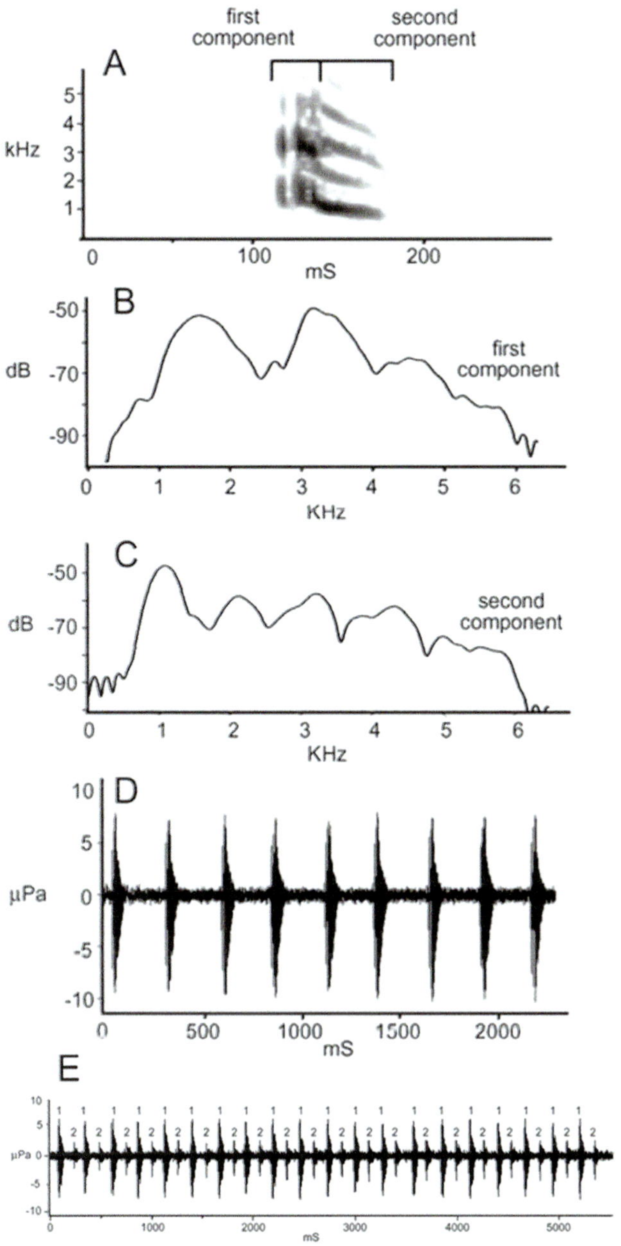

FIGURE A1.11 Calls *of Engystomops guayaco*. Cerro Masvale, Provincia del Guayas, Ecuador. A: sonogram of a single call. B: power spectrum of the first call component. C: power spectrum of the second call component. D: oscillogram of series of calls. E: oscillogram of single call. F: oscillogram of the antiphonal calls of two males. Taken from Ron *et al.* (2005).

Call type: Advertisement call. A single, short, pulsed component followed by a whine-like second component.

Sample size: Five individuals, 63 calls: QCAZ 19561(18); QCAZ 19562 (10); QCAZ 23510 (10); QCAZ 23512 (10); QCAZ 23652 (15).

Call rate:

QCAZ 19561: 221 calls per minute.
QCAZ 19562: 221 calls per minute.
QCAZ 23510: 260 calls per minute.
QCAZ 23512: 259 calls per minute.
QCAZ 23652: 260 calls per minute.

Call duration:

QCAZ 19561: $\bar{x}$ = 68.2; range = 56.7–76.0 ms.
QCAZ 19562: $\bar{x}$ = 70.6; range = 67.5–72.8 ms.
QCAZ 23510: $\bar{x}$ = 52.8; range = 49.1–54.9 ms.
QCAZ 23512: $\bar{x}$ = 65.3; range = 61.9–68.8 ms.
QCAZ 23652: $\bar{x}$ = 69.4; range = 64.6–74.9 ms.

Interval between calls:

QCAZ 19561: $\bar{x}$ = 203; range = 161.9–405.1 ms.
QCAZ 19562: $\bar{x}$ = 201; range = 188–224 ms.
QCAZ 23510: $\bar{x}$ = 178.3; range = 154.1–246.6 ms.
QCAZ 23512: $\bar{x}$ = 166.3; range = 153.7–188.1 ms.
QCAZ 23652: $\bar{x}$ = 190.5; range = 150.6–270.5 ms.

Rise time of the first component:

QCAZ 19561: $\bar{x}$ = 15.1; range = 9.1–20.8 ms.
QCAZ 19562: $\bar{x}$ = 17.7; range = 16.5–18.8 ms.
QCAZ 23510: $\bar{x}$ = 14.3; range = 8.5–15.6 ms.
QCAZ 23512: $\bar{x}$ = 19.0; range = 10.4–23.5 ms.
QCAZ 23652: $\bar{x}$ = 17.9; range = 10.5–24.4 ms.

Dominant frequency of the first component:

QCAZ 19561: $\bar{x}$ = 1.372; range = 1.281–1.491 kHz.
QCAZ 19562: $\bar{x}$ = 3.200; range = 3.149–3.273 kHz.
QCAZ 23510: $\bar{x}$ = 3.237; range = 3.168–3.341 kHz.
QCAZ 23512: $\bar{x}$ = 3.359; range = 3.274–3.442 kHz.
QCAZ 23652: $\bar{x}$ = 3.237; range = 3.004–3.325 kHz.

Fundamental frequency of the first component:

QCAZ 19561: $\bar{x}$ = 1.372; range = 1.281–1.491 kHz.
QCAZ 19562: $\bar{x}$ = 1.577; range = 1.394–1.679 kHz.
QCAZ 23510: $\bar{x}$ = 1.789; range = 1.744–1.880 kHz.
QCAZ 23512: $\bar{x}$ = 1.511; range = 1.472–1.703 kHz.
QCAZ 23652: $\bar{x}$ = 1.631; range = 1.634–1.907 kHz.

Fundamental frequency of the second component:

QCAZ 19561: $\bar{x}$ = 1.151; range = 1.125–1.179 kHz.
QCAZ 19562: $\bar{x}$ = 1.079; range = 1.060–1.098 kHz.
QCAZ 23510: $\bar{x}$ = 1.066; range = 1.064–1.097 kHz.
QCAZ 23512: $\bar{x}$ = 1.091; range = 1.070–1.120 kHz.
QCAZ 23652: $\bar{x}$ = 1.048; range = 1.007–1.084 kHz.

Harmonics: Although both notes have harmonic structures, the harmonics are clearly defined only in the second component (up to six). The first

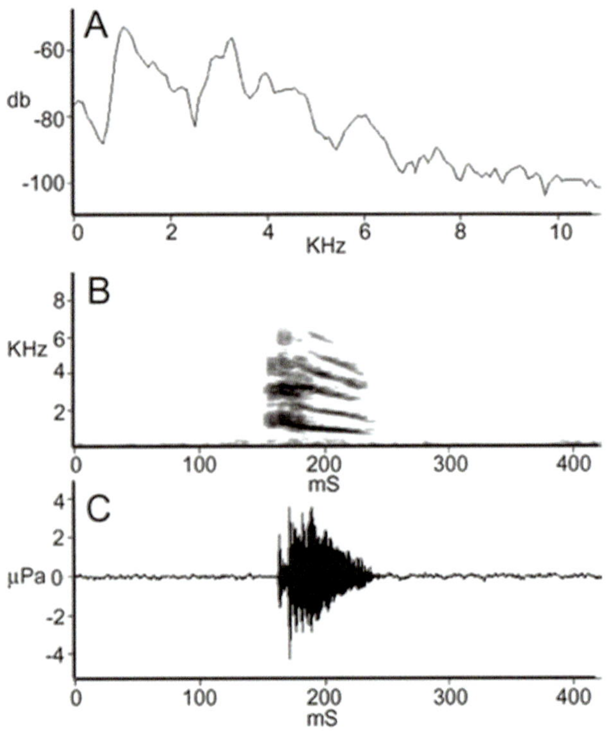

FIGURE A1.12 Call of *Engystomops montubio*. Puerto Rico, Provincia Manabí, Ecuador. A: power spectrum. B: sonogram. C: oscillogram of the advertisement call. The power spectra were measured along the entire duration of the calls. Taken from Ron *et al.* (2004).

harmonic of the whine-like note is a nearly pure tone with a descending frequency sweep.

Remarks: Calls were analyzed using Canary1.2.1 software at a sampling frequency of 22.1 kHz and a frequency grid resolution FTP = 8192. The antiphonal call was also described.

Engystomops montubio

Source: Ron *et al.* (2004). Figure A1.12.

Voucher specimen and locality: QCAZ 19517, QCAZ 19524: Puerto Rico, Provincia Manabí, Ecuador, 30 m. Collected by David C. Cannatella, Luis A. Coloma, A. Holloway, and Santiago R. Ron.

Date and time: 17 February 2002, 20:25 h (QCAZ 19517); 18 February 2002, 00:05 h (QCAZ 19524).

Temperature: Water = 27.4 °C, air = 24.2 °C.

Recording equipment: Sony WM-D6C analog tape recorder. Sennheiser ME-67 directional microphone.

Call behavior: A male (QCAZ 19517) was recorded while calling partly submerged in 2 cm of water in a roadside pool.

Call type: Advertisement call. A single, short, pulsed note that begins with two to five amplitude-modulated pulses, followed by a nearly pure tone with a downward frequency sweep.

Sample size: Two individuals.

Notes per call: One.

Call duration:

QCAZ 19517: $\bar{x}$ = 82; range = 79–86 ms.

QCAZ 19524: $\bar{x}$ = 68; range = 63–76 ms.

Call rate:

QCAZ 19517: $\bar{x}$ = 218.2 notes/min.

QCAZ 19524: $\bar{x}$ = 227.3 notes/min.

Interval between calls:

QCAZ 19517: $\bar{x}$ = 193; range = 176–216 ms.

QCAZ 19524: $\bar{x}$ = 196; range = 138–314 ms.

Rise time:

QCAZ 19517: $\bar{x}$ = 14; range = 8–24 ms.

QCAZ 19524: $\bar{x}$ = 14; range = 10–29 ms.

No. of harmonics:

QCAZ 19517: $\bar{x}$ = 5.5; range = 5–6.

QCAZ 19524: $\bar{x}$ = 4.0.

Fundamental frequency of first component:

QCAZ 19517: $\bar{x}$ = 1.369; range = 1.308–1.448 kHz (n = 5).

QCAZ 19524: $\bar{x}$ = 1.336; range = 1.114–1.550 kHz.

Fundamental frequency of second component:

QCAZ 19517: $\bar{x}$ = 1.081; range = 1.071–1.087 kHz.

QCAZ 19524: $\bar{x}$ = 0.977; range = 0.931–1.001 kHz.

No. of pulses of first component:

QCAZ 19517: $\bar{x}$ = 2.92; range = 2–5.

QCAZ 19524: $\bar{x}$ = 2.40; range = 2–3.

Remarks: Calls were analyzed using Canary1.2.1 software at a sampling frequency of 22.1 kHz and a frequency grid resolution FTP = 8192.

Engystomops petersi

Source: Guerra and Ron (2008). Figure A1.13.

Voucher specimen and locality: 5 km northeast of Puyo, Provincia Pastaza, Ecuador.

Temperature: 21–24 °C.

Recording equipment: Sony Digital Audio TCD-D8 tape recorder and a Sony Walkman WM-D6C professional stereo cassette recorder. Sennheiser SE 66–67 directional microphone.

Call type: Advertisement. A call with two components: a short, low-amplitude "prefix" followed by a longer, frequency-modulated "whine."

Sample size: Six individuals.

Call duration: $\bar{x}$ = 285.05; range = 241.93–321.99 ms.

Prefix: $\bar{x}$ = 32.35; range = 26.53–41.30 ms.

Whine: $\bar{x}$ = 252.70; range = 212.09–281.56 ms.

Call shape: $\bar{x}$ = 76.68; range = 62.08–93.19.

Fall time: $\bar{x}$ = 263.16; range = 225.76–295.15 ms.

Final frequency: $\bar{x}$ = 372.72; range = 344.53–409.10 Hz.

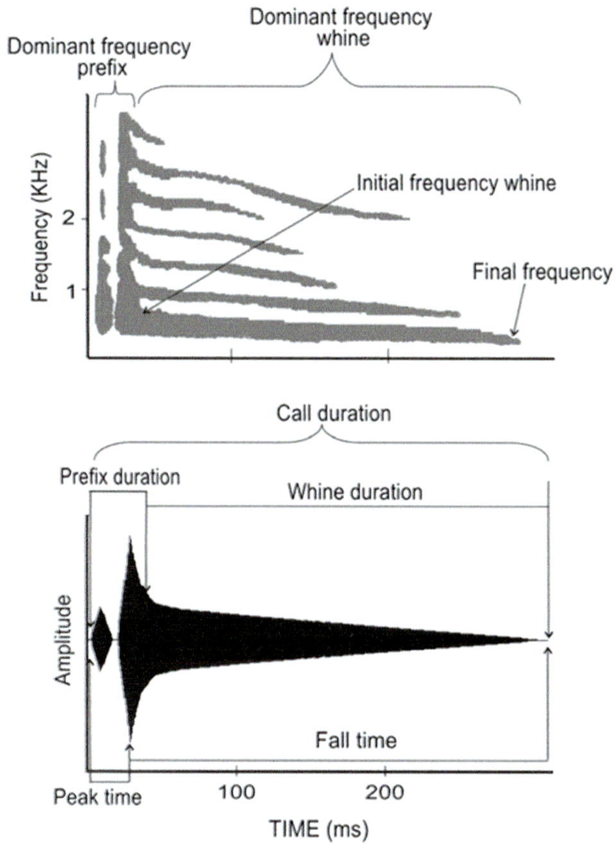

FIGURE A1.13 Spectrogram above and oscillogram below of the call of *Engystomops petersi*. 5 km northeast of Puyo, Provincia Pastaza, Ecuador. Taken from Guerra and Ron (2008).

Rise time: $\bar{x}$ = 21.90; range = 16.16–26.84 ms.

Notes/call: Two.

Dominant frequency: $\bar{x}$ = 627.65; range = 524.87–877.48 Hz.

Prefix: $\bar{x}$ = 718.75; range = 602.93–829.03 Hz.

Whine: $\bar{x}$ = 517.15; range = 473.73–544.48 Hz.

Whine initial frequency: $\bar{x}$ = 559.67; range = 516.80–592.16 Hz.

Whine frequency decrease: $\bar{x}$ = 186.95; range = 150.73–226.10 Hz.

Remarks: Calls were digitized with Cool Edit pro 2.0 (Syntrillium Software Corporation) (sampling rate = 22050 Hz, resolution =16 bits). Simple calls were analyzed using the sound analysis software Canary 1.2.4. A fast Fourier transform of 1024 points was used. *Call shape*: rise time/call duration. *Fall time*: time from the point of maximum amplitude of the call to the end of the call. *Final frequency*: frequency at the end of the first harmonic. *Whine frequency decrease*: difference in frequency between the initial frequency and final frequency of the first harmonic of the whine. *Whine initial frequency*: frequency at the beginning of the fundamental frequency of the whine.

Engystomops pustulatus

Source: Ron *et al.* (2010). Figure A1.14.

Voucher specimen and locality: No vouchers explicitly indicated: Arenillas, El Empalme y Huaquillas, Provincia El Oro, Ecuador. No vouchers explicitly indicated: La Maná, Provincia Cotopaxi, Ecuador. No vouchers explicitly indicated: Patricia Pilar, Provincia Los Ríos, Ecuador. No vouchers explicitly indicated: Puerto Rico, Provincia Manabí, Ecuador.

Recording equipment: Sony WM-D6C analog tape recorder, SONY™ MZ-NH1. Sennheiser ME-67 directional microphone.

Call type: Advertisement call. A call with two obligatory and one facultative component with harmonic structure.

Call behavior: Males call while floating on standing water.

Sample size: Six localities, 39 individuals.

Temperature: $\bar{x}$ = 27.1 ± 1.13 °C (n = 36).

Call duration: $\bar{x}$ = 0.572 ± 0.089; range = 0.384–0.750 s.

Duration first component: $\bar{x}$ = 0.150 ± 0.024; range = 0.099–0.207 s.

Duration second component: $\bar{x}$ = 0.381 ± 0.088; range = 0.150–0.555 s.

Duration third component: $\bar{x}$ = 0.116 ± 0.079; range: 0.046–0.325 s.

Notes per call: One.

Call frequency:

Frequency of the first spectral peak: $\bar{x}$ = 870 ± 55; range = 747–988 Hz.

Frequency of the second spectral peak: $\bar{x}$ = 1790 ± 113; range = 1572–2067 Hz.

First component:

Frequency of the first spectral peak: $\bar{x}$ = 925 ± 60; range = 797–1076 Hz.

Frequency of the second spectral peak: $\bar{x}$ = 1767 ± 125; range = 1552–2080 Hz.

Second component:

Frequency of the first spectral peak: $\bar{x}$ = 824 ± 57; range = 700–973 Hz.

Frequency of the second spectral peak: $\bar{x}$ = 1763 ± 115; range = 1492–2024 Hz.

Third component:

Fundamental frequency: $\bar{x}$ = 458 ± 28; range = 422–506 Hz.

Dominant frequency: $\bar{x}$ = 2544 ± 133; range = 2250–2695 Hz.

Remarks: Calls were analyzed using Raven 1.2.1 software at a sampling frequency of 22.1 kHz and a frequency resolution of 21.5 Hz. The first component is characterized by weak amplitude modulation. The second component is a whine-like note without amplitude modulation. A facultative third component (exclusive of La Maná population) is similar to a "chuck" but is less discreet

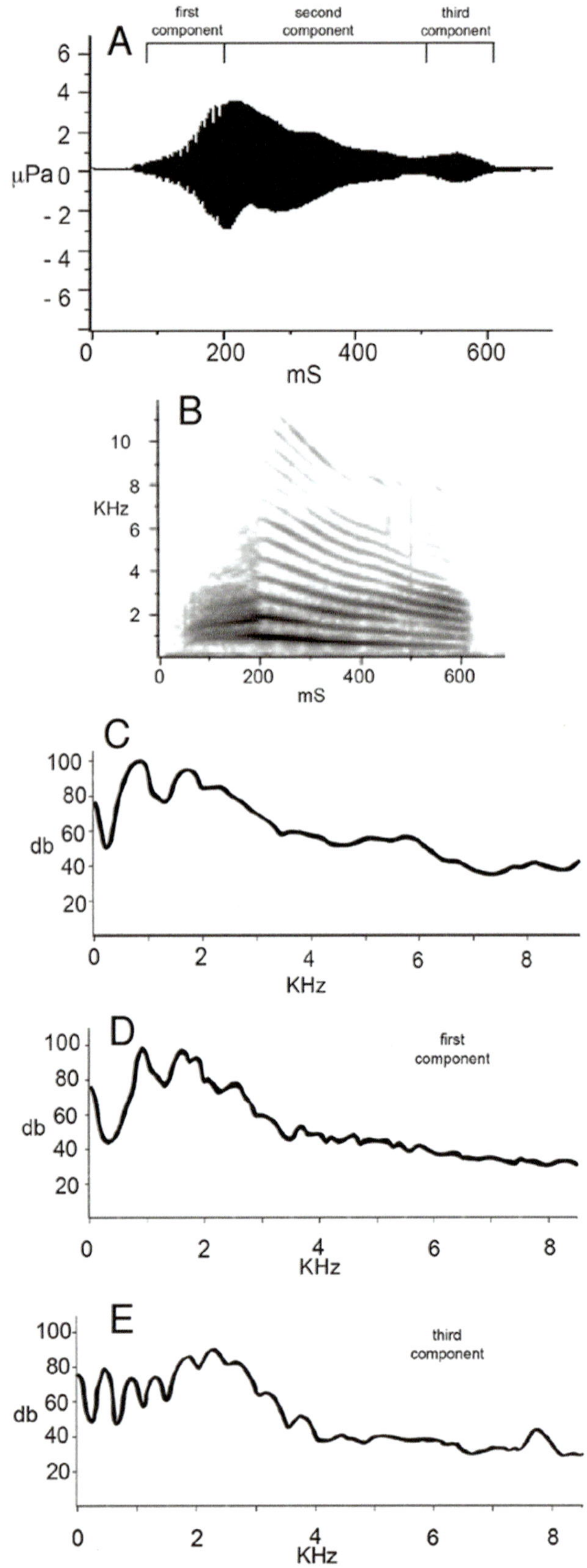

FIGURE A1.14 Calls of *Engystomops pustulatus*. El Empalme, Provincia Guayas, Ecuador; A: oscillogram. B: spectrogram. C: power spectra of complete call. D: power spectra of the first component. E: power spectra of the third component. Taken from Ron *et al.* (2010). Reproduced with permission from the copyright holder © Magnolia Press.

and has less acoustic complexity. The third component is characterized by a gradual increase in amplitude and a switch in dominant frequency from the first to the fifth or sixth harmonic. The transition between the second and third component is gradual. Ron *et al.* (2010) made comparisons between populations from Puerto Rico, La Maná, Arenillas, and Huaquillas. They found statistically significant differences between La Maná and the other populations. They also found high acoustic distinctiveness between populations of *E. pustulatus*.

Engystomops puyango

Source: Ron *et al.* (2004, 2010). Figure A1.15.

Voucher specimen and locality: Bosque Protector Puyango, Provincia El Oro, Ecuador, 321m. Collected by Santiago R. Ron, Elicio E. Tapia, Ítalo G. Tapia, Giovanna Romero, Carolina Proaño, and Mónica Guerra.

Date and time: 19 February 2004; 7 January 2005; 13–20 February 2005.

Temperature: Water = 25.9 °C (n = 9).

Recording equipment: Sony WM-D6C analog tape recorder, SONY™ MZ-NH1. Sennheiser ME-67 directional microphone.

Call type: Advertisement. A call with two obligatory components and one facultative component with harmonic structure.

Call behavior: Males call while floating on standing water.

Sample size: 10 individuals.

Call duration: $\bar{x}$ = 0.366 ± 0.049; range = 0.287–0.438 s.

- **Duration first component:** $\bar{x}$ = 0.125 ± 0.014; range = 0.102–0.142 s.
- **Duration second component:** $\bar{x}$ = 0.129 ± 0.015; range = 0.102–0.149 s.
- **Duration third component:** $\bar{x}$ = 0.111 ± 0.038; range = 0.050–0.186 s.

Notes per call: One.

Call frequency:

- **Frequency of the first spectral peak:** $\bar{x}$ = 859 ± 106; range = 683–990 Hz.
- **Frequency of the second spectral peak:** $\bar{x}$ = 2019 ± 38; range = 1970–2093 Hz.

First component:

- **Frequency of the first spectral peak:** $\bar{x}$ = 968 ± 18; range = 939–995 Hz.
- **Frequency of the second spectral peak:** $\bar{x}$ = 2017 ± 38; range = 1950–2097 Hz.

Second component:

- **Frequency of the first spectral peak:** $\bar{x}$ = 774 ± 46; range = 677–831 Hz.
- **Frequency of the second spectral peak:** $\bar{x}$ = 1807 ± 204; range = 1267–1959 Hz.

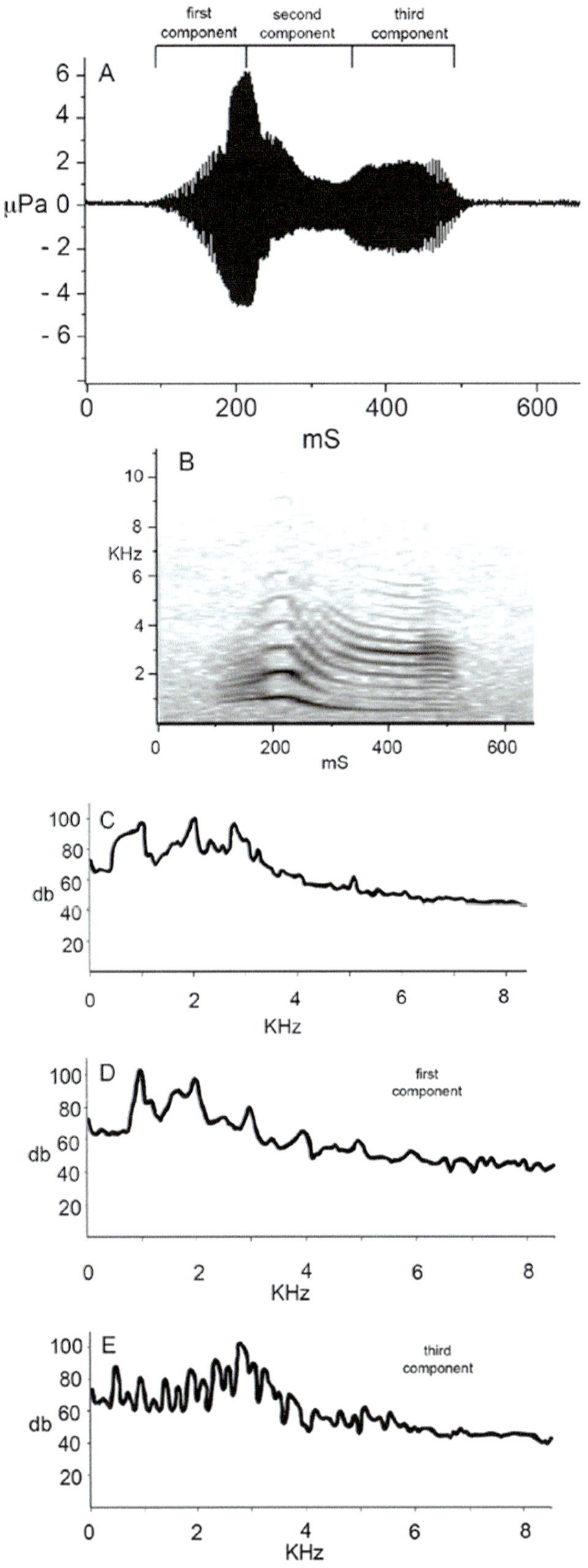

FIGURE A1.15 Calls of *Engystomops puyango*. Bosque Protector Puyango, Provincia El Oro, Ecuador. A and F: oscillograms. B and G: spectrograms. C and H: power spectra of complete calls. D and I: power spectra of the first component. E and J: power spectra of the third component. Taken from Ron *et al.* (2010). Reproduced with permission from the copyright holder © Magnolia Press.

Third component:

Fundamental frequency: $\bar{x}$ = 509 ± 25; range = 474–549 Hz.

Dominant frequency: $\bar{x}$ = 2851 ± 135; range = 2605–3083 Hz.

Remarks: Calls were analyzed using software Raven 1.2.1 software at a sampling frequency of 22.1 kHz and a frequency resolution of 21.5 Hz. The first component is characterized by a slight increase in frequency with decreasing amplitude modulation. The second component is a wine-like note. A facultative third component can be added after the whine and is characterized by a gradual increase in amplitude and a switch in dominant frequency from the first to the fifth or sixth harmonic.

Engystomops randi

Source: Ron *et al.* (2004). Figure A1.16.

Voucher specimen and locality: QCAZ 19752–23462: Cerro Masvale, Provincia Guayas, Ecuador, 92 m. Collected by David C. Cannatella, Luis A. Coloma, A. Holloway, and Santiago R. Ron.

Date and time: 20 February 2002, 20:25 h; 21 March 2003, 20:25 h.

Temperature: Water = 25.4 °C, air = 24.8 °C (QCAZ 19752); water = 27.9 °C, air = 27.5 °C (QCAZ 23462).

Recording equipment: Sony WM-D6C analog tape recorder. Sennheiser ME-67 directional microphone.

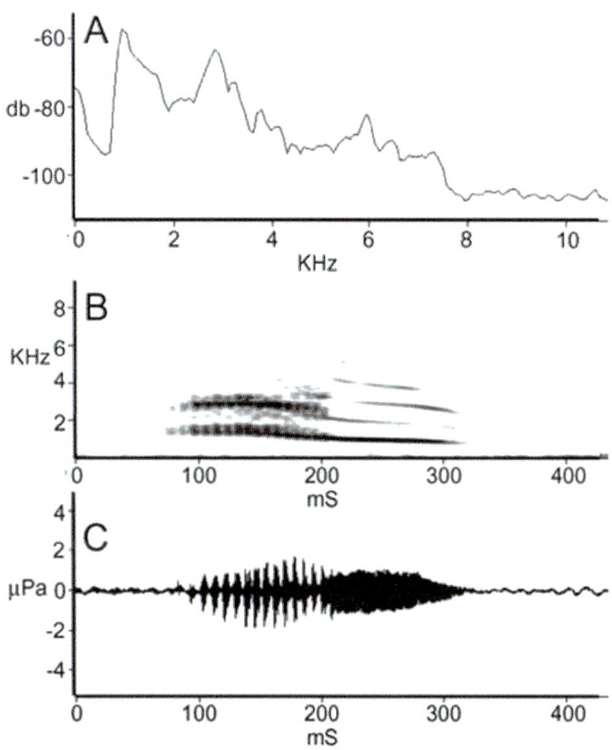

FIGURE A1.16 Advertisement call of *Engystomops randi*. Cerro Masvale, Provincia Guayas, Ecuador. A: power spectrum. B: sonogram. C: oscillogram. Taken from Ron *et al.* (2004).

Call behavior: It was recorded while calling next to a ditch, partly submerged in 0.005 m of water, among vegetation.

Call type: Advertisement call. A single, short call with an amplitude-modulated pulse component followed by a second component that is a pure whine-like tone with a downward frequency sweep.

Sample size: One individual.

Notes per call: One.

Call duration:

QCAZ 19752: $\bar{x}$ = 241; range = 231–247 ms.

QCAZ 23462: $\bar{x}$ = 246; range = 225–267 ms.

Call rate:

QCAZ 19752: $\bar{x}$ = 114.3 notes/min.

QCAZ 23462: $\bar{x}$ = 120.5 notes/min.

Interval between calls:

QCAZ 19752: $\bar{x}$ = 284; range = 232–355 ms.

QCAZ 23462: $\bar{x}$ =252; range = 202–340 ms.

Rise time:

QCAZ 19752: $\bar{x}$ = 68; range = 56–93 ms.

QCAZ 23462: $\bar{x}$ = 75; range = 57–89 ms.

No. of harmonics:

QCAZ 19752: $\bar{x}$ = 4.0.

QCAZ 23462: $\bar{x}$ = 5.1; range = 5–6.

Fundamental frequency of first component:

QCAZ 19752: $\bar{x}$ = 1.314; range = 1.141–1.351 Hz (n = 9).

QCAZ 23462: $\bar{x}$ = 1.358; range = 1.305–1.434 Hz.

Fundamental frequency of second component:

QCAZ 19752: $\bar{x}$ = 0.997; range = 0.985–1.006 kHz.

QCAZ 23462: $\bar{x}$ = 0.925; range = 0.915–0.936 kHz.

No. of pulses of first component:

QCAZ 19752: $\bar{x}$ =14.53; range = 14–15.

QCAZ 23462: $\bar{x}$ =13.00; range = 11–15.

Remarks: Calls were analyzed using Canary1.2.1 software at a sampling frequency of 22.1 kHz and a frequency grid resolution FTP = 8192. Calls described by Ryan and Rand (2001) as belonging to *E. pustulatus* actually belong to *E. randi* (from Pasaje, Provincia El Oro).

Leptodactylus discodactylus

Source: Straughan and Heyer (1976), Heyer (1997). Figure A1.17 taken from de Sá *et al.* (2014).

Voucher specimen and locality: LACM 92001: Limoncocha, Provincia Orellana, Ecuador. Collected by Ronald Heyer. USNM Tape 18 Cut 1.

Date and time: 9 July 1971, 20:00–20:34 h.

Temperature: Water = 23.6 °C, air = 23.4 °C.

Call type: Advertisement call. Series of notes repeated frequently.

Sample size: One individual, 14 calls.

Call rate: 60 calls per minute.

Call duration: $\bar{x}$ = 0.07 s.

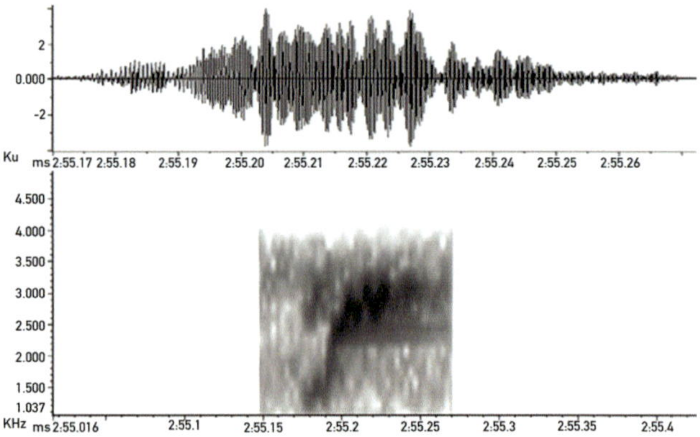

FIGURE A1.17 Advertisement call of *Leptodactylus discodactylus.* Limoncocha, Provincia Orellana, Ecuador (recording USNM 18). Taken from de Sá *et al.* (2014).

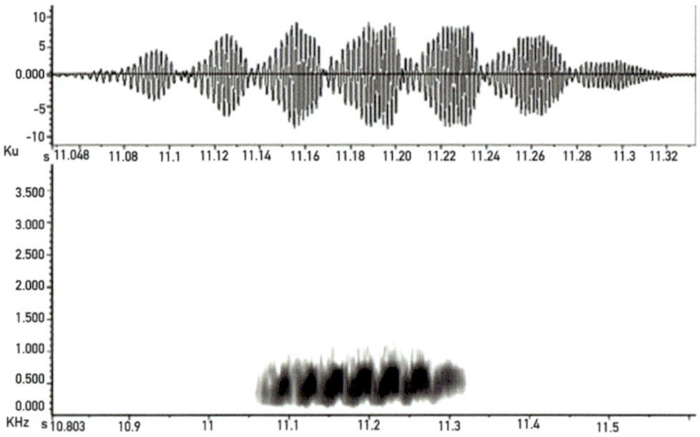

FIGURE A1.18 Advertisement call of *Leptodactylus knudseni.* Tambopata, Departamento Madre de Dios, Peru. Taken from de Sá *et al.* (2014).

Notes per call: One.

Pulses per call: 14–19 pulses per call. The note is initiated by a slow rising pulse of 0.011–0.015 s, which is poorly tuned and ill-defined in frequency.

Fundamental frequency: 1160–3190 Hz.

Dominant frequency: 2580 ± 131 to 3407 ± 142 Hz.

Remarks: Calls were analyzed by Heyer (1997) using Canary1.2 software on a Power Macintosh 8500 computer. Calls were digitized for analysis at a sample rate of 22,050 Hz and a sample size of 16 bits. Recordings had considerable noise. The filter around option was used for determining some parameters for USNM Tape 18 Cut 1 (filtered around 500–4500 Hz). Duellman (1978) describes a call from the same site as follows:

A recording of *Leptodactylus discodactylus* made at Limoncocha, Ecuador, generously was provided by W. R. Heyer. The call consists of a series of short notes "wheeet, wheeet, wheeet" produced at a rate of 44 notes per minute. The dominant frequency of the notes is at about 3000 Hz.

Leptodactylus knudseni

Source: Heyer (2005). Figure A1.18 taken from de Sá *et al.* (2014).
Voucher specimen and locality: Four unvouchered individuals: Departamentos Amazonas, Madre de Dios, Peru. Estados Amazonas, Roraima, Pará, Brazil. French Guiana.
Temperature: 23–26 °C.
Call type: Advertisement call. A single, short note. Call frequency modulated, rising whoop.
Sample size: 58 calls.
Call rate: 16–66 calls per minute.
Call duration: 0.16–0.43 s.
Notes per call: One.
Pulses per call: 6–14 pulses per call.
Pulse rate: $\bar{x}$ = ~38 pulses per second.
Fundamental frequency: 260–540 Hz.
Dominant frequency: 340–690 Hz.
Harmonics: Harmonic structure present.
Remarks: Calls were analyzed using Canary 1.2 following the terminology in Heyer *et al.* (1990). The calls were digitized at a sample rate of 22,050 Hz, sample size of 17 bits, at an input speed of 1×. The audiospectrogram analyses used settings of analysis resolution filter bandwidth 349.70 Hz and frame length 256 points; grid resolution of time 5.805 ms, overlap 50%, frequency 86.13 Hz, FFT size 256 points; window function Hamming; amplitude logarithmic; clipping level –80 dB; display style smooth.

Leptodactylus labrosus

Source: de Carvalho and Ron (2011). Figure A1.19.
Voucher specimen and locality: QCAZCC179: Río Cuaque, Provincia Manabí (0.014° S, 80.0738° W, 12 m.a.s.l.), Ecuador. No voucher: 20 km west of El Carmen, road to Pedernales, Provincia Santo Domingo de los Tsáchilas (0.28118° S, 79.61756° W, 239 m.a.s.l.), Ecuador. No voucher: 10 km north of Zapotillo, Provincia Loja (4.31198° S, 80.21690° W, 231 m), Ecuador.

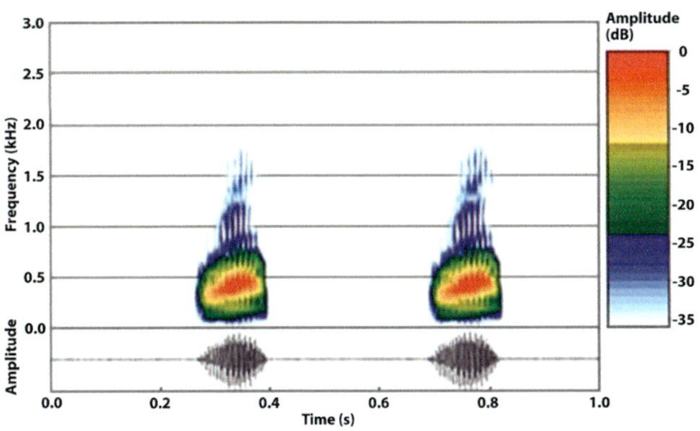

FIGURE A1.19 Audiospectrogram (above) and corresponding oscillogram (below) of a sequence of two advertisement calls of *Leptodactylus labrosus*. Río Cuaque, Provincia Manabí, Ecuador. Taken from de Carvalho and Ron (2011).

Date and time: 8 Jan 2003 (Río Cuaque). Not mentioned for other localities.
Recording equipment: Sony WM-D6C analog tape recorder or a Sony MZ-NH1 digital recorder set at a 44,100 Hz sample rate and 16 bits of resolution. Sennheiser K6/ME67 directional microphone.
Call behavior: Males call buried in underground chambers in open areas adjacent to seasonal lentic water bodies.
Call type 1: Advertisement call. A single note with slight frequency modulation.
- **Sample size:** Three individuals, 227 calls.
- **Call rate:** 118–135 calls per minute.
- **Call duration:** 0.064–0.133 s.
- **Notes per call:** One.
- **Pulses:** Absent or barely defined (completely overlaid).
- **Interval between calls:** $\bar{x}$ = 332 ± 37; range = 0.186–0.697 s.
- **Fundamental frequency:** $\bar{x}$ = 451 ± 124; range = 358–726 Hz.
- **Dominant frequency:** $\bar{x}$ = 451 ± 124; range = 358–726 Hz.

Call type 2: In the recording from Zapotillo (seven call samples). Because two males call simultaneously in the recording (background vocalization of a second male), de Carvalho and Ron (2011) suggest that these may be aggressive calls. The second type of call is broadcasted either alone (one out of seven call samples) or in groups of 2–4 simultaneous calls. It has a pulsed structure with 2–8 well-defined pulses.
Remarks: Sound analyses were carried out using Soundruler (v. 0.9.6.0); the spectral trait of dominant frequency was determined using Audacity software (v. 1.3.13 Beta); sound figures were obtained through the Seewave (v. 1.5.9), R (v. 2.12.1) package. Settings for sound analyses and figures were Hanning window function with 85% overlap and 512 points of resolution (FFT).

Leptodactylus leptodactyloides

Source: Heyer (1994). Figure A1.20.
Voucher specimen and locality: Not collected: Limoncocha, Provincia Sucumbíos, Ecuador. Collected by W. Ronald Heyer. USNM Tape 17, cut 1; USNM Tape 16, cut 2
Date and time: June–July 1971. 22:00 h (USNM Tape 17. cut 1); 15 June 1971. 20:00–22:00 h (USNM Tape 16, cut 2).
Temperature: Air = 24.6 °C (USNM Tape 17. cut 1); water = 24.6 °C (USNM Tape 16, cut 2).

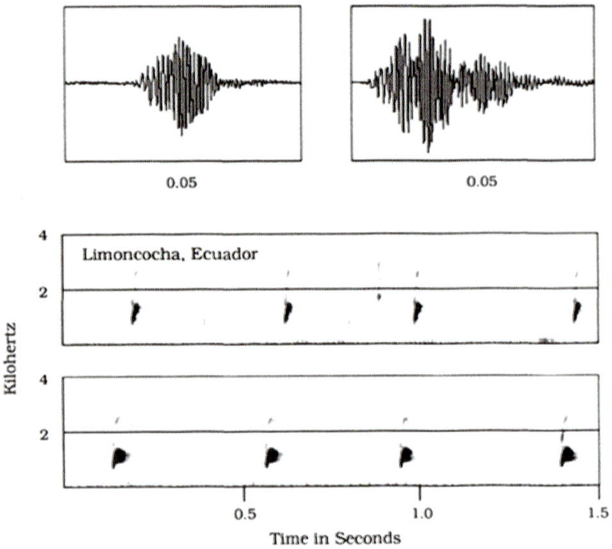

FIGURE A1.20 Advertisement call of *Leptodactylus leptodactyloides*. Limoncocha, Provincia Orellana, Ecuador. Waveform on left and upper audiospectrogram from USNM Tape 17, cut 1. Waveform on right and lower audiospectrogram from USNM Tape 16, cut 2. Taken from Heyer (1994).

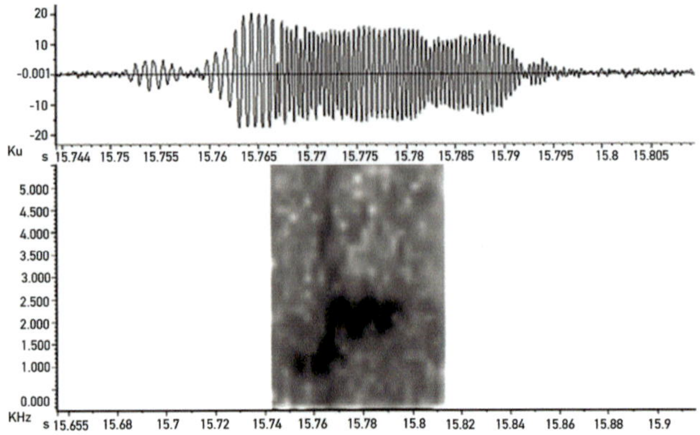

FIGURE A1.21 Advertisement call of *Leptodactylus melanonotus*. Guanacaste, Costa Rica. Recording USNM 83. Taken from de Sá *et al.* (2014).

Call type: Advertisement call. Rising frequency modulation throughout call.
Sample size: Four individuals.
Call rate: 0.3–3.3 calls per second.
Call duration: 0.01–0.04 second.
Notes per call: One.
Pulses: Calls non-pulsed or with 3–5 partial pulses.
Dominant frequency: 650–1600 Hz. Maximum energy between 1100–1300 Hz.
Harmonics: Harmonic structure present or absent.
Remarks: Heyer (1994) provided data for recordings of one individual from Rio Branco, Acre, Brazil, and four from Limoncocha, Provincia Orellana, Ecuador. He summarizes data as follows: Call rate of 0.3–3.3 calls per s; call duration 0.01–0.04 s; calls non-pulsed or with 3–5 partial pulses; calls frequency modulated with very fast rise times, dominant frequency range 650–1600 Hz, maximum energy 1100–1300 Hz; harmonic structure present or absent.

Leptodactylus melanonotus

Source: Straughan and Heyer (1976). Figure A1.21 taken from de Sá *et al.* (2014).
Voucher specimen and locality: Unvouchered: Five locations within the province of Guanacaste, Costa Rica. Specimens and tapes deposited in the Natural History Museum of Los Angeles County.
Temperature: 25 ± 1 °C.
Recording equipment: Uher series 4000 portable tape recorders. Sennheiser Microphone.
Call type: Advertisement call. A single, explosive, short yet compound note repeated rapidly but irregularly. Weak falling frequency modulations.
Sample size: 25 individuals.
Call rate: 150–160 calls per minute.
Call duration: 0.03–0.04 s.
Interval between calls: $\bar{x}$ = 0.2 s.
Notes per call: One compound note.
Fundamental frequency:
First part: 1100–2600 Hz (slight peak in power at 2200 Hz).
Second part: 2400 drops smoothly to 2000 Hz.
Remarks: Audiospectrograms were made with Kay Electronics Sonograph Models 6061 or 6061B employing the 80–8000 Hz frequency scale and narrow-band filter (45 Hz) using the Uher tape recorder for playback. To examine the internal call structure with greater resolution, the tape-recorded signal was fed into a Tektronics 502 dual-beam oscilloscope along with a 1 kHz time base and photographed continuously with a Nihon Kodin camera at film speeds of 0.01 m/s.

Leptodactylus mystaceus

Source: Heyer *et al.* (1996). Figure A1.22.
Voucher specimen and locality: LACM 92111: Limoncocha, Provincia Orellana, Ecuador, 260 m. Recorded by W. Ronald Heyer. USNM Tape 18 (cut 3).
Date and time: 15 July 1971, 20:30–21:00 h.
Temperature: Air = 22.9 °C.
Call type: Advertisement call. Rising frequency modulations throughout call.
Sample size: One individual, 40 calls.
Call duration: 0.15–0.25 s.
Call rate: 1.0–2.0 calls per second.
Notes per call: One pulsed note.
Pulses per note: 11–17 pulses per note.
Dominant frequency: 560–1760 Hz.

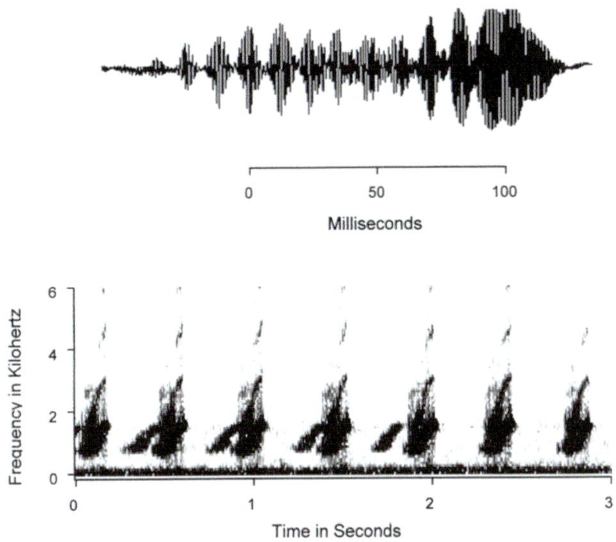

FIGURE A1.22 Call of *Leptodactylus mystaceus*. Limoncocha, Provincia Sucumbíos, Ecuador. Audiospectrogram (below) and waveform (above) of advertisement USNM Tape 16, cut 4. Taken from Heyer *et al.* (1996).

Remarks: Advertisement calls were analyzed using a Kay Digital Sona-Graph model 7800, Uniscan II equipment, or Canary software on a Macintosh IIci computer.

Heyer *et al.* (1996) described and visualized the call variation from several populations from Venezuela, Brazil, Ecuador, and Peru of the *Leptodactylus mystaceus* complex. They also provided additional data of unvouchered specimens and LACM 92111 from Limoncocha, Provincia Orellana, Ecuador, recorded by W. Ronald Heyer: USNM Tape 16 (cut 3), 20 June 1971, 22.9 °C air temperature; USNM Tape 16 (cut 4), 20 June 1971, 23.2 °C air temperature; USNM Tape 18 (cut 2), 15 July 1971, about 20:00 h, 22.9 °C air temperature; USNM Tape 18 (cut 3), 15 July 1971, 20:30–21:00 h, 22.9 °C air temperature, voucher specimen LACM 92111.

Leptodactylus pentadactylus

Source: Heyer (2005). Figure A1.23 taken from de Sá *et al.* (2014).

Voucher specimen and locality: Not collected: Meta, Colombia; Madre de Dios, Peru; Amazonas, Brazil.

Temperature: Meta: 23.2 °C; Amazonas: 26.0 °C.

Call type: Advertisement call. A rising whoop at least in the first half of the call.

Sample size: One individual, seven calls (Meta). Three individuals, six calls (Madre de Dios). Two individuals, seven calls (Amazonas). Number of frogs inferred from (Heyer, 2005).

Call rate: 4–37 calls per minute.

Call duration: 0.18–0.40 s.

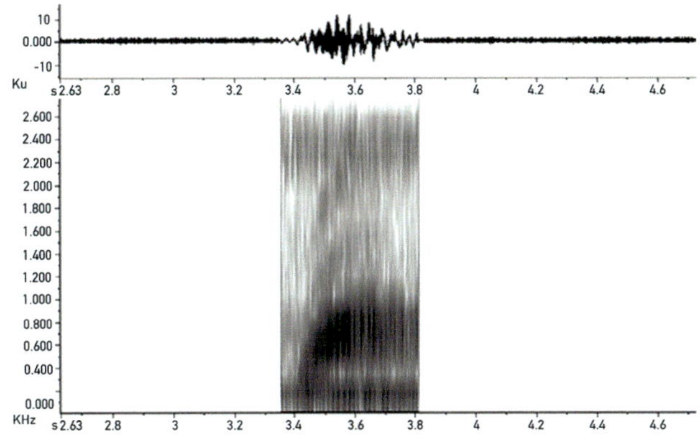

FIGURE A1.23 Advertisement call of *Leptodactylus pentadactylus*. Nova Empreza, Rio Juruá Departamento Amazonas, Brazil. Recording USNM 251 (deposited at Macaulay Library). Taken from de Sá *et al.* (2014).

Notes per call: One pulsed note.

Pulses per call: 12–18 pulses per call.

Fundamental frequency: 680–1030 Hz.

Dominant frequency: 680–1030 Hz.

Remarks: Calls were analyzed using Canary 1.2 following the terminology in Heyer *et al.* (1990). The calls were digitized at a sample rate of 22,050 Hz, sample size of 17 bits, at an input speed of 1×. The audiospectrogram analyses used settings of analysis resolution filter bandwidth 349.70 Hz and frame length 256 points; grid resolution of time 5.805 ms, overlap 50%, frequency 86.13 Hz, FFT size 256 points; window function Hamming; amplitude logarithmic; clipping level –80 dB; display style smooth. Dominant frequency was determined using the spectrum analysis in Canary with settings of analysis resolution filter bandwidth 349.70 Hz, FFT size 256 points; window function Hamming; amplitude logarithmic; clipping level –79.79 dB.

Leptodactylus peritoaktites

Source: Heyer (2005). Figure A1.24 taken from de Sá *et al.* (2014).

Voucher specimen and locality: Not collected: Provincia Pichincha, Ecuador. Tape USNM tape 98, cut 5; tape 103, cut 10.

Call type: Advertisement call. Rising whoop.

Sample size: Two individuals, 4–5 calls each frog.

Call rate: 34–37 calls per minute.

Call duration: 0.20–0.30 s.

Notes per call: One pulsed note.

Pulses per call: 5–8 pulses per call.

Pulse rate: $\bar{x}$ = 23–27 pulses per second.

Fundamental frequency: Mean initial frequency among individuals about 550–690 Hz.

Dominant frequency: ca. 860 Hz.

Remarks: Calls were analyzed using Canary 1.2 following the terminology in Heyer *et al.* (1990). The

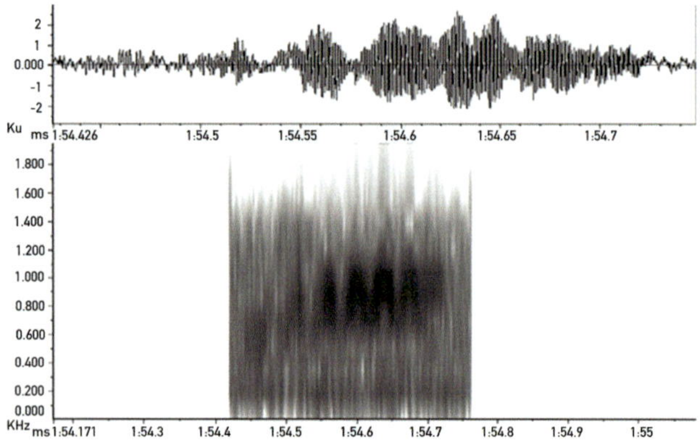

FIGURE A1.24 Advertisement call of *Leptodactylus peritoaktites.* Centro Científico Río Palenque, Provincia Los Ríos, Ecuador. Recording USNM 98 (deposited at Macaulay Library). Taken from de Sá *et al.* (2014).

calls were digitized at a sample rate of 22,050 Hz, sample size of 17 bits, at an input speed of 1×. The audiospectrogram analyses used settings of analysis resolution filter bandwidth 349.70 Hz and frame length 256 points; grid resolution of time 5.805 ms, overlap 50%, frequency 86.13 Hz, FFT size 256 points; window function Hamming; amplitude logarithmic; clipping level –80 dB; display style smooth. Dominant frequency was determined using the spectrum analysis from Canary with settings of analysis resolution filter bandwidth 349.70 Hz, FFT size 256 points; window function Hamming; amplitude logarithmic; clipping level –79.79 dB.

LEPTODACTYLUS PETERSII

Source: Heyer (1994). Figure A1.25 taken from de Sá *et al.* (2014).

Voucher specimen and locality: ZUEC 551: Cruzeiro do Sul, Acre, Brazil. Collected by Adão J. Cardoso. Recording ACJ 49.

Date and time: 5 December 1983, 19:50 h.

Temperature: Air and water 24 °C.

Call type: There are apparently two major types of advertisement call.

Call type 1:

Sample size: One individual.

Call rate: 2.9–4.2 calls per second.

Call duration: 0.04–0.05 s.

Notes per call: One pulsatile note.

Pulses per note: 3–4 partial pulses per note.

Frequency: 700–1200 Hz, maximum energy 750–800 Hz.

Harmonics: Weakly to strongly developed.

Call type 2:

Call rate: 0.6–1.3 calls per second.

Call duration: 0.03–0.05 second.

Notes per call: Two juxtaposed notes.

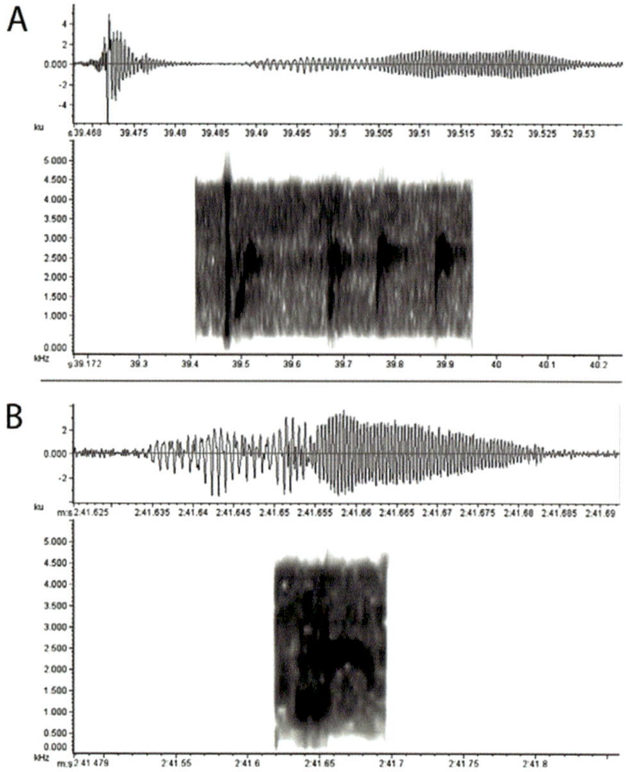

FIGURE A1.25 Advertisement call of *Leptodactylus petersii.* Cruzeiro do Sul, Acre, Brazil. A: call type 1. B: call type 2. Taken from de Sá *et al.* (2014).

Pulses per note: First note: 2–4; second note: not noticeably pulsed.

Broadcast frequency: First note: 800–1600 Hz. Second note 1800–2800 Hz.

Harmonics: First note with harmonic structure; second note apparently without harmonic structure.

Remarks: Heyer (1994) also provided data for recordings of a call type 2 call from Tambopata, Peru (USNM Tape 207, cut 19; presumed advertisement call, recorded from specimen USNM 307122, 12 Jan 1989, 23:00 h, 24.9 °C air, by Reginald B. Cocroft); from the Biological Dynamics of Forest Fragments research site north of Manaus, Amazonas, Brazil (USNM Tape 163, cut 3. Recorded 9 Apr 1987, 22:30 h, by Barbara L. Zimmerman); from IPEAN, Para, Brazil (KU Reel 81, tape 1001; recorded 8 July 1970, 23:05 h, 29 °C air, by Martha L. Crump); and from Ilha de Maraca, Roraima, Brazil (MZUSP 62430, USNM Tape 213, cut 10; recorded September–October 1985, 23:50 h, 23.5 °C air, by Celso M. de Carvalho). Heyer (1994) summarizes data: Calls of two juxtaposed notes. Dominant frequency of first note 800–1600 Hz, of second note 1800–2800 Hz; call duration 0.03–0.05 s; first note of 2–4 partial pulses, second note not noticeably pulsed;

first note frequency modulated upward with a fast rise time, second note often frequency modulated downward; call rate 0.6–1.3 s; first note with harmonic structure, second note apparently without harmonic structure.

Leptodactylus rhodomystax

Source: Zimmerman and Bogart (1984). Figure A1.26 taken from de Sá *et al.* (2014).

Voucher specimen and locality: INPA-WWF no numbers provided: Manaus reserves, Brazil. Representative individuals deposited at Museu de Zoologia da Universidade de Sao Paulo (M.Z.U.S.P.).

Date and time: November 1983–May 1984; December 1984–April 1985.

Temperature: 22–25 °C.

Recording equipment: Uher 4000 monaural tape recorder at a tape speed of 9.5 cm/s. Uher M517 microphone.

Call type: Advertisement call. Rising frequency modulations for most of call.

Sample size: Five individuals, one chorus.

Call rate: 0.23 calls per second.

Call duration: 0.12–0.23 s.

Notes per call: A single pulsed note.

Fundamental frequency: 3700–5400 Hz.

Dominant frequency: 3700–5400 Hz.

Harmonics: Harmonic structure present.

Remarks: Calls were analyzed with a Kay Electronics sonograph model 7030A at normal tape-recording speed and narrow (45 Hz) and/or wide (300 Hz) filter bandwidth setting. Sonograms were measured with a Science Accessory Corporation (SAC) GP-7 sonic digitizer interfaced to an IBM Personal Computer system.

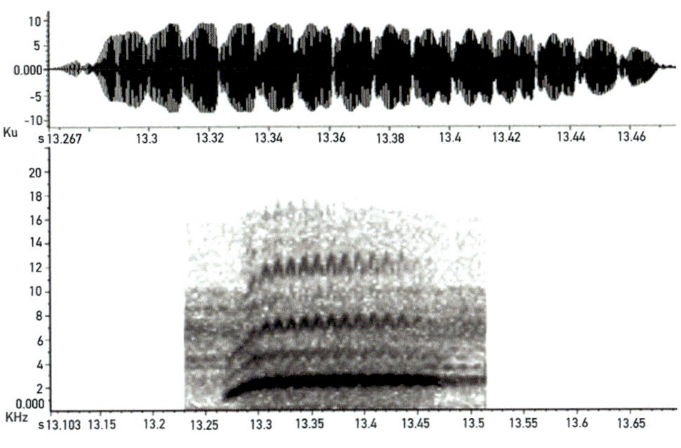

FIGURE A1.26 Advertisement of *Leptodactylus rhodomystax*. Altamira, Río Juruá, Amazonas, Brazil. Recording USNM 255, 27.7°C (deposited at Macaulay Library). Taken from de Sá *et al.* (2014).

Leptodactylus stenodema

Source: Heyer (1979). Figure A1.27 taken from Heyer (1979) and Lescure and Marty (2000).

Voucher specimen and locality: UTA A-3833: Vaupés, Colombia.

Call type: Advertisement call.

Call duration: Range = 0.26–0.36 s.

Call rate: 30 calls per min.

Notes per call: A single, multi-pulsed note.

Fundamental frequency: 760–900 Hz.

Dominant frequency: 760–900 Hz.

Remarks: Modest rising frequency modulations throughout call. de Sá *et al.* (2014) compile call data from Heyer (1979) and Lescure and Marty (2000).

Leptodactylus wagneri

Source: de Carvalho *et al.* (2022). Figure A1.28.

Voucher specimen and locality: QCAZ no voucher: Near Pacto Sumaco, Provincia Napo Ecuador. Recorded by Santiago R. Ron. MNHN Field # AF4856: Ichigkat, Eastern Cordillera, Peru. Recorded by Antoine Fouquet.

Call type: A single, partly pulsed note.

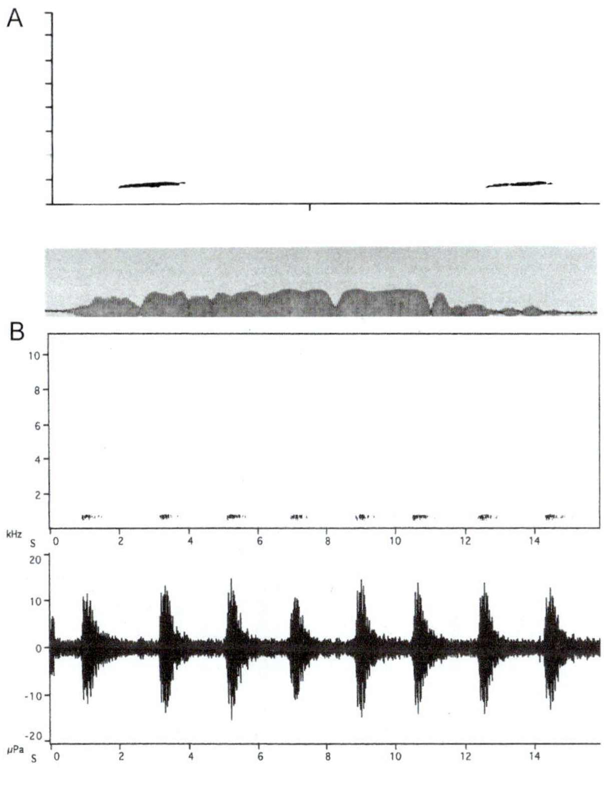

FIGURE A1.27 Advertisement call of *Leptodactylus stenodema*. A: sonogram and strip chart of the mating call. Wacará, Vaupés, Colombia. Taken from Heyer (1979). B: spectrogram and oscillogram. Guyane. Taken from Lescure and Marty (2000).

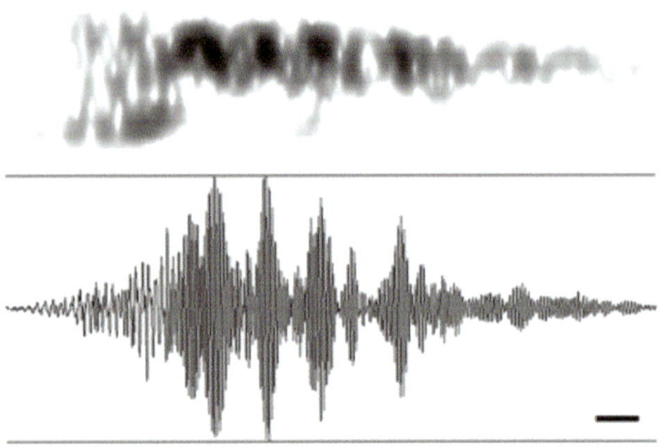

FIGURE A1.28 Call of *Leptodactylus wagneri*. Either from Pacto Sumaco, Ecuador, or Ichigkat, Peru. Taken from de Carvalho *et al.* (2022).

Sample size: 12 calls of one male from Ecuador, nine calls of a second male from Peru.
Note duration: 64–136 (103.2 ± 20.1) ms.
Rise time: 31–61% (46.4 ± 0.8) of note duration
Note rate: 50–55 (52.7 ± 3.8) per minute.
Number of pulses: 6–11 (7.9 ± 1.5) incomplete pulses.
Pulse rate: 63–118 (86.4 ± 1.9) per second.
Pulse duration: 9–18 (13.1 ± 0.6) ms.
Dominant frequency: 1572–1898 (1722.1 ± 22.9) Hz.
Fundamental harmonic: Same as dominant.
Frequency upsweep: Modest to pronounced ranging from 656 to 1172 (962.0 ± 20.6) Hz.

Lithodytes lineatus 1

Source: Marquez *et al.* (1995). Figure A1.29.
Voucher specimen and locality: Puerto Almacén, Santa Cruz, Bolivia.
Date: 1987 to 1990.
Sampling frequency: 44.1 kHz.
Resolution: 16 bits.
Recording equipment: Sony WM D6C or a Sanyo M 1120 tape recorder. Sennheiser Me 80 directional microphone.
Call type: Advertisement calls.
Call behavior: It could only be heard at night in the absence of vocalizations of other anuran species.
Sample size: 13 calls.
Call rate: $\bar{x}$ = 66.2 ±7; range = 55.4–77.9 call per min.
Notes per call: One.
Note duration: $\bar{x}$ = 155.7 ± 7.9; range = 145.8–177.1 ms.
Note rate: $\bar{x}$ = 66.2 ±7; range = 55.4–77.9 note per min.
Fundamental frequency: $\bar{x}$ = 1355.8 ± 7.3; range = 1352.9–1373.1 Hz.

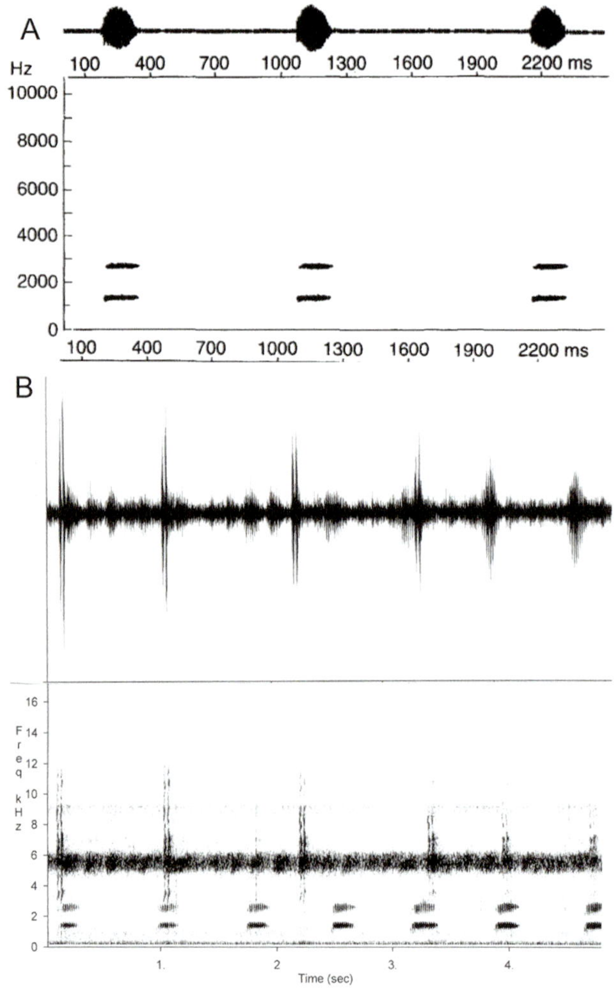

FIGURE A1.29 Oscillograms and spectrograms of the advertisement call of *Lithodytes lineatus*. A: Puerto Almacén, Santa Cruz, Bolivia. Taken from Marquez *et al.* (1995). B: male calling from the interior of an active nest of *Atta cephalotes*. Tambopata National Reserve, Peru. Taken from Schlütter *et al.* (2009).

Dominant frequency: $\bar{x}$ = 2698.6 ± 12.8; range = 2685.6–2726.0 Hz.
Change in dominant frequency: $\bar{x}$ = 16.8 ± 29.7; range = 20.2–60.6 Hz.
Remarks: Recordings were processed with an Apple Macintosh-based digital signal analysis system. Signalyze software was used to obtain numerical information and to generate audiospectrograms and oscillograms. Frequency information obtained through fast Fourier transform (FFT) (width, 1024 points).

Lithodytes lineatus 2

Source: Schlütter *et al.* (2009). Figure A1.29.
Voucher specimen and locality: Not collected: Explorer's Inn Lodge, Tambopata National

Reserve, Departamento Madre de Dios, Peru. Recorded by one of the authors.
Date and time: November/December 2005.
Recording equipment: SONY-D6C Tape recorder. Sennheiser slender microphone.
Sample size: Four individuals.
Call type: Advertisement call. Continuous series of short, repeated whistles.
Call behavior: Four males called synchronously from the interior of a large nest of leaf cutter ants (*Atta cephalotes*).
Note rate: 80–90 notes per minute.
Note duration: About 0.12 s.
Dominant frequency: 2600 Hz.
Harmonics: 1300, 2600, and 3900 Hz.
Remarks: Acoustic analyses were performed with Cool Edit software, and calls were digitized at a sampling rate of 44.1 kHz (Hamming windowing function set at a resolution of 1024 bands).

FAMILY CERATOPHRYIDAE

Ceratophrys cornuta

Source: Duellman (2005). Figure A1.30.
Voucher specimen and locality: Bolivia and Tambopata, Peru. USNM Tape 268/8.
Date and time: November 1986 and 1988. At night.
Temperature: 23.6 °C.
Call type: Advertisement. A loud "baaaaa."
Call behavior: Males call while sitting on the ground or in shallow water at the edges of ponds.
Call duration: ~1 s.
Notes per call: One.
Note duration: 0.5 s.
Note rate: 30 notes per min.
Pulse rate: 91.8 pulses per second.
Fundamental and dominant frequency: 2175–2300 Hz.
Harmonics: 4321 Hz.

FAMILY HEMIPHRACTIDAE

Gastrotheca cornuta

Source: Duellman (2015).

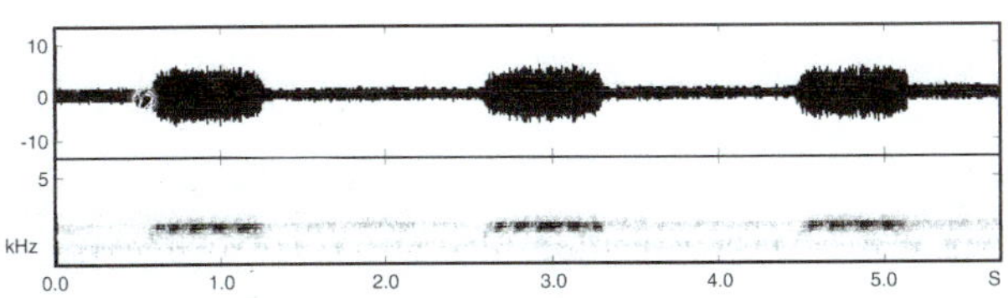

FIGURE A1.30 Oscillogram and spectrogram of the advertisement call of *Ceratophrys cornuta*. Tambopata, Peru. Taken from Duellman (2005).

Voucher specimen and locality: Río Claro, Bocas del toro, Panama. Estación Biológica Río Palenque, Provincia Los Ríos, Ecuador. KU tape No. 595, KU tape No. 732.
Date and time: 23 May 1966, 21:00 h (No. 595); 31 March 1975, 20:30 h (No. 732.).
Temperature: 18.3 °C (No. 595), 23.5 °C (No. 732).
Sampling frequency: 44.1 or 48 kHz.
Resolution: 16 bits.
Sample size: Two.
Call type: The call is a "bop."
Note duration: $\bar{x}$ = 0.08 s.
Interval between notes: $\bar{x}$ = 0.60 s.
Dominant frequency: $\bar{x}$ = 800 Hz.
Harmonics: Three harmonics are distinguishable: 800, 1600, and 2400 Hz.
Remarks: Although usually only one note is produced, sometimes, a frog produces two or three notes in quick succession. Intervals between notes can take several minutes.

Gastrotheca cuencana

Source: Carvajal-Endara *et al.* (2019). Figure A1.31.
Voucher specimen and locality: Three individuals, no vouchers: Cuenca and 3 from Río Mazán, Provincia Azuay, Ecuador. Two individuals, no vouchers: Papaloma de la Nube, Provincia Cañar. FonoZoo ID: 11170–77.
Date and time: 3 August 1986 (22:00 h), 12 August 1986, and 8 September 1986 at Río Mazán. 14 August 2008 (14:00–14:30 h) at Papaloma de la Nube. 8 June 2011 (21:00–23:58 h) at Cuenca.
Temperature: 10.3 °C at Papaloma de la Nube, 13.8–16 °C at Cuenca.
Humidity: 84.2% RH.
Recording equipment: Sony MZ-NH1 recorder and a Sennheiser K6-ME66 microphone. An Olympus LS-10S recorder and a Sennheiser ME66 microphone. An Olympus LS-11 recorder and a RØDE NTG2 microphone. An Olympus LS-14 recorder and a Sennheiser ME66 microphone.
Sampling frequency: 44.1 or 48 kHz.
Resolution: 16 bits.
Sample size: Nine calls. Eight individuals.
Call type: A simple call, composed of a series of short, single-pulsed notes, emitted at regular intervals.
Notes per call: 3–19 notes.
Note duration: $\bar{x}$ = 0.016 ± 0.007 s.
Interval between notes: $\bar{x}$ = 2.783 ± 1.004 s.
Dominant frequency: $\bar{x}$ = 1147.9 ± 99.536 Hz.
90% bandwidth: Range = 1018.6–1269.4 Hz.
Harmonics: Three.
Remarks: Acoustic analysis was conducted using Raven Pro 1.4 software (http://www.birds.cornell.edu/raven). Temporal parameters (oscillograms) and spectral parameters (spectrograms) were

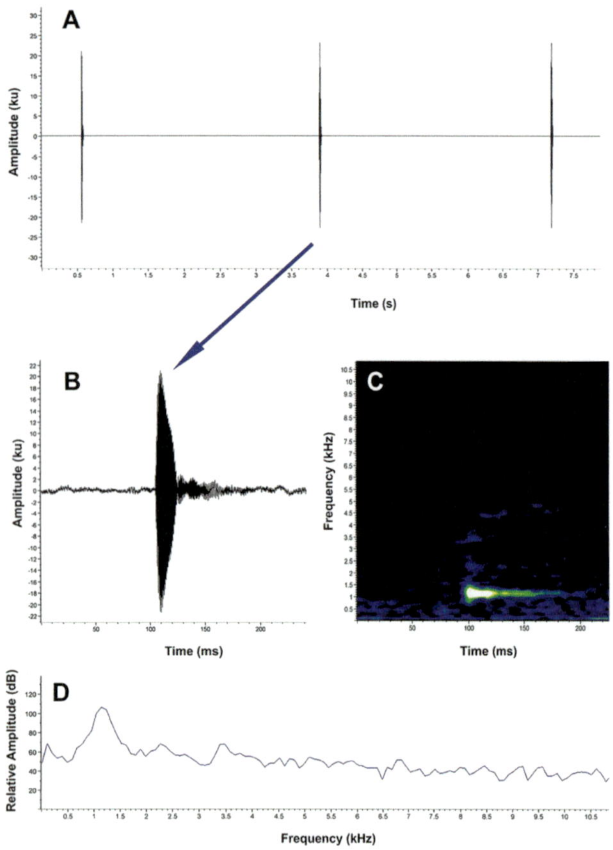

FIGURE A1.31 Advertisement calls of *Gastrotheca cuencana*. Either one of Cuenca or Río Mazán, or Papaloma de la Nube, Azuay and Cañar provinces, Ecuador. A: oscillogram of a three-note section. B: oscillogram of a single note. C: spectrogram of a single note. D: power spectrum of a single note. Taken from Carvajal-Endara *et al.* (2019). Reproduced with permission from the copyright holder © Magnolia Press.

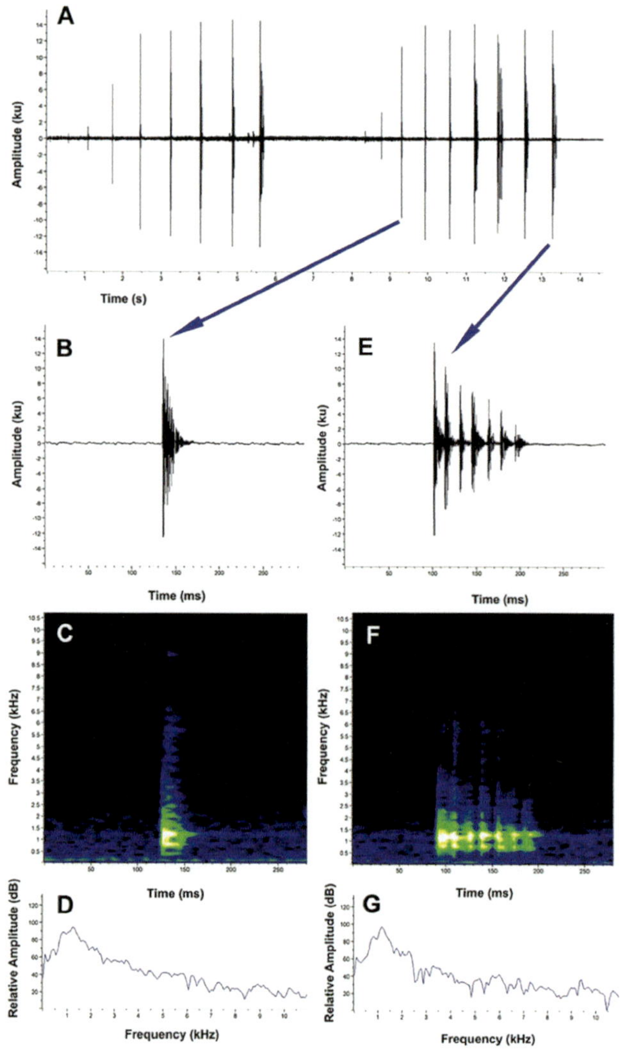

FIGURE A1.32 Calls of *Gastrotheca elicioi*. Abra de Zamora, Loja-Catamayo road, parque Universitario de Educación Ambiental y Recreación, quebrada El Salado, Provincia Loja. A: oscillogram of two consecutive calls. B: oscillogram of a single short note. C: spectrogram of a single short note. D: power spectrum of a single short note. D: oscillogram of a single long note. F: spectrogram of a single long note. G: power spectrum of a single long note. Taken from Carvajal-Endara *et al.* (2019). Reproduced with permission from the copyright holder © Magnolia Press.

measured and obtained through Hanning window function, DFT: 512 samples, 3 dB filter bandwidth: 124 Hz, and 50% overlap.

Gastrotheca elicioi

Source: Carvajal-Endara *et al.* (2019). Figure A1.32.

Voucher specimen and locality: Abra de Zamora, Loja-Catamayo road, Parque Universitario de Educación Ambiental y Recreación (PUEAR), quebrada El Salado, Loja. CJ 4308: Facilities of Centro Jambatu in San Rafael, Quito Provincia Pichincha, Ecuador.

Date and time: 20 March 2014, 20:00–22:00 h. 27 January 2016, 22:20 h. 30 November 2016, 17:58–19:09 h. 26 December 2017, 21:21 h.

Temperature: 13.1–18.5 °C.

Humidity: 74.5–82%.

Recording equipment: Sony MZ-NH1 recorder and a Sennheiser K6-ME66 microphone. An Olympus LS-10S recorder and a Sennheiser ME66 microphone. An Olympus LS-11 recorder and a RØDE NTG2 microphone. An Olympus LS-14 recorder and a Sennheiser ME66 microphone.

Sampling frequency: 44.1 or 48 kHz.

Resolution: 16 bits.

Sample size: Seven males.

Call type: A complex call, composed of up to six, short, single-pulsed notes, and followed by up to seven longer pulsed notes. It is a series of short notes, which become gradually longer, pulsated notes by the end of the call.

Long-note duration: $\bar{x}$ = 0.188 ± 0.150 s.

Pulses per note: Mean 3.47 ± 1.695 distinct pulses, partly fused, without silent intervals (amplitude modulation close but less than 100%). The amplitude of the long note decreases gradually toward the end.

Short-note duration: $\bar{x}$ = 0.027 ± 0.014 s.
Interval between notes: $\bar{x}$ = 0.452 ± 0.125 s.
Dominant frequency: $\bar{x}$ = 1249.4 ± 169.862 Hz.
90% bandwidth: Range = 807.1–1416.8 Hz.
Harmonics: Not clearly recognizable.
Remarks: Acoustic analysis was conducted using Raven Pro 1.4 software (http://www.birds.cornell.edu/raven). Temporal parameters (oscillograms) and spectral parameters (spectrograms) were measured and obtained through Hanning window function, DFT: 512 samples, 3 dB filter bandwidth: 124 Hz, and 50% overlap. The fundamental frequency and harmonics are not clearly recognizable.

Gastrotheca espeletia

Source: Sinsch and Juraske (2006). Figure A1.33.
Voucher specimen and locality: San Gabriel, Provincia Carchi, Ecuador.
Sample size: 50 calls.
Call type: Long, pulsed calls (advertisement) and series of short pulse groups. Some of the calls were irregularly interrupted by the drop-out of pulses but still formed part of a single, long pulse group.
Call duration: 807 ± 66 ms.
Interval between calls: 2225 ± 528 ms.
Call period: 3254 ± 550 ms.
Call ratio: 0.39 ± 0.05.
Pulses per call: 50 ± 2.2.
Pulse rate: 71 ± 1.2 Hz.
Dominant frequency: 1.26 ± 0.03 kHz.
Remarks: Calls were analyzed using a MEDAV Spectro 2000 signal analysis system. Terminology for call characteristics follows Schneider and Sinsch (1992).

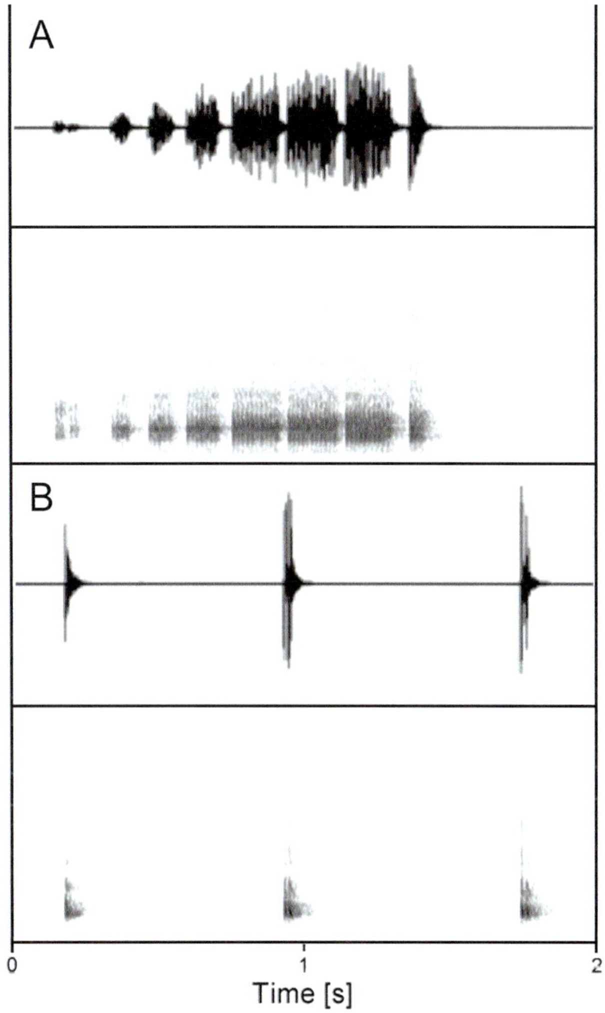

FIGURE A1.33 Calls of *Gastrotheca espeletia*. San Gabriel, Provincia Carchi, Ecuador. A: oscillogram and spectrogram of the advertisement call with several interruptions (18.9°C). B: oscillogram and spectrogram of the short pulse groups (17.2°C). Taken from Sinsch and Juraske (2006).

Gastrotheca litonedis

Source: Carvajal-Endara *et al.* (2019). Figure A1.34.
Voucher specimen and locality: QCAZ 49974, QCAZ 49977, CJ 1404: San Gerardo, San Fernando, Laguna de Busa, Provincia Azuay, Ecuador.
Date and time: 1 December 2010, 16:12 h, 17:20 h, 20:24 h–22:23 h. 9 June 2011, 22:07–23:00 h.
Temperature: 8.3–8.8 °C, 13.4–16.1 °C.
Humidity: 96–98% RH.
Recording equipment: Sony MZ-NH1 recorder and a Sennheiser K6-ME66 microphone. An Olympus LS-10S recorder and a Sennheiser ME66 microphone. An Olympus LS-11 recorder and a RØDE NTG2 microphone. An Olympus LS-14 recorder and a Sennheiser ME66 microphone.
Sampling frequency: 44.1 or 48 kHz.
Resolution: 16 bits.
Sample size: Nine individuals.
Call type: A simple call composed of a series of short, single-pulsed notes emitted at regular intervals.
Notes per call: 1–21 notes.
Note duration: $\bar{x}$ = 0.075 ± 0.025 s.
Interval between notes: $\bar{x}$ = 1.768 ± 0.286 s.
Dominant frequency: $\bar{x}$ = 1224.0 ± 66.415 Hz.
90% bandwidth: Range = 1114.7–1308.1 Hz.
Harmonics: Not clearly recognizable.
Remarks: Acoustic analysis was conducted using Raven Pro 1.4 software (http://www.birds.cornell.edu/raven). Temporal parameters (oscillograms) and spectral parameters (spectrograms) were measured and obtained through Hanning window function, DFT: 512 samples, 3 dB filter bandwidth: 124 Hz, and 50% overlap.

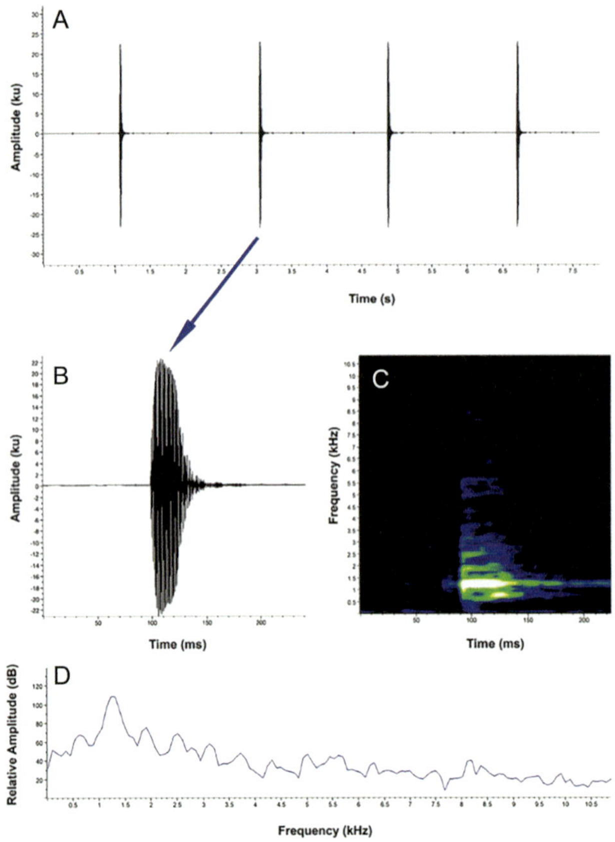

FIGURE A1.34 Calls of *Gastrotheca litonedis*. San Gerardo, San Fernando, Laguna de Busa, Provincia Azuay, Ecuador. A: oscillogram of a four notes section of the call. B: oscillogram of a single note. C: spectrogram of a single note. D: power spectrum of a single note. Taken from Carvajal-Endara *et al.* (2019). Reproduced with permission from the copyright holder © Magnolia Press.

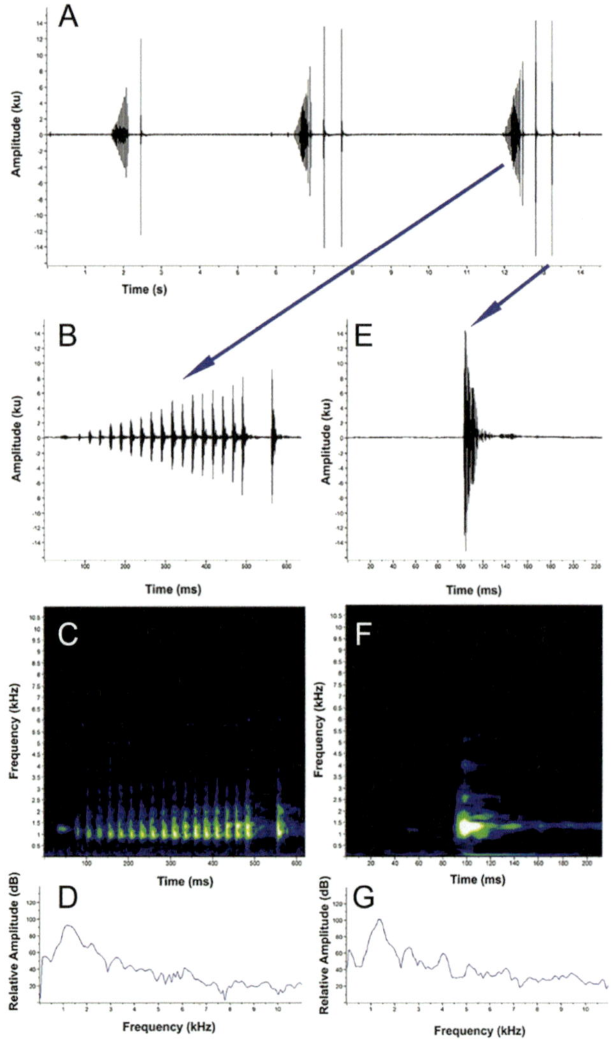

FIGURE A1.35 Calls of *Gastrotheca lojana*. San Fernando, Laguna de Busa, Azuay, Oña, Provincia Loja, Ecuador. A: long note. B: spectrogram of a single long note. C: power spectrum of a single long note. D: oscillogram of a single short note. D: spectrogram of a single short note. F: power spectrum of a single short note. Taken from Carvajal-Endara *et al.* (2019). Reproduced with permission from the copyright holder © Magnolia Press.

GASTROTHECA LOJANA

Source: Carvajal-Endara *et al.* (2019). Figure A1.35.

Voucher specimen and locality: QCAZ 42725, CJ 4300, CJ 4302: San Fernando, Laguna de Busa, Azuay, Oña, Provincia Loja, Ecuador.

Date and time: 24 October 2007, 20:19 h. 26 January 2016, 17:55–21:50 h.

Temperature: 10.8–14.6 °C.

Recording equipment: Sony MZ-NH1 recorder and a Sennheiser K6-ME66 microphone. An Olympus LS-10S recorder and a Sennheiser ME66 microphone. An Olympus LS-11 recorder and a RØDE NTG2 microphone. An Olympus LS-14 recorder and a Sennheiser ME66 microphone.

Sampling frequency: 44.1 or 48 kHz.

Resolution: 16 bits.

Sample size: Five individuals.

Call type: A complex call composed of one long, pulsed note and followed (or not) by one or two short, single-pulsed notes.

Long-note duration: $\bar{x} = 0.562 \pm 0.125$ s.

Pulses per note: $\bar{x} = 17.59 \pm 2.418$.

Short-note duration: $\bar{x} = 0.064 \pm 0.027$ s.

Interval between notes: $\bar{x} = 0.356 \pm 0.107$ s.

Dominant frequency: $\bar{x} = 1118.1 \pm 126.872$ Hz.

90% bandwidth: Range = 986.1–1348.4 Hz.

Harmonics: 5–7 harmonics in the short notes.

Remarks: Acoustic analysis was conducted using Raven Pro 1.4 software (http://www.birds.cornell.edu/raven). Temporal parameters (oscillograms) and spectral parameters (spectrograms) were measured and obtained through Hanning window function, DFT: 512 samples, 3 dB filter bandwidth: 124 Hz, and 50% overlap.

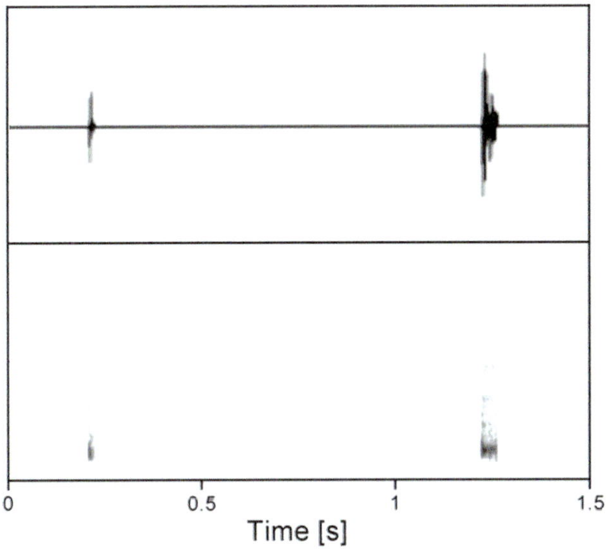

FIGURE A1.36 Oscillogram and spectrogram of the advertisement call of *Gastrotheca orophylax* (12.5°C). San Gabriel, Provincia Carchi, Ecuador. Taken from Sinsch and Juraske (2006).

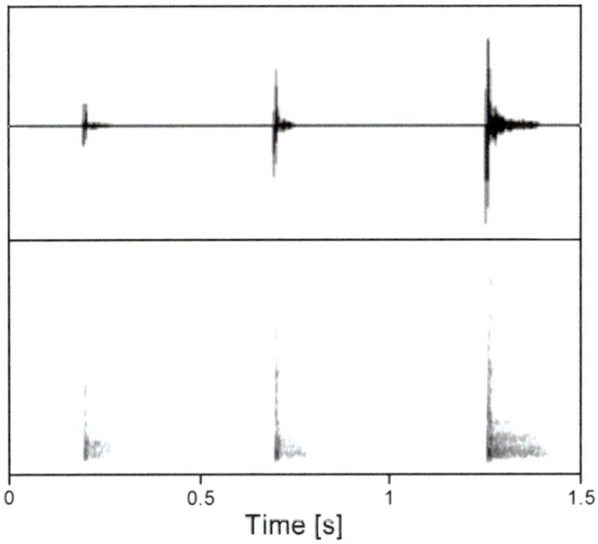

FIGURE A1.37 Oscillogram and spectrogram of the advertisement call of *Gastrotheca plumbea*. Pilaló, Provincia Cotopaxi, Ecuador. Taken from Sinsch and Juraske (2006).

Gastrotheca orophylax

Source: Sinsch and Juraske (2006). Figure A1.36.
Voucher specimen and locality: San Gabriel, Provincia Carchi, Ecuador.
Sample size: 42 calls.
Call type: Series of short pulse groups.
Temperature: Air – 11.5–12.8 °C.
Call duration: 2465 ± 419 ms.
Pulse groups per call: 3.7 ± 0.4.
Pulse group duration: 9.2 ± 0.5 ms.
Pulse group interval: 913 ± 34 ms.
Pulse group period: 921 ± 234 ms.
Pulse group ratio: 0.01 ± 0.0005.
Dominant frequency: 0.99 ± 0.1 kHz.
Remarks: Calls were analyzed using a MEDAV Spectro 2000 signal analysis system. Terminology for call characteristics follows Schneider and Sinsch (1992).

Gastrotheca plumbea

Source: Sinsch and Juraske (2006). Figure A1.37.
Voucher specimen and locality: Pilaló, Provincia Cotopaxi, Ecuador.
Sample size: 24 calls.
Call type: Series of short pulse groups. All pulse groups consisted of a single pulse.
Call duration: 1644 ± 248 ms.
Pulse groups per call: 5.6 ± 0.8.
Pulse group duration: 15 ± 0.8 ms.
Pulse group interval: 337 ± 20 ms.
Pulse group period: 352 ± 20 ms.
Pulse group ratio: 0.05 ± 0.003.
Dominant frequency: 0.97 ± 0.02 kHz.
Remarks: Calls were analyzed using a MEDAV Spectro 2000 signal analysis system. Terminology for call characteristics follows Schneider and Sinsch (1992).

Gastrotheca pseustes

Source: Carvajal-Endara *et al.* (2019). Figure A1.38.
Voucher specimen and locality: No vouchers: Bosque Washapamba, Cerro de Arcos, Provincia Loja, Ecuador; Via Urdaneta-Tutupali, Provincia Azuay, Ecuador.
Date and time: 01 April 2016, 20:02 h. 4 August 2016, 22:01–22:10 h. 9 July 2017, 20:26 h. 22 July 2017, 20:57 h.
Temperature: 6–10 °C.
Humidity: 48, 72.5, 86.5, 90, 93.5% RH.
Recording equipment: Sony MZ-NH1 recorder and a Sennheiser K6-ME66 microphone. An Olympus LS-10S recorder and a Sennheiser ME66 microphone. An Olympus LS-11 recorder and a RØDE NTG2 microphone. An Olympus LS-14 recorder and a Sennheiser ME66 microphone.
Sampling frequency: 44.1 or 48 kHz.
Resolution: 16 bits.
Sample size: Four individuals.
Call type: A complex call composed of one to five long, pulsed notes followed (or not) by one to six short, single-pulsed notes.
Long-note duration: $\bar{x}$ = 0.878 ± 0.188 s.
Pulses per note: $\bar{x}$ = 32.72 ± 6.867.
Short-note duration: $\bar{x}$ = 0.080 ± 0.046 s.
Interval between notes: $\bar{x}$ = 0.795 ± 0.502 s.
Dominant frequency: $\bar{x}$ = 1359.0 ± 37.981 Hz.
90% bandwidth: Range = 1043.2–1483.4 Hz.
Harmonics: 6–7.

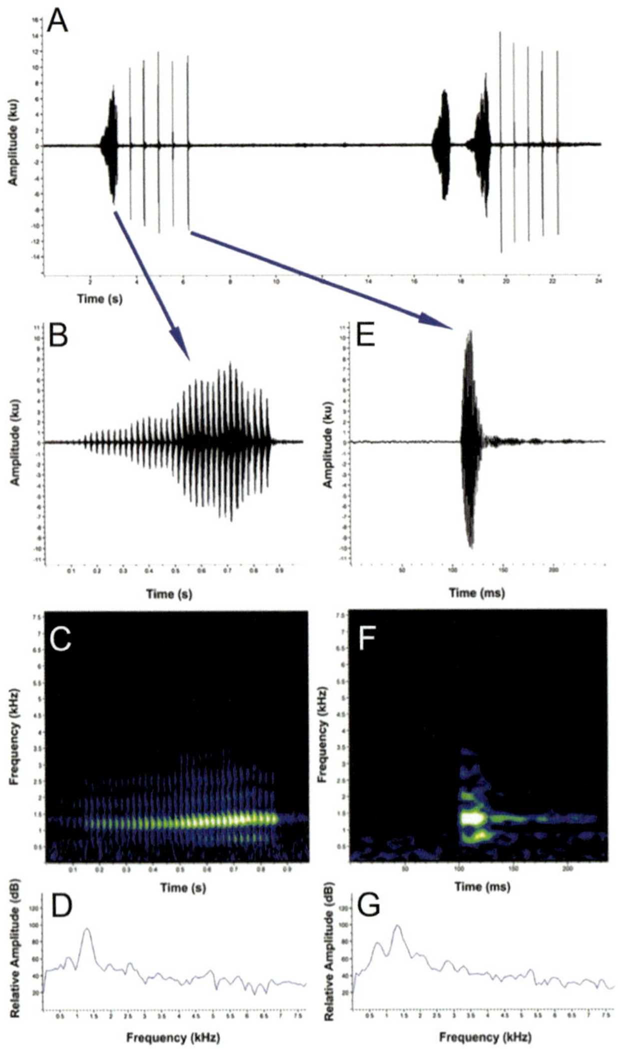

FIGURE A1.38 Calls of *Gastrotheca pseustes*. Bosque Washapamba, Cerro de Arcos, Provincia Loja, Ecuador; Road Urdaneta-Tutupali, Provincia Azuay, Ecuador. A: oscillogram of two consecutive calls, with the first call composed of one long and five short notes and the next call composed of two long and five short notes. B: oscillogram of a single long note. C: spectrogram of a single long note. D: power spectrum of a single long note. D: oscillogram of a single short note. F: spectrogram of a single short note. G: power spectrum of a single short note. Taken from Carvajal-Endara *et al.* (2019). Reproduced with permission from the copyright holder © Magnolia Press.

Remarks: Acoustic analysis was conducted using Raven Pro 1.4 software (http://www.birds.cornell.edu/raven). Temporal parameters (oscillograms) and spectral parameters (spectrograms) were measured and obtained through Hanning window function, DFT: 512 samples, 3 dB filter bandwidth: 124 Hz, and 50% overlap.

Gastrotheca riobambae

Source: Sinsch and Juraske (2006). Figure A1.39.

Voucher specimen and locality: No vouchers: Chambo, Riobamba, Provincia Chimborazo, Ecuador.

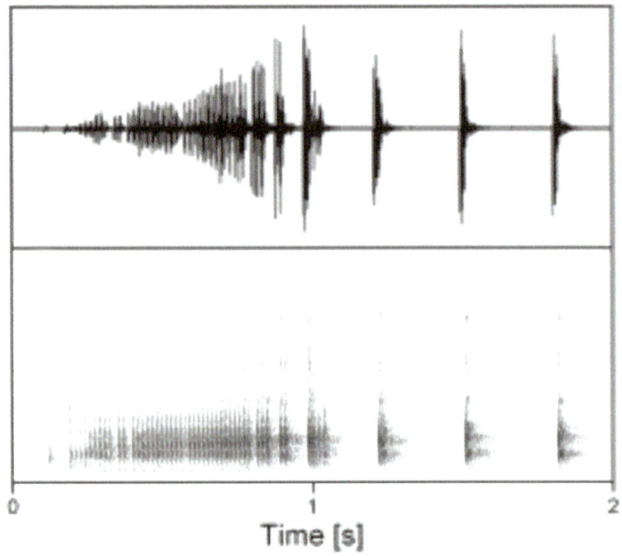

FIGURE A1.39 Oscillogram and spectrogram of the call of *Gastrotheca riobambae* followed by short pulse groups (20.0°C). Chambo, Riobamba, Provincia Chimborazo, Ecuador. Taken from Sinsch and Juraske (2006).

Sample size: 193 calls.

Call type: Long, pulsed calls (advertisement) and series of short pulse groups. Some of the calls were irregularly interrupted by the drop-out of pulses but still formed part of a single, long pulse group.

Call duration: 814 ± 16 ms.

Interval between calls: 2419 ± 90 ms.

Call period: 3232 ± 96 ms.

Call ratio: 0.36 ± 0.01.

Pulses per call: 48 ± 1.1.

Pulse rate: 61 ± 0.6 Hz.

Dominant frequency: 1.48 ± 0.02 kHz.

Remarks: Calls were analyzed using a MEDAV Spectro 2000 signal analysis system. Terminology for call characteristics follows Schneider and Sinsch (1992).

Gastrotheca testudinea

Source: Carvajal-Endara *et al.* (2019). Figure A1.40.

Voucher specimen and locality: No vouchers: Zuñag, Provincia Morona Santiago, Ecuador.

Date and time: 7 December 2012, 1:30 h.

Temperature: 10.8 °C.

Recording equipment: Sony MZ-NH1 recorder and a Sennheiser K6-ME66 microphone. An Olympus LS-10S recorder and a Sennheiser ME66 microphone. An Olympus LS-11 recorder and a RØDE NTG2 microphone. An Olympus LS-14 recorder and a Sennheiser ME66 microphone.

Sampling frequency: 44.1 or 48 kHz.

Resolution: 16 bits.

Sample size: Six individuals.

Call type: A complex call composed of one long, pulsed note followed (or not) by one to three short, single-pulsed notes.

Long-note duration: $\bar{x}$ = 0.746 ± 0.154 s.

Pulses per note: $\bar{x}$ = 18.88 ± 2.870.

Short-note duration: $\bar{x}$ = 0.080 ± 0.040 s.

Interval between notes: $\bar{x}$ = 0.452 ± 0.069 s.

Dominant frequency: $\bar{x}$ = 1069.1 ± 86.448 Hz.

90% bandwidth: Range = 800.5–1205.9 Hz.

Remarks: Acoustic analysis was conducted using Raven Pro 1.4 software (http://www.birds.cornell.edu/raven). Temporal parameters (oscillograms) and spectral parameters (spectrograms) were measured and obtained through Hanning window function, DFT: 512 samples, 3 dB filter bandwidth: 124 Hz, and 50% overlap.

Gastrotheca yacuri

Source: Carvajal-Endara *et al.* (2019). Figure A1.41.

Voucher specimen and locality: CJ 7822: El Salado, Provincia Loja, Ecuador.

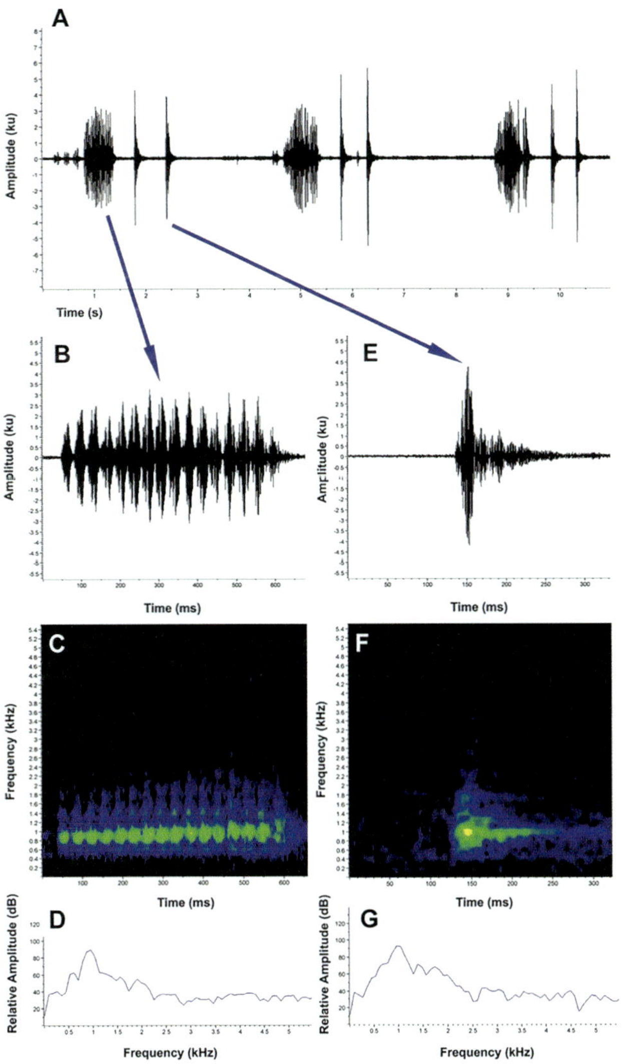

FIGURE A1.40 Calls of *Gastrotheca testudinea*. Zuñag, Provincia Morona Santiago, Ecuador. A: oscillogram of three consecutive calls, each of them composed of one long note and two short notes. B: oscillogram of a single long note. C: spectrogram of a single long note. D: power spectrum of a single long note. D: oscillogram of a single short note. F: spectrogram of a single short note. G: power spectrum of a single short note. Taken from Carvajal-Endara *et al.* (2019). Reproduced with permission from the copyright holder © Magnolia Press.

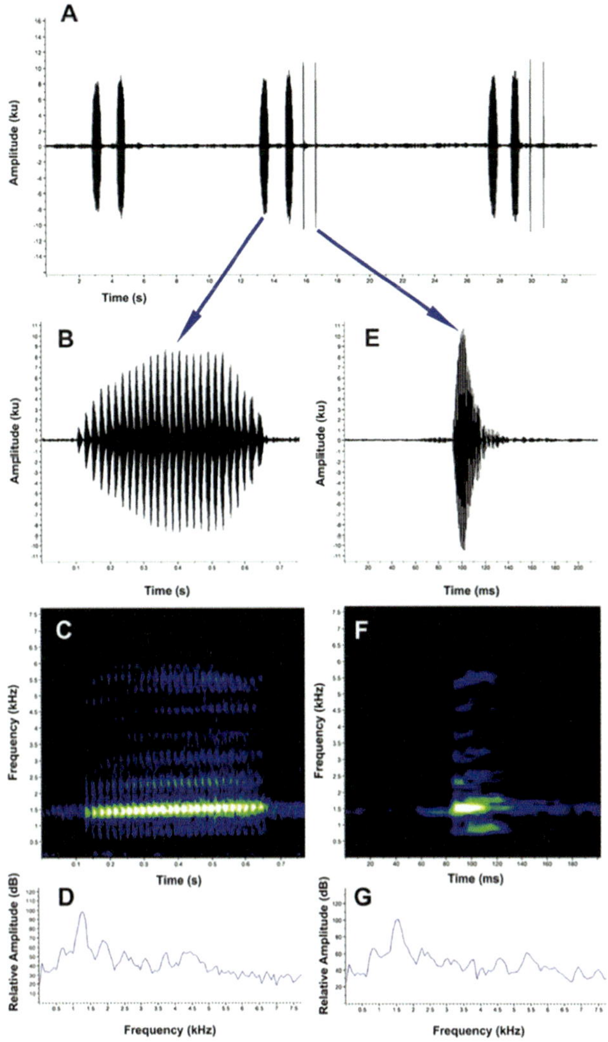

FIGURE A1.41 Calls of *Gastrotheca yacuri*. El Salado, Provincia Loja, Ecuador. A: oscillogram of three consecutive calls, with the first call composed of two long notes and the next two calls composed of two long and two short notes. B: oscillogram of a single long note. C: spectrogram of a single long note. D: power spectrum of a single long note. D: oscillogram of a single short note. F: spectrogram of a single short note. G: power spectrum of a single short note. Taken from Carvajal-Endara *et al.* (2019). Reproduced with permission from the copyright holder © Magnolia Press.

Date and time: 9 April 2016, 01:48 h. 29 July 2016, 21:50 h. 30 July 2016, 21:40 h.
Temperature: 8.5–10.5 °C.
Humidity: 93, 96.5% RH.
Recording equipment: Sony MZ-NH1 recorder and a Sennheiser K6-ME66 microphone. An Olympus LS-10S recorder and a Sennheiser ME66 microphone. An Olympus LS-11 recorder and a RØDE NTG2 microphone. An Olympus LS-14 recorder and a Sennheiser ME66 microphone.
Sampling frequency: 44.1 or 48 kHz.
Resolution: 16 bits.
Sample size: Four individuals.
Call type: A complex call, composed of one to three (usually two) long pulsed notes and followed (or not) by one to three (usually two) short, single-pulsed notes.
Long-note duration: $\bar{x} = 0.634 \pm 0.091$ s.
Pulses per note: $\bar{x} = 32.24 \pm 7.504$.
Short-note duration: $\bar{x} = 0.072 \pm 0.020$ s.
Interval between notes: $\bar{x} = 0.549 \pm 0.214$ s.
Dominant frequency: $\bar{x} = 1496.7 \pm 38.643$ Hz.
90% bandwidth: Range = 1366.7–1637.9 Hz.
Harmonics: 5–7.
Remarks: Acoustic analysis was conducted using Raven Pro 1.4 software (http://www.birds.cornell.edu/raven). Temporal parameters (oscillograms) and spectral parameters (spectrograms) were measured and obtained through Hanning window function, DFT: 512 samples, 3 dB filter bandwidth: 124 Hz, and 50% overlap.

FAMILY ELEUTHERODACTYLIDAE

Adelophryne adiastola

Source: Heyer (1977; as *Phyzelaphryne miriamae*). Figure A1.42.
Voucher specimen and locality: A4940–4943 (one of these specimens): Yapima, Departamento Vaupés, Colombia. University of Texas at Arlington.
Date and time: 21 April 1976.
Temperature: Air = 22.0 °C.
Sample size: Probably one male.
Notes per call: 1–10 notes.
Note duration: Note diphasic. First phase: ~0.009 s. Second phase: ~0.014 s.
Interval between phases of note: ~0.001 s.
Fundamental frequency: First phase = 1600 Hz; second phase = 1650 Hz.
Dominant frequency: Range = 3200–3700 Hz.

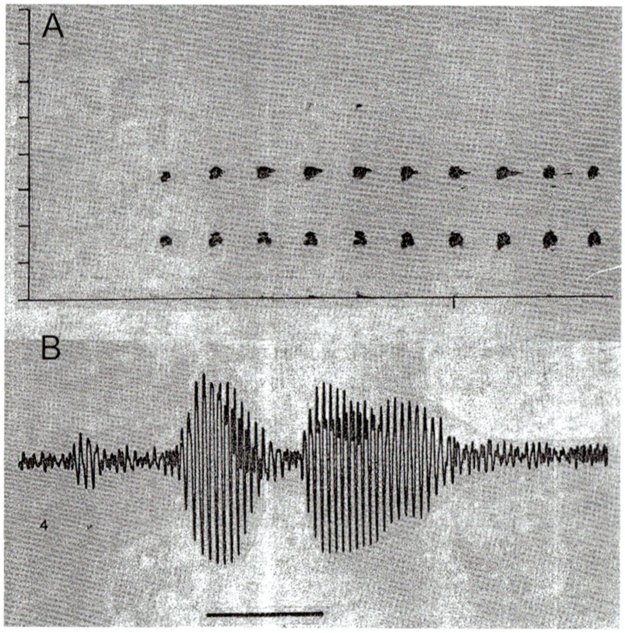

FIGURE A1.42 Sonogram and oscillogram of mating call of *Adelophryne adiastola*. Yapima, Departamento Vaupés, Colombia. A: vertical scale marks at 1 kHz intervals. Horizontal scale mark indicates 1 second. B: line indicates 0.01 second. Taken from Heyer (1977).

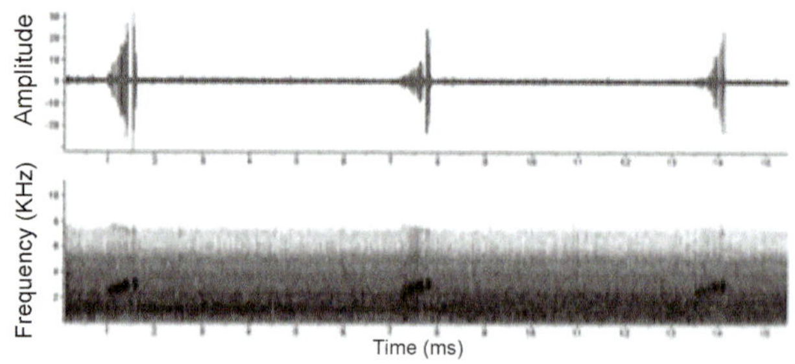

FIGURE A1.43 Call of *Centrolene ballux*. Reserva Las Gralarias, Provincia Pichincha, Ecuador. Taken from Guayasamin *et al.* (2020b).

FAMILY CENTROLENIDAE

Centrolene ballux

Source: Márquez *et al.* (1996). Figure A1.43 taken from Guayasamin *et al.* (2020b).
Voucher specimen and locality: No voucher: Las Palmeras, Reserva Las Gralarias, Provincia Pichincha, Ecuador.
Call type: Sporadically call. Single short note.
Sample size: Two individuals.
Call duration: 0.3285–0.4204 s (n = 2).
Pulses per note/call: 7–9 pulses, the first 4–5 repeated at short, regular intervals. The last three pulses are emitted at longer intervals. Mean pulses per second = 21.0 ± 0.1, range 21.3–21.4.
Dominant frequency: $\bar{x} = 4833 \pm 14$ Hz.
Remarks: The notes are frequency modulated (dominant frequency at the beginning of note is lower than at its end). Recordings obtained at Reserva Las Gralarias, Ecuador, match this description.

Centrolene buckleyi

Source: Guayasamin *et al.* (2020b). Figure A1.44.
Voucher specimen and locality: MZUTI 763: Oyacachi, Provincia Napo, Ecuador.
Call type: Calls are produced in series. Single, highly pulsed note.
Sample size: 10 individuals.
Call rate: $\bar{x}$ = 53 ± 9.9; range = 45–67 (n = 10).
Call duration: $\bar{x}$ = 0.264 ± 13.7; range = 0.248–0.288 s (n = 10).
Dominant frequency: Initial dominant frequency $\bar{x}$ = 2893 ± 38 Hz (n = 10). Final dominant frequency $\bar{x}$ = 3229 ± 12 (n = 10).
Harmonics: 5385–6393 Hz.
Remarks: Two descriptions of the call of *Centrolene buckleyi* are available in the literature (Guayasamin *et al.*, 2006b; Bolívar-García *et al.*, 1999). The differences between these calls, combined with the non-monophyly of *C. buckleyi*, clearly show that *Centrolene buckleyi*, as currently recognized, represents a species complex. Although the call presented herein is not from the neotype locality, genetically, it clusters together with samples from the type locality; thus, they correspond to the same species.

Centrolene condor

Source: Almendáriz and Batallas (2012). Figure A1.45.
Voucher specimen and locality: EPN 11343: 28.6 mm SVL, Paquisha Alto, Provincia Zamora Chinchipe.
Date: 26 March 2008.
Temperature: 8 °C.
Sample size: One individual, three calls.
Recording equipment: Panasonic RR-US450 digital recorder. Sennheiser K6-C unidirectional microphone.
Call duration: $\bar{x}$ = 668 ± 27.8; range = 650–700 ms.
Interval between calls: $\bar{x}$ = 5590 ± 11; range = 5579–5600 ms.
Notes per call: One.
Interval between notes: $\bar{x}$ = 85 ± 28.7; range = 66–118 ms.
Note duration:
First note: $\bar{x}$ = 468 ± 38; range = 435–510 ms.
Second note: $\bar{x}$ = 101 ± 1; range = 100–102 ms.
Pulses per note/call:
First note: $\bar{x}$ = 11 ± 1; range = 10–12 pulses per note.
Second note: Two pulses.
Pulse duration: $\bar{x}$ = 15 ± 3.2; range = 9–24 ms.
Interval between pulses: $\bar{x}$ = 27 ± 3.8; range = 19–34 ms.
Dominant frequency: $\bar{x}$ = 2.78; range = 2.62–2.97 kHz.
Remarks: The recording was made one and a half meters away from the specimen; a second specimen was located five meters away, and an additional four individuals were heard in a diameter of 60 meters. The time values were calculated with the timer of the Adobe Audition 3.0 program in milliseconds and the dominant frequency with the

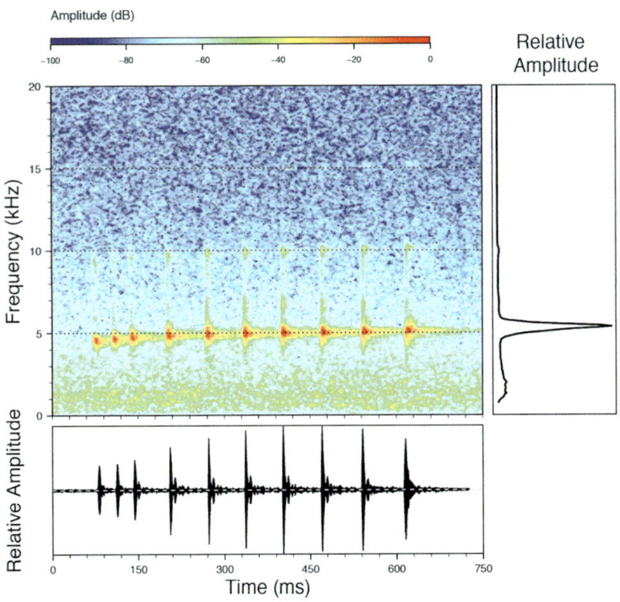

FIGURE A1.44 Call of *Centrolene buckleyi*. Paramo wetland ca. Oyacachi, 3012 m, Provincia Napo, Ecuador. Taken from Guayasamin *et al.* (2020b).

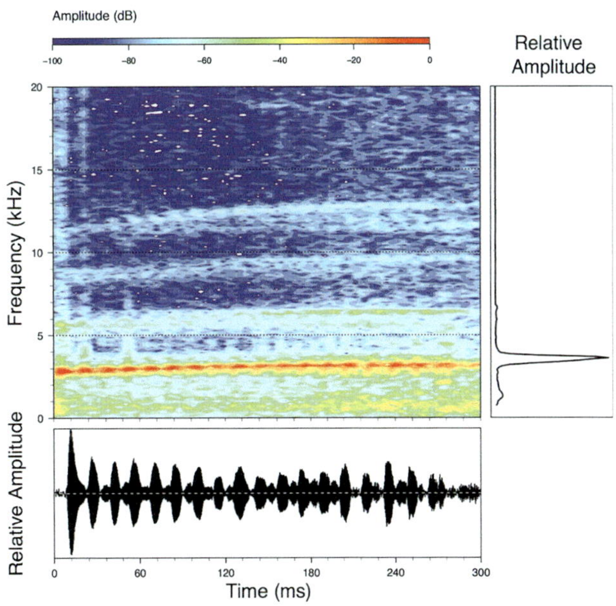

FIGURE A1.45 Oscillogram and spectrogram of the call of *Centrolene condor*. Paquisha Alto, Provincia Zamora Chinchipe, Ecuador. Taken from Almendáriz and Batallas (2012).

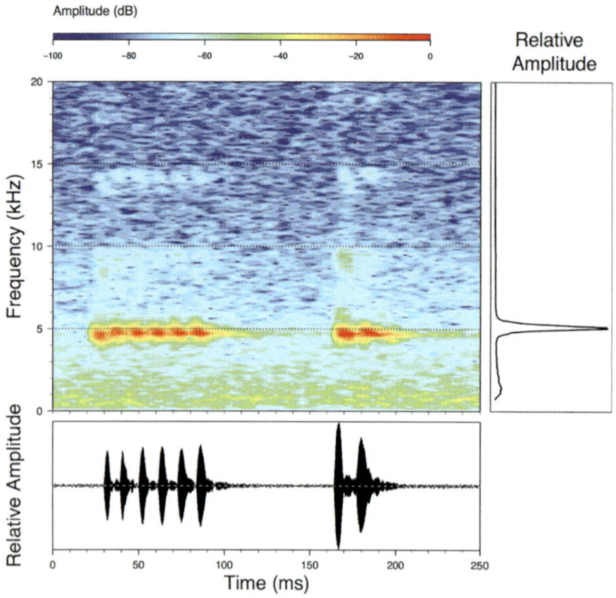

FIGURE A1.46 Call of *Centrolene heloderma*. Reserva Las Gralarias, Provincia Pichincha, Ecuador. Note that each call has two distinctive pulsed notes. Taken from Guayasamin *et al.* (2020b).

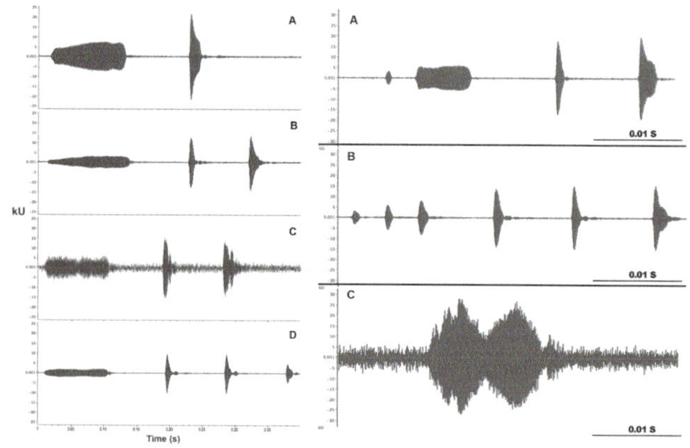

FIGURE A1.47 Calls of *Centrolene lynchi*. Reserva Las Gralarias, Provincia Pichincha, Ecuador. Left: oscillograms, variation in the advertisement call. A: call with a tonal note followed by a single, one-pulse note. B: call with a tonal note followed by two notes with one pulse each. C: call with a tonal note and two notes composed of two pulses each. D: call with a tonal note followed by three notes composed of one pulse each. Right: additional calls. A: call with the pre-tonal (left most) note. B: call that lacks a tonal note and consists of only six peaked notes. C: combat call produced during male–male, upside-down hanging grappling combat. Taken from Dautel *et al.* (2011).

Fourier calculation was given in kilohertz. Raven 1.4 software was used to plot the oscillogram and spectrogram.

Centrolene heloderma

Source: Guayasamin *et al.* (2020b). Figure A1.46.
Voucher specimen and locality: Río Alambí, 2390 m, and Reserva Las Gralarias, 1822–1864 m, Provincia Pichincha, Ecuador. LBE-C-001, 012, 015.
Call type: The typical advertisement call is relatively short.
Sample size: 58 notes contained within 29 calls from three individuals.
Call duration: $\bar{x}$ = 161 ms ± 15.4; range = 133–188 ms.
Notes per call: Two.
Pulses per note/call: $\bar{x}$ = 3.741 ± 1.906; range = 2–7.
Note duration:
First note: $\bar{x}$ = 60.2 ms ± 9.89; range = 38–76 ms.
Second note: $\bar{x}$ = 27.8 ms ± 6.283; range = 17–44ms.
Interval between notes: $\bar{x}$ = 72.9 ms ± 9.5; range = 50–99 ms.
Pulses within a note: $\bar{x}$ = 83 ± 15; range 45.5–117.6 pulses per second.
Relative peak time: $\bar{x}$ = 0.273 ± 0.242; range = 0.032–0.8697 (%).
Fundamental frequency:
Lower limit: $\bar{x}$ = 4513 ± 138; range = 4221–4651 Hz.
Higher limit: $\bar{x}$ = 4875 ± 104; range = 4651–5082 Hz.
Dominant frequency: $\bar{x}$ = 4682 ± 104; range = 4393–4823 Hz.

Centrolene lynchi

Source: Dautel *et al.* (2011). Figure A1.47.
Voucher specimen and locality: QCAZ 40191–2, 40194: Reserva Las Gralarias, Provincia Pichincha, Ecuador.
Call type: The typical advertisement call is relatively short and consists of a tonal note followed by 1–3 peaked notes.
Combat call:
Date: 5 March 2010.
Temperature: 20.7 °C.
Humidity: 100%.
Recording equipment: Sony HDR-XR520V video recorder.
Sample size: Eight males, five calls.
Call duration: $\bar{x}$ = 150.4 ± 6.7 ms.
Notes per call: One.
Pulses peaked note: Two.
Peaked frequency: $\bar{x}$ = 4892.3 ± 38.5; range = 4823.4–4909.6 Hz.
Remarks: To extract the sound from the video, the authors employed iMovie 8.0.5. This file was extracted in a reduced-quality AIFF format. They analyzed the sound clip using Raven 1.2.1 software for Mac OSX (Cornell Lab of Ornithology).

Advertisement call:

Date: 27 February and 13–14 April 2010.

Temperature: 19.5 °C, 19.9 °C, and 20.6 °C, respectively.

Humidity: 100%.

Recording equipment: Olympus linear PCM recorder LS-10 and microphone.

Sample size: Seven males, 48 calls.

Call duration: 293.5 ± 27.5 ms.

Note duration:

Tonal note: 100.4 ± 20.2 ms.

Pulsed note: 18.5 ± 4.7 ms.

Interval between notes: 75.6 ± 9.5 ms.

Notes per call: Range = 1–3 ± 2.

Pulses per note: Range = 1–3 ± 2.

Dominant frequency:

Tonal note: 5296.4 ± 58.1 Hz.

Pulsed note: 5264.4 ± 72.5 Hz.

Remarks: Individuals were recorded for 5 min at a distance of 60–400 cm, and the audiospectogram and oscillogram for each 5 min segment was produced and analyzed using Raven 1.2.1 software for MacOSX (Cornell Lab of Ornithology). The typical advertisement call is relatively short and consists of a tonal note followed by 1–3 peaked notes. The tonal notes usually is the first note of the call and is defined by flat, non-peaked amplitude sustained throughout the duration of the note. Peaked notes usually are the second through last notes in each call. Peaked notes are characterized by a clear amplitude peak and often are pulsed.

Centrolene peristicta

Source: Salgado and Guayasamin (2018). Figure A1.48 taken from Guayasamin *et al.* (2020b).

Voucher specimen and locality: QCAZ 40189–90, MZUTI 419: Reserva Las Gralarias, Provincia Pichincha, Ecuador.

Date: 15 December 2009–16 April 2010.

Temperature: $\bar{x}$ = 19.2 ± 1.09; range = 16.2–21.3 °C (n = 76).

Recording equipment: Olympus LS-10 digital recorder (Olympus Imaging America Inc. USA). Sennheiser K6-ME67 directional microphone (Sennheiser Germany).

Call type: Advertisement call consists of one short and pulsed note that resembles the sound of a cricket chirp.

Sample size: 28 individuals.

Call rate: Two to 20 calls per minute.

Note duration: $\bar{x}$ = 0.036 ± 0.087 s; range = 0.062 + 1.07 (n = 35).

Fundamental and dominant frequency: $\bar{x}$ = 6877.9 ± 177.2; range = 6470.7–7278.2 Hz (n = 35).

Harmonics: One (3235.4–6198.4 Hz) or two (6470.7– 7278.2 Hz).

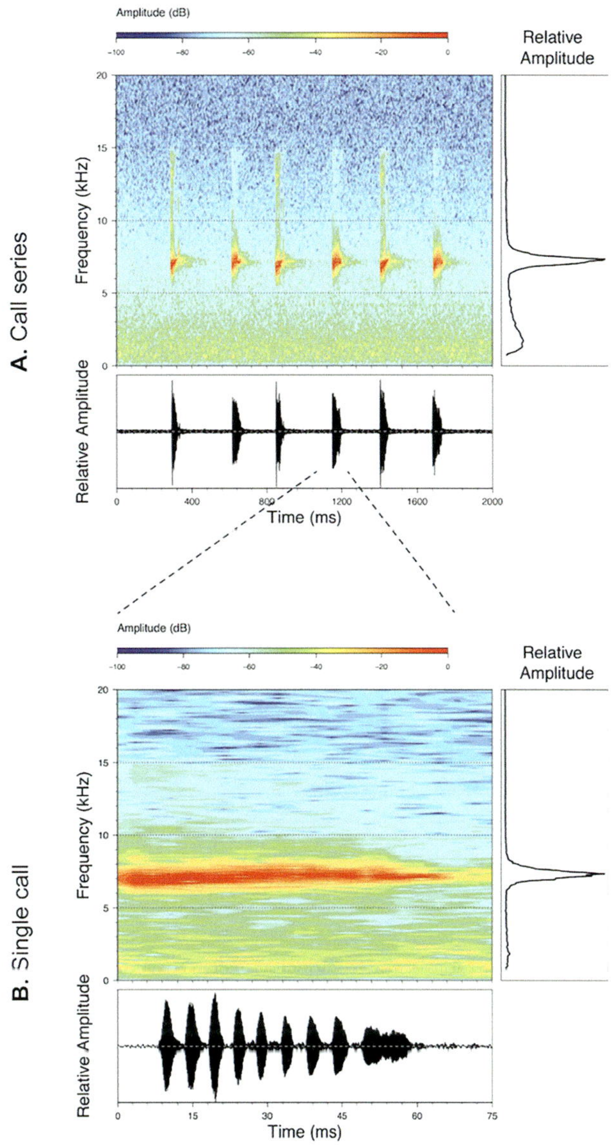

FIGURE A1.48 Call of *Centrolene peristicta*. Reserva Las Gralarias, 1800 m, Provincia Pichincha, Ecuador. A: series of calls. B: single, pulsed call. Taken from Guayasamin *et al.* (2020b).

Remarks: Raven Pro software for Mac OS X Version 1.2 (Bioacoustics Research Program, 2008) was used to analyze 10 randomly chosen advertisement calls per recorded individual. The spectrogram had a 4096 fast Fourier Transform (FFT) and a frequency resolution of 10.8 Hz. The number of pulses per note in each call was measured using SoundRuler software, Version 0.9.6.1. No frequency modulation was observed.

Centrolene sanchezi

Source: Guayasamin *et al.* (2020b). Figure A1.49.

Voucher specimen and locality: Reserva Yanayacu, Provincia Napo, Ecuador. LBE-C-023.

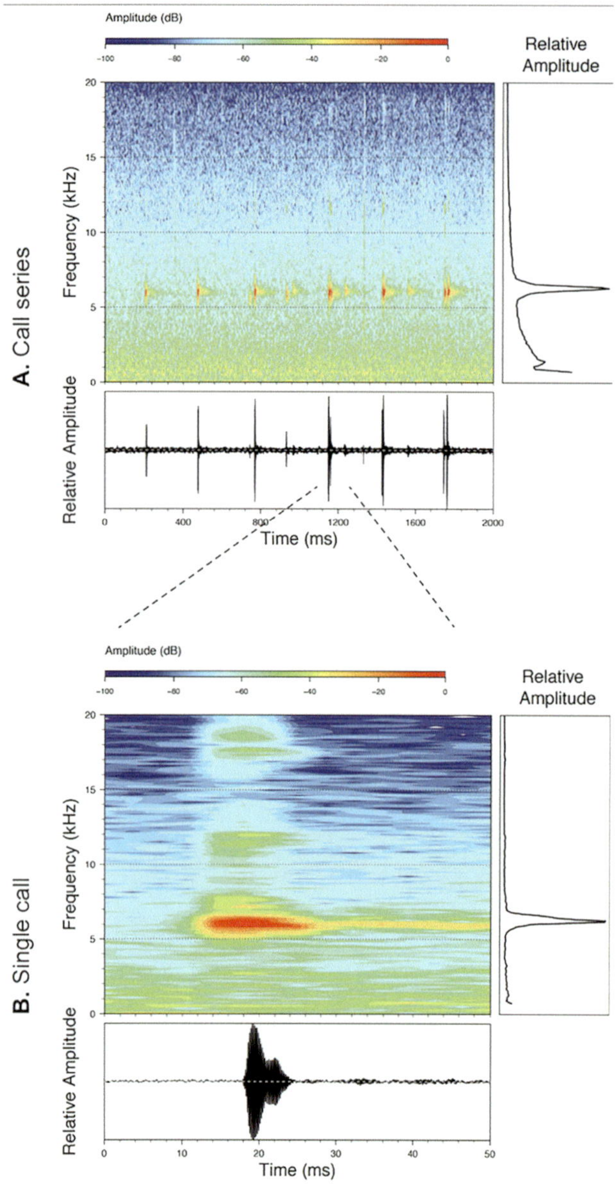

FIGURE A1.49 Call of *Centrolene sanchezi*. Reserva Yanayacu, 2150 m, Provincia Napo, Ecuador. A: series of calls. B: single, pulsed call. Taken form Guayasamin *et al.* (2020b).

Date: February 2013.

Sample size: 63 notes contained within 11 calls from one individual.

Call duration: $\bar{x}$ = 1462.6 ± 639.5; range = 618–3085 ms.

Notes per call: One or two.

Note rate: $\bar{x}$ = 5.7 ± 1.6; range = 3–9 notes per call.

Note duration: $\bar{x}$ = 11 ± 2.8; range = 6–21 ms.

Pulses per note: 1–2 amplitude peaks; $\bar{x}$ = 1.4 ± 0.5.

Pulse rate: $\bar{x}$ = 122 ± 29; range = 83–200 pulses per second.

Relative peak time: $\bar{x}$ = 0.216 ± 0.076; range = 0.0987–0.5497 (%).

Fundamental frequency:

Lower limit: $\bar{x}$ = 5714 ± 222; range = 4688–6000 Hz.

Higher limit: $\bar{x}$ = 6268 ± 144; range = 5906–6656 Hz.

Dominant frequency: $\bar{x}$ = 5996 ± 131; range = 5719–6188 Hz.

Remarks: Notes sometimes occur in pairs, and each call is very short. Notes have their peak amplitude in the first 50% of the note.

Chimerella mariaelenae 1

Source: Batallas and Brito (2016). Figure A1.50.

Voucher specimen and locality: MEPN 14514–15, 14519: Laguna Cormorán, 1750 m, Sardinayacu lake complex, Sangay National Park, Provincia Morona Santiago, Ecuador.

Date: December 2010–February 2011.

Recording equipment: Olympus WS-802 digital recorder connected to a Sennheiser K6-C modular system. Sennheiser head microphone (ME 66).

Sampling frequency: 44.1 kHz.

Resolution: 16 bits.

Sample size: Three individuals, 10 calls.

Call behavior: Vocalizations are very similar to trills, emitted from bush and tree vegetation.

Call type: Advertisement. Features frequency-modulated calls.

Call duration: $\bar{x}$ = 751.70 ± 44.98; range = 668–808 ms.

Call rate: 2–3 calls per minute.

Interval between calls: $\bar{x}$ = 32714.67 ± 4538.85; range = 27914–42940 ms.

Notes per call: Three.

Note duration: $\bar{x}$ = 71.80 ± 15.10; range = 54–116 ms.

Interval between notes: $\bar{x}$ = 257.70 ± 39.44; range = 191–325 ms.

Dominant frequency: $\bar{x}$ = 6.41 ± 0.17; range = 6.15–6.63 kHz.

Remarks: The Adobe Audition CS6 software was used to carry out the digitization, noise filters, and temporal analyses; in turn, the Raven 1.4 software was used for the diagrams and spectral analysis, using Hann's window at 80% overlap and 512 points of fast Fourier transform (FFT) resolution.

Chimerella mariaelenae 2

Source: Guayasamin *et al.* (2020b). Figure A1.50.

Voucher specimen and locality: Nearby Cascada San Rafael in Quebrada Pangayacu, Provincia Napo, Ecuador. LBE-C-021.

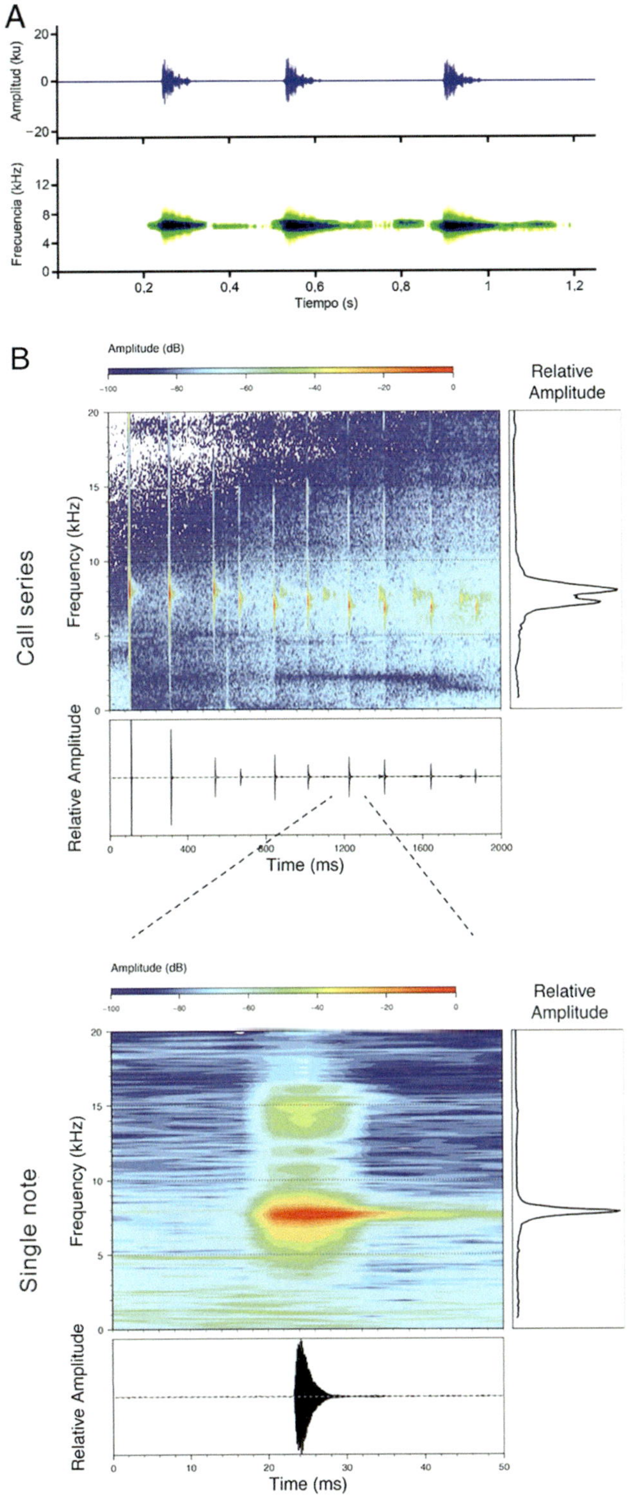

FIGURE A1.50 Call of *Chimaerella mariaelenae*. A: Laguna Cormorán, Sardinayacu, Parque Nacional Sangay, Provincia Morona Santiago, Ecuador. Taken from Batallas and Brito (2016). B: Pangayaku creek, 929 m, Provincia Napo, Ecuador. Top: series of calls. Bottom: single note. Taken from Guayasamin *et al.* (2020b).

Date: August 2012.
Sample size: 26 notes contained within six calls from one individual.
Call duration: $\bar{x}$ = 679 ± 623.3; range = 231–1761 ms.
Note rate: $\bar{x}$ = 4.3 ± 3.4; range = 2–10 notes.
Note duration: $\bar{x}$ = 6 ± 0.9; range = 4–7 ms.
Fundamental frequency:
Lower limit: $\bar{x}$ = 7222 ± 387; range = 6460–7752 Hz.
Higher limit: $\bar{x}$ = 7828 ± 398; range = 7063–8441 Hz.
Dominant frequency: $\bar{x}$ = 7510 ± 408; range = 6718–8010 Hz.
Remarks: The advertisement call is variable from being short to moderate in length. The typical call has two notes, but some calls can have up to 10.

Cochranella granulosa

Source: Kubicki (2007).
Voucher specimen and locality: Finca Los Diamantes, Guápiles, Provincia Limón, Costa Rica.
Call duration: The mean duration for a three-note call is 0.8–0.9 s.
Notes per call: Range 3–5 notes.
Note duration: Usually, the first note is longer (average duration = 0.3 s) than the subsequent notes (average duration = 0.15 s).
Interval between notes: 0.15 s.
Pulse duration: 0.003 s.
Interval between pulses: 0.02 s.
Fundamental frequency: 3700–4500 Hz (Guayasamin *et al.*, 2020b).
Remarks: The call has been described from Costa Rica and Panama by Kubicki (2007), Savage (1967), and Ibañez *et al.* (1999b). The call consists of a rapid, high-pitched pulsed trill, usually composed of three notes: "rit-rit-rit." Males often call in a chorus, with one male initiating the call and then being answered by several males.

Cochranella mache

Source: Ortega-Andrade *et al.* (2013).
Voucher specimen and locality: No voucher: Rompe frente rivulet, Reserva Biológica Bilsa, Provincia Esmeraldas, Ecuador.
Date and time: 9 May 2008, 22:54 h.
Sample size: One male.
Notes per call: The call consists of two pulses.
Call rate: 1.46 calls per minute.
Interval between calls: ~40.8 s.
Note duration: $\bar{x}$ = 0.038 ± 0.008; range = 0.029–0.049 s.
Interval between notes: Range = 0.107–0.130 s.
Fundamental frequency: Range = 5139.5–6058.8 Hz.
Dominant frequency: $\bar{x}$ = 5410.2 ± 17.9; range = 5383–5426 Hz.

Remarks: The recorded male was calling from the upper side of a dead *Heliconia* leaf, about 2.5 m above the ground, horizontally separated from the stream by about 3 m. According to Jaime Culebras (pers. comm.), this call provided by Ortega Ortega-Andrade *et al.* (2013) was erroneously assigned to *C. mache*.

Espadarana audax

Source: Guayasamin *et al.* (2020b). Figure A1.51.
Voucher specimen and locality: MZUTI 1492–93: Río Hollín, Provincia Napo, Ecuador.
Sample size: 14 notes from four calls from two individuals.

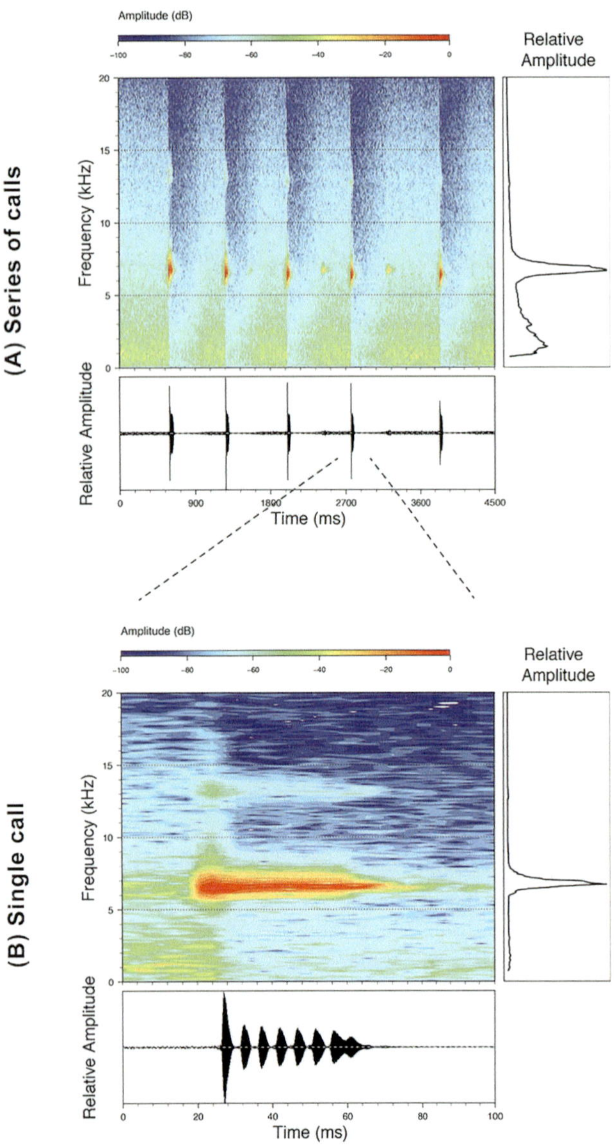

FIGURE A1.51 Calls of *Espadarana audax*. Río Hollín, 1038 m, Provincia Napo, Ecuador. A: call series. B: single call. Taken from Guayasamin *et al.* (2020b).

Call rate: $\bar{x}$ = 4.3 ± 0.6; range = 4–5 calls per call series.
Call duration: $\bar{x}$ = 2711 ms ± 497.8; range = 2398–3285 ms.
Notes per call: One.
Note duration: $\bar{x}$ = 38 ± 8.2; range = 26–53 ms.
Pulses per note: $\bar{x}$ = 7.4 ± 1.7; range = 5–11 pulse-notes/call.
Pulse rate: $\bar{x}$ = 189 ± 10; range = 171–208 pulses per second.
Relative peak time: $\bar{x}$ = 0.043 ± 0.02; range = 0.0144–0.0722 (%).
Fundamental frequency:
Lower limit: $\bar{x}$ = 5925 ± 374; range = 5254–6288 Hz.
Higher limit: $\bar{x}$ = 6608 ± 297; range = 6115–7063 Hz.
Dominant frequency: $\bar{x}$ = 6146 ± 368; range = 5426–6718 Hz.
Remarks: Notes have their peak amplitude in the first 50% of the note, where the first note usually has the highest amplitude.

Espadarana callistomma

Source: Guayasamin *et al.* (2020b). Figure A1.52.
Voucher specimen and locality: Reserva Otokiki, Provincia Esmeraldas, Ecuador. LBE-C-022.
Date: 17 May 2010.
Call duration: $\bar{x}$ = 0.38 ± 0.066; range = 0.28–0.44 s (n = 10).
Notes per call: $\bar{x}$ = 3.7 ± 0.483; range = 3–4 pulsed notes (n = 10).

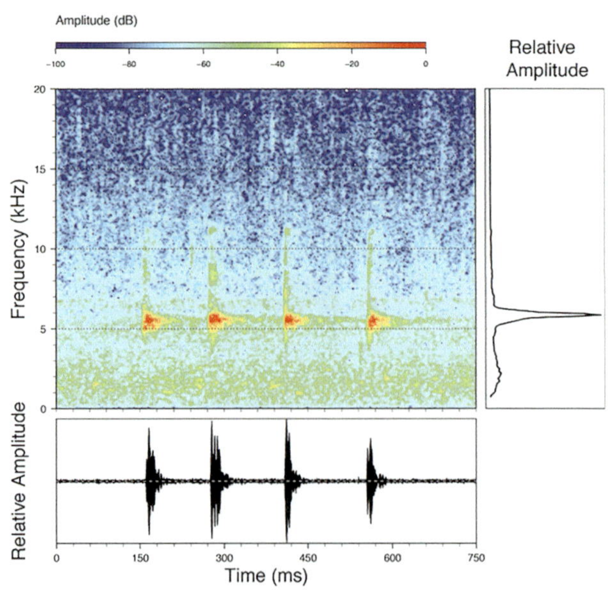

FIGURE A1.52 Call of *Espadarana callistomma*. Reserva Otokiki, 706 m, Provincia Esmeraldas, Ecuador. Taken from Guayasamin *et al.* (2020b).

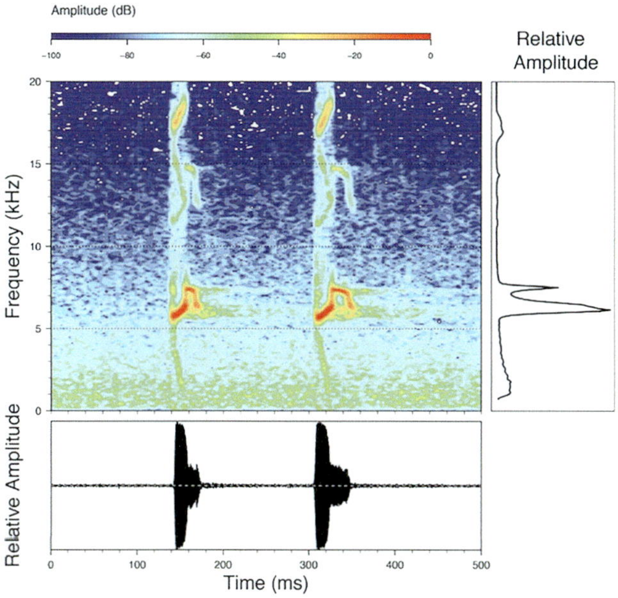

FIGURE A1.53 Call of *Espadarana prosoblepon* from Cordillera de Chontilla, 908 m, Provincia Pichincha, Ecuador. Taken from Guayasamin *et al.* (2020b).

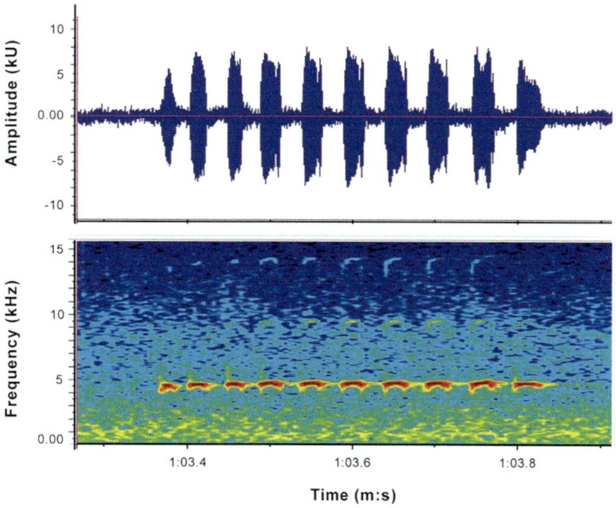

FIGURE A1.54 Call of the holotype of *Hyalinobatrachium adespinosai*. San Jacinto River, 1795 m, Provincia Tungurahua, Ecuador). Air temperature = 18°C. Taken from Guayasamin *et al.* (2020b).

Note rate: $\bar{x}$ = 9.76 ± 0.596; range = 9.1–11.0 notes per second (n = 10).

Note duration: $\bar{x}$ = 0.03 ± 0.007; range = 0.014–0.04 s (n = 10).

Dominant frequency: 5343–5812 Hz (n = 10).

Remarks: The call is not frequently modulated.

Espadarana prosoblepon

Source: Guayasamin *et al.* (2020b). Figure A1.53.

Voucher specimen and locality: MZUTI 601: Río Sune Chico (908 m), Cordillera Chontilla, Pachijal-Mashpi, Provincia Pichincha, Ecuador. Recorded by Ítalo G. Tapia.

Date: 4 July 2012.

Temperature: 18.8 °C.

Interval between calls: 4–7 min.

Call duration: $\bar{x}$ = 0.212 ± 0.007; range = 0.207–0.223 s (n = 4).

Notes per call: Two pulsed notes.

Note duration: $\bar{x}$ = 0.040 ± 0.008; range = 0.032–0.054 s (n = 8).

Interval between notes: $\bar{x}$ = 0.134 ± 0.004; range = 0.130–0.137 s (n = 4).

Dominant frequency:

At the beginning of each note: $\bar{x}$ = 5759 ± 56.1; range = 5712–5825 Hz (n = 4).

At the end of the note: $\bar{x}$ = 6252 ± 95.7; range = 6196–6394 Hz (n = 4).

Harmonics: Frequency modulated, showing complex spectral patterns.

Remarks: The call of *Espadarana prosoblepon* from Costa Rica described by Jacobson (1985) and Kubicki (2007) shows some differences from the Ecuadorian call described here (see details in Guayasamin *et al.* (2020b).

Hyalinobatrachium adespinosai

Source: Guayasamin *et al.* (2020b). Figure A1.54.

Voucher specimen and locality: ZSFQ 1648: San Jacinto River, Provincia Tungurahua, Ecuador.

Temperature: 18 °C.

Sample size: Nine individuals.

Call duration: $\bar{x}$ = 0.38 ± 0.017; range = 0.38–0.44 s.

Interval between calls: $\bar{x}$ = 4.58 ± 2.3; range = 2.0–11.0 s.

Fundamental and dominant frequency: $\bar{x}$ = 4855 ± 152; range = 4645–5203 Hz.

Harmonics: First harmonic: 9336–9754 Hz; second harmonic: 14159–1444 Hz.

Remarks: The call has a striking resemblance to the chirp of a cricket. Each call is composed of a single and high-pitched pulsed note. No frequency modulation.

Hyalinobatrachium chirripoi

Source: Guayasamin *et al.* (2020b). Figure A1.55.

Voucher specimen and locality: Reserva Itapoa, Provincia Esmeraldas, Ecuador. LBE-C-019.

Call behavior: Most males call from the undersides of leaves, but some have been observed calling from the upper sides.

Call type: Advertisement call composed of a single note. Notes are strongly pulsed. The call is a high-pitched, insect-like buzz.

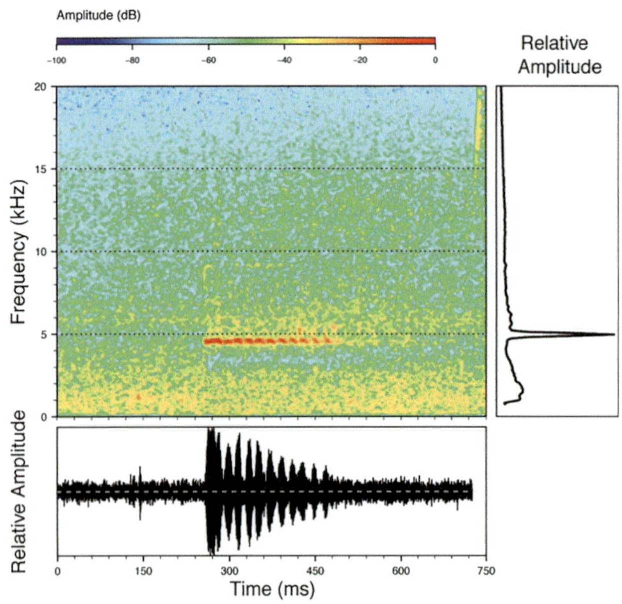

FIGURE A1.55 Call of *Hyalinobatrachium chirripoi*. Reserva Itapoa, 321 m, Provincia Esmeraldas, Ecuador. Taken from Guayasamin *et al.* (2020b).

Sample size: One individual, two notes.
Note duration: $\bar{x}$ = 250 ± 28.3; range = 230–270 ms.
Pulses per note: $\bar{x}$ = 15 ± 1.4; range = 14–16.
Note relative peak time: $\bar{x}$ = 0.063 ± 0.001; range = 0.0623–0.0638 (%).
Pulse rate: $\bar{x}$ = 60 ± 1; range = 59–61 pulses per second.
Fundamental frequency:
Lower limit: $\bar{x}$ = 2584 ± 974; range = 1895–3273 Hz.
Higher limit: $\bar{x}$ = 4866 ± 305; range = 4651–5082 Hz.
Dominant frequency: $\bar{x}$ = 4565 ± 0; range = 4565 Hz.

Hyalinobatrachium mashpi

Source: Guayasamin *et al.* (2022). Figure A1.56.
Voucher specimen and locality: CJ 11642, Reserva Mashpi, Provincia Pichincha, Ecuador. LBE-C-051. CJ 11645, Reserva Tayra, Provincia Pichincha, Ecuador. LBE-C-052.
Recording equipment: Tascam DR-05.
Call type: Advertisement call. Single, pulsed (amplitude modulated) note.
Sample size: Two individuals. Four calls (CJ 11645), eight calls (CJ 11642).
Note (call) duration: $\bar{x}$ = 0.425 ± 0.027 SD, n = 12; range = 0.373–0.461 s.
Time between calls: $\bar{x}$ = 12.80 ± 2.166 SD, n = 10.

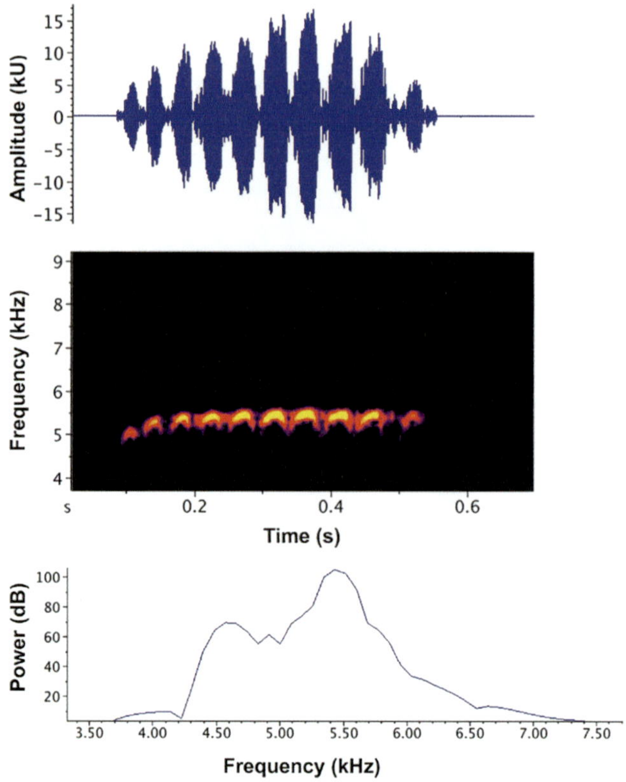

FIGURE A1.56 Call of *Hyalinobatrachium mashpi*. Either Reserva Mashpi or Reserva Tayra, Provincia Pichincha, Ecuador. Taken from Guayasamin *et al.* (2022).

Peak frequency: $\bar{x}$ = 5.38 kHz ± 0.12 SD, n = 12; range = 5.25–5.6 kHz. **Maximum frequency**: $\bar{x}$ = 5.38 kHz ± 0.11 SD, n = 12; range = 5.46–5.81 kHz. **Minimum frequency:** $\bar{x}$ = 4.79 kHz ± 0.10 SD, n = 12; range = 4.62–4.92 kHz.
Pulses per note: 8–10 pulses per note.
Remarks: All vocalizations were recorded in WAV format with a sampling rate of 44.1 kHz/s with 16 bits/sample. Measurements and definitions of acoustic variables follow Köhler *et al.* (2017).

Hyalinobatrachium munozorum

Source: Guayasamin *et al.* (2020b). Figure A1.57.
Voucher specimen and locality: Río Bigal, Provincia Orellana, Ecuador. Recorded by Morely Read. LBE-C-020.
Notes per call: The call is composed of a single note.
Sample size: 18 notes from one individual.
Note duration: $\bar{x}$ = 170 ± 9.9; range = 145–178 ms.
Pulses per note/call: $\bar{x}$ = 4.5 ± 1.6; range = 1–7 low-amplitude peaks.
Relative note peak time: $\bar{x}$ = 0.448 ± 0.123; range = 0.1699–0.661.

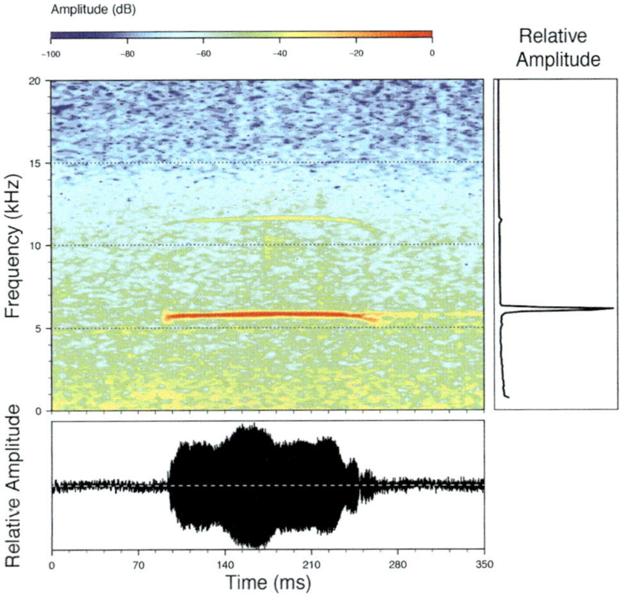

FIGURE A1.57 Call of *Hyalinobatrachium munozorum*. Río Bigal, 930 m, Provincia Orellana, Ecuador. Taken from Guayasamin *et al.* (2020b).

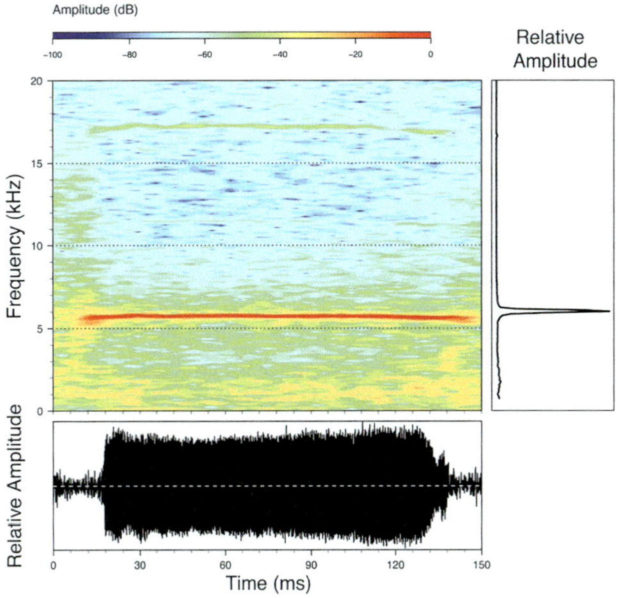

FIGURE A1.58 Call of *Hyalinobatrachium pellucidum*. Río Azuela, 1740 m, Provincia Napo. Taken from Guayasamin *et al.* (2020b).

Fundamental frequency:
- **Lower limit:** $\bar{x}$ = 5708 ± 63; range = 5625–5812 Hz.
- **Higher limit:** $\bar{x}$ = 5906 ± 64; range = 5812–6000 Hz.

Dominant frequency: $\bar{x}$ = 5812 ± 64; range = 5719–5906 Hz.

Hyalinobatrachium pellucidum

Source: Guayasamin *et al.* (2020b). Figure A1.58.
Voucher specimen and locality: USNM 286708: Río Azuela, 1740 m, Provincia Napo, Ecuador. Call recorded by Roy McDiarmid.
Call type: The advertisement call is relatively short and has a single note per call.
Sample size: 36 notes from one individual.
Note duration: $\bar{x}$ = 128 ± 8.7; range = 98–140 ms.
Fundamental frequency:
- **Lower limit:** $\bar{x}$ = 5381 ± 678; range = 5163–5685 Hz.
- **Higher limit:** $\bar{x}$ = 5778 ± 56; range = 5685–5943 Hz.

Dominant frequency: $\bar{x}$ = 5690 ± 58; range = 5599–5857 Hz.

Hyalinobatrachium tatayoi

Source: Castroviejo-Fisher *et al.* (2007). Figure A1.59.
Voucher specimen and locality: MHNLS 17174, 17183, and no voucher specimens: Tokuko, 301 m, Estado de Zulia, Venezuela.
Date: 12 July 2005.
Temperature: Air: 26.9 °C. Water: 24.9 °C.

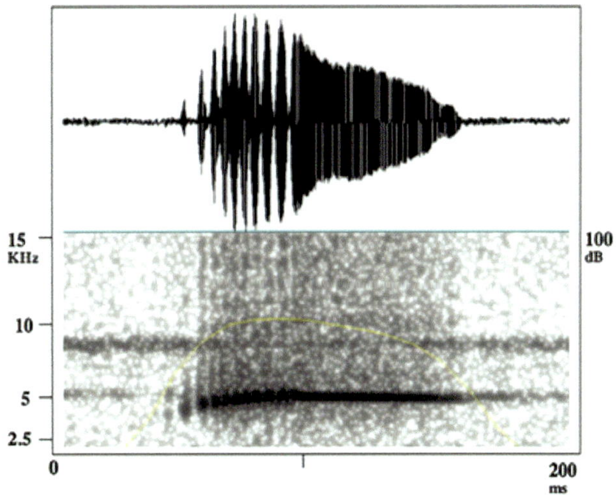

FIGURE A1.59 Oscillogram (top) and spectrogram (bottom) of the call of the holotype of *Hyalinobatrachium tatayoi*. Tokuko, 301 m, Estado de Zulia, Venezuela. Taken from Castroviejo-Fisher *et al.* (2007). Reproduced with permission from the copyright holder © Magnolia Press.

Recording equipment: Sony WM D6C tape recorder. Sennheiser Me 80 directional microphone.
Call type: Advertisement call. Single, high-pitched, half-pulsed, half-modulated note.
Call behavior: The advertisement call was emitted by individuals located on the underside of leaves. A second type of call was emitted by specimens located on the upper side of leaves.
Sample size: 20 calls.
Call rate: 8.8 calls per minute.

Call duration: $\bar{x}$ = 0.135 ± 0.016; range = 0.113–0.158 s.

Dominant frequency: $\bar{x}$ = 4655.8 ± 61.7; range = 4561.7–4728.6 Hz.

Remarks: Recordings were processed on an Apple Macintosh computer. The sounds were digitalized and edited at a sampling frequency of 44.1 kHz and 16 bits of resolution with a Delta 66 digitalizing board and Peak 3.2 software. All calls were edited with Audacity 1.2.2. Praat 4.2.22 software was used to obtain numerical information and to generate audio-spectrograms and oscillograms. Frequency information was obtained through fast Fourier transform (FFT) (width 1024 points). Digitized calls were deposited in the Fonoteca Zoológica of the Museo Nacional de Ciencias Naturales (Madrid, Spain).

Hyalinobatrachium yaku

Source: Guayasamin *et al.* (2017). Figure A1.60 taken from Guayasamin *et al.* (2020b).

Voucher specimen and locality: MZUTI 500: Kallana, Provincia Pastaza, Ecuador.

Call type: Advertisement call. Single and high-pitched tonal note with no frequency or amplitude modulation.

Call rate: Nine calls per minute.

Call duration: $\bar{x}$ = 0.3 ± 0.03; range = 0.27–0.4 s.

Interval between calls: $\bar{x}$ = 7.1 ± 1.1; range = 5.3–8.9 s.

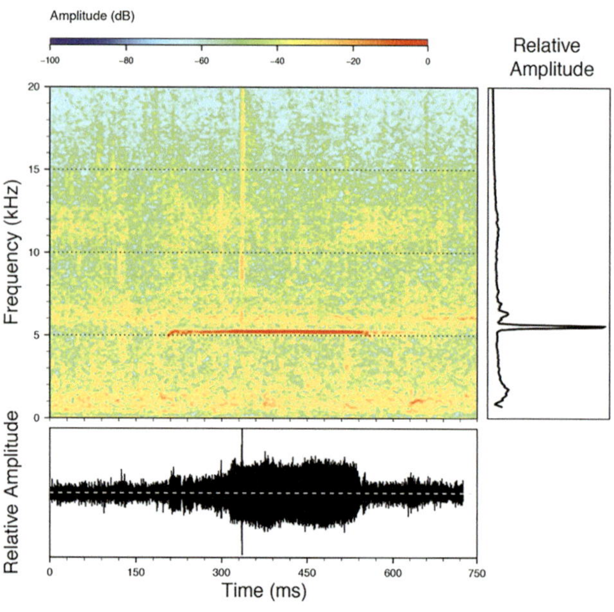

FIGURE A1.60 Call of the holotype of *Hyalinobatrachium yaku* from Kallana, 325 m, Provincia Pastaza, Ecuador. Taken from Guayasamin *et al.* (2020b).

Dominant frequency: $\bar{x}$ = 5283.8 ± 35.0; range = 5219.3–5329.6 Hz.

Lower frequency: $\bar{x}$ = 5264.6 ± 34.6; range = 5207.3–5314.8 Hz.

Upper frequency: $\bar{x}$ = 5299.1 ± 34.1; range = 5236.5–5340.5 Hz.

Remarks: Sound recordings were made with a TASCAM DR-05 Portable Digital Recorder. The calls were recorded in WAV format with a sampling rate of 44.1 kHz/second with 16 bits/sample.

Nymphargus grandisonae

Source: Hutter *et al.* (2013). Figures A1.61, A1.62.

Voucher specimen and locality:

Advertisement call: Reserva Las Gralarias, 2000–2150 m, Provincia Pichincha, Ecuador.

Encounter call: MZUTI 430–431: Reserva Las Gralarias, 2000–2150 m, Provincia Pichincha, Ecuador. MHN-UC 0250–0251: El Águila village (1833–2100 m), municipality of Manizales, Departamento Caldas, Colombia.

Date: 11 February to 14 June 2011.

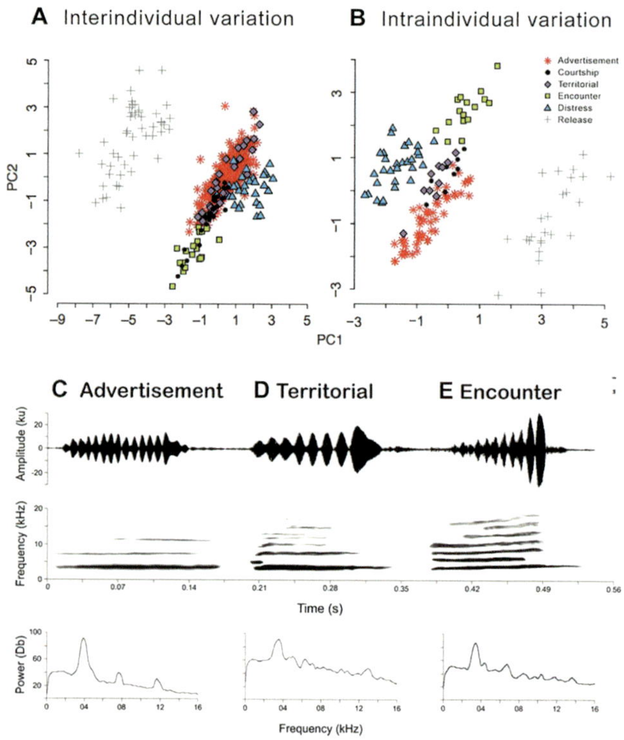

FIGURE A1.61 Call types in *Nymphargus grandisonae*. Reserva Las Gralarias, Provincia Pichincha, Ecuador, and El Águila village, municipality of Manizales, Departamento Caldas, Colombia. A: principal component analyses (PCA) of call variables for different types of calls of all individuals. B: PCA plot showing call variation within an individual. C–E: oscillogram, spectrogram, and power spectrum of different types of calls. Taken and modified from Hutter *et al.* (2013).

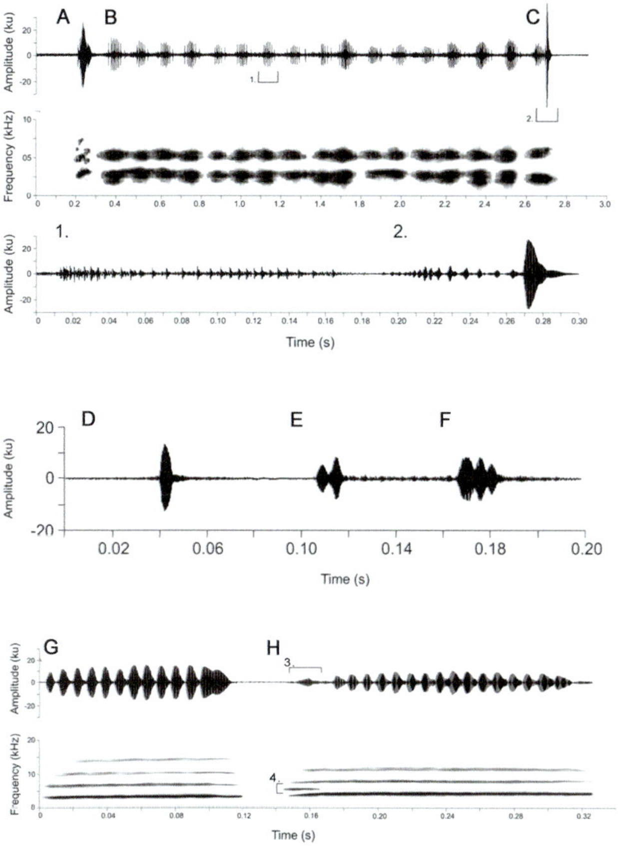

FIGURE A1.62 Oscillogram and audiospectrogram of the release call of *Nymphargus grandisonae*. Reserva Las Gralarias, Provincia Pichincha, Ecuador, and El Águila village, municipality of Manizales, Departamento Caldas, Colombia. A: first note peep; B: intermediate notes; and C: final peep. (1) Enlarged view of an intermediate note and (2) of the final peep. D–F: pulse variation of distress calls. G: oscillogram and audiospectrogram of the advertisement call. H: courtship calls. (3) Modified pulses and (4) the additional frequency component of the courtship call. Taken from Hutter *et al.* (2013).

Temperature: Between 13 and 15 °C.

Recording equipment: Olympus LS10 Linear PCM Recorder. Sennheiser M8 directional microphone.

Call type: This is the description of six different call types: advertisement, territorial, encounter, release, distress, and courtship.

Advertisement call: Call intended to advertise a male's location to other nearby males in addition to attracting females.

Sample size: 417 calls from 22 individuals.

Notes per call: One.

Call duration: $\bar{x}$ = 0.115 ± 0.018; range = 0.056–0.158 s.

Pulse rate: $\bar{x}$ = 143.2 ± 18.8; range = 107.9–320.8 pulses per second.

Dominant frequency: $\bar{x}$ = 3588 ± 189.6; range = 3101–4048.2 Hz.

Frequency modulation: $\bar{x}$ = 143.8 ± 107.9; range = 0–516.8 Hz.

Lower dominant frequency: $\bar{x}$ = 3158 ± 184; range = 2694–3645 Hz.

Higher dominant frequency: $\bar{x}$ = 4039 ± 210; range = 3441–4651 Hz.

Pulse amplitude peak ratio (first): $\bar{x}$ = 0.545 ± 0.203; range = 0.102–1.0.

Pulse amplitude peak ratio (middle): $\bar{x}$ = 0.812 ± 0.153; range = 0.403–1.0.

Pulse amplitude peak ratio (last): $\bar{x}$ = 0.616 ± 0.224; range = 0.165–1.3.

First harmonic: $\bar{x}$ = 7179 ± 409; range = 5857–8613 Hz.

Second harmonic: $\bar{x}$ = 10697 ± 557; range = 8648–13394 Hz.

Third harmonic: $\bar{x}$ = 14396 ± 627; range = 12489–15762 Hz.

Territorial call: Modification of the advertisement call, emitted by a male territory holder in response to the presence of another invading male; invading male may or may not offer a response call; emitted at long range.

Sample size: 30 calls from three individuals.

Notes per call: One.

Call duration: $\bar{x}$ = 0.110 ± 0.015; range = 0.076–0.141 s.

Pulse rate: $\bar{x}$ = 137.3 ± 17.1; range = 105.3–190.9 pulses per second.

Dominant frequency: $\bar{x}$ = 3353 ± 101; range = 3100–3445.3 Hz.

Frequency modulation: $\bar{x}$ = 212.5 ± 116.9; range = 0–430.7 Hz.

Lower dominant frequency: $\bar{x}$ = 2959 ± 118; range = 2700–3109 Hz.

Higher dominant frequency: $\bar{x}$ = 3735 ± 136; range = 3527–4000 Hz.

Pulse amplitude peak ratio (first): $\bar{x}$ = 0.509 ± 0.188; range = 0.261–0.963.

Pulse amplitude peak ratio (middle): $\bar{x}$ = 0.771 ± 0.994; range = 0.595–0.289.

Pulse amplitude peak ratio (last): $\bar{x}$ = 0.595 ± 0.289; range = 0.080–1.0.

First harmonic: $\bar{x}$ = 6523.5 ± 296; range = 6029.3–7149 Hz.

Second harmonic: $\bar{x}$ = 10304.4 ± 100.2; range = 10,077.5–10,508.2 Hz.

Third harmonic: $\bar{x}$ = 13,532 ± 125; range = 13,350–13,695 Hz.

Other frequency: $\bar{x}$ = 4657 ± 210; range = 4392–5168 Hz.

Encounter call: Further modification of the advertisement call produced during close-range antagonist interactions between males prior to or during combat; emitted at short range.

Sample size: 19 calls from two individuals.

Notes per call: One.
Call duration: $\bar{x}$ = 0.115 ± 0.016; range = 0.091–0.148 s.
Pulse rate: $\bar{x}$ = 131.4 ± 12.3; range = 115.7–169.8 pulses per second.
Dominant frequency: $\bar{x}$ = 3440 ± 142; range = 3186–3617 Hz.
Frequency modulation: $\bar{x}$ = 639.2 ± 135.6; range = 344.5–861.3 Hz.
Lower dominant frequency: $\bar{x}$ = 2950 ± 117; range = 2700–3109 Hz.
Higher dominant frequency: $\bar{x}$ = 3831 ± 165; range = 3527–4069 Hz.
Pulse amplitude peak ratio (first): $\bar{x}$ = 0.238 ± 0.156; range = 0.063–0.510.
Pulse amplitude peak ratio (middle): $\bar{x}$ = 0.628 ± 0.216; range = 0.273–0.939.
Pulse amplitude peak ratio (last): $\bar{x}$ = 0.932 ± 0.083; range = 0.746–1.0.
First harmonic: $\bar{x}$ = 6700 ± 234; range = 6201–7019 Hz.
Second harmonic: $\bar{x}$ = 10182 ± 382; range = 9431–10723 Hz.
Third harmonic: $\bar{x}$ = 13184 ± 649; range = 12165–14341 Hz.
Other frequency: $\bar{x}$ = 4750 ± 265; range = 4048–4995 Hz.

Release call: Occurred during combat behavior, when the males grasped each other in various positions.
Sample size: 48 calls from two individuals.
Notes per call: $\bar{x}$ = 18.2 ± 5.3; range = 9.0–34.0.
Call duration: $\bar{x}$ = 3.3 ± 0.877; range = 1.7–5.4 s.
Pulse rate: $\bar{x}$ = 280.1 ± 75.3; range = 152.5–454.0 pulses per second.
Dominant frequency: $\bar{x}$ = 3267 ± 235; range = 2756–3531 Hz.
Frequency modulation: $\bar{x}$ = 228.4 ± 147.6; range = 0–602.9 Hz.
Lower dominant frequency: $\bar{x}$ = 2676 ± 307; range = 1882–3109 Hz.
Higher dominant frequency: $\bar{x}$ = 3846 ± 227; range = 3255–4341 Hz.
Pulse amplitude peak ratio (first): $\bar{x}$ = 0.555 ± 0.336; range = 0.113–1.0.
Pulse amplitude peak ratio (last): $\bar{x}$ = 0.958 ± 0.145; range = 0.100–1.0.
First harmonic: $\bar{x}$ = 4959 ± 380; range = 4254–5512 Hz.

Distress call: A male emitted this as a response to pain from another male during a fight.
Sample size: 30 calls from one individual.
Notes per call: One.
Call duration: $\bar{x}$ = 0.020 ± 0.007; range = 0.010–0.037 s.
Pulse rate: $\bar{x}$ = 145.3 ± 66.8; range = 55.6–333.3 pulses per second.
Dominant frequency: $\bar{x}$ = 3080 ± 198; range = 2756–3445 Hz.
Lower dominant frequency: $\bar{x}$ = 2783 ± 182; range = 2536–3109 Hz.
Higher dominant frequency: $\bar{x}$ = 3397 ± 217; range = 3074–3798 Hz.
Pulse amplitude peak ratio (first): $\bar{x}$ = 0.786 ± 0.219; range = 0.271–1.1.
Pulse amplitude peak ratio (middle): $\bar{x}$ = 0.892 ± 0.174; range = 0.477–1.0.
Pulse amplitude peak ratio (last): $\bar{x}$ = 0.785 ± 0.220; range = 0.281–1.0.
First harmonic: $\bar{x}$ = 5735 ± 501; range = 4737–6546 Hz.
Second harmonic: $\bar{x}$ = 8494 ± 884; range = 6373–9216 Hz.
Third harmonic: $\bar{x}$ = 12058 ± 498; range = 11369–12489 Hz.

Courtship call: Communicative call emitted at close range (0.5–1 m) in the presence of a female.
Sample size: 27 calls from three individuals.
Notes per call: One.
Call duration: $\bar{x}$ = 0.120 ± 0.023, range = 0.082–0.170 s.
Pulse rate: $\bar{x}$ = 140.2 ± 12.1; range = 111.8–168.8 pulses per second.
Dominant frequency: $\bar{x}$ = 3541 ± 206; range = 3186–3876 Hz.
Frequency modulation: $\bar{x}$ = 468.9 ± 198.9; range = 258.4–861.3 Hz.
Lower dominant frequency: $\bar{x}$ = 3075 ± 218; range = 2615–3407 Hz.
Higher dominant frequency: $\bar{x}$ = 4003 ± 197; range = 3720–4372 Hz.
Pulse amplitude peak ratio (first): $\bar{x}$ = 0.202 ± 0.091; range = 0.076–0.484.
Pulse amplitude peak ratio (middle): $\bar{x}$ = 0.815 ± 0.148; range = 0.532–1.0.
Pulse amplitude peak ratio (last): $\bar{x}$ = 0.613 ± 0.231; range = 0.226–1.0.
First harmonic: $\bar{x}$ = 7056 ± 377; range = 6503–7752 Hz.
Second harmonic: $\bar{x}$ = 10537 ± 719; range = 9431–11714 Hz.
Third harmonic: $\bar{x}$ = 14211 ± 1929; range = 10508–15590 Hz.
Other frequency: $\bar{x}$ = 4929 ± 515; range = 3962–6115 Hz.

Remarks: The vocalizations were recorded in PCM format at a sampling rate of 44 Hz. We recorded calls in the field 50–300 cm from calling males. Recordings were processed with Raven Pro 1.4 software on Mac OS X. Oscillograms (waveforms), audiospectrograms, and power spectra

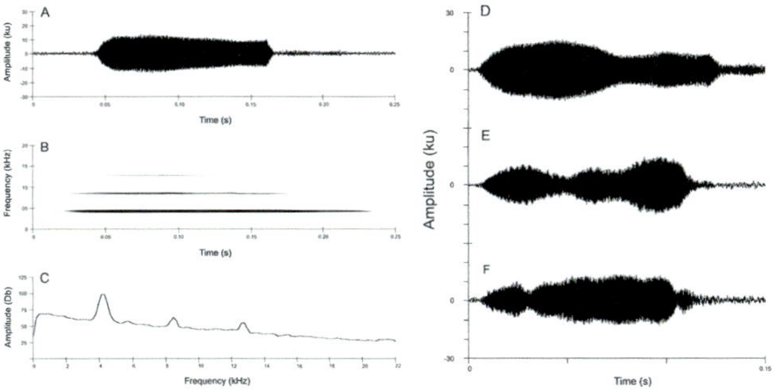

FIGURE A1.63 Advertisement call of *Nymphargus griffithsi*. Reserva Las Gralarias, Provincia Pichincha, Ecuador. A: oscillogram. B: audiospectrogram. C: power spectrum. Inter- and intraspecific waveform variation: D: MZUTI 100; E: MZUTI 098: F: MZUTI 101. Taken from Hutter and Guayasamin (2012). Reproduced with permission from the copyright holder © Magnolia Press.

(fast Fourier transform) were analyzed for various temporal and spectral call components. The fast Fourier transform size was set to 512 and frequency grid resolution to 86.1 Hz.

NYMPHARGUS GRIFFITHSI

Source: Hutter and Guayasamin (2012). Figure A1.63.
Voucher specimen and locality: Reserva Las Gralarias, Provincia Pichincha, Ecuador 1825–2400 m, Heloderma Creek, 2175–2225 m, Hercules Giant Tree Frog Creek, 2175 m, and Five Frog Creek, 2150 m.
Call type: Advertisement call. Tonal call, single note.
Sample size: 48 vocalizations from five male.
Date: 9 March 2011, 15 March 2011, 17 March 2011, and 14 April 2011.
Temperature: 13–15 °C.
Recording equipment: Olympus L10 linear PCM recorder. Sennheiser Me 80 directional microphone.
Call rate: $\bar{x}$ = 1.6 ± 0.4; range = 1.1–1.9 calls per minute.
Call duration: $\bar{x}$ = 0.122 ± 0.009; range = 0.103–0.148 s.
Notes per call: One.
Pulses per call: $\bar{x}$ = 2.4 ± 0.5; range = 1 (tonal note = one pulse)–3.
Call rise time: $\bar{x}$ = 0.066 ± 0.020; range = 0.013–0.104 s.
Dominant frequency: $\bar{x}$ = 4107.4 ± 105.5; range = 3789.8–4306.6 Hz.
Frequency modulation: $\bar{x}$ = 199.3 ± 116.2; range = 0–344.6 Hz.
Initial frequency of the fundamental frequency: $\bar{x}$ = 3290.6 ± 225.1; range = 2272.0–3777.0 Hz.
Lower frequency of the fundamental frequency: $\bar{x}$ = 3672.3 ± 115.3; range = 3400.4–3826.3 Hz.
Upper frequency of the fundamental frequency: $\bar{x}$ = 4520.5 ± 131.3; range = 4160.2–4702.2 Hz.
First harmonic: $\bar{x}$ = 8071.8 ± 268.9; range = 6546.1–8613.3 Hz.
Second harmonic: $\bar{x}$ = 12,673.8 ± 780.3; range = 7579.7–15,503.9 Hz.
Third harmonic: $\bar{x}$ = 16,437.0 ± 232.5; range = 16,020.7–17,054.3 Hz.
Remarks: Recordings were processed using the Cornell Lab of Ornithology's Raven Pro 1.4 software on Mac OS X. Oscillograms (waveforms), audiospectrograms, and power spectra (fast Fourier transform [FFT]) for various temporal and spectral components were analyzed.

NYMPHARGUS LASGRALARIAS

Source: Hutter and Guayasamin (2012). Figure A1.64.
Voucher specimen and locality: Reserva Las Gralarias, Provincia Pichincha, Ecuador: (1825–2400 m "Kathy's Creek, 2000 m"), "Heloderma Creek (2175–2225 m)," and "Five Frog Creek (2150 m)."
Call type: Advertisement call. Reminiscent of a short "ticking" or "clicking" noise. Calls consists of a short, pulsed note, and emitted in series.
Sample size: 119 vocalizations from six males.
Date: 17 March 2011, 5 April 2011, 17 April 2011, and 1 July 2011.
Temperature: 13–15 °C.
Recording equipment: Olympus L10 linear PCM recorder. Sennheiser Me 80 directional microphone.
Call rate: $\bar{x}$ = 5.5 ± 2.7; range = 2.0–9.9 call per minute.
Call duration: $\bar{x}$ = 0.026 ± 0.006; range = 0.016–0.044 s.
Notes per call: One.
Pulses per call: $\bar{x}$ = 1.5 ± 0.6; range = 1–3.
Call rise time: $\bar{x}$ = 0.007 ± 0.003; range = 0.002–0.013 s.
Dominant frequency: $\bar{x}$ = 3691.4 ± 131.9; range = 3445.3–3962.2 Hz.
Frequency modulation: $\bar{x}$ = 21.7 ± 69.3; range = –172.3–172.3 Hz.
Initial frequency of the fundamental frequency: $\bar{x}$ = 3063.6 ± 162.5; range = 2561.0–3441.0 Hz.
Lower frequency of the fundamental frequency: $\bar{x}$ = 3236.3 ± 168.7; range = 2939.4–4145.2 Hz.
Upper frequency of the fundamental frequency: $\bar{x}$ = 4139.8 ± 139.7; range = 3887.7–4473.4 Hz.
First harmonic: $\bar{x}$ = 7298.9 ± 305.8; range = 6546.1–8096.5 Hz.
Second harmonic: $\bar{x}$ = 11034.2 ± 478.3; range = 9991.4–12058.6 Hz.
Third harmonic: $\bar{x}$ = 14317.1 ± 245.5; range = 13781.2–14928.0 Hz.
Remarks: Recordings were processed using the Cornell Lab of Ornithology's Raven Pro 1.4 software

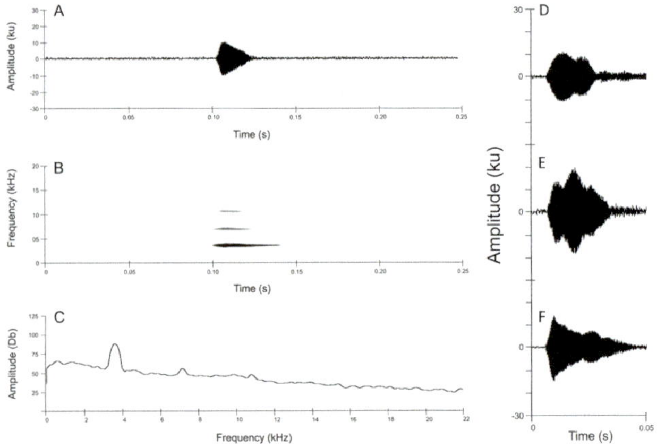

FIGURE A1.64 Advertisement call of *Nymphargus lasgralarias*. Reserva Las Gralarias, Provincia Pichincha, Ecuador. A: oscillogram. B: audiospectrogram. C: power spectrum. Inter- and intraspecific waveform variation: D: MZUTI 096; E: MZUTI 092: F; MZUTI 18839. Taken and modified from Hutter and Guayasamin (2012). Reproduced with permission from the copyright holder © Magnolia Press.

on Mac OS X. Oscillograms (waveforms), audio spectrograms, and power spectra (fast Fourier transform [FFT]) for various temporal and spectral components were analyzed.

Nymphargus mariae

Source: Guayasamin *et al.* (2020b).
Voucher specimen and locality: Stream tributary of Río Lliquino, Provincia de Pastaza, Ecuador. Call recorded by Diego Paucar.
Date: 22 February 2008.
Call behavior: Males call from the upper surfaces of leaves.
Call type: The call consists of a single, non-frequency-modulated note.
Sample size: One male.
Call duration: Range = 0.015–0.018 s (n = 3).
Interval between calls: Range = 1.733–1.940 s.
Dominant frequency: Range = 3234–4299 Hz.
Harmonics: First harmonic is visible at 7178–7928 Hz, and a second harmonic is visible at 10,767–11,714 Hz.

Nymphargus manduriacu

Source: Guayasamin *et al.* (2019a). Figure A1.65.
Voucher specimen and locality: Reserva Río Manduriacu, Provincia Imababura, Ecuador. LBE-C-042.
Recording equipment: Olympus LS-10 linear PCM field recorder. Sennheiser K6–ME 66 unidirectional microphone.
Call type: Each call is a high-pitched "chirp" that consists of a single note.

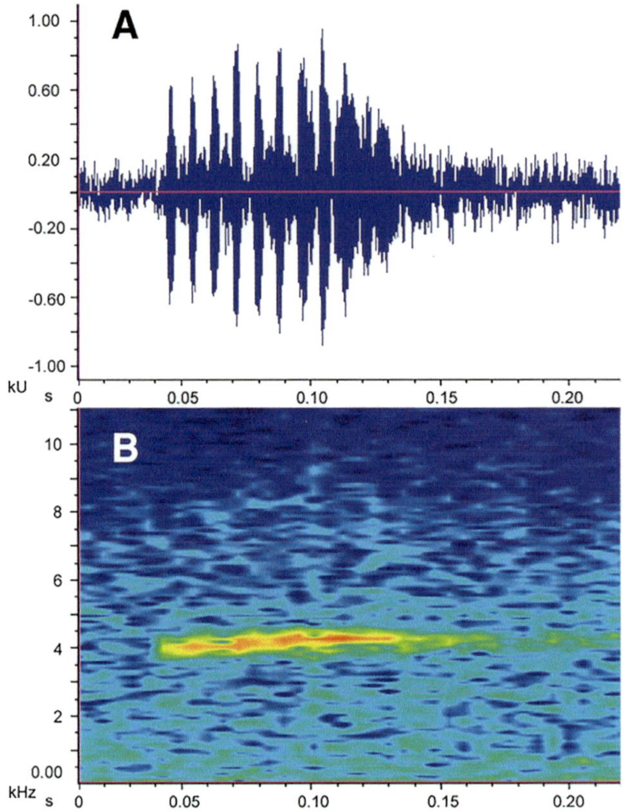

FIGURE A1.65 Call of *Nymphargus manduriacu*. Reserva Río Manduriacu. A: oscillogram. B: audiospectrogram. Taken from Guayasamin *et al.* (2019a).

Call duration: $\bar{x}$ = 0.10 ± 0.007; range = 0.093–0.118 s (n = 10).
Interval between calls: $\bar{x}$ = 5.72 ± 1.82; range = 3.9–8.6 s.
Pulses per note: $\bar{x}$ = 10.33 ± 1.366; range = 8–12.
Dominant frequency: $\bar{x}$ = 4268 ± 118.3; range = 4052–4447 Hz.
Remarks: The calls were recorded in WAV format with a sampling rate of 44.1 kHz/second with 16 bits/sample. In each call, there is slight increase in the dominant frequency with time.

Nymphargus siren

Source: Guayasamin *et al.* (2020b). Figure A1.66.
Voucher specimen and locality: MZUTI 765, 775: Oyacachi and El Chaco, Provincia Napo, Ecuador.
Temperature: 15 °C.
Call type: Advertisement call is short and composed by a single note.
Sample size: 14 notes from two individuals.
Call duration: $\bar{x}$ = 24 ± 7; range = 20–42 ms.
Relative note peak time: $\bar{x}$ = 0.123 ± 0.032; range = 0.0595–0.1751.

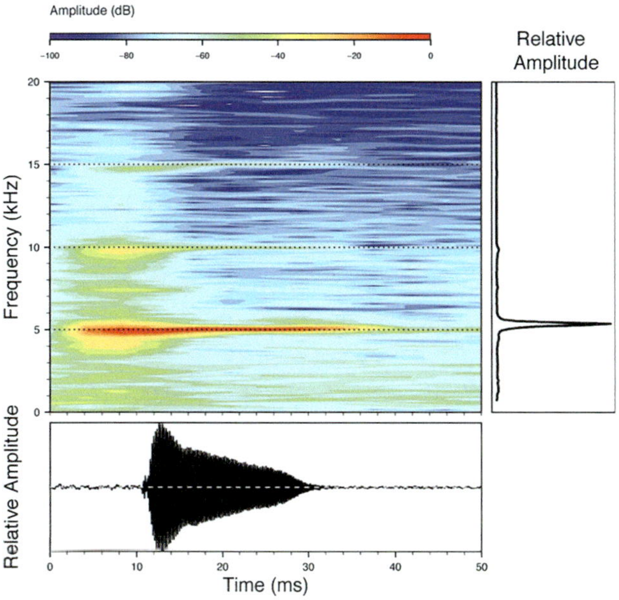

FIGURE A1.66 Call of *Nymphargus siren*. Trail between Oyacachi and El Chaco, 1645 m, Provincia Napo, Ecuador. Taken from Guayasamin *et al.* (2020b).

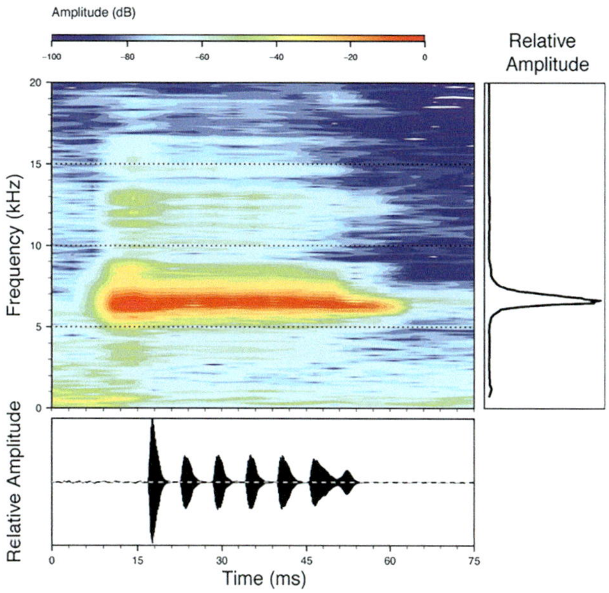

FIGURE A1.67 Call of *Rulyrana flavopunctata*. Quebrada Pangayaku, 1645 m, Provincia Napo, Ecuador. Taken from Guayasamin *et al.* (2020b).

Pulses per notes: $\bar{x}$ = 2.2 ± 0.6; range = 1–3.
Pulses within a note rate: $\bar{x}$ = 86 ± 23; range = 24–130 pulses per second.
Fundamental frequency:
Lower limit: $\bar{x}$ = 4842 ± 331= range = 4651–5943 Hz.
Higher limit: $\bar{x}$ = 5088 ± 318; range = 4823–6115 Hz.
Dominant frequency: $\bar{x}$ = 4977 ± 323; range = 4737–6029 Hz.

Rulyrana flavopunctata

Source: Guayasamin *et al.* (2020b). Figure A1.67.
Voucher specimen and locality: MZUTI 1476, 1684: Quebrada Pangayaku, Provincia Napo, Ecuador.
Temperature: 19.4 °C.
Call type: Each call is composed by a single short note.
Sample size: 85 notes contained within 13 calls from two individuals.
Call duration: $\bar{x}$ = 4 ± 1; range = 3–8 ms.
Pulses within a note rate: $\bar{x}$ = 262 ± 55; range = 125–333 pulses per second.
Fundamental frequency:
Lower limit: $\bar{x}$ = 6648 ± 447; range = 5685–7149 Hz.
Higher limit: $\bar{x}$ = 7382 ± 491; range = 6115–7924 Hz.
Dominant frequency: $\bar{x}$ = 6931 ± 472; range = 5857–7580 Hz.
Remarks: The first pulse of each note has the highest amplitude.

Sachatamia ilex

Source: Guayasamin *et al.* (2020b).
Voucher specimen and locality: Reserva Otokiki, 706 m, Provincia Esmeraldas, Ecuador. Call recorded by Elicio E. Tapia.
Date: 17 May 2010.
Call behavior: Males call from the upper surfaces of leaves.
Call type: The call consists of a single, pulsed note.
Sample size: Six calls from one male.
Call rate: 10 calls in a five-minute interval.
Note duration: $\bar{x}$ = 0.08 ± 0.033; range = 0.077–0.086 s.
Dominant frequency:
Beginning of note: $\bar{x}$ = 6640 ± 150.2; range = 6562–6937 Hz.
Tonal section of note: $\bar{x}$ = 7422 ± 38.4; range = 7406–7500 Hz.
End of note: $\bar{x}$ = 6718 ± 128.0; range = 6562–6843 Hz.
Remarks: The note ends with a series of pulses, the amplitude of which is conspicuously lower than the pulses at the beginning of the note. The maximum value of the dominant frequency is always during the first pulsed section of the note.

Sachatamia orejuela

Source: Brunner and Guayasamin (2020). Figure A1.68.
Voucher specimen and locality: Mashpi Reserve, 950 m, Provincia Pichincha, Ecuador.
Date: 6 April 2019, 19:15 and 22:15 h.

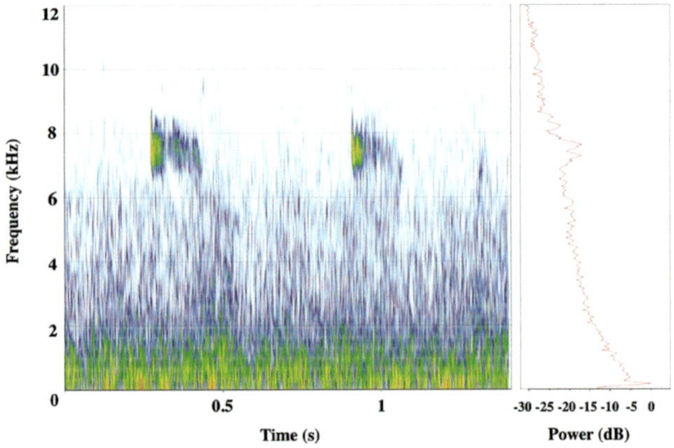

FIGURE A1.68 Spectrogram and power spectrum of the *Sachatamia orejuela* advertisement call. Reserva Mashpi, Provincia Pichincha, Ecuador. Taken from Brunner and Guayasamin (2020).

Recording equipment: Roland R-26 digital audio recorder. Sennheiser ME67 directional microphone.

Call behavior: One individual calling from a concave rock face next to a rushing waterfall.

Call type: Advertisement call. It is high pitched, even for a glassfrog, generally emitted as a single "peep."

Sample size: One individual.

Call duration: $\bar{x}$ = 0.167 ± 0.018; range = 0.137–0.189 s (n = 13).

Inter-call/note duration: $\bar{x}$ = 32.4 ± 12.9; range = 12.1– 57.0 s (n = 21).

Interval between notes: $\bar{x}$ = 0.455–0.548 s (n = 2).

Dominant frequency: $\bar{x}$ = 7.433 ± 0.1203; range = 7.219–7.688 kHz (n = 13 notes, two individuals).

Remarks: Call parameters were analyzed in Raven Pro 1.5 software (Center for Conservation Bioacoustics, Cornell Lab of Ornithology, 2014) without filters or transformations. The spectrogram was generated with a window length of 128 samples. The power spectrum was generated with a window length of 1024 samples and a 44.1 kHz sampling rate.

Teratohyla amelie

Source: Guayasamin *et al.* (2020b). Figure A1.69.

Voucher specimen and locality: QCAZ 38779: Tributary of Río Lliquino, 350 m, Provincia Pastaza, Ecuador. Recorded by Diego Paucar.

Date: 15 May 2008.

Call type: Each call consists of a single pulsed note.

Sample size: One male.

Call duration: $\bar{x}$ = 0.013 ± 0.001; range = 0.011–0.014 s (n = 10).

Interval between calls: $\bar{x}$ = 17.8 ± 7.55; range = 10–30 s (n = 8).

Pulses per note/call: $\bar{x}$ = 3.5 ± 0.515; range = 3–4 pulses (n = 10).

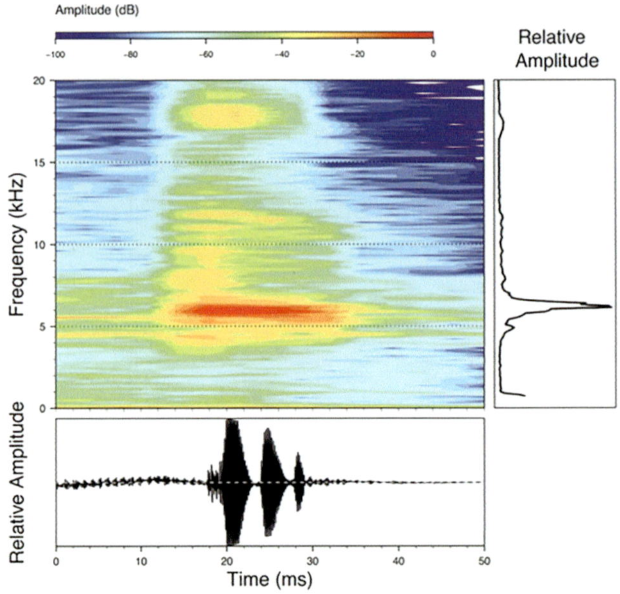

FIGURE A1.69 Call of *Teratohyla amelie*. Tributary of Río Lliquino, Provincia Pastaza, Ecuador. Taken from Guayasamin *et al.* (2020b).

Dominant frequency: Range = 4984–7085 Hz (n = 10).

Maximum frequency: Range = 5906–6023 Hz.

Harmonics: A first harmonic is visible at 10,923–12,474 Hz (n = 10); a second harmonic is visible at 16,815–19,120 Hz (n = 10).

Teratohyla midas

Source: Guayasamin *et al.* (2020b). Figure A1.70.

Voucher specimen and locality: Boanamo community, 240 m, Provincia Orellana, Ecuador. Recorded by Morley Read. LBE-C-036.

Call type: Calls are emitted in series. Each call is short and composed by a single note.

Sample size: One male.

Call rate: 16 calls in a three-minute period.

Interval between calls: $\bar{x}$ = 13.02 ± 1.732; range = 10.2–15.1 s.

Note duration: $\bar{x}$ = 0.082 ± 0.007; range = 0.073–0.092 s.

Pulses per note: $\bar{x}$ = 3.14 ± 0.378; range = 3–4 pulses.

Dominant frequency: $\bar{x}$ = 7285 ± 60.4; range = 7210–7390 Hz.

Remarks: Calls are similar to those reported by Twomey *et al.* (2014) for Peruvian populations of the species, although they report a dominant frequency at 6760–7060 Hz.

Teratohyla pulverata

Source: Kubicki (2007). Figure A1.71 taken from Ibañez *et al.* (1999b).

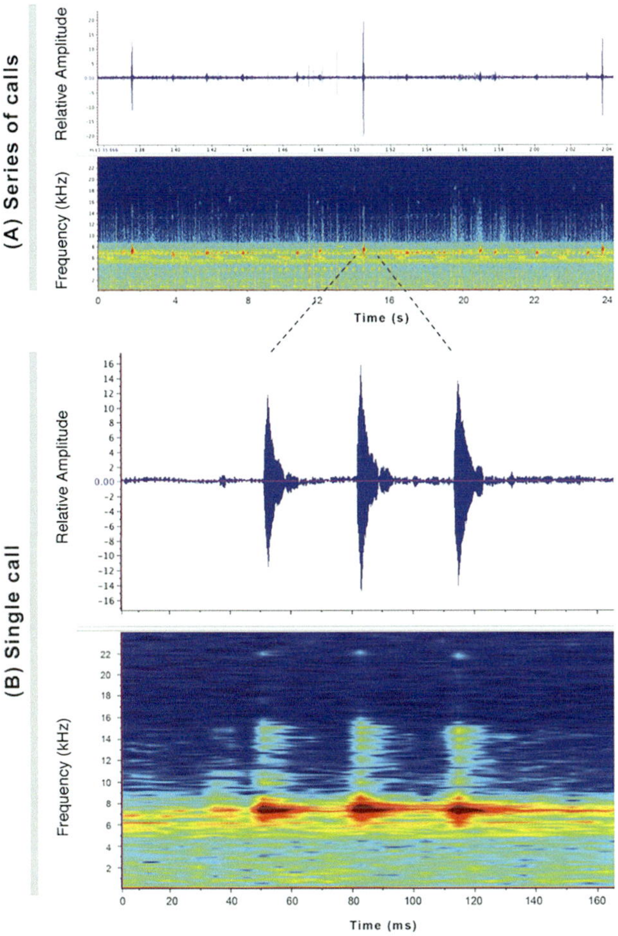

FIGURE A1.70 Calls of *Teratohyla midas*. Boanamo community, 240 m, Provincia Orellana, Ecuador. A: call series fragment showing three calls. B: single call with three pulses. Taken from Guayasamin *et al.* (2020b).

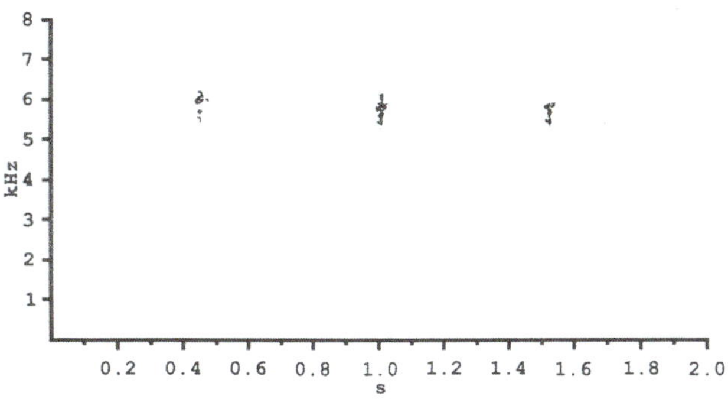

FIGURE A1.71 Spectrogram of the call of *Teratohyla pulverata*. Monumento Natural Barro Colorado, Parque Nacional Soberanía, and adjacent areas, Panama. Taken from Ibañez *et al.* (1999b).

Voucher specimen and locality: Not provided: Costa Rica.

Call behavior: Males call from the upper surface of leaves, often preferring trees and shrubs overhanging the stream 2–8 m.

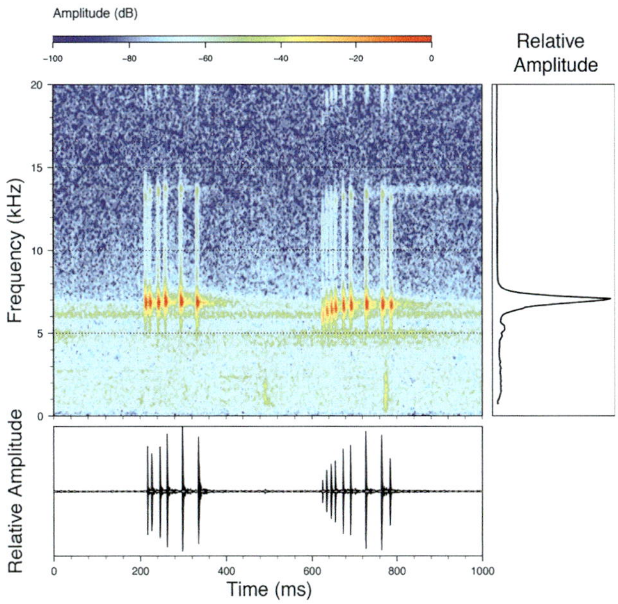

FIGURE A1.72 Call of *Teratohyla spinosa*. Reserva Otokiki, 706 m, Provincia Esmeraldas, Ecuador. Taken from Guayasamin *et al.* (2020b).

Call type: The call of is a quick high-pitched "tik," normally repeated several times with roughly a second pause between calls.

Note duration: $\bar{x}$ = 0.05 s.

Interval between notes: Range = 0.5–0.8 s.

Dominant frequency: 5.6 kHz.

Remarks: Ibañez *et al.* (1999b) provide a spectrogram of the call from Monumento Natural Barro Colorado, Parque Nacional Soberanía, and adjacent areas, Panama.

Teratohyla spinosa

Source: Guayasamin *et al.* (2020b). Figure A1.72.

Voucher specimen and locality: Reserva Otokiki, 706 m, Provincia Esmeraldas, Ecuador. Recorded by Juan Manuel Guayasamin. LBE-C-002.

Call type: Series of harsh "creep-creeps" followed by a considerable pause (Ibañez *et al.*, 1999b).

Call behavior: Males call from the upper surfaces of leaves.

Sample size: One individual.

Interval between calls: Range = 12–35 s.

Note rate: $\bar{x}$ = 1.92 ± 0.515; range = 1–3 notes (n = 12).

Call with one note duration: Range = 0.124–0.139 s (n = 2).

Call with two notes duration: $\bar{x}$ = 0.545 ± 0.067; range = 0.431–0.594 s (n = 9).

Call with three notes duration: 0.793 s.

Pulses per note: $\bar{x}$ = 7.87 ± 2.096; range = 5–15 pulses per note (n = 23).

Dominant frequency: $\bar{x}$ = 6664 ± 129.2; range = 6469–6937 Hz (n = 12).

Harmonics:

First harmonic: $\bar{x}$ = 13351 ± 305.6; range = 12843–13687 Hz (n = 12).

Second harmonic: $\bar{x}$ = 20039 ± 9885; range = 19312–20812 Hz (n = 8).

Vitreorana ritae

Source: Guayasamin *et al.* (2020b). Figure A1.73.

Voucher specimen and locality: Boanamo Community and Shiripuno Lodge, Provincia Orellana, Ecuador. Recorded by Morley Read. LBE-C-039.

Call behavior: In central Amazonia of Brazil, males of *Vitreorana ritae* were observed calling on vegetation above streams, about 0.5–3 m high on leaves of aquatic or stream-side plants.

Call type: The typical advertisement call is relatively short.

Sample size: 31 notes contained within 10 calls from two individuals.

Call duration: $\bar{x}$ = 685.2 ± 280.1; range = 288–1095 ms.

Note rate: $\bar{x}$ = 3.3 ± 1.1; range = 2–5 notes per call.

Note duration: $\bar{x}$ = 76 ± 16.3; range = 61–122 ms.

Pulses per notes: $\bar{x}$ = 13.7 ± 2; range = 10–19.

Pulses within a note rate: $\bar{x}$ = 171 ± 24; range = 131–224 pulses per second.

Note relative peak time: $\bar{x}$ = 0.719 ± 0.106; range = 0.4425–0.8815 (%).

Fundamental frequency:

Lower limit: $\bar{x}$ = 4605 ± 164; range = 4312–4875 Hz.

Higher limit: $\bar{x}$ = 5284 ± 229; range = 4969–6029 Hz.

Dominant frequency: $\bar{x}$ = 5044 ± 182; range = 4688–5344 Hz.

Remarks: The call described for the Ecuadorian populations matches the general information provided by Zimmerman and Bogart (1984) for Brazilian populations.

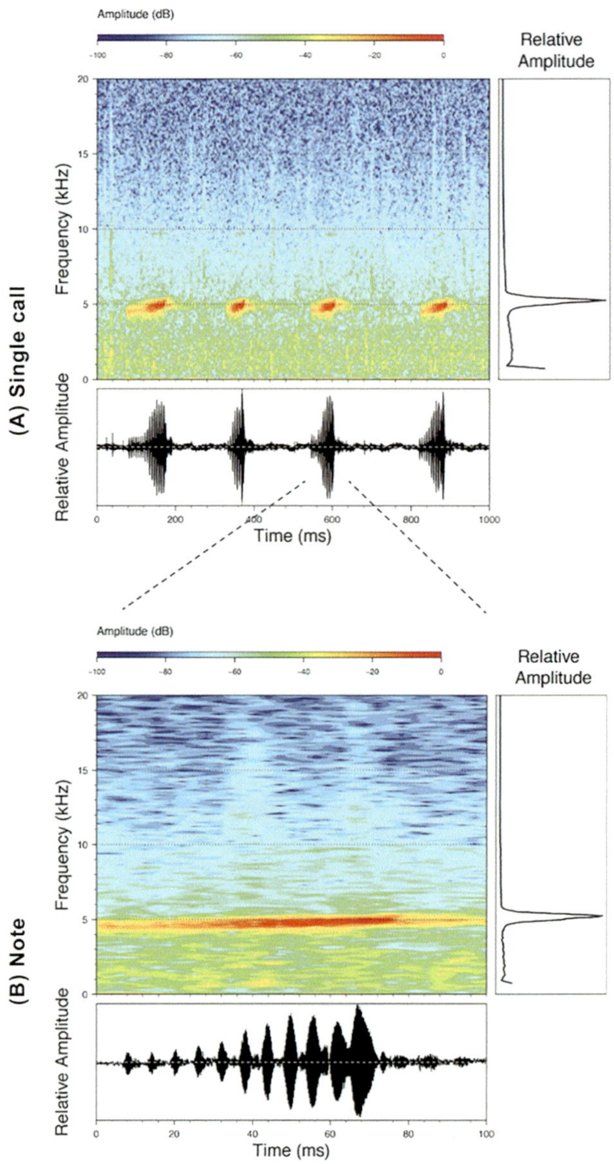

FIGURE A1.73 Calls of *Vitreorana ritae*. Boanamo Community, 240 m, Provincia Orellana, Ecuador. A: a single call with four notes. B: a single note. Taken from Guayasamin *et al.* (2020b).

Appendix 2: Addendum

As of 31 December 2022, we ceased incorporating accounts of newly published species. The same applies to data on species published since 2023, with a few exceptions noted where applicable. This allowed us to prepare final tables, maps, biogeographic information, and Red List analyses. Nonetheless, from January 2023 to the January 2025, approximately 50 scientific publications including Ecuadorian amphibians have been released. Among them, 15 publications include the description of 27 new species (three Bufonidae, six Centrolenidae, fourteen Craugastoridae, two Hylidae, and two Caeciliidae) (Cisneros-Heredia *et al.*, 2023; Székely *et al.*, 2023a, 2023b; Reyes Puig *et al.*, 2023, 2024; Sánchez-Nivicela *et al.*, 2023; Menéndez-Guerrero *et al.*, 2024; Rivera-Correa *et al.*, 2024; Ron *et al.*, 2024; Franco-Mena *et al.*, 2024; Brito-Zapata *et al.*, 2024; Bock *et al.*, 2014; Plewnia *et al.*, 2024; Ortega *et al.* 2025; Coloma *et al.* in Lötters *et al.*, 2025), the synonymy of *Epipedobates darwinwallacei* under *E. espinosai* (López-Hervas *et al.*, 2024), and the removal of *Rhinella horribilis* from the Ecuadorian fauna (Menéndez-Guerrero *et al.*, 2024). Several of these publications also include new or updated hypotheses on the phylogenetic relationships of several taxa. These additions and changes expand the number of Ecuadorian species to 680. Although we did not integrate these publications into the corresponding chapters, texts, and analyses, we provide here accounts (in alphabetical order within Anura and Gymnophiona) of 20 out of 27 newly described species. We include species published between January 2023 and January 2025 (with submissions accepted until November 6, 2024).

ANURA

Bufonidae

ATELOPUS COLOMAI PLEWNIA *ET AL.*, 2024

Coloma's Harlequin Frog/Rana arlequín de Coloma

Atelopus colomai Plewnia *et al.*, 2024: 239. **Holotype.**—CJ 12198 from Río Pucayacu, North of Diez de Agosto (1°22'43" S, 77°51'53" W, 951 m), Provincia Pastaza, Ecuador.

Etymology.—The specific epithet is a patronym for Luis A. Coloma for his contributions to the study and protection of harlequin frogs for decades.

Identification.—Figures A2.1, A2.2. Adults of this species attain a maximum SVL of 33.7 mm in females and of 25.6 in males. The body is slender, with the body width-to-SVL ratio being approximately 0.261 in males and 0.260 in females. The species has long legs, with the tibia length-to-SVL ratio being 0.433 in males and 0.443 in females. The snout is acuminate in dorsal view and protrudes beyond the apex of the lower jaw. The tympanic membrane is absent, and the tympanic annulus is not visible, but the columella and ostia pharyngea are present. The phalangeal formula of the hand is 1-2-3-3, and there is no webbing on the hand. The finger I is short, with the thumb-to-hand ratio being 0.448 in males and 0.425 in females. In males, the finger I bears keratinized dark brown spiculae, forming a nuptial pad. The phalangeal formula of the foot is 2-2-3-4-3, and the webbing in toes is well developed. The skin is smooth, but the dorsal and lateral surfaces are covered with dense, well-defined minute spiculae, giving the skin a velvety appearance on these surfaces. The ventral skin is smooth to areolate. The vertebral column is conspicuous, but neural processes are absent.

The dorsum is black with a few asymmetric yellow-green circles, dots, or lines. Dorsolaterally, there is a yellow-green foam-like reticulation forming a band from above the eye to the groin, with numerous black, rounded, irregular spots smaller than the diameter of the eye. Laterally, a black band extends from the tip of the snout almost to the groin, sometimes interrupted by yellow-green lines. The limbs and the dorsal surfaces of the hands and feet display yellow-green foam-like reticulation with black spots, varying in size. The thumb and, occasionally, the tips of the fingers and toes are orange to vermillion red. In males, the throat and venter are cream-colored, becoming yellowish posteriorly, sometimes with a few tiny black dots on the throat and lateral or posterior vent. In females, the throat and venter are yellow with numerous black dots. In both sexes, the palmar and plantar surfaces, and an oval blotch on the ventral surface of the cloaca and adjacent thigh areas are vermillion red, surrounded by small black dots. The iris is black with a golden-yellow ring surrounding the horizontally oriented pupil.

Comparisons.—*Atelopus harlequin, A. histrionicus, A. palmatus, A. planispina, A. pulcher, A. spumarius*. *Atelopus colomai* can be distinguished from *A. pulcher* by smaller male SVL (23.2 ± 1.46 mm, vs. 27.27 ± 1.07 mm, in *A. pulcher*), female ventral coloration in life (yellow with numerous black dots and a vermillion red-shaped blotch on ventral surface of cloaca and adjacent areas of thighs vs. vermillion red with irregular dark spots in *A. pulcher*). Tadpoles of *A. colomai* differ from those of *A. pulcher* in having the lower beak about twice as long as the upper (vs. one third to one fourth as long) and the nostrils positioned at about height of eye (vs. below height of eye). It differs from *A. harlequin, A. spumarius*, and *A. histrionicus* by having a long call (2049 ± 220 ms vs. 680 ± 30 ms in *A. harlequin*, 922 ± 24 ms in *A. spumarius sensu stricto*, 807 ± 80 ms in *A. histrionicus*). Further, *A. colomai* differs from *A. harlequin*

FIGURE A2.1 *Atelopus colomai.* Left: CJ 13037, female, 32.1 mm SVL, Puyo, Provincia Pastaza. Right: CJ 12198, holotype, male, 22.4 mm SVL. Río Pucayacu, Provincia Pastaza. SGS.

FIGURE A2.2 Morphological variation in *Atelopus colomai.* Top to bottom, left to right: CJ 12425, paratopotype, female, 30.4 mm SVL; CJ 12409, paratopotype, female, 30.8 mm SVL; CJ 12406, paratopotype, female, 28.5 mm SVL; CJ 12200, paratopotype, female; CJ 12408, paratopotype female; CJ 12405, paratopotype, female. Río Pucayacu, Provincia Pastaza. CJ 12711, male, 26.9 mm SVL, CJ 12712, male, 26.4 mm SVL; CJ 12716, male, 25.6 mm SVL. Puyo, Provincia Pastaza. SGS.

in dorsal pattern (reticulation pattern not densely developed on the dorsum and not broadly fused with dorsolateral bands vs. dark brown, with numerous dense yellowish-gold irregular circles or lines, fused with dorsolateral yellowish-golden reticulation in *A. harlequin*), female ventral pattern (presence of dots vs. absence in *A. harlequin*), male SVL (23.2 ± 1.46 vs. 20.6 ± 0.51 in *A. harlequin*), and by having more webbing on foot (basal in *A. harlequin*). Atelopus colomai male SVL is also smaller than in *A. histrionicus* (25.0 ± 0.67). Atelopus colomai can be distinguished from *A. palmatus* and *A. planispina* by coloration and pattern in life (dorsally black with few yellow-green asymmetric circles, dots or lines; dorsolaterally yellow-green foam-like reticulation forming a band from above eye to groin with numerous black rounded irregular spots; males below cream, posteriorly becoming yellowish, sometimes with few tiny black dots and a vermillion red ovally-shaped blotch on ventral surface of cloaca and adjacent areas of thighs, females below yellow with numerous black dots and a vermillion red ovally-shaped blotch on ventral surface of cloaca and adjacent areas of thighs vs. green blotches on reddish brown dorsum, absence of reticulation; ventrally unicolored orange in females of *A. palmatus* and whitish to reddish in *A. planispina*) and skin texture (smooth, covered in minute spiculae vs. presence of small spiculae on flanks, upper arm and from eye fading toward groin). Tadpoles of *A. colomai* differ from those of *A. palmatus* in exhibiting a golden tail pattern in life (vs. absence) and in having the upper beak shorter than the lower (vs. longer).

Natural history.—Figures A2.3, A2.4. The species inhabits *terra firme* lowland rain forest and can be found in close proximity to slow-moving clear water streams and rivers. Adult males were heard calling throughout the day with an activity peak after rainfalls in the morning and late afternoon, as observed in August and September. Reproductive males defend territories along the streams where entering males are approached by the resident and pursued, accompanied by intense calling, observed in the same months. Males were found perching on elevated positions such as branches, fallen trees, and rocks in the lower understory up to 1.5 m above ground during the day. At night, males were inactive, resting on leaves up to 2 m above ground. Females were found at night only, sleeping on leaves or in bromeliad axils 1–5 m above ground. When disturbed, specimens of *Atelopus colomai* present their colorful palmar and plantar surfaces while slowly walking away. At the type locality, amplectant pairs were observed in March, April, July–October, and December. Tadpoles were collected in July and

FIGURE A2.3 *Atelopus colomai.* Top: CJ 12198, holotype, male, hand showing the nuptial pad in finger I. SGS. Middle: amplectant pair, CJ (sc 414), male, CJ (sc 415), female. Bottom: adult resting in its microhabitat at night, 11:38 h, 10 February 2014. Río Pucayacu, Reserva Otoyacu, Provincia Pastaza. LAC.

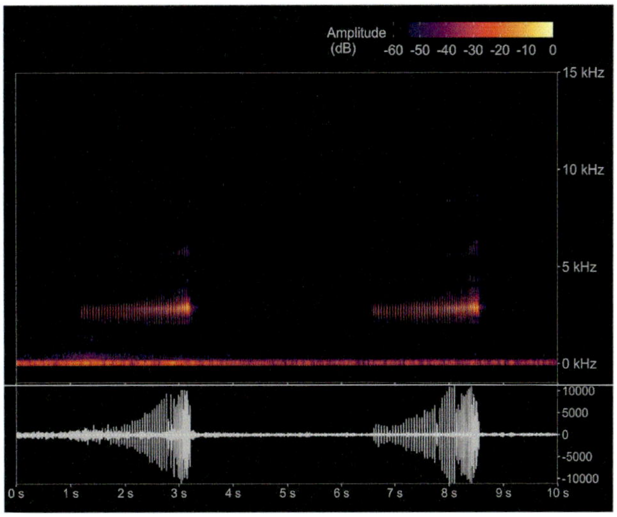

FIGURE A2.4 Oscillogram and spectrogram of two pulsed calls recorded from the holotype of *Atelopus colomai.* Taken from Plewnia *et al.* (2024).

metamorphs were observed in September suggesting that reproduction may take place year-round.

The species produces long pulsed calls for an *Atelopus.* A call recording (contains 11 pulsed calls with duration 2049 ± 220 ms (1747–2488 ms). Intervals between calls last 3601 ± 1791 ms (1791–6928 ms). Dominant frequency is 2848 ± 14 Hz (2821–2877 Hz). Amplitude is generally ascending over the course of the call being interrupted by 2–4 temporal decreases until it reaches its maximum followed by a subsequent decrease over the last 3–10 pulses. Call energy is also present in the frequency components related to the pulse rate, which shows weak modulation and a minimal increase toward the end of the call. Each call consists of 52 ± 6 pulses (42–60) that show a length of 7 ± 2 ms (3–4 ms) each. Pulses slightly increase in length over the course of each call and are not modulated in frequency.

Tadpoles.—Figure A2.5. The following description is based on a tadpole reared under laboratory conditions at CJ from parental stock collected at Río Pucayacu, Provincia Pastaza. In Stage 25, the body is depressed in lateral view, flattened ventrally, longer than wide, and widest between the eye and the spiracle. The snout is rounded in dorsal view. The eyes are large, directed laterally to anterolaterally, and positioned dorsolaterally, not visible in ventral view. The nares are ovoid, directed anterodorsally, and located closer to the eye than to the anterior margin of the snout. They are positioned at the height of the lower corner of the eye. The abdominal sucker is circular to ovoid, anteriorly bordering the oral disc and extending posteriorly to the height of the spiracle opening. The spiracle is sinistral, cylindrical, and a short tube not attached to the body wall, positioned ventrolaterally and oriented posteriorly. The anal tube is medial and tubular. The caudal musculature is massive, tapering rapidly in the distal half of the rounded tail. The caudal fins are low, with the dorsal fin slightly rounded, highest in the distal half of the tail, and not extending onto the body. The margin of the ventral fin is straight. The oral disc is ventral, almost as wide as the body, and bordered anteriorly and laterally by a single row of short, rounded papillae. A submarginal papilla is present on each side of the lower labium (P-3). The labial tooth row formula is 2/3, with all tooth rows approximately equal in length. The upper beak of the jaw sheaths is narrow

FIGURE A2.5 Ontogenetic changes in an individual of *Atelopus colomai* from tadpole to subadult, reared under laboratory conditions at CJ from parental stock collected at the type locality. Top to bottom, left to right: CJ (sc 10997), stage 25, lateral, dorsal and ventral views, TL = 8.3 mm; stage 41, TL = 14.5 mm; Stage 42, TL = 14.1 mm; Stage 46, 71 days after stage 25, 6.8 mm SVL; subadult 244 days after stage 46, 18 mm SVL. SGS. Río Pucayacu, Provincia Pastaza. SGS.

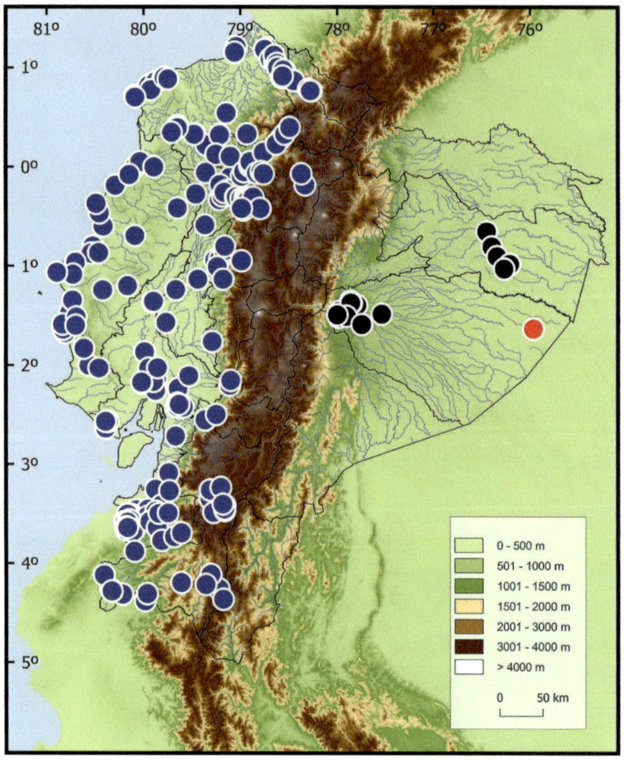

FIGURE A2.6 Localities of known occurrence of *Atelopus colomai* (black dots), *A. harlequin* (red dot), and *Rhinella bella* (blue dots) in Ecuador.

and nearly transverse, while the lower beak is broadly V-shaped and about twice as long as the upper beak.

The dorsum, flanks, and caudal musculature are black to dark brown with dense, irregular, pale brown to golden markings consisting of aggregated minute dots. These markings extend marginally onto the transparent tail fin. Below, the proximal half of the body is whitish translucent with black dots, which are densest below the suctorial disc, and golden dots appear toward the flanks. The distal half of the body is black with scattered golden dots toward the flanks and tail. The tail is cream-colored with black and golden dots. The pupil is black, surrounded by a cupreous to golden ring. Toward metamorphosis, the pattern and coloration remain stable, although the golden markings decrease in the later stages. The limbs turn orange at Stage 42.

Distribution.—Figure A2.6. *Atelopus colomai,* an endemic species to Ecuador, is known from the Amazonian lowlands of Ecuador in Orellana and Pastaza provinces. Its elevational range is 240–1020 m in an area of extent of occurrence of about 7100 km^2.

This diurnal, terrestrial, and arboreal species inhabits tropical rainforests and partially intervened areas in the evergreen forest of lowlands of Napo-Curaray, the evergreen forest of lowlands of Tigre-Pastaza, and the evergreen montane foothill forest of north of Cordillera Oriental of the Andes ecosystems within the Napo-Curaray, the North of Cordillera Oriental of the Andes, and the Tigre-Pastaza biogeographic sectors, where the average annual rainfall is 3287–4335 mm and the average annual temperature is 20.9–25.2 °C.

Conservation.—Plewnia *et al.* (2024) suggest listing *Atelopus colomai* as Endangered under criteria B1 and B2a,b of the IUCN Red List of Threatened Species as (1) it is known to persist only at six threat-defined localities, (2) all of which are severely fragmented, and (3) all of which are threatened by ongoing deforestation, agriculture, road construction, and pollution (observed September 2021; August–September 2022, June–August 2023). *Atelopus colomai* can be locally abundant with up to 21 individuals per person*hour found in nocturnal visual encounter surveys at the type locality in September 2022. Northwest of El Triunfo, where some of the paratypes were collected, the species is less common with five individuals found in approx. 20 person*hours in 2021. Tarvin *et al.* (2014) observed 356 individually identified specimens through the course of a two-year capture-recapture study in a population along Río Pucayacu, close to the type locality. Monitoring studies conducted between 1994 and 2012 in Yasuní National Park (Read, 2018) identified this species as rare. Only five individuals were discovered at two locations within the park. Including historic localities, the area of extent of occurrence is roughly 7100 km^2 and the area of occupation is less than 200 km^2 in recently confirmed localities. At the type locality, an almost complete deforestation of remaining habitats occurred in late

2022. Fieldwork in August 2023 revealed the presence of few individuals in a remaining 10 m wide fragment of forest along one side of the stream. Deforestation has also been concluded as the reason for declines in another population in the Pucayacu river by Tarvin *et al.* (2014). In the remaining four sites, there is a constant reduction in the extent and quality of remaining habitat. Tarvin *et al.* (2014) conducted a comprehensive *Bd* sampling, documenting the pathogens' absence in *Atelopus* but its presence in syntopic anurans in their study site. *Batrachochytrium dendrobatidis* (*Bd*) has been repeatedly detected in *A. colomai* (Lötters *et al.*, 2023). However, no declines related to *Bd* were noticed, similar to other lowland *Atelopus* species (Flechas *et al.*, 2012, 2015; Ballestas *et al.*, 2021). *Atelopus colomai* is further likely to be threatened by climate change in the near future, which might both impact the species habitat as well as interfer and exacerbate existing threats such as disease (Lötters *et al.*, 2023). The species from the western Amazon are among the harlequin frogs likely to be most affected by climate change (Lötters *et al.*, 2023).

It has been recorded in the public reserve Parque Nacional Yasuní in the Reserva de la Biósfera Yasuní and the private reserve Reserva Otoyaku. *Atelopus colomai* is currently kept in a survival-assurance colony at CJ, where it has already been successfully bred to produce the first generation of descendants. A backup colony is held at Centro de Conservación de Anfibios, Bioparque Amaru, Cuenca.

Remarks.—Before Plewnia *et al.* (2024) described, *Atelopus colomai* populations were classified as *A. spumarius* or *A.* sp. *(spumarius complex)*. See Chapter 3 of Volume II. According to Lötters *et al.* (2024), *A. colomai* is sister to *A. histrionicus*. Comparative adult osteology is provided in the publication by Plewnia *et al.* (2024).

According to Womack *et al.* (2018), hearing among *Atelopus* species is relatively consistent, suggesting that the partial middle ear present in *Atelopus colomai* (vs. *A. elegans* that lacks a middle ear) does not provide a significant hearing advantage. Their findings indicate that *Atelopus* hearing is unlikely to be facilitated by vibrations of the skin overlying the typical tympanic membrane region or the body lung wall, leaving the extratympanic hearing pathways in *Atelopus* unresolved.

Key references.—(Duellman and Lynch, 1969) (tadpole); (Tarvin *et al.*, 2014) (population, natural history); (Womack *et al.*, 2018) (comparative hearing morphology); (Plewnia *et al.*, 2024) (description, tadpole, call, osteology).

ATELOPUS HARLEQUIN COLOMA *ET AL.* IN LÖTTERS *ET AL.*, 2025

CLOWN HARLEQUIN FROG/RANA ARLEQUÍN PAYASO

Atelopus harlequin Coloma *et al.* in Lötters *et al.*, 2025: nnn. **Holotype.**—MUBI 2448 from km 74 on road from Iquitos to Nauta along a tributary of the Río Itaya (4°17'08" S, 73°31'45" W, ca. 100 m), Departamento Loreto, Peru.

Etymology.—The specific name refers to the widely used common name "harlequin" for frogs of the genus *Atelopus* that possess striking color patterns and occasionally make awkward moves.

Identification.—Figures A2.7, A2.8. Adults of this species attain a maximum SVL of 25.7 mm in females and of 21.2 mm in males. The body is slender, with a body width-to-SVL ratio of approximately 0.250 in males and 0.272 in the female. The species has long legs, with a tibia length-to-SVL ratio of 0.479 in males and 0.490 in the female. The snout is acuminate and protrudes beyond the apex of the lower jaw. The tympanic membrane is absent, and the tympanic annulus is not visible, but the columella and ostia pharyngea are present. The phalangeal formula of the hand

FIGURE A2.7 *Atelopus harlequin*. Left: female. Right: male. Near the type locality, Iquitos-Nauta road, Departamento Loreto, Peru. Jaime Culebras.

FIGURE A2.8 *Atelopus harlequin*. Left: QCAZ 9448, male, 20.7 mm SVL. Right: QCAZ 9544, male, 19.3 mm SVL. Lorocachi, Provincia Pastaza. LAC.

is 1-2-3-3, and there is no webbing on the hand. The finger I is short, with the thumb-to-hand ratio being 0.432 in males and 0.400 in the female. In adult males, the finger I has a keratinized brownish nuptial pad. The phalangeal formula of the foot is 2-2-3-4-3, and the webbing of the toes is basal. The skin is smooth, with the dorsal and lateral surfaces covered in dense but poorly distinct minute spiculae. The ventral skin is smooth to slightly areolate, while the cloaca and adjacent thighs are areolate. The vertebral column is conspicuous, but neural processes are absent.

The dorsum is dark brown with numerous dense yellowish-gold irregular circles or lines. In some specimens, these form a continuous foam-like reticulation pattern across the dorsum. On the head, a dark brown blotch extends from one eyelid to the other, connected centrally to a line running to the tip of the snout, forming an irregular T-shaped mark in some specimens. The dorsal pattern is fused with a dorsolateral yellowish-golden reticulation, creating a band with numerous dark brown irregular spots smaller than the diameter of the eye. Laterally, a dark brown band extends from the tip of the snout almost to the groin, sometimes interrupted posteriorly by a few yellowish-golden lines. The limbs and dorsal surfaces of the hands and feet display a yellowish-golden foam-like reticulation with dark brown spots, varying in size. The thumb and, occasionally, the tips of the fingers and toes are orange to bright red dorsally. In both sexes, the throat, chest, and venter are white to cream, with some specimens showing a few scattered small brown dots. The palmar and plantar surfaces, and a triangular blotch below the cloaca and adjacent ventral surfaces of the thighs are bright red. The iris is black, with a broad golden ring surrounding the horizontally oriented pupil.

Comparisons.—*Atelopus colomai, A. histrionicus, A. manauensis, A. spumarius*. *Atelopus harlequin* differs from *A. manauensis* in dorsal pattern in life (dark brown, with numerous dense yellowish-gold irregular circles or lines, fused with dorsolateral yellowish-golden reticulation vs. few irregular circles, lines, and blotches, not broadly fused with the foam-like reticulated dorsolateral band). It differs from *A. spumarius*, *A. colomai*, and *A. histrionicus* by call duration (680 ± 30 ms vs. 922 ± 24 ms in *A. spumarius sensu stricto*; 2,049 ± 220 ms in *A. colomai*; 807 ± 80 ms in *A. histrionicus*). Further, *A. harlequin* differs from these species in dorsal pattern (reticulation pattern not densely developed on the dorsum and not broadly fused with dorsolateral bands in *A. spumarius sensu stricto*, *A. colomai*, and *A. histrionicus*), female ventral pattern (absence in the new species vs. presence of dots) and male SVL (20.6 ± 0.51 vs. 22.2 and 23.4 in *A. spumarius sensu stricto*; 23.2 ± 1.46 in *A. colomai*; 25.0 ± 0.67 in *A. histrionicus*). *Atelopus harlequin* also differs from *A. colomai* by having less foot webbing (extensive in *A. colomai*). *Atelopus harlequin* can be distinguished from *A. pulcher* by shorter calls (680 ± 30 ms vs. 1,200 ± 100 ms). *Atelopus harlequin* further differs from *A. pulcher* by smaller male SVL (20.6 ± 0.51 vs. 27.27 ± 1.07) and male coloration in life (dorsally dark brown with numerous dense yellowish-gold irregular circles or lines, sometimes forming a continuous foam-like reticulation pattern on dorsum; dorsolateral yellowish-golden reticulation forming a band with numerous dark brown irregular spots vs. dark brown dorsum with green spots, circles, and blotches and a green broad dorsolateral band, rarely interrupted by small dark brown dots; in females, bright red venter with dark brown spots and a dark brown line on chest in some specimens).

Natural history.—Figure A2.9. The species inhabits *terra firme* lowland rain forest and can be found in close proximity to small streams. The holotype was collected during the day in disturbed forest in the leaf litter. This diurnal species is in proximity to slow-moving streams with sandy substrate that were approximately 1–3 m wide and 30 cm deep. During the day, adult males are mostly found actively walking on the forest floor, close to the stream.

The species produces pulsed calls lasting approximately 680 milliseconds, with a frequency range of 3450–4050 Hz. Pulsed calls from a population close to Nauta (Peru, Loreto) were described based on five calls (recorded in a collection bag at 26 °C) with an average call duration of 680 ± 30 ms (670–710 ms). Frequency at the beginning of the call is 3,578 ± 98 Hz (3,450–3,690 Hz) and at the end of the call is 3,900 ± 100 Hz (3,770–4,050 Hz). Each call consists of 23.4 ± 2.4 pulses (20–27), at a pulse rate of 23.4 ± 2.2 (27.5–30.6) at the beginning of the call and 61.3 ± 7.9 (52.2–71.4) at its end. In addition, vocalizations from the type locality on the Iquitos-Nauta road, Peru, were done with an iPhone 6s. Distance to recorder was 0.05 m. A 16,462 ms recording was available for analysis. It consists of 12 pure tone short calls and 9 pulsed short calls. Pure tone short calls have duration of 16 ± 4 ms (12–25 ms) and a dominant frequency of 3,270 ± 244 Hz (2,724–3,553 Hz). Frequency shows harmonics and is modulated with a

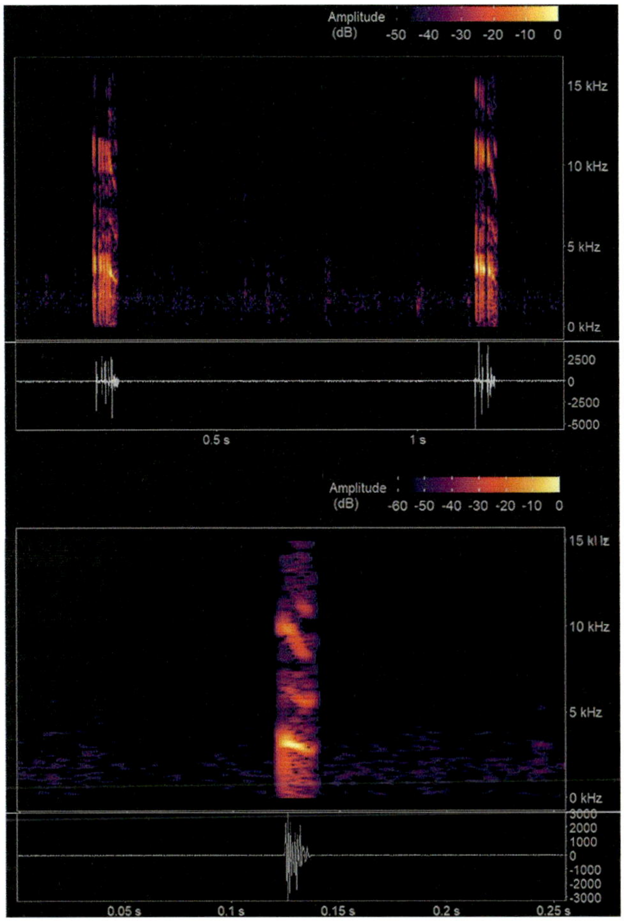

FIGURE A2.9 Oscillograms and spectrograms of calls of *Atelopus harlequin*. Top: pure tone short call. Bottom: pulsed short call. Close to Nauta on the Iquitos-Nauta road, Loreto, Peru (specimens not collected). Taken from Coloma *et al.* in Lötters *et al.*, 2025.

decrease toward the end. Pulsed short calls have call duration of 44 ± 8 ms (26–55 ms) and a dominant frequency of 3,495 ± 77 Hz (3,388–3,612 Hz). Amplitude peaks with the first pulse and descends over the course of the call. Frequency shows harmonics and is modulated in the last pulse with a decrease toward the end. Pulsed short calls consist of 3 ± 1 (2–5) pulses that show a length of 8 ± 7 ms (2–24 ms) each. The last pulse is the longest in all pulsed short calls.

Tadpoles.—Unknown.

Distribution.—Figure A2.6. *Atelopus harlequin* is known from the lowlands of eastern Ecuador to Amazonian Peru, between the Río Napo (to its right) and the Río Marañon (to its left) and west of the Amazon River at ca. 100–150 m. In Ecuador, it is known only from one locality in Provincia Pastaza, near the Peruvian border, at 195 m of elevation.

This species inhabits tropical rainforests in the Flooded Forest of Alluvial Plain of Rivers of Andean and Amazonian Cordilleras Origin ecosystem within the Tigre-Pastaza biogeographic sector, where the average annual rainfall is 3227 mm and the average annual temperature is 25.5 °C.

Conservation.—Coloma *et al.* (2025) suggest listing *Atelopus harlequin* as Data Deficient as its population status and its distributional range remain little understood. The species' EOO is 38,420 km^2 and AOO is 40 km^2. While most parts of the species' geographic range are covered in undisturbed primary forest with few human settlements, the populations on the easternmost edge of the species range along the road from Iquitos to Nauta (including its type locality) are highly threatened by habitat loss through large-scale deforestation, agriculture, and the construction of roads. Three populations along the Iquitos-Nauta road of the species persisted into the 2010s, but could not be confirmed in 2022 with almost no suitable habitat remaining. In a population a few kilometers further westwards, the species remained abundant despite ongoing deforestation in close proximity to the stream. Most other populations (including the one recorded from Ecuador) have not been visited in years hampering a thorough assessment of the species' conservation status. Presence and potential impact of *Batrachochytrium dendrobatidis* remain unknown. *Atelopus harlequin* is further likely to be threatened by climate change in the near future, which might both impact the species habitat as well as interfer and exacerbate existing threats such as disease (Lötters *et al.*, 2023). The species from the western Amazon are among the harlequin frogs likely to be most affected by climate change (Lötters *et al.*, 2023). In Ecuador, it has been recorded in the Reserva de la Biosfera Yasuní.

Remarks.—Before Coloma *et al.* (2025) described *Atelopus harlequin,* populations were classified as *Atelopus spumarius* or *Atelopus* sp. (*spumarius* complex). See Chapter 3 in Volume II. According to Lötters *et al.* (2024), *A. harlequin* is sister to all other members in the *flavescens-spumarius* lineage. Comparative adult osteology is provided in the original publication.

Key references.—Asquith and Altig (1987) (call, as *Atelopus spumarius*) (Coloma *et al.* in Lötters *et al.*, 2025) (description, call, osteology).

RHINELLA BELLA MENÉNDEZ-GUERRERO *ET AL.*, 2024

Beautiful Toad/Sapo Bello

Rhinella bella Menéndez-Guerrero *et al.*, 2024: 13.
Holotype.—QCAZ 23305 from between Palmas and Balsas (c. 2.003°S, 80.562°W, 72 m), Provincia Guayas, Ecuador.

Etymology.—The specific epithet is derived from the Latin word *bella*, meaning beautiful.

Identification.—(see Remarks). The maximum known SVL is 120 mm in males and 139.4 mm in females. The snout is short, subacuminate in dorsal view, and truncated to rounded in profile. The bony knob at the angle of the jaws is absent. The cephalic crests are low and thick with borders that have keratinized spicules and are continuous with the supraorbital, supratympanic, canthal, rostral, and preorbital crests. The infraorbital crests are reduced, and the preorbital crests are absent. The heel does not reach the posterior margin of the eye when the hind limbs are depressed. Vertebral apophyses are absent. The dorsal skin has medium and round tubercles, usually with a keratinized spicule at the tip, with or without large scattered warts, most of which are pointed. A thin middorsal line from the snout to the groin is present. The tympanic membrane and tympanic annulus are distinct. The parotoid glands are relatively large, elongate posteriorly, and relatively prominent in dorsal view, with visible small pores and a keratinized spicule at the posterior tip. The width of the parotoid glands varies from 9.17 to 14.3% of the SVL, and the height of the parotoid glands can vary from 15.91 to 22.4% of the SVL. The upper eyelid has keratinized warts. The palms lack rudimentary basal webbing between the fingers, and the lateral fringes are absent. There are small, rounded knobs at the tips of the digits, with finger lengths in the order of 3 > 4 > 2 > 1. The inner and outer metatarsal tubercles are conical and protrude distally, with the outer tubercles usually being similar in size than the inner tubercles. Subarticular tubercles are present, and supernumerary tubercles with keratinized spicules at the tip are distributed linearly on the toes on the plantar surface. The forelimbs and forearms are robust with low and rounded keratinized warts. Supernumerary tubercles are rounded with different sizes and are distributed irregularly on the palm. The palmar tubercle is larger than the thenar tubercle, and both are oval. The foot has basal webbing between the toes, and lateral fringes are present. There are small, rounded knobs at the tips of the digits, with toe lengths in the order of 4 > 5 > 3 > 2 > 1. Nuptial pads are present.

The dorsum of juveniles and females has various shades of brown mottled with irregular chocolate and beige blotches and large dark brown scapular blotches. Usually, there is a narrow beige middorsal stripe. Adult males are uniform cinnamon brown above. Juveniles have transverse dark bars on the upper limb surfaces. The venter is cream mottled with dusky pigment. The iris is brown.

Comparisons.—*Rhinella amabilis*, *R. caeruleostictus*, *R. horribilis*, *R. marina*. On the Pacific side of the Andes, the presence of cranial crests distinguishes *R. bella* from *R. amabilis*. *Rhinella bella* superficially resembles *Rhaebo caeruleostictus*, but it differs by having an externally visible tympanum (not visible in *R. caeruleostictus*).

Rhinella bella differs from populations of *R. horribilis* (from Central America) by having a smaller body size, along with a proportionally larger femur, wider head, and larger tibia. *Rhinella bella* is distinguishable from *R. horribilis* by a depressed area in the quadratojugal–squamosal contact region and the elongation of the frontoparietal bones at their most distant region. *Rhinella bella* differs in environmental envelope from *R. horribilis* and *R. marina* (and is separated by high genetic distances from both species as well). The advertisement calls of *R. bella* are significantly shorter, with fewer notes and a shorter rise time than those of *R. horribilis* and *R. marina*. The number of notes per call of *R. bella* (midrange = 23; range 6–40) is also lower than that reported in other species of the *R. marina* group [e.g., *R. arenarum* (midrange = 84, range 61–107); *R. marina* (midrange = 176.5, range 18–335); *R. poeppigii* (midrange = 27.5, range 10–45); and *R. diptycha* (midrange = 36.5, range 33–40]. Furthermore, *R. bella* produces calls with a higher dominant frequency than *R. horribilis* and *R. marina*.

Natural history.—See Remarks. Figure A2.10. Individuals have been found in artificial open areas, including agricultural fields, grasslands, and urban and suburban habitats. It occurs infrequently in undisturbed forests. Reproduction takes place in ponds, rivers, and swamps. Reproductive activity (calling males or amplectant pairs) has been observed mainly in April, June, and July. Eggs are deposited in strings in water. Tadpoles are aquatic and have a high resistance to heat, up to 42.5 °C.

The advertisement call of 17 males recorded at six localities in western Ecuador is a short trill with a mean duration of 1.51 s (range 0.94–2.56 s), a mean call rise time of 0.94 s (range 0.65–1.79 s), and a mean dominant frequency of 803.05 Hz (range 668.00–961.83 Hz). The call consists of

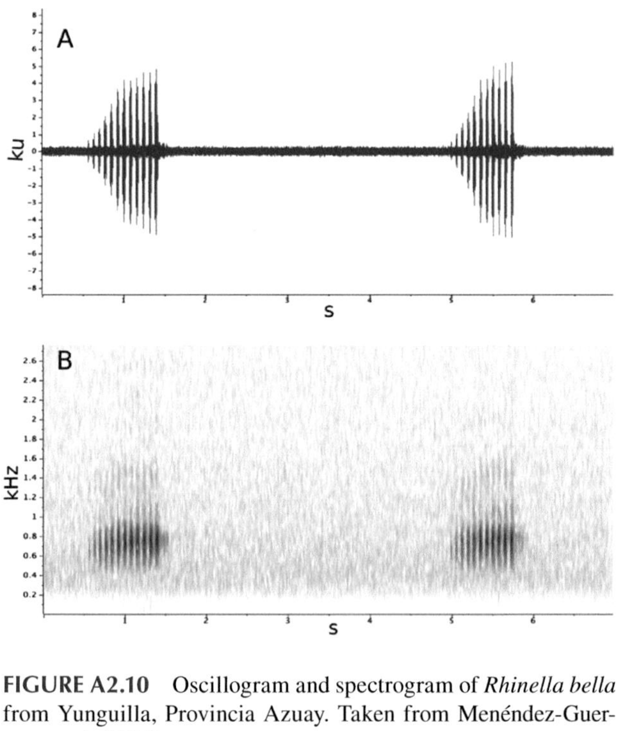

FIGURE A2.10 Oscillogram and spectrogram of *Rhinella bella* from Yunguilla, Provincia Azuay. Taken from Menéndez-Guerrero *et al.* (2024).

10–34 notes with 2–5 pulses per note. The mean pulse rate is 91.31 pulses per second (range 66.95–113.08 pulses per second), and the mean call repetition rate is 8.57 calls per minute (range 5–14 calls per minute).

Tadpoles.—See Remarks.

Distribution.—Figure A2.6. *Rhinella bella* is known from western Ecuador to extreme northwestern Peru. In Ecuador, it occurs in Esmeraldas, Carchi, Imbabura, Pichincha, Santo Domingo de los Tsáchilas, Manabí, Cotopaxi, Los Ríos, Guayas, Bolívar, Chimborazo, Cañar, Santa Elena, Azuay, El Oro, and Loja provinces. Its elevational range is 0–2606 m in an area of extent of occurrence of about 126,654 km^2.

This mostly nocturnal, terrestrial species inhabits tropical rain and dry forest, premontane and montane habitats, and intervened areas around human settlements in 19 ecosystems (Volume II, Table 4, as *Rhinella horribilis*) within nine biogeographic sectors (Volume II, Table 0.5, as *R. horribilis*), where the average annual rainfall is 231–3141 mm and the average annual temperature is 12.7–25.9 °C.

Conservation.—*Rhinella bella* is suggested to be included in the IUCN Red List as Least Concern (Menéndez-Guerrero *et al.*, 2024), in view of its wide distribution, tolerance of a broad range of habitats, presumed large population, and because it is unlikely to be declining to qualify for listing in a more threatened category. It is a very abundant species, and its range is increasing. Overall, there are no significant threats to this very adaptable, invasive species. In Ecuador, it has been recorded in the public reserves Refugio de Vida Silvestre Pacoche, Reserva Ecológica Arenillas, Reserva Ecológica Mache Chindul, Reserva Ecológica Manglares Cayapas Mataje, Reserva Ecológica Manglares Churute, and Parque Nacional Machalilla, the protected forests of Asociación Agrícola Carchi-Imbabura, Bosque Petrificado del Puyango, Chillanes Bucay, La Ceiba, Comuna Loma Alta, Loma Alta y Ampliación, Milpe Pachijal, Río Arenillas Presa Tahuín, Toachi Pilatón, and some private reserves (e.g., Reserva Otokiki, Reserva Otongachi, Reserva Río Manduriacu, Reserva Tesoro Escondido, Centro Científico Río Palenque, Bosque Integral Otonga, Estación Biológica Bilsa).

Remarks.—Before Menéndez-Guerrero *et al.* (2024) described *Rhinella bella*, populations from western Ecuador were classified as *R. horribilis*. See account and figures of *R. horribilis* in Volume II. According to Menéndez-Guerrero *et al.* (2024), *R. bella* is in the *R. marina* group and it is the sister taxon of *R. horribilis*. It was distinguished from *R. horribilis* from Central America (with which it had previously been confused) and *R. marina*, on the basis of morphometrics, advertisement call, and molecular markers. Nonetheless, the taxonomic split of *R. horribilis* remains uncertain until further studies integrate a more complete geographic sampling and specimens from intermediate areas in Central America and the Colombian Chocoan region.

Rhinella bella have been presumably used in traditional medicine and magical practices by ancient cultures in Peru and Ecuador (See Amphibian Iconography in Ecuadorian archaeology, in Chapter 1 of Volume I). Schmeda-Hirschmann *et al.* (2020) studied the parotoid gland secretions of Peruvian populations of *Rhinella bella* (as *R. horribilis*). They found 55 compounds composed of alkaloids, bufadienolides, and argininyl diacids derivatives. They inhibited the proliferation and migration of A549 lung cancer cells.

Key references.—Freile *et al.*, 2020 (account, as *Rhinella horribilis*); Armijos-Ojeda *et al.*, 2021 (account, as *Rhinella horribilis*); Menéndez-Guerrero *et al.*, 2024 (descritpion, call, phylogeny).

CENTROLENIDAE

CENTROLENE CAMPOSI CISNEROS-HEREDIA *ET AL.*, 2023

Campos' Glassfrog/Rana de Cristal de Campos

Centrolene camposi Cisneros-Heredia *et al.*, 2023: 6.
Holotype.—DHMECN 11407 from La Enramada (3.1628°S; 79.5886°W, 2950 m), Provincia Azuay, Ecuador.

Etymology.—The specific epithet honors Felipe Campos Yánez, an Ecuadorian zoologist and conservationist.

Identification.—Figure A2.11. The maximum known SVL is 31.2 mm in males, females are unknown. The snout is rounded to subacuminate in dorsal view, sloping in lateral view. The dentigerous process of the vomer is absent. The tympanic annulus is barely visible, with the lower three-fourths visible, and the tympanic membrane is colored like the dorsal skin, with the supratympanic fold present and low. The dorsal skin is shagreen with dispersed, low, rounded warts, and microspicules and spicules are present (at least in males). The ventral skin is granular, with the subcloacal area enameled, strongly granular, with two large subcloacal warts and with an enameled cloacal sheath. The parietal peritoneum is white, with iridophores covering two-thirds of the parietal peritoneum; the pericardium is covered by iridophores, with all other visceral peritonea clear. The liver is lobed (five lobes), and the hepatic peritoneum is clear (lacking an iridophore layer). Adult males have a projecting humeral spine. There is basal webbing between fingers I and II and moderate webbing between fingers II and IV; toe webbing is extensive. There is a low, enameled metacarpal fold continuing with an elevated, thick, enameled ulnar fold; elevated, low, enameled metatarsal and tarsal folds; and a low tarsal fringe on the inner tarsus. Nuptial excrescences are type I; the prepollex is concealed; finger

FIGURE A2.11 *Centrolene camposi*. DHMECN 111407, holotype, male, 29.1 mm SVL, La Enramada, Provincia Azuay. Juan Carlos Sánchez-Nivicela.

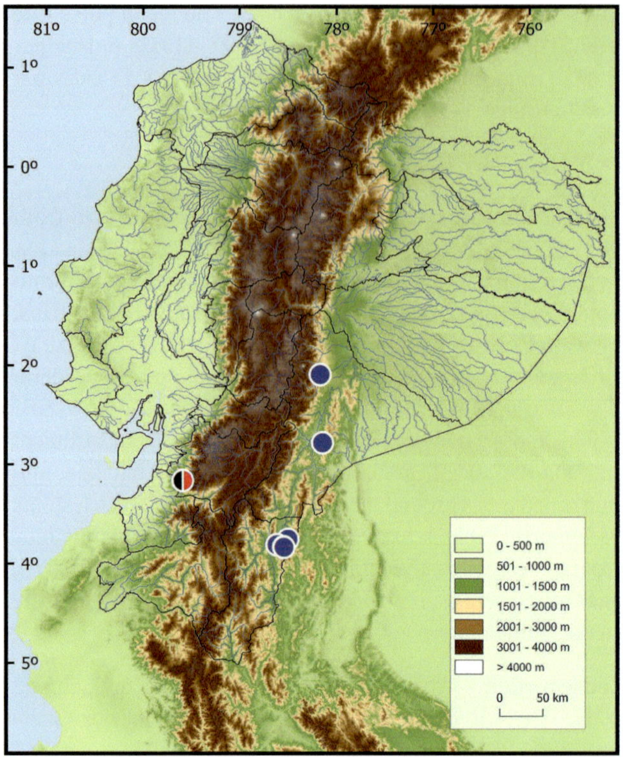

FIGURE A2.12 Localities of known occurrence of *Centolene camposi* (half-black dot), *C. ericsmithi* (half-red dot), and *C. zarza* (blue dots) in Ecuador.

I is shorter than finger II. The diameter of the eye is larger than the width of the disc on finger III.

The dorsum is bright green, with a thick yellowish-white labial stripe continuing into a faint yellowish line between the lip and the anterior one-fourth of the body. The flanks are yellowish-green as well as the hidden surfaces of limbs and digits. The metacarpal, ulnar, metatarsal, and tarsal folds are enameled. The bones are green. The iris has a white background; it is flesh-colored toward the center, with fine brown reticulations. Melanophores are present on the dorsal surfaces of the hands and feet and at the bases of finger IV, toe IV, and toe V.

Comparisons.—*Centrolene buckleyi*, *C. condor*, *C. ericsmithi*, *C. heloderma*. Characters of *C. camposi* in parentheses. *Centrolene buckleyi* differs by having low or absent ulnar and tarsal folds (elevated and thick ulnar and tarsal folds). *Centrolene heloderma* differs by having pustular dorsal skin (shagreen with dispersed warts), the tympanic annulus completely visible (tympanic annulus barely visible), a gray–lavender dorsum in preservative (lavender), an outer tarsal fold with low white tubercles (enameled fold without tubercles), and a humeral spine distinctly projecting from the arm (humeral spine curved toward arm). *Centrolene ericsmithi* differs from *C. camposi* by its round snout in lateral view (sloping), thin yellowish labial stripe (thick, white labial stripe), row of white tubercles between the lip and arm insertion (white tubercles absent), yellowish line between the arm insertion and groin (faint white line between the lip and anterior one-fourth of the body), and warts and spicules on the dorsum with the same color as the surrounding dorsal surfaces (warts and spicules on the dorsum lighter than the surrounding dorsal surfaces). *Centrolene condor* differs by having a green dorsum with abundant yellowish-white flecks and abundant dark flecks (uniformly bright green dorsum, sometimes with dark flecks); a cream–yellow iris with fine dark reticulations (white background, flesh-colored toward the center, fine brown reticulations); and vomerine teeth present (absent).

Natural history.—Individuals have been found in a steep creek. Males call from the upper sides of leaves.

Tadpoles.—Unknown.

Distribution.—Figure A2.12. *Centrolene camposi*, an endemic species to Ecuador, is known only from the type locality (La Enramada) in Provincia Azuay, near the border with Provincia El Oro, on the southwestern slopes of the Cordillera Occidental of the Andes of Ecuador at an elevation of 2950 m.

This mostly nocturnal, arboreal species inhabits a highly intervened evergreen montane ecosystem with patches of forests characterized by trees greater than 15 m in height and densely loaded with epiphytes such as bromeliads, mosses, and orchids within the Cordillera Occidental of the Andes biogeographic sector, where the average annual rainfall is 699 mm and the average annual temperature is 11.7 °C.

Conservation.—*Centrolene camposi* is herein categorized as Data Deficient. It is only known from one locality, and there is little information about the population of this species.

Remarks.—According to Ron *et al.* (2024), *Centrolene kutuku* is the sister species of *C. camposi*. This biogeographic pattern of sister species on opposite sides of the Andes is unusual. According to Cisneros-Heredia *et al.* (2023), trans-Andean distributions might have been possible until more recent geological periods. This unusual biogeographic pattern suggests a unique combination of topography and environmental history in the Andes of southern Ecuador.

Key references.—Cisneros-Heredia *et al.*, 2023 (description, phylogeny).

CENTROLENE ELISAE FRANCO-MENA *ET AL.*, 2024

ELISA'S GLASSFROG/RANA DE CRISTAL DE ELISA

Centrolene elisae Franco-Mena *et al.*, 2024: 21. **Holotype.**—MZUTI 84 from Las Caucheras (0.6133°S, 77.8974°W; 2187–2191 m), Provincia Napo, Ecuador.

Etymology.—The specific epithet honors Elisa Bonaccorso in recognition of her contributions to bird systematics and biogeography, conservation biology, batrachology, and her passionate commitment to the education of the next generation of scientists. This is also a recognition for her luminous presence in the life of Juan Manuel Guayasamin.

Identification.—Figure A2.13. The maximum known SVL is 25.3 mm in males and 27.2 mm in a female. The snout is rounded in dorsal profile and inclined in lateral profile. Adult males possess humeral spines and vocal slits. Webbing is absent between fingers I and II, and webbing is basal between fingers II and III, with the outer fingers showing webbing. Webbing on the foot is extensive. The ulnar fold is low, the inner tarsal fold is short, and the outer tarsal fold is low. The prepollex is concealed, and nuptial excrescences are present, classified as type I. Finally, toe I is shorter than toe II.

FIGURE A2.13 *Centrolene elisae.* Left: ZSFQ 5367, paratype, female, 27.2 mm SVL. Reserva Biológica Yanayacu, Provincia Napo. Mateo Vega-Yánez and Daniela Franco-Mena. Right: male. La Cueva, unión del Río Mulatos y Río Langoa, Provincia Napo. LAC.

The dorsal surfaces are dark green, adorned with small to minute white spots. The upper flanks are sharply demarcated laterally from the lower white flanks. The throat and most of the venter appear pale green. The parietal peritoneum is yellowish-white. A whitish-yellow labial line runs along the lips. The ventrolateral borders of the arms and tarsus are white. Small white spots are located posterior to the cloacal opening, corresponding to pericloacal warts. The bones are green. The iris shows a copper–white and gray coloration with thin black reticulation and a horizontal brown stripe. The digits and discs are green with yellowish interdigital webbing.

Comparisons.—*Centrolene buckleyi, C. marcoreyesi. Centrolene elisae* is morphologically similar to *C. buckleyi sensu stricto* and *C. marcoreyesi*. Characters of *C. elisae* are in parentheses. *Centrolene buckleyi* is larger (maximum SVL 34.7 mm) than *C. elisae* (maximum SVL 25.3 mm). In *C. buckleyi*, the snout is slightly sloping to sloping (rounded), the texture of dorsal skin is shagreen with or without small warts (shagreen covered with minute spinules and spots), the dorsal coloration is bright to dark green; some individuals have scattered olive-green spots on the dorsum (dark green with small to minute white spots). *Centrolene marcoreyesi* is larger (maximum SVL 27.0 mm) than C. *elisae* (maximum SVL 25.3 mm). In *C. marcoreyesi*, the snout is sloping (rounded), the texture of dorsal skin is shagreen with dispersed low warts (shagreen covered with minute spinules and spots), and the dorsal coloration is green with whitish spots (dark green with small to minute white spots). Additionally, *C. elisae* has a well-defined two-note advertisement call that structurally differs from the one-note calls exhibited by closely related species.

Natural history.—Figures A2.14, A2.15. *Centrolene elisae* is a nocturnal species found on vegetation along streams. At Las Caucheras, six individuals were found on the leaves of shrubs and ferns in a paddock near a small stream and a river, approximately 20 to 250 cm above ground level; five individuals were calling. At Río Pucayacu, a calling male was found on vegetation 2 m above stream level. In Reserva Chamanapamba, one individual was calling, perched on a fern leaf 230 cm above ground level near a small stream. The streams where the species was recorded were between 100 and 150 cm wide in primary and secondary forests. At Yanayacu Biological Station, intensive inventories for three years (2002–2004) resulted in only three individuals of *C. elisae* suggesting that this species is rare; also near Yanayacu

FIGURE A2.14 *Centrolene elisae*. Top and bottom left: adults in their microhabitats at night. Reserva Biológica Yanayacu, Provincia Napo. Mateo Vega-Yánez and Daniel Franco-Mena. Bottom right: clutch of embryos. Reserva Chamanapamba.

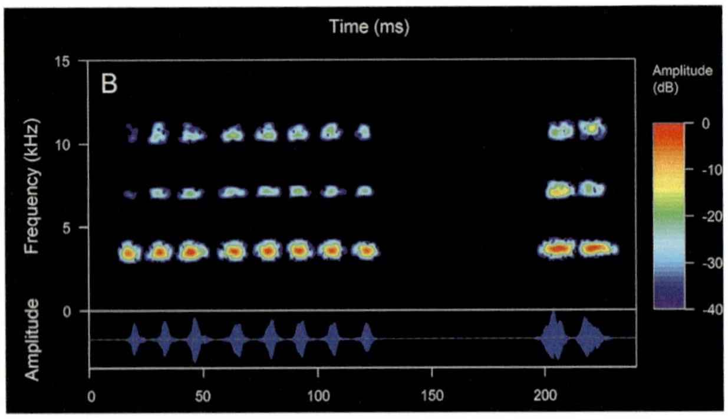

FIGURE A2.15 Oscillogram and spectrogram of *Centrolene elisae*. ZSFQ 5369, paratype, Reserva Biológica Yanayacu, Provincia Napo. Prepared by Diego Batallas-Revelo. Taken from Franco-Mena *et al.* (2024).

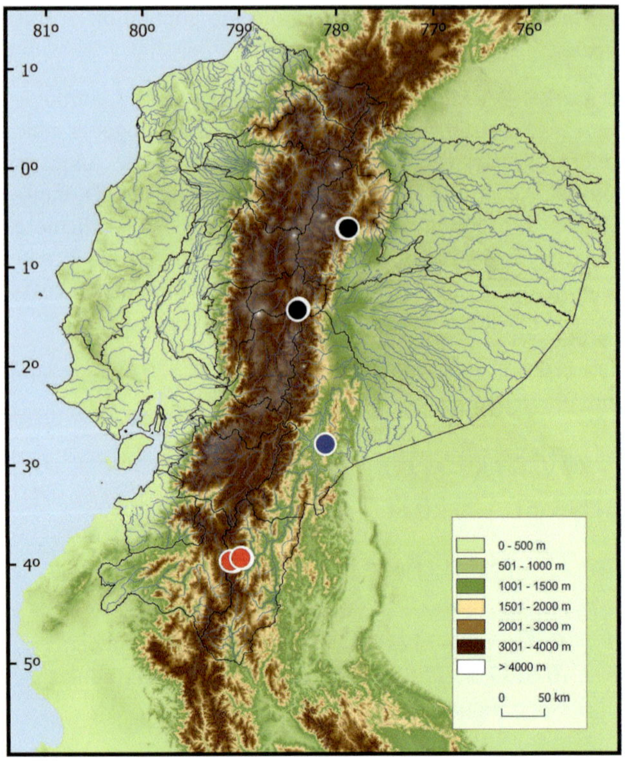

FIGURE A2.16 Localities of known occurrence of *Centrolene elisae* (black dots), *C. marcoreyesi* (red dots), and *C. kutuku* (blue dot) in Ecuador.

reserve, the area for cattle ranching and agriculture continues growing, reducing the habitat of the species. At Reserva Chamanapamba, three egg clutches with embryos in Gosner stage 22 were found. The egg clutches were attached to the upper side of a leaf ~170 cm above a small stream. The first clutch contained 28 embryos, the second clutch contained 21, and the third clutch contained 47 embryos; an adult male was observed near the eggs.

A call from Yanayacu Biological Station (0.6150S, 77.88189W; 2118 m), Napo province, Ecuador, was recorded by Daniela Franco-Mena on 11 March 2023 at 23:19 h, 10 °C. *Centrolene elisae* has a "tri" type advertisement call, composed by two pulsed notes. The calls are characterized by a mean duration of 0.203 s, an inter-call interval of 13.7–15.0 s, and a call rate of 4.1 calls/min. The first note is much longer than the second one and has a mean duration of 0.102 s, consisting on average of 12.3 pulses with a mean pulse duration of 7.9 ms and a mean pulse rate of 121.9 pulses/s; the second note has a mean duration of 0.045 s and consists of 2–3 pulses. The mean inter-note interval is 0.056 s, and the note rate is about 6.6 notes/s. The mean dominant frequency of the calls is 3.7 kHz, with a mean 90% bandwidth of 3.5–3.8 kHz. The first, longer note exhibits a very slight ascending frequency modulation, with an initial dominant frequency at 3.4–3.8 kHz (3.6 ± 0.2, n = 3) that increases to a 3.6–3.9 kHz (3.7 ± 0.1, n = 3) final dominant frequency. The fundamental frequency is not recognizable, but four harmonics are visible. The vocalizations of interacting males are composed of two notes but are different from the regular advertisement call, mostly by the much shorter inter-call interval (the calls were emitted more frequently).

Tadpoles.—Unknown.

Distribution.—Figure A2.16. *Centrolene elisae*, an endemic species to Ecuador, is known from the Cordillera Oriental of the Andes from Napo and Tungurahua provinces. Its elevational range is 2100–2586 m in an area of extent of occurrence of about 121 km^2.

This mostly nocturnal, arboreal species inhabits natural forests and intervened areas in the evergreen montane forests of the northern Cordillera Oriental of the Andes within the northern Cordillera Oriental of the Andes biogeographic sector, where the average annual rainfall is 849–2286 mm and the average annual temperature is 12.9–16.1 °C.

Conservation.—*Centrolene elisae* was categorized as Endangered by Franco-Mena *et al.* (2024) following B1b(ii)c IUCN criteria. *Centrolene elisae* is known from only four localities on the Amazonian slopes of the Andes: Las Caucheras, Yanayacu Biological Station (Provincia Napo Province), Chamanapamba Reserve, and Río Pucayaku in Nelson Palacios Reserve (Provincia Tungurahua). It occurs in the Reserva de la Biósfera Sumaco. The current estimated extent of occurrence for the species is < 5000 km^2. This area has suffered deforestation and fragmentation because of agriculture and cattle farming.

Remarks.—*Centrolene elisae* has been previously confused with *C. buckleyi*. It forms part of the *C. buckleyi* complex, with an unresolved polytomy with *C. venezuelense*, a potential new species from Colombia, and another one from northern Ecuador. Franco-Mena *et al.* (2024) described and provided CT images of osteological features of a male.

Key references.—(Amador *et al.*, 2018) (phylogeny, systematics, as *Centrolene buckleyi* [Ca2]); (Franco-Mena *et al.*, 2024) (description, phylogeny).

CENTROLENE ERICSMITHI CISNEROS-HEREDIA *ET AL.*, 2023

SMITH'S GLASSFROG/RANA DE CRISTAL DE SMITH

Centolene ericsmithi Cisneros-Heredia *et al.*, 2023: 14. **Holotype.**—DHMECN 11406 from La Enramada (3.1628°S; 79.5886°W, 2950 m), Provincia Azuay, Ecuador.

Etymology.—The specific epithet honors Eric Nelson Smith, a U.S. herpetologist and Curator of the amphibian and reptile collections at The University of Texas at Arlington. Eric is a prodigious collector who has described more than 60 new species of amphibians and reptiles from the most remote corners of the planet for almost three decades.

Identification.—Figure A2.17. The maximum known SVL is 27.3 mm in a male; females are unknown. The snout is round in dorsal and lateral views. The dentigerous process of the vomer is absent. The tympanic annulus is barely visible, with the lower three-fourths visible, the tympanic membrane is colored like the dorsal skin, and the supratympanic fold is present and low. The dorsal skin is shagreen with dispersed spicules and covered by microspicules, while the ventral skin is granular. The subcloacal area is enameled and strongly granular, with two slightly larger subcloacal warts and an enameled cloacal sheath. The parietal peritoneum is white, with iridophores covering half of the parietal peritoneum, and the pericardium is covered by iridophores, while all other visceral peritonea are clear. The liver is lobed, and the hepatic peritoneum is clear, lacking the iridophore layer. Adult males have a projecting humeral spine. There is basal webbing between fingers I and III and moderate webbing between fingers III and IV, and the toe webbing is moderate. The enameled metacarpal area continues without a fold, featuring a low, slightly elevated, enameled ulnar fold, along with low, enameled metatarsal and tarsal folds and a low tarsal fringe on the inner tarsus. Nuptial excrescences are of type I, and the prepollex is concealed. Finger I is shorter than finger II, and the diameter of the eye is larger than the width of the disc on finger III.

FIGURE A2.17 *Centrolene ericsmithi*. DHMECN 111406, holotype, male, 27.3 mm SVL, La Enramada, Provincia Azuay. Juan Carlos Sánchez-Nivicela.

The dorsum is bright green dorsum with a thin yellowish labial stripe continuing with a row of white tubercles toward the arm insertion and a yellowish line between the arm insertion and the groin. The metacarpal area and the ulnar, metatarsal, and tarsal folds are enameled. The bones are green. The iris has a flesh-colored background with fine brown reticulations. There are a few melanophores present on the dorsal surfaces of the hands and feet and at the bases of finger IV, toe IV, and toe V.

Comparisons.—*Centrolene buckleyi*, *C. camposi*, *C. heloderma*, *C. lynchi*. *Centrolene ericsmithi* differs from *C. buckleyi* (characters of *C. ericsmithi* in parentheses) by having a sloping snout (round), moderately heavy supratympanic fold (low), absent outer tarsal fold (present), and iris with a horizontal brown stripe (brown stripe absent). *Centrolene camposi* differs by having a sloping snout in lateral view (round); a thick, white labial stripe (thin yellowish labial stripe); no row of white tubercles between the lip and arm insertion (present); a faint white line between the lip and the anterior one-fourth of the body (yellowish line between arm insertion and groin), and warts and spicules on the dorsum lighter than the surrounding dorsal surfaces (warts and spicules on the dorsum with the same color as surrounding dorsal surfaces). *Centrolene heloderma* differs by having pustular dorsal skin (shagreen with dispersed spicules), the tympanic annulus completely visible (tympanic annulus barely evident), a gray–lavender dorsum in preservative (lavender), an outer tarsal fold with low white tubercles (enameled fold without tubercles), and a humeral spine distinctly projected from the arm (humeral spine curved toward arm). *Centrolene lynchi* differs by having the snout truncate to sloping in lateral view (round); shagreen dorsal skin in males and females; males with low, white warts, spicules, and spiculated warts on the sides of the head (dorsal skin shagreen with dispersed spicules); a dull green dorsum with minute yellowish-white warts and small diffuse black spots (green dorsum); absent tarsal folds (present); type II nuptial pads (type I); and a humeral spine distinctly projected from the arm (humeral spine curved toward arm). Molecular analyses clearly differentiate *C. ericsmithi* from morphologically similar species found in the Andes.

Natural history.—Individuals have been found vocalizing in March 2015. Males call from the upper side of leaves, although the advertisement call is unknown.

Tadpoles.—Unknown.

Distribution.—See Figure A2.12. *Centrolene ericsmithi*, an endemic species to Ecuador, is known only from the type locality (La Enramada) in Provincia Azuay, near the border with Provincia El Oro, on the southwestern slopes of the Cordillera Occidental of the Andes of Ecuador at an elevation of 2950 m.

This mostly nocturnal, arboreal species inhabits a mostly intervened evergreen montane ecosystem with patches of forests characterized by trees greater than 15 m in height and densely loaded with epiphytes such as bromeliads, mosses, and orchids within the Cordillera Occidental of the Andes biogeographic sector, where the average annual rainfall is 699 mm and the average annual temperature is 11.7 °C.

Conservation.—*Centrolene ericsmithi* is herein categorized as Data Deficient. It is only known from one locality, and there is little information about the population of this species.

Remarks.—According to Cisneros-Heredia *et al.* (2023), *Centrolene ericsmithi is* the sister species of *C. sabini* (from Peru) + *C.* sp. (from southeastern Ecuador).

Key references.—Cisneros-Heredia *et al.*, 2023 (description, phylogeny).

CENTROLENE KUTUKU RON *ET AL.*, 2024

Kutukú Glassfrog/Rana de Cristal de Kutukú

Centrolene kutuku Ron *et al.*, 2024: 932. **Holotype.**—QCAZ-A 71386 from 22 km ESE of Santiago de Méndez (2.7901°S, 78.1266°W, 2264 m), sector Puchimi, Cordillera del Kutukú, Provincia Morona Santiago, Ecuador.

Etymology.—The specific epithet refers to the Cordillera Kutukú, the sub-Andean Cordillera where the new species was discovered. The name originates from the Shuar language, meaning paramo with stunted vegetation.

Identification.—The maximum known SVL is 21.9 mm in males; females are unknown. The snout is slightly subacuminate to round in dorsal view and slopes in lateral view. The dentigerous process of the vomer is present, but vomerine teeth are absent. The tympanic annulus is evident and almost completely visible, except for the uppermost border. The tympanic membrane matches the color of the surrounding skin, and the supratympanic fold is present. The dorsal skin is microspiculate, with abundant tubercles on the head and body. The ventral skin is granular, and the subcloacal area features two large subcloacal warts. Elevated, abundant, enameled pericloacal warts are present, but no other cloacal ornamentation is observed. The parietal peritoneum is white, with iridophores covering half of the ventral parietal peritoneum. The pericardium is covered by iridophores, while all other visceral peritonea remain translucent. The liver is lobed, and the hepatic peritoneum is translucent. Adult males have short humeral spines that barely protrude through the skin. There is no webbing between fingers I and II, with webbing present between other fingers. Toe webbing is extensive. A dermal fringe runs along the postaxial edge of finger IV and the hand, accompanied by a slightly enameled ulnar fold. A dermal fringe also appears along the postaxial edge of toe V, with slightly enameled metatarsal and tarsal folds, and warts on the heel are absent. The nuptial pad is unpigmented and classified as type I, while the prepollex is concealed. Finger II is slightly longer than finger I, and the eye diameter is larger than the width of the disc on finger III.

The dorsal surfaces are green with light green dots, and the bones are green. The flanks show a contrasting dorsal–ventral transition in color. The iris is cream with gray reticulations. Melanophores are absent on finger IV, but scattered melanophores are present along toes IV and V.

Comparisons.—*Centrolene camposi*, *C. ericsmithi*, *C. heloderma*, *C. zarza*. Characters of *C. kutuku* are in parentheses. *Centrolene kutuku* differs from all other glassfrogs, except *C. heloderma*, by having humeral spines in males, a green dorsum with light dots and without dark marks, a dorsum with abundant tubercles, and translucent visceral peritonea (except for the pericardium). *Centrolene kutuku* is similar to *C. camposi*, *C. ericsmithi*, *C. heloderma*, and *C. zarza* in having humeral spines in males and elevated dermal ornamentations (i.e., dorsal tubercles or warts) and lacking dorsal dark-colored marks. *Centrolene camposi* differs from *C. kutuku* by having a sloping snout in lateral view (rounded), a barely visible tympanic annulus (completely visible), a coloration in life with a green dorsum with thick yellowish-white labial stripe continuing into a faint yellowish lateral line and yellowish-green flanks (green dorsum with light green dots), and a larger body size in males (SVL 29.1–31.2 mm in *C. camposi* vs. 21.5–21.9 mm in *C. kutuku*). *Centrolene ericsmithi* differs from *C. kutuku* by the having tympanic annulus barely visible (almost completely visible), dorsal skin with dispersed spicules (with abundant tubercles), a coloration in life of a green dorsum with a thin yellowish labial stripe continuing into a row of white lateral tubercles and a yellowish lateral line (green dorsum with light green dots), and a larger body size in males (SVL 27.3 mm in *C. ericsmithi* vs. 21.5–21.9 mm in *C. kutuku*). *Centrolene heloderma* differs by having pustular dorsal skin (with abundant tubercles), a gray–lavender dorsum in preservative (lavender), and an outer tarsal fold with low white tubercles (enameled fold without tubercles). *Centrolene zarza* differs by having a rounded snout with elevated warts at the tip in dorsal view and sloping in lateral view (round, without elevated warts at the tip in dorsal view and rounded in lateral view); shagreen dorsal skin with scattered warts of varying size and abundant, elevated, enameled warts on the head and flanks (abundant tubercles, lacking enameled warts); a green dorsum with yellowish-green dots and enameled warts (green dorsum with light green dots); and larger body size in males (SVL 23.2–26.2 mm in *C. zarza* vs. 21.5–21.9 mm in *C. kutuku*).

Natural history.—A male was calling over a leaf 200 cm above the ground on the edge of a small spring flowing into a black-water ravine in a primary forest. Another male was on a leaf, 300 cm above the ground, next to a black-water stream. Both individuals were found in primary forest.

Tadpoles.—Unknown.

Distribution.—Figure A2.16. *Centrolene kutuku*, an endemic species to Ecuador, is known only from the type locality (22 km east–southeast of Santiago de Méndez) in Provincia Morona Santiago on the western slopes of the Cordillera Kutukú in the southeastern Andes of Ecuador at an elevation of 2255–2264 m.

This mostly nocturnal, arboreal species inhabits the evergreen montane forests on sandstone plateaus of the Cordillera del Cóndor within the Amazonian Cordilleras biogeographic sector, where the average annual rainfall is 1701 mm and the average annual temperature is 16.8 °C.

Conservation.—*Centrolene kutuku* is categorized as Data Deficient (Ron *et al.*, 2024). It is only known from one locality, and information about the population of this species is lacking. The locality is not in a protected area, and by 2020, there were agricultural lands at a distance of 2.7 km. Because collections in Kutukú have been limited, the distribution of the species is likely to be larger.

Remarks.—According to Ron *et al.* (2024), *Centrolene kutuku* is the sister species of *C. camposi*. This biogeographic pattern of sister species on opposite sides of the Andes is unusual. Divergence times are 2.7 Mya separating *C. camposi* and *C. kutuku*. *Centrolene camposi* should result from recent dispersal across the Andes, from the Amazon Basin to the Pacific Basin, a putative example of peripatric speciation. Dispersal should have been possible due to the relatively low elevations of the Andes in southern Ecuador.

Key references.—(Ron *et al.*, 2024) (description, phylogeny).

CENTROLENE MARCOREYESI FRANCO-MENA *ET AL.*, 2024

Marco Reyes' Glassfrog/Rana de Cristal de Marco Reyes

Centrolene marcoreyesi Franco-Mena *et al.*, 2024: 21. **Holotype.**—ZSFQ 4418 from Estación Científica San Francisco (3.971667°S, 79.079167°W; 1840 m), Provincia Zamora Chinchipe, Ecuador.

Etymology.—The specific epithet is a tribute to Marco M. Reyes-Puig (1978–2013), a notable herpetologist from the herpetology division of the Museo Ecuatoriano de Ciencias Naturales (now Instituto Nacional de Biodiversidad). Marco was the original collector of this new species on a field campaign in Zamora Chinchipe. The authors of this species description honor his work and memory as a brother (Juan Pablo Reyes-Puig), sister (Carolina Reyes-Puig), and friend.

Identification.—Figure A2.18. The maximum known SVL is 27.0 mm in males; females are unknown. The snout is rounded in dorsal profile and sloping in lateral profile. Adult males possess humeral spines, a vocal sac, and vocal slits. Webbing is absent between the inner finger and finger II but is reduced to moderate between the outer fingers, with webbing between fingers III and IV. Webbing on the feet is extensive. The inner and outer ulnar and tarsal folds are conspicuous. The prepollex is concealed, and nuptial excrescences are present. Toe I is shorter than toe II.

The dorsal surfaces of the body, arms, and limbs appear green and are covered with numerous whitish spots of various sizes. The upper lip shows a yellowish-white hue. The anterior two-thirds of the venter are yellowish-white, while the posterior third is translucent. The fingers, toes, and membranes are yellowish-green, and the bones are green. The iris displays a white tone with a slight lavender hue and fine brown reticulations.

Comparisons.—*Centrolene buckleyi*, *C. elisae*. *Centrolene marcoreyesi* is morphologically similar to *C. buckleyi sensu stricto* and *C. elisae*. Characters of *C. marcoreyesi* are in parentheses. *Centrolene buckleyi* is larger (maximum SVL 34.7 mm) than *C. marcoreyesi* (maximum SVL 27.0 mm). In *C. buckleyi*, the snout is slightly sloping to sloping

FIGURE A2.18 *Centrolene marcoreyesi*. Left: CJ 11363, male. SGS. Right: CJ 11364, male (note the humeral spines), 31.4 mm SVL. DAL. Guarumales, Provincia Zamora Chinchipe.

FIGURE A2.19 *Centrolene marcoreyesi*. Male taking care of embryos. Guarumales, Provincia Zamora Chinchipe. Jaime Culebras.

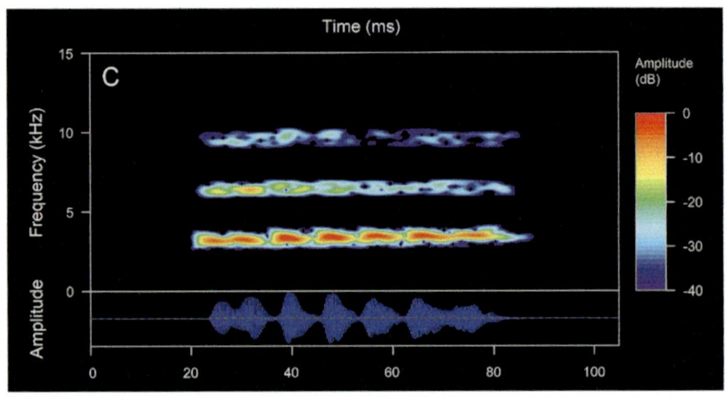

FIGURE A2.20 Oscillogram and spectrogram of *Centrolene marcoreyesi*. MUTPL 271/FUTPL-A 140, paratype, Abra de Zamora, Provincia Zamora Chinchipe. Prepared by Diego Batallas-Revelo. Taken from Franco-Mena *et al*. (2024).

(sloping), the texture of the dorsal skin of males is shagreen with or without small warts (shagreen with dispersed low warts), the dorsal coloration is bright to dark green, and some individuals have scattered olive-green spots on the dorsum (green with whitish spots). *Centrolene marcoreyesi* is larger (maximum SVL 27.0 mm) than *C. elisae* (maximum SVL 25.3 mm). In *C. elisae*, the snout is rounded (sloping), the texture of the dorsal skin of males is shagreen covered with minute spinules and spots (shagreen with dispersed low warts), and the dorsal coloration is dark green with small to minute white spots (green with whitish spots). Additionally, *C. elisae* has a well-defined two-note advertisement call that structurally differs from the one-note calls exhibited by closely related species.

Natural history.—Figures A2.19, A2.20. Individuals have been found at night in small streams on herbaceous vegetation. In Abra de Zamora, Provincia Zamora Chinchipe, several individuals were calling near small streams. The encountered males were calling from the upper surfaces of leaves, about 1 m from the water surface. In Parque Nacional Podocarpus, Provincia Zamora Chinchipe, several males were observed calling from about 2 m on leaves over fast-flowing streams. In Guarumales, Provincia Zamora Chinchipe, four individuals were found calling near a river, approximately 150 to 350 cm above the ground level; two individuals were found on leaves about 100–200 cm above ground level; another two individuals were observed on fern leaves. A male was observed calling near a clutch with 19 eggs (one of them dead). A male was observed calling from a fern leaf at 150 cm above ground level; the streams where the species is present are variable in width, between 2 and 8 m. Males were found in primary and secondary forests and on the edge of pastures.

Recordings were made by William E. Duellman on 8 March 1975 in Abra de Zamora, Provincia Zamora Chinchipe, Ecuador, between 21:45 and 21:55 h, at 11 °C, and by Paul Székely on 29 April 2017, between 20:49 and 21:08 h, at 14.5 °C. *Centrolene marcoreyesi* has a "tri" type of advertisement call composed by one pulsed note. The calls are characterized by a mean duration of 0.085 s, a mean inter-call interval of 34.1 s, and a mean call rate of 1.8 calls/min. Each call (note) consists on average of 8.1 pulses, with a mean pulse duration of 9.3 ms and a mean pulse rate of 103.6 pulses/s. The mean dominant frequency of the notes is 3.3 kHz, with a mean 90% bandwidth of

3.0–3.4 kHz. The call exhibits a slight ascending frequency modulation, with an initial dominant frequency at 2.8–3.3 kHz (3.1 ± 0.2, n = 35) that increases to a 3–3.6 kHz (3.3 ± 0.2, n = 35) final dominant frequency. The fundamental frequency is not recognizable, but up to three harmonics are visible. The two males, recorded in 2017, emitted two consecutive calls, besides the regular ones. These notes had the same duration, structure, and frequency as the regular calls, but the inter-call interval was significantly shorter, 0.8–1.7 s (1.1 ± 0.3, n = 10). It is almost certain that this behavior was triggered by the nearby presence of a competitive male, as the distance between the two recorded individuals was smaller than half a meter. It is worth mentioning that we observed some differences in physical and spectral parameters between the calls recorded at the same locality in 1975 and the ones from 2017. These differences are attributed to differences in recording conditions (especially the temperature, which in 2017 was higher) and/or a combination of general weather conditions and calling behavior of the males.

Tadpoles.—Unknown.

Distribution.—Figure A2.16. *Centrolene marcoreyesi*, an endemic species to Ecuador, is known on the southeastern slopes of Cordillera Oriental of the Andes in Provincia Zamora Chinchipe. Its elevational range is 1840–2190 m in an area of extent of occurrence of about 14 km^2.

This mostly nocturnal, arboreal species inhabits intervened areas and natural forests in the evergreen low montane forests of the southern Cordillera Oriental of the Andes ecosystem within the southern Cordillera Oriental of the Andes biogeographic sector, where the average annual rainfall is 750–942 mm and the average annual temperature is 15.6–17.5 °C.

Conservation.—*Centrolene marcoreyesi* is categorized as Endangered following IUCN criteria B1a,b(i, iii) (Franco-Mena *et al.*, 2024). *Centrolene marcoreyesi* has an estimated extent of occurrence < 100 km^2. Even if some of the known localities of *C. marcoreyesi* are inside protected areas, and as such benefit from conservation measures, this species is threatened by the degradation of its habitats, especially due to cattle farming, the introduction of invasive exotic species, and illegal and legal mining.

It occurs in the protected forests of Corazón de Oro and Cuencas que Forman los Ríos: San Francisco, San Ramón y Sabanilla and within the Podocarpus-El Cóndor biosphere reserve. It also occurs in some private reserves (e.g., Estación Científica San Francisco).

Remarks.—*Centrolene marcoreyesi* was previously confused with *Centrolene buckleyi*. According to Franco-Mena *et al.* (2024), *C. marcoreyesi* is the sister to *C. sabini*, from southeastern Peru, in the *C. buckleyi* complex, although support for this relationship is moderate. Franco-Mena *et al.* (2024) described and provided CT images of osteological features of a male.

Key references.—(Amador *et al.*, 2018) (phylogeny, systematics, as *Centrolene buckleyi* [Ca1]); (Franco-Mena *et al.*, 2024) (description, phylogeny, osteology).

CENTROLENE ZARZA SZÉKELY *ET AL.*, 2023A

Zarza Glassfrog/Rana de Cristal de Zarza

Centrolene zarza Székely *et al.*, 2023a: 60. **Holotype**.—MUTPL 932 from quebrada Las Mariposas (3.8341°S, 78.5458°W; 1434 m, datum WGS84), Refugio de Vida Silvestre El Zarza, Ecuador, Provincia Zamora Chinchipe, Ecuador.

Etymology.—The specific epithet is a noun in apposition and refers to the species' type locality, Refugio de Vida Silvestre El Zarza. This relatively small wildlife refuge conserves an impressive wealth of biodiversity, with countless species of plants and birds, more than 50 species of amphibians and reptiles, and several emblematic mammals, like the Amazonian tapir, jaguar, oncilla, and spectacled bear. It is surrounded by active mining concessions and thus fulfills an important role as a conservation island for the region, with an urgent need to expand connectivity between the reserve and neighboring conservation areas.

Identification.—Figure A2.21. The maximum known SVL is 26.2 mm in males and 27.0 mm in females. The snout is rounded in dorsal view and sloping in profile. The nostrils are slightly elevated, producing a depression in the internarial area, and the canthus rostralis is not evident in dorsal view, being rounded in cross section. The dentigerous processes of the vomers are ovoid, in a transverse row between the choanae, and separated medially by a distance slightly lower than the width of the processes, with each

FIGURE A2.21 *Centrolene zarza*. Left: MUTPL 932, holotype, female, 25.5 mm SVL. Right: male. Refugio de Vida Silvestre El Zarza, Provincia Zamora Chinchipe. Paul J. Székely.

process bearing 3–5 teeth. The tympanic annulus and tympanic membrane are evident but with coloration similar to that of the surrounding skin; the tympanum is large, with a diameter approximately 46% of the eye diameter, and there is a weak supratympanic fold present, slightly concealing the upper margin of the tympanum. The dorsal skin is shagreen with elevated warts, some enameled, corresponding to white spots, while the ventral skin is coarsely areolate. The ventral surfaces of the thighs below the vent have a pair of large, round, flat tubercles, and the cloacal region is bordered ventrally by many enameled, white warts. Half or more of the upper parietal peritoneum is covered by iridophores, but iridophores are absent on all visceral peritonea, including the pericardium. The liver is lobed and lacks iridophores. Adult males have small projecting humeral spines, round vocal slits, and a large subgular vocal sac. Webbing is absent between fingers I and II, basal between II and III, and moderate between the outer fingers, while the webbing between the toes is moderate. The outer edge of the forearms and tarsus has a row of enameled warts that often continue into the external edges of finger IV and/or toe V, with the fingers and toes having broad lateral fringes. Unpigmented type I nuptial pads are present in males, and the prepollex is concealed. Finger I is shorter than finger II, and the diameter of the eye is approximately twice as wide as the disc on finger III.

The dorsum is light green with many white or whitish elevated spots and flecks of various sizes, and the bones are green. The iris is white or yellowish-white with thick and thin black reticulations, rounded points on the upper and lower sides of the iris, and no circumpupillary ring. The fingers and toes are yellowish, usually lacking melanophores on the dorsal surfaces except for finger IV and toes IV and V.

Comparisons.—*Centrolene ballux*, *C. buckleyi*, *C. condor*, *C. pipilata*, *C. sanchezi*, *Cochranella resplendens*, *Nymphargus posadae*. *Centrolene zarza* is similar to species that have a green dorsum with white spots and/or flecks. *Centrolene condor* has a very different general habitus with a more robust body, smaller eyes, a less evident tympanum, and dark bluish-black/brown flecks and points along with white flecks (vs. slender body, larger eyes, evident tympanic annulus, and tympanic membrane and only white spots and/or flecks in *C. zarza*). *Centrolene pipilata* has a dorsum with yellowish-white flecks and diffuse dark green/black marks and a distinct prepollex (vs. only white spots and/or flecks and concealed prepollex). *Centrolene sanchezi* has a smaller body size and the presence of white warts in an area that extends from below the eye to the insertion of the arm (vs. absence of the white warts). Specimens of the *C. buckleyi* species complex that have white spots have less evident markings, dentigerous processes of the vomers without teeth, a less evident tympanum, and condition V1 of the visceral peritonea and also live at much higher elevations of 2050–3070 m (vs. dentigerous processes of vomers with teeth and condition V0 of the visceral peritonea). Somewhat similar species of the genus that live on the Pacific versant of the Ecuadorian Andes are *C. ballux*, which has a minute body size, a distinct prepollex, and lacks teeth on the dentigerous processes of the vomers (vs. medium body size, concealed prepollex, and dentigerous processes of the vomers with teeth), and *C. heloderma*, which has a unique pustular dorsum, condition V1 of the visceral peritonea, and vomers lacking teeth (vs. shagreen dorsum with elevated warts, condition V0 of the visceral peritonea, and vomers with teeth). *Cochranella resplendens* has iridophores on the pericardium and peritonea covering the intestines and stomach, and the males lack humeral spines (vs. iridophores absent on all visceral peritonea and males with humeral spines). *Nymphargus posadae* has almost indistinguishable tympanum, condition P2 and V1, and the males lack humeral spines (vs. evident tympanum, condition P3 and V0, and males with humeral spines).

Natural history.—Figures A2.22, A2.23. Individuals have been found during the night on the upper surfaces of leaves of vegetation bordering two small streams. Calling males were heard during January, March, June, October, and November, but intense activity (many males calling from lower heights, 50 cm to 3 m above the streams) was recorded between October and March. The female paratype was encountered near a calling male at ~4 m high over the stream. Two egg clutches in Gosner stage 19 were collected

FIGURE A2.22 *Centrolene zarza*. Top left: MUTPL 1051, paratype, female in its microhabitat at night. Top right: male, in its microhabitat at night. Bottom left: MUTPL-T22, egg clutches attached to a leaf (10 June 2021). Middle right: hatchlings in Gosner stage 20 (10 June 2021). Bottom right: hatchlings in Gosner stage 22 (13 June 2021). Refugio de Vida Silvestre El Zarza, Provincia Zamora Chinchipe. Paul J. Székely.

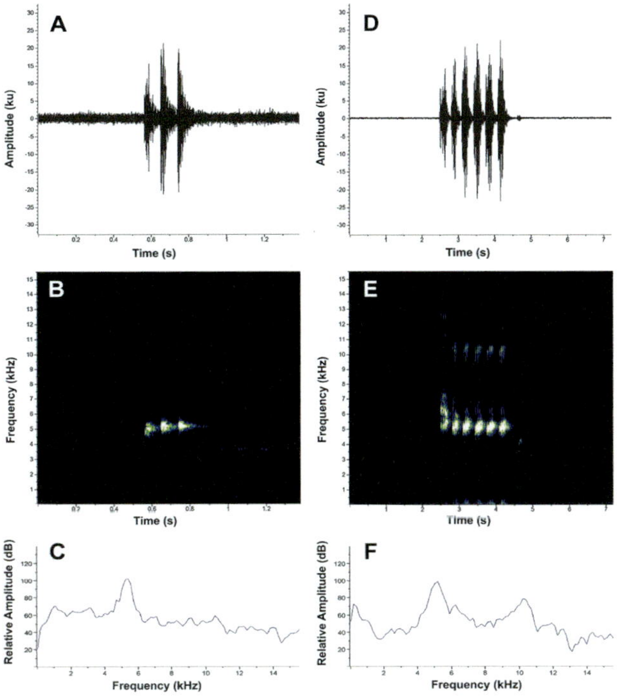

FIGURE A2.23 Calls of *Centrolene zarza*. Visual representation of the advertisement call (FUTPL-A 263; A–C) and courtship call (paratype MUTPL 933, FUTPL-A 261; D–F). A: oscillogram of a single-note call with three pulses. B: spectrogram of a single-note call. C: power spectrum of a single-note call. D: oscillogram of a multi-note call with six notes. E: spectrogram of a multi-note call. F: power spectrum of a multi-note call. Taken from Székely *et al.* (2023a).

FIGURE A2.24 Ontogenetic changes in *Centrolene zarza* from hatchlings to tadpole. Top to bottom. MUTPL-T22, lateral and dorsal views. Stage 24 (26 June 2021). Stage 25 (5 August 2021). Stage 26 (23 August 2021). Stage 31 (20 November 2021).

from the type locality (3.8379°S, 78.5418°W; 1460 m a.s.l.) on 9 June 2021. Both egg clutches were attached to the upper side of a leaf ~3 m above the stream. The clutches contained 13 and 33 embryos; no adults were observed guarding the eggs or in the near proximity. On 3 September 2022, the calls of several males were recorded from Refugio de Vida Silvestre El Zarza, quebrada "Las Mariposas." The males were calling from above, several meters high, in the vegetation bordering a small stream. The advertisement call of *C. zarza* is characterized by a high pitched, pulsed, single note, with every call/note featuring three clearly distinguishable pulses. The calls had a duration of 0.242–0.318 s (0.268 ± 0.02, n = 17), with a pulse duration of 0.072–0.159 s (0.089 ± 0.02, n = 51) and an average pulse rate of 11.8 pulses/s. The mean dominant frequency of the calls was 5309.8 Hz, with a mean 90% bandwidth of 5041.3–5487.2 Hz. The fundamental frequency is not recognizable, and no harmonics are visible. The call of a paratype was recorded on 14 October 2020 in the laboratory. The animal was encountered in the type locality on a leaf at a height of 1.5 m near the stream while vocalizing. The male was left in the same plastic bag in which it arrived from the field and had in its proximity, in a different bag, a female (the holotype). The male vocalized almost all night, but the call had a different structure from the typical advertisement call heard and recorded in the field. This call was identified as a courtship call, being composed not only of the single-note calls but mainly of multi-note calls (usually of five notes/call, but up to six notes/call). The notes had the same structure as the single-note calls, with the typical three pulses. The multi-note calls had a duration of 0.581 to 1.905 s (1.300 ± 0.45, n = 18), depending on the number of notes/calls. The individual notes had a duration similar to the ones from the single-note calls of the advertisement calls, just slightly longer: 0.244–0.378 s (0.304 ± 0.03, n = 80). The inter-call interval varied from 40.2 to 596.9 s (255.5 ± 171.23, n = 18), and the call rate was ~0.23 calls/min, or rather, ~14 calls/hour. The frequencies were slightly lower than the ones from the advertisement calls, with the mean dominant frequency of 5127.4 Hz and a mean 90% bandwidth of 4914.7–5411.2 Hz. The fundamental frequency was not recognizable, but 2–3 harmonics were visible.

Tadpoles.—Figures A2.24, A2.25. This description is based on larvae from Refugio de Vida Silvestre El Zarza, Provincia Zamora Chinchipe, in Gosner stages 26 and 31. For the tadpoles in stage 31, the total length was 26.9–34.7 mm (31.9 ± 2.67, n = 6), and the body length was 8.2–10.4 mm (9.5 ± 0.73, n = 6), the body length being ~30% of the total length.

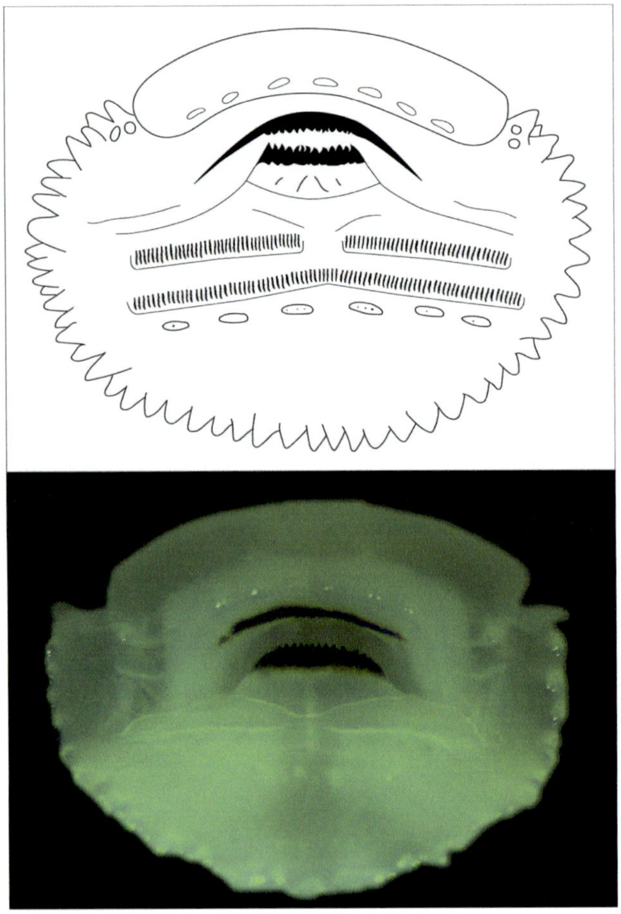

FIGURE A2.25 Oral disc of the tadpole of *Centrolene zarza*. MUTPL-T22, Top: stage 26. Bottom: stage 31. Refugio de Vida Silvestre El Zarza, Provincia Zamora Chinchipe. Paul J. Székely.

The body is elongate, oval, depressed, and wider than high; the snout is rounded in dorsal view and sloped and rounded in lateral view. The eyes are located on the dorsal surface of the head and are C-shaped (at least until stage 31). The nostrils are positioned dorsally, protruding, with very small narial apertures oriented anteriorly. The spiracle is short, single, and sinistral, located in the posterolateral region of the body; the spiracular opening is slightly below the body axis, oriented posteriorly and upward (dorsoposterior orientation). The vent tube is situated medially, short, and attached to the ventral fin (caudal), with a dextral opening directed posteroventrally. The tail is long, ~2.4× the length of the body, with a subacute tip. The dorsal fin originates near the midlength of the tail; the myotomes of the tail musculature are weakly visible in the first half of the tail length. The oral disc is large (oral disc width ~65% of body width), not emarginated, located near the tip of the snout, directed anteroventrally, and protruding ventrally but not laterally, beyond the body. The marginal papillae are uniserial, large, and distributed around the oral disc (~43–47 papillae); a large part of the anterior (upper) margin of the labium lacks papillae and instead has an involuted fold with a smooth surface, but with a row of submarginal, flattened, papilla-like ridges in proximity to the upper jaw sheath. The upper jaw sheath is broadly arched, slender (with less than half the depth of the upper jaw cartilages keratinized), and with a serrated edge; the lower jaw sheath is slightly U-shaped, slender (with less than half the depth of the lower jaw cartilages keratinized), and with a serrated edge. The labial tooth row formula (LTRF) is 0/2(1), P-1 with a medial gap, and with a row of submarginal, flattened, papilla-like ridges (composed of seven or eight papillae) on the posterior (lower) labium in stage 26. However, in stage 31, all the tooth rows were lost, and only the tooth (dermal) ridges were visible beside the submarginal ridge of the papillae. It is not clear if the loss of the tooth rows is a natural process throughout metamorphosis or is caused by disease or other factors. The general coloration varied in the different developmental stages. Hatchlings in stage 24 were slightly pinkish but with a special green coloration of the abdomen. The tadpoles had a more evident pinkish coloration in stage 25 and became almost red by stage 26. However, after a couple of months, by stage 31, they had lost the reddish coloration and had the body just slightly pink and the tail almost brown.

Distribution.—See Figure A2.10. *Centrolene zarza*, an endemic species to Ecuador, is known on the southeastern slopes of the Cordillera Oriental and the Cordillera de Kutukú of the Andes of Ecuador in Morona Santiago and Zamora Chinchipe provinces. Its elevational range is 1430–1905 m in an area of extent of occurrence of about 2750 km^2.

This mostly nocturnal, arboreal species inhabits the evergreen riparian shrub of the Cordillera Oriental of the Andes and evergreen low montane forests of the Cordilleras Cóndor-Kutukú ecosystems within the Amazonian Cordilleras biogeographic sector, where the average annual rainfall is 1288–2176 mm and the average annual temperature is 18.6–20.7 °C.

Conservation.—*Centrolene zarza* is categorized as Endangered (Ron *et al.*, 2024) following the B1ab(iii)+B2ab(iii) IUCN criteria. It occurs in the public reserves Refugio de Vida Silvestre El Zarza and Parque Nacional Sangay and in the protected forests Cordillera Kutukú y Shaimi. The localities in the Cóndor Cordillera are under intense mining pressure, and the expansion of the agricultural frontier impacts the locality in the Kutukú Cordillera. According to our calculations, the species has an extent of occurrence of 2750 km^2. The area of occupancy is 24 km^2, according to Ron *et al.* (2024). Its habitats could be affected in the near future by mining activities (both legal and illegal), as the wildlife refuge is surrounded by active mining concessions. The main threats for *C. zarza* are habitat loss and contamination. The whole southern sector of the Cordillera del Cóndor is threatened by imminent human colonization and settlement, agriculture, and cattle raising, as well as gold and copper mining. The situation of Refugio de Vida Silvestre El Zarza, in this context, is of particular concern.

For now, the wildlife refuge acts like a conservation island, being surrounded by large mining concessions and with active mining activities close to its northern borders. In recent years, signs of illegal mining activities were recorded inside the refuge. These activities are conducted especially in the streams of the reserve, and could affect, particularly but not exclusively, the survival of the glassfrogs that live in the wildlife refuge, especially due to water contamination. Another threat for the survival of *C. zarza* could be chytridiomycosis, the infectious fungal disease caused by *Batrachochytrium dendrobatidis* (*Bd*).

Remarks.—According to Ron *et al.* (2024), *Centrolene zarza* is the sister species of *C. condor* + (*C. kutuku* + *C. camposi*). It is imperative to obtain molecular data on specimens from the type locality of *C. condor* in order to clarify its placement in the phylogeny.

Key references.—(Székely *et al.*, 2023a) (description, call, tadpole, phylogeny); (Ron *et al.*, 2024) (morphology, distribution, conservation).

FIGURE A2.26 Morphological variation in *Phyllonastes arutam*. Top: ZSFQ 1882, holotype, female, 20.3 mm SVL. Middle: ZSFQ 1875, paratype, female, 17.3 mm SVL. Bottom: ZSFQ 6234, paratype, female, 18.0 mm SVL. Comunidad Río Blanco, Paquisha, Provincia Zamora Chinchipe. Carolina Reyes-Puig and David Brito-Zapata.

CRAUGASTORIDAE

PHYLLONASTES ARUTAM BRITO-ZAPATA *ET AL.*, 2024

Arútam Rain Frog/Cutín de Arútam

Noblella arutam Brito-Zapata *et al.*, 2024: 7. **Holotype.**—ZSFQ 1882 from Comunidad de Río Blanco (3.9086°S, 78.4892°W, 1850 m), Paquisha, Provincia Zamora Chinchipe, Ecuador.

Phyllonastes arutam Alessandro Catenazzi (Frost, 2024, personal communication, 5 October 2024).

Etymology.—The specific epithet refers to Arútam, the spiritual entity symbolizing divine power in the world of the Shuar. The Shuar people are native to western Amazonia, spanning Ecuador and Peru. Several Shuar communities reside in the Cordillera del Cóndor and its immediate surroundings, where this new frog species also inhabits. Arútam is believed to dwell in waterfalls, rivers, mountains, rocks, trees, and certain plants and animals. This deep association highlights the profound connection between nature and the Shuar people. This name is to pay a tribute to the Shuar people living near the type locality of *Phyllonastes arutam*, their ancestral lands and culture, acknowledging their tireless efforts and commitment to nature protection.

Identification.—Figure A2.26. The snout-vent length (SVL) of adult males ranges from 13.8 to 15.1 mm ($\bar{x}$ = 14.6 ± 0.4 mm), while the SVL of adult females ranges from 17.3 to 20.3 mm ($\bar{x}$ = 18.6 ± 1.3 mm). The skin on the dorsum is shagreen, with scattered low tubercles that are more evident on the flanks. The skin on the venter is smooth. A discoidal fold is present, and the thoracic fold is slightly defined. The tympanic annulus and membrane are defined and visible externally, while the supratympanic fold is slightly defined. The snout is subacuminate in dorsal view and rounded in lateral view. The upper eyelids bear one or two inconspicuous tubercles. Vomerine teeth are absent. Vocal slits and an external vocal sac are present, while nuptial pads are absent. The fingers are not expanded distally. The fingertips are acuminate, more evident on finger III, and lack papillae. Finger I is smaller than finger II. The distal phalanges are strongly T-shaped, and the phalangeal formula of the hands is 2-2-3-3. Minute supernumerary palmar tubercles are present and slightly visible. Ulnar tubercles are low and inconspicuous. Subarticular tubercles are rounded, and circumferential grooves are absent. The heel has one small subconical tubercle, and the tarsus has one elongated and subconical tubercle. Two prominent metatarsal tubercles are present: the inner tubercle is ovoid, and the outer tubercle is subacuminate. The toes are slightly expanded and rounded distally, with papillae present on all toes, more evident on toes II to IV. Toe V is shorter than toe III. Small, white supernumerary plantar tubercles are present, and circumferential grooves are visible. The phalangeal formula of the feet is 2-2-3-4-3.

The dorsum is light brown, grayish-brown, or dark brown, with V-shaped inverted or scattered irregular darker marks, sometimes accompanied by an inconspicuous mid-dorsal cream stripe extending from the scapular region to the cloaca. Black suprainguinal marks are present. A black or dark brown lateral band runs from the tip of the snout to the midbody, with scattered white dots. The groins are yellow to orange with minute brown dots. The venter is yellow, dotted with white and brown.

Comparisons.—*Phyllonastes lochites, P. myrmecoides, P. naturetrekii, P. personinus. Phyllonastes lochites, P. naturetrekii*, and *P. personinus* are found on the eastern slopes in the Andes of Ecuador, while *P. myrmecoides* inhabits the Amazonian lowlands. *Phyllonastes arutam* differs from *P. naturetrekii* (characters in parentheses) by the presence of three phalanges on Finger IV (two phalanges), skin on venter smooth (granular), nuptial pads absent (present), toe tips with papillae (papillae absent), larger SVL, males 13.8–15.1 mm, females 17.3–20.3 mm (males 11.2–12.1 mm, females 13.1–14.4 mm). *Phyllonastes arutam* is diagnosed from *P. personinus* (characters in parentheses) by having black suprainguinal marks (no suprainguinal marks), dorsal skin shagreen with scattered low tubercles more evident in the flanks (smooth with few low pustules, most dense on posterior half of body), snout subacuminated in dorsal view (rounded). *Phyllonastes arutam* differs from *P. lochites sensu stricto* and *P. myrmecoides sensu stricto* (characters in parentheses) by having the snout subacuminated in dorsal view (short and truncated in dorsal view), and by having ulnar tubercles (absent). Additionally, *Phyllonastes arutam* differs from *P. lochites sensu stricto* by having discoidal fold (absent), and from *P. myrmecoides sensu stricto* by having three phalanges on finger IV (two phalanges).

Natural history.—Most frogs were found active at night on the ground, amidst leaf litter, and some were found in the leaf litter but in the morning. The stomach content of one specimen contained three ant-mimicking rove beetles (Staphylinidae: Pselaphinae), two aquatic fly larvae (Diptera: Nematocera), and eight ants (Formicidae: Myrmicinae).

Distribution.—Figure A2.27. *Phyllonastes arutam*, an endemic species to Ecuador, is known only from the type locality (the Río Blanco community and surrounding areas) in Cordillera del Cóndor in Provincia Zamora Chinchipe. Its elevational range is 1720–1890 m in an area of extent of occurrence of about 2 km^2.

This mostly nocturnal species inhabits the Evergreen Low Montane Forest of Cordilleras Cóndor-Kutukú and the Evergreen Montane Forest of Cordilleras Cóndor-Kutukú ecosystems within the Amazonian Cordilleras biogeographic sector, where the average annual rainfall is 1341–1379 mm and the average annual temperature is 17.9–18.9 °C.

Conservation.—Brito-Zapata *et al.* (2024) propose that *Phyllonastes arutam* should be classified as Critically Endangered (CR), based on criteria B1ab(iii) +2ab(iii) based on the following assessment of its extinction risk: (1) Extent of occurrence (EOO) of 1.70 km^2, within the threshold for Critically Endangered (<100 km^2), criterion B1. (2) Area of occupancy AOO of 8 km^2, also within the threshold for Critically Endangered (<10 km^2), criterion B2. (3) Known from a single locality. (4) Continuing decline inferred of area of occupancy, extent of occurrence, and

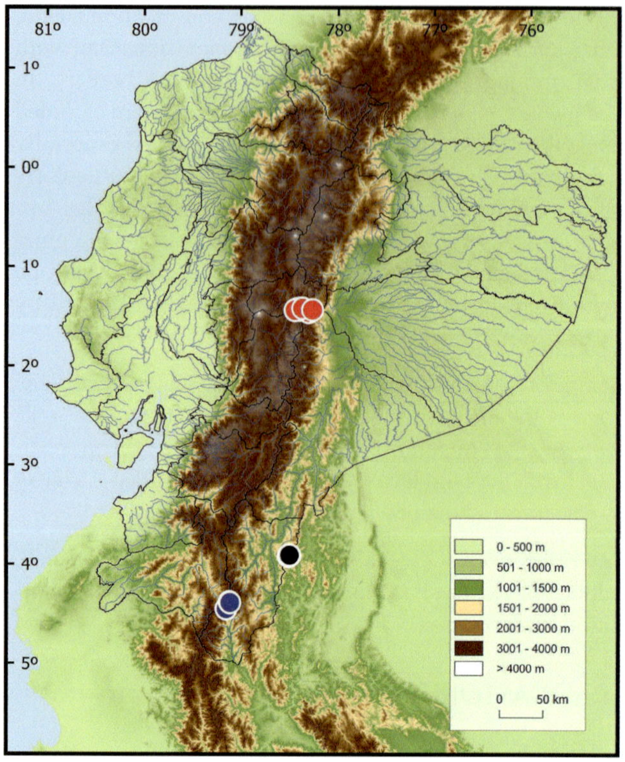

FIGURE A2.27 Localities of known occurrence of *Phyllonastes arutam* (black dot), *Pristimantis donnelsoni* (red dots), and *P. paladines* (blue dots) in Ecuador.

quality of habitat due to impacts derived from large-scale mining. The population in the type locality appears locally abundant. All known records are from surrounding areas of the Río Blanco community in the Cordillera del Cóndor, which we considered as a single locality due to their proximity and the fact that they are within an area with the same threat. The main threat to this locality is habitat degradation or loss caused by mining activities because it is within a large-scale mining concession. Other anthropogenic activities, such as agriculture and livestock farming, have been identified as threats to the native ecosystems of the Cordillera del Cóndor. However, emigration of people from the Río Blanco community has significantly increased in recent years, leading to reduction in its impacts. It has not been recorded in the public reserves, but in the Protected Forests of Cordillera del Cóndor.

Remarks.—*Phyllonastes arutam* is sister to *P. myrmecoides*, according to the original publication, where comparative morphology, osteology, and molecular markers (16S mtDNA) were detailed. According to Alejandro Catenazzi in Frost (2024), the transfer from *Noblella* to *Phyllonastes* is based on morphological evidence and phylogenetic placement in the original publication, compared to the tree provided by von May *et al.* (2024). This change affects all the species of *Noblella* previously recorded from Ecuador, which are now reclassified under the genus *Phyllonastes.*

Key references.—(Brito-Zapata *et al.,* 2024) (description).

PRISTIMANTIS DONNELSONI REYES-PUIG *ET AL.*, 2023

NELSON PALACIOS' RAINFROG/CUTÍN DE NELSON PALACIOS

Pristimantis donnelsoni Reyes-Puig *et al.*, 2023: 260. **Holotype.**—DHMECN 14701 from Reserva Chamana, eastern slopes of Tungurahua Volcano, Ulba, Baños (1.4272°S, 78.3967°W; 3028 m), Provincia Tungurahua, Ecuador.

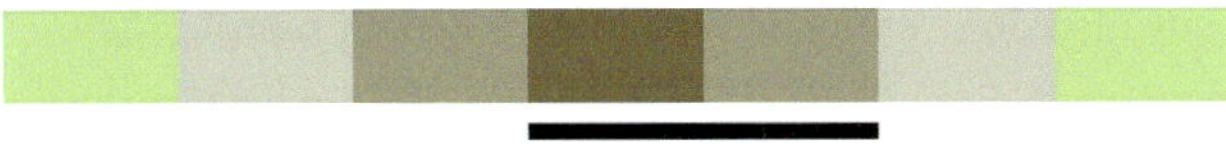

Etymology.—The specific epithet is a noun in the genitive case and a patronym for Don Nelson Palacios. The authors named this new species after him as a tribute to his friendship and collaboration in the first collections of new species described during the last decade at several additional important localities within the upper basin of the Pastaza River.

Identification.—Figure A2.28. The maximum known SVL is 16.4 mm in males and 19.7 mm in females. The skin on the dorsum is shagreen, with an evident occipital fold. The dorsolateral folds are low in banded individuals or weakly defined in those with uniform or irregular dorsal color patterns. The skin on the venter is coarsely areolate. The tympanic annulus and tympanic membrane are present, with a prominent, sexually dimorphic tympanum. The tympanum in males is 7% of the SVL, and in females, it is 6% of the SVL. The snout is subacuminate in dorsal view (with a pointed tip) and rounded in lateral view. The upper eyelid has several small, rounded to subconical tubercles, and the upper eyelid in males is 56% of the interorbital distance (in females, 59% of the interorbital distance). Cranial crests are absent. The odontophore processes of the vomers are partially concealed and oblique in outline, with each process bearing several small, poorly defined teeth. The sphenethmoids have an acuminated anterior border dorsally and a ventrally anteriorly rounded, acuminate, projected border. The posterior border is horizontal and articulates with the frontoparietals. The posterior border of the frontoparietal presents an enlarged quadrangular process posteriorly in dorsal view. The zygomatic ramus of the squamosal is short and rounded in dorsal view. The cultriform process of the parasphenoid reaches the posterior level of the vomers and has an acuminate anterior border. Males have vocal slits and a subgular vocal sac but lack nuptial pads. Finger I is shorter than finger II, and the discs on fingers II–IV are rounded. The fingers have slightly visible lateral fringes. The ulnar tubercles are small and low. The heel has small tubercles, and the outer edge of the tarsus has small tubercles, which are more accentuated in males. The inner metatarsal tubercle is oval and elongated, about twice the size of the rounded outer metatarsal tubercle. The toes lack lateral fringes, and webbing is absent. Toe V is slightly longer than toe III, with discs slightly expanded.

FIGURE A2.28 Morphological variation in *Pristimantis donnelsoni*. Top left: DHMECN 16175, paratype, female, 13.7 mm SVL. Top right: DHMECN 13321, paratype, female, 15.5 mm SVL. Bottom left: female. Bottom right: DHMECN 14701, holotype, female, 15.4 mm SVL. Cerro Candelaria. Provincia Tungurahua, Ecuador. Juan Pablo Reyes Puig.

The dorsum is brown in several shades, with extreme dark yellow or green tones. Females are generally darker than males, who usually have a banded pattern and banded limbs. The venter is light brown with dark brown flecks. The hidden surfaces of the groin, thighs, and armpits are pinkish to reddish, and the chest and throat present light brown tones.

Comparisons.—*Pristimantis festae*, *P. gladiator*, *P. kayi*, *P. leoni*, *P. sirnigeli*. *Pristimantis donnelsoni* is most similar to *P. gladiator* and *P. festae*, but it is small. Males of *P. gladiator* measure 15.8 mm vs. *P. donnelsoni* 13.8 mm; females in *P. gladiator* are 20.2 mm vs. *P. donnelsoni* 17.1 mm; males in *P. festae* are 16.4 mm vs. *P. donnelsoni* 13.8 mm; females in *P. festae* are 20.7 mm vs. *P. donnelsoni* 17.1 mm. Additionally, *P. donnelsoni* differs in having reddish and pinkish marks on the groin and ventral surfaces in several tones of brown, unlike *P. gladiator*, which features dark orange marks on a dark brown venter. *Pristimantis donnelsoni* has a subacuminate snout in comparison to *P. gladiator*, which has an acuminate snout. On the other hand, the snout of *P. festae* is rounded in dorsal view and subacuminate in *P. donnelsoni*. *Pristimantis donnelsoni* is similar to the closely related *P. kayi*. Although they are extremely similar, differences in the tympanum diameter are evident: TD in *P. donnelsoni* 0.9 mm vs. *P. kayi* 1.0 mm. *Pristimantis leoni* and *P. sirnigeli* are found at the Western Andean Cordillera, but they lack distinctive concealed reddish flash marks on the groin that are characteristic of *P. donnelsoni*.

Natural history.—Figure A2.29. Individuals were found sitting on herbs or on leaf litter on the forest floor, with dense ferns and small herbaceous plants, always near ground level (< 30 cm above ground). Calling males are active during

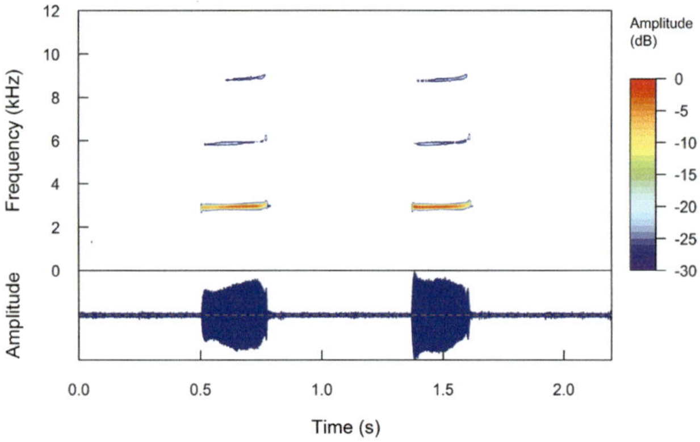

FIGURE A2.29 Oscillogram and spectrogram of *Pristimantis donnelsoni*. DHMECN 4774, paratype, Pondoa, Baños de Agua Santa, Provincia Tungurahua, Ecuador. Taken from Reyes-Puig *et al*. (2023).

the day and sporadically call at night. A frog was recorded on 21 March 2007 at 19:00 h. The calling male was sitting on herbaceous plants on the forest floor. The advertisement call has two notes, with a note duration of 236–293 ms and intervals of 522–620 ms. The call comprises tonal sounds of constant frequency, with a slight upward modulation at the end of the call. The dominant frequency is 2.93 kHz, with two partial harmonics: the first one ranging from 5.86 to 5.94 kHz and the second one ranging from 8.79 to 8.96 kHz. The call duration ranges from 1052 to 1136 ms, with intervals ranging from 4531 to 5985 ms, emitted at a rate ranging from 8.43 to 10.60 calls/minute.

Distribution.—Figure A2.27. *Pristimantis donnelsoni*, an endemic species to Ecuador, is known only from the southern mountains of the Pastaza drainage of the Cordillera Oriental of the Andes in Provincia Tungurahua. Its elevational range is 2485–3800 m in an area of extent of occurrence of about 47 km^2.

This mostly nocturnal, arboreal species inhabits the evergreen high montane forests of the northern Cordillera Oriental of the Andes and the evergreen montane forests of the northern Cordillera Oriental of the Andes ecosystems, within the northern Cordillera Oriental of the Andes biogeographic sector, where the average annual rainfall is 756–1048 mm and the average annual temperature is 10.8–14.9 °C.

Conservation.—*Pristimantis donnelsoni* is herein categorized as Endangered following B1ab(iii) IUCN criteria because: (1) its extent of occurrence is estimated to be less than 5000 km^2 (47 km^2), (2) its distribution is severely fragmented, and (3) there is a continuing decline in the extent and quality of its habitat. The major threat is habitat destruction and degradation due to the expansion of agricultural and farming activities, fires, landslides, volcanic eruptions, and pollution resulting from pesticide use. It has been recorded in seven nearby localities in the southern mountains of the Pastaza drainage, Tungurahua Province, Ecuador. Three localities are located on the Tungurahua Volcano, whereas the others correspond to nearby protected areas (Reserva Chamana and Reserva Cerro Candelaria, Finca Palmonte, and Bosque Protector Hacienda Guamag), all at elevations of 2950–3800 m. Further surveys are expected to extend its range into nearby public protected areas.

Remarks.—According to Reyes-Puig *et al.* (2023), *Pristimantis donnelsoni* is the sister species of *P. kayi* + *P. festae*, in the *P. gladiator* complex of the *P. myersi* group.

They also described and illustrated the skull morphology.

Key references.—(Reyes-Puig *et al.*, 2023) (description, phylogeny).

PRISTIMANTIS KAYI GUAYASAMIN *ET AL.*

Kay Rainfrog/Cutín de Kay

Pristimantis kayi Guayasamin *et al. in* Reyes-Puig *et al.*, 2023: 275. **Holotype.**—ZSFQ 0775 from Volcán Sumaco, Sumaco National Park (0.5696°S, 77.5939°W; 2322 m), Provincia Napo, Ecuador).

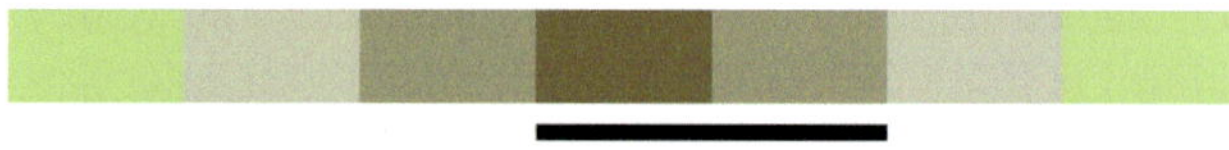

Etymology.—The specific epithet is a patronym for Andreas Kay, a German physicist and biologist who spent much of his life documenting and exploring biodiversity, contributing substantially to the conservation of Ecuadorian forests. Andreas was part of the team that created Fundación EcoMinga's Drácula Reserva in northwestern Ecuador. Additionally, his work shed light on understanding the diversity and endemism patterns of orchids in the Llanganates-Sangay Ecological Corridor, where he passed away in 2019. Some of his photographs can be seen here: https://www.flickr.com/photos/andreaskay/albums/.

Identification.—Figures A2.30, A2.31. The maximum known SVL is 15.6 mm in males and 20.2 mm in females. The skin on the dorsum is shagreen with small, scattered, rounded warts, and the upper flanks have numerous low warts; dorsolateral and W-shaped occipital folds are present. The tympanic membrane and tympanic annulus are well differentiated, and the tympanic annulus is not sexually dimorphic; the tympanum in males and females is 7% of the SVL. The snout is rounded in dorsal view and protruding in lateral view. The upper eyelid usually bears one conical or subconical tubercle and many low tubercles; the upper eyelid is about 63% of the interorbital distance in males and 59% of the interorbital distance in females, with cranial crests absent. The dentigerous processes of the vomers are evident, each process bearing 2–3 teeth. The sphenethmoid, in its dorsal view, has a short, blunt anterior border and is short and blunt in ventral view, and its posterior border presents an oblique articulation with the frontoparietals.

FIGURE A2.30 *Pristimantis kayi*. Left: ZSFQ 775, holotype, female, 19.7 mm SVL. Sumaco, Provincia Napo. Right: male. Sumaco, Provincia Napo. José Vieira-Fernandes.

FIGURE A2.31 Morphological variation in *Pristimantis kayi*. Top left: female, Sumaco, Provincia Napo. José Vieira-Fernandes. Top right and middle left: females. Middle right: DHMECN 15233, female. Bottom: males. Reserva Ecológica Machay, Río Verde, Provincia Tungurahua, Ecuador. Juan Pablo Reyes-Puig.

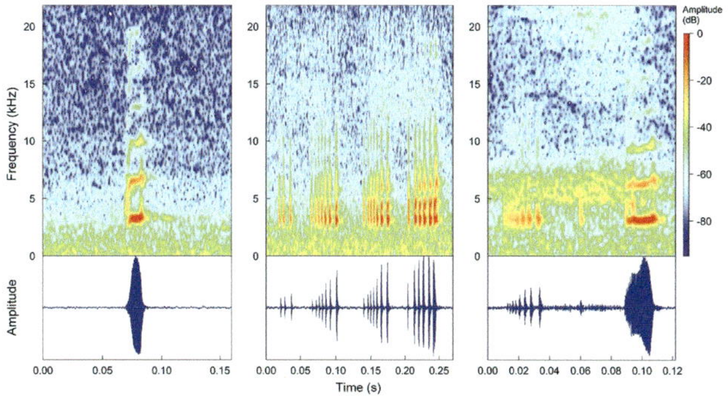

FIGURE A2.32 Calls of *Pristimantis kayi*. MZUTI 852. Cordillera de los Guacamayos, Provincia Napo, Ecuador. Left: single-note call. Middle: pulsed note calls. Right: complex calls (non-pulsed and pulsed notes). Taken from Reyes-Puig *et al.* (2023).

The posterior border of the frontoparietal has an irregular border not projected in dorsal view, and the zygomatic ramus of the squamosal is elongated anteriorly in dorsal view; the cultriform processes of the parasphenoid reach the posterior level of the vomers, with a rounded anterior border. Males have vocal slits, but nuptial pads are absent. Finger I is shorter than finger II, and the disc on finger I is not expanded, while the discs on fingers II–IV are slightly expanded. The fingers have thin lateral fringes. Two or three low ulnar tubercles are present. The heel has a low to conical tubercle, and the outer edge of the tarsus has low conical tubercles. The inner metatarsal tubercle is oval, about 1.5–2× the length of the round outer metatarsal tubercle. The toes have thin lateral fringes and lack webbing; toe V is slightly longer than toe III, with discs slightly expanded.

The dorsum is brown with darker and lighter markings; the venter is light gray to black, with or without orange to red spots of different sizes. The groin has orange to red spots that are more conspicuous in females than in males.

Comparisons.—*Pristimantis gladiator*, *P. festae*, *P. donnelsoni*. *Pristimantis kayi* differs from *P. gladiator* and *P. festae* by being smaller. Additionally, *P. kayi* differs from *P. festae* by having orange to red spots on the groin (large, white to reddish spots in *P. festae*) and slightly expanded discs on the outer fingers (discs not expanded in *P. festae*). *Pristimantis kayi* is similar to the allopatric *P. donnelsoni*; however, the diameter of the tympanum in males is larger in *P. kayi* compared to that in *P. donnelsoni* and the other closely related species (i.e., *P. festae* and *P. gladiator*). In *P. kayi*, the parasphenoid is much more developed than in *P. donnelsoni*, with its lateral alary processes almost reaching the squamosal; also, the palatine bones are shorter in *P. kayi* than in *P. donnelsoni*. Finally, the zygomatic ramus of the squamosal is much longer in *P. kayi* than in *P. donnelsoni*.

Natural history.—Figure A2.32. The species has been found in primary and disturbed Andean forests. Calling males were heard during the day and night. During the night, frogs were found on the forest floor or perched on low vegetation (always below 60 cm from ground level), including shrubs, grasses (*Neurolepis* sp.), ferns, and *Selaginella*. The call from Cordillera de los Guacamayos, Provincia Napo, Ecuador was described. The male was calling from leaf litter at ground level. The air temperature was 12.2 °C. The call is composed of several elements, ranging from single notes to pulsed notes. It is composed of 1–7 notes, with notes having a mean duration of 36.0 ± 18.3 ms and a mean interval between notes of 43.8 ± 36.2 ms, emitted at a mean rate of 15.7 ± 6.1 notes/second. The notes are composed of 2–13 pulses, with pulses having a mean duration of 3.9 ± 1.1 ms and a mean interval between pulses of 4.2 ± 2.8 ms, emitted at a mean rate of 202.2 ± 119.7 pulses/second. The dominant frequency is 3.19 ± 0.05 kHz, with 2–6 partial harmonics in the spectrogram, and the call duration is 165.1 ± 130.1 ms. The call of *Pristimantis kayi* does not present stereotyped notes. Due to their structural elements (e.g., notes/pulses), notes can be classified into three different types: single-note calls, pulsed-note calls, and complex calls, which contain single (non-pulsed) and pulsed notes in its structure.

Distribution.—Figure A2.33. *Pristimantis kayi*, an endemic species to Ecuador, is known from the Cordillera de Guacamayos, Volcán Sumaco, and the eastern slopes of Cordillera Oriental of the Andes from Napo and Tungurahua provinces. Its elevational range is 2190–3448 m in an area of extent of occurrence of about 1919 km^2.

This mostly nocturnal, arboreal species inhabits montane habitats and intervened areas in four ecosystems: evergreen high montane forests of the northern Cordillera Oriental of the Andes, evergreen low montane forests of the northern Cordillera Oriental of the Andes, evergreen montane forests of the northern Cordillera Oriental of the Andes, and herbazal of Páramo. These ecosystems are within the northern Cordillera Oriental of the Andes and Páramo biogeographic sectors, where the average annual rainfall is 892–3246 mm and the average annual temperature is 10.6–18.1 °C.

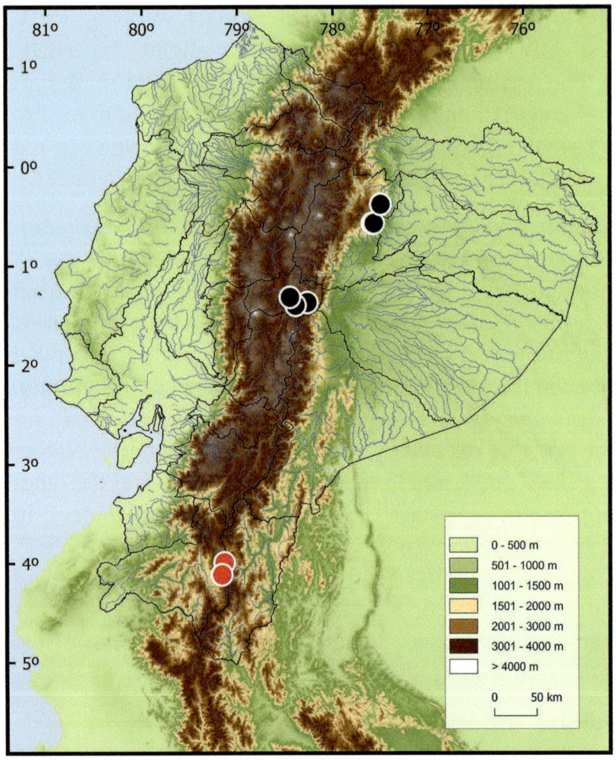

FIGURE A2.33 Localities of known occurrence of *Pristimantis kayi* (black dots) and *P. sagedunneae* (red dots) in Ecuador.

Conservation.—*Pristimantis kayi* is herein categorized as Endangered following B1ab(iii) IUCN criteria because: (1) its extent of occurrence is estimated to be less than 5000 km^2 (1919 km^2), (2) its distribution is severely fragmented, and (3) there is a continuing decline in the extent and quality of its habitat. The major threat is habitat destruction and degradation due to the expansion of agricultural and farming activities, fires, landslides, and pollution resulting from pesticide use. It has been recorded in the public area Parque Nacional Sumaco Napo-Galeras in the Reserva de la Biósfera Sumaco. It occurs in Cordillera de Guacamayos, and in some private reserves (Reserva Machay, Reserva Naturetrek Vizcaya, and Reserva Leito).

Remarks.—According to Reyes-Puig *et al.* (2023), *Pristimantis kayi* is the sister species of *P. festae*, in the *P. gladiator* complex of the *P. myersi* group. Lynch and Duellman (1980) reported *P. festae* from the northern versant of the Tungurahua Volcano, based on overall similarity. Reyes-Puig *et al.* (2022) reported these records as an undescribed species (corroborated by the molecular phylogeny produced by Franco-Mena *et al.*, 2023) and highlighted the importance of microhabitat variation to shape the high diversity and endemism of the area. Reyes-Puig *et al.* (2023) also described and illustrated the skull morphology and emphasize that skull osteological characters represent a promising area of research as a way to resolve the taxonomy of closely related species. The main key bones to discriminate species are the sphenethmoid, frontoparietals, squamosal, and parasphenoid.

Key references.—(Reyes-Puig *et al.*, 2023) (description).

PRISTIMANTIS NORMAEWINGAE REYES-PUIG, YÁNEZ-MUÑOZ, LIBKE, VINUEZA, AND CARRIÓN-OLMEDO

Norma Ewing Rainfrog/Cutín de Norma Ewing

Pristimantis normaewingae Reyes-Puig *et al.*, 2024: 60.
Holotype.—DHMECN 18409 from Cerro Candelaria (1°27.410'S, 78°18.192'W, 3042 m), Provincia Tungurahua, Ecuador.

Etymology.—The specific epithet is a matronym that honors Norma Ewing, a passionate conservationist and animal rescue volunteer from the United States. Her belief that all

FIGURE A2.34 Morphological variation in *Pristimantis normaewingae*. Top left: DHMECN 18400, paratype, male. Top right: DHMECN 18404, paratype, male. Bottom left: adult. Bottom right: DHMECN 1719, paratype, juvenile. Cerro Candelaria, Provincia Tungurahua, Ecuador. Juan Pablo Reyes-Puig.

species, and especially frogs, are interconnected is rooted in her early childhood surrounded by nature living and working on a farm. She has always cared for animals of all shapes and sizes. As a volunteer with animal rescues, she was admired for her ability to nurture orphaned and injured animals.

Identification.—Figure A2.34. The maximum known SVL is 30.6 mm in females 22.3 mm in males. The skin on the dorsum is shagreen with small, low tubercles. The skin on the throat is areolate, and the skin on the belly is strongly areolate, bearing scattered tubercles. Discoidal and thoracic folds are present. Dorsolateral folds are fragmented; instead, a dorsolateral row of enlarged tubercles is present on the posterior back, more evident in males. The tympanic membrane is absent, and the tympanic annulus is present but small (on average, 40.99% of the eye diameter). A supratympanic fold is present with low tubercles. A pair of prootic and occipital tubercles are aligned with several subconical tubercles, forming a W in the scapular region. Several subconical and rounded postrictal tubercles are present. The snout is short and subacuminate in dorsal view and rounded in lateral view, with four tubercles aligned with the canthal region forming a cross in dorsal view. The upper eyelid has one conical tubercle and a few lower tubercles. Cranial crests are absent. The dentigerous processes of the vomers are present, oblique, and separated. Vocal slits and nuptial pads are absent. Finger I is shorter than finger II, and the discs of the digits are expanded to broadly expanded and truncate. The fingers have narrow lateral fringes, and all fingers have elongated and thin hyperdistal tubercles. Low and rounded ulnar tubercles are more conspicuous in males, with a low ulnar fold evident ventrally. The heel bears a low subconical or rounded tubercle, and there is a row of subconical or rounded tubercles along the external tarsus. The toes have broad lateral fringes, with basal webbing absent, and all toes have elongated hyperdistal tubercles. Toe V is much longer than toe III. Toe III surpasses the distal border of the distal subarticular tubercle of toe V and reaches the distal border of the penultimate tubercle of toe IV, while toe V reaches toward the outermost border of the distal tubercle of toe IV.

The dorsum has irregular marks mixed with a banded pattern in different tones of brown, yellow, and green. Dark canthal, subocular, and supratympanic stripes are present. Dark, oval or irregular middorsal marks are delineated by pale lines and surrounded by irregular contrasting marks. The flanks have brown or greenish-brown tones, and dark incomplete oval marks in the sacral region with thin light borders are present on each side. The ventral surfaces of the belly, throat, and limbs are marmorated, with dark brown and cream or gray marbling. In males, light reddish or pink flecks are slightly evident on the palmar and plantar surfaces and on parts of the ventral portions of the limbs. In females, reddish or pink tones are absent on such ventral surfaces. The iris is light blue with a wide copper horizontal medial band and black reticulations, and there is a light pale sclera.

There is sexual dimorphism in the coloration and tuberculation. In dorsal view, females range from dark brown to pale brown with scapular and sacral marks that are less conspicuous than those of males. The iris coloration is pale gray with a medial horizontal line that is reddish-brown and finely darkly reticulated, and the sclera is white. Males have distinctive dorsal and flank coloration, ranging from reddish-brown to orange-brown. The posterior surface of the head and the scapular region range from light brown to yellowish-green lines surrounding marks on the flanks, which range from white to pale ochre. Sacral marks range from reddish to orange or dark brown. Iris variation in males includes bluish-gray with a coppery or faint reddish-brown medial horizontal line. Ventrally, both males and females have a grayish-brown background with dark brown blotches, yellow–cream, or grayish-cream. Juveniles exhibit pale or dark and light brown, with the interorbital region being dark brown or featuring an interorbital line that is orange–brown. Males show elevated and conical eyelid, interorbital, and scapular tubercles. Females only have small, rounded, and conical tubercles. Some individuals show plasticity in the state of tubercle texture under different environmental conditions, ranging from elevated and spiny to a relaxed, low tuberculated state; additionally, preservation could reduce the protuberance of tubercles.

Comparisons.—*Pristimantis anaiae*, *P. glendae*, *P. kunam*, *P. maryanneae*, *P. resistencia*, *P. venegasi*. *Pristimantis normaewingae* can be distinguished from species of the *Pristimantis anaiae* species group by having incomplete oval marks in the sacral region, a row of enlarged dorsolateral tubercles, and a light blue iris with a wide copper medial horizontal band with black reticulations, while the other species present different iris patterns, complete oval marks in the sacral region, and no dorsolateral row of enlarged tubercles, as in *P. anaiae*, *P. resistencia*, *P. glendae* and *P. venegasi*, with the exception of *P. kunam* and *P. maryanneae*, which respectively present larger and smaller dorsolateral rows of enlarged tubercles. *Pristimantis kunam* is the smallest species of the group (14.7 mm SVL). *Pristimantis normaewingae*'s interspecific SVL variation overlaps with the other species of the group, and it has less conspicuous ulnar tubercles and a light brown venter, in comparison with the conspicuous conical ulnar and tarsal tubercles, dark greenish body coloration with dark flecks, and bronze iris with black reticulations of *P. kunam*. *Prisitmantis maryanneae* presents less tuberculated skin, no tympanum, a small and inconspicuous row of dorsolateral tubercles, and paler ventral surfaces than *P. normaewingae*, which has contrasting body and iris coloration, the tympanum hidden beneath the skin with an evident tympanic annulus, and more tuberculated skin.

Natural history.—The female holotype was found on a shrub branch covered with moss 170 cm above the ground.

Distribution.—Figure A2.35. *Pristimantis normaewingae*, an endemic species to Ecuador, is known only from the type locality (Cerro Candelaria) in Provincia Tungurahua on the eastern slopes of the Cordillera Oriental of the Andes of Ecuador. Its elevational range is 2550–3150 m in an area of extent of occurrence of about 2 km^2.

This mostly nocturnal, arboreal species inhabits montane habitats in the evergreen low montane forests of the northern Cordillera Oriental of the Andes and evergreen montane forests of the northern Cordillera Oriental of the Andes ecosystems within the northern Cordillera Oriental of the Andes biogeographic sector, where the average annual rainfall is 973–1966 mm and the average annual temperature is 14.3–17.3 °C.

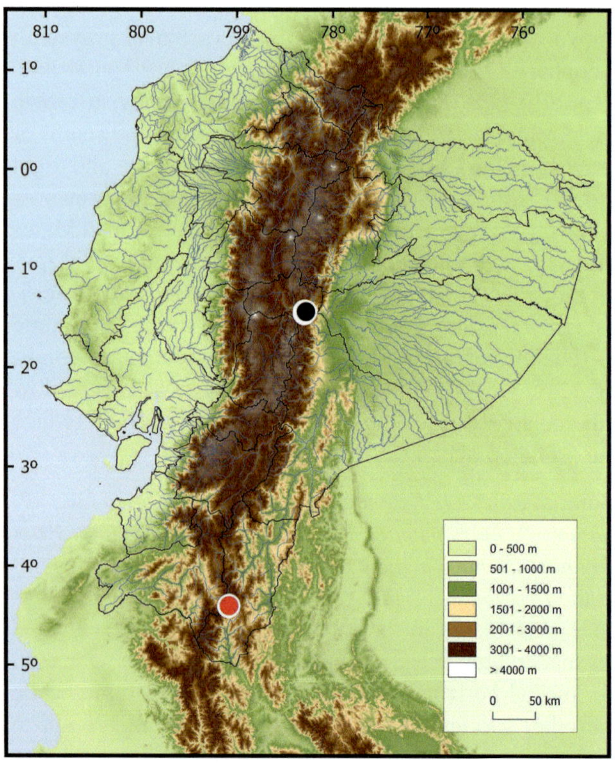

FIGURE A2.35 Localities of known occurrence of *Pristimantis normaewingae* (black dots) and *P. numbala* (red dot) in Ecuador.

Conservation.—*Pristimantis normaewingae* is categorized as Data Deficient (Reyes-Puig *et al.*, 2024) because of the lack of records, evidence of population status, and geographic distribution. The only known population is protected within the private reserve Cerro Candelaria.

Remarks.—According to Reyes-Puig *et al.* (2024), *Pristimantis normaewingae* is the sister species of *P. kunam*, in the *P. anaiae* species group.

Key references.—Reyes-Puig *et al.*, 2024 (description, phylogeny).

PRISTIMANTIS NUMBALA SZÉKELY *ET AL.*, 2023B

NUMBALA RAINFROG/CUTÍN DE NUMBALA

Pristimantis numbala Székely *et al.*, 2023b: 60. **Holotype.**—MUTPL 1190 from Reserva Numbala (4.4065°S, 79.0885°W, 2875 m), Provincia Zamora Chinchipe, Ecuador.

Etymology.—The specific epithet is a noun in apposition and refers to Reserva Numbala, an important private protected area managed by the NGO Naturaleza y Cultura Internacional. The reserve was established in 2006, with the main aim of conserving the last remnants of romerillo (*Retrophyllum rospigliosii* and *Prumnopitys harmsiana*, Podocarpaceae) forests, which are represented here by some of the last, giant trees of these species. The reserve protects 1800 ha of subparamo and montane cloud forest and is home to an important diversity of birds, amphibians, mammals, and plants. It is located between the two isolated extensions of the southern part of Parque Nacional Podocarpus, guaranteeing the connectivity needed for the preservation of the biological diversity of the national park and its area of influence.

Identification.—Figures A2.36, A2.37. The maximum known SVL is 23.2 mm in females; males are unknown. The skin on the dorsum is shagreen with some scattered tubercles (a feature more evident in life); the venter is areolate. A discoidal fold is present; dorsolateral folds are absent. The flanks usually have longitudinal lateral folds on the anterior half; a low middorsal fold is present (a feature more evident in life). The tympanic annulus and tympanic membrane are absent, but the supratympanic fold is present. The snout is rounded to broadly rounded in dorsal view and rounded in profile; the canthus rostralis is concave in dorsal view and rounded in profile. The upper eyelid bears one larger tubercle and several small tubercles (a feature more

FIGURE A2.36 Morphological variation in *Pristimantis numbala*. Top: MUTPL 1190, holotype, female, 22.3 mm SVL. Middle: MUTPL 1178, paratype, female, 23.3 mm SVL. Bottom: MUTPL 1179, paratype, female, 20.4 mm SVL. Reserva Numbala, Provincia Zamora Chinchipe. Paul J. Székely.

FIGURE A2.37 *Pristimantis numbala*. Top: juvenile. Bottom: MUTPL 1179, paratype, female, groin and thigh. Reserva Numbala, Provincia Zamora Chinchipe. Paul J. Székely.

evident in life); its width is about 83% of the interorbital distance; cranial crests are absent. The dentigerous processes of the vomers are evident, oblique, ovoid, separated medially by a distance equal to or slightly smaller than the width of the processes; each process bears four to six teeth. The condition of the vocal sacs, vocal slits, and nuptial pads is unknown. Finger I is shorter than finger II; the discs on the fingers are broadly expanded and truncate, with circumferential grooves present. The fingers bear lateral fringes; the subarticular tubercles are prominent; hyperdistal subarticular tubercles are present; supernumerary palmar tubercles are present; the palmar tubercle is usually not divided or only partially divided into a larger (inner) and a smaller (outer) tubercle; the thenar tubercle is elliptical and larger than the inner palmar tubercle. Ulnar tubercles are present (a feature more evident in life). The heel has either no larger tubercle or one larger tubercle and several small tubercles (a feature more evident in life); the outer edge of the tarsus may have a row of small tubercles (a feature more evident in life); the inner tarsal fold may be absent or present. The inner metatarsal tubercle is broadly ovoid, about 23 to 33% of the size of the subconical (in profile) outer metatarsal tubercle; the subarticular tubercles are prominent; hyperdistal subarticular tubercles are present; supernumerary plantar tubercles are present. The toes bear lateral fringes and have basal webbing; Toe V is much longer than toe III; the discs on the toes are broadly expanded and truncate, about the same size or slightly smaller than those on the fingers, with circumferential grooves present.

The dorsum is reddish-brown with dark brown bars and X markings on the back. The flanks are gray, pinkish-gray, or green, with wide, dark transverse bars. The dorsal surfaces of the arms and hind limbs have dark transverse bars; large areas of the dorsal surfaces of the thighs, groin, and concealed limb surfaces are reddish-brown. The head has black interorbital markings, dark labial bars, and black supratympanic stripes. The venter, throat, and ventral surfaces of the hind limbs and arms are pinkish-white or pinkish-gray with blackish markings. The iris is bronze with fine black reticulations and a median, wide, dark red horizontal streak.

Comparisons.—*Pristimantis colodactylus*, *P. sagedunneae*, *P. orestes*, *P. paladines*, *P. samaniegoi*, *P. matildae*. *Pristimantis numbala* is most similar to the closely related species *P. sagedunneae* and *P. paladines*. *Pristimantis numbala* differs from *P. sagedunneae* by the rounded to broadly rounded snout (vs. subacuminate to rounded snout in *P. sagedunneae*); the relative length of the toes, with toe V much longer than toe III (vs. toe V longer than toe III); the reddish-brown coloration on the dorsal surfaces of the thighs, groin, and concealed limb surfaces (vs. intense red coloration); and frequent green present in the dorsal coloration (vs. lack of green in the dorsal coloration). *Pristimantis numbala* differs from *P. paladines* by the rounded to broadly rounded snout (vs. acuminate to subacuminate snout in *P. paladines*), shagreen dorsum (vs. tuberculate dorsum), low middorsal fold (vs. tall middorsal fold that has usually several larger tubercles scattered along its length), and the relative length of the toes, with toe V much longer than toe III (vs. toe V longer than toe III). *Pristimantis numbala* differs from *P. samaniegoi* by its longer snout (Eye Nostril Distance 2.0 mm vs. 1.8 mm), longer fingers (finger III length 6.1 mm vs. finger III length 5.1 mm), longer toes (toe IV length 10.0 mm vs. toe IV length 8.3 mm), and broadly expanded discs on fingers and toes (vs. slightly expanded), and from *P. colodactylus* and *P. matildae* by the different coloration, longer fingers (finger III length 6.1 mm vs. finger III length 4.4 mm in *P. colodactylus*), longer toes (toe IV length 10.0 mm vs. toe IV length 7.5 mm in *P. colodactylus* and 9.8 mm in *P. matildae*), and by the typical reddish-brown coloration present on the dorsal surfaces of thighs, groin, and concealed limb surfaces (vs. concealed coloration missing in *P. colodactylus* and *P. matildae*).

Natural history.—Females were encountered during the night, perching on moss-covered branches or leaves, from the soil up to 1 m high, in habitats with shrubby vegetation.

Distribution.—See Figure A2.35. *Pristimantis numbala*, an endemic species to Ecuador, is known only from the southern highlands of the Cordillera Oriental of the Andes in Provincia Zamora Chinchipe at an elevation of 2875 m.

This mostly nocturnal, arboreal species inhabits the evergreen montane forests of the southern Cordillera Oriental of the Andes ecosystem within the southern Cordillera Oriental of the Andes biogeographic sector, where the average annual rainfall is 1290 mm and the average annual temperature is 13.4 °C.

Conservation.—*Pristimantis numbala* is categorized as Data Deficient (Székely *et al.*, 2023b) because of the lack sufficient data to make an assessment. *Pristimantis numbala* is known only from Reserva Numbala and its immediate proximity.

Remarks.—According to Székely *et al.* (2023b), *Pristimantis numbala* is the sister species of *P. paladines*, in the *P. colodactylus* subgroup of the *P. orestes* group.

Key references.—Székely *et al.*, (2023b) (description).

PRISTIMANTIS PALADINES SZÉKELY *ET AL.*, 2023B

PALADINES RAIN FROG/CUTÍN DE PALADINES

Pristimantis paladines Székely *et al.*, 2023b: 52. **Holotype.**—MUTPL 1159 from Parque Nacional Podocarpus in the vicinity of Reserva Tapichalaca (4.44728S, 79.14598W; 2880 m), Provincia Zamora Chinchipe, Ecuador.

Etymology.—The specific epithet honors the Paladines family from the city of Loja, in particular Félix Humberto Paladines Paladines (1938–2022), for his valuable contribution to the academic and cultural fields and for safeguarding the history and identity of southern Ecuadorian people. In addition, it constitutes recognition of the remarkable work carried out by his children, Renzo, Bruno, Pedro, and María Gabriela, who created the nongovernmental organization Naturaleza y Cultura Internacional, a consolidated organization dedicated to the protection of wildlife in Latin America, where it has managed to protect, to date, more than 8.5 million hectares of tropical forest in various countries.

Identification.—Figures A2.38, A2.39, A2.40. The maximum known SVL is 18.3 mm in males and 21.4 mm in females. The skin on the dorsum is tuberculate with numerous scattered large tubercles (in life, the tuberculated texture of the skin is more evident); the skin on the venter is areolate; a discoidal fold is present; dorsolateral folds are absent; the flanks usually have longitudinal lateral folds on the anterior half; the middorsal fold is evident and tall, usually with several larger tubercles scattered along its length (a feature more evident in life). The tympanic annulus and tympanic membrane are absent, but the supratympanic fold is present. The snout is acuminate to subacuminate in dorsal view and rounded in profile; the canthus rostralis is weakly concave to straight in dorsal view and rounded in profile. The upper eyelid bears two to three large tubercles and several small tubercles (a feature more evident in life); its width is about 72% of the

FIGURE A2.38 *Pristimantis paladines*. Left: holotype, female, 20.6 mm SVL. Parque Nacional Podocarpus, Provincia Zamora Chinchipe. Right: male, Cerro Toledo, Provincia Loja. Paul J. Székely.

FIGURE A2.39 Morphological variation in *Pristimantis paladines*. Left: females. Reserva Tapichalaca, Provincia Zamora Chinchipe. Top right: MUTPL 125, paratype, male, 16.7 mm SVL. Middle and bottom right: males. Cerro Toledo, Parque Nacional Podocarpus, Provincia Loja. Paul J. Székely.

FIGURE A2.40 Morphological details of *Pristimantis paladines*. Top: MUTPL 1107, paratype, male, 18 mm SVL. Note color change of the same individual. Bottom: groin and thighs of two males. Cerro Toledo, Parque Nacional Podocarpus, Provincia Loja. Paul J. Székely.

interorbital distance in females and 81% in males; cranial crests are absent. The dentigerous processes of the vomers are inconspicuous, oblique, ovoid, or triangular, separated medially by a distance smaller than the width of the processes; each process bears two to five teeth. Males lack vocal sacs, vocal slits, and nuptial pads. Finger I is shorter than finger II; the discs on the fingers are broadly expanded and truncate, with circumferential grooves present. The fingers bear lateral fringes; the subarticular tubercles are prominent; hyperdistal subarticular tubercles are present; supernumerary palmar tubercles are present; the palmar tubercle is usually not divided or only partially divided into a larger (inner) and a smaller (outer) tubercle; the thenar tubercle is elliptical and larger than the inner palmar tubercle. Ulnar tubercles are present (a feature more evident in life). The heel has two to three large tubercles and several small tubercles (a feature more evident in life); the outer edge of the tarsus has a row of large, conical tubercles (a feature more evident in life); the inner tarsal fold is present. The inner metatarsal tubercle is broadly ovoid, about 23 to 33% the size of the subconical (in profile) outer metatarsal tubercle; the subarticular tubercles are prominent; hyperdistal subarticular tubercles are present; supernumerary plantar tubercles are present. The toes bear broad lateral fringes and have basal webbing; Toe V is longer than toe III; the discs on the toes are broadly expanded and truncate, slightly smaller than those on the fingers, with circumferential grooves present. In life, the dorsum is various shades of brown, reddish-brown, greenish-brown, green, or reddish-orange with dark, blackish bars, with or without X or W markings on the back; the flanks, dorsal surfaces of the arms, and hind limbs have dark transverse bars; large areas of the dorsal surfaces of the thighs, groin, concealed limb surfaces, and sometimes the dorsum are orange or reddish-brown; the head usually has dark, blackish interorbital markings, dark labial bars, and dark supratympanic stripes; the venter and ventral surfaces of the hindlimbs and arms are pinkish-white or pinkish-gray, with or without dark markings on the throat; the iris is bronze with fine black reticulations and a median, wide, dark red horizontal streak.

Comparisons.—*Pristimantis colodactylus*, *P. sagedunneae*, *P. numbala*, *P. samaniegoi*, *P. muranunka*, *P. matildae*, *P. mutabilis*. *Pristimantis paladines* differs from *P. samaniegoi* by its longer snout (END 1.9 mm vs. END 1.8 mm), longer fingers (finger III length 5.4 mm vs. finger III length 5.1 mm), longer toes (toe IV length 9.7 mm vs. toe IV length 8.3 mm), broadly expanded discs on the fingers and toes (vs. slightly expanded), and by the different coloration pattern with dark, blackish bars, with or without X or W markings on the back and with large orange or reddish-brown areas on the dorsal surfaces of the thighs, groin, concealed limb surfaces, and sometimes the dorsum (vs. dorsum, flanks, dorsal surfaces of hind limbs, and arms dark brown with various white, irregular spots and a dark gray venter in females). Within the *P. colodactylus* subgroup, *P. paladines* can be distinguished from *P. muranunka* by the lack of a tympanic annulus and tympanic membrane (present in *P. muranunka*) and from *P. colodactylus* and *P. matildae* by having a non-compressed body and head (vs. compressed), longer fingers (finger III length 5.4 mm vs. finger III length 4.4 mm in *P. colodactylus*), longer toes (toe IV length 9.7 mm vs. toe IV length 7.5 mm in *P. colodactylus*), and by the orange or reddish-brown coloration present on the dorsal surfaces of the thighs, groin, and concealed limb surfaces (concealed coloration missing in *P. colodactylus* and *P. matildae*). *Pristimantis paladines* differs from the closely related *P. sagedunneae* by its smaller size; tuberculate dorsum (shagreen in *P. sagedunneae*); tall middorsal fold that has usually several larger tubercles scattered along its length (vs. low middorsal fold, without tubercles); orange or reddish-brown coloration on the dorsal surfaces of the thighs, groin, and concealed limb surfaces (vs. intense red coloration); and frequent green present in the dorsal coloration (vs. lack of green in the dorsal coloration). *Pristimantis paladines* differs from the also closely related *P. numbala* by the acuminate to subacuminate snout (vs. rounded to broadly rounded snout in *P. numbala*), tuberculate dorsum (vs. shagreen dorsum), tall middorsal fold that usually has several larger tubercles scattered along its length (vs. low middorsal fold, without tubercles), and shorter relative length of the toes, with toe V longer than toe III (vs. toe V much longer than toe III).

Natural history.—Individuals have been found at night perching on moss-covered branches or leaves or on moss-covered walls near the road, sometimes next to small streams, at heights ranging from the soil level up to 1.5 m, in habitats with shrubby vegetation.

Distribution.—See Figure A2.27. *Pristimantis paladines*, an endemic species to Ecuador, is known only from the southern highlands of the Cordillera Oriental of the Andes in Loja and Zamora Chinchipe provinces. Its elevational range is 2800–3100 m in an area of extent of occurrence of about 10 km^2.

This mostly nocturnal, arboreal species inhabits the evergreen shrub and herbazal of Páramo and evergreen montane forests of the southern Cordillera Oriental of the Andes ecosystems within the Paramo and southern Cordillera Oriental of the Andes biogeographic sectors, where the average annual rainfall is 1183–128 mm and the average annual temperature is 12.0–13.4 °C.

Conservation.—*Pristimantis paladines* is categorized as Near Threatened (Székely *et al.*, 2023b) following the IUCN criteria, due to its very small extent of occurrence (EOO). It is a common and locally abundant species known currently from only three close localities, from an estimated area of less than 20 km^2. All known localities are inside national or private protected areas, and its habitat currently does not face any major threats.

Remarks.—According to Székely *et al.* (2023b), *Pristimantis paladines* is the sister species of *P. numbala*, in the *P. colodactylus* subgroup of the *P. orestes* group.

Key references.—(Székely *et al.*, 2023b) (description).

PRISTIMANTIS SAGEDUNNEAE SZÉKELY *ET AL.*, 2023B

Sage Dunne's Rain Frog/Cutín de Sage Dunne

Pristimantis sagedunneae Székely *et al.*, 2023b: 45. **Holotype.**—MUTPL 500 from Abra de Zamora (3.9922°S, 79.1455°W, 2812 m), Provincia Loja, Ecuador.

Etymology.—The specific epithet honors Sage Dunne in recognition of her passion for Andean wildlife and her family's invaluable support of conservation work in Ecuador. Of particular importance is their contribution to amphibian conservation in the Sangay-Podocarpus connectivity corridor, Ecuador's first ecological corridor, which protects 567,067 ha of high-elevation paramo grasslands and cloud forest ecosystems, as well as chains of lakes and wetlands, with unique biological diversity and endemism.

Identification.—Figures A2.41, A2.42. The maximum known SVL is 24.3 mm in females; males are unknown. The skin on the dorsum is shagreen with some scattered tubercles (a feature more evident in life). The skin on the venter is areolate. A discoidal fold is present; dorsolateral folds are absent; the flanks usually have longitudinal lateral folds on the anterior half, and a low middorsal fold is present. The tympanic annulus and tympanic membrane are absent, but the supratympanic fold is present. The snout is subacuminate to rounded in dorsal view and rounded in profile; the canthus rostralis is concave in dorsal view and rounded in profile. The upper eyelid bears one larger

FIGURE A2.41 *Pristimantis sagedunneae.* Left: MUTPL 500, holotype, female, 21.4 mm SVL. Abra de Zamora, Provincia Loja. Right: MUTPL 597, paratype, female, 21.7 mm SVL. Cajanuma, Parque Nacional Podocarpus, Provincia Loja. Paul J. Székely.

FIGURE A2.42 Morphological variation and details of *Pristimantis sagedunneae*. Top left: female. Bottom left: MUTPL 1096, paratype, juvenile, 14.7 mm SVL. Top right: MUTPL 1095, paratype, female, 23.8 mm SVL. Bottom right: MUTPL 1095, groin and thigh. Abra de Zamora, Provincia Loja. Paul J. Székely.

tubercle and several small tubercles (a feature more evident in life), and its width is about 68% of the interorbital distance; cranial crests are absent. The dentigerous processes of the vomers are concealed in the buccal mucosa, each process bearing three to five teeth. The condition of the vocal sacs, vocal slits, and nuptial pads is unknown. Finger I is shorter than finger II. The discs on the fingers are broadly expanded and truncate, with circumferential grooves present. The fingers bear lateral fringe. The subarticular tubercles are prominent; hyperdistal subarticular tubercles are present; supernumerary palmar tubercles are present. The palmar tubercle is divided into a larger (inner) and a smaller (outer) tubercle; the thenar tubercle is elliptical, the same size as the inner palmar tubercle. Ulnar tubercles are present (a feature more evident in life). The heel has one to three large tubercles and several small tubercles (a feature more evident in life); the outer edge of the tarsus has a row of small tubercles (a feature more evident in life); the inner tarsal fold is present. The inner metatarsal tubercle is broadly ovoid, about 33 to 43% the size of the subconical (in profile) outer metatarsal tubercle. The subarticular tubercles are prominent; hyperdistal subarticular tubercles are present; supernumerary plantar tubercles are present. The toes bear broad lateral fringes and have basal webbing; toe V is longer than toe III; the discs on the toes are broadly expanded and truncate, about the same size as those on the fingers, with circumferential grooves present.

The dorsum is brownish-gray with blackish or dark brown bars that conjoin in X- or XXX-like shapes on the middle of the back; the flanks, dorsal surfaces of the arms, and hindlimbs have dark transverse bars. Large areas of the dorsal surfaces of the thighs, groin, and concealed limb surfaces are red or intense red. The head has a wide, black-bordered, light interorbital bar; dark labial bars; and dark supratympanic stripes. The venter and ventral surfaces of the hindlimbs and arms are pinkish-white or pinkish-gray, with or without dark markings on the throat. The iris is bronze with black reticulations and a median, wide, dark red horizontal streak.

Comparisons.—*Pristimantis colodactylus*, *P. matildae*, *P. muranunka*, *P. numbala*, *P. orestes*, *P. paladines*, *P. samaniegoi*. From the 17 species that are currently included in the *P. orestes* group, only six species lack tympanic annuli and membranes: *P. samaniegoi*, *P. colodactylus*, *P. matildae*, *P. paladines*, *P. numbala*, and *P. sagedunneae*. However, *P. sagedunneae* differs from *P. samaniegoi* by its longer snout (END 2.1 mm vs. END 1.8 mm), longer fingers (finger III length 6.3 mm vs. finger III length 5.1 mm), longer toes (toe IV length 10.1 mm vs. toe IV length 8.3 mm), broadly expanded discs on the fingers and toes (vs. slightly expanded), and different coloration pattern, with blackish or dark brown bars that on the middle of the back conjoin in X- or XXX-like shapes and with large areas of the dorsal surfaces of the thighs, groin, and concealed limb surfaces being red or intense red (vs. dorsum, flanks, dorsal surfaces of hindlimbs and arms dark brown with various white, irregular spots and a dark gray venter in females). Within the *P. colodactylus* subgroup, *P. sagedunneae* can be distinguished from *P. muranunka* by the lack of a tympanic annulus and tympanic membrane (vs. present in *P. muranunka*) and from *P. colodactylus* and *P. matildae* by having a non-compressed body and head (vs. compressed), longer fingers (finger III length 6.3 mm vs. finger III length 4.4 mm in *P. colodactylus* and 6.1 mm in *P. matildae*), longer toes (toe IV length 10.1 mm vs. toe IV length 7.5 mm in *P. colodactylus* and 9.8 mm in *P. matildae*), and by the typical red coloration present on the dorsal surfaces of the thighs, groin, and concealed limb surfaces (vs. concealed coloration missing in *P. colodactylus* and *P. matildae*). *Pristimantis sagedunneae* can be distinguished from *P. paladines* by its larger size; shagreen dorsum (vs. tuberculate in *P. paladines*); low middorsal fold (vs. elevated or tall middorsal fold that has usually several larger tubercles scattered along its length); intense red coloration on the dorsal surfaces of thighs, groin, and concealed limb surfaces (vs. orange or reddish-brown coloration); and the lack of green in the dorsal coloration (vs. frequently green present in the dorsal coloration). From *P. numbala*, *P. sagedunneae* can be distinguished by the subacuminate to rounded snout (vs. rounded to broadly rounded snout in *P. numbala*); the relative length of the toes, with toe V longer than toe III (vs. toe V much longer than toe III); the intense red coloration on the dorsal surfaces of the thighs, groin, and concealed limb surfaces (vs. reddish-brown coloration); and the lack of green in the dorsal coloration (vs. green frequently present in the dorsal coloration).

Natural history.—Females have been found during the night, perching on moss-covered branches or leaves from 40 cm up to 2 m high in habitats with shrubby vegetation. No individual was encountered inside bromeliads.

Distribution.—See Figure A2.33. *Pristimantis sagedunneae*, an endemic species to Ecuador, is known only in the highlands of the Cordillera Oriental of the Andes in the Abra de Zamora and about 13 km to the south, in the Cajanuma sector of Loja and Zamora Chinchipe provinces. Its elevational range is 2812–3000 m in an area of extent of occurrence of about 10 km^2.

This mostly nocturnal, arboreal species inhabits the evergreen high montane forests of the southern Cordillera Oriental of the Andes, evergreen montane forest of the southern Cordillera Oriental of the Andes, and herbazal of Páramo ecosystems within the Páramo and southern Cordillera Oriental of the Andes biogeographic sectors, where the average annual rainfall is 1113–1141 mm and the average annual temperature is 12.4–13.1 °C.

Conservation.—*Pristimantis sagedunneae* is categorized as Endangered (Székely *et al.*, 2023b) following the B1ab(iii,iv,v) + 2ab(iii,iv,v) International Union for Conservation of Nature criteria (IUCN, 2001) because: (1) its extent of occurrence (EOO) and area of occupancy (AOO) are estimated to be less than 10 km^2; (2) it is known from only a few individuals and only from two localities; and (3) its habitats could be affected in the near future, as they are situated in the vicinity of densely populated areas. Although both localities are inside a nationally protected area, Parque Nacional Podocarpus, this is a rare species, or at least with a very low detectability. Between 2016 and 2022, only seven specimens were found, and during only four days out of more than 50 days of fieldwork in the area. All the specimens from Abra de Zamora were encountered in only one very small area (less than 50 m^2) and in Cajanuma only on the Los Miradores trail.

Remarks.—According to Székely *et al.* (2023b), *Pristimantis sagedunneae* is the sister species of *P. paladines* + *P. numbala*, in the *P. colodactylus* subgroup of the *P. orestes* group.

Key references.—(Székely *et al.*, 2023b) (description).

HYLOSCIRTUS ARCANUS RIVERA-CORREA *ET AL.*, 2024

Hidden Stream Treefrog/Rana Torrentícola Misteriosa

Hyloscirtus arcanus Rivera-Correa *et al.*, 2024: 104. **Holotype.**—KU 202730 from 23.3 km WSW Plan de Milagro, (-3.2075, -78.5450; 2350 m a.s.l), Provincia Morona Santiago, Ecuador.

Etymology.—The specific epithet is derived from the Latin adjective *arcanus*, meaning "hidden" or "secret," in reference to the fact that this species remained concealed in a museum and the literature under an erroneous taxonomic identity for 40 years.

Identification.—Figure A2.43. The maximum known SVL is 67.8 mm in males and 70.8 mm in a female. The snout is rounded in dorsal view, truncate in profile. The tympanum and tympanic annulus are inconspicuous in three males and partially visible in a female. The vomerine odontophore features two small processes that are in contact with each other, each carrying four teeth. The vocal slits of males are longitudinal, originating lateral to the tongue, extending to the corner of the mouth. The vocal sac is large, single, median, subgular, and evident externally. The males have subtly more extensive webbing than the female. The foot webbing is extensive. Males exhibit large, curved, osseous, spine-shaped prepollices and strongly hypertrophied forelimbs. The prepollex is slightly smaller in a female. The female forelimbs are robust, but they do not present the remarkable hypertrophy observed in males. Males possess a large, thick supracloacal flap and supratympanic fold. Additionally, adult males have spicules on the snout, canthus rostralis, eyelid, supratympanic fold, and shank. Spicules are absent in a female. The nuptial pad is absent. The dorsum varies from pale (female) to dark gray (males), while the belly of a female is dark gray, light gray in males. The concealed surfaces of the limbs appear bluish-gray. The digital discs are orange–red to bright red. The iris is a dull olive bronze.

Comparisons.—*Hyloscirtus condor*, *H. diabolus*, *H. hillisi*, *H. lindae*, *H. tapichalaca*. *Hyloscirtus arcanus* resembles *H. lindae* in having red discs on the fingers and toes. Males of both species differ in shank dorsal texture, being smooth in *H. lindae* and spiculate in *H. arcanus*. *Hyloscirtus lindae* and most species of the *H. larinopygion* group differ from *H. arcanus* by exhibiting a blade-shaped prepollex and non-hypertrophied forelimbs. *Hyloscirtus arcanus* is comparable to four species in the *H. larinopygion* group that possess highly hypertrophied forelimbs

FIGURE A2.43 *Hyloscirtus arcanus.* KU 202731, paratype, male, 67.8 mm SVL, 23.3 km WSW of Plan de Milagro, Provincia Morona Santiago. David M. Hillis.

in adult males and an enlarged, curved, spine-shaped prepollex. These species are H. *tapichalaca*, *H. condor*, *H. diabolus*, and *H. hillisi*. *Hyloscirtus tapichalaca*: axillary and inguinal regions with pink–orange marks (absent in *H. arcanus*); thighs and shanks with brown transverse bands (absent in *H. arcanus*); concealed surfaces of limbs brown (concealed surfaces of limbs bluish-gray in *H. arcanus*); lateral and lower margin of cloaca, knee, elbow, and outer margin of forearm white (white absent on cloaca and limbs in *H. arcanus*); digital discs white (orange–red to bright red digital discs in life or grayish–orange in preserved specimens of *H. arcanus*); iris yellow–gold with fine black reticulations (dull olive bronze in *H. arcanus*); vomerine odontophore processes separate (contacting medially in *H. arcanus*); dorsal spicules absent (present in *H. arcanus*); gular and abdominal region dark gray (pale gray in *H. arcanus*); and calcar short and triangular (transverse fringe in *H. arcanus*). *Hyloscirtus condor*: dorsum dark brown with contrasting yellow spots (gray without spots in *H. arcanus*); digital discs light brown, yellow–brown, or brown with abundant yellow blotches (orange–red to bright red digital discs in *H. arcanus*); iris golden with fine tan reticulations (dull olive bronze in *H. arcanus*); vomerine odontophore processes separate (contacting medially in *H. arcanus*); dorsal spicules in males absent (present in *H. arcanus*). *Hyloscirtus diabolus*: dorsum dark brown with abundant small yellow marks or yellow blotches (gray without spots in *H. arcanus*); concealed surfaces of the thighs and shanks dark brown (bluish-gray in *H. arcanus*); digital discs dark brown, with or without yellow spots (orange–red to bright red digital discs of *H. arcanus*); iris dark red (dull olive bronze in *H. arcanus*); vomerine odontophore processes separate (contacting medially in *H. arcanus*); and dorsal spicules in males absent (present in *H. arcanus*). *Hyloscirtus hillisi*: dorsum dark brown with contrasting round orange marks (gray without spots in *H. arcanus*), digital discs dark brown (orange–red to bright red in *H. arcanus*), iris bronze or yellowish with brown reticulations (dull olive bronze in *H. arcanus*), and fleshy calcar absent (present in *H. arcanus*).

Natural history.—Specimens were found at night on 4 March 1984 on tree branches in the spray zone of a waterfall. The male was calling from a cavity beneath moss; the call was a whistle with a three-note pattern and was low-pitched. The female was on the branch of a bush. An additional calling male (not collected) was observed higher up the same waterfall.

The skin on the tip of the prepollex presents scars in the males but not in the female. In two males and one female, there are scars on the dorsum, suggesting either combat or a complex breeding behavior. The oviducts of the female are full of large, unpigmented oocytes (oocyte diameter 2.28–2.97 mm; n = 3).

Tadpoles.—Unknown.

Distribution.—Figure A2.44. *Hyloscirtus arcanus*, an endemic species to Ecuador, is known only from two nearby localities (airline distance between localities < 2 km) in Provincia Morona Santiago on the southeastern slopes of the Cordillera Oriental of the Andes of Ecuador at an elevation of 2350–2425 m.

This mostly nocturnal, arboreal species inhabits a mostly intervened area within the Amazonian Cordilleras biogeographic sector, where the average annual rainfall is 1257–1359 mm and the average annual temperature is 18.3–19.4 °C.

Conservation.—*Hyloscirtus arcanus* is herein categorized as Data Deficient. There is little information, and further surveys are necessary to assess its conservation status. The type specimens were collected in March 1984.

Remarks.—According to Rivera-Correa *et al.* (2024), *Hyloscirtus arcanus* is the sister species of *H. tapichalaca* in the *H. larinopygion* group. It was previously confused with *H. lindae*. Rivera-Correa *et al.* (2024) described and discussed both the prepollex anatomy and arm morphology.

Key references.—(Rivera-Correa *et al.*, 2024) (description, phylogeny).

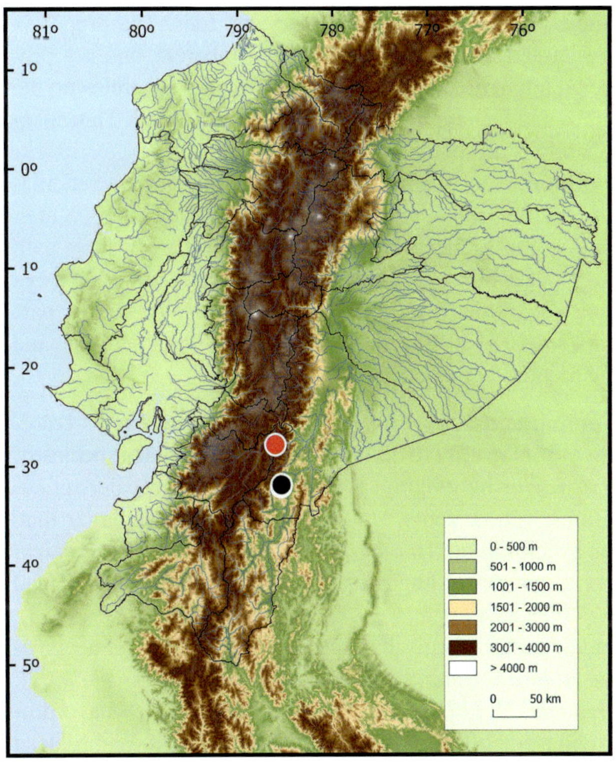

FIGURE A2.44 Localities of known occurrence of *Hyloscirtus arcanus* (black dot) and *H. tolkieni* (red dot) in Ecuador.

HYLOSCIRTUS TOLKIENI SÁNCHEZ-NIVICELA *ET AL.*, 2023

TOLKIEN'S STREAM TREEFROG/RANA TORRENTÍCOLA DE TOLKIEN

Hyloscirtus tolkieni Sánchez-Nivicela *et al.*, 2023: 79. **Holotype.**—ZSFQ-4142 from Puente de Piedra (3627 m; 3190 m), Parque Nacional Río Negro-Sopladora, Provincia Morona Santiago, Ecuador.

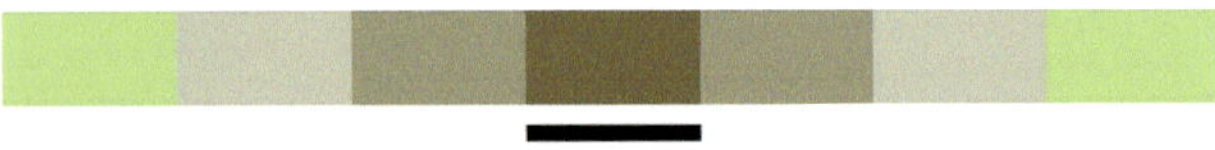

Etymology.—The specific epithet honors the writer, poet, philologist, and academic John Ronald Reuel Tolkien (J. R. R. Tolkien, 1892–1973), creator of Middle Earth and author of fantasy works like "The Hobbit" and "The Lord of the Rings." The amazing colors of the new species evoke the magnificent creatures that seem to only exist in fantasy worlds.

Identification.—Figure A2.45. The maximum known SVL is 64.9 mm in a female; males are unknown. The snout is truncate in dorsal and lateral view. The head is 7% wider than long. The tympanic membrane and annulus are evident, partially covered by the supratympanic fold. The dentigerous process of the vomer is slightly separated, with 9–13 vomerine teeth. The forearm is robust and slightly thicker than the arm. The discs are slightly expanded. Broad dermal fringes are present on the fingers and toes. The prepollex is enlarged, hidden under the thenar tubercle, and not projected into a prepollical spine. Subarticular tubercles on the hands and feet are rounded and poorly projected. The calcar tubercle is present.

The dorsum is grayish-green, with paler-hued reticulation, yellow spots, and black speckles. The throat, venter, flanks, and hidden surfaces of the limbs are yellow with large black blotches and spots. The fingers, toes, and webbing are yellow with black bars and spots. The iris is pale pink with a black periphery. The sclera is grayish-blue, and the nictitating membrane is yellow.

Comparisons.—*Hyloscirtus tolkieni* is unique in its coloration; it is readily distinguishable from all other species of *Hyloscirtus* by its grayish-green dorsum with paler reticulations, yellow spots, and black speckles. It is larger than the species in the *H. bogotensis* group. *Hyloscirtus tolkieni* has a non-protruding prepollex and narrower head (Head Width/Head Length = 1.07), more vomerine teeth (9–13), and thinner forearms than species of the southern lineage of the *H. larinopygion* species group (including *H. condor*, *H. diabolus*, *H. hillisi*, and *H. tapichalaca*), which have protruding prepollical spines, wider heads (Head Width/Head Length ≥ 1.10), fewer vomerine teeth (2–6), and hypertrophied forearms. Also, all species of the southern lineage of the *H. larinopygion* species group are darkly colored dorsally and ventrally. *Hyloscirtus tolkieni* differs from species of the *H. armatus* species group by the absence of clusters of keratinized spines on the prepollex and the proximal ventral surface of the humerus (present in *H. armatus* and *H. charazani*), non-expanded prepollex (expanded in *H. armatus* and *H. charazani*), robust but not hypertrophied forearms (hypertrophied in *H. armatus* and *H. charazani*), and absence of a skin fold in the proximoventral portion of upper arm (present in *H. armatus*, *H. charazani*, and *H. chlorostea*).

Natural history.—The female holotype was active at night at 20:30 h amid tree branches, about 5 m above the ground and 8 m from the nearest stream.

Distribution.—See Figure A2.44. *Hyloscirtus tolkieni*, an endemic species to Ecuador, is known only from the type locality (Parque Nacional Río Negro-Sopladora), Provincia Morona Santiago, on the southeastern slopes of the Cordillera Oriental of the Andes of Ecuador at an elevation of 3190 m.

This mostly nocturnal, arboreal species inhabits the evergreen shrub and herbazal of Páramo ecosystem within the Páramo biogeographic sector, where the average annual rainfall is 1194 mm and the average annual temperature is 10.8 °C.

Conservation.—*Hyloscirtus tolkieni* is categorized as Data Deficient (Sánchez-Nivicela *et al.*, 2023) in the absence of sufficient information to evaluate the conservation status and until more data is obtained. The type locality of *H. tolkieni* is officially protected as part of the Parque Nacional Río Negro-Sopladora (previously known as Bosque Protector Cuenca del Río Paute), a protected area created in 2018 where little habitat loss has occurred. Large, forested areas remain unstudied in the national park, and the species may have a wider distribution beyond the immediate surrounding of its type locality. Nonetheless, Sánchez-Nivicela *et al.*

FIGURE A2.45 *Hyloscirtus tolkieni*. ZSFQ 4142, holotype, female, 64.9 mm SVL, Puente de Piedra, Parque Nacional Río Negro-Sopladora, Provincia Morona Santiago, Ecuador. Juan C. Sánchez-Nivicela.

(2023) report that surveys were carried out over 13 effective days, and researchers could not detect additional individuals of *H. tolkieni*, despite focalized searches. They suggest that urgent research and monitoring actions should be established to study its life history and ecology, population size, and trends; survey new sites where additional populations may exist; and evaluate if threats are impacting its long-term conservation, such as invasive species, emerging diseases, or climate change.

Remarks.—Its phylogenetic relationships are unknown. According to Sánchez-Nivicela *et al.* (2023), it is in the *Hyloscirtus larinopygion* group.

Key references.—(Sánchez-Nivicela *et al*., 2023) (description).

GYMNOPHIONA

Caeciliidae

CAECILIA TESORO BOCK *ET AL.*, 2024

Tesoro's Caecilian/Ilulo Tesoro

Caecilia tesoro Bock *et al*., 2024: 222. **Holotype.**—QCAZ-A 78501 from Tesoro Escondido Reserve, N 0.54033, W 79.14432, 328 m, Provincia Esmeraldas, Ecuador.

Etymology.—The specific epithet *tesoro* means "treasure" in Spanish, thereby referring to the rich biodiversity of the Chocó lowland forests, and the name of the reserve with the type locality of the species, and thus honoring the Ecuadorian Non-governmental organization Tesoro Escondido (www.tesororeserve.org), which protects this outstanding forest.

Identification.—Figure A2.46. This robust and large *Caecilia* has a body length of up to 790 mm. The larger specimens display a unique color pattern with a grey head and posterior body, while the remaining body is yellowish-cream to caramel. The smaller individuals being almost uniformly dark grey, with or without light spotting or ventrolateral light grey bands. The species also lacks a terminal shield. The eyes are visible and are not covered by bone. It exhibits 117–121 primary folds (coefficient of variation [CV] = 1.58), 14–18 secondary folds (CV = 34.5), an absence of complete secondary folds, and 124–126 vertebrae.

A paratype shows a bicolor pattern consisting of dark grey dorsum and light grey ventrolateral bands with a sharp color transition along mid-flank. The coloration of the holotype is characteristic for its size and/or sex and darker coloration may be characteristic of juvenile and subadult individuals.

Comparisons.—*Caecilia tesoro* differs from other western Ecuadorian and south-western Colombian *Caecilia* species as follows: adults larger (> 570 mm) than *C. leucocephala, C. perdita, C. subterminalis, C. tenuissima, C. truncata* sp. n., and *C. volcani* (all < 505 mm); body more slender (TL/BW: 35–40) than (most) *C. truncata* and *C. volcani* (all in-between 22–43); body more robust (smaller TL/BW ratio) than *C. nigricans* (most), *C. pachynema*, *C. perdita* (some), *C. tenuissima*, and *C. wilkinsoni* (> 44); less primary annular grooves (117–121) than *C. nigricans, C. pachynema, C. perdita, C. subterminalis, C. tenuissima*, and *C. wilkinsoni* (all > 133); more secondary annular groves (14–18) than *C. pachynema, C. tenuissima,* and *C. truncata* (all < 11); less secondary annular groves than *C. leucocephala, C. nigricans, C. perdita, C. volcani* (most), and *C. wilkinsoni* (> 27); the absence of complete secondary annular grooves, these are present in at least some or most: *C. guntheri, C. leucocephala, C. nigricans, C. perdita, C. tenuissima, C. volcani,* and *C. wilkinsoni*; the absence of a terminal shield distinguishes *C. tesoro* from *C. leucocephala, C. nigricans, C. pachynema, C. perdita, C. subterminalis*, and *C. tenuissima*. Finally, large *C. tesoro* have a unique colour pattern, not shared by any other described *Caecilia* species (see earlier).

Natural history.—This species inhabits a range of environments, from old-growth forests to artificial open areas.

FIGURE A2.46 *Caecilia tesoro*. QCAZ-A 78501, holotype, female, 790 mm SVL, Reserva Tesoro Escondido, Provincia Esmeraldas. Note the irregular darker markings (presumably scars). Sarah Bock.

The holotype was observed moving in the late afternoon on muddy ground on a steep slope within a regenerating cacao plantation, following heavy rainfall. This plantation is located near degraded habitats and old-growth forest, with the latter being less than 200 m away. On 23 June 2023, at 4:40 PM, an individual was observed in leaf litter within a nearby primary forest, approximately 1 km from the type locality, on a montane ridge. In the Tesoro Escondido Reserve, two additional individuals, each approximately 800 mm in total length, were observed in primary forest during the day (at 11:30 AM and 5:00 PM) after heavy rains. One of these was collected near a road in a human-populated area, while the other was moving through leaf litter in old-growth forest. Three specimens were collected during the core rainy season, specifically in January, February, and April. Additional observations occurred in June and November, while one paratype was collected in August during a drier season. This suggests the species may be active on the surface primarily after heavy rains.

Distribution.—Figure A2.47. *Caecilia tesoro,* an endemic species to Ecuador, is known from the northwestern lowlands of the Ecuadorian Pacific Coast in Esmeraldas, Manabí, Carchi, and Pichincha provinces. Its elevational range is 300–1420 m in an area of extent of occurrence of about 6100 km^2.

This species inhabits natural forests and intervened areas in the Evergreen Forest of Lowlands of Ecuatorial Chocó

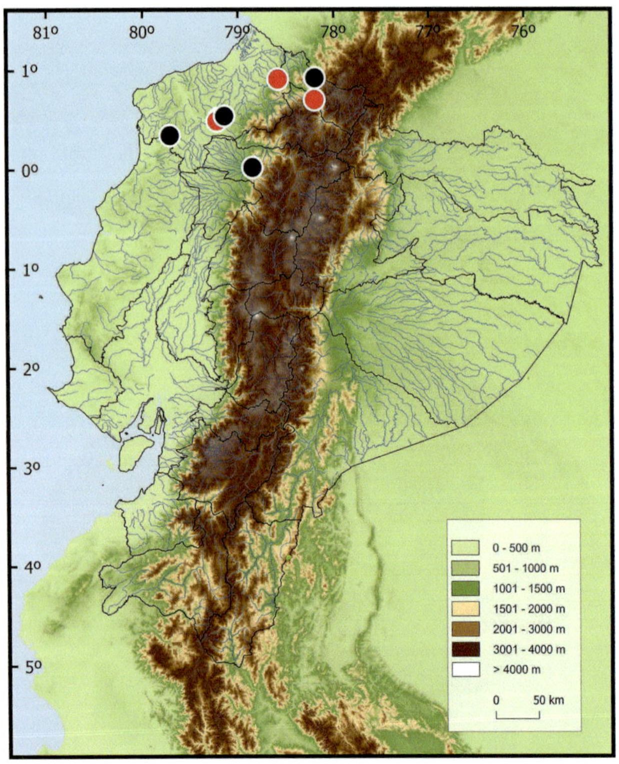

FIGURE A2.47 Localities of known occurrence of *Caecilia tesoro* (black dots) and *C. truncata* (red dots) in Ecuador.

and the Evergreen Low Montane Forest of Cordillera Costera of Ecuatorial Chocó ecosystems within the Cordillera Costera of Chocó and the Cordillera Occidental of the Andes biogeographic sectors, where the average annual rainfall is 1802–2888 mm and the average annual temperature is 19.7–23.9 °C.

Conservation.—*Caecilia tesoro* is suggested to be included in the IUCN Red List as Data Defficient (Bock *et al.*, 2024), in view of the scant known records.

It has been recorded in the public reserves Reserva Ecológica Mache Chindul, the Reserva de la Biósfera Chocó Andino de Pichincha, and some private reserves (e.g., Reserva Tesoro Escondido, Reserva Canadé, Reserva, Estación Biológica Bilsa).

Remarks.—According to the molecular phylogenetic analysis by Bock *et al.* (2024), *Caecilia tesoro* is sister to *C. truncata* + *C. volcani.*

Key references.—(Bock *et al.*, 2024) (description, osteology).

CAECILIA TRUNCATA BOCK *ET AL.*, 2024

TRUNCATED CAECILIAN/ILULO TRUNCADO

Caecilia truncata Bock *et al.*, 2024: 228. **Holotype.**—QCAZ-A 27774 from Río Carolina (N 0.7045, W 78.2011, 500 m), El Placer, Alto Tambo, Provincia Esmeraldas, Ecuador.

Etymology.—The species epithet is the Latin adjective *truncata* referring to the blunt shape of the snout.

Identification.—Figure A2.48. This small but robust species, reaching up to 482 mm in length, has a short snout that appears truncated in dorsal view. The eye is visible and not covered by bone. It exhibits 107 to 123 primary folds, with a coefficient of variation of 5.79, and 5 to 8 incomplete secondary folds, with a coefficient of variation of 19.0. A terminal shield is present, and the species has 128 to 129 vertebrae.

The dorsum and venter are dark grey, with lighter grey flanks, head, and neck, although the lighter areas are not always distinct. Cream coloration is evident in the areas surrounding the snout, tentacles, and nostrils. The entire lower mandible of a small individual is uniform, yellow creamish.

Comparisons.—From other western Ecuadorian and southwestern Colombian *Caecilia* species, *C. truncata* differs as follows: *C. truncata* has a more truncated head shape than all other species; it is smaller (< 500 mm) than *C. guntheri, C. nigricans*, *C. pachynema*, *C. tesoro*, and *C. wilkinsoni*

FIGURE A2.48 Morphological variation and details of *Caecilia truncata*. Top: QCAZ-A 79740, Reserva Tesoro Escondido, Provincia Esmeraldas. Second row to bottom: QCAZ-A 78979, paratype, juvenile, 387 mm SVL, note head shape, eye, and cloacal details, Reserva Canandé, Provincia Esmeraldas. Mark-Oliver Rödel.

(> 650 mm); it has a more robust body shape (BL/BW: 23–43) than *C. pachynema, C. tenuissima,* and *C. wilkinsoni* (all > 44, up to 108); it has less primary annular grooves (107–123) than *C. nigricans, C. pachynema, C. perdita, C. subterminalis, C. tenuissima*, and *C. wilkinsoni* (> 133); it has less secondary annular grooves (5–8) than *C. leucocephala, C. nigricans, C. perdita, C. subterminalis, C. tesoro, C. volcani*, and *C. wilkinsoni*; it has a terminal shield in contrast to *C. guntheri, C. tesoro, C. volcani,* and *C. wilkinsoni*; it shows an almost uniform greyish-blue color pattern, as typical for most *Caecilia*, but very different to *C. pachynema* and large *C. tesoro*.

Natural history.—The species occurs in old-growth and secondary forest. An individual was found above ground, near a stream of about 5 m width, next to a road. In the Reserva Tesoro Escondido and Reserva Canandé, 11 animals were observed during the night and during very heavy rain, moving on the muddy ground of paths through good standing forest. One was collected at about midnight in the Canandé Reserve on 2 December 2023. Very close to this locality, a further individual was observed on 8 April 2024 at 19:43; temperature: 22.4 °C. Five animals have been seen on two consecutive nights in the Reserva Tesoro Escondido, one on 20 March 2023 and the other four on 21 March 2023.

Distribution.—*Caecilia truncata,* an endemic species to Ecuador, is known from the northwestern lowlands of the Ecuadorian Pacific Coast in Provincia Esmeraldas. Its elevational range is 486–700 m in an area of extent of occurrence of about 1890 km^2.

This species inhabits natural forests and intervened areas in the Evergreen Montane Foothill Forest of Cordillera Occidental of the Andes ecosystem within the Cordillera Occidental of the Andes biogeographic sector, where the average annual rainfall is 1587–2490 mm and the average annual temperature is 20.7–23.6 °C.

Conservation.—*Caecilia truncata* is suggested to be included in the IUCN Red List as Data Defficient (Bock *et al.*, 2024), in view of the scant known records. It has been recorded in some private reserves (e.g., Reserva Tesoro Escondido, Reserva Canandé).

Remarks.—According to the molecular phylogenetic analysis by Bock *et al.* (2024), *Caecilia truncata* is sister to *C. volcani* from Panama and Costa Rica.

Key references.—(Bock *et al.*, 2024) (description, osteology).

Literature Cited

Amador, L., A. Parada, G. D'Elía, and J. M. Guayasamin. 2018. Uncovering hidden specific diversity of Andean glassfrogs of the *Centrolene buckleyi* species complex (Anura: Centrolenidae). *PeerJ* 6: 1–20, e5856.

Armijos-Ojeda, D., D. Székely, P. Székely, D. Cogălniceanu, D. F. Cisneros-Heredia, L. Ordóñez-Delgado, A. Escudero, and C. I. Espinosa. 2021. Amphibians of the equatorial seasonally dry forests of Ecuador and Peru. *ZooKeys* 1063: 23–48.

Ballestas, O., M. Lampo, and D. Rodríguez. 2021. Living with the pathogenic chytrid fungus: exploring mechanisms of coexistence in the harlequin toad *Atelopus cruciger*. *PLoS ONE* 16: e0254439.

Bock, S., T. E. Arroba-López, H. L. Velez-Giler, V. Moreira, M. L. M. Wiedebusch, K. Neira-Salamea, M. Wilkinson, N. Fuchs, M. Schönleitner, M.-O. Rödel, and S. R. Ron. 2024. Two new species of *Caecilia* (Gymnophiona: Caeciliidae) from the Ecuadorian humid Chocó. *Salamandra* 60: 209–236.

Brito-Zapata, D., J. D. Chávez-Reyes, M. D. Pallo-Robles, J. C. Carrión-Olmedo, D. F. Cisneros-Heredia, and C. Reyes-Puig. 2024. A new species of frog of the genus *Noblella* Barbour, 190 (Amphibia: Strabomantidae) from the Cordillera del Cóndor, Ecuador. *PeerJ* 12(e17939): 1–29.

Cisneros-Heredia, D. F., M. H. Yánez-Muñoz, J. C. Sánchez-Nivicela, and S. R. Ron. 2023. Two new syntopic species of glassfrogs (Amphibia: Centrolenidae: *Centrolene*) from the southwestern Andes of Ecuador. *PeerJ* 11: 1–27.

Flechas, S. V., C. Sarmiento, and A. Amézquita. 2012. *Bd* on the beach: high prevalence of *Batrachochytrium dendrobatidis* in the lowland forest of Gorgona Island (Colombia, South America). *EcoHealth* 9: 298–302.

Flechas, S. V., V. T. Vredenburg, and A. Amézquita. 2015. Infection prevalence in three lowland species of harlequin toads from the threatened genus Atelopus. *Herpetological Review* 46: 528–532.

Franco-Mena, D., I. De la Riva, M. A. Vega-Yánez, P. Székely, L. Amador, D. Batallas-Revelo, J. P. Reyes-Puig, D. F. Cisneros-Heredia, K. Venegas-Valencia, S. P. Galeano, J. Culebras, and J. M. Guayasamin. 2024. Simplifying the *Centrolene buckleyi* complex (Amphibia: Anura: Centrolenidae): a taxonomic review and description of two new species. *PeerJ* 12(e17712): 1–45.

Freile, J. F., L. A. Coloma, A. Terán-Valdez, D. Acosta-López, and E. E. Tapia. 2020. *Anfibios de Junín. Centro Jambatu. Publicación Miscelánea 1*. Quito: Imprenta Mariscal.

Frost, D. R. 2024. *Amphibian Species of the World: An Online Reference. Version 6.2*. New York: American Museum of Natural History. https://amphibiansoftheworld.amnh.org/index.php Accessed on 26 November 2024.

IUCN. 2001. *IUCN Red List Categories and Criteria: Version 3.1. IUCN Species Survival Commission*. Gland, Switzerland and Cambridge, UK: IUCN.

López-Hervas, K., J. C. Santos, S. R. Ron, M. Betancourth-Cundar, D. C. Cannatella, and R. D. Tarvin. 2024. Deep divergences among inconspicuously colored clades of *Epipedobates* poison frogs. *Molecular Phylogenetics and Evolution* 195: 108065.

Lötters, S., A. Plewnia, A. Catenazzi, K. Neam, A. R. Acosta-Galvis, Y. Alarcon Vela, J. P. Allen, J. O. Alfaro Segundo, A. de Lourdes Almendáriz Cabezas, G. Alvarado Barboza, K. R. Alves-Silva, M. Anganoy-Criollo, E. Arbeláez Ortiz, J. D. Arpi Lojano, A. Arteaga, O. Ballestas, D. Barrera Moscoso, J. D. Barros-Castañeda, A. Batista, M. H. Bernal, E. Betancourt, Y. O. da Cunha Bitar, P. Böning, L. Bravo-Valencia, J. F. Cáceres Andrade, *et al.* 2023. Ongoing harlequin toad declines suggest the amphibian extinction crisis is still an emergency. *Communications Earth and Environment* 4: 412.

Lynch, J. D., and W. E. Duellman. 1980. The *Eleutherodactylus* of the Amazonian slopes of the Ecuadorian Andes (Anura: Leptodactylidae). *Miscellaneous Publication* 69: 1–86.

Menéndez-Guerrero, P. A., S. P. L. dos Santos, M.-J. Salazar-Nicholls, D. M. Green, and S. R. Ron. 2024. Cryptic diversity in toads of the *Rhinella marina* species group (Anura, Bufonidae) with a subjectively beautiful new species from Western Ecuador. *Zoological Journal of the Linnean Society* 197: 1–26.

Plewnia, A., A. Terán-Valdez, J. Culebras, R. Boistel, D. J. Paluh, A. B. Quezada Riera, C. H. Heine, J. P. Reyes-Puig, D. Salazar-Valenzuela, J. M. Guayasamin, and S. Lötters. 2024. A new species of harlequin toad (Bufonidae: *Atelopus*) from Amazonian Ecuador. *Salamandra* 60: 237–253.

Reyes-Puig, J. P., C. Reyes-Puig, D. Franco-Mena, L. Jost, and M. H. Yánez-Muñoz. 2022. Strong differentiation between amphibian communities on two adjacent mountains in the Upper Rio Pastaza watershed of Ecuador, with descriptions of two new species of terrestrial frogs. *ZooKeys* 2022: 35–87.

Reyes-Puig, J. P., M. Urgilés-Merchán, D. Franco-Mena, J. M. Guayasamin, D. Batallas, C. Reyes-Puig. 2023. Two new species of terrestrial frogs of the *Pristimantis gladiator* complex (Anura, Strabomantidae) from the Ecuadorian Andes, with insights on their biogeography and skull morphology. *ZooKeys* 1180: 257–293.

Rivera-Correa, M., S. R. Ron, I. Nunes, K. Araujo-Vieira, P. D. P. Pinheiro, and T. Grant. 2024. Forty years later: a new Andean stream treefrog of the genus *Hyloscirtus* (Anura: Hylidae) from Ecuador, with comments on arm hypertrophy in the *H. larinopygion* group. *Zootaxa* 5474: 101–126.

Ron, S. R., D. García, D. Brito-Zapata, C. Reyes-Puig, E. Figueroa-Coronel, and D. F. Cisneros-Heredia. 2024. A new glassfrog of the genus *Centrolene* (Amphibia: Centrolenidae) from the Subandean Kutukú Cordillera, eastern Ecuador. *Zoosystematics and Evolution* 100: 923–939.

Sánchez-Nivicela, J. C., J. M. Falcón, and D. F. Cisneros-Heredia. 2023. A new stream treefrog of the genus *Hyloscirtus* (Amphibia, Hylidae) from the Río Negro-Sopladora National Park, Ecuador. *PeerJ* 1141: 75–92.

Székely, P., M. Córdova-Díaz, D. Hualpa-Vega, S. Hualpa-Vega, and D. Székely. 2023a. A new glassfrog species of the genus *Centrolene* (Amphibia: Anura: Centrolenidae) from Cordillera del Cóndor, southern Ecuador. *ZooKeys* 1149: 53–84.

Székely, P., D. Székely, D. Armijos-Ojeda, S. Hualpa-Vega, and J. Vörös. 2023b. Molecular and morphological assessment of rain frogs in the *Pristimantis orestes* species group (Amphibia: Anura: Strabomantidae) with the description of three new cryptic species from southern Ecuador. *Herpetological Monographs* 37: 41–69.

Tarvin, R. D., P. Peña, and S. R. Ron. 2014. Changes in population size and survival in *Atelopus spumarius* (Anura: Bufonidae) are not correlated with chytrid prevalence. *Journal of Herpetology* 48: 291–197.

von May, R., M. I. Diaz, A. Ttito, R. Santa-Cruz Farfan, and A. Catenazzi. 2024. The rediscovery of *Noblella peruviana* after more than 115 years helps resolve the molecular phylogeny and taxonomy of *Noblella* (Amphibia, Anura, Strabomantidae). *Diversity* 16(613): 1–15.

References

Acosta-Galvis, A. R. 2000. Ranas, Salamandras y Caecilias (Tetrapoda: Anphibia) de Colombia. *Biota Colombiana* 1: 289–319.

Acosta-Galvis, A. R., and D. L. Gutiérrez-Lamus. 2012. A new species of salamander (*Bolitoglossa*: Plethodontidae) from Cordillera Oriental of the Colombian Andes. *Papéis Avulsos de Zoologia* 52: 201–218.

Aguayo, R., and M. B. Harvey. 2006. A new glassfrog of the *Cochranella granulosa* group (Anura: Centrolenidae) from a Bolivian cloud forest. *Herpetologica* 62: 323–330.

Aichinger, M. 1987. Annual activity patterns of anurans in a seasonal neotropical environment. *Oecologia* 71: 583–592.

Almeida-Reinoso, D., and L. A. Coloma. 2005. *Rulyrana orejuela* (Orejuela Glass Frog). Predation. *Herpetological Review* 43: 126.

Almendáriz, A. 1987. Contribución al conocimiento de la herpetofauna centroriental Ecuatoriana. *Politécnica* 12: 77–133.

Almendáriz, A., and D. Batallas. 2012. Nuevos datos sobre la distribución, historia natural y el canto de *Centrolene condor* Cisneros-Heredia y Morales-Mite 2008 (Amphibia: Anura: Centrolenidae). *Politécnica* 30: 42–53.

Almendáriz, A., and D. Cisneros-Heredia. 2005. Nuevos datos sobre la distribución e historia natural de *Gastrotheca longipes* (Boulenger, 1882), una rana marsupial amazónica poco conocida (Amphibia: Anura: Hylidae). *Politécnica* 26: 20–27.

Almendáriz, A., D. F. Cisneros-Heredia, and M. Wilkinson. 2004a. *Chthonerpeton onorei*. The IUCN Red List of Threatened Species 2004: e.T59539A11960580. http://dx.doi.org/10.2305/IUCN.UK.2004.RLTS.T59539A11960580.en

Almendáriz, A., D. F. Cisneros-Heredia, K.-H. Jungfer, L. A. Coloma, and S. R. Ron. 2004b. *Bolitoglossa equatoriana*. The IUCN Red List of Threatened Species 2004: e.T59160A11892289. http://dx.doi.org/10.2305/IUCN.UK.2004.RLTS.T59160A11892289.en

Amador, L., A. Parada, G. D'Elía, and J. M. Guayasamin. 2018. Uncovering hidden specific diversity of Andean glassfrogs of the *Centrolene buckleyi* species complex (Anura: Centrolenidae). *PeerJ* 6: e5856.

Andersson, L. G. 1945. Batrachians from east Ecuador collected 1937, 1938 by Wm Clarke-MacIntyre and Rolf Blomberg. *Arkiv för Zoologi* 37: 1–88.

Angulo, A., C. Azevedo-Ramos, L. A. Coloma, and S. Ron. 2004a. *Edalorhina perezi*. The IUCN Red List of Threatened Species 2004: e.T56385A11458591. https://dx.doi.org/10.2305/IUCN.UK.2004.RLTS.T56385A11458591.en

Angulo, A., C. Azevedo-Ramos, L. A. Coloma, S. R. Ron, M. Hoogmoed, C. Gascon, M. J. Icochea, and D. F. Cisneros-Heredia. 2016. *Pithecopus palliatus*. The IUCN Red List of Threatened Species 2016: e.T55859A107299731. https://dx.doi.org/10.2305/IUCN.UK.2016-3.RLTS.T55859A107299731.en

Angulo, A., R. B. Cocroft, and S. Reichle. 2003. Species identity in the genus *Adenomera* (Anura: Leptodactylidae) in southeastern Peru. *Herpetologica* 59: 490–504.

Angulo, A., L. A. Coloma, S. R. Ron, M. Hoogmoed, F. Castro, J. V. Rueda-Almonacid, D. F. Cisneros-Heredia, and M. J. Icochea. 2004b. *Cruziohyla craspedopus* (errata version published in 2016). The IUCN Red List of Threatened Species 2004: e.T55291A86444116. https://dx.doi.org/10.2305/IUCN.UK.2004.RLTS.T55291A11275223.en.

Angulo, A., J. Icochea, F. Castro, and J. V. Rueda. 2004c. *Adelophryne adiastola*. The IUCN Red List of Threatened Species 2004: e.T56299A11446212. http://dx.doi.org/10.2305/IUCN.UK.2004.RLTS.T56299A11446212.en

Araújo, K. C., A. Pansonato, R. H. Oliveira, D. H. Moraisa, V. T. Carvalho, and R. W. Ávila. 2018. Advertisement call and new distribution records from Brazil of *Teratohyla midas* (Lynch & Duellman, 1973) (Anura: Centrolenidae). *Check List* 14: 303–308.

Arbeláez Ortiz, E. 2014. Amaru Zoo promotes the rescue, knowledge and protection of threatened amphibians of Southern Ecuador. *AArk Newsletter* 26: 14–15.

Arcila-Pérez, L., J. Rios-Soto, S. O. Montilla, C. Londoño-Guarnizo, and F. Vargas Salinas. 2017. Vocalization and natural history in populations of a glassfrog assigned to *Nymphargus griffithsi* in the Central Andes of Colombia. *Herpetological Review* 48: 275–280.

Arredondo Salgar, J. C. 2007. *Caecilia thompsoni* (Thompson´s caecilian). Maximum body size. *Herpetological Review* 38: 444–445.

Arroba, L. T. 2017. Sistemática del género *Caecilia* (Gymonphiona: Caeciliidae) de la vertiente del Pacífico de Ecuador. Tesis de Licenciatura, Pontificia Universidad Católica del Ecuador.

Arteaga, A., L. Bustamante, and J. M. Guayasamin. 2013. *The Amphibians and Reptiles of Mindo: Life in the Cloudforest*. Quito: Scientific Publication Series, Universidad Tecnológica Indoamérica.

Asquith, A.and R. Altig. 1987. *Atelopus spumarius*. Vocalization. *Herpetological Review*, 18: 32–33.

Auber-Thomay, M., L. A. Coloma, and G. Onore. 1990. Elevage d'une rainette marsupiale arboricole des forets nuageuses equatoriennes, *Gastrotheca plumbea* (Boulenger). *Revue Française dí Aquariologie* 17: 57–62.

Ayarzagüena, J. 1992. Los centrolénidos de la Guayana Venezolana. *Publicaciones de la Asociacion de Amigos de Doñana* 1: 6–48.

Azevedo-Ramos, C., R. Reynolds, E. La Marca, L. A. Coloma, and S. Ron. 2010. *Phyllomedusa vaillantii*. The IUCN Red List of Threatened Species 2010: e.T55868A11383140. https://dx.doi.org/10.2305/IUCN.UK.2010-2.RLTS.T55868A11383140.en

Bardua, C., M. Wilkinson, D. J. Gower, E. Sherratt, and A. Goswami. 2019. Morphological evolution and modularity of the caecilian skull. *BMC Evolutionary Biology* 19: 1–23.

Barnett, J. B., C. Michalis, H. M. Anderson, B. L. McEwen, J. Yeager, J. N. Pruitt, N. E. Scott-Samuel, and I. C. Cuthill. 2020. Imperfect transparency and camouflage in glass frogs. *Proceedings of the National Academy of Sciences* 117: 12885–12890.

Barrera Rodríguez, M., and P. M. Ruiz Carranza. 1989. Una nueva especie del género *Centrolenella* Noble 1920 (Amphibia: Anura: Centrolenidae) de la Cordillera Occidental de Colombia. *Trianea* 3: 77–84.

Barrio-Amorós, C. L. 1999. "1998". Systematics and biogeography of the amphibians (Amphibia) of Venezuela. *Acta Biológica Venezuelica* 18: 1–93.

Barrio-Amorós, C. L. 2009. Distribución y aspectos de la historia natural de las ranas lémur (Hylidae: Phyllomedusinae) en Venezuela. *Memoria de la Fundación La Salle de Ciencias Naturales* 171: 19–46.

Barrio Amoros, C. L., and C. Brewer-Carias. 2008. Herpetological results of the 2002 expedition to Sarisariñama, a Tepui in Venezuelan Guayana, with the description of five new species. *Zootaxa* 1942: 1–68.

Barrio-Amorós, C. L., F. J. M. Rojas-Runjaic, and J. C. Señaris. 2019. Catalogue of the amphibians of Venezuela: illustrated and annotated species list, distribution, and conservation. *Amphibian and Reptile Conservation* 13: 1–198.

Batallas, D., and J. Brito. 2016. Análisis bioacústico de las vocalizaciones de seis especies de anuros de la laguna Cormorán, complejo lacustre de Sardinayacu, Parque Nacional Sangay, Ecuador. *Revista Mexicana de Biodiversidad* 87: 1292–1300.

Batista, A., A. Hertz, K. Mebert, G. Köhler, S. Lotzkat, M. Ponce, and M. Vesely. 2014. Two new fringe-limbed frogs of the genus *Ecnomiohyla*. *Zootaxa* 3826: 449–473.

Baum, D. A., and S. Smith. 2013. *Tree Thinking. An Introduction to Phylogenetic Biology. Roberts and Company Publishers.* Greenwood Village: Roberts.

Beirne, C., and A. Whitworth. 2011. *Frogs of the Yachana Reserve*. Global Vision International. https://www.researchgate.net/publication/236504364_Frogs_of_the_Yachana_Reserve.

Bernarde, P. S., and M. N. de C. Kokubum. 2009. Seasonality, age structure and reproduction of *Leptodactylus (Lithodytes) lineatus* (Anura, Leptodactylidae) in Rondônia state, southwestern Amazon, Brazil. *Iheringia—Serie Zoologia* 99: 368–372.

Berthold, A. A. 1859. Einige neue Reptilien des akad. zoolog. Museums in Göttingen. Nachrichten von der Georg-Augusts-Universität und der Königl. *Gesellschaft der Wissenschaften zu Göttingen* 179–181.

Blackburn, D. C., and W. E. Duellman. 2013. Brazilian marsupial frogs are diphyletic (Anura: Hemiphractidae: *Gastrotheca*). *Molecular Phylogenetics and Evolution* 68: 709–714.

Blackburn, D. C., and D. B. Wake. 2011. Class amphibia gray, 1825 1,2,3 amphibia. In *Animal Biodiversity: An Outline of Higher-Level Classification and Survey of Taxonomic Richness*, ed. Z.-Q. Zhang, 39–55. Auckland: Zootaxa.

Block, J. E., S. L. Unser, J. K. Mooney, and E. R. Wild. 2003. *Agalychnis craspedopus* (Amazon leaf frog). Reproduction. *Herpetological Review* 34: 134–135.

Boettger, O. 1892. *Katalog der Batrachier-Sammlung im Museum der Senckenbergischen Naturforschenden Gesellschaft in Frankfurt am Main*. Frankfurt: Senckenbergischen Naturforschenden Gesellschaft.

Boettger, O. 1893. Ein neuer Laubfrosh aus Costa Rica. *Natur und Museum* 1893: 251–252.

Bolívar, W., L. A. Coloma, S. Ron, D. Cisneros-Heredia, and E. Wild. 2004. *Nymphargus grandisonae*. The IUCN Red List of Threatened Species 2004: e.T54914A11222520. https://dx.doi.org/10.2305/IUCN.UK.2004.RLTS.T54914A11222520.en

Bolívar-García, W., T. Grant, and L. Osorio. 1999. Combat behavior in *Centrolene buckleyi* and other centrolenid frogs. *Alytes* 16: 3–4.

Bossuyt, F., and K. Roelants. 2009. Anura. In *The Timetree of Life*, eds. S. B. Hedges, and S. Kumar, 357–364. New York, USA: Oxford University Press.

Boul, K. E., W. C. Funk, C. R. Darst, D. C. Cannatella, and M. J. Ryan. 2007. Sexual selection drives speciation in an Amazonian frog. *Proceedings of the Royal Society B: Biological Sciences* 274: 399–406.

Boulenger, G. A. 1882a. *Catalogue of the Batrachia Salientia s. Ecaudata in the Collection of the British Museum*, 2nd ed. London: Taylor and Francis.

Boulenger, G. A. 1882b. *Catalogue of the Batrachia Gradientia S. Caudata and Batrachia Apoda in the Collection of the British Museum (Natural History)*. London: Printed by order of the Trustees.

Boulenger, G. A. 1883. Description of a new genus of Coeciliae. *Annals and Magazine of Natural History Series* 5: 202–203.

Boulenger, G. A. 1884. "1883". On a collection of frogs from Yurimaguas, Huallaga river, Northern Peru. *Proceedings of the Zoological Society of London* 1883: 635–638.

Boulenger, G. A. 1898. An account of the reptiles and batrachians collected by Mr. W. F. H. Rosenberg in Western Ecuador. *Proceedings of the Zoological Society of London* 107–126.

Boulenger, G. A. 1902. Descriptions of new batrachians and reptiles from north Western Ecuador. *Annals and Magazine of Natural History* 9: 51–57.

Boulenger, G. A. 1903. Remarks on a specimen of *Ceratohyla bubalus* Espada. *Proceedings of the Zoological Society of London* 2: 115–116.

Boulenger, G. A. 1913. On a collection of batrachians and reptiles made by Dr. H. G. F. Spurrell, F. Z. S., in the Choco, Colombia. *Proceedings of the Zoological Society of London* 1913: 1019–1038.

Brame Jr., A. H., and D. B. Wake. 1962. A new plethodontid salamander (genus *Bolitoglossa*) from Venezuela with redescription of the Ecudorian *B. palmata* (Werner). *Copeia* 1962: 170–177.

Brame Jr., A. H., and D. B. Wake. 1963. The salamanders of South America. *Los Angeles County Museum Contributions in Science* 69: 1–72.

Brame Jr., A. H., and D. B. Wake. 1972. New species of salamanders (genus *Bolitoglossa*) from Colombia, Ecuador and Panama. *Los Angeles County Museum Contributions in Science* 219: 1–34.

Brattstrom, B. H., and R. M. Yarnell. 1968. Aggressive behavior in two species of leptodactylid frogs. *Herpetologica* 24: 222–228.

Bravo Valencia, L., and J. Delia. 2016. Maternal care in a glassfrog: care function and commitment to offspring in *Ikakogi tayrona*. *Behavioral Ecology and Sociobiology* 70: 41–48.

Brenner, W. 1960. *Gastrotheca marsupiata*, ein Laubfrosch mit hochinteressanter *Brutfliege. Aquarium und Terrarium Kunde* 7: 210–213.

Brito, J., A. Almendáriz, and D. Batallas. 2014. *Phyllomedusa ecuatoriana* Cannatella (Amphibia: Hylidae): variación, descripción del renacuajo, vocalización y anotaciones sobre la historia natural. *Papéis Avulsos de Zoologia* 54: 419–433.

Brocchi, P. 1881. *Mission scientifique au Mexique et dans l'Amérique Centrale: ouvrage publié par ordre de S.M. l'Empereur et par les soins du Ministre de l'instruction publique*. Paris: Ministère de l'Instruction Publique.

Brunner, R. M., and J. M. Guayasamin. 2020. Nocturnal visual displays and call description of the cascade specialist glassfrog *Sachatamia orejuela*. *Behaviour* 157: 1257–1268.

Bustamante, M. R., D. F. Cisneros-Heredia, M. H. Yánez-Muñoz, H. M. Ortega-Andrade, and J. M. Guayasamin. 2007. Amphibia, Centrolenidae, *Cochranella pulverata*, *Hyalinobatrachium aureoguttatum*: distribution extension, Ecuador. *Check List* 3: 271.

Bustamante, M. R., S. R. Ron, and L. A. Coloma. 2005. Cambios en la diversidad en siete comunidades de anuros en los Andes de Ecuador. *Biotropica* 37: 180–189.

Caldwell, J. P. 1994. Natural history and survival of eggs and early larval stages of *Agalychnis calcarifer* (Anura: Hylidae). *Herpetological Natural History* 2: 57–66.

Caldwell, J. P., and D. B. Shepard. 2005. Frogs of the lower Cristalino river area. In *Amphibians and Reptiles of the Lower Cristalino River Region of the Southern Amazon,* eds. L. J. Vitt, J. P. Caldwell, G. R. Colli, F. G. R. França, and D. B. Shepard. Norman: Sam Noble Oklahoma Museum of Natural History.

Cannatella, D. C. 1980. A review of the *Phyllomedusa buckleyi* group (Anura: Hylidae). *Occasional Papers of the Museum of Natural History University of Kansas* 87: 1–40.

Cannatella, D. C. 1982. Leaf-Frogs of the *Phyllomedusa perinesos* Group (Anura: Hylidae). *Copeia* 1982: 501–513.

Cannatella, D. C., and W. E. Duellman. 1984. Leptodactylid frogs of the *Physalaemus pustulosus* group. *Copeia* 1984: 902–921.

Carroll, R. L. 2009. The rise of amphibians: 365 million years of evolution. *Journal of Verterbrate Paleontology* 31: 1406–1407.

Carvajal-Endara, S., L. A. Coloma, M. A. Morales-Mite, J. M. Guayasamin, P. Székely, and W. E. Duellman. 2019. Phylogenetic systematics, ecology, and conservation of marsupial frogs (Anura: Hemiphractidae) from the Andes of southern Ecuador, with descriptions of four new biphasic species. *Zootaxa* 4562: 1–102.

Castro, F., J. V. Rueda, W. Bolívar, R. A. Estupiñan, and L. A. Almendáriz. 2004a. *Rulyrana flavopunctata*. The IUCN Red List of Threatened Species 2004: e.T54961A11219097. https://dx.doi.org/10.2305/IUCN.UK.2004.RLTS.T54961A11219097.en

Castro, F., J. V. Rueda, W. Bolívar, R. A. Estupiñán, L. A. Coloma, S. Ron, and A. Almendáriz. 2004b. *Agalychnis buckleyi*. The IUCN Red List of Threatened Species 2004: e.T55843A11379402. https://dx.doi.org/10.2305/IUCN.UK.2004.RLTS.T55843A11379402.en

Castro, F., Rueda-Almonacid, J. V., Bolívar, W., Estupinan, R., Almendáriz, A. and Wilkinson, M. 2004c. *Microcaecilia albiceps* (errata version published in 2016). The IUCN Red List of Threatened Species 2004: e.T59570A86148920. http://dx.doi.org/10.2305/IUCN.UK.2004.RLTS.T59570A11964765.en

Castroviejo-Fisher, S., J. Ayarzagüena, and C. Vilà. 2007. A new species of *Hyalinobatrachium* (Centrolenidae: Anura) from Serranía de Perijá, Venezuela. *Zootaxa* 51–62.

Castroviejo-Fisher, S., J. M. Guayasamin, A. Gonzalez-Voyer, and C. Vilà. 2014. Neotropical diversification seen through glassfrogs. *Journal of Biogeography* 41: 66–80.

Castroviejo-Fisher, S., J. Moravec, J. Aparicio, M. Guerrero-Reinhard, and G. Calderón. 2011. DNA taxonomy reveals two new species records of *Hyalinobatrachium* (Anura: Centrolenidae) for Bolivia. *Zootaxa* 2798: 64–68.

Castroviejo-Fisher, S., J. Padial, J. C. Chaparro, R. Aguayo, and I. de la Riva. 2009. A new species of *Hyalinobatrachium* (Anura: Centrolenidae) from the Amazonian slopes of the central Andes, with comments on the diversity of the genus in the area. *Zootaxa* 2143: 24–44.

Castroviejo-Fisher, S., J. Padial, I. de La Riva, J. Pombal Jr., H. Da Silva, F. Rojas-Runjaic, E. Medina-Méndez, and D. R. Frost. 2015. Phylogenetic systematics of egg-brooding frogs (Anura: Hemiphractidae) and the evolution of direct development. *Zootaxa* 4004: 1–75.

Chávez, G., P. J. Venegas, and A. Lescano. 2011. Two new records in the diet of *Ceratophrys cornuta* Linneaus, 1758 (Anura: Ceratophryidae). *Herpetology Notes* 4: 285–286.

Chippindale, P. T., R. M. Bonett, A. S. Baldwin, and J. J. Wiens. 2004. Phylogenetic evidence for a major reversal of life-history evolution in plethodontid salamanders. *Evolution* 58: 2809–2822.

Chondrocranial morphology of *Leptodactylus* larvae (Leptodactylidae : Leptodactylinae): Its utility in phylogenetic reconstruction. *Journal of Morphology* 238: 287–305.

Cintra, C. E. D., H. L. R. da Silva, and N. J. da Silva. 2014. First record of *Lithodytes lineatus* (Schneider, 1799) (Anura: Leptodactylidae) in the state of Tocantins, ecotone zone Amazon-Cerrado biomes, with notes on its natural history. *Herpetology Notes* 7: 179–184.

Cisneros-Heredia, D. F. 2005. Amphibia, Centrolenidae, *Centrolene peristictum*, *Centrolene prosoblepon*, *Cochranella cochranae*, *Cochranella midas*, *Cochranella resplendens*, *Cochranella spinosa*, *Hyalinobatrachium munozorum*: range extensions and new provincial records. *Check List* 1: 18–23.

Cisneros-Heredia, D. F. 2006. Distribution and ecology of the Western Ecuador frog *Leptodactylus labrosus* (Amphibia: Anura: Leptodactylidae). *Zoological Research* 3: 225–234.

Cisneros-Heredia, D. F. 2007a. A new species of glassfrog of the genus *Centrolene* from the foothills of Cordillera Oriental of Ecuador. *Herpetozoa* 20: 27–34.

Cisneros-Heredia, D. F. 2007b. Amphibia, Caudata, Plethodontidae: *Bolitoglossa equatoriana* and *Bolitoglossa biseriata*: range extensions, new provincial records from Ecuador, and natural history. *Check List* 2: 64–67.

Cisneros-Heredia, D. F. 2008. Habitat loss and climate change impacts on Neotropical anurans: implications for in-situ conservation. A case study with glassfrogs from eastern Ecuador (Amphibia: Anura: Centrolenidae). MSc. Dissertation, King's College.

Cisneros-Heredia, D. F. 2013. *Centrolene ritae* Lutz is a senior synonym of *Cochranella oyampiensis* Lescure and *Cochranella ametarsia* Flores (Anura: Centrolenidae). *Avances en Ciencias e Ingenierías, Sección B* 5: B1–B4.

Cisneros-Heredia, D. F., A. Almendáriz, and M. Wilkinson. 2004a. *Caecilia crassisquama*. The IUCN Red List of Threatened Species 2004: e.T59509A11952112. http://dx.doi.org/10.2305/IUCN.UK.2004.RLTS.T59509A11952112.en

Cisneros-Heredia, D. F., A. Almendáriz, and M. Wilkinson. 2004b. *Caecilia subterminalis*. The IUCN Red List of Threatened Species 2004: e.T59529A11958051. http://dx.doi.org/10.2305/IUCN.UK.2004.RLTS.T59529A11958051.en

Cisneros-Heredia, D. F., J. Delia, M. Yánez-Muñoz, and H. M. Ortega-Andrade. 2008a. Natural history and intraspecific variation of the Ecuadorian blue glassfrog *Cochranella mache* (Guayasamin and Bonaccorso, 2004). *Herpetozoa* 21: 57–66.

Cisneros-Heredia, D. F., and J. M. Guayasamin. 2014. Notes on the taxonomy of some glassfrogs from the Andes of Peru and Ecuador (Amphibia: Centrolenidae). *Papéis Avulsos de Zoologia* 54: 161–168.

Cisneros-Heredia, D. F., and R. W. McDiarmid. 2006. A new species of the genus *Centrolene* (Amphibia: Anura: Centrolenidae) from Ecuador with comments on the taxonomy and biogeography of glassfrogs. *Zootaxa* 32: 1–32.

Cisneros-Heredia, D. F., and R. W. McDiarmid. 2007. Revision of the characters of Centrolenidae (Amphibia: Anura: Athesphatanura), with comments on its taxonomy and the description of new taxa of glassfrogs. *Zootaxa* 1572: 1–82.

Cisneros-Heredia, D. F., and P. Meza-Ramos. 2007. An enigmatic new species of glassfrog (Amphibia: Anura: Centrolenidae) from the Amazonian Andean slopes of Ecuador. *Zootaxa* 41: 33–41.

Cisneros-Heredia, D. F., and M. Morales-Mite. 2008. A new species of glassfrog from the elfin forests of the Cordillera del Cóndor, southeastern Ecuador. *Wildlife Conservation* 21: 49–56.

Cisneros-Heredia, D. F., P. J. Venegas, M. Rada, and R. Schulte. 2008b. A new species of glassfrog (Anura: Centrolenidae) from the foothill Andean forests of Ecuador and Peru. *Herpetologica* 64: 341–353.

Cisneros-Heredia, D. F., and M. H. Yánez-Muñoz. 2007. A new species of glassfrog (Centrolenidae) from the southern Andean foothills on the west Ecuadorian region. *South American Journal of Herpetology* 2: 1–10.

Cochran, D. M., and C. J. Goin. 1970. Frogs of Colombia. *United States National Museum Bulletin* 288: xii + 655.

Cole, C. J., C. R. Townsend, R. P. Reynolds, R. D. MacCulloch, and A. Lathrop. 2013. Amphibians and reptiles of Guyana, South America: illustrated keys, annotated species accounts, and a biogeographic synopsis. *Proceedings of the Biological Society of Washington* 125: 317–578.

Coloma, L. A., S. Ron, and C. Azevedo-Ramos. 2004a. *Hemiphractus scutatus*. The IUCN Red List of Threatened Species 2004: e.T55371A11299534. https://dx.doi.org/10.2305/IUCN.UK.2004.RLTS.T55371A11299534.en

Coloma, L. A., S. Ron, C. Azevedo-Ramos, and E. La Marca. 2004b. *Leptodactylus leptodactyloides*. The IUCN Red List of Threatened Species 2004: e.T57140A11590413. http://dx.doi.org/10.2305/IUCN.UK.2004.RLTS.T57140A11590413.en

Coloma, L. A., S. Ron, D. Cisneros-Heredia, C. Azevedo-Ramos, F. Castro, J. V. Rueda, J. Icochea, R. Heyer, and I. de la Riva. 2004c. *Leptodactylus discodactylus*. The IUCN Red List of Threatened Species 2004: e.T57373A11628605. http://dx.doi.org/10.2305/IUCN.UK.2004.RLTS.T57373A11628605.en

Coloma, L. A., S. R. Ron, E. La Marca, M. Hoogmoed, F. Castro, J. Lynch, and M. Wilkinson. 2004d. *Caecilia tentaculata*. The IUCN Red List of Threatened Species 2004: e.T59530A11958955. http://dx.doi.org/10.2305/IUCN.UK.2004.RLTS.T59530A11958955.en

Coloma, L. A., S. R. Ron, K.-H. Jungfer, T. Grant, D. F. Cisneros-Heredia, F. Solís, R. Ibáñez, *et al.* 2008. *Gastrotheca cornuta*. The IUCN Red List of Threatened Species 2008: e.T55329A11294031. http://dx.doi.org/10.2305/IUCN.UK.2008.RLTS.T55329A11294031.en

Coloma, L. A., S. Ron, C. Azevedo-Ramos, R. Heyer, and C. Gascon. 2010. *Leptodactylus stenodema*. The UICN Red List of Threatened Species 2010: e.T57168A11594741.en

Consejo Metropolitano de Quito. 2012. Resolución No. C 349. https://www7.quito.gob.ec/mdmq_ordenanzas/Resoluciones%20de%20Concejo/A%C3%B1o%20%202012/RC-2012-349%20-%20FAUNA%20EMBLEM%C3%81TICA%20DE%20QUITO%20-%20DECLARATORIA.pdf

Cope, E. D. 1862. Catalogues of the reptiles obtained during the explorations of the Parana, Vermejo and Uraguay rivers, by Capt. Thos. J. Page, USN; and of those procured by Lieut. N. Michler, US Top. Eng., Commander of the expedition conducting the survey of the Atrato R. *Proceedings of the Academy of Natural Sciences of Philadelphia* 14: 346–359 + 594.

Cope, E. D. 1864. Contributions to the herpetology of tropical America. *Proceedings of the Academy of Natural Sciences of Philadelphia* 16: 166–181.

Cope, E. D. 1866. *On the Structure and Distribution of the Genera of the Arciferous Anura" and "On the Families of the Raniform Anura*. Philadelphia: Journal of the Academy of Natural Sciences of Philadelphia.

Cope, E. D. 1868. An examination of the Reptilia and Batrachia obtained by the Orton expedition to Equador and the upper Amazon, with notes on other species. *Proceedings of the Academy of Natural Sciences of Philadelphia* 20: 96–140.

Cope, E. D. 1876. "1875" On the Batrachia and Reptilia of Costa Rica. *Journal of the Academy of Natural Sciences of Philadelphia* 2: 93–154.

Cordeiro-Duarte, A. C., T. M. Sanaiotti, E. L. C. Duarte, and O. Pereira. 2002. Geographic distribution. *Hyalinobatrachium nouraguensis* (Glassfrog). *Herpetological Review* 33: 220.

Cossio, R., and E. Medina-Barcenas. 2020. Acoustic repertoire and calling behavior of the gliding treefrog, *Agalychnis spurrelli* (Anura: Hylidae). *South American Journal of Herpetology* 17: 71–78.

Crawford, A. J., C. Cruz, E. Griffith, H. Ross, R. Ibanez, K. R. Lips, A. C. Driskell, E. Bermingham, and P. Crump. 2013. DNA barcoding applied to *ex situ* tropical amphibian conservation programme reveals cryptic diversity in captive populations. *Molecular Ecology Resources* 13: 1005–1018.

Crump, M. L. 1974. Reproductive strategies in a tropical anuran community. *University of Kansas Publications, Museum of Natural History, Miscellaneous Publications* 61: 1–68.

Crump, M. L. 1978. "1977". Intrapopulation and interspecific variation of "standard" morphological characters of four closely related South American salamanders (*Bolitoglossa*), with description of habitat preferences. *Herpetologica* 33: 415–426.

Cuatrecasas, J. 1979. Comparación biogeográfica de páramos entre varias cordilleras. In *El Media Ambiente Páramo*, ed. M. Salgado–Labouriau, 69–99. Mérida, Venezuela: Centro de Estudios Advanzados.

Daudin, F.-M. 1803. 7 *"An. XI". Histoire Naturelle, Générale et Particulière des Reptiles; Ouvrage Faisant suit à l'Histoire Naturelle Générale et Particulière, Composée par Leclerc de Buffon; et Rédigée par C. S. Sonnini, Membre de Plusieurs Sociétés Savantes*. Paris: F. Dufart.

Dautel, N., A. L. S. Maldonado, R. Abuza, H. Imba, K. Griffin, and J. M. Guayasamin. 2011. Cantos de anúncio e de combate de Centrolene lynchi (Anura: Centrolenidae), com notas sobre seu comportamento de combate e reprodutivo. *Phyllomedusa* 10: 31–43.

Deckert, K. 1963. Die Paarung des Beutelfrosches (*Gastrotheca marsupiata*). *Zoologische Garten (NF)* 28: 12–17.

de Carvalho, T. R., A. Angulo, D. A. Barrera, C. Aguilar-Puntriano, and C. F. Haddad. 2020. Hiding in plain sight: a fourth new cryptic species of the *Adenomera andreae* clade (Anura: Leptodactylidae) from southwestern Amazonia. *Herpetologica* 76: 304–314.

de Carvalho, T. R., A. Fouquet, M. L. Lyra, A. A. Giaretta, C. E. Costa-Campos, M. T Rodrigues, C. F. B. Haddad, and S. R. Ron. 2022. Species diversity and systematics of the *Leptodactylus melanonotus* group (Anura, Leptodactylidae): review of diagnostic traits and a new species from the Eastern Guiana Shield. *Systematics and Biodiversity* 20: 1–31.

de Carvalho, T. R., A. A. Giaretta, N. M. Maciel, D. A. Barrera, C. Aguilar-Puntriano, C. F. B. Haddad, M. N. C. Kokubum, M. Menin, and A. Angulo. 2019. On the uncertain taxonomic identity of *Adenomera hylaedactyla* (Cope, 1868) and the composite type series of *A. andreae* (Müller, 1923) (Anura, Leptodactylidae). *Copeia* 107: 708–723.

de Carvalho, T. R., and S. Ron. 2011. Advertisement call of *Leptodactylus labrosus* Jiménez de la Espada, 1875 (Anura, Leptodactylidae): an unusual advertisement call within the *L. fuscus* group. *Herpetology Notes* 4: 325–326.

De la Riva, I. 1993. Ecología de una comunidad neotropical de anfibios durante la estación lluviosa. Tesis Doctoral, Universidad Complutense de Madrid.

de Lima Barros, A., J. L. López-Lozano, and A. P. Lima. 2016. The frog *Lithodytes lineatus* (Anura: Leptodactylidae) uses chemical recognition to live in colonies of leaf-cutting ants of the genus *Atta* (Hymenoptera: Formicidae). *Behavioral Ecology and Sociobiology* 70: 2195–2201.

de Sá, R. O., T. Grant, A. Camargo, W. R. Heyer, M. L. Ponssa, and E. Stanley. 2014. Systematics of the Neotropical genus *Leptodactylus* Fitzinger, 1826 (Anura: Leptodactylidae): phylogeny, the relevance of non-molecular evidence, and species accounts. *South American Journal of Herpetology* 9: S1–S100.

del Pino, E. M. 1975. Adaptaciones reproductivas para la vida terestre del sapo marsupial *Gastrotheca riobambae* (Fowler), Anura, Hylidae. *Revista de la Pontificia Universidad Católica del Ecuador* 3: 119–140.

del Pino, E. M. 1977. Núcleos múltiples en el oocito de algunos sapos marsupiales (Hylidae). *Revista de la Universidad Católica, Quito* 5: 21–30.

del Pino, E. M. 1980. Morphology of the pouch and incubatory integument in marsupial frogs (Hylidae). *Copeia* 1980: 10–17.

del Pino, E. M. 1983. Progesterone induces incubatory changes in the brooding pouch of the frog *Gastrotheca riobambae* (Fowler). *Journal of Experimental Zoology* 227: 159–163.

del Pino, E. M. 1989. Modifications of oogenesis and development in marsupial frogs. *Development* 107: 169–188.

del Pino, E. M. 1996. The expression of Brachyry (T) during gastrulation in the marsupial frog *Gastrotheca riobambae*. *Developmental Biology* 177: 64–72.

del Pino, E. M. 1998. Neural development in the marsupial frog *Gastrotheca riobambae. The International Journal of Developmental Biology* 42: 723–731.

del Pino, E. M., and I. Alcocer. 1994. Urea is necessary for the culture of embryos of the marsupial frog *Gastrotheca riobambae,* and is tolerated by embryos of the aquatic frog *Xenopus laevis*. *Development, Growth & Differentiation* 36: 73–80.

del Pino, E. M., and R. P. Elinson. 1983. A novel development pattern for frogs: gastrulation produces an embryonic disk. *Nature* 306: 589–591.

del Pino, E. M., and B. Escobar. 1981. Embryonic stages of *Gastrotheca riobambae* (Fowler) during maternal incubation and comparison of development with that of other egg-brooding hylid frogs. *Journal of Morphology* 167: 277–295.

del Pino, E. M., M. L. Galarza, C. M. de Albuja, and A. A. Humphries, Jr. 1975. The maternal pouch and development in the marsupial frog *Gastrotheca riobambae* (Fowler). *Biological Bulletin* 149: 480–491.

del Pino, E. M., and A. A. Humphries, Jr. 1978. Multiple nuclei during early oogenesis in *Flectonotus pygmaeus* and other marsupial frogs. *Biological Bulletin* 154: 198–212.

del Pino, E. M., and S. Loor-Vela. 1990. The early cleavage of the marsupial frog *Gastrotheca riobambae*. *Development* 110: 781–789.

del Pino, E. M., and G. Sánchez. 1977. Ovarian structure of the marsupial frog *Gastrotheca riobambae* (Fowler). *Journal of Morphology* 153: 153–162.

del Pino, E. M., H. Steinbeisser, A. Hofmann, C. Dreyer, M. Campos, and M. F. Trendelenburg. 1986. Oogenesis in the egg-brooding frog *Gastrotheca riobambae* produces large oocytes with fewer nucleoli and low RNA content in comparison to *Xenopus laevis*. *Differentiation* 32: 24–33.

del Pino, E. M., M. Venegas-Ferrín, A. Romero-Carvajal, P. Montenegro-Larrea, N. Sáenz-Ponce, I. M. Moya, I. Alarcón, N. Sudou, S. Yamamoto, and M. Taira. 2007. A comparative analysis of frog early development. *Proceedings of the National Academy of Sciences* 104: 11882–11888.

Delia, J., L. Bravo-Valencia, and K. M. Warkentin. 2017. Patterns of parental care in Neotropical glassfrogs: fieldwork alters hypotheses of sex-role evolution. *Journal of Evolutionary Biology* 30: 898–914.

Delia, J., D. F. Cisneros-Heredia, J. Whitney, and R. Murrieta-Galindo. 2010. Observations on the reproductive behavior of a Neotropical glassfrog, *Hyalinobatrachium fleischmanni* (Anura: Centrolenidae). *South American Journal of Herpetology* 5: 1–12.

Diaz Ricaurte, J., E. C. G. Molina, and R. Díaz-Morales. 2016. *Teratohyla midas* (Santa Cecilia Cochran Frog). Reproductive behavior. *Herpetological Review* 47: 12–13.

do Nascimento, F. A. C., R. O. de Sá, and P. C. de A. García. 2021. Tadpole of the Amazonia frog *Edalorhina perezi* (Anura: Leptodactylidae) with description of oral internal and chondrocranial morphology. *Journal of Morphology* 282: 115–126.

Donnelly, M. A., C. Guyer, D. M. Krempels, and H. E. Braker. 1987. The tadpole of *Agalychnis calcarifer* (Anura: Hylidae). Copeia 1987: 247–250.

Dubeux, M. J. M., U. Gonçalves, M. E. Ramos, A. H. Melo, A. da Silva, and T. Mott. 2020. New records of anuran amphibians from the Pedra Talhada biological reserve, a seasonal semideciduous Atlantic forest fragment in northeastern Brazil. *Herpetology Notes* 13: 997–1002.

Dubois, A., A. Ohler, and R. A. Pyron. 2021. New concepts and methods for phylogenetic taxonomy and nomenclature in zoology, exemplified by a new ranked cladonomy of recent amphibians (Lissamphibia). *Megataxa* 5: 1–738.

Duellman, W. E. 1967. Additional studies of chromosomes of anuran amphibians. *Systematic Zoology* 16: 38–43.

Duellman, W. E. 1970. *The Hylid Frogs of Middle America. Monograph of the Museum of Natural History*. Lawrence: The University of Kansas.

Duellman, W. E. 1973. Descriptions of new hylid frogs from Colombia and Ecuador. *Herpetologica* 29: 219–227.

Duellman, W. E. 1974a. A reassessment of the taxonomic status of some neotropical hylid frogs. *Occasional Papers of the Museum of Natural History, the University of Kansas* 1–27.

Duellman, W. E. 1974b. Taxonomic notes on *Phyllomedusa* (Anura: Hylidae) from the upper amazon basin. *Herpetologica* 30: 105–112.

Duellman, W. E. 1974c. A systematic review of the marsupial frogs (Hylidae: *Gastrotheca*) of the Andes Ecuador. *Occasional Papers of the Museum of Natural History, The University of Kansas* 22: 1–27.

Duellman, W. E. 1978. *The Biology of an Equatorial Herpetofauna in Amazonian Ecuador. University of Kansas Museum of Natural History, Miscellaneous Publications*, Vol. 65. Lawrence: University of Kansas.

Duellman, W. E. 1980. The identity of *Centrolenella grandisonae* Cochran and Goin (Anura: Centrolenidae). *Transactions of the Kansas Academy of Science* 83: 26–32.

Duellman, W. E. 1981. Three new species of centrolenid frogs from the Pacific versant of Ecuador and Columbia. *Occasional*

Papers of the Museum of Hatural History, The University of Kansas 88: 1–9.

Duellman, W. E. 1983. A new species of marsupial frog (Hylidae: *Gastrotheca*) from Colombia and Ecuador. *Copeia* 1983: 868–874.

Duellman, W. E. 1995. Temporal fluctuations in abundances of anuran amphibians in a seasonal Amazonian rainforest. *Journal of Herpetology* 29: 13–21.

Duellman, W. E. 2001. *The Hylid Frogs of Middle America*, Expanded ed. Ithaca: SSAR.

Duellman, W. E. 2005. *Cusco Amazónico, the Lives of Amphibian and Reptiles in an Amazonian Rainforest*. Ithaca: Cornell University Press.

Duellman, W. E. 2015. *Marsupial Frogs: Gastrotheca and Allied Genera*. Baltimore: JHU Press.

Duellman, W. E., and P. A. Burrowes. 1989. New species of frogs, *Centrolenella*, from the Pacific versant of Ecuador and Southern Colombia. *Occasional Papers of the Museum of Natural History, the University of Kansas* 132: 1–14.

Duellman, W. E., and G. Chavéz. 2010. Reproduction in the marsupial frog *Gastrotheca testudinea* (Anura: Hemiphractidae). *Herpetology Notes* 3: 87–90.

Duellman, W. E., and C. J. Cole. 1965. Studies of chromosomes of some anuran amphibians (Hylidae and Centrolenidae). *Systematic Zoology* 14: 139–143.

Duellman, W. E., and T. H. Fritts. 1972. A taxonomic review of the southern Andean marsupial frogs (Hylidae: *Gastrotheca*). *Occasional Papers of the Museum of Natural History, the University of Kansas* 9: 1–37.

Duellman, W. E., and D. M. Hillis. 1987. Marsupial frogs (Anura: Hylidae: *Gastrotheca*) of the Ecuadorian Andes: resolution of taxonomic problems and phylogenetic relationships. *Herpetologica* 43: 141–173.

Duellman, W. E., and J. D. Lynch. 1969. Descriptions of *Atelopus* tadpoles and their relevance to atelopodid classification. *Herpetologica* 25: 231–240.

Duellman, W. E., and M. Lizana. 1994. Biology of a sit-and-wait predator, the leptodactylid frog *Ceratophrys cornuta*. *Herpetologica* 50: 51–64.

Duellman, W. E., and J. D. Lynch. 1988. Anuran amphibians from the Cordillera de Cucutú, Ecuador. *Proceedings of the Academy of Natural Sciences of Philadelphia* 140: 125–142.

Duellman, W. E., A. B. Marion, and S. B. Hedges. 2016. Phylogenetics, classification, and biogeography of the treefrogs (Amphibia: Anura: Arboranae). *Zootaxa* 4104: 1–109.

Duellman, W. E., L. R. Maxson, and C. A. Jesiolowski. 1988. Evolution of marsupial frogs (Hylidae: Hemiphactinae): immunological evidence. *Copeia* 1988: 527–545.

Duellman, W. E., and J. R. Mendelson III. 1995. Amphibians and reptiles from northern departamento Loreto, Perú: taxonomy and biogeography. *The University of Kansas Science Bulletin* 55: 329–376.

Duellman, W. E., and V. R. Morales. 1990. Variation, distribution, and life history of *Edalorhina perezi* (Amphibia, Anura, Leptodactylidae). *Studies on Neotropical Fauna and Environment* 25: 19–30.

Duellman, W. E., and R. E. Pyles. 1980. A new marsupial frog (Hylidae: Gastrotheca) from the Andes of Ecuador. *Occasional Papers of the Museum of Natural History, the University of Kansas* 84: 1–13.

Duellman, W. E., and P. M. Ruiz-Carranza. 1986. Ontogenetic polychromatism in marsupial frogs (Anura: Hylidae). *Caldasia* 15, 617–627.

Duellman, W. E., and A. W. Salas. 1991. Annotated checklist of the amphibians and reptiles of Cuzco Amazónico, Peru. *Occasional Papers, Museum of Natural History, University of Kansas* 143: 1–13.

Duellman, W. E., and A. H. Savitzky. 1976. Aggressive behavior in a centrolenid frog, with comments on territoriality in anurans. *Herpetologica* 32: 401–404.

Duellman, W. E., and R. Schulte. 1993. New species of centrolenid frogs from northern Peru. *Occasional Papers of the Museum of Natural History. The University of Kansas* 155: 1–33.

Duellman, W. E., and C. A. Toft. 1979. Anurans from Serrania de Sira, Amazonian, Peru: taxonomy and biogeography. *Herpetologica* 35: 60–70.

Duellman, W. E., and L. Trueb. 1994. *Biology of Amphibians*. Baltimore: The John Hopkins University Press.

Duellman, W. E., and P. J. Venegas. 2016. Diversity of marsupial frogs (Anura: Hemiphractidae: *Gastrotheca*) in the northern Cordillera Central, Peru, with the descriptions of two new species. *Phyllomedusa* 15: 103–117.

Duellman, W. E., and E. R. Wild. 1993. Anuran amphibians from the Cordillera de Huancabamba, northern Peru: systematics, ecology, and biogeography. *Occasional Papers of the Museum of Natural History, University of Kansas* 157: 1–53.

Duméril, A. M. C., and G. Bibron. 1841. *Erpétologie Générale ou Histoire Naturelle Compléte des Reptiles. Tome 8. Comprenant l'Histoire Générale des Batraciens, et la description des cinquante-duex generes et des cent soixante-trois espéces des duex premiers sous-ordres: Les péromèles qui n'ont*. Paris: Librairie Encyclopédique de Roret.

Duméril, A. M. C., G. Bibron, and A. H. A. Duméril. 1854. *Erpétologie Genérale ou Histoire Naturelle Complète des Reptiles*. Paris: Librairie Enclyclopedique de Roret.

Dunn, E. R. 1931. New frogs from Panama and Costa Rica. *Occasional Papers of the Boston Society of Natural History* 5: 385–401.

Dunn, E. R. 1942a. A new species of frog (*Eleutherodactylus*) from Costa Rica. *Notulae Naturae of the Academy of Natural Sciences of Philadelphia* 104: 1–2.

Dunn, E. R. 1942b. The American caecilians. *Bulletin of the Museum of Comparative Zoology. Cambridge, Massachusetts* 91: 437–540.

Dunn, E. R. 1949. Notes on the South American frog genus *Edalorhina*. *American Museum Novitates* 1416: 1–10.

Echevarria, L. Y., I. de la Riva, P. J. Venegas, F. J. M. Rojas-Runjaic, R. Dias, R., and S. Castroviejo-Fisher. 2021. Total evidence and sensitivity phylogenetic analyses of egg-brooding frogs (Anura: Hemiphractidae). *Cladistics* 37: 375–401.

Elmer, K. R., R. M. Bonett, D. B. Wake, and S. C. Lougheed. 2013. Early Miocene origin and cryptic diversification of South American salamanders. *BMC Evolutionary Biology* 13: 1–16.

Ernst, R., T. Konrad, and K. E. Linsenmair. 2007. The impacts of selective logging on three sympatric species of *Leptodactylus* in a central Guyana rainforest. *Amphibia-Reptilia* 28: 51–64.

Ernst, R., M.-O. Rödel, and D. Arjoon. 2005. On the cutting edge—the anuran fauna of the Mabura Hill forest reserve, central Guyana. *Salamandra* 41: 179–194.

Fabrezi, M. 2006. Morphological evolution of Ceratophryinae (Anura, Neobatrachia). *Journal of Zoological Systematics and Evolutionary Research* 44: 153–166.

Faivovich, J., D. P. Ferraro, N. G. Basso, C. F. B. Haddad, M. T. Rodrigues, W. C. Wheeler, and E. O. Lavilla. 2012. A phylogenetic analysis of *Pleurodema* (Anura: Leptodactylidae: Leiuperinae) based on mitochondrial and nuclear gene sequences, with comments on the evolution of anuran foam nests. *Cladistics* 28: 460–482.

Faivovich, J., C. F. B. Haddad, D. Baêta, K.-H. Jungfer, G. F. R. Álvares, R. A. Brandão, C. Sheil, L. S. Barrientos, C. L. Barrio-Amorós, C. A. G. Cruz, and W. C. Wheeler. 2010. The phylogenetic relationships of the charismatic poster frogs, Phyllomedusinae (Anura, Hylidae). *Cladistics* 26: 227–261.

Faivovich, J., C. F. B. Haddad, P. C. A. García, D. R. Frost, and J. A. Campbell. 2005. Systematic review of the frog family Hylidae, with special reference to Hylinae: phylogenetic analysis and taxonomic revision. *Bulletin of the American Museum of Natural History* 294: 1–240.

Faivovich, J., L. Nicoli, B. L. Blotto, M. O. Pereyra, D. Baldo, J. S. Barrionuevo, M. Fabrezi, E. R. Wild, and C. F. B. Haddad. 2014. Big, bad, and beautiful: phylogenetic relationships of the horned frogs (Anura: Ceratophryidae). *South American Journal of Herpetology* 9: 207–227.

Felsenstein, J. 2004. *Inferring Phylogenies*. Sunderland: Sinauer Associates.

Feng, Y.-J., D. Blackburn, D. Liang, D. Hillis, D. Wake, D. Cannatella, and P. Zhang. 2017. Phylogenomics reveals rapid, simultaneous diversification of three major clades of Gondwanan frogs at the Cretaceous–Paleogene boundary. *Proceedings of the National Academy of Sciences* 114: E5864–E5870.

Ferraro, D. P., M. O. Pereyra, P. E. Topa, and J. Faivovich. 2020. Evolution of macroglands and defensive mechanisms in Leiuperinae (Anura: Leptodactylidae). *Zoological Journal of the Linnean Society* 193: 388–412.

Fick, S. E., and R. J. Hijmans. 2017. WorldClim 2: new 1-km spatial resolution climate surfaces for global land areas. *International Journal of Climatology* 37: 4302–4315.

Fitzgerald, K. T., L. J. Guillete Jr., and D. Duval. 1979. Notes on birth, development and care of *Gastrotheca riobambae* tadpoles in the laboratory (Amphibia, Anura, Hylidae). *Journal of Herpetology* 13: 457–460.

Fitzinger, L. 1843. *Systema Reptilium. Fasciculus Primus Amblyglossae*. Wien: Braumüller et Seidel.

Fletcher-Lazo, G. 2002. Taxonomía, distribución e historia natural de los ápodos (Amphibia: Gymnophiona) del occidente ecuatoriano. Tesis de Licenciatura, Pontificia Universidad Católica del Ecuador.

Flores, G. 1985. A new *Centrolenella* (Anura) from Ecuador, with comments on nuptial pads and prepollical spines in *Centrolenella*. *Journal of Herpetology* 19: 313–320.

Flores, G. 1987. A new *Centrolenella* from the Amazonian lowlands of Colombia. *Journal of Herpetology* 21: 185–190.

Flores, G., and R. W. McDiarmid. 1989. Two new species of South American *Centrolenella* (Anura: Centrolenidae) related to *C. mariae*. *Herpetologica* 45: 401–411.

Fouquet, A., A. Gilles, M. Vences, C. Marty, M. Blanc, and N. J. Gemmell. 2007. Underestimation of species richness in neotropical frogs revealed by mtDNA analyses. *PLoS One* 2: e1109.

Fouquet, A., D. Loebmann, S. Castroviejo-Fisher, J. M. Padial, V. G. D. Orrico, M. L. Lyra, I. J. Roberto, P. J. R. Kok, C. F. B. Haddad, and M. T. Rodrigues. 2012. From Amazonia to the Atlantic forest: molecular phylogeny of Phyzelaphryninae frogs reveals unexpected diversity and a striking biogeographic pattern emphasizing conservation challenges. *Molecular Phylogenetics and Evolution* 65: 547–561.

Fouquet, A., C. Santana Cassini, C. Fernando Baptista Haddad, N. Pech, and M. Trefaut Rodrigues. 2013. Species delimitation, patterns of diversification and historical biogeography of the Neotropical frog genus *Adenomera* (Anura, Leptodactylidae). *Journal of Biogeography* 41: 855–870.

Fouquette, Jr., M. J. 1960. Call structure in frogs of the family Leptodactylidae. *Texas Journal of Science* 12: 201–215.

Fowler, H. W. 1913. Amphibians and reptiles from Ecuador, Venezuela, and Yucatan. *Proceedings of the Academy of Natural Sciences of Philadelphia* 65: 153–176.

Franco-Mena, D., J. M. Guayasamin, D. Andrade-Brito, M. H. Yánez-Muñoz, and F. J. M. Rojas-Runjaic. 2023. Unveiling the evolutionary relationships and the high cryptic diversity in Andean rainfrogs (Craugastoridae: *Pristimantis myersi* Group). *PeerJ* 11: e14715.

Frost, D. R. 2022. *Amphibian Species of the World: an Online Reference. Version 6.1*. New York, USA: American Museum of Natural History. http://research.amnh.org/vz/herpetology/amphibia/index.php.

Frost, D. R. 2023. *Amphibian Species of the World: An Online Reference. Version 6.1*. New York, USA: American Museum of Natural History. http://research.amnh.org/herpetology/amphibia/index.html.

Frost, D. R., T. Grant, J. Faivovich, R. H. Bain, A. Haas, C. F. B. Haddad, R. O. de Sá, *et al.* 2006. The amphibian tree of life. *Bulletin of the American Museum of Natural History* 297: 1–370.

Funk, W. C., J. P. Caldwell, C. E. Peden, J. M. Padial, I. de la Riva, and D. C. Cannatella. 2007. Tests of biogeographic hypotheses for diversification in the Amazonian forest frog, *Physalaemus petersi*. *Molecular Phylogenetics and Evolution* 44: 825–837.

Funk, W. C., M. Caminer, and S. R. Ron. 2012. High levels of cryptic species diversity uncovered in Amazonian frogs. *Proceedings of the Royal Society B: Biological Sciences* 279: 1806–1814.

Funk, W. C., D. C. Cannatella, and M. J. Ryan. 2009. Genetic divergence is more tightly related to call variation than landscape features in the Amazonian frogs *Physalaemus petersi* and *P. freibergi*. *Journal of Evolutionary Biology* 22: 1839–1853.

Funk, W. C., G. Fletcher-Lazo, F. Nogales-Sornosa, and D. Almeida-Reinoso. 2004. First description of a clutch and nest site for the genus *Caecilia* (Gymnophiona: Caeciliidae). *Herpetological Review* 35: 128–130.

Funkhouser, A. 1957. A review of the neotropical tree frogs of the genus *Phyllomedusa*. *Occasional Papers of the Natural History Museum of Stanford University* 5: 1–90.

Gagliardi-Urrutia, L. A. G., C. R. García Dávila, A. F. Jaramillo-Martínez, O. Rojas-Padilla, E. J. Ríos-Alva, R. Aguilar-Manihuari, P. E. Pérez-Peña, *et al.* 2002. *Anfibios de Loreto*. Lima: Ministerio del Ambiente and Instituto de Investigaciones.

Gagliardo, R., E. Griffith, R. Hill, H. Ross, J. R. Mendelson III, E. Timpe, and B. Wilson. 2010. Observations on the captive reproduction of the horned marsupial frog *Gastrotheca cornuta* (Boulenger, 1898). *Herpetological Review* 41: 52–58.

García-París, M., and D. B. Wake. 2000. Molecular phylogenetic analysis of relationships of the tropical salamander genera *Oedipina* and *Nototriton,* with descriptions of a new genus and three new species. *Copeia* 2000: 42–70.

Gascon, C., S. C. Lougheed, and J. P. Bogart. 1998. Patterns of genetic population differentiation in four species of Amazonian frogs: a test of the riverine barrier hypothesis. *Biotropica* 30: 104–119.

Giaretta, A. A., F. G. de Freitas, M. M. Antoniazzi, and C. Jared. 2011. Tadpole buccal secretory glands as new support for a Neotropical clade of frogs. *Zootaxa* 44: 38–44.

Goin, C. J. 1961. Three new centrolenid frogs from Ecuador. *Zoologischer Anzeiger* 166: 95–104.

Goin, C. J. 1963. A new centrolenid frog from Venezuela. *Acta Biologica Venezuelica Caracas* 3: 283–286.

Gomez-Mestre, I., and K. M. Warkentin. 2007. To hatch and hatch not: similar selective trade-offs but different responses to egg predators in two closely related, syntopic treefrogs. *Behavioral Ecology* 153: 197–206.

Gomez-Mestre, I., J. J. Wiens, and K. M. Warkentin. 2008. Evolution of adaptive plasticity: risk-sensitive hatching in neotropical leaf-breeding treefrogs. *Ecological Monographs* 78: 205–224.

Goodman, D. E., and C. J. Goin. 1970. The habitat of *Centrolene geckoideum* in Ecuador. *Herpetologica* 26: 276.

Grant, T., G. W. Bolivar, and F. Castro. 1998. The advertisement call of *Centrolene geckoideum*. *Journal of Herpetology* 32: 452–455.

Gray, A. R. 1997. Observations on the biology of *Agalychnis spurrelli* from the Caribbean lowlands of Costa Rica. Observaciones sobre la biología de *Agalychnis spurrelli* de las tierras bajas del Caribe de Costa Rica. *Journal of the International Herpetological Society* 22: 61–70.

Gray, A. R. 2018. Review of the genus *Cruziohyla* (Anura: Phyllomedusidae), with description of a new species. *Zootaxa* 4450: 401.

Gray, A. R., and L. A. Coloma. 2022. Description of the tadpole of *Agalychnis hulli* (Anura: Hylidae). Phyllomedusa. *Belo Horizonte* 21: 141–149.

Gray, A. R., K. Taupp, L. Denès, F. Elsner-Gearing, and D. Bewick. 2021. Description of the tadpole of *Cruziohyla calcarifer* (Boulenger, 1902) (Amphibia, Anura, Phyllomedusidae). *Herpetological Journal* 31: 170–176.

Gray, J. E. 1850. *Catalogue of the Specimens of Amphibia in the Collection of the British Museum Part II. Batrachia Gradientia, etc.* London: Trustees of the British Museum.

Greeney, H. F., R. Gelis, and W. C. Funk. 2008. Predation on caecilians (*Caecilia orientalis*) by barred hawks (*Leucopternis princeps*) depends on rainfall. *Herpetological Review* 39: 162–164.

Greer, B. J., and K. D. Wells. 1980. Territorial and reproductive behavior of the tropical American frog *Centrolenella fleischmanni*. *Herpetologica* 36: 318–326.

Grenard, S. 1958. Life history of *Gastrotheca marsupiata* spp. *Herpetologica* 14: 151–152.

Guayasamin, J. M. 2013. A new yellow species of glassfrog (Centrolenidae: *Nymphargus*) from the Amazonian slopes of the Ecuadorian Andes. *Zootaxa* 3652: 193–200.

Guayasamin, J. M. 2020. *Nymphargus lindae*. In *Glassfrogs of Ecuador: diversity, evolution, & conservation,* eds. J. M. Guayasamin, D. F. Cisneros-Heredia, P. Peña, R. W. McDiarmid, and C. R. Hutter. *Diversity* 12: 184–186.

Guayasamin, J. M., and E. Bonaccorso. 2004. A new species of glass frog (Centrolenidae: *Cochranella*) from the lowlands of Northwestern Ecuador, with comments on the Cochranella granulosa group. *Herpetologica* 60: 485–494.

Guayasamin, J. M., R. M. Brunner, A. Valencia-Aguilar, D. Franco-Mena, E. Ringler, A. M. Armijos, C. Morochz, L. Bustamante, R. J. Maynard, and J. Culebras. 2022. Two new glassfrogs (Centrolenidae: *Hyalinobatrachium*) from Ecuador, with comments on the endangered biodiversity of the Andes. *PeerJ* 10: e13109.

Guayasamin, J. M., S. Castroviejo-Fisher, L. Trueb, J. Ayarzagüena, M. Rada, and C. Vilà. 2009. Phylogenetic systematics of glassfrogs (Amphibia: Centrolenidae) and their sister taxon *Allophryne ruthveni*. *Zootaxa* 2100: 1–97.

Guayasamin, J. M., D. F. Cisneros-Heredia, M. Yánez-Muñoz, and M. Bustamante. 2006a. Amphibia, Centrolenidae, *Centrolene ilex*, *Centrolene litorale*, *Centrolene medemi*, *Cochranella albomaculata*, *Cochranella ametarsia*: range extensions and new country records Juan. *Check List* 2: 70–75.

Guayasamin, J. M., M. R. Bustamante, D. Almeida-Reinoso, and W. C. Funk. 2006b. Glass frogs (Centrolenidae) of Yanayacu biological station, Ecuador, with the description of a new species and comments on centrolenid systematics. *Zoological Journal of the Linnean Society* 147: 489–513.

Guayasamin, J. M., D. F. Cisneros-Heredia, and S. Castroviejo-Fisher. 2008a. Taxonomic identity of *Cochranella petersi* Goin, 1961 and *Centrolenella ametarsia* Flores, 1987. *Zootaxa* 34: 25–34.

Guayasamin, J. M., S. Castroviejo-Fisher, J. Ayarzagüena, L. Trueb, and C. Vilà. 2008b. Phylogenetic relationships of glassfrogs (Centrolenidae) based on mitochondrial and nuclear genes. *Molecular Phylogenetics and Evolution* 48: 574–595.

Guayasamin, J. M., D. F. Cisneros-Heredia, R. J. Maynard, R. L. Lynch, J. Culebras, and P. S. Hamilton. 2017. A marvelous new glassfrog (Centrolenidae: *Hyalinobatrachium*) from Amazonian Ecuador. *ZooKeys* 2017: 1–20.

Guayasamin, J. M., D. F. Cisneros-Heredia, J. Vieira, S. Kohn, G. Gavilanes, R. L. Lynch, P. S. Hamilton, and R. J. Maynard. 2019a. A new glassfrog (Centrolenidae) from the Chocó-Andean Río Manduriacu reserve, Ecuador, endangered by mining. *PeerJ* 7: e6400.

Guayasamin, J. M., J. Vieira, R. E. Glor, and C. R. Hutter. 2019b. A new glassfrog (Centrolenidae: *Hyalinobatrachium*) from the Topo river basin, Amazonian slopes of the Andes of Ecuador. *Amphibian & Reptile Conservation* 13: 133–144.

Guayasamin, J. M., D. F. Cisneros-Heredia, R. W. McDiarmid, and C. R. Hutter. 2020a. *Nymphargus humboldti*. In *Glassfrogs of Ecuador: diversity, evolution, & conservation,* eds. J. M. Guayasamin, D. F. Cisneros-Heredia, P. Peña, R. W. McDiarmid, and C. R. Hutter. *Diversity* 12: 175–178.

Guayasamin, J. M., D. F. Cisneros-Heredia, R. W. McDiarmid, P. Peña, and C. R. Hutter. 2020b. Glassfrogs of Ecuador: diversity, evolution, and conservation. *Diversity* 12: 222.

Guayasamin, J. M., and W. C. Funk. 2009. The amphibian community at Yanayacu biological station, Ecuador, with a comparison of vertical microhabitat use among *Pristimantis species* and the description of a new species of the *Pristimantis myersi* group. *Zootaxa* 66: 41–66.

Guayasamin, J. M., and C. R. Hutter. 2020. *Nymphargus colomai*. In *Glassfrogs of Ecuador: diversity, evolution, & conservation,* eds. J. M. Guayasamin, D. F. Cisneros-Heredia, P. Peña, R. W. McDiarmid, and C. R. Hutter. *Diversity* 12: 160–163.

Guayasamin, J. M., Á. M. Mendoza, A. V. Longo, K. R. Zamudio, and E. Bonaccorso. 2014. High prevalence of *Batrachochytrium dendrobatidis* in an Andean frog community (Reserva Las Gralarias, Ecuador). *Amphibian & Reptile Conservation* 8: 33–44.

Guayasamin, J., and S. North. 2009. Amphibia, Centrolenidae, *Hyalinobatrachium iaspidiense*: distribution extension. *Check List* 5: 526–529.

Guayasamin, J. M., E. Tapia, S. Aldas, and J. Deichmann. 2011. Anfibios y Reptiles de los Tepuyes de la Cuenca Alta del Río Nangaritza, Cordillera del Cóndor. In *Evaluación Ecológica Rápida de la Biodiversidad de los Tepuyes de la Cuenca Alta del Río Nangaritza, Cordillera del Cóndor, Ecuador*, eds. J. M. Guayasamin, and E. Bonaccorso, 56–61. Quito, Ecuador: Conservación Internacional.

Guayasamin, J. M., and L. Trueb. 2007. A new species of glass frog (Anura: Centrolenidae) from the lowlands of Northwestern

Ecuador, with comments on centrolenid osteology. *Zootaxa* 1447: 27–45.

Guayasamin, J. M., S. Castroviejo-Fisher, L. Trueb, J. Ayarzagüena, M. Rada, and C. Vilà. 2009. Phylogenetic systematics of Glassfrogs (Amphibia: Centrolenidae) and their sister taxon *Allophryne ruthveni*. *Zootaxa* 2100: 1–97.

Guerra, M. A., and S. R. Ron. 2008. Mate choice and courtship signal differentiation promotes speciation in an Amazonian frog. *Behavioral Ecology* 19: 1128–1135.

Guerrero-Vargas, J. A., E. R. Coral-Plaza, M. P. Rivas-Pava, A. K. Meneses, and S. Ayerbe González. 2007. Cátologo de los anfibios de la colección de referencia del Museo de Historia Natural de la Universidad del Cauca. Taller Editorial Universidad del Cauca, Popayán.

Guevara-Molina, S. C., and M. Rada. 2018. *Sachatamia orejuela*. *Catálogo de anfibios y reptiles de Colombia* 4: 68–72.

Günther, A. C. 1858. On the systematic arrangement of the tailless batrachians and the structure of *Rhinophrynus dorsalis*. *Proceedings of the Zoological Society of London* 1858: 339–352.

Günther, A. C. 1859. Second list of cold-blooded Vertebrata collected by Mr. Fraser in the Andes of Western Ecuador. *Proceedings of the Zoological Society of London* 1859: 402–422.

Gutiérrez-C, P. D. A., and L. S. Barrientos. 2007. Geographic distribution: *Gastrotheca dendronastes* (Río Calima marsupial frog). *Herp Review* 38: 474–475.

Haas, A. 1996. Das larvale cranium von *Gastrotheca riobambae* und seine metamorpose (Amphibia, Anura, Hylidae). *Verhandlungen des Naturwissenschaftlichen Vereins in Hamburg* 36: 3–162.

Hallowell, E. 1861. "1860". Report upon the reptilia of the North Pacific exploring expedition, under command of Capt. John Rogers, U. S. N. *Proceedings of the Academy of Natural Sciences of Philadelphia* 12: 480–510.

Harrison Matthews, L. 1958. "1957". Viviparity in *Gastrotheca* (Amphibia: Anura) and some considerations on the evolution of viviparity. *Bulletin de la Société Zoologique de France* 82: 317–320.

Hartert, E. 1898. On a collection of birds from North-Western Ecuador, collected by Mr. W. F. H. Rosenberg. *Novitates Zoologicae* 5: 477–505.

Hass, C. A., R. A. Nussbaum, and L. R. Maxson. 1993. Immunological insights into the evolutionary history of caecilians (Amphibia: Gymnophiona): relationships of the Seychellean caecilians and a preliminary report on family-level relationships. *Herpetological Monographs* 7: 56–63.

Hayes, M. P. 1991. *A Study of Clutch Attendance in the Neotropical Frog Centrolenella fleischmanni (Anura: Centrolenidae)*. Dissertations Publishing, University of Miami.

Hayes, M. P., and P. H. Starrett. 1980. Notes on a collection of centrolenid frogs from the Colombian Choco. *Bulletin of the Southern California Academy of Sciences* 79: 89–96.

Hedges, S. B., W. E. Duellman, and M. P. Heinicke. 2008. New world direct-developing frogs (Anura: Terrana): molecular phylogeny, classification, biogeography, and conservation. *Zootaxa* 1737: 1–181.

Heinicke, M. P., W. E. Duellman, L. Trueb, D. B. Means, R. D. Macculloch, and S. B. Hedges. 2009. A new frog family (Anura: Terrarana) from South America and an expanded direct-developing clade revealed by molecular phylogeny. *Zootaxa* 2211: 1–35.

Hero, J.-M. 1990. An illustrated key to tadpoles occurring in the central Amazona rainforest, Manaus, Amazonas, Brasil. *Amazoniana* 11: 201–262.

Herrboldt, M. A., M. A. Steffen, C. N. McGouran, and R. M. Bonett. 2021. Pheromone gene diversification and the evolution of courtship glands in plethodontid salamanders. *Journal of Molecular Evolution* 89: 576–587.

Hershkovitz, P. 1938. A new caecilian from Ecuador. *Occasional Papers of the Museum of Zoology, University of Michigan* 370: 1–3.

Hervas, F., K. R. Elmer, D. B. Wake, and S. R. Ron. 2016. Cryptic diversity and biogeography of South American salamanders. Tesis de Maestría, Pontificia Universidad Católica del Ecuador.

Heyer, M. M., W. R. Heyer, and O. R. de Sa. 2011. *Leptodactylus pentadactylus*. *Catalogue of American Amphibians and Reptiles* 887: 1–48.

Heyer, R., C. Azevedo-Ramos, L. A. Coloma, and S. R. Ron. 2008. *Leptodactylus pentadactylus*. The IUCN Red List of Threatened Species 2008: e.T57154A11593184. http://dx.doi.org/10.2305/IUCN.UK.2008.RLTS.T57154A11593184.en

Heyer, R., and D. Cisneros-Heredia. 2008. *Leptodactylus rhodomerus*. The IUCN Red List of Threatened Species 2008: e.T136050A4242168. http://dx.doi.org/10.2305/IUCN.UK.2008.RLTS.T136050A4242168.en

Heyer, R., L. A. Coloma, S. Ron, C. Azevedo-Ramos, E. La Marca, and J. Hardy. 2010. *Leptodactylus knudseni*. The IUCN Red List of Threatened Species 2010: e.T57135A11589628. http://dx.doi.org/10.2305/IUCN.UK.2010-2.RLTS.T57135A11589628.en

Heyer, R., and M. T. Rodrigues. 2010a. *Leptodactylus mystaceus*. The IUCN Red List of Threatened Species 2010: e.T57146A11591629. http://dx.doi.org/10.2305/IUCN.UK.2010-2.RLTS.T57146A11591629.en

Heyer, R., and M. T. Rodrigues. 2010b. *Leptodactylus petersii*. The IUCN Red List of Threatened Species 2010: e.T57155A11593480 http://dx.doi.org/10.2305/IUCN.UK.2010-2.RLTS.T57155A11593480.en

Heyer, W. R. 1969. Studies on the genus *Leptodactylus* (Amphibia, Leptodactylidae) III. A redefinition of the genus *Leptodactylus* and a description of a new genus of leptodacylid frogs. *Los Angeles County Museum Contributions in Science* 155: 1–14.

Heyer, W. R. 1970a. Studies on the frogs of the genus *Leptodactylus* (Amphibia: Leptodactylidae). VI. Biosystematics of the *melanonotus* group. *Los Angeles County Museum Contributions in Science* 191: 1–48.

Heyer, W. R. 1970b. Studies on the genus *Leptodactylus* (Amphibia: Leptodactylidae). II. Diagnosis and distribution of the *Leptodactylus* of Costa Rica. *Revista de Biologia Tropical* 16: 171–205.

Heyer, W. R. 1972. The status of *Leptodactylus pumilio* Boulenger (Amphibia, Leptodactylidae) and the description of a new species of *Leptodactylus* from Ecuador. *Contributions in Science* 231: 1–8.

Heyer, W. R. 1973. Systematics of the *Marmoratus* group of the frog genus *Leptodactylus* (Amphibia, Leptodactylidae). *Natural History Museum of Los Angeles County Contributions in Science* 251: 1–50.

Heyer, W. R. 1974. Relationships of the *marmoratus* species group (Amphibia, Leptodactylidae) within the subfamily Leptodactylinae. *Natural History Museum of Los Angeles County Contributions in Science* 253: 1–46.

Heyer, W. R. 1977. Taxonomic notes on frogs from the Madeira and Purus rivers, Brasil. *Papéis Avulsos de Zoologia. São Paulo* 31: 141–162.

Heyer, W. R. 1978. Systematics of the *fuscus* group of the frog genus *Leptodactylus* (Amphibia, Leptodactylidae). *Natural*

History Museum of Los Angeles County Science Bulletin 29: 1–85.

Heyer, W. R. 1979. Systematics of the *pentadactylus* species group of the frog genus *Leptodactylus* (Amphibia: Leptodactylidae. *Smithsonian Contributions to Zoology* 301: 1–43.

Heyer, W. R. 1994. Variation within the *Leptodactylus podicipinus-wagneri* complex of frogs (Amphibia: Leptodactylidae). *Smithsonian Contributions to Zoology* 546: 1–124.

Heyer, W. R. 1997. Geographic variation in the frog genus *Vanzolinius* (Anura: Leptodactylidae). *Proceedings of the Biological Society of Washington* 110: 338–365.

Heyer, W. R. 1998. The relationships of *Leptodactylus diedrus* (Anura, Leptodactylidae). *Alytes* 16: 1–24.

Heyer, W. R. 2005. Variation and taxonomic clarification of the large species of the *Leptodactylus pentadactylus* species group (Amphibia: Leptodactylidae) from Middle America, Northern South America, and Amazonia. *Arquivos de Zoologia (São Paulo)* 37: 269–348.

Heyer, W. R., and M. S. Bellin. 1973. Ecological notes on five sympatric *Leptodactylus* (Amphibia, Leptodactylidae) from Ecuador. *Herpetologica* 29: 66–72.

Heyer, W. R., and M. J. Diment. 1974. The karyotype of *Vanzolinius discodactylus* and comments on usefulness of karyotypes in determining relationships in the *Leptodactylus*-complex (Amphibia, Leptodactylidae). *Proceedings of the Biological Society of Washington* 87: 327–336.

Heyer, W. R., J. M. García-López, and A. J. Cardoso. 1996. Advertisement call variation in the *Leptodactylus mystaceus* species complex (Amphibia: Leptodactylidae) with description of new sibling species. *Amphibia-Reptilia* 17: 7–31.

Heyer, W. R., and M. M. Heyer. 2006. *Leptodactylus knudseni. Catalogue of American Amphibians and Reptiles (CAAR)* 807: 1–2.

Heyer, W. R., and L. R. Maxson. 1982. Neotropical frog biogeography: paradigms and problems. *American Zoologist* 22: 397–410.

Heyer, W. R., A. S. Rand, C. A. G. da Cruz, O. L. Peixoto, and C. E. Nelson. 1990. Frogs of Boracéia. *Arquivos de Zoologia Sao Paulo* 31: 231–410.

Heyer, W. R., and J. Peters. 1971. The frog genus *Leptodactylus* in Ecuador. *Proceedings of the Biological Society of Washington* 84: 163–170.

Heyer, W. R., and P. A. Silverstone. 1969. The larva of the frog *Leptodactylus hylaedactylus* (Leptodactylidae). *Fieldiana—Zoology* 51: 141–145.

Hill, R. L., K. G. Martin, E. Stanley, and J. R. I. Mendelson. 2018. A taxonomic review of the genus *Hemiphractus* (Anura: Hemiphractidae) in Panama: description of two new species, resurrection of *Hemiphractus panamensis* (Stejneger, 1917), and discussion of *Hemiphractus fasciatus* Peters, 1862. *Zootaxa* 4429: 495–512.

Hime, P. M., A. R. Lemmon, E. C. Lemmon, E. Prendini, J. M. Brown, R. C. Thomson, J. D. Kratovil, B. P. Noonan, R. A. Pyron, and P. L. Peloso. 2021. Phylogenomics reveals ancient gene tree discordance in the amphibian tree of life. *Systematic Biology* 70: 49–66.

Hödl, W. 1990. Reproductive diversity in Amazonian lowland frogs. In *Biology and Physiology of Amphibians*, ed. W. Hanke, 41–60. Stuttgart, New York: Gustav Fisher Verlag.

Hoffmann, H. 2004. Description of the previously unknown tadpole of *Hyalinobatrachium pulveratum* (Anura: Centrolenidae). *Revista de Biología Tropical* 52: 219–228.

Hoffmann, H. 2010. The glass frog tadpoles of Costa Rica (Anura: Centrolenidae): a study of morphology. *Abhandlungen der Senckenberg Gesellschaft für Naturforschung* 567: 1–78.

Hoogmoed, M. S. 1967. Mating and early development of Gastrotheca marsupiata (Duméril and Bibron) in captivity (Hylidae, Anura, Amphibia). *British Journal of Herpetology* 4: 1–7.

Hoogmoed, M. S. 1986. Ergänzende Beobachtungen an *Lithodytes lineatus* (Schneider, 1799) (Salientia: Leptodactylidae). *Salamandra* 22: 215–217.

Hoogmoed, M. S., and J. E. Cadle. 1990. Natural history and distribution of *Agalychnis craspedopus* (Funkhouser, 1957) (Amphibia: Anura: Hylidae). *Zoologische Mededelingen Leiden* 65: 129–142.

Hoogmoed, M. S., and J. Lescure. 1984. A new genus and two new species of minute leptodactylid frogs from northern South America, with comments upon *Phyzelaphryne* (Amphibia: Anura: Leptodactylidae). *Zoologische Mededelingen* 58: 85–113.

Hutter, C. R., S. Esobar-Lasso, J. A. Rojas-Morales, P. D. A. Gutiérrez-Cárdenas, H. Imba, and J. M. Guayasamin. 2013. The territoriality, vocalizations and aggressive interactions of the red-spotted glassfrog, *Nymphargus grandisonae,* Cochran and Goin, 1970 (Anura: Centrolenidae). *Journal of Natural History* 47: 3011–3032.

Hutter, C. R., and J. M. Guayasamin. 2012. A new cryptic species of glassfrog (Centrolenidae: *Nymphargus*) from Reserva Las Gralarias, Ecuador. *Zootaxa* 21: 1–21.

Ibañez, R., F. Jaramillo, and C. A. Jaramillo. 1999a. Ampliación del ámbito de distribución y descripción del renacuajo de la rana de cristal *Hyalinobatrachium aureoguttatum* (Anura: Centrolenidae). *Revista de la Academia Colombiana de Ciencias Exactas, Fisicas y Naturales* 23: 293–298.

Ibañez, R., S. Rand, and C. A. Jaramillo. 1999b. *Los anfibios del Monumento Natural Barro Colorado, Parque Nacional Soberanía y áreas adyacentes*. Santa Fe de Bogotá: Fundación Natura, Círculo Herpetológico de Panamá, Smithsonian Tropical Research Institute.

Icochea, J., L. A. Coloma, S. Ron, K.-H. Jungfer, F. Castro, and J. V. Rueda. 2004. *Hemiphractus helioi*. The IUCN Red List of Threatened Species 2004: e.T55368A11299186. https://dx.doi.org/10.2305/IUCN.UK.2004.RLTS.T55368A11299186.en

IUCN Species Survival Commission. 2012. *Guidelines for Application of IUCN Red List Criteria at Regional and National Levels*. Gland, Switzerland and Cambridge: Version 4.0, IUCN.

IUCN SSC Amphibian Specialist Group. 2014. *Siphonops annulatus*. The IUCN Red List of Threatened Species 2014: e.T59593A43784684. http://dx.doi.org/10.2305/IUCN.UK.2014-1.RLTS.T59593A43784684.en

IUCN SSC Amphibian Specialist Group. 2017. *Cochranella megista*. The IUCN Red List of Threatened Species 2017: e.T54970A85872984. https://dx.doi.org/10.2305/IUCN.UK.2017-2.RLTS.T54970A85872984.en

IUCN SSC Amphibian Specialist Group. 2018a. *Gastrotheca longipes*. The IUCN Red List of Threatened Species 2018: e.T55343A61397127. http://dx.doi.org/10.2305/IUCN.UK.2018-1.RLTS.T55343A61397127.en

IUCN SSC Amphibian Specialist Group. 2018b. *Gastrotheca orophylax*. The IUCN Red List of Threatened Species 2018: e.T55349A85898791. http://dx.doi.org/10.2305/IUCN.UK.2018-2.RLTS.T55349A85898791.en

IUCN SSC Amphibian Specialist Group. 2018c. *Hemiphractus proboscideus*. The IUCN Red List of Threatened Species 2018:

e.T55370A85898013.https://dx.doi.org/10.2305/IUCN.UK.2018-1.RLTS.T55370A85898013.en
IUCN SSC Amphibian Specialist Group. 2018d. *Engystomops petersi*. The IUCN Red List of Threatened Species 2018: e.T57270A85893490.https://dx.doi.org/10.2305/IUCN.UK.2018-1.RLTS.T57270A85893490.en
IUCN SSC Amphibian Specialist Group. 2018e. *Leptodactylus labrosus*. The IUCN Red List of Threatened Species 2018: e.T57136A3056172.en
IUCN SSC Amphibian Specialist Group. 2018f. *Agalychnis hulli*. The IUCN Red List of Threatened Species2018: e.T55852A61405955.https://dx.doi.org/10.2305/IUCN.UK.2018-1.RLTS.T55852A61405955.en
IUCN SSC Amphibian Specialist Group. 2018g. *Bolitoglossa palmata*. The IUCN Red List of Threatened Species 2018: e.T59191A85861483. http://dx.doi.org/10.2305/IUCN.UK.2018-1.RLTS.T59191A85861483.en
IUCN SSC Amphibian Specialist Group. 2018h. *Caecilia albiventris*. The IUCN Red List of Threatened Species 2018: e.T59502A99288206.en
IUCN SSC Amphibian Specialist Group. 2018i. *Caecilia attenuata*. The IUCN Red List of Threatened Species 2018: e.T59505A89213981. http://dx.doi.org/10.2305/IUCN.UK.2018-2.RLTS.T59505A89213981.en
IUCN SSC Amphibian Specialist Group. 2018j. *Caecilia dissossea*. The IUCN Red List of Threatened Species 2018: e.T59511A61415092. http://dx.doi.org/10.2305/IUCN.UK.2018-1.RLTS.T59511A61415092.en
IUCN SSC Amphibian Specialist Group. 2018k. *Caecilia orientalis*. The IUCN Red List of Threatened Species 2018: e.T59523A61415454.https://dx.doi.org/10.2305/IUCN.UK.2018-1.RLTS.T59523A61415454.en
IUCN SSC Amphibian Specialist Group. 2018l. *Oscaecilia bassleri*. The IUCN Red List of Threatened Species 2018: e.T59578A89214198. http://dx.doi.org/10.2305/IUCN.UK.2018-1.RLTS.T59578A89214198.en
IUCN SSC Amphibian Specialist Group. 2018m. *Epicrionops petersi*. The IUCN Red List of Threatened Species 2018: e.T59646A89214589.https://dx.doi.org/10.2305/IUCN.UK.2018-1.RLTS.T59646A89214589.en
IUCN SSC Amphibian Specialist Group. 2018n. *Chimerella mariaelenae*. The IUCN Red List of Threatened Species 2018: e.T135968A516500. http://dx.doi.org/10.2305/IUCN.UK.2018-1.RLTS.T135968A516500.en
IUCN SSC Amphibian Specialist Group. 2018o. *Cochranella resplendens*. The IUCN Red List of Threatened Species 2018: e.T54985A54458254. https://dx.doi.org/10.2305/IUCN.UK.2018-1.RLTS.T54985A54458254.en
IUCN SSC Amphibian Specialist Group. 2018p. *Espadarana audax*. The IUCN Red List of Threatened Species 2018: e.T78163103A78162747. http://dx.doi.org/10.2305/IUCN.UK.2018-1.RLTS.T78163103A78162747.en
IUCN SSC Amphibian Specialist Group. 2018q. *Nymphargus griffithsi*. The IUCN Red List of Threatened Species 2018: e.T54965A85873493.https://dx.doi.org/10.2305/IUCN.UK.2018-1.RLTS.T54965A85873493.en
IUCN SSC Amphibian Specialist Group. 2018r. *Nymphargus posadae*. The IUCN Red List of Threatened Species 2018: e.T54981A85874598.https://dx.doi.org/10.2305/IUCN.UK.2018-1.RLTS.T54981A85874598.en
IUCN SSC Amphibian Specialist Group. 2018s. *Ceratophrys stolzmanni*. The IUCN Red List of Threatened Species 2018: e.T56341A89204905. http://dx.doi.org/10.2305/IUCN.UK.2018-1.RLTS.T56341A89204905.en Accessed on 23 November 2020.
IUCN SSC Amphibian Specialist Group. 2019a. *Diasporus gularis*. The IUCN Red List of Threatened Species 2019: e.T56640A85863703.https://dx.doi.org/10.2305/IUCN.UK.2019-2.RLTS.T56640A85863703.en
IUCN SSC Amphibian Specialist Group. 2019b. *Gastrotheca angustifrons*. The IUCN Red List of Threatened Species 2019: e.T55322A85896740. http://dx.doi.org/10.2305/IUCN.UK.2019-1.RLTS.T55322A85896740.en
IUCN SSC Amphibian Specialist Group. 2019c. *Gastrotheca dendronastes*. The IUCN Red List of Threatened Species 2019: e.T55330A85899214. http://dx.doi.org/10.2305/IUCN.UK.2019-1.RLTS.T55330A85899214.en
IUCN SSC Amphibian Specialist Group. 2019d. *Gastrotheca espeletia*. The IUCN Red List of Threatened Species 2019: e.T55332A85899399. http://dx.doi.org/10.2305/IUCN.UK.2019-1.RLTS.T55332A85899399.en
IUCN SSC Amphibian Specialist Group. 2019e. *Gastrotheca guentheri*. The IUCN Red List of Threatened Species 2019: e.T55338A85898956. http://dx.doi.org/10.2305/IUCN.UK.2019-1.RLTS.T55338A85898956.en
IUCN SSC Amphibian Specialist Group. 2019f. *Gastrotheca weinlandii*. The IUCN Red List of Threatened Species 2019: e.T55364A85898313.https://dx.doi.org/10.2305/IUCN.UK.2019-3.RLTS.T55364A85898313.en
IUCN SSC Amphibian Specialist Group. 2019g. *Leptodactylus ventrimaculatus*. The IUCN Red List of Threatened Species 2019: e.T57173A85888740.en
IUCN SSC Amphibian Specialist Group. 2019h. *Agalychnis psilopygion*. The IUCN Red List of Threatened Species 2019: e.T55861A85904690.https://dx.doi.org/10.2305/IUCN.UK.2019-3.RLTS.T55861A85904690.en
IUCN SSC Amphibian Specialist Group. 2019i. *Callimedusa ecuatoriana*. The IUCN Red List of Threatened Species 2019: e.T55851A98648763.https://dx.doi.org/10.2305/IUCN.UK.2019-3.RLTS.T55851A98648763.en
IUCN SSC Amphibian Specialist Group. 2019j. *Callimedusa perinesos*. The IUCN Red List of Threatened Species 2019: e.T55860A85905113.https://dx.doi.org/10.2305/IUCN.UK.2019-3.RLTS.T55860A85905113.en
IUCN SSC Amphibian Specialist Group. 2019k. *Caecilia bokermanni*. The IUCN Red List of Threatened Species 2019: e.T59506A49348109.https://dx.doi.org/10.2305/IUCN.UK.2019-3.RLTS.T59506A49348109.en
IUCN SSC Amphibian Specialist Group. 2019l. *Caecilia guntheri*. The IUCN Red List of Threatened Species 2019: e.T59515A49352194.en
IUCN SSC Amphibian Specialist Group. 2019m. *Epicrionops bicolor*. The IUCN Red List of Threatened Species 2019: e.T59639A85905497.https://dx.doi.org/10.2305/IUCN.UK.2019-2.RLTS.T59639A85905497.en
IUCN SSC Amphibian Specialist Group. 2019n. *Centrolene ballux*. The IUCN Red List of Threatened Species 2019: e.T54907A85877221.https://dx.doi.org/10.2305/IUCN.UK.2019-1.RLTS.T54907A85877221.en
IUCN SSC Amphibian Specialist Group. 2019o. *Centrolene geckoideum*. The IUCN Red List of Threatened Species 2019: e.T54911A85877414.https://dx.doi.org/10.2305/IUCN.UK.2019-2.RLTS.T54911A85877414.en
IUCN SSC Amphibian Specialist Group. 2019p. *Centrolene heloderma*. The IUCN Red List of Threatened Species 2019: e.T54916A85878225. http://dx.doi.org/10.2305/IUCN.UK.2019-1.RLTS.T54916A85878225.en

IUCN SSC Amphibian Specialist Group. 2019q. *Centrolene huilense*. The IUCN Red List of Threatened Species 2019: e.T54918A49363441.en

IUCN SSC Amphibian Specialist Group. 2019r. *Centrolene medemi*. The IUCN Red List of Threatened Species 2019: e.T54926A49364088. http://dx.doi.org/10.2305/IUCN.UK.2019-2.RLTS.T54926A49364088.en

IUCN SSC Amphibian Specialist Group. 2019s. *Centrolene peristictum*. The IUCN Red List of Threatened Species 2019: e.T54931A85879425. http://dx.doi.org/10.2305/IUCN.UK.2019-2.RLTS.T54931A85879425.en

IUCN SSC Amphibian Specialist Group. 2019t. *Centrolene pipilata*. The IUCN Red List of Threatened Species 2019: e.T54933A98643479.https://dx.doi.org/10.2305/IUCN.UK.2019-3.RLTS.T54933A98643479.en

IUCN SSC Amphibian Specialist Group. 2019u. *Cochranella litoralis*. The IUCN Red List of Threatened Species 2019: e.T54923A49367704. http://dx.doi.org/10.2305/IUCN.UK.2019-2.RLTS.T54923A49367704.en

IUCN SSC Amphibian Specialist Group. 2019v. *Espadarana callistomma*. The IUCN Red List of Threatened Species 2019: e.T135738A85908349. http://dx.doi.org/10.2305/IUCN.UK.2019-2.RLTS.T135738A85908349.en

IUCN SSC Amphibian Specialist Group. 2019w. *Nymphargus garciae*. The IUCN Red List of Threatened Species 2019: e.T54962A85873982.https://dx.doi.org/10.2305/IUCN.UK.2019-2.RLTS.T54962A85873982.en

IUCN SSC Amphibian Specialist Group. 2019x. *Sachatamia orejuela*. The IUCN Red List of Threatened Species 2019: e.T54976A49367802.https://dx.doi.org/10.2305/IUCN.UK.2019-2.RLTS.T54976A49367802.en

IUCN SSC Amphibian Specialist Group. 2019y. *Teratohyla amelie*. The IUCN Red List of Threatened Species 2019: e.T135771A89216740.https://dx.doi.org/10.2305/IUCN.UK.2019-3.RLTS.T135771A89216740.en

IUCN SSC Amphibian Specialist Group. 2020a. *Bolitoglossa medemi*. The IUCN Red List of Threatened Species 2020: e.T59178A54353739.https://dx.doi.org/10.2305/IUCN.UK.2020-3.RLTS.T59178A54353739.en

IUCN SSC Amphibian Specialist Group. 2020b. *Cochranella granulosa*. The IUCN Red List of Threatened Species 2020: e.T54964A3021552. https://dx.doi.org/10.2305/IUCN.UK.2020-1.RLTS.T54964A3021552.en

IUCN SSC Amphibian Specialist Group. 2020c. *Cochranella mache*. The IUCN Red List of Threatened Species 2020: e.T61764A85908989.https://dx.doi.org/10.2305/IUCN.UK.2020-3.RLTS.T61764A85908989.en

IUCN SSC Amphibian Specialist Group. 2020d. *Espadarana prosoblepon*. The IUCN Red List of Threatened Species 2020: e.T78163669A54342487. https://dx.doi.org/10.2305/IUCN.UK.2020-3.RLTS.T88380390A176743936.en

IUCN SSC Amphibian Specialist Group. 2020e. *Hyalinobatrachium aureoguttatum*. The IUCN Red List of Threatened Species 2020: e.T55003A54342927.en

IUCN SSC Amphibian Specialist Group. 2020f. *Hyalinobatrachium valerioi*. The IUCN Red List of Threatened Species 2020: e.T55036A3022651.en

IUCN SSC Amphibian Specialist Group. 2020g. *Nymphargus balionotus*. The IUCN Red List of Threatened Species 2020: e.T54948A85872712.https://dx.doi.org/10.2305/IUCN.UK.2020-3.RLTS.T54948A85872712.en

IUCN SSC Amphibian Specialist Group. 2020h. *Nymphargus Mariae (Amended Version of 2018 Assessment)*. The IUCN Red List of Threatened Species 2020: e.T88380390A176743936. https://dx.doi.org/10.2305/IUCN.UK.2020-3.RLTS.T88380390A176743936.en

IUCN SSC Amphibian Specialist Group. 2020i. *Sachatamia albomaculata*. The IUCN Red List of Threatened Species 2020:e.T54944A54342657.https://dx.doi.org/10.2305/IUCN.UK.2020-2.RLTS.T54944A54342657.en

IUCN SSC Amphibian Specialist Group. 2020j. *Sachatamia ilex*. The IUCN Red List of Threatened Species 2020: e.T54920A3021199.en

IUCN SSC Amphibian Specialist Group. 2020k. *Teratohyla pulverata*. The IUCN Red List of Threatened Species 2020: e.T55030A3022505.en

IUCN SSC Amphibian Specialist Group. 2020l. *Teratohyla spinosa*. The IUCN Red List of Threatened Species 2020: e.T54996A54342777.en

IUCN SSC Amphibian Specialist Group. 2020m. *Gastrotheca andaquiensis (Amended Version of 2019 Assessment)*. The IUCN Red List of Threatened Species 2020: e.T55321A176390035. https://dx.doi.org/10.2305/IUCN.UK.2020-3.RLTS.T55321A176390035.en

IUCN SSC Amphibian Specialist Group. 2020n. *Gastrotheca testudinea*. The IUCN Red List of Threatened Species 2020: e.T55361A89203902.https://dx.doi.org/10.2305/IUCN.UK.2020-1.RLTS.T55361A89203902.en Accessed on 13 December 2023.

IUCN SSC Amphibian Specialist Group. 2020o. *Leptodactylus melanonotus*. The IUCN Red List of Threatened Species 2020: e.T57144A53968833.en

IUCN SSC Amphibian Specialist Group. 2020p. *Cruziohyla calcarifer*. The IUCN Red List of Threatened Species 2020: e.T151594717A3027869. https://dx.doi.org/10.2305/IUCN.UK.2020-3.RLTS.T151594717A3027869.en

IUCN SSC Amphibian Specialist Group. 2022a. *Gastrotheca cuencana*. The IUCN Red List of Threatened Species 2022: e.T150084092A150084095.en

IUCN SSC Amphibian Specialist Group. 2022b. *Hemiphractus bubalus*. The IUCN Red List of Threatened Species 2022: e.T55366A85898639.https://dx.doi.org/10.2305/IUCN.UK.2022-2.RLTS.T55366A85898639.en

IUCN SSC Amphibian Specialist Group. 2022c. *Hemiphractus fasciatus*. The IUCN Red List of Threatened Species 2022: e.T154046889A54345645.en

IUCN SSC Amphibian Specialist Group. 2022d. *Bolitoglossa chica*. The IUCN Red List of Threatened Species 2022: e.T59146A85858293.https://dx.doi.org/10.2305/IUCN.UK.2022-1.RLTS.T59146A85858293.en

IUCN SSC Amphibian Specialist Group. 2022e. *Centrolene charapita*. The IUCN Red List of Threatened Species 2022: e.T78457348A89226077.en

IUCN SSC Amphibian Specialist Group. 2022f. *Centrolene lynchi*. The IUCN Red List of Threatened Species 2022: e.T188227841A188229453.en

IUCN SSC Amphibian Specialist Group. 2022g. *Centrolene sanchezi*. The IUCN Red List of Threatened Species 2022: e.T195897278A195940041.en

IUCN SSC Amphibian Specialist Group. 2022h. *Hyalinobatrachium chirripoi*. The IUCN Red List of Threatened Species 2022: e.T55323986A54343070.en

IUCN SSC Amphibian Specialist Group. 2022i. *Hyalinobatrachium iaspidiense*. The IUCN Red List of Threatened Species 2022: e.T201029146A89198728.en

IUCN SSC Amphibian Specialist Group. 2022j. *Hyalinobatrachium munozorum*. The IUCN Red List of Threatened Species 2022: e.T55023A3022224.en

IUCN SSC Amphibian Specialist Group. 2022k. *Hyalinobatrachium pellucidum*. The IUCN Red List of Threatened Species 2022: e.T47255219A89225317.en

IUCN SSC Amphibian Specialist Group. 2022l. *Hyalinobatrachium tatayoi*. The IUCN Red List of Threatened Species 2022: e.T201028107A201028339.en

IUCN SSC Amphibian Specialist Group. 2022m. *Hyalinobatrachium yaku*. The IUCN Red List of Threatened Species 2022: e.T188228424A197515963.en

IUCN SSC Amphibian Specialist Group. 2022n. *Nymphargus grandisonae*. The IUCN Red List of Threatened Species 2022: e.T54914A85878476.en

IUCN SSC Amphibian Specialist Group. 2022o. *Nymphargus megacheirus*. The IUCN Red List of Threatened Species 2022: e.T54969A85873062.en

IUCN SSC Amphibian Specialist Group. 2022p. *Nymphargus siren*. The IUCN Red List of Threatened Species 2022: e.T54992A85874078.en

IUCN SSC Amphibian Specialist Group. 2022q. *Rulyrana flavopunctata*. The IUCN Red List of Threatened Species 2022: e.T54961A3021440.en

IUCN SSC Amphibian Specialist Group. 2022r. *Rulyrana mcdiarmidi*. The IUCN Red List of Threatened Species 2022: e.T158475A89221761.en

IUCN SSC Amphibian Specialist Group. 2022s. *Vitreorana ritae*. The IUCN Red List of Threatened Species 2022: e.T55324252A88248274. https://dx.doi.org/10.2305/IUCN.UK.2022-2.RLTS.T55324252A88248274.en

Jacobson, S. K. 1985. Reproductive behavior and male mating success in two species of glass frogs (Centrolenidae). *Herpetologica* 41: 396–404.

Jetz, W., and R. A. Pyron. 2018. The interplay of past diversification and evolutionary isolation with present imperilment across the amphibian tree of life. *Nature, Ecology & Evolution* 2. 850–858.

Jiménez-Casalombo, G. B., A. P. Guarderas-Valverde, and L. A. Coloma. 2023. Abundancia de la rana marsupial *Gastrotheca riobambae* (Fowler, 1913) (Hemiphractidae, Anura) en un gradiente de urbanización en la Provincia Pichincha, Ecuador. *Revista Latinoamericana de Herpetología* 6: 45–70.

Jiménez de la Espada, M. 1870. Faunae neotropicalis species quaedam nondum cognitae. *Jornal de Sciencias Mathematicas Physicas e Naturaes* III: 54–65.

Jiménez de la Espada, M. 1872. Nuevos batracios Americanos. *Anales de la Sociedad Española de Historia Natural, Madrid* 1: 84–88.

Jiménez de la Espada, M. 1875. *Vertebrados del viaje al Pacifico verificado de 1862 a 1865 por una comision de naturalistas enviada por el Gobierno Español. Batracios.* Madrid: Miguel Ginesta.

Jimenez-Robles, O., I. de la Riva, and J. Guayasamin. 2015. *Rulyrana flavopunctata* (Yellow-spotted Cochran frog). Reproduction. *Herpetological Review* 46: 238–239.

Jungfer, K.-H., W. Bolívar, B. Kubicki, F. Bolaños, G. Chaves, F. Solís, R. Ibáñez, J. Savage, C. Jaramillo, and Q. Fuenmayor. 2008. *Agalychnis spurrelli*. The IUCN Red List of Threatened Species 2008: e.T55295A11276251.en

Kästle, W. 1963. Zur Brutpflege des Beutelfrösches (*Gastrotheca marsupiata*). *Das Aquarium und Terrarium Zeitschrift* 16: 53–56.

Keller, E. 1994. SignalyzeTM Version 3.12. Infosignal Inc., Lausanne, Switzerland.

Kenny, J. S. 1968. Early development and larval natural history of *Phyllomedusa trinitatis* Mertens. *Caribbean Journal of Science* 8: 35–45.

Köhler, G. 2001. *Anfibios y reptiles de Nicaragua*. Offenbach, Germany: Herpeton.

Köhler, G. 2011. *Amphibians of Central America*. Offenbach: Herpeton.

Köhler, J. 2000. *Amphibian Diversity in Bolivia: A Study with Special Reference to Montane Forest Regions. Bonner Zoologische Monographien No. 48*. Bonn: Zoologisches Forschungsinstitut und Museum Alexander Koenig.

Köhler, J., M. Jansen, A. Rodriguez, P. J. Kok, L. F. Toledo, M. Emmrich, F. Glaw, C. F. B. Haddad, M.-O. Rödel, and M. Vences. 2017. The use of bioacoustics in anuran taxonomy: theory, terminology, methods and recommendations for best practice. *Zootaxa* 4251: 1–124.

Kok, P. J. R., and S. Castroviejo-Fisher. 2008. Glassfrogs (Anura: Centrolenidae) of Kaieteur national park, Guyana, with notes on the distribution and taxonomy of some species of the family in the Guiana Shield. *Zootaxa* 1689: 25–53.

Kok, P. J. R., and M. Kalamandeen. 2008. *Introduction to the Taxonomy of the Amphibians of Kaieteur National Park, Guyana. Abc Taxa: A Series of Manuals Dedicated to Capacity Building in Taxonomy and Collection Management*, Vol. 5. Belgium: Belgian Development Corporation, Brussels.

Kokubum, M. N. D., and A. A. Giaretta. 2005. Reproductive ecology and behaviour of a species of *Adenomera* (Anura, Leptodactylinae) with endotrophic tadpoles: systematic implications. *Journal of Natural History* 39: 1745–1758.

Krynak, K. L., D. G. Wessels, S. M. Imba, J. A. Lyons, and J. M. Guayasamin. 2018. Newly discovered population of bumpy glassfrog, *Centrolene heloderma* (Duellman, 1981), with discussion of threats to population persistence. *Check List* 14: 261–265.

Kubicki, B. 2004a. *Ranas de hoja de Costa Rica. Leaf-frogs of Costa Rica*. Costa Rica: Editorial Santo Domingo de Heredia/Editorial INBio.

Kubicki, B. 2004b. Rediscovery of *Hyalinobatrachium chirripoi* (Anura: Centrolenidae) in southeastern Costa Rica. *Revista de Biología Tropical* 52: 214–218.

Kubicki, B. 2007. *Ranas de vidrio de Costa Rica/Glass Frogs of Costa Rica*. Santo Domingo de Heredia: Editorial INBio.

Kubicki, B., F. Bolaños, G. Chaves, F. Solís, R. Ibáñez, L. A. Coloma, S. R. Ron, E. Wild, D. F. Cisneros-Heredia, and J. Renjifo. 2010. *Espadarana prosoblepon*. The IUCN Red List of Threatened Species 2010: e.T54934A11228804. http://dx.doi.org/10.2305/IUCN.UK.2010-2.RLTS.T54934A11228804.en

La Marca, E., C. Azevedo-Ramos, L. A. Coloma, and S. R. Ron. 2004a. *Adenomera andreae*. The IUCN Red List of Threatened Species 2004: e.T56304A11453385. https://dx.doi.org/10.2305/IUCN.UK.2004.RLTS.T56304A11453385.en

La Marca, E., C. Azevedo-Ramos, L. A. Coloma, S. Ron, and C. L. Barrio Amorós. 2004b. *Phyllomedusa tarsius*. The IUCN Red List of Threatened Species 2004: e.T55864A11382243. https://dx.doi.org/10.2305/IUCN.UK.2004.RLTS.T55864A11382243.en

La Marca, E., L. A. Coloma, S. Ron, and C. Azevedo-Ramos. 2004c. *Ceratophrys cornuta*. The IUCN Red List of Threatened Species 2004: e.T56337A11464093. https://dx.doi.org/10.2305/IUCN.UK.2004.RLTS.T56337A11464093.en

La Marca, E., C. Azevedo-Ramos, L. A. Coloma, and S. R. Ron. 2016. *Callimedusa tomopterna*. The IUCN Red List of Threatened Species 2016: e.T55866A107289222. https://dx.doi.org/10.2305/IUCN.UK.2016-3.RLTS.T55866A107289222.en

La Marca, E., L. A. Coloma, S. Ron, C. Azevedo-Ramos, D. Silvano, and J. Hardy. 2010a. *Adenomera hylaedactyla*. The IUCN Red List of Threatened Species 2010: e.T56309A11455170. https://dx.doi.org/10.2305/IUCN.UK.2010-2.RLTS.T56309A11455170.en

La Marca, E., C. Azevedo-Ramos, L. A. Coloma, S. Ron, and R. Heyer. 2010b. *Leptodactylus rhodomystax*. The IUCN Red List of Threatened Species 2010: e.T57160A11578109. http://dx.doi.org/10.2305/IUCN.UK.2010-2.RLTS.T57160A11578109.en

La Marca, E., C. Azevedo-Ramos, L. A. Coloma, S. Ron, and J. Hardy. 2010c. *Lithodytes lineatus*. The IUCN Red List of Threatened Species 2010: e.T57177A11580398. https://dx.doi.org/10.2305/IUCN.UK.2010-2.RLTS.T57177A11580398.en

Lamar, W. W., and E. R. Wild. 1995. Comments on the natural history of *Lithodytes lineatus* (Anura: Leptodactylidae), with a description of the tadpole. *Apterygodon vittatus* 3: 135–142.

Larson, A. 1984. Neontological inferences of evolutionary pattern and process in the salamander family Plethodontidae. *Fossils Paleoherpetology Classification Evolution* 17: 119–217.

Laurenti, J. N. 1768. *Specimen Medicum, Exhibens Synopsin Reptilium emendatum cum experimentis circa venena at antidota reptilium austriacorum. Series: Museum of Comparative Zoology*. Viennae: Joan Thomae.

Lee, J. C. 1996. *The Amphibians and Reptiles of the Yucatan Peninsula*. Ithaca: Cornell University Press.

Lee, J. C. 2000. *A Field Guide to the Amphibians and Reptiles of the Maya World: The Lowlands of Mexico, Northern Guatemala, and Belize*. Ithaca: Cornell University Press.

Lescure, J. 1975a. Contribution a l'etude des amphibiens de guyane francaise. V. Reproduction de *Phyllomedusa tomopterna* (Cope) (Hylidae). *Bulletin de la Societe Zoologique de France* 100: 117–125.

Lescure, J. 1975b. Contribution a l'étude des amphibiens de Guyane Francaise. V. Les Centrolenidae. *Bulletin de la Société Zoologique de France* 100: 385–394.

Lescure, J., and C. Marty. 2000. *Atlas des Amphibiens de Guyane. Collection Patrimoines Naturels*, Vol. 45. Paris: Muséum National d'Histoire Naturelle.

Lescure, J., V. Marty, C. Marty, F. Starace, M. Auber-Thomay, and F. Letellier. 1995. Contribution à l'étude des Amphibiens de Guyane française. X. Les *Phyllomedusa* (Anura, Hylidae). *Revue Française d'Aquariologie* 22: 35–50.

Lima, A. P., W. E. Magnusson, M. Menin, L. Erdtmann, D. J. Rodrigues, C. Keller, and W. Hödl. 2006. *Guide to the Frogs of reserva Adolpho Ducke*. Manaus: Instituto Nacional de Pesquisas da Amazonia-INPA.

Lima, A. P., W. E. Magnusson, M. Menin, L. K. Erdtmann, D. J. Rodrigues, C. Keller, and W. Hödl. 2012. *Guia de Sapos da Reserva Adolpho Ducke, Amazônia Central/Guide to the Frogs of Reserva Adolpho Ducke, Central Amazonia*. 2nd ed, 187. Manaus: Editora INPA.

Lutz, A. 1930. Segunda memoria sobre especies brasileiras do genero Leptodactylus, incluindo outras alliadas/Second paper on Brasilian and some closely related species of the genus Leptodactylus. Memórias do Instituto Oswaldo Cruz. Rio de Janeiro 23: 1–34.

Lutz, B. 1954. Anfibios anuros do Distrito Federal/The frogs of the Federal district of Brazil. *Memórias do Instituto Oswaldo Cruz, Rio de Janeiro* 52: 155–197 (Portuguese), 219–238 (English).

Lutz, B., and G. R. Kloss. 1952. Anfíbios anuros do alto Solimões e Rio Negro. Apontamento sôbre algumas formas e suas vicariantes. *Memórias do Instituto Oswaldo Cruz, Rio de Janeiro* 50: 625–678.

Lynch, J. D. 1971. Evolutionary relationships, osteology, and zoogeography of leptodactyloid frogs. Miscellaneous publication. *Museum of Natural History, University of Kansas* 53: 1–238.

Lynch, J. D. 1976. The species groups of the South American frogs of the genus *Eleutherodactylus* (Leptodactylidae). *Occasional Papers of the Museum of Natural History, University of Kansas* 61: 1–24.

Lynch, J. D. 1982. Relationships of the frogs of the genus *Ceratophrys* (Leptodactylidae) and their bearing on hypothesis of Pleistocene forest refugia in South America and punctuated equilibria. *Systematic Zoology* 31: 166–179.

Lynch, J. D. 1986. New species of minute leptodactylid frogs from the Andes of Ecuador and Peru. *Journal of Herpetology* 20: 423–431.

Lynch, J. D. 1999. Una aproximación a las culebras ciegas de Colombia (Amphibia: Gymnophiona). *Revista Academia Colombiana de Ciencias* XXIII: 317–337.

Lynch, J. D. 2006. The Amphibian fauna in the Villacencio region of eastern Colombia. *Caldasia* 28, 135–155.

Lynch, J. D., and W. E. Duellman. 1973. A review of the centrolenid frogs of Ecuador, with descriptions of new species. *Occasional Papers of The Museum of Natural History, University of Kansas* 16: 1–66.

Lynch, J. D., and W. E. Duellman. 1997. Frogs of the genus *Eleutherodactylus* (Leptodactylidae) in Western Ecuador: systematics, ecology, and biogeography. *University of Kansas, Natural History Museum, Special Publication* 23: 1–236.

Lynch, J. D., J. Renjifo, D. F. Cisneros-Heredia, F. Solís, R. Ibáñez, C. Jaramillo, Q. Fuenmayor, M. Wilkinson, and M. Wake. 2010. *Caecilia nigricans* (errata version published in 2016). The IUCN Red List of Threatened Species 2010: e.T59521A86243710. http://dx.doi.org/10.2305/IUCN.UK.2010-2.RLTS.T59521A11955460.en

Lynch, J. D., P. M. Ruiz, and J. V. Rueda. 1983. Notes on the distribution and reproductive biology of *Centrolene geckoideum* Jiménez De la Espada in Colombia and Ecuador (Amphibia: Centrolenidae). *Studies on Neotropical Fauna and Environment* 18: 239–244.

Lynch, J. D., F. Solís, R. Ibáñez, C. Jaramillo, Q. Fuenmayor, L. A. Coloma, S. R. Ron, J. Renjifo, and M. Wilkinson. 2008. *Caecilia leucocephala*. The IUCN Red List of Threatened Species 2008: e.T59518A11954373. http://dx.doi.org/10.2305/IUCN.UK.2008.RLTS.T59518A11954373.en

Lynch, J. D., and A. M. Suárez-Mayorga. 2004. Anfibios en el Chocó biogeográfico. *Colombia Diversidad Biótica* IV: 633–667.

Maciel, A. O., H. C. Costa, L. de O. Drummond, J. O. Gomes, and A. D'Angiolella. 2013. Rediscovery of *Siphonops annulatus* (Mikan, 1820) (Amphibia: Gymnophiona: Siphonopidae) in the state of Pará, Brazil, with an updated geographic distribution map, and notes on size and variation. *Check List* 9: 106–110.

Maciel, A. O., and M. S. Hoogmoed. 2011. Taxonomy and distribution of caecilian amphibians (Gymnophiona) of Brazilian Amazonia, with a key to their identification. *Zootaxa* 53: 1–53.

Maciel, A. O., M. I. C. Sampaio, M. S. Hoogmoed, and H. Schneider. 2017. Phylogenetic relationships of the largest lungless tetrapod (Gymnophiona, *Atretochoana*) and the evolution of lunglessness in caecilians. *Zoologica Scripta* 46: 255–263.

Maddin, H. C., A. P. Russell, and J. S. Anderson. 2012. Phylogenetic implications of the morphology of the braincase of caecilian amphibians (Gymnophiona). *Zoological Journal of the Linnean Society* 166: 160–201.

Manzano, A. L. 2010. Prevalencia de quitridiomicosis en la población larvaria de *Gastrotheca riobambae* del Parque Metropolitano de Quito. Tesis de Licenciatura, Escuela Politécnica del Ejército.

Márquez, R., and I. de la Riva. 1996. Advertisement calls of three glass frogs from the Andean forests (Amphibia: Anura: Centrolenidae). *Herpetological Journal* 6: 97–99.

Marquez, R., I. de la Riva, and J. Bosch. 1995. Advertisement calls of Bolivian Leptodactylidae (Amphibia, Anura). *Journal of Zoology* 237: 313–336.

Marquis, R. J., M. A. Donnelly, and C. Guyer. 1986. Aggregations of calling males of *Agalychnis calcarifer* Boulenger (Anura: Hylidae) in a Costa Rican lowland wet forest. *Biotropica* 18: 173–175.

Martín-Torrijos, L., J. V. Sandoval-Sierra, J. Muñoz, J. Diéguez-Uribeondo, J. Bosch, and J. M. Guayasamin. 2016. Rainbow trout (*Oncorhynchus mykiss*) threaten Andean amphibians. *Neotropical Biodiversity* 2: 26–36.

Mauro, D. S., D. J. Gower, J. A. Cotton, R. Zardoya, M. Wilkinson, and T. Massingham. 2012. Experimental design in phylogenetics: testing predictions from expected information. *Systematic Biology* 61: 661–674.

McCaffery, R., and K. Lips. 2013. Survival and abundance in males of the glass frog *Espadarana* (*Centrolene*) *prosoblepon* in Central Panama. *Journal of Herpetology* 47: 162–168.

McCranie, J. R., and J. H. Townsend. 2011. Description of a new species of worm salamander (Caudata, Plethodontidae, *Oedipina*) in the subgenus *Oedopinola* from the central portion of the Cordillera Nombre de Dios, Honduras. *Zootaxa* 68: 59–68.

McCranie, J. R., D. R. Vieites, and D. B. Wake. 2008. Description of a new divergent lineage and three new species of Honduran salamanders of the genus *Oedipina* (Caudata, Plethodontidae) *Zootaxa* 17: 1–17.

McCranie, J. R., and L. D. Wilson. 2002. *The Amphibians of Honduras*. Salt Lake City: Society for the Study of Amphibians and Reptiles.

McDiarmid, R. W. 1975. Glass frog romance along a tropical stream. *Terra* 13: 14–18.

McDiarmid, R. W. 1978. Evolution of parental care in frogs. In *The Development of Behavior: Comparative and Evolutionary Aspects*, eds. G. M. Burghart, and M. Bekoff, 127–147. New York: Garland STPM Press.

McDiarmid, R. W., and K. Adler. 1974. Notes on the territorial and vocal behaviour of neotropical frogs of the genus *Centrolenella*. *Herpetologica* 30: 75–78.

Mendoza, A. M., W. Bolívar-García, E. Vázquez-Domínguez, R. Ibáñez, and G. Parra Olea. 2019. The role of Central American barriers in shaping the evolutionary history of the northernmost glassfrog, *Hyalinobatrachium fleischmanni* (Anura: Centrolenidae). *PeerJ* 7: e6115.

Mendoza-Henao, A. M., E. Arias, J. H. Townsend, and G. Parra-Olea. 2020. Phylogeny-based species delimitation and integrative taxonomic revision of the *Hyalinobatrachium fleischmanni* species complex, with resurrection of *H. viridissimum* (Taylor, 1942). *Systematics and Biodiversity* 18: 464–484.

Mendoza-Henao, A. M., K. R. Zamudio, J. M. Guayasamin, M. Escalona, and G. Parra-Olea. 2023. Environment rather than character displacement explains call evolution in glassfrogs. *Evolution* 77: 355–369.

Meneghelli, D., M. R. Messias, and P. R. M. Sampaio. 2011. Amphibia, anura, hylidae, *Cruziohyla craspedopus* (Funkhouser, 1957): Distribution extension in southwestern Amazonia, state of Rondônia, Brazil. *Check List* 7: 811–812.

Menin, M., A. P. Lima, and D. J. Rodrigues. 2009. The tadpole of *Vitreorana oyampiensis* (Anura: Centrolenidae) in Central Amazonia, Brazil. *Zootaxa* 2203: 65–68.

Menin, M., A. P. Lima, and D. J. Rodrigues. 2010. The tadpole of *Leptodactylus pentadactylus* (Anura: Leptodactylidae) from Central Amazonia. *Zootaxa* 68: 65–68.

Menin, M, and D. J. Rodrigues. 2013. The terrestrial tadpole of *Leptodactylus andreae* (Anura, Leptodactylidae), a member of the *Leptodactylus marmoratus* species group. *Zootaxa* 3683: 92–94.

Mercadal, I. 1988. Sobre la validez de *Ceratophrys testudo* Andersson, 1945 (Amphibia, Ceratophryidae). *Amphibia-Reptilia* 9: 1–6.

Minh, B. Q., H. A. Schmidt, O. Chernomor, D. Schrempf, M. D. Woodhams, A. Von Haeseler, and R. Lanfear. 2020. IQ-TREE 2: new models and efficient methods for phylogenetic inference in the genomic era. *Molecular Biology and Evolution* 37: 1530–1534.

Ministerio del Ambiente del Ecuador. 2013. *Sistema de Clasificación de los Ecosistemas del Ecuador Continental*. Quito: Subsecretaría de Patrimonio Natural.

Moraes, L. J. C. de L., and D. Pavan. 2018. On the occurrence of *Hemiphractus scutatus* (Spix, 1824) (Anura: Hemiphractidae) in eastern Amazonia. *Amphibian & Reptile Conservation* 12: 5–14.

Morales-Mite, M. 2013. *Herpetofauna en áreas prioritarias para la conservación*. Quito: Serie de Publicaciones del Museo Ecuatoriano de Ciencias Naturales. Fundación para la Conservación Jocotoco, Fundación Ecominga.

Motta, J., M. Menin, A. P. Almeida, T. Hrbek, and I. Pires Farias. 2018. When the unknown lives next door: a study of central Amazonian anurofauna. *Zootaxa* 4438: 79–104.

Mueller, R. L., J. R. Macey, M. Jaekel, D. B. Wake, and J. L. Boore. 2004. Morphological homoplasy, life history evolution, and historical biogeography of plethodontid salamanders inferred from complete mitochondrial genomes. *Proceedings of the National Academy of Sciences* 101: 13820–13825.

Mueses-Cisneros, J. J., V. Moreno-Quintero, and B. Cepeda-Quilindo. 2007. Amphibia, Caudata, Plethodontidae, *Bolitoglossa medemi*: distribution extension. *Check List* 3: 353–354.

Müller, L. 1923. Neue oder seltene Reptilien und Batrachier der Zoologischen Sammlung des bayrischen Staates. *Zoologischer Anzeiger* 57: 145–156.

Muñoz-Saravia, A., and C. R. Aguayo-Vedia. 2009. Sobre la presencia de *Vitreorana oyampiensis* (Lescure, 1975) (Amphibia: Centrolenidae) en Bolivia. *Cuadernos de Herpetología* 23: 97–99.

Murphy, J. C. 1997. *Amphibians and Reptiles of Trinidad and Tobago*. Florida: Krieger Publishing Company.

Myers, C. W. 1966. The distribution and behaviour of a tropical horned frog, *Cerathyla panamensis* Stejneger. *Herpetologica* 22: 68–71.

Myers, C. W., and M. A. Donnelly. 1997. *A Tepui Herpetofauna on a Granitic Mountain (Tamacuari) in the Borderland between Venezuela and Brazil: Report from the Phipps Tapirapecó Expedition*. New York: American Museum Novitates.

Myers, C. W., and W. E. Duellman. 1982. A new species of *Hyla* from Cerro Colorado, and other tree frog records and geographical notes from Western Panama. *American Museum Novitates* 2752: 1–32.

Narváez, A. E., and S. R. Ron. 2013. Feeding habits of *Engystomops pustulatus* (Anura: Leptodactylidae) in Western Ecuador. *South American Journal of Herpetology* 8: 161–167.

Narváez, A. E., and S. R. Ron. 2016. Spawning behaviour of *Engystomops pustulatus* (Anura: Leptodactylidae). *Journal of Natural History* 1858: 1–9.

Nascimento, L. B., U. Caramaschi, and C. A. G. Cruz. 2005. Taxonomic review of the species group of the genus *Physalameus* Fitzinger, 1826 with revalidation of the genera *Engystomops* Jimenez-de-la-Espada, 1872 and *Eupemphix* Steindachner, 1836 (Amphibia, Anura, Leptodactylidae). *Arquivos do Museu Nacional, Rio de Janeiro* 63: 297–320.

Nieden, F. 1923. Anura I. Subordo Aglossa und Phanerglossa sectio 1 Arcifera. *Das Tierreich* 46: 1–584.

Noble, G. K. 1926. An anlysis of the remarkable cases of distribution among the Amphibia, with descriptions of new genera. *American Museum Novitates* 1–24.

Núñez, D., T. Riera-Vite, F. Orellana, and J. Brito. 2015. Upper altitudinal and southern geographic range extensions for *Cruziohyla craspedopus* (Anura: Hylidae) (Funkhouser, 1957) in Ecuador. *Check List* 11: 1–5.

Nussbaum, R. A. 1977. Rhinatrematidae: a new family of caecilians (Amphibia: Gymnophiona). *Occasional Papers of the Museum of Zoology, University of Michigan* 682: 1–30.

Nussbaum, R. A. 1979. The taxonomic status of the caecilian genus *Uraeotyphlus* Peters. *Occasional Papers of the Museum of Zoology, University of Michigan* 687: 1–20.

Nussbaum, R. A. 1986. *Chthonerpeton onorei,* a new caecilian (Amphibia: Gymnophiona: Typhlonectidae) from Ecuador. *Revue Suisse de Zoologie* 93: 911–918.

Nussbaum, R. A. 1991. Cytotaxonomy of caecilians. In *Amphibian Cytogenetics and Evolution*, eds. D. M. Green, and S. K. Sessions. San Diego: Academic Press.

Nussbaum, R. A., and M. Wilkinson. 1989. On the classification and phylogeny of caecilians (Amphibia: Gymnophiona), a critical review. *Herpetological Monographs* 3: 1–42.

Obst, F. J., K. Richter, and U. Jacob. 1988. *The Completely Illustrated Atlas of Reptiles and Amphibians for the Terrarium.* Neptune City: TFH.

Oeser, R. 1959. Neue Frösche und Fröschnachzuchten. *Das Aquarium und Terrarium Zeitschrift* 12: 184–186.

Oppel, M. 1811. *Die ordnungen, familien und gattungen der reptilien als prodrom einer naturgeschichte derselben.* München: In Commission bey J. Lindauer.

Ortega-Andrade, H. M. 2008. *Agalychnis spurrelli* Boulenger (Anura, Hylidae): variación, distribución y sinonimia. *Papéis Avulsos de Zoologia (São Paulo)* 48: 103–117.

Ortega-Andrade, H. M. 2009. Amphibia, Anura, Eleutherodactylidae, *Adelophryne adiastola* Hoogmoed and Lescure, 1984: first countries records and distribution extension from Ecuador and Brazil. *Check List* 5: 139.

Ortega-Andrade, H. M., P. Meza-Ramos, D. F. Cisneros-Heredia, and M. Yánez-Muñoz. 2010a. Los anfibios y reptiles del Chocó Esmeraldeño. In *Serie Herpetofauna del Ecuador: El Chocó Esmeraldeño,* eds. M. Altamirano-Benavides, D. F. Cisneros-Heredia, and P. Meza-Ramos, 95–252. Quito: Museo Ecuatoriano de Ciencias Naturales.

Ortega-Andrade, M. H., J. Bermingham, C. Aulestia, and C. Paucar. 2010b. Herpetofauna of the Bilsa biological station, province of Esmeraldas, Ecuador. *Check List* 6: 119–154.

Ortega-Andrade, H. M., M. Rodes Blanco, D. F. Cisneros-Heredia, N. Guerra Arévalo, K. G. López de Vargas-Machuca, and J. C. Sánchez-Nivicela, D. Armijos-Ojeda, *et al*. 2021. Red list assessment of amphibian species of Ecuador: a multidimensional approach for their conservation. *PLoS One* 16: e0251027.

Ortega-Andrade, H. M., O. Rojas-Soto, and C. Paucar. 2013. Novel data on the ecology of *Cochranella mache* (Anura: Centrolenidae) and the importance of protected areas for this critically endangered glassfrog in the Neotropics. *PLoS One* 8: 1–13.

Ortega-Andrade, H. M., C. Tobar-Suárez, and M. Arellano Mónica. 2011. Tamaño poblacional, uso del hábitat y relaciones interespecíficas de *Agalychnis spurrelli* (Anura: Hylidae) en un bosque húmedo tropical remanente del noroccidente de Ecuador. *Papéis Avulsos de Zoologia* 51: 1–19.

Ortega, J. A., D. F. Cisneros-Heredia, J. D. Camper, A. Romero-Carvajal, L. Negrete, and S. R Ron. 2025. Systematics of minute strabomantid frogs allocated to the genus *Noblella* (Amphibia: Anura) with description of a new genus, seven new species, and insights into historical biogeography. *Zoological Journal of the Linnean Society* 203: zlae162.

Ortiz, D. A., D. Almeida-Reinoso, and L. A. Coloma. 2013. Notes on husbandry, reproduction and development in the Pacific horned frog *Ceratophrys stolzmanni* (Anura: Ceratophryidae), with comments on its amplexus. *International Zoo Yearbook* 47: 151–162.

Orton, G. L. 1951. The tadpole of *Leptodactylus melanonotus* (Hallowell). *Copeia* 1951: 62–66.

Orton, G. L. 1953. The systematics of vertebrate larvae. *Systematic Zoology* 2: 63–75.

Ospina-Sarria, J. J., G. W. Bolívar, and J. Mendez-Narvaez. 2010. Amphibia, Anura, Centrolenidae, *Espadarana callistomma* (Guayasamin and Trueb, 2007): first country records from Colombia. *Check List* 6: 244–245.

Ospina-Sarria, J. J., G. W. Bolívar, J. Mendez-Narvaez, and C. Burbano-Yandi. 2011. The tadpole of *Nymphargus grandisonae* (Anura: Centrolenidae) from Valle del Cauca, Colombia. *South American Journal of Herpetology* 6: 79–86.

Ouboter, P. E., and R. Jairam. 2012. *Amphibians of Suriname.* Leiden: Brill.

Padial, J. M., T. Grant, and D. R. Frost. 2014. Molecular systematics of terraranas (Anura: Brachycephaloidea) with an assessment of the effects of alignment and optimality criteria. *Zootaxa* 3825: 1–132.

Paluh, D. J., E. L. Stanley, and D. C. Blackburn. 2019. First dietary record of *Gastrotheca guentheri* (Boulenger, 1882), the lone anuran with true mandibular teeth. *Herpetology Notes* 12: 699–700.

Parker, H. W. 1932. Some new or rare reptiles and amphibians from Southern Ecuador. *The Annals and Magazine of Natural History* 9: 21–26.

Parmelee, J. R. 1999. Trophic ecology of a tropical anuran assemblage. *Scientific Papers Natural History Museum University of Kansas* 11: 1–59.

Parra-Olea, G., M. García-Paris, and D. B. Wake. 2004. Molecular diversification of salamanders of the tropical American genus *Bolitoglossa* (Caudata: Plethodontidae) and its evolutionary and biogeographical implications. *Biological Journal of the Linnean Society* 81: 325–346.

Pérez, O. D., N. B. Lai, D. Buckley, E. M. del Pino, and M. H. Wake. 2009. The morphology of prehatching embryos of *Caecilia orientalis* (Amphibia: Gymnophiona: Caeciliidae). *Journal of Morphology* 270: 1492–1502.

Peri, S. I. 1993. La validéz de *Ceratophrys testudo* Andersson, 1945 (Leptodactylidae, Ceratophryinae): aspectos morfológicos de su reconocimiento. *Alytes* 11: 107–116.

Peters, J. A. 1955. Herpetological type localities in Ecuador. *Revista Ecuatoriana de Entomología y Parasitología* 2: 335–352.

Peters, W. C. H. 1862a. Eine neue Gattung von Laubfröschen, *Plectromantis,* aus Ecuador. *Monatsberichte der königlich Akademie der Wissenschaften zu Berlin* 1862: 232–233.

Peters, W. C. H. 1862b. Über die Batrachier-Gattung *Hemiphractus. Monatsberichte der königlich Akademie der Wissenschaften zu Berlin* 1862: 144–152.

Peters, W. C. H. 1873a. Über eine neue Schildkrötenart, *Cinosternon Effeldtii* und einige andere neue oder weniger bekannte Amphibien. *Monatsberichte der königlich Akademie der Wissenschaften zu Berlin* 1873: 603–618.

Peters, W. C. H. 1873b. Über eine, zwei neue Gattungen enthaltende, Sammlung von Batrachiern des Hrn. Dr. O. Wucherer aus Bahia, so wie über einige neue oder weniger bekannte Saurier. *Monatsberichte der königlich Akademie der Wissenschaften zu Berlin* 1872: 768–776.

Peters, W. C. H. 1880. "1879". Über die Eintheilung der Caecilien und insbesondere über die Gattungen Rhinatrema und Gymnopis. *Monatsberichte der Königlichen Preussische Akademie des Wissenschaften zu Berlin* 1879: 924–945.

Peters, J. A. 1967. The generic allocation of the frog *Ceratophrys stolzmanni* Steindachner, with the description of a new subspecies from Ecuador. *Proceedings of the Biological Society of Washington* 80: 105–112.

Pinto, R. M. C., and M. Menin. 2017. Aspects of the natural history of *Leptodactylus knudseni* Heyer, 1972 (Anura: Leptodactylidae) in a pristine forest in Central Amazonia, Brazil, with comments on ontogenetic variation of its tadpoles. *Journal of Natural History* 51: 1–12.

Portik, D. M., J. W. Streicher, and J. J. Wiens. 2023. Frog phylogeny: a time-calibrated, species-level tree based on hundreds of loci and 5,242 species. *Molecular Phylogenetics and Evolution* 188: 107907.

Puschendorf, R., B. Kubicki, M. Ryan, and C. Vaughan. 2004. *Cochranella albomaculata* (NCN). Reproduction. *Herpetological Review* 35: 52.

Pyburn, W. F., and W. R. Heyer. 1975. Identity and call of the frog, *Leptodactylus stenodema. Copeia* 585–587.

Pyke, D. A., and J. M. Ray. 2006. Aggresive behavior by the amazon horned frog Ceratophrys cornuta. *Bulletin of the Maryland Herpetological Society* 42: 135–136.

Pyron, A. R., and Wiens, J. J. 2011. A large-scale phylogeny of Amphibia including over 2800 species, and a revised classification of extant frogs, salamanders, and caecilians. *Molecular Phylogenetics and Evolution* 61: 543–583.

Pyron, R. A. 2014. Biogeographic analysis reveals ancient continental vicariance and recent oceanic dispersal in amphibians. *Systematic Biology* 63: 779–797.

Quiguango-Ubillús, A., and L. A. Coloma. 2008. Notes on behaviour, communication and reproduction in captive *Hyloxalus toachi* (Anura: Dendrobatidae), an endangered Ecuadorian frog. *International Zoo Yearbook* 42: 78–89.

Rada, M., P. H. Dos Santos Dias, J. L. Pérez-Gonzalez, M. Anganoy-Criollo, L. A. Rueda-Solano, E. María Alejandra Pinto, L. M. Quintero, F. Vargas-Salinas, and T. Grant. 2019. The poverty of adult morphology: bioacoustics, genetics, and internal tadpole morphology reveal a new species of glassfrog (Anura: Centrolenidae). *PloS One* 14: e0215349.

Rada, M., and J. M. Guayasamin. 2008. Redescripción de *Cochranella megista* (Rivero, 1985) y ampliación de la distribución de nueve ranas de cristal (Anura: Centrolenidae) en Colombia. *Papéis Avulsos de Zoologia* 48: 89–101.

Raffaëlli, J. 2007. *Les Urodèles du Monde*. Plumelec: Penclen Édition.

Raffaëlli, J. 2013. *Les Urodèles du Monde*, Deuxième éd. Plumelec: Penclen Édition.

Raffaëlli, J. 2022. *Salamanders & Newts of the World*. Plumelec: Penclen Édition.

Ramírez, J. S. M., and B. M. L. Rodríguez. 2011. Estado poblacional y relaciones ecológicas de *Gastrotheca riobambae* (Anura: Hemiphractidae) en dos localidades del Volcán Pasochoa, Pichincha—Ecuador. *Grupo de Biodiversidad IASA. Boletín Técnico 10. Serie Zoológica* 7: 69–97.

Read, M. 2018. *Monitoreo biológico Yasuní. Anfibios y reptiles*. Quito: Ecuambiente Consulting Group.

Regös, J., and A. Schlüter. 1984. Erste Ergebnisse zur Fortpflanzungsbiologie von *Lithodytes lineatus* (Schneider, 1799). *Salamandra* 20: 252–261.

Reinhardt, J., and C. F. Lütken. 1862. "1861". Bidgrad til Kundskab om Brasiliens Padder og Krybdyr. *Videnskabelige Meddelelser fra den Naturhistoriske Forening i Kjøbenhavn* 3: 143–242.

Reyes-Puig, C., D. B. Wake, R. Kotharambath, J. W. Streicher, C. Koch, D. F. Cisneros-Heredia, M. H. Yánez-Muñoz, and S. Ron. 2020. Two extremely rare new species of fossorial salamanders of the genus *Oedipina* (Plethodontidae) from Northwestern Ecuador. *PeerJ* 8: e9934.

Reyes-Puig, J. P., M. H. Yánez-Muñoz, Z. Libke, P. Vinueza, and J. C. Carrión-Olmedo. 2024. Phylogenetic diversity of the Pristimantis anaiae species group, with a description of a new species of Pristimantis (Anura, Strabomantidae) from Cerro Candelaria, Tungurahua, Ecuador. Evolutionary Systematics 8: 101–118.

Rivera-Correa, M., and P. D. Gutiérrez-Cárdenas. 2006. Amphibia, Caudata, Plethodontidae, *Bolitoglossa medemi*: distribution extension. *Check List* 2: 77.

Rivero, J. A. 1961. Salientia of Venezuela. *Bulletin of the Museum of Comparative Zoology at Harvard College* 126: 1–207.

Rivero, J. A. 1968. Los centrolénidos de Venezuela (Amphibia, Salientia). *Memoria. Sociedad de Ciencias Naturales La Salle. Caracas* 28: 301–334.

Rivero, J. A. 1985. Nuevos centrolénidos de Colombia y Venezuela. *Brenesia* 23: 335–373.

Rodrigues, D. D. J., M. D. M. Lima, and R. A. Kawashita-Ribeiro. 2011. Amphibia, Anura, Hylidae, *Cruziohyla craspedopus* (Funkhouser, 1957): distribution extension, new state record and distribution map in Brazil. *Check List* 7: 149–150.

Rodrigues, D. D. J., M. Menin, and A. P. Lima. 2007. Redescription of the tadpole of *Leptodactylus rhodomystax* (Anura: Leptodactylidae) with natural history notes. *Zootaxa* 67: 61–67.

Rodríguez, L. O., and W. E. Duellman. 1994. *Guide to the Frogs of the Iquitos Region, Amazonian Peru*. Lawrence: The University of Kansas Natural History Museum.

Roelants, K., D. J. Gower, M. Wilkinson, S. P. Loader, S. D. Biju, K. Guillaume, L. Moriau, and F. Bossuyt. 2007. Global patterns of diversification in the history of modern amphibians. *Proceedings of the National Academy of Sciences* 104: 887–892.

Röhr, D. L., F. Camurugi, G. B. Paterno, M. Gehara, F. A. Juncá, G. F. R. Álvares, R. A. Brandão, and A. A. Garda. 2020. Variability in anuran advertisement call: a multi-level study with 15 species of monkey tree frogs (Anura, Phyllomedusidae). *Canadian Journal of Zoology* 98: 495–504.

Romero-Carvajal, A., N. Sáenz-Ponce, M. Venegas-Ferrín, D. Almeida-Reinoso, C. Lee, J. Bond, M. J. Ryan, J. B. Wallingford, and E. M. Del Pino. 2009. Embryogenesis and laboratory maintenance of the foam-nesting túngara frogs, genus Engystomops (= *Physalaemus*). *Developmental Dynamics* 238: 1444–1454.

Ron, S. R., and A. Merino-Viteri. 2000. Amphibian declines in Ecuador: overview and first report of chytridiomycosis from South America. *Froglog* 42: 2–3.

Ron, S. R. 2008. The evolution of female mate choice for complex calls in túngara frogs. *Animal Behaviour* 76: 1783–1794.

Ron, S. R., A. Almendáriz, and D. C. Cannatella. 2013. The *Phyllomedusa perinesos* group (Anura: Hylidae) is derived from a Miocene Amazonian lineage. *Zootaxa* 3741: 289–294.

Ron, S. R., D. C. Cannatella, and L. A. Coloma. 2004. Two new species of *Physalaemus* (Anura: Leptodactylidae) from Western Ecuador. *Herpetologica* 60: 261–275.

Ron, S. R., L. A. Coloma, and D. C. Cannatella. 2005. A new, cryptic species of *Physalaemus* (Anura: Leptodactylidae) from Western Ecuador with comments on the call structure of the *P. pustulosus* species group. *Herpetologica* 61: 178–198.

Ron, S. R., W. E. Duellman, L. A. Coloma, and M. R. Bustamante. 2003. Population decline of the Jambato toad *Atelopus ignescens* (Anura: Bufonidae) in the Andes of Ecuador. *Journal of Herpetology* 37: 116–126.

Ron, S. R., A. Narváez, and G. E. Romero. 2014. Reproduction and spawning behavior in the frog, *Engystomops pustulatus* (Shreve 1941). *Amphibian & Reptile Conservation* 8: 25–32.

Ron, S. R., J. C. Santos, and D. C. Cannatella, 2006. Phylogeny of the tungara frog genus *Engystomops* (= *Physalaemus pustulosus* species group; Anura: Leptodactylidae). *Molecular Phylogenetics and Evolution* 39: 392–403.

Ron, S. R., E. Toral, M. Rivera, and A. Terán-Valdez. 2010. A new species of *Engystomops* (Anura: Leiuperidae) from Southwestern Ecuador. *Zootaxa* 49: 25–49.

Rueda-Almonacid, J. V. 1994. Estudio anatómico y relaciones sistemáticas de *Centrolene geckoideum* (Salientia: Anura: Centrolenidae). *Trianea* 5: 133–187.

Ruiz-Carranza, P. M., M. C. Ardila-Robayo, and J. D. Lynch. 1996. Lista actualizada de la fauna de Amphibia de Colombia. *Revista de la Academia Colombiana de Ciencias Exactas, Físicas y Naturales* 20: 365–415.

Ruiz-Carranza, P. M., and J. I. Hernández-Camacho. 1976. *Gastrotheca andaquiensis*, nueva especie de la Cordillera Oriental de Colombia (Amphibia, Anura). *Caldasia* 11: 149–158.

Ruiz-Carranza, P. M., J. I. Hernández-Camacho, and M. C. Ardila-Robayo. 1986. Una nueva especie colombiana del género *Centrolene* Jiménez de la Espada 1872 (Amphibia: Anura) y redefinición del género. *Caldasia* 15: 431–444.

Ruiz-Carranza, P. M., and J. D. Lynch. 1991a. Ranas Centrolenidae de Colombia I. Propuesta de nueva clasificación genérica. *Lozania* 57: 1–60.

Ruiz-Carranza, P. M., and J. D. Lynch. 1991b. Ranas Centrolenidae de Colombia II. Nuevas especies de *Centrolene* de la Cordillera Oriental y Sierra Nevada de Santa Marta. *Lozania* 58: 1–26.

Ruiz-Carranza, P. M., and J. D. Lynch. 1995a. Ranas Centolenidae de Colombia VIII: Cuatro nuevas especies de *Centrolene* de la Cordillera Central. *Lozania* 65: 1–16.

Ruiz-Carranza, P. M., and J. D. Lynch. 1995b. Ranas centrolenidae de Colombia V. Cuatro nuevas especies de *Cochranella* de la Cordillera Central. *Revista de la Academia Colombiana de Ciencias Exactas, Fisicas y Naturales* 1–24.

Ruiz-Carranza, P. M., and J. D. Lynch. 1996. Ranas centrolenidae de Colombia. IX. Dos nuevas especies del suroeste de Colombia. *Lozania* 68: 1–11.

Ruiz-Carranza, P. M., and J. D. Lynch. 1998. Ranas Centrolenidae de Colombia XI. Nuevas especies de ranas de cristal del género *Hyalinobatrachium*. *Revista de la Academia Colombiana de Ciencias Exactas, Fisicas y Naturales* 22: 571–586.

Ryan, M. J., and R. C. Drewes. 1990. Vocal morphology of the *Physalaemus pustulosus* species group (Leptodactylidae): morphological response to sexual selection for complex calls. *Biological Journal of the Linnean Society* 40: 37–52.

Ryan, M. J., and A. S. Rand. 1993a. Phylogenetic patterns of behavioral mate recognition systems in the *Physalaemus pustulosus* species group (Anura: Leptodactylidae): the role ofancestral and derived characters and sensory exploitation. In *Evolutionary Patterns and Processes. Linnaen Society Symposium Series*, eds. D. Edward, and D. Lees, 000–000. London, UK: Academic Press. In press.

Ryan, M. J., and A. S. Rand. 1993b. Species recognition and sexual selection as a unitary problem in animal communication. *Evolution* 47: 647–657.

Ryan, M. J., and A. S. Rand. 2001. Feature weighting in signal recognition and discrimination by túngara frogs. *Anuran Communication* 86–101.

Sáenz-Ponce, N., J.-D. Santillana-Ortiz, and E. M. del Pino. 2012. The gastrocoel roof plate in embryos of different frogs. *Differentiation* 83: S62–S66.

Salazar-Nicholls, M. J., and E. M. del Pino. 2015. Early development of the glass frogs *Hyalinobatrachium fleischmanni* and *Espadarana callistomma* (Anura: Centrolenidae) from cleavage to tadpole hatching. *Amphibian & Reptile Conservation* 8: 89–106.

Salgado, A. L., and J. M. Guayasamin. 2018. Parental care and reproductive behavior of the minute dappled glassfrog (Centrolenidae: *Centrolene peristictum*). *South American Journal of Herpetology* 13: 211–219.

San Mauro, D., D. J. Gower, H. Müller, S. P. Loader, R. Zardoya, R. A. Nussbaum, and M. Wilkinson. 2014. Life-history evolution and mitogenomic phylogeny of caecilian amphibians. *Molecular Phylogenetics and Evolution* 73: 177–189.

San Mauro, D., D. J. Gower, O. V. Oommen, M. Wilkinson, and R. Zardoya. 2004. Phylogeny of caecilian amphibians (Gymnophiona) based on complete mitochondrial genomes and nuclear RAG1. *Molecular Phylogenetics and Evolution* 33: 413–427.

Sánchez-Carvajal, M. J., G. C. Reyes-Ortega, D. F. Cisneros-Heredia, and H. M. Ortega-Andrade. 2021. Rediscovery of Laura's glassfrog *Nymphargus laurae* (Anura: Centrolenidae) with new data on its morphology, colouration, phylogenetic position and conservation in Ecuador. *PeerJ* 9: e12644.

Santillana-Ortiz, J. -D., and E. M. del Pino. 2017. Características morfológicas de la gástrula en*Ceratophrys stolzmanni* (Anura: Ceratophryidae). *Revista Ecuatoriana de medicina y ciencias biológicas* 30: 50–56.

Savage, J. M. 1967. A new tree-frog (Centrolenidae) from Costa Rica. *Copeia* 1967: 325–331.

Savage, J. M. 2002. *The Amphibians and Reptiles of Costa Rica: A Herpetofauna between Two Continents, between Two Seas.* Chicago: The University of Chicago Press.

Savage, J. M., and P. H. Starrett. 1967. A new fringe-limbed treefrog (family Centrolenidae) from lower Central America. *Copeia* 1967: 604–609.

Schiesari, L. C., D. de C. Rossa-Feres, M. Menin, and W. Hödl. 2022. Tadpoles of Central Amazonia (Amphibia: Anura). *Zootaxa* 5223: 1–149.

Schlüter, A. 1980. Bio-akustische untersuchungen an leptodactyliden in einem begrenzten gebiet des tropischen regenwaldes von Peru. *Salamandra* 16: 227–247.

Schlüter, A. 1990. Reproduction and tadpole of *Edalorhina perezi* (Amphibia, Leptodactylidae). *Studies on Neotropical Fauna and Environment* 25: 49–56.

Schlüter, A., and J. Regos. 1996. The tadpole of *Lithodytes lineatus*; with note on the frogs resistance to leaf-cutting ants (Amphibia: Leptodactylidae). *Stuttgarter Beitraege zur Naturkunde Serie A (Biologie)* 536: 1–4.

Schlütter, A., P. Löttker, and K. Mebert. 2009. Use of an active nest of the leaf cutter ant *Atta cephalotes* (Hymenoptera: Formicidae) as a breeding site of *Lithodytes lineatus* (Anura: Leptodactylidae). *Herpetology Notes* 2: 101–105.

Schlütter, A., and J. Regös. 1981. *Lithodytes lineatus* (Schneider, 1799) (Amphibia: Leptodactylidae) as a dweller in nests of the leaf cutting ant *Atta cephalotes* (Linnaeus, 1758) (Hymenoptera: Attini). *Amphibia-Reptilia* 2: 117–121.

Schmeda-Hirschmann, G., J. P. de Andrade, M. R. Soto-Vasquez, P. A. A. Alvarado-García, C. Palominos, S. Fuentes-Retamal, M. Mellado, P. Correa, and F. A. Urra. 2020. The parotoid gland secretion from peruvian toad *Rhinella horribilis* (Wiegmann, 1833): chemical composition and effect on the proliferation and migration of lung cancer cells. *Toxins* 12: 608.

Schmid, M., C. Steinlein, J. P. Bogart, T. Haaf, L. Nanda, E. del Pino, W. E. Duellman, and S. B. Hedges. 2013. The hemiphractid frogs: phylogeny, embryology, life history, and cytogenetics. *Cytogenetics and Genome Research* 138: 69–384.

Schmid, M., C. Steinlein, T. Haaf, W. Feichtinger, M. Guttenbach, J. P. Bogart, S. L. Gruber, S. Kasahara, W. Kakampuy, E. del Pino, A.-B. Carrillo, A. Romero-Carvajal, M. J. Mahony, M. King, W. E. Duellman, and S. B. Hedges. 2018. *The Arboranan Frogs. Evolution, Biology, and Cytogenetics. Cytogenetics and Genome Research*, Vol. 155, eds. M. Schmid, J. P. Bogart, and S. B. Hedges, 1–325. Basel, Switzerland: S. Karger Medical and Scientific Publishers.

Schneider, H., and U. Sinsch. 1992. Mating call variation in lake frogs referred to as *Rana ridibunda* Pallas, 1771: taxonomic implications. *Journal of Zoological Systematics and Evolutionary Research* 30: 297–315.

Schneider, J. G. 1799. *Historiae Amphibiorum Naturalis et Literariae. Fasciculus Primus. Continens Ranas, Calamitas, Bufones, Salamandras et Hydros in Genera et Species Descriptos Notisque suis Distinctos.* Jena: Sumtibus Friederici Frommanni.

Scott Jr., N. J., and A. Starrett. 1974. An unusual breeding aggregation of frogs, with notes on the ecology of *Agalychnis spurelli* (Anura: Hylidae). *Bulletin of the Southern California Academy of Sciences* 73: 86–94.

Selmi, M. G., R. Brizzi, and E. Bigliardi. 1997. Sperm morphology of salamandrids (Amphibia, Urodela): implications for phylogeny and fertilization biology. *Tissue and Cell* 29: 651–664.

Señaris, C., and J. Ayarzagüena. 2005. *Revisión taxonómica de la Familia Centrolenidae (Amphibia; Anura) de Venezuela.* Sevilla: Publicaciones del Comité Español del Programa Hombre y Biosfera—Red IberoMaB de la UNESCO.

Sessions, S. K. 2008. Evolutionary cytogenetics in salamanders. *Chromosome Research* 16: 183–201.

Sever, D. M. 1994. Comparative anatomy and phylogeny of the cloacae of salamanders (Amphibia: Caudata). VII. Plethodontidae. *Herpetological Monographs* 8: 276–337.

Sheil, C. A., and J. R. Mendelson III. 2001. A new species of *Hemiphractus* (Anura: Hylidae: Hemiphractinae), and a redescription of *H. johnsoni*. *Herpetologica* 57: 189–202.

Shen, X. -X., D. Liang, M.-Y. Chen, R.-L. Mao, D. B. Wake, and P. Zhang. 2016. Enlarged multilocus data set provides surprisingly younger time of origin for the Plethodontidae, the largest family of salamanders. *Systematic Biology* 65: 66–81.

Shreve, B. 1941. Notes on Ecuadorian and Peruvian reptiles and amphibians with description of new forms. *Proceedings of the New England Zoological Club* 18: 71–83.

Siavichay-Pesántez, F., G. Maldonado-Cedeño, and D. Mejía-Coronel. 2016a. *Anfibios Urbanos de Cuenca.* Cuenca, Ecuador: Editorial Don Bosco, Centro Gráfico Salesiano.

Siavichay-Pesántez, F., G. Maldonado-Cedeño, D. Mejía-Coronel, J. F. Webster, N. Torres, and K. Costa. 2016b. *Plan de Manejo para la Conservación de los Anfibios Urbanos de Cuenca.* Cuenca: Municipalidad de Cuenca.

Silvano, D., C. Azevedo-Ramos, L. A. Coloma, S. Ron, R. Heyer, and C. Gascon. 2004. *Leptodactylus wagneri.* The IUCN Red List of Threatened Species 2004: e.T57175A11579804. http://dx.doi.org/10.2305/IUCN.UK.2004.RLTS.T57175A11579804.en

Sinovas, P., and B. Price. 2015. *Ecuador´s Wildlife Trade. Unep-Wcmc.* Ministry of the Environment of Ecuador and the German Development Cooperation (GIZ). UNEP-WCMC. https://www2.unep-wcmc.org//system/comfy/cms/files/files/000/000/808/original/Ecuador%C2%B4s_wildlife_trade.pdf

Sinsch, U., and N. Juraske. 2006. Advertisement calls of hemiphractine marsupial frogs, I: *Gastrotheca marsupiata* group. In *Herpetologia Bonnensis 2: Proceedings of the 13th Congress of the Societas Europaea Herpetologica*, eds. M. Vences, T. Köhler, T. Ziegler, and W. Böhme, 145–148. Societas Europaea Herpetologica.

Solís, F., R. Ibáñez, G. Chavez, J. Savage, C. Jaramillo, Q. Fuenmayor, F. Castro, T. Grant, E. Wild, B. Kubicki, and G. Köhler. 2010. *Sachatamia ilex.* The IUCN Red List of Threatened Species 2010: e.T54920A11224601. https://dx.doi.org/10.2305/IUCN.UK.2010-2.RLTS.T54920A11224601.en

Solís, F., R. Ibáñez, D. B. Wake, W. Bolívar, and J. M. Rengifo. 2008. *Bolitoglossa medemi.* The IUCN Red List of Threatened Species 2008: e.T59178A11881425. http://dx.doi.org/10.2305/IUCN.UK.2008.RLTS.T59178A11881425.en

Sorokin, A., and E. Steigerwald. 2017. Interspecific combat between *Nymphargus* aff. *grandisonae* and *Espadarana prosoblepon*. *Herpetology Notes* 10: 283–285.

Spannhof, I., and L. Spannhof. 1971. Beobachtungen zur Brutbiologie und Larvenentwicklung von *Gastrotheca marsupiata. Wiss. Zeit. Univ. Rostock Math. Naturwiss. Reihe* 20: 97–104.

Spix, J. B. von. 1824. *Animalia nova sive species novae Testudinum et Ranarum, quas in itinere per Brasiliam annis MDCCCXVII—MDCCCXX jussu et auspiciis Maximiliani Josephi I. Bavariae Regis.* Seraph, Hübschmanni, Monachii.

Starrett, P. H. 1960. Descriptions of tadpoles of Middle American frogs. *Miscellaneous Publications, Museum of Zoology, University of Michigan* 110: 1–37.

Steindachner, F. 1864. Batrachologische Mittheilungen. *Verhandlungen der K. K. Zoologisch-botanischen Gesellschaft in Wien* 14: 1–50.

Steindachner, F. 1892. Über zwei noch unbeschriebene Nototrema-Arten aus Ecuador und Bolivia. *Sitzungsberichten der kaiserl. Akademie der Wissenschaften in Wien* 101: 1–6.

Stejneger, L. 1911. Descriptions of three new batrachians from Costa Rica and Panama. *Proceedings of the United States National Museum* 41: 285–288.

Straughan, I. R., and W. R. Heyer. 1976. A functional analysis of the mating calls of the neotropical frog genera of the

Leptodactylus complex (Amphibia, Leptodactylidae). *Papeeis Avulsos de Zoologia* 29: 221–245.

Struck, T. H. 2007. Data congruence, paedomorphosis and salamanders. *Frontiers in Zoology* 4: 1–27.

Stuart, S. N., M. Hoffman, J. S. Chanson, N. A. Cox, R. J. Berridge, P. Ramani, and B. E. Young. 2008. *Threatened Amphibians of the World.* Barcelona: Lynx Editions.

Suárez-Mayorga, A. M. 1999. Lista preliminar de la fauna Amphibia presente en el transecto la Montañita-Alto de Gabinete, Caquetá, Colombia. *Revista de la Academia Colombiana de Ciencias Exactas, Físicas y Naturales* 23: 395–405.

Sunyer, J., and G. Köhler. 2007. New country and departmental records of herpetofauna in Nicaragua. *Salamandra* 43: 57–62.

Sunyer, J., G. Paíz, M. Dehling, and G. Köhler. 2009. A collection of amphibians from Río San Juan, Southeastern Nicaragua. *Herpetology Notes* 2: 189–202.

Székely, P., D. Armijos-Ojeda, L. Ordóñez-Delgado, D. Székely, and D. Cogălniceanu. 2016. Amphibia, Anura, Bufonidae, *Rhaebo ecuadorensis* Mueses-Cisneros, Cisneros-Heredia & McDiarmid, 2012, and Anura, Hylidae, *Phyllomedusa tarsius* (cope, 1868): range extensions and first records for Zamora-Chinchipe province, Ecuador. *Check List* 12: 2–6.

Székely, D., P. Székely, M. Denoël, and D. Cogalniceanu. 2016a. Phenotypic plasticity in a fossorial toad from a Pacific xeric forest (Ecuador). In *XIV Congreso Luso-Español de Herpetología.*

Székely, P., D. Székely, D. Armijos-Ojeda, A. Jara-Guerrero, and D. Cogălniceanu. 2016b. Anfibios de un bosque seco tropical: Reserva Ecológica Arenillas, Ecuador. *Revista Ecosistemas* 25: 24–34.

Székely, D., M. Denoël, P. Székely, and D. Cogălniceanu. 2017. Pond drying cues and their effects on growth and metamorphosis in a fast developing amphibian. *Journal of Zoology* 303: 129–135.

Székely, D., D. Cogălniceanu, P. Székely, and M. Denoël. 2018a. Dryness affects burrowing depth in a semi-fossorial amphibian. *Journal of Arid Environments* 155: 79–81.

Székely, D., P. Székely, M. Denoël, and D. Cogălniceanu. 2018b. Random size-assortative mating despite size-dependent fecundity in a Neotropical amphibian with explosive reproduction. *Ethology* 124: 218–226.

Székely, D., P. Székely, F. Stănescu, D. Cogălniceanu, and U. Sinsch. 2018c. Breed fast, die young: Demography of a poorly known fossorial frog from the xeric Neotropics. *Salamandra* 54: 37–44.

Székely, D., F. P. Gaona, P. Székely, and D. Cog. 2019. What does a Pacman eat? Macrophagy and necrophagy in a generalist predator (*Ceratophrys stolzmanni*). *PeerJ* 7: e6406.

Székely, D., D. Cogălniceanu, P. Székely, and M. Denoël. 2020a. Adult—Juvenile interactions and temporal niche partitioning between life-stages in a tropical amphibian. *PloS one* 15: 1–13.

Székely, P., J. S. Eguiguren, L. Ordóñez-Delgado, D. Armijos-Ojeda, and D. Székely. 2020b. Fifty years after: a taxonomic revision of the amphibian species from the Ecuadorian biodiversity hotspot Abra de Zamora, with description of two new *Pristimantis* species. *PloS one* 15: 1–59.

Taboada, C., J. Delia, M. Chen, C. Ma, X. Peng, X. Zhu, L. Jiang, T. Vu, Q. Zhou, and J. Yao. 2022. Glassfrogs conceal blood in their liver to maintain transparency. *Science* 378: 1315–1320.

Targueta, C. P., M. Rivera, and L. B. Lourenço. 2012. Karyotypic differentiation via 2n reduction and a finding of a case of triploidy in anurans of the genus *Engystomops* (Anura, Leiuperidae). *Genetica* 139: 1339–1347.

Tarvin, R. D., P. Pe.a, and S. R. Ron. 2014. Changes in population size and survival in *Atelopus spumarius* (Anura: Bufonidae) are not correlated with chytrid prevalence. *Journal of Herpetology* 48: 291–297.

Taylor, E. H. 1944. The genera of plethodont salamanders in Mexico, Pt. 1. *The University of Kansas Science Bulletin* 30(1): 189–232.

Taylor, E. H. 1949. Costa Rican frogs of the genera *Centrolene* and *Centrolenella. The University of Kansas Science Bulletin* XXXIII: 257–270.

Taylor, E. H. 1951. Two new genera and a new family of Tropical American frogs. *Proceedings of the Biological Society of Washington* 64: 33–40.

Taylor, E. H. 1958. Notes on Costa Rican Centrolenidae with descriptions of new forms. *The University of Kansas Science Bulletin* 39: 41–68.

Taylor, E. H. 1968. *The Caecilians of the World: A Taxonomic Review*. Lawrence: University of Kansas Press.

Taylor, E. H. 1972. Squamation in Caecilians, with an atlas of scales. *The UNiversity of Kansas Science Bulletin* XLIX: 989–1164.

Taylor, E. H. 1973. A caecilian miscellany. *University of Kansas Science Bulletin* 50: 187–231.

Taylor, E. H., and J. A. Peters. 1974. The caecilians of Ecuador. *The University of. Kansas Science Bulletin* 50: 333–346.

Terán Valdez, A., and J. M. Guayasamin. 2014. The tadpole of the glassfrog *Chimerella mariaelenae* (Anura: Centrolenidae). *CienciAmérica: Revista de divulgación científica de la Universidad Tecnológica Indoamérica* 3: 17–22.

Terán-Valdez, A., J. M. Guayasamin, and L. A. Coloma, 2009. Description of the tadpole of *Cochranella resplendens* and redescription of the tadpole of *Hyalinobatrachium aureoguttatum* (Anura, Centrolenidae). *Phyllomedusa* 8: 105–124.

Toft, C. A., and W. E. Duellman. 1979. Anurans of the lower Rio Llullapichis, Amazonian Peru: a preliminary analysis of community structure. *Herpetologica* 35: 71–77.

Toro-Sánchez, T., and M. H. Bernal-Bautista. 2015. The advertisement call of *Diasporus gularis* and *D. tinker* from the Pacific region of Colombia. *South American Journal of Herpetology* 10: 116–120.

Torres-Sánchez, M., D. J. Gower, D. Alvarez-Ponce, C. J. Creevey, M. Wilkinson, and D. San Mauro. 2019. What lies beneath? Molecular evolution during the radiation of caecilian amphibians. *BMC Genomics* 20: 1–13.

Trageser, S. J., R. J. Maynard, J. Culebras, S. Kohn, A. Quezada, and J. M. Guayasamin. 2021. Phylogenetic position of the glassfrog *"Cochranella" megista* (Anura: Centrolenidae) and first records for Ecuador. *Phyllomedusa* 20: 27–35.

Trueb, L. 1974. Systematic relationships of neotropical horned frogs, genus *Hemiphractus* (Anura: Hylidae). *Occasional Papers of the Museum of Natural History, the University of Kansas* 60: 1–60.

Tschudi, von J. J. 1838. *Classification der Batrachier, mit Berucksichtigung der Fossilen Thiere dieser Abtheilung der Reptilien*. Petitpierre: Neuchatel.

Turrell, C., B. Crnobrna, and M. Smith-bessen. 2016. Monitoring a population of *Cruziohyla craspedopus* (Funkhouser, 1957) using an artificial breeding habitat. *Amphibian & Reptile Conservation* 10: 1–6.

Twomey, E. M., and J. L. Brown. 2005. Range extension: *Cochranella resplendens. Herpetological Review* 1.

Twomey, E. M., J. Delia, and S. Castroviejo-Fisher. 2014. A review of Northern Peruvian glassfrogs (Centrolenidae), with the description of four new remarkable species. *Zootaxa* 3851: 1–87.

Vaillant, L. 1911. Cheloniens et batraciens urodèle, recuillis par M. le Dr. Rivet. In *Mission du Service Géographique de l'Armée pour la Mesure d'un Arc de Méridien Equatorial en Amérique du Sud. Vol. 9 (Zoologique). Part 2 (Reptiles, Poisson, Batraciens)*, ed. France. Service Géographique de l'Armé, 45–60. Paris: Gauthier-Villars, Imprimeur Libraire du Bureau des Longitudes, de l'Ecole Polytechnique.

Valencia, J., E. Toral, M. Morales, R. Betancourt, and A. Barahona. 2008. *Guía de campo de anfibios del Ecuador*. Quito, Ecuador: Fundación. Maxigraf S. A.

Valencia-Aguilar, A., F. Castro-Herrera, and M. P. Ramírez-Pinilla. 2012. Microhabitats for oviposition and male clutch attendance in *Hyalinobatrachium aureoguttatum* (Anura: Centrolenidae). *Copeia* 2012: 722–731.

Vanegas, J., V. Ramírez-Castaño, and E. C. G. Molina. 2014. *Nymphargus grandisonae* (Cochran & Goin 1970). *Catálogo de Anfibios y Reptiles de Colombia* 2: 51–55.

Vargas-Salinas, F. 2007. *Agalychnis spurrelli* (Clutch size). *Herpetological Review* 38: 177–178.

Vargas-Salinas, F., L. M. E. Bolaños, and B. H. Berrio. 2000. Notas sobre la ecología reproductiva de *Agalychnis spurrelli* (Anura Hylidae) en una población de Anchicayá, Pacifico colombiano. *Revista de la Academia Colombiana de Ciencias Exactas* 24: 85–99.

Vargas-Salinas, F., and P. D. Gutiérrez-Cárdenas. 2005. Cambios morfológicos y mortalidad en embriones y renacuajos de *Agalychnis spurrelli* (Hylidae: Phyllomedusinae). *Actualidades Biológicas* 27: 189–202.

Vargas-Salinas, F., and O. L. Torres-Suárez. 2013. *Agalychnis spurrelli* (Boulenger 1913). *Catálogo de Anfibios y Reptiles de Colombia. Medellín* 1: 19–25.

Verhaagh, M. 1993. Beobachtungen zur fortpflanzungs- und ernährungsbiologie von *Edalorhina perezi*. *Salamandra* 29: 231–239.

Vieites, D. R., S. N. Román, M. H. Wake, and D. B. Wake. 2011. A multigenic perspective on phylogenetic relationships in the largest family of salamanders, the Plethodontidae. *Molecular Phylogenetics and Evolution* 59: 623–635.

Vigle, G. O. 2008. The amphibians and reptiles of the Estación Biológica Jatun Sacha in the lowland rainforest of Amazonian Ecuador: a 20-year record. *Breviora* 514: 1–130.

Villacampa, J., S. Serrano-Rojas, and A. Whitworth. 2017. *Amphibians of the Manu Learning Centre and Other Areas of the Manu Region*. Cusco: The Crees Foundation.

Vitt, L. J., and J. P. Caldwell. 2014. *Herpetology. An Introductory Biology of Amphibians and Reptiles*, 4th ed. Amsterdam: Elsevier.

Vockenhuber, E. A., W. Hodl, and U. Karpfen. 2008. Reproductive behavior of the glass frog *Hyalinobatrachium valerioi* (Anura: Centrolenidae) at the tropical stream Quebrada Negra (La Gamba, Costa Rica). *Stapfia* 88: 335–348.

von May, R., A. Catenazzi, A. Angulo, J. L. Brown, J. Carrillo, G. Chávez, J. H. Córdova, *et al.* 2008. Current state of conservation knowledge on threatened amphibian species in Perú. *Tropical Conservation Science* 1: 376–396.

Wagler, J. 1828. Auszüge aus einem Systema Amphibiorum. *Isis von Oken* 740–744.

Wagler, J. 1830. *Natürliches System der Amphibien, mit vorangehender Classification der Säugthiere und Vögel*. München: J. G. Cotta'scchen Buchhandlung.

Wake, D. B. 1966. Comparative osteology and evolution of the lungless salamander, family Plethodontidae. *Memoirs of the Southern California Academy of Sciences* 4: 1–111.

Wake, D. B. 2012. Article taxonomy of salamanders of the family Plethodontidae (Amphibia: Caudata). *Zootaxa* 3484: 75–82.

Wake, D. B., and A. H. Brame Jr. 1963. The status of the plethodontid salamander genera *Bolitoglossa* and *Magnadigita*. *Copeia* 2: 382–387.

Wake, M. H. 1989. Metamorphosis of the hyobranchial apparatus in *Epicrionops* (Amphibia: Gymnophiona: Rhinatrematidae): replacement of bone by cartilage. *Annales des Sciences Naturelles Zoologie et Biologie Animale* 10: 171–182.

Walker, J. 1957. The breeding habits of the frog *Gastrotheca marsupiatum*. *British Journal of Herpetology* 2: 85–86.

Wassersug, R. J., and W. E. Duellman. 1984. Oral structures and their development in egg-brooding hylid frog embryos and larvae: evolutionary and ecological implications. *Journal of Morphology* 182: 1–37.

Weiss, H. 1960. Die Zucht von *Gastrotheca marsupiata*. *Das Aquarium und Terrarium Zeitschrift* 13: 121–122.

Werner, F. 1897. Über einige noch unbeschriebene Reptilien und Batrachier. *Zoologischer Anzeiger* 20: 261–267.

Wiens, J. J. 2011. Re-evolution of lost mandibular teeth in frogs after more than 200 million years, and re-evaluating Dollo's law. *Evolution* 65: 1283–1296.

Wiens, J. J., R. M. Bonett, and P. T. Chippindale. 2005. Ontogeny discombobulates phylogeny: paedomorphosis and higher-level salamander relationships. *Systematic Biology* 54: 91–110.

Wiens, J. J., C. A. Kuczynski, W. E. Duellman, and T. W. Reeder. 2007. Loss and re-evolution of complex life cycles in marsupial frogs: does ancestral trait reconstruction mislead? *Evolution* 61: 1886–1899.

Wild, E. R. 1994. Two new species of centrolenid frogs from the Amazonian slope of the Cordillera Oriental, Ecuador. *Journal of Herpetology* 28: 299–310.

Wild, E. R. 1996. Natural history and resource use of four Amazonian tadpole assemblages. *Occasional Papers of the Museum of Natural History University of Kansas* 176: 1–59.

Wild, E. R. 1997. Description of the adult skeleton and developmental osteology of the hyperossified horned frog, *Ceratophrys cornuta* (Anura: Leptodactylidae). *Journal of Morphology* 232: 169–206.

Wiley, O., and B. S. Lieberman. 2011. *Phylogenetics: The Theory and Practice of Phylogenetic Systematics*. Hoboken, New Jersey: John Wiley & Sons.

Wilkinson, M. 1989. On the status of *Nectocaecilia fasciata* Taylor, with a discussion of the phylogeny of the Typhlonectidae (Amphibia: Gymnophiona). *Herpetologica* 45: 23–36.

Wilkinson, M. 1996. Resolution of the taxonomic status of *Nectocaecilia haydee* (Roze) and a revised key to the genera of the Typhlonectidae (Amphibia: Gymnophiona). *Journal of Herpetology* 30: 413–415.

Wilkinson, M. 1997. Characters, congruence and quality: a study of neuroanatomical and traditional data in caecilian phylogeny. *Biological Reviews* 72: 423–470.

Wilkinson, M., P. J. R. Kok, F. Ahmed, and D. J. Gower. 2014. *Caecilita* Wake & Donnelly, 2010 (Amphibia: Gymnophiona) is not lungless: implications for taxonomy and for understanding the evolution of lunglessness. *Zootaxa* 3779: 383–388.

Wilkinson, M., A. Kupfer, R. Marques-Porto, H. Jeffkins, M. M. Antoniazzi, and C. Jared. 2008. One hundred million years of skin feeding? Extended parental care in a Neotropical caecilian (Amphibia: Gymnophiona). *Biology Letters* 4: 358–361.

Wilkinson, M., J. Measey, E. La Marca, L. A. Coloma, S. R. Ron, and F. Castro. 2010. *Potomotyphlus kaupii*. The IUCN Red List of Threatened Species 2010: e.T59589A11955756. http://dx.doi.org/10.2305/IUCN.UK.2010-2.RLTS.T59589A11955756.en

Wilkinson, M., and R. A. Nussbaum. 1999. Evolutionary relationships of the lungless caecilian *Atretochoana eiselti* (Amphibia: Gymnophiona: Typhlonectidae). *Zoological Journal of the Linnean Society* 126: 191–223.

Wilkinson, M., and R. A. Nussbaum. 2006. Caecilian phylogeny and classification. In *Reproductive Biology and Phylogeny of Gymnophiona (Caecilians),* ed. J.-M. Exbrayat, 39–78. Enfield: Science Publishers.

Wilkinson, M., R. P. Reynolds, and J. F. Jacobs. 2021. A new genus and species of rhinatrematid caecilian (Amphibia: Gymnophiona: Rhinatrematidae) from Ecuador. *Herpetological Journal* 31: 27–34.

Wilkinson, M., D. San Mauro, E. Sherratt, and D. J. Gower. 2011. A nine-family classification of caecilians (Amphibia: Gymnophiona). *Zootaxa* 2874: 41–64.

Woodhams, D. C., L. A. Rollins-Smith, C. Carey, L. Reinert, M. J. Tyler, and R. A. Alford. 2006. Population trends associated with skin peptide defenses against chytridiomycosis in Australian frogs. *Oecologia* 146: 531–540.

Wright, E. S. 2016. Using DECIPHER v2. 0 to analyze big biological sequence data in R. *R Journal* 8: 352–359.

Yánez-Muñoz, M. H., D. Batallas, D. Franco-Mena, P. A. Meza-Ramos, L. A. Oyagata, C. Paucar, J. P. Reyes-Puig, A. Rodríguez, M. A. Urgilés-Merchán, and M. Vega-Yánez. 2020. *Anfibios en los Ecosistemas Andino-Tropicales de la provincia del Carchi.* Quito: Serie de Publicaciones del Gobierno Autónomo Descentralizado Provincial del Carchi y el Instituto Nacional de Biodiversidad.

Yánez-Muñoz, M. H., and D. F. Cisneros-Heredia. 2008. Amphibia, Anura, Centrolenidae: Cochranella orejuela, first country records from Ecuador. *Check List* 4: 50–54.

Yánez-Muñoz, M. H., P. Pérez-Peña, and D. Cisneros-Heredia. 2009. New country records of *Hyalinobatrachium iaspidiense* (Amphibia, Anura, Centrolenidae) from the Amazonian lowlands of Ecuador and Peru. *Herpetology Notes* 2: 49–52.

Zhang, P., and D. B. Wake. 2009. Higher-level salamander relationships and divergence dates inferred from complete mitochondrial genomes. *Molecular Phylogenetics and Evolution* 53: 492–508.

Zheng, Y., R. Peng, M. Kuro-o, and X. Zeng. 2011. Exploring patterns and extent of bias in estimating divergence time from mitochondrial DNA sequence data in a particular lineage: a case study of salamanders (Order Caudata). *Molecular Biology and Evolution* 28: 2521–2535.

Zimmerman, B. L. 1983. A comparison of structural features of calls of open and forest habitat frog species in the Central Amazon. *Herpetologica* 39: 235–246.

Zimmerman, B. L., and J. Bogart. 1984. Vocalizations of primary forest frog species in the Central Amazon. *Acta Amazonica* 14: 473–520.

Index

Note: Page numbers in *italics* indicate a figure on the corresponding page.

A

B

C

D

E

F

G

H

J

K

L

M

R

U

V

W

Y

Z